Modeling Demographic Processes In Marked Populations

Environmental and Ecological Statistics

The Springer series **Environmental and Ecological Statistics** is devoted to the cross-disciplinary subject area of environmental and ecological statistics discussing important topics and themes in statistical ecology, environmental statistics, and relevant risk analysis. Emphasis is focused on applied mathematical statistics, statistical methodology, data interpretation and improvement for future use, with a view to advance statistics for environment, ecology, and environmental health, and to advance environmental theory and practice using valid statistics.

Each volume in the **Environmental and Ecological Statistics** series is based on the appropriateness of the statistical methodology to the particular environmental and ecological problem area, within the context of contemporary environmental issues and the associated statistical tools, concepts, and methods.

David L. Thomson • Evan G. Cooch
Michael J. Conroy
Editors

Modeling Demographic Processes In Marked Populations

Springer

Editors
David L. Thomson
Max Planck Institute for
Demographic Research
Konrad-Zuse Strasse 1
D18057 ROSTOCK
Germany

Netherlands Institute of Ecology
Centre for Terrestrial Ecology
Boterhoeksestraat 48
Postbus 40
6666 ZG Heteren
Netherlands

School of Biological Sciences
University of Hong Kong
Pokfulam
Hong Kong
thomson@demogr.mpg.de

Evan G. Cooch
Department of Natural Resources
Cornell University
Ithaca, NY
USA
evan.cooch@cornell.edu

Michael J. Conroy
D.B. Warnell School of
Forestry and Natural Resources
University of Georgia
Athens, GA
USA
mconroy@uga.edu

ISBN: 978-0-387-78150-1 e-ISBN: 978-0-387-78151-8
DOI 10.1007/978-0-387-78151-8

Library of Congress Control Number: 2008920621

Printed on acid-free paper

springer.com

Foreword

Demography can be considered the key to understanding much of biology. It is the demographic processes of birth and death which govern the spread of populations through environments and the spread of genes through populations. An understanding of demography can yield not only an understanding of population size and population change, it can help us to understand the form and function of life histories; when organisms mature, when they breed, and when they die. Demographic insights allow us to see how populations function, how they interact with their changing environment, and how they adapt.

The analysis of demographic processes in free-living organisms is however no simple task and involves considerable challenges in observation and analysis. Some 20 years ago, there was a concerted effort to promote inter-disciplinary collaboration between biologists and statisticians to address these challenges and thereby to further our understanding of demographic processes in natural populations. Although many diverse organisms can be studied in the wild, birds have proved particularly amenable with large numbers being marked and followed by large networks of observers. It was no coincidence then that the European Union for Bird Ringing (EURING) played a leading role in these initiatives, teaming up in the mid-1980s with the Mathematical Ecology Group of the Biometric Society, and the British Ecological Society, to bring together experts from diverse fields to address the challenges in hand. Twenty years on, progress has been considerable and we now have significant insights into demographic processes thanks to the wide range of quantitative tools and systematically collected datasets which have been built up over this period.

The biological questions and the methodological challenges are however by no means settled, indeed the field continues to progress at an ever accelerating pace. In 2003, a group of just under 100 scientists met to discuss and identify the key areas of development in which ongoing research effort should be focused. As listed in the Contents section, the group identified five areas defined by biological applications and five areas defined by statistical approaches including the issue of software with which to implement state-of-the-art analyses. Experts in each of these areas then took the lead in assembling authoritative contributions, with one or two overview- or perspectives- papers prepared by leading figures, and three to five primary research papers which reported the most significant new findings. A further open-forum was created for notable contributions which lay outside the ten targeted areas. Authors came together to discuss their contributions at a meeting hosted by the University of Otago at the beginning of 2007.

This field continues to move rapidly, but we hope this resulting volume will stand as a definitive compilation on the state-of-the-field at the present time, and that it will steer the further development of the field over the years ahead. As reflected in this volume, we anticipate increasing emphasis on integrated approaches which combine multiple sources of information and an increasing emphasis on Bayesian approaches. In terms of biological applications, it has traditionally been the field of wildlife management which has provided the impetus for developing modern approaches, but increasingly we see the activities of evolutionary biologists and biodemographers as a driver of growth in this field. Modeling demographic processes in marked populations is a truly interdisciplinary endeavour, and we look forward to continued fruitful dialogue not just between biologists and statisticians but between these different fields of biology which are conceptually similar and which share the same need for sound quantitative approaches to demographic analysis.

This volume has been a team effort, and as well as crediting all the work of the authors themselves and the associate editors listed in the Contents section, we would like to acknowledge Prof. Richard Barker and his team for their hard work and kind hospitality in hosting a successful meeting of contributors in Dunedin. All contributions have benefited from the expert input of at least two referees, and we would of course like to thank Prof. G.P. Patil, Manjula Jude (Project Manager at Integra Software Services), Lindy Paul and the team at Springer for facilitating the publication of this volume.

David L. Thomson
Evan G. Cooch
Michael J. Conroy

Contents

Contributors

Ray T. Alisauskas
Environment Canada, Prairie and Northern Wildlife Research Centre, 115 Perimeter Road, Saskatoon, Saskatchewan, Canada S7N 0X4, e-mail: ray.alisauskas@ec.gc.ca

Mathew W. Alldredge
NC Cooperative Fish and Wildlife Research Unit, Department of Zoology, Campus Box 7617, North Carolina State University, Raleigh, NC 27695, USA

David R. Anderson
U.S. Geological Survey, Colorado Cooperative Fish and Wildlife Research Unit, Colorado State University, Fort Collins, CO 80523, USA, e-mail: anderson@warnercnr.colostate.edu

Doug P. Armstrong
Wildlife Ecology Group, Institute of Natural Resources, Massey University, Private Bag 11222, Palmerston North, New Zealand, e-mail: d.p.armstrong@massey.ac.nz

Lise M. Aubry
Max Planck Institute for Demographic Research, Konrad-Zuse Str. 1, D-18057 Rostock, Germany; Laboratoire Evolution et Diversité Biologique, Université P. Sabatier, Toulouse, France, e-mail: Aubry@demogr.mpg.de

Larissa L. Bailey
U.S. Geological Survey, Patuxent Wildlife Research Center, 12100 Beech Forest Road, Laurel, MD 20708-4017, USA, e-mail: lbailey@usgs.gov

Stephen R. Baillie
British Trust for Ornithology, The Nunnery, Thetford, Norfolk IP24 2PU, UK, e-mail: stephen.baillie@bto.org

Dawn E. Balmer
British Trust for Ornithology, The Nunnery, Thetford, Norfolk IP24 2PU, UK

Christophe Barbraud
Centre d'Etudes Biologiques de Chizé/CNRS, UPR 1934, 79360 Villiers en Bois, France

Richard J. Barker

Department of Mathematics and Statistics, University of Otago, P.O. Box 56, Dunedin, New Zealand, e-mail: rbarker@maths.otago.ac.nz

Panagiotis Besbeas

Institute of Mathematics, Statistics and Actuarial Science, University of Kent, Canterbury, Kent CT2 7NF, England

Simon J. Bonner

Department of Statistics and Actuarial Science, Simon Fraser University, Burnaby, BC, Canada V5A 1S6, e-mail: sbonner@stat.sfu.ca

David L. Borchers

Research Unit for Wildlife Population Assessment, The Observatory, Buchanan Gardens, University of St. Andrews, Fife, KY16 9LZ, Scotland, e-mail: dlb@mcs.st-and.ac.uk

Rachel Borysiewicz

Institute of Mathematics, Statistics and Actuarial Science, University of Kent, Canterbury, Kent CT2 7NF, England

Thomas Bregnballe

VIBI, NERI, Denmark

Jerome Brewster

NC Cooperative Fish and Wildlife Research Unit, Department. of Zoology, Campus Box 7617, North Carolina State University, Raleigh, NC 27695, USA

Stephen P. Brooks

Statistical Laboratory, Centre for Mathematical Sciences, Wilberforce Road, Cambridge CB3 0WB, UK

Kenneth P. Burnham

U.S. Geological Survey, Colorado Cooperative Fish and Wildlife Research Unit, Colorado State University, Fort Collins, CO 80523, USA

Andrea E. Byrom

Landcare Research, P.O. Box 40, Lincoln 7640, Canterbury, New Zealand, e-mail: byroma@landcareresearch.co.nz

Emmanuelle Cam

Laboratoire Evolution et Diversité Biologique, Bâtiment 4 R3 – B2, 118 Route de Narbonne, Université Paul Sabatier Toulouse III, 31062 Toulouse Cedex 09, France, e-mail: emmacam@cict.fr

Hal Caswell

Biology Department MS-34, Woods Hole Oceanographic Institution, Woods Hole, MA 02543, USA, e-mail: hcaswell@whoi.edu

Vanessa M. Cave
Centre for Research into Ecological and Environmental Modelling, University of St. Andrews, The Observatory, St. Andrews KY16 9LZ, UK, e-mail: vanessa@mcs.st-and.ac.uk

Wendell O. Challenger
Department of Statistics and Actuarial Sciences, Simon Fraser University, Burnaby, BC, Canada V5A 1S6, e-mail: wchallen@stat.sfu.ca

Rémi Choquet
Centre d'Ecologie Fonctionnelle et Evolutive, UMR 5175, CNRS, 1919 Route de Mende, F34293 Montpellier cedex 5, France, e-mail: choquet@cefe.cnrs.fr

Don R. Church
Department of Biology, University of Virginia, Gilmer Hall, Charlottesville, VA 22903, USA; Conservation International, Regional Programs Division, Crystal Drive, Suite 500, Arlington, VA 22202, USA

Jaime Collazo
United States Geological Survey, North Carolina Cooperative Fish and Wildlife Research Unit, North Carolina State University, Raleigh, NC 27695, USA

Paul B. Conn
Colorado Cooperative Fish & Wildlife Research Unit, Department of Fishery, Wildlife, and Conservation Biology, Colorado State University, Fort Collins, CO 80523, USA, e-mail: pconn@warnercnr.colostate.edu

Michael J. Conroy
USGS, Georgia Cooperative Fish and Wildlife Research Unit, D. B. Warnell School of Forestry and Natural Resources, University of Georgia, Athens, GA 30602, USA, e-mail: mconroy@uga.edu

Sarah J. Converse
Colorado Cooperative Fish and Wildlife Research Unit, Patuxent Wildlife Research Center, Laurel, MD, USA, e-mail: sconverse@usgs.gov

Evan G. Cooch
Department of Natural Resources, Cornell University, Ithaca, NY 14853, USA, e-mail: evan.cooch@cornell.edu

R. Scott Davidson
Wildlife Ecology Group, Institute of Natural Resources, Massey University, Private Bag 11222, Palmerston North, New Zealand

Paul F. Doherty
Department of Fish, Wildlife, and Conservation Biology; Colorado State University, Fort Collins, CO, USA

Kiel L. Drake

University of Saskatchewan, 128 Ash Street, Saskatoon, Saskatchewan, Canada S7J 0J8, e-mail: kieldrake@sasktel.net

Brian P. Dreher

Michigan State University, Department of Fisheries and Wildlife, 13 Natural Resources Building, East Lansing, MI 48824, USA; Colorado Division of Wildlife, 4255 Sinton Road, Colorado Springs, CO 80907, USA

Holly H. Edwards

Florida Fish and Wildlife Conservation Commission, Fish and Wildlife Research Institute, 100 8th Avenue SE, St. Petersburg, FL 33701, USA

Murray G. Efford

Department of Zoology, University of Otago, P.O. Box 56, Dunedin 9010, New Zealand, e-mail: murray.efford@otago.ac.nz; murray.efford@stonebow.otago.ac.nz

Torbjørn Ergon

Centre for Ecological and Evolutionary Synthesis, Department of Biology, University of Oslo, P.O. Box 1066, Blindern, N-0316 Oslo, Norway, e-mail: torbjorn.ergon@bio.uio.no

J. Paige Eveson

CSIRO Marine and Atmospheric Research, GPO Box 1538, Hobart, Tasmania 7001 Australia, e-mail: paige.eveson@csiro.au

Rachel M. Fewster

Department of Statistics, University of Auckland, Private Bag 92019, Auckland, New Zealand, e-mail: r.fewster@auckland.ac.nz

Daniel Fink

Cornell Laboratory of Ornithology, 159 Sapsucker Woods Rd, Ithaca, NY 14850, USA, e-mail: df36@cornell.edu

David Fletcher

Department of Mathematics and Statistics, University of Otago, Dunedin, New Zealand, e-mail: dfletcher@maths.otago.ac.nz

Christopher J. Fonnesbeck

Florida Fish and Wildlife Conservation Commission, Fish and Wildlife Research Institute, 100 8th Avenue SE, St. Petersburg, FL 33701, USA, e-mail: Chris.Fonnesbeck@MyFWC.com

David A. Fournier

Otter Research Ltd, PO Box 2040, Station A, Sidney, B.C. V8L 3S3, Canada

Charles M. Francis
Canadian Wildlife Service, National Wildlife Research Centre, Environment Canada, Ottawa, ON, Canada K1A 0H3, e-mail: charles.francis@ec.gc.ca

Stephen N. Freeman
British Trust for Ornithology, The Nunnery, Thetford, Norfolk IP24 2PU, UK; Centre for Ecology and Hydrology, Maclean Building, Benson Lane, Crowmarsh Gifford, Wallingford, Oxfordshire OX108BB, UK, e-mail: sfreeman@ceh.ac.uk

Jean-Michel Gaillard
Universite Claude Bernard, Lyon, France

Olivier Gimenez
Centre d'Ecologie Fonctionnelle et Evolutive/CNRS, UMR 5175; 1919, Route de Mende, 34293 Montpellier Cedex 5, France, e-mail: olivier.gimenez@cefe.cnrs.fr

Vladimir Grosbois
Laboratoire Biométrie et Biologie Evolutive, UMR 5558, Bat. 711, Université Claude Bernard Lyon 1, 43 Boulevard du 11 novembre 1918, F-69622 Villeurbanne Cedex, France

Yann Guédon
Centre de cooperation Internationale en Recherche Agronomique pour le Developpement, Montpellier, France

Viviane Henaux
CEFE, CNRS, Montpellier, France

James E. Hines
USGS Patuxent Wildlife Research Center, Laurel, MD, USA

Wesley Hochachka
Cornell Laboratory of Ornithology, 159 Sapsucker Woods Rd, Ithaca, NY 14850, USA

Simon D. Hoyle
Oceanic Fisheries Programme, Secretariat of the Pacific Community, B.P. D5 Noumea, New Caledonia

Christine M. Hunter
Department of Biology and Wildlife, Institute of Arctic Biology, University of Alaska, P.O. Box 757000, Fairbanks, AK 99775, USA, e-mail: christine.hunter@uaf.edu

Lara E. Jamieson
The Statistical Laboratory – CMS, University of Cambridge, Wilberforce Road, Cambridge CB3 0WB, UK

Niclas Jonzén

Department of Theoretical Ecology, Ecology Building, Lund University, SE-223, 62 Lund, Sweden, e-mail: niclas.jonzen@teorekol.lu.se

William L. Kendall

U.S. Geological Survey, Patuxent Wildlife Research Center, 12100 Beech Forest Road, Laurel, MD 20708-4017, USA, e-mail: wkendall@usgs.gov

Marc Kéry

Swiss Ornithological Institute, 6204 Sempach, Switzerland, e-mail: marc.kery@vogelwarte.ch

Ruth King

Centre for Research into Ecological and Environmental Modelling, The Observatory, Buchanan Gardens, University of St. Andrews, St. Andrews, Fife KY16 9LZ, Scotland

Jonas Knape

Department of Theoretical Ecology, Ecology Building, Lund University, SE-223, 62 Lund, Sweden, e-mail: jonas.knape@teorekol.lu.se

Jeffrey L. Laake

National Marine Mammal Laboratory, Alaska Fisheries Science Center, National Marine Fisheries Service, 7600 Sand Point Way, NE, Seattle, WA 98115-0070, USA, e-mail: jeff.laake@noaa.gov

Catherine A. Langtimm

U.S. Geological Survey, Florida Integrated Science Center, 2201 NW 40th Terrace, Gainesville, FL 32605, USA, e-mail: clangtimm@usgs.gov

Geoff M. Laslett

CSIRO Mathematical and Information Sciences, Private Bag 33, Clayton South, Victoria 3169, Australia

Jean-Dominique Lebreton

CEFE UMR 5175, CNRS, 1919 Route de Mende, 34 293 Montpellier cedex 5, France, e-mail: jean-dominique.lebreton@cefe.cnrs

Mark S. Lindberg

Department of Biology and Wildlife and Institute of Arctic Biology, University of Alaska Fairbanks, Fairbanks, AK 99775, USA

William A. Link

U.S. Geological Survey, Patuxent Wildlife Research Center, 12100 Beech Forest Road, Laurel, MD 20708, USA, e-mail: wlink@usgs.gov

Paul M. Lukacs

Colorado Cooperative Fish and Wildlife Research Unit, 1484 Campus Delivery, Colorado State University, Fort Collins, CO 80523, USA; Colorado Division of Wildlife, 317 W. Prospect Rd., Fort Collins, CO 80526, USA, e-mail: Paul.Lukacs@state.co.us

James Lyons

United States Fish and Wildlife Service, Division of Migratory Bird Management, 11510 American Holly Drive, Laurel, MD 20708, USA

Hamish E. MacInnes

University of Auckland, Auckland, New Zealand

Mark N. Maunder

Inter-American Tropical Tuna Commission, 8604 La Jolla Shores Drive, La Jolla, CA, 92037-1508, USA, e-mail: mmaunder@iattc.org

Brett T. McClintock

Department of Fish, Wildlife, and Conservation Biology, Department of Statistics, Colorado State University, Fort Collins, CO 80523, USA, e-mail: btm119@warnercnr.colostate.edu; USGS Patuxent Wildlife Research Center, 12100 Beech Forest Road, Laurel, MD 20708, USA, e-mail:bmcclintock@usgs.gov

Steven D. Miller

University of Auckland, Auckland, New Zealand, e-mail: smiller@stat.auckland.ac.nz

Jean-Yves Monnat

Penn ar Run Izella, Goulien, Cap Sizun, France

Byron J.T. Morgan

Institute of Mathematics, Statistics and Actuarial Science, University of Kent, Canterbury, Kent CT2 7NF, England, e-mail: B.J.T.Morgan@kent.ac.uk

Maura B. Naughton

US Fish and Wildlife Service, Portland, OR, USA

James D. Nichols

USGS, Patuxent Wildlife Research Centre, 11510 American Holly Drive, Laurel, MD 20708-4017, USA, e-mail: jim_nichols@usgs.gov

David L. Otis

Iowa Cooperative Fish and Wildlife Research Unit, Department of Natural Resource Ecology and Management, Iowa State University, Ames IA, 50011, USA, e-mail: dotis@iastate.edu

Krishna Pacifici
NC Cooperative Fish and Wildlife Research Unit, Dept. of Zoology, Campus Box 7617, North Carolina State University, Raleigh, NC 27695, USA

Richard A. Parker
Centre for Research into Ecological and Environmental Modelling, University of St. Andrews, The Observatory, St. Andrews KY16 9LZ, Scotland

Nathalie J. Patenaude
School of Biological Sciences, University of Auckland, Private Bag 92019, Auckland, New Zealand

Shirley Pledger
School of Mathematics, Statistics and Computer Science, Victoria University of Wellington, P.O. Box 600, Wellington 6140, New Zealand, e-mail: shirley.pledger@vuw.ac.nz

Tom Polacheck
CSIRO Marine and Atmospheric Research, GPO Box 1538, Hobart, Tasmania 7001 Australia

Kenneth H. Pollock
Zoology, Biomathematics, and Statistics, Campus Box 7617, North Carolina State University, Raleigh, NC 27695, USA

Roger Pradel
CEFE, 1919 Route de Mende, 34293 Montpellier cedex, France, e-mail: roger.pradel@cefe.cnrs.fr

Moira A. Pryde
Department of Conservation, P.O. Box 13 049, Christchurch, New Zealand

John E. Reynolds
Mote Marine Laboratory, 1600 Ken Thompson Parkway, Sarasota, FL 34236, USA

Martin S. Ridout
Institute of Mathematics, Statistics and Actuarial Science, University of Kent, Canterbury, Kent CT2 7NF, England

Jay Rotella
Ecology Department, Montana State University, Bozeman, MT 59717, USA
e-mail: rotella@montana.edu

Lauriane Rouan
Centre d'Ecologie Fonctionnelle et Evolutive, UMR 5175, CNRS, 1919 Route de Mende, F34293 Montpellier cedex 5, France

J. Andrew Royle
U.S. Geological Survey, Patuxent Wildlife Research Center, 12100 Beech Forest Road, Laurel, MD 20708, USA

James C. Russell

Department of Statistics, University of Auckland, Private Bag 92019, Auckland, New Zealand, e-mail: j.russell@auckland.ac.nz

Pertti Saurola

Finnish Museum of Natural History, P.O. Box 17, FIN-00014 University of Helsinki, Helsinki, Finland

Michael Schaub

Division of Conservation Biology, Zoological Institute, University of Bern, Baltzerstrasse 6, CH-3012 Bern, Switzerland; Swiss Ornithological Institute, CH-6204 Sempach, Switzerland

Joel A. Schmutz

Alaska Science Center, U.S. Geological Survey, 1011 East Tudor Road, Anchorage, AK 99503, USA, e-mail: jschmutz@usgs.gov

Matthew R. Schofield

Department of Mathematics and Statistics, University of Otago, P.O. Box 56, Dunedin, New Zealand, e-mail: mschofield@maths.otago.ac.nz

Carl James Schwarz

Department of Statistics and Actuarial Science, Simon Fraser University, 8888 University Drive, Burnaby, BC, Canada V5A 1S6, e-mail: cschwarz@stat.sfu.ca

Kim T. Scribner

Michigan State University, Department of Fisheries and Wildlife, 13 Natural Resources Building, East Lansing, MI 48824, USA

Juan Carlos Senar

Evolutionary and Behavioural Ecology Associate Research Unit, CSIC, Museu de Ciències Naturals, E-08003 Barcelona, Spain

Theodore R. Simons

USGS, NC Cooperative Fish and Wildlife Research Unit, Department of Zoology, Campus Box 7617, North Carolina State University, Raleigh, NC 27695, USA, e-mail: tsimons@ncsu.edu

Hans J. Skaug

Department of Mathematics, University of Bergen, 5008 Bergen, Norway

Martin Sköld

Department of Economics, Statistics and Informatics, Örebro University, SE-701, 82 Örebro, Sweden, e-mail: martin.skold@esi.oru.se

Leonid Sokolov

Biological Station Rybachy, Zoological Institute, Russian Academy of Sciences, Moscow, Russia, e-mail: leonid-sokolov@mail.ru

Len Thomas
Centre for Research into Ecological and Environmental Modelling, The Observatory, Buchanan Gardens, University of St. Andrews, St. Andrews, Fife KY16 9LZ, Scotland

David L. Thomson
Max Planck Institute for Demographic Research, Konrad-Zuse Strasse 1, D18057 ROSTOCK, Germany; Netherlands Institute of Ecology, Centre for Terrestrial Ecology, Boterhoeksestraat 48, Postbus 40, 6666 ZG Heteren, Netherlands; School of Biological Sciences, University of Hong Kong, Pokfulam, Hong Kong, e-mail: thomson@demogr.mpg.de

Kasper Thorup
Zoological Museum, University of Copenhagen, Universitetsparken 15, DK-2100 Copenhagen, Denmark, e-mail: kthorup@snm.ku.dk

John M. Wettroth
Maxim Integrated Products, Cary, NC 27519, USA

Gary C. White
Department of Fish, Wildlife, and Conservation Biology, Colorado State University, Fort Collins, CO 80523, USA, e-mail: gwhite@warnercnr.colostate.edu

Scott R. Winterstein
Michigan State University, Department of Fisheries and Wildlife, 13 Natural Resources Building, East Lansing, MI 48824, USA

Nigel G. Yoccoz
Department of Biology, University of Tromsø, 9037 Tromsø, Norway, e-mail: Nigel.Yoccoz@ib.uit.no

Section I
Population Dynamics – Growth, Density-Dependence and Decomposing λ

Paul Doherty and David Fletcher

Bayesian Hierarchical Models for Inference About Population Growth

Richard J. Barker, Matthew R. Schofield, Doug P. Armstrong, and R. Scott Davidson

Abstract Mark recapture models have long been used for estimating wildlife population parameters. Typically, the data are summarized in terms of parameters that are interpreted in the context of an implicit demographic model for describing population dynamics. Usually, this demographic model plays little or no role in the mark-recapture model. Bayesian hierarchical models (BHM) offer a way to explicitly include demographic models in an analysis. We argue that such an approach should have wide appeal to ecologists as it allows inference to focus on ecological models of interest rather than obtaining a parsimonious depiction of the sampling process. We discuss the use of BHM's for modeling mark-recapture data with a focus on models describing density-dependent growth.

1 Introduction

Ecologists interested in population dynamics of wildlife populations typically work with two kinds of models: demographic models, in which predicted population trajectories are obtained conditional on parameter values and statistical models in which parameter estimates are obtained using data sampled from the study population.

Demographic models may be matrix- or individual-based and can be deterministic or stochastic (Williams et al. 2002). Whether population models are deterministic or stochastic they depend on parameters. Formally, we can write our model as $\mathcal{F}(Z;\lambda)$ where Z represents the output and λ represents demographic parameters. In using demographic models our interest lies in predicting future population behavior, usually in terms of summaries of Z such as extinction rate or equilibrium population size.

Statistical models are used to summarize data. We can formalize a statistical model as $\mathcal{F}(Y;\pi)$ where Y represents data and π parameters involved in describing

R.J. Barker (✉)
Department of Mathematics and Statistics, University of Otago, P. O. Box 56, Dunedin, New Zealand
e-mail: rbarker@maths.otago.ac.nz

D.L. Thomson et al. (eds.), *Modeling Demographic Processes in Marked Populations*, Environmental and Ecological Statistics 3, DOI 10.1007/978-0-387-78151-8_1,

the sampling process. Note that π might include some or all of the demographic parameters in λ. Once the data have been collected they are fixed; we use the statistical model to describe the mechanism by which we imagine our data were generated. We use our data and the model to obtain information about the parameters. In the model, Y is regarded as random outcome sampled from $\mathcal{F}(Y;\pi)$ with the parameters π fixed at (usually) unknown values.

In the context of the EURING technical meetings, the mark-recapture model is an important class of statistical model. A strong tradition of the EURING technical meetings has been the stimulation of the development of mark-recapture analysis, to the extent that one of the main aims of these meetings has been to establish mark-recapture as one of the standard methodologies in ecology and conservation biology (Senar et al. 2004b). The past 20 years have seen the development of models appropriate for analyzing almost every conceivable type of mark-recapture data, and the development of powerful software such as MARK (White and Burnham 1999), MSURGE (Choquet et al. 2004) and POPAN (Arnason and Schwarz 2002). Obviously, the development of mark-recapture modeling is not an end in itself. The value of mark-recapture models lies in their application. "I note that although EURING conferences have focused on estimation issues, it is important to recall that estimation is not a 'stand-alone' activity or an inherently useful endeavor and attains value primarily in the context of larger processes, such as science or management" (Nichols 2004).

Applications of mark-recapture models have tended to focus on estimation as a means for summarizing status of populations, for example the MAPS program in North America (Tautin et al. 1999), or the interpretation of vital rates and factors influencing these e.g. (Catchpole et al. 1999; Conroy et al. 2002; Reed 2004).

Implicit to a mark-recapture model is a demographic model that describes the population dynamics of the study population, at least in part. For example, in the Cormack–Jolly–Seber model (Cormack 1964; Jolly 1965; Seber 1965), the number of marked survivors alive at occasion $i + 1$ is a binomial random variable with index being the number of marked animals in the population immediately after sample i and probability the survival rate for interval $[i, i + 1)$. The implied demographic model has tended to play little part in subsequent analysis, although the most recent EURING conference (Senar et al. 2004a) saw the appearance of a number of papers that focused on assessment of population dynamics based on mark-recapture data (e.g., Caswell and Fujiwara 2004; Francies and Saurola 2004; Gauthier and Lebreton 2004; Brooks et al. 2004).

Caswell and Fujiwara (2004) stressed the potential benefits of explicitly incorporating demographic models into a mark-recapture analysis by "... making the estimation of demographic models a goal at the outset of a mark-recapture study". A key advantage to integrating demographic and statistical models, of which mark-recapture models are an example, is that it allows full expression of uncertainties.

Typically, demographic modelers have used fixed values for parameters in their models. For example, Francis and Saurola (2004) used a deterministic model based on mean parameter values to construct a deterministic predator-prey model to make predictions about tawny owl *Strix aluco* demographics. While this approach allows general predictions to be made, it does not provide any measure of prediction uncertainty associated with the fact that parameters must be estimated. Conditioning on a set of parameter estimates allows one to consider the implications of this particular set of parameter values (Caswell and Fujiwara 2004). To assess the influence of parameter uncertainty Caswell and Fujiwara (2004) discusses use of "perturbation analysis" based on derivatives of population summaries, or functions of these, such as sensitivities and elasticities.

An alternative to expression of parameter uncertainty using calculus and sensitivities is to use probability distributions. The use of probability distributions to describe uncertainty is a defining feature of Bayesian inference. In Bayesian inference prior probability distributions for parameters are combined with models for data to construct posterior distributions for parameters and posterior predictive distributions for predicted values. These posterior distributions express the uncertainty that we have about parameters and associated predictions after the data have been collected. Instead of focusing on the implications of a particular set of parameter estimates for projected population growth, the Bayesian approach allows us to consider a range of plausible parameter values with the contribution of any particular combination weighted by its posterior density.

The use of posterior distributions to summarize knowledge about parameters is convenient if interest is in exploring the demographic consequences of certain choices of parameter values in demographic simulations. Predictions under the demographic model can be made by sampling plausible values for parameters from the posterior distributions generated by the statistical model. Alternatively, we can combine the demographic and sampling models to obtain a fully integrated analysis.

A specific advantage of Bayesian inference procedures is that Bayesian models are naturally hierarchical. Hierarchical models have several levels of variability. In a Bayesian model we have data which depends on parameters that are themselves drawn from a distribution that also has parameters. The term hyperparameters is often used to describe parameters for distributions of parameters. Bayesian hierarchical models offer a way to formally integrate statistical models for estimating parameters with simulation models for predicting the likely future behavior of populations based on sample data. Mark-recapture models are also naturally hierarchical in that parameters such as survival probability and abundance are often modeled as random variables by demographers. Moreover, ecologists are often interested in relationships among parameters, such as density-dependent survival or recruitment, that have major implications for predicted population trajectories. Hierarchical mark-recapture models offer a way to model all sources of data as well as to model relationships among parameters (Link and Barker 2004) in a way that allows all posterior uncertainty, including uncertainty about predictions, to be expressed using probability distributions.

In this paper we use a case study of North Island saddlebacks (*Philesturnus rufusater*) to illustrate the use of Bayesian hierarchical modeling to predict population trajectories based on a density-dependent population model. Our emphasis is on accounting for parameter uncertainty conditional on the model. Methodology for multi-model inference in a Bayesian setting has been well covered elsewhere (Brooks et al. 2004).

2 North Island Saddlebacks of Mokoia Island

The North Island saddleback is a member of the wattlebird family (Callaeidae), a family of birds endemic to New Zealand. The Callaeidae comprise two or three (depending on taxonomic fashion) extant species. By the end of the nineteenth century saddlebacks had become locally extinct from the mainland of New Zealand: a single remnant population survived on one island off the coast of the North Island (Armstrong et al. 2005) and a single remnant population survived on one island off the coast of the South Island. Since the early 1960 s translocations have been used to re-establish populations including some mainland populations.

Armstrong et al. (2005) used mark-resighting analysis and counts of fledged and unmarked birds to monitor the outcome of a translocation to Mokoia Island in Lake Rotorua. Thirty six adult birds were released on the 135-ha island in 1992 following rat eradication. Mokoia Island is 2.1 km from the nearest shore, a distance believed to be beyond the flying range of saddlebacks. The translocated population is thus believed to be closed to emigration.

Armstrong et al. (2005) were interested in predicting future population growth of the Mokoia Island saddleback population to assess future population viability and to devise strategies for translocating birds from re-established populations to places elsewhere. In particular, they were interested in evidence for density-dependent population dynamics as the presence of density dependence can have a major stabilizing influence on dynamics. Re-introductions are useful for studying density dependent growth because founding populations are usually established at low population densities and with a relatively high level of resource availability.

In their analysis, Armstrong et al. (2005) used a stochastic matrix model to project the likely future trajectory of the Mokia Island saddleback population. Their model tracked the number of males and females in the population at the start of each breeding season and used estimates of survival and fledging rates from analyses of their banding data and from fledgling surveys. Because Armstrong et al. (2005) conditioned on these parameter estimates, their projection model does not account for all the uncertainties in the analysis. Also, Armstrong et al. (2005) treated abundance estimates obtained in their analysis as fixed, using these to assess density-dependent effects on survival and juvenile production rates. In addition to unmodeled uncertainty there is also some sampling correlation unaccounted for between the abundance and survival probability estimates.

With multiple sources of data and with interest in modeling parameters such as survival and production rate as a function of population size, this problem is ideally suited to an approach based on fitting a hierarchical model using Bayesian model fitting procedures.

3 Data and Models

3.1 Mark-Resighting Data

The 36 founder birds were color-marked before release. Mark-resighting data were obtained from 24 surveys carried out between June 1992 and December 1997 at approximately 3-month intervals: in March, June, September, and December. The marked population comprised the 36 founding birds and 245 nestlings that were banded during the five breeding seasons that took place during the study. Nestlings were banded either in December or March according to when they were born. In October 1996 an aerial poison drop was used to try and eradicate mice from the island. An additional mark-resighting survey was carried out in November 1996 to assess the effect of the poison drop on the saddlebacks. During the mark-resighting surveys counts of all unmarked birds were also recorded.

To analyze the mark-recapture data we followed Armstrong et al. (2005) and fitted a simple 2-age model in which juveniles became adults after 9 months. The logit of the 3-month survival probabilities were modeled as a linear function of effects due to the age of the bird (juvenile/adult), poison drop, and the number of breeding pairs. We did not consider a sex-specific model as the analysis of the mark-resighting data by Armstrong et al. (2005) found negligible support for sex-specific survival.

To account for unexplained variation in three month survival probabilities over time we included a normal $N(0, \sigma^2)$ random effect. Because the intervals were not exactly 3-months, interval-specific survival probabilities were adjusted for the length of the interval.

Let y_j denote the length of time between survey j and survey $j+1$ (3 months = 1.0), β_0 the overall mean 3-month survival probability on the logit scale, $Z_{1i} = 1$ for if individual i is juvenile and 0 if it is adult, $Z_{2j} = 1$ if there was a poison drop in the 3-month interval starting at the time of survey j and zero otherwise, and N_j denotes the number of breeding pairs in the breeding season associated with survey j, then:

$$\phi_{ij} = S_{ij}^{y_j}$$

and

$$\text{logit}(S_{ij}) \sim N(\beta_0 + \beta_1 Z_{1i} + \beta_2 Z_{2j} + \beta_3 Z_{1i} Z_{2j} + \beta_4 \ln(N_j), \sigma^2).$$

where ϕ_{ij} is the interval specific survival probability and S_{ij} the 3-month survival probability. Detection probabilities ($p_{ij}; i = 2 \ldots 25, j = 1$ for juveniles, $j = 2$ for adults) we modeled as age- and time-specific fixed effects.

For model fitting we used the complete data likelihood (Schofield and Barker 2008) which is proportional to:

$$[X|d, p, R][d|S, R][S|\beta, Z\sigma^2]$$

where X denotes the mark-recapture set, d the vector of times of death for each individual, R a vector indicating when birds were first released, p the detection probabilities, and S the survival probabilities. We use the notation $[Y]$ to denote the probability (density) function of the random variable Y. The completion step involves the model $[d|S]$ for the times of death for each bird. These are interval-censored, known up to the sample interval, and also left censored by the end of the study.

3.2 Production of Young

To assess the number of young produced by breeding pairs, all known nests were monitored in the first 3 years and a sample of 30–35 pairs from known nests in the last 2 years. Chicks were banded between 10 and 21 days after hatching and counted as fledged if observed in the nest at least 14 days after hatching (chicks fledge at about 28 days).

Pairs from unknown nests were also monitored and the number of fledglings counted. Because fledglings can die before their parents are detected, these counts would have underestimated the number of fledglings produced. Data from pairs that used nestboxes and that were also detected in the post-fledgling survey were used to estimate the joint probability of the parents being found and the fledglings surviving until their parents were found. These data were used to adjust the number of fledglings produced by pairs nesting at unknown sites by modeling f_{ij}^{obs}, the observed number of fledged young for pair i from an unknown nest site in year j, as a binomial random variable with index f_{ij}, the true number of young fledged by pair j and probability ξ_j.

For all pairs, we modeled f_{ij} as a Poisson $\mathcal{P}(\lambda_{ij})$ random variable with mean depending on the number of breeding pairs in September of year j (N_j), the combined age of the pair (A_{ij} = the number of adults in the pair), the length of time the pair had been established P_i, and a random pair effect. That is,

$$\ln(\lambda_{ij}) = \alpha_{0i} + \alpha_1 N_j + \alpha_2 A_{ij} + \alpha_3 P_i$$

where $\alpha_{0i} \sim N(0, \sigma_f^2)$. Because there was a known lower bound on the numbers of birds fledged from the fledgling survey we censored the total number of birds fledged at this lower bound.

To allocate new fledglings as additions to the December or April population we modeled the number fledged in December as a binomial random variable, with index given by the total number fledged and with probability π_{Dec} assumed to be the same each year. Information on δ was obtained by modeling the observed number of birds fledged in December each year conditional on the number observed fledging in December or April that year. Sex for each individual was modeled as a Bernoulli random variable with parameter π_{sex}. The sex-ratio of fledged birds was assumed to be the same as the sex-ratio of marked fledglings.

3.3 Sightings of Unmarked Birds

Sex- and age-specific counts of unmarked birds in sample j were modeled as binomial random variables with indices given by the sex- and age-specific number of birds in the population and probability equal to the detection probability in the mark-resight model for sample j. That is, we assumed that marked and unmarked birds were seen at the same rate. The binomial indices are unknown, but were determined by the (model-based) predicted number of unmarked fledglings surviving until the time of sample j.

4 Model Fitting

The model was fitted to the 4 sets of data jointly using WinBUGS (Spiegelhalter et al. 2002) and priors as given in Table 1. Except for density dependence, we used uninformative priors. We restricted density-dependent effects to be negative.

A sample of size 100,000 was drawn from the joint posterior distribution of all parameters and predicted quantities (see below) after discarding a burn-in sample of size 10,000.

For the marked subset of the population the interval-censored time of death (i.e., time of death is known only up to the period in which the bird died) was predicted as part of the data completion step. We also predicted the numbers of unmarked survivors at each sample assuming that all juveniles (marked or unmarked) had the same survival probability and all adults (marked or unmarked) had the same survival probability. From the predicted number of marked and unmarked survivors we could predict the numbers of birds at the time of each sample classified according to age and sex.

Table 1 Priors used for parameters in the joint model for saddleback mark-recapture and count data

Parameter	Prior
Mark-resight model	
β_0	Normal $N(0, 10{,}000)$
β_1	Normal $N(0, 10{,}000)$
β_2	Normal $N(0, 10{,}000)$
β_3	Normal $N(0, 10{,}000)$
β_4	Uniform $U(-5, 0)$
σ^2	Inverse-gamma $IG(0.001, 0.001)$
p_{ij}	Beta $Be(1, 1)$ ($i = 2, \ldots, 25; j = 1,2$)
Model for number of fledglings	
α_1	Normal $N(0, 10, 000)$
α_2	Normal $N(0, 10, 000)$
α_3	Normal $N(0, 10, 000)$
σ_f^2	Inverse-gamma $IG(0.001, 0.001)$
π_{Dec}	Beta $Be(1, 1)$
π_{sex}	Beta $Be(1, 1)$

To assess the likely trajectory of the saddleback population on Mokoia Island we also predicted the numbers of saddlebacks present in September for 100 years following the last survey on the island in December 1997. For these projections we ignored the effect of pair age and year of establishment, effectively assigning all pairs the average value for pair age and year of establishment. We did this to facilitate model fitting using WinBUGs which does not allow the reversible-jump (Green 1995) step needed to properly implement such an individual-based model.

To assess the likely effect of regular harvesting we also examined the population trajectory when varying numbers of birds are removed each year from 1999 onwards. Dimond and Armstrong (2007) suggested that about 30 birds could be safely removed each year to establish populations elsewhere. The number of males that were removed each year was modeled as a binomial random variable with index 30 and probability π_{sex}. We also looked at the effects on extinction rate when birds were removed at 3-yearly intervals.

5 Results

Predicted abundances during the study were in close agreement with those of Armstrong et al. (2005), although our credibility intervals tend to be wider (Fig. 1). The saddleback population on Mokoia Island quickly increased from the 14 females

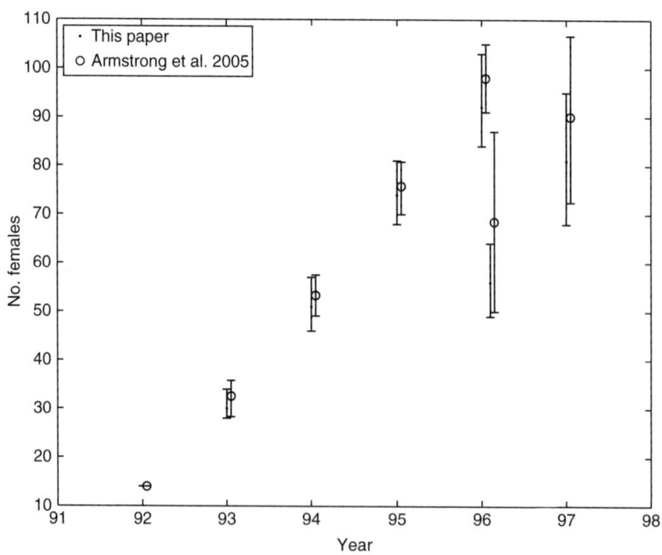

Fig. 1 Predicted number of adult female saddlebacks on Mokoia Island between September 1992 and September 1997. The error bars indicate the range of the central 95% credible interval. The estimates (and confidence intervals) of Armstrong et al. (2005) are included for reference

released in 1992 to around 90–100 females in September 1996, dropped sharply following the poison drop in October 1996 but then quickly recovered.

The acute effect of the poison operation is evident in the estimates of the survival effects (Fig. 2). Both adult and juvenile birds died as a result of the poison drop but the positive interaction between age and the poison effect (β_3) suggests that the poison drop effect was not as strong for juveniles as it was for adults. Juvenile birds survived at a lower rate than adults, with a 95% credibility interval (CI) of (0.88, 0.93) for non-poison intervals dropping to (0.36, 0.84) for the poison interval compared to (0.98, 0.99) for adults during non-poison intervals dropping to (0.63, 0.80) for the poison interval.

Our analysis also agreed with that of Armstrong et al. (2005) in finding strong evidence that production rates were higher for older birds and on territories established in the first three years (Fig. 3), as well as varying between pairs for unexplained reasons (95% CI for $\sigma_f = 0.03, 0.33$)

Our analysis confirmed that of Armstrong et al. (2005) in finding strong evidence of density dependent juvenile survival and per-capita production of young as well as evidence supporting density-dependent survival of adult saddlebacks (Fig. 4). This

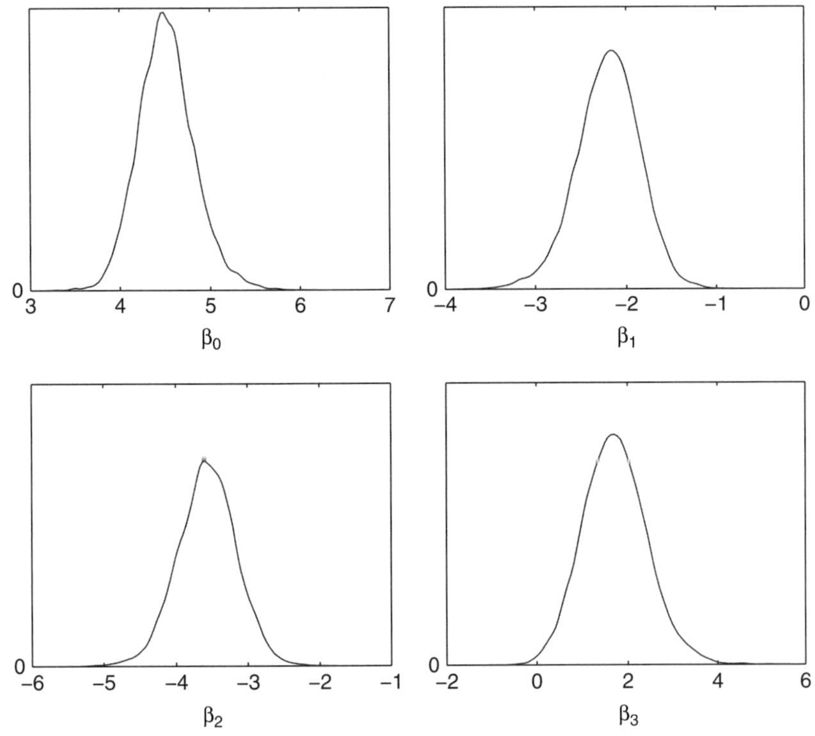

Fig. 2 Posterior density plots for the logit of expected 3-month survival probability for adult birds in a non-poison period (β_0), the juvenile effect on survival (β_1), the effect of the poison drop on adults (β_2) and the interaction between the age effect and the poison effect (β_3)

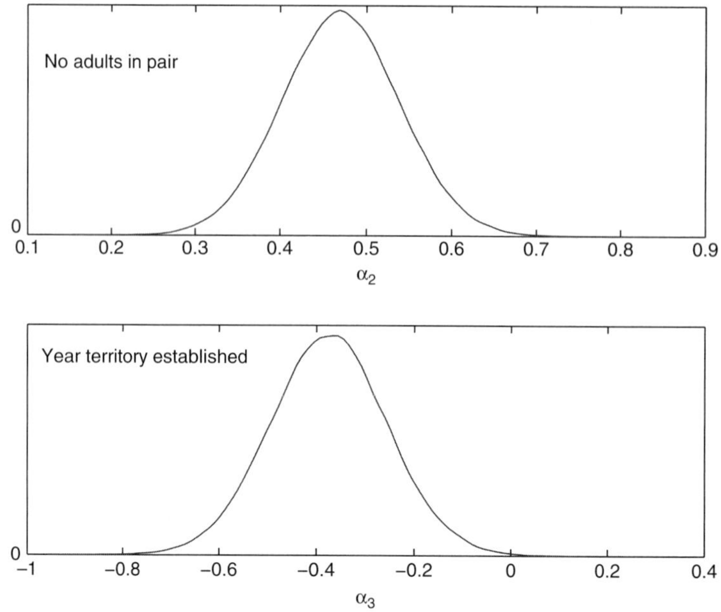

Fig. 3 Posterior density plots for the effects of α_2, the effect of the standardized number of birds of age 1^+ in the pair, and α_3, the effect of whether or not the pair were on a territory established in the first year of the study

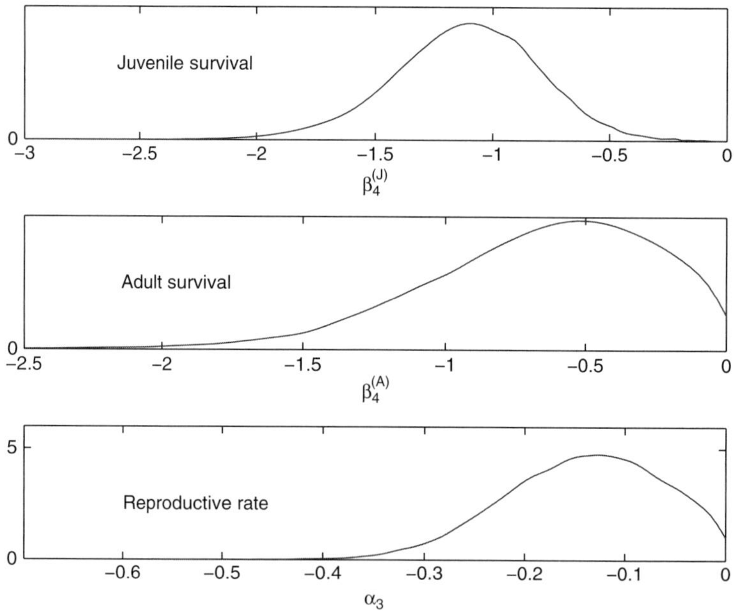

Fig. 4 Posterior density plots for the effect of log population size on juvenile survival rate ($\beta_4^{(J)}$), adult survival rate ($\beta_4^{(A)}$) and per-capita juvenile production rate (α_3)

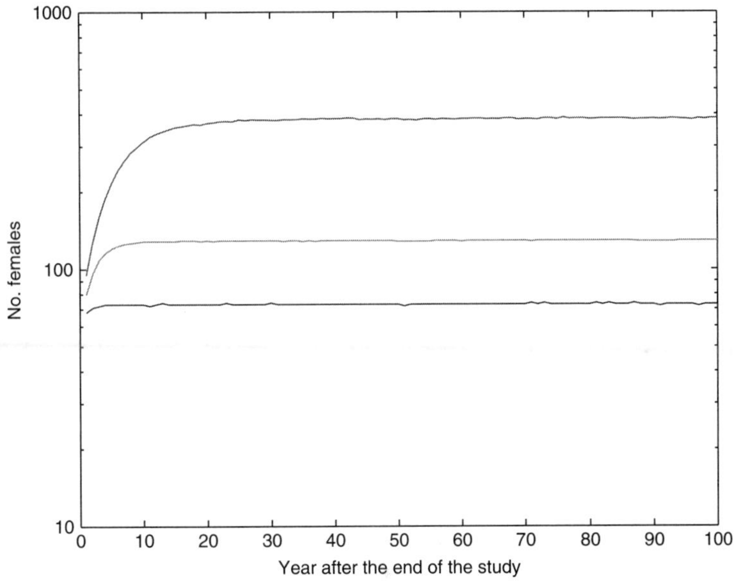

Fig. 5 Predicted number of adult female saddlebacks on Mokoia Island for the 100 years following the end of the study in 1997 and September 1997. The lower of the three lines is the 2.5% percentile, the middle the median and the upper the 97.5% percentile

density-dependence has important consequences for likely future population trajectories. Our analysis suggests that the population would quickly reach an equilibrium of around 130 females after about 10–15 years (Fig. 5), with the 2.5% value around 70 females and the 97.5% percentile around 400 females. The first 10 years of this trajectory shows that the projected development of the population follows a similar trend to that observed between 1992 and 1997 (Fig. 6).

None of our simulated population trajectories dropped below 10 females suggesting that the risk of extinction of the re-established population is small, all other things being equal. However, extinction occurred in around 5% of simulations when the population was subjected to a regular harvesting of 30 birds per year increasing to around 30% if 40 birds were removed (Fig. 7). If only 20 birds were removed each year the rate of extinction would be near zero. If simulations were based on parameter estimates (we used the median of the posterior sample) rather than the full sample then extinction risk was predicted to be near zero if fewer than 40 birds were removed each year but rapidly increased to near one if 50 birds or more were removed each year. The bias evident in the predicted extinction risk function arises because the effects of density dependence may be much weaker or stronger than indicated by the parameter estimates. If weaker, then even moderate removals will enhance extinction risk but if stronger then the population could sustain a higher level of harvesting. Note that if removals are carried out at 3-yearly intervals a much higher level of harvest could be sustained; our simulations indicated that up to 80 birds could be removed while maintaining extinction risk below 5% (Fig. 8).

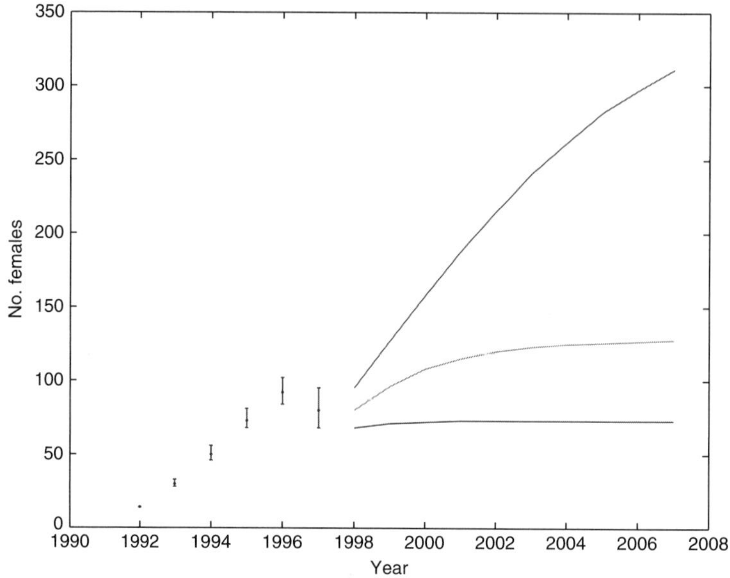

Fig. 6 Predicted number of adult female saddlebacks on Mokoia Island for the 10 years following the end of the study in 1997 with the estimates from 1992 to 1997 included. The error bars indicate the range of the central 95% credible interval. For the population projection the lower of the three lines is the 2.5% percentile, the middle the median and the upper the 97.5% percentile

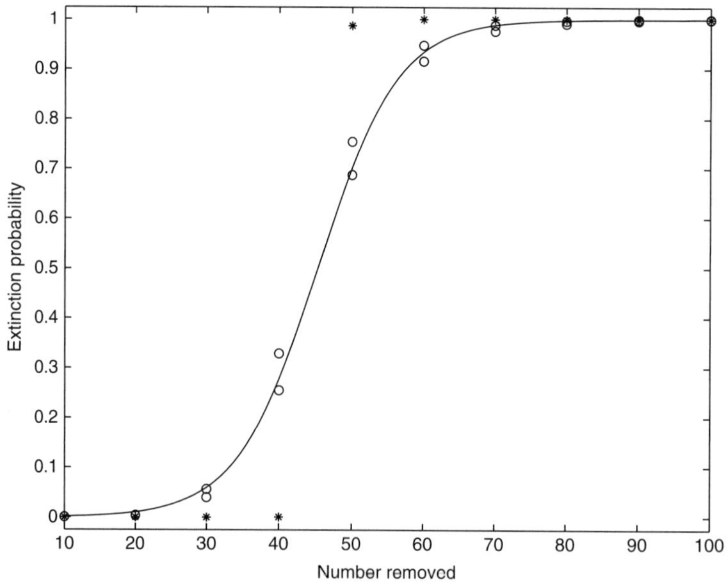

Fig. 7 Predicted rate of extinction when varying numbers of birds are removed each year from 1999 onwards (*circles* and *fitted line*) compared to rate of extinction predicted conditioning on the median values from the posterior simulations (*asterisks*)

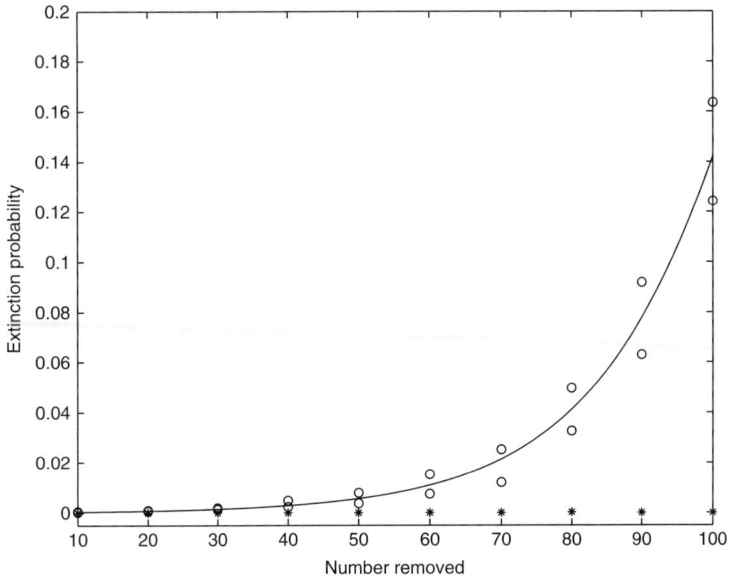

Fig. 8 Predicted rate of extinction when varying numbers of birds are removed every three year from 1999 onwards (*circles* and *fitted line*) compared to rate of extinction predicted conditioning on the median values from the posterior simulations (*asterisks*)

6 Discussion

In any modeling or inference problem there are a number of components about which we are uncertain, including values for the parameters in the model and often the exact form of the model itself. In addition, we might also be uncertain about relationships that exist among our parameters or between the parameters and external covariates. Inferences about population vital rates and subsequent population growth, must account for sampling variation in estimates that are based on sample data (Williams et al. 2002).

The analysis we have outlined for the Mokoia Island saddleback population, in which all sources of data are modeled simultaneously and population projections are considered as data-based predictions, is necessary if all uncertainties about parameters are to be expressed in population models. Although our analysis confirmed the essential findings of Armstrong et al. (2005) an important difference is that we did not condition on point estimates of effects on survival and production rate in our population projections and we did not treat abundance estimates as known. The effect of this will be an appropriate increase in uncertainty about the eventual equilibrium level of the saddleback population and the rate at which it approaches this value.

Because the banding study was discontinued in 1997, there are no reliable data to indicate how the saddleback population has fared since 1997, although Armstrong et al. (2005) reported that a survey of the island carried out in 2002 recorded 177 birds. As this was the raw count the true population size is likely well in excess of 177 birds. This suggests that our projected equilibrium value of around 130 females, and with no fewer than 70 females, is a reasonable prediction.

There are a number of ways in which we could improve our analysis. One useful step would be to base the population predictions on a fully individual-based model. Although we modeled the fate of marked individuals using an individual model we did not do so for the unmarked individuals. As noted in the model description section, implementation of such an individual-based model requires a reversible-jump step, something presently beyond the capability of WinBUGS. The consequence of ignoring individual effects is that apparent posterior uncertainty is understated. An alternative approach would be to refit the model leaving out the effects of pair-age and the time that the territory was established and allowing for these to be accounted for through the random pair effect.

In our analysis we included no model selection. In most applications there is likely to be uncertainty about which effects to include in the model and how to best express relationships among variables. The correct way to include this additional level of uncertainty would be through multi-model inference. Effective multi-model inference using hierarchical models will almost certainly require algorithms such as reversible-jump McMC (e.g., Brooks et al. 2004) to generate posterior predictions that are suitably averaged across the models considered.

References

Armstrong D, Davidson RS, Perrott JK, Roygard J, Buchanan L (2005) Density-dependent population growth in a reintroduced population of North Island saddlebacks. Journal of Animal Ecology 74:160–170.

Arnason AN, Schwarz CJ (2002) POPAN-6: exploring convergence and estimate properties with SIMULATE. Journal of Applied Statistics 29:649–668.

Brooks SP, King R, Morgan BJT (2004) A Bayesian approach to combining animal abundance and demographic data. Animal Biodiversity and Conservation 27.1:515–529.

Caswell H, Fujiwara M (2004) Beyond survival estimation: mark-recapture, matrix population models, and population dynamics. Animal Biodiversity and Conservation 27.1: 471–488.

Catchpole EA, Morgan BJT, Freeman SN, Peach WJ (1999) Modelling the survival of British lapwings *Vanellus vanellus*using ring-recovery data and weather covariates. Bird Study 46: S5–S13.

Choquet R, Reboulet, A-M, Pradel R, Gimenez O, Lebreton JD (2004) M-SURGE: new software specifically designed for multistate capture–recapture models. Animal Biodiversity and Conservation 27.1:207–215.

Conroy MJ, Senar JC, Domèenech J (2002) Analysis of individual- and time-specific covariate effects on survival of *Serinus serinus* in north-eastern Spain. Journal of Applied Statistics 29:125–142.

Cormack RM, (1964) Estimates of survival from the sighting of marked animals. Biometrika 51:429–438.

Dimond WJ, Armstrong DP (2007) Adaptive harvesting of source populations fortranslocation: a case study with New Zealand robins," Conservation Biology 21:114–124

Francis CM, Saurola P (2004) Estimating components of variance in demographic parameters of tawny owls, *Strix aluco*. Animal Biodiversity and Conservation 27.1:489–502.

Gauthier G, Lebreton JD (2004) Population models for greater snow geese: a comparison of different appraoches to assess potential impacts of harvest. Animal Biodiversity and Conservation 27.1:503–514.

Green PJ (1995) Reversible jump Markov chain Monte Carlo computation and Bayesian model determination. Biometrika 82:711–732.

Jolly G (1965) Explicit estimates from capture–recapture data with both death and immigration-stochastic model. Biometrika 52.

Link WA, Barker RJ (2004) Hierarchical mark-recapture models: a framework for inference about demographic processes. Animal Biodiversity and Conservation 27.1:441–449.

Nichols JD (2004) Evolution of quantitative methods for the study and management of avian population: on the importance of individual contributions. Animal Biodiversity and Conservation 27.1:3–19.

Reed ET, Gauthier G, Giroux J-F (2004) Effects of spring conditions on breeding propensity of greater snow goose females. Animal Biodiversity and Conservation 27.1:35–46.

Schofield MR, Barker RJ (2008) A further step toward the mother-of-all-models: flexibility and functionality in the modeling of capture–recapture Data. In: Thomson DL, Cooch EG, Conroy MJ (eds.) Modeling Demographic Processes in Marked Populations. Environmental and Ecological Statistics, Springer, New York, Vol. 3, pp. 677–690.

Seber GAF (1965) A note on the multiple-recapture census. Biometrika 52:249–259.

Senar JC, Dhondt A, Conroy MJ (2004a) Proceedings of the International Conference Euring 2003, Deutschland." in Animal Biodiversity and Conservation 27.1.

Senar JC, Dhondt AA, Conroy MJ (2004b) The quantitative study of marked individuals in ecology, evolution and conservation biology: a foreward to the EURING 2003 conference. Animal Biodiversity and Conservation 27.1:1–2.

Spiegelhalter DJ, Best NG, Carlin BP, van der Linde A (2002) Bayesian measures of model complexity and fit. Journal of the Royal Statistical Society Series B 64:583–639.

Tautin J, Metras L, Smith G (1999) Large-scale studies of marked birds in North America. Bird Study 46:S271–S278.

White GC, Burnham KP, (1999) Program MARK: survival estimation from populations of marked animals. Bird Study 46:S120–S139.

Williams B, Nichols J, Conroy M (2002) Analysis and management of animal populations. Academic Press, New York.

Assessing Density-Dependence: Where Are We Left?

Jean-Dominique Lebreton

Abstract The history of density dependence started in 1798 with Malthus' sentence: *population, when unchecked, increases in a geometrical ratio*. The famous controversy between Lack, Andrewartha and Birch and others in the 1950s and 1960s remained largely unsolved: while the impossibility of long term exponential growth required density-dependence, density-independent environmental variation in vital rates was often dominant in empirical studies. Fifty years later, where are we left? I revisit first the representation of density-dependence in dynamical models, whether deterministic or stochastic, and I emphasize the lack of theory for the simultaneous occurrence of density-dependence and environmental variation. I then review approaches to detect and measure the intensity of density-dependence, in two steps: based on population size estimates and in demographic parameter analyses. I discuss then how the question of density-dependence could be efficiently revisited, taking advantage of progress in our understanding of spatio-temporal dynamics, statistical procedures, access to individual characteristics, and possibilities of experimental approaches.

Keywords Population dynamics · Density-dependence · State-space modeling

1 Introduction

The history of density-dependence started with Malthus (1798) famous sentence: *population, when unchecked, increases in a geometrical ratio*. Deterministic population models with density-dependence (Verhulst 1838; Nicholson and Bailey 1935) were the major landmarks of a long period that could be called the embryonic

J.-D. Lebreton (✉)
CEFE UMR 5175, CNRS, 1919 Route de Mende, 34 293 Montpellier cedex 5, France
e-mail: jean-dominique.lebreton@cefe.cnrs

D.L. Thomson et al. (eds.), *Modeling Demographic Processes in Marked Populations*, Environmental and Ecological Statistics 3, DOI 10.1007/978-0-387-78151-8_2,

growth of the subject. Then, in the 1950s and 1960s, Lack (1954), Andrewartha and Birch (1954) and others (Wynne-Edwards 1962) discussed the relative role of density-dependent and density-independent variation in vital rates, in a controversy that remained largely unsolved: while the impossibility of long term exponential growth required density-dependence, density-independent environmental variation in vital rates was often dominant in empirical studies. Parallel discussions in the very different context of fishery research (Ricker 1954; Beverton and Holt 1957) lead also to attribute a prominent role to density-dependence, which is still at the core of the stock-recruitment relationship paradigm (Haddon 2001, Chapter 9).

Until recently, the subject of density-dependence seemed then to be avoided or hardly touched upon in the literature, as if a moratorium had been imposed on it, although recent discussions (Berryman 2002; Berryman et al. 2003; Berryman 2004) in particular on "local density dependence" (Murdoch 1994; Rodenhouse et al. 1997) seem to announce a revival. This would be fortunate, as density-dependence is critical to many different issues in population biology, notably selection regimes and the diversification of life histories (in particular with r–K selection; Boyce 1984) in evolutionary biology, and pest and quarry species management (in particular with the issue of harvest compensation; Burnham and Anderson 1984) in applied population ecology.

Fifty years after the Lack–Andrewartha–Birch controversy, where are we left? I attempt here, with unavoidable personal biases, to revisit one major aspect of density-dependence, namely methods of empirical assessment of density-dependence. The huge evolution of empirical population models, both statistical (Lebreton et al. 1992; Williams et al. 2002) and dynamical (Tuljapurkar 1990; Caswell 2001), certainly opens new possibilities of efficiently addressing the empirical assessment of density-dependence as well as many other questions raised by density-dependence.

As a preliminary step to set up the scene, I revisit first briefly the representation of density-dependence in dynamical models and I emphasize the need of considering simultaneously density-dependence and environmental variation (Section 2). I then review approaches to detect and measure the intensity of density-dependence based on population size estimates (Section 3). We will see that this apparently easy task is still frequently the subject of mistakes. I show then how state space models can be used in density-dependence assessment, despite open statistical questions. I go on with methods for assessing density-dependence in demographic parameter analyses (Section 4). The latter two subjects are affected in strikingly different ways by uncertainty in population size estimates.

I review then (Section 5) the potential of state-space models and discuss how the biological questions related to density-dependence could then be efficiently revisited, by taking advantage of progress in our understanding of spatiotemporal dynamics, statistical procedures, access to individual characteristics, and possibilities of experimental approaches.

I use throughout this paper a simple model to emphasize underlying ideas.

2 Density-Dependence and Stochasticity in Dynamical Models

By density dependence in a model population, what is usually meant is a variation in growth rate with population size. A general form of discrete time density-dependent model is:

$$N_{t+1} = g\left(N_t\right) N_t,$$

assuming for simplicity at this stage that N_t is a scalar. For the sake of simplicity too, I restrict my attention to "direct" density-dependence, i.e., I consider $g\left(N\right)$ as a monotonous decreasing function of N. Inverse density-dependence is inherent in the Allee effect (Courchamp et al. 1999). Among many different models, a straightforward one is the deterministic (discrete time) Gompertz model (DG) (May et al. 1974; Hassell 1975; Royama 1992, p. 48):

$$N_{t+1} = \lambda N_t^{-b} N_t, \tag{1}$$

which has the advantage of translating into a linear model for log population size $X_t = Ln\left(N_t\right)$ as (denoting $Ln\left(\lambda\right)$ as r):

$$x_{t+1} = r + \left(1 - b\right) x_t \tag{2}$$

or

$$z_{t+1} = x_{t+1} - x_t = r - bx_t \tag{3}$$

In passing, one should note that $r = Ln\left(\lambda\right)$ is the growth rate for $N = 1$, while the growth rate for $N = 0$ is infinite (Royama 1992, p. 50). This slight artifact will not interfere with our use of this model, as we will always assume $N\left(0\right) > 1$. From (2), one can easily show that if $0 < b < 2$, which we will assume in all what follows, the model population size converges asymptotically to a stable equilibrium $K = \exp\left(r/b\right)$ (Fig. 1).

By density-dependence in a natural population, what is generally meant is a decrease in individual demographic performance induced by an increase in density, generally local density (through a number of different potential mechanisms: increase in time spent in agonistic interactions, in time spent traveling farther away to collect food, decrease in chances of getting a proper nest site, etc.). However, animals or plants do not measure their density; hence their demographic performance is at most sensitive to some unknown amount of interactions between individuals or of accessible resource per individual: population size cannot be the actual determinant of demographic performance. This ambiguity has led to a lot of confusion around the concepts of regulation (i.e. density-dependence in model populations) and environment-mediated limitation (i.e. density dependence in natural populations) (Berryman 2004). Even in extreme cases, demographic traits

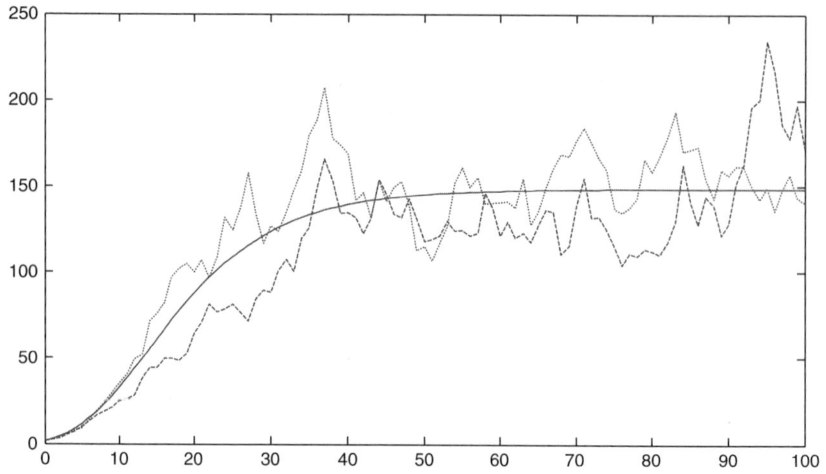

Fig. 1 Trajectory of the deterministic discrete time Gompertz model (*continuous line*) $N_{t+1} = e^r N_t^{-b} N_t$, under $r = 0.5$ and $b = 0.1$, and two trajectories (*dotted lines*) of the stochastic counterpart $N_{t+1} = e^{r+\varepsilon_t} N_t^{-b} N_t$, under $\sqrt{\operatorname{var}(\varepsilon_t)} = \sigma = 0.1$

are only linked in a statistical fashion to density. For instance 93% of variation in productivity of common terns in the North-East US is explained by change in population size (Nisbet, oral comm.). This fact immediately has three consequences:

- Density or population size, even if known without error, are only a proxy for what affects demographic performance.
- When modeling the effect over time in the same population, variation in the environment has to be accounted for as a potential modifier of the effect of density.
- Density-dependence cannot be easily compared across space, as there is no reason to assume the same population size or density in two different sites implies the same level of competitive interactions or of access to resource, and in turn the same level of demographic performance.

Another immediate consequence with direct impact on the topic of this paper is that, whatever their theoretical and conceptual interest, deterministic models of density dependence will be misleading for empirical purposes.

A natural stochastic (discrete time) Gompertz (SG) model is then:

$$x_{t+1} = r + (1 - b)\, x_t + \varepsilon_t \tag{4}$$

or

$$z_{t+1} = x_{t+1} - x_t = r - bx_t + \varepsilon_t \tag{5}$$

in which one assumes $E(\varepsilon_t) = 0$ and $\text{var}(\varepsilon_t) = \sigma^2$ with at this stage no specific assumption on the autocorrelation spectrum of ε_t. This model has wide applicability, even if a more sophisticated functional form for density dependence (Royama 1992, p. 50 ff) might be often preferable: assuming a linear relationship on an appropriate scale is a common simplification for a preliminary study of the relationship between two variables, especially when stochasticity is strong enough to mask details of a potentially more complex relationship. Because of its simplicity, this model has indeed been used repeatedly in the literature on density dependence modeling and assessment (Maelzer 1970; Slade 1977; Dennis and Taper 1994; Dennis et al. 2006). We will come back to the issue of functional form of density-dependence later.

The SG model can also be written as:

$$x_{t+1} - r/b = (1 - b)(x_t - r/b) + \varepsilon_t \tag{6}$$

Thus, if the series ε_t is not autocorrelated, i.e. is a white noise, and if, as in the DG model, $0 < b < 2$, the log-population size varies around its ergodic expectation $U = r/b$ according to an autoregressive process (Fig. 1), with a typical decreasing autocorrelation function directly depending on the value of b. If the series ε_t is autocorrelated, the autocorrelation function of the log-population size will be modified, possibly to a large extent, compared with that of an autoregressive process. In the absence of density-dependence, the model becomes a random walk model with drift r (Hamilton 1994, p. 436), a stochastic process which is no more stationary:

$$x_{t+1} = r + x_t + \varepsilon_t \tag{7}$$

In all cases, as emphasized by Bulmer (1975), it is clear that the information on density-dependence is in the autocorrelation spectrum of the population size series: the empirical assessment of density-dependence pertains to time-series analysis and raises some tricky issues. We will see that this point has often not been taken seriously enough in the literature.

The lack of relevance of deterministic models to density-dependence in natural populations extends partly to chaos in discrete time logistic models. Bifurcations in the deterministic discrete time Ricker model (May 1976) have indeed as counterparts in the stochastic version of the model sharp changes in the autocorrelation function (Texier 1996). A remaining key property of chaotic behavior is the sensitivity of the trajectory to slight changes in initial values (Hastings et al. 1993), which bears a relationship to the predictability of such systems.

Another topic concerns the behavior induced by stochastic components in density-dependent models. A first basic point is that the expected population size in a density-dependent stochastic model is not given by the population size of its deterministic counterpart because of the nonlinearity inherent in density-dependence (Dennis and Taper 1994). There is relatively little in the literature on scalar density-dependent random environment (DDRE) models, in particular in relation to empirical data, with the notable exception of Lande et al. (2003, chapter 1). If demographic stochasticity is also considered, there is even less (Gosselin and Lebreton

2000; Gosselin 2001). There is literally nothing on DDRE models for structured populations, although the effect of random environment on exponential growth models in the case of structured populations has been thoroughly investigated (Tuljapurkar 1989; Tuljapurkar 1990). There is thus a clear need for future research in this area.

3 Detecting and Measuring Density-Dependence from Counts

Population monitoring nearly always incorporates a census or some other kind of population size estimation. As a consequence, there is a great interest for being able to test for density dependence based on time series of population size estimates. However, one should realize that when investigating density-dependence, a population process, based on population size estimates, a result of the many mechanisms inherent in population dynamics, one is seeking information on process based on pattern. It is thus not surprising that many difficulties arise (Murdoch 1994, p. 275 ff).

The shortness of time series often precludes any serious testing power, but this state of affairs is changing with long term programs. Sound procedures, even with low power, may also be useful in meta-analyses of density-dependence, in which results from several populations or different pieces of information from the same population are combined into a single test or statistic. Bias becomes then critical: a small bias, even if negligible in a single study, is contrasted with a much lower standard error, with extremely deleterious consequences, if present in all components of an overall statistic. One may suspect this is the case in Brook and Bradshaw (2006).

We just saw that the information on density dependence based on a series of population size estimates is contained in the autocorrelation spectrum of the series. In an ideal world, under the SG model (6), b is linked to the first order autocorrelation ρ_1 (Box and Jenkins 1976, p. 176) as $b = 1 - \rho_1$. In the real world, any departure from the basic model that modifies the autocorrelation function must be taken into account. This critical statement has several key consequences:

(1) Naïve methods to estimate b should be avoided. Time series methods are needed and must be used.
(2) Uncertainty in population size will modify the autocorrelation and have extremely deleterious effects on the detection of density dependence and must be taken into account.
(3) Environmental variability must be taken into account, with a particular care if there is a risk of autocorrelation in the environment, again because it would modify the autocorrelation function of the series.
(4) Population structure such as age-structure must be taken into account, since it is equivalent to a delay, with, again, consequences on autocorrelation.

Although these warnings have been made repeatedly in the literature, as early as the 1970s (Saint-Amant 1970; Ito 1972; Slade 1977), deleterious consequences of naïve approaches are not yet fully realized. These four points above are developed below.

3.1 Some Time Series Difficulties

Under the SG model, a maximum likelihood estimate (MLE) of b can be obtained (Hamilton 1994, p. 118 ff). A preliminary estimate of b can be derived from that of the estimated first order autocorrelation $\hat{\rho}_1$ (Box and Jenkins 1976, p. 176) as $\hat{b} = 1 - \hat{\rho}_1$. The resulting estimate is always >0, and cannot thus be used to test for H_0 $b = 0$. Hamilton (1994, p. 123) recommends using maximum likelihood conditional on the first observation x_1. This avoids specifying a distribution for x_1, which is then considered as ancillary for the estimation of parameters of interest (Cox and Hinkley 1974, p. 35). The likelihood reduces to the product of the normal densities of x_i conditional on x_{i-1}. ML estimation amounts then to using ordinary linear regression of x_i w.r.t. x_{i-1}; the MLE of b is derived from that of the regression slope c, as $\hat{b} = 1 - \hat{c}$, i.e., apart a trivial transformation, the ordinary regression estimate is the MLE under the SG model: however, this does not imply the MLE benefits from the usual properties of the regression estimates, and even not of those of MLEs under standard conditions (Dennis and Taper 1994), for two reasons:

- First, several assumptions of ordinary regression are not met; e.g. the dependent variable values in the regression are not independent;
- Second, the model is a non-stationary stochastic process, and ML theory for dependent observations fails then to provide strong results.

As a consequence, under H_0 $b = 0$, despite $E(x_{t+1}/x_t) = r + x_t$ there is no reason for the expectation of the estimated slope to be equal to 1. One may easily see, for instance by noticing the full reversibility over time of the model, that the estimate of slope is biased below 1, and in turn that of b is biased towards positive values, i.e. detects too often density-dependence. It is only asymptotically that one will have $E(\hat{b}) = 0$, because $\mathrm{var}(x_t) \to \infty$.

Figure 2 illustrates this bias by presenting simulation results from the random walk model $x_{t+1} = r + x_t + \varepsilon_t$ with $r = 0.1$ and $\sqrt{\mathrm{var}(\varepsilon_t)} = \sigma = 0.1$. The usual regression slope is biased towards 0, pointing to the SG model with $b > 0$, i.e. to density dependence. In this case, over 1000 replicates over 20 time steps, the average slope estimate is 0.9831 ± 0.0477. The situation worsens when $\mathrm{var}(x_t) = \sigma^2$ increases: 1000 simulated slopes under $\sqrt{\mathrm{var}(\varepsilon_t)} = \sigma = 0.2$ lead to an average estimated slope equal to 0.9208 ± 0.1213. The existence of this bias in the absence of uncertainty on population size estimates goes still commonly unrecognized even in recent reviews (Freckleton et al. 2006). An approximate expression of the bias is developed in Appendix, reformulating a result first obtained, to my knowledge, by Saint-Amant (1970). It is only asymptotically that $E(\hat{b}) = 0$, because $\mathrm{var}(x_t) \to \infty$ (approximate proof in Appendix). A t-test statistic, denoted

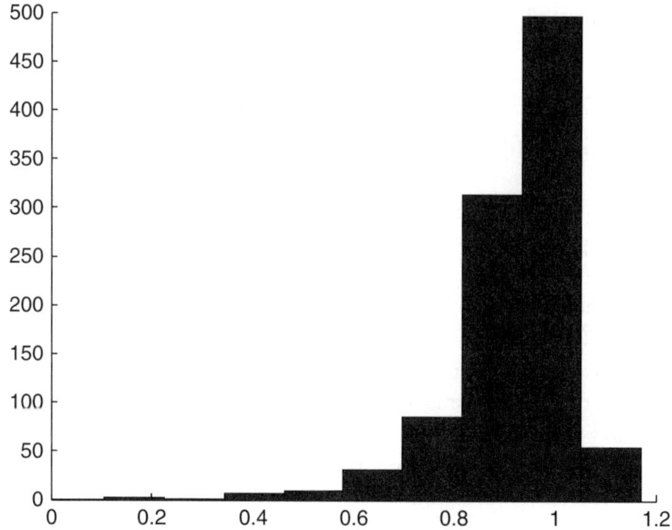

Fig. 2 Bias of standard regression when testing for density dependence based on the discrete time Gompertz model for log population size $x_{t+1} = r + (1 - b)x_t + \varepsilon_t$. The slope of the usual regression of x_{t+1} vs x_t under H_0 $b = 0$ is biased below 1 for finite sample sizes, implying this approach will reject H_0, i.e. detect density-dependence too often. (1000 simulations under with $r = 0.1$ and $\sqrt{\mathrm{var}(\varepsilon_t)} = \sigma = 0.1$)

as t_b for testing H_0 $b = 0$ is readily derived from the usual t-test statistic for the slope. As shown in Appendix, t_b remains asymptotically biased, and a simple approximation of this bias is:

$$E(t_b) \approx \sqrt{\frac{3}{2}}\sqrt{\frac{n-1}{n+1}}\frac{1}{\sqrt{1 + \frac{n-1}{2}\frac{r^2}{\sigma^2}}}$$

Simulations show that the bias corrected t-test statistic, obtained by substituting r and σ^2 in the approximate expression of the bias with their estimates (Appendix), give satisfying results (selection of results in Figs. 3, 4, and 5). In all cases, the uncorrected test is severely biased, while the bias correction brings the P-level fairly close to the nominal level, although it will reject slightly too often for $r = 0$ (Fig. 3). The results seem satisfying enough to make the bias-corrected t-test a good competitor of the parametric bootstrap procedure proposed by (Dennis and Taper 1994), while the empirical recommendation of using the reduced major axis proposed by Saint-Amant (1970) also makes sense given the reversibility of the process, but would deserve further checks.

Altogether, the review above is mainly useful to emphasize the hidden tricks of an apparently innocuous statistical exercise. Its real world applicability is obviously low, as one expects always uncertainty in population size estimates to be present.

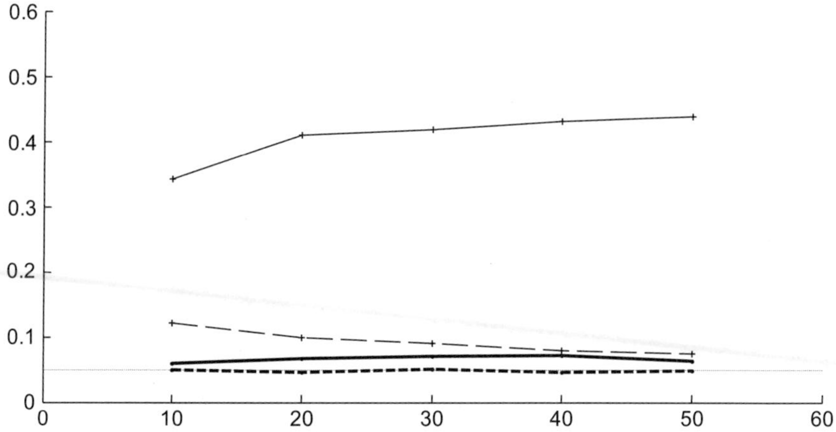

Fig. 3 Simulated P-level (5000 replicates) of one-tailed t-tests of density-dependence by regression in the model $x_{t+1} = r + (1 - b)\,x_t + \varepsilon_t$, var $(\varepsilon_t) = \sigma^2$, as a function of the length of the time series, under a nominal P-level equal to 0.05. The bias-corrected test uses the bias correction in Appendix. The results shown here are for $r = 0$. Uncorrected t-test: thin plain line, crosses, $\sigma = 0.1$; thin dotted line, crosses, $\sigma = 0.2$ Bias-corrected t-test: thick plain line, dots, $\sigma = 0.1$; Thick dotted line, dots, $\sigma = 0.2$

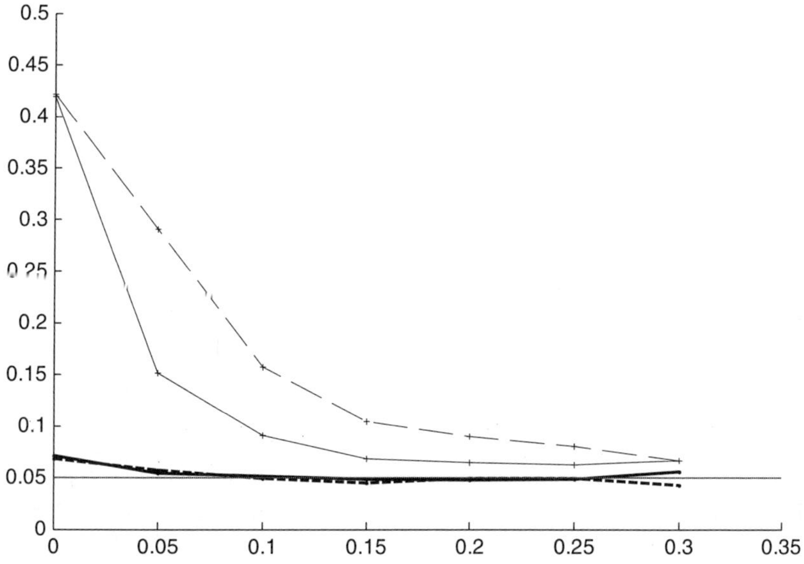

Fig. 4 Simulated P-level (5000 replicates) of one-tailed t-tests of density-dependence by regression in the model $x_{t+1} = r + (1 - b)\,x_t + \varepsilon_t$, var $(\varepsilon_t) = \sigma^2$, over $n = 30$ points, as a function of the growth rate r, under a nominal P-level equal to 0.05. The bias-corrected test uses the bias correction in Appendix. Uncorrected t-test: thin plain line, crosses, $\sigma = 0.1$; thin dotted line, crosses, $\sigma = 0.2$. Bias-corrected t-test: thick plain line, dots, $\sigma = 0.1$; thick dotted line, dots, $\sigma = 0.2$

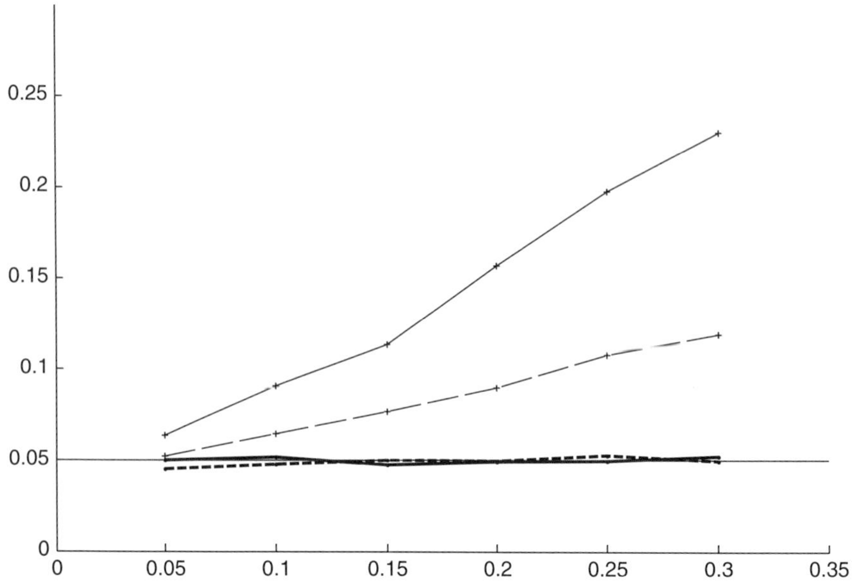

Fig. 5 Simulated P-level (5000 replicates) of one-tailed t-tests of density-dependence by regression in the model $x_{t+1} = r + (1 - b)x_t + \varepsilon_t$, $\mathrm{var}(\varepsilon_t) = \sigma^2$, over $n = 30$ points, as a function of the standard error σ, under a nominal P-level equal to 0.05. The bias-corrected test uses the bias correction in Appendix. Uncorrected t-test: thin plain line, crosses, $r = 0.1$; thin dotted line, crosses, $r = 0.2$. Bias-corrected t-test: thick plain line, dots, $r = 0.1$; thick dotted line, dots, $r = 0.2$

3.2 The Effect of Uncertainty in Population Size Estimates

When the SG model

$$x_{t+1} = r + (1 - b)x_t + \varepsilon_t \tag{8}$$

is put in relation with data, the estimated population size is linked to N_t with some measurement error. A straightforward approach is to consider that log-population size $x_t = \log(N_t)$ is estimated without bias, as y_t. Then:

$$y_{t+1} = x_{t+1} + \eta_{t+1} \tag{9}$$

One assumes the random terms x_t are independently and identically distributed (iid), that the random terms η_t are also iid, and that the two series are independent of each other. One further assumes they have zero means ($E(\varepsilon_t) = 0$ and $E(\eta_t) = 0$) and constant variances $\sigma^2 = \mathrm{var}(\varepsilon_t)$ and $\upsilon^2 = \mathrm{var}(\eta_t)$. In accordance with the generally skewed distributions of population sizes, one can reasonably assume normal distributions for ε_t and η_t. A constant bias in population size estimation on a log scale, corresponding to a constant over- or underestimation, e.g. for the latter because of incomplete detection, would change little to what follows.

Rearranging (8) and (9) as a one dimension time-series, as

$$y_{t+1} = r + (1 - b)\, y_t + \varepsilon_t + (b - 1)\, \eta_t + \eta_{t+1} \tag{10}$$

clearly shows that the random component $\varepsilon_t + (b - 1)\,\eta_t + \eta_{t+1}$ is necessarily autocorrelated when the original random terms ε_t and η_t are not. As a consequence, the autocorrelation function of the observed data y_t can be markedly different from that of the original autoregressive process. Bulmer (1975) proposed the model formed of Equations (8) and (9) as a minimal model for density-dependence in presence of population size uncertainty. He recognized the role of this uncertainty in determining largely the first order autocorrelation and proposed an ad hoc statistic close to the difference between the second order and the first order autocorrelation to test for density-dependence. However, as noted by Lebreton (1989), the state equation (8) and the observation equation (9), with parameters $r, b, \sigma^2, \upsilon^2$ constitute a linear state-space model amenable to the Kalman filter. The Kalman filter theory for state space models (Harvey 1989) is powerful and will provide deeper and stronger results than Bulmer's treatment. Moreover, although this is rarely stated, standard asymptotic stochastic process results will not be useful in the context of detection of density dependence. In relation with this remark, there is no need to assume $r = 0$, or to centre or detrend population size to come closer to usual time-series assumptions. In fact over a finite time window $t = 1, 2..., T$, the population can well grow, as expressed by the shift parameter r. Stenseth et al. (2003) and Jamieson and Brooks (2002, 2004) proposed indeed state space models, treated by Bayesian approaches to test for density-dependence. Kalman filtering has been proposed independently by several authors in contexts close to ours (Visser and Molenaar 1988, Ennola et al. 1998, Besbeas et al. 2002, Ives et al. 2003). Dennis et al. (2006) developed the likelihood approach for the model above based on Kalman filtering. The key idea is that the linearity of the state and observation equations and the assumption of normal distributions for ε_t and η_t ensure that x_t and y_t are normally distributed. It is thus sufficient to derive their expectation and variance to determine their entire distribution and obtain the likelihood $LogL\left(y_1, y_2, ..., y_n / r, b, \sigma^2, \upsilon^2\right)$. The Kalman filter (Harvey 1989) can indeed be viewed as a set of recurrence relationships for deriving the expectations and variances of x_t and y_t, and producing in turn the likelihood of the observations.

In the examples that follow, we used Matlab® programs for Kalman filtering (Besbeas et al. 2002) to calculate the likelihood and we minimized the deviance $-2\, LogL\left(y_1, y_2, ..., y_n / r, b, \sigma^2, \upsilon^2\right)$.

Reaching the minimum of the deviance $Dev\left(\hat{r}, \hat{b}, \hat{\sigma}^2, \hat{\upsilon}^2\right) = \min\left[-2\, LogL\left(y_1, y_2, ..., y_n / r, b, \sigma^2, \upsilon^2\right)\right]$ provides maximum likelihood estimates of the parameters r, b, σ^2, and υ^2, and of their covariance matrix Σ. In turn, one can test in a naïve fashion the null hypothesis $H_0\ b = 0$ against $H_1\ b > 0$ in two different ways:

1. Based on a Wald test, the statistic $z_W = (\hat{b} - 1) \Big/ \sqrt{\hat{\mathrm{var}}(\hat{b})}$ assumed to asymptotically follow under H_0 a distribution N(0,1);

2. After having fitted the model under the constraint $b = 0$ (with constrained estimates denoted as $\tilde{r}$, etc.) based on a Likelihood Ratio Test (LRT), assuming that under H_0, $Dev\left(\tilde{r}, 1, \tilde{\sigma}^2, \tilde{\upsilon}^2\right) - Dev\left(\hat{r}, \hat{b}, \hat{\sigma}^2, \hat{\upsilon}^2\right)$ asymptotically follows a χ^2_1 distribution.

Equivalently, one can use $z_{LRT} = \text{sgn}\left(\hat{b} - 1\right) \text{x} \sqrt{\text{Dev}\left(\tilde{r}, 1, \tilde{\sigma}^2, \tilde{\upsilon}^2\right) - Dev\left(\hat{r}, \hat{b}, \hat{\sigma}^2, \hat{\upsilon}^2\right)}$ and refer it to a distribution N(0,1).

However, even more than in the SG without population size uncertainty, standard MLEs results do not necessarily apply given the non stationarity of the stochastic process in the state equation under H_0 $b = 0$. Moreover, Dennis et al. (2006) review a number of practical and statistical difficulties, such as multiple modes in the likelihood, and discuss alternative estimation methods. They emphasize how much the statistical difficulties met with this relatively simple model have been overlooked in the literature on state-space models.

As a tentative illustration, Table 1 presents two examples, on the Cormorant (Brengballe, pers. comm.) and the greater snow goose (Gauthier et al. 2007), respectively. Both examples were checked for global convergence by using multiple random initial values for the iterative minimization of the deviance. The snow goose results seem reliable, in particular with an acceptable order of magnitude for the population size coefficient of variation (0.06 vs ~0.12–0.17 in field evaluation and state space modeling, respectively; Gauthier et al. 2007). They point as expected to density independence, in a context where, obviously, one expects the naïve test procedure to be biased in favor of density-dependence. While the Cormorant results point as expected to density-dependence, with the warning just given, the results seem more suspicious, with an estimated 0 variance for population size estimation, and an inadequate functional form for density-dependence. A non linear state-space model would then be advisable (De Valpine and Hastings 2002). In both cases, the

Table 1 Statistical analysis of density-dependence for number of breeding pairs of Cormorant *Phalacrocorax carbo* in Europe (Bregnballe, pers. comm.), and total spring count of Greater snow goose *Chen caerulescens caerulescens* (Gauthier et al. 2007). The analysis is based on Kalman filter likelihood estimation (Dennis et al. 2006) in the stochastic discrete time Gompertz model proposed by Bulmer (1975). See text for explanations. The statistical results presented are subject to some unknown biases as commented in the text

Parameter	Method and or statistic	Cormorant	Greater snow Goose
r	KF estimate ± se	0.6046 ± 0.0913	0.3273 ± 0.3077
r	Regression estimate ± se	0.6161 ± 0.1326	0.5044 ± 0.5255
1-b	KF estimate ± se	0.9489 ± 0.0097	0.9784 ± 0.0243
1-b	Bulmer's estimate $^+$	1.0811	1.0861
1-b	Regression estimate ± se	0.9478 ± 0.0141	0.9643 ± 0.0414
Test of H_0 *b=0*	One-sided KF Wald test	5.2710 $p < 0.01$	0.8898 NS
$\sqrt{\text{var}(\varepsilon_t)}$	Process se	0.1000	0.1049
$\sqrt{\text{var}(\eta_t)}$	Population size uncertainty	0.0000	0.0603
Test of H_0 *b=0*	One-sided KF LRT test	3.3520 $p < 0.01$	0.6509 NS
Test of H_0 *b=0*	One-sided naive z-test statistic	3.7131 $p< 0.01$	0.8625 NS
Test of H_0 *b=0*	One-sided Bulmer's Wald test	<0 NS	<0 NS
Test of H_0 *b=0*	Bulmer overall statistic R*	−0.0540 NS	−0.0167 NS

estimation of growth rate was far too high, despite a strong imprecision, and suggest biases in the estimates. As in the absence of population size uncertainty, a parametric bootstrap determination of the distribution of $\hat{b}$ under H_0 $b = 0$ could be easily set up (Dennis and Taper, 2002) and an exploration of the bias (Appendix) might be feasible too. Further progress pending, the warnings by Dennis et al. (2006) should be kept in mind.

Moreover, again, not too much can be expected in terms of inference on process based on pattern. It is not surprising that uncertainty in counts is extremely deleterious, as one is then investigating process based on a blurred pattern. Its effect is however often misunderstood and underrated. For instance Lande et al. (2003) neglect uncertainty, on the grounds that they use only precise estimates of population size. They fail to recognize that even reasonably precise estimates will be nevertheless affected by some uncertainty, and this problem jeopardizes a larger part of their conclusions. The same problem is present throughout the analyses of Brook and Bradshaw (2006).

3.3 *The Effect of Environmental Variability*

Negative autocorrelation in environmental variables may be fairly common. It is well known for Beech mast crop, a major determinant of the dynamics of the Whytham wood Great Tit population (Perrins 1970). Simultaneous analyses of density and environmental covariates a thus badly needed, unless one takes the risk of confusion between density-dependence and environmental variation. Inserting environmental covariates in such models is also a way of decreasing the residual variance and of increasing power.

Absence of autocorrelation in environmental covariates is critical to the test procedures presented above. Hence, we recommend use of relevant covariates with state equation generalised as $x_{t+1} = r + bx_t + cz_t + \varepsilon_t$. Density-dependence and environmental variation are viewed then as main additive effects, a logical first step even if some interaction, representing for instance increased density-dependence under poor environmental conditions, could be modelled in case of need.

An analysis of the Great Tit data based on a density-dependent branching process with environmental covariates is developed by Lebreton (1990). Similarly, Dennis and Otten (2000) analyze the effect of density-dependence and rainfall on population change in a Fox population. In both cases, however, the authors assume there is no uncertainty in population size estimates.

Stenseth et al. (2003) provide an example of a state space model considering simultaneously density-dependence and environmental covariates.

3.4 *The Effect of Population Structure*

When population parameters differ between various categories of individuals such as age or developmental stage, models based on a scalar population size become inappropriate and one generally resorts to matrix models. When the models above are used, it is clear that the part of the autocorrelation of total population size over time is generated by the structure. For instance, if population size becomes exceptionally large at some point in time because of a high fecundity in a given year, it will remain so for several years if survival is rather constant. While there is not problem in using a population vector in the state equation (Ives et al. 2003, Gauthier et al. 2007), this links tests of density dependence with more general state space models of population dynamics. This point is discussed later.

3.5 Conclusion

As a preliminary conclusion at this stage, two points are fairly clear:

(1) The fact that naïve methods, e.g. regression neglecting population size uncertainty (Lande et al. 2003), or ad hoc treatments of uncertainty (McGhee and Berkson 2007), are still commonly used and remain commonly undetected by referees is very annoying, and everything must be done to progress on that point.
(2) State space models are quite promising, whether as an efficient way to handle Bulmer's minimal model, or in more realistic generalized versions (e.g., with environmental covariates), or various functional forms (De Valpine and Hastings 2002), even if some statistical difficulties have still to be solved (Dennis et al. 2006).

4 Detecting and Measuring the Effect of Density-Dependence on Traits

Investigating a relationship between a demographic trait, such as a survival probability and population size does not raise all the difficulties just reviewed. This is because the information on the demographic trait, for instance Capture-Mark-Reencounter (CMR) (Thomson et al. 2008) data for estimating survival, are in general independent from the population size estimates, at least if population size is not estimated from the same capture–recapture experiment. The overall framework is that of regression, even if, as is now commonly the case for CMR models, the regression equation is embedded in a probabilistic model (North and Morgan 1979; Clobert and Lebreton 1985; Lebreton et al. 1992).

If population size could be assumed as known without uncertainty, the usual constrained model used to relate survival to any covariate in CMR analysis could then be used with success. Lebreton et al. (1992) provide an example about the Roe deer *Capreolus capreolus*. Frederiksen and Bregnballe (2000) and Hénaux et al. (2007) discuss a thorough analysis of density-dependence in survival of the Cormorant *Phalacrocorax carbo*.

However, as discussed in introduction, and in the previous section, uncertainty in population size estimates is the rule, and moreover, population size is only a proxy to a real unknown causal factor for change in demographic performance with "density". One has then to handle an "error-in-variable" problem. The effect well known in usual linear regression is to shrink the slope estimate towards zero. The usual regression test is thus conservative, i.e. tends to detect not often enough the alternative hypothesis H_1, i.e. density dependence. While the estimate of slope is biased and should thus be looked at with caution, the bias in the resulting test just amounts to a decrease in power. The same slope bias towards 0 is present for more refined regression models: logistic regression, regression embedded in a CMR model (Crespin et al. 2006).

Barker et al. (2002) adapted a simple bias correction to the regression models used in CMR. Their elegant solution provides a bias corrected estimate of slope that also improves the power of the test. One shortcoming is that one has to know beforehand an estimate of the uncertainty on population size. Moreover, they were only able to apply their correction to the ordinary least square estimate, which is less efficient than a weighted least square one would be.

It is clear also that environmental covariates, in particular autocorrelated ones, can as above have effects confounded with density-dependence; the regression models embedded in CMR models can easily handle multiple regression and this does not raise specific difficulty, out of the confounding usually induced by correlation between dependent variables. Barker's et al. approach would require some simple adaptations to handle that case.

Nevertheless, it is striking that uncertainty in population size estimates, and the fact that population size itself is a proxy to some latent variable, are currently handled in a very different fashion whether one is investigating density-dependence based on population size estimates or at the level of a demographic trait.

5 Discussion: The Potential of Integrated Modelling

Can one expand on the state space approach proposed above to handle Bulmer's minimal model for density-dependence, and obtain a general framework for assessing density-dependence? Several of our earlier comments encourage doing so, using state space models as good candidates for this general framework:

(1) Environmental covariates can easily be incorporated into state space models, expanding in a straightforward fashion over Bulmer's model (Lebreton 1990; Dennis and Otten 2000).
(2) State space models can easily be extended to consider in the state equation a vector of population size (Besbeas et al. 2002; Ives et al. 2003).
(3) State space models can cover more general functional relationships than the log-linear Gompertz representation of density-dependence. The reasonable price to pay is that one has to switch from the explicit and straightforward Kalman filtering to more sophisticated algorithms, such as numerical integration (De Valpine and Hastings 2002) or Bayesian algorithms for stochastic integration.

(4) They can simultaneously represent structure and sophisticated functional forms of density dependence, with density-dependent Leslie matrix models (Allen 1989).

The gap between the treatment of density-dependence based on population size estimates and that based on demographic traits can even be filled. The critical point here is the possibility of representing CMR models as state space models (Doris 2005). This approach has been proposed (Gimenez et al. 2007, 2008) to combine in a same state space model a Leslie matrix and a CMR model, avoiding the normal approximation used by Besbeas et al. (2002) to assemble in a simple fashion the CMR likelihood and the population size Kalman filter based likelihood. The resulting state space model can be quite general, provided again algorithms more sophisticated than Kalman filtering are used.

It seems then quite feasible to combine in a state space model one part based on a density-dependent matrix model (Allen 1989), and another part based on a density dependent CMR model; the dependence in the latter part would be on some of the component of the state vector from the former part. The uncertainty in population size would then be handled in the exact same fashion whether one works with population size or with a demographic trait.

Another advantage of such an approach would be to bridge the gap between pattern and process in statistical investigation of density-dependence.

Under the conditions discussed above, state space models seem thus to have the potential for totally renewing density-dependence assessment, and in turn, modeling. One key consequence of using state space models in the fashion just described would be that various pieces of information would be assembled in an overall diagnostic. This would partly bridge the gap between pattern and process. Another advantage would be that statistical models and dynamical models would be intimately linked. In this respect, density-dependent models with a random component, for accounting for environmental variation and for the fact that population size is always a proxy for what is called density-dependence in a biological population, should become the rule. Further research on the behavior of such models is needed and a number of statistical difficulties have to be closely examined.

A critical aspect is nevertheless that most density-dependence studies remain observational. Experimental approaches, even if they are at a shorter term than the population series considered in long term programs and in the type of modeling reviewed here, could be combined with the other information to move even farther away from pattern towards process. Nest box bird species appear as an ideal material in this respect since their density and numbers can be easily manipulated. Such experiments would moreover link easily with studies of individuals quality if based on sites in which capture–recapture study have been used for several years before the experiment starts.

Altogether, given the complexity of population numerical mechanisms and the subtleties of statistical models for density-dependence, it seems that purely biologically-oriented analyses of density-dependence (e.g., Newton 1998), will have strong shortcomings. Symmetrically, pure statistical analyses would remain too strongly pattern oriented, and be at risk of missing important mechanisms. Altogether, it seems that the scene is nearly ready for a renewal of investigations on a series of questions on density-dependence, whether they concern fundamental or applied population ecology. To cite again only two contrasted questions, the role of density on selective pressure, and its relation to the diversification of life history (Boyce 1984), and the role of potential compensation between harvesting and underlying natural mortality in exploited populations (Lebreton 2005) remain quite open and critical.

Acknowledgments I thank Brian Dennis who pointed out errors and provided thoughtful comments, and thus greatly helped to improve the manuscript.

Appendix

We consider the random walk model with drift r,

$$x_{t+1} = r + x_t + \varepsilon_t \tag{10}$$

which is a particular case of the density-dependent Gompertz model, linearized by considering logarithms of population size:

$$x_{t+1} = r + (1 - b)\,x_t + \varepsilon_t \tag{11}$$

Model (11) reduces to model (10) under H_0 $b = 0$. The alternative hypothesis is H_1 $b > 0$.

Assume data $(x_1, x_2, ..., x_n, x_{n+1})$ originating from model (10) or (11) have been observed over times $t = 1, 2, ..., n$ (i.e. $t + 1 = 2, 3, ..., n + 1$). All probability calculations are conditional on x_1. We denote as $\bar{x}_n$ and $\bar{x}_{n+1}$ the un-shifted and shifted empirical means, respectively:

$$\bar{x}_n = \frac{1}{n}\sum_{t=1}^{n} x_1 \text{ and } \bar{x}_{n+1} = \frac{1}{n}\sum_{t=1}^{n} x_{t+1}$$

related by $\bar{x}_{n+1} = \bar{x} + \frac{1}{n}(x_{n+1} - x_1)$

The ordinary least square (OLS) estimate of the slope of the regression of x_{t+1} w.r.t. x_t is:

$$\hat{a} = \frac{\sum_{t=1}^{n} (x_{t+1} - \bar{x}_{n+1})(x_t - \bar{x}_n)}{\sum_{t=1}^{n} (x_t - \bar{x}_n)^2}.$$

Under the normality assumption for ε_t, the likelihood of model (11), conditional on x_1, reduces to OLS (Hamilton, 1994, p. 118 ff). Hence $\hat{b} = 1 - \hat{a}$ is the MLE of b, conditional on x_1, under model (11).

Under model (11),

$$x_{t+1} - \bar{x}_{n+1} = r + x_t + \varepsilon_t - bx_t - \bar{x}_n - \frac{1}{n}(x_{n+1} - x_1)$$

Hence:

$$\sum_{t=1}^{n}(x_{t+1} - \bar{x}_{n+1})(x_t - \bar{x}_n) = \sum_{t=1}^{n}(x_t - \bar{x}_n)^2 - b\sum_{t=1}^{n} x_t(x_t - \bar{x}_n) + \sum_{t=1}^{n} \varepsilon_t(x_t - \bar{x}_n)$$
$$+(r - \frac{1}{n}(x_{n+1} - x_1))\sum_{t=1}^{n}(x_t - \bar{x}_n)$$

in which the last term is equal to 0 by definition of $\bar{x}_n$

Then, using

$$\sum_{t=1}^{n} x_t\,(x_t - \bar{x}_n) = \sum_{t=1}^{n}(x_t - \bar{x}_n)^2,\ \hat{b} = \frac{-\sum\limits_{t=1}^{n} \varepsilon_t\,(x_t - \bar{x}_n)}{\sum\limits_{t=1}^{n}(x_t - \bar{x}_n)^2} + b$$

Hence,

$$E\left(\hat{b}\right) - b = E\left(\frac{-\sum\limits_{t=1}^{n} \varepsilon_t\,(x_t - \bar{x}_n)}{\sum\limits_{t=1}^{n}(x_t - \bar{x}_n)^2}\right),$$

of which a first order approximation is $E\left(\hat{b}\right) - b \approx \dfrac{E\left(-\sum\limits_{t=1}^{n} \varepsilon_t\,(x_t - \bar{x}_n)\right)}{E\left(\sum\limits_{t=1}^{n}(x_t - \bar{x}_n)^2\right)}$.

The independence of x_t and ε_t, and $E\,(\varepsilon_t) = 0$, lead then to

$$E\left(\hat{b}\right) - b \approx \frac{E\left(\sum\limits_{t=1}^{n} \varepsilon_t \bar{x}_n\right)}{E\left(\sum\limits_{t=1}^{n}(x_t - \bar{x}_n)^2\right)} = \frac{C}{D}.$$

Based on $C = E\left(\sum_{t=1}^{n}\varepsilon_t\bar{x}_n\right) = \frac{1}{n}E\left(\sum_{t=1}^{n}\varepsilon_t\sum_{j=1}^{n}x_j\right)$, $E\left(\varepsilon_t x_j\right) = E\left(\varepsilon_t^2\right) = \sigma^2$ for $j > t$ and $E\left(\varepsilon_t x_j\right) = 0$ for $j \leq t$, one obtains then

$$C = \frac{1}{n}\frac{n(n-1)}{2}\sigma^2 = (n-1)\frac{\sigma^2}{2}.$$

As a consequence, $\hat{b}$ is positively biased, with

$$E\left(\hat{b}\right) - b \approx \frac{(n-1)}{2}\frac{\sigma^2}{E\left(\sum_{t=1}^{n}(x_t - \bar{x}_n)^2\right)} \tag{12}$$

This first part is a reformulation in our notation and context of the results by Saint-Amant (1970).

We will see that under H_0 $b = 0$, $E\left(\sum_{t=1}^{n}(x_t - \bar{x}_n)^2\right)$, grows as $n^{3/2}$, which implies the bias asymptotically vanishes. Thus, although the standard properties of Maximum Likelihood estimates do not necessarily hold because of the non-stationarity of the random walk model (10), $\hat{b}$ is asymptotically unbiased.

However, the bias on $\hat{b}$ has key consequences on the usual statistic for a test of slope.

The statistic for testing for H_0 $b = 0$, is

$$t_b = \frac{\hat{b}}{\sqrt{\hat{\sigma}^2 \Big/ \sum_{t=1}^{n}(x_t - \bar{x}_n)^2}},$$

or, asymptotically,

$$t_b \approx \hat{b}\sqrt{\frac{E\sum_{t=1}^{n}(x_t - \bar{x}_n)^2}{\sigma^2}}.$$

Hence, under H_0 $b = 0$, using the approximate expression (12) for $E\left(\hat{b}\right)$:

$$E\left(t_b\right) \approx \frac{n-1}{2}\sqrt{\frac{\sigma^2}{E\left(\sum_{t=1}^{n}(x_t - \bar{x}_n)^2\right)}}$$

The calculation of $E\left(\sum_{t=1}^{n}(x_t - \bar{x}_n)^2\right)$ is a bit more involved.

First,

$$E\left(\sum_{t=1}^{n}(x_t-\bar{x}_n)^2\right)=\sum_{t=1}^{n}E\left(x_t^2\right)-nE\left(\bar{x}_n^2\right)=\sum_{t=1}^{n}\left(\mathrm{var}\left(x_t\right)+E\left(x_t\right)^2\right)$$
$$-n\left(\mathrm{var}\left(\bar{x}_n\right)+E\left(\bar{x}_n\right)^2\right).$$

Then, from $x_t = (t-1)r + x_1 + \varepsilon_1 + \varepsilon_2 + \ldots + \varepsilon_{t-1}$, one gets the following intermediate results:

$$\bar{x}_n=\frac{n-1}{2}r+x_1+\frac{1}{n-1}\left((n-1)\,\varepsilon_1+(n-2)\,\varepsilon_1\ldots+\varepsilon_{n-1}\right)$$

$$E\left(\bar{x}_n\right)=\frac{n-1}{2}r+x_1 \text{ and } E\left(\bar{x}_n\right)^2=\left(\frac{n-1}{2}r+x_1\right)^2=\frac{(n-1)^2}{4}r^2$$
$$+(n-1)\,rx_1+x_1^2$$

$$\mathrm{var}\left(\bar{x}_n\right)=\frac{1}{n^2}\left(\sum_{t=1}^{n-1}t^2\right)\sigma^2,\text{ in which }\left(\sum_{t=1}^{n-1}t^2\right)$$
$$=\frac{(n-1)n(2n-1)}{6},\text{ denoted as } S(n-1)$$

$$E(x_1)=(t-1)r+x_1,\;E(x_t)^2=\left((t-1)r+x_1\right)^2=(t-1)^2r^2+2(t-1)rx_1+x_1^2,$$

$$\mathrm{var}\left(x_t\right)=(t-1)\,\sigma^2,\;\textit{and}\;\sum_{t=1}^{n}\mathrm{var}\left(x_t\right)=\frac{n\,(n-1)}{2}\sigma^2$$

Then

$$E\left(\sum_{t=1}^{n}(x_t-\bar{x}_n)^2\right)=\frac{n\,(n-1)}{2}\sigma^2+S\,(n-1)\,r^2+rx_1n\,(n-1)+nx_1^2$$
$$-\frac{n}{n^2}S\,(n-1)\,\sigma^2$$
$$-n\left(\frac{(n-1)^2}{4}r^2+(n-1)\,rx_1+x_1^2\right)$$
$$=\left(\frac{3\,n^2-3\,n}{6}-\frac{n}{n^2}\frac{n\,(n-1)\,(2\,n-1)}{6}\right)\sigma^2$$
$$+\,n\,(n-1)\left(\frac{2\,n-1}{6}-\frac{n-1}{4}\right)r^2$$
$$=\frac{(n+1)\,(n-1)}{6}\left(\sigma^2+\frac{n-1}{2}r^2\right)$$

In turn,

$$E\left(t_b\right) \approx \sqrt{\frac{3}{2}}\sqrt{\frac{n-1}{n+1}}\frac{1}{\sqrt{1+\frac{n-1}{2}\frac{r^2}{\sigma^2}}}$$

As, under H_0, r can be estimated without bias as $\hat{r} = \frac{1}{n}\left(x_{n+1} - x_1\right) = \bar{x}_{n+1} - \bar{x}_n$ and σ^2 can be estimated as

$$\hat{\sigma}^2 = \frac{1}{n-1}\sum_{t=1}^{n}\left(x_{t+1} - x_t - \hat{r}\right)^2 = \frac{1}{n-1}\sum_{t=1}^{n}\left(x_{t+1} - \bar{x}_{n+1} - \left(x_t - \bar{x}_n\right)\right)^2,$$

the approximate bias of the t-test can thus be estimated by substituting the estimates in the formula above.

References

Allen LJS (1989) A density-dependent Leslie matrix model. Mathematical Biosciences 95(2): 179–187.

Andrewartha HG, Birch LC (1954) The distribution and abundance of animals. Chicago, University of Chicago Press.

Barker R, Fletcher D, Scofield P (2002) Measuring density dependence in survival from mark-recapture data. Journal of Applied Statistics 29:305–313.

Berryman AA (2002) Population: a central concept for ecology?. Oikos 97(3):439–442.

Berryman AA (2004) Limiting factors and population regulation. Oikos 105(3):667–670.

Berryman AA, Lima M, Hawkins BA (2003) Population regulation, emergent properties, and a requiem for density dependence. Oikos 100(3):636–636.

Besbeas P, Freeman SN, Morgan BJT, Catchpole EA (2002) Integrating mark-recapture-recovery and census data to estimate animal abundance and demographic parameters. Biometrics 58(3): 540–547.

Beverton RJH, Holt SJ (1957) On the dynamics of exploited fish populations. London, Her Majesty's Stationery Office.

Box GE, Jenkins GM (1976) Time series analysis forecasting and control. Oakland, CA, Holden-Day.

Boyce MS (1984) Restitution of r- and K-selection as a model of density-dependent natural selection. Annual Review of Ecology and Systematics 15:427–447.

Brook BW, Bradshaw CJA (2006) Strength of evidence for density dependence in abundance time series of 1198 species. Ecology 87(6):1445–1451.

Bulmer MG (1975) The statistical analysis of density-dependence. Biometrics 31:901–911.

Burnham KP, Anderson DR (1984) Test of compensatory vs. additive hypotheses of mortality in Mallards. Ecology 63(1):105–112.

Caswell H (2001) Matrix population models. Sunderland, MA, Sinauer.

Clobert J, Lebreton J-D (1985) Dépendance de facteurs de milieu dans les estimations de taux de survie par capture-recapture. Biometrics 41:1031–1037.

Courchamp F, Clutton-Brock T, Grenfell B (1999) Inverse density dependence and the Allee effect. Trends in Ecology and Evolution 14(10):405–410.

Cox DR, Hinkley DV (1974) Theoretical statistics. London, Chapman and Hall.

Crespin L, Harris MP, Lebreton J-D, Wanless S (2006) Increased adult mortality and reduced breeding success with age in a population of common guillemot Uria aalge using marked birds of unknown age. Journal of Avian Biology 37(3):273–282.

Dennis B, Otten MRM (2000) Joint effects of density dependence and rainfall on abundance of San Joaquin kit fox. Journal of Wildlife Management 64(2):388–400.

Dennis B, Ponciano JM, Lele SR, Taper ML, Staples DF (2006) Estimating density dependence, process noise, and observation error. Ecological Monographs 76(3):323–341.

Dennis B, Taper ML (1994) Density-dependence in time-series observations of natural populations – Estimation and testing. Ecological Monographs 64(2):205–224.

De Valpine P, Hastings A (2002) Fitting population models incorporating process noise and observation error. Ecological Monographs 72(1):57–76.

Doris B (2005) Modélisation intégrée en dynamique des populations. Application des méthodes de filtrage. Montpellier, CEFE & University Montpellier II:67p.

Ennola K, Sarvala J, Dévai G (1998) Modelling zooplancton population dynamics with the extended Kalman Filetring technique. Ecological Modelling 110:135–149.

Freckleton RP, Watkinson AR, Green RE, Sutherland WJ (2006) Census error and the detection of density dependence. Journal of Animal Ecology 75(4):837–851.

Frederiksen M, Bregnballe T (2000) Evidence for density-dependent survival in adult cormorants from a combined analysis of recoveries and resightings. Journal of Animal Ecology 69(5): 737–752.

Gauthier G, Besbeas P, Morgan BJT, Lebreton J-D (2007) Population growth in Snow Geese: a modeling approach integrating demographic and survey information. Ecology 88:1420–1429.

Gimenez O, Rossi V, Choquet R, Dehais C, Doris B, Varella H, Vila J-P, Pradel R (2007) State-space modelling of data on marked individuals. Ecological Modelling 206:431–438.

Gimenez O, Bonner S, King R, Parker RA, Brooks SP, Jamieson LE, Grosbois V, Morgan BJT, Thomas L (2008) WinBUGS for population ecologists: Bayesian modeling using Markov Chain Monte Carlo methods. In: Thomson DL, Cooch EG, Conroy MJ (eds.) Modeling Demographic Processes in Marked Populations. Environmental and Ecological Statistics, Springer, New York, Vol. 3, pp. 883–916.

Gosselin F (2001) Asymptotic behavior of absorbing Markov chains conditional on non-absorption for applications to conservation biology. Annals of Applied Probability 11: 261–284.

Gosselin F, Lebreton J-D (2000) The potential of branching processes as a modeling tool for conservation biology. Quantitative methods in conservation biology. S. Ferson and M. Burgman. New-York, Springer: 199–225.

Haddon M (2001) Modelling and quantitative methods in fisheries. Boca Raton, FL, Chapman & Hall/CRC.

Hamilton JD (1994) Time series analysis. Princeton, NJ, Princeton University Press.

Harvey AC (1989) Forecasting, structural time series models and the Kalman filter. Cambridge, Cambridge University Press.

Hassell MP (1975) Density-dependence in single-species populations. Journal of Animal Ecology 44(1):283–295.

Hastings A, Hom CL, Ellner S, Turchin P, Godfray JC (1993) Chaos in ecology: is mother nature a strange attractor?. Annual Review of Ecology and Systematics 24:1–33.

Henaux V, Bregnballe T, Lebreton J-D (2007) Dispersal and recruitment during population growth in a colonial bird, the great cormorant Phalacrocorax carbo sinensis. Journal of Avian Biology 38(1):44–57.

Ito Y (1972) Methods for determining density-dependence by means of regression. Oecologia 10(4):347–372.

Ives AR, Dennis B, Cottingham KL, Carpenter SR (2003) Estimating community stability and ecological interactions from time-series data. Ecological Monographs 73(2):301–330.

Jamieson LE, Brooks SP (2002) State space models for density dependence in population ecology. Cambridge, UK, University of Cambridge.

Jamieson LE, Brooks SP (2004) Density dependence in north american ducks. Animal Biodiversity and Conservation 27.1:113–128.

Lack D (1954) The natural regulation of animal numbers. Oxford, Oxford University Press.

Lande R, Engen S, Sæther B-E (2003) Stochastic population dynamics in ecology and conservation. Oxford, Oxford University Press.

Lebreton J-D (1989) Statistical methodology for the study of animal populations. Bulletin of the International Statistical Institute 53(1):267–282.

Lebreton J-D (1990) Modelling density-dependence, environmental variability, and demographic stochasticity from population counts: an example about Wytham Great tits. Population biology of passerine birds an integrated approach. Blondel J, Gosler A, Lebreton J-D, McCleery R. Berlin, Springer 24: 89–102.

Lebreton, J-D (2005) Dynamical and statistical models for exploited populations. Australian and New Zealand Journal of Statistics 47(1):49–63.

Lebreton J-D, Burnham KP, Clobert J, Anderson DR (1992) Modeling survival and testing biological hypotheses using marked animals: a unified approach with case studies. Ecological Monographs 62:67–118.

Maelzer DA (1970) Regression of Log Nn+1 on Log Nn as a test of density dependences – an exercise with computer-constructed density-independent populations. Ecology 51(5): 810–822.

Malthus TR (1798) An essay on the principle of population. Population, evolution, and birth control: a collage of controversial ideas (1964). G. Hardin. San Francisco, Freeman: 4–16.

May RM (1976) Simple mathematical-models with very complicated dynamics. Nature 261(5560):459–467.

May RM, Conway GR, Hassell MP, Southwood TR (1974) Time delays, density-dependence and single-species oscillations. Journal of Animal Ecology 43(3):747–770.

McGhee JD, Berkson JM (2007) Estimation of a non-linear density-dependence paramter for wild turkey. Journal of Wildlife Management 71(3):706–712.

Murdoch WW (1994) Population regulation in theory and practice. Ecology 75(2):271–287.

Newton I (1998) Population limitation in birds. London, Academic Press.

Nicholson AJ, Bailey VA (1935) The balance of animal populaitons, part I. Proceedings of the Zoological Society of London 1:551–598.

North PM, Morgan BJT (1979) Modelling heron survival using weather data. Biometrics 35:67–681.

Perrins CM (1970) Population studies of the Great Tit, Parus major. Proceedings of the Advanced study Institute on Dynamics Number 6 in Population, Oosterbeek.

Ricker WE (1954) Stock and recruitment. Journal of the Fisheries Research Board of Canada 11;559–623.

Rodenhouse N, Sherry TW, Holmes RT (1997) Site-dependent regulation of population size: a new synthesis. Ecology 78:2025–2042.

Royama T (1992) Analytical population dynamics. London, Chapman & Hall.

Saint-Amant JLS (1970) Detection of regulation in animal populations. Ecology 51(5): 823–828.

Slade NA (1977) Statistical detection of density dependence from a series of sequential censuses. Ecology 58(5):1094–1102.

Stenseth NC, Viljugrein H, Saitoh T, Hansen TF, Kittilsen MO, Bølviken E, Glöckner F (2003) Seasonality, density dependence, and population cycles in Hokkaido voles. Proceedings of the National Academy of Sciences USA 100(20):11478–11483.

Texier R (1996) Contrepartie stochastique du modèle déterministe de Ricker en biologie des populaitons. Montpellier, CEFE, rapport de stage de magistère: 24.

Thomson DL, Conroy MJ, Anderson DR, Burnham KP, Cooch EG, Francis CM, Lebreton J-D, Lindberg MS, Morgan BJT, Otis DL, White GC (2008) Standardising terminology and notation for the analysis of demographic processes in marked populations. In: Thomson DL,

Cooch EG, Conroy MJ (eds.) Modeling Demographic Processes in Marked Populations. Environmental and Ecological Statistics, Springer, New York, Vol. 3, pp. 1099–1106.
Tuljapurkar S (1989) An uncertain life: demography in random environments. Theoretical Population Biology 35:227–294.
Tuljapurkar S (1990) Population dynamics in variable environments. New York, Springer.
Verhulst PE (1838) Notice sur la loi que suit la population dans son accroissement. Corresp. Math. Phys. 10:113–121.
Visser H, Molenaar J (1988) Kalman filter analysis in dendroclimatology. Biometrics 44(4): 29–940.
Williams BK, Nichols JD, Conroy MJ (2002) Analysis and management of animal populations. San Diego, Academic Press.
Wynne-Edwards VC (1962) Animal dispersion in relation to social behaviour. Edinburgh, Oliver & Boyd.

The Efficient Semiparametric Regression Modeling of Capture–Recapture Data: Assessing the Impact of Climate on Survival of Two Antarctic Seabird Species

Olivier Gimenez and Christophe Barbraud

Abstract A nonparametric approach has recently been proposed for estimating survival in capture–recapture models, which uses penalized splines to achieve flexibility in exploring the relationships with environmental covariates. However, this method is highly time-consuming because it is implemented through a fully Bayesian approach using Markov chain Monte Carlo simulations. To cope with this issue, we developed a two-step approach in which the existing method is used in conjunction with a multivariate normal approximation to the capture–recapture data likelihood. The ability of our approach to capture various nonlinearities in demographic parameters was validated by carrying out a simulation study. Two examples dealing with Snow petrel and Emperor penguin capture–recapture data sets were also considered to illustrate our procedure, including the relationship between survival rate, population size and climatic covariates.

Keywords Auxiliary variables · Bayesian inference · Bivariate smoothing · Computational efficiency · Demographic rates · Environmental covariates · Interactions · Multivariate normal approximation · Penalized-splines · WinBUGS

1 Introduction

Climate change, specifically global warming, is projected to accelerate in the next century (IPCC 2001). Consequences of this on the functioning of ecosystems are at present difficult to predict, and the study of climatic fluctuations on populations is a major topic in ecology (Hughes 2000; McCarty 2001; Stenseth et al. 2002). Recent investigations show that global warming affects some animal populations, through changes in their physiology, phenology, distribution and demography (Hughes 2000; Walther et al. 2002; Root et al. 2003; Walther et al. 2005). The vast majority of studies assume that the potential effects of both climate and population

O. Gimenez (✉)
Centre d'Ecologie Fonctionnelle et Evolutive/CNRS, UMR 5175, 1919 Route de Mende, 34293 Montpellier Cedex 5, France
e-mail: olivier.gimenez@cefe.cnrs.fr

D.L. Thomson et al. (eds.), *Modeling Demographic Processes in Marked Populations*, Environmental and Ecological Statistics 3, DOI 10.1007/978-0-387-78151-8_3,

density on demographic parameters are linear. However, there is strong evidence that environmental factors may affect population dynamics in more complex ways. For instance, using a global proxy to describe climatic conditions (such as the North Atlantic Oscillation) may induce nonlinear relationships as a consequence of similar nonlinear relations between the proxy and local climatic variables (Mysterud et al. 2001). Empirical data that can be used to investigate the effects of climate change on populations is increasing. Yet, at present there is insufficient modeling methodology to investigate nonlinear relationships between environmental covariates and demographic rates, and to create reliable predictions concerning the impact that the anticipated changes might have on populations.

In this paper, we focus on a new nonparametric approach which has recently been developed to model flexible nonlinear relationships between environmental covariates and demographic rates assessed using capture–recapture/recovery models (Gimenez et al. 2006a). In the spirit of Generalized Additive Models (Hastie and Tibshirani 1990), the shape of the relationship is determined by the data without making any prior assumption regarding its form, using penalized splines (P-splines; Ruppert et al. 2003). However, the whole approach is implemented in a Bayesian framework using MCMC algorithms, and our experience shows that the model fitting process may be highly time-consuming, which can be an obstacle to model selection and to comparative analyses of the response of several species' population dynamics to environmental factors.

Here, we propose to overcome this difficulty by the use of multivariate normal approximation to the capture–recapture model likelihood in a first step (Lebreton et al. 1995; Besbeas et al. 2003). This approximation is then used in a second step in conjunction with a Bayesian approach using MCMC methods in order to implement the P-splines. This combination allows purpose-built programs (e.g. M-SURGE, Choquet et al. 2005; or MARK, White and Burnham 1999) to be used for analyzing capture–recapture data with maximum flexibility and results in a considerable reduction in the computational burden. To validate the ability of our approach to capture various nonlinearities in demographic parameters, we carry out a simulation study. Two examples are also considered to illustrate our approach, including the relationship between survival rate, population size and climatic covariates. Using this new approach we reanalyzed two capture–recapture data sets of Antarctic seabirds, for which previous analyses have investigated (and found) linear relationships between survival and environmental covariates (Jenouvrier et al. 2005). For the Snow petrel (*Pagodroma nivea*), we analyzed the nonlinear relationships between sex-specific adult survival and the Southern Oscillation Index (SOI). For the Emperor penguin (*Aptenodytes forsteri*), we investigated nonlinear relationships between sex specific adult-survival, sea ice extent and population size.

2 Efficient Nonparametric Regression in Capture–Recapture Modeling

In this section, we introduce our approach following two steps. First, the data are analyzed using standard capture–recapture models in a Frequentist framework.

The survival parameter estimates and the associated estimated variance–covariance matrix are then used to approximate the likelihood of the model best supported by the data (*Step 1*). This allows us to adopt a Bayesian approach using MCMC algorithms to implement the nonparametric approach using P-splines (*Step 2*).

2.1 Step 1: Handling the Capture–Recapture Data

We used standard capture-recapture models (Lebreton et al. 1992) to get maximum likelihood estimates (MLEs) for the probability ϕ_i that an individual survives to occasion $i + 1$ given that it is alive at time i, and for the probability p_j that an individual is recaptured at time j. All models were fitted using program M-SURGE (Choquet et al. 2005), but program MARK could have been used instead (White and Burnham 1999).

Using program U-CARE (Choquet et al. 2003), we assessed the fit of the most general time-dependent CJS model to determine whether it provided an adequate description of the data. In both examples (see the Section 2.4), we detected a trap-dependence effect on capture (Pradel 1993), meaning that capture probability at occasion $j + 1$ was different for individuals captured at occasion j than for individuals not captured at occasion j. Such a trap-dependent effect in long-lived species is common and partly reflects heterogeneity in the quality of individuals in a population. For emperor penguins and snow petrels, trap-dependence was at least partly caused by heterogeneity between individuals in their capacity to breed at the colony every year and therefore to be captured. Consequently, we used a multistate capture–recapture model to cope with this departure from the null hypothesis that the CJS fits the data (Gimenez et al. 2003). We distinguished two states whether a capture occurred on the prior occasion (say state A) or not (say state B). In practice, we considered a separate formulation (i.e., the transition probabilities are split into survival and movement probabilities – see Hestbeck et al. 1991). The survival probabilities were time-dependent while the capture probabilities in the states A and B were set constant and fixed to 1 and 0 respectively, and the transition probabilities were state- and time-dependent. By using this formulation, the transition probabilities between states A and A were the capture probabilities given a capture on the prior occasion, and the transition probabilities between states B and A were the capture probabilities given no capture on the prior occasion. See Gimenez et al. (2003) for further details. If any lack of fit remained, we applied a correction to the estimates and their estimated variance–covariance based on the calculation of the coefficient of overdispersion (Lebreton et al. 1992).

As is seen above, we conducted modeling in two steps (Lebreton et al. 1992). We first focused on a model that described the nuisance parameter – i.e., the capture probabilities – in the most parsimonious way, while survival remained time-dependent. Then, preserving the most parsimonious structure of the nuisance parameters, we worked out the survival probabilities using P-splines. Note that for simplicity, we analyzed males and females separately for both data sets (see Section 2.4).

We now turn to the approximation of the capture–recapture likelihood, which will be denoted $L(\phi, p)$. Lebreton et al. (1995) and Besbeas et al. (2003) proposed to use a multivariate normal to approximate the function $L(\phi, p)$. More precisely, the maximum likelihood estimates of the parameters *on the logit scale*, $\hat{\theta}$, and the associated estimated variance–covariance matrix, $\hat{\Sigma}$, both obtained from fitting an appropriate capture–recapture model (see above), are used to approximate the log-likelihood as:

$$2 \log \{L(\phi, p)\} = \text{constant} - \left(\hat{\theta} - \theta\right)^T \hat{\Sigma}^{-1} \left(\hat{\theta} - \theta\right). \quad (1)$$

Note that Besbeas et al. (2003) showed that it is only necessary to make the approximation for the parameters of interest, which are the survival probabilities in our case. Obviously, using a multivariate normal distribution in place of the usual product of multinomial distributions (where cells are complex nonlinear functions of the survival and recapture probabilities) results in a much simpler form for the likelihood $L(\phi, p)$, which in turn greatly speeds up the Bayesian fitting process using MCMC algorithms.

Nevertheless, the use of Eq. (1) may be made difficult by numerical issues. Indeed, some parameters may be estimated close to or on a boundary (0 or 1 as we are dealing with probabilities), resulting in the impossibility to properly quantify the variability associated to the MLEs using standard methods. Technically, the dispersion matrix $\hat{\Sigma}$ is ill-conditioned which prevents us from obtaining its inverse as required in Eq. (1). We circumvent this issue by neglecting the covariances, and considering the diagonal $\hat{\Sigma}$ matrix of the estimated variances with off-diagonal terms all zeros. Still, calculating variances for boundary estimates remains problematic. One option is to use profile-likelihood intervals (Gimenez et al. 2005), the problem being that this approach does not formally provide a point estimate nor a standard error. In this paper, we decided to assign a large variance (10,000) to those para-meters estimated close to or on the boundary, thus affecting relative negligible weights to the corresponding MLEs (see Eq. (1)). This ad-hoc procedure was used in the Section 2.4 only.

2.2 Step 2: Semiparametric Modeling of the Survival

2.2.1 Univariate Smoothing

We consider the following regression model for the survival probability ϕ_i:

$$\text{logit}(\phi_i) = \log\left(\frac{\phi_i}{1 - \phi_i}\right) = f(x_i) + \varepsilon_i, \quad (2)$$

where x_i is the value of the covariate applying between occasions i and $i + 1$, f is a smooth function and ε_i are i.i.d. random effects $N\left(0, \sigma_\varepsilon^2\right)$. The function f specifies a nonparametric flexible relationship between the survival probability and the covariate that allows nonlinear environmental trends to be detected. Following Gimenez et al. (2006a), we used a truncated polynomial basis to handle *f*:

$$f(x) = \beta_0 + \beta_1 x + \ldots + \beta_P x^P + \sum_{k=1}^{K} b_k (x - \kappa_k)_+^P, \tag{3}$$

where x is the covariate, $\beta_0, \beta_1, \ldots, \beta_P, b_1, \ldots, b_K$ are regression coefficients to be estimated, $P \geq 1$ is the degree of the spline, $(u)_+^P = u^p$ if $u \geq 0$ and 0 otherwise, and $\kappa_1 < \kappa_2 < \ldots < \kappa_K$ are fixed knots. We considered $K = \min\left(\frac{1}{4}I, 35\right)$ knots to ensure the desired flexibility, and let k_k be the sample quantile of x's corresponding to probability $\frac{k}{K+1}$. To avoid overfitting, we penalized the b's by assuming that the coefficients of $(x - \kappa_k)_+^P$ are normally distributed random variables with mean 0 and a certain variance σ_b^2 to be estimated. This is the reason why this approach is referred to as penalized splines (Ruppert et al. 2003). For further details see Gimenez et al. (2006a) and references therein.

2.2.2 Bivariate Smoothing

To incorporate the interaction between two continuous environmental covariates, we opted for bivariate smoothing using thin-plate splines (Green and Silverman 1994). The main challenge here was to achieve the ideal balance between roughness and smoothness, which is controlled by a parameter δ usually referred to as the smoothing parameter. We considered the restricted maximum likelihood (REML) criterion to choose this amount of smoothing using the data (Searle et al. 1992), which allows the whole modeling exercise to be easily implemented in a mixed model framework (Ruppert et al. 2003; Crainiceanu et al. 2005; Gimenez et al. 2006a). Specifically, we consider a nonparametric model for the survival with respect to environmental covariates as follows:

$$\text{logit}(\phi_i) = f(\mathbf{x_i}) + \varepsilon_i \tag{4}$$

where $\mathbf{x}_i = \left(x_i^1, x_i^2\right)^{\mathrm{T}}$ is the value of the vector of two covariates $\mathbf{x}^1$ and $\mathbf{x}^2$ for year i, T denotes transpose, ε_i are i.i.d $N\left(0, \sigma_\varepsilon^2\right)$ and f is a smooth function. Because they have good numerical properties, we used radial basis functions to handle f (Ruppert et al. 2003):

$$f(\mathbf{x}) = \mathbf{X}\mathbf{b} + \mathbf{Z_K}\mathbf{v}, \tag{5}$$

where $\left\{1, x_i^1, x_i^2\right\}$ is the ith row of matrix $\mathbf{X}$, $\{C(\|\mathbf{x_i} - \boldsymbol{\kappa}_1\|), \ldots, C(\|\mathbf{x_i} - \boldsymbol{\kappa}_K\|)\}$ is the ith row of matrix $\mathbf{Z_K}$, the $\boldsymbol{\kappa}_k$'s are bi-dimensional vectors of fixed knots, the function $C(\|\mathbf{r}\|) = \|\mathbf{r}\|^2 \log \|\mathbf{r}\|$ with $\|\mathbf{r}\| = \sqrt{\mathbf{r}^{\mathrm{T}}\mathbf{r}}$ handles the nonlinear structure of the survival surface, $\mathbf{b} = (b_1, b_2, b_3)^{\mathrm{T}}$ and $\mathbf{v} = (v_1, \ldots, v_K)^{\mathrm{T}}$ are vectors of fixed and random regression parameters respectively to be estimated with $\text{Cov}(v) = \sigma_u^2 \mathbf{\Omega}_K^{-1}$

where $\mathbf{\Omega}_K$ has (k, k')th element $C(\|\boldsymbol{\kappa}_k - \boldsymbol{\kappa}_{k'}\|)$. Using the re-parameterization $u = \mathbf{\Omega}_K^{1/2} v$ and defining $Z = Z_K \mathbf{\Omega}_K^{-1/2}$, Eq. (5) becomes equivalent to

$$f(x) = \mathbf{Xb} + \mathbf{Zu}, \tag{6}$$

where $\mathbf{u}$ is assumed to be normally distributed, independent from $\boldsymbol{\varepsilon}$, with $\mathrm{Cov}(\mathbf{u}) = \sigma_u^2 \mathbf{I}_K$. It can be shown that the optimal amount of smoothing using the REML criterion is given by $\delta = \sigma_u^2/\sigma_\varepsilon^2$, which turns out to be also the case in the univariate smoothing (Ruppert et al. 2003). To choose the number and the location of the knots, we considered $K = \max\{20, \min(I/4, 150)\}$ knots as suggested by Ruppert et al. (2003) and used the space-filling algorithm of Nychka and Saltzman (1998) to select the location of these knots. This algorithm automatically places knots in regions with high density of observed values while maximizing the average spacing between knots of those regions. Finally, to plot the fitness surface, we obtained contours and perspectives views by generating a 30×30 grid of predicted values.

2.2.3 Bayesian Inference

Vague prior distributions were provided for all parameters. Specifically, we chose uniform distributions on [0,1] for the detection probabilities, normal distributions with mean 0 and variances 1,000 for the β 's and normal distributions with mean 0 and variances σ_u^2, σ_b^2 and σ_ε^2 for the u 's, b 's and ε 's respectively. The priors for the hyperparameters σ_u^2, σ_b^2 and σ_ε^2 were chosen as inverse-gamma with both parameters equal to 0.001. We generated two chains of length 100,000, discarding the first 50,000 as burn-in. Convergence was assessed using the Gelman and Rubin statistic which compares the within to the between variability of chains started at different and dispersed initial values (Gelman 1996). All covariates were standardized to improve convergence. The simulations were performed using WinBUGS (Spiegelhalter et al. 2003). The R (Ihaka and Gentleman 1996) package R2WinBUGS (Sturtz et al. 2005) was used to call WinBUGS and export results in R. To implement the space-filling algorithm, we used the R package FIELDS (Fields Development Team 2006).

Whenever needed, we used the Deviance Information Criterion (DIC; Spiegelhalter et al. 2002) to discriminate between candidate models: the smaller the DIC value, the better the model. We acknowledge that the DIC is somewhat controversial in the statistical literature, and should be used with caution (see Spiegelhalter et al. 2002 and Celeux et al. 2006 and the discussion papers following these two papers). The R and WinBUGS codes are available on request from the first author.

2.3 Simulation Study

We conducted a simulation study to investigate the performance of our approach, in particular to check that the use of the approximation for the capture–recapture likelihood did not affect the estimation of parameters. Following Gimenez et al. (2006a), we considered two scenarios with different forms for the underlying nonlinear regression function f of Eq. (3). Study 1 used the regression function $f(x) = 2.2$ if $x \leq -0.06$ and $f(x) = 2.08 - 2x$ otherwise to represent a threshold effect. The x's were equally spaced on $[-1.5; 1.5]$. Study 2 used the regression function $f(x) = 1.5g\left(\frac{x-0.35}{0.15}\right) - g\left(\frac{x-0.6}{0.1}\right)$ where $g(x) = \frac{1}{\sqrt{2\pi}} \exp\left(-\frac{x^2}{2}\right)$ to represent complex non-linear patterns. The x's were equally spaced on $[0; 1]$. For both studies, we simulated 100 capture–recapture data sets covering 26 sampling occasions, so that 25 survival probabilities had to be estimated. We considered 50, 100 and 250 newly marked individuals per occasion and two levels of variability with $\sigma_\varepsilon^2 = 0.02$ or $\sigma_\varepsilon^2 = 0.1$. The capture probability was set constant and equal to 0.7.

For each data set, we applied our approach in two steps, first fitting a capture–recapture model with time-dependent survival probabilities and constant recapture probabilities, second using the MLEs and the variance–covariance matrix to approximate the capture–recapture likelihood of this model in order to implement the P-splines in a Bayesian framework using MCMC algorithms. Details on the practical implementation can be found in the Section 2. For each x value, we computed the median along with a 95% confidence interval for the posterior medians of f and then back-transformed in order to compare the estimated survival curve to its true counterpart. The results are displayed in Figs. 1 and 2, showing that our two-step approach does a good job in capturing the nonlinearities in the survival vs. covariate relationship. For a fixed number of newly released individuals, the greater the variance the lower the precision (both Figs. 1 and 2, left column – low variability vs. right column – high variability), the difference being clearer for Study 1. When the sample size increases, the precision gets better (both Figs. 1 and 2, going down – 50, 100 and 250 newly released individuals), although for high variability the gain was not substantial (right column in both Figs. 1 and 2). Overall, as noted by Gimenez et al. (2006a), the relationship in Study 1 was more precisely estimated than that of Study 2.

2.4 Applications

2.4.1 Snow Petrels

As a first example, we analyzed the data used in Gimenez et al. (2006a) to illustrate the full Bayesian implementation of the semiparametric modeling of survival probabilities. The data were obtained in a 40–year study on individually marked Snow petrels, nesting at Petrels Island, Terre Adélie, from 1963 to 2000 (see also Barbraud et al. 2000; Jenouvrier et al. 2005). We considered the Southern Oscillation Index (*SOI*) as a proxy of the overall climate condition, available from the Climatic Research Unit (http://www.cru.uea.ac.uk/cru/data/soi.htm). In total, we considered 563 female and 561 male capture histories (more than in Gimenez et al. 2006a who were limited by the computational burden).

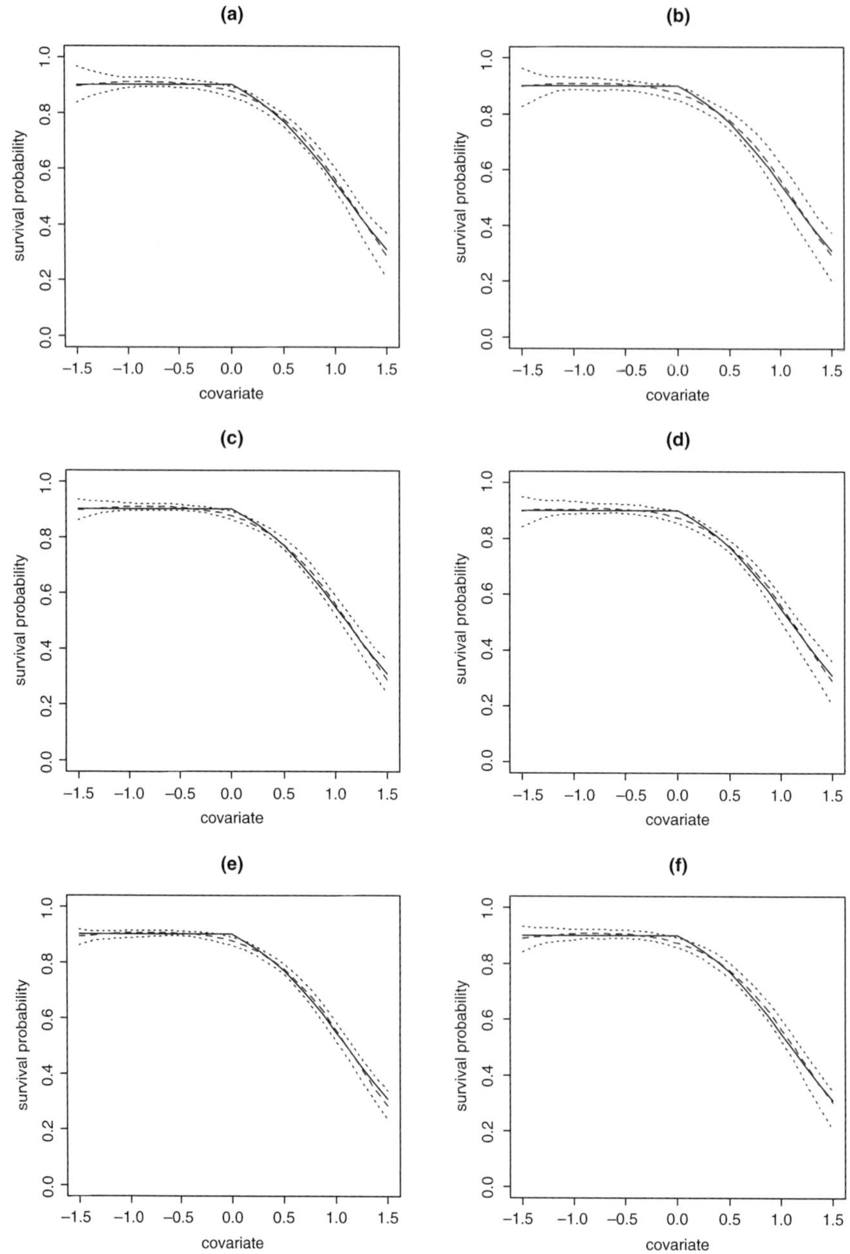

Fig. 1 Performance of the nonparametric approach for estimating nonlinearities in the survival probability – Study 1 (see the Section 2.3 for details). We used 100 simulated capture–recapture data sets with 50, 100 and 250 newly released individuals per occasion (from top to bottom resp.) and two levels of variability, $\sigma_\varepsilon^2 = 0.02$ or $\sigma_\varepsilon^2 = 0.1$ (from left to right resp.). The solid line is the true regression function, the dashed line is the median of the 100 estimated posterior medians and the dotted lines indicate the associated 95% confidence interval

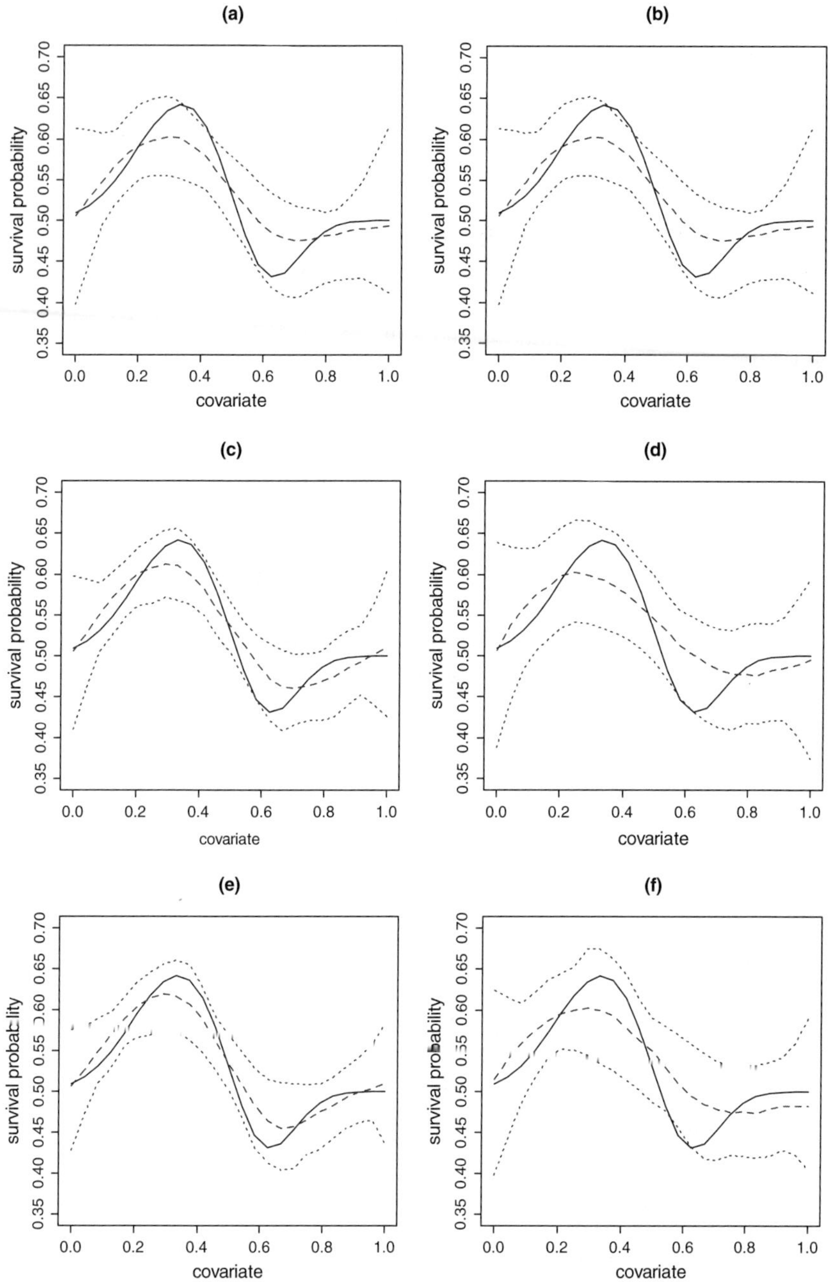

Fig. 2 Performance of the nonparametric approach for estimating nonlinearities in the survival probability – Study 2 (see the Section 2.3 for details). We used 100 simulated capture–recapture data sets with 50, 100 and 250 newly released individuals per occasion (from top to bottom resp.) and two levels of variability, $\sigma_\varepsilon^2 = 0.02$ or $\sigma_\varepsilon^2 = 0.1$ (from left to right resp.). The solid line is the true regression function, the dashed line is the median of the 100 estimated posterior medians and the dotted lines indicate the associated 95% confidence interval

Step 1. After removing the first capture to remove heterogeneity, the CJS model still poorly fitted the data for both females and males (females:χ^2_{167} (349.98) $<$ 0.01; males:χ^2_{169} (472.20) $<$ 0.01), and a closer inspection of the results revealed that a large part of the CJS χ^2 statistic was explained by a trap-dependence effect (females: χ^2_{34} (175.74) $<$ 0.01; males: χ^2_{34} (248.697) $<$ 0.01). The goodness-of-fit for the model with trap-dependence was still significant (females:χ^2_{133} (174.239) $<$ 0.01; males:χ^2_{135} (223.503) $<$ 0.01) so we used a lack-of-fit coefficient for further analyses (females: 1.3; males: 1.7). Time-dependent survival probability estimates and the estimated variance-covariance were then obtained for both sexes using M-SURGE (Choquet et al. 2005).

Step 2. First, because sex differences in the survival probabilities were found before, we considered a model with an additive effect of both SEX and SIE factors. This was achieved by extending the nonparametric approach introduced above to allow a predictor to enter the model linearly (we will refer to semiparametric modeling when both linear and nonlinear effects appear in a model). To do so, we wrote:

$$\text{logit}\left(\phi_i^l\right) = \beta_0 + \gamma\text{SEX} + \beta_1\text{SOI}_\text{i} + \sum_{k=1}^{K} b_k\left(\text{SOI}_i - \kappa_k\right) + \varepsilon_i, \tag{7}$$

where ϕ_i^l is the survival probability between occasion i and $i+1$ for l = males (SEX = 0) or l = females (SEX = 1). Interestingly, only little adjustments to the modeling introduced in Section 2.2.1 are needed to specify the model defined by Eq. (7) (see Gimenez et al. 2006a). We also fitted a model with an interaction effect between the SEX and the SOI factors. It basically consists of considering different smooth functions according to the SEX qualitative variable (Coull et al. 2001). Table 1 shows that the model with an additive effect of both covariates is preferred to the model with interaction.

Finally we considered two further models corresponding to two biological hypotheses. First, we were interested in assessing the significance of the SEX effect, so we fitted a model without the SEX effect, while keeping the nonparametric feature of the model. This model performs better than the two models having the SEX effect (Table 1). This was also confirmed by the 95% posterior credible interval [−0.49;0.15] of the parameter γ which contains 0. Second, we were interested in testing for the presence of nonlinearities in the survival probability. One way to answer this question was to fit a model with a linear effect of the SOI covariate

Table 1 Models fitted to the Snow petrel data. DIC is the deviance information criterion, and pD the number of effective parameters. ΔDIC is the difference between the DIC of a model and the DIC for the minimum DIC model. The model best fitting the data is in bold font

Model	DIC	pD	ΔDIC
Additive effect of SEX and SOI	1129.29	1062.99	604.74
Interaction effect of SEX and SOI	1644.82	1595.38	1120.27
SOI effect only (no SEX effect)	679.27	607.34	154.72
Linear effect of SOI (no SEX effect)	**524.55**	**446.49**	**0**

upon the survival probability, and to compare with its nonparametric counterpart. To do so, we used:

$$\operatorname{logit}\left(\phi_i\right) = \beta_0 + \beta_1 \mathrm{SOI_i} + \varepsilon_i. \tag{8}$$

As already noted by Gimenez et al. (2006a), the relationship between the climatic covariate SOI and the Snow petrel survival seems to be linear (Table 1). The graphical representation of the two latter models tends to confirm this result (Fig. 3) During negative SOI, characteristic of El Niño episodes, cooler waters in the western part of the tropical Pacific and southern Australia down to the Ross Sea region seem to favor enhanced productivity in this oligotrophic area (Wilson and Adamec 2002). Therefore these oceanographic conditions may increase the food availability for snow petrels and reduce their mortality risk associated with starvation. However, the effect of SOI on adult survival is small, with only a 1–2% difference in survival between negative and positive SOI conditions, which might explain the linear relationship between survival and SOI. We will go back to the issue of formally detecting nonlinearities in Section 2.5.

In this section, we have considered an interaction between a discrete variable SEX and a continuous variable SOI. In the next section, we consider an interaction between two continuous variables using bivariate smoothing (Ruppert et al. 2003).

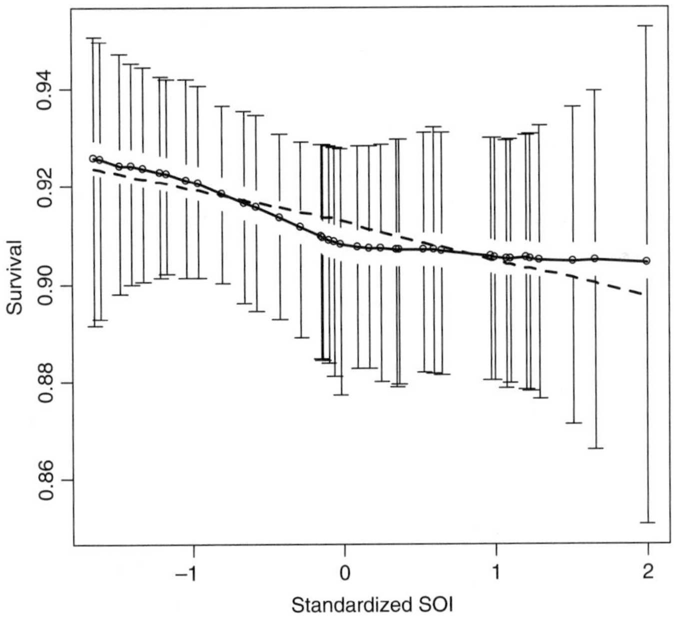

Fig. 3 Annual variations in Snow petrel survival as a function of the standardized SOI using a nonparametric model. Medians (*solid line*) with 95% pointwise credible intervals (*vertical solid lines*) are shown, along with the estimated linear effect (*dotted line*)

2.4.2 Emperor Penguins

As a second example, we analyzed data on the emperor penguin which consist of data from a long-term study on marked individuals, nesting at Petrels Island, Terre Adélie, from 1962 to 2002 (see Barbraud and Weimerskirch 2001; Jenouvrier et al. 2005). We considered the Sea Ice Extent (*SIE*) as the distance from the colony to the limit of a 15% or higher sea ice concentration, which was obtained at longitude 140°E using the sea ice data available from the Antarctic CRC and Australian Antarctic Division Climate Data Sets (http://www.antcrc.utas.edu.au/~jacka/seaice_C_html). We also considered the number of breeding pairs (*POPSIZE*). In total, we considered 382 female and 331 male capture histories.

Step 1. After removing the two first captures to remove heterogeneity, the CJS model still poorly fitted the data for both females and males (females:χ^2_{85} (182.05) < 0.01; males:χ^2_{79} (198.12) < 0.01), and a closer inspection of the results revealed that a large part of the CJS χ^2 statistic was explained by a trap-dependence effect (females: χ^2_{27} (112.07) < 0.01; males: χ^2_{26} (131.85) < 0.01). The goodness-of-fit for the model with trap-dependence indicated that the fit was satisfactory (females:χ^2_{58} (69.98) $= 0.135$; males:χ^2_{53} (66.28) $= 0.104$). Time-dependent survival probabilities estimates and the estimated variance–covariance were then obtained for both sexes using M-SURGE (Choquet et al. 2005).

Step 2. The results of the bivariate smoothing for male and female Emperor penguins are given in Fig. 4. Overall, females survive better than males, which is in agreement with previous studies (Barbraud and Weimerskirch 2001; Jenouvrier et al. 2005). Now if we look into the relationship between survival and the interaction of the SIE and POPSIZE effects, interesting patterns emerge. Strategies differ by sex. While the survival optimum for both males and females is reached for average values of SIE, there is a marked difference regarding POPSIZE: females prefer very high POPSIZE while males survive better with relatively low POPSIZE. These differences may be interpreted in the light of the contrasting breeding strategies of males and females. After their 3.5 months fast incubating the egg, emaciated males return to sea for feeding and density dependent processes may affect their survival chances through competition for food when POPSIZE is high. This should be particularly accentuated when food resources are scarce, i.e., when sea ice extent is low. During the entire incubation, females are absent from the colony, feeding within the pack ice and below the fast ice. Males at the colony face very harsh climatic conditions and it has been shown that they also form huddles to save energy (Ancel et al. 1997). Therefore, we hypothesize that when the population is large it might be easier to find congeners and to form huddles than when the population is small, which may increase their chances of survival. However, we note that we could not formally assess sex differences since the two data sets were analyzed separately. Interestingly, it is relatively easy to get a picture of the precision associated with the survival surface as a by-product of the use of the MCMC procedure (Fig. 1, right column). Having a visualization of the precision helps us in determining to what extent the patterns we detected are supported by the data. In the present example, the standard deviations are low, except for extreme values of both covariates (Fig. 1, right column).

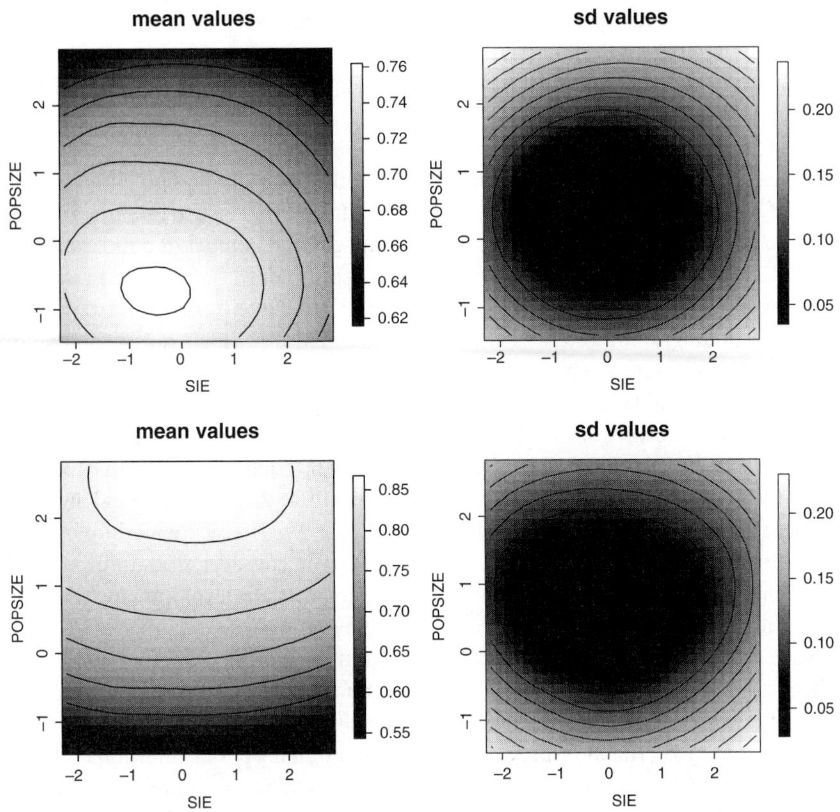

Fig. 4 Bayesian thin-plate spline visualization of the survival surface for the Emperor penguin as a function of the two external covariates sea-ice extent (SIE) and number of breeding pairs (POPSIZE). Posterior mean survival probabilities (*left column*) and associated posterior standard deviations (*right column*) are provided for males (*top*) and females (*bottom*)

2.5 Discussion

In this paper, we have used a combination of the Frequentist and the Bayesian approaches to implement semiparametric modeling of survival probabilities as a function of environmental covariates using capture–recapture data. Instead of opposing the two frameworks and forcing one to make a choice between the two, we have utilized the merits of each of the two approaches: the Frequentist approach was used to handle the capture–recapture data using specialist programs like M-SURGE (Choquet et al. 2005) or MARK (White and Burnham 1999) which allows flexible fitting of complex models including age, cohort and/or site effects; the Bayesian approach was used to avoid making any prior assumption regarding the form of the relationship between the survival and the covariates, while taking benefit of the automatic adjustment of the amount of smoothing in the P-splines. Besides, the combination allows the computational burden to be substantially reduced. For example, it took about 25 hours to fit the semiparametric model of Eq. (7) to the

Snow petrel data with the full Bayesian approach (Gimenez et al. 2006a), while only 5 minutes were required to obtain the MLEs with the estimated variance–covariance matrix and to fit the semiparametric model of Eq. (7) using the multinormal likelihood approximation.

Apart from the gain in time of calculation, the use of a normal approximation to the capture–recapture data likelihood has another appealing application. We can think of using the information published in the literature to investigate the impact of climatic conditions on demographic rates, in the general context of a meta-analysis. The MLEs and the associated standard errors could indeed be extracted from relevant papers and then used to form a likelihood, which in turn, could be used to relate the demographic rates to climatic conditions, for which measurements are often freely available from the Internet. Maximum flexibility in describing those relationships would be assured by the use of the approach advocated here.

Although the results of the simulations are encouraging, the fact that we did not detect a sex effect in the Snow petrel analysis is in contradiction with a previous study (Jenouvrier et al. 2005), although no sex differences were found in earlier studies (Chastel et al. 1993; Barbraud et al. 2000). Possible explanations are very small differences in survival and/or a loss of power caused by assuming that the covariances are all zeros (see Section 2.1). Pending further developments, extensive simulations are needed to assess the loss of precision when standard errors are used in place of the whole estimated variance–covariance matrix.

Regarding the Emperor penguin example, our analysis should be considered as a preliminary step towards a more comprehensive study. We envisage that model selection will be a crucial issue, as we would like to incorporate additional climatic variables (e.g., SOI and SEX) to POPSIZE and SIE, making the number of scenarios numerous. Besides, determining whether nonlinearities are required in the model still needs to be properly addressed. A Reversible-Jump MCMC procedure is a promising solution to that aim (Bonner et al. this volume).

Finally, so far we have considered environmental covariates only, i.e., variables with values changing over time. A semiparametric approach to incorporate individual covariates, i.e., variables with values changing at the individual level, has recently been proposed to assess natural selection on a single quantitative trait (e.g. body mass: Gimenez et al. 2006b) as well as estimating and visualizing fitness surfaces (Gimenez et al. submitted) using capture–recapture data. There is high interest in considering both types of covariates in a model (e.g. Coulson et al. 2001), and the normal approximation might be useful to reduce the computational burden.

References

Ancel A, Visser M, Handrich Y, Masman D, Le Maho Y (1997) Energy saving in huddling penguins. Nature 385:304–305.

Barbraud C, Weimerskirch H, Guinet C, Jouventin P (2000) Effect of sea-ice extent on adult survival of an Antarctic top predator: the snow petrel Pagodroma nivea. Oecologia 125: 483–488.

Barbraud C, Weimerskirch H (2001) Emperor penguins and climate change. Nature 411: 183–186.

Besbeas P, Lebreton J-D, Morgan BJT (2003) The efficient integration of abundance and demographic data. Applied Statistics 52:95–102.

Celeux G, Forbesy F, Robert CP, Titterington DM (2006) Deviance information criteria for missing data models. Bayesian Analysis 1:651–674.

Chastel O, Weimerskirch H, Jouventin P (1993) High variability in demographic parameters in the Snow petrel Pagodroma nivea; a 27–year study. Oecologia 94:278–285.

Choquet R, Reboulet AM, Pradel R, Gimenez O, Lebreton JD (2003) U-CARE user's guide, Version 2.0. Mimeographed document CEFE/CNRS, Montpellier.

Choquet R, Reboulet AM, Pradel R, Gimenez O, Lebreton JD (2005) M-SURGE: new software specifically designed for multistate capture–recapture models. Animal Biodiversity and Conservation 27:207–215.

Coull BA, Ruppert D, Wand MP (2001) Simple incorporation of interactions into additive models. Biometrics 57:539–545.

Coulson T, Catchpole EA, Albon SD, Morgan BJT, Pemberton JM, Clutton-Brock TH, Crawley MJ, Grenfell BT (2001) Age, sex, density, winter weather, and population crashes in Soay sheep. Science 292:1528–1531.

Crainiceanu C, Ruppert D, Wand MP (2005) Bayesian analysis for penalized spline regression using WinBUGS. Journal of Statistical Software 14(14).

Fields Development Team (2006) fields: Tools for Spatial Data. National Center for Atmospheric Research, Boulder, CO. http://www.cgd.ucar.edu/Software/Fields.

Gelman A (1996) Inference and monitoring convergence. Pp. 131-143 in Gilks WR, Richardson S, Spiegelhalter DJ (eds.) Markov Chain Monte Carlo in Practice. Chapman and Hall, London.

Gimenez O, Choquet R, Lebreton J-D (2003) Parameter redundancy in multistate capture–recapture models. Biometrical Journal 45:704–722.

Gimenez O, Choquet R, Lamor L, Scofield P, Fletcher D, Lebreton et, J-D, Pradel R (2005) Efficient profile-likelihood confidence intervals for capture–recapture models. Journal of Agricultural, Biological, and Environmental Statistics 10:184–196.

Gimenez O, Crainiceanu C, Barbraud C, Jenouvrier S, Morgan BJT (2006a) Semiparametric regression in capture–recapture Modelling. Biometrics 62:691–698.

Gimenez O, Covas R, Brown CR, Anderson MD, Bomberger Brown M, Lenormand T (2006b) Nonparametric estimation of natural selection on a quantitative trait using capture-mark-recapture data. Evolution. 60:460–466.

Gimenez O, Grégoire A, Lenormand T, Submitted. Estimating and visualizing fitness surfaces using mark-recapture data.

Green PJ, Silverman BW (1994) Nonparametric Regression and Generalized Linear Models: A Roughness Penalty Approach. Chapman and Hall, New York.

Hastie T, Tibshirani R (1990) Generalized additive models. Chapman and Hall, London.

Hestbeck JB, Nichols JD, Malecki RA (1991) Estimates of movement and site fidelity using mark-resight data of wintering Canada geese. Ecology 72:523–533.

Hughes L (2000) Biological consequences of global warming; is the signal already apparent? Trends in Ecology and Evolution 15:56–61.

Ihaka SP, Gentleman EA (1996) R: a language for data analysis and graphics. Journal of Computational and Graphical Statistics 5:299–314.

IPCC (2001) Climate change; a synthesis report. A contribution of Working Groups I, II and III to the third assessment report of the IPCC. Cambridge University Press, Cambridge.

Jenouvrier S, Barbraud C, Weimerskirch H (2005) Long-term contrasted responses to climate of two Antarctic seabird species. Ecology 86:2889–2903.

Lebreton JD, Burnham KP, Clobert J, Anderson DR (1992) Modeling survival and testing biological hypotheses using marked animals: a unified approach with case-studies. Ecological Monographs 62:67–118.

Lebreton J-D, Morgan BJT, Pradel R, Freeman SN (1995) A simultaneous survival rate analysis of dead recovery and live recapture data. Biometrics 51:1418–1428.

McCarty JP (2001) Ecological consequences of recent climate change. Biological Conservation 15: 320–331.

Mysterud ACSN, Yoccoz NG, Langvatn R, Steinheim G (2001) Nonlinear effects of large-scale climatic variability on wild and domestic herbivores. Nature 410:1096–1099.

Nychka D, Saltzman N (1998) Design of Air-Quality Monitoring Networks in Case Studies in Environmental Statistics, Lecture Notes in Statistics. Nychka D, Cox L, Piegorsch W (eds.) Springer Verlag, New York.

Pradel R (1993) Flexibility in survival analysis from recapture data: handling trap-dependence. Pages 29–37 in J.D. Lebreton and North PM (eds.) Marked Individuals in the Study of Bird Population. Birkhäuser, Basel.

Root TL, Price JT, Hall KR, Schneider SH, Rosensweig C, Pounds JA (2003) Fingerprints of global warming on wild animals and plants. Nature 421:57–60.

Ruppert D, Wand MP, Carroll RJ (2003) Semiparametric Regression. Cambridge University Press, Cambridge.

Searle SR, Casella G, McCulloch EC (1992) Variance Components. Wiley. New York.

Spiegelhalter DJ, Best NG, Carlin BP, van der Lind A (2002) Bayesian measures of complexity and fit. Journal of the Royal Statistical Society, Series B 64:583–639.

Spiegelhalter D, Thomas A, Best N, Lunn D (2003) WinBUGS User Manual. Version 1.4 (http://www.mrc-bsu.cam.ac.uk/bugs). Technical report, Medical Research Council Biostatistics Unit. Cambridge.

Stenseth N, Mysterud A, Ottersen G, Hurrell JW, Chan K-S, Lima M (2002) Ecological effects of climate fluctuations. Science 297:1292–1296.

Sturtz S, Ligges U, Gelman A (2005) R2WinBUGS: a package for running WinBUGS from R. Journal of Statistical Software 12:1–16.

Walther G-R, Post E, Convey P, Menzel A, Parmesan C, Beebee TJC, Fromentin J-M, Heogh-Guldberg O, Bairlein F (2002) Ecological responses to recent climate change. Nature 416:389–395.

Walther G-R, Berger S, Sykes MT (2005) An ecological 'footprint' of climate change. Proceedings of the Royal Society of London B 272:1427–1432.

White GC, Burnham KP (1999) Program MARK: survival estimation from populations of marked animals. Bird Study 46:120–139.

Wilson C, Adamec D (2002) A global view of bio-physical coupling from SeaWiFS and TOPEX satellite data, 1997–2001. Geophysical Research Letters 29:1–4.

Multivariate State Space Modelling of Bird Migration Count Data

Jonas Knape, Niclas Jonzén, Martin Sköld, and Leonid Sokolov

Abstract We analyse 54 year long time series data on the numbers of common redstart (*Phoenicurus phoenicurus*), common whitethroat (*Sylvia communis*), garden warbler (*Sylvia borin*) and lesser whitethroat (*Sylvia curruca*) trapped in spring and autumn at Ottenby Bird Observatory, Sweden. The Ottenby time series could potentially serve as a reference on how much information on population change is available in count data on migrating birds. To investigate this, we combine spring and autumn data in a Bayesian state-space model trying to separate demographic signals and observation noise. The spring data are assumed to be a measure of the breeding population size, whereas the autumn data measure the population size after reproduction. At the demographic level we include seasonal density dependence and model winter dynamics as a function of precipitation in the Sahel region, south of the Sahara desert, where these species are known to spend the winter. Results show that the large fluctuations in the data restrict what conclusions can be drawn about the dynamics of the species. Annual catches are highly correlated between species and we show that a likely explanation for this is that trapping numbers are strongly dependent on local weather conditions. A comparative analysis of a related data set from the Courish Spit, Russia, gives rather different dynamics which may be caused by low information in the two data sets, but also by distinct populations passing Ottenby and the Courish Spit. This highlights the difficulty of validating results of the analyses when abundance indices derived by other methods or from other populations do not agree.

Keywords Trapping data · State space models · Migration · Bird · Seasonal

1 Introduction

Populations of organisms living in seasonal environments are exposed to different conditions during different parts of the demographic cycle (Fretwell 1972). For

J. Knape (✉)
Department of Theoretical Ecology, Ecology Building, Lund University, SE-223 62 Lund, Sweden
e-mail: jonas.knape@teorekol.lu.se

D.L. Thomson et al. (eds.), *Modeling Demographic Processes in Marked Populations*, Environmental and Ecological Statistics 3, DOI 10.1007/978-0-387-78151-8_4,

migratory organisms such as many bird species, the reproductive success may be mostly influenced by the conditions on the breeding grounds, whereas mortality is probably highest during migration and wintering (Sillett and Holmes 2002). Thus, different parts of the seasonal demographic cycle are affected by the conditions at geographically and environmentally distinct locations. Changes in the environment on wintering grounds, along the route of migration or on the breeding grounds may then have different implications for the dynamics of the species (Saether et al. 2004). Population dynamics of migratory birds is therefore interesting from an ecological perspective but just because of the long distances they travel, collecting appropriate data for analysing their dynamics is difficult.

Currently there is no general method for locating the same individuals or populations at the breeding and wintering grounds and analyses of population dynamics through the full seasonal cycle are usually restricted to count data or to mark-recapture data at either wintering or breeding grounds (but see Webster et al. 2002). Mark-recapture analyses of long distance migrants are often hard due to the possibility of birds returning to sites outside of the study area but can sometimes be used to estimate, e.g., effects of weather conditions on survival (e.g. Peach et al. 1991). Traditionally, data from counts at breeding locations such as the North American Breeding Bird Survey and the Common Bird Census in the United Kingdom have been used to compute indices of population sizes (e.g. James et al. 1996). Analyses of this type of data require care since the data often suffer from variation related to sources at different scales and levels, e.g., differences in skill between observers and differences in detection probabilities between types of habitats, and of biases due to biased selections of surveyed habitats (Thomas 1996; Nichols et al. 2008). Although some recent analyses have tried to take the most serious sources of variation in breeding bird survey data into account (e.g. Link and Sauer 2002), it would be helpful if other types of data could be used to confirm conclusions drawn from analyses of point count data (Dunn and Hussel 1995).

A complementary method for monitoring populations of migratory birds is to use visual counts or trapping numbers of birds at fixed locations during migration. Many bird observatories have data from standardised annual or even biannual catches of passerine birds during the periods of migration. Trapping data from bird observatories have recently been used to study phenology shifts in relation to climate change (e.g. Jonzén et al. 2006) and to estimate population trends and dynamics of passerine birds (e.g. Sokolov et al. 2001; Jonzén et al. 2002; Berthold et al. 2004). However, because of the high between year variation typically present in such data, the use of trapping data as indices of population size has been criticised (Svensson 1978). The day-to-day variation in trapping numbers is high and is influenced by local weather conditions. There are a number of studies analysing daily variation in migration count data with the aim of retrieving population abundance indices (e.g. Dunn et al. 1997; Francis and Hussel 1998). Most of these studies regress daily counts or log counts on sets of weather and time dependent variables and from this derive annual abundance indices. Here we take a different approach and analyse seasonal total trapping numbers using state-space modelling techniques. Thus, instead of accounting for weather effects by estimating adjusted annual indices we deal with the problem of noisy data by integrating the noise in seasonal totals as part of the model.

We analyse trapping numbers from the Ottenby Bird Observatory (Sweden) and from the Courish Spit (Russia) on common redstart (*Phoenicurus phoenicurus*), common whitethroat (*Sylvia communis*), garden warbler (*Sylvia borin*) and lesser whitethroat (*Sylvia curruca*). These species are thought to spend part of the winter in the Sahel area south of the Sahara desert. Previous studies of migrants that winter in the Sahel area have shown that winter survival is dependent on the amount of rainfall in the Sahel area (Peach et al. 1991; Szep 1995). Particularly, a severe drought in the Sahel in the late 1960s and early 1970s (Hulme 1992; Nicholson et al. 1998) is thought to have been the cause of a crash reported for UK populations of common whitethroat (Baillie and Peach 1992) and Hjort and Lindholm (1978) found a strong relationship between the water level in Lake Chad and the number of whitethroats caught at Ottenby the following autumn. With 30 more years of data, we try to verify the influence of conditions at the wintering grounds on the dynamics of this species and compare it to the effects on three other Sahel migrants. We do this using a state-space modelling approach to explicitly deal with the problem of extracting a dynamical process from data in the presence of sampling error. In order to try to determine the relevance of the derived abundance indices as measures of population change we compare an analysis of Ottenby data to an analysis of data from the Courish Spit and to indices from the Swedish Breeding Bird Survey.

2 Materials and Methods

State-space models (Durbin and Koopman 2001) are becoming a standard tool among ecologists working on models of population dynamics (Buckland et al. 2004; Jamieson and Brooks 2004), and have been been extensively used in fisheries stock assessment (e.g. Millar and Meyer 2000). Sampling error is a common feature of data from surveys on wild animal populations and a state-space approach to analysing population dynamics time series data therefore seems natural. For the Ottenby time series on annual catches, there are reasons to believe that a large portion of the variation in trapping data on migratory birds is related to varying external conditions during migration and not to real changes in population sizes (Svensson 1978). This is further supported by the tendency for high between species correlations in total annual catches. The high correlations may be caused by the fact that the species experience similar external conditions during their migration. Because of these potential problems we model multivariate observation disturbances within the state space framework.

Since we are interested in comparing the population dynamics between breeding and wintering seasons we construct a model with two simple dynamical components, one for the breeding season and one for the wintering and migration seasons. Both spring and autumn trapping numbers are inputs for this model which will henceforth be referred to as a seasonal model. The model will allow us to ask questions about how strong forces of density dependence are during summer and winter respectively (Stenseth et al. 2003) and what effect conditions on the wintering grounds have on the winter season population dynamics. We further evaluate if there is any gain in terms of improved parameter estimates and abundance indices in including both spring and autumn data in the same model and in modelling correlated observation disturbances. The results from fitting the model are therefore compared to results from fitting models with uncorrelated observation disturbances and with models where spring and autumn data are included separately (referred to as non-seasonal models).

For all models we make the assumption that the disturbance terms in the process part of the state-space model are independent between species. This assumption may not be entirely satisfying since species having similar life-histories and geographical distributions may well have correlated dynamics even when covariates suspected to influence the dynamics are included in the model. On the other hand, we expect errors in observations to be large and potentially influence similar species in a similar manner.

To get an idea about the validity of the population abundance indices that are derived from the state part of the models as measures of larger scale population change we compare our results with patterns reported for other European populations. We also compare population indices estimated from our model with indices derived with the same analysis of similar data from the bird station at the Courish Spit and with indices from the Swedish Bird Survey which are computed using another type of data (see the Data section below). Both the Courish Spit and the Swedish Breeding Bird Survey may however cover populations distinct from the ones passing Ottenby and therefore comparisons between these indices are not very informative unless results agree.

2.1 Data

Ottenby Bird Observatory (56°12′N, 16°24′E) is situated at the southernmost point of Öland, a 137 km long island ca. 10 km off the coast of south-eastern Sweden. The trapping area in the observatory garden is 1.2 ha and contains most of the higher vegetation within the nearest 2 km, and therefore attracts migratory birds. Birds have been caught at Ottenby in funnel traps of Helgoland-type (Bub 1991) since the first year of trapping in 1946. Since 1960 birds have also been caught in mist nets and to avoid a potential increase in trapping numbers due to the increase in the number of traps we only use data between 1960 and 2005. The start of spring trapping varied considerably between 1952 and 1979, whereas from 1980 onwards, the spring trapping started on March 15 and ended on June 15. The spring passage of the species analysed in this paper is mainly in May, which has been well covered in all years except for 1966 and 1967 when there were no spring trappings. These years are treated as missing data points. The spring data we use is the total number of birds caught per year between March 15 and June 15 in the Helgoland traps and in the mist nets. The autumn trapping season starts on July 25 and ends on November 15. In some years the season ended before November 15, but very few birds of the species studied in this paper are trapped after mid October. By using the total number of individuals trapped per year between July 25 and October 25 during 1960–2005 we include 99.9% of all trapped individuals of the species studied here. Both juveniles and adult birds are caught in autumn but there is an over representation of juveniles for most species. Since age classification were not complete for all years, both adults and juveniles are included in our data. For more details about the trapping conditions, see Stervander et al. (2005).

Data from the Courish Spit consist of the number of birds caught in two traps between 1977 and 2005. The project was carried out by the Biological Station Rybachy of the Zoological Institute, Russian Academy of Sciences (Sokolov et al. 2000). Index values from the Swedish Breeding Bird Survey (SBBS) (Lindström and Svensson 2005) are available from 1975. These indices are derived from point counts along routes freely chosen by observers and are therefore potentially subject to biases such as habitat bias, differences in skill between observers, etc. (Thomas 1996). Annual Sahel rainfall indices were obtained from the web-page of the Joint Institute for the Study of the Atmosphere and the Ocean, http://jisao.washington.edu/data_sets/sahel, and are computed as the mean of monthly rainfall indices from June through October. The annual indices were standardised for the period 1950–2004.

2.2 Models

A (rather) general definition of a multivariate linear Gaussian state-space model with covariates can be given as

$$\begin{aligned} \mathbf{y}_t &= Z_t \mathbf{x}_t + \boldsymbol{\epsilon}_t, \quad && \boldsymbol{\epsilon}_t \sim N(0, \Omega_t) \\ \mathbf{x}_{t+1} &= T_t \mathbf{x}_t + W_t \boldsymbol{c} + \boldsymbol{\eta}_t, \quad && \boldsymbol{\eta}_t \sim N(0, \Sigma_t) \end{aligned} \tag{1}$$

for $t = 1, 2, \ldots, n$, where all $\boldsymbol{\epsilon}_t$ and $\boldsymbol{\eta}_t$ are independent (the parameters of the $N(\boldsymbol{\mu}, \Sigma)$-distribution denote the mean vector and variance matrix respectively, vectors are denoted by bold face and matrices by capital letters). The first state vector, $\mathbf{x}_1$, also needs to be defined to complete the model specification. This can be done in various ways, and in our models described later in this section the initial vector is treated as a parameter with an informative prior. An interpretation of the model is that the vectors $\mathbf{y}_t$ represent the data which are noisy observations of linear transformations (Z_t) of hidden state vectors $\mathbf{x}_t$ which need not be of the same size as the vectors of observations. The hidden state is a linear normal stochastic process with autoregression coefficient matrix T_t. The matrix W_t contains covariates for the transition from t to $t + 1$ and their (linear) effect on the process is measured by the regression coefficients in the vector $\mathbf{c}$. Depending on the setting, the elements of the matrices of the model may either be completely specified or may depend on unknown parameters.

All our models of the bird observatory data are special cases of the more general model defined above. To find out if anything is gained by using both spring and autumn data in the same model we used both a seasonal and a non-seasonal version of the state space model. For the non-seasonal model, the data are arranged so that the vectors $\mathbf{y}_t$ contain the log of the total seasonal trapping numbers in year t in either spring or autumn for the four species. The system is modelled on the log scale in line with common practise in studies on population dynamics. The log transformation also had the effect of making the data appear more Gaussian. The non-seasonal model is a simplified version of the model in (1):

$$\begin{aligned} \mathbf{y}_t &= \mathbf{x}_t + \boldsymbol{\epsilon}_t, \quad \boldsymbol{\epsilon}_t \sim N(0, \Omega) \\ \mathbf{x}_{t+1} &= \mathbf{a} + B\mathbf{x}_t + r_t\mathbf{c} + \boldsymbol{\eta}_t, \quad \boldsymbol{\eta}_t \sim N(0, \Sigma). \end{aligned} \tag{2}$$

The vectors $\mathbf{a}$ and $\mathbf{c}$ of length 4 contain parameters a_i and c_i respectively on position i. The autoregression coefficient matrix B is diagonal with parameter b_i on position (i, i). All the regression parameters a_i, b_i and c_i are different between species so that there are no shared parameters between species in the deterministic part of the model. The hidden state vector $\mathbf{x}_t$ of length 4 should be interpreted as the logarithm of population indices for the species in year t. The scalars r_t are indices of mean Sahel rainfall during the wet season (June–October) in year t. The amount of rainfall here serves as a surrogate for availability of food and water for the populations during the winter, which in turn might affect winter survival (Peach et al. 1991). The process disturbance variance matrix Σ was constrained to be diagonal with entries σ_i^2 on the diagonal. We consider two models for the observation disturbance variance matrix Ω. In the first Ω is allowed to be non-diagonal and all elements of the matrix are estimated. In the second model Ω is constrained to be diagonal, meaning that we have a set of four independent models for the species.

Based on the above definition, the process part of the model for species i can be written as:

$$x_{it+1} = a_i + b_i x_{it} + c_i r_t + \eta_{it},$$

where subscripts i refer to element i of the vectors. Thus the processes are $AR(1)$-processes with covariates and since the model is defined for the log of the data, this can be seen as a Gompertz model for the population dynamics (see e.g. Royama 1992).

The quantities $1 - b_i$ in the Gompertz model can be interpreted as measures of density dependence in growth. The log-linearity of the process guarantees that the coefficients b_i are invariant to multiplying the population process $\exp(x_i)$ by a constant. More specifically, if

$$N_{t+1} = \exp(x_{t+1}) = \exp(a + bx_t + \eta_t) = N_t^b \exp(a + \eta_t),$$

and the population size is rescaled to an index $M_t = kN_t$, then

$$M_{t+1} = kN_{t+1} = (kN_t)^b \exp(a + (1 - b)\ln k + \eta_t) = M_t^b \exp(a' + \eta_t),$$

where $a' = a+(1-b)\ln k$. Hence, the parameter b can be interpreted as a measure of density dependence regardless of the value of the constant of proportionality implied by the index interpretation of $\exp(x_i)$. The parameter a on the other hand depends on the constant and is thus of little interest to us. A critical assumption of the model is that k, the proportionality constant linking the trapping numbers to the "true" population size is constant through time. In fact, the same interpretations of the model parameters hold if the assumption is lightened by letting $\ln k$ be independent and identically distributed according to a normal distribution since the terms then can be seen as a part of the disturbance terms η_t. If however k is not independent over time or if k depends on population size it might well affect estimates of e.g. density dependence and abundance indices.

For the seasonal model, both spring and autumn trapping numbers are included simultaneously. We let $\mathbf{y}_t^a$ be vectors containing the log of autumn trapping numbers for the four species and $\mathbf{y}_t^s$ be vectors containing the log of the spring trapping numbers. Using sub- and superscripts b and w referring to breeding and winter season respectively and sub- and superscripts a and s referring to autumn and spring, the model is defined as:

$$\begin{aligned} \mathbf{y}_t^a &= \mathbf{x}_t^a + \boldsymbol{\epsilon}_t^a, \quad \boldsymbol{\epsilon}_t^a \sim N(0, \Omega_a) \\ \mathbf{x}_t^a &= \mathbf{a}_b + B_b\mathbf{x}_t^s + \boldsymbol{\eta}_t^b, \quad \boldsymbol{\eta}_t^b \sim N(0, \Sigma_b) \end{aligned} \tag{3}$$

$$\begin{aligned} \mathbf{y}_t^s &= \mathbf{x}_t^s + \boldsymbol{\epsilon}_t^s, \quad \boldsymbol{\epsilon}_t^s \sim N(0, \Omega_s) \\ \mathbf{x}_t^s &= \mathbf{a}_w + B_w\mathbf{x}_{t-1}^a + r_{t-1}\mathbf{c}_w + \boldsymbol{\eta}_t^w, \quad \boldsymbol{\eta}_t^w \sim N(0, \Sigma_w). \end{aligned} \tag{4}$$

In the same way as for the seasonal model, the vectors $\mathbf{a}_b$, $\mathbf{a}_w$ and $\mathbf{c}_w$ contain species specific parameters and the matrices B_b and B_w are diagonal with species specific autoregressive parameters for measuring seasonal density dependence. Again, two versions of observation disturbance variance matrices are considered, in the first these are non-diagonal and in the second they are diagonal. The process disturbance variance matrices are diagonal. This model is also included in the general definition in (1), but here the index t refers to year.

Similarly to the non-seasonal model, $\exp(x_t^s)$ and $\exp(x_t^a)$, should be interpreted as indices of spring and autumn population sizes respectively. However, since it may well be that different populations or parts of populations pass the observatories in spring and in autumn respectively, the spring and autumn indices may not share the same constant of proportionality to the "true" population size. In the same way as above, the parameters b_w and b_s are invariant to multiplying the population time series by a constant.

To try to validate the assumption of correlated observation disturbances we compared the estimated correlation matrices of the observation disturbances to a heuristic estimate calculated from the amount of overlap in migration between the species. The sum of daily catches over the years from 1950 to 2005 for each species was divided by the total number of catches for the species. The heuristic estimate

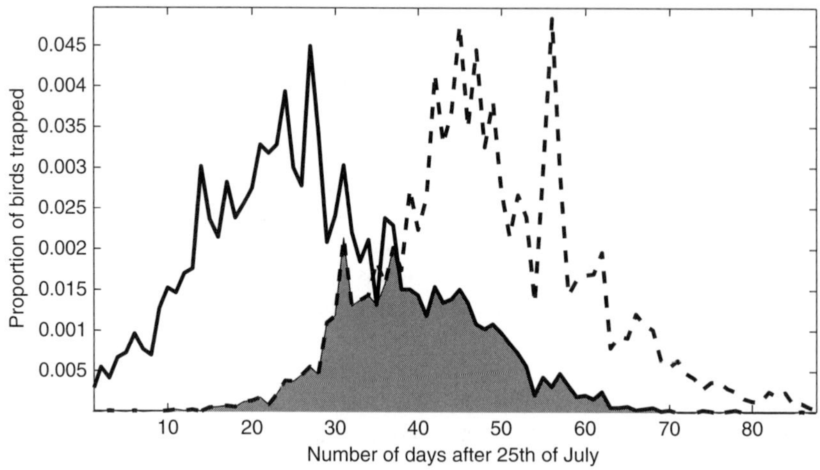

Fig. 1 Autumn phenology curves for whitethroat (*solid line*) and redstart (*dashed line*) and the amount of overlap in migration period (*shaded area*)

of the correlation was then computed as the area under the overlap of the curves (Fig. 1). The correlation of the observation disturbances between species i and species j was estimated as the posterior mean of element (i, j) of the Ω matrix divided by the square root of the product of the posterior means of the elements (i, i) and (j, j) of the same matrix.

2.3 Priors

The model was fitted to data from 1960 to 2005 for Ottenby and from 1977 to 2005 for the Courish Spit. The state vectors of the seasonal model were initialised by putting normal priors on x_{i1}^s with means equal to the mean of the log of spring data between 1950 and 1959 and with prior standard deviations set to 1.5 times the empirical standard deviations over these periods. Analogous priors were used for the initial states of the non-seasonal models. Courish Spit data between 1957 and 1976 was used as prior information for the initial state of the model of the Courish Spit data.

For the other parameters we used vague priors (given below) since there was no obvious a priori information available. To improve convergence of the Gibbs sampler (see Section 2.5), the regression was centred around $m_x = 5$, i.e. the model in (2) was reparametrised as

$$\mathbf{x}_{t+1} = \mathbf{a}' + B'(\mathbf{x}_t - m_x\mathbf{1}) + r_t\mathbf{c} + \boldsymbol{\eta}_t,$$

where $\mathbf{1}$ is a vector of ones. This parametrisation gives the same interpretation of B' as of B. The components of the regression parameter vectors $\mathbf{a}'$ and $\mathbf{c}$ were then given independent $N(0, 100)$-prior distributions. When $|b_i'| > 1$, the model is non-stationary and there is no reason to expect extreme non-stationarities of the population indices. The autoregression parameters b'_i were therefore given slightly more informative $N(0, 10)$-priors. The stationarity argument does not translate directly to the seasonal model in 3, but to simplify comparisons between the models we used the same priors for the corresponding parameters a'_b, a'_w, b'_b and b'_w.

For the variances of the observation and state disturbances, the information in the data on separating the two can be low (Dennis et al. 2006). Therefore, if there is no prior information on the relative size of these it is desirable to give them similar priors. The variances of the state disturbances, σ_i^2, were given conditionally conjugate improper inverse gamma distributions with shape parameter 0 and inverse scale parameter 0.01, $IG(0, 0.01)$ (Fig. 2). In the models with diagonal observation error variance matrices, Ω, the elements on the diagonal were also given $IG(0, 0.01)$-priors. When the matrices Ω, Ω_s and Ω_a were allowed to be non-diagonal we gave them improper inverse Wishart priors with 3 degrees of freedom and scale matrix $0.02I$ where I is the identity matrix. Since the marginal distribution of the elements on the diagonal of a matrix having an inverse Wishart distribution with scale matrix V of size $p \times p$ and ν degrees of freedom is an $IG((\nu - p + 1)/2, V_{ii}/2)$-distribution, the marginal prior distributions for the elements on the diagonal then also correspond to $IG(0, 0.01)$-distributions.

We analysed sensitivity to priors by changing the prior distribution for the σ_i^2 parameters to an improper $IG(-0.5, 0.001)$ distribution and at the same time

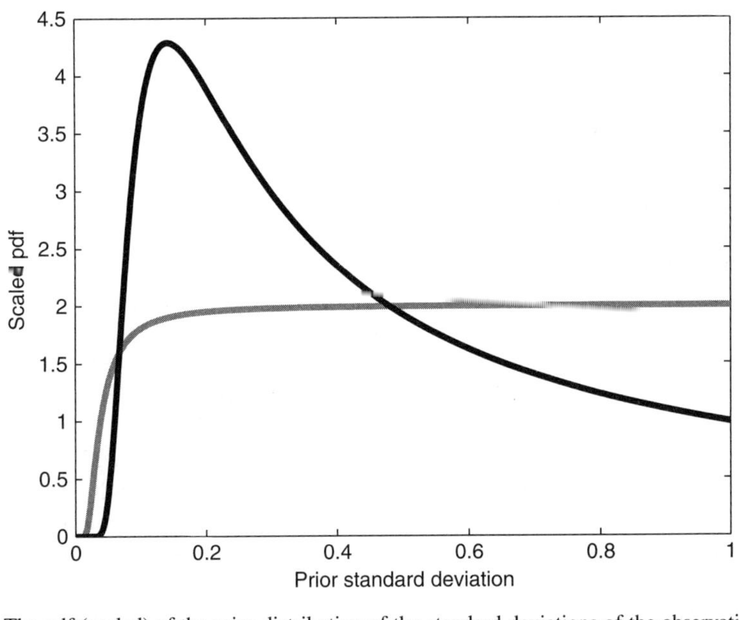

Fig. 2 The pdf (scaled) of the prior distribution of the standard deviations of the observation and process disturbances (*black line*) and of the alternative prior used for sensitivity analysis (*grey line*). The priors on the standard deviations correspond to $IG(0, 0.01)$ and $IG(-0.5, 0.001)$ prior distributions on the variances respectively

changing the parameters of the inverse Wishart distribution to 2.5 degrees of freedom and scale matrix equal to $0.002I$. In this way the priors on the variances of the process disturbances and on the variances of the observation disturbances are kept identical. The $IG(-0.05, 0.001)$ distribution on the variances is close to a uniform distribution on the standard deviations except that it has low support for small values (Fig. 2). It also lacks the peak of the $IG(0, 0.01)$ probability density function.

2.4 Goodness of Fit

Bayesian p-values build on some measure of discrepancy between the model and data and is the posterior probability that a replicate data set yields a larger value of this measure. We used the deviance, i.e.

$$D(y, \theta) = -2 \log f_{Y|\theta}(y)$$

where $f_{Y|\theta}(y)$ is the likelihood of the data given the vector θ containing the regression parameters and the parameters of the variance matrices of the observation and process errors, as a measure of discrepancy. The likelihood was computed using the Kalman filter (see e.g. Durbin and Koopman 2001).

Goodness of fit was also checked by analysing the residuals of the models. Residuals in the state space model, $\hat{\epsilon}$ and $\hat{\eta}$, can be defined as the expected value of ϵ and η given the data and given the parameters θ equal to their posterior marginal mean. The residuals were analysed by computing their correlation, autocorrelation and by qq-plots.

2.5 Fitting the Models

The model was fit by implementing a Gibbs sampler in the program Matlab. In each iteration of the sampler, all the state vectors $\mathbf{x}_t^s$ and $\mathbf{x}_t^a$ were updated simultaneously using the Kalman simulation smoother of Durbin and Koopman (2002). All the regression parameters a, b and c for all species and both seasons were updated as a block according to their multivariate normal conditional posterior. The inverse Wishart prior is conditionally conjugate for the observation variance matrices Ω. When these matrices were allowed to be non-diagonal, they were therefore updated by simulating a draw from the inverse Wishart posterior. For the diagonal variance matrices, each diagonal element was updated with a draw from the inverse gamma conditional posterior. The Gibbs sampler for the non-seasonal models was constructed in a similar manner but is even more simple since the regression parameters a, b and c then are the same for each time step of the state space model.

The sampler was run with a single chain for half a million iterations where the first 20,000 iterations were discarded as burn. Every 20th value of the output was then used as a draw from the posterior. Convergence and mixing of the MCMC's were investigated by looking at trace plots and autocorrelation functions of the thinned chains. Visual inspection revealed no sign of poor mixing and all autocorrelations of the thinned chains were below 0.15 at lag 5. All chains seemed to have converged after just a few iterations.

Starting values of the MCMC were chosen at least a small distance away from the expected region of high posterior density. Specifically, the state vectors were initialised to 1 for all species, all the regression parameters, a, b and c were initially set to 2 and all variance matrices were initialised as identity matrices.

Although we have not done so, we believe all our models could be implemented and fit in e.g. the program WinBUGS if the priors on the variance matrices are changed to proper ones. The non-seasonal model with uncorrelated observation disturbances is especially simple and parameter estimates can be obtained using maximum likelihood or REML methods (see e.g. Dennis et al. 2006). These methods can probably also be used for estimating parameters of at least some of our more complex models.

3 Results

Unless otherwise stated, the results below refer to the Ottenby data. Tests always refer to the informal test of whether or not 95% credible intervals contain the value of the null hypothesis. Estimates of abundance indices from the non-seasonal model on autumn catches with observation disturbances allowed to be correlated across species are shown in Fig. 3. A comparison with estimates from spring catches (Fig. 4) shows that on a coarse (long term) scale, the indices for the two data sets have similar tendencies with sample correlations 0.8, 0.6, 0.3 and 0.5 for redstart, whitethroat, garden warbler and lesser whitethroat respectively. (Note that the sample correlations should be interpreted with care as the indices are autocorrelated.) The estimated whitethroat indices show declines in the early 1970 and 1980s which roughly coincide with droughts in the Sahel area. Declines in numbers of whitetroats following these droughts have been reported in the UK. A decline by the time of the first drought was reported for redstarts in the UK (Gibbons et al. 1993) and the indices derived here decline at about the time of the start of the drought in the late 1960s but this result is weaker than for the whitethroat indices. Any trends in the garden warbler and lesser whitethroats indices are less clear although there was a drop in autumn catches of lesser whitethroats in the early 1970s and a sudden drop in both spring and autumn catches of garden warblers around 1990. The regression coefficient for Sahel rainfall, c, is only significantly larger than zero in the model of whitethroat autumn data (Table 1). Estimates of parameters representing density dependence, b, all had wide credible intervals that don't allow us to make any comparisons between species or seasons (Table 1). However, the credible intervals of b for whitethroat and garden warbler are well separated from one. Since b equal to one represents density independence this could be an indication of some degree of density dependence, but because of the wide credible intervals we avoid drawing any firm conclusions.

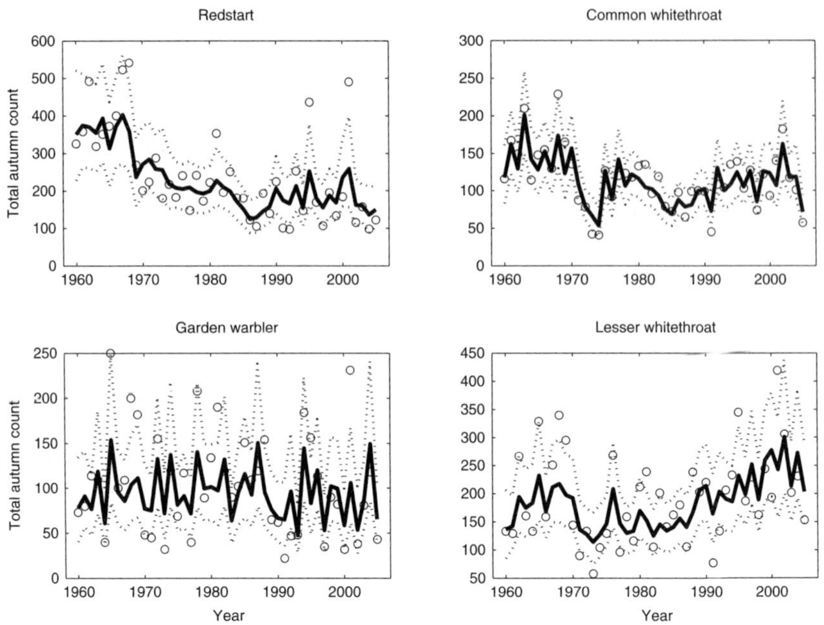

Fig. 3 Posterior mean abundance indices from the non-seasonal model of Ottenby autumn data (*black lines*) with 95% credibilty bands (*dotted lines*). The circles denote the observed data

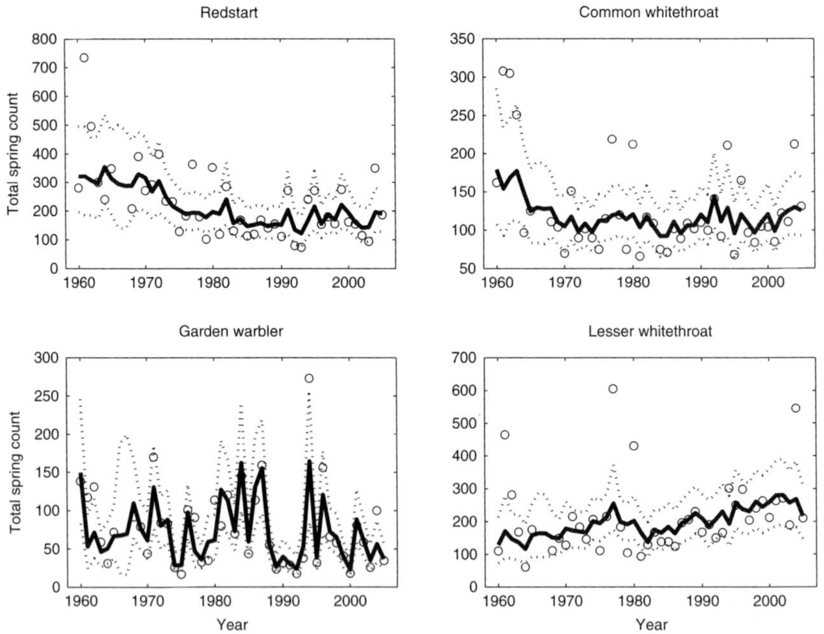

Fig. 4 Posterior mean abundance indices from Ottenby spring data (*black lines*) with 95% credibility bands (*dotted lines*). The circles denote the observed data

Table 1 Parameter estimates and 95% credibility intervals for the non-seasonal models applied to autumn and spring data at Ottenby and to autumn data at the Courish Spit. Estimates of the square root of the elements on the diagonal of the observation disturbance variance matrix Ω are denoted ω_i. Indices refer to redstart, r, whitethroat, w, garden warbler, g, and lesser whitethroat l. All models in the table have correlated observation disturbances

Parameter	Ottenby spring		Ottenby autumn		Courish spit autumn	
p-value	0.62		0.51		0.51	
b_r	0.58	(−0.06, 0.96)	0.63	(0.20, 0.94)	0.24	(−0.45, 0.76)
b_w	0.25	(−0.63, 0.82)	0.11	(−0.27, 0.45)	0.35	(−0.17, 0.78)
b_g	0.13	(−0.26, 0.51)	−0.30	(−0.90, 0.22)	0.02	(−0.52, 0.54)
b_l	0.57	(−0.27, 0.98)	0.43	(−0.11, 0.93)	0.65	(0.11, 1.07)
c_r	0.08	(−0.05, 0.22)	0.10	(−0.03, 0.24)	−0.19	(−0.48, 0.08)
c_w	0.07	(−0.05, 0.20)	0.27	(0.15, 0.39)	0.05	(−0.21, 0.32)
c_g	−0.17	(−0.44, 0.09)	−0.01	(−0.26, 0.25)	−0.18	(−0.50, 0.12)
c_l	0.02	(−0.10, 0.17)	0.13	(−0.01, 0.28)	−0.13	(−0.47, 0.19)
σ_r	0.22	(0.07, 0.40)	0.24	(0.08, 0.41)	0.37	(0.23, 0.58)
σ_w	0.16	(0.07, 0.28)	0.20	(0.09, 0.29)	0.42	(0.22, 0.62)
σ_g	0.57	(0.41, 0.75)	0.36	(0.12, 0.55)	0.41	(0.13, 0.66)
σ_l	0.19	(0.08, 0.32)	0.22	(0.10, 0.34)	0.41	(0.13, 0.66)
ω_r	0.39	(0.25, 0.52)	0.32	(0.19, 0.42)	0.41	(0.17, 0.69)
ω_w	0.34	(0.24, 0.45)	0.23	(0.13, 0.33)	0.35	(0.14, 0.61)
ω_g	0.39	(0.19, 0.62)	0.48	(0.30, 0.68)	0.50	(0.24, 0.78)
ω_l	0.42	(0.30, 0.56)	0.34	(0.23, 0.47)	0.45	(0.18, 0.76)

When combining spring and autumn data in the seasonal model, the derived indices appear more similar to the indices from the non-seasonal model of autumn data than to the indices from the non-seasonal model of spring data (Fig. 5). This indicates that the information in the spring data is less than the information in the autumn data in agreement with a belief that spring ringing figures at Ottenby are more dependent on local weather than autumn figures (Hjort and Lindholm 1978). Estimates of the regression coefficients on standardised Sahel rainfall, c_i, are qualitatively similar between the seasonal and the non seasonal models with a positive effect for whitethroat (Table 2). There is however a stronger indication of a positive effect of Sahel rainfall for redstart in the seasonal model even though it is barely significantly larger than zero. As for density dependence the credible intervals are still very wide and not much can be said about differences between seasons. The whitethroat estimates of b are however lower in the winter season than in the breeding season although this is not significant at the 95% level. We therefore leave it as a hypothesis that whitethroats are more strongly regulated by density dependence in the period between leaving and arriving at the breeding grounds than in the period spent at the actual breeding grounds. As an indication of whether or not the combined model improved abundance indices we summed the lengths of the 95% credible intervals of the log abundance indices x_{it} for each species across time for spring and autumn indices separately. This was done for both the seasonal and the non-seasonal models. We then computed the percent reduction of these summed totals for the seasonal model compared to the non-seasonal models. The total lengths of the log spring index credible intervals were reduced by 10, 15, 13 and 9 percent

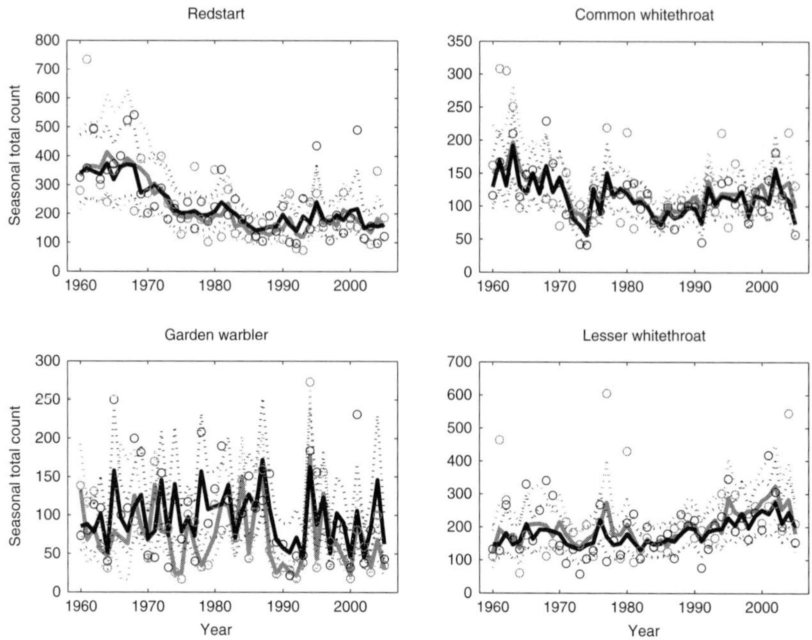

Fig. 5 Posterior mean abundance indices from the seasonal model of Ottenby data with 95% credibility bands (*dotted lines*). Grey lines denote spring indices and black lines denote autumn indices

for redstart, whitethroat, garden warbler and lesser whitethroat respectively. The analogous reductions in autumn were 6, 1, 9 and 4 percent. Hence in this sense the seasonal model performes slightly better than the non-seasonal models. The fact that the reduction is larger for the spring indices also supports the conclusion above that the spring data are less informative.

The reduction in total length (both spring and autumn) of log abundance credible intervals when moving from the seasonal model with independent observation disturbances to the seasonal model with correlated observation disturbances are 9% for redstart, 4% for whitethroat, 30% for garden warbler and 22% for lesser whitethroat.

The agreement between SBBS-indices and indices from the non-seasonal model of spring and autumn data is weak (Fig. 6). Sample correlations between SBBS-indices and autumn indices from the non seasonal model were 0.4, 0.2, −0.1 and 0.1 for redstart, whitethroat, garden warbler and lesser whitethroat. For indices from the non-seasonal model of spring data these correlations were 0.5, 0.3, 0.1 and 0.2. For the redstart, both the SBBS and our indices indicate a decline in the early 1980s but that is much more marked in the former. For the whitethroat no decline at all at this point is seen in the SBBS indices. A noticeable feature is that a sudden sharp decline in whitethroats in 1991 occurs in both autumn data at Ottenby and in the SBBS-indices and is further consistent with a drop in the British CBC-indices (Gibbons et al. 1993).

Table 2 Parameter estimates with 95% credible intervals for the seasonal model with correlated and uncorrelated* observation disturbances applied to data from Ottenby. First indices refer to breeding season b, winter season w, spring observation s and autumn observation a. Second indices refer to species as in Table 1.

	Seasonal model		Seasonal model*		Prior sensitivity	
p-value	0.60		0.68		0.66	
b_{br}	0.72	(0.39, 1.13)	0.88	(0.38, 1.50)	0.68	(0.33, 1.13)
b_{bw}	1.05	(0.56, 1.79)	1.33	(0.65, 2.38)	1.08	(0.52, 1.99)
b_{bg}	0.28	(0.01, 0.54)	0.46	(−0.12, 1.50)	0.27	(0.01, 0.53)
b_{bl}	0.62	(0.19, 1.14)	0.57	(−0.39, 1.98)	0.55	(0.15, 1.04)
b_{wr}	0.92	(0.42, 1.49)	0.81	(0.36, 1.41)	0.83	(0.32, 1.48)
b_{ww}	0.20	(−0.13, 0.60)	0.22	(−0.04, 0.57)	0.16	(−0.15, 0.55)
b_{wg}	−0.03	(−0.73, 0.65)	0.26	(−0.83, 1.89)	−0.01	(−0.60, 0.54)
b_{wl}	0.68	(0.10, 1.55)	0.14	(−1.25, 1.67)	0.70	(0.08, 1.76)
c_{wr}	0.13	(0.00, 0.27)	0.09	(−0.02, 0.23)	0.14	(0.00, 0.31)
c_{ww}	0.19	(0.09, 0.31)	0.17	(0.07, 0.29)	0.19	(0.08, 0.31)
c_{wg}	−0.10	(−0.34, 0.15)	−0.12	(−0.37, 0.12)	−0.10	(−0.35, 0.16)
c_{wl}	0.10	(−0.03, 0.24)	0.09	(−0.08, 0.25)	0.10	(−0.04, 0.25)
σ_{br}	0.18	(0.07, 0.35)	0.18	(0.06, 0.40)	0.21	(0.04, 0.41)
σ_{bw}	0.16	(0.07, 0.25)	0.17	(0.07, 0.29)	0.17	(0.05, 0.27)
σ_{bg}	0.40	(0.17, 0.54)	0.31	(0.08, 0.62)	0.43	(0.27, 0.56)
σ_{bl}	0.16	(0.07, 0.28)	0.23	(0.07, 0.44)	0.17	(0.05, 0.30)
σ_{wr}	0.18	(0.07, 0.35)	0.15	(0.06, 0.33)	0.21	(0.04, 0.41)
σ_{ww}	0.14	(0.07, 0.24)	0.13	(0.06, 0.22)	0.15	(0.05, 0.26)
σ_{wg}	0.61	(0.46, 0.79)	0.49	(0.12, 0.79)	0.63	(0.49, 0.81)
σ_{wl}	0.20	(0.09, 0.31)	0.22	(0.07, 0.47)	0.23	(0.08, 0.34)
ω_{sr}	0.38	(0.26, 0.50)	0.39	(0.27, 0.51)	0.37	(0.22, 0.51)
ω_{sw}	0.35	(0.27, 0.45)	0.36	(0.28, 0.46)	0.36	(0.27, 0.46)
ω_{sg}	0.31	(0.13, 0.53)	0.42	(0.09, 0.75)	0.30	(0.10, 0.53)
ω_{sl}	0.41	(0.30, 0.53)	0.38	(0.12, 0.55)	0.40	(0.28, 0.53)
ω_{ar}	0.33	(0.21, 0.44)	0.32	(0.15, 0.45)	0.31	(0.16, 0.44)
ω_{aw}	0.23	(0.14, 0.34)	0.18	(0.07, 0.31)	0.22	(0.13, 0.33)
ω_{ag}	0.41	(0.24, 0.62)	0.43	(0.10, 0.68)	0.39	(0.21, 0.60)
ω_{al}	0.36	(0.25, 0.47)	0.31	(0.09, 0.49)	0.36	(0.25, 0.49)

Estimated abundance indices for the Courish Spit data show a quite different picture than the Ottenby estimates (Fig. 7). There is e.g. a decreasing trend in the lesser whitethroat and a drop in the number of whitethroats in the mid 1990s. No clear effect of Sahel rainfall is found for the Courish Spit data (Table 1). Credibility intervals for the parameter estimates are in most cases too wide to allow for comparisons with estimates from Ottenby data but, except for the garden warbler, estimates of state disturbance variances are less precise for the Courish Spit data.

A comparison between the heuristic correlation estimate and the estimate from fitting the seasonal model with correlated observation disturbances (Table 3) reveals, especially in autumn, a close agreement between the two. The estimates from the model are in general higher than the heuristic estimates, but sample correlations between the off diagonal correlation estimates were 0.99 in autumn and 0.76 in

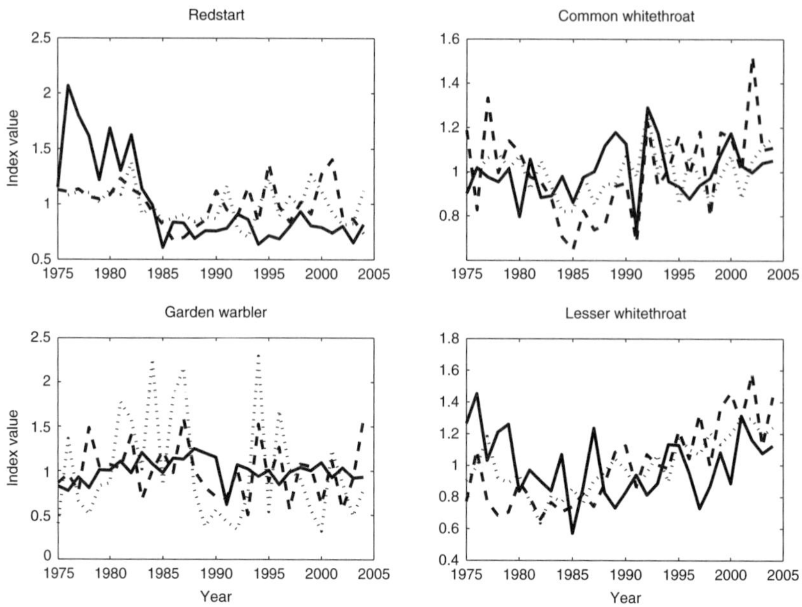

Fig. 6 Indices from the Swedish breeding bird survey (*solid line*) and from the non-seasonal model of data from Ottenby in spring (*dotted line*) and autumn (*dashed line*). The indices are scaled to have mean equal to one for the given time period

spring. A good estimate of the overlap in trapping season between the species can thus be computed from just total annual catches.

P-values did not indicate signs of bad fit for any of the models (Tables 1 and 2). Analysis of residuals showed that the model with correlated observation disturbances in total had somewhat less correlation and autocorrelation of the residuals than the model with independent observation disturbances. Also, the observation disturbance residuals of the seasonal model show that the fit is worse for spring than for autumn data (Fig. 8). None of the autocorrelations of the residuals at lag 1 were larger than 0.2 but correlations between residuals for different species were in some cases larger than expected. This was true for all of the models we considered.

Table 3 Estimates of correlations in observation disturbances from the seasonal model and the heuristic estimate. Upper right triangles show estimates from spring data and lower left triangles from autumn data

Model estimate					Heuristic estimate			
	r	w	g	l	r	w	g	l
r	1.00	0.60	0.64	0.84	1.00	0.67	0.50	0.84
w	0.51	1.00	0.79	0.83	0.39	1.00	0.77	0.79
g	0.76	0.73	1.00	0.76	0.69	0.62	1.00	0.62
l	0.66	0.84	0.85	1.00	0.57	0.77	0.82	1.00

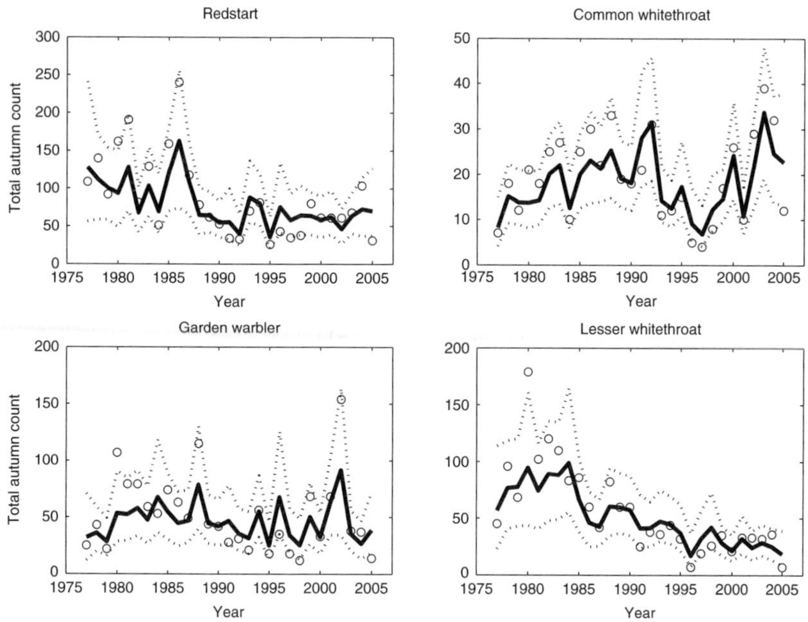

Fig. 7 Posterior mean abundance indices from autumn data at the Courish Spit spring (*black lines*) with 95% credibility bands (*dotted lines*). The circles denote the observed data

A reason for this might for example be that the correlations or variances of the disturbances are not constant through time as we assume in our models.

Residual analysis for the Courish Spit data also show a worse fit than Ottenby data for at least whitethroat and lesser whitethroat, which are the species that are caught in lowest numbers. Using the log of total counts is not very appropriate when counts are small and an overdispersed Poisson model of observations could have been a better alternative here.

4 Discussion

The linear dynamics derived from our state space model of trapping data is presumably a mix of "true" variation in population abundance, of weather dynamics or other external forces that influence migration patterns and of trapping probabilities and possibly also of dynamic changes in migratory routes. The relative influence of these processes determines the amount of information available in the data and the relevance of the data as indicators of population size. It is however hard to assess this amount of information unless there is a close agreement between analyses of various kinds of data at several locations. Different methods of recording and analysing data may give rise to different kinds of bias and geographically (or temporally) separated populations may experience different conditions that cause differences in

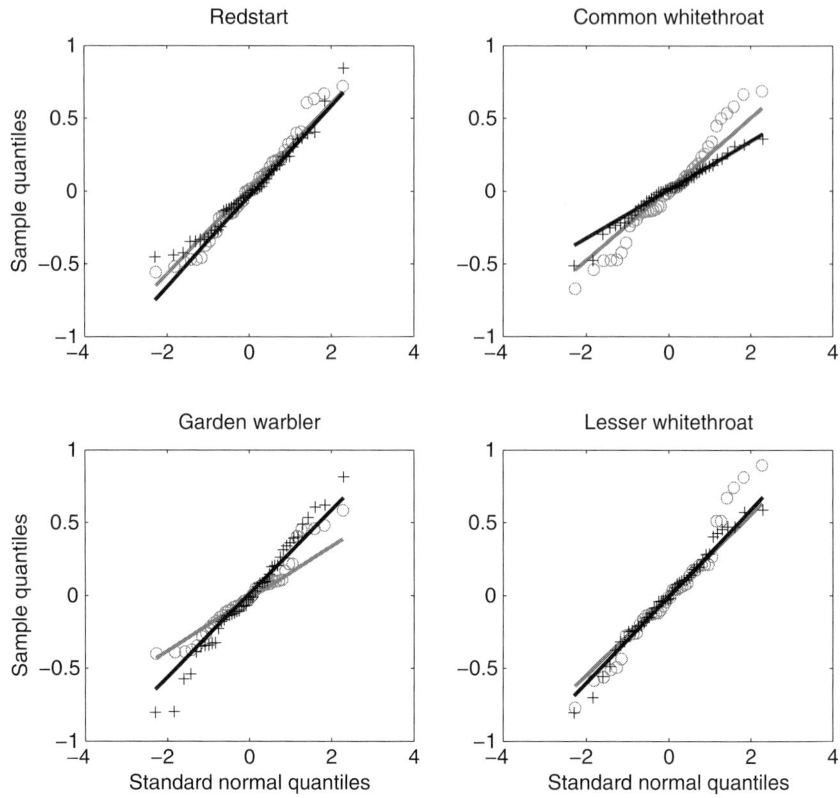

Fig. 8 Qq-plots of observation disturbance residuals for spring (*grey circles*) and autumn (*black crosses*) for the seasonal model with correlated observation disturbances

dynamics. The rough agreement between spring and autumn indices from Ottenby may thus indicate that there is a relation to changes in abundance but let alone can not exclude the possibility of e.g. dynamical changes in catching probabilities. The lack of agreement between indices from Ottenby, the Courish Spit and the SBBS on the other hand can not be taken as more than an indication that some of the indices are not very precise as there may well be differences in both breeding and wintering locations of the populations passing the stations. What can be done using statistical analyses of ringing figures is to set limitations to what information can be extracted from the data. This is exemplified by the fact that autoregressive parameters in our model could not be estimated with any reasonable degree of precision.

The good agreement between the heuristic estimates of correlations in observations and the matrices estimated from the model (Table 3) indicates that the assumption of correlated observation disturbances and independent population dynamics is reasonable. The relatively high variance of these disturbances in turn show that catches are highly dependent on extrinsic factors which has been hypothesised before (Svensson 1978). Ignoring effects of correlated measurement errors could cause the dynamics between species that migrate during the same time period to appear overly coherent, if for no other reason, simply because the sample size is overestimated. There is not much sign of such an effect in this analysis but it is important to be aware of the correlations in the data, not only when analysing population dynamics or population sizes but in all analyses relying on ringing figures (e.g. analyses of phenology). We expect that the high correlation in catches is not a special feature of the Ottenby and Courish Spit data but rather is common in counts of populations during migration.

Despite the high variance in observation disturbances, there are still hints that there is some valuable information in the data, at least for some species. There is support for a positive effect of Sahel rainfall on between year fluctuations in abundance indices for whitethroats and weak support for the same effect on redstarts. Visual inspection of the abundance indices for redstart and whitethroat show declines following the Sahel droughts and give some support for the possibility of picking up clear population trends in the data for some species. When the purpose is to produce visual population trends, the state-space modelling approach could be used to produce more smoothed estimates than the ones given here. For example, a local linear trend model (Durbin and Koopman 2001) could be used instead of the autoregressive model.

Our analyses give some support to the view that spring catches at Ottenby are less informative about population sizes than autumn catches (Hjort and Lindholm 1978). The location of Ottenby at the southern tip of Öland may influence the dynamics of spring and autumn catches differently. In autumn, migrating birds may use Öland as a lead line on their southward migration whereas no such lead line is available for birds passing Ottenby in spring (Stervander et al. 2005). This can lead to spring catches being more dependent on local weather conditions (Hjort and Lindholm 1978).

Due to the high variance in observation disturbances, we believe that ringing figures from bird stations are not very suitable for picking up even drastic changes in population abundances. However, long term ringing figures on migrating birds from bird stations with carefully standardised trapping methods might in some cases be useful in recovering long term trends and biological information but any conclusions from such analyses need to be confirmed by independent data.

Acknowledgments The authors would like to thank Martin Stervander for his help on information about the standardisation of the Ottenby data and Emma Ådahl and Fredrik Haas for fruitful discussions. Len Thomas and Charles Francis gave comments on the manuscript that greatly helped in improving the paper. We also thank everyone who has been involved in collecting and managing data through the years. J.K. and N.J. are financially supported by the Swedish Research Council. This is contribution no 224 from Ottenby Bird Observatory.

References

Baillie SR, Peach WJ (1992) Population limitation in Palaearctic-African migrant passerines. IBIS 134(suppl. 1):120–132.

Berthold P (2004) Use of mist nets for monitoring landbird fall population trends, and comparison with other methods. Studies in Avian Biology 29:112–115.

Bub H (1991) Bird trapping and bird banding – a handbook for trapping methods all over the world. Cornell University Press, Hong Kong.

Buckland ST, Newman KB, Thomas L, Koesters NB (2004) State-space models for the dynamics of wild animal populations. Ecological Modelling 171:157–175.

Dennis B, Ponciano JM, Lele SR, Taper ML, Staples DF (2006) Estimating density dependence, process noise and observation error. Ecological Monographs 76:323–341.

Dunn EH, Hussel DJT (1995) Using migration counts to monitor landbird populations: review and evaluation of current status. Current Ornithology 12:43–88.

Dunn EH, Hussel DJT, Adams RJ (1997) Monitoring songbird population change with autumn mist netting. Journal of Wildlife Management 61:389–396.

Durbin J, Koopman SJ (2001) Time Series Analysis by State Space Methods. Oxford University Press, New York.

Durbin J, Koopman SJ (2002) A simple and efficient simulation smoother for state space time series analysis. Biometrika 89:603–616.

Francis CM, Hussel DJT (1998) Changes in numbers of land birds counted in migration at Long Point bird observatory, 1961–1997. Bird Populations 4:37–66.

Fretwell SD (1972) Monographs in population biology 5. Populations in a seasonal environment. Princeton University Press, Princeton, New Jersey.

Gibbons DW, Reid JB, Chapman RA (1993) The new atlas of breeding birds in Britain and Ireland: 1988–1991. T. & A.D. Poyser, London.

Hjort C, Lindholm CG (1978) Annual bird ringing totals and population fluctuations. OIKOS 30:387–392.

Hulme M (1992) Rainfall changes in Africa: 1931–1960 to 1961–1990. International Journal of Climatology 12:685–699.

James FC, McCulloch CE, Wiedenfeld DA (1996) New approaches to the analysis of population trends in land birds. Ecology 77:13–27.

Jamieson LE, Brooks SP (2004) Density dependence in North American ducks. Animal Biodiversity and Conservation 27:113–128.

Jonzén N, Hedenström A, Hjort C, Lindström, Å, Lundberg P, Andersson A (2002) Climate patterns and the stochastic dynamics of migratory birds. OIKOS 97:329–336.

Jonzén N, Lindén A, Ergon T, Knudsen E, Vik JO, Rubolini D, Piacentini D, Brinch C, Spina F, Karlsson L, Stervander M, Andersson A, Waldenström J, Lehikoinen JA, Edvardsen E, Solvang R, Stenseth NC (2006) Rapid advance of spring arrival dates in long-distance migratory birds. Science 312:1959–1961.

Lindström Å, Svensson SE (2005) Övervakning av fåglarnas populationsutveckling. Årsrapport för 2004. Ekologiska institutionen, Lunds universitet, Lund, Sweden.

Link WA, Sauer JR (2002) A hierarchical analysis of population change with application to cerulean warblers. Ecology 83:2832–2840.

Millar RB, Meyer R (2000) Non-linear state space modelling of fisheries biomass dynamics by using Metropolis-Hastings within-Gibbs sampling. Journal of the Royal Statistical Society Series C 49:327–342.

Nichols JD, Thomas L, Conn PB (2008) Inferences about landbird abundance from count data: recent advances and future directions. In: Thomson DL, Cooch EG, Conroy MJ (eds.) Modeling Demographic Processes in Marked Populations. Environmental and Ecological Statistics, Springer, New York, Vol. 3, pp. 201–236.

Nicholson SE, Tucker CJ, Ba MB (1998) Desertification, drought and surface vegetation: an example from the West African Sahel. Bulletin of the American Meteorological Society 79:2837–2849.

Peach W, Baillie S, Underhill L (1991) Survival of British sedge warblers, *Acrocephalus schoenobaenus*, in relation to west African rainfall. IBIS 133:300–305.

Royama T (1992) Analytical population dynamics. Chapman &; Hall, London.

Saether BE, Sutherland WJ, Engen S (2004) Climate influences on avian population dynamics. In: Moller AP, Fielder W, Berthold, P. (eds) Advances in Ecological Research: 35, Academic Press, London, pp. 185–209.

Sillett TS, Holmes RT (2002) Variation in survivorship of a migratory songbird throughout its annual cycle. Journal of Animal Ecology 71:296–308.

Sokolov LV, Yefremov YD, Markovets MY, Shapoval AP, Shumakov ME (2000) Monitoring of numbers in passage populations of passerines over 40 years (1958–1999) on the Courish Spit of the Baltic Sea. Avian Ecology and Behaviour 4:31–53.

Sokolov LV, Baumanis A, Leivits A, Poluda AM, Yefremov YD, Markovets MY, Morozov YG, Shapoval AP (2001) Comparative analysis of long-term monitoring data on numbers of passerines in nine European countries in the second half of the 20th century. Avian Ecology and Behaviour 7:41–74.

Stenseth NC, Viljugrein H, Saitoh T, Hansen TF, Kittilsen MO, Bolviken E, Glockner F (2003) Seasonality, density dependence, and population cycles in Hokkaido voles. Proceedings of the National Academy of Sciences of the United States of America 100:11478–11483.

Stervander M, Lindström Å, Jonzén N, Andersson A (2005) Timing of spring migration in birds: long-term trends, North Atlantic Oscillation and the significance of different migration routes. Journal of Avian Biology 36:210–221.

Svensson SE (1978) Efficiency of two methods for monitoring bird population levels breeding bird censuses contra counts of migrating birds. OIKOS 30:373–386.

Szep T (1995) Survival rates of Hungarian sand martins and their relationship with Sahel rainfall. Journal of Applied Statistics 22:891–904.

Thomas L (1996) Monitoring long-term population change: Why are there so many analysis methods? Ecology 77:49–58.

Webster MS, Marra PP, Haig SM, Bensch S, Holmes RT (2002) Links between worlds: unraveling migratory connectivity. Trends in Ecology and Evolution 17:76–83.

Section II
Evolutionary Ecology

Jim Nichols and Torbjørn Ergon

Contribution of Capture-Mark-Recapture Modeling to Studies of Evolution by Natural Selection

Emmanuelle Cam

Abstract Capture-Mark-Recapture (CMR) modeling is one of the most commonly used estimation methods in population ecology of wild animals. Until recently, much of the emphasis of this method was on the estimation of abundance and survival probability. Despite common interest in estimation of such demographic parameters, evolutionary ecologists have often been more critical of CMR estimation methods than wildlife biologists, mostly because the available models did not allow investigators to address what is at the heart of evolutionary ecology. Evolutionary ecology aims at explaining biological diversity: studies in this area of research necessarily involve assessment of variation in traits among individuals, including fitness components. The main limitation of early CMR models was the inability to handle *states* among which individuals move in a stochastic manner throughout life (e.g., breeding activity and number of offspring raised, locations, physiological states, etc.). Several important advances have enhanced ecologists' ability to address evolutionary hypotheses using CMR data; namely multistate models and models with individual covariates.

Recently, methodological advances have allowed investigators to handle random effects models. This is bringing CMR models close to modern statistical models (Generalized linear mixed models) whose use is rapidly increasing in quantitative genetics. In quantitative genetics, the *animal model* aims at disentangling sources of phenotypic variation to draw inferences about heritability of any type of trait (morphological, demographic, behavioral, physiological traits). The *animal model* partitions variation in the trait of interest using variance components. Understanding evolution by natural selection and predicting its pace and direction requires understanding of the genetic and environmental influences on a trait. Phenotypic characteristics such as morphological or life-history traits (i.e. demographic parameters such as number of offspring raised and survival probability) are likely to be influenced by a large number of genes, the genetic basis of which can be quantified via statistical inferences based on similarities among relatives in a population. The

---

E. Cam (✉)
Laboratoire Evolution et Diversité Biologique, Bâtiment 4 R3 – B2, 118 Route de Narbonne, Université Paul Sabatier Toulouse III, 31062 Toulouse Cedex 09, France
e-mail: emmacam@cict.fr

D.L. Thomson et al. (eds.), *Modeling Demographic Processes in Marked Populations*, Environmental and Ecological Statistics 3, DOI 10.1007/978-0-387-78151-8_5,

extent of evolutionary responses in a quantitative trait is assumed to be proportional to the force of natural selection and heritability of a trait. Estimating the genetic basis of quantitative traits can be tricky for wild animal populations in natural environments: environmental variation often obscures the underlying evolutionary patterns. However, this genetic basis of traits is at the heart of natural selection, and recently there has been increased interest in applying the *animal model* to natural populations to understand their evolutionary dynamics. Such models have been applied to estimation of heritability in life history traits, either in the rare study populations where detection probability is close to 1, or without considering the probability of detecting animals that are alive and present in the study area (recapture or resighting probability). Applications of the *animal model* to demographic parameters (fitness components) such as survival, breeding probability or to lifetime reproductive success in wild animal populations where detection probability is < 1 require trans-disciplinary efforts; this is necessary to address evolutionary processes in such populations.

Keywords Capture-mark-recapture · Dispersal · Evolution · Fitness functions · Heritability · Life history theory

1 Introduction

Approaches to estimating demographic parameters using capture-mark-recapture data while accounting for incomplete detection of individuals by investigators during sampling sessions have tremendously diversified over the past 20 years (CMR estimation models; reviewed in Williams et al. 2002). Early efforts in development of CMR models have been directed mostly to estimation of abundance and survival probability, but recent advances now allow investigators to estimate other population vital rates such as breeding and recruitment probability, or movement probability among units of fragmented populations, temporary emigration, etc. (e.g., Spendelow et al. 1995, 2002; Pradel 1996; Pradel and Lebreton 1999; Oro and Pradel 2000; Schwarz and Arnason 2000, 2001; Schwarz and Stobo 2000; Nichols et al. 2000; Lindberg et al. 2001; Kendall and Bjorkland 2001; Lebreton et al. 2003; Reed et al. 2004; Barbraud and Weimerskirch 2005; Cam et al. 2005; Crespin et al. 2006; Hadley et al. 2006; Martin et al. 2006). In other words, CMR models allow estimation of the main parameters governing demographic processes. In addition, CMR models are now increasingly used to address community vital rates (e.g., species extinction or colonization probability; Nichols et al. 1998 a, b; Williams et al. 2002), and vital rates specific to large-scale features of species distribution (e.g., site occupancy models, MacKenzie et al. 2006). Recent technical advances relevant to the particular field of evolutionary ecology are extensively explained in Conroy (2008, this volume).

Recent CMR models aimed at estimating demographic parameters have two important features: (1) an increased variety of population vital rates can be estimated (provided appropriate sampling design), and (2) an increased degree of stratification of populations can be achieved (i.e., classes in which individuals stay permanently or temporarily, or individual attributes) and stratum-specific vital rates can be estimated. This has partly alleviated an old tension between biologists and statisticians, the former blaming the latter for designing estimation methods corresponding to unrealistically simple populations or biological systems, the latter doubting that appropriate data could ever be collected to match the requirements of complex models (i.e., appropriate sampling design and large sample sizes in all strata within populations). Moreover, development of software programs and documentation has considerably facilitated access to CMR estimation techniques for biologists (some examples of software used in ecology, in alphabetical order: CAPTURE: White et al. 1978; Rexstad and Burnham 1991, MARK: White and Burnham 1999; Cooch and White 2007, MSURGE: Choquet et al. 2003, MSSURVIV: Hines 1994, POPAN: Arnason et al. 1998, SURGE: Clobert and Lebreton 1985; Lebreton and Clobert 1986, SURPH: Smith et al. 1994, SURVIV: White 1983). From the viewpoint of an evolutionary ecologist, an enormous stride has been made with the development of multistate models (e.g., Nichols and Kendall 1995). This was a first step towards accommodation of a pervasive property of the history of individuals in long-lived species: individuals change *state* (location, social or physiological state, etc.) in a stochastic manner.

CMR field techniques have long been popular among biologists (especially in studies of birds, small and large mammals, or fish). They have been widely used to estimate demographic parameters, but the corresponding statistical approaches disentangling sampling processes and demographic ones have not always been used (Martin et al. 1995). Not all the fields of ecology have taken full advantage of the potential that new CMR estimation models offer. The Proceedings of the 1994 EURING conference held at Patuxent Wildlife Research Center (Laurel, MD, U.S.A.) include a paper entitled: "Capture–recapture and evolutionary ecology: a difficult wedding?" (Clobert 1995; see also Clobert 2002). Despite the regular presence of researchers involved in evolutionary ecology studies at EURING meetings and the recent advances in CMR methodology applied to evolutionary studies, analytical tools that estimate demographic parameters of wild animal populations while accounting for imperfect detection of individuals do not seem as widely used in this field as in other fields of ecology, especially wildlife ecology and conservation biology. This probably partly results from differences in history and educational practices.

Although mathematics and statistics play an important part in education for evolutionary ecologists (e.g., Charlesworth 1994; Lynch and Walsh 1998; Caswell 2001), as far as estimation is concerned (e.g., genetic parameters; Lynch and Walsh 1998) analytical tools used typically do not account for the sampling processes that are specific to studies of wild animal populations, more precisely, incomplete detection of individuals (Martin et al. 1995). This may not be a problem for traits whose phenotypic values are independent of detection probability, but the assumption that the sample of captured or resighted individuals and the sample of undetected individuals have identical features is unlikely to be met for demographic parameters like breeding probability after recruitment (e.g., Nichols et al. 1994), age-specific recruitment probability (e.g., Viallefont et al. 1995 a, b; Spendelow et al. 2002), breeding success probability or clutch size (e.g., Yoccoz et al. 2002).

Efforts to carefully design sampling protocols specific to wild animal populations and to master the corresponding statistical analyses (e.g., Williams et al. 2002) are probably more common in education programs offered to students in wildlife ecology and conservation biology than to evolutionary ecologists. There is no international review available on the topic, but when one scans web sites detailing course sequences offered in undergraduate and graduate programs organized by evolutionary ecology departments, classes on CMR modeling hardly ever appear. Such classes appear in several wildlife ecology programs offered in internationally known universities. Because it is necessary to have accurate estimates of demographic parameters to assess the state of populations, design efficient management plans and make decisions, not accounting for sampling "biases" that have long been known (e.g., Cormack 1964; Jolly 1965, 1993; Seber 1965; Lebreton et al. 1992) may be considered more "irresponsible" in wildlife ecology educational programs (Anderson et al. 2003) than in others. The quantitative content of university programs for wildlife ecology may be considered as insufficient by quantitative wildlife ecologists, but there is some pressure in specialized scientific journals to develop quantitative skills in education (e.g., Kendall and Gould 2002; Seber and Schwarz 2002; Anderson et al. 2003).

Because wildlife ecologists focus on wild animal populations, they are constantly faced with the challenge of using or designing appropriate approaches for analysis. This is not true for evolutionary ecologists. The discipline differs from that of wildlife ecology in that there are strong historic relationships with some other disciplines that tend to utilize experimental study systems and molecular approaches to address the genetic basis of evolution (e.g., evolutionary biology, molecular evolution, phylogeography, population, quantitative and developmental genetics, systematics, etc.; e.g., Freeman and Herron 2000). Students in evolutionary ecology devote a large proportion of time to these fields and may not be introduced to the specificity of the data and analytical tools required to conduct demographic studies in wild animal populations (e.g., non-detection of marked individuals that are alive and present in the study area). Nevertheless, comparative methods play an important part in evolutionary ecology (e.g., life history evolution and evolution of morphological traits; Promislow et al. 1992; Bennett and Owens 2002; Liker and Székely 2005) and such studies are unlikely to be possible without using data from wild animal populations, especially for long-lived species. Several comparative studies have ignored the distinction between studies that have estimated demographic parameters while accounting for incomplete detection of individuals, and those that have not, even in situations where survival probability was the focal trait (e.g., Owens and Bennett 1994; Liker and Székely 2005). Use of estimates of demographic parameters ignoring incomplete detection of individuals may lead to erroneous conclusions.

Almost 15 years later, the answer to the question: "Capture–recapture and evolutionary ecology: a difficult wedding?" (Clobert 1995) may not be unanimous in the EURING meeting audience. Despite the slowness of integration of CMR estimation models in evolutionary ecology, the range of questions relevant to evolutionary ecology addressed in studies that have used appropriate CMR estimation techniques is large. In addition, recent advances in development of CMR models may open new opportunities for evolutionary ecologists to use empirical data from wild animal populations to address novel questions. In the following Section (2), I will address the specificity of evolutionary ecology as a discipline. If evolutionary ecology has generally been viewed as relevant to basic research exclusively, several researchers are now advocating consideration of the theoretical framework of (micro-) evolution to address questions about ecological processes and the consequences of human influence on wild animal populations and their habitat (e.g., Ferrière et al. 2004; Reznick et al. 2004). In Section 3, I will define natural selection, one of the corner stones of evolutionary ecology. In this section, I will also describe how the specific question of the evolution of demographic parameters is usually addressed. Understanding how natural selection is addressed in wild animal populations is necessary to assess the features of existing CMR studies in evolutionary ecology, which I will do in Section 4. Concerning the evolution of demographic parameters by natural selection, up to now CMR estimation techniques have been used to address only part of the prerequisites for natural selection. A central prerequisite, heritability, has not been addressed. Recent advances in estimation methods used in quantitative genetics and in methods aiming at estimating demographic parameters from CMR data have some common features that should be useful to address heritability. In Section 5, I will describe the method used in quantitative genetics with data from wild animal populations to address heritability of traits (the *animal model*), and I will offer suggestions concerning the type of development needed in CMR estimation models to address heritability in demographic traits.

2 Evolutionary Ecology: Historical Background

There have been numerous attempts to classify the different disciplines of biology "to deal with the enormous range of phenomena brought together under the heading of biology" (Mayr 1997, p. 111). According to Mayr (1997), classification of disciplines according to the type of question asked in research is one of the most logical classification systems. There are three main questions: "What?", "How?", and "Why?" However, the first question, "What?", is shared by all biological disciplines. Description (establishment of a "solid factual basis") is the first step in any branch of biology. However, "Answers to the 'What?' questions alone failed to produce a satisfactory solution to the problem of how to classify the subdivisions of biology" (Mayr 1997, p. 115). In addition, it is impossible to conduct any descriptive work without identifying the object to describe. Identification of the objects on which scientific research focuses in different biological disciplines is possible because there are specific theoretical bodies. Description cannot be conducted without referring to a theoretical background because no "factual basis" emerges ex nihilo, independently of the hypotheses and theories. Consequently, answers to the "What?" question depend on identification of biological disciplines using other criteria.

Answers to the "How?", and Why?" questions provide a more efficient basis for classification of biological disciplines – a distinction that appeared in 1870 in debates among biologists. Biology has considerably diversified since then, but a fundamental distinction between disciplines has survived, even if this classification system has its problems. The main distinction between these two types of questions lies in the type of causes invoked to explain biological phenomena. Proximate causes allow investigators to explain the functioning of biological entities "here" and "now" (answers to the "How?" question). Conversely, ultimate causes allow investigators to explain observed phenomena in the light of the history of life and evolutionary theory (answers to the "Why?" question). "Why?" questions usually relate to adaptation or organic diversity (Mayr 1997, p. 118); "ultimate causes attempt to explain why an organism is the way it is, as a product of evolution". [...]. However, "no biological phenomenon is fully explained until both proximate and ultimate causes are illuminated". "One of the special properties of the living world is that it has these two sets of causations." (p. 67).

Mayr (1997) also described ecology as the most heterogeneous and comprehensive field of biology, and one that is difficult to assign to one single type of question ("How?", or "Why?"): both types of questions are addressed. Ecologists focusing on ultimate causes are called *evolutionary ecologists* (Fox et al. 2001). According to Fox et al. (2001), "Evolutionary ecology and ecology share the goals of describing variation in natural systems and discovering its functional basis. Within this common framework, evolutionary biologists emphasize historical and lineage-dependent processes and hence often incorporate phylogenetic reconstructions and genetic models in their analyses. Ecologists, while recognizant of historical processes, tend to explain variation in terms of contemporary effects of biotic and abiotic environmental factors. Evolutionary ecology spans the two disciplines and incorporates the full range of techniques and approaches from both"

The apparent difference between the timescales invoked in ecology and evolutionary ecology may lead to the conclusion that the dichotomy is natural. However, this traditional dichotomy may have become an obstacle to our understanding of ecological phenomena, and this may have consequences on our ability to design efficient conservation plans. There is growing evidence that evolutionary responses to environmental changes can be so fast that researchers are able to witness them both in the laboratory and in the wild (Ferrière et al. 2004; Frankham and Kingsolver 2004; Reznick et al. 2004). Hendry and Kinnison (1999) suggested that rapid microevolution is the norm in contemporary populations confronted with environmental change. According to Saccheri and Hanski (2006), "there is a growing acceptance that the traditional dichotomy between ecological and evolutionary timescales is a false one".

3 Evolution by Natural Selection

3.1 Natural Selection

"Evolution may be defined as any net directional change or any cumulative change in the characteristics of organisms or populations over many generations [...]" and "may occur as a result of natural selection, genetic drift, or both" (Endler 1986, p. 5). According to Fairbairn and Reeve (2001, p. 30),

> Natural selection is notoriously difficult to define. In the broadest sense, the process of natural selection has been defined by the following deductive argument:
> If there is:
> (i) variation in some attribute or trait among biological entities (phenotypic variation),
> (ii) a consistent relationship between the trait and fitness (a fitness function), and
> (iii) descent with heritability for the trait (i.e., the variation in the trait must have a genetic component),
>
> Then the trait distribution will change:
>
> (I) within generations more than expected from ontogeny alone, and
> (II) across generations "in a predictable way" until an equilibrium is reached.
>
> This definition is true to Darwin's original description of natural selection, and was adopted by Endler (1986) in his review of selection in natural populations. However, in addition to being rather cumbersome, the deductive argument is flawed because conclusion (I) does not require premise (iii) and holds for any fitness difference caused by differences in survival (i.e., differences in fecundity alone will not cause within-generation changes in trait distributions) In constructing a more concise and logically consistent definition of natural selection, most authors (e.g., Lande and Arnold 1983; Futuyma 1998) prefer to distinguish the process of natural selection occurring within generations (premise i and ii) from the evolutionary consequences of that selection (premise iii and conclusion II)

Researchers working in different areas have used different definitions of fitness (Endler 1986, p. 38). In life history theory, the fitness concept currently relies on invasibility: the possibility of a rare mutant strategy to replace the strategy played predominantly in the population (Metz et al. 1992). However, as emphasized by Brommer et al. (2002), invasibility is not readily measured in natural populations, and many empirical studies focus on other measures of evolutionary success. When focusing on selection at the level of individual organisms, fitness "in its most general sense is success in contributing descendants to the next generation" (Fairbairn and Reeve 2001, p. 31). The definition of fitness by Endler (1986, p. 39) highlights the direct relevance of CMR estimation models to evolutionary ecology: "Fitness is the degree of demographic difference among phenotypes", or a measure of the degree of the following condition for natural selection: "a consistent relationship between [a] trait and mating ability, fertilizing ability, fertility, fecundity, and or survivorship". Natural selection is based on demographic processes, and estimation of demographic parameters is a key point in some approaches to detecting natural selection (Endler 1986).

3.2 Evolution of Demographic Parameters by Natural Selection

Demographic parameters (i.e., age at maturity, number of offspring produced, longevity, age-specific reproductive investment and mortality schedule, etc.) are not only involved in the evolution of morphological, behavioral, or physiological traits by natural selection, but they are themselves subjected to natural selection (Roff

1992; Stearns 1992). The field of life history evolution focuses on the evolution of demographic parameters, a class of traits also called *fitness components* (Stearns 1992)

Studies of the evolution of demographic parameters do not necessarily address all the above Premises: some studies do not require Premise (ii) because they do not involve other classes of traits than fitness components themselves. Such studies involve two fitness components (or more), and Premise (i) concerns several components simultaneously. In life history theory, it is assumed that there are relationships among traits and that natural selection operates on complexes of traits: "Age-specific survival and fecundity are not free to independently evolve, but are constrained by physiological and ecological trade-offs" (Tatar 2001). As emphasized by Clobert (1995), Viallefont et al. (1995a) and Cooch et al. (2002), reproductive costs and trade-offs between life history traits are central to the theory of life history evolution (Reznick et al. 2000). The basis of physiological trade-offs (Stearns 1992) is the following: because individuals have access to limited resources, resources allocated to one trait are assumed not to be allocated to another (*Principle of allocation*, Levins 1968). According to Reznick et al. (2000), "[...] it became convenient to think of the life history as being similar to a pie divided into slices, each slice being devoted to a different function, such as growth, maintenance, storage or reproduction. Because the pie is of fixed size, increasing the size of a given slice necessarily decreases the size of another slice". For example, individuals that raised two offspring to independence in a given breeding occasion may not be able to invest the same amount of energy in their own maintenance functions as individuals that raised only one offspring, and the former may incur survival costs. Trade-offs may also have a behavioral basis when reproductive activity is associated with increased mortality risk because of predators or fights with conspecifics. Last, there are intergenerational trade-offs linking parental allocation of resources to reproduction and offspring fitness (e.g., offspring size at birth; Stearns 1992).

Trade-offs also play a central part in one of the evolutionary theories of senescence: antagonistic pleiotropy (Williams et al. 2006). Antagonistic pleiotropy assumes that "improvements early in life are purchased at a cost to later-age fitness components" (Williams et al. 2006). Several hypotheses have been put forward concerning the mechanisms responsible for pleiotropy (Tatar 2001), but one of them is based on physiological trade-offs: "For instance, the allocation of lipid to current egg production may preclude its use in cell or mitochondria cell membrane turnover. Natural selection favors genotypes that maximize fitness within such sets of constraints" (Tatar 2001, p. 131). Studies of antagonistic pleiotropy may address covariation between age at maturity and at last reproduction. Here, Premise (ii) isn't relevant.

However, traits other than fitness components may be taken into account in studies of life history evolution, even if the evolution of such traits isn't the central topic of these studies. For instance, variation in fitness components such as fecundity (e.g., number of eggs in fish) may be intrinsically linked to variation in morphological traits such as body size (*n.b.*: here *fecundity* describes the actual reproductive performance; see Caswell 2001, p. 10, for an alternative definition from human demography). Consequently, studies of the evolution of demographic parameters by natural selection may address Premise (ii): a consistent relationship between a (non-demographic) trait (e.g., morphology) and fitness. However, in studies of life history evolution involving morphological traits, the latter are usually addressed as part of a trio including two fitness components. In the above example (fecundity and body size), in species with delayed maturity, changes in age at maturity may result in changes in body size, which in turn may result in changes in fecundity (Roff 2001). In other words, body size is involved in the trade-off between development time (age at maturity) and fecundity.

4 Features of Existing CMR Studies in Evolutionary Ecology

CMR estimation models have been used in evolutionary ecology studies addressing evolution of "non-demographic traits". For that class of traits (i.e., morphological, physiological, behavioral traits) CMR estimation methods have been used to address Premise (ii), i.e., to estimate fitness components and assess fitness functions (e.g., Gimenez et al. 2006). Apart from estimation of fitness components, CMR estimation models may be considered to address other Premises for natural selection and Deductions I and II. For morphological, behavioral or physiological traits, the question of whether samples are representative in studies addressing Premises (i) and (iii) in the above definition of natural selection, or Deductions I and II, should probably receive attention. If detected and undetected individuals have different characteristics with respect to the trait of interest (e.g., if there is a relationship between size or color and detection probability), then use of CMR estimation models may be necessary to assess the distribution of trait values in populations (in this case closed-population models may be useful to estimate the frequency of trait values). In most studies, the observed distribution of phenotypic trait values in samples is assumed to reflect the distribution of trait values in populations in an exhaustive and consistent manner (i.e., samples are assumed to be representative of (sub-) populations). In addition, multitrait *animal models* (see Section "Multitrait Models") or random coefficients *animal models* (Schaeffer 2004) can be used to address fitness functions based on breeding values (a possible approach to Premise (ii)) and heritability (Premise (iii)) of traits changing over life (e.g., body size in some mammal or snake species, egg size in some bird species). In this case, repeated data from individuals are needed. For continuous traits, this may raise the issue of missing individual covariates when individuals are not recaptured or resighted (Bonner and Schwarz 2004), an area where methodological development is needed. In most studies, data are assumed to be missing at random (e.g., missing data in growth curves).

Concerning the evolution of demographic parameters themselves, Premise (i) (phenotypic variation and covariation between parameters) has been addressed using CMR estimation models, as well as Premise (ii) when morphological or physiological traits are assumed to be involved in the trade-offs between demographic parameters. I am not aware of studies that have addressed Premise (iii) using CMR estimation models (Premise (iii) is the third condition for evolution of demographic parameters by natural selection: heritability). Interestingly, CMR estimation models required to address Deduction I (within-generation selection) have been developed (Burnham and Rexstad 1993; Pledger and Schwarz 2002; Royle and Link 2002; Royle 2007), but I am not aware of studies that have addressed the evolutionary consequences of selection (Deduction II) for demographic parameters using CMR estimation techniques.

Despite the relatively "limited" scope of the evolutionary ecology studies that have used appropriate CMR estimation models (several aspects of natural selection have not been addressed), such studies have required considerable efforts from biologists and statisticians: details of the technical advances are reviewed in Conroy (2008, this volume). In this section, I will focus on some evolutionary ecology studies that have accounted for incomplete detection of animals to estimate demographic parameters. This is not an exhaustive review; my goal is to provide examples illustrating the diversity of topics addressed.

4.1 Life History Evolution

4.1.1 Trade-offs Between Life History Traits

Trade-offs are one of the topics that have been addressed in the largest proportion of CMR studies in evolutionary ecology (e.g., Nichols et al. 1994; Nichols and Kendall 1995; Viallefont et al. 1995a, b; Cam et al. 1998; Yoccoz et al. 2002; Barbraud and Weimerskirch 2005). In a large number cases, the studied species exhibited a small range of reproductive investment levels (e.g., non-breeding versus a single egg produced in birds or a single young in mammals, or only two or three young raised; McElligott et al. 2002; Barbraud and Weimerskich 2005). In such long-term observational studies that have retrospectively used data collected over long periods of time, trade-offs have been addressed using few discrete categories of reproductive investment, as opposed to continuous trade-off functions (Reekie et al. 2002). Technical development in CMR estimation methodology in the 1990's has considerably broadened the scope of studies aimed at detecting trade-offs. First, in iteroparous species, the proportion of individuals breeding more than once is larger than 0, which implies that individuals may change breeding *state* (i.e., breeding activity or success; Nichols et al. 1994) over time and raise a different number of offspring to independence on different breeding occasions. For this reason, development of multistate models has played a central part in studies of trade-offs (Arnason 1973; Brownie et al. 1993; Nichols and Kendall 1995; Williams et al. 2002; Conroy 2008 this volume). Several versions of multistate models have been designed to estimate state-specific transition probabilities in situations where there are unobservable states or where individuals are sometimes misclassified (e.g., when individuals are erroneously considered as nonbreeders in a given sampling occasion; Kendall and Nichols 2002; Kendall et al. 2003; Kendall 2004; Nichols et al. 2004). These tools allow investigators to accommodate situations that are common in empirical studies (i.e., state uncertainty). In addition, development of models allowing use of different sources of data provides a means of improving estimates of local survival probability by estimating permanent emigration (Burnham 1993). Indeed, unless an additional source of data is available, estimates of local survival incorporate permanent emigration out of the study area. The ability to disentangle the factors influencing "true" local survival and those influencing dispersal is important for studies of trade-offs between survival and other life history traits. Unfortunately, studies in evolutionary ecology mostly use live captures/resightings of marked animals, partly because study systems focusing on hunted species may be considered as "artificial". However, there may be situations where selective hunting may correspond to carefully designed experimental systems that have played a major role in development of theory in evolutionary ecology (e.g., Mertz 1975).

Before multistate models became standard, other approaches were possible (reviewed in Viallefont et al. 1995a); for example one may compare survival probability in individuals assumed to have identical reproductive history (e.g., first time breeders with no prior experience). That is, groups of individuals are defined on the basis of the number of offspring raised in the first breeding occasion, and group-specific survival probabilities over the first year of reproductive life are compared. Alternatively, experiments may be conducted by randomly assigning "identical" individuals (i.e., same age, same year, same environmental condition, same prior reproductive history etc.) to treatments (increased/decreased clutch or brood size). The underlying assumptions are that no important factor influencing survival probability has been missed (left uncontrolled), and that the controlled factors correctly reflect the factors involved *in natura* in the studied process (the factors have been correctly identified). In some instances, recapture probability in the year following the first breeding attempt has been compared to subsequent recapture probability to address possible experience-related reproductive costs in terms of future reproduction, assuming that recapture probability reflects breeding probability to some extent (Viallefont et al. 1995b). Last, temporary emigration has also been used as an indicator of breeding probability (e.g., Kendall and Nichols 1995; Schmidt et al. 2002; Frétey et al. 2004).

Progressively, more complex definitions of state have been used to address the question of trade offs more thoroughly. More complex multistate models incorporating individual covariates have also been developed. Indeed, because individuals vary in their ability to obtain resources, or because they live in environments with different resource availability (Stearns 1992), reproductive costs may not be identical in every individual; i.e., some individuals may be able to invest more in reproduction without incurring as large costs as others, depending on their state. Here, state may correspond to different things depending on the organism studied (e.g., body size, parasite load, immunological state, social dominance, experience, etc.). State is assumed to reflect a hierarchy among individuals in their ability to acquire resources or to use them, or simply a baseline efficiency of functions (e.g., maintenance, reproduction, etc.). Detection of reproductive costs from observational data requires comparison of fitness between individuals having different condition or social rank but identical reproductive investment (reproductive activity and success).

Using states combining information from reproduction and other individual characteristics (e.g., experience), multistate models allow investigators to address trade-offs in heterogeneous populations. Development of models including individual covariates (e.g., body mass or size; Bonner and Schwarz 2004) has played a large part in emergence of studies addressing state-specific reproductive costs and reproductive tactics (e.g., Barbraud and Weimerskirch 2005).

Moreover, temporal environmental variation (i.e., variation in resource availability or density of conspecifics) may lead to time-specific variation in reproductive costs within categories of individuals in the same state in different years (Orzack and Tuljapurkar 2001). Assuming that individuals are able to assess resource availability in time, individuals may adjust the amount of energy allocated to functions according to their state; i.e., there may be *individual optimization* (Van-Noordwijk and de Jong 1986; Pettifor et al. 1988, 2001; Tinbergen and Both 1999). If individual optimization occurs, only experimental approaches (e.g., Yoccoz et al. 2002) may allow detection of trade-offs. The difficulty in designing appropriate experiments to address trade-offs has been discussed in Cooch et al. (2002, p. 35): "Trade-offs within an individual must be true (Tuomi et al. 1983; Emlen 1984); if an individual is forced to expend greater energy on one activity, then this necessarily reduces the amount of energy available for another activity. However, this does not necessarily mean that trade-offs occur among individuals. This is important, since natural selection operates on the additive genetic covariance among individuals, not correlations within individuals". If trade-offs are necessarily based on physiological or behavioral mechanisms operating at the individual level (trade-offs "within" individuals), addressing relationships between fitness components by comparing different individuals assigned to different experimental treatments (or naturally exhibiting different levels of reproductive investment) is addressing relationships between components expressed "among" individuals (trade-offs "among" individuals). Comparing different individuals should allow inferences about within-individual trade-offs if investigators can be sure that individuals are strictly identical with respect to all traits except the ones involved in the trade-off itself, but failure to fulfill this condition is thought to be a major reason for failure to detect trade-offs.

The process of individual optimization is assumed to lead individuals to make decisions according to their current state. State may change over time, but there may also be permanent differences among individuals. CMR studies have contributed to identification of permanent differences among individuals. For example, several studies have provided evidence of permanent differences in fitness components (e.g., survival probability) among individuals according to morphological traits reflecting relative body conditions (e.g., Barbraud and Weimerskirch 2005; Blums et al. 2005). Similarly, long-lasting cohort effects have been identified (e.g., Cam et al. 2005), as well as a permanent influence of conditions during development on several life history traits (e.g., Cam et al. 2003). Blums et al. (2005) have used relative time of nesting to account for individual differences in "quality". Experimental studies have provided contrasting results concerning the hypothesis of individual optimization in wild animal populations (e.g., Tinbergen and Sanz 2002; Török et al. 2004). However, in all cases tests of this hypothesis require high levels of stratification of the data according to state (which may be assigned to individuals in the framework of an experiment) and year or environmental conditions (biotic and abiotic).

Last, some of the physiological and behavioral mechanisms underlying intra- and inter-generational trade-offs between life history traits have been investigated using CMR estimation models. For instance, in an experimental study, Reed et al. (2006) have addressed the relationship between survival probability and manipulated testosterone level in free-living dark-eyed juncos (*Junco hyemalis carolinensis*). Testosterone-treated males increased levels of activity and home range size and had elevated levels of stress hormones. They exhibited increased ability to attract females (increased ability to produce extra-pair offspring), but produced smaller offspring with lower postfledging survival. In addition testosterone-treated adult males had increased detectability and susceptibility to predation, which led to lower adult survival.

4.1.2 Level at Which Natural Selection Operates and Estimation of Demographic Parameters

Obviously, modern CMR estimation models allow evolutionary ecologists to address a large range of questions directly relevant to fitness functions, selection, and adaptation. However, the ultimate goal of evolutionary ecologists is to address differences in demographic parameters at the level at which natural selection operates, which is often identified as the individual level (Endler 1986). Mayr (1997) identified a reason why evolutionary ecologists constantly press statisticians to develop complex CMR models allowing investigators to stratify populations according to large numbers of criteria (e.g., multistate models, models with time-varying individual covariates; Pollock 2002). This is an irresolute tension between the conceptual foundations of evolutionary ecology and population ecology which was extensively discussed in Cooch et al. (2002). The community of researchers involved in development and use of CMR estimation models is mostly composed of researchers focusing on wildlife ecology and conservation biology, or of statisticians who often develop models to answer questions from these same fields of ecology. Historically, these fields have been dominated by concepts from population ecology, which "can be tracked back to a school of mathematical demographers interested in the growth of populations and the factors controlling it" (Mayr 1997, p. 211). However, the population concept specific to population ecology is different from that of evolutionary ecology. "The population concept adopted by most mathematical population ecologists was basically typological, in that it neglected the genetic variation among the individuals of a population. Their 'populations' were not populations in any genetic or evolutionary sense but were what mathematicians refer to as sets. The crucial aspect of the population concept to have emerged in evolutionary biology, by contrast, is the genetic uniqueness of the composing individuals. This kind of 'population thinking' is in sharp contrast with the typological thinking of essentialism. In ecology, the genetic uniqueness of the individuals of a population is usually ignored" (Mayr 1997, p. 211).

Ideally, evolutionary ecologists would like to apply this concept of genetic "uniqueness" of individuals to demographic parameters as well. This is because "the ultimate context for estimation is the degree to which selection and the fitness differences upon which selection operates translates into evolutionary change" (Cooch et al. 2002). Indeed, the individual level is assumed to be the relevant level of selection in many studies (Endler 1986). However, as Nichols (2002, pp. 49–50) pointed out, "[...] the larger the number of strata, the fewer individuals in each stratum, and the more difficult it will be to estimate stratum-specific survival probability. [...] increasing stratification will yield a single individual in each stratum, with the corresponding estimation problem analogous to that of being asked to estimate the probability of heads from a single flip of a loaded coin. [...] some form of aggregation is necessary for the conduct of science. [...]. If we view an individual organism's fate or behaviour at any point in space and time as a unique event not capable of informing us about the likelihood of the event for other individuals or points in space and time, then generalization and prediction become impossible. The task of the biologist then involves simply recording and describing these unique events and possibly developing a posteriori stories to explain them. Although such descriptive work might be interesting, it is not consistent with most definitions of science".

Stratification in large numbers of discrete categories and limited sample size makes statistical inference impossible. However, evolutionary ecologists are familiar with an approach assuming that phenotypic traits of individuals in a population are characterized by a distribution. The distribution is assessed using random individual effects models, also called *frailty* models in human demography (Vaupel and Yashin 1985a, b; Service et al. 2000; Cam et al. 2002a; Link et al. 2002a, b; Service 2004; Wintrebert et al. 2004; Fox et al. 2006). Mixed models (Fahrmeir and Tutz 1994) are extensively used in quantitative genetics to address the genetic basis of phenotypic values of quantitative traits in populations (i.e., as opposed to "qualitative traits" such as gender); more precisely to assess their variance in populations (Lynch and Walsh 1998). Mixed models are commonly used in human demography to address senescence (e.g., Yashin et al. 2001; Service 2000, 2004), and they are also used in studies of behavior (e.g., Hernández-Lloreda et al. 2003). The motivations for the use of mixed models in these different fields have common points: (i) incorporation in statistical models of terms accounting for heterogeneity among individuals in the focal trait, (ii) the possibility for dependence of individuals for trait values (e.g., incorporation of a particular variance–covariance matrix for random effects), and (iii) assessment of the influence of specific covariates (fixed effects) on the trait while accounting for specific variance–covariance structures for random effects.

The Process of Natural Selection: Within-Generation Mortality Selection

Several long term studies have provided evidence that wild animal populations are demographically heterogeneous (e.g., Fox et al. 2006): it has been suggested that populations are composed of groups of individuals with a permanent hierarchy in fitness components among individuals. Whenever measurable individual characteristics can be used to account for individual heterogeneity in a satisfactory manner, individual covariates may be used (e.g., Blums et al. 2005). However, studies of survival in humans have provided evidence that measurable covariates are not always sufficient to account for heterogeneity among individuals (Hougaard 1991). In this case the hierarchy among individuals can be accounted for in discrete time survival models by incorporating an individual random effect with mean 0 and a variance term accounting for the distribution of individual survival probability around the mean (Cam et al. 2002a, Link et al. 2002a, b). The mean survival probability may depend on fixed effects such as age, sex, year, birth cohort, etc. These models assume that there are differences in *underlying*, or *latent* survival among individuals (Cooch et al. 2002). The same approach can be used to model underlying differences in breeding probability or breeding success probability among individuals.

Under the *good genes hypothesis*, some individuals are assumed to have a higher breeding success probability than others, or higher survival probability, or both (e.g., Curio 1983; Cam et al. 2002a; Link et al. 2002a, b). This is likely to have consequences for studies of senescence or any class of age effect on demographic parameters (Curio 1983; Vaupel and Yashin 1985a, b). In heterogeneous populations, one might expect an age-related change in the composition of the population. The *selection hypothesis* accounts for the progressive concentration of individuals with higher intrinsic survival probability in older age classes (Endler 1986), and if there is a positive correlation between breeding success probability and survival at the individual level, the progressive concentration of individuals with higher success probability in older age classes (Cam et al. 2002a; Barbraud and Weimerskirch 2005; Beauplet et al. 2006). This within-generation phenotypic selection process corresponds to Deduction I in the definition of natural selection (see Section 3.1). Within-generation phenotypic selection is not a sufficient condition for natural selection, but this process may explain some within-generation changes in survival or reproductive parameters detectable in heterogeneous populations (e.g., Forslund and Pärt 1995; Service 2004). More generally, assessment of phenotypic variation in fitness components among individuals within populations (e.g., Fox et al. 2006) is at the heart of studies of life history evolution by natural selection and is relevant to studies of the evolution of traits other than age-specific life histories (Mazer and Damuth 2001).

Senescence has been detected in a fair number of wild animal populations using CMR estimation models (e.g., Nichols et al. 1997; Festa-Bianchet et al. 1999; Bryant and Reznick 2004; Gaillard et al. 2004), and is common in captive birds and mammals (Ricklefs 2000; Ricklefs and Scheuerlin 2001). Senescence has motivated an enormous number of studies in humans, probably because of the economical and sociological implications of the phenomenon. In addition, senescence is one of the most challenging paradoxes from a fundamental perspective: "Senescence is an intriguing problem for evolutionary theory: can natural selection favour an age-specific decline in fitness?" (Bennett and Owens 2002). Not all authors agree on the occurrence of senescence in wild vertebrates. According to Williams (1992) "Both birds and mammals have life cycles that should make them similarly vulnerable to the evolution of senescence, but there is little evidence that senescence effects birds at all. Where data on avian age structures are most abundant, it usually appears that mortality rates of young adults prevails through life. This conspicuously violates expectation from theory" [of Hamilton 1966]. However, as emphasized by van de Pol and Verhulst (2006); "Phenotypic traits can change as a result of within-individual changes (phenotypic plasticity) and between-individual changes, as selection may favour some individuals over others. When quantifying how population values of phenotypic traits change over time or differ between groups of individuals, it is therefore important to realize that both within–and between–individual process might be underlying causal mechanisms". Individual heterogeneity may mask senescence and patterns of change in life history traits over life, or may hamper quantification of the rate of change in fitness components with age (Service 2004; van de Pol and Verhulst 2006).

Development of *frailty* CMR models

Until now, estimating individual variation in life history traits without using observable covariates, and estimating age-specific variation in life history traits (e.g., survival probability) while accounting for individual heterogeneity in underlying survival probability were difficult because methods were not designed to handle incomplete detection of individuals. The very first CMR estimation models developed to address heterogeneity in survival probability were developed in the 1970s. The development of models accounting for individual heterogeneity in parameters was motivated by the issue of heterogeneity in detection probability among individuals. The importance of such heterogeneity in wild animal populations has long been acknowledged (e.g., Carothers 1973; Gilbert 1973; Pollock et al. 1990; Norris and Pollock 1996; Pledger and Efford 1998; Pollock 2002; Link 2004): for example, failure to account for such heterogeneity may result in biased estimates of survival probability in open population models or of population size in both closed and open populations models. Efforts to account for heterogeneity in detection probability have triggered development of models accounting for individual heterogeneity in other parameters (e.g., survival; Burnham and Rexstad 1993; Pradel et al. 1995; Burnham and White 2002; Pledger and Schwarz 2002; Royle 2007).

Recently developed CMR estimation models allow consideration of heterogeneity in survival via random individual effects (Royle 2007). The state-space formulation of the Cormack–Jolly–Seber model proposed by Royle (2007) offers flexible means of extending the model to account for the specificity of different study systems and sampling schemes, and address different biological hypotheses. Briefly, the model accounts for the individual *state* on a given sampling occasion (e.g., dead or alive), and is specified using two distinct models: one for the process of interest (i.e., the survival process over a given time interval, partly unobservable), and one for the observations (i.e., whether the individual was captured/resighted on a given occasion). The observation process depends on recapture/resighting probability, and is conditional on the *latent* survival process (i.e., survival probability). Survival probability can be modelled as a function of covariates such as year, age, environmental covariates, individual measurable characteristics (e.g., body size etc.), and an individual random effect if one has reasons to suspect additional heterogeneity in survival probability (e.g. Royle 2007). With this innovation, frailty (Yashin et al. 2001) CMR models can be developed. Parameters in this class of models can be estimated using a Bayesian approach (e.g., using WinBUGS; Spiegelhalter et al. 1996). In addition, it is theoretically possible to design a state-space formulation of the multistate Arnason-Schwarz model (Arnason 1973; Dupuis 1995) with frailty, by using an additional (partially unobservable) transition process conditional on survival (i.e., a process accounting for the probability of being in a given stratum in a given occasion, and transition probability among strata in consecutive sampling occasions).

4.1.3 Individual Fitness and Population Growth Rate

On a related topic, Link et al. (2002a) used correlated latent survival, breeding and probability of raising 1 or 2 offspring to independence to estimate individual fitness. As emphasized above, estimation of fitness is at the heart of studies of evolutionary change by natural selection. Many empirical studies of selection have used fitness components to address fitness functions, but there is increased interest in developing estimates of "total fitness" (not components only; e.g. Coulson et al. 2006). Ideally, one may want to estimate an "individual growth rate" measuring the capacity of a given phenotype to be propagated into future generations. Because the growth rate of a genotype depends on the timing of production of viable offspring during life (Brommer et al. 2002), McGraw and Caswell (1996) suggested using an individual-specific Leslie matrix to estimate fitness. Link et al. (2002b) assessed the performance of the growth rate estimated using individual-specific Leslie matrices as an estimator of individual fitness. They defined latent fitness as the "latent individual growth rate", which corresponds to the latent survival characterizing the individual, as well as the individual breeding probability, and probability of producing a given number of offspring. They concluded that individual capture–recapture history data (i.e., one realization of the stochastic process defined by latent life history traits, McGraw and Caswell 1996) result in realized fitness that isn't consistent with latent fitness, and advocated a model-based approach to estimating fitness.

Interestingly, the distribution of individual demographic parameters (i.e., latent parameters) in populations has received much attention in another field, namely, applied population dynamics and conservation biology. Indeed, Conner and White (1999), Kendall and Fox (2001, 2003), and Fox and Kendall (2002) have provided evidence that certain forms of demographic heterogeneity substantially influence population persistence, a question that is particularly relevant to small populations. Development of CMR estimation models allowing investigators to estimate the distribution of individual life history traits and the possible covariation among latent traits may help develop an empirical basis for investigations of population persistence.

4.2 Evolution of Morphological and Behavioral Traits

CMR studies have also contributed to investigations of covariation between morphological or behavioral traits and fitness components. Just like for life history evolution, most CMR studies have addressed fitness functions and have drawn inferences about the possible consequences of these functions in terms of natural selection. However, I am not aware of studies that have addressed fitness functions of physiological traits using CMR estimation models, apart from the study by Reed et al. (2006, see above) whose aim was to investigate the physiological mechanisms underlying trade-offs between life history traits in juncos. The fitness costs incurred by testosterone-treated males suggest that high-testosterone phenotypes have selective disadvantages *in natura*. The body of studies listed below may seem eclectic compared to studies of life history evolution for two reasons. First, whether a trait is under selective pressures, or not, strongly depends on the study system and the type of organism concerned. Second, there are only a few studies that have used CMR models and have focused on some classes of fitness functions (e.g., fitness functions of behavioral traits).

4.2.1 Morphology

Several studies of birds have used CMR estimation models with individual covariates to address selection on body size (wing length), mass, condition in juveniles (e.g., lesser snow goose, Cooch et al. 2002) or adults (tufted duck *Aythya fuligula*, common pochard *Aythya ferina*, and Northern shoveler *Anas clypeata,* Blums et al. 2005), or both (serins *Serinus serinus* Conroy et al. 2002). Body condition can be viewed "as the size of the individual's energy reserves relative to its body size" (Blums et al. 2005). As migration is demanding in terms of energy, Blums et al. (2005) predicted a positive relationship between survival probability and body condition, but they also considered the possibility for costs associated with very high mass, as did Conroy et al. (2002). Specifically, they considered non-monotonic functions of body condition for survival probability. Both Cooch (2002) and Conroy et al. (2002) found evidence that the relationship between body mass and survival probability varied according to other covariates. Cooch (2002) found evidence of a positive relationship between survival probability and body mass in late-hatched young only. Conroy et al. (2002) found evidence of a negative influence of body mass in years with low density of competitors (siskins; *Carduelis spinus*). That is, in serins the shape of the fitness function varied with environmental conditions (biotic conditions). In lesser snow geese the fitness function differed according to the value of another trait (hatching date), which suggests that selection on body mass cannot be understood without considering the covariance between several traits (e.g., Lande and Arnold 1983; Houle 1991; Pigliucci 2006). The ability to use models including several covariates, both individual and time-specific environmental covariates proved important. Similarly, Wikelski and Trillmich (1997) addressed sex-specific relationships between survival probability, fertility, and body size in Iguanas (*Amblyrhynchus cristatus;* see also Laurie and Brown 1990) and suggested that balanced selective forces shaped body size in this species: sexual selection favoring large sizes in males in a lek-mating species, but natural selection penalizing large individuals in years with lower resource availability. CMR estimation models have also been used to address micro-evolutionary processes in morphological traits in taxa that are not often mentioned in EURING meetings. For example, Kingsolver and Smith (1995) have addressed wing pattern traits in a butterfly species (*Pontia occidentalis*). They found evidence of a negative influence of mean grey level of the dorsal wing and of ventral hind wings on daily survival. They suggested that this relationship resulted from the influence of color on thermoregulation ability.

Fitness functions based on estimates of survival probability have also been used to address the balance between sexual and natural selection in the wild. For example, Gregoire et al. (2004) have addressed the relationship between bill color in European blackbirds (*Turdus merula*) and survival probability. Theory of sexual selection assumes that there are advantages associated with exaggerated sexual characters in males, more precisely a larger breeding success probability (Andersson 1994). However, there may also be costs associated with secondary sexual traits, such as energetic costs of producing ornaments, increased detectability by predators or intra-specific competition (survival costs). Gregoire et al. (2004) found evidence of stabilizing selection on bill color using models with individual covariates. However, there are several non-exclusive hypotheses concerning the relationship between ornament expression and survival probability. Ornamental traits are assumed to have evolved through mate choice: in birds, individuals with the most showy feathers for example are assumed to be "higher-quality" individuals because they can afford to display costly adornments. In long-lived species, re-mating with the same mate has been shown to have advantages; loss of the mate through mortality may be very costly. Ornaments are assumed to serve as viability indicators: individuals may benefit from choosing a "higher-quality" mate with high survival ability. In this view, one may expect a positive relationship between survival probability and ornament size. Jones et al. (2002) have addressed the relationship between sexually selected feather ornaments and survival probability in crested auklets (*Aethia cristatella*), but have not found evidence of such a relationship. Here again, ultrastructural models were used.

For quantitative traits (as defined in Conner and Hartl 2004, p. 97), the shape of the relationship between fitness and trait values provides insight into the type of phenotypic selection (Conner and Hartl 2004). Directional selection is characterized by a linear fitness function, stabilizing selection by a quadratic function where fitness is highest at some intermediate value of the phenotype, and disruptive selection by a quadratic function where fitness is lowest at some intermediate value of the phenotype. However, as emphasized by Gimenez et al. (2006), the shape of the fitness function estimated using empirical data may not be quadratic, and more complicated forms of selection can occur (Conner and Hartl 2004). There is an analogy between the need for development of relevant fitness functions to address natural selection and the need for development of relevant forms for reproductive or survival functions in optimal control solutions of problems in population dynamics (Runge and Johnson 2002). Gimenez et al. (2006) have developed a nonparametric approach to fitting cubic splines within a CMR framework to address the relationship between body mass and survival in sociable weavers (*Philetairus socius*). Model parameters were estimated using a Bayesian approach in WinBUGS (Spiegelhalter et al. 1996). They found evidence that the fitness function is not symmetric, which suggests that body mass may not be under stabilizing selection. The technical development in Gimenez et al. (2006) reflects development in quantitative genetics where cubic splines and locally weighted least squares are used to assess the shape of fitness functions (Conner and Hartl 2004).

4.2.2 Behavior

Very few CMR studies have addressed fitness functions of behavioral traits other than those involved in dispersal and breeding activities. The studies I am aware of have investigated the influence of discrete behavioral traits on fitness components. For example, Webb (2006) addressed the consequences of tail autotomy in gekos (*Oedura lesueurii*) on survival probability. Some animals autotomize their tails, which is thought to facilitate escape from predators. Tail autotomy may increase the likelihood of surviving a predator's attack, however, this may have costs including: reduced growth, loss of energy reserves, decreased mating success, loss of social status, and decreased probability of survival during subsequent encounters with predators. Results did not provide evidence that spontaneous tail autotomy influences survival of juvenile geckos.

In a completely different framework, Cam et al. (2002b) have addressed fitness functions of behavior before recruitment: age-specific survival and recruitment probability, and breeding success probability in the first breeding occasion and subsequent occasions. Squatters are individuals present on nesting sites they don't own, containing chicks, when the owners are absent (e.g., during foraging trips at sea). Squatters may be aggressive and even kill the chicks, exhibit territorial behavior and coordination behavior with another squatter of the opposite sex. It has been suggested that squatting is part of behavioral maturation and territory acquisition and may influence age-specific recruitment probability. Results provided evidence that squatters have a higher age-specific local survival and recruitment probability than non-squatters in age-classes where squatting is represented, and a higher breeding success probability than non-squatters at the same age (Cam et al. 2002b). In addition, the relationship between initial breeding success probability and subsequent success probability was addressed using random intercept models (i.e., frailty models): individuals with high initial breeding success probability consistently have higher subsequent success probability. Consequently, it may be relevant to use squatting status before recruitment as a measurable covariate to account for permanent differences among individuals over life (i.e., as an observable criterion to classify individuals in "quality" classes).

4.3 Coevolution

Few CMR studies have addressed the evolution of morphological and life history traits within the framework of coevolution. Although morphology and life histories have been treated above, studies of coevolution are rare and are worth identifying separately. For example, Benkman et al. (2005) addressed bill size and survival probability in red crossbills (*Loxia curvirostra*). Their hypothesis to explain the difference in mean bill size between two populations was local infestation by the scaly-leg mite (*Knemidokoptes jamaicensis*) which favored local selection of smaller-bill birds. Indeed, large-billed males were more likely to exhibit symptoms of ectoparasitic mites. The authors found evidence that infestation by mites was associated with lower survival probability and caused directional selection against larger-billed individuals. In a recent review of dispersal and parasitism, Boulinier et al. (2001) deplored the weakness of the empirical basis in this area of research.

Concerning life history evolution, Dugger and Blums (2001) addressed brood parasitism in ducks using several fitness components – breeding success, recruitment in offspring, and adult survival probability. They conducted an experimental study by adding eggs and ducklings to clutches and broods, and also analyzed a larger observational data set. Their objective was to compare fitness components of parasitized and nonparasitized female common pochard (*Aythya ferina*) and tufted ducks (*Aythya fuligula*). They found that addition of small numbers of eggs to host nests (i.e., simulated parasitism) did not influence host clutch size, host hatching success, or nest success for either species. Parasitism by large numbers of eggs did not influence nest success in pochards, but it did in tufted ducks nests (numbers of eggs = 6 or more). Recruitment probability did not differ between parasitized and nonparasitized nests for either species, and parasitism had no negative effect on adult survival. Dugger and Blums (2001) concluded that moderate levels of parasitism do not have a negative influence on host fitness in these species.

4.4 Evolution of Sex-Ratio

A topic that has received much attention in studies of human populations is evolution of sex-ratio and its variation, either at birth or in the adult segment of the population (i.e., secondary sex-ratio). This topic has also received much attention in studies of animals (e.g., Nager et al. 1999, Weimerskirch et al. 2005), but in many cases without accounting for incomplete detection of individuals. Sex ratio theory is based on the idea that if the fitness benefits of producing males or females vary with environmental or social conditions, parents should adjust the sex ratio of offspring in a way that maximizes their own fitness. For example, if maternal condition influences survival probability in male and female offspring, the mother should produce offspring whose sex ratio maximizes the mother's fitness.

Empirical tests of hypotheses about adjustment of offspring sex ratio according to environmental and social conditions are scarce. Uller et al. (2004) addressed the influence of pre-natal sex-ratio on offspring survival and adult reproductive parameters in common lizards (*Lacerta vivipara*). In viviparous animals, sex ratio in-utero may influence the characteristics of offspring through exposure to sex-specific steroids in-utero and hormonal interactions between offspring. Evidence from studies in mammals suggests that both sexes are negatively affected by opposite-sexed siblings. Uller et al. (2004) provided evidence of a long-lasting influence of early conditions on fitness components, more precisely of an influence of pre-natal sex ratio on female fecundity, but not on survival probability. Age at maturity was also influenced by pre-natal sex-ratio. Prenatal sex-ratio may be maladaptive: females from male-biased clutches have lower fecundity and mature earlier than females from female-biased clutches. The fitness return for the mother may not be compromised because negative effects on the underrepresented sex could be counteracted by positive effects on the overrepresented sex. Thus, evolutionary consequences of pre-natal sex ratio on secondary sex ratio are still poorly understood.

4.5 Movement Among Locations and Habitat Selection Studies

Development of multistate models (Arnason 1973; Hestbeck et al. 1991; Brownie et al. 1993) also gave an enormous stride to studies of movement, migration and dispersal using data from wild marked animals (Bennetts et al. 2001; Kendall and Nichols 2004). This is one of the topics that have received attention in a large proportion of CMR studies relevant to evolutionary ecology. In a paper focusing on use of multistate models in evolutionary ecology, Nichols and Kendall (1995) laid the foundations of many subsequent studies of movement in subdivided populations (e.g., Spendelow et al. 1995; Senar et al. 2002; Blums et al. 2003, Skvarla et al. 2004). They basically explained in detail the relationship between model parameterization and classical hypotheses put forward in the literature on dispersal. One class of hypotheses considered corresponds to models of gene flow in systems of subdivided populations (e.g., influence of distance among locations on movement probability; Skvarla et al. 2004). In evolutionary ecology, because CMR studies mostly focus on vertebrates (i.e., mobile animals dispersing actively), movement among locations has mostly been addressed within the framework of habitat selection theory (e.g., Fretwell and Lucas 1970).

In long-lived species, it is natural to assume that individuals have to make decisions concerning breeding sites several times during life. When the individual's perspective is considered, the evolution of dispersal can be addressed within the framework of "habitat selection", whose broad scope encompasses both the decision of leaving a site and the choice of a new one (Ronce et al. 2001). Environmental conditions are likely to vary over space and time; for this reason, fixed dispersal strategies are unlikely to be favoured by natural selection (e.g., Ronce et al. 2001). Dispersal can be viewed as a decision making problem (i.e., "to stay or to leave?";. e.g., Danchin et al. 1998; Doligez et al. 1999; Brown et al. 2000; Serrano et al. 2001). It has been hypothesized that decisions are state-specific (i.e., depend on the individual state, such as condition, previous breeding success, breeding habitat, other environmental factors; e.g., Danchin et al. 1998). Recent syntheses about dispersal highlighted the growing attention to questions of individual plasticity and condition-dependant dispersal (Danchin et al. 2001; Ims and Hjermann 2001; Ronce et al. 2001; Serrano et al. 2001; Serrano and Tella 2003).

A key question is how individuals make decisions concerning fidelity to the previous breeding site, or if they decide to move, selection of a new one. One of the main predictions of the "ideal free habitat selection theory" (Fretwell and Lucas 1970) is that natural selection should favour dispersal tactics where moving leads to increased realized fitness (i.e., habitat selection should be shaped by fitness maximization; Holt and Barfield 2001). Densities in the various locations are expected to change as well as realized fitness in each habitat, and eventually realized fitness is equilibrated. This is why studies based on this theoretical framework do not always address "realized" fitness functions: fitness is not assumed to vary according to behavioral decisions in a systematic manner. The form of the function depends on the state of the study system (e.g., a sub-divided population), whether it is at evolutionary equilibrium, whether fitness is density-dependent, etc. Because of the numerous assumptions this theory relies on (e.g., individuals have perfect knowledge of their environment, there is no cost of moving, etc.; Holt and Barfield 2001), the scenario leading to the "ideal free distribution" of individuals in space developed by Fretwell and Lucas (1970) is unlikely to be observed in the wild (Nichols and Kendall 1995). However, the seminal idea that habitat selection is shaped by fitness maximization leads to some specific predictions that have been tested in several CMR studies of habitat selection. The hypothesis that fitness maximization shapes habitat selection tactics leads to the question of how individuals can assess fitness prospects in different potential locations (Danchin et al. 1998).

For example, it has been hypothesized that individuals use their own breeding success and the success of conspecifics as cues to assess expected location-specific fitness (Danchin et al. 1998; Doligez et al. 1999; Brown et al. 2000; Serrano et al. 2001), and that the decision regarding the location where they will breed in year $t + 1$ is made based on evidence from year t. Serrano et al. (2005) suggested that colony size also contributes to determine fitness prospects. Several CMR studies have addressed breeding habitat selection and movement within this framework (e.g., Doligez et al. 2002, 2004; Cam et al. 2004a; Serrano et al. 2003, 2005). In a different vein, Brown et al. (2005) have addressed how intrinsic individual characteristics (more precisely, steroid hormone level and its influence on competitive ability) influenced movement probability among colonies and colony choice in cliff swallows (*Petrochelidon pyrrhonota*)

Furthermore, theories of habitat selection have been invoked to address life history traits other than movement probability per se. Indeed, because the quality of the breeding habitat is likely to influence individual fitness, natural selection may favor habitat selection tactics involving decisions about "when to breed". It has been suggested that the two decisions "where to breed" and "when to breed" are "two sides of the same coin" (Ens et al. 1995). Habitat selection tactics and age-specific recruitment probability have been addressed in several studies based on CMR data (e.g., Oro and Pradel 2000; Frederiksen and Bregnballe 2001). The evolution of dispersal has also been addressed outside the framework of habitat selection theory. More specifically, it is sometimes assumed that parents produce offspring with fixed dispersal strategies (e.g., philopatric versus dispersing offspring). Within this framework, Hamilton and May (1997) have suggested that in species with senescent decline in survival, whether parents should produce philopatric versus resident offspring should depend on their age. Few CMR studies have addressed hypotheses about age-specific variation in reproductive investment or reproductive performance, and dispersal (e.g, natal dispersal and senescence; Ronce et al. 1998).

Last, dispersal evolution theories or habitat selection theories make assumptions about whether dispersal is costly (e.g., there may be mortality costs associated with movement, costs of settling because of competition with conspecifics, or costs associated with reproduction in unfamiliar environments). Massot et al. (1994) have studied settlement ability using experimentally translocated common lizards (*Lacerta vivipara*). They compared individuals that survived after the introduction with those of non-manipulated populations. Results provided evidence that translocated individuals had a lower survival probability after being transferred to their new habitat, except juveniles. Adults may thus incur costs associated with unfamiliarity with the new habitat. Selected individuals had particular features in terms of body mass and size. In addition, surviving transplanted males have the same characteristics as transients or immigrants in natural populations (body mass and size); they may thus have been transients or immigrants in their own population of origin. However, this did not hold in females.

5 Development of CMR Estimation Models to Address the Genetic Basis of life History Traits in Wild Animal Populations

A condition for natural selection is heritability in the focal trait (Premise (iii), Section 4.1.). Until now, evolutionary ecologists have used CMR estimation models to address the relationship between demographic parameters themselves (i.e., covariation in life history traits) or between demographic parameters and morphological traits (i.e., fitness functions), which has permitted them to gain insights into Premise (ii) for natural selection. However, studies of Premise (iii) have concerned either "non-demographic" traits (morphological, physiological, behavioral traits), or demographic traits estimated without accounting for imperfect detection of individuals by investigators. As far as the evolution of demographic parameters themselves is concerned, the genetic basis of these traits has not been addressed using CMR estimation techniques. Here I suggest that recent methodological development concerning both quantitative genetics models of estimation of additive genetic variance of traits and CMR models of estimation of demographic parameters theoretically allows integration of the two fields.

5.1 Features of Current Knowledge of Heritability of Demographic Parameters in Wild Animal Populations

Without using CMR estimation models, several studies have provided evidence that life history traits (fitness components such as age of first breeding, lifetime reproductive success, clutch size or litter size for example) exhibit low heritability compared to other traits (morphological, physiological or behavioral traits), but the studies in question have also provided evidence that such heritability levels cannot be ignored, and that they vary among species for the same trait and among populations of a single species (Stearns 1992; Matos et al. 1997; Kruuk et al. 2000; Réale and Festa-Bianchet 2000; Réale et al. 2003; Sheldon et al. 2003; Charmantier et al. 2006a, b; Pigliucci 2006). Comparison of results obtained using different statistical techniques to address heritability and additive genetic variance of traits (more precisely parent-offspring regression versus the *animal model*; Section 5.2.2) have provided evidence that the question of heritability of life-history traits should be re-addressed using the most modern techniques (e.g., Kruuk et al. 2001). Moreover, several researchers have pointed out that despite the success of quantitative genetics theory in domesticated animal and plant breeding, very few studies of natural populations have provided evidence of micro-evolutionary changes in heritable traits in response to selection in the presence of directional selection (Kruuk 2004). Two hypotheses have been put forward to explain this: (1) the approaches to estimation of the amount of genetic variation transmitted from parents to offspring we have used so far lead to biased estimates of variation or of the strength of selection (more precisely they overestimate the additive genetic variance or the directional selection differential, (see below)), and (2) the genetic basis of different traits should not be addressed separately (Lande and Arnold 1983; Houle 1991). Researchers in quantitative genetics have developed approaches to estimating genetic parameters that may partly solve these problems (Lynch and Walsh 1998).

Importantly, the problem of imperfect detection of individuals by investigators has been overlooked. Until now, in wild animal populations, quantitative genetics studies focusing on demographic parameters have been conducted using observed values of life-history traits (e.g. Sheldon et al. 2003), which is reasonable only in situations where detection probability of marked individuals alive and present in the study area is close to 1. Such situations have long been known to be rare in wild animal populations (Lebreton et al. 1992; Clobert 1995; Martin et al. 1995). In the vast majority of populations, estimation of survival or dispersal probability, age-specific recruitment probability (one definition of which is the probability of making a transition between state "pre-breeder" and state "breeder" at a given age), and adult breeding probability (in species exhibiting intermittent breeding), all require use of estimation models explicitly incorporating detection probability. Unless this probability is equal to 1, the age at which the first breeding event was recorded cannot be assumed to be a reliable measure of age of recruitment, and the age at which the last breeding event was observed cannot be assumed to be a reliable measure of age of last reproduction, observed breeding events cannot be assumed to account for all the breeding attempts in an individual's life, and individuals not recaptured or not resighted in a given year cannot be assumed to be dead that year. In addition, estimation of breeding success probability in a given breeding occasion, the probability of laying a clutch of a given size, or giving birth to a litter of a given size, may require CMR estimation models accounting for state-specific detection probability (i.e., multistate models, Nichols et al. 1994; Nichols and Kendall 1995; Williams et al. 2002). Besides, even if heritability of survival probability per se is not of interest, in several classes of CMR models for open populations, several of the aforementioned demographic parameters are conditional on survival probability over one or several time intervals between sampling occasions. Therefore, the issue of estimation of survival probability in situations where detection probability is <1 cannot be ignored.

5.2 Mixed CMR Models to Address Heritability of Demographic Parameters

Lynch and Walsh (1998, p. 50) summarized the problem of analysis of the genetic basis of quantitative traits as follows: "inferences concerning the genetic basis of quantitative traits can be extracted from phenotypic measures of resemblance between relatives". Obviously, to use CMR statistical techniques to estimate parameters relevant to quantitative genetics, CMR data from marked *relatives* are needed. Long-term monitoring programs of marked individuals in wild animal populations have often led to such data. The principles of quantitative genetics are general; they have been widely used in animal and plant breeding and are valid in wild organism populations as well. However, "because the systems of mating and evolutionary forces found in natural populations are generally quite different than the controlled programs imposed on domesticated species, study of the inheritance of quantitative traits in natural populations presents a number of challenges" (Lynch and Walsh 1998, p. 5). The distinctive feature of the data sets from wild animal populations is that they correspond to complex pedigrees; there is a variety of degrees of relatedness between individuals. In domesticated animals, investigators design experiments to address the genetic basis of phenotype. In wild animal populations fathers of individuals often cannot be identified unless molecular approaches are used (Thomas et al. 2002), especially in species without paternal care or species where extra-pair paternity is common. Most long-term monitoring programs are purely observational if the target species is protected or is (locally) of conservation concern. Consequently, specific sources of phenotypic variation in populations have seldom been addressed using experimental approaches. Experiments have sometimes been conducted (e.g., cross-fostering, Wiggins 1989), but designs specifically relevant to quantitative genetics are rare and such experiments are usually short term.

One of the main obstacles encountered by researchers using CMR estimation methods to address heritability in demographic parameters is obtaining "one" estimate ("one measurement") of the focal trait per individual (even the mean value over the lifetime). Early efforts to assess heritability in quantitative traits (morphological traits or life history traits) have relied on estimation of the slope of parent-offspring regression (Lynch and Walsh 1998), which requires one measurement of the trait in parents (or one parent) and one measurement in offspring. This was done either by estimating the mean value of the trait over the parent's lifetime and the mean value of the offspring trait over its lifetime, or by taking one single measurement of the parent and the offspring. Approaches to estimation of demographic parameters almost inevitably involve some degree of aggregation of data (Cooch et al. 2002; Nichols 2002), and it is not possible to estimate quantities like "the individual survival probability" using data from a single individual. Development of models with individual covariates (Skalski et al. 1993) allows investigators to achieve high levels of stratification of populations (Cooch et al. 2002) and to estimate individual-specific demographic parameters (e.g., an "individual survival probability"), provided relevant measurable covariates are available. However, the distribution of the focal trait (the individual survival probability) in the population may not be accounted for in a satisfactory manner by the relationship between survival probability and a measurable individual covariate (e.g., body size, laying date in birds, etc.).

Recent developments in CMR estimation methods share a common feature with the most recent techniques to estimate heritability (Fry 1992; Lynch and Walsh 1998; Kruuk 2004; Schaeffer 2004), namely, use of mixed models (including both fixed and random effects). More specifically, the feature shared by the *animal model* in quantitative genetics and some CMR estimation models is that part of the variation in the focal trait is accounted for by random effects. As emphasized in Section 4.1.2, in human demography, random effects have long been used to account for individual heterogeneity in mortality risk. Mixed effects models may not provide investigators with "one estimate of survival probability per individual", but they provide an estimated distribution of the demographic parameter in the population, that is, an estimated variance of the focal trait among individuals (after accounting for relevant fixed effects: sex, age, location, etc.). Similarly, according to Kruuk (2004) "One of the major recent changes in the study of quantitative genetics of natural populations has been the use of mixed models, in particular the form of mixed models known as the 'animal model', for the estimation of variance components."

The following section is largely inspired by Lynch and Walsh (1998), and Conner and Hartl (2004): it is intended for readers without background in quantitative genetics. This material is needed to understand the points shared by recently developed CMR and quantitative genetics estimation techniques. All the topics addressed here are extensively covered in two quantitative genetics "bibles" (Falconer and Mackay 1996; Lynch and Walsh 1998).

5.2.1 A Very Short Introduction to Quantitative Genetics Theory

The *phenotypic value* of an individual (the measurement of a given quantitative trait for an individual: morphological, demographic, physiological, or behavioral trait), z, is determined by the individual genotype and the environment. Quantitative genetics "focuses on the phenotype, usually without knowing the genotype underlying the traits" (Conner and Hartl 2004, p. 3). The traits "are encoded by a large number of genetic loci, and for practical reasons, the individual loci are generally unobservable" (Lynch and Walsh 1998, p. 4). The phenotypic value is assumed to be the sum of the total effects of all loci on the trait, G, the *genotypic value* and an *environmental deviation* E. That is,

$$z = G + E \tag{1}$$

The genotypic value is the phenotype produced by a given genotype averaged across environments. The environmental deviation is the difference between the phenotypic and the genotypic values caused by the environment (temperature, prey availability, rainfall etc., Conner and Hartl 2004, p. 101). The mean and the variance (σ_p^2) characterizing the distribution of individual phenotypic values are properties of populations; there is a distribution of individual genotypic values and of individual environmental deviations. The distribution of environmental deviations is generally assumed to be normal with mean = 0. One central goal of quantitative genetics is partitioning the phenotypic variance σ_p^2 into genetic and non-genetic components:

$$\sigma_p^2 = \sigma_G^2 + \sigma_E^2 \tag{2}$$

where σ_G^2 is the genotypic variance and σ_E^2 is the environmental variance.

The evolutionary response of a trait to selection is a function of the intensity of selection and the fraction of the phenotypic variance attributable to certain genetic effects (Lynch and Walsh 1998). More specifically, one draws a distinction between *additive genetic effects* (the effects of each allele in the genotype adds to determine the total effect on the phenotype) and interactions between alleles at the same locus, *dominance,* or at different loci, *epistasis*. Hence, the genotypic variance σ_G^2 can be partitioned into σ_A^2, the additive genetic variance, σ_D^2 the dominance variance, and σ_{EPI}^2, the epistatic variance (or interaction). That is,

$$\sigma_{\mathrm{G}}^2 = \sigma_A^2 + \sigma_G^2 + \sigma_{EPI}^2 \tag{3}$$

The additive genetic variance is the most important for sexually reproducing species because only the additive effects of genes are transmitted directly from parents to offspring; information on other sources of genetic variation (e.g., linkage disequilibrium, polyploidy, etc.) can be found in Lynch and Walsh (1998).

The directional selection differential, S, is the within-generation difference between the mean phenotype after an episode of selection (but before reproduction) and the mean before selection. Because of direct transmission of additive effects, σ_A^2 is most important in determining changes in mean phenotypic values across generations in sexual species. It is also the easiest of the genetic components of variance to estimate using resemblance between relatives: resemblance is caused primarily by additive variation (Lynch and Walsh 1998). Change in mean phenotypic values across generations is the definition of phenotypic evolution (Deduction II). The degree to which the mean phenotype after selection μ_s deviates from the mean before selection μ_0 depends on survival probability and reproduction (fitness) of individuals with different phenotypes. Under specific assumptions concerning the (un-) importance of genotype $\times$ environment covariance and interaction, if the regression of the offspring phenotype on that of its average parent is linear with slope β, a change in the parental mean phenotype induces an expected change in the mean phenotype across generations equal to:

$$\Delta\mu = \mu_0 - \mu_s = \beta^* S. \tag{4}$$

where μ_0 is the mean phenotype of the offspring of the selected parents. This equation is the *breeder's equation*. "It combines information on the forces of selection (S) with that on inheritance (β) to yield a predictive equation for evolutionary change across generations. If β is zero, no matter how large S is, the response to selection across generations is zero" (Lynch and Walsh 1998, p. 47).

In sexually reproducing species, genotypes are not passed on from parents to offspring, but are created anew in each offspring by combining an allele from each parent at each locus. "The breeding value can be defined as the effect of an individual's genes on the value of the trait in its offspring; this effect is caused by the additive effects of genes – it is sometimes called 'additive genotype' and has variance" σ_A^2 (Conner and Hartl 2004, p. 111). Heritability is the proportion of the phenotypic variance that is due to genetic causes. Broad-sense heritability is defined as follows:

$$H^2 = \frac{\sigma_G^2}{\sigma_P^2} \tag{5}$$

However, because the genotypic value includes genetic components (e.g., epistatis) that do not contribute to resemblance between relatives as much as the additive genetic component, we usually define the narrow-sense heritability:

$$h^2 = \frac{\sigma_A^2}{\sigma_P^2} \tag{6}$$

In parent-offspring regression, h^2 can be estimated using the slope of the regression.

5.2.2 The Classical Parent-Offspring Regression and the Animal Model

Parent-offspring regressions involve measurements from individuals with specific degrees of relatedness. This commonly used technique to estimate heritability has advantages that are not restricted to practical considerations (such data are usually available and the computations are done using classical least squares regression): parent offspring regression is not affected by dominance or linkage (loci close together on the chromosome are said to be genetically linked, which makes recombination between loci during meiosis rarer; see Lynch and Walsh 1998 for details, p. 537). However, in monitoring programs of wild animal populations, information from individuals with different degrees of relatedness is available. The so called *animal model* allows investigators to make full use of the available information. The fact remains that the degree of relatedness of the individuals included in the analysis must be known, and the corresponding information is used in a model where phenotypic values are expressed as a function of fitness, breeding values, environmental effects, etc.

In addition, in parent-offspring regression, measurements from individuals were either averaged over life, or a single observation was retained for analysis. For some traits, lack of variation over life may be a reasonable assumption (e.g., body size after sexual maturity in species whose growth takes place before reproduction), but not for all traits (e.g., clutch size in birds) or all species (e.g., body size in snakes). Last, genotype × environment covariance and interactions are assumed to be negligible, as well as permanent maternal effects if data from the sole mother are available (permanent maternal effects occur when the phenotype of the offspring is influenced by the phenotype of the mother, which may be caused either by genetic or environmental effects; Mousseau and Fox 1998). One of the advantages of the *animal model* is to allow investigators to take such effects into account explicitly in analyses of phenotypes to address variance components and heritability. Studies that have used both parent-offspring regression and the *animal model* with the same data set to address heritability have often provided evidence of discrepancies in results (e.g., Kruuk et al. 2001), which may result from the fact that variance components accounting for a larger number of sources of variation in phenotypic values can be included in analyses with the *animal model* compared to parent-offspring regression (Kruuk 2004).

5.2.3 Estimation of Heritability and Breeding Values: The Animal Model

General Formulation of the Animal Model

For the sake of simplicity, here I will assume that phenotypic values of a trait (morphological, physiological, behavioral, or demographic) are normally distributed, as in the case in many quantitative genetics applications (when focusing on traits such as body mass, size, or laying dates in birds for example; e.g., Kruuk et al. 2000; Wilson et al. 2005). However, the statistical theory for estimation of variance components and prediction of random effects in mixed models exists for variables with other distributions (e.g., Bernoulli, Poisson; Fahrmeir and Tutz 1994; Matos et al. 1997; Lynch and Walsh 1998, p. 745, 779), which may be more useful for addressing heritability in traits such as survival probability or breeding success probability. Besides, estimation per se is beyond the scope of this paper: several approaches have been developed in quantitative genetics (namely, REML and Bayesian methods based on Markov Chain Monte Carlo simulations; Blasco 2001), which are well suited for complex pedigrees with unbalanced data, as is usually the case in long-term monitoring programs of wild marked animal populations. These methods can be implemented within the framework of CMR estimation models (e.g., Dupuis 1995; Dupuis et al. 2002; Vounatsou and Smith 1995; Royle and Link 2002; Brooks et al. 2002, 2004; King and Brooks 2002, 2004; Link and Barker 2004; Otis and White 2004; Royle 2007; Royle and Kéry 2007). However, to estimate heritability in fitness components, more flexible mixed CMR models than existing models are needed.

In the general case (i.e., where the phenotype is determined by genetic and environmental effects plus interactions between them), the phenotype of the *k*th individual of the *i*th genotype exposed to the *j*th environmental effect can be described as a linear function of four components (Lynch and Walsh 1998).

$$z_{ijk} = G_i + I_{ij} + E_j + e_{ijk} \tag{7}$$

where G_i is the genotypic value, which may be a function of the population mean phenotype (e.g., z may depend on age, year of birth, gender for example, which can be accounted for using additional fixed effects, and year using either fixed or random effects, depending on the study). G_i includes the additive genetic effects (breeding value) and possible genetic components such as dominance and epistasis (different gene effects on the phenotype). E is the environmental effect on the phenotype, I is the genotype × environment interaction effect on the phenotype, and *e* is the residual deviation.

Accordingly, the total phenotypic variance of a population can be written as:

$$\sigma_P^2 = \sigma_G^2 + \sigma_I^2 + 2\sigma_{G,E} + \sigma_E^2 + \sigma_e^2 \tag{8}$$

where $\sigma_{G,E}$ is the genotype × environment covariance. σ_I^2 is the genotype × environment interaction, which corresponds to the variation in the phenotypic response of specific genotypes to different environments. $\sigma_{G,E}$ is the physical association of specific genotypes with environments: if the genotypes are randomly distributed with respect to environment, $\sigma_{G,E}$ is zero. σ_G^2 may be further decomposed according to additive genetic variance (σ_A^2) and types of gene effects on the phenotype (see above). A source of environmental variation of particular interest in evolutionary ecology is maternal effects (for additional information, see Lynch and Walsh 1998).

A distinctive feature of the *animal model* is that random effects are used to account for the additive genetic variance (breeding values), and that information on the degree of relatedness of the individuals included in the analysis is used to estimate σ_A^2. Other random effects may be used to account for other components of the genotypic value (e.g., genotype × environment covariation), and also to account for sources of variation in the phenotypic values other than genetic effects (e.g., environmental effects). The general formulation of the *animal model* is the following. Consider a n × 1 column vector y with n observed phenotypic values. The model assumes that y can be described as a linear model with a $p \times 1$ vector of p levels of fixed effects (β), a $q \times 1$ vector of q levels of random effects (u), and an n × 1 vector of random, residual terms (e). The first element of vector β *is* generally the population mean. Importantly, the elements of the vector u are usually genetic effects, including additive genetic effects (i.e., breeding values). The residual deviations are assumed to be independent of random genetic effects.

$$y = X\beta + Zu + e \tag{9}$$

X and Z are design matrices whose elements are equal to 0 or 1 depending on whether the effect influences the individual's phenotype. The expectation of y is:

$$E\begin{pmatrix} y \\ u \\ e \end{pmatrix} = \begin{pmatrix} X\beta \\ 0 \\ 0 \end{pmatrix} \tag{10}$$

and the variance–covariance structure of y is:

$$V\begin{pmatrix} u \\ e \end{pmatrix} = \begin{pmatrix} G & 0 \\ 0 & R \end{pmatrix} \tag{11}$$

G is the variance–covariance matrix for random effects other than residual terms, and R is the variance–covariance matrix of residuals. The square matrices G and R are assumed to be non-singular and positive definite. V is usually expressed as follows:

$$\mathrm{V} = \mathrm{V(y)} = \mathrm{ZGZ'} + \mathrm{R} \tag{12}$$

In many applications, residual terms are assumed to be independently and identically distributed with mean 0 and variance σ_e^2. Therefore, $\mathrm{R} = \mathrm{I}\sigma_e^2$. In situations where the phenotype is assumed to be completely influenced by fixed effects (e.g., age), and observations are independent (e.g., there is only one observation per individual), V is equal to R. However, the distinctive feature of genetic analysis is that V is generally not diagonal. By definition, individuals with some degree of relatedness share part of their genes and the main objective of quantitative genetics analysis is to address the genetic basis of phenotype using resemblance between relatives. Hence, it is hypothesized that the phenotypic values of relatives are not independent and that part of the dependence is caused by additive genetic effects: addressing this hypothesis is central to quantitative genetics.

Incorporation of Dependency Among Relatives in CMR Models

The matrix G describes the covariance among random effects. Assuming that the only random effects (u) in the model are the additive genetic effects, G corresponds to the covariance in additive genetic effects among relatives. It can be shown that the covariance between two relatives i and j is given by $2\Theta_{ij}\sigma_A^2$, where Θ_{ij} is the coefficient of coancestry (Falconer and Mackay 1996, Lynch and Walsh 1998). It is the probability that an allele drawn at random from individual i will be identical by descent to an allele drawn at random from individual j. For example, this probability is 0.25 for parent and offspring, so that the additive genetic covariance between them is $0.5\,\sigma_A^2$ (Kruuk 2004). The matrix including (twice the) coefficients of coancestry must be built before the analysis (it is often called the *Numerator Relationship Matrix*, A), according to the specific data set in hand and the corresponding pedigree. Several pieces of software or routines have been designed to build it (e.g., Kruuk 2004, Saxton 2004, Kinghorn and Kinghorn 2007). This matrix is used to specify the variance–covariance structure of u: G. R and G have to be modified according to the design of the study and the question addressed. For example, their structure may account for repeated measures from the same individual through an additional random effect reflecting permanent environmental effects on all observations from the same individual, and through non-independence of residual terms (Kruuk 2004). Similarly, maternal effects or common environmental effects can be accounted for by an additional random effect on all offspring of a given mother, or on all individuals sharing the same environment during development, respectively.

Recently developed CMR models designed to estimate demographic parameters allow consideration of both fixed and random effects. Importantly, development of individual-level models to accommodate individual covariates (Skalski et al. 1993; Royle 2008) required specification of the likelihood for each individual capture–recapture history; this is also required to accommodate individual random effects (Royle 2007). Incorporation of a user-defined design matrix and variance–covariance matrix for random effects is what remains to be made possible to address the genetic basis of demographic parameters estimated using CMR estimation models accounting for incomplete detection of individuals. It is important to note that in quantitative genetics, the assumption of non-independence among individuals is central: the main objective of analyses is assessment of the contribution of common genetic material to resemblance among relatives. Therefore, development of flexible tools to specify the variance-covariance matrix of random effects will probably greatly influence the success of the efforts to address heritability in demographic parameters.

Multitrait Models

The *animal model* has been used in quantitative genetics to conduct multivariate analyses of life history traits (Charmantier et al. 2006a). The "multivariate breeder's equation" (Pigliucci 2006) allows consideration of pleiotropic effects such as antagonistic pleiotropy invoked in evolutionary theories of aging (e.g., Tatar 2001). In theory, the state-space formulation of the Arnason–Schwarz model for example allows simultaneous estimation of several life history traits. Several individual random effects can be used (Yashin et al. 2002; Cam et al. 2004b) to address the correlation in latent life history traits.

Other Methodological Challenges

In addition to the technical difficulties associated with the specification of a user-defined variance–covariance matrix for random effects in CMR estimation models, two other issues will require additional efforts. First if estimation is done within the Bayesian framework, according to some researchers, how to conduct model selection is unclear for models with random effects (e.g., Spiegelhalter et al. 2002). Some computer-intensive methods (Reversible Jump Markov chain Monte Carlo simulations; Green 1995) have been proposed to explore variable dimension statistical models, but may be difficult to implement in a flexible manner in standard software programs (Brooks et al. 2002). In addition, there has been a strong emphasis on evaluation of the fit of models using information criteria to perform model selection (e.g., Lebreton et al. 1992; White 2002; Choquet et al. 2003; Pradel et al. 2003). Here again, how to assess the fit of models is not straightforward.

Breeding Values and Selection Studies

As emphasized above, studies involving CMR models to estimate demographic parameters have addressed the question of the evolution of morphological or physiological traits using fitness functions: the relationship between the phenotypic value of a (morphological, physiological, behavioral) trait and fitness components. Kruuk (2004) pointed out that one may find evidence of such a relationship even in situations where there is no relationship between the genetic basis of the trait and fitness. This may occur when there is an environment-induced covariance between the trait and fitness (variation in environmental conditions are associated with joint variation in trait values and fitness). To detect such a phenomenon, the *animal model* may be used to predict individual breeding values (i.e., prediction of an individual random effect accounting for σ_A^2, the variance of the additive genotype). Comparisons between two ways of assessing selection gradients provide insight into the above covariance: fitness functions obtained using breeding values, and using phenotypic values.

5.3 Concluding Remarks

As emphasized by Lynch and Walsh (1998), evolutionary biology has considerably been influenced by quantitative genetics, but the need for statistical tools (more specifically, mixed models) to analyze complex pedigrees in wild animal populations is currently one of the motivations for statisticians developing methods to estimate relevant quantities in quantitative genetics. The material introduced above suffers from the simplifying assumptions early quantitative genetics suffered from, but the current machinery of the field and of statistics can handle more complex situations likely to be relevant to wild animal populations, such as genotype × environment interactions, maternal, or family effects (e.g., Massot and Clobert 2000). In addition, multivariate phenotypes and pleiotropic effects can be addressed, and a few studies conducted using empirical data from wild animal populations where detection probability of individuals is high have provided evidence of additive genetic variation in life history traits, and of evolutionary trade-offs and opposing directional selection on traits (e.g., Charmantier et al. 2006a).

One should keep in mind, however, that if fitness functions (common in CMR evolutionary ecology studies) are not sufficient to address evolutionary change in traits, quantitative genetics has its limitations as well. Pigliucci (2006, p. 5) recently pointed out that heritability is a "local measure, meaning that it can, and often does, change with changes in the population's gene frequencies and environments encountered. [...] Evolution de facto changes gene frequencies. [...] Heritabilities do not provide a useful measure of the long-term capability of traits to respond to selection". Quantitative genetics is successful at making short-term predictions, mostly qualitative predictions, but in its current state evolutionary biology theory is unable to predict long-term evolutionary change in traits.

In addition, there is a long tradition of experimentation in quantitative genetics, which uses creation of "artificial sets of offspring derived from carefully designed crosses among parents sampled at random from a natural population" (Pigliucci 2006, p. 9). Experimentations of that type may be possible in some wild animal populations, but long-term studies (e.g., $\approx$ 30 years) resulting in complex pedigrees haven't been designed that way. This implies that some inferences (about maternal effects for example) may not be possible with data from wild animal populations. In addition, complex pedigrees may lead to situations where there are small sample sizes to estimate some variance components (Quinn et al. 2006). However, Pigliucci (2006, p. 9) also questioned whether an artificially created set of genotypes should be used to draw inference about genetic parameters in the natural population, because "it is vanishingly unlikely that the individuals in the population in question would ever cross in even approximately the same pattern as required by statistics tests and laboratory experiments". In other words, the corresponding estimated genetic parameters may not be thought of as the parameters in the natural population.

Nevertheless, limitations of the inferences that can be drawn from quantitative genetics parameter estimates should not overshadow the weakness of our knowledge of heritability of life history traits and of their genetic basis in wild animal populations. As emphasized above, the emergence of new statistical tools calls for studies of the genetic basis of these traits (Kruuk 2004). In addition, the rare studies where detection probability is likely to be high (although not formally estimated) and that have used *animal models* with values of life history traits directly observed (e.g., the mute swan, *Cygnus olor*, population in Abbotsbury, Dorset, U.K.) have provided stimulating results (Charmantier et al. 2006a, b). Last, Houle (1992) pointed out that many previous inferences about the potential for evolution of life history traits compared to other traits have been drawn using narrow-sense heritability, h^2 (see above). Lower heritability of life history traits compared to others has been interpreted as evidence of lower genetic variation in demographic parameters than in other traits. However, Houle (1992) argued that heritability is not appropriate for comparative studies of genetic variation in traits and proposed a dimensionless criterion for this purpose. This also calls for new studies of the genetic basis of life history traits in wild animal populations.

6 Additional Topics

"Evolutionary ecologists consider both historical and contemporary influences on patterns of variation and study variation at all levels, from within-individual variation (e.g., ontogenetic, behavioral) to variation among communities or major taxonomic groups" (Fox et al. 2001, Preface). The enormous range of questions potentially relevant to evolutionary ecology is reflected in the explosion of studies that have used CMR estimation models, and of the scope of such models. In this paper, for the sake of conciseness and homogeneity, several important topics and CMR models have not been addressed. One of the reasons for this choice is that overall, studies relevant to evolutionary ecology that have used these approaches are still rare. However, as for models developed earlier, evolutionary ecologists may only grab these tools once they are widely used elsewhere and are sufficiently developed and flexible to allow them to address novel questions relevant to current evolutionary theory, concepts and methods. There is no conceptual reason why such models should not be used by evolutionary ecologists in the near future.

(i) CMR approaches to estimation of vital rates specific to ecological communities (e.g., Nichols et al. 1998a, b; Dupuis and Joachim 2006; Kéry and Royle 2008, this volume). These approaches use the species ID as the individual mark in population modelling. A few studies have opened the way to questions undoubtedly anchored in evolutionary ecology. For example, Doherty et al. (2003) have addressed the relationship between sexual selection and local extinction probability in bird communities. As Stenseth and Saetre (2003, p. 5576) emphasized: "community ecology and evolutionary ecology have in the past, to a large extent, been moving along separate paths. [. . .]. However, CMR estimation methods have already been used to draw inferences about variation in species diversity at long evolutionary time scales (Nichols and Pollock 1983). Stenseth and Saetre (203, p. 5577) also emphasized that "Doherty et al. (2003) provide an excellent demonstration of the potential power of using long-term ecological monitoring data to address key problems in community ecology and evolution"

(ii) Occupancy estimation models (MacKenzie et al. 2006). This recent book by MacKenzie and colleagues focuses on estimation methods to address "occupancy in ecological investigations", either occupancy of sampling units by one species, or by several. In other words, these methods are relevant not only to ecological communities (see above), but also to all the studies using "presence–absence data" over space and time for a given species. The number of questions relevant to evolutionary ecology that can be addressed using these methods is very large (e.g., metapopulation dynamics, changes in geographical range, epidemiology). This is an area of active research to develop models to estimate spatio-temporal variation in occupancy probability, but to date mostly methodological work as been done (e.g., MacKenzie and Kendall 2002; MacKenzie et al. 2002, 2004; MacKenzie and Bailey 2004; MacKenzie and Nichols 2004; Royle et al. 2005).

(iii) Noninvasive genetic sampling. Despite the intensive use of molecular markers in evolutionary ecology (e.g., Conner and Hartl 2004), capture–recapture analysis of DNA-based data has received little attention in this field. Recent development in capture–recapture theory designed for molecular markers may open the way to new studies (Lukacs and Burnham 2005; Petit and Valière 2006).

Acknowledgments I thank Jim Nichols and Torbjørn Ergon for the invitation to present this paper, and Mike Conroy for sharing his material with me before the meeting. Two anonymous referees helped improve the paper in a substantial manner. I am grateful to Jean-Yves Monnat for the numerous discussions about the contribution of CMR models to the long-term study of kittiwakes in France, Matthieu Authier for stimulating discussions about estimation of quantitative genetics parameters using data from wild animal populations, and Christophe Thebaud for sharing some key papers on the topic.

References

Andersson M (1994) Sexual selection. Princeton University Press, Princeton, New Jerey, U.S.A.

Anderson DR, Cooch EG, Gutiérriez RJ, Krebs CJ, Lindberg MS, Pollock KH, Ribic CA, Shenk TM (2003) Rigorous science: suggestions on how to raise the bar. Wildlife Society Bulletin 31: 1–10.

Arnason AN (1973) The estimation of population size, migration rates and survival in a stratified population. Research in Population Ecology 15: 1–8.

Arnason AN, Schwarz CJ, Boyer G (1998) POPAN 5: a data maintenance and analysis system for mark–recapture data. Scientific Report, Department of Computer Science, University of Manitoba, Winnipeg. Available at http://www.cs.umanitoba.c/~popan

Barbraud C, Weimerskirch H (2005) Environmental conditions and breeding experience affect costs of reproduction in blue petrels. Ecology 86: 682–692.

Beauplet G, Barbraud C, Dabbin W, Küssener C, Guinet, C (2006) Age–specific survival and reproductive performances in fur seals: evidence of senescence and individual quality. Oikos 112: 430–441.

Benkman CW, Colquitt JS, Gould WR, Fetz T, Keenan PC, Santisteban L (2005) Can selection by an ectoparasite drive a population of crossbills from its adaptive peak? Evolution 59: 2025–2032.

Bennett PM, Owens IPF (2002) Evolutionary ecology of birds. Oxford University Press, Oxford.

Bennetts RE, Nichols JD, Lebreton J–D, Pradel R, Hines JE, Kitchens WM (2001) Methods for estimating dispersal probabilities and related parameters using marked animals. In: Clobert J, Danchin E, Dhondt A, Nichols JD (eds), Disperal, Oxford University Press, Oxford, U.K., pp 5–17.

Blasco A (2001) The Bayesian controversy in animal breeding. Journal of Animal Science 79: 2023–2046.

Blums P, Nichols JD, Lindberg MS, Hines JE, Mednis A (2003) Estimating natal dispersal movement rates of female European ducks with multistate modeling. Journal of Animal Ecology. 72: 1027–1042.

Blums P, Nichols JD, Hines JE, Lindberg MS, Mednis A (2005) Individual quality, survival variation and patterns of phenotypic selection on body conditions and timing of nesting in birds. Oecologia 143: 365–376.

Bonner SJ, Schwarz CJ (2004) Continuous time–dependent individual covariates and the Cormark–Jolly–Seber model. Animal Biodiversity and Conservation 27: 149–155.

Boulinier T, McCoy KD, Sorci G (2001) Dispersal and parasitism. In: Clobert J, Danchin E, Dhondt A, Nichols JD (eds), Disperal, Oxford University Press, Oxford, U.K., pp 169–179.

Brommer JE, Merillä J, Kokko H (2002) Reproductive timing and individual fitness. Ecology Letters 5: 802–810.

Brooks SP, Catchpole EA, Morgan BJT, Harris M (2002) Bayesian methods for analysing ringing data. Journal of Applied Statistics 29 (Special Issue): 187–206.

Brooks SP, King R, Morgan BJT (2004) A Bayesian approach to combining animal abundance and demographic data. Animal Biodiversity and Conservation 27: 515–529.

Brown CR, Bromberger M, Danchin E (2000) The effect of conspecific reproductive success on colony choice in cliff swallows. Journal of Animal Ecology 69: 133–142.

Brown CR, Bromberger M, Brown, Raouf SA, Smith LC, Wingfield JC (2005) Steroid hormone levels are related to choice of colony size in cliff swallows. Ecology 86: 2904–2915.

Brownie C, Hines JE, Nichols JD, Pollock KH, Hestbeck JB (1993) Capture–recapture studies for multiple strata including non–Markovian transitions. Biometrics 49: 1173–1187.

Bryant MJ, Reznick D (2004) Comparative studies of senescence in natural populations of guppies. American Naturalist 163: 55–68.

Burnham KP (1993) A theory for combined analysis of ring recovery and recapture data. In: Lebreton J-D, North PM (eds) The use of marked individuals in the study of bird population dynamics, Birkhauser, Basel, pp 199–213.

Burnham KP, Rexstad EA (1993) Modeling heterogeneity in survival rates of banded waterfowl. Biometrics 49: 1194–1208.

Burnham KP, White GC (2002) Evaluation of some random effects methodology applicable to bird ringing data. Journal Applied Statistics 29 (Special Issue): 245–264.

Cam E, Hines JE, Monnat J-Y, Nichols JD, Danchin E (1998) Are nonbreeders prudent parents? The Kittiwake model. Ecology 79: 2917–2930.

Cam E, Link WA, Cooch EG, Monnat J-Y, Danchin E (2002a) Individual covariation in life history traits: seeing the trees despite the forest. American Naturalist 159: 96–105.

Cam E, Cadiou B, Hines JE, Monnat J-Y (2002b) Influence of behavioural tactics on recruitment and reproductive trajectory in the kittiwake. Journal of Applied Statistics 29 (Special Issue): 163–186.

Cam E, Monnat J-Y, Hines JE (2003) Long-term consequences of early conditions in the kittiwake. Journal of Animal Ecology 72: 411–424.

Cam E, Oro D, Pradel R, Jimenez J (2004a) Assessment of hypotheses about dispersal in a long-lived seabird using multistate capture-recapture models. Journal of Animal Ecology 73: 723–736.

Cam E, Monnat J-Y, Royle JA (2004b) Dispersal and individual quality in a long-lived species. Oikos 106: 386–398.

Cam E, Cooch EG, Monnat J-Y (2005) Earlier recruitment or earlier death? On the assumption of homogeneous survival in recruitment studies. Ecological Monographs 75: 419–434.

Carothers AD (1973) Capture–Recapture methods applied to a population with known parameters. Journal of Animal Ecology 42: 125–146

Caswell H (2001) Matrix population models. Construction, analysis, and interpretation. Sinauer Associates, Sunderland, Massachusetts.

Charmantier A, Perrins C, McCleery RH, Cheldon BC (2006a) Quantitative genetics of age at reproduction in wild swans: support for antagonistic pleiotropy models of senescence. Proceedings of the National Academy of Sciences of the U.S.A. 103: 6587–6592.

Charmantier A, Perrins C, McCleery RH, Cheldon BC (2006b) Evolutionary response to selection on clutch size in a long-term study of the mute swan. American Naturalist 167: 453–565.

Charlesworth B (1994) Evolution in age-structured populations. Cambridge University Press, Cambridge.

Choquet R, Reboulet AM, Pradel R, Gimenez O, Lebreton J-D (2003) U-Care user's guide, Mimeographed document, CEFE/CNRS, Montpellier, France. Available at: ftp://ftp.cefe.cnrs-mop.fr/biom/Soft-CR/

Clobert J, Lebreton JD (1985) Dependance de facteurs du milieu dans les estimations de survie par capture–recapture. Biometrics 41: 1031–1037. SURGE is available at :http://ftp.cefe.cnrs.fr/biom/Archives/

Clobert J (1995) Capture–recapture and evolutionary ecology: a difficult wedding? Journal of Applied Statistics 22: 989–1008.

Clobert J (2002) Capture–recapture and evolutionary ecology: further comments. Journal of Applied Statistics 29 (Special Issue): 53–56.

Conner MM, White GC (1999) Effects of individual heterogeneity in estimating the persistence of small populations. Natural Resource Modeling 12: 109–127.

Conner JK, Hartl DL (2004) A primer of ecological genetics. Sinauer Associates, Inc. Publishers, Sunderland, Massachusetts.

Conroy MJ, Senar JC, Domènech J (2002) Analysis of individual- and time-specific covariate effects on survival of *Serinus serinus*in north-east Spain. Journal of applied Statistics 29 (Special Issue): 125–142.

Conroy MJ (2008) Some contributions of capture–recapture to evolutionary ecology and population modeling. In: Thomson DL, Cooch EG, Conroy MJ (eds.) Modeling Demographic Processes in Marked Populations. Environmental and Ecological Statistics, Springer, New York, Vol. 3, pp. 131–156.

Cooch EG, Lank DB, Rockwell RF, Cooke F (1999) Body size and age of recruitment in snow geese *Anser c. caerulescens*. Bird Study (supplement) 46: 112–119.

Cooch EG, Cam E, Link WA (2002) Occam's shadow: levels of analysis in evolutionary ecology – where to next? Journal of Applied Statistics 29 (Special Issue): 19–48.

Cooch EG, White GW (2007) Using MARK – a gentle introduction. Available at ttp:// www.phidot.org/software/

Cormack RM (1964) Estimates of survival from the sighting of marked animals. Biometrics 51: 429–438.

Coulson T, Benton TG, Lunberg P, Dall SRX, Kendall BE, Gaillard J–M (2006) Estimating individual contributions to population growth: evolutionary fitness in ecological time. Proceedings of the Royal Society London B: 273: 547–555.

Crespin L, Harris MP, Lebreton J-D, Frederiksen M, Wanless S (2006) Recruitment to a seabird population depends on environmental factors and on population size. Journal of Animal Ecology 75: 228–238.

Curio E (1983) Why do young birds reproduce less well? Ibis 125: 400–404.

Danchin E, Boulinier T, Massot M (1998) Conspecific reproductive success and breeding habitat selection: implications for the evolution of coloniality. Ecology 79: 2415–2428.

Danchin E, Heg D, Doligez B (2001) Public information and breeding habitat selection. In: Clobert J, Danchin E, Dhondt A, Nichols JD (eds), Dispersal, Oxford University Press, Oxford UK pp 243–258.

Doligez B, Danchin E, Clobert J, Gustafsson L (1999) The use of conspecific reproductive success for breeding habitat selection in a non-colonial, hole-nesting species, the collared flycatcher. Journal of Animal Ecology 68: 1193–1206.

Doligez B, Clobert J, Pettifor RA, Rowcliffe M, Gustafsson L, Perrins CM, McCleery RH (2002) Costs of reproduction: assessing responses to brood size manipulation on life-history and behavioural traits using multi-state capture-recapture models. Journal of Applied Statistics 29 (Special Issue): 407–423.

Doligez B, Pärt T, Danchin E, Clobert J, Gustafsson L (2004) Availability and use of public information and conspecific density for settlement decisions in the collared flycatcher. Journal of Animal Ecology 73: 75–87.

Doherty Jr. PF, Sorci G, Royle JA, Hines JE, Nichols JD, Boulinier T (2003) Sexual selection affects local extinction and turnover in bird communities. Proceedings of the National Academy of Science U.S.A.: 100: 5858–5862.

Dugger BD, Blums P (2001) Effect of conspecific brood parasitism on host fitness for tufted duck and common pochard. The Auk 118: 717–726.

Dupuis JA (1995) Bayesian estimation of movement and survival probabilities from capture–recapture data. Biometrika 82: 761–772.

Dupuis JA (2002) Prior distributions for stratified capture–recapture models. Journal of Applied Statistics 29 (Special Issue): 225–237.

Dupuis JA, Badia J, Maublanc M, Bon R (2002) Survival and spatial fidelity of mouflons (*Ovis gmelini*): a Bayesian analysis of age-dependant capture–recapture model. Journal of Agricultural, Biological and Environmental Statistics 7: 277–298.

Emlen JM (1984) Population biology – the coevolution of population dynamics and behaviour. Macmillan, New York, U.S.A.

Dupuis JA, Joachim J (2006) Bayesian estimation of species richness from quadrat sampling data in the presence of prior information. Biometrics 62: 706–712.

Endler JA (1986) Natural selection in the wild. Princeton, University Press, Princeton, New Jerey, U.S.A.

Ens JB, Weissing FJ, Drent R (1995) The despotic distribution and deferred maturity: two sides of the same coin. The American Naturalist 146: 625–650

Fahrmeir L, Tutz G (1994) Multivariate statistics modelling based on generalized linear models. Springer-Verlag, New-York, Inc., New York.

Fairbairn DJ, Reeve JP (2001) Natural Selection. In: Fox CW, Roff DA, Fairbairn DJ (eds) Evolutionary Ecology. Oxford University Press, Oxford, pp 29–43.

Falconer DS, Mackay TFC (1996) Introduction to quantitative genetics, 4th edition, Longman, Harlow, U.K.

Ferrière R, Dieckmann U, Couvet D (2004) Introduction. In: Ferrière R, Dieckmann U, Couvet D (eds) Evolutionary conservation biology. Cambridge University Press, Cambridge, pp 1–14.

Festa-Bianchet M, Gaillard J-M, Jorgenson JT, Jullien J-M, Loison A (1999) Age-specific survival in five populations of ungulates: evidence of senescence. Ecology 80: 2539–2554.

Fox CW, Roff DA, Fairbain DA (2001) Preface. In: Fox CW, Roff DA, Fairbairn DJ (eds), Evolutionary Ecology, Oxford University Press, Oxford.

Fox GA, Kendall BE (2002) Demographic stochasticity and the variance reduction effect. Ecology 83: 1928–1934.

Fox GA, Kendall BE, Fitzpatrick JW, Woolfenden GE (2006) Consequences of heterogeneity in survival probability in a population of Florida scrub-jay. Journal of Animal Ecology 75: 921–927.

Forslund P, Pärt T (1995) Age and reproduction in birds – hypotheses and tests. Trends in Ecology and Evolution 10: 374–378.

Frankham R, Kingsolver J (2004) Responses to environmental change: adaptation or extinction. In: Ferrière R, Dieckmann U, Couvet D (eds) evolutionary conservation biology. Cambridge University Press, Cambridge, pp. 85–100.

Freeman S, Herron JC (2000) Evolutionary analysis. Prentice Hall 2nd edition. Prentice Hall, Inc. Upper Saddle River, New Jerey, U.S.A.

Frederiksen M, Bregnballe T (2001) Conspecific reproductive success affects age of recruitment in a great cormorant, *Phalacrocorax sinensis*, colony. Proceedings of the Royal Society of Lonbon B: 268: 1519–1526.

Frétey T, Cam E, Le Garff B, Monnat J-Y (2004) Adult survival and temporary emigration in the common toad. Canadian Journal of Zoology 82: 859–872

Fretwell SD, Lucas HL (1970) On territorial behaviour and other factors influencing habitat distribution in birds. I. Theoretical development. Acta Biotheoretica 19: 16–36.

Fry JD (1992) The mixed-model analysis of variance applied to quantitative genetics: biological meaning of parameters. Evolution 46: 540–550.

Futuyma DJ (1998) Evolutionary Biology. Sinauer. Sunderland, Massachusetts, U.S.A.

Gaillard, J-M, Viallefont A, Loison A, Festa-Bianchet M (2004) Assessing senescence patterns in populations of large mammals. Animal Biodiversity and Conservation 27:47–58.

Gilbert RO (1973) Approximation of the bias in the Jolly–Seber capture–recapture model. Biometrics 29: 501–526.

Gimenez O, Covas R, Brown CR, Anderson MD, Brown MB, Lenormand T (2006) Nonparametric estimation of natural selection on a quantitative trait using mark-recapture data. Evolution 60: 460–466.

Gregoire A, Preault M, Cezilly F, Wood MJ, Pradel R, Faivre B (2004) Stabilizing selection on the early expression of a secondary sexual trait in a passerine bird. Journal of Evolutionary Biology 17: 1152–1156.

Green P (1995) Reversible jump Markoc chain Monte Carlo computation and Bayesian model determination. Biometrika 82: 711–732.

Hadley GL, Rotella JJ, Garrott RA, Nichols JD (2006) Variation in probability of first reproduction of Weddell seals. Journal of Animal Ecology 75: 1058–1070.

Hendry AP, Kinnison MT (1999) The pace of modern life: measuring rates of contemporary microevolution. Evolution 53: 1637–1653.

Hernández-Lloreda MV, Colmenarez F, Arias RM (2003) Application of hierarchical linear modeling to the study of trajectories in behavioural development. Animal Behaviour 66: 607–613.

Hestbeck JB, Nichols JD, Malecki RA (1991) Estimates of movement and site fidelity using mark-resight data of wintering Canada geese. Ecology 72: 523–533.

Hines JE (1994) MSSURVIV Users Manual. National Biological Service, Patuxent Wildlife Research Center, Laurel, MD 0708–4017. Available at: http://www.mbr-pwrc.usgs.gov/software.html

Holt RD, Barfield M (2001) On the relationship between the ideal free distribution and the evolution of dispersal. In: Clobert J, Danchin E, Dhondt A, Nichols JD (eds), Dispersal, Oxford University Press, Oxford UK pp 83–95.

Houle D (1991) Genetic covariance of fitness correlates: what genetic correlations are made of and why it matters. Evolution 45: 630–648.

Houle D (1992) Comparing evolvability and variability of quantitative traits. Genetics 130: 195–204.

Hougaard P (1991) Modeling heterogeneity in survival data. Journal of Applied Probability 28: 695–701.

Ims RA, Hjermann DØ (2001) Condition-dependent dispersal. In: Clobert J, Danchin E, Dhondt A, Nichols JD (eds) Dispersal, Oxford University Press, Oxford UK pp 203–216.

Jolly GM (1965) Explicit estimates from capture–recapture data with both death and immigration-stochastic model. Biometrika 52: 225–247.

Jolly GM (1993) Instinctive statistics. In: Lebreton J-D, North PM (eds) Marked individuals in the study of bird populations. Birkhäuser Verlag Basel, Switzerland, pp 1–5.

Jones IL, Hunter FM, Robertson GJ, Fraser G (2002) Natural variation in the sexually selected feather ornaments of crested auklets (*Aethia cristatella*) does not predict future survival. Behavioral Ecology 15: 332–337.

Kendall WL, Nichols JD (1995) On the use of secondary capture–recapture samples to estimate temporary emigration and breeding proportions. Journal of Applied Statistics 22: 751–762.

Kendall BE, Fox GA (2001) Variation among individuals and reduced demographic stochasticity. Conservation Biology 16: 109–116.

Kendall BE, Fox GA (2003) Unstructured individual variation and demographic stochasticity. Conservation Biology 17: 1170–1172.

Kendall WL, Bjorkland R (2001) Using open robust design models to estimate temporary emigration from capture–recapture data. Biometrics 57: 1113–1122

Kendall WL, Gould WR (2002) An appeal to undergraduate wildlife programs: send scientists to learn statistics. Wildlife Society Bulletin 30: 623–627.

Kendall WL, Nichols LD (2002) Estimating state-transition probabilities for unobservable states using capture–recapture/resighting data. Ecology 83: 3276–3284.

Kendall WL, Hines JE, Nichols JD (2003) Adjusting multi-state capture–recapture models for mis-classification bias: manatee breeding proportions. Ecology 84: 1058–1066.

Kendall WL (2004) Copying with unobservable and mis-classified states in capture-recapture studies. Invited paper. Animal Biodiversity and Conservation 27: 97–107.

Kendall WL, Nichols JD (2004) On the estimation of dispersal and movement of birds. The Condor 106: 720–731.

Kéry M, Royle JA (2008) Inference about species richness and community structure using species-specific occupancy model in the national swiss breeding bird survey MHB. In: Thomson DL, Cooch EG, Conroy MJ (eds.) Modeling Demographic Processes in Marked Populations. Environmental and Ecological Statistics, Springer, New York, Vol. 3, pp. 639–656.

King R, Brooks SP (2002) Bayesian model discrimination for multiple strata capture–recapture data. Biometrika 89: 785–806.

King R, Brooks SP (2004) Bayesian analysis of the Hector's dolphin data. Animal Biodiversity and Conservation 27: 343–354.

Kinghorn B, Kinghorn S (2007) Pedigree. Available at http://www.personal.une.edu.au/~bkinghor/pedegree.htm

Kingsolver JG, Smith SG (1995) Estimating selection on quantitative traits using capture–recapture data. Evolution 49: 384–388.

Kruuk LEB, Clutton-Brock TH, Slate J, Pemberton JM, Brotherstone S, Guinness FE (2000) Heritability of fitness in a wild mammal population. Proceedings of the National Academy of Sciences of the U.S.A. 97: 698–703.

Kruuk LEB, Merilä J, Sheldon BC (2001) Phenotypic selection on a heritable size trait revisited. American Naturalist 158: 557–571.

Kruuk LEB (2004) Estimating genetic parameters in natural populations using the 'animal model'. Philosophical Transactions of the Royal Society London B. 359: 873–890.

Lande R, Arnold SJ (1983) The measurement of selection on correlated characters. Evolution 37: 1210–1226.

Laurie WA, Brown D (1990) Population biology of marine iguanas (*Amblyrhynchus cristatus*) II. Changes in annual survival rates and the effects of size, sex, age and fecundity in a population crash. Journal of Animal Ecology 59: 529–544.

Lebreton J-D, Clobert J (1986) User's Manual for program SURGE, Montpellier, France: CEPE/CNRS.

Lebreton J-D, Burnham KP, Clobert J, Anderson DR (1992) Modeling survival and testing biological hypotheses using marked animals: a unified approach with case studies. Ecological Monographs 62: 67–118.

Lebreton J-D, Hines JE, Pradel R, Nichols JD, Spendelow JA (2003) Estimation by capture–recapture of recruitment and dispersal over several sites. Oikos 101: 253–264.

Liker A, Székely T (2005) Mortality costs of sexual selection and parental care in natural populations of birds. Evolution 59: 890–897.

Levins R (1968) Evolution in changing environments. Princeton University Press, Princeton, New Jersey, U.S.A.

Lindberg MS, Kendall WL, Hines JE, Anderson MG (2001) Combining band recovery data and Pollock's robust design to model temporary and permanent emigration. Biometrics 57: 273–281.

Link WA, Cooch EG, Cam E (2002a) Model-based estimation of individual fitness. Journal of Applied Statistics 29 (Special Issue): 207–224.

Link WA, Cam E, Nichols JD (2002b) Of BUGS and birds. An introduction to Markov Chain Monte Carlo. Journal of Wildlife Management 66: 277–291.

Link WA (2004) Individual heterogeneity and identifiability in capture–recapture models. Animal Biodiversity and Conservation 27: 87–91.

Link WA, Barker RJ (2004) Hierarchical mark–recapture models: a framework for inference about demographic processes. Animal Biodiversity and Conservation 27: 441–449.

Lukacs PL, Burnham KP (2005) Review of capture–recapture methods applicable to noninvasive genetic sampling. Molecular Ecology 14: 3909–3919.

Lynch M, Walsh B (1998) Genetics and analysis of quantitative traits. Sinauer Associates Inc. Publishers. Sunderland, Massachusetts, U.S.A.

MacKenzie DI, Kendall WK (2002) How should detection probability be incorporated into estimates of relative abundance? Ecology 83: 2387–2393.

MacKenzie DI, Nichols JD, Lachman GB, Droege S, Royle JA, Lagtimm CA (2002) Estimating site occupancy rates when detection probabilities are less than one. Ecology 83: 2248–2255.

MacKenzie DI, Bailey LL (2004) Assessing the fit of site occupancy models. Journal of Agricultural, Biological, and Environmental Statistics 9: 300–318.

MacKenzie DI, Bailey LL, Nichols JD (2004) Investigating species co–occurrence patterns when species are detected imperfectly. Journal of Animal Ecology 73: 546–555.

MacKenzie DI, Nichols JD (2004) Occupancy as a surrogate for abundance estimation. Animal Biodiversity and Conservation 27: 461–467.

MacKenzie DI, Nichols JD, Royle AJ, Pollock KH, BaileyLL, Hines JE (2006) Occupancy estimation and modeling. Inferring patterns and dynamics of species occurrence. Academic Press, Elsevier Inc. Burlington, Massachusetts, U.S.A.

Martin TE, Clobert J, Anderson DR (1995) Return rates in studies of life history evolution: are biases large? Journal of Applied Statistics 22: 863–875.

Martin J, Nichols JD, Kitchens WM, Hines JE (2006) Multiscale patterns of movement in fragmented landscape and consequences on demography of the snail kite in Florida. Journal of Animal Ecology 75: 527–539.

Massot M, Clobert J, Lecompte J, Barbault R (1994) Incumbent advantage in common lizards and their colonizing ability. Journal of Animal Ecology 63: 431–440.

Massot M, Clobert J (2000) Processes at the origin of similarities in dispersal behaviour among siblings. Journal of Evolutionary Biology 13: 707–719.

Matos CAP, Thomas DL, Gianola D, Tempelman RJ, Young LD (1997) Genetic analysis of discrete reproductive traits in sheep using linear and nonlinear models: I. Estimation of genetic parameters. Journal of Animal Science 75: 76–87.

Mayr E (1997) This is biology: the science of the living world. The Belknap Press of Harvard University Press, Cambridge, Massachussetts, U.S.A.

Mazer SJ, Damuth J (2001) Nature and causes of variation. Evolutionary significance of variation. In: Fox CW, Roff DA, Fairbairn DJ (eds) Evolutionary Ecology. Oxford University Press, Oxford, pp 3–28.

McElligott AG, Altwegg R, Hayden TJ (2002) Age-specific survival and reproductive probabilities: evidence for senescence in male fallow deer (*Dama dama*). Proceedings: Biological Sciences 269 : 1129–1137.

McGraw JB, Caswell H (1996) Estimation of individual fitness from life–history data. American Naturalist 147: 47–64.

Mertz DB (1975) Senescent decline in flour beetle straints selected for early adult fitness. Physiological Zoölogy 48: 1–23.

Metz JAJ, Nisbet RM, Geritz SAH (1992) How should we define 'fitness' for general ecological scenarios? Trends in Ecology and Evolution 7: 198–202.

Mousseau TA, Fox CW (1998) The adaptive significance of maternal effects. Trends in Ecology and Evolution 13: 403–407.

Nager RG, Monaghan P, Griffiths R, Houston DC, Dawson R (1999) Experimental demonstration that offspring sex ratio varies with maternal condition. Proceedings of the National Academy of Science U. S. A. 96: 570–573.

Nichols JD, Pollock KH (1983) Estimating taxonomic diversity, extinction rates and speciation from fossil data using capture–recapture models. Paleobiology 9: 150–163.

Nichols JD, Hines JE, Pollock KH, Hinz RL, Link WA (1994) Estimating breeding proportions and testing hypotheses about costs of reproduction with capture–recapture data. Ecology 75: 2052–2065.

Nichols JD, Kendall WL (1995) The use of multi-state capture–recapture models to address questions in evolutionary ecology. Journal of Applied Statistics 22: 835–846.

Nichols JD, Hines JE, Blums P (1997) Tests for senescent decline in annual survival probabilities of Common Pochards, *Aythya ferina*. Ecology 78: 1009–1018.

Nichols JD, Boulinier T, Hines JE, Pollock KH, Sauer JR (1998a) Estimating rates of local extinction colonization and turnover in animal communities. Ecological Applications 8: 1213–1225.

Nichols JD, Boulinier T, Hines JE, Pollock KH, Sauer JR (1998b) Inference methods for spatial variation in species richness and community composition when not all species are detected. Conservation Biology 12: 1390–1398.

Nichols JD, Hines JE, Lebreton J-D, Pradel R (2000) The relative contribution of demographic components to population growth: a direct estimation approach based on reverse-time capture–recapture. Ecology 81: 3362–3376.

Nichols JD (2002) Discussion comments on: "Occam's shadow: levels of analysis in evolutionary ecology – where to next?" by Cooch, Cam and Link. Journal of Applied Statistics 29 (Special Issue): 49–52.

Nichols JD, Kendall WL, Hines JE, Spendelow JA (2004) Estimation of sex-specific survival from capture–recapture data when sex is not always known. Ecology 85: 3192–3201.

Norris JL, Pollock KH (1996) Nonparametric MLE under two closed capture–recapture models with heterogeneity. Biometrics 52:639–649.

Oro D, Pradel R (2000) Determinant of local recruitment in a growing colony of Audouin's gull. Journal of Animal Ecology 69: 119–132.

Orzack SH, Tuljapurkar S (2001) Reproductive effort in variable environments, or environmental variation is for the birds. Ecology 82: 2659–2665.

Otis DL, White GC (2004) Evaluation of ultrastuctural and random effects band recovery models for estimating relationships between survival and harvest rates in exploited populations. Animal Biodiversity and Conservation 27: 157–173.

Owens IPF, Bennett PM (1994) Mortality costs of parental care and sexual dimorphism in birds. Proceedings of the Royal Society London B: Biological Sciences 257: 1–8.

Petit E, Valière N (2006) Estimating population size with non-invasive capture–recapture data. Conservation Biology 20: 1062–1073.

Pettifor RE, Perrins CM, McCleery RH (1988) Individual optimization of clutch size in great tits. Nature 336: 160–162.

Pettifor RE, Perrins CM, McCleery RH (2001) Individual optimization of fitness: variation in reproductive output, including clutch size, mean nestling mass and offspring recruitment, in manipulated broods of great tits *Parus major*. Journal of Animal Ecology 70: 62–79.

Pigliucci M (2006) Genetic variance–covariance matrices: a critique of the evolutionary quantitative genetics research program. Biology and Philosophy 21: 1–23.

Pledger S, Efford M (1998) Correction of bias due to heterogeneous capture probability in capture–recapture studies of open populations. Biometrics 54: *898.

Pledger S, Schwarz CJ (2002) Modelling heterogeneity of survival as a random effect using finite mixtures. Journal of Applied Statistics 29 (Special Issue): 315–327.

Pollock KH, Nichols JD, Brownie C, Hines JE (1990) Statistical inference for capture–recapture experiments. Wildlife Monographs 107: 1–97.

Pollock KP (2002) The use of auxiliary variables in capture–recapture modeling: an overview. Journal of Applied Statistics 29 (Special Issue): 85–102.

Pradel R, Cooch EG, Cooke F (1995) Transient animals in a resident population of snow geese: local emigration or heterogeneity? Journal of Applied Statistics 22: 695–710.

Pradel R (1996) Utilization of capture-mark-recapture for the study of recruitment and population growth rate. Biometrics 52: 703–709.

Pradel R, Lebreton JD (1999) Comparison of different approaches to the study of local recruitment of breeders. Bird Study 46 (supplement): 74–81.

Pradel R, Wintrebert CMA, Gimenez O (2003) A proposal for a goodness-of-fit test to the Arnason–Schwarz multisite capture–recapture model. Biometrics 59: 43–53.

Promislow DEL, Montgomerie R, Martin TE (1992) Mortality costs of sexual dimorphism in birds. Proceedings of the Royal Society London B: Biological Sciences 250: 143–150.

Quinn JL, Charmantier A, Garant D, Scheldon BC (2006) Data depth, data completeness and their influence on quantitative genetics estimation. Journal of Evolutionary Biology 19: 994–1002.

Réale D, Festa-Bianchet M (2000) Quantitative genetics of life-history traits in a long-lived wild mammal. Heredity 85: 593–603.

Réale D, Berteaux D, McAdam AG, Boutin S (2003) Lifetime selection on heritable life-history traits in a natural population of red squirrels. Evolution 57: 2416–2423.

Reed ET, Gautier G, Giroux JF (2004) Effects of spring conditions on breeding propensity of greater snow goose females. Animal Biodiversity and Conservation 27: 35–46.

Reed WL, Clark ME, Parker PG, Raouf SA, Arguedas N, Monk DS, Snajdr E, Nolan Jr V, Ketterson ED (2006) Physiological effects on demography: a long-term experimental study of testosterone's effects on fitness. American Naturalist 167: 667–683.

Reekie EG, Budge S, Baltzer JL (2002) The shape of the trade-off function between reproduction and future performance in *Plantago major*and *Plantago rugelii.* Canadian Journal of Botany 80: 140–150.

Rexstad E, Burnham KP (1991) User's guide for interactive program CAPTURE. Colorado Cooperative Fish and Wildlife Research Unit, Fort Collins, Colorado, U. S. A.

Reznick D, Nunney L, Tessier A (2000) Big houses, big cars, superfleas and the cost of reproduction. Trends in Ecology and Evolution 15: 421–425.

Reznick D, Rodd H, Nunnery L (2004) Empirical evidence for rapid evolution. In: Ferrière R, Dieckmann U, Couvet D (eds) Evolutionary conservation biology. Cambridge University Press, Cambridge, pp 101–118.

Ricklefs RE (2000) Intrinsic aging-related mortality in birds. Journal of Avian biology 31: 103–111.

Ricklefs RE, Scheuerlein A (2001) Comparison of age-related mortality among birds and mammals. Experimental Gerontology 36: 845–857.

Roff DA (1992) The evolution of life histories. Chapman and Hall, New York, U.S.A.

Roff DA (2001) Age and size at maturity. In: Fox CW, Roff DA, Fairbairn DJ (eds), Evolutionary Ecology. Oxford University Press, Oxford, pp 99–112.

Ronce O, Clobert J, Massot M (1998) Natal dispersal and senescence. Proceedings of the National Academy of Science U. S. A. 95: 600–605.

Ronce O, Oliviery I, Clobert J, Danchin E (2001) Perspectives on the study of dispersal evolution. In: Clobert J, Danchin E, Dhondt A, Nichols JD (eds) Dispersal. Oxford University Press, Oxford UK pp 341–357.

Royle JA, Link WA (2002) Random effects and schrinkage estimation in capture–recapture models. Journal of Applied Statistics 29 (Special Issue): 329–351.

Royle JA, Nichols JD, Kéry M (2005) Modeling occurrence and abundance of species when detection is imperfect. Oikos 110: 353–359.

Royle JA (2007) Modeling individual effects in the Cormack-Jolly-Seber model: a state-space formulation. Biometrics 63: 568–578.

Royle JA (2008) Analysis of capture-recapture models with individual covariates using data augmentation. Biometrics. 2008 Apr 15. [Epub ahead of print].

Royle JA, Kéry M (2007) Analysis of multinomial models with unknown index using data augmentation. Journal of Computational and Graphical Statistics 16: 67–85.

Runge MC, Johnson FA (2002) The importance of functional form in optimal control solutions of problems in population dynamics. Ecology 83:1357–1371.

Saccheri I, Hanski I (2006) Natural selection and population dynamics. Trends in Ecology and Evolution 21: 341–346.

Saxton AM (2004) Genetic analysis of complex traits using SAS. SAS Institute Inc. Cary, North Carolina, U.S.A.

Schaeffer LR (2004) Application of random regression models in animal breeding. Livestock Production Science 86: 35–45.

Schmidt BR, Schaub M, Anholt BR (2002) Why you should use capture–recapture methods when estimating survival and breeding probabilities: on bias, temporary emigration, overdispersion and common toads. Amphibia-Reptilia 23: 375–388

Schwarz CJ, Arnason AN (2000) Estimation of age-specific breeding probabilities from capture–recapture data. Biometrics 56: 59–61.

Schwarz CJ, Stobo WT (2000) Estimation of juvenile survival, adult survival, and age-specific pupping probabilities for the female grey seal (*Halichoerus gryprus*) on Sable Island from capture–recapture data. Canadian Journal of Fisheries and Aquatic Sciences 57: 247–253.

Schwarz CJ, Arnason AN (2001) Comment on Schwarz and Arnason: estimation of age-specific breeding probabilities from capture–recapture data – authors reply. Biometrics 57: 976.

Seber GAF (1965) A note on the multiple recapture census. Biometrika 52: 249–259.

Seber GAF, Schwarz CJ (2002) Capture–recapture: before and after EURING 2000. Journal of Applied Statistics 29 (Special Issue): 5–18.

Serrano D, Tella JL, Forero MG, Donázar JA (2001) Factors affecting breeding dispersal in the facultatively colonial lesser kestrel: individual experience vs. conspecific cues. Journal of Animal Ecology 70: 568–578.

Serrano D, Tella JL (2003) Dispersal within a spatially structured population of lesser kestrels: the role of spatial isolation and conspecific attraction. Journal of Animal Ecology 72: 400–410.

Serrano D, Oro D, Ursúa E, Tella JL (2005) Colony size selection determines adult survival and dispersal preferences: Allee effects in a colonial bird. American Naturalist 166: 22–31.

Senar JC, Conroy MJ, Borras A (2002) Asymetric exchange between populations differing in habitat quality: a metapopulation analysis of the Citril Finch. Journal of Applied Statistics 29 (Special Issue): 425–442.

Service PM (2000) Heterogeneity in individual mortality risk and its importance for evolutionary studies of senescence. The American Naturalist 156:1–13.

Service PM (2004) Demographic heterogeneity explains age-specific patterns of genetic variance in mortality rates. Experimental Gerontology 39: 25–30.

Sheldon BC, Kruuk LEB, Merilä J (2003) Natural selection and inheritance of breeding time and clutch size in collared flycatcher. Evolution 57: 406–420.

Skalski JR, Hoffman A, Smith SG (1993) Testing the significance of individual- and cohort-level covariates in animal survival studies. In: Lebreton JD, North PM (eds) Marked individuals in the study of bird population. Birkhauser Verlag, Basel, Switzerland, pp 9–29.

Skvarla JL, Nichols JD, Hines JE, Waser PM (2004) Modeling interpopulation dispersal by banner-tailed kangaroo rats. Ecology 85: 2737–2746.

Stenseth NC, Saetre G-P (2003) Sexual selection forms the structure and dynamics of ecological communities. Proceedings of the National Academy of Science U.S.A. 100: 5576–557.

Smith SG, Skalski JM, Schlethe JW, Hoffman A, Cassen V (1994) SURPH. Statistical survival analysis of fish and wildlife tagging studies. Center for Quantitative Sciences, University of Washington, Seattle, Wahington, U.S.A.

Stearns SC (1992) The evolution of life histories. Oxford University Press, New York, U.S.A.

Spiegelhalter DJ, Thomas A, Best NG, Gilks WR (1996) BUGS 1.4. Bayesian inference using Gibbs sampling Manual http://www.mrc-bsu.cam.ac.uk/bugs/.

Spiegelhalter DJ, Best NG, Carlin BP, van der Linde A (2002) Bayesian measures of model complexity and fit. Journal of the RoyalStatistical Society, Series B. 64: 583–639.

Spendelow JA, Nichols JD, Nisbet ICT, Hays H, Cormons GD, Burger J, Safina C, Hines JE, Gochfeld M (1995) Estimating annual survival and movement rates of adults within a metapopulation of roseate terns. Ecology 76: 2415–2428.

Spendelow JA, Nichols JD, Hines JE, Lebreton J-D, Pradel R (2002) Modelling postfledging survival and age-specific breeding probabilities in species with delayed maturity: a case study of Roseate terns at Falkner island, Connecticut. Journal of Applied Statistics 29 (Special Issue): 385–405.

Tatar M (2001) Senescence. In: Fox CW, Roff DA, Fairbairn DJ (eds) Evolutionary ecology. Oxford University Press. Oxford, pp 128–141.

Tinbergen JM, Sanz JJ (2002) Strong evidence for selection for larger brood size in a great tit population. Behavioral Ecology 15: 525–533.

Tinbergen JM, Both C (1999) Is clutch size individually optimized? Behavioral Ecology 10: 504–509.

Thomas SC, Coltman DW, Pemberton JM (2002) The use of marker-based relationship information to estimate the heritability of body weight in a natural population: a cautionary tale. Journal of Evolutionary Biology 15: 92–99.

Török J, Hegvi G, Tóth L, Könczey R (2004) Unpredictable food supply modifies costs of reproduction and hampers individual optimization. Oecologia 141: 432–443.

Tuomi J, Hakala T, Haukioja E (1983) Alternative concept of reproductive effort, costs of reproduction and selection in life-history evolution. American Zoologist 23: 25–34.

Uller T, Massot M, Richard M, Lecompte J, Clobert J (2004) Long-lasting fitness consequences of prenatal sex-ratio in a viviparous lizard. Evolution 58: 2511–2516.

Van de Pol M, Verhulst S (2006) Age-dependent traits: a new statistical model to separate within- and between-individual effects. American Naturalist 167: 766–773

Van-Noordwijk AJ, de Jong G (1986) Acquisition and allocation of resources: their influence on variation in life-history tactics. The American Naturalist 128: 137–142.

Vaupel JW, Yashin AI (1985a) Heterogeneity's ruses: some surprising effects of selection on population dynamics. The American Statistician 39: 176–185.

Vaupel JW, Yashin AI (1985b) The deviant dynamics of death in heterogeneous populations. In: Brandon Tuma N (ed) Sociological methodology. Jossey-Bass, Inc., Publishers. San Fransisco, California, U.S.A. pp 180–211

Viallefont A, Cooch EG, Cooke F (1995a) Estimation of trade-offs with capture–recapture models: a case study on the lesser snow goose. Journal of Applied Statistics 22: 847–861.

Viallefont A, Cooke F, Lebreton J-D (1995b) Age-specific costs of first-time breeding. Auk 112: 67–76.

Vounatsou P, Smith AFM (1995) Bayesian analysis of ring-recovery data via Markov Chain Monte Carlo simulation. Biometrics 51: 687–708.

Webb JK (2006) Effects of tail autotomy on survival, growth and territory occupation in free-ranging juvenile geckos (*Oedura lesueurii*). Austral Ecology 31: 432–440.

Weimerskirch H, Lallemand J, Martin J (2005) Population sex ratio in a monogamous long-lived bird, the wandering albatross. Journal of Animal Ecology 74: 285–291.

White GC, Burnham KP, Otis DL, Anderson DR (1978) Users Manual for Program CAPTURE, Utah State Univ. Press, Logan, Utah. http://www.mbr-pwrc.usgs.gov/software.html

White GC (1983) Numerical estimation of survival from recovery and biotelemetry data. Journal of Wildlife Management 47: 716–728.

White GC, Burnham KP (1999) Program MARK: survival estimation from populations of marked animals. Bird Study 46 (supplement): 120–138.

White GC (2002) Discussion comments on: the use of auxiliary variables in capture–recapture modeling. An overview. Journal of Applied Statistics 29 (Special Issue) 103–106.

Wiggins DA (1989) Heritability of body size in cross-fostered tree swallow broods. Evolution 43: 1808–1811.

Wikelski M, Trillmich F (1997) Body size and sexual dimorphism in marine iguanas fluctuate as a result of opposing natural and sexual selection: an island comparison. Evolution 51: 922–936.

Wilson AJ, Kruuk LEB, Coltman DW (2005) Ontogenetic patterns in heritable variation for body size: using random regression models in a wild ungulate population. American Naturalist 166: 177–192.

Williams BK, Nichols JD, Conroy MJ (2002) Analysis and management of animal populations. Academic Press, San Diego, California, U.S.A.

Wintrebert CMA, Zwinderman AH, Cam E, Pradel R, van Houwelingen JC (2004) Joint modelling of breeding and survival in the kittiwake using frailty models. Ecological modelling 181: 203–213.

Williams GC (1992) Natural selection. Oxford University Press, Oxford.

Williams PD, Day T, Fletcher Q, Rowe L (2006) The shaping of senescence in the wild. Trends in Ecology and Evolution 21: 458–463.

Yashin AI, Iachine IA, Begun AZ, Vaupel JW (2001) Hidden frailty: myths and reality. Working paper. University of southern Denmark, Department of Statistics. ISBN 87-90700 41 4. Available at: http://demografie.de

Yoccoz NG, Erikstad KE, Bustnes JO, Hanssen SA, Tveraa T (2002) Costs of reproduction in common eiders (*Someteria mollissima*): an assessment of relationships between reproduction effort and future survival and reproduction based on observational and experimental studies. Journal of Applied Statistics 29 (Special Issue): 57–64.

Application of Capture–Recapture to Addressing Questions in Evolutionary Ecology

Michael J. Conroy

Abstract Capture–recapture (CR) is one of the most commonly used methods in quantitative ecology. Until recently, much of the emphasis of CR was on the estimation of abundance and vital rates, especially survival rates. Here, I discuss several important advances that have enhanced ecologists' ability to address questions in evolutionary ecology. Generalizations of CR methodology to include group and covariate effects have allowed direct, empirical modeling of the influence of extrinsic and intrinsic factors on demographic rates. Advances in sampling design and software now allow CR modeling to address questions such as dispersal and natal fidelity, tradeoffs between reproductive effort and survival, senescence, and variability in demographic rates in relation to individual traits, among others. Furthermore, complex ecological and evolutionary questions seem to be especially amenable to a paradigm of multiple alternative (vs. single null) hypotheses, which is consistent both with information-theoretic and Bayesian approaches to inference.

Previous CR approaches have emphasized the estimation of averages of demographic parameters for individuals grouped into classes (age, sex, size or other attributes), but evolutionary questions tend to emphasize individual variability, with fitness "parameters" best characterized by frequency distributions. Bayesian approaches are particularly appropriate for modeling individual, temporal, spatial, and other components of variation via random effects models. Finally, Bayesian methods and conditional/hierarchical modeling allow for ready construction of complex models of life history from a variety of data sources. I present selected examples to illustrate each of these major points.

M.J. Conroy (✉)
USGS, Georgia Cooperative Fish and Wildlife Research Unit, D. B. Warnell School of Forestry and Natural Resources, University of Georgia, Athens, GA 30602, USA
e-mail: mconroy@uga.edu

D.L. Thomson et al. (eds.), *Modeling Demographic Processes in Marked Populations*, Environmental and Ecological Statistics 3, DOI 10.1007/978-0-387-78151-8_6,

1 Introduction

Capture–recapture (CR), in which animals are captured, marked with a tag or other device, and recaptured at some future time, is one of the most common techniques used in ecological studies. CR may be employed simply to estimate population size or vital rates, or more interestingly, to investigate a wide array of questions about variability (over space, time, and among individuals) in these parameters (Williams et al. 2002).

The rich nature of CR data now makes it possible, given appropriate experimental designs and sufficient data, to address a wide range of evolutionary questions, including

- Modeling of factors influencing demographic rates.
- Analysis of individual heterogeneity in fitness.
- Metapopulation dynamics.
- Alternative reproductive strategies, including survival/reproduction tradeoffs, breeding propensity and age at first reproduction, senescence, and fidelity to breeding areas.

By definition, because CR allows an investigator to follow individual animals or groups of animals through time, it apparently provides a rich source of information on individual behavior and fates. It also is possible to determine the "state" of animals at various points in time, where "state" can refer to any attribute about the individual, such as the age, physical condition, breeding status, or the location of the animal in space.

However, this richness is balanced by the fact that CR data are based on *recaptures*, so if animals are marked and released, but no animals are ever encountered, no information is provided to the study. Also, unless care is taken in modeling, there will be confounding between the parameters of biological interest, and nuisance parameters related to the sampling process. Finally, statistical models must balance the desire of evolutionary biologists to exploit the rich information in CR studies (seemingly calling for complex models), with the fact that in most studies, data are relatively limited (calling for simpler models).

2 Historical Advances

2.1 Separation of Encounter from Survival and Other Parameters – The CJS Model

As noted earlier, CR studies fundamentally depend upon recaptures. However, in order to model recaptures, we must also model the events that lead to animals *not* being recaptured. It is obvious what events lead to a recapture: for instance, if an animal is marked and released at time t, in order to be recaptured at t+1, it must

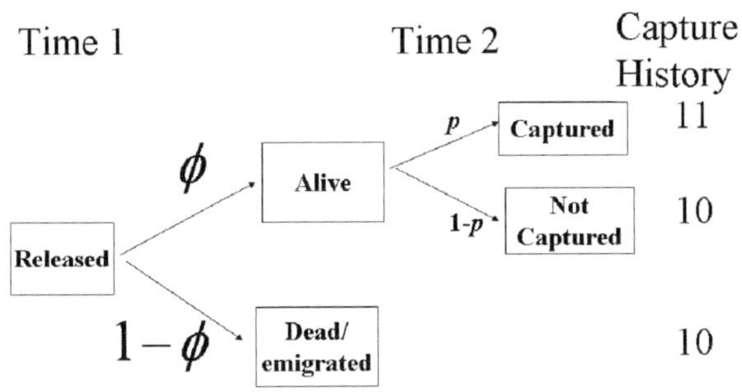

Fig. 1 Events leading to capture histories under the Cormack–Jolly–Seber model

(1) survive the interval [*t*, *t*+1], and (2) be recaptured at time *t*+1. By contrast, there are three ways in a simple CR study that an animal can fail to be recaptured at *t*+1: (1) it died between *t* and *t*+1, (2) it left the study area between *t* and *t*+1, and (3) it was alive and present on the study area, but was not recaptured at *t*+1 (Fig. 1). Failure to separate these events in the parameterization of a statistical model leads to biases in estimates, and confounding of inferences. For example, the "traditional" approach of return analysis, still used by some ecologists, which considers the proportion of animals marked and released that are recaptured or resighted in a natal area, is at best a crude index, and confounds fidelity with mortality and nuisance sampling parameters. Because these factors likely vary over time and space, there is no assurance that the return rate is even a relative index to fidelity, survival, or other biological parameter. Unfortunately, there is still a tendency among ecologists and evolutionary biologists to ignore sampling intensity; Clobert (1995) lamented that fewer than 10% of evolutionary studies until that time considered sampling intensities (capture probabilities). This percentage has doubtless increased in the intervening period, but it is interesting to note that of 45 citations of Clobert (1995), only 6 do not involve either Clobert as a coauthor, or one of the other regular attendees at EURING (e.g., Barker, Cam, Lebreton, Hines, Nichols, Schwarz, to name a few). This by no means proves that evolutionary ecologists are ignoring sampling intensities, but perhaps does suggest that use of CMR or methods in evolutionary studies is still confined to a relatively small group of scientists.

Fortunately, a statistical tool has existed for 40 years that allows at least for valid separation of sampling intensity from the biological processes of interest. Independently developed by Cormack (1964), Jolly (1965), and Seber (1965), the Cormack–Jolly–Seber (CJS) model has been an important mainstay of empirical population analysis. The CJS model separates the biological and sampling events by defining two parameters:

- ϕ_i, the probability that an animal alive at sampling occasion *i* is alive and on the study area at sampling occasion *i*+1,
- p_{i+1}, the probability that a marked animal that is present at *i*+1 is recaptured.

Note that the first of these parameters still combines the event of "surviving" with that of "remaining on the study area", and for that reason is sometimes termed *apparent survival*; I will return to this point later. Note too that the second parameter, recapture, is only defined at the second and subsequent capture occasions. The reason for this is that CJS models only the fate of *marked* animals, and at the first capture period, no animals are marked.

The importance of separating sampling probabilities (p) from the parameters of the process of interest (in this case, ϕ), is made clear by way of a simple example. Suppose that one marks and releases 1,000 animals and subsequently observes 250 in a recapture sample. Under the CJS model, the expected value of this outcome is $1{,}000 \times \phi \times p$, and thus 250 is consistent with any combination of ϕ and p where this quantity equals 250, for instance, $\phi = 0.5$ and $p = 0.5$, or $\phi = 0.25$ and $p = 1.0$. Because there are an infinite number of such combinations, the observation "250 recaptures out of 1,000 releases" provides no information on the parameter of interest (ϕ), unless p can be estimated and "removed" as a nuisance parameter.

Besides the important fact that CJS allowed for separate estimation of survival (or at least, apparent survival) from sampling intensity, several other advances have occurred more recently. Notable among these were the development of flexible approaches for modeling the effects of group factors (such as sex and geographic area), time effects, and age dependencies, as part of an expanded CJS approach (e.g., Lebreton et al. 1992). However, the basic CJS model remained somewhat limited in its ability to allow modeling of continuous factors, whether these be temporal (e.g., environmental) or individual effects. Secondly, survival and permanent emigration remained confounded in the CJS apparent survival parameter. Both of these issues were addressed by more recent advances, discussed in the second section.

2.2 Use of Multiple Alternative Hypotheses and Information Theory

As noted earlier, and observed by others (e.g., Cooch et al. 2002) there is a tension between the desire of evolutionary biologists to extract detailed information from CR data, and that of statisticians to work toward models no more complicated than are supported by data. In particular, evolutionary biologists are often interested in the variation of traits among individuals, so that models that reduce this variation to a single parameter are not of much interest. However, science is also presumably guided by parsimony (Occam's Razor – but see Cooch et al. 2002), and by the periodic feedback of information from observational or experimental studies. Thus, CR modeling is best viewed as collaboration between the biological and statistical sciences. Fortunately, important advances in both fields are helpful for this problem.

Paralleling the development of CR modeling in the 1960s–1980s, but obviously of broader scientific relevance, were two trends: one in scientific philosophy, the other in statistical inference. The first of these was not new, dating at least to Chamberlin (1897), who advocated an approach based on multiple working hypotheses.

Under this approach, rather than seeking to prove (or falsify) individual hypotheses, scientists would entertain multiple, plausible explanations, each with a connection to an underlying body of theory. Science would proceed by the collection of data, just as in single-hypothesis testing, but the emphasis is on accumulating weights of evidence across the alternatives, rather than in rejecting or supporting single hypotheses.

More recently, a similar paradigm has emerged in statistics, wherein emphasis is on inference across multiple alternative (but plausible) models, rather than in statistical null hypothesis testing. This paradigm is supported by important developments in information theory, including the development of Akaike's Information Criterion (AIC; Akaike 1973) and the synthesis by Burnham and Anderson (2002) on multimodel inference and model selection. Link and Barker (2006) recently suggested an alternative based on a weighted Bayesian Information Criterion (BIC), and others (e.g., King and Brooks 2001) have advocated reversible jump Markov chain Monte Carlo (MCMC), also under the Bayesian paradigm. Regardless of the approach, emphasis now can be on parameter estimation, prediction, and weights of evidence under alternative models, rather than on artificially constructed null hypotheses. Below, I cite examples where this general approach has been effectively applied to addressing ecological and evolutionary questions using CR.

2.3 Experimental Approaches

Although this paper emphasizes statistical modeling for observational studies, experimentation, when feasible, will often provide stronger inferences. Under experimentation, subjects (individuals, populations, or other units) are assigned, ideally at random, to treatments corresponding to the environmental or other factors under study. Properly designed studies can provide inferences about causation that are stronger than those provided by observational studies (William et al. 2002). Of course, experiments can and often should take advantage of statistical modeling tools such as CR, including the advances outlined below. Recent examples of studies to address evolutionary questions that have combined observational and experimental approaches with statistical modeling (including CR) include Yoccoz et al. (2002) and Keyser (2003).

3 Recent Advances

Here I focus on more recent advances that have greatly enhanced the power of CR modeling to address ecological and evolutionary questions. My coverage is by no means exhaustive, and is somewhat biased toward studies in avian and mammalian population dynamics. Nonetheless, these, and the examples cited, should give readers a good sense of the power of CR for addressing important questions.

3.1 Modeling of Environmental and Individual Covariates

The basic CJS model, and extensions up through the early 1990s, permitted the modeling of many group and temporal effects on survival and other parameters. However, certain types of variation, including temporal and individual variation in parameters, could only be crudely modeled. Under modifications of the basic CJS model (Pollock et al. 1990), in fact, there are only 2 alternative expressions for time variation in apparent survival, ϕ:

- Survival is time varying but otherwise unstructured, so that $k-1$ survival parameters are estimated for k capture occasions.
- Survival is constant over all time periods, so that a single survival parameter is estimated.

Neither of these models of survival variation is likely to be of interest to evolutionary biologists. Indeed, the second of these must be false a priori, since it implies no variation in this fitness parameter over time; it would only be selected as the appropriate statistical model if the observational evidence is insufficient to allow time-specific estimation. Almost always, we are interested not in *whether* survival (or other parameters) vary over time but *how* and *why* they vary. In particular, is there an environmental, biotic, or other factor that drives (or at least is correlated with) temporal variation in survival or other parameters of interest? Are there individual traits that mediate this variation? One approach to investigate this question would be to estimate parameters under a time-specific model (the general CJS model), and then plot (or conduct a formal regression analysis) these estimates against temporal variation in a potential predictor variable. However, this approach is inefficient, and potentially biased. The preferred approach is to reparameterize the likelihood in terms of the hypothesized relationship. This is more efficient statistically, and also allows direct comparison of alternative biological models (e.g., using AIC).

For example, suppose that one is interested in modeling the relationship between annual survival ϕ_t and a covariate predictor X_t (for example, winter temperature). A model that specifies this relationship is

$$\phi = \frac{\exp(\beta_0 + \beta X_t)}{[1 + \exp(\beta_0 + \beta X_t)]}$$

equivalently

$$\text{In}\left[\frac{\phi_t}{(1-\phi_t)}\right] = \beta_0 + \beta X_t$$

where the logit transform is used to assure that predicted survival stays on the interval (0,1). If we now introduce this into the CJS model we see that the observation "captured at occasion 1, next recaptured at occasion 2" is modeled as

$$\Pr(\{11\}) = \phi_1 p_2.$$

However, the covariate model now replaces ϕ_1 in the likelihood so that we have

$$\Pr(\{11\}) = \frac{\exp(\beta_0 + \beta X_t)}{\left[1 + \exp(\beta_0 + \beta X_t)\right]} p_2.$$

This is repeated for each of the ϕ_i in the model, with the effect that instead of having $k-1$ survival probabilities to estimate, we have only the two parameters β_0 and β. We have also effectively inserted an interesting model between the general CJS model ($\phi_t, t = 1, ..., k-1$) and the "null" model ($\phi_t = \phi, t = 1, ..., k-1$), which can also be obtained from the above expression simply by specifying that β=0, yielding

$$\Pr(\{11\}) = \frac{\exp(\beta_0)}{\left[1 + \exp(\beta_0)\right]} p_2 = \phi p_2.$$

Our "interesting model" can be generalized to allow for multiple predictors and curvilinear relationships by

$$\ln\left[\frac{\phi_t}{(1-\phi_t)}\right] = \beta_0 + \beta' X_t$$

where X and β are vectors of predictors and parameters, respectively. For example,

$$\ln\left[\frac{\phi_t}{(1-\phi_t)}\right] = \beta_0 + \beta_1 X_t + \beta_2 X_t^2$$

specifies a quadratic (on the logit scale) relationship between survival and the factor X_t.

An early, and simple, time-specific covariate analysis was conducted on European dippers (*Cinclus cinclus*) by Lebreton et al. (1992). Here, interest focused on the impact of flood events on annual survival. Lebreton et al. (1992) created a time-specific indicator variable (X_t=1 for flood years, and X_t=0 for non flood years) and incorporated this relationship into the CJS likelihood. They found a negative relationship between flood events and survival, with predicted survival 0.469 in flood years and 0.607 in non-flood years. The Lebreton et al. (1992) model, although a simple approach effectively equivalent to grouping the data into two groups of years and estimating survival for each group, is readily extendable to continuous temporal covariates, such as temperature, or biotic variables such as density of conspecifics or competitors (thus allowing direct estimation of density dependence or competition).

The above approach, however, does not really take into account the fact that many environmental or other factors vary at levels of resolution different than the individuals (the actual units of observation in the study). In the above, for instance, factors varying over time (e.g., flood effects) are treated as fixed effects, and essentially the study is replicated across these effects. There are at least 2 problems with this: first, it is a form of "pseudoreplication" (Hurlbert 1984), second, it does not properly deal with the random, hierarchical nature of the "year" effects. Later, I will return to this example, using a hierarchical modeling approach that more appropriately accounts for these random year effects.

Modeling of individual covariate effects is based on a similar approach, but the statistical implementation is a bit different. Here, survival is modeled at the level of an individual animal i, with attributes X_i that are (potentially) unique to that animal. Thus the fundamental model is

$$\ln\left[\frac{\phi_i}{(1-\phi_i)}\right] = \beta_0 + \beta' X_i$$

for each of the $i = 1, \ldots, n$ animals in the study. Typically, implementation of individual covariate models is based on modeling capture histories as multiple Bernoulli trials, rather than as multinomials (as in CJS) (Skalski et al. 1993, Smith et al. 1994).

I illustrate covariate analysis with a study of Serin finches (*Serinus serinus*), ringed in Spain during 1985–2000 (Conroy et al. 2002). We ringed birds during two periods (October–March, April–September) in order to estimate survival over 6-month periods. In addition to a series of time-, age-, and sex-specific CJS models, we modeled survival in relation to both temporal (minimum temperature, days $< 0\,^\circ$C, and presence of siskins [*Carduelis spinus*], a competitor, during winter, and rain and maximum temperature during summer) and individual (g body mass and mm wing length, standardized within age–sex group) covariates, measured at the first capture of individuals. Overall, we found weak support (as judged by AIC)

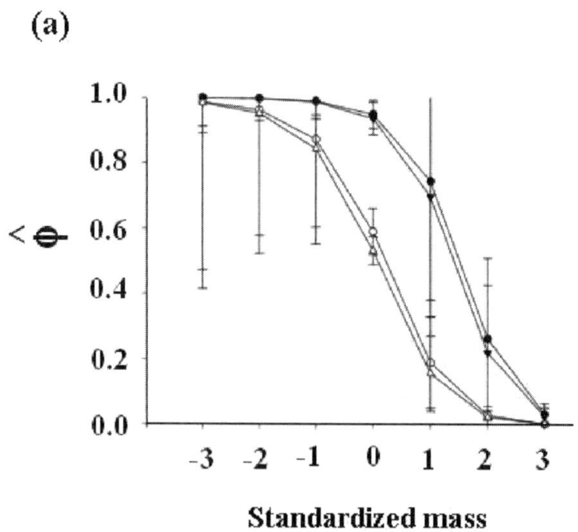

Fig. 2 Predictive model of the relationship between body mass and 6-month survival for Serins in northeastern Spain. Prediction with below-average (*top*), average (*middle*), and above-average number of competing Siskins present

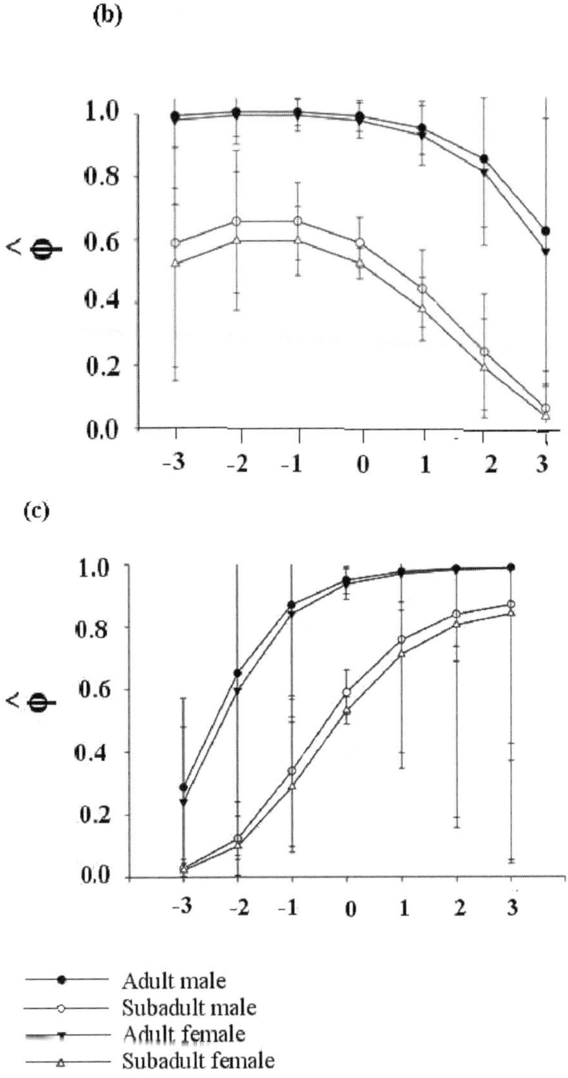

Fig. 2 (continued)

for the temporal covariates but stronger support for individual covariates. I focus here on the estimated (via AIC averaging across alternative models based on AIC weight) quadratic relationship between body mass and 6-month survival (Fig. 2). Interestingly, this body mass-survival relationship seemed to be influenced by the presence of Siskins, with high numbers of this competitor essentially reversing the predictive relationship. Our interpretation of these results was that they (1) provide support for a theoretical view of a fitness tradeoff between starvation and predation risk, since lighter-than-average birds had best predicted survival, but (2) this tradeoff disappears in favor of a strategy toward maximizing mass, when the competitor is present in large numbers.

Blums et al. (2005) used a similar approach to identify individual traits correlated with fitness components, finding quadratic relationships between standardized body size and clutch initiation date as traits varying among individuals, and individual survival probability. Analyses such as these, particularly when buttressed by evidence that individual traits such as body size, coloration, or breeding propensity are heritable, can provide powerful, additional evidence for evolutionary studies.

3.2 Modeling of Heterogeneity via Random Effects

The identification of individual traits correlated with fitness is, as seen above, a powerful tool in evolutionary studies. However, even in the absence of correlated traits, it can be important to characterize the manner in which individual animals vary with respect to survival or other fitness components, or how these components vary over time. An alternative approach is via random effects models, in which the emphasis is on estimation of parameters of an underlying stochastic model that specifies the nature of variation in the component among individuals, or over space, time, or other dimension (e.g., Royle and Link 2002). Many times, interest will focus more on how variable fitness components are, rather than average fitness (Fig. 3a). Alternatively, it may be possible to identify groups within the population having two or more heritable traits, and to identify "fitness profiles" for each (Fig. 3b). Obviously, as illustrated by this last case, there is no bright line between "fixed" (identifiable group or covariate trait) and "random" effects, and which approach is used will depend upon the nature of the hypotheses and data (see also Cam et al. (2002) and references therein). Finally, the above discussion emphasizes random effect modeling where the goal is inference about parameters of interest to ecology and evolution, such as components of variation in fitness. Random effects models can also be useful for modeling "nuisance" variation that often occurs in ecological studies, such as spatial or temporal dependencies, although such variation may itself be of primary interest. Finally, while much of evolutionary ecology is focused on individual variation, modeling variation among demes, populations, or larger aggregations of animals is also of great importance. Proper modeling at all these levels should take into account previous cautions about pseudoreplication, and in most cases should incorporate hierarchical modeling.

3.3 Modeling of Movement and State Transition

As noted earlier, the basic CJS model, while properly separating sampling intensity from demographic processes, cannot distinguish between the event that an animal died between two sampling occasions, and the event that it emigrated from the study area (Fig. 4). If emigration is permanent, there is of course no possibility of future recapture, and this event is treated identically to mortality in CJS.

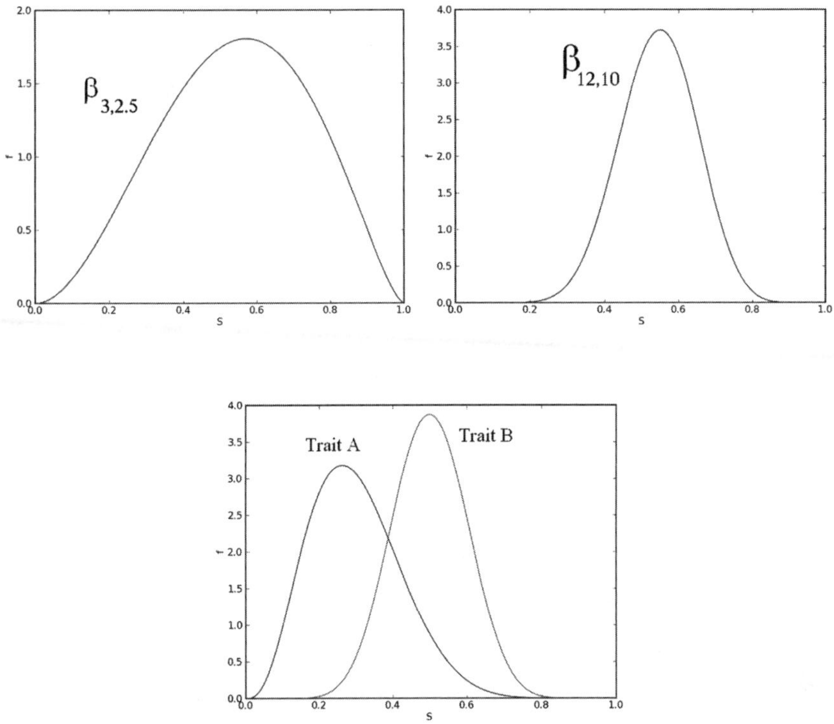

Fig. 3 Random effects models for a fitness component, e.g., survival probability. (*Top*) Survival as a random process from a beta(3, 2.5) distribution (*left*) and from a beta (12,10) distribution (*right*), providing distributions with identical mean survival but very different variances. (*Bottom*) Survival distribution for 2 heritable traits, with higher probability mass under trait B, providing evidence in favor of selection for that trait

Permanent emigration is confounded with survival, however, *temporary* emigration is confounded with recapture, and more complex modeling, such as that provided under the Robust Design (below), is required to separate these effects.

The problem of emigration is a serious one for CR studies, particularly for mobile species such as birds, for which it is virtually impossible to establish a spatial recapture design that would assure that all animals are subject to recapture. Burnham (1993) developed an approach for combined analysis of ring recovery and CR data that, under certain assumptions, allows for separating demographic survival from permanent emigration (confounded under CJS models) and thus enables the estimation of fidelity rates. Barker (1997) expanded this idea to joint modeling of recaptures, resightings, and dead recoveries of marked animals. The Burnham–Barker models depend on the critical assumption that, unlike recaptures, recoveries can potentially occur throughout the range of the species. Blums et al. (2002) applied this model successfully for estimating breeding site fidelity in European ducks, and it has been applied as well to fidelity in North American

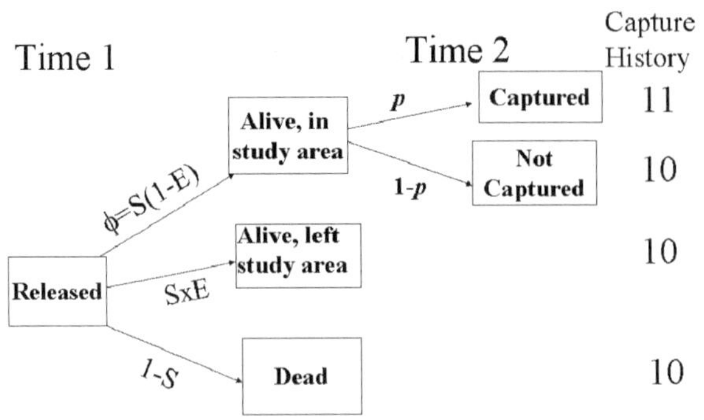

Fig. 4 Problem of apparent survival under conventional CJS model. Animals alive at time 1 may be alive and present on the study area at time 2; alive but absent from the study area; or dead. Capture histories (10) of dead, alive and present but not captured, and alive but absent animals are indistinguishable

ducks and geese (Alisauskas and Lindberg 2002; Doherty et al. 2002; Zimpfer and Conroy 2006).

A more general approach involves the use of multi-state (multi-stratum) models, in which animals are captured, released, and recaptured in two or more "states". Again, "state" refers to any attribute about the individual, such as the age, physical condition, breeding status, or spatial location. These data are summarized in a manner analogous to the capture history format already seen, but where numerical or other codes now signify the state at each capture. For instance, "20310223" signifies an animal that was released in state 2 at occasion 1, recaptured in state 3 at occasion 3, state 1 at 4, state 2 at 6 and 7, state 3 at 8, and not captured at occasions 2 or 5. Multi-state models require definition of many new parameters, notably state-specific transition and recapture probabilities. Transition probabilities are particularly interesting and allow for complex modeling of many attributes. State transition can potentially occur among any of the states, and is reversible (i.e., animals can return to states that they previously occupied). For the 3-state example, this results in a matrix of transition probabilities (ϕ_t^{rs} – the probability that an animal alive and in state r at sampling period t is alive and in state s at period t+1)

$$\begin{bmatrix} \phi^{11} & \phi^{12} & \phi^{31} \\ \phi^{21} & \phi^{22} & \phi^{23} \\ \phi^{31} & \phi^{32} & \phi^{33} \end{bmatrix}.$$

These transition models generally are Markovian, meaning that transition depends only on the state of the animal at t, and not at previous times (Arnason 1972, 1973; Brownie et al. 1993; Schwarz et al. 1993; Williams et al. 2002) but have been extended to non-Markovian (memory) transitions as well (Hestbeck et al. 1991; Brownie et al. 1993; Williams et al. 2002). Also, while in general certain kinds of transition, such as growth, are reversible (an animal can move from state r to state s and back to state r), others, such as stage development, are not (e.g., once an animal transitions to an adult stage it cannot transition back to neonate). Finally, not all state transitions are stochastic: some (such as age-transition) are deterministic, so that transition occurs on a fixed schedule (e.g., for some organisms, 1 year after birth animals transition with certainty to an adult or breeding state). These cases can easily be accommodated by the general state transition model, so that, for instance

$$\begin{bmatrix} \phi^{11} & \phi^{12} & \phi^{13} \\ 0 & 0 & 1 \\ 0 & 0 & \phi^{33} \end{bmatrix}$$

signifies that animals in state 2 transition to state 3 with certainty, those in state 3 never transition to states 1 or 2, but remain in state 3 with probability ϕ^{33}.

Multi-state CR models allow separate estimation of state-specific transition and recapture probabilities, under assumptions analogous to those of ordinary CJS models. Under certain assumptions, the transition probability ϕ_t^{rs} has additional biological significance. In order to transition from state r to state s over $[t, t{+}1]$, the animal must of course survive the interval. Under the assumption that survival over the interval $[t, t{+}1]$ is dependent only on the state at time t, ϕ_t^{rs} can be re-written as

$$\phi_t^{rs} = S_t^{\,r}\Psi_t^{rs},$$

where S_t^r is survival dependent on the originating state r, and Ψ_t^{rs} is the probability of moving from state r to state s, conditioned on survival. Under the assumption that there are j=1,..,k mutually exclusive states to which the animal can transition if it survives, state-specific survival S_t^r, can be computed as $S_t^r = \sum_{s=1}^{k} \phi_t^{rs}$. For instance, in the 3-state example, $S_t^1 = \phi_t^{11} + \phi_t^{12} + \phi_t^{13}$. Finally, marked animals that survive the interval $[t, t{+}1]$ and transition to one of the states, may or may not be recaptured, and recapture probabilities may depend upon the state the animal is in at the next capture period. Thus, for instance, the probability that an animal that is marked and released in state 2 at time t, transitions to and is captured in state 3 and $t{+}1$, would be

$$\phi_t^{23} p_{t+1}^{3}.$$

Figure 5 shows how multi-state modeling works for a simple example involving only 2 states, and should be compared to the much simpler modeling of fates under ordinary CJS (Fig. 1).

Multi-state modeling has been effectively used to model fidelity to breeding areas and other types of movement questions (e.g., Lindberg et al. 1998). I illustrate multi-state CR modeling with a metapopulation example from Senar et al. (2002). Citril finches (*Serinus citrinella*) were ringed in two habitats in the pre-Pyrenees of northeast Spain, separated by approximately 5 km. One of these habitats

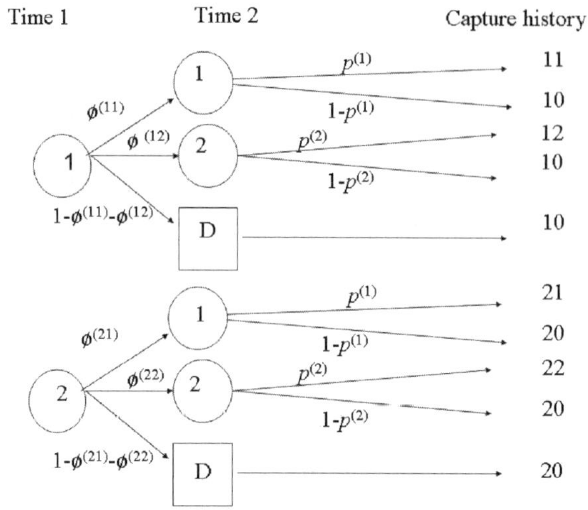

Fig. 5 Multistate CR model, with captures in two strata (1, 2), and state-specific transition ($\phi^{(i1)}$, $\phi^{(i2)}$, i=1, 2) or death ($1 - \phi^{(i1)} - \phi^{(i2)}$, i = 1, 2) to the second occasion, where animals still alive and present in one of the states are captured with state-specific recapture probability $p^{(i)}$, i=1, 2

(La Bofia, "B"), has relatively xeric conditions, resulting in low productivity of pine seeds that are food for the finches, resulting in birds with poor body condition and lower reproductive output. The other habitat (La Vansa, "V"), is more mesic, with higher productivity of pine seeds, better body condition, and higher reproductive output. Senar et al. (2002) captured and ringed birds in both habitats between 1991 and 1999 to estimate habitat-specific survival and probability of movement between habitats. A model containing age and habitat effects for both survival and movement provided over one-half of the total weight of evidence, as judged by AIC; other models containing area effects for movement but not for survival constituted most of the remainder. Model-averaged estimates (Fig. 6) indicate that survival was higher in the better habitats. Additionally, they suggest asymmetry of movement,

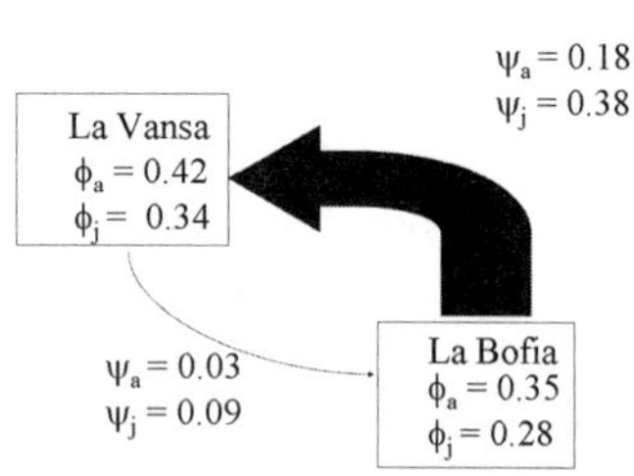

Fig. 6 Asymmetric survival (ϕ_a, ϕ_j) and movement ψ_a, ψ_j probabilities for adult and juvenile, respectively, Citril Finches captured and released in two environmentally different locations in the pre-Pyrenees of northeast Spain

with higher rates of movement from Bofia to Vansa than reverse. The authors interpreted these results both as supporting microgeographic adaptation in finches, as well as supporting a "source-pool" model of movement in favor of alternatives. Finally, the results are especially interesting, in light of evidence that morphological differences in finches appear to be correlated with genetic differences between the 2 sites (Senar et al. 2006).

In another example of multi-state modeling, Senar and Conroy (2004) used multi-state models to estimate transition and survival rates of Serins that either exhibited or did not exhibit the symptoms of avian pox. In this study, state transition rates had interpretation in terms of infection and recovery rates, and estimates provided information for modeling the spread of diseases in wild populations. Senar and Conroy (2004) also found evidence that symptomatic and asymptomatic birds had different recapture rates, which strongly suggests that studies of disease prevalence need to take into account the unequal probabilities of encountering these different states in samples.

Finally, evolutionary biologists are often interested in addressing questions about life history strategies, such as fitness tradeoffs under alternative reproductive strategies (e.g., Brown and Brown 1998; Cam et al. 2003). Multi-state models have been effective tools for addressing these types of hypotheses, by casting reproduction status (or other life history attributes) as states among which animals can stochastically move. For many organisms, reproductive effort affects behavior, and consequently the probability that animals are reencountered in a CR study. Because sampling is often restricted to sites where breeding animals are known to occur this potentially creates situations where certain classes of animals (e.g., non-breeders) are virtually undetectable. Under certain designs (e.g., variations of the Robust Design, discussed below) multi-state models can be used to allow modeling of transition between an observable state (breeding) and an unobservable one (non-breeding). Bailey et al. (1995) applied such an approach to pond-breeding amphibians, and were able to address questions such as whether breeding activity has fitness effects on survival. Likewise, Cam and Monnat (2000) applied MSM to test for the influences of age and habitat quality on both survival and breeding individual heterogeneity in kittiwakes (*Rissa tridactlya*). Cam and Monnat (2000) were able not only to estimate reproductive state, but to model the interaction between age and correlation between fitness components. Tradeoffs between dispersal and reproductive effort have also been investigated by several authors using multi-state models (e.g., Grosbois and Tavecchia 2003; Lebreton et al. 2003; Danchin and Cam 2002). Other examples of multi-state models applied to questions of reproductive strategy and fitness tradeoffs include Yoccoz et al. (2002); Reed et al. (2003); Orell and Belda 2002; Cam et al. (2004), and Rivalan et al. (2005).

3.4 Modeling the Components of Population Growth

Several important classes of problems revolve around the question of the relative contribution of different sources to population growth. Specifically, many questions of theoretical and applied population ecology relate to the relative contribution of age-specific reproduction and survival rates. Estimates of relative contributions will allow prediction about the relative consequences of hypothetical changes in each type of parameter on population growth, and in turn, can be helpful in directing conservation actions.

Estimation of components of population growth has been greatly facilitated by a novel approach to CR modeling, based on viewing capture histories in reverse time order (Pollock et al. 1974; Pradel 1996; Nichols et al. 1994; Nichols and Kendall 1995; Nichols et al. 2000; Nichols and Hines 2002). In this approach, analysis is conditioned on the *last* capture of animals, and modeling is with respect to events that led to the animal's capture history. If we consider just two capture occasions, the capture histories of these animals are all of "1" type. For open populations, a marked animal that is recaptured at occasion 2 obviously must have been in the population at a previous time, and survived from occasion 1 to occasion 2, and could not have been a new recruit into the population. Those animals either may have been captured (11) or not captured (01) at occasion 1. Animals that were not in the population at 1 but are captured at 2, and so are recruits, also will have history "01", and so based on only a single recapture sample these outcomes cannot be distinguished (Fig. 7). This is much like the situation in forward time, in which for the history "10" we cannot distinguish the outcomes "died or emigrated" from "survived but not recaptured" (Fig. 4).

With several (3 or more) capture occasions, reverse-time modeling can be used to estimate the parameter γ_{i+1}, the probability that a member of the population N_{i+1} is a survivor from the previous period, *i*. This in turn can be used to estimate the two demographic components (survival, recruitment) of growth, $\lambda_i = N_{i+1}/N_i$, as the weighted average of abundance

$$E(\lambda_i) = \frac{\gamma_{i+1}N_{i+1} + (1-\gamma_{i+1})N_{i+1}}{N_i}.$$

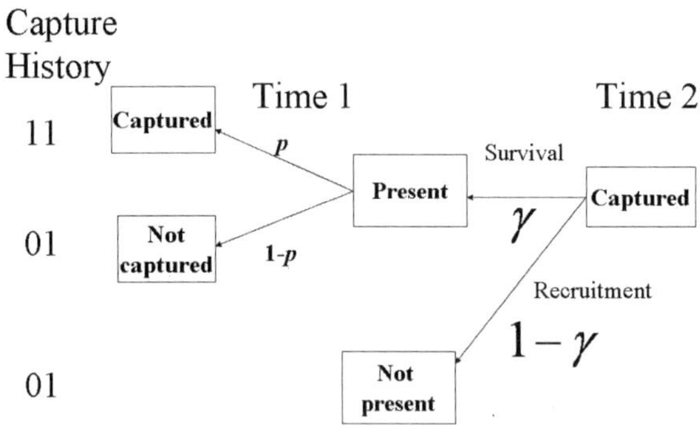

Fig. 7 Reverse-time CJS model showing modeling of survival and recruitment components leading to population growth

The parameter γ_{i+1} also can be used to provide an estimate of the proportional change in growth rate from a proportional change in one of the components, a quantity similar to elasticity (Caswell 2001). For example if survival was increased by an amount α between i and $i+1$, $(1+\alpha)S_i$, then we can obtain the corresponding proportional change in growth as $\alpha\gamma_{i+1}$. Likewise, for a proportional change α in recruitment, the corresponding proportional change in growth would be $\alpha(1-\gamma_{i+1})$. Reverse-time modeling also can be combined with multi-state modeling to allow for estimation of the relative contributions to growth from individual states, such as animals in different behavioral or physiological conditions, or occupying different habitat patches.

The Robust Design (RD; Pollock 1982; Williams et al. 2002; Runge et al. 2006) is another advance useful in this type of analysis. The RD is a two-stage CR design involving primary sampling periods between which the population is assumed to be demographically open, and secondary periods over which the population is assumed closed (Fig. 8). The RD was originally developed to allow for robust modeling of capture probabilities (best done in a closed model) with estimation of survival, recruitment, and other demographic parameters. Originally the RD was used in conjunction with closed CR models to estimate abundance based on the secondary periods; open (CJS) models to estimate survival; and a combination of estimates to estimate recruitment. More recently, the RD has been used in conjunction with multiple-age CR to allow separate estimation of in situ reproduction from immigration, and with reverse-time modeling to allow estimation of components of recruitment from multiple sites, breeding propensity, and other parameters.

A study by Nichols et al. (2000) nicely illustrates the combination of reverse-time modeling with the RD. Meadow voles (*Microtus pennsylvanicus*) were captured and released in grids in two areas during primary trapping periods every 2 months during 1991–1993. During each primary period, trapping occurred on 2–6 consecutive

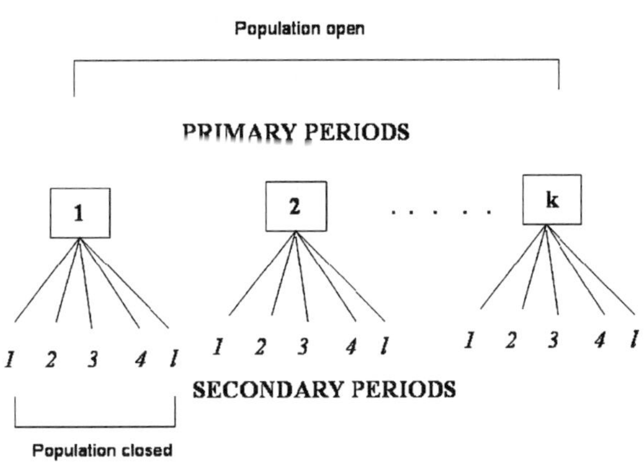

Fig. 8 Pollock's robust design, with combination of primary sampling periods between which population is assumed demographically open, and secondary periods over which population is assumed closed

days. Thus, the data were analyzed according to the RD, with monthly primary periods and daily secondary periods. Estimates of growth rate (λ_i) and proportion of growth contributed by survival and movement and recruitment for each area (γ_i) were estimated using reverse-time modeling. These estimates can be used for making two kinds of inference about population growth. First, as mentioned earlier, the estimates of γ_i (for the moment, not referenced to area) can be used to address questions in evolutionary biology or resource management, such as "how much change in population growth rate can be expected by a specified change in survival or recruitment?" This can be illustrated with the specific case of period 9 growth by examining the estimates $\hat{\lambda}_9 = 0.59$, estimating growth between primary periods 9 and 10, and $\hat{\gamma}_{10} = 0.70$, the estimated proportion of the population at period 10 derived from survival. To address the question "what would population growth rate have been, if survival had increased by 5%" we would use these estimates to obtain

$$\hat{\lambda}_9^* = \hat{\lambda}_9 \left[1 + \alpha \hat{\gamma}_{10}\right] = 0.59(1 + 0.035) = 0.61,$$

that is, population growth rate would have increased from 0.59 to 0.61 with a 5% increase in survival. By comparison, the change in growth due to a 5% change in recruitment would have been

$$\hat{\lambda}_9^* = \hat{\lambda}_9 \left[1 + \alpha(1 - \hat{\gamma}_{10})\right] = 0.59(1 + 0.015) = 0.599.$$

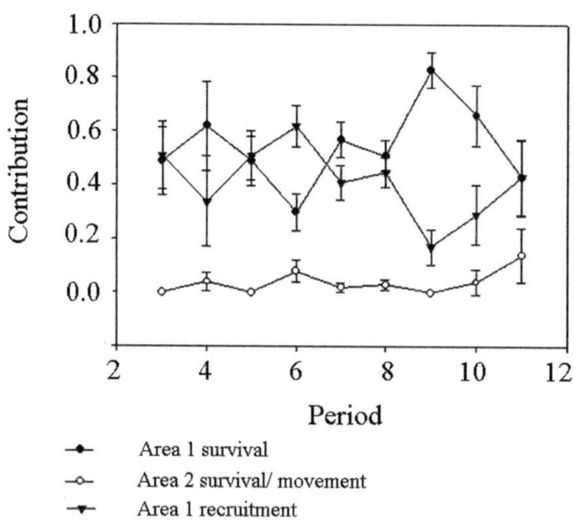

Fig. 9 Contribution of Area 1 survival, Area 1 recruitment, and survival and recruitment from Area 2, to population growth in Area 1, for meadow voles captured and recaptured in Maryland, USA. "Recruitment" in Area 1 includes in situ recruitment, plus immigration from outside the study system

Because of the use of multi-state modeling and the RD, a richer analysis is possible, allowing for inferences on the time- and area-specific variation in the proportion of population growth for each area derived from three sources: (1) survival from animals alive on the area at time t, (2) recruitment (which includes both in situ reproduction, as well as immigration from areas outside the study system) for this area, and (3) survival and movement from other areas. This is illustrated from the vantage point of Area 1 in Fig. 9, from which at least 2 major conclusions can be drawn. First, the relative contribution of area 1 survival and recruitment to area 1 population growth is much larger than the movement into area 1 from other areas. Second, these relative contributions vary over time, raising the possibility that site-specific factors are interacting with temporal factors, such as environmental conditions.

4 Other Advances

The rapid advance in CR modeling has been greatly aided by several other developments, some of which arose directly as a result of efforts in CR modeling, and others which were developed more-or-less for other reasons.

4.1 Bayesian Modeling of Random and Hierarchical Effects

As noted earlier, many important ecological questions require modeling of the components of individual, spatial, and temporal heterogeneity, and random effects models can be especially useful. In other cases, data structures, ecological questions, or both, dictate hierarchical relationships among model parameters. For example, individual variation in survival probability may be expressed by

$$\phi_i, i = 1, \ldots, N$$

where there are N individuals in the population under study. Rather than estimating each of the individual effects (which, in any case, would be impossible), they may instead be modeled hierarchically, for instance, as random variables drawn from a beta distribution

$$\phi_i \sim Beta(\alpha, \beta)$$

with parameters α and β. In this example, a complex relationship (individual variation in survival rates) is summarized in a very parsimonious way, with just 2 parameters. Link and Barker (2004) provide a very useful overview of hierarchical modeling in the context of demographic studies.

Both types of problems, while sometimes tractable by conventional, maximum likelihood approaches, are often much easier to analyze in a Bayesian framework. In Bayesian analysis, parameters (such as p of the binomial distribution) are thought of as "random outcomes" (i.e., are not fixed, but are uncertain). This is in contrast to "classical" (frequentist) analysis in which parameters are thought of as unknown constants. The difference largely revolves around differing philosophical viewpoints about the nature of observations. Frequentists interpret study results in terms of repeatability, and talk about things like "the number of times the confidence interval is predicted to include the parameter p in 100 future experiments." Bayesians wish to make probability statement about parameters, based on *the current* study, while potentially incorporating knowledge about the parameter from previous studies or other sources (prior information). Formally, inference about the parameter θ is based on a *posterior* distribution. By Bayes' Theorem

$$P(\theta|x)P(x) = P(x|\theta)P(\theta)$$

where x is the sample data and $P(\theta)$ is the prior distribution of θ; $P(x|\theta)$ is the probability of the data, under the assumed statistical model and value for θ, and is related to the familiar statistical likelihood. For a given sample, $P(x)$ is a constant, and so the above formulation reduces to

$$P(\theta|x) \propto P(x|\theta)P(\theta).$$

In other words, the posterior distribution of θ is proportional to the likelihood times the prior distribution. Given the posterior distribution, one obtains inference on θ by the usual approach of averaging, computing variances, quantiles, etc. For some combinations of priors and posteriors the resulting posterior is analytically tractable, i.e., one can compute these statistics directly from the data in a single step, much like maximum likelihood estimation. In many others, the statistical likelihood and priors cannot be put together to form a solvable posterior. MCMC gets around this problem by generating samples from a distribution that should, given certain conditions are met, converge on the posterior distribution. There are various ways of doing this; the method employed in MCMC is known as Metropolis–Hastings. Basically, the procedure "proposes" values of θ, which are used to compute a value for the likelihood times the prior. This latter value is then evaluated via an acceptance criterion which is compared to a random uniform deviate, and if accepted, the proposed value is kept in the sample. Given sufficient numbers of MCMC samples the algorithm converges on values for θ that are essentially indistinguishable from random samples from the posterior. Finally, these samples (after convergence or "burn-in") are then used to compute means, variances, quantiles, and other statistics. Papers such as Brooks et al. (2000), Link et al. (2002), Royle and Link (2002), Fonnesbeck and Conroy (2004) and Conroy et al. (2005) provide further discussion and examples of Bayesian analysis of CR and tag-recovery data.

Current versions of Program MARK (White and Burnham 1999) now allow MCMC analysis of simple hierarchical models, which can be particularly useful for properly dealing with random effects in model parameters (Burnham and White 2002). I used the MCMC procedure in MARK to re-analyze the dipper example (Lebreton et al. 1992), modeling variation in survival over the flood and non-flood years via a logit-normal distribution (i.e., a *hyperdistribution*) of the random time effects . After appropriate back-transformation, the new analysis indicates a mean survival of 0.609 (0.555–0.660 95% CI) during non-flood years, and 0.473 (0.418–0.528 95% CI), which is similar to the results of the earlier, simple analysis, but now properly accounts for random, inter-year variability in survival during flood and non-flood years in the calculation of confidence intervals.

4.2 Integrated Parameter Modeling

Ecologists frequently collect several different types of data to address similar or even identical ecological questions. For example, it is not uncommon for physical recaptures, radiotelemetry, and resighting of marked animals to be simultaneously used to estimate survival. CR and other data may be used in conjunction with abundance surveys, so that one potentially is both predicting abundance (via a demographic model) and observing abundance independently. It makes sense both ecologically and statistically to use multiple data sources in an integrated fashion in modeling. In some cases (e.g., Burnham 1993; Powell et al. 2000; Kendall et al. 2006) it is possible to include the different types of data in a common statistical likelihood, in other cases the resulting likelihood would be intractable for analysis. Thus, integrated modeling is a natural realm for the application of the Bayesian methods just discussed (see for example Fonnesbeck and Conroy 2004).

4.3 Innovative Marking and Recapture

Most ecologists think of CR data as arising from a traditional "capture-mark-recapture" study, in which animals are physically captured, marked, and released, and then potentially recaptured at some future occasions. Perhaps the majority of CR studies fall into this category, but increasingly, novel methods are being used for capture, recapture, or both. These approaches may eliminate the need for physical recapture, or even initial marking, with obvious benefits in terms of the removal of concerns of affecting animal behavior, expense, trap mortality, etc. Additionally, some "marking" approaches permit the acquisition of additional information about animals that would otherwise not be obtained.

For purposes here, we may consider marking methods to be either active or passive. *Active* methods require the physical capture of animals at an initial occasion, but may or may not require physical recapture. These methods obviously include conventional CR, where reencounters are by physical recapture, but also designs in which reencounters are via re-sighting of color tags or other visible markers, radiotelemetry, or other passive methods (e.g., Barker 1997). These types of data can be analyzed by themselves to estimate parameters, or potentially combined with conventional CR data. The key here is recognizing that the different types of reencounters involve different physical processes, and thereby should be modeled by different parameters (Williams et al. 2002). In some instances (e.g., radiotelemetry) reencounters are potentially continuous through time, both providing rich sources of data for addressing detailed questions, as well as challenges in deciding on the appropriate level of detail to use in analysis.

By contrast, *passive* markers do not require the initial capture of animals, in that they are based on unique markers that individual animals exhibit and can be detected remotely. For example, camera trapping and CR modeling have been used to estimate abundance of tigers (*Panthera tigris*), relying on the unique signature of striping that each individual tiger possesses (Karanth and Nichols 1998). Genetic markers also are increasingly being used to estimate abundance and other parameters, based on samples of hair, scat, or other tissues not necessarily obtained through physical capture. These methods have the obvious advantage of avoidance of capture and handling of individual animals, which may affect animals' behavior or fitness in ways that could cause the sampled (i.e., captured and marked) population to differ from the population of interest (Williams et al. 2002).

4.4 Computer Software

Early CR analysis involved relatively simple data structures and models, and in many cases, solutions by hand or desk calculator were feasible. Today's CR analyses involve complex data structures and modeling, and generally require the use of high-speed computers. Fortunately, both computer and software technologies have seen rapid advances in the last two decades, and all the analyses described here (and many more) can be performed on desktop computers of moderate capacity. Modern CR software has greatly facilitated the task of handling data and analyses. Ideally, CR software should permit

- Management of data and analyses in a common framework.
- Handling of model data types and structures.
- Rapid construction of and evaluation of alternative models, including goodness of fit.
- Model selection and multiple-model inference using information criteria.

Program MARK (White and Burnham 1999) provides all these capabilities and more, and is freely available. Other programs that provide some or all of these features are MSSURGE (Choquet et al. 2004), POPAN (Arnason and Schwarz 1999), SURGE (Pradel and Lebreton 1991), SURPH (Smith et al. 1994), and U-CARE (Choquet et al. 2003).

5 Summary

This review, though selective, demonstrates that modern CR methods are a powerful tool for exploring evolutionary hypotheses. This modeling framework is now sufficiently general to allow for models that are motivated by underlying theory, in comparison to historical CR modeling, which was constrained to address a few relatively narrow questions. Successful empirical modeling depends, of course, on appropriate sampling or experimental designs. Again, novel designs, in conjunction with appropriate statistical models, have now greatly expanded the nature of questions that can be addressed by CR. Thus, today's CR methodology allows, in addition to estimation of abundance and demographic parameters, investigation of factors influencing variation in these parameters, analyses of evolutionary trade-offs (e.g., survival versus reproductive effort, timing of reproduction), investigation of metapopulation dynamics, and exploration of the sensitivity of populations to perturbations.

A key to the successful application of CR analysis to the study evolutionary questions is the proper separation of the processes leading to the observations. This separation is at the root of all CR analysis, going back to the relatively simple CJS model, in which it was recognized that recapture events require animals to both survive over some interval of time, and be captured at a subsequent sampling occasion. Separation of these events into constituent parameters allows for inferences that are not confounded by the sampling process, and that can lead to unambiguous inferences about the processes of interest (survival, reproduction, fidelity, etc.). A second key is the direct incorporation of biological hypotheses as part of the statistical likelihood. This, in turn, allows the analysis within a scientific framework that admits to plausible, a priori alternative explanations for any ecological phenomenon, and incorporates these as alternative models, each to be challenged by the data.

Acknowledgments I thank Emmanuelle Cam, Torbjorn Ergon, Jim Nichols , Juan Carlos Senar, and Nigel Yoccoz for helpful comments on earlier drafts. The Georgia Cooperative Fish and Wildlife Research Unit is jointly sponsored by USGS, the University of Georgia, the Georgia Department of Natural Resources, the U.S. Fish and Wildlife Service, and the Wildlife Management Institute.

References

Akaike H (1973) Information theory and an extension of the maximum likelihood principle. In: Petran BN, Csaaki F (eds), Second International Symposium on Information Theory. Akdèemiai Kiadi, Budapest, Hungary, pp 267–281.

Alisauskas RT, Lindberg MS (2002) Effects of neckbands on survival and fidelity of white-fronted and Canada geese captured as non-breeding adults. Journal of Applied Statistics 29:521–537.

Arnason AN (1972) Parameter estimates from mark-recapture experiments on two populations subject to migration and death. Research on Population Ecology. 15:1–8.

Arnason AN (1973) The estimation of population size, migration rates, and survival in a stratified population. Research on Population Ecology 13:97–113.

Arnason AN, Schwarz CJ (1999) Using POPAN-5 to analyse banding data. Bird Study 46: S157–S168.

Bailey LL, Kendall WL, Church DR, Wilbur HM (1995) Estimating survival and breeding probability for pond-breeding amphibians: a modified robust design. Ecology 85:2456–2466.

Barker RJ (1997) Joint modeling of live-recaptures tag-resight and tag-recovery data. Biometrics 53:666–677.

Blums P, Nichols JD, Hines JE, Lindberg MS, Mednis A (2005) Individual quality survival variation and patterns of phenotypic selection on body condition and timing of nesting in birds. Oecologia 143:365–376.

Blums P, Nichols JD, Hines JE, Mednis A (2002) Sources of variation in survival and breeding site fidelity in three species of European ducks. Journal of Animal Ecology 71:438–450.

Brooks SP, Catchpole EA, Morgan BJT, Barry SC (2000) On the Bayesian analysis of ring-recovery data. Biometrics 56:951–956.

Brown CR, Brown MB (1998) Fitness components associated with alternative reproductive tactics in cliff swallows. Behavioral Ecology 9:158–171.

Brownie C, Hines JE, Nichols JD, Pollock KH, Hestbeck JB (1993) Capture–recapture studies for multiple state including non-Markovian transition probabilities. Biometrics 49: 1173–1187.

Burnham KP, Anderson DR (2002) Model selection and multi-model inference. Springer, New York.

Burnham KP (1993) A theory for combined analysis of ring recovery and recapture data. In: Lebreton JD, North PM (eds), The study of bird population dynamics using marked individuals. Birkhauser-Verlag, Berlin, pp 199–213.

Burnham KP, White GC (2002) Evaluation of some random effects methodology applicable to bird ringing data. Journal of Applied Statistics 29:245–264.

Cam E, Monnat JY (2000) Apparent inferiority of first-time breeders in the Kittiwake: the role of heterogeneity among age classes. Journal of Animal Ecology 69:380–394.

Cam E, Link WA, Cooch EG, Monnat J-Y, Danchin E (2002) Individual covariation in life history traits: seeing the trees despite the forest. American Naturalist 159:96–105.

Cam E, Monnat JY, Hines JE (2003) Long-term fitness consequences of early conditions in the kittiwake. Journal of Animal Ecology 72:411–424.

Cam E, Monnat JY, Royle JA (2004) Dispersal and individual quality in a long lived species. OIKOS 106:386–398.

Caswell H (2001) Matrix population models:construction analysis and interpretation. Sinauer, Sunderland, MA.

Chamberlin TC (1897) The method of multiple working hypotheses. Journal of Geology 5: 837–848.

Choquet R, Reboulet AM, Pradel R, Gimenez O, Lebreton JD (2003) User's manual for U-Care. Mimeographed document, CEFE/CNRS, Montpellier, France.

Choquet R, Reboulet AM, Pradel R, Gimenez O, Lebreton JD (2004) M-SURGE: new software specifically designed for multistate capture–recapture models. Animal Biodiversity and Conservation 27:207–215.

Clobert J (1995) Capture–recapture and evolutionary ecology: a difficult wedding? Journal of Applied Statistics 22: 989–1008.

Conroy MJ, Fonnesbeck CJ, Zimpfer NL (2005) Modeling regional waterfowl harvest rates using Markov chain Monte Carlo. Journal of Wildlife Management 69:77–90.

Conroy MJ, Senar JC, Domènech J (2002) Analysis of individual- and time-specific covariate effects on survival of *Serinus serinus* in north-eastern Spain. Journal of Applied Statistics 29:125–142.

Cooch EG, Cam E, Link W (2002) Occam's shadow: levels of analysis in evolutionary ecology—where to next? Journal of Applied Statistics 29:19–48.

Cormack RM (1964) Estimates of survival from the sighting of marked animals. Biometrika 51:429–438.

Danchin E, Cam E (2002) Can non-breeding be a cost of breeding dispersal? Behavioral Ecology and Sociobiology 51:153–163.

Doherty PF, Nichols JD, Tautin J, Voelzer JF, Smith GW, Benning DS, Bentley VR, Bidwell JK, Bollinger KS, Brazda AR, Buelna EK, Goldsberry JR, King RJ, Roetker FH, Solberg JW, Thorpe PP, Wortham JS (2002) Sources of variation in breeding-ground fidelity of mallards (*Anas platyrhynchos*). Behavioral Ecology 13:543–550.

Fonnesbeck CJ, Conroy MJ (2004) Application of integrated Bayesian modeling and Markov chain Monte Carlo methods to the conservation of a harvested species. Animal Biodiversity and Conservation 27:267–281.

Grosbois V, Tavecchia G (2003) Modeling dispersal with capture-recapture data: disentangling decisions of leaving and settlement. Ecology 84:1225–1236.

Hestbeck JB, Nichols JD, Malecki RA (1991) Estimates of movement and site fidelity using mark-resight data of wintering Canada geese. Ecology 72:523–533.

Hurlbert SH (1984) Pseudoreplication and the design of ecological field experiments. Ecological Monographs 54:187–211.

Jolly GM (1965) Explicit estimates from capture–recapture data with both death and immigration – stochastic model. Biometrika 52:225–247.

Karanth KU, Nichols JD (1998) Estimation of tiger densities in India using photographic capture and recaptures. Ecology 79:2852–2862.

Kendall WL, Conn PB, Hines JE (2006) Combining multistate capture–recapture data with tag recoveries to estimate demographic parameters. Ecology 87:169–177.

Keyser AJ (2003) Life history, demography, and individual variation in western bluebirds. PhD Disseration, University of Georgia, Athens, USA.

King R, Brooks SP (2001) On the Bayesian analysis of population size. Biometrika 88:317–336.

Lebreton JD, Burnham KP, Clobert J, Anderson DR (1992) Modeling survival and testing biological hypotheses using marked animals – a unified approach with case-studies. Ecological Monographs 62:67–118.

Lebreton JD, Hines JE, Pradel R, Nichols JD, Spendelow JA (2003) Estimation by capture–recapture of recruitment and dispersal over several sites. OIKOS 101:253–264.

Lindberg MS, Sedinger JS, Derksen DV, Rockwell RF (1998) Natal and breeding philopatry in a Black Brant, *Branta bernicla* nigricans, metapopulation. Ecology 79:1893.

Link WA, Cam E, Nichols JD, Cooch EG (2002) Of BUGS and birds: Markov chain Monte Carlo for hierarchical modeling in wildlife research. Journal of Wildlife Management 66: 277–291.

Link WA, Barker RJ (2004) Hierarchical mark-recapture models: a framework for inferences about demographic processes. Animal Biodiversity and Conservation 27:441–449.

Link WA, Barker RJ (2006) Model weights and the foundations of multimodel inference. Ecology 87:2626–2635.

Nichols JD, Kendall WA (1995) The use of multi-state capture-recapture models to address questions in evolutionary ecology. Journal of Applied Statistics 22:835–846.

Nichols JD, Hines JE, Pollock KH, Hinz RL, Link WA (1994) Estimating breeding proportions and testing hypotheses about costs of reproduction with capture–recapture data. Ecology 75: 2052–2065.

Nichols JD, Hines JE, Lebreton J-D, Pradel R (2000) Estimation of contributions to population growth: a reverse-time capture recapture approach. Ecology 81:3362–3376.

Nichols JD, Hines JE (2002) Approaches for the direct estimation of lambda, and demographic contributions to lambda, using capture–recapture data. Journal of Applied Statistics 29: 539–568.

Orell M, Belda EJ (2002) Delayed cost of reproduction and senescence in the willow tit *Parus montanus*. Journal of Animal Ecology 71:55–64.

Pollock KH (1982) A capture–recapture design robust to unequal probability of capture. Journal of Wildlife Management 46:757–760.

Pollock KH, Nichols JD, Brownie C, Hines JE (1990) Statistical inference for capture–recapture Wildife Monographs 107:70 pp.

Pollock KH, Solomon DL, Robson DS (1974) Tests for mortality and recruitment in a K-sample tag-recapture experiment. Biometrics 30:77–87.

Powell LA, Conroy MJ, Hines JE, Nichols JD, Krementz DG (2000) Simultaneous us of mark-recaptuer and radio telemetry to estimate survival, movement, and capture rates. Journal of Wildlife Management 64:302–313.

Pradel R (1996) Utilization of capture-mark-recapture for the study of recruitment and population growth rate. Biometrics 52:703–709.

Pradel R, Lebreton J-D (1991) Users manual for program SURGE version 4.1. C.E.F.E. C.N.R.S Montpelier France.

Reed ET, Gauthier G, Pradel R, Lebreton J-D (2003) Age and environmental conditions affect recruitment in greater snow geese. Ecology 84:219–230.

Rivalan P, Prevot-Julliard AC, Choquet R, Pradel R, Jacquemin B, Girondot M (2005) Trade-off between current reproductive effort and delay to next reproduction in the leatherback sea turtle. Oecologia 145:564–574.

Royle A, Link WA (2002) Random effects and shrinkage estimation in capture–recapture models. Journal of Applied Statistics 29: 329–351.

Runge JP, Runge MC, Nichols JD (2006) The role of local populations within a landscape context: defining and classifying sources and sinks. American Naturalist 167:925–938.

Schwarz CJ, Schweigert JF, Arnason AN (1993) Estimating migration rates using tag recovery data. Biometrics 49:177–193.

Seber GAF (1965) A note on the multiple-recapture census. Biometrika 52:249–259.

Senar JC, Borras A, Cabrera J, Cabrera T, Björklund M (2006) Local differentiaion in the presence of gene flow in the citril finch *Serinus citronella*. Biology Letters 11:85–87.

Senar JC, Conroy MJ (2004) Multi-state analysis of the impacts of avian pox on a population of Serins (*Serinus serinus*): the importance of estimating recapture rates. Animal Biodiversity and Conservation 27:133–146.

Senar JC, Conroy MJ, Borras A (2002) Assymetric exchange between populations differing in habitat quality: a metapopulation study on the citril finch. Journal of Applied Statistics 29: 425–441.

Skalski JR, Hoffman A, Smith SG (1993) Testing the significance of individual and cohort-level covariates in animals survival studies. The study of bird population dynamics using marked individuals. (J.-D. Lebreton & North PM, editors) Birkhauser Verlag, New York.

Smith SG, Skalski JR, Schlecte W, Hoffman A, Cassen V (1994) SURPH.1 Manual. Statistical survival analysis for fish and wildlife tagging studies. Bonneville Power Administration, Portland, OR.

White GC, Burnham, KP (1999) Program MARK: survival rate estimation from both live and dead encounters. Bird Study 46:S120–S139.

Williams BK, Nichols JD, Conroy MJ (2002) Analysis and management of animal populations. Elsevier-Academic, Amsertdam.

Yoccoz NG, Erikstad KE, Butnes JO, Hannsen SA, Tveraa, T (2002) Costs of reproduction in common eiders (Somateria mollissima): an assessment of relationships between reproductive effort and future survival and reproduction based on observational and experimental studies. Journal of Applied Statistics 29:57–64.

Zimpfer NL, Conroy, MJ (2006) Modeling movement and fidelity of American black ducks. Journal of Wildlife Management 70:1770–1777.

Estimating Reproductive Costs with Multi-State Mark-Recapture Models, Multiple Observable States, and Temporary Emigration

Jay Rotella

Abstract Multi-state mark-recapture models have seen increased use in recent years for studies of reproductive costs. When individuals in both breeding and non-breeding states can be observed, multi-state models can be used to directly estimate reproductive costs by comparing state-specific estimates of survival and breeding probabilities. The method assumes that each state that an animal occupies is observable, an assumption that is violated if some animals are absent for one or more breeding seasons and are thus, unobservable due to temporary emigration. Previous research on the case of a single observable state and a single unobservable state has shown that non-random (Markovian) temporary emigration can, if not accounted for, bias estimates of survival. Here, simulation is used to study effects of non-random (Markovian) temporary emigration on estimates of survival and breeding probabilities for the case of two observable states and one unobservable state. Results clearly show that temporary emigration can cause estimates of survival and breeding probability to be biased if the unobservable state is ignored. Bias was either positive or negative depending on circumstances, and was sometimes severe (percent relative bias was as high as 67% for estimates of breeding probability). Accordingly, the strengths and limitations of including an unobservable state in analyses are also considered. For some situations, simply including an unobservable state will be an adequate solution. But, for those studies particularly interested in temporal variation in costs of reproduction, it will be necessary to collect other information to avoid problems of parameter constraints. Additional information can consist of data from sub-sampling during primary sampling occasions, radio telemetry, or ring recoveries.

1 Introduction

An organism's lifetime reproductive output determines its fitness. To maximize output, individuals must optimize their life-history decisions because producing

J. Rotella (✉)
Ecology Department, Montana State University, Bozeman, MT 59717, USA
e-mail: rotella@montana.edu

D.L. Thomson et al. (eds.), *Modeling Demographic Processes in Marked Populations*, Environmental and Ecological Statistics 3, DOI 10.1007/978-0-387-78151-8_7,

and caring for offspring in one year may decrease energy available for subsequent reproductive opportunities and reduce longevity. Thus, organisms are expected to have to balance current reproduction against future survival and reproduction (Williams 1966). The trade-off between current reproductive effort and future reproductive value is known as the cost of reproduction. The trade-off is expressed through decreased survival, future probability of reproduction, and/or offspring quality and is hypothesized to be an important factor shaping life-history strategy for many species (Roff 1992; Stearns 1992). Accordingly, evolutionary ecologists have devoted a great deal of effort to empirically measuring reproductive costs.

In testing for reproductive costs, it is common to be interested in whether breeding at time t negatively affects an individual's probability of surviving from time t to time $t+1$ or its probability of breeding at time $t+1$, although delayed costs can certainly be of interest as well (Nichols et al. 1994; Clobert 1995). To estimate possible effects, survival and breeding probabilities are often compared between breeders and non-breeders. As shown by Nichols et al. (1994), mark-recapture methods provide a useful approach for comparing these probabilities and allow one to deal with possible differences in the probabilities of individually marked breeders and non-breeders appearing in samples. In particular, one can use a multistate modeling approach (Darroch 1961; Arnason 1972, 1973; Brownie et al. 1993; Schwarz et al. 1993) to estimate reproductive costs while accounting for possible time- and age-specific variation in sampling probabilities across K successive potential breeding periods (Nichols et al. 1994). When applying multistate modeling to questions of costs of reproduction when breeders and non-breeders can be observed, it is common to define each observed individual as being in either the breeder or non-breeder state for each sampling occasion (breeding period), and to estimate and compare state-specific probabilities of survival (probability of being alive at time $t+1$ given being alive and in a particular state at time t) and breeding (probability of being in the breeder state at time $t+1$ given being in a particular state at time t and conditional on being alive at time $t+1$). Transitions between the breeder and non-breeder state are typically treated as a first-order Markov process (state at time $t+1$ depends only on state at time t) (Brownie et al. 1993).

Since Nichols et al. (1994) first described the multistate modeling approach for studying reproductive costs, the details and benefits of the approach have been further discussed (Clobert 1995; Nichols and Kendall 1995; Viallefont et al. 1995; Boulinier et al. 1997; Doligez et al. 2002; Lebreton and Pradel 2002; Williams et al. 2002; Sandercock 2006), and the multistate modeling approach has been used to assess costs of reproduction in numerous studies of diverse taxa (e.g., Viallefont et al. 1995; Cam et al. 1998; Sandercock et al. 2000; McElligot et al. 2002; Yoccoz et al. 2002; Rivalan et al. 2005; Barbraud and Weimerskirch 2005; Tavecchia et al. 2005; Beauplet et al. 2006; Hadley et al. 2007). Although there are certainly other approaches to studying costs of reproduction (e.g., Reznick 1985; 1992; Golet et al. 2005), results from multistate modeling can provide rigorous empirical estimates of potential costs of reproduction under certain circumstances and have contributed much to our understanding of life history evolution.

As discussed by Nichols et al. (1994), multistate modeling is especially applicable to studies directed at phenotypic correlations, but the approach can also be used in some field, experimental-manipulation studies and may be useful for dealing with possible genotypic variation in capture or observation probabilities in genetic field studies. However, it is important to note that multistate models based on live encounter data yield survival estimates that are commonly referred to as "apparent survival" rates in the mark-recapture literature because they combine the probability of survival and the probability of not permanently emigrating from the study area, i.e., apparent survival differs from true survival to an extent that depends on the level of permanent emigration. This has potentially important implications if breeders and non-breeders have different rates of permanent emigration: differences in apparent survival between animals in different breeding states could be due to variation in site fidelity and not true survival. Accordingly, the multistate modeling approach to estimating costs of reproduction is most appropriate for species in which site fidelity is high or for which permanent emigration occurs at similar rates for breeders and non-breeders.

When using multistate modeling to compare survival and breeding probabilities for breeders and non-breeders, one assumes that each state that an animal occupies is observable (for details of all assumptions, see Nichols et al. 1994; Williams et al. 2002). As reviewed by Kendall (2004), this assumption fails if some members of the population are unavailable for capture or detection when sampling occurs, i.e., heterogeneity in detection probability exists such that some individuals are in an unobservable state (probability of detection is zero). Failure of this assumption can lead to biased estimates.

In multistate modeling of reproductive costs, some individuals may temporarily emigrate and be absent for one or more breeding seasons. For example, in some species, both breeders and non-breeders may be present and observable on breeding sites, but some non-breeders may use non-breeding habitat and be unobservable (for an overview of the widespread nature of temporary emigration see Schaub et al. 2004). Estimates obtained from multistate modeling that ignores temporary emigration can be biased in the presence of some forms of temporary emigration (Kendall et al. 1997; Fujiwara and Caswell 2002; Kendall and Nichols 2002; Schaub et al. 2004), and consequently, estimates of costs of reproduction will also be biased. In particular, it is well known that survival estimates can be biased if the probability of temporary emigration depends on an individual's state during the previous occasion, i.e., is non-random or Markovian (Kendall et al. 1997).

The existence of an unobservable state can be accommodated by incorporating it into multistate models. For the simplest case where individuals occur in either one observable state or one unobservable state (e.g., only breeders present on study sites; all non-breeders in an unobservable state; Reed et al. 2003), the effects of Markovian temporary emigration on parameter estimates have been well studied, methods for incorporating the unobservable state in mark-recapture modeling have been developed, parameter redundancies are quite well understood, suggestions for obtaining useful additional information have been provided, and comprehensive guidelines exist (e.g., Lebreton et al. 1999; Pradel and Lebreton 1999; Fujiwara and Caswell 2002;

Kendall and Nichols 2002; Schaub et al. 2004; Choquet et al. 2004; Kendall 2004). Of particular note, Schaub et al. (2004) used simulation and computer algebra to comprehensively evaluate the performance of the traditional Cormack–Jolly–Seber model (Cormack 1964; Jolly 1965; Seber 1965) and a multistate model that included one observable state and one unobservable state when Markovian temporary emigration occurred. Schaub et al. (2004) found that estimates of survival from the Cormack–Jolly–Seber model were biased in the presence of Markovian temporary emigration, whereas estimates from the multistate model were not. They concluded that when survival and recapture probabilities are high the multistate model works well when Markovian temporary emigration occurs and individual states can be well described with one observable and one unobservable state.

However, for situations with multiple observable states and at least one unobservable state, the complexity is much greater due to the increased number of states, possible transitions between the various states, and the increased number of parameters. For example, Kéry et al. (2005), in an investigation of parameter identifiability for a variety of multistate model for perennial plants with two observable states and one unobservable state, found that most models had identifiable parameters and that some models allowed for estimation of state-specific survival rates, including that of the unobservable state. However, they also found that several models contained no identifiable parameters, which highlights the additional complexity that arises when working with an unobservable state and multiple observable states. At this time, it is difficult to know how well information from the single-observable-state situation applies to the multiple-observable-state situation (Kendall 2004). Of particular relevance to studies of costs of reproduction, information from studies of single-observable states provides no information about how temporary emigration may bias estimates of transitions between observable states, which include breeding probabilities for breeders and for non-breeders. Consequently, it is not clear how one should proceed with multistate analyses of breeders and non-breeders if temporary emigration is suspected. If one includes the unobservable state, then additional constraints such as time constancy must be placed on demographic parameters (Kendall and Nichols 2002; Kendall 2004; Schaub et al. 2004) unless additional information about detection probability is available (Kendall 2004).

As clearly articulated by Kendall (2004:100), "To be forced to assume a priori that parameters are equal over time or group is unsatisfactory. In fact, testing that hypothesis might be of interest." Alternatively, one can choose to ignore the temporary emigration and risk having some level of bias in the resulting estimates. Analysis choices have been recently considered for a situation with multiple observable states and a single unobservable state in orchids (Kéry et al. 2005). However, to date, I am aware of only two papers that have discussed these choices with respect to estimates of reproductive costs based on multistate analyses of breeders and non-breeders. Beauplet et al. (2006) chose to incorporate an unobservable state at the cost of additional parameter constraints and numerical convergence issues. In contrast, Hadley et al. (2007) chose to ignore the unobservable state in their primary analyses after preliminary analyses provided evidence that temporary emigration was not having important effects on estimates of reproductive costs for their situation. Until better information becomes available on how best to proceed in such situations, researchers will continue to face difficult decisions when designing studies of reproductive costs and analyzing resulting data.

Given the rapid increase in the use of multistate modeling to study reproductive costs in recent years and the possibility that temporary emigration has been, or will be, an issue in at least some multistate studies, the objectives of this paper are to stimulate thinking about how to handle temporary emigration in multistate modeling of reproductive costs by (1) using simulated data to illustrate how temporary emigration, if ignored, can cause important levels of bias in estimates of survival, breeding probability, and reproductive costs, (2) using simulated data to show that the inclusion of an unobservable state into multistate models imposes modeling constraints that can limit one's ability to fully estimate reproductive costs, and (3) reviewing suggestions for additional information that can be collected to improve future multistate comparisons of survival and breeding probabilities in breeders and non-breeders. The results have implications that reach beyond multistate modeling of reproductive costs as they are also relevant to a variety of other studies containing multiple observable states or sites and at least one unobservable state or site (e.g., dispersal among sites when some sites are not monitored, Lebreton et al. 2003; demography of perennial plants with observable vegetative and flowering states and unobservable below-ground rhizomes, Kéry et al. 2005).

2 Methods

To assess the effects of Markovian temporary emigration on estimates of survival and breeding probabilities for breeders and non-breeders, I conducted simulations using M-SURGE (Choquet et al. 2004) and an expected-values approach for a variety of multistate modeling scenarios that might be encountered. Choquet et al. (2005) and Devineau et al. (2006) provide details regarding the process for generating expected data using M-SURGE. Devineau et al. (2006) provide additional details about the expected-values approach, its benefits, and how it relates to Monte Carlo simulation. For the simulations done here, all analyses considered one group with three states: breeders (B), non-breeders in an observable state (N), and non-breeders in an unobservable state (U). Individuals in state N and U were assigned the same reference parameters except that detection probability (p) was zero for those in state U. Because it is a fairly common real-world scenario, I created data for a situation where only breeders could be captured but in which field-readable markers could be observed on breeders and non-breeders.

For the multistate model with states B, N, and U, the transition matrix and associated vectors of survival and capture probabilities were

$$\begin{bmatrix} \psi^{BB} & 1-\psi^{BB}-\psi^{BU} & \psi^{BU} \\ \psi^{NB} & 1-\psi^{NB}-\psi^{NU} & \psi^{NU} \\ \psi^{UB} & 1-\psi^{UB}-\psi^{UU} & \psi^{UU} \end{bmatrix}_t \begin{bmatrix} S^B \\ S^N \\ S^N \end{bmatrix}_t \begin{bmatrix} p^B \\ p^N \\ 0 \end{bmatrix}_t,$$

where ψ_t^{rs} is the probability that an individual which is still alive and present in the study population at the end of period t will move from state r to state s; S_t^r is the probability that an individual in state r survives and remains in the study population through period t, where non-breeders are assigned the same survival rate on a given occasion regardless of whether they are in state N or U; and p_t^r is the probability that a marked individual alive in state r at time t is captured or observed (zero for those in state U). The matrix and vector subscript t denotes that time dependence was possible: in the actual modeling, parameters were held constant for some scenarios and allowed to vary with time in others (see below).

The key steps in the process used here were to (1) define a set of reference parameter values, (2) obtain the expected data (i.e., expected values for encounter histories) using the Arnason–Schwarz model (Arnason 1972, 1973; Schwarz et al. 1993), (3) obtain maximum-likelihood estimates of parameters for models of interest (see below); and (4) derive relevant measures of bias and precision. Expected data were generated based on (1) the reference values (see below), (2) the number of breeders released on the first occasion (n = 10,000), and (3) the population growth rate for the number of breeders between successive occasions. I released a large number of breeders to minimize possible rounding errors that may have been associated with working with fractional numbers of animals that result from using expected data in the encounter histories (trials with even larger numbers of released individuals indicated that the number used was adequate to avoid rounding errors; smaller numbers may have sufficed). The population growth rate for breeders was set to one between all successive occasions. Thus, as breeders died or transitioned to other states, an adequate number of breeders was injected into the population to maintain 10,000 breeders on each occasion, and all new individuals were captured and released. Thus, there was a staggered entry of newly marked individuals in state B, and there were resightings of individuals in states B and N. The reference parameters determined the actual survival, transition, and detection rates for animals in each state on each occasion (see below).

For multistate modeling, it may not be possible to estimate all parameters because some may be aliased and not separately identifiable, and the parameter redundancy is not always intuitive (Gimenez et al. 2003, 2004). Therefore, parameter redundancy was evaluated for each model considered here using the numerical version of the Catchpole, Morgan, and Freeman approach as implemented in M SURGE and described by Choquet et al. (2005). After checking that all parameters were identifiable for a given model, parameter estimates were compared to the reference parameter values and, where applicable, between competing models applied to a given dataset. Absolute bias was computed as the difference between a given parameter estimate and the true underlying reference parameter value, and percent relative bias was the absolute bias divided by the reference parameter value and multiplied by 100. Percent coefficient of variation was calculated for each parameter as the estimated standard error divided by the parameter estimate and multiplied by 100.

The scenarios that were simulated were not meant to encompass the extremely broad range of sampling scenarios that might be encountered in actual studies. Rather, they were chosen to illustrate some estimation problems that can arise if Markovian temporary emigration is not dealt with properly and to motivate future work on the problem. The situations were limited to scenarios where only breeders were captured and where breeders were always resighted at a higher rate than were non-breeders. However, within these constraints, and as explained in more detail below, I did bracket conditions such that reproductive costs to survival, future fecundity, or both were large, small, or absent and for which recapture rates were high or low. The actual parameter values used were arbitrarily chosen. As detailed below, I considered scenarios in which parameters were constant over time or time-varying.

2.1 *Simulations to Evaluate Effects of Ignoring Markovian Temporary Emigration when Estimating Reproductive Costs*

Based on published information regarding factors affecting bias in survival rates estimated with multistate models of one observable and one unobservable state in the presence of temporary emigration (e.g., Kendall 2004; Schaub et al. 2004), I chose to consider simulations with (1) Markovian temporary emigration, (2) high versus low capture probabilities $(p^{\mathrm{B}} = 0.9$ or $0.3; p^{\mathrm{N}} = p^{\mathrm{B}} - 0.2)$, (3) high versus low survival rates ($S^B = 0.9$ to 0.2, $S^N = 0.9$ or 0.3), (4) presence or absence of reproductive costs to survival rate ($S^B = S^N \times 0.667$ or $S^B = S^N$), and (5) presence or absence of reproductive costs to breeding probability ($\psi^{BB} = \psi^{NB} \times 0.445$ or $\psi^{BB} = \psi^{NB}$) (Table 1). In all scenarios, the temporary emigration rate was higher for breeders ($\psi^{BU} = 0.4$) than for non-breeders ($\psi^{NU} = 0.1$), which could occur if (1) breeders have a propensity to move to an alternate habitat for replenishing body reserves in the year after a breeding attempt such that $\psi^{BU} > \psi^{NU}$, but (2) non-breeders can also gain benefits from being present on breeding sites, e.g., for evaluating site quality, such that ψ^{BU} and $\psi^{NU} < 1$. In all simulations, parameters were constant across time and age, and data were generated for eight occasions.

For each scenario, two models were run. First, the true generating model (a model that included state U) was used to estimate parameters from the expected data. Results from the generating model were checked to ensure that the model converged on the reference parameters and that all parameters were estimable. Second, a simplified version of the generating model (included states B and N but ignored state U and held parameters constant across time and age) was used and the resulting estimates were evaluated for absolute bias and precision (percent coefficient of variation), with emphasis on the estimates of S^B, S^N, ψ^{BB}, and ψ^{NB}, i.e., those used to estimate costs of reproduction.

2.2 *Simulation of Time-Varying Reproductive Costs in the Presence of Markovian Temporary Emigration*

When no additional information about movements, capture probabilities, or survival rates is available, parameter estimation for a model that includes states B, N, and U is only possible if at least one of the following constraints is applied: the order of Markovian transition probabilities is reduced to make them random, partial determinism is imposed on transition probabilities (e.g., probabilities of transition

Table 1 Absolute bias and coefficient of variation (%CV) for estimates of survival and breeding probabilities for breeding and non-breeding individuals for different MSMR modeling scenarios using an expected-values approach in M-SURGE where data were generated in the presence of Markovian temporary emigration and analyzed with a model that ignored temporary emigration

Reference parameter values						S^B		S^N		ψ^{BB}		ψ^{NB}	
p^B	p^N	S^B	S^N	ψ^{BB}	ψ^{NB}	Bias	%CV	Bias	%CV	Bias	%CV	Bias	%CV
0.9	0.7	0.6	0.9	0.267[1]	0.6[1]	−0.01	0.5	0.00	0.3	−0.02	1.0	−0.07	0.6
0.3	0.1	0.6	0.9	0.267[1]	0.6[1]	−0.03	3.0	−0.06	2.8	−0.10	17.3	−0.25	17.3
0.9	0.7	0.2	0.3	0.267[1]	0.6[1]	−0.01	2.2	0.03	2.4	−0.02	2.6	−0.11	2.7
0.3	0.1	0.2	0.3	0.267[1]	0.6[1]	−0.02	15.4	−0.04	6.0	−0.18	15.7	−0.40	14.9
0.9	0.7	0.9	0.9	0.267[1]	0.6[1]	−0.01	0.3	0.01	0.2	−0.02	0.8	−0.09	0.5
0.3	0.1	0.9	0.9	0.267[1]	0.6[1]	−0.02	2.1	0.01	1.2	−0.08	10.7	−0.21	11.5
0.9	0.7	0.3	0.3	0.267[1]	0.6[1]	−0.01	1.8	0.03	2.0	−0.02	2.2	−0.12	2.2
0.3	0.1	0.3	0.3	0.267[1]	0.6[1]	−0.02	12.9	0.00	7.2	−0.17	102.2	−0.40	102.5
0.9	0.7	0.6	0.9	0.4[2]	0.4[2]	0.09	0.4	0.10	0.0	0.35	0.6	0.12	1.2
0.3	0.1	0.6	0.9	0.4[2]	0.4[2]	0.13	1.1	0.10	0.0	0.43	2.2	0.17	4.2
0.9	0.7	0.2	0.3	0.4[2]	0.4[2]	0.03	2.4	0.13	3.1	0.33	2.9	0.06	5.6
0.3	0.1	0.2	0.3	0.4[2]	0.4[2]	0.05	6.2	0.13	13.2	0.43	7.0	0.10	16.1
0.9	0.7	0.9	0.9	0.4[2]	0.4[2]	0.00	0.2	0.04	0.3	0.32	0.6	0.11	1.3
0.3	0.1	0.9	0.9	0.4[2]	0.4[2]	0.00	0.7	0.07	1.8	0.39	2.4	0.14	4.4
0.9	0.7	0.3	0.3	0.4[2]	0.4[2]	0.01	1.4	0.10	2.6	0.28	2.6	0.02	4.8
0.3	0.1	0.3	0.3	0.4[2]	0.4[2]	NA[3]	NA	NA	NA	NA	NA	NA	NA

[1]Reproductive costs to fecundity present: probabilities for other transitions were: $\psi^{BN}= 0.333$, $\psi^{BU}= 0.4$; $\psi^{NN}= 0.3$, $\psi^{NU}= 0.1$.
[2]Reproductive costs to fecundity absent: probabilities for other transitions were: $\psi^{BN}= 0.2$, $\psi^{BU}= 0.4$; $\psi^{NN}= 0.5$, $\psi^{NU}= 0.1$.
[3]Estimates from the generating model did not converge to reference parameter values so no results reported for scenario.

from non-breeder to breeder are assumed to be 0 before a certain minimum age and all animals are assumed to begin breeding by a certain age), or parameters are constrained to be constant over time or to follow a temporal trend (Kendall and Nichols 2002; Kendall 2004). For some species Markovian temporary emigration is expected and partial determinism in transitions is not. For example, breeders in year *t* may be more likely to be absent from the breeding site in year *t+1* than are non-breeders in year *t*, but the organisms do not follow a set temporal pattern of breeding and non-breeding. In such cases, one is only left the option of making some parameters constant.

To illustrate some of the potential problems with this solution, I generated expected data for the following scenario. Survival and breeding probabilities varied by state and environmental conditions (S^B = 0.8 [good year] or 0.6 [bad year], S^N = 0.8 [all years], ψ^{BB} = 0.6 [good year] or 0.3 [bad year], and ψ^{NB}= 0.7 [good year] or 0.5 [bad year]). Temporary emigration was Markovian in bad years (ψ^{BU}= 0.1 [good year] or 0.5 [bad year], and ψ^{NU}= 0.1 [good year] or 0.3 [bad year]). The simulated study encompassed six breeding seasons, and conditions for survival and breeding probabilities were good, bad, good, bad, and good, respectively. Capture probabilities varied by occasion around different state-specific means (values of p^B had a mean of 0.7 and were drawn from a uniform distribution bounded between 0.6 and 0.8, whereas values of p^N had a mean of 0.5 and were drawn from a uniform distribution bounded between 0.4 and 0.6). This scenario was chosen to illustrate performance limitations of the multistate model with multiple observable states and a single unobservable state for a mildly challenging situation. Specifically, survival and detection rates were moderate to high, and temporary emigration was Markovian in only some years; under such circumstances, the multistate model as applied to a one observable and one unobservable state performs quite well (Schaub et al. 2004).

A single model was used to estimate parameters from the simulated data (S^B, S^N, p^B, and p^N allowed to vary by time; ψ^{BB} and ψ^{NB} constant through time). This model assumes that researchers were aware that parameters might vary by time but were not aware of the true underlying pattern of good and bad years and the effects of those changing conditions on parameters. Results from the model were evaluated for absolute bias and percent coefficient of variation, again with emphasis on the estimates of S^B, S^N, ψ^{BB}, and ψ^{NB}.

3 Results

3.1 *Effects of Ignoring Markovian Temporary Emigration when Estimating Reproductive Costs*

For time-constant reference parameters and Markovian temporary emigration, parameters in the models evaluated were separately identifiable, and estimates obtained from the generating model converged on reference parameters for all but one of the scenarios simulated (Table 1). However, convergence was typically not achieved unless repeated random initial values were used, and local minima were apparent in the output, especially when reference values of p^B and p^N were low. When a model that ignored temporary emigration (i.e., a model that included states B and N but not U) was employed, the resulting estimates of survival and breeding probabilities typically had low coefficients of variation (%CV $\leq$7% for 24 of 30 estimates) (Table 1). However, the level of bias varied quite substantially depending on the simulation scenario and ranged from strongly negative to strongly positive. When costs of reproduction to breeding probability were present, bias was low for survival estimates and higher for estimates of breeding probability (average % relative bias = −4.4, 0.2, −28.2, and −34.2 for estimates of S^B, S^N, ψ^{BB}, and ψ^{NB}, respectively), any bias tended to be negative, and absolute bias was always greatest for estimates of ψ^{NB}. In contrast, when costs of reproduction to breeding probability were absent, bias tended to increase (average % relative bias = 11.4, 22.2, 90.3, and 25.7 for estimates of S^B, S^N, ψ^{BB}, and ψ^{NB}, respectively), was always positive, and was always highest for estimates of ψ^{BB}.

Given the bias in estimates of S^B, S^N, ψ^{BB}, and ψ^{NB}, estimates of reproductive costs were also typically biased: sometimes at low levels and sometimes severely so (Table 2). Estimates of reproductive costs to survival were typically over-estimated, whereas estimated costs to breeding probability were always negatively biased. Absolute bias was always greater for estimates of costs to breeding probability. Of particular note, when true ψ^{BB} was equal to ψ^{NB}, estimates indicated that ψ^{BB} was greater than ψ^{NB} by 0.22–0.33 (absolute difference between the rates).

3.2 Difficulties of Estimating Time-Varying Reproductive Costs in Presence of Markovian Temporary Emigration

Results from the simulation with full time-varying parameters, which could be separately identified for the data and model evaluated, illustrate some of the limitations that can arise when one is forced to impose time constraints on some parameters. Even for a scenario with moderately high values for survival rate (>0.6) and capture probability (>0.4), random and low values of temporary emigration in good years (0.1), and Markovian temporary emigration only in bad years, the time-varying multistate model that considered states B, N, and U was not able to fully characterize several key features of the reference parameters. As this is but one possible scenario out of a broad set of circumstances that could be considered, the results are kept brief and only a few highlights are mentioned.

First, the time constraints that were imposed on breeding probabilities, made it impossible to identify the large changes in breeding probabilities between good and bad years, which simply emphasizes a known problem when considering an unobservable state in models with time-varying parameters. Second, in one of the good years (reference values: $S^B = S^N$), time-varying estimates of survival rate were higher for breeders than for non-breeders (estimated difference = 0.04). The model did perform reasonably well in some respects, however. Time-invariant estimates were

Table 2 Absolute bias in estimates of reproductive costs to survival and breeding probability estimated for breeding and non-breeding individuals for different MSMR modeling scenarios using an expected-values approach in M-SURGE where data were generated in the presence of Markovian temporary emigration and analyzed with a model that ignored temporary emigration

Reference parameter values						Costs of reproduction to survival probability			Costs of reproduction to breeding probability		
p^B	p^N	S^B	S^N	ψ^{BB}	ψ^{NB}	Actual	Estimate	Bias	Actual	Estimate	Bias
0.9	0.7	0.6	0.9	0.267[1]	0.6[1]	0.30	0.31	0.01	0.33	0.28	−0.05
0.3	0.1	0.6	0.9	0.267[1]	0.6[1]	0.30	0.27	−0.03	0.33	0.19	−0.14
0.9	0.7	0.2	0.3	0.267[1]	0.6[1]	0.10	0.14	0.04	0.33	0.24	−0.09
0.3	0.1	0.2	0.3	0.267[1]	0.6[1]	0.10	0.09	−0.01	0.33	0.11	−0.22
0.9	0.7	0.9	0.9	0.267[1]	0.6[1]	0.00	0.02	0.02	0.33	0.27	−0.06
0.3	0.1	0.9	0.9	0.267[1]	0.6[1]	0.00	0.02	0.02	0.33	0.20	−0.13
0.9	0.7	0.3	0.3	0.267[1]	0.6[1]	0.00	0.04	0.04	0.33	0.24	−0.09
0.3	0.1	0.3	0.3	0.267[1]	0.6[1]	0.00	0.02	0.02	0.33	0.10	−0.23
0.9	0.7	0.6	0.9	0.4[2]	0.4[2]	0.30	0.31	0.01	0.00	−0.22	−0.22
0.3	0.1	0.6	0.9	0.4[2]	0.4[2]	0.30	0.27	−0.03	0.00	−0.26	−0.26
0.9	0.7	0.2	0.3	0.4[2]	0.4[2]	0.10	0.20	0.10	0.00	−0.26	−0.26
0.3	0.1	0.2	0.3	0.4[2]	0.4[2]	0.10	0.18	0.08	0.00	−0.33	−0.33
0.9	0.7	0.9	0.9	0.4[2]	0.4[2]	0.00	0.04	0.04	0.00	−0.22	−0.22
0.3	0.1	0.9	0.9	0.4[2]	0.4[2]	0.00	0.07	0.07	0.00	−0.25	−0.25
0.9	0.7	0.3	0.3	0.4[2]	0.4[2]	0.00	0.09	0.09	0.00	−0.27	−0.27
0.3	0.1	0.3	0.3	0.4[2]	0.4[2]	NA[3]	NA	NA	NA	NA	NA

[1]Reproductive costs to fecundity present: probabilities for other transitions were: $\psi^{BN} = 0.333$, $\psi^{BU} = 0.4$; $\psi^{NN} = 0.3$, $\psi^{NU} = 0.1$.
[2]Reproductive costs to fecundity absent probabilities for other transitions were: $\psi^{BN} = 0.2$, $\psi^{BU} = 0.4$; $\psi^{NN} = 0.5$, $\psi^{NU} = 0.1$.
[3]Estimates from the generating model did not converge to reference parameter values so no results reported for scenario.

quite accurate for ψ^{BB} (estimate = 0.46, true average = 0.48) and ψ^{NB} (estimate = 0.62, true average = 0.62) and accurately estimated the average cost of reproduction to breeding probability over the study as being 0.14. Finally, the model typically produced estimates of survival rates with low bias (range in absolute bias: S^{B} = −0.01 to −0.02; S^{N} = −0.07 to +0.02) such that estimates of costs of reproduction to survival had low bias in all but one year.

4 Discussion

For the special application of multistate modeling that was investigated here, i.e., comparing survival and breeding probabilities between breeders and non-breeders, the results obtained make it clear that estimated costs of reproduction, especially costs of reproduction to breeding probability, can be badly biased when temporary emigration occurs but is not properly accounted for. For some scenarios evaluated, estimates of costs of reproduction were only slightly biased. However, for other situations, biased estimates of costs of reproduction to both survival and breeding probability could lead to misleading conclusions. For example, when reproductive costs to breeding probability were absent, and Markovian temporary emigration was present but ignored, estimates of breeding probability were biased high for both breeders and non-breeders, and the bias was greater for breeders. Thus, it appeared that breeding probability was higher for breeders than non-breeders. Such a result is of great interest in studies of life-history evolution as it relates to important questions regarding effects of heterogeneity in individual quality (e.g., Cam et al. 1998; Wintrebert et al. 2005). It would be useful to have similar evaluations regarding permanent emigration to aid decision-making for investigations in which breeders, non-breeders, or both might permanently leave the study area.

Given the importance of studying reproductive costs and the potential for biased results if temporary emigration is not properly incorporated into analyses, it is important to consider the available analysis options. In some situations, researchers might choose to use multistate models without an unobservable state because temporary emigration is thought to be a non-issue in their study. In such cases, it would be helpful if authors would provide justification for ignoring possible temporary emigration in analyses based on the biology, the sampling situation, and any available supporting data. It may be less clear how best to proceed in some studies because (1) less may be known about possible levels of temporary emigration and whether or not it may be Markovian and (2) there is currently no specific goodness-of-fit test for detecting temporary emigration. For such studies, it might be useful to compare results obtained from analyses conducted with and without an observable state and to then consider how best to proceed (e.g., Hadley et al. 2007). For still others, it may be known that temporary emigration occurs and an unobservable state may be incorporated in analyses (e.g., Beauplet et al. 2006).

When an unobservable state is included in analyses, several options are available. The simplest approach is to include the unobservable state in the model (with zero capture probability) using recently described methods (Lebreton et al. 1999; Pradel and Lebreton 1999; Kendall and Nichols 2002; Kendall 2004; Schaub et al. 2004) and readily available software (White and Burnham 1999; Choquet et al. 2004). When an unobservable state is considered, numerical and statistical problems inherent to multistate models are increased, and it will be important to use recently developed software to help ensure convergence (Lebreton and Pradel 2002; Choquet et al. 2004). As was found in the simulation work reported on here, convergence was not always easy to achieve even when working with the generating model or a close approximation and knowing the values of reference parameters. With the multistate approach, parameter identifiability will be a problem for some models and datasets and be important to evaluate (Gimenez et al. 2003, 2004; Kéry et al. 2005).

When deciding whether to address temporary emigration by simply including an unobservable state, researchers should also carefully consider whether or not the parameter redundancy problems and limitations of the approach will prevent them from asking key questions of interest. As thoroughly explained by Kendall (2004) and shown here with a simple simulation with time-varying reference parameters, the approach will be inadequate if hypotheses of interest involve time-varying reproductive costs and other solutions to handling parameter redundancy (such as partial determinism in state transitions) are inappropriate. As recently discussed by Tavecchia et al. (2005), there are excellent reasons to be interested in time variation in reproductive costs because trade-offs may vary in stochastic environments, thus affecting optimal reproductive strategies.

The results presented here further emphasize previous recommendations to properly evaluate the magnitude and nature of temporary emigration in mark-recapture studies (e.g., Kendall et al. 1997; Fujiwara and Caswell 2002; Kendall and Nichols 2002; Kendall 2004; Schaub et al. 2004). Also, as had been shown previously for the case of a single observable and a single unobservable state (Schaub et al. 2004), it is possible to obtain relatively unbiased estimates of survival even when Markovian temporary emigration occurs if capture probabilities and survival rates are high.

Kendall (2004) and White et al. (2006) provide excellent reviews of options for gaining greater flexibility in multistate modeling in the presence of an unobservable state. The key idea is that various types of additional information can be used to make additional parameters estimable. If sub-sampling is done within each breeding season, the robust design (Pollock 1982) can be used, which allows estimation of time-varying probabilities (Kendall 2004). This approach still requires the assumption that the survival probability for individuals in the unobservable state is equal to that for one of the observable states. For the case of breeder and non-breeder states, it may be reasonable in some situations to assume that unobservable individuals are non-breeders and therefore, have the same survival rate as do observable non-breeders. However, in other studies such an assumption may be unreasonable or a question of biological interest. In such cases, further additional information will be needed. Simply put, it will be necessary to sample unobservable individuals through radio telemetry, ring recoveries, or combinations of approaches. Examples of innovative approaches for estimating temporary emigration and combining sources of information now exist (e.g., Lindberg et al. 2001; Bailey et al. 2004; Barker et al. 2005) and should be valuable to future studies of reproductive costs.

An excellent set of analysis options and software now exist for estimating temporary emigration. With thoughtful study design and analysis, future comparisons of survival and breeding probabilities for breeders and non-breeders should provide valuable information regarding reproductive costs regardless of Markovian temporary emigration. In fact, estimates of temporary emigration rates may provide insights into the strategies used to avoid incurring reproductive costs. In planning future studies or choosing among analysis options, simulations, which can now be readily conducted in software such as M-SURGE (Choquet et al. 2004) and MARK (White and Burnham 1999), should prove useful (Schaub et al. 2004; Devineau et al. 2006).

Acknowledgments I thank James Nichols and William Kendall for helpful discussions on the topic, and Rémi Choquet for assistance with conducting simulations in M-SURGE. I also thank Brett Sandercock and an anonymous reviewer for helpful comments on an earlier version of the manuscript. This work was supported by the National Science Foundation, Division of Polar Programs (grant no. OPP-0225110 to R. A. Garrott, J. J. Rotella and D. B. Siniff).

References

Arnason AN (1972) Parameter estimates from mark–recapture experiments on two populations subject to migration and death. Researches on Population Ecology 13:97–113.

Arnason AN (1973) The estimation of population size, migration rates, and survival in a stratified population. Researches in Population Ecology 15:1–8.

Bailey LL, Kendall WL, Church DR, Wilbur HM (2004) Estimating survival and breeding probability for pond–breeding amphibians: a modified robust design. Ecology 85:2456–2466.

Barbraud C, Weimerskirch B (2005) Environmental conditions and breeding experience affect costs of reproduction in blue petrels. Ecology 86:682–692.

Beauplet G, Barbraud C, Dabin W, Küssener C, Guinet C (2006) Age-specific survival and reproductive performances in fur seals: evidence of senescence and individual quality. Oikos 112:430–441.

Barker RJ, White GC, McDougall M (2005) Movement of paradise shelduck between molt sites: a joing multistate-dead recovery mark-recapture model. Journal of Wildlife Management 69:1194–1201.

Boulinier T, Sorci G, Clobert J, Danchin E (1997) An experimental study of the costs of reproduction in the Kittiwake *Rissa tridactyla:* comment. Ecology 78:1284–1287.

Brownie C, Hines JE, Nichols JD, Pollock KH, Hestbeck JB (1993) Capture–recapture studies for multiple strata including non–Markovian transitions. Biometrics 49:1173–1187.

Cam E, Hines JE, Monnat JY, Nichols JD, Danchin E (1998) Are adult nonbreeders prudent parents? The Kittiwake model. Ecology 79:2917–2930.

Choquet R, Reboulet A-M, Pradel R, Gimenez O, Lebreton J–D (2004) M–SURGE: an integrated software for multi state recapture models. Animal Biodiversity and Conservation 27.1: 207–215.

Choquet R, Reboulet A-M, Pradel R, Gimenez O, Lebreton J–D (2005) M-Surge 1.7 user's manual. Centre d'Ecologie Fonctionnelle et Evolutive, Montpellier, France.

Clobert J (1995) Capture–recapture and evolutionary biology: a difficult wedding. Journal of Applied Statistics 22:989–1008.

Cormack RM (1964) Estimates of survival from the sighting of marked animals. Biometrika 51:429–438.

Darroch JN (1961) The two-sample capture–recapture census when tagging and sampling are stratified. Biometrika 48: 241–260.

Devineau O, Choquet R, Lebreton J-D (2006) Planning capture–recapture studies: straightforward precision, bias, and power calculations. Wildlife Society Bulletin 34:1028–1035.

Doligez B, Clobert J, Pettifor RA, Rowcliffe M, Gustafsson L, Perrins CM, McCleery RH (2002) Costs of reproduction: assessing responses to brood size manipulation on life-history and behavioural traits using multi-state capture–recapture models. Journal of Applied Statistics 29:407–423.

Fujiwara M, Caswell H (2002) Temporary emigration in mark-recapture analysis. Ecology 83: 3266–3275.

Gimenez O, Choquet R, Lebreton J-D (2003) Parameter redundancy in multistate capture–recapture models. Biometrical Journal 45:704–722.

Gimenez O, Viallefont A, Catchpole EA, Choquet R, Morgan BJT (2004) Methods for investigating parameter redundancy. Animal Biodiversity and Conservation 27:561–572.

Golet GH, Schmutz JA, Irons DB, Estes JA (2005) Determinants of reproductive costs in the long-lived Black-legged Kittiwake: a multiyear experiment. Ecological Monographs 74:353–372.

Hadley GL, Rotella JJ, Garrott RA (2007) Evaluation of reproductive costs for Weddell Seals in Erebus Bay, Antarctica. Journal of Animal Ecology 76:448–458.

Jolly GM (1965) Explicit estimates from capture–recapture data with both death and immigration-stochastic model. Biometrika 52:225–247.

Kendall WL (2004) Coping with unobservable and mis-classified states in capture–recapture studies. Animal Biodiversity and Conservation 27:97–107.

Kendall WL, Nichols JD (2002) Estimating state-transition probabilities for unobservable states using capture–recapture/resighting data. Ecology 83:3276–3284.

Kendall WL, Nichols JD, Hines JE (1997) Estimating temporary emigration using capture–recapture data with Pollock's robust design. Ecology 78:563–578.

Kéry M, Gregg KB, Schaub M (2005) Demographic estimation methods for plants with unobservable life-states. Oikos 108:2–307.

Lebreton J-D, Almeras T, Pradel R (1999) Competing events, mixtures of information and multi stratum recapture models. Bird Study 46:S39–S46.

Lebreton J-D, Hines JE, Pradel R, Nichols JD, Spendelow JA (2003) Estimation by capture–recapture of recruitment and dispersal over several sites. Oikos 101:253–264.

Lebreton J-D, Pradel R (2002) Multistate recapture models: modeling incomplete individual histories. In: Morgan BJT and Thomson DL (eds) Statistical analysis of data from marked bird populations. Journal of Applied Statistics 29:353–369.

Lindberg MS, Kendall WL, Hines JE, Anderson MG (2001) Combining band recovery data and Pollock's robust design to model temporary and permanent emigration. Biometrics 57:73–282.

McElligot AG, Altwegg R, Hayden TJ (2002) Age-specific survival and reproductive probabilities: evidence for senescence in male fallow deer (*Dama dama*). Proceeding of the Royal Society of London Series B 269:1129–1137.

Nichols JD, Hines JE, Pollock KH, Hinz RL, Link WA (1994) Estimating breeding proportions and testing hypotheses about costs of reproduction with capture–recapture data. Ecology 75: 2052–2065.

Nichols JD, Kendall WL (1995) The use of multi-state capture–recapture models to address questions in evolutionary ecology. Journal of Applied Statistics 22: 835–846.

Pollock KH (1982) A capture–recapture design robust to unequal probability of capture. Journal of Wildlife Management 46:757–760.

Pradel R, Lebreton J-D (1999) Comparison of different approaches to study the local recruitment of breeders. Bird Study 46:S74–S81.

Reed ET, Gauthier G, Pradel R, Lebreton J-D (2003) Age and environmental conditions affect recruitment in greater snow geese. Ecology 84:219–230.

Reznick D (1985) Costs of reproduction: an evaluation of the empirical evidence. Oikos 44: 257–267.

Reznick D (1992) Measuring the costs of reproduction. Trends in Ecology and Evolution 7:42–45.

Rivalan P, Prevot-Julliard AC, Choquet R, Pradel R, Jacquemin B, Girondot M (2005) Trade-off between current reproductive effort and delay to next reproduction in the leatherback sea turtle. Oecologia 145:564–574.

Roff DA (1992) The evolution of life histories: theory and analysis. Chapman & Hall, New York.

Sandercock BK (2006) Estimation of demographic parameters from live-encounter data: a summary review. Journal of Wildlife Management 70:1504–1520.

Sandercock BK, Beissinger SR, Stoleson SH, Melland RR, Hughes CR (2000) Survival rates of a neotropical parrot: implications for latitudinal comparisons of avian demography. Ecology 81:1351–1370.

Schaub M, Gimenez O, Schmidt BR, Pradel R (2004) Estimating survival and temporary emigration in the multistate capture–recapture framework. Ecology 85: 2107–2113

Schwarz CJ, Schweigert JF, Arnason AN (1993) Estimating migration rates using tag-recovery data. Biometrics 49:177–193.

Seber GAF (1965) A note on the multiple-recapture census. Biometrika 52:249–259.

Stearns SC (1992) The evolution of life histories. Oxford University Press, New York.

Tavecchia G, Coulson T, Morgan BJT, Pemberton JM, Pilkington JC, Gulland FMD, Clutton-Brock TH (2005) Predictors of reproductive cost in female Soay sheep. Journal of Animal Ecology 74:201–213.

Viallefont A, Cooch EG, Cooke F (1995) Estimation of trade-offs with capture– recapture models: a case study on the lesser snow goose. Journal of Applied Statistics 22: 847–862.

White GC, Burnham KP (1999) Program MARK for survival estimation. Bird Study 46: S120–S139.

White GC, Kendall WL, Barker RJ (2006) Multistate survival models and their extensions in Program MARK. Journal of Wildlife Management 70:1521–1529.

Williams GC (1966) Natural selection, the cost of reproduction and a refinement of Lack's principle./American Naturalist 100:687–690.

Williams BK, Nichols JD, Conroy MJ (2002) Analysis and management of animal populations. Academic Press, New York, USA.

Wintrebert CMA, Zwinderman AH, Cam E, Pradel R, van Houwelingen JC (2005) Joint modelling of breeding and survival in the kittiwake using frailty models. Ecological Modelling 181: 203–213

Yoccoz NG, Erikstad KE, Bustnes JO, Hanssen SA, Tveraa T (2002) Costs of reproduction in common eiders (*Somateria mollissima*): an assessment of relationships between reproductive effort and future survival and reproduction based on observational and experimental studies. Journal of Applied Statistics 29:57–64.

Estimating Latent Time of Maturation and Survival Costs of Reproduction in Continuous Time from Capture–Recapture Data

Torbjørn Ergon, Nigel G. Yoccoz, and James D. Nichols

Abstract In many species, age or time of maturation and survival costs of reproduction may vary substantially within and among populations. We present a capture-mark-recapture model to estimate the latent individual trait distribution of time of maturation (or other irreversible transitions) as well as survival differences associated with the two states (representing costs of reproduction). Maturation can take place at any point in continuous time, and mortality hazard rates for each reproductive state may vary according to continuous functions over time. Although we explicitly model individual heterogeneity in age/time of maturation, we make the simplifying assumption that death hazard rates do not vary among individuals within groups of animals. However, the estimates of the maturation distribution are fairly robust against individual heterogeneity in survival as long as there is no individual level correlation between mortality hazards and latent time of maturation. We apply the model to biweekly capture–recapture data of overwintering field voles (*Microtus agrestis*) in cyclically fluctuating populations to estimate time of maturation and survival costs of reproduction. Results show that onset of seasonal reproduction is particularly late and survival costs of reproduction are particularly large in declining populations.

Keywords Capture-mark-recapture · Latent traits · Life-history theory · Maximum likelihood · Multi-state/multi-strata · Continuous time · Hazard rates · Heterogeneity · Maturation · Cost of reproduction · Disease infection dynamics

1 Introduction

Age and timing of reproduction are important components of fitness and population dynamics. Causes and consequences of variation in age at first reproduction have for long been central topics in ecological and evolutionary theory (e.g., Cole 1954;

T. Ergon (✉)
Centre for Ecological and Evolutionary Synthesis, Department of Biology (now at Program for Integrative Biology), University of Oslo, P. O. Box 1066, Blindern, N-0316 Oslo, Norway
e-mail: torbjorn.ergon@bio.uio.no

D.L. Thomson et al. (eds.), *Modeling Demographic Processes in Marked Populations*, Environmental and Ecological Statistics 3, DOI 10.1007/978-0-387-78151-8_8,

Stearns 1992; Gaillard et al. 2005), but also timing of reproduction within seasons is of prime interest. In multivoltine species, such as many small mammals, early onset of seasonal reproduction enables more generations to be completed within the season (Fairbairn 1977; Lambin and Yoccoz 2001; Ergon 2007), and reproductive success and viability of offspring in univoltine birds and mammals are often strongly related to the date of reproduction within the season (Lack 1996; Clutton-Brock et al. 1987; Hochachka 1990; Winkler and Allan 1996). The latter is particularly relevant for match-mismatch theory and changing phenologies due to climate change (Visser et al. 1998; Visser and Holleman 2001; Stenseth et al. 2002; Both et al. 2006).

It is of general interest to characterize the full distribution of age or time of reproduction in a population, and not just the mean value. The variances of trait values are central for evolutionary theory, and life-history traits such as age and timing of reproduction may evolve in response to differential selection of different parts of the trait distributions. Furthermore, to study response to selection or trade-offs between different life-history traits, it is the *latent* trait values that are of interest (i.e., the propensities existing at birth, e.g. Link et al. 2002). The distribution of latent maturation times is the distribution that potentially could be observed if, hypothetically, all individuals survived until maturation. In contrast, the realized distribution applies only to the subset of the population that survives until maturation, and is hence filtered by mortality. In a sense, the latent distribution is the true trait distribution that is independent of survival. For example, a group of individuals in a given environment may be characterized with a latent distribution for time of maturation, but the individual trait value is not observable for all members of the group because some individuals may die before they mature. Because individuals with a late latent maturation time will have a higher probability of dying before maturation (unless individuals with a late latent maturation also survive better), the distribution of the realized maturation times will be biased downwards compared to the distribution of the latent trait values – the higher the mortality rate is, the stronger this bias will be. This type of 'right censoring' is commonplace in, e.g., medicine where some study subjects may die or be removed from the study before an effect is observed (e.g. Cox 1972). In such cases, statistical modeling must take into account that observations of reproductive events might be censored due to the death of individuals; we might know that a given individual has not reproduced up to a given age or time, but death prevented recording the actual age or time of reproduction. Common models incorporating such censoring or competing risks (e.g. Pintilie 2006) are however not readily applicable for wildlife studies due to the usual inability in such studies to detect all breeding and non-breeding individuals alive in the population, and it is not possible to know the exact time that an individual matured or was censored out of the population due to natural mortality. When detection probability is less than one, we do not even know *if* an animal was censored.

As with other important life history characteristics, such as survival rate, capture–recapture modeling provides a natural means of estimating quantities of interest in the presence of probabilities of capture or detection less than one. Our goal in this paper is to include both processes relating to capture probability and censoring due to natural mortality in order to provide unbiased and efficient estimates of the latent distribution of age or time at reproduction. As animals may have different survival rates before and after reproduction, we can also use the modeling to estimate costs of reproduction even when reproduction takes place at unknown points in continuous time. The model can also be applied to estimate the latent timing of other irreversible state transitions and the survival costs associated with these transitions. It may for example be applied in capture–recapture studies of disease or parasitism dynamics to obtain unbiased estimates of the underlying infection rates and survival costs of infection (see Discussion).

Although we explicitly model individual heterogeneity in the latent time of state transition, we make the simplifying assumption that mortality hazard rates do not vary among individuals belonging to a given group. The implications of this assumption are investigated in a simulation study. Finally, we apply the model to a case study on life-history correlations in a cyclic population of field voles.

2 The Model

We primarily consider a study design in which there are many sampling occasions within a reproductive season and where state transition can take place at any point in continuous time, not just at given points between sampling occasions. For example, a population of individually marked animals may be monitored with repeated sampling occasions from winter, when no individuals are reproducing, through spring until most individuals have either initiated reproduction or died (see the Case Study below). In principle, the data could also cover several years as long as latent time of state transition can be described with a probability distribution.

The aim of the modeling is to characterize the distribution of latent times of state transition in a population. This distribution may, for example, be represented by a normal distribution function with a population mean and variance describing individual heterogeneity in the population. The distribution can also be made discrete, for example at yearly intervals, to estimate the latent age at first time reproduction in long-lived species (e.g. Cam et al. 2005). An alternative perspective is to focus on the transition hazard rates that correspond to the distribution of latent times of state transition. In capture–recapture studies of disease or parasitism dynamics, our modeling will then yield estimates of infection rates and lethality that are not biased by background mortality (see Discussion).

The basic structure of our model is based on a special case of a multi-state capture–recapture model (Arnason 1972, 1973; Hestbeck et al. 1991; Brownie et al. 1993; Schwarz et al. 1993) in which there are only two states and transitions are allowed in only one direction, from the immature to mature state. Such models have been much used to address questions in evolutionary ecology and in other applications where animals can change state or location (see Cam (2008), Conroy (2008) and Rotella (2008) of this volume). In these multi-state models, transition probabilities are defined as the joint probability of surviving and moving from one state to another over a time interval, $\Phi = \Pr(\text{'Survive'} \cap \text{'Move'})$. These transition probabilities can be partitioned into survival probabilities and movement probabilities conditional on survival, $\Phi = \Pr(\text{'Survive'}) \times \Pr(\text{'Move'}|\text{'Survive'})$. To facilitate estimation, it is usually assumed that movement takes place at a given point in time within the intervals, usually at the very end of the intervals so that the survival probabilities are independent of the terminal state of the interval (Hestbeck et al. 1991; Brownie et al. 1993). While this assumption may be reasonable in some applications, it is clearly not in others. Joe and Pollock (2002) relaxed this assumption by treating the time of movement within intervals as a random variable with a given distribution, $g(t)$. Thus, denoting survival rate before movement in interval i as S_i^A and survival rate after movement as S_i^B, they obtained

$$\Phi_i = \int_0^1 (S_i^A)^t \psi_i (S_i^B)^{(1-t)} g(t)dt$$

where ψ_i is the probability of movement given survival (Joe and Pollock 2002, Eq. 1). Note that the probability distribution $g(t)$ is conditional on knowing that state transition took place during the given interval. Hence, $\int_0^1 g(t)dt = 1$ (e.g., $g(t)$ is a uniform or a beta distribution).

Our approach is similar to the approach of Joe and Pollock (2002). However, both the movement probabilities and the probability distributions for times of state transition for each interval depend on a distribution for latent times of state transition, $f(t)$, which span the entire study period. The parameters of $f(t)$ are included in the parameter vector estimated by maximum likelihood. In this way we are able to partition the overall transition probability ($\Phi = \Pr(\text{'Survive'} \cap \text{'Mature'})$) into the probability that latent maturation is located within the interval unconditional on whether the individual survives the interval or not (Pr('Mature')) and survival conditional on maturation (Pr('Survive'|'Mature')), rather than survival (Pr('Survive')) and maturation conditional on survival (Pr('Mature'|'Survive')). We also model survival probabilities by the use of continuous mortality hazard functions over time or age. This allows easy implementation of proportional hazard models and models addressing questions related to longevity and senescence (see Gaillard et al. (2004)).

Our model is formulated in detail below. Sections 2.1 and 2.2 introduce the multi-state m-arrays (e.g. Brownie et al. 1993) and apply generally for multi-state models with two states where transitions can only take place in one direction – these sections are provided for completeness. Section 2.3 describes our modeling of the cell probabilities corresponding to the m-array, and the rest of Section 2 is devoted to discussing maturation and mortality hazard functions and to describing the computer implementation and numerical optimization methods that we have used. In our modeling, we make the simplifying assumption that mortality hazard rates do not vary among individuals belonging to a given group. The implications of this assumption are investigated in a simulation study presented in Section 3. Finally, we apply the model to a case study on life-history correlations in a cyclic population of field voles (Section 4), before moving on to a general Discussion about the modeling (Section 5).

2.1 Data Structure

The data are assumed to come from a capture-recapture study (Lebreton et al. 1992; Williams et al. 2002) of individually marked animals where it is recorded (without error) whether the individuals are immature (state A) or mature (state B) every time they are captured. Individuals may belong to one of several groups (e.g., sites or cohorts).

We assume data from each group of animals to be summarized in the standard multi-state m-array of Brownie et al. (1993), shown for four capture sessions in Table 1. Here, m_{ij}^{XY} is the number of the animals of state $X(A$ or $B)$ observed alive in session i that were *first* recaptured in session j when being in state $Y(A$ or $B)$, and $m_{i\dagger}^{X}$ is the number of the animals that were never recaptured. Note that state transition from mature (state B) to immature (state A) is not possible, and that index j is always greater than or equal to index $i+1$.

In addition, a vector with the times of the capture sessions, $\mathbf{t} = [t_1, t_2, \ldots, t_n]$, must be given for each group.

2.2 Likelihood Function

Each of the m_{ij}^{XY}-elements in the m-array (Table 1) has a corresponding cell probability, π_{ij}^{XY}, which is the probability that an individual of state X released in session i will be recaptured for the first time in state Y and session j;

$$\pi_{ij}^{XY} = \Pr(\text{first recaptured in state } Y \text{and session } j \mid \text{released in state } X \text{ and session } i).$$

The cell probabilities for each row in the above m-array sum to one. Hence, the cell probabilities corresponding to the last column of the array may be set to

Table 1 Multi-state m-array with 4 capture occasions and 2 states (A = 'immature', B = 'mature'). There is one m-array for each group of animals

		Session of first recapture (j) State of recapture						
Session of release (i)	State of release	2 A	2 B	3 A	3 B	4 A	4 B	never
1	A	m_{12}^{AA}	m_{12}^{AB}	m_{13}^{AA}	m_{13}^{AB}	m_{14}^{AA}	m_{14}^{AB}	$m_{1\dagger}^{A}$
1	B		m_{12}^{BB}		m_{13}^{BB}		m_{14}^{BB}	$m_{1\dagger}^{B}$
2	A			m_{23}^{AA}	m_{23}^{AB}	m_{24}^{AA}	m_{24}^{AB}	$m_{2\dagger}^{A}$
2	B				m_{23}^{BB}		m_{24}^{BB}	$m_{2\dagger}^{B}$
3	A					m_{34}^{AA}	m_{34}^{AB}	$m_{3\dagger}^{A}$
3	B						m_{34}^{BB}	$m_{3\dagger}^{B}$

one minus the other cell probabilities of the row; $\pi_{i\dagger}^{A} = 1 - \sum_j (\pi_{ij}^{AA} + \pi_{ij}^{AB})$ and $\pi_{i\dagger}^{B} = 1 - \sum_j \pi_{ij}^{BB}$.

Assuming independence among individuals, the elements of each row of the m-arrays follow a multinomial distribution, and the likelihood function to be maximized is a product of multinomials. Since the multinomial coefficients are independent of the cell probabilities, it is sufficient to maximize the log-likelihood function

$$\ell(\boldsymbol{\pi}; \mathbf{m}) = \mathbf{m}' \log(\boldsymbol{\pi}),$$

where vectors $\mathbf{m}$ and $\boldsymbol{\pi}$ contain the elements of the m-arrays and the corresponding cell-probabilitics.

2.3 Cell Probabilities

The description so far applies also for previous formulations of multi-state models in the special case where there are only two states (A and B) and where transitions are only allowed in one direction (from A to B). However, our model differs in the way the constraints on the cell probabilities are parameterized. We start by defining a continuous probability density function $f(t)$ for the latent distribution for time of maturation. This function describes the trait distribution among all animals in the study population that are ever seen as immature. The function *does not* only apply for individuals that survive until maturation – it is independent of survival. Thus, the probability of having a latent time of maturation (trait value) before time t_i is $\int_{-\infty}^{t_i} f(t)dt$, and the probability of having a latent time of maturation in the interval t_i to t_j given that the individual has not matured before t_i is

$$\alpha_{ij} = \frac{\int_{t_i}^{t_j} f(t)dt}{1 - \int_{-\infty}^{t_i} f(t)dt}. \tag{1}$$

Note that this probability is not conditional on the individual surviving the interval, or even surviving until maturation. In this way we can partition the probabilities of surviving an interval and at the same time maturing (Pr('Survive' $\cap$ 'Mature')) into products of the probability of having the latent time of maturation within the interval (α = Pr('Mature')) and the probability of surviving given that the animal matured during the interval (ω = Pr('Survive'|'Mature')). The latter probability is not conditional on maturation taking place at a given point in time within the interval, but is instead found by integrating over the probability distribution for time of maturation determined by $f(t)$.

Since we only deal with state transitions in one direction between two states, we do not use the matrix formulation used by Brownie et al. (1993) and by Joe and Pollock (2002). Instead we incorporate state dependent capture probabilities into the transition probabilities by the use of logical index functions.

To write the cell probabilities of the m-array, π_{ij}^{XY}, we introduce the following notation:

$\phi_{s_1 \to s_2}^X$ = Probability of surviving from time s_1 to time s_2 given that the individual was alive at time s_1 and in state X during the entire time-interval.
ϕ_{ij}^X = Short-hand for $\phi_{t_i \to t_j}^X$.
p_k^X = Probability of capture of an individual at session k given that it was alive and in state X.
α_{ij} = Probability of having latent time of maturation in the interval between session i and j given that the individual was immature at session i.
ω_{ij} = Probability of surviving *and* not being captured between session i and j given that the animal matured between session i and j.

The cell probabilities are then

$$\pi_{ij}^{AA} = (1 - \alpha_{ij})\phi_{ij}^A p_j^A \prod_{\forall k: t_i < t_k < t_j} (1 - p_k^A), \tag{2}$$

$$\pi_{ij}^{BB} = \phi_{ij}^B p_j^B \prod_{\forall k: t_i < t_k < t_j} (1 - p_k^B), \tag{3}$$

and

$$\pi_{ij}^{AB} = \alpha_{ij}\omega_{ij} p_j^B. \tag{4}$$

In these expressions, the conditions $\forall k : t_i < t_k < t_j$ indicate that the product should involve all capture probabilities representing occasions between occasion i and occasion j.

In the expression for π_{ij}^{AB}, both α_{ij} and ω_{ij}, relate to the probability density function $f(t)$ for the latent time of maturation T, given by $f(t)dt = \Pr(t < T < t + dt)$. The expression for α_{ij} is given in Eq. 1 above, while $\omega_{ij}$is found by integrating the *joint* probability of surviving and not being captured during the interval, conditional on time of maturation ($T = t$), over the probability distribution for T given that $t_i < T < t_j$,

$$\omega_{ij} = \int_{t_i}^{t_j} \phi_{t_i \to t}^A \phi_{t \to t_j}^B \prod_{\forall k: t_i < t_k < t_j} \left\{ (1 - p_k^A)^{I(t > t_k)} (1 - p_k^B)^{I(t \le t_k)} \right\} \frac{f(t)}{\int_{t_i}^{t_j} f(t)dt} dt, \tag{5}$$

where the function I('expression')is a logical function that takes the value 1 if the argument is true, and 0 if it is false, and where $\frac{f(t)}{\int_{t_i}^{t_j} f(t)dt}$ is the probability density function for time of maturation given that maturation occurred between t_i and t_j.

Hence, by expansion of Eq. 4 we get

$$\pi_{ij}^{AB} = \frac{1}{1 - \int_{-\infty}^{t_i} f(t)dt} p_j^B \int_{t_i}^{t_j} \phi_{t_i \to t}^A \phi_{t \to t_j}^B \prod_{\forall k: t_i < t_k < t_j} \left\{ (1 - p_k^A)^{I(t > t_k)} (1 - p_k^B)^{I(t \le t_k)} \right\} f(t)dt. \quad (6)$$

The integral in the above expression cannot be solved analytically when using reasonable functions for survival and maturation (see below). Some numerical solutions are outlined in Section 2.5 below.

The functions $f(t)$, $\phi_{s_1 \to s_2}^A$ and $\phi_{s_1 \to s_2}^B$ depend on hyperparameters, and various functional forms are discussed below. The main aim of the modeling is to estimate these hyperparameters. In principle, it should also be possible to estimate an 'instantaneous' cost of reproduction by including such a parameter in ω_{ij} (Eq. 5). However, we do not consider such a model in this paper.

2.4 Maturation and Survival Functions

Several functional forms for the latent distribution of time of maturation (or, generally, any irreversible state transition), $f(t)$, and survival probabilities, $\phi_{s_1 \to s_2}^X$, are possible. The choice depends on the application. When modeling time of maturation or reproduction (e.g., time of first reproduction after the winter season), it may be reasonable to use a symmetric distribution for $f(t)$, such as the normal distribution or logistic distribution, but skewed, or even bimodal, distributions may also be applied. When modeling age (rather than time) of state transition, it may be necessary to use a probability distribution with a lower bound of zero. If the model is used to estimate the temporal distribution of individual infection by diseases or parasites (or the corresponding infection hazard rates), exponential, Weibull or gamma distributions may be natural candidates. In principle, the distribution can also be made discrete, for example at yearly intervals, to estimate the latent age at first time reproduction in long-lived species (e.g. Cam et al. 2005).

Continuous time variation in survival rates is best modeled by using mortality hazard functions, $h(t)$, defined as $h(t)dt = \Pr\,(t < \text{time of death} < t + dt|\text{time of death} > t)$. The probability of an individual of state X surviving from time s_1 to s_2 is then simply the exponent of the negative cumulative hazard during the time interval,

$$\phi_{s_1 \to s_2}^X = e^{-\int_{s_1}^{s_2} h^X(t)dt}.$$

The state-specific hazard rates $h^X(t)$ should be ≥ 0 for all values of t and the cumulative hazard rates should become infinite over infinite time (e.g. not of the form $e^{-\beta^2 t}$ which integrates to $1/\beta^2$ over $t = 0 \ldots \infty$).

When modeling hazard rates as a function of age, and if effects of environmental variability are negligible, it may be reasonable to assume hazard rates that increase or decrease monotonically with age. In such cases one may use a hazard function of the form $h(a_t) = \alpha\lambda a_t^{\alpha-1}$, where a_t is age at time t. This hazard function corresponds to survival function $\phi_{a_{s1} \to a_{s2}} = e^{-\lambda(a_{s2}^{\alpha} - a_{s1}^{\alpha})}$, and age at death will follow a Weibull distribution (exponential distribution in the special case when $\alpha = 1$ and the hazard rate is constant).

In other applications, survival will be largely influenced by random environmental events, as is the case for, e.g., small rodents that are heavily predated. One will then need a flexible hazard function that guarantees positive hazards, and it is desirable that the function can be integrated to find an analytical expression for the survival probabilities. In the Case Study on field voles below we used a hazard function composed as a sum of Gaussian curves parameterized as

$$h(t) = e^{a_1 - b_1^2(t-c_1)^2} + e^{a_2 - b_2^2(t-c_2)^2} + \cdots + e^{a_n - b_n^2(t-c_n)^2}. \tag{7}$$

Survival probabilities corresponding to this hazard function can be found analytically by the use of the erf-function of the cumulative normal. Note that $h(t)$ in this case does *not* integrate to infinity over infinite time. The function can nevertheless be used as a curve-fitting tool as long as extrapolations are not made. Proportional hazard rates among groups and states can be specified by using the same b- and c-parameters for different groups while constraining the a-parameters to additive differences among terms and groups (exponents of common components of the a-parameters can be factored out of the sum). One way to use this function is to fix the c-parameters to regular time-points over the study period (including the end points) and further constrain the b-parameters to be the same for all terms. Less constrained models can, in our experience, lead to strong correlations between the parameters and hence cause problems when fitting the model (this is especially the case when there is little temporal overlap in the presence of the two states in the population – see the Case Study below).

2.5 Implementation and Numerical Optimization

The model was implemented in Matlab® 7.1 in a way that the probability density functions for time of state-transition, $f(t)$, and mortality hazard functions, $h(t)$, can readily be replaced and where the state, group and occasion specific hyperparameters can be constrained by design matrices and link-functions. Optimization of the log-likelihood was done by using the quasi-Newton gradient method implemented in the 'fminunc' function of the Matlab Optimization Toolbox Version 3.0.3. The variance–covariance matrix of the maximum likelihood estimates was estimated as the inverse of the finite-difference approximation to the Hessian matrix at the optimum. The Matlab code be will shared upon request to the first author.

The integral in the expression for π_{ij}^{AB} in Eq. 6 cannot be solved analytically when using reasonable functions for survival and maturation. Several numerical integration (quadrature) methods are implemented in programming environments such as Matlab® and R (http://www.r-project.org/). The Matlab function 'quadl' performed well in our implementation. However, the optimization can sometimes be made more efficient by discretizing Eq. 6 directly. By dividing the interval between capture session i and j into M short sections (e.g. days) in such a way that none of the capture sessions are included within the sections (captures only at section borders) and breaking down ω_{ij} (Eq. 5) into a sum of probabilities conditional on maturation during each of the time sections, and further assuming that maturation can only take place at specific points within each time section (e.g. at a given time of the day), we get

$$\pi_{ij}^{AB} = \frac{1}{1-\int_{-\infty}^{t_i} f(t)dt} p_j^B \sum_{m=1}^{M} \left\{ \phi_{t_i \to \tau_m}^{A} \phi_{\tau_m \to t_j}^{B} \prod_{\forall k: t_i < t_k < \tau_m} \{(1-p_k^A)\} \prod_{\forall k: \tau_m \leq t_k < t_j} \{(1-p_k^B)\} \int_{t_m}^{t_{m+1}} f(t)dt \right\}, \tag{8}$$

where t_m and t_{m+1} are the beginning and the end points of time section m, and where $t_m \leq \tau_m \leq t_{m+1}$ (e.g., $\tau_m = (t_m + t_{m+1})/2$).

In the simulation study presented below we used a Normal distribution for the individual latent time of maturation, $f(t)$, and time-independent mortality hazard rates, $h(t)$ (implemented as a Weibull hazard and fixing the shape parameter, α, to 1). For the Case Study we used a logistic distribution for $f(t)$ and the sum of Gaussian terms described above (Eq. 7) for $h(t)$. The denominator in the expression of α_{ij} (Eq. 1) and in the expression of π_{ij}^{AB} (Eqs. 6 and 8) can potentially become numerically zero when the cumulative $f(t)$ approaches one (more likely to happen for the Normal distribution which has lighter tails than the logistic distribution). This problem can be avoided by constraining the upper limit of the cumulative to one minus a very small number (e.g., $1 - 1e^{-12}$).

The recapture probabilities, p's, can be state, group and occasion specific and were constrained to the $\langle 0, 1 \rangle$ interval by a logistic link-function. An identity link was used for the group specific mean latent maturation-time parameters, while the standard deviation of the maturation distribution was forced to be positive with a log-link (hence modeling multiplicative effects). The hazard function in Eq. 7 is always positive, but the b-parameters were still forced to be positive with a log-link to avoid symmetry around $b = 0$. A log-link was also used on the parameters of the Weibull mortality hazard function. The parameters of the mortality hazard functions can be state and group specific, and the hazard rates of the different states and groups can be constrained to be proportional (i.e., parallel over time on a log-scale).

3 Individual Heterogeneity and Simulation Results

In the formulation of this model we have specifically addressed individual heterogeneity in latent time of maturation (or any irreversible state transition). However, we have made the simplifying assumption that mortality hazard rates *within* groups do not vary among individuals. This is a potentially critical condition that may not be met. Individual heterogeneity is ubiquitous in animal populations due to variation in individual quality and the resources they possess (Łomnicki 1988). Furthermore, correlations in individual life-history traits exist due to life-history strategies and trade-offs (Roff 2002). For example, animals that reproduce early may face a higher mortality hazard due to the costs of accumulating resources, or, alternatively, animals in good condition may be able to both reproduce early and avoid mortality risks.

To investigate the assumption of individual homogeneity within groups we fitted the model to simulated data with variable degrees of individual heterogeneity in mortality hazard rates and correlations between individual mortality hazard and latent time of state transition. Individual combinations of log(mortality hazard) and time of maturation were drawn from multivariate normal distributions, and capture histories were simulated based on Bernoulli trials of survival and capture. We used 11 capture occasions at regular intervals along a time-line from -1 to 1. Capture probability was set to 0.5 for all animals at all occasions, mean time of maturation was 0, and the standard deviation in time of maturation was 0.2 . Median hazard rate in the immature stage was 1 (corresponding to an expected lifetime of 1), and the individual hazard rate doubled when the animal reached the mature stage. In different runs of the simulations we varied the standard deviation of log(hazard rate) from 0 to 1. A standard deviation in log-hazard of 1 means that the 97.5% quantile of the distribution is 50 times higher than the 2.5% quantile, which is a very substantial heterogeneity.

Results of simulations where individual hazard rate and latent time of maturation were uncorrelated are presented in Fig. 1. The bias in the estimators of mean ($\widehat{\mu}$) and standard deviation ($\widehat{\sigma}$) of the maturation distribution is very moderate. Even with the highest degree of heterogeneity, the bias in $\widehat{\sigma}$ is less than 2% and the bias in $\widehat{\mu}$ is less than 6% of the true value of σ. There is however a very substantial bias in the estimated hazard ratios when heterogeneity increases. When the standard deviation of log-hazard rate is one, the hazard in the mature stage relative to the immature stage is underestimated by 56% (the mature stage is estimated to have 11% *lower* hazard rate than the immature stage even though the data were simulated with a doubling of the hazard rate after maturity). This bias is due to the fact that individuals with a low survivability (high hazard rate) will have a higher probability of dying before they mature, and since the hazard estimates of mature animals are based on only individuals that are alive at maturity, the survival cost of reproduction for the total population will be underestimated.

There was also a slight upward bias in the estimates of mortality hazard rates at the immature stage. However, when data were simulated with no difference in the mortality hazard rates of the two stages and fitted with the corresponding model (not shown), there was a *downward* bias in the estimates of the overall hazard rate, and no detectable bias in $\widehat{\mu}$ and $\widehat{\sigma}$. The observation that the estimates of hazard rates *underestimate* even median hazard rate when adding heterogeneity may seem odd given that log-normal heterogeneity *increases* mean hazard rate relative to the median. However, log-normal heterogeneity in hazard rates will *increase* mean survival and

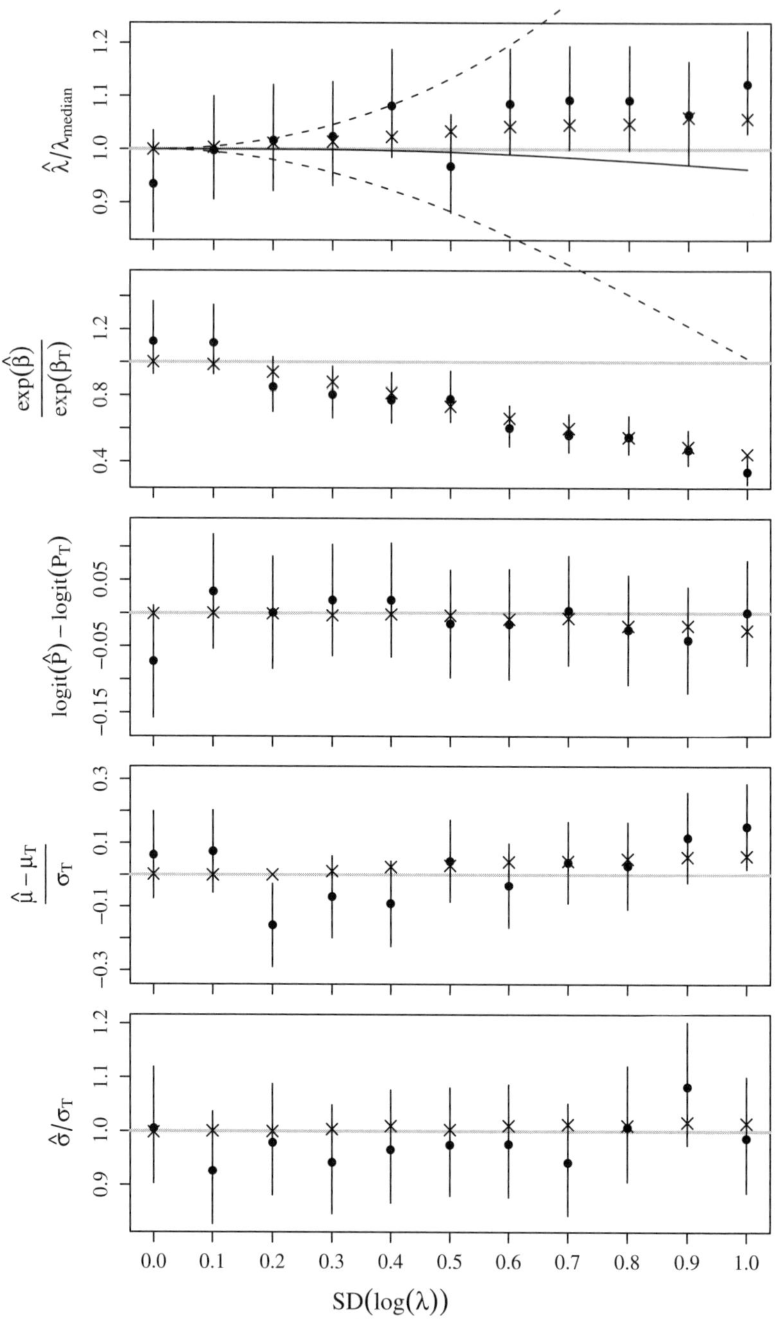

Fig. 1 Bias in the parameter estimates (*y*-axes) from the model fitted to simulated data with various degrees of log-normal heterogeneity in the mortality hazard rates (*x*-axis). Points with 95% confidence bars are from simulations with 1000 individuals released at the initial occasion

life-expectancy even though mean hazard rate also increases (see text of Fig. 1). Also note that, in the presence of heterogeneity in hazard rates, the hazard rates for both states will be biased downwards for the same reason that the hazard ratio representing cost of reproduction was underestimated: a cohort will be more and more overrepresented by high quality individuals (those with low mortality hazards) as time goes by (see Cam et al. (2002); Zens and Peart (2003); Cam et al. (2005)). Hence, if only studying population level hazard rates it may look as if survival rate increases after reproduction even when reproduction has a very substantial survival cost for the individuals, as is seen in this simulation study.

Note that biased estimation of survival (or mortality hazards) due to individual heterogeneity in mortality hazards is not specific to the model presented here. Any method for estimation of survival rates will overestimate survival if individual heterogeneity is not fully accounted for in the analysis (Zens and Peart 2003). However, the simulations show that, although we get severe bias in estimates of survival parameters, we retain fairly robust estimation of the latent maturation distribution even when there is rather extreme individual heterogeneity in mortality hazards.

Even though we obtain fairly robust estimation of the latent time of maturation distribution in the presence of high heterogeneity in the mortality hazard rates, the situation is very different when individual hazard rate and latent time of maturation are correlated (Fig. 2). When individuals that mature late tend to have a lower hazard rate (i.e., a negative correlation), the estimated mean time of maturation will be positively biased, while the bias in the hazard-rate estimators are reduced compared to the case when the parameters are uncorrelated. When there is a positive correlation between the parameters, the situation is the opposite: a negative bias in the estimators of mean time of maturation, and increased bias in the estimators of hazard rates. Standard deviation of the latent maturation distribution is negatively biased when the correlation between the parameters is strong, but is quite accurately estimated when the correlation between the parameters is within ± 0.5.

Fig. 1 (continued) (one simulation per point), while the crosses are from simulations with 500,000 initial releases (details in text). The panels from top to bottom represent: (1) Mortality hazard rate of immatures (λ); (2) Hazard ratio ($e^{\beta} = \lambda^{\text{mature}}/\lambda^{\text{immature}}$) (3) Recapture probability (p); (4) Mean latent time of state transition (μ); and (5) Standard deviation in individual latent time of state transition (σ). Parameter estimates $\widehat{\beta}$, $\widehat{p}$, $\widehat{\mu}$ and $\widehat{\sigma}$ are expressed in relation to the true fixed value used in the simulations (subscripts 'T'), while $\widehat{\lambda}$ is expressed in relation to the expected median of the simulated hazard rates. The upper dashed line in the top panel shows the mean expectations of the simulated hazard rates ($E[\lambda] = \exp(\mu + \sigma^2/2)$, where μ and σ are the mean and standard deviation of $\log(\lambda)$), which increases to 1.65 at SD $(\log(\lambda)) = 1$. The solid black line shows the hazard rate corresponding to the mean expectation of survival rate, $-\log(E[\exp(-\lambda)])$, (found numerically). The lower dashed line shows the hazard rate corresponding to the mean lifetime expectancy, $E[\lambda^{-1}]^{-1} = \exp(\mu - \sigma^2/2)$

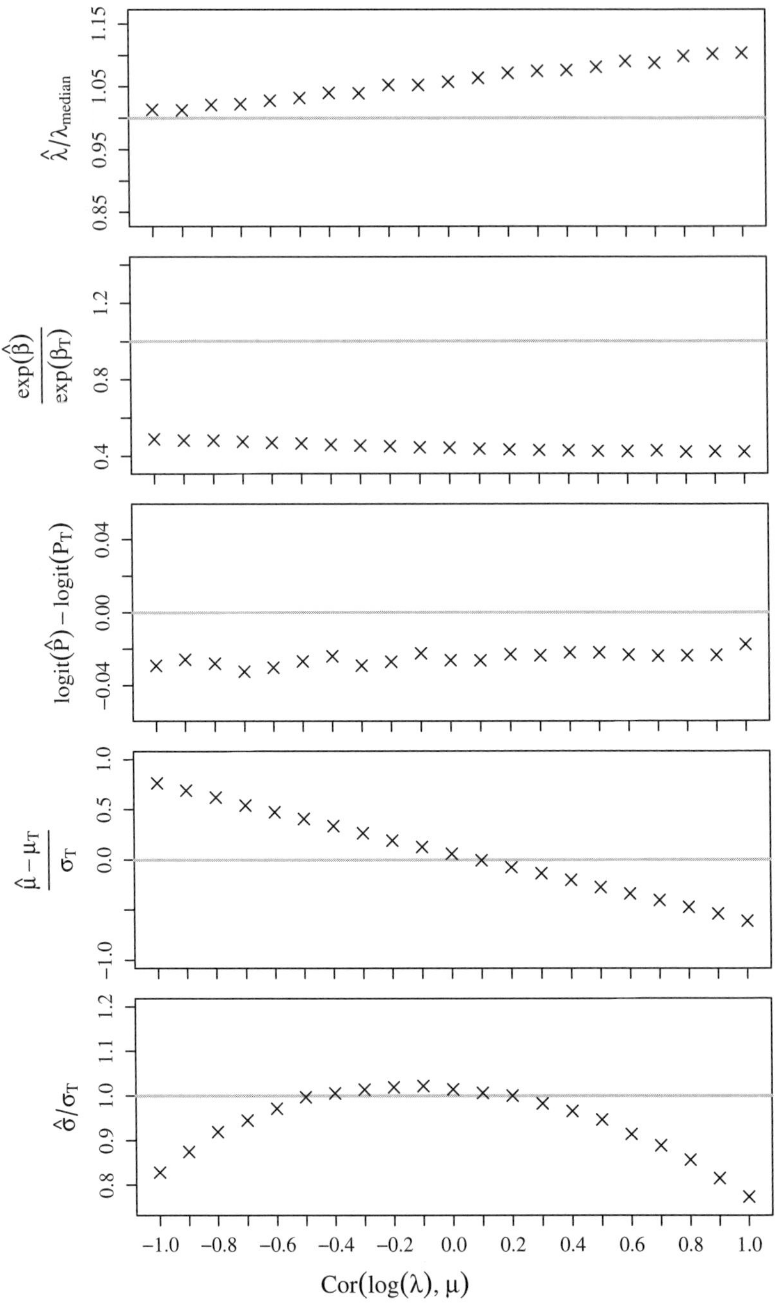

Fig. 2 Simulation results when individual mortality hazard rate and latent time of maturation were correlated to different degrees and in different directions (x-axes). The simulations were based on the most extreme degree of heterogeneity among the simulations presented in Fig. 1 (i.e., SD(log-hazard) = 1), and individual values for log(mortality hazard) and latent time of maturation were drawn from a multivariate normal distribution. See text of Fig. 1 for explanation

4 Case Study

We here illustrate the use of the model with a case study on timing of spring reproduction and associated survival costs in field voles (*Microtus agrestis*) from a cyclically fluctuating population in northern England. Onset of spring reproduction in this study area varies by more than 2 months among years and locations, and this variation is strongly correlated with population densities in the previous spring, but not with present densities (Ergon 2003). Capture–recapture data with records of individual reproductive state were collected at 2-week intervals from four study sites following a translocation experiment (Ergon et al. 2001) in order to study the effects of the present and past environment of the individuals on life-history traits among the overwintering animals. Analyses of these data are also presented by Ergon et al. (2004) and Ergon (2007).

In the analysis presented here we will apply our model to estimate latent distribution of time of first parturition in the season and survival costs of reproduction at the four study sites. It is of particular interest to look at the correlation between mean maturation date and survival costs of reproduction across the four sites; higher survival costs where reproduction commences late in the season could indicate correlated effects of the environment on both timing and costs of reproduction, whereas a higher survival cost at the sites where reproduction takes place early in the season could indicate an adaptive life-history trade-off as a response to fluctuations in the environment (Ergon 2007). An analysis of the data by the use of traditional multi-state models (see Discussion) is presented by Ergon (2007). There are however a couple of differences in the data. First, data from an additional trapping occasion in the beginning of the study are now included. This trapping occasion was excluded in the previous analysis because the length of the first interval varies among sites, and since transition probabilities are not scaled according to the length of the intervals in the traditional multi-state model (but transitions are, unrealistically, assumed to always occur instantaneously at the end of an interval between trapping sessions) it was difficult to make use of these data in the previous analysis. Secondly, to simplify the presentation, we now only include females in the analysis. We also ignore the history of the individuals prior to translocation, as this has been shown to be unimportant (Ergon et al. 2001). The aim here is primarily to show how inferences can be made from fitting the model to data, and a thorough analysis is beyond the scope of this paper.

The data include records of 354 individuals (from 45 to 115 within each site) at 7 trapping occasions, and the effective sample size (number of releases) was 763. The data, summarized in multi-state m-arrays are given in the Appendix.

4.1 Choice of Model Constraints and Model Selection

The data cover the entire spring season in which the voles commence reproduction after the winter; in the beginning of the study all animals were non-reproductive, and by the end of the study all animals known to be alive had started to reproduce.

In the beginning of the study, the populations within sites were fairly homogeneous; most individuals were animals that were born late in the previous breeding season and had suspended growth and delayed maturation before the winter. Age or cohort effects are thus not of any concern. Survival may, however, vary quite unpredictably over time because the voles are heavily predated by mobile predators (mustelids, birds of prey and foxes), that may occupy the study area for a longer or shorter period of time. Accordingly, it is desirable to use a rather flexible model for the time-variation in the hazard rate, and there may be support for models with different time-effects at the different sites (a 'time × site' interaction). We used the hazard function composed as a sum of Gaussian terms described above (Eq. 7), and we also considered models with constant hazard rates over time.

The trapping followed a standardized protocol and previous analysis has shown that recapture probability mainly varied according to reproductive state (Ergon 2007). We thus only considered models with a 'state' effect on the recapture probabilities. The main aim of the analysis was to study the association between mean latent time of parturition and survival costs or reproduction among the study sites. Hence a 'state × site' interaction effect on mortality hazard and mean time of state transition was included in all models. The 'time × site' and 'state × site' effects on mortality hazard was made additive on a log-scale, so that the hazard rates of the two states within sites will remain proportional over time. For simplicity, we only present models including the same within-site variance in the latent maturation distribution (models with different variances of the distribution at the different sites were not supported by the data). Model statistics for 5 candidate models are presented in Table 2.

4.2 Results

Ranking the candidate models according to their AIC_c model-selection criteria (Burnham and Anderson 2002) revealed that two different models were supported

Table 2 Model statistics for 5 models with different mortality hazard functions fitted to the vole data. The distributions of individual latent maturation dates within sites were modelled as logistic distributions with a different mean for each site (4 parameters) and a common variance (1 parameter). All models include a different recapture probability for each state (2 parameters), but recapture probabilities are assumed to not differ among sites. The label 'time(G2)' means a hazard function constructed as a sum of two Gaussian curves (Eq. 7) with the same width (b-parameters) where the means (c-parameters) are fixed to each of the endpoints of the study period (the heights of the curves (a-parameters) vary freely). Models labeled 'time(G3)' include an additional Gaussian curve at the centre point of the study period. All models have a proportional difference between mortality hazard rates of immature and mature animals within sites. See text for the meaning of the other labeling

Hazard model	Total no. parameters	Deviance	Dev. DF	Δ AIC_c
state × site	15	368.86	82	0.00
state × site + time(G2) × site	20	358.40	77	0.03
state × site + time(G2)	17	366.84	80	2.16
state × site + time(G3)	18	365.99	79	3.41
state × site + time(G3) × site	24	355.12	73	5.25

equally well by the data (Table 2). One of the models had a constant hazard rate within sites and states (15 parameters in total), and the other more complex model (20 parameters) had a different hazard curve over time for each of the four sites. Estimates of hazard ratios, recapture probabilities, and the means and standard deviations of the latent maturation distributions based on both these models are given in Table 3. The estimates of recapture probabilities and maturation distributions are virtually identical for the two models. The estimates of the hazard ratios of the two states are however rather different. The model with constant hazard rates shows a large difference between the sites in the hazard ratios representing cost of reproduction. In the more complex model, the confidence intervals are too wide to support any conclusions. The discrepancies can be understood from looking at plots of the estimated hazard curves (Fig. 3); there is only a rather short period of time when animals of both states are present in the same population. Hence, it is difficult to separate the effects of 'state' and 'time' (part of the 'state' effect may be absorbed by the 'time' effect or vice versa). The problem of separating these two effects is also evident from the sampling correlations between the log of the hazard ratios and the height of the second Gaussian component relative to the height of the first component; this correlation was respectively −0.95, −0.90, −0.57, and −0.89 for site A, B, C, and D. This illustrates a general potential problem when estimating survival costs associated with state transition if there is little temporal overlap between the animals of the two states in the population and when there is little a priori knowledge about the shape of the temporal change in hazard rates.

Plotting the estimated hazard ratios from the model with time-independent hazard rates against the estimates of the mean time of latent maturation (Fig. 4) shows that estimated cost of reproduction is higher in the sites where reproduc-

Table 3 Estimates from the two top (AIC*c*-best) models in Table 2. Differences in mortality hazard rates of animals in the immature and mature state (the cost of reproduction) are expressed as hazard ratios (hazard rate of the mature state divided by hazard rate of the immature state), which is the same as the ratio of log(survival) over an interval. Values within brackets show 95% confidence intervals

	Hazard model	
	state × site	state × site + time(G2) × site
Survival cost of reproduction (hazard ratio)		
Site A	2.43 [1.05, 5.65]	8.56 [0.73, 99.92]
Site B	0.35 [0.14, 0.89]	0.82 [0.10, 6.64]
Site C	1.79 [0.82, 3.91]	0.74 [0.26, 2.09]
Site D	0.55 [0.22, 1.40]	0.65 [0.09, 4.66]
Recapture probability		
Immature	0.80 [0.73, 0.85]	0.80 [0.73, 0.85]
Mature	0.87 [0.80, 0.92]	0.87 [0.80, 0.92]
Mean latent parturition date (day of year)		
Site A	113.2 [108.8, 117.6]	112.8 [108.3, 117.3]
Site B	93.9 [90.2, 97.7]	93.9 [89.9, 97.9]
Site C	109.5 [106.2, 112.8]	109.9 [106.6, 113.2]
Site D	97.5 [94.0, 101.0]	97.5 [93.9, 101.1]
SD latent parturition date (days)		
Within all sites	8.5 [6.8, 10.5]	8.5 [6.8,10.6]

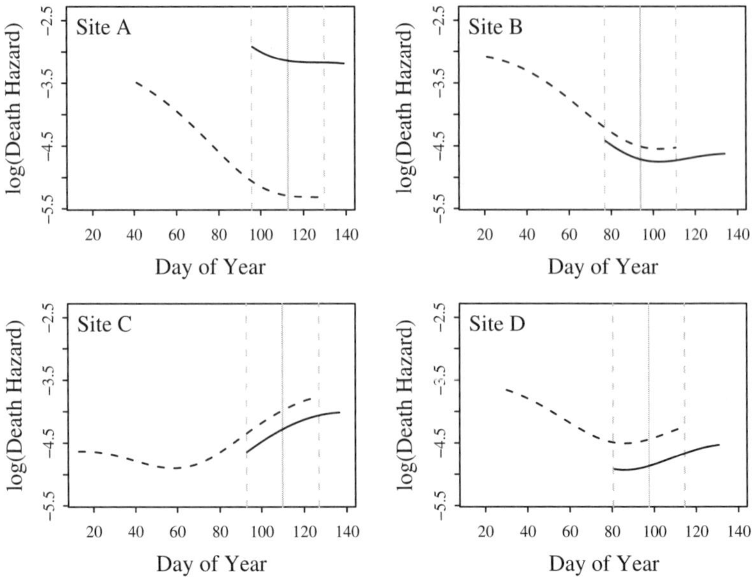

Fig. 3 Mortality hazard rates on a one-day time-scale according to model 'state × site + time(G2) × site' plotted on a log-scale as a function of day of year for each of the four sites. Dashed curves show the hazard rates for the immature state, and solid curves represent the mature state. Grey vertical lines show the mean ± 2SD of the estimated distribution for latent time of maturation. The curves for the immature state are plotted from the first trapping session to the 97.5% quantile of the maturation distribution, and the curves for the mature state are plotted from the 2.5% quantile of the distribution to the last trapping session at the site

tion commenced late, which indicates correlated effects of the environment on both timing and costs of reproduction. That the voles in site B and D show a *decrease* in the hazard rates after parturition is not surprising when considering the fact that mortality is confounded with dispersal; voles may disperse prior to reproducing, but they will obviously not leave the site while they are nursing young in the nest – illustrating yet another potential problem in estimating survival cost of reproduction. Finally, we should also keep in mind the lesson from the simulation study above, that individual heterogeneity in mortality hazard rates can lead to substantial bias in the estimates of cost of reproduction.

5 Discussion

Age at first reproduction has long been known to be an important life history component with substantive consequences for fitness (Cole 1954; Roff 1992; Stearns 1992). Age at first reproduction is also a component of several so-called life history invariants (Charnov 1993), and predictions about age at maturity 'probably represent the most successful empirical area of evolutionary life history theory' (Charnov

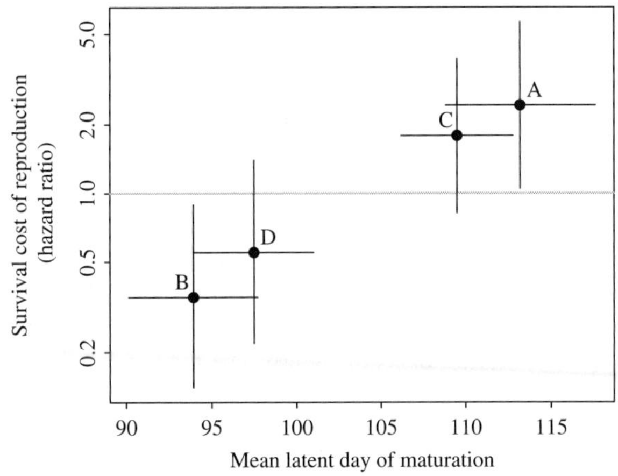

Fig. 4 Survival costs of reproduction plotted as hazard ratios (y-axis has a logarithmic scale) against the mean of the estimated distribution for latent maturation date (x-axis) at each of the study sites. Error bars show 95% confidence intervals of the estimates. Estimates were obtained from the 'state × site' model of Table 2. The estimates of monthly (30 days) survival rates of immature/mature voles (95% c.i.) from this model were: Site A: 0.63 (0.50, 0.74)/0.32 (0.10, 0.57); Site B: 0.50 (0.36, 0.63)/0.79 (0.57, 0.90); Site C: 0.75 (0.66, 0.81)/0.59 (0.37, 0.76); Site D: 0.63 (0.51, 0.73)/0.77 (0.57, 0.89)

1989, p. 237). Changes in survival probability that might accompany an animal's transition from an immature to a reproductively mature state are also relevant to theory about life history trade-offs and costs of reproduction (Roff 1992; Stearns 1992). Despite the central role of age at maturity in ecological and evolutionary theory, empirical work lags behind theory, and approaches for estimation of age at maturity for natural animal populations in the face of imperfect detection have been relatively recent.

Clobert et al. (1994) developed a capture-recapture approach to inferences about age at first reproduction for species that are available for detection as young at a breeding colony and are then unobservable until they return to breed for the first time. Subsequent recognition that this problem fits naturally within the framework of multi-state models (Arnason 1972, 1973; Brownie et al. 1993; Schwarz et al. 1993) led to the direct estimation and modeling of age-specific probabilities of transition between pre-breeding and breeding states (e.g., Lebreton et al. 1999, 2003; Pradel and Lebreton 1999; Fujiwara and Caswell 2002; Spendelow et al. 2002; Kendall et al. 2003; Crespin et al. 2006; Hadley et al. 2006).

This work using multi-state models as the basis for inference about age at first reproduction deals primarily with organisms that reproduce, at most, once a year, and that can be sampled during the reproductive season each year. Existing multi-state modeling approaches do not permit inference about the precise timing of pre-breeder to breeder transition during the interval between sample occasions and do not account for the different state-specific survival rates that may apply during intervals of transition. Conroy et al. (1996) and Hestbeck (1995) noted the potential for biased inferences about survival arising because of unknown times of transition. Joe and Pollock (2002) considered multi-state models in which the probability distribution of the time of state transition within an interval was known (e.g., uniform or beta distributions), and found that these models performed fairly well when evaluated in terms of relative bias of survival, transition and capture probability estimators.

In this paper, we use a latent variable modeling approach (e.g., Skrondal and Rabe-Hesketh 2004) on multi-state capture-recapture data to address questions about age at first reproduction in organisms for which the reproductive season cannot be readily viewed as a single discrete period (e.g., Fairbairn 1977; Ergon et al. 2001; and Ergon 2007). Our approach focuses on the latent probability distribution for time at maturation or, equivalently, the hazard function associated with this distribution. This probability distribution is not conditional on survival, whereas realized (i.e., observable) times of maturation represent functions of survival and conditional probabilities of maturation. For studies of responses to selection, life-history trade-offs and phenotypic correlations (e.g. Roff 2002), estimates of the latent trait distributions are generally more relevant than estimates of the realized distribution. Also in studies of density-dependence in timing of seasonal reproduction (e.g., Smith et al. 2006), it will be more informative to study density dependence on the latent times of reproduction than the observable times which will be influenced by (perhaps density-dependent) variation in survival.

Using this model to analyze variation in maturation and survival rates in field voles, we found evidence that the survival cost of reproduction (estimated as mortality hazard ratios of mature vs. immature voles) was higher at sites for which maturation was later, suggestive of correlated effects of environmental conditions on both time of maturation and survival, which is consistent with previous analyses of the study system (Ergon et al. 2001; Ergon et al. 2004; Ergon 2007). The survival differences between immature and mature voles were relatively large, again arguing for models such as these that permit the partitioning of between-sample intervals based on time of maturation.

In our example analysis, it was difficult to discriminate between models with time-constant versus time-varying mortality hazard rates. This difficulty was largely a result of synchronous maturation, such that most voles were immature for one set of time intervals and mature for another set of intervals, with few sample intervals containing sufficient numbers of both mature and immature voles. This seasonality in our example leads to the recommendation that sampling should be most intense during periods for which numbers of animals can be found in both immature and mature states. Increased synchronization of maturation should lead to an increase in the focus of sampling intensity on transition periods, especially if large time variation in mortality hazard rates can be expected. In the case of highly synchronous maturation, it will be difficult to separate temporal trends in survival from true costs of reproduction.

The simulation study provided evidence of the sensitivity of estimates of reproductive costs to heterogeneity in individual mortality hazards. Estimation of the latent time of maturation distribution was found to be fairly robust to heterogeneity of hazard rates. However, when there were strong correlations between individual mortality hazard and time of maturation (e.g., individuals that mature late tend to have lower mortality hazard rate), then estimates of the maturation distribution tend to be biased.

These latter observations about potential bias in the face of correlated mortality hazards and time at maturation lead to clear recommendations for future work. Our models assume independence of the two demographic processes, transition to reproductive state and survival. Individual quality can lead to positive correlations between survival and reproductive parameters estimated at the population level (Cam et al. 2002; Weladji et al. 2006), even if reproductive costs are expected to lead to negative correlations within individuals. Indeed, it is a major empirical question to determine the relative importance of these two factors, and how they can be affected by environmental variability (see Ergon (2007)). One major future development will be to incorporate such correlation in the model.

The focus of this paper has been on the transition from immature to mature, but we see applications for modeling of other ecological processes as well. For example, in disease ecology, the state transition from uninfected to infected and the mortality hazard ratio for these two states (i.e., lethality) are of primary importance (e.g. Begon et al. 2002; Telfer et al. 2002; Burthe et al. 2006; Oli et al. 2006). Because infection will frequently produce large differences in state-specific mortality hazards, bias resulting from failure to deal adequately with transition times would be expected to be substantial. Indeed failure to re-observe animals that become infected during the interval between sample periods (because of death) will result in underestimation of the lethality as well as the rate of infection (especially if lethality is high). The model presented in this paper will remedy these problems and yield estimates of infection rates and lethality that are not biased by background mortality. However, it should be noted that when changing the focus from the latent distribution for time of state-transition to the corresponding transition hazard rates, it is assumed that the variation in time of transition (infection) is solely due to stochasticity and there is no individual heterogeneity in the infection hazards (susceptibility). Hence, individual heterogeneity in the infection susceptibility (other than what can be accounted for by group variables, age or time) is not accounted for in the current model and may lead to biases.

Acknowledgments We thank Evan Cooch, Rolf Ergon, Roger Pradel and Jean-Dominique Lebreton for comments that helped us improve earlier versions of the manuscript.

Appendix – Case Study Data

The data used in the Case Study, represented as multi-state m-arrays (See Table 1), are given in Tables 4–7 below. Times of trapping session at each site, expressed as day of year, are given in the caption of each table. These should be rescaled to have absolute values less than ~ 1 to avoid numerical problems when using some mortality hazard functions. Note that site A and site C have 7 trapping sessions, while site B and site D have 6 trapping sessions.

Table 4 Multi-state m-array Site A. Trapping sessions on days 1:41, 2:69, 3:84, 4:98, 5:112, 6:127, 7:139

		Session of first recapture (j) State of recapture												
Session of	State of	2		3		4		5		6		7		never
release (i)	release	A	B	A	B	A	B	A	B	A	B	A	B	
1	A	18	0	1	0	0	0	0	0	0	0	0	0	16
1	B	0	0	0	0	0	0	0	0	0	0	0	0	0
2	A			14	0	1	0	0	2	0	0	0	0	7
2	B			0	0	0	0	0	0	0	0	0	0	0
3	A					13	0	1	1	0	0	0	0	0
3	B					0	0	0	0	0	0	0	0	0
4	A							8	4	0	0	0	0	3
4	B							0	0	0	0	0	0	0
5	A									0	5	0	2	2
5	B									0	4	0	0	3
6	A											0	0	0
6	B											0	3	7

Table 5 Multi-state m-array Site B. Trapping sessions on days 1:21, 2:76, 3:92, 4:106, 5:119, 6:134

		Session of first recapture (j) State of recapture										
Session of	State of	2		3		4		5		6		never
release (i)	release	A	B	A	B	A	B	A	B	A	B	
1	A	4	0	0	1	0	0	0	0	0	0	23
1	B	0	0	0	0	0	0	0	0	0	0	0
2	A			13	8	0	1	0	0	0	0	7
2	B			0	0	0	0	0	0	0	0	0
3	A					2	13	0	2	0	0	1
3	B					0	10	0	1	0	0	2
4	A							0	3	0	0	0
4	B							0	28	0	0	2
5	A									0	0	0
5	B									0	27	13

Table 6 Multi-state m-array Site C. Trapping sessions on days 1:13, 2:67, 3:82, 4:96, 5:110, 6:124, 7:137

		Session of first recapture (j) State of recapture												
Session of	State of	2		3		4		5		6		7		never
release (i)	release	A	B	A	B	A	B	A	B	A	B	A	B	
1	A	26	0	4	0	0	0	0	0	0	0	0	0	18
1	B	0	0	0	0	0	0	0	0	0	0	0	0	0
2	A			28	0	3	0	0	0	1	1	1	1	10
2	B			0	0	0	0	0	1	0	0	0	0	0
3	A					28	1	1	1	0	0	0	0	5
3	B					0	0	0	0	0	0	0	0	0
4	A							9	14	0	3	0	0	9
4	B							0	1	0	0	0	0	0
5	A									0	11	0	0	1
5	B									0	15	0	0	3
6	A											0	1	1
6	B											0	21	14

Table 7 Multi-state m-array Site D. Trapping sessions on days 1:30, 2:74, 3:89, 4:103, 5:117, 6:131

		Session of first recapture (j) State of recapture										
Session of release (i)	State of release	2 *A*	2 *B*	3 *A*	3 *B*	4 *A*	4 *B*	5 *A*	5 *B*	6 *A*	6 *B*	never
1	*A*	12	0	3	1	0	0	0	0	0	0	22
1	*B*	0	0	0	0	0	0	0	0	0	0	0
2	*A*			19	2	2	1	0	3	0	0	5
2	*B*			0	0	0	0	0	0	0	0	0
3	*A*					3	20	0	0	0	1	4
3	*B*					0	3	0	0	0	0	0
4	*A*							2	2	0	2	4
4	*B*							0	22	0	3	5
5	*A*									0	4	1
5	*B*									0	29	7

References

Arnason AN (1972) Parameter estimates from mark-recapture experiments on two populations subject to migration and death. Researches on Population Ecology 13:97–113.

Arnason AN (1973) The estimation of population size, migration rates and survival in a stratified population. Researches on Population Ecology 15:1–8.

Begon M, Bennett M, Bowers RG, French NP, Hazel SM, Turner, J (2002) A clarification of transmission terms in host-microparasite models: numbers, densities and areas. Epidemiology and Infection 129(1):147–153.

Both C, Bouwhuis S, Lessells CM, Visser ME (2006) Climate change and population declines in a long-distance migratory bird. Nature 441:81–83.

Brownie C, Hines JE, Nichols JD, Pollock KH, Hestbeck, JB (1993) Capture–recapture studies for multiple strata including non-Markovian transitions. Biometrics 49:1173–1187.

Burnham KP, Anderson DR (2002) Model selection and inference: a practical information-theoretic approach. New York, Springer-Verlag.

Burthe S, Telfer S, Lambin X, Bennett M, Carslake D, Smith A, Begon M (2006) Cowpox virus infection in natural field vole *Microtus agrestis* populations: delayed density dependence and individual risk. Journal of Animal Ecology 75(6):1416–1425.

Cam E (2008) Contribution of capture-mark-recapture modeling to studies of evolution by natural selection. In: Thomson DL, Cooch EG, Conroy MJ (eds.) Modeling Demographic Processes in Marked Populations. Environmental and Ecological Statistics, Springer, New York, Vol. 3, pp. 83–130.

Cam E, Link WA, Cooch EG, Monnat JY, Danchin E (2002) Individual covariation in life-history traits: Seeing the trees despite the forest. The American Naturalist 159:96–105.

Cam E, Cooch EG, and Monnat JY (2005) Earlier recruitment or earlier death? On the assumption of equal survival in recruitment studies. Ecological Monographs 75:419–434.

Charnov EL (1989) Natural selection on age at maturity in shrimp. Evolutionary Ecology 3: 236–239.

Charnov EL (1993) Life history invariants. Oxford, Oxford University Press, UK.

Clobert J, Lebreton JD, Allainé D, and Gaillard, JM (1994) The estimation of age-specific breeding probabilities from recaptures or resightings in vertebrate populations. II. Longitudinal models. Biometrics 50:375–387.

Cole LC (1954) The population consequences of life history phenomena. Quarterly Review of Biology 29:103–137.

Conroy MJ (2008) Application of capture–recapture to addressing questions in evolutionary ecology. In: Thomson DL, Cooch EG, Conroy MJ (eds.) Modeling Demographic Processes in Marked Populations. Environmental and Ecological Statistics, Springer, New York, Vol. 3, pp. 131–156.

Conroy MJ, Anderson JE, Rathbun SL, and Krementz, DG (1996) Statistical inference on patch-specific survival and movement rates from marked animals. Environmental and Ecological Statistics 3:99–116.

Cox DR (1972) Regression models and life tables. Journal of the Royal Statistical Society B 74:187–220.

Crespin L, Harris MP, Lebreton JD, Frederiksen M, Wanless S (2006) Recruitment to a seabird population depends on environmental factors and on population size. Journal of Animal Ecology 75:228–238.

Ergon T (2003) Fluctuating life-history traits in overwintering field voles (*Microtus agrestis*). PhD thesis. Faculty of Mathematics and Natural Sciences, Norway: University of Oslo.

Ergon T, (2007) Optimal onset of seasonal reproduction in stochastic environments: when should overwintering small rodents start breeding? Ecoscience 14(3):330–346.

Ergon T, Lambin X, and Stenseth NC (2001) Life-history traits of voles in a fluctuating population respond to the immediate environment. Nature 411:1043–1045.

Ergon T Speakman JR, Scantlebury M, Cavanagh R, Lambin X (2004) Optimal body size and energy expenditure during winter: why are voles smaller in declining populations? American Naturalist 163(3):442–457.

Fairbairn DJ (1977) Why breed early? A study of reproductive tactics in Peromyscus. Canadian Journal of Zoology 55:862–871.

Fujiwara M, Caswell H (2002) Estimating population projection matrices from multi-stage mark-recapture data. Ecology 83:3257–3265.

Gaillard JM, Loison A, Viallefont A, Festa-Bianchet M (2004) Assessing senescence patterns in populations of large mammals. Animal Biodiversity and Conservation 27:47–58.

Gaillard JM, Yoccoz NG, Lebreton JD, Bonenfant C, Devillard S, Loison A, Pontier D, Allainé D (2005) Generation time: a reliable metric to measure life-history variation among mammalian populations. American Naturalist 166:119–123.

Hadley GL, Rotella JJ, Garrott RA, Nichols J (2006) Variation in probability of first reproduction of Weddell seals. Journal of Animal Ecology 75:1058–1070.

Hestbeck JB (1995) Bias in transition-specific survival and movement probabilities estimated using capture-recapture data. Journal of Applied Statistics 22:737–750.

Hestbeck JB, Nichols JD, Malecki RA (1991) Estimates of movement and site fidelity using mark-resight data of wintering Canada geese. Ecology 72:523–533.

Hochachka W (1990) Seasonal decline in reproductive performance of song sparrows. Ecology 71:1279–1288.

Joe M, Pollock KH (2002) Separation of survival and movement rates in multi-state tag-return and capture–recapture models. Journal of Applied Statistics 29:373–384.

Kendall WL, Hines JE, Nichols JD (2003) Adjusting multistate capture–recapture models for misclassification bias: Manatee breeding proportions. Ecology 84:1058–1066.

Lambin X, Yoccoz NG (2001) Adaptive precocial reproduction in voles: reproductive costs and multivoltine life-history strategies in seasonal environments. Journal of Animal Ecology 70:191–200.

Lebreton JD, Burnham KP, Clobert J, Anderson DR (1992) Modelling survival and testing biological hypotheses using marked animals: a unified approach with case studies. Ecological Monographs 62:67–118.

Lebreton JD, Almeras T, Pradel R (1999) Competing events, mixtures of information and multistratum recapture models. Bird Study 46:39–46.

Lebreton JD, Hines JE, Pradel R, Nichols JD, Spendelow JA (2003) Estimation by capture-recapture of recruitment and dispersal over several sites. OIKOS 101:253–264.

Łomnicki A (1988) Population ecology of individuals. Princeton, NJ: Princeton University Press.

Oli MK, Venkataraman M, Klein PA, Wendland LD, Brown MB (2006) Population dynamics of infectious diseases: a discrete time model. Ecological Modelling 198(1–2):83–194.

Pintilie MM (2006) Competing risks : a practical perspective. Chichester: John Wiley & Sons, Ltd.

Pradel R, Lebreton JD (1999) Comparison of different approaches to the study of local recruitment of breeders. Bird Study 46:74–81.

Roff DA (1992) The evolution of life histories. London: Chapman and Hall.

Roff DA (2002) Life history evolution. Sunderandl, MA: Sinauer Associates, Inc.

Rotella J (2008) Estimating reproductive costs with multi-state mark-recapture models, multiple observable states, and temporary emigration. In: Thomson DL, Cooch EG, Conroy MJ (eds.) Modeling Demographic Processes in Marked Populations. Environmental and Ecological Statistics, Springer, New York, Vol. 3, pp. 157–172.

Schwarz CJ, Arnason AN (2000) Estimation of age-specific breeding probabilities from capture-recapture data. Biometrics 56(1):59–64.

Schwarz CJ, Schweigert JF, Arnason AN (1993) Estimating migration rates using tag-recovery data. Biometrics 49:177–193.

Skrondal A, Rabe-Hesketh S (2004) Generalized latent variable modeling. London: Chapman & Hall.

Smith MJ, White A, Lambin X, Sherratt JA, Begon M (2006) Delayed density-dependent season length alone can lead to rodent population cycles. American Naturalist 167:695–704.

Spendelow JA, Nichols JD, Hines JE, Lebreton JD, Pradel R (2002) Modelling postfledging survival and age-specific breeding probabilities in species with delayed maturity: a case study of Roseate Terns at Falkner Island, Connecticut. Journal of Applied Statistics 29:385–405.

Stearns SC (1992) The evolution of life histories. Oxford: Oxford University Press.

Stenseth NC, Mysterud A, Ottersen G, Hurrell JW, Chan KS, Lima M (2002) Ecological effects of climate fluctuations. Science 297:1292–1296.

Telfer S, Bennett M, Bown K, Cavanagh R, Crespin L, Hazel S, Jones T, Begon, M (2002) The effects of cowpox virus on survival in natural rodent populations: increases and decreases. Journal of Animal Ecology 71(4):558–568.

Visser ME, Holleman LJM (2001) Warmer springs disrupt the synchrony of oak and winter moth phenology. Proceedings of the Royal Society, London B 268:289–294.

Visser ME, van Noordwijk AJ, Tinbergen JM, Lessells CM (1998) Warmer springs lead to mistimed reproduction in great tits (*Parus major*). Proceedings of the Royal Society, London B 265:1867–1870.

Weladji RB, Gaillard JM, Yoccoz NG, Holand Ø, Mysterud A, Loison A, Nieminen M, Stenseth NC (2006) Good reindeer mothers live longer and become better in raising offspring. Proceedings of the Royal Society B-Biological Sciences 273:1239–1244.

Williams B, Nichols JD, Conroy MJ (2002) Analysis and Management of Animal Populations. Academic Press, San Diego.

Winkler DW, Allen PE (1996) The seasonal decline in tree swallow clutch size: physiological constraint or strategic adjustment? Ecology 77:922–932.

Zens MS, Peart DR (2003) Dealing with death data: individual hazards, mortality and bias. Trends in Ecology and Evolution 18:366–374.

Section III
Abundance Estimation – Direct Methods, Proxies, Occupancy Models and Point Count Data

Len Thomas and Paul Conn

Inferences About Landbird Abundance from Count Data: Recent Advances and Future Directions

James D. Nichols, Len Thomas, and Paul B. Conn

Abstract We summarize results of a November 2006 workshop dealing with recent research on the estimation of landbird abundance from count data. Our conceptual framework includes a decomposition of the probability of detecting a bird potentially exposed to sampling efforts into four separate probabilities. Primary inference methods are described and include distance sampling, multiple observers, time of detection, and repeated counts. The detection parameters estimated by these different approaches differ, leading to different interpretations of resulting estimates of density and abundance. Simultaneous use of combinations of these different inference approaches can not only lead to increased precision but also provides the ability to decompose components of the detection process. Recent efforts to test the efficacy of these different approaches using natural systems and a new bird radio test system provide sobering conclusions about the ability of observers to detect and localize birds in auditory surveys. Recent research is reported on efforts to deal with such potential sources of error as bird misclassification, measurement error, and density gradients. Methods for inference about spatial and temporal variation in avian abundance are outlined. Discussion topics include opinions about the need to estimate detection probability when drawing inference about avian abundance, methodological recommendations based on the current state of knowledge and suggestions for future research.

1 Introduction

For decades, the majority of inferences about landbird abundance and density have been based on counts conducted by investigators, either stationed at points (e.g., Blondel et al. 1970) or walking along line transects (Emlen 1971; Jarvinen and Vaisanen 1975). Counts resulting from point and transect sampling have been treated frequently as indices to abundance, in the sense that the expected counts

J.D. Nichols (✉)
U.S. Geological Survey, Patuxent Wildlife Research Center, Laurel, MD, USA
e-mail: jnichols@usgs.gov

D.L. Thomson et al. (eds.), *Modeling Demographic Processes in Marked Populations*, Environmental and Ecological Statistics 3, DOI 10.1007/978-0-387-78151-8_9,

have been assumed to represent an unknown, yet relatively constant, proportion of the sampled population. Early on, some investigators argued that the proportionality assumption is not likely to be widely met, or at least merits testing, and advocated the collection of supplemental data with counts that permit inference about the detection probabilities of individual birds and thus about true abundance and density (e.g., Ramsey and Scott 1979; Burnham et al. 1980, 1981). Debate about approaches for drawing inferences about population size and dynamics from avian count data has motivated symposia and workshops over the years (Ralph and Scott 1981; Ralph et al. 1995) and has persisted through the present time (e.g., Anderson 2001; Hutto and Young 2002, 2003; Rosenstock et al. 2002; Thompson 2002b; Ellingson and Lukacs 2003).

The past 5 years have been a period of especially active research on inference methods for avian count data. Such research has included development of new estimation methods, application of these and previous methods in investigations of relatively large scale, and serious testing of existing methods using novel experimental approaches. These developments are sufficiently recent that it has been difficult for investigators to keep up with progress that has been made. Thus, we hosted a small workshop at Patuxent Wildlife Research Center, Maryland USA, inviting 19 biometricians and avian population ecologists (see Acknowledgements) who have played large roles in the recent research. The purposes of the workshop were to obtain a synthesis of the current "state of the art" in methods for estimating landbird abundance from point count and related data, to highlight future research needs, and to determine how best to bridge the gap between statisticians and practitioners. This paper represents an effort to summarize some of the central conclusions and points of discussion from the workshop.

2 Conceptual Framework

2.1 Basic Framework

Discussions of use of count data as a basis for inference about animal populations frequently begin with 2 facts about sampling animal populations (e.g., Lancia et al. 1994, 2005; Borchers et al. 2002; Williams et al. 2002):

(1) Interest is frequently in areas that are sufficiently large that animals cannot be counted over the entire area for which inference is desired;
(2) At locations where investigators do obtain counts of animals by whatever means (counts of animals seen, heard, captured, etc.), these counts seldom include all animals at the sampled location.

Fact 1 is common to many areas of statistics, and traditional design-based sampling approaches are applicable (e.g., Cochran 1977; Thompson 2002a). These sampling approaches are designed to use data from locations at which counts are made to draw inferences about locations where counts are not made. In design-based sampling, the key to such extrapolation is to sample locations in such a manner that all locations about which inference is desired have some known, non-negligible probability of being included in the sample. These probabilities (sometimes called "coverage" probabilities) are determined by the type of design (e.g., random sampling, stratified random sampling, systematic sampling, adaptive cluster sampling), and can be computed based on a knowledge of the design (including desired sample size). Model-based sampling represents a somewhat different approach in which covariate relationships estimated from data on locations that are visited are assumed to apply to locations that are not visited. If this assumption holds, then as long as covariate information is available for all locations of interest, inferences about animal abundance can be made even from locations at which no counts are made. Although geographic variation and spatial sampling were not the primary foci of the workshop, discussion of these topics arose frequently, as their importance was clearly recognized.

Fact 2 involves detectability, and the workshop focused on approaches for dealing with this issue. Traditional discussions of detectability view counts of animals (C_i for location i) as random variables, the expectation of which can be written as the product of the true number of animals at the location at the time of the survey (N_i) and the detection probability (p_i), the probability that a member of N_i appears in C_i:

$$E(C_i) = N_i p_i \ . \tag{1}$$

In the context of the workshop, counts were usually the numbers of birds seen or heard, and discussion focused on how to translate these counts into inferences about true abundance or density.

For some purposes, estimates of true abundance or density are required, and can be obtained as:

$$\hat{N}_i = \frac{C_i}{\hat{p}_i} \ . \tag{2}$$

In many cases, we can view location i as corresponding to an individual sampling unit selected from a large area for which an abundance estimate, $\hat{N}$, is desired. In an effort to deal with facts 1 and 2 listed above, we define $p_{c,i}$ as the coverage probability of sample i within this large area and estimate N as:

$$\hat{N} = \sum_i \frac{C_i}{\hat{p}_i p_{c,i}} \ . \tag{3}$$

More frequently, inferences of interest involve not abundance itself, but ratios of abundance over space (often termed relative abundance) or time (often termed trend or rate of population change). One approach to inference about ratios of abundance is to standardize data collection procedures in hopes of obtaining similar detection probabilities for the different times or locations to be compared. If similar detection probabilities can be obtained, then ratios of the counts themselves provide reasonable estimates of ratios of abundances. Another approach is to hope that most of the relevant temporal and spatial variation in detection probability is associated with recorded covariates that have no possibility of also being associated with variation in true abundance. For example, observer identity is such a covariate and can be incorporated into analyses that use raw count data (e.g., Link and Sauer 1997). Those who do not believe it is safe to rely on standardization and covariate identification typically advocate collection of data needed to draw direct inference about detection probability and its variation. Given such data, it is possible to compare alternative models that express different hypotheses about how detection probability varies as a function of time, space, or recorded covariates. The workshop included discussion of the relative efficacy of these 2 general approaches: (1) use of raw counts with assumptions about relevant sources of variation in detection probability, versus (2) collection of data needed to draw inferences about variation in detection probabilities, using methods that also require assumptions about the detection process.

2.2 Decomposition of Coverage and Detection Probabilities

K.H. Pollock presented a conceptual framework for the workshop that extended the ideas presented above to include different components of detection (also see Pollock et al. 2002; Farnsworth et al. 2002, 2005). Specifically, he noted that detection can be broken into components associated with availability and detection given availability. The issue of availability has been discussed, mainly with respect to aquatic organisms that may be submerged at the time of the survey and thus not exposed to surface survey methods (e.g., Marsh and Sinclair 1989; Laake and Borchers 2004; Okamura et al. 2006). In an auditory survey, a bird that does not vocalize during the survey period is not available to be detected. A bird that does vocalize is available and may or may not be detected depending on the probability of detection given availability. During the course of the workshop, there was also discussion of temporary emigration and the possibility that a bird that sometimes uses a particular sampled site (i.e., the site is included in the bird's territory or home range) may not be present on the site at the time of the survey.

These ideas about geographic sampling and detection probability cause us to further subdivide the coverage and detection probabilities of Section 2.1 and view the probability that a bird in some large area of interest is actually detected during a survey within that area as the product of 4 conceptually distinct probabilities. We begin by considering all birds whose territories or home ranges lie at least partially within the large area about which inference is to be drawn. We refer to this number as superpopulation size, N^*, indicating that these birds have some probability of being exposed to sampling efforts at any given survey time. We are not necessarily interested in estimating N^*, but we simply identify members of N^* as the individuals that may appear in our sample counts. We then consider a randomly chosen bird from the superpopulation N^* and consider its probability of being counted during a survey.

One way to define coverage probability is the probability that the location of the bird is within a sampling unit (location at which a survey count is conducted) at the time of the survey. It is helpful to decompose coverage probability into two parts. The first is the probability that the bird's home range or territory at least partly overlaps a sampling unit. We denote this probability as p_s, and associate it with spatial sampling (Cochran 1977; Thompson 2002a). This probability depends on the spatial sampling design (how are sampling units selected, what are the sizes of the units) and on the size and shape of the bird's home range, or of the portion of it lying within the large area of interest. Conditional on a bird's home range overlapping at least one selected sample unit, the second probability of interest is the probability that a bird is present at a sample unit during the survey period (the time spent surveying at that unit). We refer to this as the probability of presence, p_p, as it indicates the probability that the bird is within the area exposed to sampling efforts for at least some of the survey period. For ease of extrapolation and discussion, we assume the simple case in which the range of a single bird can overlap at most a single selected sample unit, although more complicated situations are certainly possible. Decomposing coverage probability into these two components allows us to account for temporary emigration of birds from a sampling unit during the time of the survey; this occurs with probability $(1-p_p)$.

We now decompose detection probability into two parts. If a bird is located within the sample unit during the survey period, we still consider the possibility that the bird is not available for detection during that time. For example, in an auditory survey, a bird that fails to vocalize during the survey period is essentially unavailable for detection despite being present in the surveyed area. We use the term availability for this detection component and denote the associated probability as p_a. We further note that the methods we consider focus on birds for which $p_a>0$. For example, if females of some species are simply invisible in the general period during which surveys are conducted, then none of our estimation approaches will be useful for them. Finally, we consider the probability of detection given presence and availability, p_d. In an auditory survey, this probability corresponds to the observer actually hearing a bird and being able to identify the species. This component of detection probability is likely to vary as a function of such factors as observer skill and sensory (e.g., hearing) abilities, vocalization characteristics of the bird species, habitat structure, and distance from the observer.

All of these probabilities are viewed as components of the process generating the count data, and their identification and specification during the workshop hopefully led to clear thinking and discussion of the various methods used to estimate bird numbers and density. In particular, we note that most of the estimation methods discussed at the workshop incorporate some sort of parameter representing detection probability. We will use the terms "detection probability" to describe the parameters estimated by different approaches, and we reserve the notation p_d for detection conditional on presence and availability.The nature of the detection probability parameter varies among the different methods, as it can reflect only p_d, the product $p_a p_d$, or even the product $p_p p_a p_d$. Combinations of methods provide opportunities for estimating these components separately and in some cases such decomposition may be useful. We believe that it is especially important that the investigator be aware of which detection parameter applies to a particular method, as this determines the population about which inferences are being made. We discuss this further when introducing each method, below.

Note that for ease of presentation and interpretation, the above components of the detection process are listed as single parameters, implying that every individual in the population of interest shares these common parameters. In reality, at least some of these components will vary by individual, such that these parameters are best viewed as averages. In some cases, these probabilities are best viewed at the level of the individual and modeled as a function of individual-level covariates.

2.3 Closed Populations and Open Fields

Frequently, we will express interest in inferences about a closed population of size N in a known area of size A. While this conceptualization is directly applicable in some circumstances, for example an island population, it is often the case that the survey boundaries are not closed to animal movement, so that population size is not a fixed quantity but varies even over short time intervals. The inclusion of p_p as a component of the detection process in the framework presented above reflects a recognition of this reality. Further, there are some circumstances when the area of inference A is not well defined, although these are often associated with circumstances where there is a poor survey design. M.G. Efford (collaborating with D.K. Dawson) made a presentation, part of which described an alternative perspective to the "closed population" paradigm, which he called an "open field". In this perspective, detectors are located within some area of interest and the focus is on estimating animal density, defined as the local intensity of a spatial point process.

2.4 Index Methods

In certain situations, it may be possible to diagnose population trends without explicitly estimating detection probabilities, treating counts as indices to abundance. In a recent survey of 224 papers using field counts of landbirds, Rosenstock et al. (2002) found that 95% viewed the counts as indices for inference about variation in abundance. Index methods traditionally assume constant detectability, so that changes in raw counts over time or space are viewed as representing changes in the population of interest. Use of indices can also involve covariates hypothesized to influence detection probability but not true abundance or density. Such analyses require that all systematic changes in detectability can be explained and modeled with covariates (e.g., Link and Sauer 1997). At the workshop, W.A. Link and D.H. Johnson gave provocative presentations that focused on the key points of consideration when making decisions about whether to use an index approach to point counts or to instead try to model and/or estimate detection probabilities. W.A. Link maintained that covariates influencing detectability can often be controlled for, and that it was indeed possible to make inferences about population trends without explicit estimation of detection probability. Both presenters argued that index approaches to monitoring, such as the BBS, have provided valuable information on population trends, and have helped to identify species of concern whose status warrants further investigation.

Subsequent discussion of index methods largely focused on the tenability of model assumptions. Workshop participants reached consensus that the assumption of constant detectability needed for index methods was sometimes overstated. Rather, it would suffice that expected detection probability, E(p), is constant over time (e.g., Nichols et al. 2000; Yoccoz et al. 2001; Conn et al. 2004). D. H. Johnson argued that even if there is a small trend in detectability, index methods may still be sufficient to diagnose large scale changes in abundance. The same general comments apply to covariate analyses of index data, although the interpretation of E(p) is in this case changed to the expected detection probability after available covariates have been used to control for variation in detectability.

Index-based methods are inherently attractive when their assumptions are met; they require fewer data to be collected and avoid potentially problematic assumptions about the functional form of individual heterogeneity in detection probability (Link 2003). On the other hand, when index assumptions are sufficiently violated, they may lead to erroneous inferences. When trends are estimated as ratios of raw counts or as regressions of counts on time, undetected trends in detection probability can either mask changes in abundance or cause one to observe a spurious trend in abundance. Use of appropriate covariates can correct for some of the factors influencing detection probability. For instance, Link and Sauer (1998) were able to detect, and correct for, changes in the competency of bird point count participants over time. However, these approaches cannot be used to correct for unrecorded covariates that influence detection, nor for covariates that may be associated with trends in abundance. For instance, decibel level has undoubtedly increased along roads in the United States over the past several decades, but historically has not been recorded at U. S. Breeding Bird Survey (BBS) point counts. Decibel level appears to be inexorably linked with the auditory detection process (Simons et al. 2007), and so may have led to unmodeled trends in detectability in BBS analyses. Even if decibel level had been recorded, this covariate is likely related to land use and proximity of human development, variables that may influence actual bird abundance. Global warming, succession, and introduction of invasive species are some of the examples proposed by workshop participants that might simultaneously influence both abundance and detectability. If these factors have large influences on expected detection probability, index methods will not be adequate for the analysis of population trends. In these cases, methods designed to explicitly estimate detection probability are needed.

3 Basic Inference Methods

3.1 Distance Sampling

L. Thomas presented to the workshop an overview of "conventional" distance sampling (CDS) methods (Burnham et al. 1980; Buckland et al. 1993, 2001). There are two main variations: line transects, where the survey is performed from a set of randomly located lines, and point transects, where it is from a set of randomly located points. The basic idea is the same for both. Observers record the distance from the line or point to all birds detected within some truncation distance, w (which in practice may be infinity, i.e., all detections are recorded, but some finite truncation distance is almost invariably specified at the analysis stage). The sample units are therefore a set of strips (line transects) or circles (point transects) of known size. Not all birds within the sample units are detected, but a fundamental assumption of the conventional methods is that all birds at zero distance are available and detected. Intuitively, one would expect that birds become harder to detect on average with increasing distance from the line or point. The key to distance sampling is to use the distribution of the observed distances to estimate the "detection function", denoted $g(y)$ – that is the probability of detecting a bird, given it is at distance y. This function can then be used to estimate the average probability of detecting a bird given that it is within a sample unit and available for detection – i.e., p_d. Note that the conventional methods are not designed to estimate availability, p_a – if this is <1 then additional data are required. Given an estimate of detection probability, p_d, it is straightforward to estimate density using design-based methods such as equation (3), since coverage probability is known by design. Let N_s represent the total number of birds whose home ranges overlap the set of sample units surveyed by the two observers. Because distance sampling estimates p_d, abundance estimates obtained using this approach are associated with the birds present at the sample locations during the sample period and available to be detected during that time. If ranges of individual bird do not overlap multiple surveyed sample units, then $E(\hat{N}_s) \approx N_s p_p p_a$.

Assumptions of CDS used in estimating p_d are (1) animals are distributed independently of the line or point locations; (2) all birds at zero distance are detected; (3) distances are measured without error; (4) observations at a line or point take place at an instant in time, so that animal movement is negligible. Assumption 1 (independent animal distribution) is true by design if a large number of sample units are located at random within the study area, and may be violated if there is non-random sample unit placement such as surveys along roads or trails. Substantial violation can lead to substantial bias in the estimator of abundance. Assumption 2 (g(0)=1) may be violated due to "availability bias" (i.e., non-availability of a component of the population, such as female birds not vocalizing in an aural survey) or "perception bias" (birds that are available being missed at zero distance). In both cases, additional information is required to estimate the proportion missed, either through a separate survey (e.g., observations of radiomarked or colormarked birds to directly estimate availability, Section 4.3) or more complex mark-recapture distance sampling methods (Section 3.7). Assumption 3 (no measurement error) may be more easily met in visual surveys, where laser rangefinders can be used to provide accurate distances, but is more problematic in aural surveys, as we discuss later. We also discuss methods designed to correct for bias caused by measurement error in cases where the measurement error process is relatively simple and well understood. Assumption 4 (no animal movement) can also be problematic in some field situations. Responsive movement of animals, such as attraction of birds to the observer before detection, can cause substantial bias, and even random movement can cause significant bias for point transects. One reason that bias tends to be worse in point transects is that the observer is (more or less) stationary; short survey periods (perhaps 3 min) can minimize the problems, but a preferable protocol is a "snapshot" method where only known locations of birds at a pre-defined instant are recorded – see Buckland (2006) for a more detailed description. Another potential solution is to use cue-counting methods, where distances to individual cues such as song bursts are recorded, rather than individual birds. Cue production rate is estimated in a separate survey, and used to convert an estimate of the abundance of cues back to abundance of animals. If possible, estimation of cue production rate is carried out at the same time/place of the actual survey. Detailed advice on survey design and field methods is given in Buckland et al. (2001, Chapter 7) and Strindberg et al. (2004), and recommendations specifically oriented to landbird studies are given in Buckland (2006). A fifth assumption, that each bird detection is an independent event, is not as important in practice.

While not strictly assumptions, there are some additional requirements for robust estimation. First, the detection function should have a "shoulder" – i.e., the probability of detection should remain at or close to 1 initially as distance from the line or point increases. This is often referred to as the "shape criterion". Second, the detection function should be smooth. Third, the models used for g(y) should be flexible, in the sense that they can take a wide variety of plausible shapes, so that they will be a good approximation to the true detection function given a large sample size. Such models are termed "model robust". Fourth, an adequate sample size of distances is required. "Adequate" is hard to define unambiguously, since more samples are required for "difficult" detection functions (e.g., small shoulder or steep fall-off in detectability), however Buckland et al. (2001, Section 7.2.2) recommend at least 60–80 observations for line transect studies and 75–100 for point transect studies. Under these conditions, the estimators of abundance are "pooling robust", meaning that even large variations among individuals in probability of detection due to observer, habitat, etc. cause little bias in the estimate of p_{d} and hence abundance. Thomas presented simulations that demonstrated this, but which also showed when heterogeneity in detectability is extreme (e.g., singing males and cryptic females), significant bias can arise (Thomas et al. in prep.). One potential solution to extreme heterogeneity is to include the factors causing the heterogeneity as additional covariates in the detection function model (Marques et al. 2007). This approach may also offer a partial solution to the sample size requirements if rarer species can be combined with more common ones and species used as a detection function covariate (Alldredge et al. 2007b). A fifth requirement is of an adequate sample of lines or points (minimum 10–20, Buckland et al. 2001, Section 7.2.1) for reliable estimation of the spatial component of variance of the abundance estimate.

3.2 Multiple Observers

M.W. Alldredge presented the basic ideas underlying models based on multiple observers and time of detection. The multiple-observer approach requires that 2 or more observers either sample a point together or traverse a line transect together, keeping track of observer-specific detections of individual birds. The approach is adapted from work by Cook and Jacobson (1979) on estimation approaches for aerial surveys and is closely related to capture–recapture modeling of closed population data (Otis et al. 1978; Seber 1982). Field sampling by multiple observers can be treated in either of two general ways, labeled dependent and independent. Both approaches have been used with avian point count data. Our descriptions will be of point counts, although we note that both approaches can be implemented along line transects as well.

Under a dependent double-observer approach, at each point one observer is designated as "primary" and the other as "secondary". The primary observer identifies all birds detected and communicates each detection to the secondary observer. The secondary observer records these detections of the primary observer, as well as additional birds that the primary observer does not detect (e.g., Nichols et al. 2000). Observers switch roles at different points such that each observer serves as primary observer for about half the sample points. The data for a series of point counts conducted in this manner by two observers can be summarized as four sufficient statistics for each species or group of species to be analyzed together: the number of birds detected by observer i (i=1,2) when that observer was the primary observer, and the number of extra birds detected by observer i when the other observer was primary observer. These data can then be used to estimate the number of birds exposed to sampling efforts at the group of surveyed points under a general model in which detection probabilities differ between observers and among species. Reduced-parameter models can then be developed to evaluate hypotheses about the similarity of detection probabilities for observers and bird species.

The general model for the dependent double-observer approach assumes that detection probability of an observer does not vary depending on the observer's role as primary or secondary. It is assumed that detection and recording of a bird by the secondary observer does not influence the probability that the primary observer detects the bird. The approach assumes that all birds of a species found within the sampled area (frequently defined by a specified fixed radius) at the time of the sample have the same probability of being detected by an observer (the probabilities may be different for the two observers). The number of birds exposed to sampling efforts is assumed to be fixed (population closure), and this assumption may limit the sample to a very short time period (e.g., 3 min). Nichols et al. (2000) provide a more detailed discussion of field sampling methods, analysis methods, underlying model assumptions and field approaches directed at meeting model assumptions.

The independent multiple-observer approach to point counts requires that two or more observers independently record detections from the same basic field sampling point for some specified short period of time (e.g., 3 min). The observers will typically have a schematic diagram of the surveyed area (concentric circles of different radii around the sample point) such that detections of a species are recorded on the diagram with a time of detection. Immediately following the count, observers confer and compare diagrams with the purpose of matching detections of the same birds and developing detection histories for every bird detected (Alldredge et al.

2006). For example, with two independent observers, three detection histories and associated sufficient statistics can be observed: x_{11} = number of birds detected by both observers, x_{10} = number of birds detected by observer 1 and not observer 2, and x_{01} = number of birds detected by observer 2 and not observer 1. These data are then analyzed as closed model capture–recapture data, where time-specific variation in the capture–recapture context is analogous to observer-specific variation in the point count context. Observer variation in detection probability can be incorporated into models, or detection probability can be modeled as a constant for all observers. If >2 observers conduct the sampling, then finite mixture heterogeneity models (e.g., Norris and Pollock 1996; Pledger 2000) can be fit that permit variation among individual birds in their probabilities of being detected (Alldredge et al. 2006).

Assumptions underlying the independent multiple observer approach include population closure and independence of detections among observers. It is further assumed that detection histories are correct (i.e., that there are no matching errors). The double observer models assume the same detection probabilities for the different individuals of the same species within the sample area for each observer. This homogeneity assumption can be relaxed with >2 observers using finite mixture heterogeneity models (Alldredge et al. 2006). If limited-radius counts are used, it is assumed that the observer correctly determines whether each detected bird is inside or outside of the specified radius.

The detection probabilities estimated using multiple-observer approaches pertain to the conditional probability of detection, p_d, given that the bird is present in the area exposed to sampling efforts (e.g., located inside the area defined by a fixed radius) at the time of the sample (probability associated with this event is p_p) and given that it vocalizes or is otherwise available during the sample period (associated probability is p_a). The detection probability is also conditional on the initial probability that the point count sample unit is overlapped by the home range of a particular bird (associated probability p_s), but as this probability is a component of all bird detections, we will omit it from our discussions of the different detection parameters estimated by the different methods. Let N_s represent the total number of birds whose home ranges overlap the set of sample units surveyed by the two observers. Because multiple-observer estimation focuses on p_d, the abundance estimated using this approach is associated with the birds present at the sample locations during the sample period and available to be detected during that time. If ranges of individual bird do not overlap multiple surveyed sample units, then $E(\hat{N}_s) \approx N_s p_p p_a$.

3.3 Time of Detection

The time of detection approach to abundance estimation from avian point counts requires only a single observer at each sampled point. The duration of the entire point count is divided into component time intervals (e.g., a 3-minute point count might be divided into three 1-minute time intervals). The initial development of this approach focused on the time interval of first detection for each bird detected in the count (Farnsworth et al. 2002). If we let K denote the total number of time intervals in a point count, then the sufficient statistics under this approach are the numbers of birds detected for the first time in each interval, $x_1, x_2, \ldots x_K$.

Modeling of these sufficient statistics requires an abundance parameter, *N*, and detection parameters, *p*. The modeling is identical to that of removal modeling in capture–recapture literature (Otis et al. 1978; Seber 1982). Estimation is not possible with interval-specific detection parameters, and these parameters are typically assumed to be constant over time when all intervals are of equal length. Farnsworth et al. (2002) also consider the situation where the intervals are of unequal length. Let t_i represent the length of interval *i* expressed in some relevant time unit (e.g., minutes). Then the probability of a bird being detected during interval *i* can be written as: $p_i = 1-(1-p)^{t_i}$, where *p* is the probability of detection for a single unit of time. Under an equivalent continuous time formulation, let ϕ_i be the instantaneous rate of detection during interval *i*, or "Poisson detectability coefficient" (Alldredge et al. 2007a). Then the probability of detection during interval *i* is: $p_i = 1 - e^{-\phi_i t_i}$. Heterogeneity among individual birds at a sample unit can be modeled using a finite mixture (e.g., Pledger 2000) or other approach (other estimators for M_{bh} of Otis et al. 1978).

It is also possible to treat time of detection data as standard capture–recapture data, rather than as simply removal data (e.g., Alldredge et al. 2007a). For example, instead of recording the time interval of first detection, the observer records all intervals of detection for each bird. For example, with two time intervals, three detection histories and associated sufficient statistics can be observed: x_{11}= number of birds detected in both time intervals, x_{10} = number of birds detected only in the first time interval, and x_{01} = number of birds detected only in the second time interval. These data are then analyzed using standard capture–recapture models for closed populations, with time interval of detection being equivalent to a sample period in closed capture–recapture. It seems likely that the initial detection of an individual might have a different (typically smaller) detection probability than subsequent detections, in which case analysis would be based on the time intervals of first detection.

Assumptions underlying the time of detection approach include population closure and independence of detections of an individual among the different sample intervals. If time of first detection only is modeled then the latter independence assumption is no longer relevant. The time of detection approach assumes the same detection probabilities for the different individuals of the same species within any sample interval. This homogeneity assumption can be relaxed with >2 intervals using finite mixture heterogeneity models. It is assumed that birds are not double-counted (1 bird mistakenly counted as 2). The time-of-detection approach is typically applied to sample plots defined by a fixed radius (fixed distance from the point). If fixed-radius counts are used, it is assumed that the observer correctly determines whether each detected bird is inside or outside of the specified radius. A final assumption concerns the circle defined by the modeling of the availability process and thus of p_a. Farnsworth et al. (2002) initially modeled availability as a random process in the sense that each bird had an equal probability of vocalizing in any time interval. However, if the process for individual birds is Markovian, in the sense that vocalization during one interval causes the probability of vocalization in subsequent intervals to be larger or smaller, then this process should be incorporated into the modeling and estimation (e.g., see analogous situation in capture–recapture with temporary emigration, Kendall et al. 1997).

The detection probabilities estimated using time of detection approaches pertain to the product of (1) the conditional probability of being available (p_a), given presence in the sample unit at the time of the survey (associated probability p_p), and (2) the conditional probability of bird detection, p_d, given presence and availability. The detection probability estimated by this approach is also conditional on the initial probability that the point count sample unit is overlapped by the home range of a particular bird (associated probability p_s), but this term is omitted in our development as in the example for multiple observers. Let N_s once again represent the total number of birds whose home ranges overlap the set of sampled points. Because estimation based on time of detection focuses on the product $p_a\, p_d$, the abundance estimated using this approach is associated with the birds present at the sample location during the sample period but is not conditioned on availability. Thus, if ranges of individual birds do not overlap multiple surveyed sample units, then$E(\hat{N}_s) \approx N_s p_p$.

3.4 Repeated Counts

J. A. Royle outlined approaches to the use of repeated count data from the same locations as a basis for inference. In doing so, he noted that the data arising from point counts can be viewed naturally in terms of hierarchical models with two basic components, an observation component and a process component. The observation component of such models deals with survey methods and avian detection, conditional on true abundance, whereas the process component deals with the distribution of true abundance over space or survey points. Royle noted that the likelihoods for the approaches described above (distance sampling, multiple observers, time of detection) can all be viewed as multinomial observation models in at least some instances (e.g., distance sampling with data grouped by intervals). He then specified two other data types resulting from repeated point counts at the same locations, the replicate counts themselves and the reduced presence–absence (detection–nondetection) data separating 0 and positive counts.

The sampling protocol involves simple point counts (no necessary collection of ancillary data on distance, time of detection, etc.) at the same locations at multiple times. The different sampling occasions are typically close together in time (e.g., 5 counts at each point during May or perhaps the breeding season) to achieve a kind of closure in the sense that the same group of breeding birds is potentially exposed to sampling efforts at each occasion (see more complete discussion below). The data arising from such sampling are the counts for each sampling occasion. Conditional on the true abundance of birds potentially exposed to sampling efforts at a point, the counts at one location can be viewed as binomial random variables with detection probability modeled as a constant, or perhaps as a function of site- or time-specific covariates (Royle 2004). If abundances are viewed as site-specific, then the resulting likelihood contains many abundance parameters and can be difficult to maximize (e.g., Carroll and Lombard 1985).

Under the approach outlined by Royle, the conditional (on detection probability and site-specific abundances) binomials can be viewed as the observation component of the likelihood. Royle (2004; see also Kery et al. 2005) then proposed use of a reasonable density (e.g., Poisson or negative binomial, with parameters possibly modeled as functions of covariates) for the distribution of true abundances over the spatial sampling locations as the process component of the model. Estimation under the resulting hierarchical model can be accomplished numerically, although a large number of sampling locations will typically be needed to achieve adequate performance.

When the repeat count data at each location are condensed into detections (at least 1 individual of the species counted at the sampling occasion) and nondetections (species not detected at that occasion), abundance can be estimated by relying on the relationship between detection probability at the level of the sample unit ($p*$) and the detection probability of individual birds (p). This relationship is a function of abundance at location i, as $p_i* = 1-(1-p)^{N_i}$ Presence–absence detection data provide information about detection probability at the level of the sample unit, and variation in this probability over space provides information about the distribution of N_i (Royle and Nichols 2003).

Under the basic approach using replicate counts, the binomial detection parameter is actually the product$p_p p_a p_d$. In order to appear in a count, an individual bird must be present in the sample unit at the time of the sampling, must be available for detection during the time of sampling and then must be detected. Similarly, the individual bird detection parameter of models for presence–absence data reflects this product. Thus, the detection parameters estimated using the approach of Royle (2004) and also Royle and Nichols (2003) will usually be smaller than detection probabilities using other approaches as they include the detection component associated with the complement of temporary emigration; the probability that a bird is in the portion of its range exposed to sampling efforts during the survey. Similarly, the abundance estimated using these approaches includes not just the animals that are present during a single survey, but all birds with some non-negligible probability of being in the sample unit during a survey (i.e., all birds whose ranges overlap the sample unit). In the language of capture–recapture modeling (e.g., Kendall et al. 1997; Williams et al. 2002), the abundance estimates produced by the replicate count approaches represent "superpopulation" sizes (N_s) of all of the birds whose home ranges overlap the sample units.

We note that the above inclusion of p_p as a component of the detection parameter estimated using a repeated counts approach is based on the usual field application in which the replicate counts are separated in time by a period during which substantial bird movement is expected (e.g., at least 24 hours). If repeat counts can be independently conducted on an area within a much shorter time frame (e.g., every third minute for 12 min), then this approach will also be conditional on the set of birds present during this period. Our point is that the different interpretation of estimates resulting from the repeated count approaches is not based on anything inherent in the approach itself, but is instead determined by the time intervals over which it is typically applied.

The population closure assumption underlying repeated count approaches applies to the superpopulation size rather than to abundance during any survey occasion. So the number of birds with ranges overlapping the sample unit is assumed not to change over the time period of the repeat counts. The probability of detecting an individual in a given survey (representing the product $p_p p_a p_d$) is generally assumed to remain constant over survey occasions, but this assumption can likely be relaxed in various ways. The repeat count approach generally assumes the same detection probabilities for the different individuals of the same species within any survey, but this assumption can likely be relaxed using various mixture distributions for the detection parameters. It is assumed that birds are not double-counted (1 bird mistakenly counted as 2) during a survey. Although not an assumption, the estimation of superpopulation size, rather than abundance of birds in a sample unit at a snapshot in time, has implications for interpretation and use of resulting estimates. In particular, use of estimates obtained from repeated counts to estimate abundance in some larger area of interest may be more difficult. In addition, the superpopulation of birds exposed to any particular sample unit depends on bird mobility and may change with bird density, such that comparative uses of resulting estimates must be evaluated carefully.

3.5 Double Sampling

The general term "double sampling" from general sample survey statistics (e.g., Cochran 1977; Thompson 2002a) has recently been used to refer to a specific approach to estimation of avian density from count data (Bart and Earnst 2002). The approach involves extensive rapid survey methods on a typically large number of sample units and intensive surveys on a subset of these sample units. If the intensive surveys yield unbiased estimates of true abundance, then the ratio of counts from rapid surveys to estimates based on intensive surveys provides estimates of detection probabilities (for avian applications see Smith 1995). The usual approach to double sampling in wildlife surveys (e.g., Pollock et al. 2002; MacKenzie and Royle 2005) uses methods such as those described above as the intensive surveys, as these methods require ancillary data (distances to detected birds, times of detection), multiple observers, or repeat counts. The rapid surveys often involve single counts with no ancillary data.

The intensive approach described by Bart and Earnst (2002) for Alaskan shorebirds required several hours per day for about 3 weeks, with much of this time reportedly spent searching for nests and counting territorial birds. No estimator was presented for abundance on the intensive survey plots, so we did not discuss or attempt to evaluate this approach. The definition of abundance on a plot as "number of territorial males whose first nest of the season, or territory centroid for non-nesters, was within the plot" (Bart and Earnst 2002) indicates interest in a subset of the superpopulation of all birds whose ranges overlap the sample unit. The basic assumption underlying this approach is that observers on intensive survey plots end the season with an unbiased estimate of abundance on the plot.

3.6 Interpreting Estimates

Both formal and informal discussion at the workshop was devoted to the issue of interpreting estimates from these basic approaches to abundance estimation. Interpretation is dependent on which components of the detection process are the bases for conditioning and which are considered as part of the process to be estimated. All of the approaches begin by conditioning on the birds whose home ranges overlap the selected sample units, with expected value for this set of birds $E(N_s) \approx N^* p_s$, where N^* again is the total number of birds in the entire area of interest from which samples are drawn. If ranges of individual birds do not overlap multiple selected sample units, then abundance estimates based on repeat sample approaches at the selected sample units estimate N_s. Estimates based on time of detection approaches estimate the number of birds present in the selected sample units during the survey period, $N_s p_p$. Estimates based on distance sampling and multiple observers estimate the number of birds that are present in the selected sample units during the survey period and available to be detected, $N_s p_p p_a$. From an "open field" perspective, the quantity $N_s p_p$ is roughly equivalent to the number of birds with home range centers lying within sample units (assuming that observers do not influence the probability of presence, etc.). Thus, one key distinction among abundance estimates based on these four approaches is that a repeated count estimate does not apply to a known fixed area (unless the time scale for repeat surveys is very short), whereas the other three approaches yield snapshot estimates that can be associated with sample units of known area. However, estimates or assumptions about p_a will typically be needed to make inferences about absolute abundance when distance sampling or multiple observer methods are employed. Finally, recall that the above expectations apply to birds for which $p_a > 0$. If a subset of birds is simply invisible to detection efforts, then abundance estimates will of course not include this subset.

The above considerations lead to three points that deserve emphasis. First, if abundance is estimated for a set of sample units at which point counts are conducted, different estimates are expected depending on which estimation approach is used. Second, approaches based on combinations of the above methods provide an opportunity to separate components of the detection process in cases where this might be useful. Third, the different interpretations of the closure assumption for the basic methods lead to the recognition that some approaches lend themselves more readily to estimation of abundance for the entire area from which samples are drawn, than do others. For example, investigators will frequently be interested in the number of birds whose range centers are located within the specified area of interest (define this number as N). Distance sampling, multiple observers, and time of detection all condition on the birds that are actually present in the sample units during the survey period (expectation $N_s p_p$). Given application of design-based sampling protocols, estimates based on these methods can be readily used to extrapolate (based on p_s) to an estimate of overall abundance, N, for the entire area of interest. However, repeated count approaches yield estimates of the number of birds whose ranges overlap the selected sample units such that extrapolation of these estimates to estimate N would require extra information (e.g., about the average number of sample units overlapped by each bird or the actual area sampled by a point count).

3.7 Combination Approaches

Several attendees noted the possibility of using both distance sampling and either multiple observer or time of detection sampling (e.g., D.L. Borchers, M.W. Alldredge, K.H. Pollock, M. Efford). The different kinds of information resulting from such combination-method point or transect counts can be viewed in various ways. For example, from the perspective of multiple-observer and time at detection approaches, distance may be viewed as a covariate that is used to deal with heterogeneous detection probabilities (Alldredge et al. 2006; 2007a, b). From the perspective of distance sampling, use of multiple observers at the same locations and times can be used to estimate the probability of detection on the transect line and/or to test the hypothesis that this probability is 1 (e.g., in aerial surveys). In addition, multiple observers at lagged times (e.g., while traversing an oceanic line transect) can be used to estimate p_a, where the complement, $1{-}p_a$, includes individuals that are submerged. The need for a time lag between multiple observer counts in order to estimate p_a is based on the same thinking as for time of detection approaches. Indeed, Farnsworth et al. (2005) combined time at detection and distance sampling in order to estimate both p_a and density, corrected for availability.

A well-developed literature now exists for combined applications of distance sampling and double-observer approaches (Alpizar-Jara and Pollock 1996, 1999; Manly et al. 1996; Hiby and Lovell 1998; Borchers et al. 1998a, b; Borchers 1999; Laake and Borchers 2004), although applications to avian point count data have been relatively recent (Alldredge et al. 2006; Kissling and Garton 2006). It is now widely recognized that such combination approaches can be used to relax assumptions required by single component approaches and sometimes to permit separation of different components of the detection process. For example, mark-recapture distance sampling (MRDS) allows estimation of $g(y)$ without the requirement that animals be distributed independently of the transect lines or points (Laake and Borchers 2004, Section 6.3.1.1). Intuitively this is done using the proportion of animals seen ("marked") by one observer at distance y that were also seen ("recaptured") by the other observer. A similar approach can be used to deal with responsive animal movement (Laake and Borchers 2004, Section 6.3.1.2).

When detection on a point or line is not certain, MRDS methods are typically not pooling robust in the sense of CDS. However, D. L. Borchers presented some recent work at the workshop indicating that if the distribution $g(y)$ is assumed to be known (e.g. uniform), one can use data on the distribution of observed distances to relax heterogeneity assumptions (Borchers et al. 2006).

K.H. Pollock discussed the possibility of using double observers and time of detection simultaneously. Pollock noted that this approach can be viewed as a robust design (Pollock 1982; Williams et al. 2002), with the point count divided into K

time intervals and detection–nondetection data for each observer at each interval. For example, in a 3-minute point count with 1-minute intervals and 2 observers, a detection history for a bird might be: 00 11 01. During the first interval the bird was not detected by either observer. Both observers detected it during the second, and only the second observer detected the bird during the final interval. This approach would permit separate estimation of 2 detection components, p_d and p_a. Pollock also noted that whereas the initial development of the time of detection approach assumed random availability (i.e., constant p_a, Farnsworth et al. 2002, or perhaps variation in p_a associated with time or distance, Alldredge et al. in 2007a), the combination approach would permit treatment of availability as a Markov process. Under such a model (similar to Markovian temporary emigration models of Kendall et al. 1997), the availability of a bird in a specific time interval could differ depending on whether the bird was available in the previous interval. Such Markovian modeling would seem to represent a reasonable hypothesis about avian singing behavior.

4 Field Tests, Field Trials, and a Simulation Experiment

4.1 Bird Radio System

One of the primary motivations for the workshop was the recent and ongoing work on field tests of estimation methods for avian point counts. In particular, the bird radio system developed by the group at N.C. State University has provided an excellent test system with which to test estimation methods themselves as well as specific assumptions that underlie those methods (e.g., Alldredge et al. 2007c; Simons et al. 2007). Workshop presentations by T.R. Simons, M. Alldredge and K. Pacifici noted that tests with their system showed that most observers had substantial difficulties estimating distances to birds that are detected aurally, particularly at large distances (beyond 60 m). Heaping of observations at certain distances was also a problem. Double counting of individual birds was a substantial problem, especially in experiments simulating relatively complex field situations (several bird species with heterogeneous singing rates and different speaker orientations). Density estimates based on distance sampling approaches were too high and estimated detection probabilities too small when observer-estimated distances were used. However, there was speculation during discussion that with more truncation better results might be obtained at the cost of decreased sample size of detections, and hence reduced precision. Distance sampling approaches based on true distances to detected bird vocalizations performed well.

Use of the radio bird system to test double-observer and time of detection approaches led to interesting insights as well. K. Pacifici noted that multiple observer approaches sometimes seemed to provide reasonable estimates and at other times yielded biased estimates. In the case of independent observers, there were substantial difficulties with matching, in that observers found it difficult to decide whether detected birds represented a bird detected by both observers or different birds detected by each observer separately. There was some experimentation with objective decision rules for determining matches. Despite matching difficulties for independent observers, estimation results were fairly similar for independent and dependent double-observer approaches. Double counting of birds occurred fairly frequently during experiments with time of detection methods. Placement of singing birds in the wrong time interval was another common error in these tests. In cases where observers were asked to restrict data to birds within a fixed radius, it was common to include birds outside the detection radius. Finally, many observers were unable to identify the species of all detected birds and thus had a number of "unknown" observations. Overall, estimates based on time of detection approaches were biased low and frequently appeared to better estimate $N_s p_p p_a$ than $N_s p_p$. Thus, the approach did not seem to account well for birds that were present but did not happen to vocalize during the survey period.

Although it was difficult to summarize results of all the various experimental tests conducted using the bird radio system, it was clear that none of the basic estimation methods performed as well as would be desired in the common situation of many individual birds from a diverse community. Although this work is still underway, workshop participants involved in the bird radio project offered several summary conclusions and recommendations. Auditory detection of singing birds was found to be much more problematic than previously thought. Localization of sound proved to be very difficult, leading to problems in matching birds when applying multiple-observer approaches, in double counting for single observers, and in estimating distances both for distance sampling and for the purpose of identifying birds lying in and out of fixed radius sample units. Observer performance was especially poor in more complicated experiments with larger numbers of species and individual birds. Performance was better in simple experiments with a small number of species, leading to the recommendation that many of the methods discussed above would be most useful in situations where observers concentrate on a very small number of focal species. In situations where interest must focus on an entire diverse community, it was proposed that alternative approaches such as occupancy estimation and modeling (e.g., MacKenzie et al. 2006) should be considered. The rationale for this recommendation is that occupancy estimation simply requires information on detection, or not, of each species, and does not require keeping track of individual birds over time and space. Another suggestion was that more effort be allocated to the field effort with different observers being responsible for different sub-taxa.

4.2 Robust Distance Sampling Methods Study

S.T. Buckland reported results of a comparison of various distance sampling methods against results from territory mapping for four species (common chaffinch, great tit, European robin and winter wren) in a Scottish woodland and parkland study area of approximately 40 ha (Buckland 2006). The methods compared were

(1) a conventional point transect of 5 min duration; (2) a point transect survey using the "snapshot" approach, with 3 min allowed before the snapshot moment and 2 min afterwards; (3) a point transect using cue counts, where the number of song bursts were recorded using 5 min surveys, and a separate note was kept of birds where the observer was confident all cues were heard so that cue production rate could be calculated; (4) a conventional line transect survey. The habitat was more open than in the bird radio study, resulting in more frequent visual detections, or at least identification of the tree or bush from which a bird was heard to sing. A laser binocular was used to accurately measure distance in these cases. The estimated 95% confidence limits contained the values obtained by territory mapping for all species and methods with one exception: the estimate for great tit from conventional point transects was too high, likely due to animal movement. The snapshot method was found to be the most efficient of the point transect methods, in terms of precision per hour field time, but line transect sampling was more efficient than all the point transect methods. This is likely a general finding, leading to a recommendation that line transect sampling be employed rather than point transect methods where they are possible. Where this is not feasible (for example in difficult terrain, or multi-species surveys where observers may get swamped) then snapshot methods are preferred over conventional point transects. Cue counting may be particularly useful for single-species surveys, but are unlikely to be of use when the goal is to survey many species simultaneously as the observer will quickly become overwhelmed. Insufficient resources were devoted to estimating cue rates in this study, but given more resources the method offers great potential to provide reliable results where animal movement is a problem. The study also included a computer simulation component, which showed that edge effects caused by undersampling near the edge of small study areas do not necessarily cause a serious problem with the reliability of results, nor does sampling at closely spaced points, so that the same birds are heard from multiple points.

4.3 Grassland Bird Availability Study

D.R. Diefenbach reported results of a study of grassland birds in central Pennsylvania, U.S.A. designed to estimate availability, p_a (Diefenbach et al. 2007). Color-marked and radio-marked grassland sparrows were followed for relatively long periods of time (e.g., 1 hour) while observers recorded their availability with respect to auditory detections at point counts, and auditory and sight-based detections at both point counts and line transects. Overall, availability of male sparrows was low ($\hat{p}_a < 0.5$) and quite variable for periods $\leq$ 10 min. Time of the breeding season was an important source of variation with availability being greatest in late May and early June but declining to very low levels late June and July. Diefenbach et al. (2007) concluded that substantial bias and heterogeneity in abundance estimates would be obtained if this source of detection probability were not incorporated in population estimates.

4.4 Other Field Trials

P.F. Doherty recounted experiences with organizing substantial field efforts based on avian point counts and occupancy surveys in Colorado and southern California. The California work was on an endangered species, and survey efforts included many points with no birds detected and only small numbers of detections at remaining points. His Colorado example included work on design-based sample allocation, and Doherty noted substantial problems in accessing a large number of selected sites. A relatively large proportion (10%) of landowners denied access to their lands, leading to a change in scope of inference. Doherty highlighted the frequent need to deal with unanticipated practical problems that may necessitate reevaluation of methods and approaches.

4.5 A Simulation Experiment

Using his "Open Field" view of abundance, M.G. Efford presented preliminary results from a simulation study, in which the probability of detection of an animal depended primarily on distance of the animal's home range centre from the detector (Efford and Dawson in prep.) He showed that if distance has a strong effect on probability of capture then many methods show substantial negative bias in estimated density, including multiple-observer, time of detection and repeated count methods (all using 4 "capture occasions"). This was the case even when using mixture models to try to account for distance induced heterogeneity. Using distance sampling methods or adding distance as a covariate in some of the above methods produced lower bias, although it appeared to be important which model was used for the relationship between detection probability and distance.

5 Methodological Advances and Extensions

5.1 Distance Sampling with Density Gradients

T. A. Marques presented recent work extending conventional distance sampling methods to deal with situations where animal distribution is not independent of distance from the transect line or point (Marques 2008). Such methods are useful in two contexts. The first is when there are few transects located at random: although random transect placement ensures independence on average, the actual distribution of animals for a single realization may be far from the average if there are few transects. The second is where transects are not located at random, but instead follow roads, trails, rivers, etc. In these circumstances, it is not possible with conventional distance sampling to distinguish between changes in animal density with respect to distance from the transect line and changes in detectability, and hence it is not possible to use the observed distances to estimate p_d without additional data. One solution is to use double-observer methods mentioned earlier. A second related approach, for line transects, is a "crossed design", where two sets of transects are laid out perpendicular to one another; information in the along-transect distribution of observations is then used to estimate the distribution of animals perpendicular to the other transect (see Buckland et al. 2007a for an application to plant surveys). Indeed any additional survey that provides information about the distribution of observations perpendicular to the main line transect could be used. A third approach is to perform point transects along the linear feature (road, trail, etc.) and record both the distance and angle of observations. Then, under the assumption that the detection function is radially symmetric, it is possible to use differences in the distribution of observed distances in the along-road direction and perpendicular-to-road direction to simultaneously estimate detection probability and animal distribution. Marques presented a simulation study of this third method that included different effects of the linear feature (road) on animal density, and demonstrated marked reduction in bias relative to conventional methods, although the method did not appear very robust to misspecification of the model for animal density. It appears that the new method will not over-ride the advice that roadside (and other non-random) transects should be avoided whenever possible.

5.2 Measurement Error in Distance Sampling

Part of the presentation by Marques concerned the effects of measurement error on distance sampling estimators, and extensions to the conventional methods that account for measurement error (Marques 2004, 2008; Burnham et al. 2004: 11.9, Borchers et al. in prep.). Random, non-systematic measurement error can lead to bias in abundance estimates, with the magnitude of the effect depending on the type of survey (bias is generally much worse for point transects than line transects) and type of measurement error (e.g., constant variance with respect to distance from the transect vs. constant CV) and its magnitude. In general, measurement error close to the line or point has more effect than the same level of error far from the transect. Systematic errors cause larger bias than non-systematic errors. Given a model for the error process it is generally possible to estimate error process parameters jointly with detection function parameters, and so to reduce or eliminate the bias if the error process model is correct, at the expense of lower precision in the estimate of p_d. In some cases, the correction method is quite simple to implement (Marques 2004). However, the true process leading to measurement errors may be quite complex, as has been demonstrated by the bird radio studies, and the methods may not be robust to misspecification of the error model. Marques' take-home message was that distance measurement error should be minimized wherever possible in the field through the use of, e.g., training, calibration exercises, appropriate field methods (e.g., allowing observers to move around during or after point transects to better locate birds) and technological aids (e.g., laser binoculars for visual surveys).

5.3 Bird Misclassification

R. Webster presented some joint work with K.H. Pollock and T.R. Simons on dealing with misidentification of individual birds in certain approaches to estimation using point count data. He focused on the time of detection approach, although this work has clear relevance to multiple observer approaches as well. The basic problem involves the misidentification of individual birds in the different intervals, such that detection histories contain errors. For example, assume that a bird is detected in both periods of a 2-period point count. Assume that the bird is misidentified as a new bird in period 2, so that the true detection history, 1 1, gives rise to two incorrect detection histories, 1 0 and 0 1. Webster modeled this problem by introducing a correct classification probability, α. For example, if p_t denotes detection probability for period t, then the underlying probability associated with detection history 1 0 would be written as: $p_1(1 - p_2) + p_1 p_2(1 - \alpha)$, with the second additive term corresponding to a misidentification of the original bird as a different individual. Webster noted that he was experimenting with approaches to fitting such models, including use of a χ^2 loss function. Subsequent discussion (e.g., by W.A. Link, L. Thomas) focused on the possibility of accounting for the dependencies among the detection histories resulting from a single study with misidentification.

5.4 Surveys of Cryptic Species

S.T. Buckland presented three recently-developed methods based on distance sampling that are appropriate for cryptic species not surveyed well by conventional methods. The first is a "lure point transect" (Buckland et al. 2006), where lures such as playbacks of territorial songs are used to elicit a response from animals of the target species. Conventional methods are not applicable because it is likely that the animal will respond by approaching the surveyors before detection – causing a positive bias in estimates of abundance using conventional estimators. Instead, the detection function is estimated in a separate survey, where trials are conducted by two observation teams, both of which search for animals without using the lure. When one team detects an animal, the other team deploys the lure and records whether they detect a response. A set of such trials allows a binary regression to be used to determine probability of detection given distance of the animal from the lure, and therefore average detection probability. If this detection probability can be assumed to apply also to birds detected in the main survey (for which their initial location before the lure was used is not known) then it is possible to calculate abundance. This method has been used in a survey of Scottish crossbills (Buckland et al. 2006).

The second approach is a "trapping point transect", which is essentially identical to the above approach, except that a trap is used to capture animals rather than a lure. Trials can be set up in much the same manner as for lure point transects: a marked animal is located (e.g., because it is radio-collared, or because it has been released from a different trap) at the same time as the trial trap is set a known distance away, and whether the animal is caught in the trial trap or not within some fixed time interval is recorded. The methods are described in Buckland et al. (2006) and are being tried on an endangered woodrat in Florida. Note that if repeated captures of marked animals in different traps are anticipated then standard mark-recapture methods may be used for estimating animal abundance, or the more recent methods of Efford (2004) and Efford et al. (2005).

The third approach is crossed line (or strip) transects (Buckland et al. 2007a), which were mentioned in Section 5.1. They are potentially suitable for sedentary cryptic objects (such as cryptic plants or bird nests) and allow for probability of detection on the transect line to be less than 1 and for non-independence of animal distribution with respect to the transect.

6 Space-time Variation in Abundance

Although much of the workshop emphasis was on approaches for dealing with the various components of detection, the inferences of most interest to biologists, and thus the ultimate objective of our efforts, involves variation of density or abundance over space and/or time. Several workshop attendees were involved in such work and gave brief presentations on their directions.

6.1 Spatial Variation

J.A. Royle presented examples of work on spatial modeling of avian abundance and occupancy based on avian count data. Specifically, he focused on replicate counts as a basis for inference about abundance (Royle 2004) and occupancy (Royle and Nichols 2003). In the terminology of his presentation on replicate count approaches, his focus in this presentation was on the process model component rather than the binomial component dealing with observation. He noted that the Poisson distribution provides a reasonable model for spatial distribution of abundance, N_i for location *i*. Modeling can be based on the Poisson mean, λ, and linear models can be developed for $\log\lambda_i$ as a function of covariates associated with site *i*. If counts exhibit excess variation with respect to the Poisson, then the negative binomial provides an alternative model for spatial distribution, and the negative binomial mean can again be modeled as a function of site-specific covariates. Royle presented several examples in which this basic approach was used for mapping avian distribution as a function of environmental covariates (Kery et al. 2005; Royle et al. 2005).

R. Webster was similarly interested in spatial modeling (with K.H. Pollock and T.R. Simons), but used the time of detection approach as a basis for modeling. Abundance was modeled as a Poisson–lognormal mixture. He explored the use of conditional autoregressive models in which abundance at a site was modeled as a function not only of selected site covariates (e.g., elevation), but also as a weighted (by distance) function of abundances at nearby sites. He noted that the approach provided a reasonable means of investigating spatial processes and that it could be used with repeated count, as well as time of detection, data.

S.T. Buckland described the general framework of Hedley (2000), in which the location of animals is seen as a realization of a random process with some spatially-indexed intensity (the density), and the number of animals in a given area can therefore be described by an inhomogeneous Poisson process (IPP). The detection process represents a thinning of the IPP, which yields another IPP. Given a parametric form for the intensity and detection processes, a likelihood can be derived, but it is often more convenient to treat the estimation of parameters for the two processes separately as they often operate on very different scales (the detection process over a small scale – e.g., 10's or 100's of meters, while we usually wish to model spatial variation in density at much larger scales). This notion has led to the two-stage methods for spatial modeling of line transect data of Hedley and Buckland (2004) and Hedley et al. (2004). Spatial smoothing is accomplished using generalized additive modeling, and improved methods are in development to deal with problems caused by irregular topography (Wood et al. in press) and local clustering that is not explained by the large-scale smooth (Bravington et al. in prep.).

6.2 Temporal Variation

A general overview of methods for temporal inferences when detection is uncertain (but focusing on distance sampling) is given by Thomas et al. (2004). One distinction made in that paper is between empirical modeling and process (or mechanistic) modeling. Aspects of the former approach were covered in a talk by W.A. Link (in joint work with J.R. Sauer) based on work with data from the North American Breeding Bird Survey (BBS) and the Christmas Bird Count (CBC) survey. Their hierarchical modeling approach to BBS data models the logarithm of the expected count at a survey site as a linear function with stratum-specific intercept, time slope, and year effect, with observer effects based on observer identity and year (reflecting temporal changes in observer abilities), and an overdispersion parameter (e.g., Link and Sauer 2002, 2007; Sauer and Link 2002). The year, observer and overdispersion are modeled as random effects. Their modeling of CBC data includes observer effort, rather than observer identity. Such models can be fit using BUGS (Spiegelhalter et al. 1995). Time trend and annual indices to abundance can be derived from estimates under such models.

Link and Sauer (2007) noted that these continental surveys have the potential to provide information about seasonality of population change. In particular, models for the logarithm of counts from each survey include components reflecting proportional change between approximately January and June and between July and December. Such modeling permits inference about the relative amount of temporal variation associated with different seasons of the year, an inference of potential use for managers.

Link and Sauer noted current interest in estimates not only of trend, but also of population size. They noted that BBS has two primary deficiencies that preclude direct inferences about abundance: the absence of information about detection probability on surveyed routes, and the restriction of the BBS to roads. Experimental work using distance and time of detection approaches on both on- and off-road routes provided data that can be incorporated into BBS analyses, again using a hierarchical modeling approach, to obtain abundance estimates.

Process modeling was discussed by L. Thomas, who pointed out that this approach is becoming increasingly popular as recent developments in computer intensive (largely Bayesian) statistical methods allow reasonably complex, high dimensional models to be fit to biological data. Overviews of recent developments in this area are given by Newman et al. (2006) and Buckland et al. (2007b), and an example using bird (lapwing) count data is given by Besbeas et al. (2005). Thomas described his work fitting biological models of grey seal population dynamics to data on seal pup numbers, using a Bayesian fitting algorithm called particle filtering or sequential importance sampling (early work is described in Thomas et al. 2005).

7 Discussion

The workshop format emphasized participant discussion, with the result that useful discussion followed virtually every presentation. Here we attempt to capture some of the ideas expressed, especially on topics that recurred throughout the workshop. One of these topics was the issue of Section 2.2 about whether to use raw counts as indices or to instead attempt to collect ancillary data in order to try to better deal with detection probabilities when drawing inferences about abundance and its variation over time and space. It seems clear that abundance estimation requires some sort of estimate of detection probability in order to translate counts into estimates of abundance (2). However, inferences about temporal or spatial variation in abundance do not require abundance estimates. Raw counts can be used for inferences about temporal or spatial variation in cases where (1) detection probabilities over the dimension of interest (space, time) are similar or (2) the primary sources of variation in detection probability over the dimension of interest are identified, measured, and used as covariates. This latter approach is only appropriate when the primary sources of variation in detection probability do not have the potential to also be associated with variation in abundance.

An important question about inferences for variation in abundance is then: when is it best to collect ancillary data needed to draw specific inference about detection probabilities and when is it better to forego the collection and subsequent modeling of ancillary data and base inference on counts themselves? Many participants held a belief similar to that expressed in books such as Skalski and Robson (1992), Borchers et al. (2002), and Williams et al. (2002), that collection and modeling of ancillary data was wise, even if model selection has the potential to result in inferences being based on models assuming constant detection probabilities. These participants argued that collection and modeling of data on detection probability guarded against incorrect inferences about abundance that can be caused by detection probabilities that differ over the dimension of comparison. Participants who did not necessarily agree with this approach suggested that the modeling associated with the detection process was also likely to produce incorrect inferences in some cases. All agreed that if detection probability is not formally incorporated into the estimation and modeling, then design issues become more important as the only means of dealing with variation in detection probability over dimensions of interest (e.g., via standardization).

In considering this difference in opinion, we recall the early days of capture-recapture modeling, in which some investigators advocated use of these models whereas others suggested that underlying model assumptions were too restrictive and that enumeration methods were preferable. Capture–recapture proponents investigated performance of model-based estimators in the face of likely assumption violations (e.g., Carothers 1973, 1979; Gilbert 1973) and in some cases compared performance with direct enumeration approaches (e.g., Jolly and Dickson 1983; Nichols and Pollock 1983; Conn et al. 2004). These exercises led many investigators to the conclusion that capture–recapture models incorporating detection probability parameters are typically preferable to raw captures as a basis for inference, even in the face of common assumption violations. We believe that such exercises investigating the robustness of the various point count approaches to likely assumption violations, such as those reported by M.G. Efford (with D.K. Dawson), would be useful here as well. In addition to computer simulation work, test systems such as the N.C. State University bird radio system could be very useful in such work.

A conclusion emerging from the bird radio system work about which there was general agreement was the inability to prescribe an omnibus estimation approach likely to be the wise choice in all situations. Instead, the need to tailor methods to the particular study was emphasized, with such features as structural habitat type, relative likelihood of visual versus auditory detections, species richness and study objectives (e.g., time trend versus habitat variation) all identified as important determinants of method selection.

Weaknesses associated with each of the basic estimation approaches were readily identified. Many of these specific weaknesses were associated with auditory detections and the general difficulty in localizing bird sound. For example, this difficulty led to problems with determining matches of individual birds between multiple independent observers. Inability to localize sound also produces problems in distinguishing multiple observations of the same individual within a point count (e.g., double counting and phantom individuals), and such problems can affect all of the basic methods. Difficulties in determining distances at which birds are heard cause problems not only with distance sampling but with all approaches using fixed radius plots. In addition to sound localization, undetected movement in or out of the sample plot has the potential to lead to bias under all of the considered methods, causing more difficulties with some approaches than others. Simulation results presented by M.G. Efford (with D.K. Dawson) further indicated that distance-induced heterogeneity in detection probabilities had substantial potential to cause negative bias in estimators of abundance that did not explicitly account for distance (Efford and Dawson in prep.). If these approaches are used, the suggestion was made to limit the effective area surveyed at each line or point to only include individuals that have naturally "high" detection probabilities, in the sense that they are close to the observer. Distance estimation is also typically easier for closer animals.

The bird radio system test results led to the conclusion that it will likely not be possible to use point count data to draw strong inferences about populations of all species in even a moderately rich bird community. This conclusion led to the recommendation that sampling be restricted to a small number of focal species, but this recommendation was countered by the claim that sometimes the entire bird community is of interest. Subsequent discussion included the observation that many avian studies and monitoring programs are not well-conceived in the sense of having concise objectives. All participants agreed on the importance of establishing clear study objectives and ensuring that sampling was consistent with these objectives. It was noted that in cases where community-level inferences really are of primary interest, then occupancy estimation and modeling (MacKenzie et al. 2002, 2006; Dorazio and Royle 2005) could provide a methodological alternative to abundance estimation that would yield achievable results consistent with many kinds of study objectives. Some participants suggested designs in which the observers focus on 1–3 species for abundance estimation and collect occupancy data on remaining species, whereas another recommendation was to devote one person to the focal species and an additional observer to collect occupancy data for the entire community.

Another aspect of study objectives that is relevant to method selection is whether one goal is to use abundance estimates on the selected sample units to estimate abundance for the entire area from which the samples were drawn. Such extrapolation seems reasonable for many of the methods, conditional on the ability to determine whether each detected bird is inside or outside of the prescribed sample unit. However, such extrapolation is expected to be more difficult for repeated count approaches (Royle and Nichols 2003; Royle 2004). The fact that detection probability estimates under these approaches include the probability of a bird whose range overlaps the sample unit actually being present in the unit at the time of sampling (p_p), means that resulting abundance estimates should include all birds with ranges that overlap sample units (Section 3.3). Extrapolation based on such estimates should estimate the product of overall abundance and the average number of sample units covered by an individual bird's home range (Section 3.6). If sample unit size is large relative to individual home range, this sample-units-per-home-range multiplier is likely to be small, so overall abundance estimates should be fairly good. If sample units and home ranges are of similar size, then it will be more difficult to estimate overall abundance from repeat count estimates without separately estimating the average number of units overlapped by the range of an individual bird. Note that such extrapolation to an overall abundance estimate will not be required by many objectives.

Participants discussed possible changes to existing methods that might be worthy of future consideration. For example, several participants noted that some movement of observers within a plot for point counts, rather than standing at a single point, would likely increase detection probabilities of birds in a sample unit. However, it was suggested that another consequence of such observer movement would likely be a decrease in ability to keep track of individual birds, with subsequent increases in the problems of double counting and phantom birds. Observer movement might induce unwanted bird movement as well. For multiple-observer approaches, a suggestion was made to place observers at different edges of a sample plot as a means of reducing the correlation in detection probabilities of individual birds by the different observers, thus decreasing the problems associated with individual heterogeneity. Once again, though, this approach could lead to an increase in matching and other errors associated with localizing individual birds and determining which birds were detected by one or multiple observers.

Recommended durations of point counts have long been discussed and debated. The explicit recognition of a probability that a bird vocalizes and becomes available for detection in auditory surveys (p_a) leads to the natural suggestion to increase point count duration in order to increase p_a. However, increases in survey duration increase such problems as bird movement into and out of the study plot and difficulties in tracking individual birds over time. These problems can be reduced to some degree by employing a "snapshot" approach (Buckland et al. 2006).

A relatively new approach to collection of field data was that proposed by M.G. Efford (with D.K. Dawson) of using acoustic recording devices (e.g., Hobson et al. 2002) to collect bird vocalization data for subsequent use with estimation approaches. Software would then be developed to help extract the relevant information (species and individual identification) about bird vocalizations. An array of such recording devices could be used with models of the type developed for other passive detectors (e.g., Efford 2004; Efford et al. 2005; Borchers and Efford in press) to directly estimate bird density.

During discussion, Thomas speculated that it may be possible to develop a highly portable acoustic array that could be used to provide real-time localizations of received signals (such as bird vocalizations). Such a system could then be used as a field aid, to provide better estimation of distance in aural surveys than is possible using human ears. More reliable estimation of location than is possible for humans may be achieved if more than two sensors are used, or if the sensors are located farther apart than human ears.

This proposed work by Efford and Dawson on passive arrays of recording devices was among the top proposals for future work recommended by workshop participants. This work will include the recording devices themselves, computer software for resulting sound data, and estimation approaches for computing density estimates from these data. Additional work on combining data and methods was another recommendation of virtually all participants. Research on singing rates and patterns was identified as a priority research problem. For example, current time of detection approaches typically assume a random process in which all birds in the survey area exhibit some probability of vocalizing during each time interval. However, it is likely that this probability is better viewed as resulting from a Markov process in which vocalization during one interval is partially dependent on whether the bird vocalized during the previous interval(s). The likely influence of vocalizations of other birds on singing rates and patterns was also noted, potentially leading to very complicated models of the stochastic process of bird vocalization. Additional research on models of individual bird misidentification (Section 5.3) was also judged to be an important direction of future work. There is no question that double counting and phantom individuals occur and can influence all of the recommended estimation approaches, and formally accounting for the additional uncertainty associated with these errors will represent an important advance.

Finally, there were some very pragmatic recommendations for future work. The most obvious was to field test the various proposed methods using such testing platforms as the bird radio system. The objective of these tests would simply be to determine which estimation approaches "work best" under various conditions and for various objectives. Such field testing could also be used to establish an approximate relationship between estimator performance and number of species considered in the sampling. Knowledge of such a relationship could then lead to the development of rules of thumb about how many focal species can be surveyed with the expectation of drawing reasonable inferences. Recommendations also included work on survey design and allocation of effort based on survey objectives. Practical considerations about effort devoted to visiting many sample units versus estimating components of detection probability for a unit are important and may lead to recommendations about the use of double sampling (sensu Pollock et al. 2002).

Substantial research has been conducted on point counts during the last 5 years. The workshop and this paper represent an attempt to summarize key results of that work and to point to key areas of future research. We offer the opinion that the various recommendations for future research should not be used as reasons for postponing consideration of changes to existing point count programs. Although future work will certainly lead to more refined recommendations, we believe that many existing programs that use point counts can be improved based on the current state of knowledge.

Acknowledgments We are extremely grateful to the workshop participants for attending, self funded, and for sharing so freely their research findings and ideas. We are also grateful for their comments and suggestions on earlier drafts of this manuscript, which have led to a much improved paper. Participants were: Matthew W. Alldredge, David L. Borchers, Stephen T. Buckland, Paul B. Conn, Duane R. Diefenbach, Paul F. Doherty, Murray G. Efford, Douglas H. Johnson, Marc Kéry, William A. Link, Tiago A. Marques, James D. Nichols, Krishna Pacifici, Kenneth H. Pollock, J. Andrew Royle, Michael Runge, Theodore R. Simons, Len Thomas and Ray Webster. We thank Kenneth P. Burnham, Rachel M. Fewster, Marshall Howe, and Bruce Peterjohn for reviews and constructive comments.

References

Alldredge MW, Pollock KH, Simons TR, Shriner SA (2007b) Multiple species analysis of point count data: a more parsimonious modeling framework. J. Appl. Ecol. 44:281–290.

Alldredge MW, Simons TR, Pollock KH (2007c) Factors affecting aural detections of songbirds. Ecol. Appl. 17:948-955.

Alldredge MW, Pollock KH, Simons TR (2006) Estimating detection probabilities from multiple-observer point counts. Auk 123:1172–1182.

Alldredge MW, Pollock KH, Simons TR, Collazo JA, Shriner SA (2007a) Time of detection method for estimating abundance from point count surveys. Auk 124:653-664.

Alpizar-Jara R, Pollock KH (1996) A combination line transect and capture–recapture sampling model for multiple observers in aerial surveys. Environ. Ecol. Stat. 3:311–327.

Alpizar-Jara R, Pollock KH (1999) Combining line transect and capture–recapture for mark-resighting studies. Pages 99–114 *in* Garner GW, Amstrup SC, Laake JL, Manly BFJ, McDonald LL, Robertson DG (eds.) Marine mammal survey and assessment methods. Balkema, Rotterdam.

Anderson DR (2001) The need to get the basics right in wildlife field studies. Wildl. Soc. Bull. 29:1294–1297.

Bart J, Earnst S (2002) Double sampling to estimate density and population trends in birds. Auk 119:36–45.

Besbeas P, Freeman SN, Morgan BJT (2005) The potential for integrated population modeling. Aust. N.Z. J. Stat. 47:35–48.

Blondel J, Ferry C, Frochot B (1970) La methode des indices ponctuels d'abondance (IPA) ou des releves d'avifaune par "stations d'ecoute." Alauda 41:63–84.

Borchers DL (1999) Composite mark-recapture line transect surveys. Pages115–126 *in* Garner GW, Amstrup SC, Laake JL, Manly BFJ, McDonald LL, Robertson DG (eds.) Marine mammal survey and assessment methods. Balkema, Rotterdam.

Borchers DL, Buckland ST, Goedhart PW, Clarke ED, Hedley SL (1998a) Horvitz–Thompson etsimators for double-platform line transect surveys. Biometrics 54:1221–1237.

Borchers DL, Buckland ST, Zucchini W (2002) Estimating animal abundance. Springer, New York.

Borchers DL, Efford MG. Spatially explicit maximum likelihood methods for capture-recapture studies. Biometrics (in press).

Borchers DL, Laake JL, Southwell C, Paxton CGM (2006) Accommodating unmodeled heterogeneity in double-observer distance sampling surveys. Biometrics 62:372–378.

Borchers DL, Marques TA, Gunnlaugsson T, Víkingsson GA. Distance sampling with measurement errors. (in prep.).

Borchers DL, Zucchini W, Fewster RM (1998b) Mark-recapture models for line transect surveys. Biometrics 54:1207–1220.

Bravington MV, Hedley SL, Wood SN. A general approach for modelling clustered line transect data using gamma random fields. (in prep.)

Buckland ST (2006) Point transect surveys for songbirds: robust methodologies. Auk 123: 345–357.

Buckland ST, Anderson DR, Burnham KP, Laake JL (1993) Distance sampling: estimating abundance of biological populations. Chapman and Hall, London, UK.

Buckland ST, Anderson DR, Burnham KP, Laake JL, Borchers DL, Thomas L (2001) Introduction to distance sampling. Oxford University Press, Oxford, UK.

Buckland ST, Borchers DL, Johnston A, Henrys PA, Marques TA (2007a) Line transect methods for plant surveys. Biometrics 63:989–998.

Buckland ST, Newman KB, Fernández C, Thomas L, Harwood J (2007b) Embedding population dynamics models in inference. Stat. Sci. 22:44–58.

Buckland ST, Summers RW, Borchers DL, Thomas L (2006) Point transect sampling with traps or lures. J. Appl. Ecol. 43:377–384.

Burnham KP, Anderson DR, Laake JL (1980) Estimation of density from line transect sampling of biological populations. Wildl. Monogr. 72:1–202.

Burnham KP, Anderson DR, Laake JL (1981) Line transect estimation of bird population density using a Fourier series. Stud. Avian Biol. 6:466–482.

Burnham KP, Buckland ST, Laake JL, Borchers DL, Marques TA, Bishop JRB, Thomas L (2004) Further topics in distance sampling. Pages 307–392 *in* Buckland ST, Anderson DR, Burnham KP, Laake JL, Borchers DL, Thomas L, eds. Advanced distance sampling. Oxford University Press, Oxford, UK.

Carothers AD (1973) The effects of unequal catchability on Jolly-Seber estimates. Biometrics 29:79–100.

Carothers AD (1979) Quantifying unequal catchability and its effect on survival estimates in an actual population. J. Anim. Ecol. 48:863–869.

Carroll RJ, Lombard F (1985) A note on *N* estimators for the binomial distribution. J. Amer. Stat. Assoc.80:423–426.

Cochran WG (1977) Sampling techniques. Third ed. Wiley, New York, USA.

Conn PB, Bailey LL, Sauer JR (2004) Indexes as surrogates to abundance for low-abundance species. Pages 59–74 *in* Thompson WL (ed.) Sampling rare or elusive species. Island Press, Washington, DC, USA.

Cook RD, Jacobsen JO (1979) A design for estimating visibility bias in aerial surveys. Biometrics 35:735–742.

Diefenbach DR, Marshall MR, Mattice JA, Brauning DW (2007) Incorporating availability for detection in estimates of bird abundance. Auk 124:96-106.

Dorazio RM, Royle JA (2005) Estimating the size and composition of biological communities by modeling the occurrence of species. J. Amer. Stat. Assoc. 100:389–398.

Efford MG (2004) Density estimation in live-trapping studies. Oikos 106:598–610.

Efford MG, Dawson DK. The effect of distant - related heterogeneity on population size estimates from point counts (in prep.)

Efford MG, Warburton B, Coleman MC, Barker RJ (2005) A field test of two methods for density estimation. Wildl. Soc. Bull. 33:731–738.

Ellingson AR, Lukacs PM (2003) Improving methods for regional landbird monitoring: a reply to Hutto and Young. Wildl. Soc. Bull. 31:896–902.

Emlen JT (1971) Population densities of birds derived from transect counts. Auk 88:323–342.

Farnsworth GL, Nichols JD, Sauer JR, Fancy SG, Pollock KH, Shriner SA, Simons TR (2005) Statistical approaches to the analysis of point count data: a little extra information can go a long way. Pages 736–743 *in* Ralph CJ, Rich TD (eds.) Bird Conservation Implementation and Integration in the Americas: Proceedings of the 3rd International Partners in Flight Conference. Volume 2. Gen. Tech. Rep. PSW-GTR-191. Pacific Southwest Research Station, Forest Service, U.S. Dept. Agriculture: Albany, CA.

Farnsworth GL, Pollock KH, Nichols JD, Simons TR, Hines JE, Sauer JR (2002) A removal model for estimating detection probabilities from point count surveys. Auk 119: 414–425.

Gilbert RO (1973) Approximations of the bias in the Jolly-Seber capture–recapture model. Biometrics 29:501–526.

Hedley SL (2000) Modelling heterogeneity in cetacean surveys. Ph.D. thesis, Univ. St. Andrews, St. Andrews, Scotland, UK.

Hedley SL, Buckland ST (2004) Spatial models for line transect sampling. J. Agric. Biol. Environ. Stat. 9:181–199.

Hedley SL, Buckland ST, Borchers DL (2004) Spatial distance sampling models. Pages 48–70 *in* Buckland ST, Anderson DR, Burnham KP, Laake JL, Borchers DL, Thomas L (eds.) Advanced distance sampling. Oxford University Press, Oxford, UK.

Hiby L, Lovell P (1998) Using aircraft in tandem formation to estimate abundance of harbour porpoise. Biometrics 54:1280–1289.

Hobson KA, Rempel RS, Greenwood H, Turnbull B, Van Wilgenburg SL (2002) Acoustic surveys of birds using electronic recordings: new potential from an omnidirectional microphone system. Wildl. Soc. Bull. 30:709–720.

Hutto RL, Young JS (2002) Regional landbird monitoring: perspectives from the northern Rocky Mountains. Wildl. Soc. Bull. 30:738–750.

Hutto RL, Young JS (2003) On the design of monitoring programs and the use of population indices: a reply to Ellingson and Lukacs. Wildl. Soc. Bull. 31:903–910.

Jarvinen O, Vaisanen RA (1975) Estimating relative densities of breeding birds by the line transect method. Oikos 26:316–322.

Jolly GM, Dickson JM (1983) The problem of unequal catchability in mark-recapture estimation of small mammal populations. Can. J. Zool. 61:922–927.

Kendall WL, Nichols JD, Hines JE (1997) Estimating temporary emigration and breeding proportions using capture–recapture data with Pollock's robust design. Ecology 78:563–578.

Kery M, Royle JA, Schmid H (2005) Modeling avian abundance from replicated counts using binomial mixture models. Ecol. Appl. 1450–1461.

Kissling ML, Garton EO (2006) Estimating detection probability and density from point-count surveys: a combination of distance and double-observer sampling. Auk 123: 735–752.

Laake JL, Borchers DL (2004) Methods for incomplete detection at distance zero. Pages 108–189 *in* Buckland ST, Anderson DR, Burnham KP, Laake JL, Borchers DL, Thomas L (eds.) Advanced distance sampling. Oxford University Press, Oxford, UK.

Lancia RA, Kendall WL, Pollock KH, Nichols JD (2005) Estimating the number of animals in wildlife populations. Pages 106–153 *in* Braun CE ed. Research and management techniques for wildlife and habitats. The Wildlife Society, Bethesda, MD, USA.

Lancia RA, Nichols JD, Pollock KH (1994) Estimating the number of animals in wildlife populations. Pages 215–253 *in* Bookhout T (ed.) Research and management techniques for wildlife and habitats. The Wildlife Society, Bethesda, MD, USA.

Link WA (2003) Nonidentifiability of population size from capture–recapture data with heterogeneous detection probabilities. Biometrics 59:1123–1130.

Link WA, Sauer JR (1997) Estimation of population trajectories from count data. Biometrics 53:499–497.

Link WA, Sauer JR (1998) Estimating relative abundance from count data.Austrian J. Stat. 27: 83–97.

Link WA, Sauer JR (2002) A hierarchical analysis of population change with application to Cerulean Warblers. Ecology 83:2832–2840.

Link WA, Sauer JR (2007) Seasonal components of avian population change: joint analysis of two large-scale monitoring programs. Ecology 88:49–55.

Mackenzie DI, Nichols JD, Lachman GB, Droege S, Royle JA, Langtimm CA (2002) Estimating site occupancy when detection probabilities are less than one. Ecology 83: 2248–2255.

Mackenzie DI, Nichols JD, Royle JA, Pollock KH, Bailey LA, Hines JE (2006) Occupancy modeling and estimation. Academic Press, San Diego, CA, USA.

MacKenzie DI, Royle JA (2005) Designing occupancy studies: general advice and allocating survey effort. J. Appl. Ecol. 42:1105–1114.

Manly, B.F.J., McDonald LL, Garner GW (1996) Maximum likelihood estimation for the double-count method with independent observers. J. Agric. Biol. Environ. Stat. 1: 170–189.

Marques TA (2004) Predicting and correcting bias caused by measurement error in line transect sampling using multiplicative error models. Biometrics 60:757–763.

Marques TA (2008) Incorporating measurement error density gradients in distance sampling surveys. Ph.D. thesis, Univ. St. Andrews, St. Andrews, Scotland, UK.

Marques TA, Thomas L, Fancy SG, Buckland ST (2007) Improving estimates of bird density using multiple covariate distance sampling. Auk (124:1229-1245).

Marsh H, Sinclair DF (1989) Correcting for visibility bias in strip transect aerial surveys of aquatic fauna. J. Wildl. Manage. 53:1017–1024.

Newman KB, Buckland ST, Lindley ST, Thomas L, Fernández C (2006) Hidden process models for animal population dynamics. Ecol. Appl. 16:74–86.

Nichols JD, Hines JE, Sauer JR, Fallon FW, Fallon JE, Heglund PJ (2000) A double-observer approach for estimating detection probability and abundance from point counts. Auk 117: 393–408.

Nichols JD, Pollock KH (1983) Estimation methodology in contemporary small mammal capture–recapture studies. J. Mammal. 64:253–260.

Norris III JL, Pollock KH (1996) Nonparametric MLE under two closed capture–recapture models with heterogeneity. Biometrics 52:639–649.

Okamura H, Minamikawa S, Kitakado T (2006) Effect of surfacing patterns on abundance estimates of long-diving animals. Fish. Sci. 72:631–638.

Otis DL, Burnham KP, White GC, Anderson DR (1978) Statistical inference from capture data on closed animal populations. Wildl. Monogr. 62:1–35.

Pledger S (2000) Unified maximum likelihood estimates for closed capture–recapture models for mixtures. Biometrics 56:434–442.

Pollock KH (1982) A capture–recapture design robust to unequal probability of capture. J. Wildl. Manage. 46:757–760.

Pollock KH, Nichols JD, Simons TR, Farnsworth GR, Bailey LL, Sauer JR (2002) Large scale wildlife monitoring studies: statistical methods for design and analysis. Environmetrics 13: 1–15.

Ralph CJ, Sauer JR, Droege S (eds.) (1995) Monitoring bird populations by point counts. Gen. Tech., Rep. PSW-GTR-149. U.S. Forest Service Pacific Southwest Research Station, Albany, CA, USA.

Ralph CJ, Scott JM (eds.) (1981) Estimating numbers of terrestrial birds. Stud. Avian Biol. No. 6:1–630.

Ramsey FL, Scott JM (1979) Estimating population densities from variable circular plot surveys. Pages 155–181 in Cormack RM, Patil GP, Robson DS (eds.) Sampling biological populations. Statistical Ecology Series, Vol. 5, International Cooperative Publication House, Fairland, MD, USA.

Rosenstock SS, Anderson DR, Giesen KM, Leukering T, Carter MF (2002) Landbird counting techniques: current practices and an alternative. Auk 119:46–53.

Royle JA (2004) N-mixture models for estimating population size from spatially replicated counts. Biometrics 60:108–115.

Royle JA, Nichols JD (2003) Estimating abundance from repeated presence absence data or point counts. Ecology 84:777–790.

Royle JA, Nichols JD, Kery M (2005) Modeling occurrence and abundance of species when detection is imperfect. Oikos 110:353–359.

Sauer JR, Link WA (2002) Hierarchical modeling of population stability and species group attributes from survey data. Ecology 86:1743–1751.

Seber GAF (1982) The estimation of animal abundance and related parameters. Second ed. MacMillian Publication Co., Inc., New York, USA.

Simons TR, Alldredge MW, Pollock KH, Wettroth JM (2007) Experimental analysis of the auditory detection process on avian point counts. Auk 124:986-999.

Skalski JR, Robson DS (1992) Techniques for wildlife investigations. Academic Press, San Diego, USA.

Smith GW (1995) A critical review of the aerial and ground surveys of breeding waterfowl in North America. U.S. Dept. Interior, Biological Science Report 5. Washington, DC, USA.

Spiegelhalter DJ, Thomas A, Best NG, Gilks WR (1995) BUGS: Bayesian inference using Gibbs sampling. Version 0.50. MRC Biostatistics Unit, Cambridge, UK.

Strindberg S, Buckland ST, Thomas L (2004) Design of distance sampling surveys and geographic information systems. Pages 190–228 *in* Buckland ST, Anderson DR, Burnham KP, Laake JL, Borchers DL, Thomas L (eds.) Advanced distance sampling. Oxford University Press, Oxford, UK.

Thomas L, Borchers DL, Buckland ST, Hammond IE. Estimation of population size from distance sampling data with heterogenous detection probabilities. (in prep.)

Thomas L, Buckland ST, Newman KB, Harwood J (2005) A unified framework for modelling wildlife population dynamics. Aust. N. Z. J. Stat. 47:19–34.

Thomas L, Burnham KP, Buckland ST (2004) Temporal inferences from distance sampling surveys. Pages 71–107 *in* Buckland ST, Anderson DR, Burnham KP, Laake JL, Borchers DL, Thomas L (eds.) Advanced distance sampling. Oxford University Press, Oxford, UK.

Thompson SK (2002a) Sampling. Wiley, New York, USA.

Thompson WL (2002b) Towards reliable bird surveys: accounting for individuals present but not detected. Auk 119:18–25.

Williams BK, Nichols JD, Conroy MJ (2002) Analysis and management of animal populations. Academic Press, San Diego, USA.
Wood SN, Bravington MV, Hedley SL. Soap film smoothing. J. Royal Stat. Soc. Ser. B. (in press).
Yoccoz NG, Nichols JD, Boulinier T (2001) Monitoring of biological diversity in space and time. Trends Ecol. Evol. 16:446–453.

Sources of Measurement Error, Misclassification Error, and Bias in Auditory Avian Point Count Data

Theodore R. Simons, Kenneth H. Pollock, John M. Wettroth, Mathew W. Alldredge, Krishna Pacifici, and Jerome Brewster

Abstract Avian point counts vary over space and time due to actual differences in abundance, differences in detection probabilities among counts, and differences associated with measurement and misclassification errors. However, despite the substantial time, effort, and money expended counting birds in ecological research and monitoring, the validity of common survey methods remains largely untested, and there is still considerable disagreement over the importance of estimating detection probabilities associated with individual counts. Most practitioners assume that current methods for estimating detection probability are accurate, and that observer training obviates the need to account for measurement and misclassification errors in point count data. Our approach combines empirical data from field studies with field experiments using a system for simulating avian census conditions when most birds are identified by sound. Our objectives are to: identify the factors that influence detection probability on auditory point counts, quantify the bias and precision of current sampling methods, and find new applications of sampling theory and methodologies that produce practical improvements in the quality of bird census data.

We have found that factors affecting detection probabilities on auditory counts, such as ambient noise, can cause substantial biases in count data. Distance sampling data are subject to substantial measurement error due to the difficulty of estimating the distance to a sound source when visual cues are lacking. Misclassification errors are also inherent in time of detection methods due to the difficulty of accurately identifying and localizing sounds during a count. Factors affecting detection probability, measurement errors, and misclassification errors are important but often ignored components of the uncertainty associated with point-count-based abundance estimates.

T.R. Simons (✉)
USGS, NC Cooperative Fish and Wildlife Research Unit, Department of Zoology, Campus Box 7617, North Carolina State University, Raleigh, NC 27695, USA

D.L. Thomson et al. (eds.), *Modeling Demographic Processes in Marked Populations*, Environmental and Ecological Statistics 3, DOI 10.1007/978-0-387-78151-8_10,

1 Introduction

The most common method of estimating avian abundance is the point count (Ralph et al. 1995) where a single observer records all birds seen or heard at a point during a prescribed interval (usually 3-10 min) (Fig. 1). Surveys of breeding birds rely heavily on auditory detections which can comprise 70% of observations in suburban landscapes (Sauer et al. 1994), 81% in tropical forests (Scott et al. 1981), and up to 97% of observations in closed-canopy deciduous forest (DeJong and Emlen 1985; Brewster 2007).

Avian abundance estimates can vary over space and time due to actual differences in abundance, differences in detection probabilities among counts, and differences associated with measurement and misclassification errors (Nichols et al. 2008; Nichols et al. 2000; Farnsworth et al. 2002; Pollock et al. 2002; Rosenstock et al. 2002; Thompson 2002).

A general conceptual model (Marsh and Sinclair 1989; Pollock et al. 2004, 2006) for auditory count-based abundance estimates can be represented as:

$$\hat{N} = \frac{C}{p_{area}\,\hat{p}_a\hat{p}_{da}}$$

(A) (B)

Fig. 1 Common point count protocols. (**a**) A single observer standing at the center of a circular plot maps the location and estimated distance to all birds seen or heard during a prescribed time interval, usually 3–10 min. Time of detection information can be incorporated by color-coding or annotating observations to indicate the time interval of initial and subsequent detections. (**b**) Multiple observer methods employ two or more observers who collect data simultaneously. Observers can assume either dependent (Nichols et al. 2000) or independent (Alldredge et al. 2006) roles where:

$\hat{N}$ = the population estimate

C = the count statistic

p_{area} = the fraction of the area sampled

$\hat{p}_a$ = the probability that a bird is available to be counted

$\hat{p}_{da}$ = the probability that a bird is detected given that it is available.

Many factors influence the probability of detecting birds during auditory point counts. These factors include both "measurement error" factors associated with observer skill and ability, and "signal to noise ratio" factors that influence how much information about bird diversity and abundance is available to observers. Measurement error factors relate to observer skill in identifying and localizing individual birds (Kepler and Scott 1981), and hearing ability (Emlen and Dejong 1981, 1992; Sauer et al. 1994; Kendall et al. 1996; Downes 2004). Signal to noise ratio factors include the spectral qualities of songs (Schieck 1997), song volume, singing rate (Best 1981; Ralph 1981; Skirvin 1981), time of day (Sheilds 1977; Robbins 1981a; Skirvin 1981), the orientation of singing birds (toward or away from observers), (Alldredge et al. 2007c) presence of an observer (McShea and Rappole 1997), the number of species and number of individuals singing during a count (Simons et al. 2007), pairing status (Krebs et al. 1980; Johnson 1983; Cuthill and Hindmarsh 1985; Gibbs and Wenny 1993), stage of nesting cycle (Wilson and Bart 1985), vegetation structure (Diehl 1981; McShea and Rappole 1997; Simons et al. 2006; Pacifici et al. in press), topography, weather (Mayfield 1981; Robbins 1981b), temperature, humidity, and ambient noise (Simons et al. 2007). Systematic variation in any of these factors will impart a systematic bias in count data.

At least five methods of estimating detection probabilities on avian point counts are currently available (Nichols et al. 2008): distance sampling (Buckland et al. 2001), multiple-observer methods (Nichols et al. 2000; Alldredge et al. 2007a), time of detection methods (Farnsworth et al. 2002; Alldredge et al. 2007b), double sampling (Bart and Earnst 2002) and repeated count methods (Royle and Nichols 2003; Kery et al. 2005). Applications of combined methods are also possible (Kissling and Garton 2006; Alldredge et al. 2007a, b). Different methods estimate different components of the detection process. For example, distance sampling and multiple observer approaches assume that all birds on a given sample plot are available (sing during the count interval) and they estimate the probability of detection given availability. Time of detection methods provide estimates of the product of availability and detection given availability (Alldredge et al. 2007a) but they cannot separate the two components. Repeated count methods (Royle and Nichols 2003; Nichols et al. this volume) estimate the product of availability, detection given availability, and a third component of the detection process, the probability that an individual is present in the sample area.

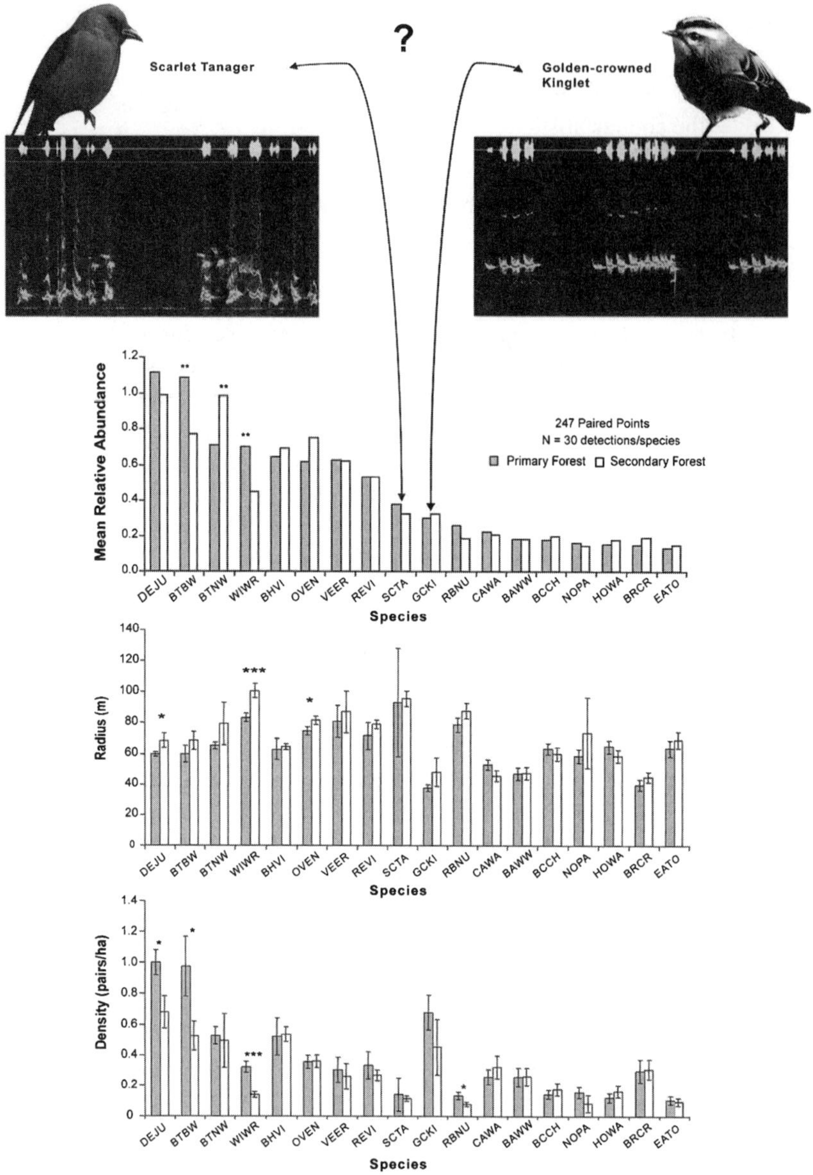

Fig. 2 Comparisons of abundance measures for 18 species of breeding birds recorded at 247 paired survey locations in primary and secondary forests in Great Smoky Mountains National Park (Simons et al. 2006). (*Top*) Mean relative abundance; counts are not adjusted for differences in de tectability. (*Middle*) Effective detection radii (EDR) estimates (calculated using Program DISTANCE, Thomas et al. 1998). (*Bottom*) Estimated densities based on effective detection radii. Error bars represent standard errors. Significant differences are denoted by asterisks (*$p < 0.05$, **$p < 0.01$, ***$p < 0.001$)

A brief example will illustrate how correcting avian point count data for variations in detection probability among species and habitats can dramatically alter abundance estimates. Figure 2 summarizes count data for 18 species of forest songbirds from 247 paired point count locations in primary and secondary southern Appalachian forests (Simons et al. 2006). Of interest is the similarity of unadjusted counts (Fig. 2a) for two very dissimilar species, the Scarlet Tanager (SCTA, *Piranga olivacea*) and the Golden-crowned Kinglet (GCKI, *Regulus satrapa*). The Scarlet Tanager is a brightly-colored, very active and vocal species that flies constantly about the forest canopy giving a loud, high-energy call. The Golden-crowned Kinglet is a small, drab, and generally inconspicuous species that forages along branches and gives a high, thin, low-energy call. The unadjusted counts for these two species suggest that their abundance is similar in primary and secondary forests. However when we examine the effective detection radii (Buckland et al. 2001) of the two species (Fig. 2b) and use this information to convert the raw counts into density estimates (Fig. 2c) dramatic differences in abundance become apparent. Accounting for differences in detection probability related to differences in the conspicuousness of theses two species results in a nearly four-fold difference in our abundance estimate. Similarly, differences in abundance between primary and secondary forest habitats become apparent once counts are adjusted for differences in detection probability between habitats. Failure to account for such differences in detection probabilities among species and habitats weakens inferences from comparative studies of avian abundance (Yoccoz et al. 2001; Pollock et al. 2002; Williams et al. 2002).

Nevertheless, there is still considerable disagreement over the importance of estimating detection probabilities associated with individual counts (Rosenstock et al. 2002). Although common survey methods are largely unvalidated, most practitioners assume that current methods for estimating detection probability are accurate, and that observer training obviates the need to account for measurement and misclassification errors in point count data. Given the substantial time, effort, and money expended conducting avian point counts to address ecological research and monitoring objectives (Bart 2005; Simons et al. 2007), validating current avian sampling methods has enormous practical importance. In this paper we first review factors that influence detection probabilities on auditory counts, and we then summarize key findings of recent field experiments aimed at understanding the factors affecting detection probabilities, and the sources and magnitude of measurement and misclassification errors inherent in several common sampling methods.

2 Approach

Our approach uses empirical data from field studies of southern Appalachian songbirds (Shriner 2001; Lichstein et al. 2002; Brewster 2007) to inform the development of new avian sampling methods (Farnsworth et al. 2002; Alldredge et al. 2007a), and to design field experiments (Simons et al. 2007) that assess the factors affecting detection probabilities on auditory counts (Alldredge et al. 2007b;

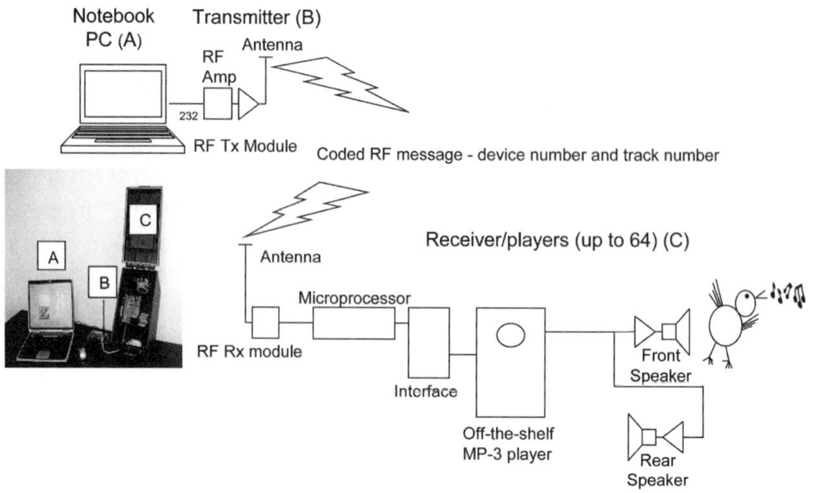

Fig. 3 System diagram of song playback system. (*A*) laptop computer and playback software, (*B*) transmitter, (*C*) portable receiver/player. See Simons et al. (2007) for specifications

Pacifici et al. in press), and the precision and accuracy of auditory avian point count methods. (Alldredge et al. 2007c, d; Alldredge et al. in press).

Field experiments are conducted using a system for simulating avian census conditions when most birds are identified by sound (Fig. 3). The system uses a

Fig. 4 Volunteers participating in a playback experiment. Observers standing at the center of the experimental plot conduct point counts simulated using up to 45 different playback devices. Players are placed at known locations (up to 200 m) and heights (up to 15 m) on the surrounding plot

laptop computer to control up to 50 amplified MP3 players placed at known locations up to 200 m around a survey point (see Simons et al. (2007) for details). To date we have simulated over 5,000 unlimited radius point counts with 50 observers (Fig. 4). The system can realistically simulate a known population of songbirds under a range of factors that affect detection probabilities. Validation experiments evaluate traditional methods for estimating detection probabilities, such as distance sampling, and new approaches that incorporate information from multiple observers, the time sequence of observations, and combined methods.

2.1 Factors Affecting Detection Probabilities

Figure 5 illustrates the number of six observers able to hear (Heard), correctly identify (Correct), and number of observers who misidentified (Wrong) calls of Black-throated Blue Warblers (BTBW, *Dendroica caerulescens*), at 25 distances between 40 and 160 m. Calls were played randomly at each distance for approximately 20 s.

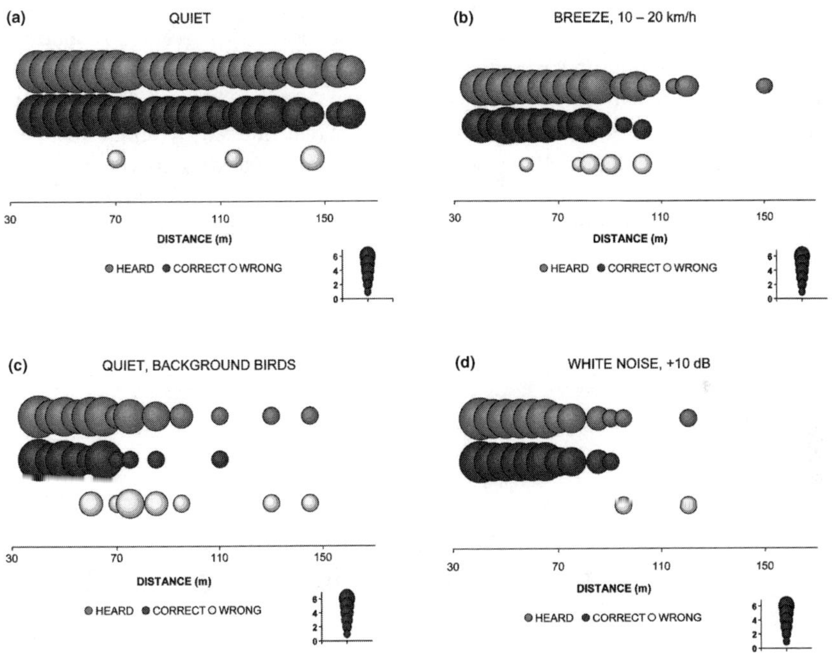

Fig. 5 Number of six observers able to hear (Heard), correctly identify (Correct), and number of observers who misidentified (Wrong) calls of Black-throated Blue Warblers at 25 distances between 40 and 160 m. Calls played randomly at each distance for approximately 20 s. Experiments were replicated under four ambient noise conditions: (**a**) quiet (mean ambient noise 40.6 dB, S.D. 4.47 dB), (**b**) breezy (10–20 km/h gusty winds, 55.4 dB, S.D. 3.87 dB), (**c**) quiet conditions with 1–3 background birds (Winter Wren, *Troglodytes troglodytes*, Yellow-throated Warbler, *Dendroica dominica*, and Ovenbird) singing 20 m behind or to either side of the observers, and (**d**) quiet conditions with white noise added (10 dB above ambient). White noise (uniform power, spectral frequency = 1.0) was played from a speaker facing the observers at a distance of 10 m

Experiments were replicated under four ambient noise conditions: (a) quiet (mean ambient noise 40.6 dB, S.D. 4.47 dB), (b) breezy (10–20 km/h gusty winds, 55.4 dB, S.D. 3.87 dB), (c) quiet conditions with three background birds Winter Wren (WIWR, *Troglodytes troglodytes)*, Yellow-throated Warbler (YTWA, *Dendroica dominica)*, and Ovenbird (OVEN, *Seiurus aurocapillus*) singing 40 m behind or to either side of the observers, and (d) quiet conditions with white noise added (10 dB above ambient). White noise (uniform power, spectral frequency = 1.0) was played from a speaker facing the observers at a distance of 10 m. Results illustrate how detection distances decline and identification errors increase with increasing levels of ambient noise. Overall, the proportion of birds heard by observers decreased by 28 ± 4.7% under breezy conditions, 41 ± 5.2% by the presence of additional background birds, and 42 ± 3.4% by the addition of 10 dB of white noise.

Temporal trends in environmental factors such as ambient noise can impart trends in count data unrelated to the true abundance of birds. To provide some context for our ambient noise experiment we asked observers to record ambient noise levels on 21 Breeding Bird Survey (Sauer et al. 2005) routes across North Carolina in 2006. Note the proportion of North Carolina BBS counts in which ambient noise levels exceed 40 dB (Fig. 6). Ambient noise experiments (Fig. 5) indicate that an increase in ambient noise from 40 to 50 dB produces a 42% average reduction in the counts of six common species. Thus, if ambient noise levels along these North Carolina routes increased by 10 dB over the past 20 years, we would expect BBS counts of species detected by ear to decline over that interval by about 40%, even if populations were stable. Because BBS counts are not adjusted for differences in detection probability, in this example there is no way of knowing if declines represent actual population declines or simply declines in detection probability due to increasing ambient noise.

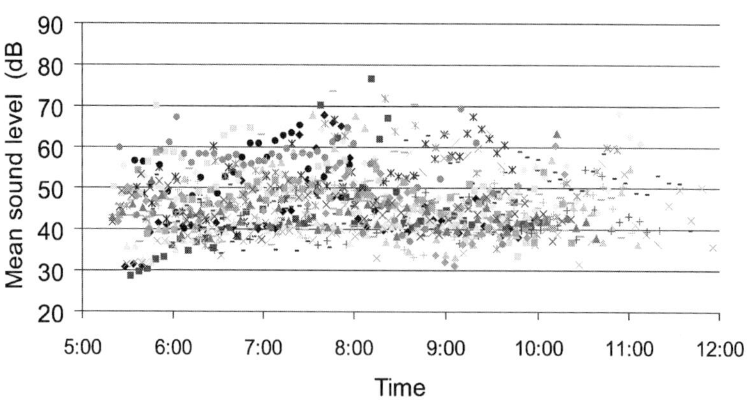

Fig. 6 Measured levels of ambient noise on 20 North Carolina Breeding Bird Survey routes in 2006. Observers conduct 50 3-min unlimited radius point counts along a 40 km route. Symbols represent the mean of three sound pressure readings measured along each route using a Martel Electronics model 325 sound level meter (accuracy ± 1.5 dB)

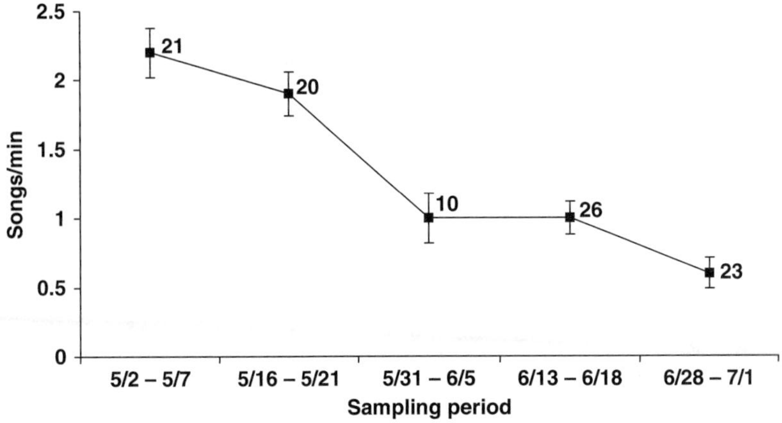

Fig. 7 Declining singing rates of Black-throated Blue Warblers (*Dendroica caerulescens*) over the breeding season in North Carolina. Sample sizes (individual birds sampled for a minimum of 30 min) and standard errors indicated for each sample period (Brewster 2007)

Trends in other factors affecting detection probabilities, such as observers or habitat conditions (Sauer et al. 1994; Norvell et al. 2003), can impart similar biases. For example, a recent analysis indicated that 76% of observers conducting Canadian Breeding Bird Survey routes are over 45 years old (Downes 2004; Simons et al. 2007). Forty-five percent of observers cited "hearing loss" as their primary reason for retiring from the survey. As with ambient noise, trends in age-related hearing loss can impart trends in count data that are unrelated to true abundance.

Singing rates of most breeding songbirds decline steadily during the breeding season. Brewster (2007) found that singing rates of southern Appalachian Black-throated Blue Warblers declined by 50% during the first month of the breeding season (Fig. 7). Temporal trends in factors such as average singing rates, that influence availability during a count, can also impart trends in count data unrelated to the true abundance. For example, there is increasing evidence that birds are breeding earlier now than in the past, presumably due to global warming (Butler 2003). Climatic trends that impart trends in the average singing rates of birds will bias abundance estimates over time unless analyses account for the temporal trends in detection probabilities.

We assessed several factors thought to influence overall detection probabilities (p_a p_{da}) on 40 experimental 3-min point counts comprised of 10 birds per count and five primary species (Black-and-white Warbler (BAWW), *Mniotilta varia*), Black-throated Blue Warbler, Black-throated Green Warbler (BTNW, *Dendroica virens*), Hooded Warbler (HOWA, *Wilsonia citrina*), and Ovenbird over a range of 15 distances (34–143 m). Songs were played at low (two songs per count) and high (13–21 songs per count) singing rates (Alldredge et al. 2007b). Detection probabilities at 100 m ranged from 0.60 (Black-and-white Warbler) to 0.83 (Hooded Warbler) at the high singing rate and 0.41 (Black-and-white Warbler) to 0.67 (Hooded Warbler) at the low singing rate (Fig. 8). Logistic regression

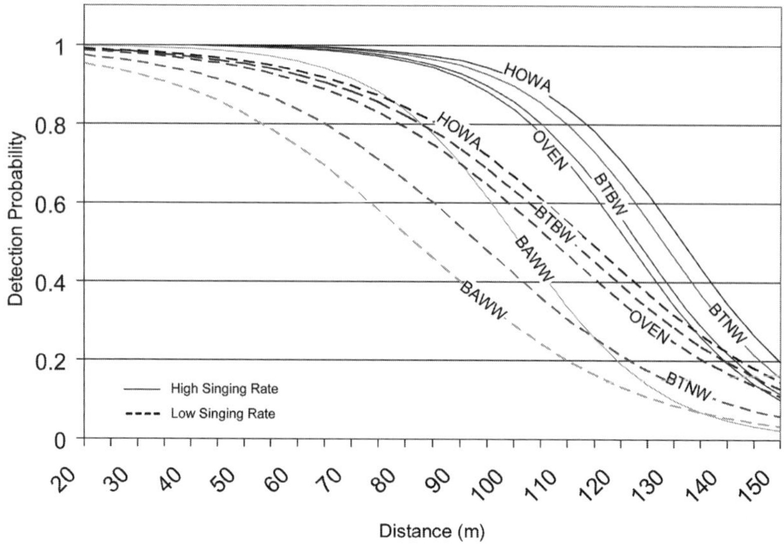

Fig. 8 Logistic regression models for a single observer illustrating the relationship between detection probability and distance for counts of five species (Black-and-white Warbler (BAWW, *Mniotilta varia*), Black-throated Blue Warbler (BTBW, *Dendroica caerulescens*), Black-throated Green Warbler (BTNW, *Dendroica virens*), Hooded Warbler (HOWA, *Wilsonia citrina*), and Ovenbird (OVEN, *Seiurus aurocapillus*) singing at high and low singing rates (Alldredge et al. 2007b). Note the consistent effect of singing rate on detection probability

analyses indicated that species, singing rate, distance, and observer were all significant factors affecting detection probabilities. Simulations of expected counts based on the best logistic model (Table 1), indicated that observers detected between 19% (190/1000 birds for the worst observer, lowest singing rate, and least detectable species) and 65% (653/1000 birds for the best observer, highest singing rate, and most detectable species) of the true population.

2.2 Evaluation of Distance Measurement Error

Detection distance is one of the most important and common auxiliary variables measured during point count surveys of avian abundance. The distance to individual birds is used to determine the effective area sampled, to determine if birds are within a fixed radius plot, and to model the detection process. In densely vegetated habitats, visual detections of birds are rare, and most estimates of detection distance are based solely on auditory cues. Distance sampling theory assumes that detection distances are measured accurately, but empirical validation of this assumption for auditory detections is lacking. We simulated avian point counts in a forested habitat to determine the error structure of distance estimates based on auditory detections (Alldredge et al. 2007c). Experiments were conducted with six experienced observers both before and after distance estimation training. Experiments

Table 1 Detection probabilities at distances from 30 to 150 m, and expected counts for a simulated population of 1,000 uniformly distributed birds, based on the logistic models for BAWW (least detectable species) and HOWA (most detectable species) using the best and worst observers and both high and low singing rates

	Worst		Best	
Distance	Low	High	Low	High
BAWW				
30 (40)	0.87	0.99	0.94	1.00
60 (120)	0.61	0.92	0.80	0.97
90 (200)	0.26	0.55	0.48	0.75
120 (280)	0.08	0.11	0.17	0.23
150 (360)	0.02	0.01	0.05	0.03
Expected Count	190	294	295	382
HOWA				
30	0.97	1.00	0.99	1.00
60	0.88	0.99	0.95	1.00
90	0.64	0.93	0.82	0.97
120	0.29	0.55	0.51	0.76
150	0.08	0.11	0.19	0.24
Expected Count	382	538	529	653

were also conducted to determine the effect of the height and orientation (toward or away from observers) of the song source on distance estimation error. Distance estimation errors for all experiments were substantial, although training did reduce errors and bias in distance estimates. Distance estimates for all observers increased for all species played between 23 and 65 m. Distance estimates did not increase for songs played at distances between 65 and 86 m, indicating observers were not able to differentiate distances among songs played within this range. The height from which songs were played had no effect on distance estimation errors. The orientation of the song source did have a large effect on distance estimation errors; observers generally doubled their estimates for songs played away from them compared to songs played directly toward them (Fig. 9).

2.3 Double-Observer Methods

Comparing simultaneous observations by two or more observers provides another measure of point count detection probabilities (Alldredge et al. 2006). The method requires that observers accurately map bird locations and match birds detected by all observers. We evaluated the accuracy and sources of measurement and classification error in double-observer counts by conducting 60 experimental 3-min point counts with six experienced observers (Alldredge et al. in press). Thirty five players were uniformly distributed with respect to area around a single point. All players were set 1m above ground at radial distances between 0 and 120m. Songs for all species were played at a sound intensity of approximately 90 dB at a distance of 1m.

Each count had exactly 12 birds of up to eight species. Six primary species; Scarlet Tanager (SCTA, *Piranga olivacea*), Acadian Flycatcher (ACFL, *Empidonax*

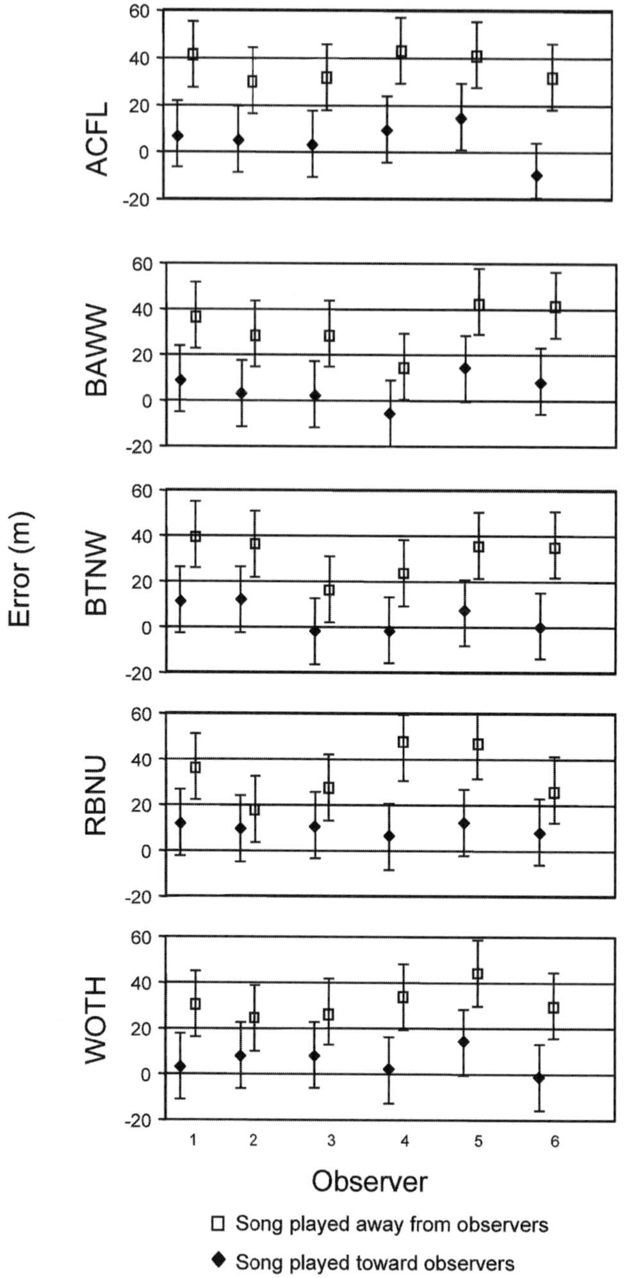

Fig. 9 Differences in distance estimation errors for songs oriented toward observers compared to those oriented away from observers. Errors for six observers averaged across three distance categories. WOTH, (Wood Thrush *Hylocichla mustelina*), RBNU, (Red-breasted Nuthatch, *Sitta Canadensis*), BTNW (Black-throated Green Warbler, *Dendroica virens*), BAWW (Black and White Warbler, *Mniotilta varia*), ACFL (Acadian Flycatcher, *Empidonax virescens*)

Table 2 Identification and matching errors for two observers conducting a double-observer point count

	Observer A		Observer B
# Birds simulated	730	–	730
# Birds recorded (% correct)	581 (72.1)	–	598 (75.3)
Birds mapped within true quadrant (%)	448 (77.1)	–	424 (70.9)
Birds double-counted (%)	52 (9.0)	–	42 (7.0)
Imagined birds (%)	3 (0.5)	–	6 (1.0)
180 degree birds (%)	4 (0.7)	–	5 (0.8)
Total observations for both	–	679	–
Observations match (%)	–	495 (72.9)	–
Observations matched in same quadrant (%)	–	432 (63.6)	–

virescens), BAWW, BTBW, BTNW, and HOWA were simulated across a range of distances to approximate a population of 100 birds uniformly distributed with respect to area. Table 2 provides typical results from a pair of observers who conducted 60 3-min double-observer point counts. A quadrant was a 90° segment of the circular plot, centered on the true location of the song (for individual observers) or the mapped location of one observer (for combined data). Overall, observers undercounted the total number of birds available, recording on average 75.5% (S.E. 1.7%) of simulated birds, and correctly matching 75% (S.E. 1.7%) of birds recorded. In contrast, counts were inflated by an average of 8% (S.E. 1.6%) due a combination of double-counting and misidentification (imagined birds) errors.

2.4 *Time of Detection Methods*

We evaluated the time of detection method (Farnsworth et al. 2002; Alldredge et al. 2007a) by conducting 60, 8-min point counts with four experienced observers (Alldredge et al. 2007d). Counts were divided into four, 2-min intervals, and observers recorded birds using multi-colored pens to distinguish time intervals. Detections of birds recorded in a previous interval were noted by underlining the initial notation in the color of the current time interval. Thirty five players were uniformly distributed with respect to area around a single point. All players were set 1m above ground at radial distances between 0 and 1m. Songs for all species were played at a sound intensity of approximately 90 dB at a distance of 1m.

Eighteen birds of 12 species were simulated on each point. Analyses were focused on eight species; ACFL, BAWW, BTBW, BTNW, HOWA, OVEN, SCTA, and YTWA. Songs of 100 total individual birds of each of these eight species were played on the 60 counts. The total simulated population size was 800 birds because not all birds in the simulated population were available on some counts. Availability of ACFL, BAWW, BTNW, SCTA, and YTWA was simulated under a Markovian process with availability during the count interval varying between 0.60 and 1.0, and singing rates varying from 2 to 8 songs per minute. Singing rates and availability for BTBW and OVEN were based on empirical field data (Brewster 2007).

Double counting, counting a single bird as more than one bird, was a significant source of error among the four experienced observers. Double counting rates ranged from 0.9 to 3.4% (S.E. 0.6%) of total observations among observers. Double counting occurred in a variety of forms. In most cases single birds were recorded as two birds throughout the count interval. This created two or more capture histories that clearly indicated a single individual tracked as two birds throughout the count. Occasionally observers mapped a bird in one location at the beginning of the count, then mapped the same bird in a new location and continued to track it at the new location for the remainder of the count. These cases produced two or more capture histories of the form xx00 for the original bird and 00xx for the double count, where x could be a either a one or a zero.

Overall 2.0–4.1% (S.E. 0.43%) of observations were recorded in the wrong time interval among the four observers. This can occur when two or more individuals of the same species sing during a count, and observers attribute a song to the wrong individual. Finally, as we found in the multiple-observer experiment, misidentification errors were rare with experienced observers. Misidentification errors ranged from 0.1 to 0.6% (S.E. 0.09%) of total observations among the four experienced observers.

3 Discussion

Many known and unknown factors influence detection probabilities on auditory point counts. These include factors such as ambient noise, habitat structure, and the singing rates of individual birds that can impart trends in abundance estimates unrelated to true abundance. Fortunately, multiple methods of estimating detection probabilities are now available, and we believe direct estimates of detection probability should accompany all analyses of avian point count data.

Unfortunately, methods that require the localization of auditory detections are subject to large measurement and misclassification errors. Our experiments illustrate the source and magnitude of those errors, although we believe errors on actual point counts are probably larger than our results suggest, because our simulations involved a relatively small number of species, our observers were highly experienced, and many sources of variability were carefully controlled in our experiments.

As the example for the Golden-crowned Kinglet and Scarlet Tanager in the introduction illustrates, distance sampling can reduce sampling bias for species with large differences in behavior, plumage, and song characteristics. Nevertheless we found that measurement error on auditory point counts is substantial, presumably because the complexities of sound attenuation and reverberation in natural environments make the localization of auditory cues very difficult. We recommend that practitioners distinguish between visual and auditory detections in their analyses of point count data, and recognize the limitations of distance sampling methods on avian point counts when detections are auditory. If distance sampling based solely on auditory detections is unavoidable, analyses should incorporate estimates of measurement error.

Classification and matching errors are important and generally ignored components of overall error rates on multiple-observer and time of detection point counts. Errors result primarily due to the difficulty of localizing auditory cues and accurately mapping bird locations (Buckland 2006). We would expect substantially larger errors on actual point counts that employ these methods due to the movement of birds during the count interval.

These findings imply that the uncertainty surrounding estimates of avian diversity and abundance based on distance, double-observer, and time of detection auditory sampling methods is much higher than is currently assumed by practitioners. Therefore conclusions based on those estimates are, in most cases, weaker than those currently reported in studies based on auditory detections. Approaches to account for this uncertainty in abundance models are clearly needed if abundance estimates are to withstand critical scrutiny.

Alternatively, it is possible that modern avian auditory sampling methods, that require observers to simultaneously track the location and singing rates of multiple individuals and multiple species, often exceed the ability of human observers. Simplified protocols, such as single species surveys, or occupancy approaches based on presence/absence data (Royle and Nichols 2003) that reduce the demands on human observers, may ultimately yield better results.

Acknowledgments We are very grateful to the many volunteers who assisted with this research: David Allen, Brady Beck, Jenna Begier, Scott Bosworth, Amy Bleckinger, Dan Boone, Marshall Brooks, Gordon Brown, Becky Browning, Sue Cameron, Susan Campbell, John Connors, Deanna Dawson, Jimmy Dodson, Barbara Dowell, Curtis Dykestra, Adam Efird, Patrick Farrell, John Finnegan, Lena Gallitano, John Gerwin, Stephanie Horton, Becky Hylton, Mark Johns, Chris Kelly, Salina Kovach, Ed Laurent, Harry Legrand, Merrill Lynch, Sarah Mabey, Jeff Marcus, Kevin Miller, Melissa Miller, Ryan Myers, Keith Pardieck, Bruce Peterjohn, Andrei Podolsky, Chan Robbins, James Sasser, Shiloh Schulte, Clyde Sorenson, Ed Swab, Chris Szell, Nathan Tarr, Kendrick Weeks, Dan Williams, and Diana Wray. C.M. Downes generously allowed us to cite her survey of Canadian Breeding Bird Survey volunteers. Electrical engineering students at NCSU: John Marsh, Marc Williams, and Michael Foster and Wendy Moore provided valuable technical assistance. Funding for this research was provided by the USGS Status and Trends Program, the US Forest Service, the US National Park Service, and the North Carolina Wildlife Resources Commission.

References

Alldredge MW, Pollock KH, Simons TR (2006) Estimating detection probabilities from multiple observer point counts. Auk 123:1172–1182.

Alldredge MW, Pollock KH, Simons TR, Collazo J, Shriner SA (2007a) Time of detection method for estimating abundance from point count surveys. Auk 124:653–664.

Alldredge MW, Simons TR, Pollock KH (2007b) Factors affecting aural detections of songbirds. Ecological Applications 17:948–955.

Alldredge MW, Simons TR, Pollock KH (2007c) An experimental evaluation of distance measurement error in avian point count surveys. Journal of Wildlife Management 71: 2759–2766.

Alldredge MW, Simons TR, Pollock KH, Pacifici K (2007d) A field evaluation of the time-of-detection method to estimate population size and density for aural avian point counts. *Avian*

Conservation and Ecology – Écologie et conservation des oiseaux 2(2):13. http://www.ace-eco.org/vol2/iss2/art13/

Alldredge MW, Pacifici K, Simons TR, Pollock KH (In press) A novel field evaluation of the effectiveness of distance sampling and double independent observer methods to estimate aural avian detection probabilities. Journal of Applied Ecology.

Bart J (2005) Monitoring the abundance of bird populations. AUK 122:15–25.

Bart J, Earnst S (2002) Double sampling to estimate density and population trends in birds. Auk 119:36–45.

Best LB (1981) Seasonal changes in detection of individual bird species. Studies in Avian Biology 6:252–261.

Brewster JP (2007) Spatial and temporal variation in the singing rates of two forest songbirds, the Ovenbird and Black-throated Blue Warbler; Implications for aural counts of songbirds. MS Thesis. Department of Zoology, North Carolina State University, Raleigh, NC.

Buckland ST (2006) Point-transect surveys for songbirds: Robust methodologies. Auk 123: 345–357.

Buckland ST, Anderson DR, Burnham KP, Laake JL, Borchers DL, Thomas L (2001) Introduction to distance sampling: Estimating abundance of biological populations. Oxford University Press, New York.

Butler C (2003) The disproportionate effect of global warming on the arrival dates of short-distance migratory birds in North America. IBIS 145:484–495.

Cuthill I, Hindmarsh A (1985) Increase in Starling song activity with removal of mate. Animal Behavior 33:326–335.

DeJong MJ, Emlen JT (1985) The shape of the auditory detection function and its implication for songbird censusing. Journal of Field Ornithology 56:213–223.

Diehl B (1981) Bird populations consist of individuals differing in many respects. Studies in Avian Biology 6:225–229.

Downes CM (2004) Canadian Wildlife Service, Environment Canada. Results of the 2004 Questionnaire for Canadian Participants in the Breeding Bird Survey. http://www.cws-scf.ec.gc.ca/nwrc-cnrf/default.asp?lang=en&n=929AA800-1

Emlen JT, DeJong MJ (1981) The application of song detection threshold distance to census operations. Studies in Avian Biology 6:346–352.

Emlen JT, DeJong MJ (1992) Counting birds: The problem of variable hearing abilities. Journal of Field Ornithology 63:26–31.

Farnsworth GL, Pollock KH, Nichols JD, Simons TR, Hines JE, Sauer JR (2002) A removal model for estimating detection probabilities from point-count surveys. Auk 119:414–425.

Gibbs JP, Wenny DG (1993) Song output as a population estimator: Effect of male pairing status. Journal of Field Ornithology 64:316–322.

Johnson LS (1983) Effect of mate loss on song performance in the Plain Titmouse. Condor 85: 378–380.

Kendall WL, Peterjohn BG, Sauer JR (1996) First-time observer effects in the North American Breeding Bird Survey. Auk 113:823–829.

Kepler CB, Scott JM (1981) Reducing bird count variability by training observers. Studies in Avian Biology 6:366–371.

Kery M, Royle JA, Schmid H (2005) Modeling avian abundance from replicated counts using binomial mixture models. Ecological Applications 15:1450–1461.

Kissling ML, Garton EO (2006) Estimating detection probability and density from point-count surveys: a combination of distance and double-observer sampling. Auk 123:735–752.

Krebs JR, Avery M, Cowie RJ (1980) Effect of removal of mate on the singing behavior of Great Tits. Animal Behavior 29:635–637.

Lichstein JW, Simons TR, Franzreb KE (2002) Landscape effects on breeding songbird abundance in managed forests. Ecological Applications 12:836–857.

Marsh H, Sinclair DF (1989) Correcting for visibility bias in strip transect aerial surveys of aquatic fauna. Journal of Wildlife Management 53:1017–1024.

Mayfield HF (1981) Problems in estimating population size through counts of singing males. Studies in Avian Biology 6:220–224.

McShea WJ, Rappole JH (1997) Variable song rates in three species of passerines and implications for estimating bird population. Journal of Field Ornithology 68:367–375.

Nichols JD, Hines JE, Sauer JR, Fallon FW, Fallon JE, Heglund PJ (2000) A double-observer approach for estimating detection probability and abundance from point counts. Auk 117: 393–408.

Nichols JD, Thomas L, Conn PB (2008) Inferences about landbird abundance from count data: Recent advances and future directions. Journal of Ecological and Environmental Statistics 3:201–236.

Pacific K, Simons TR, Pollock KH (In press) Effects of vegetation and background noise on the detection process in auditory avian point count surveys. The Auk 125.

Norvell RE, Howe FP, Parrish JR (2003) A seven-year comparison of relative abundance and distance sampling methods. Auk 120:1013–1028.

Pollock KH, Marsh H, Bailey LL, Farnsworth GL, Simons TR, Alldredge MW (2004) Separating components of detection probability in abundance estimation: An overview with diverse examples. Pages 43–58 *in* Sampling Rare and Elusive Species: Concepts, Designs and Techniques for Estimating Population Parameters (W. L. Thompson ed.). Island Press, Washington DC.

Pollock KH, Marsh HD, Lawler IR, Alldredge MW (2006) Estimating animal abundance in heterogeneous environments: An application to aerial surveys for dugongs. Journal of Wildlife Management 70:255–262.

Pollock KH, Nichols JD, Simons TR, Farnsworth GL, Bailey LL, Sauer JR (2002) Large scale wildlife monitoring studies: Statistical methods for design and analysis. Environmetrics 13:105–119.

Ralph CJ (1981) An investigation of the effects of seasonal activity levels on avian censusing. Studies in Avian Biology 6:265–270.

Ralph JC, Droege S, Sauer JR (1995) Managing and monitoring birds using point counts: standards and applications. Pages 161–168 *in* Monitoring bird populations by point counts (Ralph JC, Sauer JR, Droege S Eds.). U.S. Forest Service General Technical Report PSWGTR-149.

Robbins CS (1981a) Effect of time of day on bird activity. Studies in Avian Biology 6:275–286.

Robbins CS (1981b) Bird activity levels related to weather. Studies in Avian Biology 6:301–310.

Rosenstock SS, Anderson DR, Giesen KM, Leukering T, Carter MF (2002) Landbird counting techniques: Current practices and an alternative. Auk 119:46–53.

Royle JA, Nichols JD (2003) Estimating abundance from repeated presence-absence data or point counts. Ecology 84:777–790.

Sauer JR, Barker RJ, Geissler PH (1994) Statistical aspects of modeling population change from population size data. Pages 451–466 *in* Wildlife Toxicology and Population Modeling: Integrated Studies of Agroecosystems (R. J. Kendall and Lacher TE, Jr., eds.). CRC Press, Boca Raton, FL.

Sauer JR, Hines JE, Fallon J (2005) The North American Breeding Bird Survey, Results and Analysis 1966–2004. Version 2005.2, USGS Patuxent Wildlife Research Center, Laurel, MD http://www.mbr-wrc.usgs.gov/bbs/bbs.html.

Schieck J (1997) Biased detection of bird vocalizations affects comparisons of bird abundance among forested habitats. The Condor 99:179–190.

Scott JM, Ramsey FL, Kepler CB (1981) Distance estimation as a variable in estimating bird numbers from vocalizations. Studies in Avian Biology 6:334–340.

Sheilds WM (1977) The effect of time of day of avian census results. Auk 94:380–383.

Shriner SA (2001) Distribution of breeding birds in Great Smoky Mountains National Park. Ph.D. Dissertation, Department of Zoology, North Caroline State University, Raleigh, NC.

Simons TR, Shriner SA, Farnsworth GL (2006) Comparison of breeding bird and vegetation communities in primary and secondary forests of Great Smoky Mountains National Park. Biological Conservation 129:302–311.

Simons TR, Alldredge MW, Pollock KH, Wettroth JM (2007) Experimental analysis of the auditory detection process on avian point counts. The Auk 124:986–999.

Skirvin AA (1981) Effects of time of day and time of season on the number of observations and density estimates of breeding birds. Studies in Avian Biology 6:271–274.

Thomas L, Laake JL, Derry JF, Buckland ST, Borchers DL, Anderson DR, Burnham KP, Strindberg S, Hedley SL, Burt ML, Marques FFC, Pollard JH, Fewster RM (1998) *Distance 3.5. Release 6*. Research Unit for Wildlife Population Assessment, University of St.Andrews, UK. Available: http://www.ruwpa.st-and.ac.uk/distance/

Thompson WL (2002) Towards reliable bird surveys: Accounting for individuals present but not detected. Auk 119:18–25.

Williams BK, Nichols JD, Conroy MJ (2002) Analysis and management of animal populations. Academic Press, New York.

Wilson DM, Bart J (1985) Reliability of singing bird surveys: Effects of song phenology during the breeding season. Condor 87:69–73.

Yoccoz NG, Nichols JD, Boulinier T (2001) Monitoring of biological diversity in space and time. Trends in Ecology and Evolution 16:446–453.

Density Estimation by Spatially Explicit Capture–Recapture: Likelihood-Based Methods

Murray G. Efford, David L. Borchers, and Andrea E. Byrom

Abstract Population density is a key ecological variable, and it has recently been shown how captures on an array of traps over several closely-spaced time intervals may be modelled to provide estimates of population density (Borchers and Efford 2007). Specifics of the model depend on the properties of the traps (more generally 'detectors'). We provide a concise description of the newly developed likelihood-based methods and extend them to include 'proximity detectors' that do not restrict the movements of animals after detection. This class of detector includes passive DNA sampling and camera traps. The probability model for spatial detection histories comprises a submodel for the distribution of home-range centres (e.g. 2-D Poisson) and a detection submodel (e.g. halfnormal function of distance between a range centre and a trap). The model may be fitted by maximising either the full likelihood or the likelihood conditional on the number of animals observed. A wide variety of other effects on detection probability may be included in the likelihood using covariates or mixture models, and differences in density between sites or between times may also be modelled. We apply the method to data on stoats *Mustela erminea* in a New Zealand beech forest identified by microsatellite DNA from hair samples. The method assumes that multiple individuals may be recorded at a detector on one occasion. Formal extension to 'single-catch' traps is difficult, but in our simulations the 'multi-catch' model yielded nearly unbiased estimates of density for moderate levels of trap saturation ($\leq$ 86% traps occupied), even when animals were clustered or the traps spanned a gradient in density.

Keywords Density Estimation · Maximum likelihood · Capture–recapture · Competing risks · *Mustela erminea* · DNA · Camera Traps · Individual heterogeneity

---

M.G. Efford (✉)
Zoology Department, University of Otago, PO Box 56, Dunedin 9010, New Zealand
e-mail: murray.efford@stonebow.otago.ac.nz

D.L. Thomson et al. (eds.), *Modeling Demographic Processes in Marked Populations*, Environmental and Ecological Statistics 3, DOI 10.1007/978-0-387-78151-8_11,

1 Introduction

Trapping is a common source of capture–recapture data, but the spatial component of such data has generally been ignored. By trapping we mean sampling an animal population with traps set for a known time at known points in the habitat, often on a grid. Time is usually divided into discrete intervals ('occasions'), and new animals may be captured, marked and released on each occasion. By convention, closed-population encounter histories are coded in binary form: on each occasion an individual is either captured (1) or not captured (0) (Otis et al. 1978). A spatial encounter history also records the location of each capture. We are concerned with the estimation of population density using information in the spatial encounter history.

Previous methods for estimating population density D with arrays of traps have used the relation $\hat{D} = \hat{N}/A$, where N is the population size and A is the area occupied by the population. This is the method of choice if the biological population occupies a defined geographic area (e.g. an island) and if every member of the population is at risk of capture. More commonly, the individuals at risk of capture in traps are an ill-defined subset N_c of a larger biological population that extends indefinitely beyond the trap array. We may estimate N_c empirically from the encounter histories with conventional closed population methods (Otis et al. 1978; Chao and Huggins 2005), but this quantity bears only a vague relationship to the biological parameters of interest (N and D). While we may hypothesize the existence of an 'effective trapping area' A_c such that $D = N_c/A_c$, rigorous general methods for estimating A_c are lacking (but see White et al. 1982; Jett and Nichols 1987).

A more secure approach is to estimate density directly without recourse to the quantities N_c and A_c. The feasibility of estimating D directly from trapping data was demonstrated by Efford (2004) and Efford et al. (2004). Their method relied on a simulation of the trapping process. Here we describe a likelihood-based approach that is in some ways more general and flexible. The underlying theory was developed by Borchers and Efford (2007).

The literature on nonspatial capture–recapture has not been concerned with the trapping process although it is an important determinant of capture probability. At the simplest level, increasing the number of traps per home range will increase capture probability; more subtly, per capita capture probability will decline with increasing local density if traps are of a type that 'fill up', particularly if they are 'single-catch' traps. Such patterns result naturally from suitably formulated spatial trapping models, so long as care is taken to match the model to the process by which data were collected. The focus in Borchers and Efford (2007) was on traps that do not fill up, but which stop an animal from advancing to another trap within the same occasion (we call these 'multi-catch' traps). We extend their treatment to allow for other types of trapping process; in particular, we model 'proximity detectors' such as automatic cameras and devices that passively collect DNA samples from animals without limiting their movement. As an example, we analyse data from stoats *Mustela erminea* identified by their microsatellite DNA in hair samples. We also discuss the extension of likelihood-based methods to single-catch traps. In lieu of a likelihood function for single-catch traps, we use simulation to evaluate the performance of the multi-catch density estimator applied to data from single-catch traps.

2 Model

We wish to construct a probability model for encounter histories that include the location of each detection. Our model comprises one submodel for the distribution of animals in a region that includes the traps, and another submodel for the capture process. The capture process submodel gives the probability of catching an individual in a particular trap, given the location of its home range. We introduce these models before proceeding to the likelihood.

2.1 Distribution Submodel

We assume that for the duration of trapping the general location of each individual in the population may be summarised by the coordinates of a point that we call the animal's home range centre. Later we relate probability of detection to radial distance from this point. The density of the population is equivalent to the intensity of a spatial point process for the home range centres. In this paper we model the distribution with a homogeneous spatial Poisson process; more generally, we could use an inhomogeneous Poisson process (Borchers and Efford 2007).

2.2 Capture Submodel

A spatial model of capture probability must take into account properties of the trap or detector. We distinguish three types of detector:

- Proximity detector
- Multi-catch trap
- Single-catch trap

We order these by increasing complexity in the probability model, rather than novelty. Multi-catch traps were treated by Borchers and Efford (2007), while the likelihood given here for proximity detectors is new.

A proximity detector records the presence of an individual at or near a point, but leaves it free to visit other detectors on the same occasion. Multiple individuals may be recorded at a detector on one occasion (see Discussion for one-shot detectors). Examples are camera traps and passive DNA sampling devices such as sticky hair traps. The probability that a particular individual i with home range centre $\mathbf{X}(i)$ is recorded at detector k, located at $\mathbf{Y}(k)$, is assumed to be a function of the Euclidean distance $d_k[\mathbf{X}(i)] = |\mathbf{X}(i) - \mathbf{Y}(k)|$, and possibly also of other covariates. Here vector notation (in bold) is used for location, which might otherwise have been represented by Cartesian coordinates (x, y). We assume independence between visits to different detectors. Each occasion-specific entry in an encounter history from an array of K proximity detectors is itself a vector of length K whose elements take the value 1 for detectors at which the individual was recorded at least once and 0 otherwise.

Multi-catch traps differ from proximity detectors in that capture in one trap precludes capture of the same individual in other traps on the same occasion. As with proximity detectors, multiple individuals may be recorded at a trap on one occasion. Mist nets for birds and pitfall traps for lizards are examples of multi-catch traps used in capture–recapture studies. The probability of capture is modified by 'competition' among traps for the chance to capture an individual (multiple traps within an individual's home range may reduce its probability of capture in any particular trap). An additive competing risks hazard formulation is appropriate for trap-specific capture probability (Borchers and Efford 2007). Each occasion-specific entry in an encounter history from an array of K multi-catch traps is a single trap index k where $0 \leq k \leq K$, and $k = 0$ indicates no capture.

Single-catch traps are able to catch only one animal at a time, and capture probability is affected by the presence of other individuals that may 'compete' for traps. The majority of traps used for capture–recapture of small mammals are of this type. The encounter history has the same form as for multi-catch traps, but different histories may have the same entry on one occasion only if both are zero. Capture of an animal disables a trap and immediately reduces the capture probabilities of neighbouring animals. Simulation of the capture process is straightforward in continuous time, and a capture model may be fitted by inverse prediction (Efford 2004). A likelihood model for single-catch traps is considerably more complicated than for multi-catch traps, and remains to be developed.

3 Likelihood

The probability associated with each capture history may be treated as the product of the probability of catching an individual at least once ($p.$) and the probability of the observed history given that it includes at least one capture. Each part is conditional on the location of the individual's home range centre $\mathbf{X}$, but using the distribution submodel we may integrate over possible locations to evaluate the likelihood without knowing $\mathbf{X}$.

We start by defining a spatial analogue of detection probability $a = \int p.(\mathbf{X}; \boldsymbol{\theta})\, d\mathbf{X}$, where $\boldsymbol{\theta}$ is a vector of detection parameters and a has units of area (the parallel between a and detection probability becomes clear in the next section). For a homogeneous Poisson distribution model, the probability of observing exactly n capture histories is itself Poisson-distributed:

$$\Pr(n) = \frac{(Da)^n \exp(-Da)}{n!}. \tag{1}$$

The likelihood given n observed capture histories $\omega = (\omega_1,\ldots,\omega_n)$ is then

$$L(\boldsymbol{\theta}, D) = \Pr(n) \times \prod_{i=1}^{n} \frac{\int \Pr(\omega_i|\mathbf{X};\boldsymbol{\theta})d\mathbf{X}}{a}, \quad (2)$$

where $\Pr\{\omega_i|\mathbf{X};\boldsymbol{\theta}\}$, the probability of the capture history for a given home range location and model parameters, is defined below. The probability of being caught at least once over S occasions depends on the distances $d_k(\mathbf{X})$ to each of the K traps:

$$p.(\mathbf{X};\boldsymbol{\theta}) = 1 - \prod_s \prod_k \{1 - p_s(d_k(\mathbf{X})\,;\boldsymbol{\theta})\}. \quad (3)$$

Here p_s is analogous to the detection function in distance sampling (e.g. Buckland et al. 2001).[1] Its parameters $\boldsymbol{\theta}$ control the overall efficiency of detection and also its spatial scale, which we expect to increase with home range size. Three suitable forms for p_s are shown in Table 1. These use the independent parameters g_0 for overall efficiency of detection and σ for spatial scale.[2] The hazard function has an additional shape parameter b ($b > 0$); when b is fixed at a large value (e.g. 100) the hazard function comes to resemble a step function ($p_s(d) \approx 0$ for $d > \sigma$). Although our experience tends to favour the hazard function, we recommend that a function is selected for each dataset only after comparing the fit of alternatives.

The preceding formulation applies to all three types of detector. Differences arise in the term $\Pr\{\omega_i\,|\mathbf{X};\boldsymbol{\theta}\}$. This has the general form

$$\Pr\{\omega_i|\mathbf{X};\boldsymbol{\theta}\} = \prod_s \prod_k p_{ks}^{\delta_{iks}}\,(1 - p_{\cdot s})^{1-\delta_{i\cdot s}}, \quad (4)$$

where p_{ks} is the probability of detection at k on occasion s, $\delta_{iks} = 1$ if individual i was detected at k on occasion s and $\delta_{i\cdot s} = 1$ if $\sum_k \delta_{iks} > 0$ and $\delta_{i\cdot s} = 0$ otherwise.

Table 1 Detection functions p_s for spatially explicit capture–recapture models. d is the distance between an animal's home range centre and a detector. The parameter g_0 is common to all functions and represents the probability of detection at a single detector placed in the centre of the home range; values of the spatial scale parameter σ are not comparable between functions

Detection function		Parameters $\boldsymbol{\theta}$
Halfnormal	$p_s = g_0 \exp\left(\frac{-d^2}{2\sigma^2}\right)$	g_0, σ
Hazard rate	$p_s = g_0\left[1 - \exp\left\{-\left(d/\sigma\right)^{-b}\right\}\right]$	g_0, σ, b
Negative exponential	$p_s = g_0 \exp\left(\frac{-d}{\sigma}\right)$	g_0, σ

[1] Borchers and Efford (2007) use p_s^1 for p_s.

[2] Independence may not always be appropriate: intuitively, an animal that spreads its activity over a larger area will become less trappable at any particular place. An alternative parameterization would scale g_0 by $1/\sigma^2$, as in the pdf of a bivariate normal distribution.

For an array of proximity detectors we use

$$p_{ks} = p_s(d_k(\mathbf{X})\ ;\boldsymbol{\theta}). \tag{5}$$

Multi-catch traps 'compete' for animals and a competing risks hazard-rate form is appropriate:

$$p_{ks} = \frac{h(d_k(\mathbf{X}))}{h.(\mathbf{X})}[1 - e^{-h.(\mathbf{X})}], \tag{6}$$

where $h(d_k(\mathbf{X})) = -\ln\{1 - p_s(d_k(\mathbf{X};\boldsymbol{\theta}))\}$ and $h.(\mathbf{X}) = \sum_{k=1}^{K} h(d_k(\mathbf{X}))$.

4 Estimation

We estimate D and $\boldsymbol{\theta}$ by numerically maximising the full likelihood (1) with respect to the parameters. For maximisation we log-transform D and σ, and logit-transform g_0 to keep each within feasible ranges. Each evaluation of the likelihood requires numerical integration over the plane, once for each observed encounter history ω_i and once for the null capture history to calculate a. The speed of the algorithm used for integration is therefore critical. We have not obtained satisfactory results with standard algorithms such as the adaptive method of Genz and Malik (1980) used in some packages; our preferred method at present is to sum function values over a grid of points. Integration may be limited to a subset of the plane that contains plausible animal locations $\mathbf{X}$; the estimated density will then apply to that area of habitat. Asymptotic variances may be estimated from the inverse of the information matrix. Confidence limits for $\hat{D}$ may be estimated as $\exp[\ln(\hat{D}) \pm z_\alpha \hat{s}_D]$, where $\hat{s}_D$ is the estimated SE of $\hat{D}$ on the log scale and z_α is the appropriate normal deviate, or by profile likelihood. Software is available (Efford 2007).

An alternative procedure is to maximise the conditional likelihood (the product over capture histories in (1)) to get estimates $\hat{\boldsymbol{\theta}}$ and hence $a(\hat{\boldsymbol{\theta}})$, and to estimate $\hat{D} = n/\hat{a}$. This is advantageous when there are individual covariates $\mathbf{z}_i$ as we can then use the Horvitz-Thompson-like estimator $\hat{D} = \sum_{i=1}^{n} \hat{a}(\mathbf{z}_i)^{-1}$, which does not require the pdf of covariates to be modelled (Borchers and Efford 2007). Similar methods are used in conventional capture–recapture to estimate population size N from individual detection probabilities p_i ($\hat{N} = \sum_{i=1}^{n} \hat{p}_i^{-1}$) (Huggins 1989).

5 Extensions

5.1 Modelling Additional Variation in Detection

Our core model accounts for variation in capture probability due to the varying number and location of traps in each animal's home range. This confers a robustness that is lacking in conventional closed-population analyses of trapping data.

Other sources of variation that *are* addressed in conventional analyses (e.g. Otis et al. 1978; Chao and Huggins 2005) may readily be included (Borchers and Efford 2007). Capture probability p in conventional models is replaced in the spatial model by a vector of at least two parameters, g_0 and σ. Each conventional source of variation in p (i.e. time, response to capture, and individual heterogeneity; Otis et al. 1978) may affect either or both of g_0 and σ. For example, we can fit a model for a change after first capture in either the efficiency of detection ($M_b(g_0)$) or its spatial scale ($M_b(\sigma)$). Individual heterogeneity may be incorporated via mixture models for either parameter (e.g. 2-class finite mixture $M_{h2}(g_0)$ cf. Pledger 2000). In addition, the spatial model allows for novel sources of variation, such as dependence of capture on the type of trap or on other trap-level covariates describing the habitat at the trap site. Modelling and estimation of additional sources of variation in detection probability requires additional parameters and adjustments to the expression for p_{ks} (Eqs. (5) and (6)). We do not describe these in detail because they follow directly from current practice (Otis et al. 1978; Chao and Huggins 2005).

5.2 Variation Across Space or Time

The purpose of a capture–recapture study will often be to compare density at different places, or at different times. For convenience, we use the word ‘session’ for each sampled population, whether populations are separated by space, time or an attribute such as sex. Our ‘sessions’ have been termed ‘groups’ in other capture–recapture contexts (e.g. Williams et al. 2002, p. 426), and are loosely equivalent to primary sessions in the open-population robust design of Pollock (1982). Even when a separate density is to be estimated for each session, if data are sparse it may be efficient to estimate a common detection function across all sessions. In general, session effects may be treated as constant (pooled across sessions), as fixed effects (e.g. session-specific levels or a trend across sessions) or as random effects (yet to be implemented). Thus the utility of the method is greatly extended by a multi-session model. Alternative models may be compared by standard methods (e.g., Akaike’s Information Criterion or likelihood ratio tests).

Sessions are assumed to be demographically closed (no births, deaths, immigration or emigration), and each encounter history spans only one session. If the same individual is caught in two sessions it is artificially assigned a new identity in the second session. Under these independence assumptions it is appropriate to use a combined multi-session likelihood (the product of within-session likelihoods) to model variation between sessions.

Session-specific parameter values (levels of D and the elements of $\boldsymbol{\theta}$) may be treated as functions of session-level covariates, including time. For each evaluation of the combined likelihood we substitute the current values of D and $\boldsymbol{\theta}$ into the within-session likelihood (1), and sum the resulting log-likelihoods across sessions. The combined likelihood is maximised over all parameters, including those of the functions controlling the session-specific D and $\boldsymbol{\theta}$.

6 Example: Stoats Identified by DNA Microsatellites

Stoats *Mustela erminea* introduced to New Zealand have deleterious effects on populations of native birds, and their ecology and population management are therefore a prime concern for conservation. Capture–recapture with traps is onerous and not always successful because of low capture rates. An alternative to trapping is to register the presence of individuals from DNA in hair or dung. For stoats, a convenient sampling device is a tube with a transverse adhesive-coated rubber band that retains hairs from stoats that pass through (Duckworth et al. 2005). Here we analyse data from a pilot study in *Nothofagus fusca* forest in the Matakitaki Valley, South Island, New Zealand (172°30'E, 42°00'S). Hair sampling tubes ($K =$ 94) were placed on a 3×3 km grid with 500 m spacing between rows and 250 m spacing along rows. Tubes were baited with rabbit meat and checked daily for 7 days, starting 15 December 2001. Stoat hair samples were identified to individual using DNA microsatellites amplified by PCR from follicular tissue.[3] Six loci were amplified and the mean number of alleles was 7.3 per locus, allowing identification of individuals even in samples for which not all loci could be amplified (27%).

Table 2 Spatial encounter histories of 20 stoats identified by DNA microsatellites in hair samples collected daily on 7 parallel lines of sampling stations A–G; spacing 500 m between lines and 250 m along lines. Matakitaki, South Island, New Zealand, 15–21 December 2001. '–' indicates the stoat was not detected. There were no multiple records (>1 stoat per trap or >1 trap per stoat on any occasion)

	Occasion						
Animal	1	2	3	4	5	6	7
1	A9	–	–	–	–	–	–
2	A12	A12	–	–	–	–	–
3	–	–	C6	B5	–	–	–
4	–	–	G3	–	F3	–	–
5	–	–	–	F2	E2	–	F1
6	–	–	–	–	–	E8	–
7	–	–	F5	–	–	G7	–
8	F6	–	–	–	–	–	–
9	–	A4	–	–	–	–	–
10	–	C5	–	–	–	–	–
11	–	D4	–	–	–	–	–
12	–	–	D7	–	–	–	–
13	–	–	E5	–	–	E4	–
14	–	–	F1	–	–	G3	–
15	–	–	F9	–	–	–	–
16	–	–	G13	–	F13	G13	–
17	–	–	–	G9	–	–	–
18	–	–	–	–	–	F1	–
19	–	–	–	–	–	G9	–
20	–	–	–	–	–	–	G8

[3]We do not address here the problems of identification due to 'allelic dropout' and other difficulties when the samples contain only small amounts of DNA that is potentially degraded.

The dataset included 20 individuals of which 7 were 'recaptured' a total of 10 times (Table 2). The largest detected movement (707 m) was small relative to the usual home range size of stoats (the average home-range diameter of stoats in New Zealand is at least 1.3 km; data from King and Murphy 2005, Table 5.5), and individuals appeared to be localised within the grid (Fig. 1). No stoat was detected in more than one tube on the same day, although both the field methodology and the analysis allow for this.

We fitted a homogeneous Poisson density by maximising the full likelihood. For numerical integration we evaluated the function at 1024 evenly-distributed points in a rectangular area extending 1000 m beyond the grid. Modelling of even a single-session dataset such as this requires multiple choices: among forms for the detection function p_s, and among models for variation in the parameters of p_s in relation to previous capture and random individual variation. We did not collect data on occasion-, trap- or individual covariates (we note that sex determined from DNA would be a potentially useful covariate of g_0 and σ in this sexually dimorphic species).

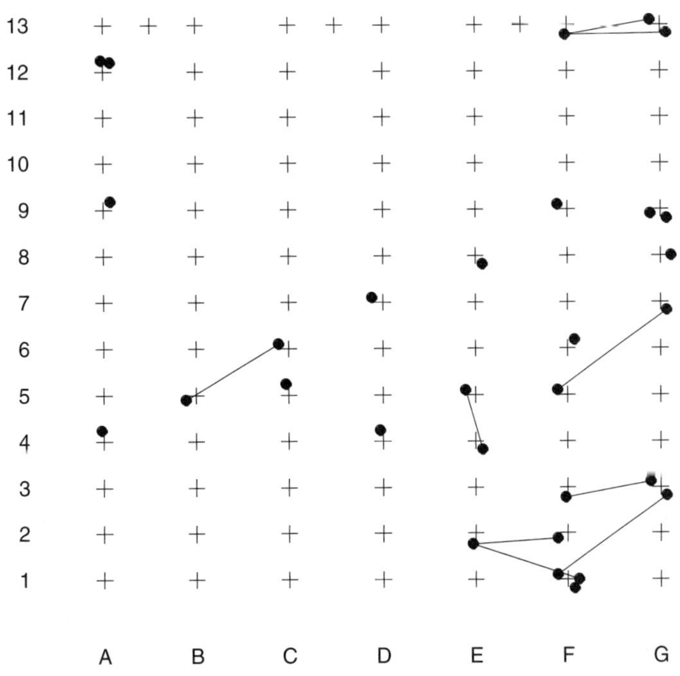

Fig. 1 Map of detections of individual stoats in red beech forest, Matakitaki Valley, South Island, New Zealand, 15–21 December 2001. Sampling stations (crosses) were on seven lines A–G; stations spaced 500 m between lines and 250 m along lines, with three additional detectors as shown. The first station on each line was on a forest-pasture edge. Lines are drawn between locations of the same individual on different occasions; locations are shifted slightly from the actual location for clarity

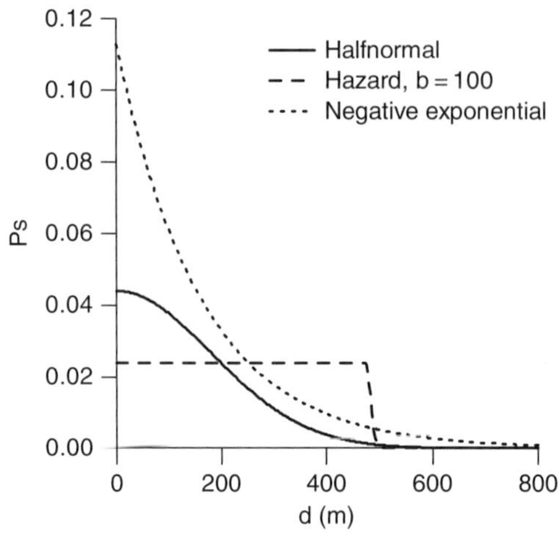

Fig. 2 Detection functions $p_s(d)$ fitted to stoat data from proximity detectors in the Matakitaki Valley, New Zealand, where d is the distance between an animal's home range centre and a detector. Density estimates (Table 3) were stable despite the considerable variation in the fitted detection functions

Strictly territorial animals may have a 'hard' edge to their range that is best represented by a step function ($p_s(d) = 0$ for $d >$ range radius), but zero values for p_s can cause problems when maximising the likelihood. Instead, we used the hazard function to emulate a step function by setting $b = 100$. Density estimates and confidence intervals were not noticeably affected by the form used for p_s (halfnormal, hazard or negative exponential; Fig. 2), and asymptotic intervals resembled profile likelihood intervals (Table 3). We also fitted models with additional parameters allowing for a response to previous capture ($M_b(g_0)$, $np = 4$) or random individual variation using a 2-class finite mixture ($M_{h2}(g_0)$, $M_{h2}(\sigma)$, $np = 6$), but these barely increased the maximised log likelihood ($\Delta LL < 0.05$) and were clearly inferior by AIC.

7 Single-Catch Traps

Single-catch traps are used very widely in studies of small mammals, and biologists will ask whether such data may be analysed with the methods described here. Competition for single-catch traps breaches the model assumption that animals are caught independently. We expect any resulting bias to be small when trap saturation (the proportion of traps occupied) is low. Trap saturation will be higher when population density is high, the intervals between trap checks are longer, traps are highly attractive or the animals are inherently very trappable.

With high trap saturation we would intuitively expect density estimates from single-catch trap data analysed with multi-catch models to be biased downwards. We conducted simulations to test this prediction for a scenario in which 100 traps on a square grid with spacing c were operated for 5 occasions. Notional home range centres were placed at expected density D in a rectangular area extending

Table 3 Comparison of models for density and spatial detection of stoats, fitted to data from hair samples collected in the Matakitaki valley, South Island, New Zealand, December 2001. LL = maximised log-likelihood, np = number of parameters. See Section 4 for calculation of asymptotic confidence intervals

							95% confidence interval for $\hat{D}$ (km^{-2})	
Detection function p_s	LL	np	ΔAIC	$\hat{g}_0 \pm$ SE	$\hat{\sigma} \pm$ SE (m)	$\hat{D} \pm$ SE (km^{-2})	Asymptotic	Profile likelihood
Hazard (shape fixed $b = 100$)	−186.9	3	0.0	0.024 ± 0.007	482 ± 12	2.6 ± 0.8	1.4–4.6	1.4–4.7
Halfnormal	−188.0	3	2.3	0.045 ± 0.019	255 ± 44	2.5 ± 0.7	1.4–4.4	1.4–4.5
Negative exponential	−188.7	3	3.6	0.113 ± 0.056	162 ± 35	2.4 ± 0.7	1.4–4.4	1.4–4.6

4c beyond the traps. Three distributions were compared. In the first ('Poisson'), centres were placed at random uniformly and independently across the area. For the second distribution ('clustered'), centres followed a Neyman-Scott distribution in which the foci of clusters were from a spatial Poisson process with intensity D/μ, and μ range centres were located around each focus according to a bivariate normal distribution with scale σ_c; the clustering parameters were set to $\mu = 5$ and $\sigma_c = c$. For the third distribution ('inhomogeneous Poisson'), centres were placed independently, but with a linear gradient in expected density from east to west, from zero at –4c from the western-most traps to 2D at +4c from the eastern-most traps (the gradient over the traps themselves was from 0.47D to 1.53D). Detection was simulated with a halfnormal function ($g_0 = 0.2$, $\sigma = c$) using the algorithm in Efford (2004, Appendix) to allow for competition between traps for animals and between animals for traps. Simulated average densities spanned the range $0.0625\sigma^{-2}$– $2\sigma^{-2}$; 100 replicate simulations were performed for each level of density. For estimation, a halfnormal detection function was fitted by maximising the conditional likelihood for multi-catch traps (see Eqs.(2), (3), (4), (6)). Trap saturation was measured by the proportion of traps occupied at the end of each occasion. Relative bias is estimated by $RB(\hat{\upsilon}) = \frac{\hat{\upsilon}-\upsilon}{\upsilon}$, where υ represents any of the parameters D, g_0 and σ.

Our simulations detected no bias in $\hat{D}$ for a Poisson distribution, even when 86% of traps were occupied (Table 4). Clustering of home range centres caused no

Table 4 Simulation of bias in density and detection parameters estimated by spatially explicit capture recapture when data are from single-catch traps and the fitted model assumes multi-catch traps. Simulations used one of three alternative distributions for animal range centers. Results are mean ± SE over 100 replicates. Density is expressed in terms of σ ($1.0\ \sigma^{-2} \times 16$ is equal to 6.25 ha^{-1} when $\sigma = 10$ m, and 0.0625 ha^{-1} when $\sigma = 100$ m). Trap saturation is the proportion of occupied traps at the end of each occasion

D ($\sigma^{-2} \times 16$)	Trap saturation	$RB(\hat{D})$	$RB(\hat{g}_0)$	$RB(\hat{\sigma})$
a. Poisson distribution				
1	0.049 ± 0.002	0.018 ± 0.031	–0.002 ± 0.032	0.017 ± 0.019
2	0.098 ± 0.002	–0.029 ± 0.022	–0.007 ± 0.025	0.010 ± 0.011
4	0.198 ± 0.003	–0.027 ± 0.017	–0.071 ± 0.016	–0.011 ± 0.007
8	0.353 ± 0.004	0.012 ± 0.012	–0.208 ± 0.010	0.002 ± 0.006
16	0.601 ± 0.004	–0.017 ± 0.009	–0.357 ± 0.006	–0.003 ± 0.005
32	0.860 ± 0.002	–0.013 ± 0.007	–0.573 ± 0.005	–0.012 ± 0.004
b. Clustered distribution of animals (Neyman–Scott distribution, $\mu = 5$ and $\sigma_c = \sigma$)				
1	0.044 ± 0.003	–0.034 ± 0.055	–0.067 ± 0.042	0.029 ± 0.019
2	0.090 ± 0.003	–0.028 ± 0.033	–0.118 ± 0.025	0.008 ± 0.012
4	0.169 ± 0.005	–0.047 ± 0.027	–0.160 ± 0.014	–0.003 ± 0.008
8	0.332 ± 0.007	–0.026 ± 0.021	–0.239 ± 0.010	0.000 ± 0.005
16	0.567 ± 0.006	–0.021 ± 0.012	–0.404 ± 0.006	0.004 ± 0.004
32	0.843 ± 0.004	–0.010 ± 0.010	–0.587 ± 0.005	–0.005 ± 0.004
c. Inhomogenous Poisson distribution of animals (east–west density gradient)				
1	0.051 ± 0.002	–0.012 ± 0.031	0.064 ± 0.038	0.007 ± 0.016
2	0.097 ± 0.002	–0.014 ± 0.025	–0.037 ± 0.022	–0.003 ± 0.010
4	0.188 ± 0.003	–0.021 ± 0.016	–0.096 ± 0.015	–0.002 ± 0.007
8	0.355 ± 0.004	0.003 ± 0.012	–0.222 ± 0.009	0.008 ± 0.005
16	0.585 ± 0.004	–0.008 ± 0.009	–0.388 ± 0.006	–0.002 ± 0.004
32	0.824 ± 0.002	–0.052 ± 0.007	–0.586 ± 0.004	0.000 ± 0.001

detectable bias in $\hat{D}$ at any level of trap saturation (Table 4b). The simulated gradient in density had a detectable effect only at the highest level of trap saturation, when the density estimates showed a 5% negative bias (Table 4c). These results are surprising. We note that $\hat{\sigma}$ also remains unbiased at high levels of trap saturation, whereas $\hat{g}_0$ becomes negatively biased. We infer that competition for traps causes a spatially homogeneous reduction in capture probability under the conditions of these simulations, and that this is adequately modelled by a multi-catch likelihood with lower g_0. We tentatively conclude that the associated estimator for D may be sufficiently robust to use for single-catch traps without further development. Extreme trap saturation should be avoided by increasing the density of traps or the frequency of trap checking, if only because the additional captures will increase precision.

8 Discussion

Many general benefits accrue from the estimation of density in a likelihood-based framework (Borchers and Efford 2007). Attention is shifted from the artificial parameters N_c and A_c to the ecologically significant parameter D. The model may be applied to any configuration of detectors, and is not restricted to compact arrays such as trapping grids (Efford et al. 2005). Differences between individuals in capture probability due to spatial location can be modelled with these methods and so do not result in unmodelled heterogeneity, the bane of conventional population estimation (there may of course be other sources of unmodelled heterogeneity). Hypotheses for variation in density over time or space may be evaluated using nested models and likelihood ratio tests. The fitted model describes the detection process and may be used in simulations to evaluate the effect of altering the study design, for example by changing the number and placement of traps.

To these general benefits we now add the ability to adapt the analysis for specific types of detector, for greater realism in modelling the detection process. In the case of proximity detectors, the model embraces the possibility of detecting an individual at multiple points on one occasion. Our results support the tentative use of a multi-catch model with data from single-catch traps if the goal is unbiased estimation of density. However, estimates of the detection parameter g_0 by this method are highly biased by trap saturation, and the method of simulation and inverse prediction (Efford 2004) should be used to fit the full process model to data from single-catch traps if it is intended to use the process estimates in simulations.

Our stoat example establishes the feasibility of applying spatially explicit capture-recapture methods to quite small data sets. Precision increases with the number of recaptures (Efford et al. 2004; M. G. Efford unpubl.), and it is generally desirable to obtain at least 20 recaptures. There was a promising robustness to the choice of detection function. Robustness of density estimates to the shape of the fitted detection function (step function vs halfnormal) was also found in simulations using inverse prediction (Efford 2004). Passive DNA sampling (e.g. Woods et al. 1999; Mills et al. 2000; Boulanger and McLellan 2001; Boulanger et al. 2004) and camera traps (e.g. Karanth and Nichols 1998; Trolle and Kéry 2003; Soisalo and Cavalcanti 2006) are used increasingly for mobile and difficult-to-trap animals such as carnivores. We expect our likelihood for proximity detectors to be widely applicable, assuming reliable identification of individuals.

The three detector types introduced so far do not exhaust the possibilities. A 'one-shot' proximity detector would become disabled once it detected an animal, but would not prevent the animal from finding another detector. Examples are a camera that does not reset itself, or hair sampling for DNA by some method that blocks collection of more than one sample per site per occasion (this might be desirable if sample mixing degrades the accuracy of identification). While there is no competition among detectors for animals, there is a sense in which animals 'compete' for access to 'one-shot' detectors. Equation (6) may possibly be adapted to allow for this by recasting it in terms of a trap-specific hazard rate.

A more difficult issue arises if the laboratory protocol is to reject all mixed DNA samples because individuals cannot be distinguished with confidence. Censoring mixed samples effectively creates a new type of detector (one which works only if fewer than two animals use it). At present we lack a satisfactory model for p_{ks} with this detector, as with single-catch traps. Field methods should be adapted to minimise the frequency of mixed samples (e.g. by frequent checking of devices). For analysis, we advise the use of simulation-based methods (e.g. Efford 2004) or cautious application of the proximity detector or multi-catch likelihoods; simulations should be used to check that the bias in the estimates is small relative to sampling error.

Other types of single-catch and multi-catch detector may remove animals permanently from the population. Used alone, these are not useful for fitting movement-based models such as we describe, because in the absence of recaptures we have no information on the scale of movements. However, such detectors may in principle be used in composite arrays with other detectors described here.

Acknowledgments We would like to thank Dianne Gleeson and Robyn Howitt of Landcare Research NZ who performed the stoat DNA analyses. Thanks also to Rachel Fewster for helpful comments on a draft.

References

Borchers DL, Efford MG (2007) Spatially explicit maximum likelihood methods for capture–recapture studies. Biometrics OnlineEarly doi:10.1111/j.1541-0420.2007.00927.x.

Boulanger J, McLellan B (2001) Closure violation in DNA-based mark–recapture estimation of grizzly bear populations. Canadian Journal of Zoology 79:642–651

Boulanger J, McLellan BN, Woods JG, Proctor MF, Strobeck C (2004) Sampling design and bias in DNA-based capture–mark–recapture population and density estimates of grizzly bears. Journal of Wildlife Management 68:457–469

Buckland ST, Anderson DR, Burnham KP, Laake JL, Borchers DL, Thomas L (2001) Introduction to distance sampling. Oxford University Press, Oxford

Chao A, Huggins RM (2005) Modern closed population models. In: Amstrup SC, McDonald TL, Manly BFJ (eds) Handbook of capture–recapture methods. Princeton University Press, Princeton, pp 58–87

Duckworth J, Byrom AE, Fisher P, Horn C (2005) Pest control: does the answer lie in new biotechnologies? In: Allen RB, Lee WG (eds) Biological invasions in New Zealand. Springer, Berlin, pp 421–434

Efford MG (2004) Density estimation in live-trapping studies. Oikos 106:598–610

Efford MG (2007) DENSITY 4.1: software for spatially explicit capture–recapture. Department of Zoology, University of Otago, Dunedin, New Zealand. http://www.otago.ac.nz/density.

Efford MG, Dawson DK, Robbins CS (2004) DENSITY: software for analyzing capture–recapture data from passive detector arrays. Animal Biodiversity and Conservation 27:217–228

Efford MG, Warburton B, Coleman MC, Barker RJ (2005) A field test of two methods for density estimation. Wildlife Society Bulletin 33:731–738

Genz AC, Malik AA (1980) Remarks on algorithm 006: an adaptive algorithm for numerical integration over an N-dimensional rectangular region. Journal of Computational and Applied mathematics 6:295–302

Huggins RM (1989) On the statistical analysis of capture experiments. Biometrika 76:133–140

Jett DA, Nichols JD (1987) A field comparison of nested grid and trapping web density estimators. Journal of Mammalogy 68:888–892

Karanth KU, Nichols JD (1998) Estimation of tiger densities in India using photographic captures and recaptures. Ecology 79:2852–2862

King CM, Murphy EC (2005) Stoat. In: King CM (ed) Handbook of New Zealand mammals. 2nd edition. Oxford University Press, South Melbourne, pp 261–287

Mills LS, Citta JJ, Lair KP, Schwartz MK, Tallmon DA (2000) Estimating animal abundance using noninvasive DNA sampling: promise and pitfalls. Ecological Applications 10:283–294

Otis DL, Burnham KP, White GC, Anderson DR (1978) Statistical inference from capture data on closed animal populations. Wildlife Monographs 62:1–135

Pledger S (2000) Unified maximum likelihood estimates for closed capture–recapture models using mixtures. Biometrics 56:434–442

Pollock KH (1982) A capture–recapture design robust to unequal probability of capture. Journal of Wildlife Management 46:752–757

Soisalo MK, Cavalcanti SMC (2006) Estimating the density of a jaguar population in the Brazilian Pantanal using camera-traps and capture–recapture sampling in combination with GPS radio-telemetry. Biological Conservation 129:487–496

Trolle M, Kéry M (2003) Estimation of ocelot density in the Pantanal using capture–recapture analysis of camera-trapping data. Journal of Mammalogy 84:607–614

White GC, Anderson DR, Burnham KP, Otis DL (1982) Capture–recapture and removal methods for sampling closed populations. Los Alamos National Laboratory, Los Alamos

Williams BK, Nichols JD, Conroy MJ (2002) Analysis and management of animal populations. Academic Press, San Diego

Woods JG, Paetkau D, Lewis D, McLellan BN, Proctor M, Strobeck C (1999) Genetic tagging free ranging black and brown bears. Wildlife Society Bulletin 27:616–627.

A Generalized Mixed Effects Model of Abundance for Mark-Resight Data When Sampling is Without Replacement

Brett T. McClintock, Gary C. White, Kenneth P. Burnham, and Moira A. Pryde

Abstract In recent years, the mark-resight method for estimating abundance when the number of marked individuals is known has become increasingly popular. By using field-readable bands that may be resighted from a distance, these techniques can be applied to many species, and are particularly useful for relatively small, closed populations. However, due to the different assumptions and general rigidity of the available estimators, researchers must often commit to a particular model without rigorous quantitative justification for model selection based on the data. Here we introduce a nonlinear logit-normal mixed effects model addressing this need for a more generalized framework. Similar to models available for mark-recapture studies, the estimator allows a wide variety of sampling conditions to be parameterized efficiently under a robust sampling design. Resighting rates may be modeled simply or with more complexity by including fixed temporal and random individual heterogeneity effects. Using information theory, the model(s) best supported by the data may be selected from the candidate models proposed. Under this generalized framework, we hope the uncertainty associated with mark-resight model selection will be reduced substantially. We compare our model to other mark-resight abundance estimators when applied to mainland New Zealand robin (*Petroica australis*) data recently collected in Eglinton Valley, Fiordland National Park and summarize its performance in simulation experiments.

Keywords Population size · Logit-normal · Program NOREMARK · Sighting probability · Mark-recapture · Bowden's estimator

---

B.T. McClintock (✉)
USGS Patuxent Wildlife Research Center, 12100 Beech Forest Road, Laurel, MD 20708, USA
e-mail:bmcclintock@usgs.gov

D.L. Thomson et al. (eds.), *Modeling Demographic Processes in Marked Populations*, Environmental and Ecological Statistics 3, DOI 10.1007/978-0-387-78151-8_12,

1 Introduction

The mark-resight method for estimating population abundance when the number of marked individuals is known (White and Shenk 2001; McClintock and White 2007) may in many circumstances be considered a reliable, cost-effective alternative to traditional mark-recapture or index methods based on counts. Mark-resight is generally most useful for estimating relatively small, closed populations, and because animals only need to be physically captured and marked once prior to resighting surveys, the method is typically less expensive and less invasive than mark-recapture. The various mark-resight estimators available include the Joint Hypergeometric Maximum Likelihood Estimator (JHE) (Bartmann et al. 1987), the Minta–Mangel Estimator (MME) (Minta and Mangel 1989), the Immigration/Emigration Joint Hypergeometric Estimator (IEJHE) (Neal et al. 1993), Bowden's Estimator (BOWE) (Bowden and Kufeld 1995), and the Beta-Binomial Estimator (BBE) (McClintock et al. 2006). These primarily differ in their sampling protocols and means of modeling variability in resighting probabilities. Temporal variation in resighting probabilities is readily handled by all of the estimators, but individual heterogeneity (where sighting probabilities vary among animals) is not. Similar to mark-recapture abundance models, individual heterogeneity has been particularly problematic and often causes biased estimates when not properly accounted for (Otis et al. 1978; Neal et al. 1993).

JHE requires the standard assumptions of mark-resight estimators for the size of a closed population: (1) geographic and demographic closure; (2) no loss of marks; (3) no errors in distinguishing marked and unmarked animals; (4) independently and identically distributed (iid) resighting probabilities for marked and unmarked animals; (5) homogeneity of resighting probabilities within an occasion; and (6) sampling without replacement within occasions (Neal et al. 1993; White and Shenk 2001). IEJHE requires the same assumptions of JHE, but geographic closure need not be met because the presence of marked animals on the area surveyed is determined explicitly (Neal et al. 1993). BOWE relaxes several assumptions of JHE by allowing temporary movement off the study area, individual heterogeneity, and sampling with replacement (Bowden and Kufeld 1995). Some study designs, such as those using camera traps or lacking a defined "occasion," may only be conducted with replacement and necessitate the use of BOWE. MME has similar assumptions to BOWE, but its performance in simulation experiments has proven inferior to the other models allowing individual heterogeneity and its use is not recommended (White 1993; White and Shenk 2001). BBE has the same assumptions of BOWE, but sampling must be without replacement. Any heterogeneity model requires that marked animals be individually identifiable, but in some cases this is not feasible and necessitates the use of JHE. If individually identifiable marks are used, both BOWE and BBE tolerate less than 100% individual identification given that the animal is identified as marked (White and Shenk 2001; Magle et al. 2007). These models also allow demographic closure to be violated via mortality independent of mark status, but abundance estimates produced when this occurs become the population residing in the study area at the beginning of the resighting period. As with IEJHE, when geographic closure is violated via temporary movement off the study area, the interpretation of abundance estimates for the heterogeneity models becomes the total population using the study area, often termed a "super population."

JHE is generally contended to be the most precise when its assumptions hold, but confidence interval coverage can fall well below the nominal 95% when individual heterogeneity is moderate to high (Neal et al. 1993; McClintock et al. 2006). BOWE performs well when individual heterogeneity is present, but is not likelihood-based and therefore lacks the benefits of likelihood theory, including information-theoretic model selection and model averaging methods. BBE successfully combines likelihood theory and the ability to model individual heterogeneity. The model may also incorporate a "robust" sampling design, which combines data from both closed and open sampling periods to estimate demographic parameters (Pollock 1982; Kendall et al. 1995). The open periods between primary sampling occasions apply to longer intervals of time where closure need not be met. Each primary sampling occasion consists of ≥ 2 secondary sampling occasions, and time intervals between these must be short enough for the assumption of closure to be satisfied. This approach has many advantages in long-term monitoring studies, including the ability to model detection probabilities similarly across time (or groups) for increased efficiency. Under the robust design, BBE has advantages over JHE in the presence of individual heterogeneity and over BOWE in cases where sighting probabilities are similar between primary occasions (McClintock et al. 2006).

When sampling is without replacement, BBE will often outperform other estimators and aid researchers in determining which model is most appropriate, but it is by no means a superlative mark-resight estimator for all situations. Due to the different assumptions and general rigidity of JHE, BOWE, and BBE, researchers must commit to a particular model based on educated guesswork without rigorous quantitative justification for model selection based on the data. Because there is no quantitative criterion to choose between these estimators, there remains a need for a more generalized framework for mark-resight abundance estimation. Similar to those available for mark-recapture studies, this framework would allow a wide variety of sampling conditions to be parameterized efficiently and provide quantitative justification for model selection regardless of the types and levels of variation encountered in the field. These parameterizations would include complex models utilizing covariates and simpler models where potential sources of variation such as individual heterogeneity may be ignored. By incorporating a more flexible model structure under a generalized framework, the uncertainty that remains in mark-resight model selection would be reduced substantially. In the following section, we introduce a model addressing this need for a more generalized framework when sampling is without replacement. We then apply the model to New Zealand robin (*Petroica australis*) data and compare its performance to the other estimators. Finally, we evaluate the performance of our model based on simulation experiments and discuss the implications for mark-resight model selection.

2 The Model

The logit-normal mixed effects mark-resight model (LNE) has the same assumptions of BBE and allows the data to be combined across t primary sampling occasions in a robust sampling design. In order to incorporate heterogeneity parameters into the model, we will assume marked individuals are individually identifiable. However, unlike BBE and BOWE, LNE does not require individually identifiable marks (although its utility is somewhat diminished without them). A known number of individuals (n_j, $j = 1, \ldots, t$) must first be marked at the beginning of interval j, and resighting data are collected during the t closed intervals consisting of k_j ($j = 1, \ldots, t$) distinct secondary resighting occasions. The data consist of resightings for marked individual s on secondary occasion i of primary occasion j (δ_{sij}) and the total number of unmarked sightings across all k_j secondary occasions of primary occasion j (T_{u_j}). The δ_{sij} are modeled as independent Bernoulli random variables, where $\delta_{sij} = 1$ if individual s is seen on secondary occasion i of primary occasion j, and $\delta_{sij} = 0$ otherwise. Individual sighting probabilities are approximated as the realization of a logit-normal random variable, where time is modeled as a fixed effect (β_{ij}) and individual heterogeneity as a random effect with mean zero and unknown variance σ_j^2. The marked individual resighting data have conditional expectation

$$\mathrm{E}(\delta_{sij} \mid \sigma_j, Z_{sj}, \beta_{ij}) = p_{sij} = \frac{1}{1 + \exp(-(\sigma_j Z_{sj} + \beta_{ij}))},$$

where $Z_{sj} \overset{iid}{\sim} N(0, 1)$. Therefore, a randomly selected individual s from primary occasion j with latent sightability Z_{sj} has the marginal probability of being seen on secondary occasion i

$$\mathrm{E}_{Z_{sj}}(p_{sij}) = p_{ij|s} = \int \frac{1}{1 + \exp(-(\sigma_j z_{sj} + \beta_{ij}))} \phi(z_{sj}) dz_{sj},$$

where $\phi(z_{sj})$ is the standard normal distribution. Time (β_{ij}) could possibly be treated as a random effect, but we chose not to investigate this approach because the number of occasions is generally too small for this to be useful. Under this framework, resighting probabilities may be modeled with no time or heterogeneity effects within secondary occasions ($\beta_{ij} = \theta_j, \sigma_j = 0$), only time effects, only heterogeneity effects, or additive time and heterogeneity effects. Across all marked individuals and secondary occasions, an unconditional likelihood function for σ_j and β_{ij} is

$$L(\sigma_j, \beta_{ij} \mid \delta_{sij}, n_j, k_j) = \prod_{s=1}^{n_j} \int \left[\prod_{i=1}^{k_j} p_{sij}^{\delta_{sij}} (1 - p_{sij})^{(1-\delta_{sij})} \right] \phi(z_{sj}) dz_{sj}. \quad (1)$$

Abundance (N) enters the equation by focusing on T_{u_j} and the number of unmarked individuals in the population ($U_j = N_j - n_j$). Using the approach validated for BBE (McClintock et al. 2006), $T_{u_j} \overset{ind}{\sim} N[\mathrm{E}(T_{u_j}), \mathrm{var}(T_{u_j})]$, the approximate likelihood function for N_j is:

$$L(N_j \mid \sigma_j, \beta_{ij}, \delta_{sij}, n_j, k_j, T_{u_j}) = \frac{1}{\sqrt{2\pi \,\mathrm{var}(T_{u_j})}} \exp\left\{\frac{-[T_{u_j} - \mathrm{E}(T_{u_j})]^2}{2\mathrm{var}(T_{u_j})}\right\}.$$

Combining the two likelihoods across the t primary occasions yields the LNE likelihood of the general form:

$$\begin{aligned} &L(\boldsymbol{N}, \boldsymbol{\sigma}, \boldsymbol{\beta} \mid \boldsymbol{\delta}, \boldsymbol{n}, \boldsymbol{k}, \boldsymbol{T_u}) \\ &= \prod_{j=1}^{t} \left\{ \prod_{s=1}^{n_j} \int \left[\prod_{i=1}^{k_j} p_{sij}^{\delta_{sij}} (1 - p_{sij})^{(1-\delta_{sij})} \right] \phi(z_{sj}) dz_{sj} \right\} \\ &\quad \times \prod_{j=1}^{t} \frac{1}{\sqrt{2\pi \,\mathrm{var}(T_{u_j})}} \exp\left\{\frac{-[T_{u_j} - \mathrm{E}(T_{u_j})]^2}{2\mathrm{var}(T_{u_j})}\right\}. \end{aligned} \quad (2)$$

For the simplest model, with no time or individual heterogeneity effects within secondary occasions,

$$\mathrm{E}(T_{u_j}) = (N_j - n_j)k_j \frac{1}{1 + \exp(-\theta_j)},$$

and

$$\mathrm{var}(T_{u_j}) = (N_j - n_j)k_j \frac{\exp(\theta_j)}{[1 + \exp(\theta_j)]^2}.$$

For the case of fixed time effects only within secondary occasions,

$$\mathrm{E}(T_{u_j}) = (N_j - n_j) \sum_{i=1}^{k_j} \frac{1}{1 + \exp(-\beta_{ij})},$$

$$\mathrm{var}(T_{u_j}) = (N_j - n_j) \sum_{i=1}^{k_j} \frac{\exp(\beta_{ij})}{[1 + \exp(\beta_{ij})]^2}.$$

The individual heterogeneity model with no time effects within secondary occasions has unconditional

$$\mathrm{E}(T_{u_j}) = (N_j - n_j)k_j\mu_j,$$

and

$$\mathrm{var}(T_{u_j}) = (N_j - n_j)k_j\left[\mu_j(1-\mu_j) + (k_j-1)(\gamma_j - \mu_j^2)\right], \tag{3}$$

where

$$\mu_j = \int \frac{1}{1+\exp(-(\sigma_j z_j + \theta_j))}\phi(z_j)dz_j,$$

$$\gamma_j = \int \left[\frac{1}{1+\exp(-(\sigma_j z_j + \theta_j))}\right]^2 \phi(z_j)dz_j,$$

and $\phi(z_j)$ is the standard normal distribution. For the heterogeneity model with fixed time effects within secondary occasions,

$$\mathrm{E}(T_{u_j}) = (N_j - n_j)\sum_{i=1}^{k_j}\mu_{ij},$$

and

$$\mathrm{var}(T_{u_j}) = (N_j - n_j)\left[\sum_{i=1}^{k_j}\mu_{ij}(1-\mu_{ij}) + \sum_{l\neq i}\sum(\gamma_{lij} - \mu_{lj}\mu_{ij})\right], \tag{4}$$

where

$$\mu_{ij} = \int \frac{1}{1+\exp(-(\sigma_j z_j + \beta_{ij}))}\phi(z_j)dz_j,$$

and

$$\gamma_{lij} = \int \frac{1}{1+\exp(-(\sigma_j z_j + \beta_{lj}))}\frac{1}{1+\exp(-(\sigma_j z_j + \beta_{ij}))}\phi(z_j)dz_j.$$

Interested readers may find the derivations of (3) and (4) in the Appendix.

LNE may incorporate the number of marked individuals that were identified as marked, but not identified to individual (ϵ_{ij}). These data enter the likelihood in (2) via $\mathrm{E}(T_{u_j})$ and $\mathrm{var}(T_{u_j})$. For the general case with fixed time effects and individual heterogeneity,

$$\mathrm{E}(T_{u_j}) = (N_j - n_j)\sum_{i=1}^{k_j}\mu'_{ij},$$

and

$$\mathrm{var}(T_{u_j}) = (N_j - n_j)\left[\sum_{i=1}^{k_j} \mu'_{ij}(1-\mu'_{ij}) + \sum\sum_{l\neq i}(\gamma'_{lij} - \mu'_{lj}\mu'_{ij})\right],$$

where

$$\mu'_{ij} = \int\left[\frac{1}{1+\exp\left(-(\sigma_j z_j + \beta_{ij})\right)} + \frac{\epsilon_{ij}}{n_j}\right]\phi(z_j)dz_j,$$

and

$$\gamma'_{lij} = \int\left[\frac{1}{1+\exp(-(\sigma_j z_j + \beta_{lj}))} + \frac{\epsilon_{lj}}{n_j}\right] \times \left[\frac{1}{1+\exp(-(\sigma_j z_j + \beta_{ij}))} + \frac{\epsilon_{ij}}{n_j}\right]\phi(z_j)dz_j.$$

Similar to BOWE and BBE, with high levels of individual heterogeneity the adjustment to incorporate unidentified marks is reliable when the proportion of unidentified marks remains < 0.10. When > 0.10, the resulting underestimates of variances can cause confidence interval coverage of N to fall as low as 88% (White and Shenk 2001; Magle et al. 2007). When individual heterogeneity is low to moderate, the proportion of unidentified marks can approach 0.20 and still achieve nominal 95% confidence interval coverage (Magle et al. 2007).

Because the integrals in (2) do not have a closed form solution, they must be solved numerically. These can be approximated with relative ease using Gaussian–Hermite quadrature (Givens and Hoeting 2005), with

$$\int\left[\prod_{i=1}^{k_i} p_{sij}^{\delta_{sij}}(1-p_{sij})^{(1-\delta_{sij})}\right]\phi(z_{sj})dz_{sj}$$
$$\approx \frac{1}{\sqrt{\pi}}\sum_{m=1}^{M} w_m \prod_{i=1}^{k_j}\left(\frac{1}{1+\exp(-(\sqrt{2}\sigma_j v_m + \beta_{ij}))}\right)^{\delta_{sij}} \left(1-\frac{1}{1+\exp(-(\sqrt{2}\sigma_j v_m + \beta_{ij}))}\right)^{1-\delta_{sij}},$$

$$\mu_{ij} \approx \frac{1}{\sqrt{\pi}}\sum_{m=1}^{M} w_m \frac{1}{1+\exp(-(\sqrt{2}\sigma_j v_m + \beta_{ij}))},$$

and

$$\gamma_{lij} \approx \frac{1}{\sqrt{\pi}} \sum_{m=1}^{M} w_m \frac{1}{1+\exp(-(\sqrt{2}\sigma_j v_m + \beta_{lj}))} \frac{1}{1+\exp(-(\sqrt{2}\sigma_j v_m + \beta_{ij}))}$$

for tabulated (v_m, w_m) pairs corresponding to M quadrature points (Abramowitz and Stegun 1964).

3 Example: New Zealand Robin

3.1 Methods

The New Zealand robin data were collected in March 2005 from $t = 2$ study sites in the Eglinton Valley of Fiordland National Park, New Zealand (44'58's, 168'01'E). The two sites, Knobs Flat and Walker Creek, consisted of 100 ha grids and were part of an exploratory investigation by the Department of Conservation on the usefulness of this technique for estimating N of endangered populations in the Chatham Islands. Prior to the resighting surveys, as many juvenile and adult birds as possible were captured within the study areas and given individually identifiable bands. Between September 2003 and March 2005, 80 and 79 birds were banded in Knobs Flat and Walker Creek, respectively. Immediately prior to collecting resighting data in March 2005, an independent visual survey was conducted to sample a known "marked" subset ($n_{kf} = 23$, $n_{wc} = 20$) of the previously banded birds. This was necessary because banded birds could have died or emigrated during the extended capture period prior to the resighting surveys. The resighting effort was divided into 7 distinct secondary occasions where the entire area of both study sites was surveyed. Secondary sampling occasions were conducted in the morning and typically required four hours each. The populations were assumed closed during the sampling intervals. On several occasions a marked or banded individual was seen more than once. However, because the extended capture period left few birds unbanded, the researchers believed they could identify double counts and satisfy the assumption of sampling without replacement. Raw estimates of $\bar{p}$ from the marked populations were 0.40 (SE = 0.04) and 0.41 (SE = 0.04) for Knobs Flat and Walker Creek, respectively. Total unmarked sightings ($T_{u,kf} = 45$, $T_{u,wc} = 54$) included previously banded birds that were not included in the marked subset.

With $t = 2$ primary occasions both consisting of 7 secondary occasions, there are 30 possible LNE parameterizations with $N_{kf} \neq N_{wc}$ (Table 1). The models range in complexity from the simplest no heterogeneity model, $\{\beta(.)\sigma(0)\}$, to the most general time and heterogeneity model, $\{\beta(t1, t2)\sigma(., .)\}$. The models were easily implemented using the nonlinear mixed-effects model (NLMIXED) maximum likelihood procedure in the SAS System for Windows (SAS Institute 2002). By default, NLMIXED computes the integrals in (1) using adaptive Gaussian quadrature.

Table 1 LNE parameterizations for β_{ij} and σ_j with $t = 2$ primary sampling occasions both consisting of k secondary sampling occasions . Combining the six β_{ij} parameterizations with the five σ_j parameterizations yields 30 possible LNE models with $N_1 \neq N_2$. The number of estimated parameters (not including N_j) in the models range from $K = 1$ for model $\{\beta(.)\sigma(0)\}$ to $K = 2k+2$ for model $\{\beta(t1, t2)\sigma(., .)\}$

Model notation	Parameterization	k
$\beta(.)$	$\beta_{i1} = \beta_{i2} = \theta$	1
$\beta(., .)$	$\beta_{i1} = \theta_1,\ \beta_{i2} = \theta_2$	2
$\beta(t1 = t2)$	$\beta_{i1} = \beta_{i2}$	k
$\beta(t1, .)$	$\beta_{i1} \neq \beta_{i2} = \theta$	$k+1$
$\beta(., t2)$	$\beta_{i2} \neq \beta_{i1} = \theta$	$k+1$
$\beta(t1, t2)$	$\beta_{i1} \neq \beta_{i2}$	$2k$
$\sigma(0)$	$\sigma_1 = \sigma_2 = 0$	0
$\sigma(.)$	$\sigma_1 = \sigma_2$	1
$\sigma(., 0)$	$\sigma_1 \neq \sigma_2 = 0$	1
$\sigma(0, .)$	$\sigma_2 \neq \sigma_1 = 0$	1
$\sigma(., .)$	$\sigma_1 \neq \sigma_2$	2

The integrals in E(T_{u_j}) and var(T_{u_j}) must be programmed separately within the procedure, and we found Gaussian–Hermite quadrature with $M = 4$ points to be an adequate approximation.

We used Akaike's Information Criterion (AIC$_c$) (Burnham and Anderson 2002) and the Bayesian Information Criterion (BIC) (Schwarz 1978) as a basis for ranking the 30 LNE models and obtaining model-averaged point estimates and unconditional variances with AIC$_c$ and BIC weights (Burnham and Anderson 2002; Link and Barker 2006). We defined the effective sample size for AIC$_c$ and BIC calculation as $n = \sum_{j=1}^{t} n_j k_j + t$. We compared the LNE model-averaged estimates to those of JHE, BOWE, and BBE. BBE estimates were also model-averaged using AIC$_c$ and BIC weights. As "equivalents" to JHE and BOWE, we also compared estimates of the most general LNE and BBE models where all parameters were estimated independently. Logarithm-transformed 95% confidence intervals for BOWE were computed as in Bowden and Kufeld (1995). Confidence intervals for LNE, JHE, and BBE were computed similarly but with the lower bound constrained to be greater than the known number of marked individuals. In comparing the performance of the models, our results focus on the precision of the estimates. Bias is also an important issue, but we were unable to quantify this property because N is unknown for these populations. However, both AIC$_c$ and BIC address the trade-off between bias and precision as a means of model selection.

3.2 Results

AIC$_c$ and BIC model rankings differed, with AIC$_c$ giving higher weights to the more complex additive models (Table 2). BIC rankings suggest mean resighting probabilities did not differ between secondary occasions or between the two study areas, but AIC$_c$ rankings provide some evidence of temporal variation in Knobs Flat resighting probabilities. The vast majority of AIC$_c$ weight (85%) was given to

Table 2 AIC_c and BIC weights for selected LNE models of New Zealand robin abundance in Knobs Flat and Walker Creek study areas in Fiordland National Park, New Zealand. Numbers of estimated parameters include N

Model	AIC_c weight	No. est. parameters	Model	BIC weight	No. est. parameters
$\beta(t1,.)\sigma(.,0)$	0.22	11	$\beta(.)\sigma(0)$	0.61	3
$\beta(t1,.)\sigma(.)$	0.14	11	$\beta(.)\sigma(.,0)$	0.15	4
$\beta(.)\sigma(.,0)$	0.09	4	$\beta(.)\sigma(.)$	0.14	4
$\beta(t1,.)\sigma(.,.)$	0.08	12	$\beta(.)\sigma(0,.)$	0.04	4
$\beta(.)\sigma(.)$	0.08	4	$\beta(.,.)\sigma(0)$	0.03	4
$\beta(t1,.)\sigma(0)$	0.06	10	$\beta(.)\sigma(.,.)$	0.01	5
$\beta(.)\sigma(0)$	0.06	3	$\beta(.,.)\sigma(.,0)$	0.01	5
$\beta(.)\sigma(.,.)$	0.04	5	$\beta(.,.)\sigma(.)$	0.01	5
...	...	...	...	...	...
$\beta(t1,t2)\sigma(.,.)$	0.00	18	$\beta(t1,.)\sigma(.,0)$	0.00	11
...	...	...	...	...	...
$\beta(.,t2)\sigma(0,.)$	0.00	11	$\beta(t1,t2)\sigma(.,.)$	0.00	18

models incorporating individual heterogeneity. BIC favored less complex models, with 36% of BIC weight given to those with heterogeneity parameters. The highest ranking BIC model estimates were therefore more precise than those of AIC_c. Estimates for the three-parameter minimum-BIC model, $\{\beta(.)\sigma(0)\}$, were $\hat{\theta} = -0.38$ (SE = 0.12), $\hat{N}_{kf} = 38.7$ (SE = 2.11), and $\hat{N}_{wc} = 38.9$ (SE = 2.37). Heterogeneity and abundance estimates for the 11-parameter minimum-AIC_c model, $\{\beta(t1,.)\sigma(.,0)\}$, were $\hat{\sigma}_{kf} = 0.79$ (SE = 0.28), $\hat{N}_{kf} = 38.7$ (SE = 2.87), and $\hat{N}_{wc} = 38.8$ (SE = 2.74). In comparing the various estimators, point estimates were very similar regardless of the method used, but precision levels did vary (Table 3). The BIC model-averaged LNE and JHE had the highest precision, but given the AIC_c and BIC evidence that individual heterogeneity may be an issue with these data, we believe JHE is underestimating the uncertainty about N and is therefore inappropriate. Model-averaged LNE and BBE results were very similar for these data because both incorporated a robust sampling design and estimated individual heterogeneity parameters. Even when compared to the "equivalent" BBE and LNE models with all parameters estimated independently, BOWE was the least precise of the estimators. Although inferences in this simple example were quite similar regardless of the model used, the model-averaged LNE or BBE appear to be the most appropriate because they were more efficient. Had there been less evidence of heterogeneity, we suspect the AIC_c model-averaged LNE would also have been more efficient than its BBE counterpart because of its ability to incorporate these parameters as deemed necessary by the data.

The use of AIC_c or BIC has received much attention in recent years (Burnham and Anderson 2004; Link and Barker 2006). Philosophical issues aside, this example provides no information on the appropriateness of AIC_c or BIC for use with these models. Further, the results from this single data set are not indicative of the expected relative performance of LNE. We therefore conducted simulation

Table 3 Abundance estimates ($\hat{N}$), percent coefficient of variation (% CV), 95% confidence intervals (CI), and percent confidence interval lengths (% CIL) for Knobs Flat (KF) and Walker Creek (WC) study areas when using the AIC$_c$ model-averaged (modAIC) LNE and BBE, BIC model-averaged (modBIC) LNE and BBE, LNE and BBE with both areas estimated independently, BOWE, and JHE. Estimators are ordered by the smallest average % CV

				95% CI		
Estimator	Study area	$\hat{N}$	% CV	Lower	Upper	% CIL
LNE modBIC	KF	38.7	5.8	34.9	43.7	22.8
	WC	38.9	6.4	34.6	44.4	25.4
JHE	KF	38.4	6.0	34.5	43.5	23.4
	WC	38.9	7.1	34.2	45.1	27.9
BBE modBIC	KF	38.7	6.4	34.6	44.2	24.9
	WC	38.8	7.1	34.1	45.0	28.1
BBE modAIC	KF	38.7	6.6	34.5	44.4	25.7
	WC	38.8	7.2	34.0	45.1	28.5
LNE modAIC	KF	38.7	6.7	34.4	44.5	26.3
	WC	38.8	7.2	34.1	45.2	28.6
BBE	KF	38.7	7.4	34.1	45.2	28.8
	WC	38.8	7.6	33.8	45.4	29.9
LNE	KF	38.7	7.4	34.1	45.2	28.9
	WC	38.8	7.6	33.8	45.4	29.9
BOWE	KF	38.7	7.7	33.0	45.4	32.2
	WC	38.7	7.9	32.8	45.6	33.1

experiments to assess the model's utility in a wide variety of sampling conditions using both AIC$_c$ and BIC.

4 Simulation Experiments

4.1 Methods

Data were generated under the assumptions of geographic and demographic closure within secondary resighting occasions, sampling without replacement, iid sighting probabilities for marked and unmarked individuals, 100% mark identification, and no error in distinguishing marked versus unmarked individuals. Individual resighting probabilities were modeled as logit-normal random variables based on an underlying population $\bar{p}$ and individual heterogeneity level (σ_{IH}), but additive temporal variation (σ_{TV}) allowed p_{sij} to vary for each secondary occasion. Because resighting probabilities were modeled using this transformation, input values for $\bar{p}$, σ_{IH}, and σ_{TV} did not back-transform identically to their original values. McClintock et al. (2006) used the same methods and categorized the realized values for the data-generating parameters. For $\bar{p}$, the categories were Low ($0.15 < \bar{p} < 0.16$), Medium ($0.30 < \bar{p} < 0.38$), and High ($\bar{p} = 0.50$). The categories for σ_{IH} and σ_{TV} were Low ($0.00 < \sigma < 0.05$), Medium ($0.10 < \sigma < 0.15$), and High ($0.16 < \sigma < 0.26$).

We first generated simulated mark-resight data for $t = 1$ primary sampling occasion. The input parameter values for generating resighting probabilities were all possible combinations of $\bar{p}$ = (Low, Medium, or High) and σ_{IH} =

σ_{TV} = (Low, Medium, or High). This limited the number of resighting probability scenarios to seven because when $\bar{p}$ = Low, only $\sigma_{IH} = \sigma_{TV}$ = Low is theoretically possible. Applying these seven resighting probability scenarios to the four sample size classes with $k = 3$ or 5 and $n = 25$ ($N = 100$) or 75 ($N = 500$) totaled 28 simulation scenarios. These scenarios ranged in sample size from smallest ($k = 3$, $n = 25$, $N = 100$, $\bar{p}$ = Low) to largest ($k = 5$, $n = 75$, $N = 500$, $\bar{p}$ = High) with the variation in $\bar{p}$ determined by the level of $\sigma_{IH} = \sigma_{TV}$.

We next generated data for $t = 2$ primary sampling occasions. With so many possible input parameters determining resighting probabilities and sample sizes, we restricted these simulations to six pseudo-randomly selected scenarios fixing $k_1 = k_2$, $n_1 = n_2$, and $N_1 = N_2$ (Table 4). We first designated "small" ($k = 3$, $n = 25$, $N = 100$, $\bar{p}$ = Low), "medium" ($k = 5$, $n = 25$, $N = 100$, $\bar{p}$ = Medium), and "large" ($k = 5$, $n = 75$, $N = 500$, $\bar{p}$ = High) samples. We then randomly assigned $\sigma_{IH(1)} = \sigma_{IH(2)}$ and $\sigma_{TV(1)} = \sigma_{TV(2)}$ from (None, Low, Medium, or High) to create three scenarios. For the other three scenarios, all values were randomly selected from k = (3, 5), n = (25, 75), $\bar{p}_j$ = (Low, Medium, High), $\sigma_{IH(j)}$ = (None, Low, Medium, High), and $\sigma_{TV(j)}$ = (None, Low, Medium, High) with N_j = 100 if $n = 25$, and $N_j = 500$ otherwise.

With 1 primary occasion, there are four possible LNE parameterizations: (1) no time or heterogeneity effects, $\{\beta(.)\sigma(0)\}$, with $K = 3$ parameters; (2) time effects only, $\{\beta(t)\sigma(0)\}$, $K = k + 2$; (3) heterogeneity only, $\{\beta(.)\sigma(.)\}$, $K = 4$; and (4) time and heterogeneity effects, $\{\beta(t)\sigma(.)\}$, $K = k + 3$. With 2 primary occasions and $k_1 = k_2$, there are 30 possible LNE parameterizations (Table 1). If $k_1 \neq k_2$, there are 25 parameterizations because constraining $\beta_{i1} = \beta_{i2}$ is no longer possible. For each of the 1000 replications within a given simulation scenario, we compared the performance of LNE with JHE, BBE, and BOWE. For LNE, we examined both AIC$_c$ and BIC model-averaged parameter estimates. For simulations with 2 primary occasions, we also examined the AIC$_c$ and BIC model-averaged parameter estimates for BBE. AIC$_c$, BIC, and confidence intervals were computed as in Section 3. Model performance was based primarily on percent confidence interval coverage of N, Bias/SE $= \mathrm{E}(\hat{N} - N)/\mathrm{SE}(\hat{N})$, percent confidence interval length (%CIL $= 100(\mathrm{UCI} - \mathrm{LCI})/N$), and root mean

Table 4 Data generating scenarios for simulation experiments with $t = 2$ primary sampling occasions. Number of secondary resighting occasions (k), marked sample size (n), and population abundance (N) were the same for both primary sampling occasions, but mean sighting probability $\bar{p}$, individual heterogeneity (σ_{IH}), and temporal variation (σ_{TV}) were allowed to vary

Scenario	k	n	N	$\bar{p}_1$	$\sigma_{IH(1)}$	$\sigma_{TV(1)}$	$\bar{p}_2$	$\sigma_{IH(2)}$	$\sigma_{TV(2)}$
A	3	25	100	Low	Low	None	Low	Low	None
B	3	75	500	Med	None	High	Low	None	Low
C	5	25	100	Med	High	Med	Med	High	Med
D	5	25	100	High	Med	None	Med	High	Med
E	5	75	500	Med	Low	High	High	None	High
F	5	75	500	High	Med	Low	High	Med	Low

squared error $\left(\text{RMSE} = \sqrt{\text{Bias}(\hat{N})^2 + \text{var}(\hat{N})}\right)$. Bonferroni intervals with family confidence coefficient $\alpha = 0.05$ (Hocking 2003) were used to simultaneously compare average estimator coverage, Bias/SE, and % CIL across scenarios. All analyses were performed using NLMIXED as described above and the Interactive Matrix Language (IML) in SAS (SAS Institute 2002).

4.2 Results

In simulations with 1 primary occasion, bias was not an appreciable problem for any of the estimators, with average Bias/SE across all 28 scenarios < 0.1 for all models (Cochran 1977) (Table 5). BOWE had the highest average point estimate for coverage and JHE had the lowest % CIL across the seven resighting probability scenarios in all four sample size classes. However, BOWE also had the highest % CIL, and JHE had the lowest coverage across all four sample size classes. No significant differences were observed between the AIC_c or BIC model-averaged LNE approaches. No significant differences in average coverage for the four sample size classes were observed between BOWE and LNE, but average % CILs were significantly lower in all sample size classes for LNE than for BOWE. Overall coverage and % CIL for BBE did not significantly differ from BOWE or LNE. When σ_{IH} = Low, no significant difference in average coverage was observed between the approaches. However, JHE and the LNE approaches had significantly smaller % CILs, and the two were not significantly different from one another. BBE tended to have slightly higher RMSEs than the other heterogeneity models, but BOWE had the highest RMSEs with the largest sample sizes. Except with the largest sample sizes, BOWE generally had slightly smaller RMSEs than the LNE approaches. This is attributable to a slight positive bias for LNE with smaller sample sizes, but because Bias/SE ratios remained small, the LNE approaches still achieved optimal coverage and % CILs.

Table 5 Average percent confidence interval coverage, percent confidence interval length (% CIL), and Bias/SE of abundance estimates for BBE, BOWE, JHE, AIC_c model-averaged (modAIC) LNE, and BIC model-averaged (modBIC) LNE across 28 simulated scenarios with $t = 1$ primary sampling occasion

	% Coverage		% CIL		Bias/SE	
Estimator	Est.	SE	Est.	SE	Est.	SE
BBE	94.1	0.14	42.4	0.11	0.06	0.01
BOWE	94.8	0.13	43.3	0.10	0.03	0.00
JHE	91.6	0.16	37.9	0.10	0.10	0.01
modAIC LNE	93.4	0.15	41.1	0.11	0.07	0.01
modBIC LNE	93.1	0.15	40.6	0.11	0.07	0.01

Table 6 Average percent confidence interval coverage, percent confidence interval length (% CIL), and Bias/SE of abundance estimates for BOWE, JHE, AIC_c model-averaged (modAIC) LNE and BBE, and BIC model-averaged (modBIC) LNE and BBE across six simulated scenarios with $t = 2$ primary sampling occasions

	% Coverage		% CIL		Bias/SE	
Estimator	Est.	SE	Est.	SE	Est.	SE
modAIC BBE	94.4	0.21	47.5	0.24	0.07	0.02
modBIC BBE	92.6	0.24	45.0	0.21	0.07	0.04
BOWE	94.5	0.21	50.1	0.23	−.01	0.01
JHE	89.8	0.27	43.7	0.26	0.13	0.03
modAIC LNE	93.7	0.22	47.1	0.25	0.07	0.02
modBIC LNE	92.0	0.25	43.8	0.21	0.06	0.04

Across the six scenarios with 2 primary occasions (Table 4), BOWE again had the highest average coverage and largest average % CILs. JHE on average had lower coverage and smaller % CILs than the other approaches. Average coverage for the AIC_c model-averaged LNE and BBE were not significantly different than BOWE, but average coverage for the BIC model-averaged LNE and BBE were significantly lower than BOWE. Both LNE and BBE model-averaged approaches produced significantly smaller % CILs than BOWE. Average Bias/SE was only > 0.1 for JHE (Table 6). The poorest performance for all approaches was in estimating N_2 of scenario D, where coverage was 87.7% (SE $=$ 1.04) for the AIC_c model-averaged BBE, 82.8% (SE $=$ 1.20) for the BIC model-averaged BBE, 92.3% (SE $=$ 0.84) for BOWE, 80.7% (SE $=$ 1.25) for JHE, 89.0% (SE $=$ 0.99) for the AIC_c model-averaged LNE, and 81.6% (SE $=$ 1.23) for the BIC model-average LNE. In this scenario, coverage was not significantly different between BOWE and the AIC_c model-averaged LNE or BBE, but all other approaches were significantly lower. When $\sigma_{IH(j)} \leq$ Low, no significant differences in average coverage or % CIL were detected between JHE and the AIC_c model-averaged LNE and BBE, but BBE had the highest point estimate for coverage (95.4%, SE $=$ 0.27) and LNE had the smallest point estimate for % CIL (53.7%, SE $=$ 0.46). BIC model-averaged LNE had significantly lower % CILs than JHE with no significant difference in coverage for these low heterogeneity scenarios. With the smallest sample size (scenario A), RMSE was largest for JHE and smallest for the BIC model-averaged BBE. With the largest sample size (scenario F), RMSE was largest for BOWE and smallest for the BIC model-averaged BBE. For the other scenarios, RMSE was generally largest for BOWE or BBE and smallest for JHE. Although average performance across all scenarios was very similar for LNE and BBE, LNE tended to be more efficient than BBE in scenarios with low levels of heterogeneity and BBE tended to be slightly more efficient when $\bar{p} =$ Low.

5 Discussion

With $t = 1$, little difference was observed in LNE performance when using AIC_c or BIC for model-averaged inference. When the number of occasions, marked individuals, and resighting probabilities were all at the lowest levels (scenario A), BOWE did perform better than the maximum likelihood models. A non-parametric model such as BOWE (whose properties are not based on asymptotic theory) may be a less biased approach with such small sample sizes, but precision is so poor that none of the estimators are particularly useful for inferences. With sample sizes suitable for producing useful levels of precision, LNE was generally a more precise estimator with no significant loss in coverage. Its higher efficiency compared to BBE and BOWE is attributable to LNE's ability to invest in estimating heterogeneity parameters as deemed necessary by the data. With low levels of heterogeneity, LNE had similar coverage and precision to JHE.

With $t = 2$, the advantages of data pooling in a robust sampling design were apparent in the increased precision of LNE and BBE. In the few scenarios with low levels of individual heterogeneity, LNE appeared to be more efficient than the other estimators, but not enough scenarios of this type were examined to detect a significant difference. However, based on these results and those from the simulations with 1 primary occasion, we expect that unlike BBE, the model-averaged LNE will be as or more efficient than JHE when heterogeneity levels are low. We also expect these advantages of LNE over the other estimators to be more pronounced with > 2 primary sampling occasions.

Although little difference was found in the use of AIC_c versus BIC with 1 primary occasion, we found a slight advantage in the use of AIC_c in some cases with 2 primary sampling occasions. The tendency of BIC to select less complicated models with small to moderate sample sizes (Burnham and Anderson 2004; Link and Barker 2006) was somewhat of a disadvantage in terms of coverage when the population mean resighting probabilities were different. Abundance estimates are particularly sensitive to biases in mean resighting probability estimates, and BIC's greater tendency to "split the difference" in estimating fewer parameters can result in underestimation of N in one primary occasion and overestimation in the other. We are not suggesting that AIC_c is not susceptible to similar problems with small marked sample sizes, but it did appear to alleviate them more than BIC. For example, in scenario D the "true" generating model had different values for all of the resighting probability input parameters, and all of the estimators failed to achieve nominal coverage in estimating N_2. As evidence of the criterion "splitting the difference," Bias/SE for the LNE model-averaged estimates of N_1 and N_2 were 0.35 and −0.26 for BIC, but were 0.11 and 0.02 for AIC_c, respectively. For BBE, these were 0.43 and −0.18 for BIC and 0.21 and 0.00 for AIC_c, respectively. Although coverage was close to nominal for N_1, coverage for N_2 using the BIC model-averaged approach was significantly lower than its AIC_c counterpart, and the problem appeared more severe for BBE than for LNE. However, this was not an appreciable problem for either approach in simulations with different population mean resighting probabilities and larger sample sizes, such as scenarios B and E.

Because the estimation of resighting probability parameters is so critical to estimates of N, we advise against the use of BIC model averaging under sampling conditions similar to those simulated in scenario D. We recommend as a general guideline that researchers carefully compare the estimates obtained via model averaging to those from the most general model where all parameters are estimated independently. If the parameter estimates (particularly N) are quite different, a moderately conservative approach would be to use $\mathrm{AIC_c}$ model averaging for inferences. The most conservative approach would be to use the general model. As indicated by the simulations with 1 primary occasion, the use of the most general LNE will typically still be more efficient than BBE and BOWE. When compared with the BIC model-averaged results for scenario D, average performances were better with the most general LNE and BBE models. Bias/SE for N_1 and N_2 were 0.02 and 0.10 for LNE, and 0.02 and 0.12 for BBE, respectively. Coverage of N_2 was 90.9% (SE $= 0.91$) and 90.7% (SE $= 0.92$) with % CILs of 57.8 (SE $= 0.69$) and 58.0 (SE $= 0.68$) for LNE and BBE, respectively. Similar to the $\mathrm{AIC_c}$ model-averaged results, these coverages were not statistically different from BOWE. Despite being slightly larger than when using $\mathrm{AIC_c}$ model averaging, % CILs for the general models were still significantly smaller than BOWE. Although we found it to be a problem with 2 primary occasions, we expect this small sample issue for BIC to be less of a concern in longer-term monitoring studies with > 2 primary occasions.

6 Conclusions

In terms of efficiency, we found LNE to be equivalent to or better than the other available mark-resight abundance estimators (with no appreciable loss in coverage) regardless of the sampling conditions. LNE provides researchers a more efficient alternative to JHE capable of incorporating a robust sampling design when individually identifiable marks are not feasible. LNE is more efficient than BOWE or BBE and equivalent to JHE when observed heterogeneity levels are low because it may ignore this variability as deemed appropriate by the data. When heterogeneity levels are high, LNE is more efficient than BOWE and equivalent to BBE because it may incorporate a robust sampling design. When sampling is without replacement, its flexible modeling framework provides quantitative justification for model selection based on the data, thereby eliminating the need to determine which of JHE, BOWE, or BBE is most appropriate based on educated guesswork. Overlooking philosophical issues, we did identify some potential advantages and disadvantages of using $\mathrm{AIC_c}$ or BIC for these models, but little difference in inferences can generally be expected between the two approaches when using model averaging. Although computationally more complicated than the other estimators, we believe the increased complexity that comes with the generalized modeling framework of LNE is justified by its increased efficiency and rigorously defendable means of mark-resight model selection. While not investigated here, the ability of LNE to incorporate environmental or individual covariates in modeling resighting probabilities may further increase its efficiency. However, when sampling must be with replacement, BOWE is still the only reliable option available for these studies. A flexible structure similar to LNE allowing sampling with replacement is still desirable and warrants further research.

Acknowledgments We thank the New Zealand Department of Conservation for collecting and allowing us the use of their data. Funding for this research was provided by NPS, USGS, and the Colorado State University Program for Interdisciplinary Mathematics, Ecology, and Statistics (PRIMES, NSF-IGERT Grant DGE-#0221595).

Appendix

If by definition the sightings of the $N - n$ unmarked individuals (any primary occasion j) are independent over individuals and conditionally (on Z_s) independent over occasions, then

$$T_u = \sum_{s=1}^{N-n} \sum_{i=1}^{k} \delta_{si} = \sum_{s=1}^{N-n} T_s.$$

Hence,

$$\text{var}(T_u) = (N-n)\text{var}\left(\sum_{i=1}^{k} \delta_{si}\right) = (N-n)\text{var}(T_s),$$

and

$$T_s = \sum_{i=1}^{k} \delta_{si}.$$

The general variance formula for any individual s is

$$\begin{aligned} \text{var}(T_s) &= \text{E}_Z\left[\text{var}(T_s \mid Z)\right] + \text{var}_Z\left[\text{E}_{T_s}(T_s \mid Z)\right] \\ &= \text{E}_Z\left[\sum_{i=1}^{k} p_{si}(1-p_{si})\right] + \text{var}_Z\left(\sum_{i=1}^{k} p_{si}\right) \\ &= \text{E}_Z\left[\sum_{i=1}^{k} p_{si}(1-p_{si})\right] + \sum_{l=1}^{k}\sum_{i=1}^{k} \text{cov}_Z(p_{sl}, p_{si}). \end{aligned} \tag{5}$$

For (3), with no fixed time effects (5) becomes

$$\begin{aligned} \text{var}(T_s) &= \text{E}_Z\left[kp_s(1-p_s)\right] + k^2\text{var}_Z(p_s) \\ &= k\mu - k\text{E}_Z(p_s^2) + k^2\left[\text{E}_Z(p_s^2) - \mu^2\right], \end{aligned}$$

where

$$\gamma = \mathrm{E}_Z(p_s^2) = \int \left[\frac{1}{1+\exp(-(\sigma z_s + \theta))} \right]^2 \phi(z_s) dz_s.$$

Hence,

$$\begin{aligned}\mathrm{var}(T_s) &= k\mu - k\gamma + k^2(\gamma - \mu^2) \\ &= k\left[\mu(1-\mu) + (k-1)(\gamma - \mu^2)\right],\end{aligned}$$

and

$$\mathrm{var}(T_u) = (N-n)k\left[\mu(1-\mu) + (k-1)(\gamma - \mu^2)\right].$$

For (4), with fixed time effects (5) becomes

$$\begin{aligned}\mathrm{var}(T_s) &= \sum_{i=1}^{k} \mathrm{E}_Z[p_{si}(1-p_{si})] + \sum_{l=1}^{k}\sum_{i=1}^{k} [\mathrm{E}_Z(p_{sl}, p_{si}) - \mu_l \mu_i] \\ &= \sum_{i=1}^{k} \mu_i(1-\mu_i) + \sum\sum_{l \neq i} (\gamma_{li} - \mu_l \mu_i),\end{aligned}$$

and

$$\mathrm{var}(T_u) = (N-n)\left[\sum_{i=1}^{k} \mu_i(1-\mu_i) + \sum\sum_{l \neq i} (\gamma_{li} - \mu_l \mu_i)\right].$$

References

Abramowitz M, Stegun I (1964) Handbook of Mathematical Functions, with Formulas, Graphs and Math Tables. Applied Math Series 55, National Bureau of Standards, U.S. Government Printing Office, Washington, DC.

Bartmann RM, White GC, Carpenter LH, Garrott RA (1987) Aerial mark-recapture estimates of confined mule deer in pinyon-juniper woodland. Journal of Wildlife Management 51:41–46.

Bowden DC, Kufeld RC (1995) Generalized mark-resight population size estimation applied to Colorado moose. Journal of Wildlife Management 59:840–851.

Burnham KP, Anderson DR (2002) Model Selection and Multi-Model Inference: A Practical Information-Theoretic Approach. 2nd edition. Springer-Verlag, New York.

Burnham KP, Anderson DR (2004) Multimodel inference: understanding AIC and BIC in model selection. Sociological Methods and Research 33:261–304.

Cochran WG (1977) Sampling Techniques. 3rd edition. Wiley, New York.

Givens GH, Hoeting JA (2005) Computational Statistics. Wiley, Hoboken, NJ.

Hocking RR (2003) Methods and Applications of Linear Models. 2nd edition. Wiley, Hoboken, NJ.

Kendall WL, Pollock KH, Brownie C (1995) A likelihood-based approach to capture–recapture estimation of demographic parameters under the robust design. Biometrics 51: 293–308.

Link WA, Barker RJ (2006) Model weights and the foundations of multimodel inference. Ecology 87:2626–2635.

Magle SM, McClintock BT, Tripp DW, White GC, Antolin MF, Crooks KR (2007) Mark-resight methodology for estimating population densities for prairie dogs 71:2067–2073.

McClintock BT, White GC (2007) Bighorn sheep abundance following a suspected pneumonia epidemic in Rocky Mountain National Park. Journal of Wildlife Management 71: 183–189.

McClintock BT, White GC, Burnham KP (2006) A robust design mark-resight abundance estimator allowing heterogeneity in resighting probabilities. Journal of Agricultural, Biological, and Environmental Statistics 11:231–248.

Minta S, Mangel M (1989) A simple population estimate based on simulation for capture-recapture and capture–resight data. Ecology 70:1738–1751.

Neal AK, White GC, Gill RB, Reed DF, Olterman JH (1993) Evaluation of mark-resight model assumptions for estimating mountain sheep numbers. Journal of Wildlife Management 57:436–450.

Otis DL, Burnham KP, White GC, Anderson DR (1978) Statistical inference from capture data on closed animal populations. Wildlife Monographs 62.

Pollock KH (1982) A capture–recapture design robust to unequal probability of capture. Journal of Wildlife Management 46:752–760.

SAS Institute (2002) SAS OnlineDoc, Version 9. SAS Institute, Inc., Cary, NC.

Schwarz G (1978) Estimating the dimension of a model. Annals of Statistics 6:461–464.

White GC (1993) Evaluation of radio tagging marking and sighting estimators of population size using Monte Carlo simulations. In Marked Individuals in the Study of Bird Populations, Lebreton JD, North PM, eds. Birkhauser-Verlag, Basel, Switzerland, pp. 91–103.

White GC, Shenk TM (2001) Population estimation with radio-marked animals. In Radio Tracking and Animal Populations, Millspaugh J, Marzluff JM eds. Academic Press, San Diego, pp. 329–350.

Evaluation of the Linkage Disequilibrium Method for Estimating Effective Population Size

James C. Russell and Rachel M. Fewster

Abstract Data on linkage disequilibrium at unlinked loci provide an estimate of the inbreeding effective population size of the parental generation of the sampled cohort. The inbreeding effective population size, N_e, is the reciprocal of the probability that two gametes, selected at random without replacement from those that produced the sampled cohort, derive from the same parent. Effective population size is an important parameter for measuring the rate of inbreeding in a population. We detail the construction of the linkage disequilibrium estimator of N_e, and evaluate its performance by simulation. We simulate populations which are dioecious and non-selfing. We use the simulations to examine the effects of several types of deviation from ideal population conditions, and of sample size, genotyping errors, number of loci typed, and polymorphic loci. We find substantial bias in the N_e estimator when there have been recent fluctuations in census population size, when the index of breeding variability is greater than one, and when the ratio of sample size to effective population size differs substantially from one. Due to high variability, estimators that have low bias for the reciprocal of N_e can present substantial bias when used as estimators of N_e itself. We consider a recent small sample size bias correction proposed for the method, and find that it improves bias in the reciprocal, but at the expense of increased bias for N_e. The improvements in the bias of the reciprocal are usually small, but are substantial when sample size is much less than N_e, while the increase in bias for N_e is often substantial. We test the method on two exhaustively sampled rat populations, and find it performs as expected from simulation. For practitioners, we recommend that resources are spent first in ensuring that the sample size is likely to be greater than the effective population size, and only then that the number of loci is increased to improve the precision of the estimate.

Keywords Burrow's composite disequilibrium measure · Effective population size · Non-ideal populations · Rats · *Rattus* · Squared correlation coefficient

---

J.C. Russell (✉)
Department of Statistics, University of Auckland, Private Bag 92019, Auckland, New Zealand
e-mail: j.russell@auckland.ac.nz

D.L. Thomson et al. (eds.), *Modeling Demographic Processes in Marked Populations*, Environmental and Ecological Statistics 3, DOI 10.1007/978-0-387-78151-8_13,

1 Introduction

1.1 Census Population Size

Population size is a fundamental parameter of interest in ecological systems. Classical statistical methods have been developed for estimating the census population size (N_c) from ecological capture data over a period of sampling (e.g. Seber 1982; Borchers et al. 2002). For population modeling, we are often interested in the total number of breeding individuals, often differentiating between breeding males and females (Caswell 2001; Buckland et al. 2004). Only the breeders will contribute to the next generation. The number of potentially breeding individuals (adults) can be determined from census data using knowledge of the age structure of the population, or by using an appropriate surrogate such as the size of animals.

1.2 Effective Population Size

From a genetic perspective, concepts of population size are related to the rates of loss of genetic variation, fixation of deleterious alleles, and inbreeding. In an infinite population without mutation, migration, or selection, allele frequencies are constant over the generations. However, for a finite population, a random process of genetic drift operates to change allele frequencies from one generation to the next (Caballero 1994). This occurs because each generation is formed by taking a finite sample of the gametes produced by the parental generation. The sampled allele frequencies will not match the parental frequencies exactly, and departures become greater as the sample size becomes smaller. The magnitude of genetic drift between generations therefore contains information about the sampled number of gametes, and hence the population size. Small populations also increase the chance of an individual possessing two copies of the same allele (homozygosity), because both copies were inherited from the same ancestor. This can cause inbreeding depression.

Measures of genetic change are not reflected directly by the census population size, but are related to the breeding population size, life history characteristics such as variable individual breeding success and biased sex ratios, and fluctuations in population size over the generations. The effective population size, N_e, is a surrogate size that is related directly to the genetic change being experienced by the population (Crow and Denniston 1988). The effective population size is the size of an ideal population that would experience the same amount of genetic change as that observed in the population under study (Wright 1931; Crow and Kimura 1970). The ideal population meets the three conditions of equal sex ratio, random mating, and constant census population size over generations (Crow and Denniston 1988; Caballero 1994). The notion of effective size provides a yardstick for understanding the rate of genetic change for any population, irrespective of its life history and other characteristics. When generations do not overlap, the effective population size estimate refers to the size of a single generation (Waples 2005).

To maintain constant population size in the ideal population, N adults must produce $2N$ gametes. The numbers of gametes contributed by adults $1, \ldots, N$ are $k_1, \ldots, k_N$, where these values sum to $2N$. Random mating in the ideal population means that, for each required gamete, the contributing adult is selected randomly and independently from the N adults available, so the joint distribution of $(k_1, \ldots, k_N)$ is Multinomial $(2N;\ 1/N, \ldots, 1/N)$ and the marginal distribution of k_i is Binomial $(2N;\ 1/N)$. This model is also termed binomial mating. In a population of constant size, the mean number of gametes contributed per adult is therefore $\mu_k = 2$, and the variance is $\sigma_k^2 = 2(1 - 1/N)$. As $N \to \infty$, the distribution tends to Poisson with $\sigma_k^2 = \mu_k = 2$. The model assumes that a gamete can unite with any other gamete, and hence the ideal population is capable of selfing (mating with oneself) and sex is not taken into account (Crow and Denniston 1988).

Effective population size (N_e) quantifies the size of an ideal population that would undergo a given level of genetic change. In the ideal population, regardless of which measure of genetic change is considered, N_e is equal to the census population size of a generation. For most real populations, however, the ideal conditions do not hold and N_e is smaller than the census population size. This is largely because real mating is not binomial and some individuals have greater breeding success than others. Frankham (1995) suggested that N_e may be as small as a tenth of the census population size for many species. It is rare for N_e to exceed the census size, but this can occur if variability in the number of gametes contributed by each parent is less than that expected from binomial chance, i.e. $\sigma_k^2 < \mu_k(1 - 1/N)$. This is termed minimal inbreeding and can be produced in managed populations (Caballero 1994).

When considering deviations from the ideal population, it is necessary to specify what measure of genetic change underlies the definition of N_e. This leads to different notions of effective population size, the most common of which are inbreeding effective size and variance effective size (Waples 2005). For inbreeding effective size, genetic change refers to the rate of increase in inbreeding per generation, while for variance effective size, genetic change refers to the variance of the change in allele frequency from one generation to the next (Crow and Denniston 1988; Caballero 1994). At small population sizes the different effective sizes can differ substantially (Crandall et al. 1999). Care must also be taken to specify what generation an estimate of effective size refers to, because a single sampled generation may yield estimates of effective size for its own generation, its parental generation, or its grandparental generation, depending upon which effective size is intended, whether the estimates are genetically based or demographically based, and whether or not the population exhibits selfing (Caballero 1994; Waples 2005).

The linkage disequilibrium method estimates the inbreeding effective size of the parental generation (Waples 2005, 2006). We define the inbreeding effective size of the parental generation to be the number N_e such that its reciprocal, $1/N_e$, is the probability that two gametes, selected at random without replacement from those occurring in the offspring, derive from the same parent. Caballero (1994) shows how this probability is related to the rate of increase in the coefficient of inbreeding per generation in the ideal population. If some parents contribute many more successful gametes than others (non-binomial mating), the probability will be inflated and N_e

will be smaller than the census size. With our definition above, we do not require the randomly selected gametes to be united in a single offspring, so the definition does not require that selfing has taken place. We use this definition to avoid confusion over whether the effective size refers to the parental generation or the grandparental generation in the case of non-selfing populations. For a non-selfing population, two gametes from a single parent cannot unite in an offspring, but two gametes from a single grandparent can, so the generation to which inbreeding N_e applies in a non-selfing population is commonly cited as the grandparental generation (Crow and Denniston 1988; Caballero 1994). However, the linkage disequilibrium method estimates its value for the parental generation (Waples 2005, 2006). We circumvent this confusion by defining N_e via a random selection of two successful gametes that is notional, rather than a selection united in an offspring. Throughout this paper, the inbreeding effective size refers to that of the parental generation.

The reciprocal probability, $1/N_e$, is a more direct driver of evolutionary processes than N_e itself (Wang 2001; Waples 2005). This has led previous authors to report bias in N_e estimators using the harmonic mean of estimated values rather than the arithmetic mean, because this reflects the bias of $1/N_e$. However, this policy has not been made explicit and could lead to confusion because most researchers focus on N_e rather than its reciprocal. In this paper, we will focus on bias in N_e itself, which is assessed by the arithmetic mean of estimated values rather than the harmonic mean.

1.3 Demographic Estimation

Effective population size can be estimated from demographic data on the total number of breeding males and females, and the mean and variance across individuals of their lifetime number of offspring that survive to reproduction (Crow and Denniston 1988, with corrections in Caballero and Hill 1992; Caballero 1994; Rockwell and Barrowclough 1995). However, these parameters are notoriously hard to estimate accurately (Waples 1991; Barrowclough and Rockwell 1993; Schwartz et al. 1998). Additionally, equations used to estimate effective population size from demographic data are not always comprehensive (Frankham 1995), because they do not simultaneously incorporate all three conditions leading to deviation from the ideal population.

1.4 Genetic Estimation

Genetic estimates of effective population size operate by measuring genetic processes that are known to be functions of N_e (Waples 1991). Genetic estimates incorporate all three conditions which lead to deviation from the ideal population (Frankham 1995). The genetic signal from N_e is strongest when the population size is small (Waples 1991), and this is where we have the most potential to estimate N_e accurately (Waples 1991; Wang 2005). For genetic estimation we assume that:

(1) mutation is negligible; (2) the alleles considered are not subject to natural or sexual selection (selectively neutral) and not linked with other loci subject to selection; (3) the samples of individuals for genetic analysis are randomly drawn from a specified population or generation; and (4) there is no immigration from neighbouring populations (Waples 1991).

Genetic estimates of effective population size are most commonly obtained from two or more temporally separated samples from a population (Waples 1989; Williamson and Slatkin 1999; Berthier et al. 2002), and represent an average of N_e over the appropriate time-scale (Waples 2005; Wang 2005). Other methods estimate historical effective population size over longer time periods using coalescent theory and mutation rates (Crandall et al. 1999; Wang 2005), or jointly with other parameters such as migration (Wang and Whitlock 2003) or mutation rate (Garza and Williamson 2001).

By contrast with the temporal and historical methods for estimating N_e, the linkage disequilibrium method gives an estimate of contemporary N_e from just one sampled generation. The single sample exhibits linkage disequilibrium from the two different processes of genetic sampling (selection of parental gametes) and statistical sampling to form the set of individuals for genetic analysis. By quantifying the level of linkage disequilibrium in the sample, an estimate of inbreeding N_e for the parental generation can be obtained (Waples 2006), which we detail below.

1.5 Our Purpose

The effective population size provides a single summary value of the contributions of breeding variability, sex ratio, and fluctuations in population size to the population biology of a species (Wang 2005). Most research in effective population size has focused on rare and endangered species (Nunney and Campbell 1993; Nunney and Elam 1994), where it is considered important to increase effective population size to raise the persistence of a population (Lande and Barrowclough 1987; Lynch and Lande 1998). Our work focuses on invasive species where reducing the census population size is the desired outcome. Invasive species are highly successful colonizers, despite initially small census population sizes and associated limited effective population sizes. This scenario is contrary to what would be expected following experiences with threatened species at small population sizes (Sax and Brown 2000).

Census and effective population sizes are both readily estimated when samples from a population can be taken repeatedly across time and space: for example, mark-recapture methods can be used for census size, and change in heterozygosity over time can be used for effective size. However ecologists commonly operate in the less ideal situation of having only one opportunity to sample a single population. This scenario is particularly the case for pest species where individuals are removed as they are encountered. Where the conservation goal is to remove all individuals as rapidly as possible (eradication) there is very little scope for long-term study of populations. Researchers then have minimal data from which to gain understanding of the population biology.

Our goal is to consider the utility of the linkage disequilibrium method for making inferences on a closed population which can only be sampled once without replacement. We present the theory underlying the linkage disequilibrium method, and use simulation to evaluate its performance. We use selectively neutral and highly variable microsatellite markers to characterise genetic diversity within a population (Selkoe and Toonen 2006). We consider only closed diploid populations with discrete generations. We also examine the performance of the method on real data of approximately known census population size, sex ratio and breeding success. Some previous work has simulated the performance of the linkage disequilibrium method (Waples 2005, 2006; England et al. 2006), and applied it to real datasets (Bartley et al. 1992). We focus on a thorough simulation of the parameters that can affect effective population size, using an ecologically plausible population. We discuss finally the utility of the method, and how this may affect the partitioning of field and laboratory work.

2 Linkage Disequilibrium

Linkage disequilibrium is the non-random association of alleles at different gene loci. Linkage disequilibrium can be produced by a number of factors, including physical linkage (the two loci are on the same chromosome), epistatic selection (alleles interacting to control fitness), genetic hitch-hiking (physical linkage with a selected locus), migration or population admixture, and random drift in finite populations (Hill 1981; Waples 1991). We are concerned only with the effect of genetic drift on linkage disequilibrium, in the absence of the other effects.

Genetic drift linkage disequilibrium is generated from the finite sampling of gametes from the parental generation. Sampling effects in a small sample mean that the sample correlation between the alleles possessed at two loci will not be zero, despite the underlying correlation or physical linkage being zero. The expected squared correlation becomes larger as the population size gets smaller, and can be shown to depend on $1/N_e$ (Sved 1971; Laurie-Ahlberg and Weir 1979; Weir and Hill 1980).

Specifically, consider two alleles A and B at loci 1 and 2 respectively, and suppose for the moment that we have data on individual gametes, as opposed to genotype data (see below). Let g be the total number of gametes whose alleles are known at both loci 1 and 2, g_A be the number of these gametes with allele A at locus 1, g_B be the number with allele B at locus 2, and g_{AB} be the number with both allele A at locus 1 and allele B at locus 2. Under random assortment, the proportion of AB gametes should be approximately the product of the proportion of A gametes and the proportion of B gametes. The linkage disequilibrium measure for alleles A and B at these loci is the difference between the observed and the expected proportions:

$$D_{AB} = \frac{g_{AB}}{g} - \frac{g_A}{g} \times \frac{g_B}{g}$$

A variety of methods can be used for estimating D_{AB} from gametic data (Weir 1996, p. 112). However, it is more common that only genotypic data are available, from which we do not know which gametes the individual's two alleles are located on. When we sample individuals with genotype AA', BB', for example, we do not know whether the alleles are arranged on the individual's two gametes as $(A,B) \mid (A',B')$ or as $(A,B') \mid (A',B)$. Additional to the disequilibrium D_{AB} within the gamete, we can define a second analogous disequilibrium D_B^A referring to opposite gametes within the same individual (Weir and Hill 1980). Neither D_{AB} nor D_B^A is observable from genotypic data, but their sum is. This suggests that we can use a composite disequilibrium measure, attributable to Dr Peter Burrows (see Cockerham and Weir 1977, p. 142, Weir 1979, p. 241, and later unattributed in Weir 1996, p. 126):

$$\Delta_{AB} = D_{AB} + D_B^A$$

The composite Δ_{AB} is known as Burrow's composite D, written as Δ in Weir (1996, p. 126), and D^* in Campton (1987, p. 184). It performs better than an alternative maximum likelihood estimator for linkage disequilibrium (Weir 1979). It is estimated directly from genotype counts as follows:

$$\hat{\Delta}_{AB} = \frac{n_{AB}}{n} - 2\hat{p}_A\hat{q}_B$$

where $n_{AB} = 2n_1 + n_2 + n_4 + n_5/2$, and n_1, ..., n_9 are genotype counts defined in Table 1, and $n = n_1 + ... + n_9$ is the total number of counts (sampled individuals). Here, $\hat{p}_A$ and $\hat{q}_B$ are the sample proportions of alleles A and B in the n individuals typed at both loci: for example $\hat{p}_A = (2n_1 + 2n_2 + 2n_3 + n_4 + n_5 + n_6)/(2n)$. A small-sample correction factor of $n/(n-1)$ should be applied to $\hat{\Delta}_{AB}$ (Weir 1979, p. 241; Campton 1987, p. 185).

It can be shown that $\hat{\Delta}_{AB} = \hat{\text{cov}}(K_A, K_B)/2$, where K_A and K_B give respectively the number of A alleles and B alleles possessed by an individual, and each take values 0, 1, or 2. The covariance is taken across individuals. We can then estimate the corresponding correlation coefficient, $r_{AB} = \text{cor}(K_A, K_B)$, as

$$\hat{r}_{AB} = \frac{\hat{\Delta}_{AB}}{\sqrt{\left\{\hat{p}_A(1-\hat{p}_A) + (\hat{h}_{AA} - \hat{p}_A^2)\right\}\left\{\hat{q}_B(1-\hat{q}_B) + (\hat{h}_{BB} - \hat{q}_B^2)\right\}}}$$

Table 1 Possible genotypes and their sample counts for alleles A and B at loci 1 and 2 respectively. $A'$and B' denote any other alleles

	$B\,B$	$B\,B'$	$B'B'$
$A\,A$	n_1	n_2	n_3
$A\,A'$	n_4	n_5	n_6
$A'A'$	n_7	n_8	n_9

where $\hat{h}_{AA}$ and $\hat{h}_{BB}$ are the observed proportions of AA and BB homozygotes in the sample of size n, for example $\hat{h}_{AA} = (n_1 + n_2 + n_3)/n$. This estimate of the correlation is the ratio of two estimators: the covariance $\hat{\text{cov}}(K_A, K_B) = 2\hat{\Delta}_{AB}$, and the square root of the product of the sample variances for K_A and K_B (Weir 1996, p. 38; equation (2) in Waples 2006).

The correlation coefficient r_{AB} has $E(r_{AB}) = 0$ for unlinked loci, where the expectation is taken over conceptual replicate populations. In finite populations, however, the correlation is likely to take non-zero values, with small populations giving the largest values, so the expectation of its square is non-zero and is a function of the effective population size, N_e. The expression for $E(r_{AB}^2)$ depends upon the mating structure and recombination fraction c in a population (Weir and Hill 1980; Weir et al. 1980), and is also affected by sample size n, because linkage disequilibrium arises from statistical sampling as well as genetic sampling. The distribution of r_{AB}^2 is not known, so the expectation $E(r_{AB}^2)$ is approximated as the ratio of the expectation of the numerator and the denominator (Weir and Hill 1980, p. 484; Laurie-Ahlberg and Weir 1979, p. 1309; Waples 2006, p. 169). For randomly mating populations, this yields

$$E(r_{AB}^2) \approx \frac{c^2 + (1-c)^2}{2N_e^I c(2-c)} + \frac{1}{n}$$

where $N_e{}^I$ is the inbreeding effective size of the parental generation. Thus $E(r_{AB}^2)$ is inversely related to both inbreeding effective population size and sample size. This expression is the same for dioecious species with random pairing and for monoecious species with or without selfing (Weir and Hill 1980). Rearranging the expression and replacing $E(r_{AB}^2)$ with $\hat{r}_{AB}^2$ gives the formula for estimating $N_e{}^I$ in the case of unlinked loci ($c = 0.5$):

$$N_e^{LD} = \frac{1}{3(\hat{r}_{AB}^2 - 1/n)} \tag{1}$$

where $N_e{}^{LD}$ is the linkage disequilibrium estimate of $N_e{}^I$, and n is the number of individuals sampled (Laurie-Ahlberg and Weir 1979; Hill 1981; Waples 1991). Equation 1 incorporates the contribution of both genetic and statistical sampling to the estimate of effective population size (Waples 2006). For species with a mating system of lifetime monogamy, the numerator becomes 2 (Weir and Hill 1980), but we do not consider this case.

The method above shows how $N_e{}^{LD}$ is obtained from data on a single pair of biallelic loci, where there are only two alleles A and A' at locus 1, and B and B' at locus 2. In many applications, there will be several loci which are polymorphic, in other words have more than two alleles. In this case we must derive an estimate $\hat{r}^2$ that combines all possible allele–allele comparisons within a single pair of loci, and additionally all possible pairs of loci. All these comparisons contain information on the underlying parameter r^2.

For a single allele–allele comparison with a single pair of loci, for example A and B above, all other alleles at the loci are binned together as with A' and B' in

Table 1 (Wang 2005). This gives a single allele–allele estimate $\hat{r}^2_{AB}$. Let x_i and x_j be the number of alleles at loci i and j respectively. We obtain $x_i \times x_j$ estimates of r^2, but only $(x_i - 1) \times (x_j - 1)$ of these estimates are independent. This can be seen by noting that (for example) $r_{AB} = \text{cor}(K_A, K_B)$, and the K_A values have to sum to 2 across the x_i different alleles at locus i, and similarly for locus j. The estimate of r^2 within the locus pair is obtained as the arithmetic mean of the $x_i \times x_j$ estimates (England et al. 2006, p. 304). If there are L loci, this produces $L(L-1)/2$ locus-pair estimates of r^2. A single estimate for r^2 across all locus pairs is gained from the weighted arithmetic mean of the estimates for each locus pair, weighted by the number of independent allelic comparisons $(x_i - 1) \times (x_j - 1)$ in each pair. This estimate for r^2 is substituted into equation (1) in place of $\hat{r}^2_{AB}$.

The appropriate sample size n to substitute into equation (1) is complicated by the possibility of missing data for some individuals at some loci, and the different numbers of estimates of r^2 contributed by the different locus pairs. Let n_{ij} be the number of sampled individuals with data available for both loci i and j. There are $(x_i - 1) \times (x_j - 1)$ independent estimates of r^2 available from this locus pair. The final value n in equation (1) is the harmonic mean of the n_{ij} values, where each n_{ij} is included $(x_i - 1) \times (x_j - 1)$ times.

For calculating confidence intervals, the distribution of r^2 is approximated by a chi-square distribution with $M = L(L-1)/2$ degrees of freedom (Hill 1981; Waples 1991). Confidence limits for r^2 are estimated with

$$(1-\alpha)\text{CI} = \left(\hat{r}^2 \times M/\chi^2_{(\alpha/2),M},\quad \hat{r}^2 \times M/\chi^2_{(1-\alpha/2),M}\right) \tag{2}$$

and confidence intervals for N_e^I are obtained from equation (2) using equation (1).

The method above assumes that loci are neutral (non-selected) and physically unlinked ($c = 0.5$). Microsatellite loci are highly suitable for the linkage disequilibrium method (Schwartz et al. 1998), because they are highly polymorphic and nearly selectively neutral, although this may be compromised by genetic hitchhiking. To avoid physical linkage of microsatellite loci, the loci should be located on different chromosomes where possible. Unfortunately, the greatest information about N_e is provided when loci have tight physical linkage (Hill 1981; Hayes et al. 2003), but this would require knowledge of the recombination fraction c, which is not usually available for natural populations (Waples 1991). As the recombination fraction decreases ($c < 0.5$), the effective population size estimate applies to more distant generations (Hill 1981; Hayes et al. 2003; Waples 2006). For unlinked loci, the linkage disequilibrium signal for N_e is determined by the random reassortment from the breeding process in the parental generation, and is greatest when the population size is small (Waples 1991, 2006).

The relationship between the estimated r^2 and N_e^{LD} takes the form of a hyperbolic curve (Fig. 1). When $\hat{r}^2$ is less than $1/n$, negative estimates of N_e^{LD} are possible. In these cases, which are most likely to arise when the sample size is small, the contribution of genetic drift to linkage disequilibrium is swamped by the contribution from statistical sampling. Because it is not possible for N_e to be negative, the conventional way of interpreting a negative N_e^{LD} is to replace it with an

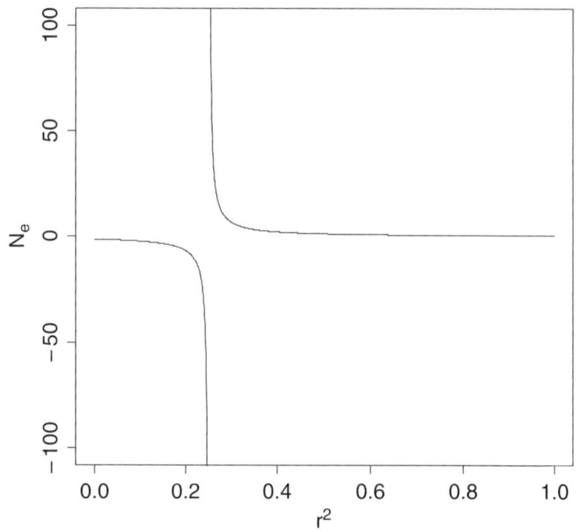

Fig. 1 Relationship between N_e^{LD} and $\hat{r}^2$ ($n = 4$ for illustration)

estimate of infinity (Waples 1991), meaning that the observed linkage disequilibrium is estimated to be entirely due to sampling and with zero contribution due to drift, as would occur in an infinite population. This scenario is a considerable disadvantage to the linkage disequilibrium method. There is also a singularity (undefined value) associated with N_e^{LD} (equation 1) when $\hat{r}^2 = 1/n$.

The linkage disequilibrium in a sample is also affected by residual disequilibria from previous generations. If a population of effective size N_e is initially drawn from an infinite population at time 0, and then remains at constant size N_e for subsequent generations, the expected value of r^2 takes a few generations to reach its equilibrium value. The rate of convergence is given by Sved (1971), and is $1-(1/4)^t$ for unlinked loci, where t is the number of generations (Waples 2005). It follows that r^2 reaches its equilibrium value after about four generations. Waples (2005) performed preliminary simulations for populations with recent increases and decreases in N_e, and confirmed that accumulated disequilibria over multiple generations can affect N_e^{LD}. Populations suffering recent declines were only affected for about one generation until N_e stabilized. For populations that had undergone recent increases, residual effects could persist for a few generations, and caused negative bias in estimates of N_e, depending upon the severity of the bottleneck and the magnitude of the subsequent increase in N_e.

England et al. (2006) simulated ideal populations and found that the N_e^{LD} estimator was robust to different distributions of allele frequencies with up to five alleles per locus. However, for small sample sizes (less than 100), they found serious negative bias in N_e^{LD} when the sample size was smaller than the true value of N_e. Waples (2006) additionally noted a less serious positive bias in N_e^{LD} when the sample size was larger than the true N_e. The conclusion that the sample size should be approximately equal to the quantity that we are trying to estimate, in order for the estimates to be unbiased, is clearly problematic. To improve the method, Waples (2006) used simulated data to derive an empirical correction factor to adjust the estimated r^2 to its correct value. This was to take account of the second order terms in $(1/N_e^I)$ and $(1/n)$ that were omitted in the original derivation of the approximation $E(r^2) \approx 1/(3N_e^I) + 1/n$. He derived two separate modified equations for $N_e{}^{LD}$, one for $n < 30$ and one for $n \geq 30$. Using biallelic simulations, he showed that the harmonic mean of the corrected N_e estimates compared well with the true values (Fig. 4 of Waples 2006; Waples personal communication). The harmonic mean was used because the method adjusted bias in r^2, which is related to the reciprocal $(1/N_e^I)$. He also found that dependencies in r^2 effectively lower the degrees of freedom for estimating var($N_e{}^{LD}$). Both England et al. (2006) and Waples (2006) stated that their simulations were only exploratory and that a thorough evaluation of the linkage disequilibrium method was still required.

The linkage disequilibrium method has seen limited application to real datasets. Bartley et al. (1992) applied the method to natural populations, though obvious errors in their equations (2) and (3), and incorrect sample variances for allele frequencies in estimating r, should all be noted. Recent studies have estimated $N_e{}^{LD}$ using the software package NEESTIMATOR (Peel et al. 2004), following Bartley et al. (1992), but do not provide confidence intervals for their estimates (Lippé et al. 2006), or compare it to inappropriate demographic estimates (Schmeller and Merilä 2007). A new software package LDNE estimates $N_e{}^{LD}$ using the bias-corrected method, removing alleles below a specified proportion, and implementing bootstrap confidence intervals (Waples and Do in press).

3 Simulation

We are concerned with populations consisting of two sexes (male and female; i.e. dioecious and diploid), two distinct generations (adults and juveniles), and a small number of offspring annually from each adult (number of offspring ≤ 15), as is common for terrestrial vertebrate species. We consider only populations with discrete breeding generations where breeding occurs once within a generation. We simulate small populations ($N < 200$) of individuals for four discrete generations after initially drawing alleles from an effectively infinite pool. This allows us to estimate N_e from genetic data on a generation once r^2 has reached its equilibrium value.

We first generate a population of individuals with L loci and A_i alleles in locus i, where $A_1, \ldots, A_L$ can be set *a priori* as constant, or drawn from a Normal distribution (rounded to integer values) to create polymorphism. Initial allele frequencies for the simulation are drawn from a Dirichlet distribution with a specified shape parameter which controls allele rarity. We use relatively common allele frequencies ($p > 0.1$) to reduce the chance of alleles being lost from the population through genetic drift over the generations.

The first generation of N_1 individuals is given alleles drawn from an infinite sampling distribution with the selected frequencies, which effectively means that

$N_0 = \infty$. Generations 1–4 have finite sizes $N_1, ..., N_4$ determined by the chosen value of N_c for the simulations. Over these four generations genetic drift occurs, which creates the desired linkage disequilibrium but can alter the number of alleles and their frequencies from those specified for the simulation.

In each generation, individuals are assigned male or female sex so that the generation's sex ratio is exactly equal to that specified by the simulation parameters. The ideal population has a 1:1 sex ratio. Males and females in the population contribute k gametes (or, equivalently, k offspring when gametes are united) to the next generation, where the value of k for any individual is drawn from a negative binomial distribution with mean μ_k and variance σ_k^2, and μ_k and σ_k^2 are specified separately for males and females. For a population of constant size and equal sex ratio, the mean must be $\mu_k = 2$ for each sex.

The ratio of offspring variance to mean is called the index of variability in breeding success: $IV = \sigma_k^2/\mu_k$ (Barrowclough and Rockwell 1993; Waples 2006). It is a key parameter for controlling departures from the ideal population. If some individuals have much greater breeding success (k) than others, the index of variability is high, and the inbreeding N_e is lowered due to an increased probability of two randomly selected gametes being derived from the same, successful, parent. We control the index of variability separately for each sex. In an infinite ideal population, $IV = 1$ and k follows a Poisson distribution. This specifies a randomly mating (promiscuous) population. For $IV > 1$, the negative binomial distribution for k has greater than Poisson variance, which is characteristic of polygamous populations where a few individuals have the most successful matings. We do not consider the unusual case where $IV < 1$.

For creating simulated populations with IV exactly equal to the value specified by the simulation parameters, we require vectors of gametes $(k_1, ..., k_M)$ for the M male adults such that the mean and variance of $(k_1, ..., k_M)$ are exactly equal to the specified values for μ_k and σ_k^2, and similarly for the female adults. We achieve this by reformulating the vectors as $(n_0, n_1, ..., n_K)$, where n_k is the number of males contributing k gametes, to a maximum allowed number of gametes $K = 15$. We have three equations for the $K + 1$ unknown integers $n_0, n_1, ..., n_K$, where these are $\sum_{k=0}^{K} n_k = M$, $\sum_{k=0}^{K} kn_k = M\mu_k$, and $\sum_{k=0}^{K} (k - \mu_k)^2 n_k = M\sigma_k^2$. We select guesses for all but three of the $n_0, n_1, ..., n_K$, and solve a matrix equation to find the remaining three values, which leads to some initial guesses being discarded because there are no solutions in integers ≥ 0. Only certain combinations of μ_k and σ_k^2 yield exact solutions in non-negative integers for M individuals. We restrict our simulations to these combinations where possible, or use closely approximate solutions otherwise when the sex ratio and index of variability are both non-ideal.

Once the gamete vectors have been generated for both sexes, gametes from each sex are united with those from the other sex at random without replacement, to create offspring from breeding pairs. Because each offspring is formed from the union of a male and female gamete, the two sexes contribute the same number of gametes, even when sex ratios differ. For unequal sex ratios, μ_k and σ_k^2 are determined separately for each sex to achieve the target population size and index of variability.

Finally, we simulate statistical sampling from the generated population by drawing a sample of n individuals at random without replacement from the final generation (N_4). The genotypes of these n individuals are inspected, and genotyping errors may be added with a specified probability to mimic real laboratory conditions (van Oosterhout et al. 2004; Hoffman and Amos 2005). We simulated two types of errors. The first is allelic drop-out, where one of two alleles for an individual is not typed. This causes the individual to appear homozygous when in fact it is heterozygous but one allele was not typed. We simulate allelic dropout for an individual by replacing one of its two alleles with the other one. The second type of error is missing data, where the individual fails to type for both alleles at a locus. This reduces the sample size for that locus across the population. All error rates are assigned across individuals × loci. For a single individual at a single locus, allelic dropout is assigned first (yes or no), then missing data, which will override allelic dropout if both are selected. We do not consider other microsatellite typing errors which change the length of the microsatellite allele due to either contamination (allele is drawn from the population frequencies) or stutter error (allele is altered by a multiple of the repeat unit). We assume that the error rates are independent and multiplicative. Reported error rates from studies may be conservative, since missing data errors can mask allelic dropout, and both can mask typing errors.

The inbreeding effective population size at time $t-1$ for the non-selfing populations such as in our simulation is approximated from demographic parameters as:

$$\frac{1}{N_e^I} = \frac{\mu_k - 1 + \sigma_k^2/\mu_k}{N_{t-1}\mu_k - 2} \tag{3}$$

(equations 2, 2′ and 2″ in Crow and Denniston 1988; equation (23) in Caballero 1994). Here, the overall μ_k and σ_k^2 for both sexes combined are given by $\mu_k = 2m\mu_m = 2f\mu_f$ and $\sigma_k^2 = m\sigma_m^2 + f\sigma_f^2 + mf(\mu_m - \mu_f)^2$, where m and f are the proportions of adult males and females, μ_m and μ_f are the mean number of progeny of adult males and females, and σ_m^2 and σ_f^2 are the variances in the number of progeny of adult males and females. N_{t-1} is the total number of adults in the parental generation (time $t-1$), given by $N_m + N_f$.

Because inbreeding is slightly retarded in non-selfing populations, $N_e{}^I$ is slightly less than N even in an ideal population without selfing. Our estimate of $N_e{}^{LD}$ is for a non-selfing population, and so we do not adjust $N_e{}^I$ so that our effective population size is exactly equal to our census population size under ideal conditions (Caballero and Hill 1992; Waples 2006). Equation 3 is thus an appropriate true value for comparison to our simulation $N_e{}^{LD}$ estimates when linkage disequilibrium is generated from the non-selfing parental population.

Using this model, we simulate across a range of ecologically realistic parameter values (Table 2), and consider how each parameter influences our estimates of the mean $N_e{}^{LD}$ and 95% confidence intervals, compared to the true comprehensive demographic $N_e{}^I$ calculated from equation (3).

The default simulation values involve eight loci with five alleles per locus at approximately equal allele frequencies, and the entire generation is sampled to estimate effective population size ($n = N$). First we consider ideal populations of different census population sizes. We then determine how the linkage disequilibrium estimate is affected by deviation from each of the three ideal population conditions:

Table 2 Simulation parameters and values. Bold indicates default values that are used unless the simulation specifies that the associated parameter is to be varied. The number of alleles per locus is an integer generated by rounding a Normal(s,v) variate with mean s and variance v given in the table

	Parameter	Values
Population properties	Census population size	10, **50**, 100 and 200
	Index of variability	1, 2, **3**, 4 and 5
	Sex ratio	**1:1**, 1:1.5
Sample properties	Proportion sampled	**1**, 0.5, 0.2
	Number of loci	**8**, 16, 24
	Allele numbers	**N(5,0)**, N(2,0), N(10,0), N(5,1)
	Sequencing errors	Dropout (1%)
		Missing (5%)

constant population size; equal sex ratio; and random mating with $IV = 1$. To investigate the effect of non-constant population size, we run the simulation for a further four generations after r^2 has stabilized in generation 4, and allow N_e to change over generations 5–8 before using the generation 8 data to calculate the estimate N_e^{LD}. Table 3 shows the patterns of population change that we simulate, which we label increasing, decreasing, fluctuating up (where 'up' refers to the last change from generation 7 to 8), and fluctuating down. The populations are ideal in all characteristics except for the changes in population size.

The sequences of population sizes chosen in Table 3 are dictated by our requirement that the mean number of gametes per individual is exactly equal to the variance, to ensure that $IV = 1$, and that the new population size (half the mean number of gametes, times the old population size) is an integer. Sequences of population sizes meeting these requirements are unusual and hard to generate. We restrict our simulations to these exact sequences so that the performance of the N_e^{LD} estimator is not confounded by the unknown effect of using an approximate IV.

After investigating the impact of the three departures from the ideal population, we then test how N_e^{LD} is affected by sampling properties. We select ecologically realistic population parameters, for which $N_e^I = N_{t-1}/2$, and vary the proportion of the population sampled; the number of loci typed; the number of alleles occurring per locus (distributed Normally and rounded to 0 d.p.); and the presence of genotyping errors.

Multiple paternity (offspring within a litter sired by multiple fathers) can impact on the effective population size (Sugg and Chesser 1994), but we do not model this

Table 3 Simulation parameters for changing population sizes in an otherwise ideal population

Population change	$N_1, \ldots N_4$	N_5	N_6	N_7	N_8
Increasing	12	24	36	48	60
Decreasing	64	48	36	24	12
Fluctuating up	36	48	36	24	36
Fluctuating down	36	24	36	48	36

separately because its only effect would be to alter σ_k^2 (and hence IV) contingent on the model for multiple paternity. Dominance multiple paternity where the most successful male breeders also sire other litters increases σ_k^2, while sneaky multiple paternity where unsuccessful male breeders sire other litters reduces σ_k^2, and random multiple paternity should not alter σ_k^2.

4 Results

We simulated over a range of N_e values from 6 to 199, where $N_e = 6$ occurs with census size $N = 10$, sex ratio 1.5:1, and $IV = 2$; and $N_e = 199$ occurs with $N = 200$, sex ratio 1:1, and $IV = 1$. For most values of census size N, we allowed IV to vary from 1 to 5, but for $N = 10$ we only used IV of 1 and 2, because exact solutions for larger IV were not possible. Substantial loss of alleles due to genetic drift over the four generations of equilibration was only problematic for populations with $N = 10$. We refer to the original linkage disequilibrium method as the standard method (SM), and the small sample bias correction introduced by Waples (2006) as the Waples adjustment (WA). WA results are discussed but not shown.

4.1 Population Properties

For both methods, N_e^{LD} has a right-skewed distribution (Fig. 2; only showing SM). The reciprocal $1/N_e^{LD}$, which estimates the probability that two randomly selected successful gametes derive from the same parent, is distributed approximately Normally. For the ideal population ($IV = 1$), both the arithmetic mean and the harmonic mean of N_e^{LD} are almost equal to the true value of N_e, for both the SM and WA methods. Because most researchers currently focus on estimators for N_e rather than its reciprocal, we focus on the bias of N_e^{LD} rather than the bias of $1/N_e^{LD}$ and therefore report arithmetic means of our results rather than harmonic means. Estimates of infinity are omitted when calculating the arithmetic mean for each simulation, which is justifiable in the sense that these estimates would be discarded by practitioners when they occur in practice. Omitting these poor results will enhance the apparent performance of the N_e^{LD} estimator to some extent, but in practice they only occurred in one of our simulations (the biallelic plot A2 in Fig. 6 below). Infinite values are recorded on the boxplots when they occur within the central 95% of the distribution of simulated N_e^{LD} estimates, which again only occurred once.

In the ideal populations, where sample size $n \approx N_e$, a positive bias in N_e^{LD} is present which increases with census population size (Fig. 3; upper left plot). This is consistent with the suggestion that the N_e^{LD} method should work best in small populations, where the genetic signal is strongest (Waples 1991, 2006). The SM performs better than the WA (graphs not shown). For each of the 10,000 simulations in every boxplot, 95% confidence intervals are calculated using the chi-squared approximation in equations (2) and (1). Except for very small populations, uninformative upper

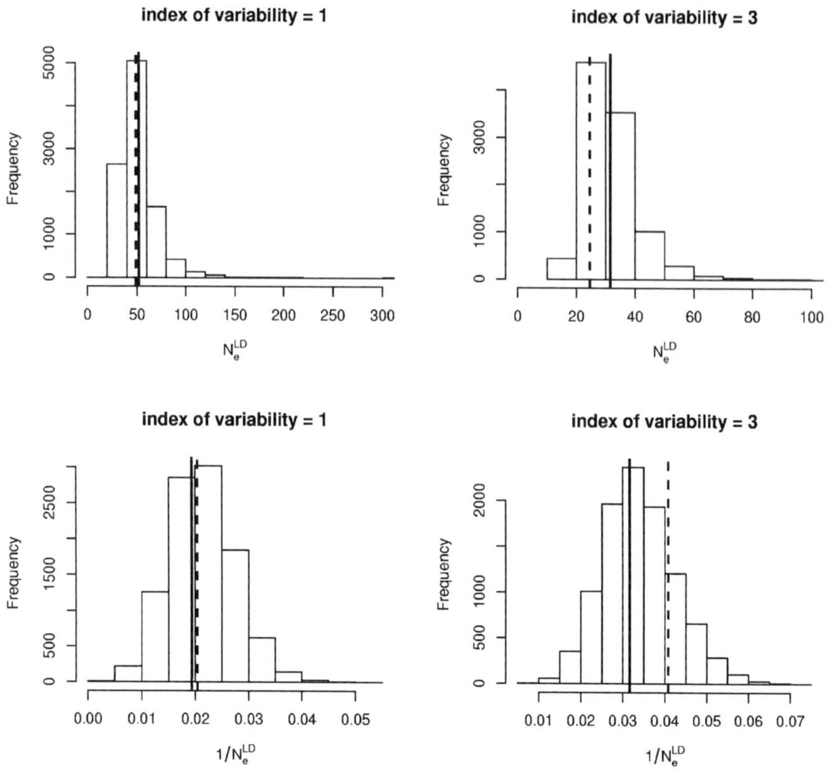

Fig. 2 Distribution of linkage disequilibrium SM estimates from 10,000 simulations with $N = 50$, equal sex ratio and index of variability 1 (an ideal population), and index of variability 3. Bold lines are means, dotted lines are true values. The entire generation is sampled

95% confidence interval estimates of infinity are ubiquitous when the SM is applied to ideal populations. Precision is constant in ratio to the mean for increasing census population sizes. Changing the sex ratio from the ideal 1:1 to the non-ideal 1.5:1 has very little effect on the calculation of N_e by equation (3), and also has very little effect on the bias or precision of either method (Fig. 3).

Figure 4 shows the impact of increasing the index of variability, simultaneously for both sexes, for different population sizes. The bias of both methods increases with increasing *IV* (Fig. 4), in the sense of the true value lying below the 25% quantile of the estimator distribution, and the bias of the WA becomes more similar to that of the SM. For populations with identical N_e the bias is greater at higher indices of variability, but precision remains similar. Both biases are considerable at high indices of variability, with the true value lying at about or below the 25% quantile of the estimator distribution, but the SM still has less bias than the WA at our maximum index of variability (5). High indices of variability did substantially decrease the number of upper 95% confidence interval estimates of infinity, beyond that expected from a decrease in N_e alone.

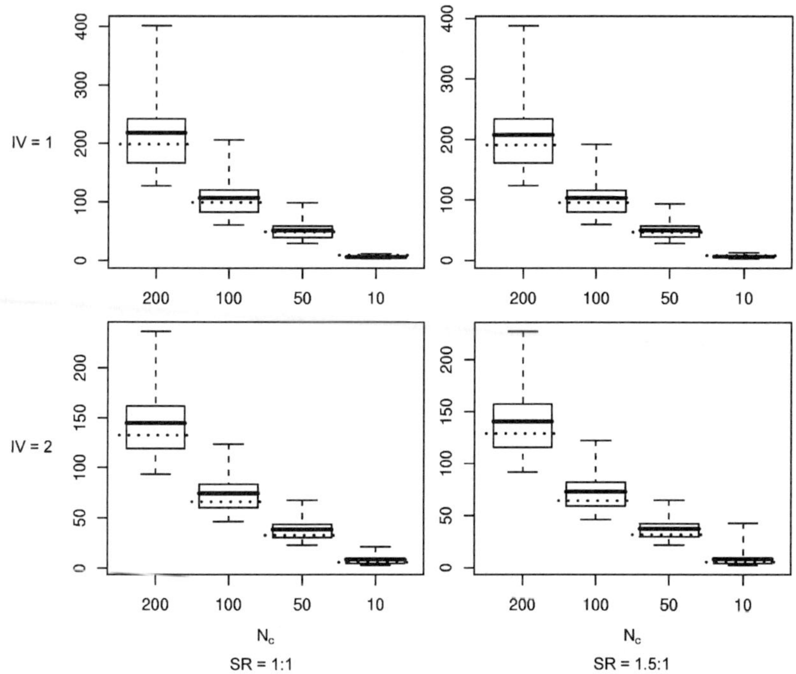

Fig. 3 Boxplots of the SM N_e^{LD} estimator for sex ratios 1:1 and 1.5:1, and indices of variability 1 and 2 for populations $N = 200$, 100, 50 and 10 from 10,000 simulations. Boxplots show 2.5, 25%, mean, 75 and 97.5% quantiles of the estimator distribution. Dotted lines are true N_e values

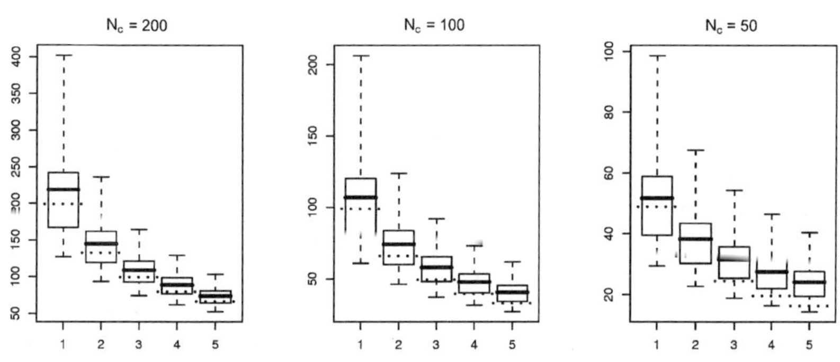

Fig. 4 Boxplots of the SM N_e^{LD} estimator for sex ratio 1:1 and indices of variability 1–5 for populations $N = 200$, 100 and 50 from 10,000 simulations. Boxplots show 2.5, 25%, mean, 75 and 97.5% quantiles of the estimator distribution. Dotted lines are true N_e values

If males and females are given different values of *IV*, the resulting bias is greater than the bias obtained if the same overall average *IV* is used and is equal for both sexes (results not shown). If the *IV* differs between sexes and the sex ratio is also

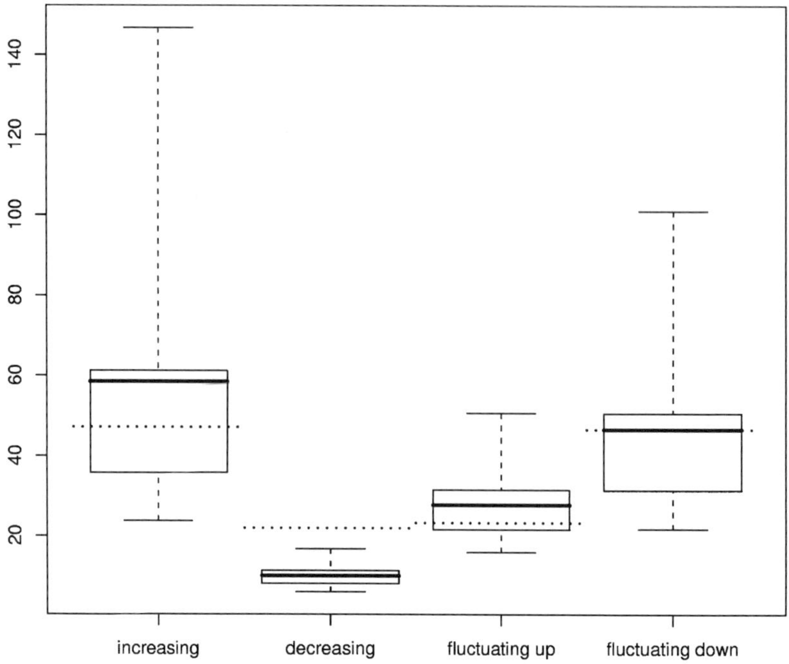

Fig. 5 Boxplots of the SM N_e^{LD} estimator when generations $N_5, \ldots, N_8$ are steadily increasing, steadily decreasing, and fluctuating with a final increase and final decrease, from 10,000 simulations. Boxplots show 2.5, 25%, mean, 75 and 97.5% quantiles of the estimator distribution. Dotted lines are true N_e values

deviated from 1:1, we see an additional, but small, positive bias again. Precision is similar to that in Fig. 4 when both sexes are given the average of the two *IV* values.

Figure 5 shows the results of changing the census population size over generations 5–8 in populations that are otherwise ideal, using the population size sequences in Table 3. For both SM and WA methods, the N_e^{LD} estimate is affected differently depending on the form of the population change. The N_e^{LD} estimate from generation 8 will be distorted by residual linkage disequilibrium from roughly the previous four generations (Sved 1971). For a systematically increasing population N_e^{LD} was positively biased, and imprecise (Fig. 5). For a systematically decreasing population N_e^{LD} was negatively biased but very precise. For the 'fluctuating up' population where the final fluctuation was upwards, N_e^{LD} was also slightly positively biased, but not to the extent of the 'increasing' population. For the 'fluctuating down' population where the final fluctuation was downwards, N_e^{LD} showed no bias. Waples (2005) also found that decreases in population size had less severe effects than increases. Bias reflected the persistence of directional population change, while precision reflected the final census population size as expected from previous results. In all cases the SM performed better than the WA, which was highly positively biased for both cases of decreasing populations. The N_e^{LD} for systematically increasing and decreasing populations was in fact more closely aligned with the census population size of the offspring generation 8 rather than that of the parental generation 7, which would correspond to the variance effective population size rather than the inbreeding effective size. Further investigation beyond that of our limited population sequences is warranted for this ecologically important scenario.

N_e^{LD} estimates of infinity are a concern for inference from the linkage disequilibrium method, but in our simulations they only occurred once (biallelic case: plot A2, Fig. 6 below). They occur because the N_e^{LD} estimator is sensitive to variation in r^2 (Fig. 1). The estimator is constructed by replacing the expectation $E(r^2)$ by the sample value $\hat{r}^2$ in equation (1), so a large variance in $\hat{r}^2$ could draw the estimator into the part of the parameter space left of the singularity in Fig. 1, which would be impossible for the true expectation $E(r^2)$. The result is a negative estimate of N_e^{LD},

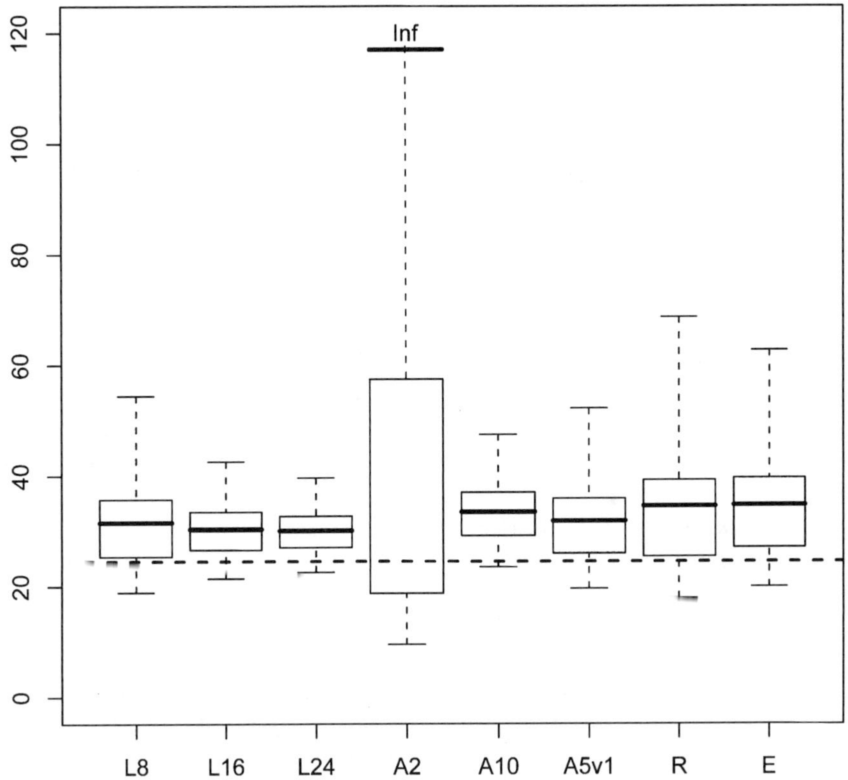

Fig. 6 Boxplot of the SM N_e^{LD} estimator for $N = 50$, sex ratio 1:1 and index of variability 3 from 10,000 simulations (except A2 = 20,000 simulations). Boxplots show 2.5, 25%, mean, 75 and 97.5% quantiles of the estimator distribution. The dotted line is the true value $N_e = 25$. L8 = standard reference sample with 8 loci and 5 alleles; L16 = 16 loci; L24 = 24 loci; A2 = biallelic loci, i.e. number of alleles per locus is Normal(2,0); A10 = polymorphic loci with $N(10,0)$ alleles per locus; A5v1 = $N(5,1)$ alleles per locus; R = rare alleles in initial population ($p < 0.1$); E = errors in genetic sequencing at an individual rate of allelic dropout = 0.01 and missing data = 0.05

which has to be interpreted as infinity. A greater problem in our simulations is upper confidence limits of infinity, which are gained using the chi-squared approximation in the SM and are very common. The SM consistently overestimates the variance of N_e^{LD}. Except for very small census population sizes of $N = 10$, the 95% confidence intervals contained the true value 100% of the time, usually with uninformative upper confidence intervals of infinity. The chi-square approximation for the variance of r^2 does not appear to be appropriate for estimating suitable confidence intervals.

By contrast, coverage of the 95% confidence intervals using the WA method was usually below 95%, sometimes as low as 40% for small census population sizes. Only for large census population sizes with significant deviation from the ideal population did the WA coverage approach 100%. Waples (2006) did some investigations of confidence interval coverage and also found that for $N_e < 100$ his bias-corrected confidence intervals would have coverage well below 95% (Figs. 5 and 6 in Waples 2006). The poor coverage of the WA confidence intervals was attributable to a decrease in the estimator variance, coupled with an increase to the positive bias already present when sample size was greater than N_e. This often led to the lower confidence interval being above the true value. The WA 95% confidence interval was additionally problematic when the square root component of his adjusted N_e equations (Waples 2006) was positive for the corrected $\hat{r}^2$, but not for the lower 95% confidence interval of $\hat{r}^2$. This meant a second approximate equation was necessary to estimate the lower 95% confidence interval for N_e^{LD} in the WA method.

4.2 Sample Properties

Effects of number of loci sampled, allele numbers and rarity, and genotyping errors were investigated with the ecologically reasonable value $IV = 3$ (e.g. Heiberg et al. 2006). Increasing the number of loci sampled substantially improves the precision of the N_e^{LD} estimate for both methods (Fig. 6, plots L8, L16, L24), and reduces the number of upper confidence interval estimates of infinity in the SM. However, increasing the number of loci sampled has little effect on the bias, and confidence interval coverage remains at 100% for the SM. For the WA, confidence interval coverage decreased markedly to 40% with 24 loci. For strictly biallelic loci, 20,000 simulations were required to attain stability. Biallelic loci generated a substantial positive bias in both methods (Fig. 6, plot A2), due to the presence of many large estimate values, including infinity. The biallelic loci case was the only case where the WA performed better than the SM, which is notable because the WA method was derived using biallelic loci (Waples 2006, p. 182). In the biallelic loci case for the SM, confidence interval coverage was reduced slightly less than 95%, and the precision of the estimator was reduced to the extent that it included infinity in the central 95% of its distribution. However, the harmonic mean of N_e^{LD} was almost exactly correct in the biallelic case for the SM.

With an increased polymorphism of 10 alleles per locus (plot A10), precision was improved slightly but bias was also increased slightly over the 5 allele setting (plot L8). Normally distributed variation in the number of alleles had no effect on the estimates from either method (plot A5v1). The presence of rare alleles meant some were lost during genetic drift, creating a positive bias and decrease in precision as found for the extreme biallelic case (plot R). The presence of errors in the data at our specified rates also created a positive bias and decrease in precision (plot E).

The important effect of sample size n on the N_e^{LD} estimator was emphasised by both Waples (2006) and England et al. (2006). For the SM we found a similar pattern to that reported in Fig. 1 of Waples (2006). When $n > N_e$ (so sampling fraction $S = n/N_c$ satisfies $S > N_e/N_c$), positive bias occurs (Fig. 7). When $n > N_e$ ($S < N_e/N_c$), we obtain severe negative bias. The least bias occurs when $n < N_e$ (Fig. 7: $S = 1$ with $IV = 1$; and $S = 0.5$ with $IV = 3$). In our simulations, the WA

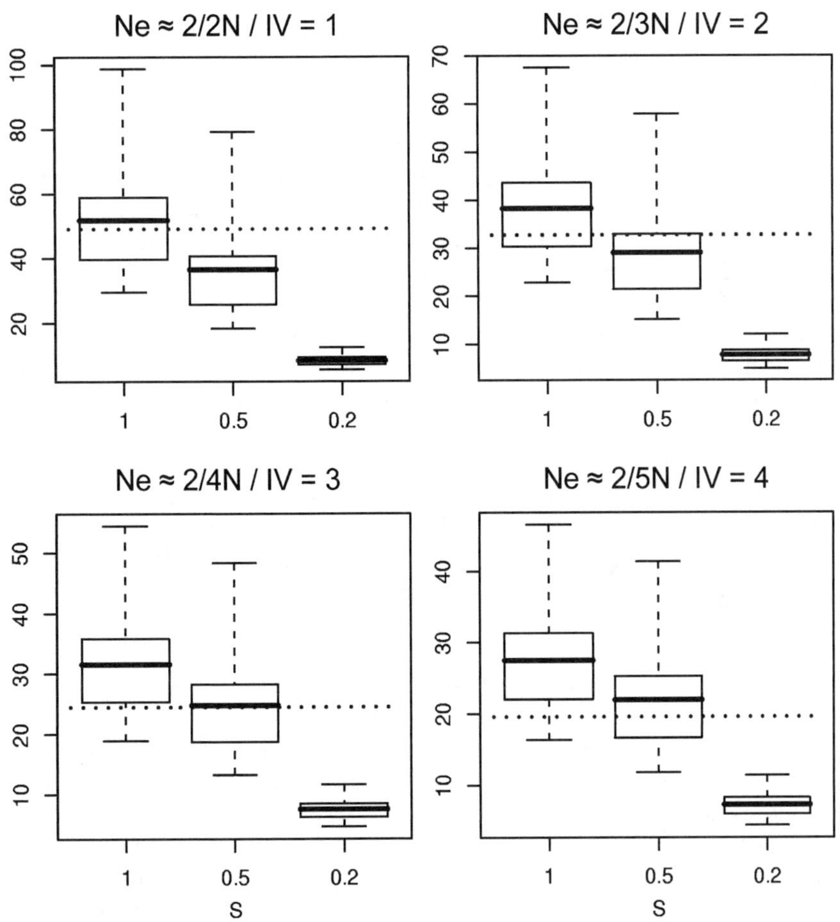

Fig. 7 Boxplots of the SM N_e^{LD} estimator for $N = 50$, sex ratio 1:1 and indices of variability 1–4 from 10,000 simulations. Boxplots show 2.5, 25%, mean, 75 and 97.5% quantiles of the estimator distribution. The dotted line is the true value N_e. $S = n/N_c$ is the proportion of the census population size sampled

removed the negative bias but created a positive bias of at least similar magnitude. The WA did however reduce the bias of the reciprocal estimator when $n < N_e$. As the sampling fraction decreased, the coverage of the WA 95% confidence interval for N_e decreased to below 50%, because the positive bias lifted the lower confidence interval above the true value. The SM became more precise at smaller sampling fractions, but the estimator distribution did not include the true value. By contrast, the WA estimator became less precise at lower sampling fractions. Increasing the number of loci typed when the sample size was very small led to the very poor interaction of large bias and high precision, because the estimator became more precise with more loci, but had confidence interval coverage below 50% due to the negative bias. Increasing the number of loci did however reduce the bias of the WA.

5 Rat Populations

We now compare demographic and linkage disequilibrium estimates of inbreeding effective population size using data collected from closed rat populations on two small (<10 ha) islands in New Zealand, with negligible migration. Random mating can be achieved as all rats can physically find each other on the islands. We assume that rats follow a breeding model with discrete generations. In the wild, rats rarely survive longer than one year (Innes 2005a, b), usually only breeding in one generation (season). Hence we assume that any adults caught can not be offspring of other adults in our sample from a previous generation. As many rats as possible were sampled prior to eradication from the islands. Rats were classified by sex and assigned as adult or juvenile based on an arbitrary weight value and breeding condition. On each island we therefore have a total population of size N, comprising two generations of rats (N_t and N_{t-1}), from which we will use the linkage disequilibrium method with a sample of size n_t juveniles to estimate the effective population size of generation $t-1$. In each case, N was approximately known due to exhaustive eradication sampling and follow-up observation.

We estimate the mean and variance of the number of progeny of adults, separately for males and females, by applying the two-stage parentage assignment software CERVUS (Marshall et al. 1998) to the juvenile genotypes. Missing data rates were estimated from the real populations, and a default mis-typing rate of 1% was assumed. We then use equation 3 to estimate an approximate demographic N_e^I for the parental generation. We assume equal catchability between age classes, and therefore estimate the total number of adults (N_{t-1}) by dividing the number of adults caught by the approximate sampling proportion of the population. Our demographic estimates are only approximate due to possible incorrect assignment of parentage and missing individuals, which will influence our estimates of mean and variance in breeding success. We have essentially ignored the statistical issues in estimating these demographic parameters (Crow and Denniston 1988), and do so merely for comparative purposes with the linkage disequilibrium estimates. We assume individuals were sampled at random so that our estimates are representative of the entire population.

5.1 Ship Rats (*Rattus rattus*) – Hawere (9.3 ha)

A total of n = 29 ship rats were genotyped from a known population of N = 31 individuals, giving a sampling proportion of S = 0.94 for the population. An arbitrary weight of 120 g was used to distinguish between adults and juveniles, giving 7 male and 11 female adults (n_{t-1} = 18), and 6 male and 5 female juveniles (n_t = 11). Parentage was assigned with 80% confidence for the fathers of 8/11 juveniles, and mothers of 10/11 juveniles. We estimated $\mu_m = 1.14$, $\sigma_m^2 = 0.48$, $\mu_f = 0.91$, and $\sigma_f^2 = 1.69$. Using the expressions below equation (3) (but using $\mu_k = m\mu_m + f\mu_f$ to allow for unequal parentage assignment for male and female adults), the overall estimates were $\mu_k = 1.00$ and $\sigma_k^2 = 1.23$. By equation (3):

$$\frac{1}{N_e^I} \approx \frac{1.00 - 1 + 1.23/1.00}{18/0.94 \times 1.00 - 2}$$

This gives our demographic estimate $N_e{}^I$ = 13.93 (73% of N_{t-1}). The $N_e{}^{LD}$ estimates from the linkage disequilibrium method were: SM = 7.52, 95%CI = [2.35, ∞]; and WA = 30.68 [17.37, 47.34].

5.2 Norway Rats (*Rattus norvegicus*) – Moturemu (5.0 ha)

A total of n = 27 Norway rats were genotyped from a population of approximately $N \approx 39$ individuals, giving a sampling proportion of S = 0.69 for the population. An arbitrary weight of 220 g was used to distinguish between adults and juveniles, giving 11 male and 5 female adults (n_{t-1} = 16), and 4 male and 7 female juveniles (n_t = 11). Most likely parentage was assigned for the fathers and mothers of all 11 offspring (ties in parentage were assigned 0.5 to each parent), giving estimates $\mu_m = 1.00$, $\sigma_m^2 = 2.40$, $\mu_f = 2.20$, and $\sigma_f^2 = 11.08$. The overall estimates were therefore $\mu_k = 1.38$ and $\sigma_k^2 = 5.42$, so by equation (3):

$$\frac{1}{N_e^I} \approx \frac{1.38 - 1 + 5.42/1.38}{16/0.69 \times 1.38 - 2}$$

therefore $N_e{}^I$ = 6.92 (30% of N_{t-1}), while the $N_e{}^{LD}$ estimates were SM = 1.92, 95%CI = [0.85, 4.39], and WA = 1.16 [0.63, 1.84].

6 Discussion

6.1 Simulations

Data on linkage disequilibrium in a population provide a method of estimating inbreeding effective population size in the parental generation, and its reciprocal, the probability that two randomly chosen successful gametes derive from the same parent. The concepts underlying the method are intuitively challenging, and this has led to some confusion in the literature regarding gametic and genotypic estimators (Waples 1991; Bartley et al. 1992), and variance and inbreeding effective population sizes (Leberg 2005). The method can be used when just one sample is available from a population, complementing alternative methods for estimating effective population size when multiple samples are available (Wang 2005).

Throughout our study, we evaluated the bias of N_e^{LD} as an estimator of N_e, rather than using the harmonic mean of N_e^{LD} estimates. The harmonic mean would be appropriate if the parameter of interest were the reciprocal, $1/N_e$, rather than N_e itself, and has been used by previous authors in assessing bias of N_e^{LD} (Waples 2006). We intentionally conducted our investigations on N_e^{LD} rather than the reciprocal estimator, because this is the parameter that researchers focus on and set out to estimate in conservation biology. The bias correction method (WA) of Waples (2006) was developed in such a way that it improves the bias of $1/N_e^{LD}$ for biallelic loci, especially when the sample size is much less than N_e, but this has the effect of exacerbating upwards bias in N_e itself in most other situations. For the standard method (SM), the bias in the arithmetic mean and harmonic mean of N_e^{LD} estimates were similar in direction and magnitude. For the Waples (2006) adjusted method (WA), the bias in the harmonic mean was generally reduced slightly below that of the SM, but the bias in the arithmetic mean was increased substantially above that of the SM. Using the arithmetic mean had the disadvantage that infinite values had to be discarded before assessing bias, but this happened only in the biallelic simulation A2 (Fig. 6).

Our simulation results suggest that the N_e^{LD} estimator performs poorly for non-ideal populations and when the sample size is either substantially greater than or less than the true value N_e, which we wish to estimate. Deviation from ideal sex ratios has little effect, but deviation from random (binomial) breeding by using index of variability $IV > 1$ leads to bias in the method. Of some concern is that the size of the census population, and recent fluctuations in it, have considerable effects on N_e^{LD} (Waples 2005). This is a problem as much of the interest in N_e^{LD} is in its application to natural populations, where usually N is unknown and changing. Non-constant population size can cause considerable bias, because N_e^{LD} is affected by residual disequilibria from previous generations. Waples (2005) simulated changes in population size and the subsequent rate of recovery in N_e^{LD} for the population at its final, stable, size. For a population that had gone through a bottleneck and then increased, the bias was considerable for several generations because the small population size during the bottleneck generated strong levels of linkage disequilibrium which took several generations to decay. In our limited simulations of non-constant population size, we found bias in the direction of the most recent population change. The bias became greater as population change persisted in a single direction.

The bias in N_e^{LD} changes direction according to whether the sample size is larger or smaller than the true value N_e, and this is a considerable drawback of the method (Hill 1981; England et al. 2006; Waples 2006). The bias correction method proposed by Waples (2006) only improves properties of the reciprocal estimator $1/N_e^{LD}$ when n is much less than N_e, but in our simulations did not perform as well as the standard method when considering bias, precision and confidence interval coverage for estimating N_e itself. England et al. (2006) suggested a way of addressing the problem that the optimal sample size is the same as the value of the unknown parameter we wish to estimate. They recommended sub-sampling the available sample to create several N_e^{LD} estimates from different sample sizes and plotting the results against sample size to investigate whether it stabilizes. Stability suggests that the correct value has been reached, because the bias from $n > N_e$ is much less than that from $n < N_e$. The sample size effect has been observed in every evaluation of the N_e^{LD} undertaken to date, and it would be wise for researchers applying the method to real data to routinely investigate results from this sub-sampling procedure.

The index of breeding variability exerts a strong influence on N_e, and another tactic for overcoming the problem of requiring $n \approx N_e$ is to produce a rudimentary estimate of the index of variability within a population. With this, it may be possible to gain a crude estimate of N_e from demographic data using equation (3), and hence an estimate of N_e/N_c and the proportion of the population that should be sampled. For example, assuming $IV = 3$, $N_e \approx N/2$, so $n > N/2$ is appropriate. Estimates of the proportion of the population sampled can be obtained from removal data and catch-effort modeling methods (Seber 1982). However, these approaches are themselves subject to considerable statistical error.

Waples (2006, p. 180) remarked that simulating ideal populations of $N = N_e$, and assuming that estimators will behave similarly if N_e is of an equivalent value under non-ideal conditions, may be a reasonable approximation; however, this was not always the case with our simulations. Different forms of deviation from the ideal population led to substantially different biases in the linkage disequilibrium estimate. Waples (2006) simulated a standard population of sample size 50, which we also used as our default value and which performs very well in most simulations. As census (and effective) population size increase, the magnitude of the positive bias, and lack of precision, in the method increases. This is consistent with the point made by Waples (1991, 2006) that the linkage disequilibrium method is most useful for small populations in which the genetic signal from linkage disequilibrium is strongest.

Although genetic drift over the four generations of burn-in for our simulations did reduce allelic diversity for small census population sizes or for populations with rare alleles ($p < 0.1$), no loci ever became monomorphic. Monomorphic loci positively bias the linkage disequilibrium method because all their pair-wise loci combinations have $r = 0$, which will lead to infinite estimates of N_e^{LD}. In general as the number of alleles decreases for a locus (reduced polymorphism, or apparent fixation), estimates of N_e^{LD} will become increasingly positively biased due to a poor ability to detect linkage disequilibrium, causing r^2 to be underestimated in some samples. With a high level of polymorphism, precision is improved and there appears to be little if any effect on bias. Researchers should be aware of the effect of polymorphisms and select appropriate loci for estimating N_e^{LD}. The presence of occasional genotyping errors had little effect on N_e^{LD} in our simulations.

Increasing the number of loci sampled improves precision substantially, but this is only desirable when bias is relatively small compared to precision, otherwise misleading estimates which are precise but highly biased are possible. Practitioners should therefore focus primarily on increasing their sample size of individuals. Once they are confident that their sample size exceeds the effective population size, increasing the number of loci sampled is a useful secondary consideration in order to give precise estimates which will be useful for inference. Increasing the number of loci sampled with an inadequate sample size can lead to highly misleading results. This recommendation is similar to advice given when estimating census population size through mark-recapture studies, where captures of new individuals provide more information than recaptures of previously caught individuals (Seber 1982; Borchers et al. 2002).

6.2 Rats

With a removal sample from a population, an estimate of census population size is possible, but this is prone to poor precision and accuracy (Borchers et al. 2002). By estimating the index of variability for all adults from trapping data and parentage assignment, and with a reliable estimate of N_e (and knowledge of any recent fluctuations in N) it may be possible to improve or corroborate estimates of N. As others have previously noted this relies heavily on the accurate estimation of the index of variability (Barrowclough and Rockwell 1993), which is problematic.

The sample sizes (parents and offspring) for our rat datasets were 29 and 27, and the total population sizes were 31 and approximately 39 respectively. Despite small sample sizes, n was very likely to be greater than N_e for both populations. Both populations had a mean number of offspring less than two, suggesting declining populations, which our simulations suggested would lead to negative bias in N_e^{LD}. Consistent with this, the standard method estimates were less than the demographic estimates in each case. The index of variability of the Norway rats was three times that of the ship rats, for which the index of variability was only a little greater than the ideal value of 1. The ship rat confidence interval was far too wide, and included infinity as expected from our simulations for populations close to $IV = 1$, while the Norway rat confidence interval was unexpectedly narrow and did not include our demographic estimate. As N_e approaches N, we expect the WA to have some positive bias with a wide confidence interval approaching 95% coverage, and this is somewhat consistent with the ship rat results. For $N_e < N$ we expect the WA confidence interval to have very poor coverage as seems likely for the Norway rat results.

For ship rats, the demographic, SM and WA methods give substantially different answers. It is not possible to make a judgement as to which is correct, as all three methods have associated problems. The N_e^{LD} estimates between 1 and 2 for Norway rats are likely to be underestimates for a population of this size. The comparison of estimates is encouraging in appearing to corroborate some of our simulation results, but also illustrates the poor results that may be obtained from the N_e^{LD} method with real data, even though the sampling fractions are large and N_e is small enough for a strong genetic signal to be expected.

6.3 Application

The effective population size provides a single statistic which simultaneously adjusts for the effects of fluctuations in population size, deviations from random mating, and unequal sex ratios when measuring genetic change in a population. As such, it can be a useful summary value in population dynamics, particularly for conservation managers (Wang 2005). It does not directly quantify the genetic diversity in a population, however, for example the number of different alleles per locus. Populations can display local adaptation and persistence with low genetic diversity and effective population size (McKay et al. 2001). The linkage disequilibrium method for estimating effective population size is particularly useful for providing information about populations that can only be sampled once. However, the methodology is still at a stage where estimates must be treated with caution.

Threatened species at small population sizes are routinely found to have low ratios of effective to census population sizes (Frankham 1995), which are implicated in their bottleneck. This creates something of a paradox, however, as introduced invading populations created from small numbers of founders also undergo a severe bottleneck and might be expected to be poorly adapted to successful establishment (Sax and Brown 2000). We would expect this to be reflected in invasive species also having a low ratio of effective to census population size. Effective population size can therefore play an important comparative role in understanding the persistence of not just threatened but also invading species (Holland 2000). Particularly, it is of interest to study whether there are certain mechanisms that help invasive species to overcome the long-term effects of severe bottlenecks and low effective population sizes.

We have performed a reasonably extensive simulation on the effect of multiple and simultaneous deviations from ideal conditions, as well as sampling properties, on the linkage disequilibrium estimate of effective population size. From this and other work (England et al. 2006; Waples 2005, 2006) we now have a reasonable understanding of the effect of multiple loci and alleles on the method, and sensitivity to allele frequencies and genotyping errors. Violation of critical assumptions needed for genetic estimation of effective population size remain to be investigated, including selected or linked markers; mutation, population admixture, migration, and sub-division; non-random sampling; and overlapping generations, which will all affect the method (Vitalis and Couvet 2001; Waples 2006). Our simulation had no spatial component, and so was biased towards 'random mating', in that it implicitly assumed all individuals could access all other individuals, which would only be possible in small closed populations. In other situations, a 'Wahlund effect' might occur where genetically similar family clusters are created. The presence of such population subdivision will affect the linkage disequilibrium estimate (Wang and Caballero 1999). More importantly, a thorough treatment of r^2 is required, considering both its accurate estimation with respect to population and sample size, and the impact of dependencies among pairwise loci comparisons. From this it may be possible to propose more robust methods of estimating N_e from data on linkage disequilibrium.

Acknowledgments Thanks to Robin Waples and Chi Do for helpful discussion on implementing the method, and to Jawad Abdelkrim for genotyping rats. Paul Conn, Paul Lukacs and an anonymous referee provided valuable feedback on the manuscript. James Russell was supported by a Top Achiever Doctoral Scholarship from the NZ Tertiary Education Commission, and Rachel Fewster was supported by a Royal Society of New Zealand Marsden grant.

References

Barrowclough GF, Rockwell RF (1993) Variance of lifetime reproductive success: estimation based on demographic data. American Naturalist 141:281–295

Bartley D, Bagley M, Gall G, Bentley B (1992) Use of linkage disequilibrium data to estimate effective size of hatchery and natural fish populations. Conservation Biology 6:365–375.

Berthier P, Beaumont MA, Cornuet J-M, Luikart G (2002) Likelihood-based estimation of the effective population size using temporal changes in allele frequencies: a genealogical approach. Genetics 160:741–751.

Borchers DL, Buckland ST, Zucchini W (2002) Estimating animal abundance: closed populations. Springer, London.

Buckland ST, Newman KB, Thomas L, Koesters NB (2004) State-space models for the dynamics of wild animal populations. Ecological Modelling 171:157–175.

Caballero A, Hill WG (1992) A note on the inbreeding effective population size. Evolution 46:1969–1972.

Caballero A (1994) Developments in the prediction of effective population size. Heredity 73: 657–679.

Campton DE (1987) Natural hybridization and introgression in fishes. In: Ryman N, Utter F (eds) Population genetics and fishery management. University of Washington Press, Seattle, pp 161–192.

Caswell H (2001) Matrix population models: construction, analysis, and interpretation. Sinauer Associates, Sunderland.

Cockerham CC, Weir BS (1977) Digenic descent measures for finite populations. Genetical Research 30:121–147.

Crandall KA, Posada D, Vasco D (1999) Effective population sizes: missing measures and missing concepts. Animal Conservation 2:17–319.

Crow JF, Kimura M (1970) An introduction to population genetics theory. Burgess, Minneapolis.

Crow JF, Denniston C (1988) Inbreeding and variance effective population numbers. Evolution 42:482–495.

England PR, Cornuet J-M, Berthier P, Tallmon DA, Luikart G (2006) Estimating effective population size from linkage disequilibrium: severe bias in small samples. Conservation Genetics 7:303–308.

Frankham R (1995) Effective population size / adult population size ratios in wildlife: a review. Genetical Research 66:95–107.

Garza JC, Williamson EG (2001) Detection of reduction in population size using data from microsatellite loci. Molecular Ecology 10:305–318.

Hayes BJ, Visscher PM, McPartlan HC, Goddard ME (2003) Novel multilocus measure of linkage disequilibrium to estimate past effective population size. Genome Research 13:635–643.

Heiberg A-C, Leirs H, Siegismund HR (2006) Reproductive success of bromadiolone-resistant rats in absence of anticoagulant pressure. Pest Management Science 62:862–871.

Hill WG (1981) Estimation of effective population size from data on linkage disequilibrium. Genetical Research 38:209–216.

Hoffman JI, Amos W (2005) Microsatellite genotyping errors: detection approaches, common sources and consequences for paternal exclusion. Molecular Ecology 14:599–612.

Holland BS (2000) Genetics of marine bioinvasions. Hydrobiologia 420:63–71.

Innes JG (2005a) Ship rat. In: King CM (ed) The handbook of New Zealand mammals. Oxford University Press, Melbourne, pp 187–203.
Innes JG (2005b) Norway rat. In: King CM (ed) The handbook of New Zealand mammals. Oxford University Press, Melbourne, pp 174–187.
Lande R, Barrowclough GF (1987) Effective population size, genetic variation, and their use in population management. In: Soulé M (ed) Viable populations for conservation. Cambridge University Press, Cambridge, pp 87–123.
Laurie-Ahlberg CC, Weir BS (1979) Allozymic variation and linkage disequilibrium in some laboratory populations of *Drosophila melanogaster*. Genetics 92:1295–1314.
Leberg P (2005) Genetic approaches for estimating the effective population size of populations. Journal of Wildlife Management 69:1385–1399.
Lippé C, Dumont P, Bernatchez L (2006) High genetic diversity and no inbreeding in the endangered copper redhorse, *Moxostoma hubbsi* (Catostomidae, Pisces): the positive side of a long generation time. Molecular Ecology 15:1769–1780.
Lynch M, Lande R (1998) The critical effective size for a genetically secure population. Animal Conservation 1:69–73.
Marshall TC, Slate J, Kruuke LEB, Pemberton JM (1998) Statistical confidence for likelihood-based paternity inference in natural populations. Molecular Ecology 7:639–655.
McKay JK, Bishop JG, Lin J-Z, Richards JH, Sala A, Mitchell-Olds T (2001) Local adaptation across a climatic gradient despite small effective population size in the rare sapphire rockcress. Proceedings of the Royal Society of London B 268:1715–1721.
Nunney L, Campbell KA (1993) Assessing minimum viable population size: demography meets population genetics. Trends in Ecology & Evolution 8:234–239.
Nunney L, Elam DR (1994) Estimating the effective population size of conserved populations. Conservation Biology 8:175–184.
Peel D, Ovenden JR, Peel SL (2004) NeEstimator: software for estimating effective population size (version 1.3). Queensland Government, Department of Primary Industries and Fisheries, Brisbane.
Rockwell RF, Barrowclough GF (1995) Effective population size and lifetime reproductive success. Conservation Biology 9:1225–1233.
Sax DF, Brown JH (2000) The paradox of invasion. Global Ecology & Biogeography 9:363–371.
Schmeller DS, Merilä J (2007) Demographic and genetic estimates of effective population and breeding size in the amphibian *Rana temporaria*. Conservation Biology 21:142–151.
Schwartz MK, Tallmon DA, Luikart G (1998) Review of DNA-based census and effective population size estimators. Animal Conservation 1:293–299.
Seber GAF (1982) The estimation of animal abundance and related parameters. The Blackburn Press, Caldwell.
Selkoe KA, Toonen RJ (2006) Microsatellites for ecologists: a practical guide to using and evaluating microsatellite markers. Ecology Letters 9:615–629.
Sugg DW, Chesser RK (1994) Effective population sizes with multiple paternity. Genetics 137:1147–1155.
Sved JA (1971) Linkage disequilibrium and homozygosity of chromosome segments in finite populations. Theoretical Population Biology 2:125–141.
Van Oosterhout C, Hutchinson WF, Wills DPM, Shipley P (2004) MICRO-CHECKER: software for identifying and correcting genotyping errors in microsatellite data. Molecular Ecology Notes 4:535–538.
Vitalis R, Couvet D (2001) Two-locus identity probabilities and identity disequilibrium in a partially selfing subdivided population. Genetical Research 77:67–81.
Wang J (2001) A pseudo-likelihood method for estimating effective population size from temporally spaced samples. Genetical Research 78:243–257.
Wang J (2005) Estimation of effective population sizes from data on genetic markers. Philosophical Transactions of the Royal Society B 360:1395–1409.

Wang J, Caballero A (1999) Developments in predicting the effective size of subdivided populations. Heredity 82:212–226.
Wang J, Whitlock MC (2003) Estimating effective population size and migration rates from genetic samples over space and time. Genetics 163:429–446.
Waples RS (1989) A generalized approach for estimating effective population size from temporal changes in allele frequency. Genetics 121:379–391.
Waples RS (1991) Genetic methods for estimating the effective size of cetacean populations. Reports of the International Whaling Commission Special Issue 13:279–300.
Waples RS (2005) Genetic estimates of contemporary effective population size: to what time periods do the estimates apply? Molecular Ecology 14:3335–3352.
Waples RS (2006) A bias correction for estimates of effective population size based on linkage disequilibrium at unlinked loci. Conservation Genetics 7:167–184.
Waples RS, Do C (in press) LDNE: a program for estimating effective population size from data on linkage disequilibrium. Molecular Ecology Resources.
Weir BS (1979) Inferences about linkage disequilibrium. Biometrics 35:235–254.
Weir BS, Hill WG (1980) Effect of mating structure on variation in linkage disequilibrium. Genetics 95:477–488.
Weir BS, Avery PJ, Hill WG (1980) Effect of mating structure on variation in inbreeding. Theoretical Population Biology 18:396–429.
Weir BS (1996) Genetic Data Analysis II: methods for discrete population genetic data. Sinauer Associates, Sunderland.
Williamson EG, Slatkin M (1999) Using maximum likelihood to estimate population size from temporal changes in allele frequencies. Genetics 152:755–761.
Wright S (1931) Evolution in mendelian populations. Genetics 16:97–159.

Section IV
Dispersal, Movement and Migration – Methods and Multi-State Models

Blandine Doligez and Thierry Boulinier

Migration and Movement – The Next Stage

Carl James Schwarz

Abstract The design and analysis of multi-state studies when the states are discrete entities is now well understood with several robust software packages (e.g. M-Surge, MARK) available. However, recent technological advances in radio and archival tags will provide very rich datasets with very fine details on movement. Current methods for the analysis of such data often discretize the data to very coarse states. This paper will review the current state of the art on the analysis of such datasets and make some (bold) forecasts of future directions for the analysis of these data.

1 Introduction

Many animals move, often over long distances. Where do they move to, how many move, which routes do they choose, and how many survive the migration or dispersal are key questions to ecologists. Sugden and Pennisi (2006) introduced a special issue of *Science* on this topic where up-to-date references can be obtained.

The earliest attempts to answer these questions used simple capture–recapture or tag-recovery studies. Animals are tagged at various locations and simple tabulations of subsequent recoveries show where the tagged animals moved. However, unless recovery rates are equal in all reporting locations, the relative numbers of recoveries are uninformative about the real underlying migration rates, or survival rates.

Mark-recapture methods have been developed for cases where detection rates are less than 100% and for differential detection probabilities (Williams et al. 2002; Barker and White 2004). These methods have become more and more advanced (e.g. multi-state and multi-event models) but still have the same basic form, discrete capture occasions on a discretized landscape.

Advanced technology will provide new, exciting sources of data on movement. For example, archival and satellite tags can record, store, and report on the location

C.J. Schwarz (✉)
Department of Statistics and Actuarial Science, Simon Fraser University, 8888 University Drive, Burnaby, BC V5A 1S6 Canada
e-mail: cschwarz@stat.sfu.ca

D.L. Thomson et al. (eds.), *Modeling Demographic Processes in Marked Populations*, Environmental and Ecological Statistics 3, DOI 10.1007/978-0-387-78151-8_14,

of animals on a daily basis. In the past, animals had to be physically handled to read tags, but newer tags can be interrogated at a distance. These new data streams can be used to construct the entire trajectory of an animal. In many cases, detection rates are essentially 1 (i.e. known fate studies) and the challenge is to summarize a vast quantity of data measured on very fine temporal and spatial scales. Additional complications such a seasonality of movement will also need to be modeled.

In this paper, I will review some of the recent work on dealing with measuring migration starting with very coarse spatial and temporal scales and finishing with examples of how known fate and known path data may be analyzed.

2 Discrete Time – Discrete State Models

The discrete state/discrete time models are based on stratified capture–recapture methods and are well developed. Much development has occurred in applications to bird migration, driven in large part by the triannual EURING conferences. These methods extend the capture–recapture methods developed by Cormack (1964), Jolly (1965), and Seber (1965) for estimating survival (Cormack–Jolly–Seber or CJS models) and abundance (Jolly–Seber or JS models) to populations that are spatially stratified. Schaefer (1951), Chapman and Junge (1956), Darroch (1961), Plante et al. (1998), and Schwarz and Taylor (1998) dealt extensively with the two sample cases. Arnason (1972, 1973) extended these methods to three sample times while Hestbeck et al. (1991), Brownie et al. (1993), Schwarz et al. (1993), and Dupuis and Schwarz (2007) completely generalized the problem. In the past 10 years, there has been an explosion of effort in applying the multi-state models not only to geographic movement, but also to any movement among discrete states of animals (e.g. Kendall and Nichols 2002; Lebreton and Pradel 2002). A related model is used for tag returns typically from exploited animal populations where the key difference is that now recoveries take place over an extended period of time between release occasions, many of the tags that are captured are not reported, and tag reporting rates are unknown. Hilborn (1990) and Schwarz et al. (1993) used this type of data to estimate migration rates in a fishery context.

In general, the population is first divided into discrete, non-overlapping states corresponding to geographic locations, stock units, weight classes or condition factors. (States defined by age are not usually considered in the same context as movement models because of the fixed, non-probabilistic movement between age classes.) Each and every animal that is captured should be readily classified into one and only one state. Without loss of generality, suppose there are K states corresponding to geographical locations numbered $1, \ldots, K$.

Releases occur at regular widely spaced intervals (say, yearly) for a total of T years in the study. At each release time point, animals are tagged and released in state s in year i. It is assumed that releases are instantaneous to ensure that animals are alive and present in state s at the time of release. Usually releases take place in all states. Each released animal requires an individually labeled tag so that the location of release and recovery can be determined.

Recaptures take place every year in every state. Tagged animals have their tag numbers recorded. Modern practice is to construct a capture history vector for each animal observed in the study. The capture history (Lebreton et al. 1992) is a vector, ω, of length T with components (ignoring losses-on-capture)

$$\omega_i = \begin{cases} 0 & \text{animal not seen at time } i \\ s & \text{animal seen in state } s \text{ at time } i \text{ and released} \end{cases}$$

For example, the history vector (3, 0, 2, 0, 1, 0) represents an animal released in state 3 in year 1, not seen in year 2, seen in state 2 in year 3, not seen in year 4, seen in year 5 in state 1, and not seen in year 6.

There are three sets of parameters needed to model these data:

ϕ_i^s Apparent survival rate, which is the probability that an animal alive in year i in state s will survive and remain in the study until year $i + 1$. Permanent migration out of the study area is indistinguishable from mortality.

ψ_i^{st} Inter-state movement rates conditional upon survival from year i to year $i +$ 1. This is the probability than an animal who survived from year i to year $i+1$ will "move" from state s to state t. $\sum_{t=1}^{K} \psi_i^{st} = 1$.

p_i^s Recapture rate. This is the probability that an animals which is alive in year i in state s will be captured. All captured animals have their tag read.

The Arnason–Schwarz model for migration does not specify the timing of the inter-state movements, which can take place any time between year i and year $i + 1$. The inter-state movement parameters are also net movement rates that are conditional upon the animal surviving for the year. Lastly, because of the Markovian assumption, movement patterns between the years is unrestricted, that is, an animals may have moved from state 1 to state 4 via state 2 or state 3 between the 2 years.

Similarly, the survival terms measure net survival "averaged" over all destination states. Because the migration rates sum to 1 over the destination states, it is possible to combine the migration and survival parameters into one parameter measuring the product of survival and movement, but there are some advantages to the current parameterization.

These parameters are used to construct a probability expression for each history conditional upon the initial release. These expressions are complex and must account for all the (hidden) paths when an animal is not seen. For example, the history vector (0, 2, 0, 2, 1, 0) has probability (assuming only two states):

$$\left[\phi_2^2\psi_2^{21}\left(1-p_3^1\right)\phi_3^1\psi_3^{12}+\phi_2^2\psi_2^{22}\left(1-p_3^2\right)\phi_3^2\psi_3^{22}\right]\times p_4^2\phi_4^2\psi_4^{21}p_5^1\chi_5^1$$

where the term in square brackets represents the two possible paths between being seen in state 2 at time 2 and state 2 at time 4 without being seen at time 3; and χ_5^1 is the probability of not being seen after time 5 in state 1. Construction of these expressions is complex (Brownie et al. 1993), but, as shown by Brownie et al. (1993) and Schwarz et al. (1993) can easily be expressed using matrix notation.

While capture–recapture models assume that captures take place at a point in time, harvest models assume that returns are from dead animals typically captured in a harvest (e.g. fishery, hunting season). The multi-state models are an extension of the Brownie et al. (1985) suite of models with the key difference being that not all tags are reported when harvested. Schwarz et al. (1993) developed the theory for the multi-state harvest models which is similar to the theory of the multi-state capture models. It is also possible to combine multiple sources of information (e.g. the live recapture and dead recovery models of Barker (1997)) and to separate survival from philopatry (e.g. Henaux et al. 2007).

Closed models assume movement among a set of states with no losses or new entrants to the population. Schwarz and Ganter (1995) developed methodology for this case.

While many of the applications of the multi-state model are extensions of the CJS models and do not estimate abundance, Dupuis and Schwarz (2007) recently developed an extension of the ordinary Jolly–Seber models that can be used to also estimate abundance. The key change is that now, like in the Jolly–Seber models, the probability of the first capture of an animal must also be modeled. For example, the history vector (0, 2, 0, 2, 1, 0) now has probability (assuming only two states):

$$\left[\beta_0^1\left(1-p_1^1\right)\phi_1^1\psi_1^{12}+\beta_0^2\left(1-p_1^2\right)\phi_1^2\psi_1^{22}+\beta_1^2\right]p_2^2\times$$

$$\left[\phi_2^2\psi_2^{21}\left(1-p_3^1\right)\phi_3^1\psi_3^{12}+\phi_2^2\psi_2^{22}\left(1-p_3^2\right)\phi_3^2\psi_3^{22}\right]\times p_4^2\phi_4^2\psi_4^{21}p_5^1\chi_5^1$$

where β_t^s is the probability that an animal in the study will enter the study population between time t and $t+1$ in state s. The term before the first capture is analogous to the term after the last capture. Pradel (1996) noticed this relationship and used a reverse capture-history approach for simple CJS models to estimate seniority rates but this has not been extended to the multi-state case. McGarvey and Feenstra (2002) used an analogous approach in the multi-state approach to estimate transition rates by conditioning upon recaptures (rather than releases), but this methodology requires further development as it requires external information about natural and other sources of mortality (e.g. fishing mortality).

In all of the above cases, the state of the animal is known with certainty whenever the animal is captured. Recently, Pradel (2005) introduced the multi-event model where the state of the animal is not directly observable; rather "events" are observable whose probabilities may depend upon the underlying state. For example, an underlying state may be breeding or non-breeding while the events may be "sitting on egg", "courting", or "not observed". An event history such as (3,0,1) has probability:

$$\sum_{\text{i,j,k}\in\text{states}}\pi_1^i b_1^{i3}\phi_1^{ij}b_2^{j0}\phi_2^{jk}b_3^{k3}$$

where π_t^s is the probability of initially being in state s at time t; b_t^{so} is the probability that an animal in state s is observed in event o at time t; and ϕ_t^{st} is the probability that an animal alive at time t moves from state s to state t. The multi-event model provides a nice way to integrate uncertainty about the underlying state based on an observed event. The multi-event model can be used for a much wider diversity of ecological problems such as the mover-stayer problem (see Pradel and Kendall 2008).

In some cases, not only is the state not observable, animals are not even present in the set of states under study, e.g. temporary emigration. There are a number of ways to deal with this problem. Some animals could be tagged with a radio tag so that they can be tracked in these non-observable states. Or, auxiliary information may be available such as returns of dead animals and models similar to Barker (1997) can be fit. Lastly, a robust design can be used to estimate the temporary emigration rates (e.g. Kendall and Nichols 2002, 2004).

Finally, age-structured models are not traditionally modeled using multi-state approaches but the multi-state approach is a natural application. In this case, because the transitions are exact, there is no need to model the transition matrix at all. Pollock's age-structured Jolly–Seber model (Pollock 1981) has been structured using the multi-state approach in Dupuis and Schwarz (2007).

The use of these models is now standard practice in ecology with well developed software such as MARK (White and Burnham 1999) and M-SURGE (Choquet et al. 2005). But there are still some outstanding issues that need to be resolved.

The main issue is how to actually fit the models. Both standard likelihood and Bayesian approaches have been developed (e.g. King et al. 2006). The key problem is the huge number of parameters often found in these models. The likelihood surface is very flat, and there are many local maxima. Multiple starting points (as done in M-SURGE) are one way to check that the final solution is a global maximum. Bayesian methods seem to explore the parameter space quite well and they rarely have problems in locating the appropriate mode of the posterior but the multiple-starting point likelihood methods of M-SURGE require less computations than a single Bayesian run. Much more algorithmic work is needed on this problem.

Model selection in likelihood methods follows the well understood AIC paradigm. This facet needs more work for the Bayesian paradigm. There are a number of ad hoc methods that attempt to mimic the AIC framework such as DIC and BIC, but these have drawbacks (Gelman et al. 2004, p. 182). A natural way to incorporate model uncertainty is to allow the Bayesian models to also explore the model space via RJMCMC methods. The DIC/BIC or RJMCMC approaches both require the pre-specification of the set of models to be explored which, despite a similar requirement when using AIC, is rarely followed. Or some sort of automatic searching through "all possible models" (e.g. King et al. 2006) is required. At the moment, the use of RJMCMC methods in capture–recapture is limited to a handful of researchers and much work is needed to make these methods more accessible to the general scientist.

A secondary issue which leads to the next section is that in many cases the states are often arbitrary. It is implicitly assumed that all animals in a state have the same chance to moving to other states regardless of where located in the state. It is possible that animals could be very close geographically, but in two different states. It is becoming increasingly possible to measure exact locations of animals so that discretization of states may not be necessary.

3 Discrete Time – Continuous Space Models

While discrete time intervals often occur naturally (e.g. yearly samples), the discretization of space is rather arbitrary. At the moment, the multi-state models assume completely separate movement patterns for animals that may be in geographically similar areas, but happen to lie on either side of the arbitrary boundary.

I think it is useful to distinguish two cases. First, the movement may be linear (such as often occurs in fisheries up and down a river), or the movement may be over a two dimensional space. The linear case has an almost exact correspondence with recent work on individual time-varying covariates (such as body mass) that can only be measured when the animal is observed.

The first attempts to model individual covariates discretized the individual covariate into discrete classes (e.g. body mass classes) and used the Arnason–Schwarz multi-state approach. For example, Nichols et al. (1992) classified body mass of voles into four weight classes and used the standard multi-state models to estimate survival and "movement" as a function of the mass category. The process of discretizing a continuous covariate into classes imposes no model for the shape of response. This may be desirable when the form of the response is unknown, especially given the power to select among models is typically low and the bias due to imposing incorrect covariate response models is unknown. However, it seems logical that the response of neighbouring covariates should be similar, so spline models may be a more suitable approach. The splines impose local smoothing but allow the response model to vary globally.

Bonner and Schwarz (2006) generalized the CJS model to allow for individual time-varying covariates. For example, here is a sample capture history and covariate information:

$$\begin{array}{cccc} 1 & 0 & 1 & 0 \\ 47 & \bullet & 42 & \bullet \end{array}$$

The animal is known to be alive at time 2 (because it was recaptured after time 2), but because the animal was not captured, the value of the covariate is unknown at time 2. Similarly, the animal was not seen after time 3, so both the alive status and the covariate are unknown at the fourth occasion.

As in the multi-state case, both the transition (change in covariate to the next sampling interval) and the survival rates may be related to the covariate. Because the covariate is unknown whenever the animal is not captured, the likelihood function must integrate over all possible values of the unmeasured covariates. For example, the probability of the above history and covariate vector is:

$$\int_{c_2} \phi_1(47)\, f(c_2\,|47)(1 - p_2(c_2))\,\phi_2(c_2) f(42\,|c_2)\, dc_2 \times p_3(42)$$

$$\times \left[1 - \int_{c_4} \phi_3(42)\, f(c_4\,|42)\, p_4(c_4)\, dc_4 \right]$$

where $p_t(c_t)$ is the probability of seeing the animal at time t given the covariate value c_t; $\phi_t(c_t)$ is the probability of surviving from t to $t + 1$ given the covariate value c at t; and $f(c_t\,|c_{t-1})$ is the probability of moving to covariate c_t at time t given the animal had covariate value c_{t-1} at time $t - 1$. The likelihood is intractable for all but the simplest cases.

Bayesian methods provide a more natural way to attack these problems using MCMC methods to numerically integrate over the latent covariates. Bonner and Schwarz (2006) modeled covariate changes between sampling occasions using a normal distribution with unknown mean and variance. The survival depended upon the covariate through a logistic function. Following the standard Bayesian paradigm, the missing values (covariates and survival status) were first imputed (conditionally upon the observed data and the current parameter steps), and then the parameters were updated based on the complete-data likelihood (which has a simple form), the observed and imputed data, and the prior distributions.

These models assume a simple global model for the relationship between the covariate and survival or capture probabilities. As noted earlier, it seems reasonable that covariate values close together should have similar effects, but the general form of the relationship may vary over the range of the covariates. Bonner et al. (2008) and other papers in this conference show how to use splines to fit non-parametric models to the survival function rather than relying on a simple parametric form. A similar method can be used to model the capture rates.

Currently, very strong assumptions are made about changes to individual covariates over time. A spline model would also seem to be a reasonable way to model these changes, but I am unaware of any work in this area.

These models have a direct correspondence with mark-recapture studies on a stream. For example, salmon may be captured and tagged with radio tags as they begin their migration up the stream. Daily walks of the banks attempt to locate the salmon. The locations of detected salmon are marked. Survival and catchability are a function of location and time. Unfortunately, simple functions for the survival and catchability as a function of location are unlikely to be useful – the survival and detection curve is likely to be very non-linear.

I think these models will be difficult to fit in a two-dimensional space because of the difficulty of fitting a two-dimensional spline for both capture and survival even with a very simple movement model. Data sparsity will be an issue as well.

4 Continuous Time – Continuous Space

When dealing with large geographic extent with animals dispersed over the entire space, movement over a large spatial area is of interest rather than interchanges between discrete stock units. Sibert et al. (1999) is a typical example where the movement of tuna through the southwest Pacific Ocean is studied.

The basic idea can be summarized by a differential equation (Sibert et al. 1999):

$$\frac{\partial N}{\partial t} = \frac{\partial}{\partial x}\left(D\frac{\partial N}{\partial x}\right) + \frac{\partial}{\partial y}\left(D\frac{\partial N}{\partial y}\right) - \frac{\partial}{\partial x}(uN) - \frac{\partial}{\partial y}(vN) - ZN$$

where N is the number of tagged fish alive at a point (x, y) at time t in the study area. The first two terms represent movement of fish through diffusion in the x and y directions. The next two terms represent directed movement in the x and y directions in response to some gradient. The generic term for these types of models is advection-diffusion. The last term represents mortality from all causes. In general, the diffusion parameter, D, the directed movement parameters, u and v, and the mortality parameter, Z, vary in space and time.

For practical purposes, the physical space is first discretized to a regular spatial and temporal grid. Sibert et al. (1999) used this model to estimate movement rates of skipjack tuna (*Katsuwonus palamis*) in the southwest Pacific Ocean (Fig. 1). The entire region was subdivided into 1-degree square blocks and time was discretized into monthly intervals. Between 1977 and 1980 over 94,000 tuna were tagged

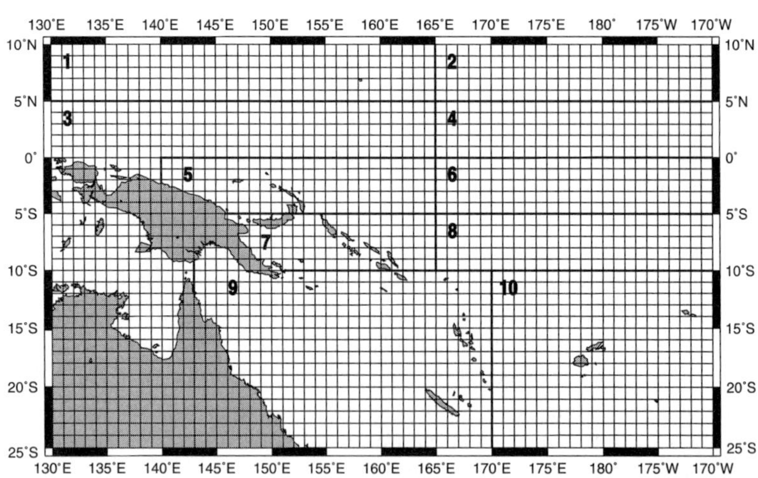

Fig. 1 Map of the southwest Pacific Ocean showing the location of the 10-region model area used for the skipjack tuna analysis. The regions are sub-divided into 1-degree blocks. Reproduced from Sibert et al. (1999, Fig. 2, p. 929)

and released, and about 5,000 tags were returned from harvested tuna throughout the grid.

The most general model has four parameters at each grid point in space and time, which is clearly too many to fit. At this point, simpler models are assumed. For example, movement parameters are equal over regions or seasons, and fishing mortality is a known function of fishing effort.

The 1-degree squares were further grouped into 10 larger regions and time was further grouped into single-, two-, or four-season blocks. Within region–season combinations, movement parameters (over the 1-degree grid) were set equal. These parameters were also held constant over years. They further assumed that natural mortality was assumed constant over time and space, that fishing mortality was a simple function of known fishing effort, and that reporting rates were 100%. Sibert et al. (1999) considered single, two, or four season models for the movement parameters with varying start points for the seasons. Figure 1 illustrates the grouping of the smaller grids into larger spatial structures – note that modeling still took place at the fine spatial scale; the larger groups of units were used to reduce the number of parameters. As an aside, I think that an alternate approach would be to implement a smoothed surface over the fine grid that would force close grids to have similar parameters.

The observed number of tag returns are related to the predicted number of tag returns using (typically) a Poisson distribution:

$$C_{ij}^t \sim \text{Poisson}\,(\tilde{C}_{ij}^t) \text{ where } \tilde{C}_{ij}^t = N_{ij}^t \left(1 - e^{-Z_{ij}^t}\right) \frac{F_{ij}^t}{Z_{ij}^t} \beta_{ij}^t$$

and

$C_{ij}^t, \tilde{C}_{ij}^t$ are the observed and expected catch at grid (i, j) at time t
N_{ij}^t is the number of tags present at grid (i, j) at time t according to the discretized solution to the differential equation;
Z_{ij}^t is the total mortality at grid (i, j) at time t;
F_{ij}^t is the fishing mortality at grid point (i, j) at time t; and
β_{ij}^t is the reporting rate at grid point (i, j) at time t.

The latter two parameters are often assumed to be known (e.g., reporting rates of 100%) or a function of known covariates (e.g., fishing mortality proportional to effort).

The likelihood function is constructed as the product of the above over all grid points in space and time. Standard numerical methods are used to maximize the likelihood function. As pointed out in Sibert et al. (1999), the discretization process needs to be carefully implemented to obtain meaningful results. I am unaware of any general software packages that implement the above models, and these must be hand constructed for each problem.

The best fitting model selected was that with two seasons (starting in March and September). It has 66 parameters – three movement parameters (diffusion D and north–south and east–west rates (u and v) for each of the 10 regions for each of two

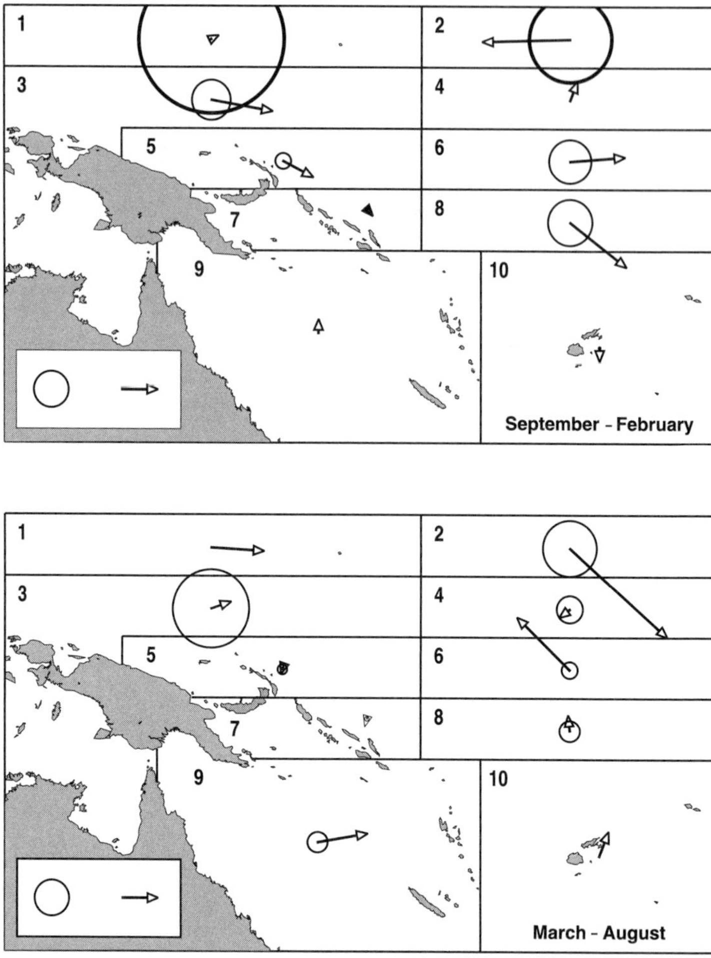

Fig. 2 Estimated movement pattern for model 2 of Sibert et al. (1999). In this model, movement differs between the two seasons (September–February in (**a**) and March–August in (**b**)). The length of the arrows is proportional to the resultant directed movement component (u, v), and the areas of the circles are proportional to the random movement component (D). The arrow in the legend inset represents directed movement of 100 Nmi·month^{-1}; the circle represents random movement of 3000 Nmi^2·month^{-1}. Reproduced from Sibert et al. (1999, Fig. 8, p. 933)

seasons plus one parameter for natural mortality plus five parameters for the catchability coefficients for the five fleets operating. Results from this model are shown in Fig. 2. There is considerable spatial variability in both the directed and random components of movement. Figure 2 also shows a consistent eastward and southward directed movement in the September–February season for five of the regions. The diffusive movements are larger for most regions in this season compared to the diffusive movements in the March–August season. Numerical estimates can be found in Sibert et al. (1999).

5 Continuous Time – Discrete States

In the previous cases, the time intervals were fixed, and movement was of interest. In some cases, the reverse is true – monitoring stations are fixed (check-points) and the time to pass the monitoring stations is of interest. Modeling is easiest when movement is essentially uni-directional (e.g. fish along a river) and interest lies in where mortality occurs in the migration, and the migration speed. Typically, a series of check-points are set up along the migration path, and animals may be detected as they pass the check-point.

Burnham et al. (1987) discuss the simplest of these models where fishes are tracked as they move downstream past a series of dams but the time taken to move between check-points is not of interest. In this case, the dams form the "sampling points", and survival between dams is of interest. These models are the exact analogue of the CJS models where states are now the sampling points, but the time taken to move between the states is not of interest (but could be measured). The analysis is identical to the standard CJS model (replacing time by check-points) with parameters p_i being the probability of detection at check-point i, and ϕ_i being the probability of survival between check-point i and $i + 1$ given that it was alive at check-point i. Despite the differences in the way time and space are handled, the analysis is identical to that of standard CJS models and so is not discussed here further.

In some cases, the path is still one-way but there are several routes from the start to the end. For example, Skalski et al. (2002) described a situation where there are multiple paths across a dam (spillway, turbine, or fish way) and the transition probabilities among these three paths is also of interest (e.g. Fig. 3). As a fish moves over the dam, it can choose among three possible routes with various probabilities. For example, the probability that a fish is released alive at R_1, detected at the Rock Island powerhouse 1, and then detected in the Wanapum pool is expressed as:

$$P(121) = S_{\text{pool}}(1 - E)(1 - G)(1 - \delta)\, p_{\text{T1}} S_{\text{T1}} \lambda$$

where the following parameters were defined:

- S_{pool} Rock Island pool survival probability;
- E probability that smolts will travel over the spillway at Rock Island Dam, i.e., spill efficiency at Rock Island Dam;
- G conditional probability of guidance to powerhouse 2, given that smolts were going to a powerhouse;
- p_{T1} powerhouse (turbine) 1 primary array detection probability;
- $p_{\text{T1}'}$ powerhouse 1 secondary array detection probability;
- p_{T2} powerhouse 2 primary array detection probability;
- $P_{\text{T2}'}$ powerhouse 2 secondary array detection probability;
- p_{s} spillway primary array detection probability;
- $p_{\text{s}'}$ spillway secondary array detection probability;
- S_{T1} powerhouse 1 survival probability at Rock Island Dam;

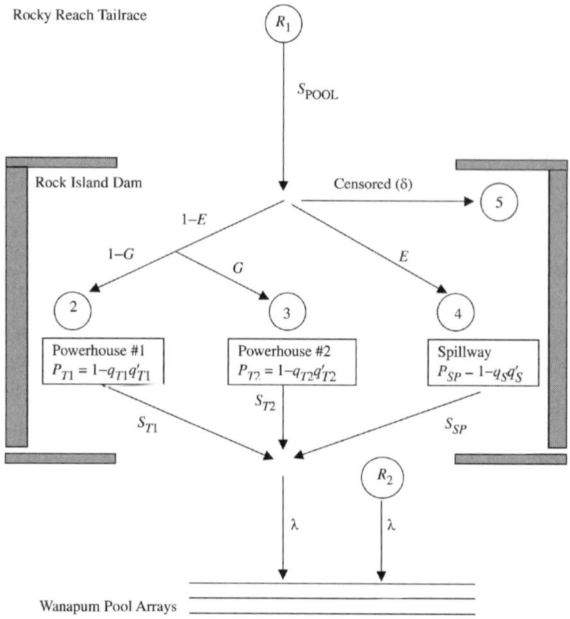

Fig. 3 Schematic example of route-specific passage and survival through the check point (the Rock Island Dam) based on releases up-stream after the previous check-point (Rocky Reach Tailrace R1) and releases downstream just after the check-point in the Rock Island Dam tailrace (R2). Parameters are defined in the text. Reproduced from Skalski et al. (2002, Fig. 2, p. 1389)

S_{T2} powerhouse 2 survival probability at Rock Island Dam;
S_{SP} spillway survival probability at Rock Island Dam;
λ joint probability of surviving and being detected at the three Wanapum pool arrays;
δ probability that a smolt is censored (i.e. lost on capture) at Rock Island Dam.

Similarly, the probability of path (141) (through the spillway) is found as:

$$P\,(141) = S_{\mathrm{pool}}\,(1-\delta)\,E p_{\mathrm{SP}} S_{\mathrm{SP}} \lambda$$

The likelihood is found as the product of the multinomial probabilities in the usual fashion.

As Skalski et al. (2002) noted, some care is needed if measurements are to be taken at too fine a scale. For example, only the products of the parameters along a path can be estimated without additional information. In their protocol, they also included multiple detection points at each of the powerhouses and spillway (similar to the robust design with multiple secondary samples at each primary period) so that information about the specific route capture probability (the p_{T1}, p_{T2}, and p_{SP}) can be determined from the secondary sampling capture histories. A standard likelihood analysis similar to the robust design gave estimates as shown in Fig. 4.

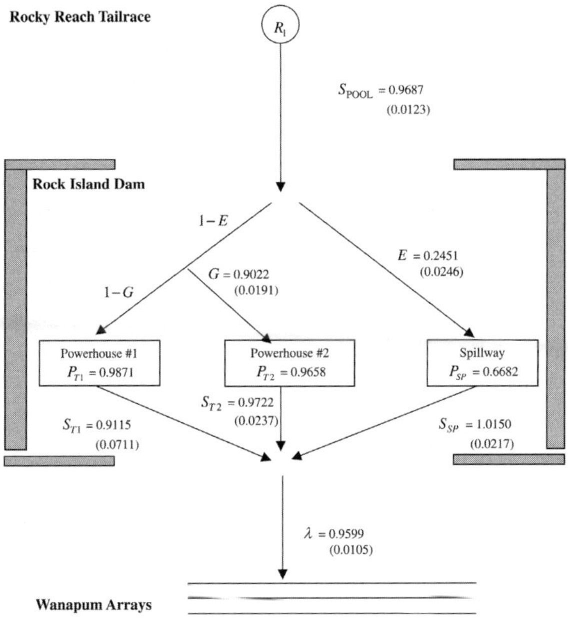

Fig. 4 Schematic representation of estimated route-specific passage and survival probabilities based on releases as outlined in Fig. 3. Reproduced from Skalski et al. (2002, Fig. 3, p. 1390)

These models have a close relationship to the multi-state models but require additional constraints placed on some of the movement parameters. For example, referring to Fig. 3, all fish must return to the common state (the tailrace) after selecting one of the three pathways past the dam, i.e. the probability of movement from power house #1 to the tailrace is 1.0.

The above models ignored the travel time between the check-points and assumed that survival between the check-points was independent of travel-time. For highly clustered groups of animals this may be appropriate, but in many migrations, differences in travel time between check-points is considerable. Hence the travel time should also be recorded for each animal as it is detected at a check-point.

Muthukumarana et al. (2008) examined the case where the timing is measured and is of interest. The context is the POST (Pacific Ocean Shelf Tracking System, http://www.postcoml.org/) where tagged fishes are released at the start of their migration, and tracking lines along the ocean floor record the tag-number and time of passage of the fish if it passes close to the detector. (Detection probabilities are still less than one.)

The typical data looks like a standard capture history but timing information is also recorded:

(1, 1, 0, 1, 0) with times (0, 6.1, •, 12.5, •)

This fish was released at $t = 0$ at location 1, took 6.1 days to travel to location 2 where it was detected, was not detected at location 3 (hence the time the fish passed the 3rd check-point is unknown and is indicated by •), was detected at location 4 at t = *12.5* days, and not detected afterwards. [This fish was released at time 0 at location 1, but fishes could be released at different times.] The fish must have passed location 3 (because it was detected at location 4), but the time of passage between locations 2 and 4 is somewhere between 6.1 and 12.5 days.

The parameters of interest are p_j – the probability of detection at location j, that is independent of when the fish arrived at location j, a survival function for the ith fish that may depend upon the inter-location travel time, e.g. $\phi_{ij} = f(t_{j+1}, t_j, \phi_j) = \phi_j^{t_{j+1}-t_j} = \phi_j^{\Delta t_j}$, and parameters ($\theta$) describing the distribution of travel times between any two check-points $g_i(\bullet|\theta)$. A common choice for the travel time distribution is an inverse Gaussian or log-normal distribution. For example, the probability of history (1, 1, 0, 1, 0) with times (0, 6.1, •, 12.5, •) can be written as:

$$g_1(6.1|\theta)\phi_1^{6.1} p_2 \times \int_{t=0}^{6.4} g_2(t|\theta)\phi_2^t (1 - p_3)\, g_3(6.4 - t|\theta)\, \phi_3^{6.4-t} dt \times p_4$$
$$\times \left(1 - \int_{t=0}^{\infty} g_4(t|\theta)\, \phi_4^t p_5 dt\right)$$

The likelihood would be written as the product of these probabilities, but is intractable because of the "missing" data (e.g. the time of passage at location 3 is unknown, but the fish was known to be alive; both the time of passage at location 5 and whether the fish is alive at time 5 are unknown). The formal likelihood would need to be integrated over all the possible times in a similar fashion as when the Arnason–Schwarz models "integrate" over all possible paths when there is movement between states but the intermediate states are unknown. However, numerical quadrature is difficult to do properly.

However, as in similar situations, Bayesian methods using MCMC methods provide a tractable solution. For convenience, think of the problem as movement of the tag past check-points to which is attached a fish that may be alive or dead. If the fish is dead, then the probability of detection at any check-point is 0.

The problem has two sets of unknown random variables – the parameters of interest, and the missing data. This missing data are the times when the tag passed the check-points when not detected and the final time of death after the last check-point where the fish was seen.

The integration is performed using MCMC methods and requires two parts in each iteration. First is the imputation of the missing data values (conditional upon the observed data and current values of the parameters). Then the parameters are updated based on the observed and imputed data and a complete data-likelihood (that has a very simple form). The sample from the posterior distribution of the parameters or missing data is found numerically.

Of more interest is that now additional questions can be answered. For example, not all fishes pass the detection lines at the same time. Are some fishes naturally slower than others or is the difference in timing purely stochastic? One way to answer this question is to ask if there is a correlation between Δ_j and Δ_k across

Table 1 Estimated posterior mean (standard deviations) for a four check-point passage model where the travel times are modeled

	Capture rate	Daily survival rate	Mean log(travel time	Covariance (upper triangle) and correlation (lower triangle) of travel times		
	p_i	ϕ_i	μ_i	Σ		
1	0.871 (0.019)	0.992 (0.000)	5.377 (0.026)	0.193(0.016)	−0.027(0.021)	−0.032(0.039)
2	0.672 (0.026)	0.974 (0.001)	4.205 (0.047)	−0.09	0.479(0.046)	0.058(0.061)
3	0.547 (0.028)	0.940 (0.002)	3.202 (0.084)	−0.07	−0.07	1.272(0.135)

fish? A positive correlation would indicate that slower (faster) fish tend to be slower (faster) across the various intervals. For example, Muthukumarana et al. (2008) assumed a multivariate normal distribution for the log-travel times:

$$\begin{bmatrix} \log(\Delta t_1) \\ \log(\Delta t_2) \\ \log(\Delta t_3) \end{bmatrix} : N\left(\begin{bmatrix} \mu_1 \\ \mu_2 \\ \mu_3 \end{bmatrix}, \begin{bmatrix} \sigma_1^2 & \rho_{12}\sigma_1\sigma_2 & \rho_{13}\sigma_1\sigma_3 \\ \rho_{12}\sigma_1\sigma_2 & \sigma_2^2 & \rho_{23}\sigma_2\sigma_3 \\ \rho_{13}\sigma_1\sigma_3 & \rho_{23}\sigma_2\sigma_3 & \sigma_3^2 \end{bmatrix} \right)$$

where ρ_{ij} refers to the correlation in travel times between intervals i and j. This again is easily modeled in the Bayesian framework. Muthukumarana et al. (2008) obtained estimates shown in Table 1. The correlations among the travel times between the check-points is very small indicating that despite wide differences in travel times among fishes, these cannot be attributed to individual fishes being "slow" or "fast" movers.

Even more complex models for travel time or survival could easily be fitted where these are a function of individual-fixed covariates (such as sex or body mass at time of release).

6 Continuous Time – Continuous Space – Known Fates

In the above examples, detection of animals was uncertain and sampling effort at the check-points is often substantial. However, modern technology will lead to very rich sets of data where it is not necessary to set up fixed check-points and essentially continuous data will be available on animal movement. For example, Argos satellite tags and archival tags with internal GPS sensors give essentially continuous data on animal positions such as seen in Fig. 5. These data are complex both in their underlying biological mechanisms and in their statistical properties.

Any pathway may consist of several different behaviors (e.g. foraging, swimming, sleeping). Is a single average movement parameter (e.g. swimming speed and direction) estimated for the entire pathway or are separate parameters estimated for

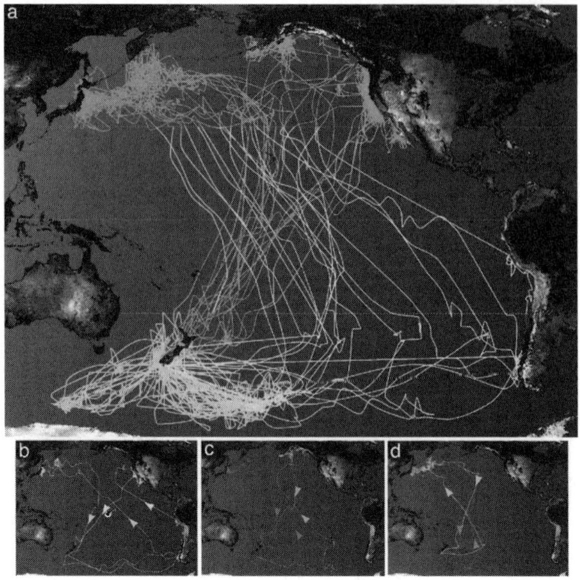

Fig. 5 Shearwater migrations originating from breeding colonies in New Zealand. Interpolated geolocation tracks of 19 sooty shearwaters during breeding and subsequent migration pathways. The 30° parallels, equator, and international dateline are indicated by dashed lines. Reproduced from Shaffer et al. (2006, Fig. 1a, p. 12800). Copyright (2006) National Academy of Sciences, USA

each behavior? In the latter case, how is it determined when one behavior ends and another begins?

Furthermore, many remote sensing devices impose complex error structures on the data that must be dealt with appropriately so that important biological variability can be separated from artificial noise. Estimation errors in position are typically non-Gaussian and vary in time; observations, while on a fine spatial scale, often occur irregularly in time.

The key difference between these types of data and classical mark-recapture models is that the concept of detection probabilities at discrete check-points or time-points is not really relevant. For all intents and purposes, this data is known-fate data where the time/location of death are known.

Several authors have proposed the use of state-space models (e.g. Newman 2000; Sibert and Fournier 2001; Brillinger et al. 2004; Buckland et al. 2004; Jonsen et al. 2005; Newman et al. 2006; Jonsen et al. 2006) for this type of data. The key feature of state-space models is that stochasticity in the underlying unobservable process is separated from uncertainty in the measurement process. This is unlike standard likelihood methods where all sources of uncertainty are integrated into one expression. Likelihood approaches will likely be intractable except for simple cases involving standard normal distribution (e.g. the Kalman filter as described by Harvey (1991)). As seen in many papers at this conference, Bayesian methodology is favored in these cases. In the Bayesian state-space methods, the unobservable (latent) process variables are first simulated, and then inference is based on a complete data-likelihood. MCMC methods integrate over the unobservable data value.

The first component of the state-space model is a description of the state process. In this case, the state process consists of the actual (unobserved) locations at regular time intervals for each animal, i.e.

$$\mathbf{x}_1, \mathbf{x}_2, \mathrm{L}, \mathbf{x}_K$$

where $\mathbf{x}_i$ is a two-dimensional location vector (latitude and longitude).

There are many different models for the state process. The simplest (and often unrealistic) is that the state process is a simple first order random walk:

$$\mathbf{x}_{t+1} = \mathbf{x}_t + \boldsymbol{\varepsilon}_t$$

where $\boldsymbol{\varepsilon}_t$ are uncorrelated random disturbances with drift (i.e. direction).

A slightly more complex (and often biologically more realistic) model uses a correlated random walk. As in many time series models, this model looks at the differences in locations rather than the actual locations, i.e. it models

$$\mathbf{d}_t = \mathbf{x}_t - \mathbf{x}_{t-1}$$

with

$$\mathbf{d}_t = \gamma \mathbf{T} \mathbf{d}_{t-1} + \boldsymbol{\varepsilon}_t^*$$

where $\mathbf{T}(\theta) = \begin{bmatrix} \cos\theta & \sin\theta \\ \sin\theta & \cos\theta \end{bmatrix}$ is a transition matrix that describes the rotational movement of the random walk (i.e. change in direction),

$$\varepsilon_t^* : N\left(\begin{bmatrix} 0 \\ 0 \end{bmatrix}, \begin{bmatrix} \sigma_{\text{lon}}^2 & \rho\sigma_{\text{lon}}\sigma_{\text{lat}} \\ \rho\sigma_{\text{lon}}\sigma_{\text{lat}} & \sigma_{\text{lat}}^2 \end{bmatrix}\right)$$

is the random walk in the difference with different variance in the latitude and longitude directions, and γ is a parameter describing the degree of autocorrelation in movement ($\gamma = 0$ corresponds to an uncorrelated random walk).

If there are several behavior modes, then a separate equation and sets of parameters can be defined for each behavior, along with a switching process with switching probabilities at each time step being described by a first order "movement" process. For example, with two behaviors, the matrix:

$$\begin{array}{c} \\ \text{mode}_1 \\ \text{mode}_2 \end{array}\begin{array}{c} \begin{array}{cc} \text{mode}_1 & \text{mode}_2 \end{array} \\ \begin{bmatrix} \psi_{11} & 1-\psi_{11} \\ 1-\psi_{22} & \psi_{22} \end{bmatrix} \end{array}$$

measures the switching rates between the two behaviours at each time step, i.e. *P* (being in mode 1 at *t* | in mode 1 at *t* – *1*) = ψ_{11}

The second part of the state-space formulation is the observational process. While the underlying state-space operates on regular time intervals, the observed locations are typically recorded at irregular points in time. A reasonable assumption is that animals move linearly between the underlying states, and so the observed location is a weighted average of the bounding state locations plus some measurement error, i.e.

$$\mathbf{y}_{t*} = (1-\alpha)\,\mathbf{x}_{t-1} + \alpha\mathbf{x}_t + \varepsilon^{obs}_{t*}$$

where *t** lies in some interval *(t – 1,t)*, $t^* = (t-1)+\alpha$, and ε^{obs}_{t*} is the observational error.

One challenge with satellite data is that large observational errors are quite common. Consequently, many researchers use a scaled *t*-distribution with small *df* to model the observational errors where both the scaling factor and the *df* are estimated from the data as well.

Jonsen et al. (2005) fit this model for individual animals. For example, Fig. 6 shows the fitted line for two behaviors (foraging and swimming) with estimates presented in Table 2 based on satellite tags from a hooded seal. There was good evidence of a separation in parameters between the two behaviors (foraging and migrating) and data points that were clearly erroneous were handled without the need for arbitrary pre-processing of the dataset.

Jonsen et al. (2006) extended these types of models in a natural way to multiple animals (turtles) using a hierarchical Bayesian framework. Now the parameters for individual animals come from a hyper-distribution on these parameters. There is an enormous amount of information that is estimated from such models at various levels. For example, consider Fig. 7. This displays the distribution of travel rates for individual animals and the distribution of the population of the individual parameters in three modes of behavior.

For the present, these models have been used for long-lived animals. However, it is not difficult to see how survival rates could be added to these models in a similar way as done by Muthukumarana et al. (2008). Similarly, these models currently ignore the physical environment, but again, extension where parameters vary depending on local typography (e.g. turning angles dependent upon closeness to mountains) are easily implemented.

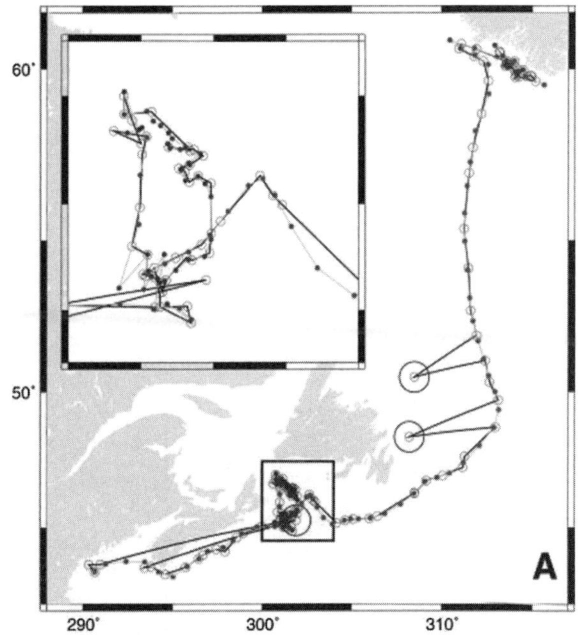

Fig. 6 Plot of hooded seal track data, with observed locations as open circles and state estimates from the DCRWS (first-difference correlated random walk switching) model as filled circles (migrating behavior) or open diamonds (foraging behavior). Erroneous detections are indicated with large circles. The black line is the straight-line path between observations and the gray line is the straight-line path between state estimates. Reproduced from Jonsen et al. (2005, Fig. 1A, p. 2876)

Table 2 Posterior medians and 95% credible intervals for parameters from a two behaviour mode, correlated random walk on the differences fit to Argos satellite data from a hooded seal. Extracted from Table 1 of Jonsen et al. (2005)

Parameter	Lower	Median	Upper
Mean turning angle θ – migrating (radians)	–0.07	0.04	0.16
Mean turning angle θ – foraging (radians)	–3.13	–2.70	3.13
Autocorrelation parameter γ – migrating	0.71	0.83	0.94
Autocorrelation parameter γ – foraging	0.07	0.62	0.94
Pr(stay in migration mode)	0.45	0.85	0.96
Pr(stay in foraging mode)	0.80	0.91	0.97
σ_{lat} (degrees)	0.04	0.06	0.09
σ_{long} (degrees)	0.09	0.12	0.20

7 Discussion

The design and analysis of the discrete-time discrete-space migration models is very mature, but the story is not finished. The theory for the analysis of these models is well known. Likelihood based methods are well developed and work well for small sets of models. Future work will likely tend more and more towards the use of

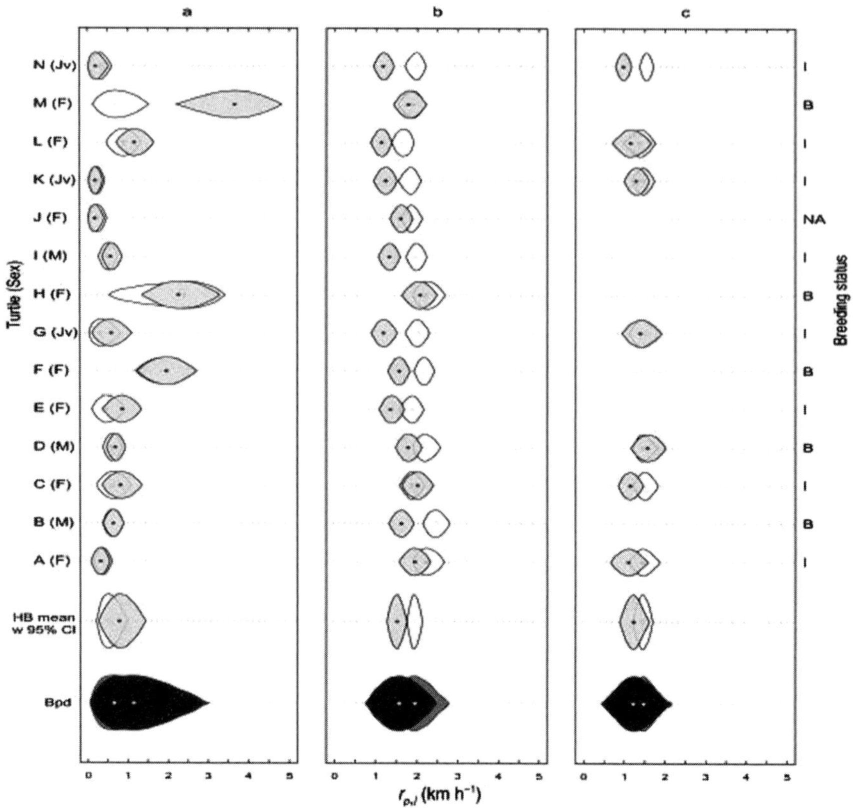

Fig. 7 Raindrop plots (Barrowman and Myers 2003) summarizing the 95% credible intervals of the posterior densities of the absolute day (*white*) and night (*grey*) travel rates of leatherback turtles during the predominant phases of their migratory cycle: (**a**) northern foraging (**b**) southward migration, and (**c**) northward migration. The point within each raindrop indicates the posterior mode. The top and bottom symmetric curves show the log(density) for the highest 95% predictive density. The width of the raindrop plot reflects variation in the parameter value while the height reflects probability density. A perfect normal distribution would have a (reflected) quadratic shape

The sex (M = male, F = female, Jv = juvenile when sex cannot be easily determined) and breeding status (B = breeding, I = interbreeding, NA = not known) of each turtle are displayed on panels (a) and (c), respectively.

The taller raindrops in the bottom two rows of the panels are hierarchical summaries, and are interpreted differently from the raindrops for the individual turtles. The "mean w 95% CI" raindrops are the estimated hierarchical mean travel rate with 95% credible intervals. The last row represents the Bayes predictive distributions (Bpd) for individual travel rates and are shaded-in darker (day = *grey*, night = *black*) to emphasize that they are different from the mean. Reproduced from Jonsen et al. (2006, Fig. 4, p. 1052).

Bayesian methods because of the intractability of the likelihood when movement rates depend upon underlying unknown states. The key problems for the future for Bayesian methods are model selection – RJMCMC methods (e.g. King et al. 2006) hold great promise for an "automated" model selection, but fitting these models is currently limited to a very small subset of researchers! Unfortunately, no amount of statistical wizardry can salvage a study with too few releases and recoveries – much more effort needs to be placed into planning these studies – there is a need for easy to use planning tool based on the lines of Xiao (1996) or Devineau et al. (2006).

The discrete time – continuous space models require substantially more work. The key problem is that space is very large and most datasets are far too sparse to fit parameters at very fine spatial scales.

In many cases, particularly with certain species of fishes, migration routes are very linear (along a river or along the continental shelf). Large scale detection arrays will soon provide large amount of data on migration speed. Of particular interest will be the relationship of migration choices (e.g. routes) or speeds upon individual covariates.

Finally, the continuous time – continuous space models will require work on fitting smooth surfaces rather than arbitrarily discretizing these models to larger scales.

All of these models work well when the states and time intervals are very coarse. Researchers are pushing towards finer and finer scales in both dimensions. In these cases, classical models where rates are either completely time dependent or completely constant over time simply are not reasonable. Random effect models are a step towards intermediate models, but these still implicitly assume some constancy over time. Parameters that are close in time or space should be similar; those further apart should be different. One way to attack this is through the use of autocorrelated random effects (e.g. along the lines of Johnson and Hoeting (2003)) or automated smoothing methods analogous to the spline methods (e.g. along the lines of Gimenez et al. (2006); or papers presented at this conference) where the data drive the complexity of the variation in the parameters over time or space.

To make predictions about the future is always dangerous – particularly if your predictions can be retrospectively examined! What do I see in my cloudy crystal ball?

My boldest prediction is that using mark-recapture with $p < 1$ to monitor migrations will be obsolete in 15 years (just in time for me to retire!).

What do I mean by this? The greatest challenge with mark-recapture is that detection probabilities are less than 1 so that when an animal is not captured, there is no information about its current state. However, technology is rapidly advancing. A special issue of *Science* (Sugden and Pennisi 2006) reported on some cutting edge technology. For example, 300 mg transmitters attached to dragonflies that allowed tracking for 12 days; the worldwide POST system will enable thousands of smolt to be tracked as they migrate along the continental shelves. With relatively modest investments in detection equipment, these signals are essentially always detected.

What are the challenges with this deluge of data? The statistical analysis of remotely sensed movement data is complicated by high serial correlation within individuals; missing data caused by equipment malfunction; irregularly spaced but finely grained data in time; complexity in the underlying behavior processes; and estimation errors that are (currently) often non-normal and ill-defined. When large collections of animals are marked, the individual parameters must be collated together.

These types of data have many similarities to standard longitudinal data for which there is a vast literature. For example, hidden Markov models (Cappé et al. 2007) are essentially multi-event models with 100% detectability; standard longitudinal survival analysis (e.g. Fitzmaurice et al. 2004) can be used for CJS models when the probability of detection is certain (i.e. $p = 1$). However, many of the longitudinal models assume regularly spaced collection points and simple covariance structures, none of which are likely to hold for these complex datasets.

Classical likelihood methods are unlikely to be practical in these contexts because of the need to integrate over many latent variables. State-space models that separate the underlying real process from the observed data seem like a natural way to proceed (Buckland et al. 2004; Newman et al. 2006).

Simple models will likely be suitable to model correlations within individuals, e.g. modeling first or second differences. Extensive experience with standard time series data have shown that simple ARMA(1,1) applied at a suitable degree of differencing can give rise to a very rich set of possible curves (Nelson 1973).

Rather than trying to regularize the collected data to force it to match a regular time-step, the approaches outlined above where the observed data are related to an underlying regular framework will likely make modeling easier. For example, modeling the distance moved on regular time intervals is easily done using log-normal (or related) distributions, but becomes much harder when the time-intervals vary within an animal.

The underlying behaviors are latent data that are rarely directly observable. However, in many cases, the underlying parameters describing movement are different between the behavior modes and MCMC methods provide a natural way to model these latent variables.

The state-space model avoids the need to remove "bad" observations by using a recursive procedure where predictions of the unobserved states are made via a process model and these predictions are refined by combining them with current observations via Bayes' rule.

Finally, Bayesian hierarchical models provide a natural way to model different levels of data from individuals, to families, to larger aggregations. By taking a hierarchical Bayes approach to modeling migratory travel rates, it is assumed that individual animals exhibit some degree of similarity in their migratory behavior. The hierarchical model is the formal representation of this assumption, and allows the estimation of not only the parameters for the individual data sets but also those at a higher level(s) in the model. These hierarchical models use information from all the data to estimate parameters at each of the levels in the hierarchy and this has the effect of reducing uncertainty in parameter estimates at the individual data set level. Hierarchical Bayes models allow quantification of all the uncertainty at all levels of the hierarchy.

What are the current stumbling blocks? First, software. While WinBUGS is a very powerful fitting tool, it still requires a high level of expertise to use with very detailed coding required. What is needed is a front-end that would take a sequence of model statements and create the underlying WinBUGS code.

Second, fitting many models is "relatively" easy, but discerning among models is still difficult. A Bayesian analogue of Akaike's information criterion (AIC), deviance information criterion (DIC) does not work with many models because the underlying likelihood is not log-concave. This situation results in negative estimates of the effective number of parameters (pD) which is clearly invalid.

Third, model fits are wonderful, but it is dangerous to become enamored with models without formal model checking. Many people underestimate the power of graphical displays to assess model adequacy – more work is needed here particularly with Bayesian models where simply looking at the numbers is not at all helpful.

References

Arnason AN (1972) Parameter estimation from mark-recapture experiments on two populations subject to migration and death. Researches in Population Ecology 13:97–113.

Arnason AN (1973) The estimation of population size, migration rates, and survival in a stratified population. Researches in Population Ecology 15:1–8.

Barker RJ (1997) Joint modeling of live-recapture, tag-resight, and tag-recovery data. Biometrics 53:657–668.

Barker RJ, White GC (2004) Towards the mother-of-all-models: customised construction of the mark-recapture likelihood function. Animal Biodiversity and Conservation 27:177–185.

Barrowman NJ, Myers RA (2003) Raindrop plots: a new way to display collections of likelihoods and distributions. The American Statistician 57: 268–274.

Bonner SJ, Schwarz CJ (2006) An extension of the Cormack–Jolly–Seber model for continuous covariates with application to *Microtus pennsylvanicus*. Biometrics 62:142–149.

Bonner SJ, Thomson DL, Schwarz CJ (2008) Time-varying covariates and semi-parametric regression in capture–recapture: an adaptive spline approach. In: Thomson DL, Cooch EG, Conroy MJ (eds.) Modeling Demographic Processes in Marked Populations. Environmental and Ecological Statistics, Springer, New York, Vol. 3, pp. 657–675.

Brillinger DR, Preisler HK, Ager AA, Kie JG (2004) An exploratory data analysis (EDA) of the paths of moving animals. Journal of Statistical Planning and Inference 122:43–63.

Brownie C, Anderson DR, Burnham KP, Robson DS (1985) Statistical inference from band recovery data: a handbook. United States Fish and Wildlife Service Resource Publication Number 15i

Brownie C, Hines JE, Nichols JD, Pollock KH, Hestbeck JB (1993) Capture–recapture studies for multiple strata including non-Markovian transition probabilities. Biometrics 49:1173–1187

Buckland ST, Newman KB, Thomas L, Koesters NB (2004) State-space models for the dynamics of wild animal populations. Ecological Modeling 171:157–175.

Burnham KP, Anderson DR, White GC, Brownie C, Pollock KH (1987) Design and analysis methods for fish survival experiments based on release-recapture. American Fisheries Society Monograph 5.

Cappé O, Moulines E, Rydén T (2007) Inference in Hidden Markov Models. Springer, New York.

Chapman DG, Junge CO (1956) The estimation of the size of a stratified animal population. Annals of Mathematical Statistics 27:375–389.

Choquet RR, Reboulet AM, Pradel R, Gimenez O, Lebreton JD (2005) M-SURGE : new software specially designed for multistate capture-recapture models. Animal Biodiversity and Conservation 27:207–215.

Cormack RM (1964) Estimates of survival from the sighting of marked animals. Biometrika 51:429–438.

Darroch JN (1961) The two-sample capture–recapture census when tagging and sampling are stratified. Biometrika 48:241–260.

Devineau O, Choquet R, Lebreton JD (2006) Planning capture–recapture studies: straightforward precision, bias, and power calculations. Wildlife Society Bulletin 34:1028–1035.

Dupuis JA, Schwarz CJ (2007) The stratified Jolly-Seber model. Biometrics 63:1015–1022. Published article online: 14-May-2007. doi: 10.1111/j.1541-0420.2007.00815.x

Fitzmaurice G, Laird N, Ware J (2004) Applied Longitudinal Analysis. Wiley-Interscience, New York.

Gelman A, Carlin JB, Stern HS, Rubin DB (2004) Bayesian Data Analysis. 2 nd edn. Chapman & Hall/CRC, Boca Raton, FL.

Gimenez O, Crainiceanu C, Barbraud C, Jenouvrier S, Morgan BJT (2006) Semiparametric regression in capture–recapture modeling. Biometrics 62:691–698.

Harvey AC (1991) Forecasting, Structural Time Series Models and the Kalman Filter. Cambridge University Press, Cambridge.

Henaux V, Bregnballe T, Lebreton JD (2007) Dispersal and recruitment during population growth in a colonial bird, the great cormorant *Phalacrocorax carbo sinensis*. Journal of Avian Biology 38:44–57.

Hestbeck JB, Nichols JD, Malecki RA (1991) Estimates of movement and site-fidelity using mark-resight data of wintering Canada geese. Ecology 72:523–533.

Hilborn R (1990) Determination of fish movement patterns from tag-recoveries using maximum likelihood estimators. Canadian Journal of Fisheries and Aquatic Sciences 47:635–643.

Johnson DS, Hoeting JA (2003) Autoregressive models for capture–recapture data: a Bayesian approach. Biometrics 59:341–350.

Jolly GM (1965) Explicit estimates from capture-recapture data with both death and immigration – Stochastic model. Biometrika 52:225–247.

Jonsen ID, Flemming JM, Myers RA (2005) Robust state-space modeling of animal movement data. Ecology 86:2874–2880.

Jonsen ID, Myers RA, James MC (2006) Robust hierarchical state-space models reveal diel variation in travel rates of migrating leatherback turtles. Journal of Animal Ecology 75: 1046–1057.

Kendall WL, Nichols JD (2002) Estimating state-transition probabilities for unobservable states using capture–recapture/resighting data. Ecology 83:3276–3284.

Kendall WL, Nichols JD (2004) Modern statistical methods for the study of dispersal and movement of marked birds. Condor 106:720–731.

King R, Brooks SP, Morgan BJT, Coulson T (2006) Factors influencing Soay sheep survival: a bayesian analysis. Biometrics 62:211–220.

Lebreton JD, Burnham KP, Clobert J, Anderson DR (1992) Modeling survival and testing biological hypotheses using marked animals. A unified approach with case studies. Ecological Monographs 62:67–118.

Lebreton JD, Pradel R (2002) Multi-stratum recapture models: modeling incomplete individual histories. Journal of Applied Statistics 29:353–369.

McGarvey R, Feenstra JE (2002) Estimating rates of fish movement from tag recoveries: conditioning by recapture. Canadian Journal of Fisheries and Aquatic Sciences 59:1054–1064.

Muthukumarana S, Schwarz CJ, Swartz TB (2008) Bayesian analysis of mark-recapture data with travel-time-dependent survival probabilities. Canadian Journal of Statistics 36:5–28.

Nelson CR (1973) Applied time series analysis for managerial forecasting. Holden-Day, San Francisco.

Newman KB (2000) Hierarchic modeling of salmon harvest and migration. Journal of Agricultural Biological and Environmental Statistics 5:430–455

Newman KB, Buckland ST, Lindley ST, Thomas L, Fernández C (2006) Hidden process models for animal population dynamics. Ecological Applications 16:74–86

Nichols JD, Sauer JR, Pollock KH, Hestbeck JB (1992) Estimating transition probabilities for stage based population projection matrices using capture recapture data. Ecology 73:306–312.

Plante N, Rivest LP, Tremblay G (1998) Stratified capture–recapture estimation of the size of a closed population. Biometrics 54:47–60.

Pollock KH (1981) Capture–recapture models allowing for age-dependent survival and capture rates. Biometrics 37:521–529.

Pradel R (1996) Utilization of capture-mark-recapture for the study of recruitment and population growth rates. Biometrics 52:371–377.

Pradel R (2005) Multievent: an extension of multistate capture–recapture models to uncertain states. Biometrics 61:442–447.

Schaefer MB (1951) Estimation of the size of animal populations by marking experiments. US Fish and Wildlife Service Fisheries Bulletin 69:191–203.

Schwarz CJ, Ganter B (1995) Estimating the movement among staging areas of the barnacle goose (*Branta leucopsis*). Journal of Applied Statistics 22:711–725.

Schwarz CJ, Schweigert J, Arnason AN (1993) Estimating migration rates using tag recovery data. Biometrics 49:177–194.

Schwarz CJ, Taylor CG (1998) The use of the stratified-Petersen estimator in fisheries management with an illustration of estimating the number of pink salmon (*Oncorhynchus gorbuscha*) that return to spawn in the Fraser River. Canadian Journal of Fisheries and Aquatic Sciences 55:281–296.

Seber GAF (1965) A note on the multiple recapture census. Biometrika 52:249–259.

Shaffer SA, Tremblay Y, Weimerskirch H, Scott D, Thompson DR, Sagar PM, Moller H, Taylor GA, Foley DG, Block BA, Costa DP (2006) Migratory shearwaters integrate oceanic resources across the Pacific Ocean in an endless summer. Proceedings of the National Academy of Sciences 103:12799–12802

Sibert J, Fournier DA (2001) Possible Models for Combining Tracking Data with Conventional Tagging Data. In: Sibert J, Nielsen J (eds) Electronic Tagging and Tracking in Marine Fisheries Reviews: Methods and Technologies in Fish Biology and Fisheries. Kluwer Academic Press, Dordrecht, pp 443–456.

Sibert JR, Hampton J, Fournier DAA, Bills PJ (1999) An advection–diffusion-reaction model for the estimation of fish movement parameters from tagging data, with application to skipjack tuna (*Katsuwonus pelamis*). Canadian Journal of Fisheries and Aquatic Science 56:925–938.

Skalski JR, Townsend R, Lady J, Giorgi AE, Stevenson JR, McDonald RD (2002) Estimating route-specific passage and survival probabilities at a hydroelectric project from smolt radiotelemetry studies. Canadian Journal of Fisheries and Aquatic Sciences 59:1385–1393.

Sugden A, Pennisi E (2006) When to go, where to stop: introduction to special issue of science on migration and movement. Science 313:775–775.

White GC, Burnham KP (1999) Program MARK: survival estimation from populations of marked animals. Bird Study 46 (suppl):s120–s139.

Williams BK, Nichols JD, Conroy MJ (2002) Analysis and Management of Animal Populations. Academic Press, New York.

Xiao Y (1996) A framework for evaluating experimental designs for estimating rates of fish movement from tag recoveries. Canadian Journal of Fisheries and Aquatic Sciences 53:1272–1280.

Stopover Duration Analysis with Departure Probability Dependent on Unknown Time Since Arrival

Shirley Pledger, Murray Efford, Kenneth Pollock, Jaime Collazo, and James Lyons

Abstract In stopover duration analysis for migratory birds, models with the probability of departure dependent upon time since arrival are useful if the birds are stopping over to replenish body fat. In capture–recapture studies, the exact time of arrival is not generally known, as a bird may not be captured soon after arrival, or it may not be captured at all. We present models which allow for the uncertain knowledge of arrival time, while providing estimates of the total number of birds stopping over, and the distribution and mean of true stopover times for the population.

Keywords Age-related survival · Capture–recapture · Jolly–Seber model · Mark-recapture · migratory birds · residence time · Schwarz–Arnason model · Stopover duration · Survival curve

1 Introduction

Many migratory bird populations stop over at predictable sites en route, replenishing body reserves before flying on. The total number of birds using the staging site is important when studying the population (see e.g. Routledge et al. 1999; Fredericksen et al. 2001; Ydenberg et al. 2004), and detection of trends in population size uses comparisons of these totals over the years. Individual residence time at stopover sites is also an important variable in the biology of migratory birds for at least three reasons. First, if there is turnover of the population during staging, with some birds leaving before others have arrived, a snapshot estimate of the number of birds will underestimate the total throughput. Second, individual residence times, together with rate of refueling, shape overall migration strategies (Ålerstam and Lindström 1990). Migrating birds typically spend more time and energy at stopover sites than aloft (Wikelski et al. 2003); the total time spent on

S. Pledger (✉)
School of Mathematics, Statistics and Computer Science, Victoria University of Wellington, P.O.Box 600, Wellington 6140, New Zealand
e-mail: shirley.pledger@vuw.ac.nz

D.L. Thomson et al. (eds.), *Modeling Demographic Processes in Marked Populations*, Environmental and Ecological Statistics 3, DOI 10.1007/978-0-387-78151-8_15,

stopovers and the number of stopover sites visited largely determine the spatiotemporal course of migration. Effective conservation and management of migratory birds therefore depends on a fundamental understanding of stopover behaviour, including time spent resting and refuelling. Finally, stopover duration is a critical component in models of optimal bird migration (Ålerstam and Lindström 1990; Ålerstam and Hedenström 1998). Testing models of bird migration and reducing parameter uncertainty requires accurate estimates of stopover duration.

In stopover duration analysis, the key problem is estimating residence time before first capture and after last encounter (recapture or resighting). If birds are individually marked and uniquely identified at a stopover site, frequent sampling during the stopover provides a record of dates when caught (a capture history) for each bird which was caught or seen at least once. A capture–recapture analysis may then be used, with "age" meaning residence time (time since arrival). A Cormack–Jolly–Seber (CJS) model (Cormack 1964; Jolly 1965; Seber 1965) provides estimates of the probabilities of "survival" (retention at the site) from one sample to the next, and the probability of recapture at each sample. In the basic CJS model, these probabilities are assumed to depend on time (sample). This model was extended to multiple age classes by Pollock (1981). Lebreton et al. (1992) produced a comprehensive framework of likelihood models based on the CJS. They extended the basic CJS model to allow for multiple groups and covariates, and allowed survival and/or capture probabilities to be constant, to depend on time, known age and/or group (e.g. sex or site). They also introduced the idea of selecting from a wide class of models using Akaike's Information Criterion (AIC, Akaike 1973). The basic CJS model has been used to estimate stopover duration (Kaiser 1995; Dinsmore and Collazo 2001; Rice et al. 2007). In these studies, the estimated daily probability of retention was used in the life expectancy formula of Seber (1982), stopover duration $= -1/\log_e$(daily retention probability). The life expectancy method is not unbiased however because the CJS model is conditional on first capture. If the "age" (time since arrival) is known at the time of first capture, CJS models provide estimates of "age-related survival", where the probability of retention at a particular sample is assumed to be related to the duration of stopover so far. If, however, exact arrival times are unknown, assuming each newly caught bird has just arrived biases the estimates of the parameters of interest. It is necessary to estimate how long the bird was in residence before its first capture. Schaub et al. (2001) used Pradel's (1996) recruitment parameters to get an overall estimate of stopover time, but see Efford (2005) and Pradel et al. (2005) for the limitations of this model.

We present new capture–recapture models which use information from each individual capture history to estimate the arrival times, and hence provide estimates of retention probabilities (which are dependent on time since arrival). The Jolly–Seber (JS) model (Jolly 1965; Seber 1965) provides estimates of the population size at each sample, and from these an estimate of the total number of birds stopping over may be obtained. Schwarz and Arnason (Schwarz and Arnason 1996; Schwarz 2001) provided a fully likelihood-based variant of the JS model, which we call

JSSA. This makes available maximum likelihood estimates of all the parameters, likelihood ratio tests and model comparisons based on AIC. The JSSA model also directly estimates N, the total number of birds stopping over, which is an advantage for our application. The likelihood framework gives profile likelihood intervals for N and other parameters (Cormack 1992).

This paper extends the JSSA model, providing a new collection of models in which capture and retention probabilities may depend on the residence time so far, even if arrival times are unknown. The models may also be applied to true births and deaths in populations with no migration, giving estimates of frailty and senescence.

Section 2 sets out the assumptions and notation of these models, Section 3 describes models in discrete time, and Section 4 introduces retention curves in continuous time. Statistical methods are in Section 5, Section 6 illustrates the models with real data, and Section 7 reports a simulation study. Evaluation and discussion are in Section 8.

2 Assumptions and Notation

2.1 Assumptions

Assumptions 1–6 are those of the JSSA model, but interpreting "birth" as arrival, "death" as departure, "survival" as retention at the site, "age" as time since arrival or residence time, and "lifetime" as stopover duration. We assume there are no actual births or deaths during the study.

1. K samples are taken at intervals which are large in relation to the time needed for the sample, so that samples may be regarded as instantaneous.
2. Arrivals and departures occur between samples, and departure is assumed to be permanent.
3. Each individual bird is uniquely and correctly identified.
4. There is a superpopulation of N birds, each available for capture on at least one sampling occasion.
5. Proportions β_0, β_1, ... β_{K-1} of the N birds enter the population and are first available for capture at times 1, 2, ... K respectively ($\sum \beta_j = 1$).
6. Capture and departure events are independent between birds and between samples, and the birds are independent in their arrival times.
7. Sampling covers all the time when birds are present.

The extra assumption 7 prevents boundary effects from biasing estimates. Early birds arriving long before the first sample and late lingerers after the last sample would have their stopover durations underestimated.

The discussion to follow also assumes the samples are equally spaced in time, although unequal intervals may be modelled by adjusting all retention probabilities to a standard time unit.

2.2 Data

The data come in the form of a $D \times K$ capture matrix **X**, where D is the number of distinct birds caught. Element x_{ij} is 1 if bird i is captured on occasion j, otherwise $x_{ij} = 0$. Row i of **X** is the **capture history** (CH_i) for bird i, $i = 1,2,\ldots,D$.

Thus there are $N - D$ birds never caught, each with capture history **0**, a K-vector of zeros. N is an unknown parameter.

2.3 Parameters

We use the JSSA N and β_j parameters, while the retention and capture parameters are extended to allow dependence on time since arrival as well as calendar time (sample). A bird which arrived a time units ago and is present at sample j is assumed to have probability p_{ja} of capture at sample j and ϕ_{ja} of retention from sample j to $j + 1$. These simplify to ϕ_j and p_j (the JSSA models) if there is no dependence on residence time, to ϕ_a and p_a if there is no calendar time dependence, and to ϕ and p if the probabilities are constant over both residence time and calendar time.

3 Models in Discrete Time

3.1 Capture Histories and Their Likelihoods

We now develop likelihoods for open population models which allow both time since arrival and calendar time to affect the capture and retention probabilities of the birds. Modelling individual capture histories in the JSSA framework enables us to allow for different possible arrival and departure times for each bird via random variables $\mathcal{B}$ and $\mathcal{D}$. If bird i is first present and available for capture at sample b_i, and is last available for capture at sample d_i before departure, we denote its **presence history** PH_i by $\{b_i, d_i\}$. The ordered pair (b_i, d_i) is an unobserved realisation of the joint distribution of $\mathcal{B}, \mathcal{D}$. If retention is related to residence time, $\mathcal{B}$ and $\mathcal{D}$ are correlated. Suppose a bird has capture history CH_i with first capture at sample f_i and the last at ℓ_i ($1 \le b_i \le f_i \le \ell_i \le d_i \le K$). Then the probability of this capture history, conditional on this presence history, is (omitting the i subscripts on b and d)

$$\text{P}(\text{CH}_i \mid \text{PH}_i = \{b, d\}) = \prod_{j=b}^{d} p_{ja}^{x_{ij}} (1 - p_{ja})^{1-x_{ij}} \tag{1}$$

where a is the time from arrival to sample j, e.g. if time is measured in days, $a = j - b + 1$ assuming "age" 1 day at arrival time. However, the JSSA model provides the probability of this presence history, allowing retention to depend on time since arrival:

$$\mathrm{P}\left(\mathrm{PH}_i = \{b, d\}\right) = \mathrm{P}\left(\mathcal{B} = b \text{ and } \mathcal{D} = d\right) = \beta_{b-1}\left(\prod_{j=b}^{d-1} \phi_{ja}\right)(1 - \phi_{da}) \quad (2)$$

where, if b=d, the empty product $\prod_{j=b}^{d-1}$ is taken to be unity.

Hence the unconditional probability of bird i's capture history with unknown arrival and departure times is found by summing $\mathrm{P}\left(\mathrm{CH}_i \mid \{b, d\}\right) \times \mathrm{P}\left(\{b, d\}\right)$ over all possible presence histories, using equations 1 and 2:

$$\mathrm{P}(\mathrm{CH}_i) = \sum_{b=1}^{f_i}\sum_{d=\ell_i}^{K}\left\{\beta_{b-1}\left(\prod_{j=b}^{d-1}\phi_{ja}\right)(1-\phi_{da})\left(\prod_{j=b}^{d} p_{ja}^{x_{ij}}(1-p_{ja})^{1-x_{ij}}\right)\right\}. \quad (3)$$

Similar reasoning gives the unconditional probability of no captures, denoted by CH_0:

$$\mathrm{P}\left(\mathrm{CH}_0\right) = \sum_{b=1}^{K}\sum_{d=b}^{K}\left\{\beta_{b-1}\left(\prod_{j=b}^{d-1}\phi_{ja}\right)(1-\phi_{da})\left(\prod_{j=b}^{d}(1-p_{ja})\right)\right\}. \quad (4)$$

The summation over possible departure times in equations 3 and 4 is an extension of the use of the parameter χ_j = probability not seen after sample j in the JS model. The JSSA model's entry parameters, β_j, allow us to do a similar summation over the entry times before the first sighting.

Let h index the different observed capture histories, with n_h being the number of birds with capture history h, and write the parameters N, β, p and ϕ as a parameter vector θ. Then a multinomial model to allocate the N birds to their capture histories gives the full likelihood of θ given the capture matrix $\mathbf{X}$ as

$$L(\theta \mid \mathbf{X}) = \frac{N!}{\Pi_h n_h!\,(N-D)!} \times \left(\prod_{i=1}^{D} L_i\right) \times L_0^{N-D} \quad (5)$$

where L_i = $\mathrm{P}(\mathrm{CH}_i)$ (equation 3) and L_0 = $\mathrm{P}(\mathrm{CH}_0)$ (equation 4). If the a subscripts are dropped from equation 5, some algebra reduces the formula to the full likelihood for the JSSA model.

3.2 Linear Logistic Models for Retention and Capture Probabilities

The full model above may be labelled $\{\beta(t), \phi(t \times a), p(t \times a)\}$, to indicate βs dependent on time (sample), while ϕ and p both allow for time and age effects in an interactive way. However, there is not enough information in capture–recapture data to estimate the interactive parameters, and so we propose simplifications of ϕ and p with fewer parameters. An **additive** or **main effects** model on the logistic scale (which is preferable to a raw scale for normality of estimators and avoidance of ϕ or p estimates outside [0,1]) could have ϕ_{ja} reparameterised as

$$\log\left(\frac{\phi_{ja}}{1-\phi_{ja}}\right) = \tau_j + \alpha_a \quad (a, j = 1, 2, \ldots (K-1)) \tag{6}$$

where τ is the time effect and α is the age effect. A constraint is needed on α, perhaps a sum-to-zero constraint ($\sum \alpha_a = 0$) or a corner-point constraint ($\alpha_1 = 0$). Similarly capture probabilities may be modelled with additive time and age effects on the logistic scale:

$$\log\left(\frac{p_{ja}}{1-p_{ja}}\right) = \nu_j + \zeta_a \quad (a, j = 1, 2, \ldots (K-1)) \tag{7}$$

Suitable notation for labelling these additive models would be $\phi(t+a)$ and $p(t+a)$.

Further simplifications could have

- $\phi(a)$ or $p(a)$, with probabilities depending on residence time but constant through calendar time,
- $\phi(t)$ or $p(t)$, with probabilities independent of residence time but varying through calendar time, or
- ϕ or p constant over changing residence time and calendar time, denoted by $\phi(.)$ and $p(.)$ in the model specification.

If all 2^K observable capture histories are actually seen, the models above are feasible, except for some minor parameter redundancy in early or late p or ϕ parameters (Catchpole and Morgan 1997). If no individual birds have a presence history with $f_i = 1$ and $l_i = K$, there is virtually no information about parameters beyond a certain maximum observed duration of stay ($M = \max\ (l_i - f_i + 1)$), and there is near-singularity of models (Catchpole et al. 2001); in this case parameters from "age" $M + 1$ onwards are not estimated. Also sparse data, with few different capture histories observed, can give substantial parameter redundancy. Schwarz and Arnason (1996) suggested various options for dealing with the two redundant parameters in the Jolly–Seber model $\{\beta(t), \phi(t), p(t)\}$, including setting $p_1 = p_K = 1$. However, as this gives an underestimate of β_0, leading on to overestimates of the later βs (Jim Nichols, pers. comm.), we have set such end parameters to the mean of the estimable ones, on a logistic scale. For example, our JS model has $\text{logit}(p_1) = \text{logit}(p_K) = \text{mean}\{\text{logit}(p_2), \ldots, \text{logit}(p_{K-1})\}$. Table 1 shows the numbers of estimable parameters in the models proposed so far.

Table 1 Numbers of independent parameters with K samples, an observed maximum stopover duration of $M \leq K$, and a large enough set of different capture histories to make the estimates feasible

	$p(.)$	$p(t)$	$p(a)$	$p(t+a)$
$\phi(.)$	$K+2$	$2K$ *	$K+M-1$	$2K+M$
$\phi(t)$	$2K$	$3K-3$ *†	$2K+M-2$ *	$3K+M-4$ *†
$\phi(a)$	$K+M$	$2K+M-2$ *	$K+2M-2$ *†	$2K+2M-4$ *†
$\phi(t+a)$	$2K+M-2$	$3K+M-5$ *†	$2K+2M-4$ *	$5K-9$ *†

All models assume $\beta(t)$. Some parameters are not estimable. Models marked * must have p_1 or its logistic equivalent assigned, and models marked † must have p_K or its logistic equivalent assigned.

3.3 Using Covariates

Covariates in calendar time may be incorporated into this scheme, as shown in Lebreton et al. (1992). For example, a time effect in the probability of capture due to weather or varying search effort could be accounted for by modifying equation 7 to

$$\log\frac{p_{ja}}{1-p_{ja}} = \zeta_a + \lambda x_j + \delta w_j$$

where x_j is search effort and w_j is a relevant weather covariate at sample j. The parameters λ and δ are logistic regression coefficients. An example for retention parameters could use a measure of weather between samples (w_j = weather between samples $j-1$ and j) as a covariate. One such modification of equation 6 is

$$\log\frac{\phi_{ja}}{1-\phi_{ja}} = \alpha_a + \delta_a w_j.$$

The different slopes (δ_a rather than δ) provide for a differential effect of severe weather conditions on retention – perhaps birds which arrived more recently are more likely to delay departure if the weather is poor.

3.4 Comparing Different Groups of Birds

The data may come from two or more populations which are separated spatially, temporally, taxonomically or sexually. These groups are modelled as in Lebreton et al. (1992) but using the full likelihoods of Sections 3.1 and 3.2. The joint likelihood is formed as the product of the individual likelihoods for each group. Comparison between groups of retention and/or capture probabilities is effected by starting with a global model allowing each group its own parameters, and then fitting submodels with various constraints on parameters. For example, we could compare the residence time-related retention probabilities of different groups, while still allowing each group its own N, β and p parameters. The constrained model would have the same retention parameters across the groups. These groups could be females and males, or populations at different locations, or different sub-species. With migratory birds, we may look for changes over the years in the total number stopping over, by comparing a constrained model with equal N over different years with a model having N fluctuating, or on a linear trend over time.

4 Retention Curves in Continuous Time

Traditional survival curve analysis and lifetime modelling provides a method of smoothing the discrete retention probabilities into a retention curve depending on the time since arrival, using fewer parameters.

Suppose a continuous random variable X for the stopover duration of an individual has distribution function $F(x)$ = probability of departure by age x, and probability density function $f(x) = \frac{d\,F(x)}{dx}$ $(x > 0)$. The retention function, probability of retention for at least x time units, is $S(x) = 1 - F(x)$, and the hazard function (instantaneous departure rate) is $h(x) = \frac{f(x)}{S(x)}$.

The retention probabilities from discrete-time capture–recapture may be constrained to lie on such curves, using

$$\phi_x = \mathrm{P}(\text{duration} > x + 1 \mid \text{duration} > x) = \frac{S(x+1)}{S(x)}. \tag{8}$$

Standard survivorship (retention) curves of Types I, II and III (see, e.g. Richter and Söndgerath 1990) may be modelled with a Weibull distribution for the stopover duration random variable X $(X > 0)$. The distribution function $F(x) = 1 - \exp\left\{-\left(\frac{x}{\gamma}\right)^{\kappa}\right\}$ has scale parameter $\gamma > 0$ and shape parameter $\kappa > 0$, and gives retention function $S(x) = 1 - F(x) = \exp\left\{-\left(\frac{x}{\gamma}\right)^{\kappa}\right\}$ and hazard rate (instantaneous departure rate) $\mu = \frac{f(x)}{S(x)} = \left(\frac{\kappa x^{\kappa-1}}{\gamma^{\kappa}}\right)$. The value of κ gives the type of retention curve, with $\kappa > 1$ for a Type I retention curve (high retention rate until near the end of the stopover, then high departure rate, a J-shaped hazard curve), $\hat{\kappa} = 1$ for Type II retention (constant hazard rate), and $\kappa < 1$ for Type III (lowest retention soon after arrival, a reverse J-shaped hazard curve). The case $\kappa = 1$ with an exponential retention curve and constant hazard rate $1/\gamma$ is implicit in all models, such as the JS, where departure probability is assumed to be unrelated to time since arrival. If the κ estimate from the data gives a rejection of H_0: $\kappa = 0$ in favour of H_A: $\kappa > 1$, there is evidence for high retention soon after arrival and lower retention later.

The connection with discrete time data is $\phi_a = \exp\left\{-\left(\frac{a+1}{\gamma}\right)^{\kappa} + \left(\frac{a}{\gamma}\right)^{\kappa}\right\}$ where ϕ_a is the probability of retention from "age" a to $a + 1$. The assumed arrival time for a bird first available for capture at sample b is midway between samples $b - 1$ and b, and for those present at the first sample it is the time of the first sample minus half the average interval between samples.

Type IV retention has highest departure rates when either recently arrived or after staying a while with high retention, a "bathtub" shaped hazard curve (Richter and Söndgerath 1990). One example is the 6-parameter Siler curve (Siler 1979), which also allows for a trend in retention over time to be tested. The retention probability from "age" a (at time t) to a +1 is

$$\phi_{at} = \exp\left[\frac{\beta_1}{\gamma_1}\left\{e^{-\gamma_1(a+1)} - e^{-\gamma_1 a}\right\} - \beta_2 a - \frac{\beta_3}{\gamma_3}\left\{e^{-\gamma_3(a+1)} - e^{-\gamma_3 a}\right\} - \beta_4 t\right]. \quad (9)$$

If the data select a bathtub curve in preference to a Weibull, there is evidence for two types of bird, transients which depart soon after arrival, and stayers with high retention before ultimate departure.

The Weibull curve may also be adapted for time trends by allowing the shape parameter to vary by calendar time, $\gamma = \gamma_0 + \gamma_1 t$. A significantly non-zero γ_1 could show if, say, later arrivals spend less time at the site.

5 Statistical Analysis

5.1 Model Comparison and Parameter Estimation

Model selection among these likelihood-based models may be done using Akaike's Information Criterion (AIC) or some variant of that (Lebreton et al. 1992; Burnham and Anderson 2002). For confirmatory studies, all the models have the usual likelihood ratio tests (χ^2 tests) available for comparing two models or for testing parameters. Maximum likelihood estimates of parameters arise from the model fitting, with estimated standard errors available from the inverse of the estimated Hessian matrix.

Following Lebreton et al. (1992), the models are fitted using the logits of the retention and capture probabilities. This gives better convergence properties, and more appropriate confidence intervals. The greater normality of the estimators on the logit scale means the associated symmetric confidence intervals ($\pm$ 1.96 standard errors) are realistic. The centres and endpoints of the logit confidence intervals are back-transformed to the [0,1] scale to give asymmetric confidence intervals for the original probabilities. Similar advantages result from using log(N) as a parameter in the optimisation, with back-transformation giving an asymmetric confidence interval for N.

Profile likelihood intervals (PLI) are also strongly recommended for interval estimation with these models (Cormack 1992). They also provide the asymmetric intervals appropriate to the data.

5.2 Stopover Duration Estimation

For specific models, estimates of mean stopover duration have been used in the past. If a model with ϕ constant has been selected, mean stopover duration may be estimated by

$$-\frac{1}{\log_e \phi}$$

(Seber 1982). If a fitted stopover duration curve (e.g. Weibull or Siler) was selected, the mean of that distribution may be used. If a discrete-time model with ϕ_a was chosen, with the final $\phi_K = 0$ (valid if sampling continued to the final departures), a probability branching diagram gives estimates of the mean and standard deviation of stopover duration.

However, for any of the models in this paper, the parameter estimates and equation 2 provide estimates of the joint distribution of arrival and departure times, $P(\mathcal{B} = b \text{ and } \mathcal{D} = d)$. To obtain the (discrete) derived distribution of stopover duration $\mathcal{S} = \mathcal{D} - \mathcal{B} + 1$, the probabilites of histories with a common duration are summed. With unequal spacing of samples the support of the distribution has irregular spacing, but fitting a retention curve in continuous time will give a distribution, mean and variance for the stopover duration.

6 Real Data Example

At the Cabo Rojo salt flats in Puerto Rico, 113 previously-banded semipalmated sandpipers *(Calidris pusilla)* were sighted over 18 weeks in 1992–3. These data represent overwintering residency rather than a short stopover.

Analysis of resightings by Rice et al. (2007) using CJS models (Lebreton et al. 1992) selected as the best models $\phi(\text{fat})p(t)$ (with a covariate of body fat, $\Delta AIC_c = 0$), and $\phi(.)p(t)$ ($\Delta AIC_c = 0.66$).

Our models differ by including estimation of arrival time, and by using first sighting information more fully. The model selected by AIC was $\{\beta(t), \phi(t + a, Weibull), p(a)\}$ (Table 2), where a Weibull model for retention has its scale parameter on a linear trend over time. This allows for later cohorts to be on a longer or shorter stopover duration, while keeping the shape of the curve constant.

In this example, the parameter estimates indicate that the later arrivals are staying longer. This model choice is being driven by a number of early arrivals being seen only once, while a large group arriving about the middle of the study were seen frequently until the end.

The distribution of stopover time was found for the best four models, giving the means and standard deviations shown in Table 3.

However, this real data set has high capture probabilities, around 0.8 per sample, leading to an estimate of N, $\hat{N} = 113 = D$, which is the number actually seen. This

Table 2 Relative AIC values for semipalmated sandpipers *(Calidris pusilla)* at Cabo Rojo

	$p(.)$	$p(t)$	$p(a)$	$p(t+a)$
$\phi(.)$	13.90	19.25	10.39	9.34
$\phi(t)$	15.62	21.79	9.43	9.96
$\phi(a)$	21.52	22.94	17.84	15.11
$\phi(t+a)$	20.43	24.85	14.53	14.62
$\phi(a, Weibull)$	15.69	20.66	11.98	9.31
$\phi(a, Siler)$	21.58	26.36	18.02	16.15
$\phi(t+a, Weibull)$	10.99	20.26	0.00	9.99
$\phi(t+a, Siler)$	20.43	24.92	15.28	14.17

Table 3 Means and standard deviations of estimated distributions of stopover times (weeks) for Cabo Rojo semipalmated sandpipers *(Calidris pusilla)*, using the best four models

Model	ΔAIC	Mean Duration	Standard Deviation
$\phi(t + a, Weibull), p(a)$	0.00	10.09	3.50
$\phi(a, Weibull), p(t + a)$	9.31	9.97	3.43
$\phi(c), p(t + a)$	9.34	10.04	3.59
$\phi(t), p(a)$	9.43	10.33	3.69

data set is not providing a good test of the value of these models for estimation of N. It is also not really necessary to distinguish true arrival time from time of first capture, as most birds were seen very soon after arrival. The simplest model, $\phi(.)p(.)$, gave stopover time estimates fairly similar to those from the four models above.

Because of the differences in the analyses, detailed comparisons with the model of Rice et al. (2007) are not meaningful.

7 A Simulation Study

A simulation study was run to evaluate the model selection procedure, estimation of N and estimation of stopover duration. Three scenaria were tried, encompassing low and high K values (either 5 or 10) and different patterns of entry probabilities, with details given in Table 4. All simulations used $n = 200$ birds and the generating model $\{\phi(a)p(.)\}$ with constant capture probability 0.4 and retention probabilities 0.9, 0.8, 0.2, 0.1, 0. This gave high retention for two intervals, followed by low retention. No birds were retained for more than four intervals (five samples).

At each replication, a population was simulated to give a capture matrix of observed birds, which was then analysed with all 16 discrete-time models. A model fit was ruled inadmissible if any parameter estimate was at the boundary of the parameter space, which happened sometimes with sparse generated data. Table 4 gives an overview of model selection and estimation of N from the simulations.

In all three scenaria, the generating model was selected by AIC more times than any other, with improvement of the proportion of times selected as K increased. Overall, the best two models were the generating model $\{\phi(a), p(.)\}$ and the model $\{\phi(.), p(a)\}$. Strong correlations between ϕ estimates and adjacent p estimates introduce "leakage" of parameters (see e.g. Sidhu et al. 2007); if we insist on constant ϕ, the failure to observe long-staying birds is attributed instead to capture probabilities, and the $p(a)$ estimates become zero from a certain "age" onwards. In this case, common sense would dictate that birds do not suddenly become uncatchable when they have stayed for a certain time – a far more reasonable explanation is that they have departed. Models with constant ϕ or ϕ dependent on time only are unrealistic

Table 4 Simulation Study Results. The generating model was $\phi(a)$, $p(.)\}$, and in each replication the 16 models were fitted. All scenaria had the same retention probabilities $\phi(a) = (0.9, 0.8, 0.2, 0.1, 0)$, and constant capture probability 0.4. Entry probabilities β were (.3,.2,.2,.2,.1) for Scenario A, (.15,.15,.1,.1,.1,.1,.1,.1,.05,.05) for Scenario B, and (.05,.1,.1,.15,.15,.15,.1,.1,.05,.05) for Scenario C. Coverage is from a 95% log-based confidence interval for n, back-transformed

Scenario	A	B	C
Number of samples, K	5	10	10
Number of birds, N	200	200	200
Entry pattern (details in caption)	low–high	low–low	high–low
Retention probabilities	.9 .8 .2 .1 0	.9 .8 .2 .1, 0	.9 .8 .2 .1, 0
Capture probability	0.4	0.4	0.4
Number of replications	100	100	100
Model selection results:			
% reps, correct model 1st choice	33	45	38
% reps, correct model 2nd choice	38	30	36
% reps, correct model 3rd choice	24	16	22
Analysis by correct model:			
Coverage for N	0.95	0.94	0.93
Average % bias of $\hat{N}$	−0.07%	−2.71%	−1.58%

for stopover duration analysis. If the unrealistic model $\{\phi(.), p(a)\}$ is excluded, the selection of $\{\phi(a), p(.)\}$ improves considerably.

The coverage of the nominal 95% confidence intervals for n (log based and back-transformed) was acceptable with analysis by the true (generating) model (Table 4), and was almost always nearer 0.95 than coverage from the other 15 analysing models.

To evaluate the estimation of stopover duration, we compare the true distribution of duration (from the generating model) with the estimated distribution from the analysing models, averaged over the 100 simulated populations. The comparisons of true and estimated stopover duration distributions are shown for Scenario B ($K = 10$) in Fig. 1. The probabilities from fitting the correct model $\{\phi(a), p(.)\}$ are much closer to the true probabilities than those from analysis by three competing models $\{\phi(.), p(.)\}$, $\{\phi(.), p(a)\}$ and $\{\phi(a), p(a)\}$. The simulated populations were also analysed using the correct $\{\phi(a), p(.)\}$ model but with a simplified likelihood model using only the birds seen at least once; no attempt was made to estimate N or allow for the unseen birds in the likelihood. Arrival and departure times were estimated only for the seen birds, and ϕ and p parameters were estimated from this incomplete data set. The estimated stopover duration probabilities were calculated, averaged over the 100 simulations, and the trace of this "seen" model is also shown in Fig. 1. The resulting overestimation of stopover duration probably occurs because the unseen birds are largely those with short stopovers, so estimation without allowing for them gives a positive bias.

The means and standard deviations of these generating and analysing distributions are in Table 5. Note the 23% overestimation of the mean stopover duration which results from considering only the seen birds.

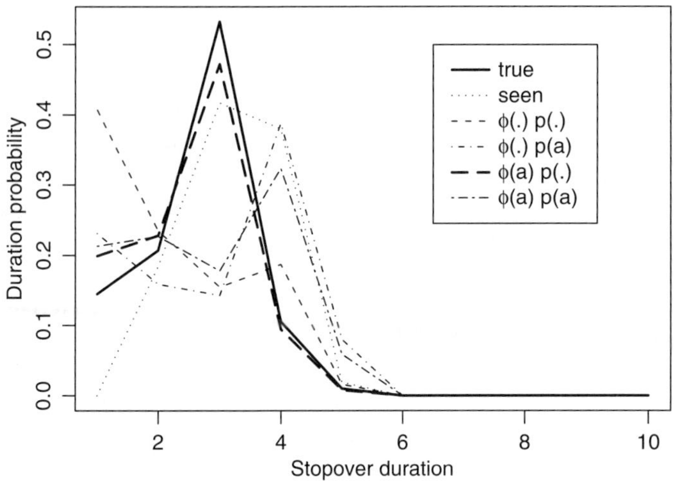

Fig. 1 True and estimated distributions of stopover duration, Scenario B. The true (generating) model is $\{\phi(a), p(.)\}$. Five analysing models were fitted for each simulated population, and the probabilities estimated from averages over the 100 simulations. Analysis by the true model gives a close fit. The "seen" model ignored the unseen birds

Table 5 Means and standard deviations of the true stopover duration distribution and the estimated distributions

Model	Mean	Standard deviation
True model		
$\{\phi(a), p(.)\}$	2.63	0.79
Analysing model:		
Seen birds only	3.23	0.59
$\{\phi(.), p(.)\}$	2.17	1.43
$\{\phi(.), p(a)\}$	2.93	1.78
$\{\phi(a), p(.)\}$	2.49	0.88
$\{\phi(a), p(a)\}$	2.79	1.59

8 Discussion

We have introduced new models, aimed at improving accuracy of estimation of the total number of birds using a stopover site, and the duration of stopover. Likelihood-based models are employed, bringing a range of benefits: AIC comparisons, likelihood ratio tests, and the estimation of the distribution of stopover duration rather than just a mean and standard deviation. The use of joint likelihoods allows for comparisons of different groups of birds, perhaps two sexes, different species, or the same species over different years. Tests may be constructed to see if, for example, there is a trend over the years of the numbers stopping over.

In the survival literature, it is known that individual heterogeneity of survival, if not allowed for in a model, has consequences for the estimation of survival parameters

(Burnham and Rexstad 1993; Pledger and Schwarz 2002; Efford 2005). In stopover applications, the birds with intrinsically longer stopover times provide the information about the upper end of the retention curve. Since they are not representative of the whole group, an increasing instantaneous departure rate for the whole population can be masked. Our models allow for heterogeneity of retention via an "age" effect, using the (unknown) true time since arrival rather than the time since first capture.

It is necessary to sample over the whole time of the stopover, to eliminate boundary effects. If the sampling starts late, and some birds have already been present a long time before a first opportunity for capture, their arrival times are underestimated. Similarly if sampling finishes too soon, birds still present at the last sample will be assumed to depart soon, when in fact they may stay much longer.

The models are performing well, as shown by a simulation study with substantial turnover in the population. Compared with existing models (with constant or time-dependent retention and capture probabilities), the new models allowing for retention to depend on residence time give a much more accurate estimation of distribution, mean and variance of stopover duration.

These models may be extended in various ways. Allowance can be made for unequal spacing of samples, using smoothing or "lifetime" curves in continuous time. Adaptations are possible to allow for some occasions which have resighting only, with no attempt to capture new birds. Also, the capture–recapture data may be combined via likelihoods with count data of unmarked birds, using joint multinomial models for the capture–recapture and Poisson models for the counts. Joint likelihoods also allow the inclusion of extra information such as some birds having known arrival and departure times, perhaps from radiotelemetry information.

These models also apply to population dynamic studies, where age and survival have their true meanings, and are not interpreted as residence time and retention. With studies which are long in relation to the lifetime of the animal, and where there is little or no migration, the distribution of lifetimes may be estimated and the detection of senescence in animals of unknown age is possible.

Acknowledgments We thank Carl Schwarz, Neil Arnason and Jim Nichols for useful discussions of the Schwarz and Arnason model, and Hal Caswell for Siler curve information.

References

Akaike H (1973) Information theory as an extension of the maximum likelihood principle. In: Petrov BN, Csaki F (eds) Second International Symposium on Information Theory. Akademiai Kiado, Budapest, Hungary.

Ålerstam T, Hedenström A (1998) The development of bird migration theory. Journal of Avian Biology 29:343–369.

Ålerstam T, Lindström, Å (1990) Optimal bird migration: the relative importance of time, energy, and safety. Pages 331–351 in Gwinner E. (eds) Bird Migration: The Physiology and Ecophysiology. Springer, Berlin, Germany.

Burnham KP, Anderson DR (2002) Model Selection and Multimodel Inference: A Practical Information-Theoretic Approach. Springer-Verlag, Berlin, Heidelberg, New York.

Burnham KP, Rexstad EA (1993) Modeling heterogeneity in survival rates of banded waterfowl. Biometrics 49:1194–1208.

Catchpole EA, Morgan BJT (1997) Detecting parameter redundancy. Biometrika 84:187–196.

Catchpole EA, Kgosi PM, Morgan BJT (2001) On the near-singularity of models for animal recovery data. Biometrics 57:720–726.

Cormack RM (1964) Estimates of survival from the sighting of marked animals. Biometrika 51:429–438.

Cormack RM (1992) Interval estimation for mark-recapture studies of closed populations. Biometrics 48:567–576.

Dinsmore SJ, Collazo JA (2001) The influence of body condition on local apparent survival of spring migrant sanderlings in coastal North Carolina. Condor 105:465–473.

Efford MG (2005) Migrating birds stop over longer than usually thought: Comment. Ecology 86:3415–3418.

Fredericksen M, Fox A, Madsen J, Colhoun K (2001) Estimating the total number of birds using a staging site. Journal of Wildlife Management 65:282–289.

Jolly GM (1965) Explicit estimates from capture-recapture data with both death and immigration – stochastic model. Biometrika 52:225–247.

Kaiser A (1995) Estimating turnover, movements and capture parameters of resting passerines in standardized capture–recapture studies. Journal of Applied Statistics 22:1039–1047.

Lebreton J-D, Burnham KP, Clobert J, Anderson DR (1992) Modelling survival and testing biological hypotheses using marked animals: A unified approach with case studies. Ecological Monographs 62:67–118.

Pledger S, Schwarz CJ (2002) Modelling heterogeneity of survival in band recovery data using mixtures. Journal of Applied Statistics 29:315–327.

Pollock KH (1981) Capture–recapture models allowing for age-dependent survival and capture rates. Biometrics 37:521–529.

Pradel R (1996) Utilization of capture-mark-recapture for the study of recruitment and population growth. Biometrics 52:703–709.

Pradel R, Schaub M, Jenni L, Lebreton J-D (2005) Migrating birds stop over longer than usually thought: Reply. Ecology 86:3418–3419.

Rice SM, Collazo JA, Alldredge MW, Harrington BA, Lewis, AR (2007) Local annual survival and seasonal residency rates of semipalmated sandpipers (Calidris pusilla) in Puerto Rico. The Auk 124:1397–1406 .

Richter O, Söndgerath, D (1990) Parameter Estimation in Ecology. VCH Verlagsgesellschaft mbH, Weinham, Germany.

Routledge RD, Smith GEJ, Sun L, Dawe N, Nygren E, Sedinger JS (1999) Estimating the size of a transient population. Biometrics 55:224–230.

Schaub M, Pradel R, Jenni L, Lebreton J-D (2001) Migrating birds stop over longer than usually thought: An improved capture–recapture analysis. Ecology 82:852–859.

Schwarz CJ (2001) The Jolly–Seber model: more than just abundance. Journal of Agricultural, Biological and Environmental Statistics 6:195–205.

Schwarz CJ, Arnason AN (1996) A general methodology for the analysis of capture–recapture experiments in open populations. Biometrics 52:860–873.

Seber GAF (1965) A note on the multiple-recapture census. Biometrika 52:249–259.

Seber GAF (1982) The Estimation of Animal Abundance and Related Parameters. Second edition. Macmillan, New York, USA.

Sidhu LA, Catchpole EA, Dann P (2007) Mark-recapture-recovery modeling and age-related survival in Little Penguins *(Eudyptula minor)*. The Auk 124:815–827.

Siler W (1979) A competing-risk model for animal mortality. Ecology 60:750–757.

Wikelski M, Tarlow EM, Raim A, Diehl RH, Larkin RP, Visser GH (2003) Costs of migration in free-flying song birds. Nature 423:704.

Ydenberg RC, Butler RW, Lank DB, Smith BD, Ireland J (2004) Western sandpipers have altered migration tactics as peregrine falcon populations have recovered. Proceedings of the Royal Society of London B 271:1263–1269.

Habitat Selection, Age-Specific Recruitment and Reproductive Success in a Long-Lived Seabird

Lise M. Aubry, Emmanuelle Cam, and Jean-Yves Monnat

Abstract Delayed recruitment (i.e. first reproduction) is a key feature of the demography of long-lived species such as seabirds. If physiological, behavioral, and environmental factors are thought to influence age at first breeding, knowledge of the fitness prospects corresponding to different recruitment tactics is needed to get insight into the evolution of delayed recruitment.

Because the age at which an individual recruits may depend on the location chosen to breed, we first investigated the relationship between habitat quality and age of first breeding in a long-lived seabird, the black-legged the Kittiwake (*Rissa tridactyla*). We used multi-state mark-recapture approaches to model the transition from non-breeding to breeding status as a function of age and habitat quality. We also investigated whether there was a relationship between age at recruitment and reproductive success in the year of recruitment. We assessed several non-exclusive hypotheses. (i) If experience plays a part in reproductive success *per se* (e.g. in the quality of parental care), or in acquisition of higher-quality breeding sites (i.e. increased competitive ability), then reproductive success should be lower for early recruits (i.e. age 3) than others. (ii) In the same vein, if delayed recruitment corresponds to a queuing tactic allowing access to higher-quality sites, then late recruits (age 6 or 7) should exhibit higher breeding success than others. Alternatively, delayed recruitment may reflect behavioral inability to access to higher-quality sites, in this case, late recruits should exhibit poorer breeding success than younger ones. (iii) Experience combined with social constraints may lead to an initial increase in breeding success with recruitment age, and a decrease in older recruits.

We found that recruitment probability was highest at intermediate ages (i.e. 5–6 years old), and that recruitment probability was maximal in habitat patches (i.e. 'cliffs') of medium quality. This may reflect harsh competition in the most productive cliffs, and avoidance of the least productive ones (i.e. where predation on eggs is high). In accordance with our predictions (i and iii), we found that the youngest

---

L.M. Aubry (✉)
Max Planck Institute for Demographic Research, Konrad-Zuse Str. 1, D-18057 Rostock, Germany
e-mail: Aubry@demogr.mpg.de

D.L. Thomson et al. (eds.), *Modeling Demographic Processes in Marked Populations*, Environmental and Ecological Statistics 3, DOI 10.1007/978-0-387-78151-8_16,

recruits experienced poor breeding success at the beginning of their reproductive life, and that breeding success was higher for birds recruiting at intermediate age. In addition, recruitment probability was best predicted by apparent habitat quality the year preceding recruitment. The latter result suggests either that habitat selection takes place the year preceding settlement and first reproduction, or that the information available to individuals at the beginning of a season is temporally autocorrelated to past productivity.

Reproductive choices and/or the constraints met during the pre-reproductive stage of life may influence age at recruitment. Our results show that there is a relationship between age of first breeding and breeding success probability. However, age of first breeding may also have substantial effects on breeding success over life. Future study should examine if reproductive success improves, shows senescent decline, or remains the same over the life course of individuals recruiting at various ages.

Keywords Age-specific recruitment · Black-legged the Kittiwake · Capture-Mark-Recapture · Habitat selection · Habitat quality · Multi-state modeling · Breeding success

1 Introduction

Age of first breeding (i.e. recruitment) in vertebrates is determined in part by age at sexual maturity, a constraint limiting flexibility in the minimum age at recruitment. However, even within the same population, a wide range of ages at first breeding is observed in many vertebrates, and many seabirds delay first reproduction well beyond physiological maturity. Accordingly, the pre-breeding segment of the population constitutes a significant part of the population. It is important to understand factors influencing the timing of recruitment to the breeding population, as the age of first breeding may have a significant impact on population dynamics and fitness (Caswell and Hastings 1980; Stearns 1992; Charlesworth 1994).

A consistent prediction from models based on life history trade-offs is that early reproduction should be favoured by natural selection (Stearns 1992; Charlesworth 1994), except under specific circumstances (i.e. population decline or fluctuating juvenile survival, Charlesworth 1994). Thus, unless one underestimates the importance and evolutionary consequences of temporal variation in juvenile survival, one might expect delayed reproduction to be rare in the wild. Interestingly, empirical observations do not always support this prediction (e.g. in birds: Viallefont et al. 1995; Pradel et al. 1997; Cooch et al. 1999; Lebreton et al. 2003). In a habitat selection framework, delaying recruitment has been suggested to allow individuals to gather information about potential breeding patches before recruitment (Boulinier and Danchin 1997). Reproductive delay might also be beneficial in terms of fitness if reproductive success increases with age, experience (i.e. skill enhancement through learning), or both (Charlesworth 1994), as long as the survival costs associated to such a delay do not exceed its benefits. Furthermore, delayed reproduction can be adaptive (i.e. 'bet-hedging strategy') in environments where reproduction is uncertain (Tuljapurkar 1990).

In the black-legged Kittiwake (*Rissa tridactyla*), a cliff-nesting seabird, reproduction can begin at age two (Cam et al. 2002b, 2003), but the bulk of recruitment takes place between 3 and 6 years of age (Danchin et al. 1991; Cam et al. 2005), and some individuals delay recruitment even longer (e.g., 15 years; Cadiou 1993). In addition to physiological constraints, a certain level of behavioural maturity is required to complete reproduction successfully (Danchin 1987a; Porter 1988). Behavioral maturity is part of the general complex of 'increasing reproductive ability with age and experience' proposed by Charlesworth (1994), and may explain delayed age of first breeding. A certain level of maturity (e.g. competitive abilities in males to gain ownership on a nest-site), that may require a relatively long learning process, is therefore mandatory before reproduction can begin (Nur 1984; Pickering 1989; Monnat et al. 1990; Danchin et al. 1998).

Environmental conditions may interact with an individual's intrinsic quality and result in a variety of recruitment tactics. Energetic constraints on reproduction are likely to depend on resource acquisition, which is determined by both resource availability (i.e. a feature of the environment) and the individual's ability to harvest them (i.e. intrinsic quality). Only individuals of high quality may be able to recruit early if resource limitation occurs (e.g. in the lesser snow goose, Viallefont et al. 1995; in the blue petrel, Barbraud and Weimerskirch 2005).

Within this framework of constraints setting limits to variation in the age at first reproduction, individuals still have 'a decision to make'. For instance, age at first reproduction may directly depend on habitat selection tactics based on optimization of expected fitness (Fretwell and Lucas 1970). Potential recruits may decide either to breed or to wait until the next breeding season based on the quality of potential breeding locations in a given year (Boulinier and Danchin 1997). For example, severe predation events on eggs or chicks in the colonies attended by pre-breeders in a given year (Cam et al. 2004a) may lead some of them to postpone recruitment until the following breeding season. In this view, habitat selection may be one of the main components of the 'selective environment' of the age at recruitment. The decisions of where and when to start breeding may actually be 'two sides of the same coin' as Ens et al. emphasized (1995), and delayed recruitment might be the outcome of a specific habitat selection strategy involving:

(1) Information gathering, in order to identify potential habitats and assess habitat quality (e.g., assessment of conspecific reproductive success and predation pressure in different colonies over time, Cadiou 1993; Boulinier and Lemel 1996; Boulinier and Danchin 1997; Danchin et al. 1998);
(2) Gaining 'ownership' on a new site via competition or by queuing for an already occupied site to become available (see Wiley and Rabenold 1984; Ens et al. 1995; Cam et al. 2002b).

Obviously, constraints related to the acquisition of a nest-site may influence the timing of recruitment, especially in colonial cliff-nesting seabirds where competition amongst individuals to acquire a nest-site is strong, and may in turn delay accession to the breeding population. In the study-population, nest-site density is stronger in higher-quality patches (i.e. cliffs) than in low-quality ones. Density is part of the environmental features that may influence individual age at recruitment. However, long-term observations (1979–2007) have shown that individuals can always 'create' new nest-sites in higher-quality patches if they have the competitive abilities to do so; thus, none of the patches are saturated in the study area. The common observation of individuals competing for specific nest-sites that are already occupied (Cadiou 1993) in higher-quality patches contradicts the idea that density dependence is the main factor influencing settlement decisions. Indeed, these individuals might be more interested in queuing for already occupied nest-sites, or in evicting previous owners of occupied sites, as in both cases they can directly observe how much this site is 'worth' (i.e. based on their conspecific's reproductive success on this particular site). The pays-offs of such behaviour might overtake the benefits associated with the creation of a new nest-site, in which case individuals have no information on its potential. Our assumption is that social constraints and competitive abilities, rather than density dependence *per se*, are the main determinants of the age at first reproduction.

The optimal age at first reproduction, if any, is likely to depend on an individual's phenotype, the environment, and their interaction. It is probably not achievable to fully understand why an individual recruits at a given age without identifying habitat characteristics (nest-site, territory, colony or breeding location) where recruitment takes place. Regrettably, these tend to be treated independently in the literature (but see Ens et al. 1995; Boulinier and Lemel 1996). To circumvent this shortfall, our main objective was to address the relationship between age at first reproduction and habitat selection. Several behavioural tactics of habitat selection characterized by different ages at first reproduction may coexist in populations. For example, age *per se* may be associated with increased behavioural maturity and competitive ability, which may in turn translate into a higher probability of acquiring a good nest-site in older, more experienced individuals. A non-exclusive hypothesis may explain a similar relationship between age and habitat quality: the queuing hypothesis (Ens et al. 1995). Higher quality individuals may acquire higher quality nest-sites if they wait for a productive site (i.e. site where the current and past reproductive success is high) to become available. Alternatively, individuals with poor competitive abilities may recruit in lower-quality sites, regardless of age, which may lead to a situation where the oldest recruits breed on low-quality sites.

We used capture-mark-recapture (i.e. CMR) multi-state models (Nichols and Kendall 1995; Nichols 1996; Cam et al. 2005) to estimate recruitment probabilities as a function of age, cohort, as well as covariates used as surrogates for habitat quality. We first examined the age at which birds recruit (whether this choice reflects an individual decision or results from constraints), and where they settle at recruitment (in terms of habitat quality) as a function of age, in order to determine whether delayed recruitment results in the acquisition of higher-quality sites within productive cliffs.

Our second objective was to determine how well individuals recruiting at various ages and in habitat patches of different qualities perform in terms of breeding success probability in the year of recruitment. This aims at assessing whether delayed recruitment is associated with higher breeding success than early breeding; more generally, we aim at assessing the importance of habitat selection on fitness prospects in the very first breeding event.

2 Methods

2.1 Estimation of Recruitment Probabilities as a Function of Habitat Quality Using CMR Multi-State Models

In 1979, a black-legged Kittiwake monitoring program was initiated in Brittany, France, and is continuing today (five colonies located in Cap Sizun a few kilometers apart from each other, 48°5'N, 4°36'W; Monnat et al. 1990). Here, we examine the capture–recapture histories of twelve birth cohorts (1986–1997) over 18 years (1986–2003), that is a total of 4030 individuals. The fieldwork covers each breeding season entirely such that observers do not miss a single reproductive event in the study area (Cam et al. 1998). It is therefore possible to identify the very first reproductive event of each individual returning to the study area (Cam et al. 2002b, 2003, 2005). We acknowledge that some pre-breeding individuals may have recruited into another population (e.g. the British Isles or Spain; Cam et al. 2002b) before breeding in Brittany, that is, we may have missed the very first breeding event. However, we believe that such cases are rare as most individuals resighted as recruits attend Brittany colonies in the years preceding recruitment, and the majority of the recruits are sexed through behaviour before first breeding. Thus, attending Brittany colonies while breeding in the British Isles is likely to be rare.

2.1.1 Habitat Quality

To address the relationship between habitat selection and recruitment probability, we first defined habitat patches as sections of a cliff delimited by topographical discontinuities (e.g. Danchin et al. 1998). In the following, we will use 'cliff' and 'habitat patch' interchangeably. Only patches hosting at least ten nest-sites were included in our study. Following the approach developed by Danchin et al. (1998), we calculated yearly 'habitat quality' (i.e. local productivity) as the percentage of nests in a success situation within each cliff (0–33% for poor quality cliffs, 33–66% for medium quality cliffs, and 66–100% for highly productive cliffs). A 'success' was defined as a nest fledging at least 1 offspring. Similar to Danchin et al. (1998), and Cam and Monnat (2000a), the performance of the focal individual was excluded from the calculation of habitat quality in order to maintain independence between measures of individual breeding success and habitat quality. Our measure of habitat (cliff) quality will be referred to as 'Cliff' in the statistical analyses.

2.1.2 Approach to Modeling

Recruitment Probability

We used the definition of recruitment given by Pradel and Lebreton (1999): the probability that a pre-breeder in year t, which survived up to year $t+1$, reproduces in year $t+1$ (i.e. transition probability from pre-breeding to a breeding state; Brownie et al., 1993). Transition probabilities from breeding to non-breeding states were fixed to zero (i.e. impossible transitions). The recapture probability of breeders is ≈ 1 in the study population (Cam et al. 1998, 2005), but previous studies have shown that recapture probabilities for pre-breeders are lower than 1 (Cam et al. 2005). Hence, estimation of transition probabilities conditional on survival (i.e. recruitment probability) requires probabilistic models incorporating recapture probability.

Multi-state (MS) models (Arnason 1973; Nichols et al. 1992, 1993; Nichols and Kendall 1995; Schwarz et al. 1993; Lebreton and Pradel 2002) are designed in such a manner that individuals can move among states (e.g., states can be geographical states, or biological states such as size classes, breeding states, etc.). ψ_t is the probability of moving among states between time t and $t+1$ (in our case, transition from non-breeding to breeding state) conditional on surviving up to time $t+1$. We used the multi-state models implemented in Program MARK (White and Burnham, 1999) to estimate recapture, survival, and transition probabilities denoted as:

- P_t^r: Recapture probability at time t for an individual in state r at time t (t = 2, 3, . . ., k)
- S_t^r: Probability of being alive at time $t+1$, for an individual alive and of state r at time t (t = 1, 2, 3, . . ., k–1)
- Ψ_t^{rs}: Transition probability from state r (non-breeder) at time t (t = 1, 2, 3, . . ., k–1) to state s (breeder) at time $t+1$, for an individual surviving between t and $t+1$.

Here, age is accounted for by inclusion of both cohort and year (for additional details see Cam et al. 2005)

Influence of Habitat Quality on Recruitment Probability

We modeled the effect of habitat quality on transition probabilities Ψ_t^{rs} (from a non-breeding state r to a breeding state s) using two different approaches. First, we assigned a covariate corresponding to the quality of the recruitment habitat to each individual. Because previous studies have provided evidence that patch quality the year preceding recruitment (t–1) influences settlement decisions in year t in both dispersers and recruits (Cadiou et al. 1994; Danchin et al. 1998; Cadiou 1999), we considered models with a covariate accounting for habitat quality the year preceding recruitment (covariate q_{t-1}), or the year of recruitment (covariate q_t). The biological hypotheses underlying a model including habitat quality in the year preceding recruitment is that recruits might be prospecting for a high-quality patch. They might make the decision of where they are going to settle and breed for the first time at least a year in advance. We also considered quadratic models including q_{t-1} squared (q^2_{t-1}), and q_t squared (q^2_t), to evaluate possible non-monotonic relationships between recruitment probabilities and habitat quality. Testing for a positive linear relationship between recruitment probability and cliff quality is testing whether recruitment probability increases as cliff quality increases. A quadratic relationship term may account for higher recruitment probabilities in habitat of medium quality and lower recruitment probabilities in habitats of poor and high quality, or conversely if the sign is switched. We also considered a model without any covariate accounting for habitat quality to address the 'null' biological hypothesis that is a lack of influence of habitat quality on recruitment probability.

Since one must assign a covariate value to each individual, we had to dispense a value to individuals that did not recruit and for which recruitment-habitat characteristics did not exist (i.e. individuals that have never reproduced, died before recruitment, or emigrated out of the study area before recruitment). Following Cooch and White's (2006) two-step solution to the 'missing-value' issue, we assigned average covariate values to individuals that did not recruit (i.e. $q_t = 0.464$, $q^2_t = 0.215$, $q_{t-1} = 0.456$, $q^2_{t-1} = 0.208$). This may artificially skew the estimate of transition probability towards these values (i.e. habitat patches of intermediate quality). To assess the importance of the bias, we compared our results (i.e. models receiving large support) to recruitment estimates obtained in a second set of analyses, where no covariates were involved, but where states were defined differently and accounted for the quality of the recruitment habitat.

The second approach assesses the effect of habitat quality on transition probabilities Ψ^{rs}_i (from non-breeding state r to breeding state s) by specifying four states. We considered (1) pre-breeders; (2) breeders recruiting in high-quality cliffs (i.e. cliffs where local productivity is between 66 and 100%); (3) breeders recruiting in cliffs of intermediate quality (i.e. local productivity between 33 and 66%); and (4) breeders recruiting in poor quality cliffs (i.e. local productivity between 0 and 33%). Because the previous approach indicated that models including habitat quality the year preceding recruitment performed better than others (according to information criteria, Burnham and Anderson, 1998, see also results), the categorical index of habitat quality used in the second approach to define the states is based on habitat quality the year preceding recruitment.

Our models included covariates (i.e. various measures of habitat quality) or breeding states accounting for habitat quality, as well as age and cohort effects on transition probabilities. We never used interaction terms between age and cohort as it would be equivalent to considering a time effect. However, additive models allowed disentangling age effects on recruitment probabilities from cohort-related effects. The latter may reflect long lasting birth-year effects on age-specific recruitment probabilities (e.g. climatic effects). We also used such additive models (cohort + age) for purely technical reasons, that is to fix some parameters to zero according to the specificity of the distribution of ages at recruitment in the different cohorts (e.g. if in the birth cohort 1992, the minimum transition from a non-breeding to a breeding state occurred between ages 3 and 4, the 'cohort + age' format allowed us to fix the parameters representing the probability of recruitment in younger age classes to zero, such as recruitment probabilities in between age 0 and 1, 1 and 2, or 2 and 3).

Model Selection

Based on prior studies of recruitment probability (Cam and Monnat 2000a; Cam et al. 2002b, 2003, 2005), and movement among colonies of black-legged kittiwakes in Brittany (Danchin and Monnat 1992; Danchin et al. 1998), we were primarily interested in hypotheses pertaining to the relationship between age-specific recruitment probability and habitat selection, conditional on age- and state-specific survival. Accordingly, we designed a general model reflecting all biological processes of interest. Because recapture probability of the youngest pre-breeders is known to be low, sample sizes within age-by-state combinations were assessed to design the most general model (e.g. Cam et al. 2002b, 2003, 2005).

Previous studies indicated that adult recapture probabilities have always been ≈ 1 in the study area (Danchin and Monnat 1992; Cam et al. 1998, 2005), hence we assumed that adult recapture probabilities were independent of time and cohort in all models: $p2(.)$ (state '2' corresponds to adults i.e., after recruitment). On the contrary, we expected recapture probabilities to vary with age amongst pre-breeders: $p1(.)$ (state '1' corresponds to pre-breeders). As the majority of pre-breeders recruit before 7 years old, we pooled data from pre-breeders of age 7 or more (i.e., recapture at age 1, 2, 3, 4, 5, 6, 7 or greater).

Previous work also indicated that apparent survival probability is lower for pre-breeders than for breeders (Cam et al. 2005), as they might be subjected to greater extrinsic causes of death than adults, or have a higher probability of permanent emigration. Therefore, we considered a 7 age-class effect on pre-breeders' survival as well. Climatic conditions experienced during early development or during the first winter at sea may affect differently each birth cohort justifying why we considered cohort variations in pre-breeders' survival (i.e. $S1(c12, a7)$, where '$c12$' stands for the cohort effect and '$a7$' for the seven age-class effect). We did not consider cohort variation in adult survival, as we were trying to limit model size (i.e. number of estimated parameters). We focussed preferentially on the parameters of interest (i.e. parameters representative of the 'pre-breeding' stage, and of 'first-time breeding' events). We then considered an age effect on adult survival (denoted as $S2(a5)$). In the case of adults, we defined only 5 age classes (i.e. $a5$ defines age classes 2, 3, 4, 5, 6, 7+, where 7+ stands for individuals aged 7 years old and more), as the minimum age to become a breeder is 2 years old.

The last set of assumptions concerns the probability of transition from a non-breeding to breeding state, Ψ. As previous studies have provided evidence of an increase in recruitment probability with age, and in a limited sense, with experience as well (Cam et al. 2002b, 2003, 2005), we considered age effects on transition probabilities. Also, we included an additive cohort effect to account for the influence of annual environmental change (climatic conditions or predation events affecting the proportion of high-quality breeding habitats available to recruits), when individuals belonging to different cohorts reach the age at which transition to the breeding state is theoretically possible. Such environmental factors may influence age-specific recruitment differently from one cohort to another. As our primary objective was to examine the influence of habitat quality on age at first reproduction, we also

included the effect of habitat quality, either by using individual covariates (approach 1), or by stratifying the data set into different states, reflecting different combinations of habitat quality and age (approach 2). The initial model accounted for an age effect on recruitment probabilities characterized by 6 age-specific transition probabilities (transition in between 1 and 2 years old, 2 and 3, 3 and 4, 4 and 5, 5 and 6, 6 and 7). Transition probabilities were denoted $\Psi 12(c12 + a6 + q_t + q_t^2 + (a6 \text{ x } q_t))$ in the first approach and $\Psi 12(c12 + a6)$ $\Psi 13(c12 + a6)$ $\Psi 14(c12 + a6)$ in the second, where 2, 3 and 4 corresponded to the three different habitat qualities (i.e. poor, medium, high) in which a bird can recruit. As transitions cannot biologically occur in the opposite direction, $\Psi 21(.)$ $\Psi 31(.)$ $\Psi 41(.)$ were fixed to zero.
For each approach, our starting model was defined as follows:

> Approach 1 (with individual covariates):
> $S1(c12 + a7)$ $S2(a5)$ $p1(a7)$ $p2(.)$ $\Psi 12(c12 + a6 + q_t + q_t^2 + (a6 \text{ x } q_t))$ $\Psi 21(.)$
>
> Approach 2 (discrete states):
> $S1(c12 + a7)$ $S2(a5)$ $S3(a5)$ $p1(a7)$ $p2(.)$ $p3(.)$ $p4(.)$ $\Psi 12(c12 + a6)$ $\Psi 13(c12 + a6)$ $\Psi 14(c12 + a6)$

We acknowledge that both global (starting) models are not saturated, even though it would be desirable to compare the performance of saturated models and less parameterized ones. A saturated model is defined as the model where the number of parameters equals the number of data points. Such a model is needed to compute the baseline deviance, which is in turn used to estimate the amount of over-dispersion in the data (Cooch and White 2006). However, we had to limit the degree of stratification of the data to make parameter estimation feasible. Furthermore, the large number of biological parameters of interest made it difficult to define a small set of alternative models defined 'a priori' (Burnham and Anderson 1998). We sequentially specified models by simplifying the starting model to test for specific biological hypotheses. Nevertheless, we acknowledge that sequential development of models might lead to different conclusions compared to considering a set of models defined *a priori*.

Unfortunately, formal goodness-of-fit tests for MS models do not allow for treatment of situations with permanent transitions (Pradel 2006). As an alternative approach to a formal goodness-of-fit test, we estimated an overdispersion parameter (i.e. $\hat{c}$) for the global model without individual covariates (approach 2 described above) using bootstrap simulations in MSSURVIV (Hines 1994).

We used Akaike's Information Criterion modified for small sample size, AICc, in the first approach (where no overdispersion parameter $\hat{c}$ can be calculated), and the qAICc modified for overdispersion in the second approach (where q stands for quasi-likelihood; Akaike 1973, see also Sakamoto et al. 1986; Lebreton et al. 1992; Burnham and Anderson 1998). We also used Akaike's weights, $w_{i,}$ to select the best models from our set of candidate MS models in both approaches 1 and 2. Only models with an Akaike weight exceeding 0.95 were systematically retained. If the weight was shared among 2 or more models, we discussed the interpretation of each of them.

2.2 *Estimation of Breeding Success in the Year of Recruitment*

2.2.1 Sample Specifications

Only individuals that survived until recruitment and recruited were considered in analyses of breeding success (1450 individuals, 5054 observations). More specifically, we addressed breeding success probability in the year of recruitment and in subsequent breeding occasions. As emphasized above, for birds recruiting in the Cap Sizun, the probability of recapture is virtually 1 after recruitment (Cam et al. 1998, 2005). Working on the sample of individuals that has recruited allows us to use simple statistical models that do not account for recapture probabilities, such as generalized linear models and mixed models. Individuals whose breeding success in the year of recruitment was unknown or uncertain were excluded from the analyses. Only individuals that fledged at least one chick up to independence were considered 'successful', others were considered to have 'failed'.

2.2.2 Generalized Linear Models (glm) and Mixed Models (glmmM)

We used generalized linear models (Agresti 1990) to address the influence of two covariates (i.e. age at first reproduction and habitat quality) on breeding success probability, a binary response variable (i.e. success *versus* failure). Sample sizes incited us to minimize the number of states in the analysis, thus we did not include different levels of failure (e.g. early failure when the chick dies at the nest or late failure when the chick died at fledging) or success (e.g. kittiwakes generally produce 1 or 2 eggs, and occasionally produce up to 3 eggs, and may fledge several chicks).

We built a series of glms (use of the logit link) accounting for cliff quality the year preceding recruitment (found to be a better predictor of the recruitment process than cliff quality the year of recruitment, see Results) and age at recruitment. Age at recruitment was treated either as a continuous or as a categorical covariate. We tested several transformations of cliff quality (i.e. proportions of successful nests within a cliff in a given year):

- the arcsine transformation, suitable for binary data summarized as proportions.
- the square root transformation, suitable for Poisson-distributed covariates where sample means are proportional to the variances of the respective samples; replacing each measure by its square root will often result in homogeneous variances (Neter et al. 1996).

We also built models including a quadratic effect and a cubic effect of age and habitat quality on breeding success probability. A quadratic effect of cliff quality on success probability would mean that maximum (or minimum) success probability is reached in cliffs of intermediate quality. Similarly, a quadratic effect of age on success probability would account for a minimum, or maximum breeding success at intermediate ages. A cubic relationship would account for a bimodal pattern in success probability as a function of covariates.

We accounted for temporal variation in breeding success (possibly resulting from environmental fluctuations; e.g. climatic conditions, predation events, food shortage, etc.) by incorporating a random effect of time (year) only to the best performing model. We modeled year as a random effect for two reasons. First, we had no motive to suspect any specific shape for the influence of year on breeding success probability (e.g. a systematic trend). Second, using a random effect to account for temporal variation in breeding success leads to fewer parameters than a fixed effect model. We viewed this as an advantage (i.e. larger sample sizes) to address the influence of covariates more relevant to hypotheses pertaining to habitat selection (e.g. habitat quality). We used the package 'glmmML' (i.e. package 'MASS', R version 2.3.1) to implement mixed models.

2.2.3 Model Selection

First, we compared pairs of models containing the same covariate but parameterized in different ways (e.g. a model containing the age at first reproduction AFR, treated as a continuous covariate, was tested against a model containing AFR treated as a factor). After retaining the best parameterization, we compared models with an additive effect or an interaction term. Each model was created to discriminate between various underlying biological hypotheses. The models selected will be discussed in the manuscript. For model comparison, we only reported model selection based on Akaike's Information Criterion AIC (Akaike 1973; Burnham and Anderson 1998), as results based on AICc were consistent with results based on AIC.

3 Results

3.1 Estimation of Recruitment Probabilities as a Function of Habitat Quality: A CMR Approach Using Multi-State Models

The estimated overdispersion parameter (i.e. variance inflation factor) for the global model was 1.94 (bootstrap procedure in MSSURVIV, 1000 simulations).

3.1.1 Analysis with Individual Covariates

The best model, '2-state-model-19', is structured as follow (see Appendix 1; Table 1):

$$S1\,(c12 + a6)\,S2(.)\,p1(a4)\,p2(.)\,\Psi12(c12 + a6 + q_{t\text{-}1} + q_{t-1}{}^{2})\,\Psi21(.)$$

This model includes a cohort effect on survival probability of pre-breeders, and survival probability at a given age varied according to birth year. Pre-breeder

Table 1 Modelling the influence of habitat quality on the recruitment process I: model selection results based on the first approach (approach with covariates, see Methods)

Model selection	AICc	Δ_i	W_i	Likelihood	NP	Deviance
M19	21374	0.00	0.77	1.00	42	21290
M24	21377	2.69	0.20	0.26	43	21290
M13	21382	7.98	0.01	0.02	43	21296
M18	21838	8.87	0.01	0.01	44	21294
M11	21384	9.93	0.00	0.01	33	21318
M7	21384	10.44	0.00	0.00	33	21318
M12	21395	20.50	0.00	0.00	50	21294
M5	21396	21.98	0.00	0.00	49	21297
M4	21397	22.51	0.00	0.00	51	21294
M3	21401	26.75	0.00	0.00	40	21320
M14	21404	30.35	0.00	0.00	57	21290
M20	21406	32.16	0.00	0.00	58	21289
M6	21408	34.17	0.00	0.00	59	21289
M10	21409	35.22	0.00	0.00	56	21297
M16	21418	44.06	0.00	0.00	43	21332
M22	21420	45.78	0.00	0.00	43	21333
M15	21428	53.50	0.00	0.00	43	21341
M21	21430	55.55	0.00	0.00	43	21343
M9	21449	75.11	0.00	0.00	45	21359
M8	21457	83.04	0.00	0.00	34	21389
M17	21643	268.88	0.00	0.00	38	21567
M23	21643	269.12	0.00	0.00	38	21567
M1	21826	452.40	0.00	0.00	60	21706
M1bis	21861	934.67	0.00	0.00	47	21766
M2	22666	1291.50	0.00	0.00	54	22557

Note: NP: number of estimated parameters; AIC: Akaike's Information Criterion = −2*log-likelihood + 2*NP; W_i = exp (−0.5*ΔAIC)/NP.

survival also varied across ages (i.e. survival probabilities between ages 0-1, 1-2, 2-3, 3-4, and 4-5 years old were significantly different).

Adult survival probability was best accounted for by a model with constant survival across ages and cohorts ($S2(.)$).

The recapture probability of pre-breeders did not vary across cohorts, but varied across ages '$p1(a4)$'. The best model retained had a 4 age-class structure (0-1, 1-2, 2-3, and 3-4 years old and more) showing a non-negligible difference in recapture probabilities across age groups.

Recapture probability of adults was ≈ 1, regardless of cohort and age class ($p2(.)$) and confirms previous findings (Danchin and Monnat 1992; Cam et al. 1998; 2005).

The probability of transition from the 'pre-breeding' to the 'breeding' state varied across cohorts and ages ($\Psi 12(c12 + a6 + q_{t-1} + q_{t-1}{}^2)$). The model selected included six age classes (transition from 1 to 2, 2 to 3, 3 to 4, 4 to 5, 5 to 6 and 6 to 7+ years old and more). Averaged across cohorts, the recruitment probability between ages 1 and 2 was close to zero (only a handful of individuals recruited at such an early age). Model selection provided support for a model where recruitment probability increased with age at first breeding up to 5 and 6 years of age, after which it declined (i.e. recruitment probability peaks for the transition occurring in between

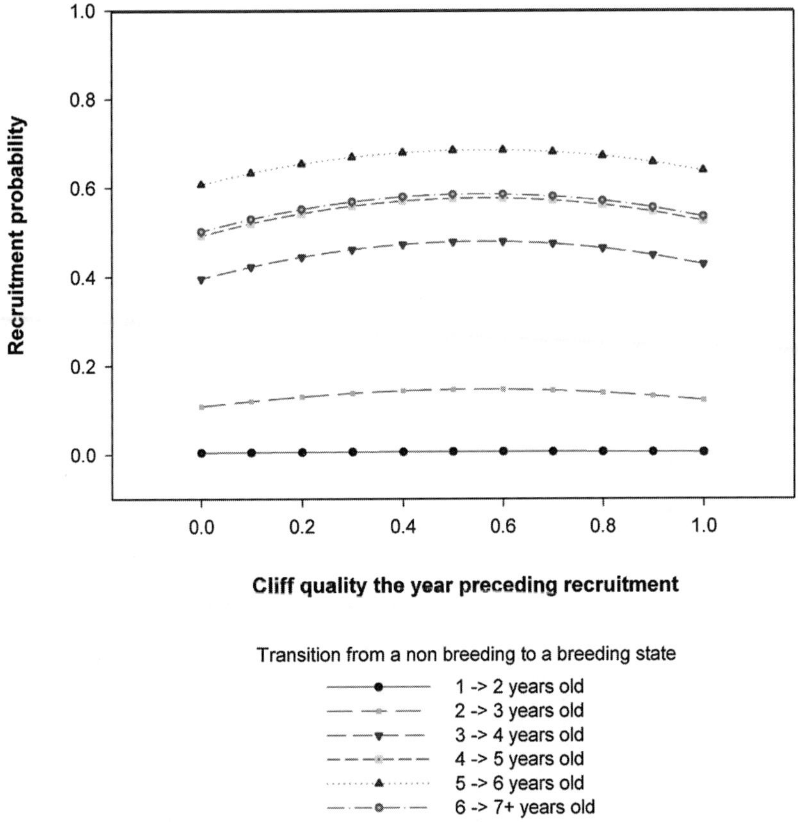

Fig. 1 Recruitment probabilities as a function of habitat quality and age at first reproduction Cliff quality (i.e. continuous covariate) was calculated the year preceding recruitment. Recruitment probabilities were estimated from the best performing multi-state model. The recruitment probabilities were averaged across cohorts (birth cohorts 1986 to 1997, followed from 1986 to 2003).

age 5 and 6; Fig. 1). The model selected also included an effect of cliff quality the year preceding recruitment (i.e. q_{t-1}), largely preferred (according to AICc) over a model with an effect of cliff quality in the recruitment year (i.e. q_t), and over a model without a covariate accounting for cliff quality. Moreover, a quadratic effect of cliff quality in the year preceding recruitment (i.e. $q_{t-1} + q_{t-1}^2$) received more support than a linear effect (Table 1). Thus, for each recruitment tactic, maximum recruitment probability occurred in cliffs of intermediate quality the year preceding recruitment (Fig. 1).

One may argue that this result does not reflect any active individual choice, but rather that the availability of habitat patches of intermediate quality exceeds that of patches of other qualities (i.e. poor and highly productive cliffs). Individuals may simply distribute themselves randomly according to habitat availability. That is true

Table 2 Time series of the proportion of poor, medium, and high quality cliffs from 1986 to 2003

Year	Cliffs of good quality (%)*	Cliffs of intermediate quality (%)**	Cliffs of poor quality (%)***
1986	29	45	26
1987	32	39	30
1988	16	35	49
1989	6	34	59
1990	11	39	50
1991	35	40	26
1992	33	35	33
1993	23	28	49
1994	10	23	67
1995	26	21	53
1996	19	19	62
1997	24	34	41
1998	7	22	7
1999	15	54	31
2000	21	62	17
2001	41	32	27
2002	20	44	36
2003	62	35	4

Note: Cliff quality was calculated the year preceding recruitment as model selection results indicated that this quantity best predicts recruitment probability.
* Proportion of cliffs in a given year containing 66–100% nests in a success situation.
** Proportion of cliffs in a given year containing 33–66% nests in a success situation.
*** Proportion of cliffs in a given year containing 0–33% nests in a success situation.

in less than half of the cases (see Table 2, years 1986, 1987, 1991, 1992, 1999, 2000 and 2002). Consequently, a higher recruitment probability in cliffs of intermediate quality cannot be interpreted as resulting exclusively from a spatially random recruitment process. In more than half of the years included in this study, individual choice and/or constraints led recruits to select habitat features different from those that would be obtained by random settlement.

3.1.2 Analysis Without Individual Covariates

The above results provided evidence that models including an effect of cliff quality in the year preceding recruitment on transition probability best fit the data. In the second approach (i.e. without individual covariates), we therefore defined three states for breeders ('2', '3', and '4', settling in poor, medium, and high quality habitat patch, respectively) according to cliff quality in the year preceding recruitment, as the model including a quadratic form of this covariate was found to perform better than models including cliff quality the year of recruitment (Appendix 2; Table 3).

Table 3 Modeling the influence of habitat quality on the recruitment process II: model selection results for the second approach (approach without covariates, see Methods)

Model selection	qAICc*	Δ_i	W_i	Model likelihood	NP	Deviance
Md21	32218	0.00	1.00	1.00	76	16355
Md12	32238	20.35	0.00	0.00	86	16355
Md8	32238	20.37	0.00	0.00	86	16355
Md7	32239	21.32	0.00	0.00	87	16354
Md9	32240	22.39	0.00	0.00	85	16360
Md15	32241	22.82	0.00	0.00	84	16362
Md6	32241	23.15	0.00	0.00	88	16354
Md14	32242	24.53	0.00	0.00	85	16362
Md20	32244	26.36	0.00	0.00	83	16368
Md4	32241	33.25	0.00	0.00	102	16336
Md3	32252	34.08	0.00	0.00	103	16335
Md1	32254	36.12	0.00	0.00	104	16335
Md5	32255	37.05	0.00	0.00	101	16342
Md11	32284	66.32	0.00	0.00	76	16422
Md19	32291	73.34	0.00	0.00	53	16475
Md16	32348	130.43	0.00	0.00	50	16539
Md13	32481	263.59	0.00	0.00	81	16609
Md18	32513	295.45	0.00	0.00	68	16667
Md10	32550	331.86	0.00	0.00	69	16702
Md17	32561	342.85	0.00	0.00	35	16781
Md2	34003	1785.50	0.00	0.00	98	18096

Note: NP: number of estimated parameters.
* We used Akaike's Information Criterion modified for sample size qAICc (where q stands for quasi-likelihood) and for an estimated overdispersion parameter $\hat{c}$ of 1.93 using bootstrap simulations (see Methods).

The best approximating model, '4-states-model-21', had the following structure:

$$S1(c12+a6)S2,3,4(a3)p1(a5)p2,3,4(.)\Psi 12(c12+a6)\Psi 13(c12+a6)\Psi 14(c12+a6)$$

This model included a cohort effect on pre-breeder survival, '$\Phi 1(c12)$', showing that birth year influences survival. In addition, pre-breeder survival probabilities changed with age, $S1(a6)$ (i.e. survival probabilities from age 0 to 1, 1 to 2, 2 to 3, 3 to 4, 4 to 5, and from 5 years old to any higher age were significantly different). Adult survival probability did not vary across cohorts or across cliff qualities in the year preceding recruitment. We did however detect an age effect on adult survival (where individuals aged 3, 4, 5 years old and more had different survival probabilities: '$S2,3,4(a3)$').

Recapture probability of pre-breeders was constant across cohorts, but varied across five age classes: 0, 1, 2, 3, 4 years and more; '$p1(a5)$'. For adults, neither cliff quality, cohort, nor age influenced recapture probabilities.

Recruitment probabilities varied according to birth cohort and age (transition probabilities from 1 to 2, 2 to 3, 3 to 4, 4 to 5, 5 to 6, 6 to 7 years old and more were significantly different). Consistent with the results obtained using the previous approach, transition probabilities were highest at intermediate ages (i.e. 5 and 6 years old; quadratic age effect on recruitment probability). In addition, age-specific

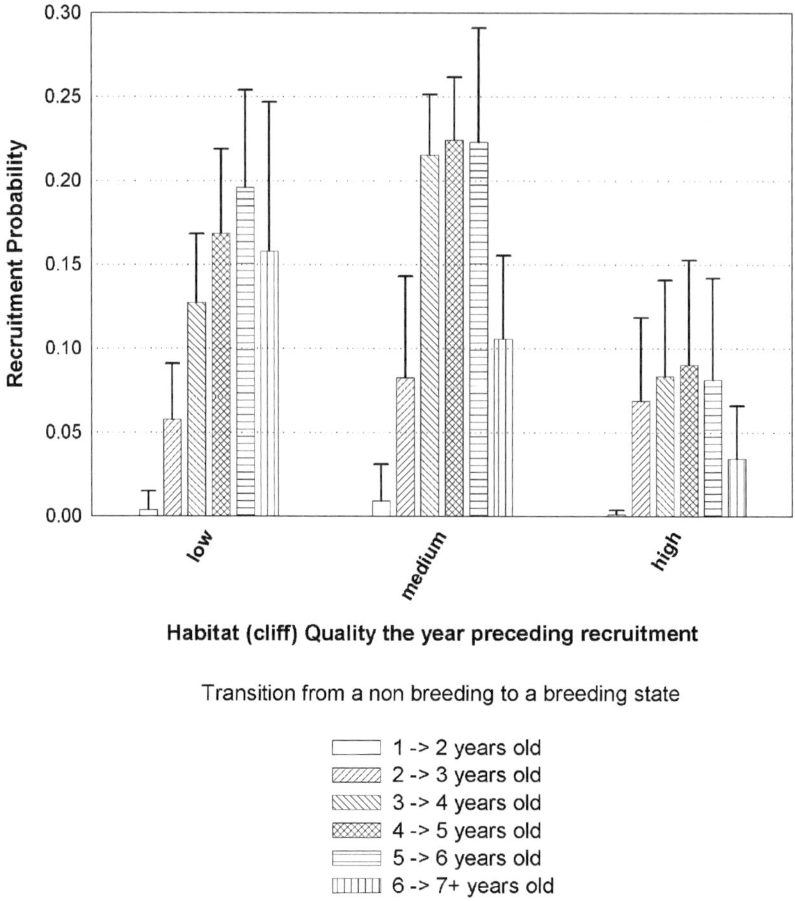

Fig. 2 Recruitment probability as a function of habitat quality and age at first reproduction Cliff quality was calculated the year preceding recruitment. Three states accounted for the quality of the recruitment habitat the year preceding recruitment: poor, medium and high quality cliffs. A fourth state accounted for the pre-breeding segment of the population. Recruitment probabilities were estimated from the multi-state model that received the most support, and were averaged across cohorts (birth cohorts 1986–1997, followed from 1986 to 2003).

recruitment probabilities varied according to cliff quality in the year preceding recruitment, and were higher in habitat patches of intermediate quality Fig. 2).

3.2 Breeding Success the Year of Recruitment

The best model contained both an effect of age at first reproduction (treated as a factor: AFR = 3, 4, 5, 6, 7 years old and more) and a quadratic effect of cliff quality, on breeding success (Table 4; Fig. 3). Breeding success probability was maximal

Table 4 Model selection results: generalized linear models and mixed model testing the effects of age at recruitment and habitat quality on breeding success in the year of recruitment

Model	NP	AIC	Δ_i	exp(–(1/2)* Δi)	W_i
AFR.cat + (Cliff)^2+ ε (time)*	**7**	**1613.0**	**0.0**	**1.000**	**0.609**
AFR.cat * (Cliff)^2	10	1619.6	6.6	0.037	0.022
AFR.cat + (Cliff)^2	*6*	*1614.0*	*1.0*	*0.606*	*0.369*
Cliff + (Cliff)^2	3	1658.4	45.4	0.0	0.0
Sqrt (Cliff)	2	1668.5	55.5	0.0	0.0
(Cliff)^2	2	1656.9	43.9	0.0	0.0
Arcsin (Cliff)	2	1661.1	48.1	0.0	0.0
Cliff	2	1659.6	137.3	0.0	0.0
AFR + (AFR)^2 + (AFR)^3	4	1733.0	120.0	0.0	0.0
AFR + (AFR)^3	3	1735.8	122.8	0.0	0.0
AFR + (AFR)^2	3	1734.5	121.5	0.0	0.0
AFR.cat	5	1733.3	120.3	0.0	0.0
AFR	2	1750.3	137.3	0.0	0.0

Note. Model selected in bold characters; Models contained above in italic character(s) are models that were not retained but that have some weight in explaining the biological process that gave rise to the data; we used a mixed model to add a random time effect 'ε (time)' to the best performing glm; '+' additive effect; '*' interaction; NP = number of estimated parameters; AIC: Akaike's Information Criterion = –2*log-likelihood + 2*NP; $W_i = \exp(-0.5*\Delta AIC)/\sum \exp(-0.5*\Delta AIC)$. Covariates: Age at first reproduction (AFR if continuous, AFR.cat if categorical, AFR^2 for a quadratic effect, AFR^3 for a cubic effect); cliff quality (Cliff if continuous, Arcsin(Cliff) if arcsinus transformed, Sqrt(Cliff) if square root transformed, and Cliff^2 for a quadratic effect).

for individuals recruiting at an intermediate age of 5 years old (Fig. 3). By including a random effect of time in this model, the AIC dropped by 1 unit, down to the value of 1613, providing slight evidence of yearly variation in breeding success over time.

4 Discussion

In this paper, we first aimed at examining whether habitat selection and age at recruitment were linked, and if so, which recruits gained the best breeding habitats: early recruits or individuals delaying recruitment. We also examined which recruitment tactic led to the highest breeding success in the year of recruitment. Overall, this paper studies (1) recruitment-habitat selection, (2) when and where recruits breed for the first time, and (3) breeding success as a function of the location and the age at which individuals recruit.

4.1 The Timing of Habitat Selection, Habitat Quality and Age-Specific Recruitment

A number of investigators (Danchin 1988b; Danchin et al. 1991, 1998; Cadiou et al. 1994) have suggested that dispersers actively select their recruitment habitat the year preceding settlement (e.g. the number of prospectors in habitat patches

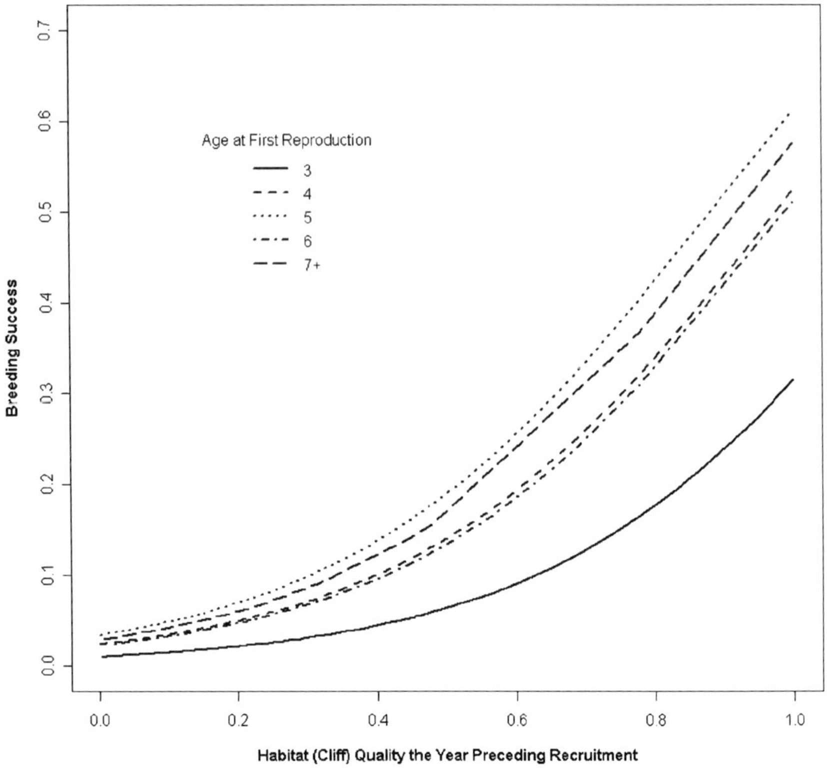

Fig. 3 Breeding success the year of recruitment as a function of cliff quality the year preceding recruitment and age at first reproduction

depends on their current productivity; Cadiou 1993). Corroborating their hypothesis, we also observed that models including cliff quality the year preceding recruitment performed better in explaining the recruitment process than models including cliff quality the year of recruitment (Appendix 1; Table 1). Two *scenarii* can be proposed. Habitat selection may take place in the year of recruitment based on information available at the beginning of a breeding season on habitat quality (i.e. information based on social activity and attendance of individuals that bred in that patch the previous year and returned to the same breeding patch). Such information may be strongly autocorrelated to local productivity in the preceding year (Boulinier et al. 1996). However, young recruits (i.e. recruiting at 3 years old) might only benefit from an imperfect knowledge of cliff quality, as they arrive on average 2 months later than individuals delaying recruitment (based on direct observations). Alternatively, settlement decisions may be made the year preceding recruitment. This implies that pre-breeders prospect for a breeding ground at least a year in advance (Danchin et al. 1991). Both *scenarii* rely on the assumption that habitat quality in a given year t is a reliable indicator of its quality in year $t+1$ (Boulinier and Lemel 1996).

If so, temporal autocorrelation in breeding success over 2 consecutive years in a given patch should allow pre-breeders to locate a higher-quality breeding location a breeding season in advance. Although we only considered habitat quality in the year of recruitment and the year preceding recruitment, recruitment probability may depend on past productivity over several consecutive years (with temporal autocorrelation of cliff quality being superior to a year).

In most age classes, recruitment probability in a given year (from 1986 to 2003) was highest in cliffs of intermediate quality the year preceding recruitment (with the exception of individuals recruiting at age 2, for which transition probabilities are not reliable as sample size is very small). According to habitat selection theory, if no constraints are operating (i.e., no competition, no dominance in social hierarchy, and if individuals have information on the range of habitats available), we might expect natural selection to favour habitat selection tactics that maximize fitness (Holt and Barfield 2001). That is, recruitment probability should be highest in the most productive cliffs, where the fitness prospects are maximal. Our results do not support this prediction, and therefore suggests the existence of constraints. High-quality cliffs may not be accessible to most recruits (i.e. the youngest). They may be constrained to breed in habitats where competition for nest-sites is lower. Our results also provided evidence that older first-time breeders recruit in habitats of lower quality than intermediate age individuals; therefore, the queuing hypothesis is unlikely to explain the pattern observed in these recruits. Features of individual quality relevant to habitat selection may involve differences in behavioural maturity, social and territorial dominance. These differences could be expressed in terms of the age at recruitment, whereby old recruits (i.e. of low intrinsic quality) can only afford to breed in low-quality patches because of competitive inferiority. However, in the case of young recruits, if they are sometimes assumed to be of high intrinsic quality (Nur 1988), we did not find evidence that this translates into access to higher-quality habitat, as they do have the advantage of an early breeding start, but still do not recruit on the best breeding-sites. These individuals may not be of lower intrinsic quality, they may simply lack competitive skills. Behavioral maturation may explain why individuals recruiting at intermediate age have access to higher-quality sites.

These results are valuable only if one assumes that recruits make an active selection of the habitat in which they will breed for the first time. One could imagine that individuals breed preferentially in intermediate quality cliffs because these are more abundant than other cliffs type (i.e. cliffs of low or high quality). However, our results suggest that we are observing the outcome of an individual choice involving active habitat selection rather than random settlement. Indeed, over all the years studied, the proportion of cliffs of intermediate quality was not larger than the proportion of cliffs of high or poor quality, as cliffs of low, high, and intermediate quality were equally available in the study area. In addition, we acknowledge that density dependant processes may influence settlement decisions. However, we believe that density dependance alone cannot explain the observed distribution of recruits according to habitat quality. Indeed, behavioural studies have provided evidence that creation of new nest-sites by pre-breeders is possible even in highly productive patches, but that this option is not usually preferred by pre-breeders: they mostly compete for occupied sites (Cadiou 1993; Cadiou et al. 1994).

Both multi-state modeling approaches showed age-related variation in recruitment probability, with highest recruitment probabilities reached at intermediate ages (transition probability in between 5 and 6 years old). Yet, a non-negligible proportion of individuals recruited earlier (i.e. recruitment probability of 0.15 at age 3, and approximately 0.40 at age 4). Age-specific variation in recruitment probability, more specifically the initial increase in recruitment probability, may partly reflect the progressive acquisition of behavioural and physiological maturity of individuals in the population. Whether individuals delaying reproduction do so because they are not sexually mature, or because of a lack of behavioural maturity (in sexually mature individuals), is beyond the scope of this paper: physiological and behavioural data are required to address this question. However, within the framework of physiological and social constraints (e.g. competition), it is possible to address whether there is scope for natural selection processes to operate on age of first breeding by evaluating and comparing fitness components associated with each age-specific recruitment tactic. We addressed whether there was a relationship between each tactic and age-specific reproductive success in the recruitment year to determine which one(s) might yield highest breeding success levels.

4.2 Breeding Success

We found evidence that birds recruiting at intermediate ages (i.e. recruiting at age 5) experienced the highest reproductive success in the year of recruitment. These results complete our findings regarding age-specific recruitment probability, where again, the highest probability of recruitment was observed at that age. Based on these results, it is tempting to suggest that the age at first reproduction has been shaped by an optimization process. Under this view, recruiting around age 5 (i.e. delaying recruitment up to intermediate ages) would be associated with fitness advantages that offset the direct costs of delayed recruitment (i.e. costs such as 'missed' breeding opportunities in comparison with individuals recruiting earlier).

One of the predictions of life history theory is that early reproduction should be favoured by natural selection in stable or increasing populations (Stearns 1992; Charlesworth 1994), except in situations where delayed reproduction is beneficial in terms of fitness. A well-known case explicitly addressed by Charlesworth (1994) is when reproductive success increases with age, experience (or both). In this case, theory suggests that younger individuals may balance the potential benefits of recruiting early (e.g. more breeding events accumulated throughout life compared to recruits delaying first reproduction), with the cost of unsuccessful breeding attempts early in life (Charlesworth 1994), as the youngest recruits lack experience and have a higher probability of breeding failure than others. Also, the time spent prospecting for a site may provide benefits in terms of information gathered on a potential breeding site, despite the costs associated with missing breeding opportunities (Boulinier and Danchin 1997). Early recruitment in this population is indeed associated with low reproductive success in the year of recruitment. Behavioural maturity and competitive abilities gained before breeding may explain why individuals recruiting at intermediate ages exhibit higher breeding success than early recruits (Nol and Smith 1987; Lunn et al. 1994).

Also, it has been suggested that heterogeneity in individual quality contributes to explain the age-specific variation in age at first breeding (Nur 1988; Curio 1983). According to this hypothesis, higher-quality individuals are assumed to be able to invest more into reproduction without incurring as large costs as lower-quality individuals; this may favour early investment into reproduction for higher-quality individuals. At this point, our results concerning breeding success probability are not consistent with this hypothesis. Overall, our results provided evidence that the youngest first-time breeders (the ones that theoretically are assumed to be of highest intrinsic quality: Nur 1988; Pyle et al. 1997), experienced the poorest breeding success probability in the recruitment year, compared to intermediate-age recruits, regardless of the quality of the recruitment habitat. However, it is possible that early recruits improve their breeding success as they age and gain experience; again, they may not be of lower intrinsic quality. Heterogeneity in quality among individuals may explain only partially our results: the decrease in recruitment probability in first-time breeders after age 5 (i.e. individuals delaying recruitment), and the fact that late breeders recruit in lower-quality habitat than birds recruiting at intermediate age. Social inferiority may prevent these individuals from beginning reproduction earlier in life. However, their breeding success probability in the year of recruitment is high: experience gained over a longer pre-breeding period may result in this pattern.

Attempting to explain the evolutionary (dis)advantages of early or delayed recruitment by addressing reproductive success in the first breeding attempt exclusively is too restrictive: first reproduction is only a snapshot of the lifetime profile of reproductive success for individuals recruiting at various ages in each habitat type. However, this first step was crucial in the understanding of the age-specific recruitment process and how it is related to habitat selection theory.

4.3 Prospects

Regarding habitat selection mechanisms, preliminary analysis conducted in the same study population (Aubry, unpublished; Bled 2006; Bled et al. in prep), suggests that it is critical to work at a much finer spatial scale to address recruitment; that is the nest-site itself, within a given cliff (e.g. it may be disadvantageous to gain ownership on a site of poor quality within the most productive cliff). Using an approach based on the quality of patches, there is no clear hierarchy among age classes in terms of access to habitat of lower, intermediate, or higher quality, but there is a relationship between habitat quality at the patch level and success probability in the year of recruitment. The shape of the relationship between age and success probability (which is highest in birds of intermediate age) cannot be explained by higher recruitment probability in higher-quality habitat: intermediate age first-time breeders do not recruit in higher-quality habitats than others (e.g. the interaction between age and quality was not retained in multi-state models with individual covariates). Heterogeneity in quality among sites within habitat patches may obscure the relationship between habitat quality, age, and individual success probability. A more detailed study of habitat selection at the site level may help understand the observed influence of age on breeding success probability. In this mobile species exhibiting breeding dispersal (Danchin and Monnat 1992; Danchin et al. 1998), it may not be possible to fully understand age-specific variation in fitness components (e.g. variation over life) without considering features of the habitat where each reproductive event takes place. But the very high degree of stratification required by such analysis of age-specific reproductive success as a function of habitat quality may be a major obstacle, and further work is needed to assess whether it is feasible using this data set.

The study of reproductive success indicates that reproductive choices and/or the constraints met during the pre-reproductive stage of life may influence age at recruitment, which may in turn have substantial effects on breeding success over life. To address the overall fitness of recruitment tactics, one may consider measures of lifetime fitness such as Lifetime Reproductive Success (Clutton-Brock 1988), or individual lambda (McGraw and Caswell 1996). This would be a first step to assess whether different tactics are associated with a different total number of viable offspring, and if there is scope for natural selection to operate on age of first breeding in a different manner than understood on the basis of breeding success in the first breeding attempt exclusively. However, the same lifetime fitness may be achieved in very different ways in terms of longevity, age-specific reproductive investment, and choices in terms of habitat selection. As for further investigations, an interesting step to take would be to look over the life course of individuals recruiting at different ages (i.e., recruitment tactics), and determine whether breeding success improves, shows senescent decline, or remains the same across ages for the different recruitment tactics identified above. Our results provided evidence that early recruits (i.e. 3 years old) start their reproductive life with a handicap, as their initial breeding success probability is the lowest (Fig. 3). It would be worth addressing whether recruits experiencing poor breeding success in the year of recruitment catch up and perform increasingly better throughout life. One may also assess whether individuals recruiting at intermediate ages (i.e. recruits of 5 years old which show the highest reproductive success in the recruitment year), are the ones performing best overall (i.e. highest fitness prospects). Last, one may determine if recruiting beyond this age leads to the lowest fitness prospects or not. Assessing fitness differences among reproductive tactics and determining the selective advantages of adopting one tactic or the other will require additional work (e.g. Evolutionary Stable Strategy modeling; Maynard Smith 1982).

Moreover, our work suggests that there may be a substantial level of individual heterogeneity in the study population (i.e. variation in age-specific recruitment tactics leading to variation in reproductive success), and highlights the need to develop multi-state models for estimating transition probabilities while properly accounting for unobserved heterogeneity in reproduction (and in survival) in cases where recapture probability is lower than 1. Multi-state CMR models allowed us to address the influence of observable covariates on recruitment probability, but measurable covariates may not account for heterogeneity in a satisfying manner. Heterogeneity in survival (e.g. frailty) has been looked at in human demography starting some 20 years ago (Vaupel and Yashin 1985), and ecologists have long been concerned with heterogeneity as well (e.g. Burnham and Rexstad 1993; Pledger and Schwarz 2002). However, developments regarding heterogeneity in both survival and reproductive success in wild animal population are only fairly recent (Burnham and Rexstad 1993; Cam et al. 2002b; Pledger and Schwarz 2002; Barbraud and Weimerskirch 2005; Crespin et al. 2006; Gordon et al. 2006; Royle 2008), and require additional efforts.

Acknowledgments We are grateful to the Conseil rale du Finistère for allowing us to work in the nature reserve of Goulien Cap Sizun (Brittany, France). The nature reserve is managed by the Société pour l'Etude et la Protection de la Nature en Bretagne. We also thank Jacques Nisser (Office National de la Chasse et de la Faune Sauvage) for his help with fieldwork, all the people who have helped collecting data over 28 years. We thank Gary White for his help in the use of the program MARK, Jim Hines for advising us on bootstrap methods via the program MSSURVIV, and Florent Bled for his useful methodological suggestions. We are very grateful to David Koons for stimulating discussions on earlier versions of the manuscript. Comments from Blandine Doligez, Thierry Boulinier, Morten Frederiksen, David Thomson, and an anonymous reviewer greatly improved the manuscript. The Max Planck Institute for Demographic Research in Rostock, Germany, funded this research.

Appendix 1

Table 5 Multi-state model pertaining to various biological hyphothesis underlying the recruitment process, while accounting for potential sources of variation in recapture and survival probabilities

Model	Recapture probability	Survival probability	Transition probability
2-state-model-1	p1(a7), p2(.)	S1(c12 + a6), S2(.)	Ψ12(c12 + a6 + q1 + q1c +,(a6*q1)), Ψ21(.)
2-state-model-1bis	p1(a7), p2(.)	S1(c12 + a6), S2(.)	Ψ12(c12 + a6), Ψ21(.)
2-state-model-2	p1(.), p2(.)	S1(c12 + a7), S2(a5)	Ψ12(c12 + a6 + q1 + q1c + (a6*q1)), Ψ21(.)
2-state-model-3	p1(a6), p2(.)	S1(c12+a7), S2(a5)	Ψ12(c12+a6+q1+q1c+(a6*q1)), Ψ21(.)
2-state-model-4	p1(a5), p2(.)	S1(c12+a7), S2(a5)	Ψ12(c12+a6+q1+q1c+(a6*q1)), Ψ21(.)
2-state-model-5	p1(a4), p2(.)	S1(c12+a7), S2(a5)	Ψ12(c12+a6+q1+q1c+(a6*q1)), Ψ21(.)
2-state-model-6	p1(a3), p2(.)	S1(c12+a7), S2(a5)	Ψ12(c12+a6+q1+q1c+(a6*q1)), Ψ21(.)
2-state-model-7	p1(a4), p2(.)	S1(c12+a7), S2(.)	Ψ12(c12+a6+q1+q1c+(a6*q1)), Ψ21(.)
2-state-model-8	p1(a4), p2(.)	S1(.), S2(.)	Ψ12(c12+a6+q1+q1c+(a6*q1)), Ψ21(.)
2-state-model-9	p1(a4), p2(.)	S1(c12), S2(.)	Ψ12(c12+a6+q1+q1c+,(a6*q1)), Ψ21(.)
2-state-model-10	p1(a4), p2(.)	S1(a7), S2(.)	Ψ12(c12+a6+q1+q1c+(a6*q1)), Ψ21(.)
2-state-model-11	p1(a4), p2(.)	S1(c12+a6), S2(.)	Ψ12(c12+a6+q1+q1c+(a6*q1)), Ψ21(.)
2-state-model-12	p1(a4), p2(.)	S1(c12+a5), S2(.)	Ψ12(c12+a6+q1+q1c+(a6*q1)), Ψ21(.)
2-state-model-13	p1(a4), p2(.)	S1(c12+a6), S2(.)	Ψ12(c12+a6+q1+q1c), Ψ21(.)
2-state-model-14	p1(a4), p2(.)	S1(c12+a7), S2(a5)	Ψ12(a6+q1+q1c), Ψ21(.)
2-state-model-15	p1(a4), p2(.)	S1(c12+a6), S2(.)	Ψ12(c12+a6+q1), Ψ21(.)
2-state-model-16	p1(a4), p2(.)	S1(c12+a6), S2(.)	Ψ12(c12+a6+q1c), Ψ21(.)
2-state-model-17	p1(a4), p2(.)	S1(.), S2(.)	Ψ12(c12+q1+q1c), Ψ21(.)
2-state-model-18	p1(a4), p2(.)	S1(c12+a6), S2(.)	Ψ12(c12+a5+q1+q1c), Ψ21(.)
2-state-model-19*	**p1 (a4) p2 (.)**	**S1 (c12+a6) S2 (.)**	**Ψ12 (c12+a6+q2+q2c) Ψ21 (.)**
2-state-model-20	p1(a4), p2(.)	S1(c12+a6), S2(.)	Ψ12(a6+q2+q2c), Ψ21(.)
2-state-model-21	p1(a4), p2(.)	S1(c12+a6), S2(.)	Ψ12(c12+a6+q2), Ψ21(.)
2-state-model-22	p1(a4), p2(.)	S1(c12+a6), S2(.)	Ψ12(c12+a6+q2c), Ψ21(.)
2-state-model-23	p1(a4), p2(.)	S1(c12+a6), S2(.)	Ψ12(c12+q2+q2c), Ψ21(.)
2-state-model-24	p1(a4), p2(.)	S1(c12+a6), S2(.)	Ψ12(c12+a5+q2+q2c), Ψ21(.)

2-state-model-19*: Model selected based on AIC weight – $w_i = 0.99$ (see Table 3).

Appendix 2

Table 6 Multi-state models based on biological hypotheses pertaining to the recruitment process (second approach without covariates, see Methods) various while accounting for potential sources of variation in recapture and survival probabilities

Model	Recapture robability	Survival probability	Transition probability
4-state-model-1	p1(a7),p2(.),p3(.),p4(.)	S1(c12+a7),S2(a5),S3(a5).S4(a5)	Ψ12(c12+a7),Ψ13(c12+a7),Ψ14(c12+a7)
4-state-model-2	p1(.),p2(.),p3(.),p4(.)	S1(c12+a7),S2(a5),S3(a5),S4(a5)	Ψ12(c12+a7),Ψ13(c12+a7),Ψ14(c12+a7)
4-state-model-3	p1(a6),p2(.),p3(.),p4(.)	S1(c12+a7),S2(a5),S3(a5),S4(a5)	Ψ12(c12+a7),Ψ13(c12+a7),Ψ14(c12+a7)
4-state-model-4	p1(a5),p2(.),p3(.),p4(.)	S1(c12+a7),S2(a5),S3(a5),S4(a5)	Ψ12(c12+a7),Ψ13(c12+a7),Ψ14(c12+a7)
4-state-model-5	p1(a4),p2(.),p3(.),p4(.)	S1(c12+a7),S2(a5),S3(a5),S4(a5)	Ψ12(c12+a7),Ψ13(c12+a7),Ψ14(c12+a7)
4-state-model-6	p1(a5),p2(.),p3(.),p4(.)	S1(c12+a7),[S2(a5),S3(a5),S4(a5)]	Ψ12(c12+a7),Ψ13(c12+a7),Ψ14(c12+a7)
4-state-model-7	p1(a5),p2(.),p3(.),p4(.)	S1(c12+a7),[S2(a4),S3(a4),S4(a4)]	Ψ12(c12+a7),Ψ13(c12+a7),Ψ14(c12+a7)
4-state-model-8	p1(a5),p2(.),p3(.),p4(.)	S1(c12+a7),[S2(a3),S3(a3),S4(a3)]	Ψ12(c12+a7),Ψ13(c12+a7),Ψ14(c12+a7)
4-state-model-9	p1(a5),p2(.),p3(.),p4(.)	S1(c12+a7),[S2(a2),S3(a2),S4(a2)]	Ψ12(c12+a7),Ψ13(c12+a7),Ψ14(c12+a7)
4-state-model-10	p1(a5),p2(.),p3(.),p4(.)	S1(.),[S2(a3),S3(a3),S4(a3)]	Ψ12(c12+a7),Ψ13(c12+a7),Ψ14(c12+a7)
4-state-model-11	p1(a5),p2(.),p3(.),p4(.)	S1(a7),[S2(a3),S3(a3),S4(a3)]	Ψ12(c12+a7), Ψ13(c12+a7),Ψ14(c12+a7)
4-state-model-12	p1(a5),p2(.),p3(.),p4(.)	S1(c12+a6),[S2(a3),S3(a3),S4(a3)]	Ψ12(c12+a7),Ψ13(c12+a7),Ψ14(c12+a7)
4-state-model-13	p1(a5),p2(.),p3(.),p4(.)	S1(c12),[S2(a3),S3(a3),S4(a3)]	Ψ12(c12+a7), Ψ13(c12+a7),Ψ14(c12+a7)
4-state-model-14	p1(a5),p2(.),p3(.),p4(.)	S1(c12+a5),[S2(a3),S3(a3),S4(a3)]	Ψ12(c12+a7),Ψ13(c12+a7),Ψ14(c12+a7)
4-state-model-15	p1(a5),p2(.),p3(.),p4(.)	S1(c12+a4),[S2(a3),S3(a3),S4(a3)]	Ψ12(c12+a7),Ψ13(c12+a7),Ψ14(c12+a7)
4-state-model-16	p1(a5),p2(.),p3(.),p4(.)	S1(c12+a6),[S2(a3),S3(a3),S4(a3)]	[Ψ12(c12+a7),Ψ13(c12+a6),Ψ14(c12+a7)]
4-state-model-17	p1(a5),p2(.),p3(.),p4(.)	S1(c12+a6),[S2(a3),S3(a3),S4(a3)]	Ψ12(.),Ψ13(.),Ψ14(.)
4-state-model-18	p1(a5),p2(.),p3(.),p4(.)	S1(c12+a6),[S2(a3),S3(a3),S4(a3)]	Ψ12(c12),Ψ13(c12),Ψ14(c12)
4-state-model-19	p1(a5),p2(.),p3(.),p4(.)	S1(c12+a6),[S2(a3),S3(a3),S4(a3)]	Ψ12(a7),Ψ13(a7),Ψ14(a7)
4-state-model-20	p1(a5),p2(.),p3(.),p4(.)	S1(c12+a6),[S2(a3),S3(a3),S4(a3)]	Ψ12(c12+a6),Ψ13(c12+a6),Ψ14(c12+a6)
4-state-model-21*	**p1(a5) [p2(.) p3(.) p4(.)]**	**S1(c12+a6) [S2(a3) S3(a3) S4(a3)]**	**Ψ13(c12+a6) Ψ13(c12+a6) Ψ14(c12+a6)**

4-state-model-21*: Model selected based on AIC weight – w_i= 0.99 (see Table 3).

References

Agresti A (1990) Categorical data analysis. John Wiley and Sons, New York.
Akaike H (1973) Information theory and an extension of the maximum likelihood principle. In Petron BN, Csaki F (eds.) International Symposium on Information Theory, 2nd edition, pp 267–281. Akademiai Kiadi, Budapest, Hungary.
Arnason AN (1973) The estimation of population size, migration rates, and survival in stratified populations. Research on Population Ecology 15:1–8.
Barbraud C, Weimerskirch H (2005) Environmental conditions and breeding experience affect costs of reproduction in blue petrels. Ecology 86, 3:682–692.
Bled F (2006) Selection de l'habitat de reproduction au sein de la falaise chez un oiseau marin longevif: Rissa tridactyla. Master 2 Biodiversité Ecologie Evolution. Université Paul Sabatier, Toulouse, France.
Boulinier T, Lemel JY (1996) Spatial and temporal variations of factors affecting breeding habitat quality in colonial birds: some consequences for dispersal and habitat selection. Acta Oecologica 17:531–552.
Boulinier T, Danchin E, Monnat J-Y, Doutrelant C, Cadiou B (1996) Timing of prospecting and the value of information in a colonial breeding bird. Journal of Avian Biology 27:252–256.
Boulinier T, Danchin E (1997) The use of conspecific reproductive success for breeding patch selection in territorial migratory species. Evolutionary Ecology 11:505–517.
Brownie C, Hines JE, Nichols JD, Pollock KH, Hestbeck JB (1993) Capture–recapture studies for multiple strata including non-Markovian transitions. Biometrics 49:1173–1187.
Burnham KP, Rexstad E (1993) Modeling heterogeneity in survival rates of banded waterfowl. Biometrics 49:1194–1208.
Burnham KP, Anderson DR (1998) Model selection and inference, a practical information-theoretic approach. Springer-Verlag, New York.
Cadiou B (1993) L'accession à la reproduction: un processus social d'ontogenèse. Cas de la mouette tridactyle (Rissa tridactyla). Thèse. Université de Rennes I .
Cadiou B (1999) Attendance of breeders and prospectors reflects the quality of colonies in the kittiwake Rissa tridactyla. IBIS 141:321–326.
Cadiou B, Monnat JY, Danchin E (1994) Prospecting in the kittiwake, Rissa tridactyla: different behavioural patterns and the role of squatting in recruitment. Animal Behaviour 47:847–856.
Cam E, Monnat JY (2000a) Apparent inferiority in first time breeders in the kittiwake: the role of heterogeneity among age-classes. Journal of Animal Ecology 69:380–394.
Cam E, Hines JE, Monnat JY, Nichols JD, Danchin E (1998) Are adult non breeders prudent parents? The Kittiwake model. Ecology 79:2917–2930.
Cam E, Link WA, Cooch EG, Monnat JY, Danchin E (2002b) Individual covariation between life-history traits: seeing the trees despite the forest. American Naturalist 159:96–105.
Cam E, Monnat JY, Hines JE (2003) Long term fitness consequences of early conditions in the kittiwake. Journal of Animal Ecology 72:411–424.
Cam E, Monnat JY, Royle JA (2004a) Dispersal and individual quality in a long-lived species. OIKOS 106:386–398.
Cam E, Cooch E, Monnat JY (2005) Earlier recruitment or earlier death? On the assumption of homogeneous survival in recruitment studies. Ecological Monographs 75:419–434.
Caswell H, Hastings A (1980) Fecundity, developmental time, and population growth rate: an analytical solution. Theoretical Population Biology 17:71–79.
Charlesworth B (1994) Evolution in age-structured populations. 2nd Edition. Cambridge University Press, Cambridge, UK.
Clutton-Brock TH (1988) Reproductive Success: Studies of Individual Variation in Contrasting Breeding Systems. The University of Chicago Press, Chicago.
Cooch EG, White GC (2006) Mark book: http://www.phidot.org/software/mark/docs/book/.
Cooch EG, Lank DB, Rockwell RF, Cooke F (1999) Body size and recruitment in snow geese. Bird Study 46:112–119.

Crespin L, Harris MP, Lebreton J-D, Frederiksen M, Wanless S (2006) Recruitment to a seabird population depends on environmental factors and on population size. Journal of Animal Ecology 75:228–238.

Curio E (1983) Why do young birds reproduce less well? IBIS 125:400–404.

Danchin E (1987a) The behaviour associated with the occupation of the breeding site in the kittiwake gull Rissa tridactyla: the social status of landing birds. Animal Behaviour 35:81–93.

Danchin E (1988b) Role of behavioural processes in the mechanisms of population regulation in colonial seabirds: Case of the Kittiwake (Rissa tridactyla). Thèse d'Etat, Université Paris VI. France.

Danchin E, Monnat JY (1992) Population dynamic modeling of two neighboring kittiwake Rissa tridactyla colonies. Ardea 80:171–180.

Danchin E, Cadiou B, Monnat JY, Rodriguez Estrella R (1991) Recruitment in long lived birds: conceptual framework and behavioral mechanisms. Proceedings of the International Ornithology Congress 20:1641–1656.

Danchin E, Boulinier T, Massot M (1998) Conspecific reproductive success and breeding habitat selection: implications for the study of coloniality. Ecology 79:2415–2428.

Ens JB, Weissing FJ, Drent R (1995) The despotic distribution and deferred maturity: two sides of the same coin. The American Naturalist 146:625–650.

Fretwell S, Lucas HJ (1970) On territorial behaviour and other factors influencing habitat distribution in birds. Acta Biotheoretica 19:1636.

Gordon AF, Kendall BR, Fitzpatrick JW, Woolfenden GE (2006) Consequences of heterogeneity in survival probability in a population of Florida scrub-jays. Journal of Animal Ecology 75: 921–927.

Hines JE (1994) MSSURVIV User's Manual. National Biological Survey, Patuxent Wildlife Research Center, Laurel, MD 20708 USGS-PWRC. Available at: http://www.mbr-pwrc.usgs.gov/Software.html.

Holt RD, Barfield M (2001) On the relationship between the ideal free distribution and the evolution of dispersal. In Clobert J, Danchin E, Dhondt A and Nichols JD (eds) Dispersal, pp 83–95. Oxford University Press, Oxford, UK.

Lebreton JD, Pradel R (2002) Multistate recapture models: modeling incomplete individual histories. In Morgan BJT and Thomson DL (eds) Statistical Analysis of Data from Marked Bird Population. Journal of Applied Statistics 29:353–369.

Lebreton JD, Burnham KP, Clobert J, Anderson DR (1992) Modeling survival and testing biological hypothesis using marked animals: a unified approach with case studies. Ecological Monographs 62:67–118.

Lebreton JD, Hines JE, Pradel R, Nichols JD, Spendelow JA (2003) Estimation by capture-recapture of recruitment and dispersal over several sites. OIKOS 101:253.

Lunn NJ, Boyd IL, Croxall JP (1994) Reproductive performance of female Antarctic fur seals: the influence of age, breeding experience, environmental variation and individual quality. Journal of Animal Ecology 63:827–840.

Maynard Smith J (1982) Evolution and the Theory of Games, 224 pp. Cambridge University Press, Cambridge.

McGraw JB, Caswell H (1996) Estimation of individual fitness from life-history data. The American Naturalist 147(1):47–64.

Monnat JY, Danchin E, Rogriguez Estrella R (1990) Evaluation de la qualité du milieu dans le cadre de la prospection et du recrutement : le squattérisme chez la mouette tridactyle. Comptes rendus de l'Académie des sciences, Paris, Série 311:391–396.

Neter M, Kutner H, Nachtsheim C, Wasserman W (1996) Applied Linear Statistical Models, 4th edition, 1408 pp. Irwin, Chicago.

Nichols JD (1996) Sources of variation in migratory movements of animal populations: statistical inference and selective review of empirical results for birds. In Rhodes OE, Chesser RK, Smith MH (eds) Population Dynamics in Ecological Space and Time, pp 147–149. University of Chicago Press, Chicago.

Nichols JD, Kendall WL (1995) The use of multi-state capture-recapture models to address questions in evolutionary ecology. Journal of Applied Statistics 22:835–840.

Nichols JD, Brownie C, Hines JE, Pollock KH, Hestbeck JB (1992) The estimation of exchanges among populations or subpopulations. In Lebreton JD, North PM (eds) Mark Individuals in the Study of Bird Populations, pp 265–280. Birkhauser-Verlag, Basel, Switzerland.

Nichols JD, Brownie C, Hines JE, Pollock KH, Hestbeck JB (1993) The estimation of exchanges among populations or subpopulations. In Lebreton JD and North PM(eds) The Use of Marked Individuals in the Study of Bird Population Dynamics: Models, Methods, and Software, pp 265–279. Birkhauser Verlag, Berlin, Germany.

Nol E, Smith JNM (1987) Effects of age and breeding experience on seasonal reproductive success in the song sparrow. Journal of Animal Ecology 56:301–313.

Nur N (1984) Increased reproductive success with age in the Californian gull: due to an increased effort or improvement of skill? OIKOS 43:407–408.

Nur N (1988) The consequences of brood size for breeding blue tits. . Measuring the cost of reproduction: survival, future fecundity, and differential dispersal. Evolution 42(2):351–362.

Pickering SPC (1989) Attendance patterns and behaviour in relation to the experience and pair-bond formation in the wandering albatross Diomeda exulans at South Georgia. IBIS 131: 183–195.

Pledger S, Schwarz CJ (2002) Modeling heterogeneity of survival as a random effect using finite mixtures. Journal of Applied Statistics 29 (Special Issue):315–327.

Porter JM (1988) Prerequisites of recruitment of kittiwakes Rissa tridactyla. IBIS 130:204–215.

Pradel R, Lebreton JD (1999) Comparison of different approaches to the study of local recruitment of breeders. Bird Study 46:74–81.

Pradel R, Hines JE, Lebreton JD, Nichols JD (1997) Capture–recapture survival models taking account of transients. Biometrics 53(1):60–72.

Pyle P, Nur N, Sydeman WJ, Emslie SD (1997) Cost of reproduction and the evolution of deferred breeding in the western gull. Behavioral Ecology 8(2):140–147.

Royle JA (2008) Modeling individual effects in the Cormack–Jolly–Seber model: a state-space formulation. Biometrics In press.

Sakamoto Y, Ishiguro M, Kitagawa G (1986) Akaike Information Criterion Statistics. KTK Scientific Publishers, Tokyo.

Schwarz CJ, Schweigert JF, Arnason AN (1993) Estimating migration rates using tag-recovery data. Biometrics 49:177–193.

Stearns SC (1992) The Evolution of Life Histories. Oxford University Press, New York.

Tuljapurkar S (1990) Delayed Reproduction and Fitness in Variable Environments. Proceedings of the National Academy of Sciences 87:1139–1143.

Vaupel JW, Yashin AI (1985) Heterogeneity's ruses: some surprising effects of selection on population dynamics. American Statistician 39:176–185.

Viallefont A, Cooke F, Lebreton JD (1995) Age specific cost of first time breeding. The Auk 112(1):67–76.

White GC, Burnham KP (1999) Program MARK – survival estimation from populations of marked animals. Bird Study 46:120–139. Available at: http://www.cnr.colostate.edu/-gwhite/mark/mark.htm.

Wiley RH, Rabenold KN (1984) The evolution of cooperative breeding by delayed reciprocity and queuing for favourable positions. Evolution 38:609–621.

Cubic Splines for Estimating the Distribution of Residence Time Using Individual Resightings Data

Rachel M. Fewster and Nathalie J. Patenaude

Abstract Residence time, or stopover duration, is of considerable interest to biologists studying migratory populations. We present a method for estimating the distribution of residence time for a population of southern right whales (*Eubaleana australis*) in the subantarctic Auckland Islands, using photo-ID resightings data from the 1998 winter breeding season. We explain how we can estimate a smooth probability distribution for residence time, by formulating a likelihood penalized for roughness in the residence distribution. The estimated residence distribution is a cubic spline that maximizes the penalized likelihood. The non-parametric approach allows complete flexibility in the shape of the distribution for residence time, and can fit distribution shapes that would be difficult to obtain using a parametric mixture distribution. We show that cubic splines give a general solution to penalized likelihood problems, and fitting the spline is an optimization problem accessible to users of standard statistical software. The methodology is quite general in its potential for fitting smooth probability distributions to data.

Keywords Auckland Islands · Cubic spline · Nonparametric estimation · Penalized likelihood · Photo-identification · Southern right whale *Eubaleana australis* · Stopover duration

1 Introduction

The southern right whale (*Eubaleana australis*) was once widely distributed across the southern hemisphere (Townsend 1935). Extensive coastal and vessel-based whaling reduced their numbers to near extinction during the 19th century (Dawbin 1986). The species was abundant in New Zealand waters, around both the mainland and the subantarctic islands (Dawbin 1986). Although the mainland New Zealand population has yet to show signs of recovery (Patenaude 2003), southern right

---

R.M. Fewster (✉)
Department of Statistics, University of Auckland, Private Bag 92019, Auckland, New Zealand
e-mail: r.fewster@auckland.ac.nz

D.L. Thomson et al. (eds), *Modeling Demographic Processes in Marked Populations*,
Environmental and Ecological Statistics 3, DOI 10.1007/978-0-387-78151-8_17,

whales have reestablished winter calving grounds further south in the subantarctic Auckland Islands, latitude 50°33′ S (Patenaude and Baker 2001).

Little is known about interactions between southern right whales from the Auckland Islands and those from other calving grounds. Significant differentiation in mitochondrial DNA (mtDNA) has been found between calving grounds from the Auckland Islands, Argentina, South Africa, and Western Australia (Patenaude et al. 2007). This suggests strong maternal fidelity among these calving grounds, despite the lack of geographic barriers to movement. However, mtDNA provides no evidence about movements of males and non-calving females among the calving grounds.

Photo-identification studies have revealed occasional records of individual whales sighted in different years in the Auckland Islands and a South Australian wintering ground, but no whale has yet been sighted at both grounds within the same year (Pirzl et al. in review). However, studies in Australian wintering ranges show that movements of hundreds of kilometres are common within a winter season, particularly among adult males and females without a calf of the year (Burnell 2001). Our current study is motivated by the possibility of within-season movements of individual whales between the Auckland Islands and other wintering locations, perhaps including wintering grounds not yet discovered. Within-season movements of males between the Auckland Islands and other calving grounds might mean there is male-mediated gene flow between stocks that are currently thought to be non-interacting, and has management implications for protection of transit waters. Additionally, population estimates of the species might be affected if the same individuals are being counted in two different locations over a winter season.

To gain insight into movement patterns within a season, we investigate the distribution of within-season residence time of whales in Auckland Island waters. If some whales remain in the Auckland Islands for a short time only, they might be making a brief stop on the way to another calving ground. Our analysis is based on photo-ID sightings of individual right whales in the Port Ross area of the Auckland Islands, New Zealand, to investigate the residence time of the animals in the waters around the islands over a single winter breeding season. Data were collected in the austral winter over the 54 days from 26th June to 18th August, 1998 (Patenaude 2002). This period was thought to cover most, although not all, of the residency period of whales in the area. The population appeared to be confined to a 20 km^2 area of shallow waters around Port Ross. Whale densities reached a maximum of about 8 whales per km^2 in the middle of the survey period.

Photographs of whales were collected from small vessels, focusing on callosity patterns, lip ridges, unusual skin pigmentation or prominent scars for individual identification. All photographs were reviewed by several independent photo-ID experts. Where possible, whales were sexed by molecular methods by collecting biopsy samples. Any whale seen associated with a newborn calf in the season was classified as a cow (female adult calving in 1998). All other whales, whether male, female, or unknown, were classified as non-cows.

In total, individual photo-ID records were collected for 34 cows and 188 non-cows. These photo records, for days $1, 2, \ldots, 54$ of the survey period, constitute 222 individual capture histories. About 62% of whales in the data were photographed once only, 25% were photographed twice, 7% three times, and 3% each of four and five times. The mean number of days between the first and last sightings was 8.0 for cows, and 6.5 for non-cows. We analyse cows and non-cows separately, because cows use the coastal waters to give birth and their behaviour with respect to residence time might differ from non-cows.

Previous work on estimating residence time from individual capture histories has used a conceptual equivalence between residence time for migrating animals, and survival time in Jolly–Seber models (Crosbie and Manly 1985; Lady and Skalski 1998; Manske and Schwartz 2000; Efford 2005; Pledger et al. 2008). Time since arrival in the residence model corresponds to age in the survivorship models, and a departure from the stopover location corresponds to a death. Lady and Skalski (1998) developed models for stream residence time of salmon from a single release of marked fish, while Manske and Schwartz (2000) allowed for multiple releases and a distribution on entry time. Pledger et al. (2008) further extended the ideas to allow 'age-dependent survival', so that the time of departure is affected by how long the animal has already stayed at the stopover location. Crosbie and Manly (1985) suggested many of these models and recommended a parametric framework for parsimonious modelling and to allow smooth distributions for residence and entry times. An alternative modelling approach was provided by Whitehead (2001), who used continuous time Markov models and diffusion models to analyse movements of sperm whales between study areas using photo-ID data.

Here we investigate a new alternative, in which we estimate the probability distribution of residence time directly. Pradel et al. (2005) commented that there is currently no statistical method to assess the distribution of residence time from empirical data. We present such a method using a penalized likelihood based on the individual capture histories. Similar to Crosbie and Manly (1985), we aim to generate a smooth distribution of residence time, but our work differs from all previous work by modelling the residence distribution directly with a smooth but non-parametric curve.

The model and likelihood are developed in Section 2. In Section 3 we show how the likelihood can be penalized to guarantee a smooth, but otherwise completely flexible, distributional shape for residence time. We show that the residence time distribution is fitted by a cubic spline that can be computed using standard software. Finally, we apply the cubic spline method to the right whale data in Section 4.

2 Model for Resightings Data

2.1 Log-Likelihood

We use the following notation for the likelihood formulation. Let T_i be a discrete random variable denoting the residence time (in days) of whale i, where

$i = 1, \ldots, n$ and T_i can take values $1, 2, \ldots, \tau$. The random variables $T_1, \ldots, T_n$ all have the same distribution, which we wish to estimate.

Let $a_t = P(T_i = t)$ for $t = 1, \ldots, \tau$. The a_t are the unknown parameters of interest, and they satisfy $\sum_{t=1}^{\tau} a_t = 1$ and $a_t \geq 0$ for all t.

We number the days of the survey period $1, 2, 3, \ldots, D$. For whales $i = 1, \ldots, n$, let f_i be the day on which whale i was first photographed, and ℓ_i be the day whale i was last photographed. We assume that whale i was present in the region at least for days $f_i, (f_i + 1), \ldots, \ell_i$, so the total length of stay is $T_i \geq \ell_i - f_i + 1$.

Let ω_{id} be an indicator such that $\omega_{id} = 1$ if whale i was photographed on day d, and 0 otherwise. Define p_d to be the probability that any whale present in the region on day d is photographed. These probabilities are allowed to vary by day, but are the same for all whales. The observations for whale i are given by the capture history $\boldsymbol{\omega}_i = (\omega_{i1}, \ldots, \omega_{iD})$. Sightings of the same whale on different days are assumed to be independent.

To link residence time with the observations, we also require a distribution for the entry day of whale i, E_i, which is the day on which whale i first entered the region. Let $q_e = P(E_i = e)$ for allowable entry days $e_L, \ldots, 1, 2, \ldots, e_U$, where the initial entry day e_L may occur before day 1 when the survey began. The nuisance parameters $q_{e_L}, \ldots, q_{e_U}$ satisfy $\sum_e q_e = 1$ and $q_e \geq 0$ for all e.

To form the log-likelihood, we assume that all whales $i = 1, \ldots, n$ are independent, and that residence time t is independent of entry day e for each whale. Let $\boldsymbol{\theta} = (a_1, \ldots, a_\tau, q_{e_L}, \ldots, q_{e_U}, p_1, \ldots, p_D)$ be the vector of all unknown parameters. Because residence time and entry day are unknown, we partition over all possible pairs (t, e) for whale i, and use the law of total probability to give the log-likelihood:

$$\log \mathcal{L}(\boldsymbol{\theta}\,; \boldsymbol{\omega}_1, \ldots, \boldsymbol{\omega}_n) = \sum_{i=1}^{n} \log \left\{ \sum_{t=1}^{\tau} \sum_{e=e_L}^{e_U} a_t\, q_e P(\boldsymbol{\omega}_i \mid t, e\,; \boldsymbol{\theta}) \right\}.$$

Now $P(\boldsymbol{\omega}_i \mid t, e) = 0$ if $t < \ell_i - f_i + 1$, the known minimum length of stay. For given $t \geq \ell_i - f_i + 1$, the possible days of entry range from $e = \ell_i - t + 1$ (for which the last sighting day ℓ_i coincides with the day of departure), to $e = f_i$ (for which the first sighting day f_i coincides with the day of entry). For all other (t, e) combinations, $P(\boldsymbol{\omega}_i \mid t, e) = 0$. Define $D_i(t, e) = \{e, \ldots, f_i - 1, \ell_i + 1, \ldots, e + t - 1\}$ to be the set of days that whale i must have remained unobserved strictly before the first sighting f_i and strictly after the last sighting ℓ_i, if it was present for t days and entered the region on day e. Then:

$$\log \mathcal{L}(\boldsymbol{\theta}; \boldsymbol{\omega}_1, \ldots \boldsymbol{\omega}_n) = \sum_{i=1}^{n} \log \left\{ \sum_{t=\ell_i - f_i + 1}^{\tau} \sum_{e=\ell_i - t + 1}^{f_i} a_t q_e \prod_{d=e}^{e+t-1} p_d^{\omega_{id}} (1 - p_d)^{1-\omega_{id}} \right\}$$

$$= \sum_{i=1}^{n} \log \left\{ \sum_{t=\ell_i - f_i + 1}^{\tau} \sum_{e=\ell_i - t + 1}^{f_i} a_t q_e \prod_{d \in D_i(t,e)} (1 - p_d) \prod_{d=f_i}^{\ell i} p_d^{\omega_{id}} (1 - p_d)^{1-\omega_{id}} \right\}$$

$$= \sum_{i=1}^{n} \sum_{d=f_i}^{\ell_i} \left\{ \omega_{id} \log(p_d) + (1 - \omega_{id}) \log(1 - p_d) \right\}$$

$$+ \sum_{i=1}^{n} \log \left\{ \sum_{t=\ell_i - f_i + 1}^{\tau} \sum_{e=\ell_i - t + 1}^{f_i} a_t q_e \prod_{d \in D_i(t,e)} (1 - p_d) \right\}, \quad (1)$$

where the first term of (1) is gained by taking out the factor $\prod_{d=f_i}^{\ell_i} p_d^{\omega_{id}} (1 - p_d)^{1-\omega_{id}}$ that is common to all allowed combinations of (t, e).

The likelihood (1) can be maximized with respect to the parameters $\boldsymbol{\theta}$, subject to the constraints that $\sum_{t=1}^{\tau} a_t = \sum_{e=e_L}^{e_U} q_e = 1$ and $0 \le a_t, q_e, p_d \le 1$ for all t, e, and d. The daily sighting probabilities $p_1, \ldots, p_D$ are estimated primarily from the first component of (1), which does not involve the other parameters. The parameters for residence time and entry time, a_t and q_e, are estimated only from the second component of (1). This component does not involve any of the sightings data from days f_i to ℓ_i for whale i. Information about $a_t = P(T_i = t)$ and $q_e = P(E_i = e)$ is gained from the number of days that the whale would have to be unobserved at the start and end of its residence period, if it entered at time e and stayed for time t. Given the sighting probabilities $p_1, \ldots, p_D$, it is likely that there are a few days on which the whale was unobserved at either end of its stay, but it is unlikely that the whale would be unobserved for very long runs of days. This is sufficient to estimate the distribution of T_i and E_i across all whales.

The likelihood (1) does not contain a component for whales never sighted. We therefore restrict all inference to the whales that were sighted at least once during the survey period. Whales seen at least once might not be representative of all whales with respect to residence time, because the whales staying the shortest times are those least likely to be seen. We investigate the effect of this restriction in Section 4.3.

2.2 Fitting Options

There are several options for using the log-likelihood (1) to estimate the distribution of residence time. The simplest option is to maximize (1) directly with respect to all the parameters $\boldsymbol{\theta} = (a_1, \ldots, a_\tau, q_{e_L}, \ldots, q_{e_U}, p_1, \ldots, p_D)$, subject to the constraints $\sum_{t=1}^{\tau} a_t = \sum_{e=e_L}^{e_U} q_e = 1$ and $0 \le a_t, q_e, p_d \le 1$ for all t, e, and d. This gives flexibility to each parameter, so that a_t does not vary smoothly with t but has the freedom to spike up and down from one value of t to the next, and similarly for q_e and p_d for values of e and d respectively. This behaviour is appropriate for the daily sightings probabilities $p_1, \ldots, p_D$, because the probabilities do not vary smoothly through time due to abrupt changes in weather and sea state. However, we might wish to estimate smooth distributions for residence time ($a_t = P(T_i = t)$) and entry time ($q_e = P(E_i = e)$).

In the case of residence time, the aim of the analysis is to identify the important qualitative features of the residence distribution, in particular to investigate evidence for transient animals. This motivates the need to estimate a smooth distribution of residence time. One possibility is to smooth the output $\hat{a}_1, \ldots, \hat{a}_\tau$ from the direct maximization, using a sum-preserving smoothing spline such as the function smooth.spline in the software R (R Development Core Team 2007). However, it is preferable to incorporate the smoothness directly into the model fitting, rather than to apply a smoother to the output from an unsmoothed model. The difference is akin to the difference between fitting a generalized additive model (GAM: Hastie and Tibshirani 1990) and smoothing the output from a generalized linear model.

It is possible to fit smooth parametric distributions to the residence distribution, for example using a mixture of distributions $\mathcal{P}_1$ and $\mathcal{P}_2$ such that $a_t = \mu\mathcal{P}_1(t)+(1-\mu)\mathcal{P}_2(t)$, where $\mathcal{P}_1$ and $\mathcal{P}_2$ represent the probability mass functions of distributions such as Poisson or discretized log-normal, and μ is a mixing parameter to be estimated. This has the disadvantage that the number of mixing distributions must be stipulated beforehand, and the final estimated distribution is constrained by the shapes of the mixing distributions.

The aim of this article is to show how we can use cubic splines to estimate a distribution for residence time which has true flexibility of shape while remaining smooth. This non-parametric approach ensures that the distribution shape is not influenced by parametric modelling choices. The procedure is motivated by the use of cubic splines in GAMs (Hastie and Tibshirani 1990), but it applies to a general likelihood such as (1), which cannot be cast into the GAM framework. In GAMs, a flexible regression line is fitted assuming a specified response distribution and link function. By contrast, here we are attempting to estimate a flexible distributional shape.

For the nuisance parameters for entry time, $q_{e_L}, \ldots, q_{e_U}$, either a smooth parametric distribution or a completely flexible distribution may be considered. Using a smooth parametric distribution has the disadvantage that choices of parametric shape for the nuisance parameters might unduly influence the shape of the estimated residence distribution. Additionally, animals might arrive into the area in bursts (for example in social groups or in certain weather conditions), so a parametric distribution might not be able to capture the true pattern of entries.

We investigate both options for entry distribution. For a flexible entry model, we allow each q_e to be estimated individually so that the model choices exert minimal influence on the estimated distribution for residence time. The entry distribution is not of primary interest, so we are not concerned about the lack of smoothness in the output. We also investigate a parametric entry model, in which q_e is obtained from a discretized beta distribution with parameters α and β to be estimated, and with the support scaled to the interval $[e_L, e_U + 1]$. Under the beta entry model,

$$q_e = \int_e^{e+1} f\left(\frac{x - e_L}{e_U + 1 - e_L}\,;\ \alpha, \beta\right) dx \quad \text{for } e = e_L, \ldots, e_U, \tag{2}$$

where $f(\cdot;\alpha,\beta)$ is the probability density function of the beta (α, β) distribution. This model is similar to an entry model used by Crosbie and Manly (1985). A third alternative, which we do not trial here, is to use cubic splines to provide a flexible but smooth distribution for entry time, as we will demonstrate for the residence time distribution.

The set of possible entry days $e_L, \ldots, e_U$ need not correspond exactly to the sightings period $1, \ldots, D$. Because we are restricting inference to the whales seen at least once between day 1 and day D, we stipulate that all whales must have entered by day D: therefore $e_U = D = 54$ in all analyses. The choice of lower boundary e_L is less straightforward, especially in our context where whales began entering the survey region well before sightings commenced. Taking into account biological opinion and trials of several different values of e_L, we use $e_L = -26$ for all analyses presented here. This choice corresponds to the point at which the probability of arrival in the first 10 days was estimated at about 1%. If e_L is moved 10 days earlier ($e_L = -36$), the probability mass for the extra 10 days is estimated at less than 1%, and conclusions about q_e, a_t, and p_d do not change. The choice of $e_L = -26$ allows whales to enter the survey region from 30th May onwards, while sightings began on 26th June. By a similar process, we fix the upper limit for the support of T_i as $\tau = 60$ days.

3 Cubic Splines for Estimating Residence Time

3.1 Penalized Likelihood

We now show how we can estimate values $a_1, \ldots, a_\tau$ for the residence time distribution, where $a_t = P(T_i = t)$, such that the shape of the distribution formed by $a_1, \ldots, a_\tau$ is smooth but is otherwise unconstrained.

Consider a smooth, continuous function $s(x)$ defined on the interval $0 \leq x \leq \tau + 1$, where (as before) τ is the maximum supported value of the discrete distribution of T_i. The values $s(t)$ at integers $t = 1, 2, \ldots, \tau$ are to be the required values $a_t = P(T_i = t)$ that determine the distribution of T_i: thus $s(t) = a_t = P(T_i = t)$ for $t = 1, 2, \ldots, \tau$. Note that T_i is a discrete random variable, and $s(x)$ is not a probability density function; rather, $s(x)$ is a continuous function that is evaluated at the discrete points $s(1), \ldots, s(\tau)$ to yield the τ discrete probabilities for T_i. This formulation is designed to produce a discrete distribution with a smooth shape. It might seem more natural to approximate T_i by a continuous random variable with probability density function $s(x)$, but the discrete formulation can readily be solved using standard approaches, and is easier to handle than the continuous approach which would involve maximizing the likelihood under the integral constraint $\int_0^{\tau+1} s(x)\, dx = 1$.

The basis of our method is to find the function $s(x)$ that maximizes an expression involving the log-likelihood together with penalty terms that ensure s conforms to desired properties. To ensure that s is smooth, we introduce a roughness penalty:

$$\lambda \int_0^{\tau+1} s''(x)^2\, dx. \tag{3}$$

Here, λ is the 'smoothing parameter'. By integrating the squared second derivative of s, we penalize curvature in s, and disallow rough or 'wiggly' functions. Large values of λ force s to be increasingly smooth. Equation (3) is the same as the penalty term for GAMs (Hastie and Tibshirani 1990:27).

Additionally, the probabilities $a_t = s(t)$ should be ≥ 0 and sum to 1, for $t = 1, \ldots, \tau$. To ensure that $a_1, \ldots, a_\tau \geq 0$, the usual procedure would be to use a parameter transformation, for example using $\log(a_1), \ldots, \log(a_\tau)$ as inputs to the likelihood maximization, because their values are unconstrained in $\mathbb{R}$. However, this is not possible in our context, because the a_t are closely connected with $s(x)$, and $s(x)$ and its derivatives are required elsewhere in the objective function (for example, in the penalty term (3)). Instead, we can impose a second penalty term to ensure positivity of $a_1, \ldots, a_\tau$. Because the penalized likelihood must be differentiable, the penalty term must also be smooth, so discontinuous step functions such as 0 for $a_t \geq 0$, 10^6 for $a_t < 0$ are not appropriate. Although it is possible to force strict positivity of a_t, we have found that the most effective strategy is to allow a small amount of leeway. Forcing strict positivity with a smooth $s(x)$ makes it difficult for genuinely unsupported values of t to yield the estimate $\hat{a}_t = 0$. We therefore propose the penalty term $\sum_{t=1}^{\tau} \exp(-10^6 a_t - 20)$, for which the penalty associated with a_t is of the order 10^{-9} when $a_t = 0$, escalating to 10^{34} when $a_t = -0.0001$.

For the condition $\sum_{t=1}^{\tau} a_t = 1$, we add a third penalty term: $K\left(\sum_{t=1}^{\tau} a_t - 1\right)^2$, where K is a large positive number. We use $K = 10^5$, which produces results for $\sum_{t=1}^{\tau} \hat{a}_t$ between 1 and 1.001 for all our analyses. Our primary interest is in distributional shape, so this approximation is adequate for our purposes.

Fulfilling constraints for the nuisance parameters is more straightforward. We ensure $0 \leq p_d \leq 1$ for $d = 1, \ldots, D$ by using input parameters $\log\{p_d/(1 - p_d)\}$, which take values in $\mathbb{R}$, so that the domain for numerical maximization is unconstrained. For the flexible entry model, we require $q_e \geq 0$ for $e = e_L, \ldots, e_U$ and $\sum_e q_e = 1$. The first constraint is satisfied by using the real-valued parameters $\log(q_e)$ for input to the maximization, and the second is satisfied by adding an extra penalty term $K(\sum_{e=e_L}^{e_U} q_e - 1)^2$, where $K = 10^5$ as before. For the beta entry model (2), only α and β need to be estimated, and the constraints $\alpha, \beta > 0$ are ensured by using the real-valued inputs $\log(\alpha)$ and $\log(\beta)$.

The final expression for the penalized negative log-likelihood is:

$$PL(\boldsymbol{\theta}) = -\log \mathcal{L}(\boldsymbol{\theta}\,; \boldsymbol{\omega}_1, \ldots, \boldsymbol{\omega}_n) + \lambda \int_0^{\tau+1} s''(x)^2\,dx + \sum_{t=1}^{\tau} \exp(-10^6 a_t - 20)$$
$$+\, 10^5 \left(\sum_{t=1}^{\tau} a_t - 1\right)^2 + 10^5 \left(\sum_{e=e_L}^{e_U} q_e - 1\right)^2, \tag{4}$$

where the last term is automatically zero when using the beta entry model (2). The parameters are $\boldsymbol{\theta} = (a_1, \ldots, a_\tau, q_{e_L}, \ldots, q_{e_U}, p_1, \ldots, p_D)$ for the flexible entry model, or $\boldsymbol{\theta} = (a_1, \ldots, a_\tau, \alpha, \beta, p_1, \ldots, p_D)$ for the beta entry model. The log-likelihood $\log \mathcal{L}(\boldsymbol{\theta}; \boldsymbol{\omega}_1, \ldots, \boldsymbol{\omega}_n)$ is given by (1), and the parameters are connected to the smooth function $s(x)$ via $a_t = s(t)$ for $t = 1, \ldots, \tau$.

Our overall problem is to find a smooth function s and additional parameters $\{q_e, p_d\}$ that minimize expression (4). The space of smooth functions is defined as the space of all functions s for which s, s' and s'' are continuous on the interval $(0, \tau + 1)$. Additionally, because $s(x)$ is an envelope for the distribution of T_i, we constrain $s(0) = s(\tau + 1) = 0$. We also add the constraints that $s''(0) = s''(\tau + 1) = 0$, so that s is linear or constant at the endpoints: these constraints are reasonable in context and are needed for the optimization (see Appendix).

If the nuisance parameters $\{q_e, p_d\}$ are temporarily considered fixed, we can rewrite $P\mathcal{L}(\boldsymbol{\theta})$ in (4) as the following expression over the function-space:

$$G(s) = \lambda \int_0^{\tau+1} s''(x)^2 dx - H\left(s(1), \ldots, s(\tau)\right), \tag{5}$$

where $H(a_1, \ldots, a_\tau)$ is a function of τ arguments that is formed from the right-hand side of (4) by excluding the term $\lambda \int_0^{\tau+1} s''(x)^2 dx$ and substituting (1) for $\log \mathcal{L}$. Our optimization problem is to minimize $G(s)$ over all functions $s : (0, \tau + 1) \to \mathbb{R}$ with s, s', and s'' continuous, and with $s(0) = s(\tau + 1) = s''(0) = s''(\tau + 1) = 0$.

Minimizing expressions of the form (5) is a standard problem in the calculus of variations (e.g. Reinsch 1967). Surprisingly, the expression is minimized by a unique function s that can be written down in closed form. The minimizing s is a cubic spline, which means it is a piecewise cubic with joins or knots at the integers $1, 2, \ldots, \tau$:

$$s(x) = a_t + b_t(x-t) + c_t(x-t)^2 + d_t(x-t)^3 \quad \text{for } t \le x < t+1\,;\ t = 0, 1, \ldots, \tau. \tag{6}$$

The coefficients $\{a_t, b_t, c_t, d_t\}$ are all readily computable. Note the consistency of notation, so that $a_t = s(t) = P(T_i = t)$ as in the rest of the manuscript.

A sketch of the derivation of (6) is given in the Appendix. It should be noted that the methodology is quite general. Any optimization of the form

$$\text{minimize} \left\{\lambda \int_a^b s''(x)^2 dx - H\Big(s(x_1), \ldots, s(x_m)\Big)\right\} \tag{7}$$

has a cubic spline solution. The cubic spline is a result of the term $\int_a^b s''(x)^2 dx$, which can be shown to guarantee a zero fourth derivative of the optimizing s (Appendix equation (18)). The third derivative of s is a step function with jumps given by the partial derivatives of H (equation (20)), and the other coefficients are determined by the conditions that s, s', and s'' are continuous, as we demonstrate below. The penalty term $\int_a^b s''(x)^2 dx$ produces outputs that accord well with visual perceptions of smoothness, so the technique is applicable to any problem in which the aim is to construct a smooth function that feeds into a likelihood at a set of discrete values.

Initially, it appears that there are four parameters a_t, b_t, c_t, and d_t to be estimated for each $t = 0, \ldots, \tau$ in order to compute the spline s. In the next section, we show how the smoothness constraints of the spline reduce this burden to just one parameter for each value of t. In particular, all coefficients may be expressed in terms of just $a_1, \ldots, a_\tau$, or alternatively (and more conveniently) in terms of $c_1, \ldots, c_\tau$. The optimal solution for s can then be found simply by minimizing the penalized log-likelihood (4) with respect to the τ parameters $c_1, \ldots, c_\tau$, and using the optimal values $\{\hat{c}_t\}$ to compute the other spline coefficients $\{\hat{a}_t, \hat{b}_t, \hat{d}_t\}$, and particularly the required parameters $\{\hat{a}_t\}$.

Overall, the estimated $\hat{\theta} = (\hat{a}_1, \ldots, \hat{a}_\tau, \hat{q}_{e_L}, \ldots, \hat{q}_{e_U}, \hat{p}_1, \ldots, \hat{p}_D)$ is obtained by minimizing (4) simultaneously with respect to the spline parameters $c_1, \ldots, c_\tau$ detailed in the next section, and the nuisance parameters $(q_{e_L}, \ldots, q_{e_U}, p_1, \ldots, p_D)$.

3.2 Computation of the Cubic Spline

Here we show how continuity conditions on the spline s enable all spline parameters $\{a_t, b_t, c_t, d_t\}$ to be expressed in terms of $\{c_t\}$ only. We emphasize that the log-likelihood is not involved in the following development, showing that the details of the spline computation are general to any problem of the form (7), regardless of the specific model, data, and log-likelihood.

The following relations are obtained by differentiating (6) repeatedly and imposing continuity conditions:

$$s''(x) \text{ is continuous} \implies d_t = \frac{c_{t+1} - c_t}{3} \text{ for } t = 0, 1, \ldots, \tau. \tag{8}$$

$$s'(x) \text{ is continuous} \implies b_t + 2c_t + 3d_t = b_{t+1} \text{ for } t = 0, 1, \ldots, \tau. \tag{9}$$

$$s(x) \text{ is continuous} \implies a_t + b_t + c_t + d_t = a_{t+1} \text{ for } t = 0, 1, \ldots, \tau. \tag{10}$$

We also have the following constraints discussed above:

$$s(0) = s(\tau + 1) = 0 \implies a_0 = a_{\tau+1} = 0. \tag{11}$$

$$s''(0) = s''(\tau + 1) = 0 \implies c_0 = c_{\tau+1} = 0. \tag{12}$$

$$s(x) \equiv 0 \text{ for } x \geq \tau + 1 \implies b_{\tau+1} = d_{\tau+1} = 0. \tag{13}$$

Conditions (11) and (13) are natural requirements of the function s enveloping the probability distribution of T_i, which is supported only between $x = 1$ and $x = \tau$. Condition (12) is required for the optimization of (5) (see Appendix).

We compute the spline as follows. Equation (10) shows that b_t can be expressed in terms of coefficients a, c, and d; substituting for d from (8) puts b_t in terms of

a and c coefficients only. Inserting these expressions into (9) gives the following equation for a and c:

$$a_{t+2} - 2a_{t+1} + a_t = \frac{1}{3}c_{t+2} + \frac{4}{3}c_{t+1} + \frac{1}{3}c_t \quad \text{for } t = 0, 1, \ldots, \tau - 1. \tag{14}$$

Equation (14) can conveniently be rewritten in matrix notation. Let $\boldsymbol{a} = (a_1, \ldots, a_\tau)$ and $\boldsymbol{c} = (c_1, \ldots, c_\tau)$. Note that $a_0 = a_{\tau+1} = c_0 = c_{\tau+1} = 0$. Let M and V be $\tau \times \tau$ matrices such that:

$$M_{ij} = \begin{cases} 1 & \text{if } i = j+1 \text{ or } i = j-1, \\ -2 & \text{if } i = j, \\ 0 & \text{otherwise.} \end{cases} \qquad V_{ij} = \begin{cases} \frac{1}{3} & \text{if } i = j+1 \text{ or } i = j-1, \\ \frac{4}{3} & \text{if } i = j, \\ 0 & \text{otherwise.} \end{cases}$$

Equation (14) is then:

$$M\boldsymbol{a} = V\boldsymbol{c} \implies \boldsymbol{a} = M^{-1}V\boldsymbol{c}. \tag{15}$$

We can show that the $\tau \times \tau$ matrix M^{-1} can be computed by a simple closed form expression:

$$M_{ij}^{-1} = \begin{cases} -\frac{(\tau+1-i)j}{\tau+1} & \text{if } i \geq j+1, \\ -\frac{(\tau+1-j)i}{\tau+1} & \text{if } i \leq j. \end{cases}$$

This makes it easiest to work with the c coefficients rather than the a coefficients, because we have a closed form for the matrix inverse M^{-1} needed to compute $\boldsymbol{a}$ from $\boldsymbol{c}$.

The development above gives a complete formulation of the cubic spline s in terms of only the second derivative coefficients $c_1, \ldots, c_\tau$. The a coefficients are expressed in terms of c by (15); the d coefficients by (8); and the b coefficients by using the previous expressions for a and d in (10).

We can also derive a useful matrix expression for the penalty term:

$$\lambda \int_0^{\tau+1} s''(x)^2\, dx = 2\lambda \boldsymbol{c}^T V \boldsymbol{c}. \tag{16}$$

All the results above apply for any problem with smoothing penalty term $\lambda \int_0^{\tau+1} s''(x)^2 dx$ and conditions (11)–(13), regardless of the log-likelihood and other penalty terms in H. The spline is estimated by minimizing equation (4) with respect to $c_1, \ldots, c_\tau$, using (16) to compute the smoothness penalty and (15) to compute the parameters $a_1, \ldots, a_\tau$ needed for the other terms of (4).

Code for the analyses is written in the programming languages C, SPLUS, and R, and likelihoods are maximized numerically using the downhill simplex routine amoeba from Press et al. (1988), with multiple restarts to ensure convergence. To be conservative, we use 40 restarts; however, observation suggests that 6–8 restarts are usually sufficient.

4 Results

4.1 Fitting Methods

Figures 1 and 2 show results from the three fitting methods for cows and non-cows. Row A shows results from maximizing the unpenalized likelihood (1) directly with respect to all parameters in $\boldsymbol{\theta}$. In each case, a basic shape is evident for the residence distribution, but the jagged output distracts from the global picture. The output from smoothing the jagged distribution is overlaid on the barplot, but is not ideal because the smooth output is not the optimal solution to an explicitly formulated likelihood or penalized likelihood.

Row B shows results from the flexible entry model, in which the residence distribution is estimated by a cubic spline, but the nuisance parameters in the entry distribution are allowed to be completely flexible. Row C shows the results from the beta entry model (2), in which the entry distribution is constrained to a beta distribution scaled onto the interval $(-26, 54)$.

The spline fits for residence time (rows B and C) demonstrate that the estimated distribution is both smooth and flexible. The smoothing parameters were $\lambda = 0.5 \times 10^6$ (0.5 M) and $\lambda = 1 \times 10^6$ (1 M), selected to give outputs that retained a good visual balance between smoothness and flexibility. Suitable choices for λ depend upon several effects, including the size of the log-likelihood term in (4) against which the smoothness penalty has to compete, the length of the interval $(0, \tau + 1)$ for the penalty in (3), and the desired smoothness in the output. For our applications with $\tau = 60$, we have found that values of λ in the range of 100–1000 times the magnitude of the log-likelihood are effective.

Especially for the flexible entry model, the estimated distributions have shapes that would not be easily captured by parametric modelling. For both cows and non-cows, the estimated distribution is broad but essentially unimodal. Short-stay peaks suggestive of a group of transient animals are absent in both cases.

The flexible entry models suggest that entry into the survey region occurred in two to three bursts, with non-cows entering earlier than cows. The beta entry model appears to be an over-simplification of the arrival process, so it seems preferable to retain complete flexibility in the nuisance parameters. The impact on the residence time distributions of using the beta entry model is relatively slight, given the level of oversimplification, but it appears to make the residence distributions less broad and more symmetric.

The sighting probabilities are relatively insensitive to the model choice, which is expected because they are estimated primarily from the first term of the log-likelihood (1), which does not involve the parameters for residence time or entry time. For both cows and non-cows, there were 17 days for which weather precluded sighting effort and $p_d = 0$. Of the remaining 37 days of the survey period, the mean estimated daily resighting probability was about 0.10 for cows (standard deviation 0.09), and 0.07 for non-cows (standard deviation 0.04), consistently across the

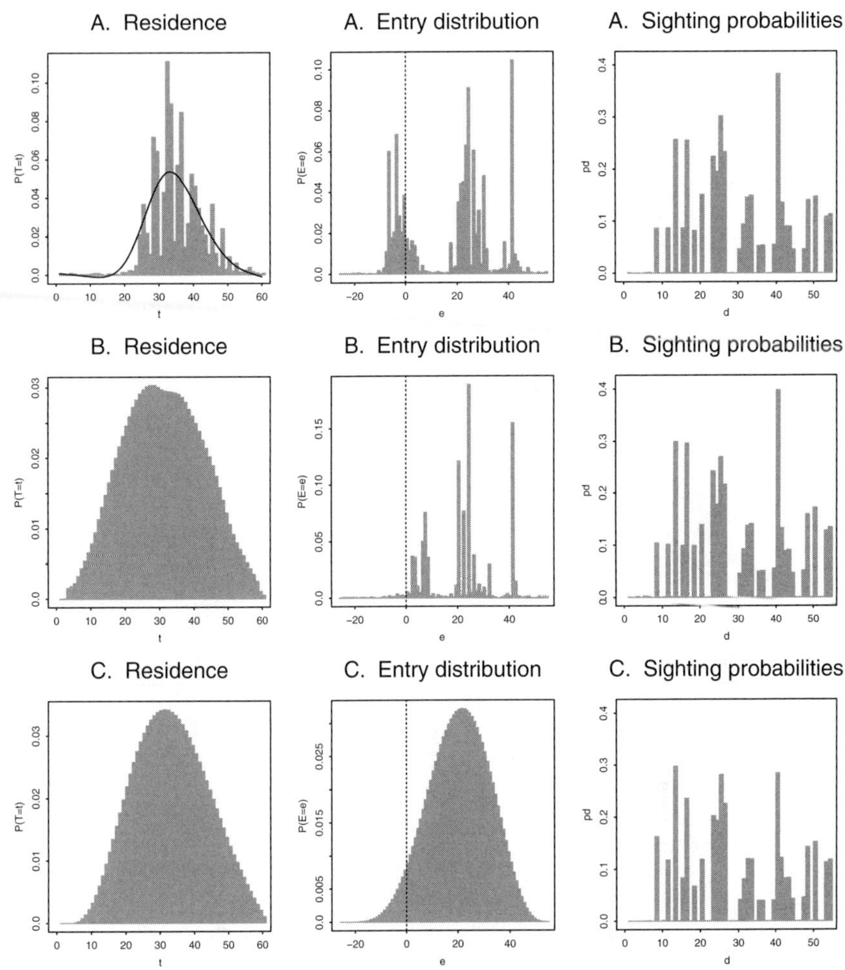

Fig. 1 Results from the three fitting methods for cows. **A**. Direct maximization of the likelihood with respect to all parameters in θ. **B**. Flexible entry model, with residence time estimated with a cubic spline. **C**. Beta entry model, with residence time estimated with a cubic spline. For each fit, plots are shown of the estimated discrete distributions of residence time and entry distribution, and estimated sightings probabilities for each day of the survey period are shown. For each spline fit, the smoothing parameter is $\lambda = 0.5 \times 10^6$ (0.5 M). For the direct maximization (case A), a smoothing spline is applied to the output for residence distribution and is shown as a curve overlaying the barplot. The estimated beta parameters for the entry distribution in case C are $\alpha = 6.2$, $\beta = 4.6$

different models. It is biologically reasonable that the sighting probabilities should be higher for cows than for non-cows, especially after calving. Cows spend more time resting at the surface, move slowly, and after calving gain extra visibility due to the presence of their newborn. Additionally, survey effort was sometimes directed specifically at cows.

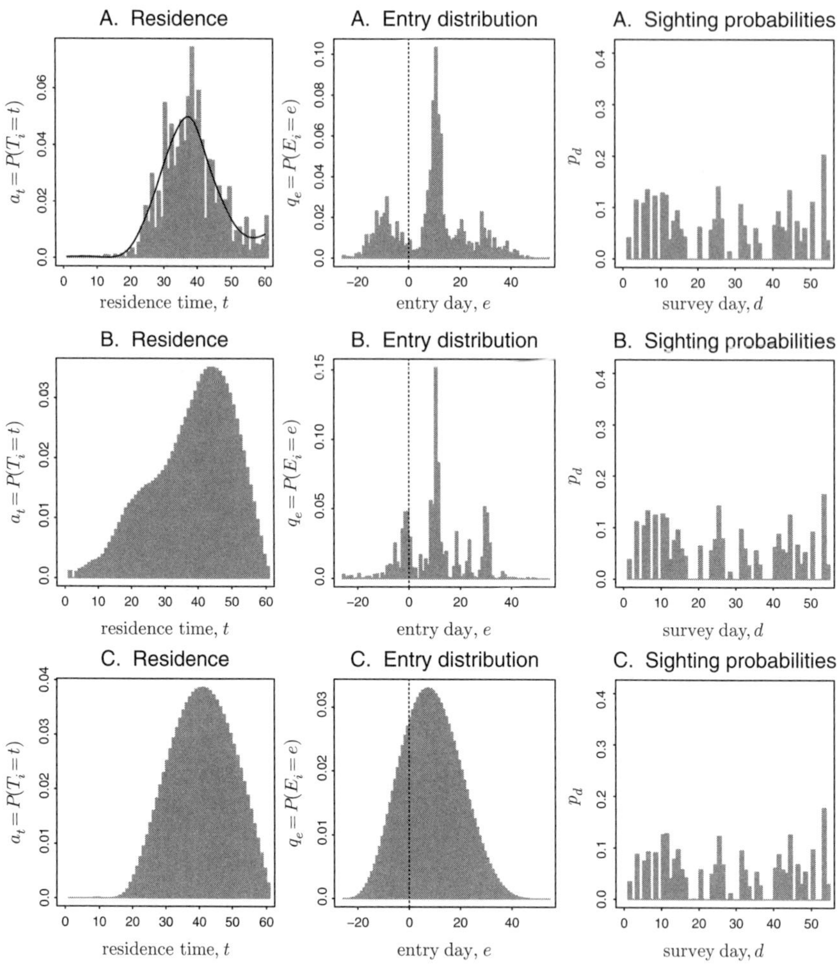

Fig. 2 Results from the three fitting methods for non-cows. Details as for Fig. 1. The smoothing parameter is $\lambda = 1 \times 10^6$ (1 M). The estimated beta parameters for the entry distribution in case C are $\alpha = 4.9$, $\beta = 6.5$

4.2 Variance and Characteristics of the Cubic Spline Fit

In principle, we could assess the variance of the cubic spline fit for residence time by obtaining pointwise confidence intervals for the parameters $a_1, \ldots, a_\tau$, for example using bootstrap resampling. However, this is problematic because the parameters are highly correlated, being constrained to be part of the same smooth curve. A bootstrap resample that shifts the entire curve to the left or to the right will create the impression of high pointwise variance in the individual parameters, when it might in fact represent biologically similar conclusions with respect to the shape and characteristics of the fitted output.

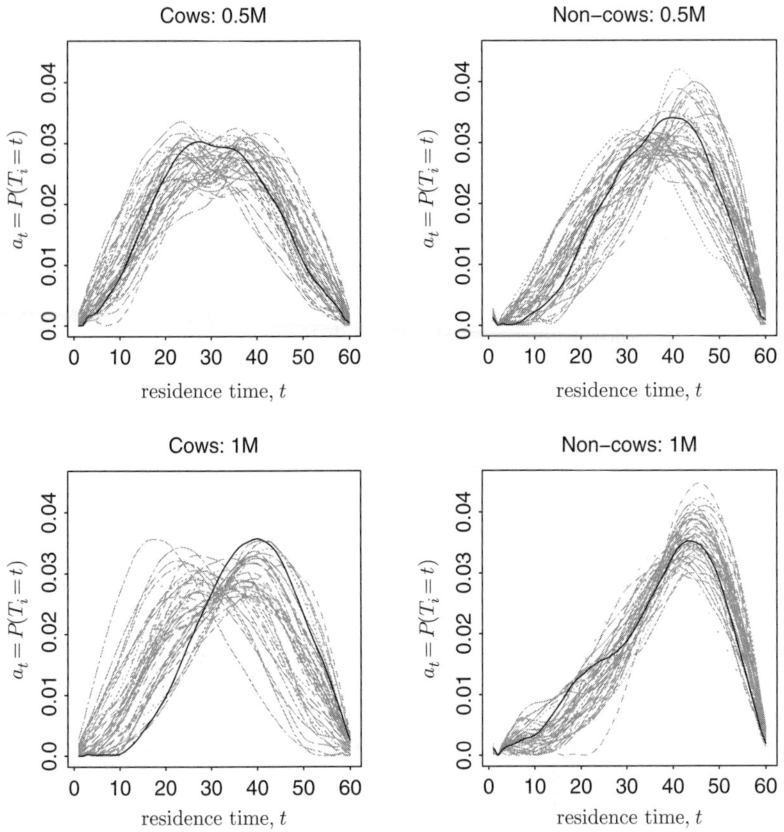

Fig. 3 Traces of the residence time distribution obtained from 40 replicates of the parametric bootstrap for cows and non-cows, for smoothing parameter $\lambda = 0.5 \times 10^6$ (0.5 M) and $\lambda = 1 \times 10^6$ (1 M). The flexible entry model is used

Figure 3 shows traces of 40 bootstrap replicates from the fitted flexible entry models. Because we are restricting inference to the animals in our data set, the parametric bootstrap is used rather than the non-parametric bootstrap. For each bootstrap replicate, a new data set of cows or non-cows is generated from the fitted model, with the same number of animals as in the original data sets. The flexible entry model is refitted to the new data, with the optimization started from a neutral (uniform) starting point for each set of parameters $\{a_t, q_e, p_d\}$. The original fitted residence distributions, which give the 'correct answer' for each of the bootstrap replicates, are shown by bold lines in Fig. 3. Figure 3 also shows the effects of using different smoothing parameters for each of the cows and non-cows data sets.

Figure 3 shows a number of points of interest. Firstly, the flexible entry model has been quite successful in reproducing the correct residence time distributions, at least in general shape. This is especially true of the larger data set of non-cows, which also has lower variance in a visual sense than the cows distributions. For non-cows, changing the smoothing parameter from $\lambda = 0.5$ to 1 M makes very little qualitative difference to the output, except for reducing variance with the smoother model. The difference in the fitted distributions is much greater for cows, but interestingly the bootstrap traces for cows with $\lambda = 1$ M reproduce much the same picture as those for $\lambda = 0.5$ M. We deduce that the two fitted distributions shown in Fig. 3 for cows are not readily distinguishable from the data set of only 34 animals.

For more formal assessment of variance, Table 1 shows the bootstrapped variance of features of the fitted residence distribution, namely the distribution mean, variance, quartiles, and position of the peak. In nearly every case, the estimation method has performed very well in fitting the residence time distribution to data drawn from the known correct model through the parametric bootstrap, with the mean of the quartiles, peak, and distribution mean from the bootstrap replicates being within one or two days of their known correct answers (given by the previous row for each quantity in Table 1). The fitting method therefore seems to be approximately unbiased in recreating the important features of the residence distribution. The exception is the cows fit with $\lambda = 1$ M, which did not perform so well. Given the heavy parametrization of the models, the flexibility of potential spline fits, and the relatively small data sets, the overall performance of the method is encouraging.

4.3 Model Validation

Monte Carlo goodness-of-fit tests for the fitted model can be conducted by comparing attributes of the data and model fit with the same attributes obtained

Table 1 Characteristics of the fitted residence time distribution for cows and non-cows under the flexible entry model, with two values of the smoothing parameter: $\lambda = 0.5 \times 10^6$ (0.5 M) and $\lambda = 1 \times 10^6$ (1 M). The characteristics given for the fitted residence time distribution are: distribution mean; distribution variance; lower quartile; median; upper quartile; interquartile range (IQR); and peak. Except for variance, all quantities are measured in days. Below each quantity are the mean and standard deviation of the quantity from 100 replicates of the parametric bootstrap

	Cows				Non-cows			
Distribution	$\lambda = 0.5$ M		$\lambda = 1$ M		$\lambda = 0.5$ M		$\lambda = 1$ M	
characteristic	mean	sd	mean	sd	mean	sd	mean	sd
Mean	31		38		37		38	
–bootstrap	31	3.0	32	3.8	37	2.6	39	2.0
Variance	130		110		120		140	
–bootstrap	150	21	150	18	140	18	130	22
Lower 25%	22		31		29		30	
–bootstrap	21	3.2	23	4.1	29	3.4	32	3.1
Median	31		39		37		40	
–bootstrap	31	3.6	32	4.3	38	3.0	40	2.1
Upper 25%	39		46		45		47	
–bootstrap	40	3.3	41	3.9	46	2.4	48	1.5
IQR	17		15		16		17	
–bootstrap	19	1.9	18	1.6	17	1.7	16	1.9
Peak	28		40		40		43	
–bootstrap	31	7.3	33	7.4	40	5.0	44	2.7

from data that were genuinely simulated from the fitted model, using the parametric bootstrap replicates above.

A check on the sightings data may be performed by generating data from the fitted models and compiling sightings tables for each simulated data set, giving the number of animals detected once only, the number detected twice, and so on. By creating these frequencies over 10,000 simulated animals, we determine the expected frequency table under the fitted model. We then conduct a chi-squared test to determine whether the real sightings tables were consistent with the frequencies generated from the fitted model.

Results from the cows fits were good ($p = 0.50$ for $\lambda = 0.5$ M, $p = 0.55$ for $\lambda = 1$ M). However, results from the non-cows fits were extremely poor ($p < 10^{-8}$ in each case). The number of non-cows in the real data that were sighted only once is far in excess of the number expected under the fitted models. Possible reasons for this bad failure of the sightings component of the non-cows model are given in the Discussion.

For an indicative test of model fit, we can compare the optimized log-likelihood of the real data against optimized log-likelihood values generated by refitting the model to the parametric bootstrap replicates, which create the true distribution of optimized log-likelihoods if the fitted model is correct. By ranking the real-data log-likelihood value amongst those from the simulated replicates, we gain a Monte Carlo p-value for the fit. The p-value is the proportion of simulated data sets with a lower (worse) optimized value than that of the real data.

Again, the results were good for cows: $p = 0.80$ for $\lambda = 0.5$ M, $p = 0.70$ for $\lambda = 1$ M. The results were extremely poor for non-cows, with the real data log-likelihood standing well away from the distribution of log-likelihood values genuinely created from the fitted model. Investigation reveals that this is a

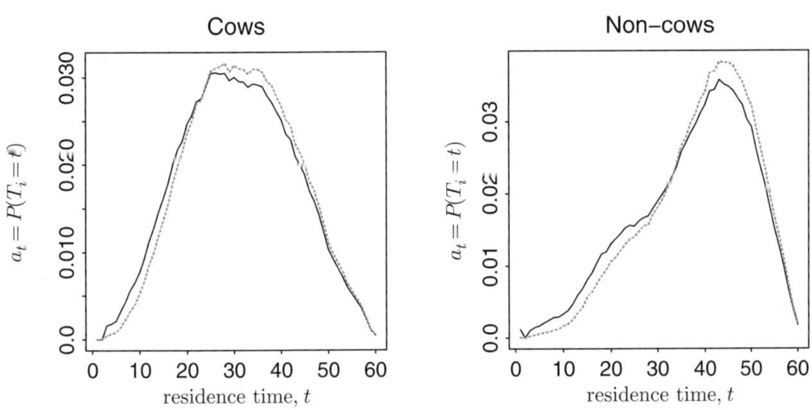

Fig. 4 Comparison of the empirical distribution of residence times for simulated data of 100,000 whales generated from the fitted models, for all animals (seen and unseen) and for animals seen at least once. The solid line shows the distribution for all animals, and the dashed line shows the distribution for animals detected at least once. The smoothing parameters for the fitted models are $\lambda = 0.5$ M for cows, and $\lambda = 1$ M for non-cows

consequence almost entirely of failure of the sightings model mentioned above. Splitting the log-likelihood (1) into components given by the first term (sightings component) and the second term (residence component), we find that the log-likelihood of the residence component is only marginally outside of the distribution created by simulation, while the large discrepancy in overall log-likelihood is caused mainly by the sightings component.

Finally, we noted in Section 2.1 that we have restricted inference to whales seen at least once, because these might not be representative of all whales with respect to residence time. In Fig. 4, we show the effect of this restriction using data simulated from the fitted models. The dashed line shows the residence time distribution for simulated whales seen at least once, while the solid line is the distribution for all simulated whales. There is a slight bias towards higher residence times among the whales seen at least once, but the effect is almost negligible. We conclude that, if the fitted models are accurate, the impact on residence estimation of restricting to the whales seen at least once will be slight.

5 Discussion

The analysis has demonstrated the flexibility of the cubic spline method for estimating the distribution of residence time. The estimated residence time distribution for both cows and non-cows covers a broad range with median of about 31 days for cows and 37 days for non-cows. In each case, the central 50% of the distribution covers a range of about 8 days above and below the median.

The results for cows appear to be a good fit, although there is high variability in the output. The high variability is an expected consequence of fitting a complex distribution to observations on only 34 cows. Of some concern to the validity of our results is the right-censoring of observations. An informal count conducted on the last day of the survey period detected at least 38 cows still present (Patenaude 2002), showing that the survey period ended before all cows had left the area. Few cows appeared to be present before the beginning of the survey period, however. To avoid bias, the survey period should cover the entire residence time at both ends, or at least cover the whole residence period of most individuals. The length of this survey was dictated by logistics in the subantarctic Auckland Islands, and it is not known whether the survey covered the whole residence period for the majority of cows or non-cows.

Our estimate of mean residence time of 31 days for cows in the Auckland Islands is much lower than the mean of 70 days (s.d. 30 days) found by Burnell and Bryden (1997) for right whale cows in South Australia using direct observation. This discrepancy might be due to right-censoring, or might genuinely reflect behavioural differences between the Australian calving grounds and the much more southerly subantarctic grounds. The Auckland Islands data were collected from mid-June to mid-August, while the Australian data were collected over a much longer period from mid-June to mid-October.

The results for non-cows did not fit well, primarily in the sightings component of the model. Far too many non-cows were sighted only once during their stay, compared with the number of sightings that would be expected from the fitted model. This in turn affected the likelihood of the fitted model, providing very poor results for likelihood-based goodness-of-fit tests. The sightings component is not affected by the fit for residence time, as can be seen from the first term of equation (1) where the sightings component covers each whale from its first sighting to its last sighting, and does not involve the residence parameters $\{a_t\}$ or the entry parameters $\{q_e\}$. The explanation for the poor fit must lie primarily with the sightings model rather than the residence model.

There are several possible explanations for failure of the sightings model. Firstly, our assumption that whales remained in the survey area between their first and last sighting may be incorrect. In their South Australian study, Burnell and Bryden (1997) reported that some non-cows left the calving grounds for a period of time during the season before returning. This would cause the effect seen with our data, where a period of absence would explain why non-cows were not sighted as often as expected. This explanation is worth following up biologically, because important information about mixing of known or unknown stocks might be gained from determining where non-cows go if they leave the Auckland Islands during the season.

Another possible explanation is that the matching of photographs was imperfect, and some resightings were logged as new records because the photographs were not identified as belonging to the same whale. This explanation is unlikely to account for the severe failure of the resighting model in our data, because photos were reviewed by several independent experts, and individual right whales are generally easily recognisable from their photographs within a season. Data on within-season error rates were not recorded, but discrepancies between experts for the more difficult task of matching right whale photographs between years were less than 3%.

A third possible explanation is that whales exhibited avoidance behaviour after the first photograph, which was sometimes accompanied by a biopsy sample. However, no evidence for avoidance after a biopsy was found in the data, using a chi-squared test on number of sightings with and without biopsy ($p = 0.73$). Like most other residence models, our model does not accommodate heterogeneous behaviour between whales. The daily sighting probabilities $p_1, \ldots, p_D$ vary according to day, but for a given day we assume that every whale present has the same probability of being photographed. It would not be easy to reformulate our method to allow for individual heterogeneity, because differential behaviours and differential residence times would be confounded. An alternative model that might perform better in the presence of individual heterogeneity is the Jolly–Seber model of Pledger et al. (2008), in which daily capture probability for each whale is allowed to depend upon the length of time the whale has already spent in the region. Their model also removes our assumption that residency time is independent of the time of arrival, which might also be responsible for bias in our results.

It is surprising that the model for residence time did not fit a low (short-stay) peak to the non-cows data, which would have attributed the large number of one-off sightings to short residence times rather than to long periods of non-detection during residence. Burnell and Bryden (1997) found a much lower mean residence period of only 20 days (s.d. 21 days) for non-cows. The long-residence fit might be explained by the severe left and right censoring of observations in the case of non-cows. An estimated minimum number of about 100 non-cows was already in the survey region before the photo-ID survey commenced, and at least 60 were still left on the last day of the survey (Patenaude 2002). Additionally, there were 17 days during the 54-day survey period during which sightings could not be made due to weather conditions. Overall, a whale could easily reside for 40 days and have only 10–15 sighting opportunities, allowing the one-off sightings to remain reasonably likely with an average sighting probability of about 7% per opportunity.

In addition to censoring of observations, failure of the assumption that non-cows remain in the survey area between their first and last sighting will also tend to cause positive bias in residence estimates. With some whales absent, resightings probabilities are likely to be underestimated, leading to overestimation of residence time because long periods of non-sightings at the start and end of the whale's stay are more easily tolerated. A final effect causing positive bias in residence time for non-cows is that some animals may have been mistakenly recorded as non-cows when they were in fact cows, because they were never photographed after giving birth.

We conclude that the results for non-cows should be treated with caution. Censoring of observations is problematic, and the assumption that non-cows remain in the survey area between their first and last sighting is probably violated. However, this conclusion is interesting in itself, and warrants further investigation. Right-censoring is also problematic for the cows data. In addition, successful estimation of sighting probabilities, entry time, and residence distribution places heavy demands on the data. For small data sets or low capture probabilities, it is likely that these effects can not be fitted reliably, illustrated by our results for cows in Fig. 3.

To test the reliability of our residence model, we conducted informal simulations with different sample sizes and levels of censoring. Using daily capture probabilities of 0.2, and with the true residence distribution given by the fit for non-cows with $\lambda = 1$ M, we found that the model reliably fitted the correct residence distribution for a sample size of only 100 whales, even with severe left and right censoring. However, if the true residence distribution was strongly bimodal, with a low peak at 7 days and a high peak at 38 days, and zero support between 19 and 25 days, detection of the low peak was very poor for samples of 100 whales, even with no censoring. With 500 whales, detection of the two peaks was good as long as at least 80% of whales had their residence period covered completely by the survey period.

For fitting the model, capture probabilities should not be fitted for days before the first animals arrived, otherwise the model will be unable to distinguish between zero capture probability for the pre-arrival days (and consequently longer residence times), and higher capture probability with no animals to capture. The simulations generally performed well for estimating entry distribution, even with severe censoring, but high data quality was needed for reliably untangling effects of sighting probabilities and residence time.

Overall, we recommend that biologists should simulate likely scenarios for residence distribution before collecting data, to ascertain the data quality needed to detect the effects of interest. With our example, the data for non-cows were almost certainly not sufficient to detect a short-stay residence peak if one were present, so it is still possible that the Auckland Islands population includes a set of transient animals. The method is promising as a way of modelling residence time directly, but it requires good data quality and the computational burden is high.

Acknowledgments We thank two anonymous reviewers and the AEs Blandine Doligez and Thierry Boulinier for their constructive comments which greatly improved earlier versions of this manuscript. We are indebted to the biologists and support staff who assisted in the collection and processing of the photo-ID data.

Appendix

We wish to minimize the expression (5) with respect to functions s satisfying $s(0) = s(\tau+1) = s''(0) = s''(\tau+1) = 0$ and with two continuous derivatives. We write the function H with τ arguments as $H(y_1, \ldots, y_\tau)$. Consider the effect of a small perturbation function $\epsilon h(x)$ on the minimizing function $s(x)$, where h is any function with two continuous derivatives, satisfying $h(0) = h(\tau+1) = 0$ so that the endpoints of s are not perturbed. At the minimizing s, we require

$$\lim_{\epsilon \to 0} \left(\frac{G(s+\epsilon h) - G(s)}{\epsilon} \right) = 0.$$

Substituting $s + \epsilon h$ into (5), and using a first-order Taylor expansion of $H\Big(s(1) + \epsilon h(1), \ldots, s(\tau) + \epsilon h(\tau)\Big)$ about $\Big(s(1), \ldots, s(\tau)\Big)$, we obtain

$$\lim_{\epsilon \to 0} \left(\frac{G(s+\epsilon h) - G(s)}{\epsilon} \right) = 2\lambda \int_0^{\tau+1} s''(x) h''(x)\, dx - \sum_{t=1}^{\tau} h(t) \left. \frac{\partial H}{\partial y_t} \right|_{y=s(1),\ldots,s(\tau)} \tag{17}$$

The minimizing s is the function that sets the right-hand side of (17) to 0, for any function h satisfying $h(0) = h(\tau+1) = 0$. Here it becomes clear why it was necessary to set the penalty terms in (4) (partly constituting H) to differentiable functions.

The first term on the right-hand side of (17) is handled by integrating by parts twice. The constraints $s''(0) = s''(\tau+1) = h(0) = h(\tau+1) = 0$ make most of the evaluated integrals vanish, leaving

$$2\lambda \int_0^{\tau+1} s''(x) h''(x)\, dx = 2\lambda \int_0^{\tau+1} h(x) s^{(4)}(x)\, dx.$$

For the second term in (17), we note that the point values $h(t)\frac{\partial H}{\partial y_t}\Big|_{y=s(1),\ldots,s(\tau)}$ can be written as the integral of the Dirac delta function, $\delta(t-x)$ defined for $-\infty < x < \infty$ such that (by definition),

$$\sum_{t=1}^{\tau} h(t) \left.\frac{\partial H}{\partial y_t}\right|_{y=s(1),\ldots,s(\tau)} = \int_0^{\tau+1} \left(\sum_{t=1}^{\tau} \delta(t-x) h(x) \left.\frac{\partial H}{\partial y_t}\right|_{y=s(1),\ldots,s(\tau)} \right) dx.$$

Substituting these in (17), and equating to 0, we get:

$$\int_0^{\tau+1} \left(2\lambda s^{(4)}(x) - \sum_{t=1}^{\tau} \delta(t-x) \left.\frac{\partial H}{\partial y_t}\right|_{y=s(1),\ldots,s(\tau)} \right) h(x)\, dx = 0.$$

Because this holds for all functions h, it follows that

$$2\lambda s^{(4)}(x) - \sum_{t=1}^{\tau} \delta(t-x) \left.\frac{\partial H}{\partial y_t}\right|_{y=s(1),\ldots,s(\tau)} \equiv 0 \text{ for all } 0 \le x \le \tau+1. \tag{18}$$

Whenever $x \notin \{1, 2, \ldots, \tau\}$, then $\delta(t-x) = 0$ by definition, so $2\lambda s^{(4)}(x) = 0$. This tells us that $s^{(3)}(x)$ is constant, with possible jumps at $x = 1, 2, \ldots, \tau$. This immediately reveals that the solution s, with zero fourth derivative, must be a cubic spline, with knots at the points $x = 1, 2, \ldots, \tau$. We may therefore write:

$$s(x) = a_t + b_t(x-t) + c_t(x-t)^2 + d_t(x-t)^3 \text{ for } t \le x < t+1. \tag{19}$$

This confirms result (6). Equation (19) may be differentiated to find s, s', s'', and s''' directly in terms of the coefficients a_t, b_t, c_t, and d_t.

To find the jump size in the third derivative $s^{(3)}$ at integer point t, we integrate (18) from $t-\epsilon$ to $t+\epsilon$ for $\epsilon \ll 1$. The integral is necessarily 0, so we obtain:

$$s^{(3)}(t+\epsilon) - s^{(3)}(t-\epsilon) = 6\,(d_t - d_{t-1}) = \frac{1}{2\lambda} \left.\frac{\partial H}{\partial y_t}\right|_{y=a_1,\ldots,a_\tau} \quad \text{for } t = 1, 2, \ldots, \tau, \tag{20}$$

where $a_1, \ldots, a_\tau$ have been substituted for $s(1), \ldots, s(\tau)$. Using (8), (15), and (20), some algebra shows that the spline coefficients $\mathbf{c}$ may be written succinctly as the solution to the equation

$$\mathbf{c} = \frac{1}{4\lambda} M^{-1} \left.\frac{\partial H}{\partial y_k}\right|_{y=M^{-1}Vc}$$

References

Burnell SR, Bryden MM (1997) Coastal residence periods and reproductive timing in southern right whales, *Eubalaena australis*. Journal of Zoology 241:613–621.

Burnell SR (2001) Aspects of the reproductive biology, movements and site fidelity of right whales off Australia. Journal of Cetacean Research and Management Special Issue 2:89–102.

Crosbie SF, Manly BFJ (1985) Parsimonious modelling of capture-mark-recapture studies. Biometrics 41:385–398.

Dawbin WH (1986) Right whales caught in waters around south eastern Australia and New Zealand during the nineteenth and early twentieth centuries. Report of the International Whaling Commission Special Issue 10:261–267.

Efford MG (2005) Migrating birds stop over longer than usually thought: Comment. Ecology 86:3415–3418.

Hastie TJ, Tibshirani RJ (1990) Generalized additive models. Chapman and Hall, London.

Lady JM, Skalski JR (1998) Estimators of stream residence time of Pacific salmon (*Oncorhynchus* spp.) based on release-recapture data. Canadian Journal of Fisheries and Aquatic Sciences 55:2580–2587.

Manske M, Schwarz CJ (2000) Estimates of stream residence time and escapement based on capture-recapture data. Canadian Journal of Fisheries and Aquatic Sciences 57:241–246.

Patenaude NJ, Baker CS (2001) Population status and habitat use of southern right whales in the sub-antarctic Auckland Islands of New Zealand. Journal of Cetacean Research and Management, Special Issue 2:111–116.

Patenaude NJ (2002) Demographic and genetic status of southern right whales at the Auckland Islands, New Zealand. PhD Thesis, University of Auckland.

Patenaude NJ (2003) Sightings of southern right whales around 'mainland' New Zealand. Science for Conservation 225.

Patenaude NJ, Portway VA, Schaeff CM, Bannister JL, Best PB, Payne RS, Rowntree VJ, Rivarola M, Baker CS (2007) Genetic differentiation and population structure among southern right whales (*Eubaleana australis*). Journal of Heredity 98:147–157.

Pirzl R, Patenaude NJ, Burnell S, Bannister J (in review) Movements of southern right whales (*Eubaleana australis*) between Australian and sub-Antarctic New Zealand populations. Marine Mammal Science.

Pledger SA, Efford MG, Pollock KH, Collazo J, Lyons J (2008) Stopover duration analysis with departure probability dependent on unknown time since arrival. In: Thomson DL, Cooch EG, Conroy MJ (eds.) Modeling Demographic Processes in Marked Populations. Environmental and Ecological Statistics, Springer, New York.

Pradel R, Schaub M, Jenni L, Lebreton J-D (2005) Migrating birds stop over longer than usually thought: Reply. Ecology 86:3418–3419.

Press WH, Teukolsky SA, Vetterling WT, Flannery BP (1988) Numerical recipes in C. Cambridge University Press, Cambridge.

R Development Core Team (2007) R: A language and environment for statistical computing. R Foundation for Statistical Computing, Vienna, Austria.

Reinsch CH (1967) Smoothing by spline functions. Numerische Mathematik 10:177–183.

Townsend CH (1935) The distribution of certain whales as shown by logbook records of American whaleships. Zoologica 19:1–50.

Whitehead H (2001) Analysis of animal movement using opportunistic individual identifications: Application to sperm whales. Ecology 82:1417–1432.

Detecting Invisible Migrants: An Application of Genetic Methods to Estimate Migration Rates

Steven D. Miller, Hamish E. MacInnes, and Rachel M. Fewster

Abstract In studies of migration, both between and within populations, it is not always feasible to use physical tags to track the movement of animals. Funding and time constraints may not allow for the trapping and tagging of a sufficiently large set of animals to expect that a reasonable number will be recaptured at a future time in another population. An alternative approach is to use genetic markers to estimate migration and population parameters of interest. This is a rapidly developing area of research, an advantage being that each captured subject has effectively been "tagged". The choice of tag however is not at the discretion of the researcher, and is a realisation of a complex array of historical events and random fluctuations. It is therefore necessary to develop methods to interpret observed genetic characteristics in order to describe inter- and intra-population movements. We present research using simulated and real-world data which evaluates the performance of one recent genetic approach to handling these sorts of problems. The collected data is of an invasive species, where it is likely the populations from which the samples were taken were recently established and therefore did not meet the usual genetic equilibrium conditions.

Keywords Genetics · Migration · Rattus

1 Introduction

Traditionally, mark-recapture studies have been used to study the demographics of populations. However, such an operation is sometimes impractical or infeasible, subject to constraints such as funding or time. It might even be impossible, for instance when the initial capture is made by kill-trapping. In such cases, methods that use genetics to infer historical migration patterns among populations might be considered as alternatives. These methods should be distinguished from methods

S.D. Miller (✉)
University of Auckland, Auckland, New Zealand
e-mail: smiller@stat.auckland.ac.nz

D.L. Thomson et al. (eds.), *Modeling Demographic Processes in Marked Populations*, Environmental and Ecological Statistics 3, DOI 10.1007/978-0-387-78151-8_18,

which make use of animals' genetic signatures within the classical mark-recapture framework (Nichols and Kendall 1995).

Using the genetics of animals is closely analogous to the physical tagging of animals. With the exception of monozygotic siblings, an animal's genetic code is unique. This means each individual has a unique tag which does not need to be applied by the researcher. Thus there is no "mark" phase, in the classical sense.

When attempting to detect migration between populations, it is necessary to take representative genetic samples from every potential source population, much as it would be necessary to tag animals in each population when using mark-recapture models. In a genetic study however, it is not necessary for an individual sampled in the source population to subsequently migrate. When a migrant has a genetic profile that is more similar to related individuals in its source population than individuals in its new population, it is not necessary to mark the migrant in its source population prior to its migration. Rather, it is sufficient to have a representative sample of the genetic profiles of individuals from the source population with which the migrant should closely associate.

It has been demonstrated that indirect methods of characterising movement between populations using genetics can be as effective as methods involving direct observation, but require a fraction of the field-time (Berry et al. 2004). Although use of genetics can substantially reduce field-time and its associated costs, these efficiencies must be balanced against the lab-time and costs necessary to extract the genetic information from the biological samples.

2 Estimating Migration Through Genetics

We were interested in calculating the levels of migration between populations of an invasive species. Because invasive species are generally short-lived, it can be difficult to complete further sampling as part of a longer-term study involving resampling tagged individuals during those individuals' lifetimes. However, because such species are necessarily fast-breeding, it is likely that a migrant will leave descendants in its destination population. This is an ideal scenario for tracking migration through genetic signals.

It is not necessary for a sampled individual to be a migrant itself in order to detect migration events. Within several generations, and assuming populations are relatively distinguishable genetically, a genetic signature carried by an immigrant will remain detectable within that immigrant's progeny. By contrast, physical tags cannot be passed from parent to offspring. However, estimates of migration based on genetics rely on migrants successfully breeding in their destination populations. If resident individuals are unwilling to breed with migrants from other populations, the genetic record will underestimate the true level of migration.

Several factors affect the detectability of ancestral migration:

- How representative was the genetic profile of the migrant of its source population?

- How similar were the genetic profiles of the source population when the migration event occurred, and when the sampling occurred?
- How distinguishable are the genetic profiles of the migrant's population of origin and destination?
- How likely is a migrant individual to breed successfully in its new population?
- What is the probability of sampling the migrant or one of its descendants in the new population?

The detection of migration will not always be certain therefore, and the chance of detection decreases over time.

Another advantage of using a genetic approach to detect migration is that every sampled individual has the potential to be identified as a migrant. Contrast this with studies involving tagged individuals, where it is impossible to discriminate between untagged individuals – they could be untagged resident individuals, untagged migrants from identified source populations, or migrants from unidentified source populations.

However, methods using genetics might not behave properly if underlying genetic models on which they rely do not hold. For example, many early methodologies rely on the assumption that the populations analysed conform to the Hardy-Weinberg Equilibrium (HWE) conditions. In practice, these conditions can never hold perfectly. Recent modifications to earlier implementations of genetic methods relax many of these assumptions.

2.1 An Individual Assignment Approach

Assignment methods are a class of methodologies that use genetics to estimate individuals' most likely source populations or ancestries (Paetkau et al. 1995; Rannala and Mountain 1997; Pritchard et al. 2000; Wilson and Rannala 2003). Many of these methods allow a diploid individual's genes to have descended from different populations, representing an ancestral migration event, even if the individual itself is a non-migrant (Rannala and Mountain 1997; Pritchard et al. 2000; Wilson and Rannala 2003). With the appropriate data, these methods are more efficient at characterising migration than other methods, particularly if migration has only occurred over the last few generations (Rousset 2001).

Once migrant ancestries have been estimated, it is possible to construct point estimates of migration rates by dividing the number of migrants in a population by the size of that population (Manel et al. 2005). This requires the population sizes to be estimated, which is often difficult. Wilson and Rannala (2003) developed a Bayesian assignment-based method to estimate rates of migration between populations without needing population sizes. Analysis of the output from the MCMC chains enables distributional characteristics of the migration rates to be evaluated (Wilson and Rannala 2003).

Assignment methods are useful for studying migration over short time-scales (Manel et al. 2005). Many contemporary approaches rely on fewer assumptions than traditional models, such as the requirement for the populations to be in HWE, but this comes at the cost of the results being applicable to only recent trends in migration (Wilson and Rannala 2003). The complexity of these models increases exponentially with the number of generations since the unobserved migration (Rannala and Mountain 1997; Pritchard et al. 2000). Although performance is better when gene-flow is low, accurate estimates are possible even when migration is frequent (Berry et al. 2004). The ability to precisely estimate migration rates when populations lack genetic diversity might be enhanced, for example by taking larger sample sizes from each population, and particularly by using highly-variable loci (Berry et al. 2004; Paetkau et al. 1995; Wilson and Rannala 2003).

Bayesian variations of assignment methods often recognise that the genetic information provided is only a sample from the population, and there will be uncertainty inherent in the population genetic profiles calculated from these samples due to sampling variation (Rannala and Mountain 1997; Pritchard et al. 2000; Wilson and Rannala 2003).

3 Data and Motivation

We have collected data for a study of the invasion dynamics of Norway rats (*Rattus norvegicus*) and ship rats (*R. rattus*) among offshore islands in New Zealand. The aim of this study is to investigate the geographical and biological characteristics affecting the rate at which rats migrate between islands. Outcomes from this research are intended to assist conservation managers during the planning stages of the restoration of protected islands. This involves the eradication of invasive species present on the islands, the reintroduction of native species, and ongoing monitoring to prevent the reinvasion of the eradicated species. Resource availability and the biology of rats rendered it infeasible to attempt to infer the migration rates of rats between islands using physical tagging.

Twelve existing Norway rat microsatellite markers for unlinked loci on separate chromosomes were selected for this study (Jacob et al. 1995). One locus failed to amplify for both species, and another locus did not produce reliable results for ship rats. Thus, samples from Norway rats were genotyped at 11 independent loci, while those from ship rats were typed at 10 loci.

The populations we were interested in were likely to have been founded recently, so the equilibrium conditions were unlikely to hold. In order to describe migration in this sort of system, assignment methods are most appropriate (Rousset 2001).

We have focused on the assignment method described by Wilson and Rannala (2003), and implemented in the program BayesAss+ (http://www.rannala.org/labpages/software.html). This approach estimates migration rates between populations, which are of primary interest for our study.

3.1 Genetic Migration Models

3.1.1 The BayesAss+ Method

BayesAss+ assumes that migration is relatively rare (Wilson and Rannala 2003). This is necessary to derive the equations for the ancestry probabilities employed in the conditional multinomial distribution of ancestries, given migration rates. If migration rates are assumed to be low, terms involving squared or higher polynomials of migration rates are deemed to be negligible, and the probability of a pair of migrants breeding in their destination population is effectively zero.

BayesAss+ describes an individual's genetic ancestry through the use of the parameters *S*, the population the individual was sampled from; *M*, the individual's ancestral population; and *t*, the individual's ancestral time. An ancestral time of *t=1* implies an individual is a direct migrant from the population specified by *M*, which means the individual was born in that population and migrated to the population it was caught in, *S*. The individual's parents' ancestries are irrelevant in this case. If *t=2*, an individual was born in the population it was sampled from, one of its parents was also born in that population, and the other was a direct migrant from the population specified by *M*. Finally, if *t=0*, the individual has no migrant ancestry, which means it was born in the population it was sampled from, to parents who were also born in that population.

Consider all individuals sampled from population *S=a*. These individuals must have one of the three ancestries described above. BayesAss+ defines m_{ji} to be the proportion of individuals in population i who migrated from population j. These are referred to as migration rates. When we refer to "external migration rates", we are talking about those values m_{ji} where $i \neq j$. That is, the proportion of individuals in population i who were not born in population i, and therefore must have migrated from an external population. "Internal migration rates" refer to values m_{ii}, the proportion of individuals in a population born in that population. Thinking in terms of genes, being born in a population is a migration event from the parents in population i to their offspring in population i, so the genes have migrated within the same population.

BayesAss+ assumes the proportion of individuals with full migrant ancestry from population b in population a, $(S = a, M = b, t = 1 : a \neq b)$, to be m_{ba}. The proportion of individuals with half migrant ancestry from population b in population a, $(S = a, M = b, t = 2 : a \neq b)$, is approximated as $2m_{ba}$. In order to ensure the ancestry proportions sum to one, the proportion of individuals in population a with no migrant ancestry, $(S = a, M = a, t = 0)$, is required to be $m_{aa} = 1 - 3\sum_{\{k \in \mathbf{I}:k \neq a\}} m_{ka}$, where $\mathbf{I}$ is the set of all populations. This requires that the sum of all external migrant proportions be less than $\frac{1}{3}$ in order to ensure the proportion of individuals with no migrant ancestry is non-negative. We will refer to this method of calculating the ancestry proportions in a population as "Method 1" from here on.

3.1.2 A New Method

We couldn't rely on migration being rare between populations of rats for our study. In order to make the migration model employed by Method 1 more flexible, we introduce the parameters M_1 and M_2, in place of M and t. M_1 represents the ancestry of one of an individual's parents and M_2 represents the ancestry of the other parent. As before, S represents the population the individual was sampled from. We will assume we cannot distinguish between the cases $(M_1 = i, M_2 = j)$ and $(M_1 = j, M_2 = i)$. For our purposes, it is reasonable to assume that an individual may only migrate once in its lifetime. We also assume that generations are non-overlapping. In this scenario, the two parents are required to move from their initial populations to a common population so that they can breed together. When born in this population, the individual in question must migrate to the population where it was captured. Let b and c be the populations from which an individual's parents originated, and let a be the population from which we sampled the individual. If m_{ji} is defined as before, then the probability of such an ancestry can be written as

$$\Pr(S = a, M_1 = b, M_2 = c) = \begin{cases} \sum_{\kappa \in \mathrm{I}} m_{bk} m_{ck} m_{ka} & \text{if } b = c; \\ \sum_{\kappa \in \mathrm{I}} 2 m_{bk} m_{ck} m_{ka} & \text{otherwise.} \end{cases} \quad (1)$$

We will refer to this method of calculating the ancestry proportions in a population as "Method 2".

A comparison of the ancestry probabilities between the two methods is shown in Table 1, where the scenario involves three possible populations.

Table 1 Comparison of the formulae for expected ancestry proportions between Method 1 and Method 2, for the case of three populations. The equations shown are only for individuals sampled from population a. Equations for pure and half ancestries from only population b are shown – analogous equations exist for ancestries involving population c.

Description	Notation	Probability
(a) Method 1		
Pure Migrant Ancestry from populations b	$(S = a, M = b, t = 1)$	m_{ba}
Mixed Migrant Ancestry from populations b and c	*Not possible*	0
Half Migrant Ancestry from population b	$(S = a, M = b, t = 2)$	$2m_{ba}$
No Migrant Ancestry	$(S = a, M = a, t = 0)$	$m_{aa} = 1 - 3(m_{ba} + m_{ca})$
(b) Method 2		
Pure Migrant Ancestry from population b	$(S = a, M_1 = b, M_2 = b)$	$m_{ba}^2 m_{aa} + m_{bb}^2 m_{ba} + m_{bc}^2 m_{ca}$
Mixed Migrant Ancestry from populations b and c	$(S = a, M_1 = b, M_2 = c)$	$2(m_{ba} m_{ca} m_{aa} + m_{bb} m_{cb} m_{ba} + m_{bc} m_{cc} m_{ca})$
Half Migrant Ancestry from population b	$(S = a, M_1 = a, M_2 = b)$	$2(m_{aa} m_{ba} m_{aa} + m_{ab} m_{bb} m_{ba} + m_{ac} m_{bc} m_{ca})$
No Migrant Ancestry	$(S = a, M_1 = a, M_2 = a)$	$m_{aa}^2 m_{aa} + m_{ab}^2 m_{ba} + m_{ac}^2 m_{ca}$

3.2 Assessment of the Genetic Models Using an Individual-Based Ecological Migration Model

We simulated three populations that had experienced constant rates of migration between each other for two generations. The simulation was designed to mimic individual behaviour, and not adhere to either migration model used by Method 1 or Method 2. In this way, we were testing the suitability of the two methods, by comparing their predicted proportions of ancestries with those derived from the neutral, ecologically-based simulation.

There were 1000 individuals in each of the populations. In the first migration step, a proportion m of these individuals were moved between each of the nine pair-wise combinations of source and destination populations. Because the proportions of migrants were symmetric, the population sizes were still 1000 at the conclusion of this step. Breeding then took place. Pairs of individuals were selected at random with replacement. This allowed individuals to be involved in multiple matings. Each pair produced one offspring by randomly selecting an allele from each parent for each locus. The resulting offspring then undertook migration in the same fashion as their parents. The offspring's true genetic ancestries were recorded for comparison with the expected proportions under each model. Method 1 and Method 2 were tested at three levels of migration: $m = 0.05$ (rare), $m = 0.15$ (frequent), and $m = 0.3$ (very frequent).

The 18 simulated ancestry proportions spread over the four types of ancestry described in Table 1 are shown in Fig. 1. The expected proportions for each ancestry

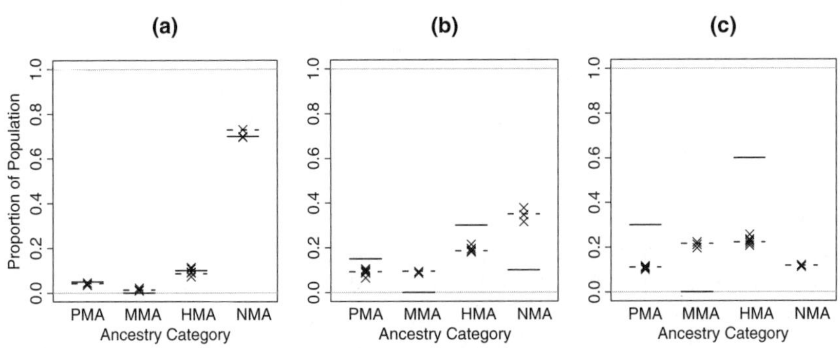

Fig. 1 Expected ancestry proportions for a three population system when (**a**) $m_{ii} = 0.9$, and $m_{ji} = 0.05$; (**b**) $m_{ii} = 0.7$, and $m_{ji} = 0.15$; and (**c**) $m_{ii} = 0.4$, and $m_{ji} = 0.3$, $j \neq i$. Crosses are the population proportions of each ancestry type calculated from simulations using a neutral, individual based movement model. Solid lines are the expected ancestry proportions according to Method 1, and dashed lines indicate the expected ancestry proportions using Method 2. PMA = Pure Migrant Ancestry, MMA = Mixed Migrant Ancestry, HMA = Half Migrant Ancestry, NMA = No Migrant Ancestry (see Table 1 for details)

type under Method 1 are indicated with solid horizontal bars, while the ancestry proportions expected by Method 2 are shown with dashed horizontal bars.

The results in Fig. 1 justify the suitability of Method 1 in cases of low migration, as the authors of BayesAss+ advise, but indicate there could be discrepancies between the expected and observed proportions at the higher rates of migration BayesAss+ can accommodate. Certainly the model used by BayesAss+ is inadequate in cases where migration rates exceed the maximum value ensuring proportions are non-negative, i.e. when the sum of the proportions of external migrants in a population exceeds $\frac{1}{3}$ (Fig. 1c). Method 2 accurately reconstructs the migrant proportions under all of the migration scenarios investigated.

3.3 Application of the Model to Simulated Data

After establishing Method 2 was suitable for estimating ancestry proportions when migration rates are known, we incorporated it into the MCMC framework used by BayesAss+ to assess how it performed when estimating migration rates from genetic data.

Samples were taken of 50 individuals from each of the populations generated according to the neutral simulation routine described above. Individuals' genotypes consisted of 10 loci. There were 10 allele types at each locus. Individuals generated from the ancestral populations were randomly assigned genotypes according to the specific allele frequencies for their population. We desired the ancestral populations to be genetically distinct. The coancestry coefficient θ (Weir and Cockerham 1984) for the ancestral populations was 0.22, with a 95% bootstrap confidence interval of (0.17, 0.28). The closer values of θ are to 0, the more genetically similar the populations. Values of θ in excess of 0.2 are considered high, suggesting populations have been separated for a long time. Individuals in the offspring generation were randomly assigned one allele from each of their parents at each locus.

Migration rates of $m = 0.05$ are considered to be relatively low. After two generations of migration at this rate and breeding, the coancestry coefficient reduced to 0.1123 (0.0871, 0.1417). Migration of $m = 0.15$ is close to the maximum migration that Method 1 can accommodate for three populations and equal external migration rates (since we require $1 - 2m > \frac{2}{3}$). Within two generations, the genetic differentiation between the populations is markedly reduced, down to just 0.0229 (0.0173, 0.0294). A migration rate of $m = 0.3$ makes the reconstruction of the ancestral allele profiles for the populations very difficult. The coancestry coefficient was effectively zero ($< 1 \times 10^{-5}$). Method 1 is not designed to cope with such a situation, but it was of interest to see the results from Method 1 under this scenario.

MCMC chains from a Bayesian framework employing Method 1 and Method 2 were run for 3,000,000 iterations, with the first 500,000 iterations discarded as burn-in. Samples were taken every 1000 iterations. Three replicate chains were run with different starting configurations to avoid reporting results drawn from locally optimal solutions.

3.4 Multiple Solutions from MCMC

During these simulations, we discovered that Method 2 was capable of finding multiple migration solutions for which the posterior probabilities were equivalent, particularly when migration was high. The flexibility of Method 2 means that there are often two or more sets of migration rates that fit the estimated ancestries equally well. This is not so much of an issue with Method 1 because of the constraints it imposes on the migration rates. These restrict the set of possible scenarios that can be considered, potentially excluding the true migrant proportions, if migration is high.

Figure 2 shows an example of the problem of equally probable parameter sets. The top row shows the traces of the posterior probability of the parameter sets from three separate MCMC chains using Method 2, initiated with different seeds. The lower three rows display the traces of the estimates for the three internal migrant proportions (m_{ii}) when $m_{ii} = 0.7$, for each of the three chains.

Examining the posterior probabilities alone, all three chains seem to have found a common solution, since the posterior probability distributions are the same across the three chains. However, the estimated migration rates switch between solutions during all three chains, and each chain finds solution sets that the others do not. By comparing the estimated allele frequencies with the known allele frequencies of the three ancestral populations, we found all three chains had calculated the frequencies

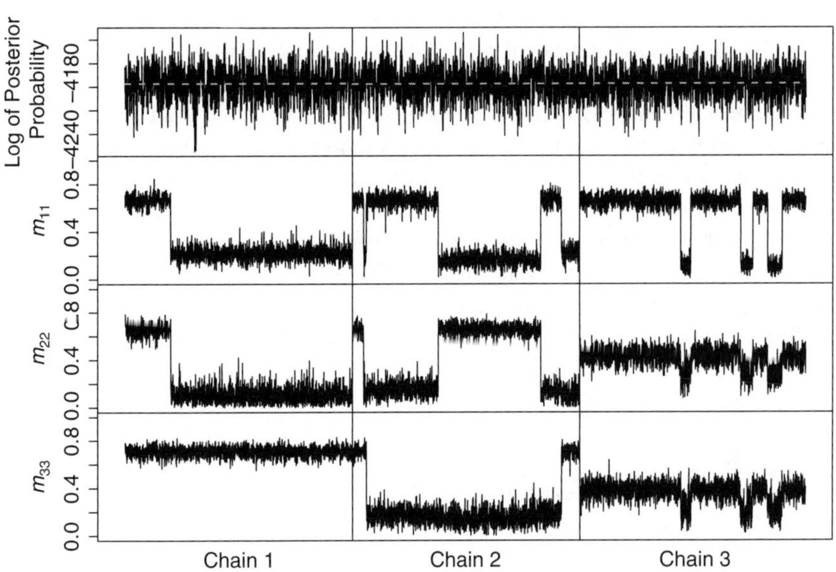

Fig. 2 Plot of the traces of the log of the posterior probability, and each of the internal migration rate estimates when $m_{ji} = 0.15$, $j \neq i$, for three separate MCMC chains. The true value of the internal migration rates was $m_{ii} = 0.7$. The third chain confused the allele frequencies for two of the populations. The three mean posterior probabilities were almost identical. These appear as three indistinguishable dashed lines in the first row

well, but that the third chain had swapped the frequencies for population 2 with those for population 3.

Although chains 1 and 2 estimated the ancestral allele frequencies correctly, four separate, similarly supported migration solutions were derived from these two chains. These sets were isolated using a clustering method. In the absence of additional information it would be necessary to treat each possible solution as equally plausible.

In practice however, often some solutions are less believable than others. For example, a small population might receive a large proportion of migrants from a neighbouring large population, but the large population is unlikely to receive the majority of its individuals from the smaller population. In this way, additional information can be used to select the most plausible set of parameters. Alternatively, such information could be incorporated into the shape of prior distributions used for the parameters before the MCMC chains are run. Implicitly, this is the approach taken by Method 1. By not allowing internal migration rates to be below $\frac{2}{3}$, the method is incorporating the prior belief that no more than $\frac{1}{3}$ of the population comprises individuals who were not born in that population.

Typical convergence diagnostics did not provide much insight for these problems. For Method 1, migration rate estimates quickly reached an equilibrium (typically within 50,000 iterations), which rarely varied for the length of the chain. Estimates also corresponded across multiple chains. These indicators confirmed consistency, not accuracy or appropriateness. In contrast, at higher migration rates Method 2 provided estimates that regularly jumped between apparent solutions. Within a chain, there was little consistency, and the three chains exhibited wildly different behaviour.

Because of the inconsistency of the behaviour across chains, and the existence of multiple solutions, particularly when using Method 2, we found it was advisable to use a number of replicate chains. The choice of three chains for the simulation study was probably too few to guarantee the discovery of all plausible parameter sets, although correct estimates of the true parameter set appeared to be recovered for each of the migration rates tested (Section 3.5). The comparison of results between chains can be used to distinguish globally optimal solutions from locally optimal solutions, and provides a greater probability of detecting all globally optimal parameter sets.

3.5 Simulation Results

Figure 3a shows the estimated migration rates from Methods 1 and 2 when the external migration rates were set to $m = 0.05$. Recall that for low external migration rates, both methods predict similar proportions of ancestries (Fig. 1a). The populations are also still relatively genetically distinct. Both methods successfully allocate allele frequencies to populations, and the estimated migration rates are similar between the methods. The interval containing 95% of all estimated

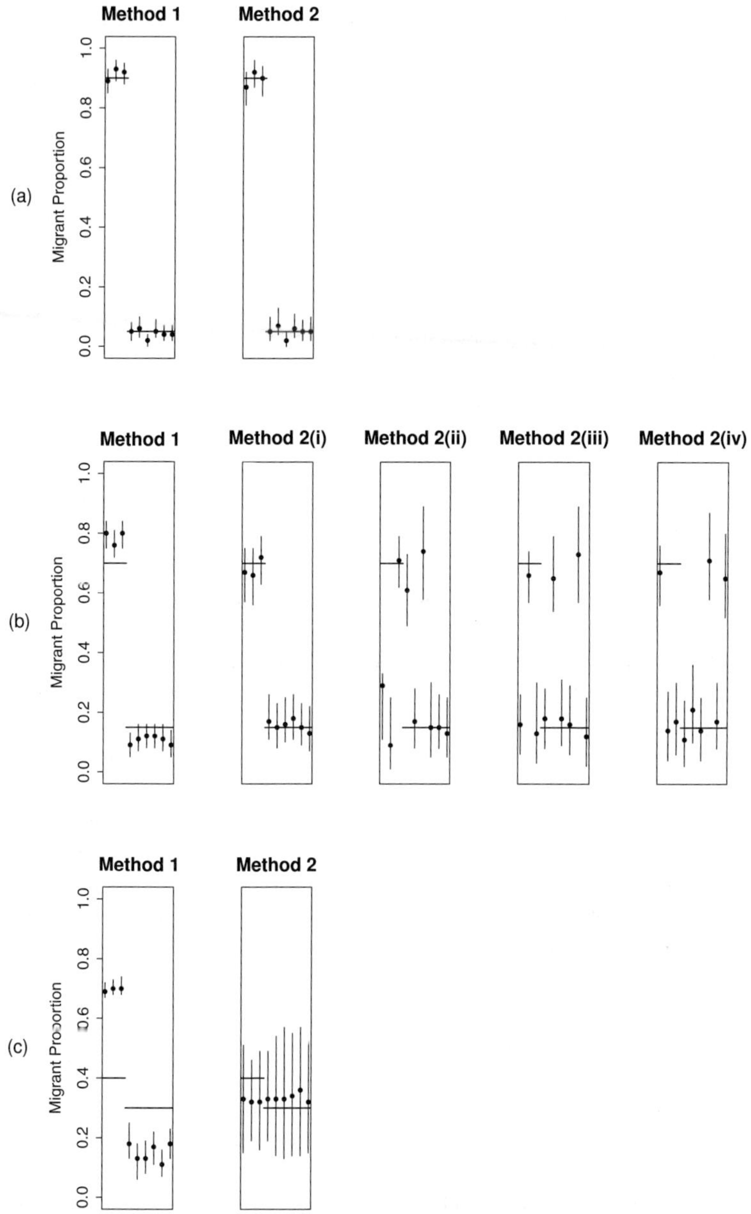

Fig. 3 Estimated migrant proportions for samples generated from a neutral model. True values are indicated with solid horizontal lines. The first three proportions in each plot are internal migrant proportions (m_{ii}) and the following six proportions are external migrant proportions ($m_{ji},\ i \neq j$). Point estimates are indicated with 95% credible intervals. Part **(a)** shows results for $m_{ii} = 0.9$, $m_{ji} = 0.05$; part **(b)** shows results for $m_{ii} = 0.7$, $m_{ji} = 0.15$; and part **(c)** shows results for $m_{ii} = 0.4$, $m_{ji} = 0.3$

migration rates contains the values of m used to generate the data in all cases, except for m_{12} under Method 1.

Figure 3b presents the migration rates estimated when the true external migration rates were $m = 0.15$. While diagnostic tests gave no indication that Method 1 had any difficulty with the similarity of the populations, in only four of the nine cases is the true migration rate within the central 95% of values estimated. This is due to the migration model adopted by Method 1 inadequately estimating the proportion of ancestries in the presence of high external migration rates (Fig. 1b). As mentioned, Method 2 found four sets of parameters that were judged equally likely from this data set (Fig. 2). Point estimates of the migration rates are close to the true values for the first parameter set, but the 95% intervals are quite wide, reflecting the uncertainty inherent in the genetic data due to the low genetic diversity between the populations.

Figure 3c shows the estimated migration rates from Methods 1 and 2 when the true migration rates were $m_{ii} = 0.4$, and $m_{ji} = 0.3$. Method 1 was not designed to cope with scenarios where more than 33% of a population is comprised of individuals with full migrant ancestry. Despite this, when the simulated populations were generated with external migration rates of $m = 0.3$ (i.e. 60% of individuals in each population were migrants), diagnostics from Method 1 did not signal any problems. However, it is impossible for the model to estimate the true value of the internal migration rates, $m_{ii} = 0.4$. Similarly, the sum of the external migration rates must be less than $\frac{1}{3}$, so both external migration rates cannot attain their true value of 0.3. Method 1 estimated equal migrant proportions from each external population, which is correct, but this requires both external migrant proportions to be underestimated. The internal migration rate is estimated to be nearly as low as possible (0.67), but is nevertheless over-estimated. Perhaps of most concern, the Markov chain realises the external migration rates are too low and internal migration rates are too high, so it seldom accepts proposals that move away from the upper and lower bounds for the external and internal migration rates respectively.

The effect is to generate values from the chain with little variability, which might be mistakenly interpreted as indicating strong support for the estimated parameters.

For Method 2, one of the three chains gave very different migration estimates from those given by the other two chains. The posterior probability distribution for this chain was centred lower than those for the other chains, suggesting this chain was drawing values from a local optimum rather than the global optimum. This chain was discarded. The other two chains estimated similar migrant proportions. All estimates were highly variable, reflecting a lack of information in the genetic data with respect to migration.

These simulations using a neutral ecological model indicate that Method 2 improves upon Method 1 in high migration scenarios. When migration is low, both methods cope fairly well with estimating migration parameters. At higher migration rates, Method 2 is capable of better estimating the correct migration parameters, but suffers from its flexibility, and will suggest a range of solutions that are as probable as each other.

By restricting consideration to only a subset of possible solutions, Method 1 is unable to even suggest the correct parameter values when migration is very high. Although Method 1 is not able to cope with such high levels of migration, it is not always obvious from the output, creating the potential for misinterpretation. At very high levels of migration, Method 2 finds it difficult to conclusively estimate the migration parameter values, but this is clearly indicated in the output, with high variability of estimates or frequent switching between alternative solutions. While such results may be of little practical use, they present a more accurate reflection of the difficulty in isolating information about population movement from the genetic data in such situations.

4 Application to Real Data

We investigated the invasion rates of rats over the island archipelago in the Bay of Islands (*Ipipiri o Tokerau*), Northland, New Zealand. There are seven main islands in this group (see Fig. 4). Samples were collected from survey lines over 2 week-long surveys, one in January (mid-summer) and the other in May (late autumn) 2005. Private landowners on several of the islands provided additional samples from their properties between surveys. Ship rats and *kiore* (*R. exulans*) were detected in small numbers, but the vast majority of captures were Norway rats. In addition, we trapped surrounding mainland areas, resulting in the addition of many ship rat samples. We included five DNA samples from Norway rats preserved in *Te Papa Tongarewa*, the Museum of New Zealand, taken during a previous survey of rodents from the islands conducted in 1984 (Moller and Tilley 1986).

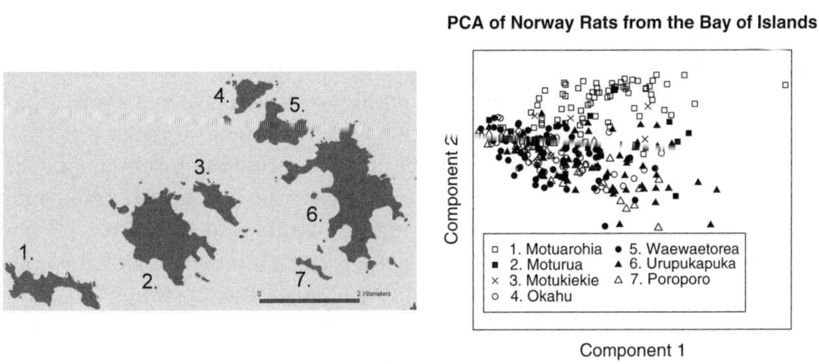

Fig. 4 Bay of Islands, Northland, New Zealand. Map of the archipelago and Principal Components plot of genetic distance between Norway rats. Individuals are grouped according to sampling populations. Substantial overlap indicates genetic similarity between groups. The first two principal components describe 89% of the variation seen in the data

4.1 Norway Rats

We analysed 310 Norway rat samples from this area, genotyped at 11 loci. A multivariate extension of the plotting method shown in Fig. 6 of (Paetkau et al. 2004) was used to explore the genetic relationships between individuals. This exploratory analysis of the data showed there was a large degree of genetic similarity between Norway rats in the Bay of Islands (Fig. 4), possibly indicating frequent migration between populations. Norway rats are known to be capable swimmers, covering distances of at least 400 m across open water between islands (Russell et al. 2005). The five historic samples did not differentiate themselves, indicating a level of consistency in the genetic make-up of the Norway rats on the islands over 20 years.

Each island was treated as a single population. Pair-wise estimates of Weir and Cockerham's coancestry coefficient θ (Weir and Cockerham 1984) were typically low. Ignoring Motuarohia (1.), all point estimates were less than 0.05 for every pair of islands except Moturua (2.) and Poroporo (7.), which was 0.06. The value of θ for pairs of islands involving Motuarohia were all greater than 0.05. Other than $\theta = 0.06$ between Motuarohia and Motukiekie (3.), all other point estimates of θ for population pairs involving Motuarohia were in the range 0.09–0.11.

An estimate of θ over all seven populations is

$$\theta = 0.06\ (0.04, 0.09)$$

The value of θ for the populations as a group is quite low. This might cause the genetic migration models to have trouble distinguishing populations. The pair-wise values of θ suggest this is likely to be the case for some pairs of islands more than others. Table 2 shows the results from the data-set analysed using Method 2 (Section 3.1).

Six random seeds were used to generate MCMC chains to estimate the parameters in an attempt to avoid interpreting a locally optimal solution and to identify differences due to attributing allele profiles to different populations. Convergence to the parameters' equilibrium distribution is guaranteed under MCMC, but this convergence may not occur within the number of iterations observed. Chains which suggest a consistent set of parameter values for the majority of the chain are probably drawing parameter values from an optimal region of the parameter space, but it might be unclear whether this is a local optimum or a global optimum. Using multiple chains increases the possibility of at least one chain finding a global optimum region. If the parameter sets suggested by some chains have higher posterior support than others, it can be concluded that the chains with lower posterior support are stuck in local optima. Our choice of six chains attempted to balance the increased probability of finding globally optimal solutions with computational burden. If we had found that few chains were suggesting the same solutions at the end of the procedure, it would have been prudent to use a larger set of chains initiated with different random seeds.

Table 2 Median immigration rates between populations of Norway rats in the Bay of Islands calculated with Method 2. 95% credible intervals from 99,000 draws from a Markov chain are indicated in brackets. Destination populations are represented by the columns, and source populations by the rows. Cell m_{ji} is the estimated proportion of migrants from population j in population i within a generation. Estimates with 95% credible intervals which do not cover zero are shown in bold

	Motuarohia (1.)	Moturua (2.)	Motukiekie (3.)	Okahu (4.)	Waewaetorea (5.)	Urupukapuka (6.)	Poroporo (7.)
Motuarohia (1.)	**0.78** **(0.71, 0.83)**	**0.09** **(0.01, 0.28)**	0.02 (0.00, 0.09)	0.01 (0.00, 0.06)	0.00 (0.00, 0.01)	0.00 (0.00, 0.02)	0.02 (0.00, 0.09)
Moturua (2.)	0.00 (0.00, 0.03)	0.04 (0.00, 0.25)	0.01 (0.00, 0.06)	0.01 (0.00, 0.08)	0.00 (0.00, 0.01)	0.00 (0.00, 0.02)	0.02 (0.00, 0.10)
Motukiekie (3.)	**0.18** **(0.12, 0.25)**	**0.32** **(0.02, 0.64)**	**0.61** **(0.35, 0.82)**	0.02 (0.00, 0.10)	0.00 (0.00, 0.02)	0.03 (0.00, 0.09)	0.02 (0.00, 0.11)
Okahu (4.)	0.00 (0.00, 0.02)	0.07 (0.00, 0.36)	0.01 (0.00, 0.07)	0.03 (0.00, 0.12)	0.00 (0.00, 0.01)	0.01 (0.00, 0.03)	0.05 (0.00, 0.32)
Waewaetorea (5.)	0.01 (0.00, 0.05)	**0.20** **(0.01, 0.63)**	**0.29** **(0.09, 0.54)**	**0.84** **(0.65, 0.94)**	**0.98** **(0.95, 0.99)**	0.02 (0.00, 0.10)	**0.41** **(0.10, 0.74)**
Urupukapuka (6.)	0.01 (0.00, 0.03)	**0.10** **(0.01, 0.32)**	0.02 (0.00, 0.09)	0.03 (0.00, 0.17)	0.01 (0.00, 0.03)	**0.91** **(0.82, 0.97)**	**0.40** **(0.10, 0.65)**
Poroporo (7.)	0.00 (0.00, 0.02)	0.05 (0.00, 0.25)	0.01 (0.00, 0.06)	0.02 (0.00, 0.11)	0.00 (0.00, 0.01)	0.00 (0.00, 0.03)	0.03 (0.00, 0.13)

Examining the posterior probabilities of the solutions suggested by the separate chains indicated three of the chains had located parameter sets that the data supported better than the solutions suggested by the other three chains. The less-well supported solutions were discarded. Two of the three chains that appeared to be performing well were calculating similar parameter estimates. The other chain suggested parameter values that we felt were unlikely, for example suggesting the majority of rats on the large island of Urupukapuka (6.) were migrants from the small neighbouring island of Poroporo (7.). On closer inspection, this chain was estimating allele profiles that matched those from the other two well-performing chains, except it had swapped the allele frequencies for Urupukapuka and Poroporo, which would explain some of the unusual estimates. We therefore decided to rely on the output from the two chains previously mentioned. These chains were run for a further 97,000,000 iterations, during which they rarely suggested any different sets of parameter values, and no parameter sets with the same or higher posterior support. The results with highest support from these two chains were used in Table 2.

The two largest eastern islands, Urupukapuka (6.) and Waewaetorea (5.) seem to consist predominantly of non-migrants. The other of the major eastern islands, Okahu (4.), is estimated to consist predominantly of immigrants from Waewaetorea (5.). We were expecting the Okahu and Waewaetorea populations to be particularly genetically indistinguishable, which should have resulted in variable estimates of the migrant and non-migrant proportions of these islands within their respective populations. There is little evidence to suggest Okahu has any significant proportion of non-migrants, and virtually no evidence to suggest Waewaetorea's population consists of any Okahu immigrants. It is possible that the strong sea current between these islands favours migration from Waewaetorea to Okahu only. Alternatively, the spatial location of traps may have influenced this result. Waewaetorea had three traplines laid across it: one on the shoreline on the south coast facing Urupukapuka (6.), one on the shoreline on the west coast facing Okahu, and a line running between these two, spanning the interior of the island. There was only one trapline on Okahu, and the majority of samples were taken close to the south-eastern shoreline facing Waewaetorea. Although the subpopulations on the facing shores of Okahu and Waewaetorea might be frequently swapping migrants, the diversity of samples from the wider spatial range across Waewaetorea might swamp this signal, instead leading to the conclusion that while the majority of rats on Okahu could fit the Waewaetorea population profile, not all Waewaetorea rats would fit well into the Okahu population. This scenario suggests that accurate spatial representation of populations is necessary to derive accurate migration estimates.

In the set of solutions chosen, Poroporo (7.) is described as consisting of a majority of migrants from Urupukapuka (6.) and Waewaetorea (5.), but there are wide credible intervals associated with these estimates, and for the proportion of Okahu (4.) migrants in Poroporo. This reflects the genetic similarity of those three populations with each other, and with the samples taken from Poroporo. That is, it is likely that Poroporo's population is dominated by migrants each generation, but it is difficult to discern which of the three other populations they are coming from. Common sense would suggest most, if not all, the migrants hail from Urupukapuka

(6.), due to its proximity. Incorporating an isolation-by-distance rationale in the BAYESASS+ framework to favour migration between neighbouring populations, or using geographical distances to inform the posterior probability through the prior distribution of migration rates would be a useful addition.

The migrant proportion estimates for Moturua (2.) were highly variable, probably resulting from the low sample size from this island. No conclusive interpretations could be made from these results.

Despite low sample sizes from Motukiekie (3.), five of the other six islands are strongly discounted as sources of migrants. This leaves Waewaetorea (5.) as a major source of migrants, and a large proportion of non-migrants. These two migrant proportion estimates are highly variable, and strongly negatively correlated. Because Motukiekie and Waewaetorea rats are genetically very similar, it is possible that the population on Motukiekie is relatively recently established from immigrants from Waewaetorea, and the program has difficulty deciding which ancestry fits these individuals best, particularly when the reference population for Motukiekie is so low (11 samples).

Based on the exploratory analysis (Fig. 4), we expected the Motuarohia (1.) population to be the most isolated population in our study. While there is very strong evidence to suggest there has been no recent migration between Motuarohia and five of the six other islands in the chain, there is surprisingly strong evidence that about 18% of individuals on Motuarohia are migrants from Motukiekie (3.). The precision of this estimate is especially surprising considering the low number of samples from Motukiekie. The density of rats on Motuarohia was very high, whereas we detected very few rats on Motukiekie. A fifth of the size of the Motuarohia population at that time was likely to be many times the size of the population on Motukiekie. It is likely there is some genetic characteristic of the Motukiekie sample that accounts for this unusual result, which we have yet to determine.

For the ship rats caught in the Bay of Islands, 32 samples were from the island archipelago, 29 coming from Urupukapuka (6.) alone. Two more samples were from Okahu (4.), and the other sample was from Moturua (2.). We classed all these rats together as island rats. We also analysed samples from 34 ship rats trapped on the surrounding mainland area of Rawhiti. Eight more samples were taken from a marina at Doves Bay, more than 15 km from the islands by sea, and more than 80 km from Rawhiti by land (see Fig. 5). These 74 ship rat samples were genotyped at 10 loci. Ship rats are relatively recent arrivals to the island chain, with none being detected on the islands in 1984 (Moller and Tilley 1986). We expected the island ship rats to be more closely related to the ship rats from Rawhiti, and the Doves Bay rats to look distinct from both these populations. Instead, exploratory analyses seem to indicate rats from the two mainland sites are more closely related to each other than to the island rats (Fig. 5). It is clear however that one of the island rats (the single rat caught on Moturua (2.)) has strong ancestral links with the mainland rats. Pair-wise values of θ were greater than 0.2 for both pairs involving the islands and each of the mainland populations. The value of θ between the two mainland populations at Doves Bay and Rawhiti was surprisingly only 0.06, despite the improbability of exchanging migrants over such a long distance.

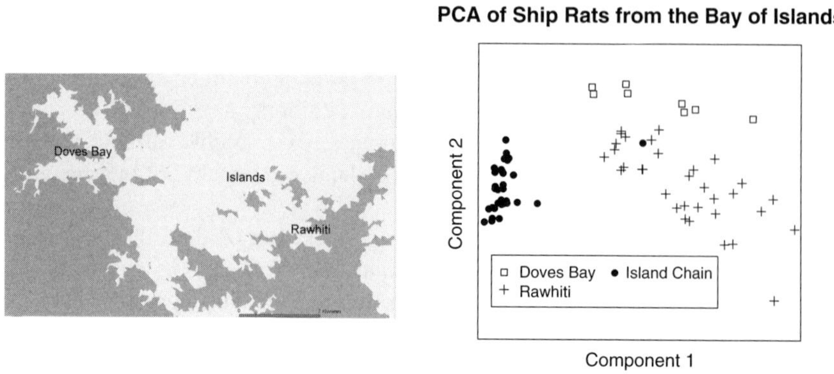

Fig. 5 Bay of Islands, Northland, New Zealand. Map of the islands and surrounding mainland, and Principal Component plot of genetic distances between ship rats. Individuals are grouped by sampling population. The first two principal components describe 98% of the variation seen in the data

The estimate of θ over all three populations is

$$\theta = 0.20\ (0.12, 0.32)$$

The θ for the combined populations is very high. The pair-wise values of θ indicate this is mostly due to the large differentiation between the island rats and the mainland rats. However, even the mainland sites show only a moderate level of similarity. The results of applying Method 2 to this sample of rats is provided in Table 3.

Although rates of convergence varied, all six chains settled on the same solution. We therefore feel confident that this is a plausible solution set.

Table 3 indicates all three populations are essentially isolated from each other. Due to the geographical distance, we were not expecting Doves Bay rats to be migrating between the other populations. Although the Rawhiti and island populations are proximate, the plot of genetic distances (Fig. 5) supports the conclusion that there is negligible gene-flow between these populations. Ship rats are reluctant swimmers (Innes 1990), so migration from the mainland to the islands is likely to

Table 3 Median immigration rates between populations of ship rats in the Bay of Islands region calculated with Method 2. 95% credible intervals from 25,000 draws from Markov chains are indicated in brackets

	Doves Bay	Rawhiti	Islands
Doves Bay	**0.93**	0.03	0.02
	(0.79, 0.99)	(0.00, 0.14)	(0.00, 0.06)
Rawhiti	0.03	**0.96**	0.01
	(0.00, 0.16)	**(0.85, 0.99)**	(0.00, 0.04)
Islands	0.02	0.01	**0.97**
	(0.00, 0.12)	(0.00, 0.04)	**(0.93, 0.99)**

stem from unintentional migration (for example, stowing aboard boats which visit the islands). Migration in this manner is likely to be infrequent, making it difficult for ship rats to establish populations in the presence of the high densities of Norway rats on most of the islands. Populations that do manage to establish are unlikely to frequently exchange migrants with the mainland population.

Although there was no evidence of frequent recent gene-flow between any of our populations, our exploratory analysis identified an island rat that showed clear mainland ancestry (Fig. 5). This rat was consistently identified as a full migrant with pure genetic ancestry from the Doves Bay population by Method 2. This might be evidence of human-assisted migration by boat, and emphasises the need for care from boats passing near eradicated islands, in order to ensure they do not unintentionally reintroduce pests.

5 Discussion

An appreciation of the rate of migration between populations is useful in many conservation applications. As an example of this, we introduced a project to characterise the migration dynamics of rats over island archipelagos in New Zealand. Interest might lie in which islands are most suitable to reintroduce and protect threatened animals and plants. This requires any pest species present to be eradicated, and the prevention of any re-establishment of these populations. Eradications are expensive, both in terms of money and time. Success must be assured.

We gave examples where the conventional means of detecting migration through direct observation might be impractical. In such situations, methods which estimate migration indirectly through genetics may be suitable.

For the purpose of this project, estimating rates of migration through genetics was the only approach practicable. However, estimates of direct migration would have been useful. Genetic estimates of migration relate only to successful migration, i.e. where a migrant manages to breed in its new population. For the purposes of eradication and preventing reinvasion, once an island is rat-free, the rate of reinvasion will solely depend on how often rats from neighbouring islands move to that island. There is no need to consider the chances of successfully being incorporated into the resident population. Prior to an eradication, the resident population might have forcibly repelled migrant rats, and so there would be little indication of migration in the genetic composition of the island's population. In the event, the levels of genetic admixture we saw in populations of Norway rats in the Bay of Islands suggests that migrants were readily accepted into their destinations' breeding populations.

Methods now exist that relax many of the unrealistic assumptions that earlier genetic methods relied on. This flexibility often comes at the cost of generality, where estimates might be applicable to only recent generations (Wilson and Rannala 2003). However, pest species often experience rapid growth and decline, meaning extant populations have often existed for only a small number of generations. This makes the application of these new flexible methods ideal.

A particular method, BayesAss+, was introduced as an example of these new flexible methodologies. Through simulations, we demonstrated that the original version of this program works well in situations where migration is low. However, at higher rates of migration, the simplification of equations used to estimate the proportions of migrants in populations becomes inaccurate. Beyond a certain level of migration, the program is unable to consider the range containing the true proportions of migrants, and potentially provides inaccurate results that indicate spurious precision.

We introduced a new model based on equations that accurately calculate these migrant proportions at any level of migration. This modification increases the program's flexibility by allowing it to consider any recent migration scenario, although this substantially increases the number of migration scenarios available. This places greater onus on users to select the most reasonable solution from the sets provided. It is also important to ensure all plausible sets of parameter estimates have been considered. One way to achieve this efficiently would be to implement Metropolis-coupled Markov chain Monte-Carlo (MCMCMC; parallel tempering) (Geyer 1991). Here, several MCMC chains are run in parallel, where the posterior distribution associated with each chain is a weighted version of the desired distribution, increasing in diffusion. During the course of the simulation, parameter values can be swapped between chains, according to the Metropolis algorithm. Examining the distribution of parameter values drawn from the chain corresponding to the unweighted posterior distribution provides the areas of the posterior density with greatest support.

There will always be situations where direct methods should be preferred over indirect methods to estimate levels of migration, and the contrasting approaches should be seen as complementary (Rousset 2001). Indirect methods which rely on genetics often struggle when populations experience large degrees of gene-flow, but in this situation, direct observation of movement might be easier. Conversely, when migration events are rare, observing an event directly might take multiple sampling occasions, but an unusual genetic signature might persist for a short time, giving the researcher a better chance of detecting the migration, if successful reproduction followed the migration. Synergy is created when both approaches inform each other, for example where a small scale physical assessment of migration is used to weight the possible solution sets available from the genetic analysis.

Even if a mark-recapture study is preferred, there is nothing lost in taking genetic samples of animals as they are captured. Such records would be useful if a tag is ever lost or becomes illegible. There is also the benefit of being able to discriminate between individuals from known populations who had not previously been tagged, and individuals from populations outside of the study. As samples are collected over generations, parentage analysis could be performed, or lineages constructed. This could aid in the calculation of other demographic parameters. The opportunity to collect such a wealth of information could only benefit the research behind catching the animals in the first place.

Acknowledgments We wish to thank Nicolas Perrin and an anonymous reviewer, and AEs Blandine Doligez and Thierry Boulinier for suggestions that greatly improved the content of this manuscript. We would also like to thank the staff and volunteers from the New Zealand Department of Conservation's Bay of Islands Area Office, the landowners and caretakers of the private residences on the islands studied and around the Rawhiti peninsula, and the teams of field-workers from the University of Auckland who helped collect the genetic samples. This work was funded by a Royal Society of New Zealand Marsden grant. Steven Miller was supported by a scholarship from the New Zealand Tertiary Education Commission.

References

Berry O, Tocher MD, Sarre SD (2004) Can assignment tests measure dispersal? Molecular Ecology 13:551–561.

Geyer GC (1991) Markov chain Monte Carlo maximum likelihood. In Computing Science and Statistics: Proceedings of the 23rd Symposium on the Interface, Keramidas EM (ed), Fairfax Station: Interface Foundation, pp. 156–163.

Innes JG (1990) Ship rat. In The Handbook of New Zealand Mammals, King CM (ed), chapter 17. Oxford University Press, Oxford, pp. 206–225.

Jacob HJ, Brown DM, Bunker RK, Daly MJ, Dzau VJ, Goodman A, Koike G, Kren V, Kurtz T, Lernmark A, Levan G, Mao Y-P, Pettersson A, Pravenec M, Simon JS, Szpirer J, Troillet MR, Winer ES, Lander ES (1995) A genetic linkage map of the laboratory rat, *rattus norvegicus*. Nature Genetics 9:63–69.

Manel S, Gaggiotti OE, Waples RS (2005) Assignment methods: matching biological questions with appropriate techniques. Trends in Ecology and Evolution 20:136–142.

Moller H, Tilley JAV (1986) Rodents and their predators in the eastern Bay of Islands. New Zealand Journal of Zoology 13:563–572.

Nichols JD, Kendall WL (1995) The use of multi-state capture–recapture models to address questions in evolutionary ecology. Journal of Applied Statistics 22:835–846.

Paetkau D, Calvert W, Stirling I, Strobeck C (1995) Microsatellite analysis of population structure in canadian polar bears. Molecular Ecology 4:347–354.

Paetkau D, Slade R, Burden M, Estoup A (2004) Genetic assignment methods for the direct, real-time estimation of migration rate: a simulation based exploration of accuracy and power. Molecular Ecology 13:55–65.

Pritchard JK, Stephens M, Donnelly P (2000) Inference of population structure using multilocus genotype data. Genetics 155:945–959.

Rannala B, Mountain JL (1997) Detecting immigration by using multilocus genotypes. Proceedings of the National Academy of Science 94:9197–9201.

Rousset F (2001) Genetic approaches to the estimation of dispersal rates. In Dispersal, Clobert J, Danchin E, Dhont AA, and J. D. Nichols (eds), chapter 2. Oxford University Press, Oxford, pp. 18–28.

Russell JC, Towns DT, Anderson SH, Clout MN (2005) Intercepting the first rat ashore. Nature 437:1107.

Weir BS, Cockerham CC (1984) Estimating f-statistics for theanalysis of population structure. Evolution 38:1358–1370.

Whitlock MC, McCauley DE (1999) Indirect measures of gene flow and migration: $F_{ST} \neq 1/(4Nm+1)$. Heredity 82:117–125.

Wilson GA, Rannala B (2003) Bayesian inference of recent migration rates using multilocus genotypes. Genetics 163:1177–1191.

Wright S (1951) The genetical structure of populations. Annuls of Eugenics 15:323–354.

Section V
Wildlife and Conservation Management

Morten Frederiksen and Mark Lindberg

Stochastic Variation in Avian Survival Rates: Life-History Predictions, Population Consequences, and the Potential Responses to Human Perturbations and Climate Change

Joel A. Schmutz

Abstract Stochastic variation in survival rates is expected to decrease long-term population growth rates. This expectation influences both life-history theory and the conservation of species. From this expectation, Pfister (1998) developed the important life-history prediction that natural selection will have minimized variability in those elements of the annual life cycle (such as adult survival rate) with high sensitivity. This prediction has not been rigorously evaluated for bird populations, in part due to statistical difficulties related to variance estimation. I here overcome these difficulties, and in an analysis of 62 populations, I confirm her prediction by showing a negative relationship between the proportional sensitivity (elasticity) of adult survival and the proportional variance (CV) of adult survival. However, several species deviated significantly from this expectation, with more process variance in survival than predicted. For instance, projecting the magnitude of process variance in annual survival for American redstarts (*Setophaga ruticilla*) for 25 years resulted in a 44% decline in abundance without assuming any change in mean survival rate. For most of these species with high process variance, recent changes in harvest, habitats, or changes in climate patterns are the likely sources of environmental variability causing this variability in survival. Because of climate change, environmental variability is increasing on regional and global scales, which is expected to increase stochasticity in vital rates of species. Increased stochasticity in survival will depress population growth rates, and this result will magnify the conservation challenges we face.

Keywords Adult survival · Elasticity · Fitness · Life history · Process variance · Sensitivity · Stochasticity · Trade-off

---

J.A. Schmutz (✉)
Alaska Science Center, U. S. Geological Survey, 4210 University Drive, Anchorage, AK 99508, USA
e-mail: jschmutz@usgs.gov

D.L. Thomson et al. (eds.), *Modeling Demographic Processes in Marked Populations*, Environmental and Ecological Statistics 3, DOI 10.1007/978-0-387-78151-8_19,

1 Introduction

Survival rates and population sizes fluctuate over time. Such stochastic variation can be the consequence of simple binomial variation among individuals (demographic stochasticity) or through variability in the environment (environmental stochasticity). Except for very small populations, only the latter contributes significantly to population variation (Lande et al. 2003). Importantly, the long-term growth rate of a population, $\lambda_{s,}$ is reduced by stochastic variation in a, an element of a projection matrix (such as adult survival rate), which Tuljapurkar (1982) identified with the following equation,

$$\log_e(\lambda_s) = \log_e(\lambda_1) - 0.5/\lambda_1{}^2 \times \mathrm{var}(a) \times S_a{}^2 \quad (1)$$

where λ_1 is the asymptotic growth rate predicted by the mean matrix, $S_a{}^2$ is the (squared) sensitivity (Caswell 2001) of matrix element a, and a term for covariances among matrix elements is ignored. This equation can be recast in terms of elasticities, which are proportional sensitivities, as follows (Morris and Doak 2004),

$$\log_e(\lambda_s) = \log_e(\lambda_1) - 0.5 \times \mathrm{CV}_a{}^2 \times E_a{}^2 \quad (2)$$

where CV is the coefficient of variation of matrix element a and E is the elasticity of a. Pfister (1998) recognized that these equations established a tradeoff between the mean and variance of a trait, such as adult survival rate. Specifically, she predicted that for trait values with greater sensitivity that a given amount of variability in these traits would have greater effect on fitness than comparable variability in other traits with lower sensitivity. Consequently, selection against variability in a trait should be stronger for those traits with high sensitivity. Such selection pressure would lead to the canalization of a trait, such as adult survival. Pfister (1998) tested and largely verified her predictions by contrasting 30 different populations of very wide taxonomic diversity.

Pfister's predictions and support for the idea that variance in a trait is negatively related to its sensitivity is of strong importance to life history theory. I chose to further examine this life history tradeoff for two principal reasons. First, Pfister's (1998) work was flawed in that she used the total variance of a trait (or matrix element), which includes both sampling and process variances, whereas only process variance is of ecological interest. The decomposition of total variance into sampling and process variances has received increasing attention in recent years (Gould and Nichols 1998; White 2000; Burnham and White 2002; Morris and Doak 2004). Using estimates of process variance, Gaillard and Yoccoz (2003) examined whether variability in survival led to canalization of survival rates with high sensitivity, through a meta-analysis of 27 populations of mammals. Doherty et al. (2004) examined the relationship between process variance and its sensitivity across a set of vital rates for one species – red-tailed tropicbirds (*Phaethon rubricauda*). In the only meta-analysis of birds, Sæther and Bakke (2000) examined 18 populations where sampling circumstances made them believe that sampling errors were negligible and that total variance could be interpreted as process variance. Given the ecological and evolutionary importance of Pfister's (1998) predictions, more evaluation of their universality is needed. Thus, with a focus on birds, I was motivated to re-examine her predictions with a large set of populations (n = 62) and where I decomposed total variance to enable use of an estimate of process variance. In this paper, I use the term stochasticity as equivalent to process variation.

My second motivation for such re-examination relates to the impacts of stochasticity on population growth and the conservation of species. Given that species' life histories are shaped by their environment (Roff 2002), it is logical to expect that increased environmental variability causes increased stochasticity in species' vital rates. Because stochastic variation in adult survival rates (or any other vital rate) reduces the long-term growth rate of a population (Tuljapurkar 1982), then increased environmental variability is expected to reduce species' abundance. Importantly, this prediction reflects just the effects of stochastic variation and connotes that populations would decline without necessitating changes in the means of vital rates. The conservation relevance of this prediction is that several lines of evidence suggest that climatic aspects of the global environment are becoming increasingly stochastic (Easterling et al. 2000; Emanuel 2005; Boyce et al. 2006), which leads to predicted increases in stochasticity in species' vital rates and reduced long-term population growth rates. While such increases in stochasticity will create conservation challenges, it is unclear how regional such effects will be or if certain life histories will be disproportionately affected.

Given these two motivations, my goal is to conduct a meta-analysis of adult survival rates of bird populations wherein I test Pfister's (1998) prediction that the sensitivity of adult survival is negatively related to its variance. Further, I combine mean survival and variance, as their product gives a prediction of the effect on population growth from stochastic variation in survival (Caswell 2001; Haridas and Tuljapurkar 2005). As population growth is diminished by stochasticity (equations (1) and (2)), the expectation is that all life histories across the r to k selection continuum (Sæther and Bakke 2000) have evolved to limit the impact of stochastic variation, and thus the product of sensitivity (squared) and variance in a trait yields the expected variance in population growth, which should be small and similar across populations (Fig. 1). I test this idea and look for what species or populations may deviate from this life history expectation. I principally focused on three species groups that reflect different life histories: marine birds, waterfowl, and passerines, which represent a continuum from long- to short-lived species and which may experience different types of anthropogenic effects (e.g., waterfowl are the only ones commonly hunted). I did not restrict the geographic extent of this meta-analysis, and thus I have included studies from areas that I categorize as North America, Europe, Southern Ocean, and Tropics (< 20° latitude).

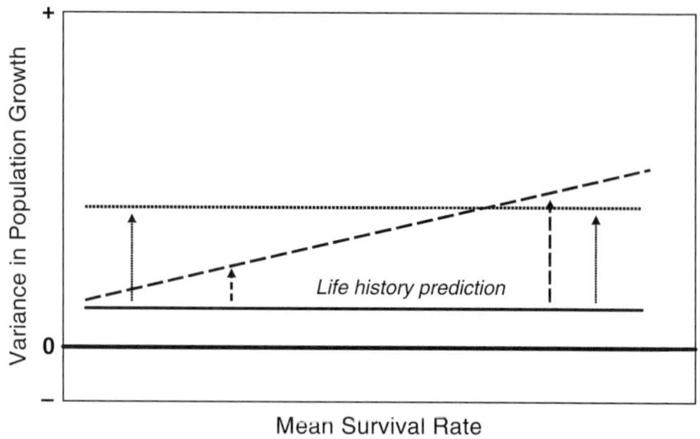

Fig. 1 All life histories are expected to have evolved low variability in population growth rate (Tuljapurkar 1982). Thus, across the range of mean survival (*S*) rates, one expects low values for the variance in population growth which is calculated as $\text{var}(\lambda) = (\lambda/S \times e)^2 \times \text{var}(S)$, where *e* is the elasticity of adult survival rate (Pfister 1998). One can also use a somewhat analogous equation with the sensitivity rather than elasticity. If all environments for all species become similarly more stochastic, we expect all species to experience similar increases in var(λ), as shown by the *dotted line*. Alternatively, if only the environments for long-lived species (e.g., marine ecosystems) became more stochastic, then we might expect a shift in var(λ) toward the *dashed line*

2 Methods

I obtained data for this analysis by locating journal articles containing annual estimates of survival and annual estimates of sampling variance (typically standard errors or confidence intervals). Specifically, I used keywords (e.g., 'survival' and species or genus name) at www.scopus.com, and restricted the search to articles published in 2000 or later. By focusing on only recent articles, I anticipated that the vast majority of papers would use modern modeling methods, which give less biased estimates than precursors (Williams et al. 2002). To further promote confidence that these selected papers presented largely unbiased estimates of survival, I identified what specific survival model (e.g., Cormack–Jolly–Seber model vs Burnham combined model), what method of goodness of fit evaluation, and what software programs were used in these analyses (Appendix). I only selected papers with annual estimates for five or more years. If estimates were given in figure form, I interpolated them to the third decimal point. Annual estimates were typically derived either from the best approximating model in an analysis or from model averaging procedures across a set of candidate models (Burnham and Anderson 2002). Also, any lack of fit was usually accounted for by use of a variance inflation factor, which increased the size of the sampling variance associated with each year's estimate.

A total of 62 populations were used in this analysis. Most populations reflected a combination of adults of both sexes, but a few were sex specific and seven pairs of populations reflected separate estimates for males and females (Appendix).

I used a variance decomposition approach outlined by Burnham et al. (1987:260) and White (2000), which uses an iterative procedure to inversely weight the contribution of individual years relative to the size of each year's sampling variance. This approach partitions the total variance into sampling and process variation, and it is this latter quantity that I then use for testing Pfister's (1998) predictions. This approach assumes there are no sampling covariances among annual estimates within a given study. Sampling covariances exist for most mark-recapture based estimates of survival, and likely for most estimates used in this study; however, sampling covariances are rarely reported. To gauge the consequence of ignoring sampling covariances when decomposing the total variance, I more closely examined three species for which sampling covariances were available: roseate terns (*Sterna dougallii*; Gould and Nichols 1998), black-capped chickadees (*Parus atricapillus*; Gould and Nichols 1998), and emperor geese (*Chen canagica*, Schmutz unpublished). Using the variance decomposition approach of Gould and Nichols (1998), which includes the contribution of sampling covariances, I determined that for these three studies sampling variances contributed 13–90 times more weight to the total measured sampling error than did sampling covariances. Thus, omission of sampling covariances in the variance decomposition approach I used for all the study populations likely resulted in negligible bias.

I input the mean estimate of annual adult survival into a matrix model, whose size was dictated by a literature-based estimate of age of breeding. I then input values for fertility and prebreeding survival to whatever values were necessary to yield a population growth rate of $\lambda = 1.0$. Although not all study populations were necessarily stable, population trend was often not known or reported. Further, the comparability of study populations was enhanced by deriving all projection matrices to predict stability ($\lambda = 1.0$). Because fertility and prebreeding survival have the same elasticity and both relate to the recruitment process only, the relative magnitudes of these two vital rates to one another is immaterial, given the constraints of age of breeding, adult survival, and stable population growth. From this matrix I calculated elasticity and sensitivity of adult survival for each population. Aside from the issue of using total variance versus process variance, another potential hindrance to examining the predicted relationship between the sensitivity of survival and its variance is the fact that high survival rates are bounded by 1.0, which may result in a statistical rather than ecological limitation to its variability. Two methods of accommodating this issue have been suggested. One is to use a variance stabilizing formula (Link and Doherty 2002), and the other is to express observed variance (or CV) as the proportion of theoretically possible variance, which for binomially distributed survival rates is greatest at 0.5 and least as survival approaches 0 and 1 (Gaillard and Yoccoz 2003; Morris and Doak 2004). I here employed both these methods, with use of an arcsine transformation for variance stabilization.

Relationships between sensitivity and variance can be equivalently explored with the proportional sensitivity (elasticity) and the proportional variance (coefficient of variation), which minimizes issues relating to scaling of vital rates (Pfister 1998). Although such scaling complications do not exist in this analysis, due to an exclusive focus on adult survival, I show all results using elasticities and CVs to maximize comparability to other studies. Following Pfister (1998) I calculated the expected variance of population growth as $\mathrm{var}(\lambda) = (\lambda/S \times e)^2 \times \mathrm{var}(S)$, where e is the elasticity of adult survival rate, S. This quantity is a theoretical expectation of variation in population growth, given the e and the observed variation in S.

I was concerned that estimates of process variance might be biased high for short time series of data (Burnham and White 2002; White et al. 2008). Thus, I used a maximum likelihood generalized linear model to examine whether the SD or CV of survival was negatively related to the number of years in a time series.

3 Results

Mean annual survival varied among populations from 0.303 for *Phylloscopus collybita* to 0.950 *Rissa tridactyla* (Appendix). Estimates of process variance in survival, expressed as a SD, varied among populations from 0 to 0.213. Across all populations, mean annual survival and SD of survival were 0.738 and 0.062, respectively. Years of data in individual time series varied from 5 to 42 years, with a mean of 12. There was little evidence that number of years in the time series affected point estimates of mean survival, its process variance, or the CV (for all three statistics, $r^2 < 0.02$ and confidence intervals on slope parameters broadly overlapped zero).

I affirmed Pfister's (1998) prediction of an inverse relationship between elasticity of survival and its variability. Inverse relationships were found when using arcsine transformations (slope parameter = -0.905, SE = 0.164, $r^2 = 0.29$, Fig. 2)

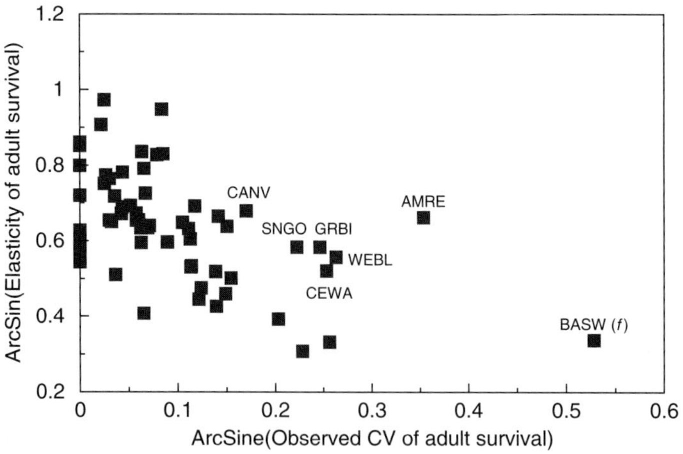

Fig. 2 Arcsine transformed elasticities of adult survival in relation to the arcsine transformed CV of adult survival for 62 populations of birds. The seven species with high var(λ) (> 0.01) are explicitly identified, where species codes are AMRE = American redstart, BASW = barn swallow (females), CANV = Canvasback, CEWA = cerulean warbler, GRBI = great bittern, SNGO = greater snow goose, and WEBL = western bluebird

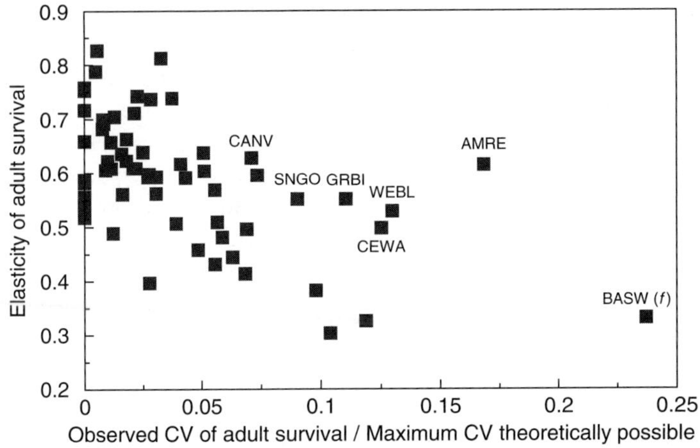

Fig. 3 Elasticities of adult survival in relation to an adjusted CV of adult survival, where the adjusted CV represents the observed CV divided by the theoretically maximum possible value of CV for that mean survival rate. The seven species with high var(λ) (> 0.01) are explicitly identified, where species codes are AMRE = American redstart, BASW = barn swallow (females), CANV = Canvasback, CEWA = cerulean warbler, GRBI = great bittern, SNGO = greater snow goose, and WEBL = western bluebird

and when using a CV adjusted by the maximum potential CV (slope parameter = -1.439, SE = 0.272, $r^2 = 0.318$, Fig. 3). As predicted, the variability in population growth, var(λ), was small (≤ 0.010) for the vast majority of populations (Fig. 4). However, for seven populations, var(λ) was one to five orders of magnitude higher.

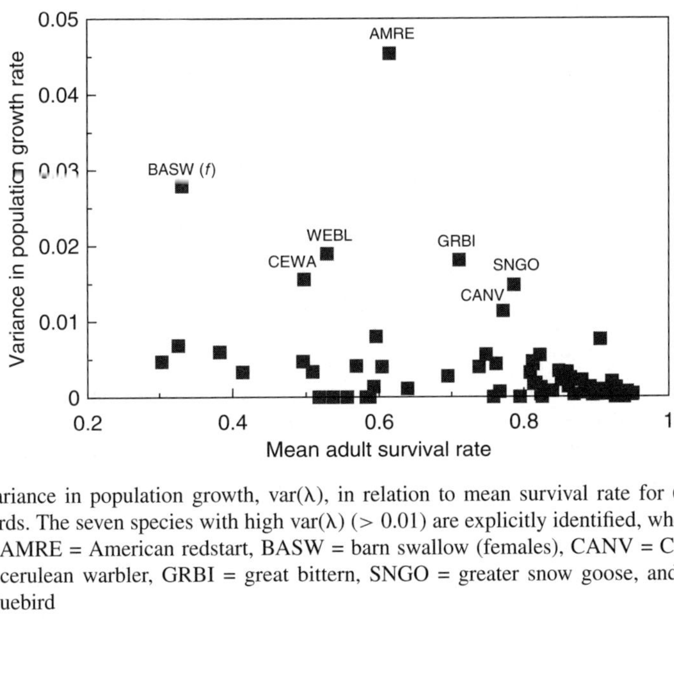

Fig. 4 Variance in population growth, var(λ), in relation to mean survival rate for 62 populations of birds. The seven species with high var(λ) (> 0.01) are explicitly identified, where species codes are AMRE = American redstart, BASW = barn swallow (females), CANV = Canvasback, CEWA = cerulean warbler, GRBI = great bittern, SNGO = greater snow goose, and WEBL = western bluebird

Although var(λ) is a function of multiple quantities (survival rate, its elasticity, and the variance of survival), these high values in var(λ) were principally a function of comparatively high point estimates of process variance in survival. The species with var(λ) > 0.010 were American redstart, barn swallows (females), western bluebird, great bittern, cerulean warbler, greater snow goose, and canvasback (Latin names in Appendix). I then input estimates of process variance for these species into equation (2), and by projecting over 25 years, obtained an estimate of population reduction over that time span solely due to stochastic variation in survival. For these species, these percent population reductions ranged from 13 to 44%.

4 Discussion

The significant population impact of changes in mean survival rates of adult, iteroparous organisms is widely recognized. Regarding long-lived birds, this impact has been identified for many taxa (Sæther and Bakke 2000). However, less well appreciated is the consequence of Tuljapurkar's (1982) approximations (equations (1) and (2)), which is that variability in survival rates without changes in mean values is alone sufficient to reduce population growth rate from that expected in an invariant environment. This occurs because the asymptotic population growth rate (equation (1)) is best measured by the geometric, not arithmetic, mean of annual population change, and temporal variability in annual population growth (caused by variability in one or more vital rates) reduces the geometric mean (Morris and Doak 2004). For the species exhibiting the most variability in survival rates in Fig. 2–4, the potential effect on population trend is not trivial. If process variance in survival of American Redstarts continued for 25 years at the magnitude estimated here, then its population size is expected to be only 56% of that in an environment with no stochasticity in survival. A notably similar population statistic is that Partners in Flight, a North American based organization concerned with the monitoring and conservation of landbirds such as American redstarts, sets its monitoring goal as the ability to detect a 50% decline in population size over 25 years (Rich et al. 2004). If one attempted to monitor the population by estimating survival rates (such as the MAPS program (DeSante et al. 1995)), and used an unbiased point estimate of survival in a deterministic matrix model to project what population growth was occurring, one may falsely conclude that the population was stable whereas in reality it had almost halved. Such a result emphasizes the need to estimate process variances of demographic parameters and use stochastic methods (Lande et al. 2003; Haridas and Tuljapurkar 2005) to monitor and model population dynamics.

Based on a foundation of life-history theory that predicted reductions in fitness due to variability in traits (Lewontin and Cohen 1969; Gillespie 1974; Tuljapurkar 1982), Pfister (1998) developed and tested the important life-history prediction that natural selection will minimize variability in vital rates (matrix elements) with high sensitivities. Her prediction has not been widely tested, in part due to statistical difficulties of doing so (Morris and Doak 2004). Nonetheless, recent rigorous analyses of mammalian (Gaillard and Yoccoz 2003) and alpine plant (Morris and Doak 2004) populations have supported her prediction. My results here for bird populations also support the prediction. However, I think the large amount of residual variation exhibited in Figs. 2–4 merits some attention, and that it may be fruitful to think about some residual variation (e.g., populations with var(λ) >0.010 in Fig. 4) as true departures from life history expectations and why that may occur. But before doing so, I will first consider potential bias in my results.

One concern with my study is that I focused only on adult survival rates and did not examine patterns of variation in other vital rates (e.g., fecundity). In the absence of temporal correlation among vital rates (matrix elements), a focus on just survival is safely unbiased. However, covariation among vital rates can influence the estimate of how much variation in any one vital rate affects population growth (Coulson et al. 2005; Doak et al. 2005). If adult survival is negatively correlated with other matrix elements, then I will likely have overestimated the effect of process variance in survival on population growth. The converse is true if adult survival is positively related to other elements. I do not know the correlational structure among vital rates in these 62 studies. However, despite the predicted and observed tradeoff between survival and reproduction at the individual level, at the population level positive correlations between survival and reproduction are much more common than negative ones (Clutton-Brock 1988). Positive correlations arise for two principal reasons. One is that phenotypic variability among individuals is large, which results in a favored phenotype in a particular environment outperforming less favored phenotypes in multiple fitness attributes (e.g., both survival and reproduction) (Clutton-Brock 1988). Second is that changes over time in an environment tend to similarly affect multiple vital rates; a 'good year' may boost both average survival and average fecundity. An example of these contrasting correlations at the individual and population level was evident in an experimental study that is one of the 62 populations shown here – black-legged kittiwakes in Alaska (Golet et al. 2004). Among unmanipulated birds, those that successfully reproduced consistently had higher survival rates than those that failed to lay eggs. However, individuals that were on the path to successful reproduction (eggs nearly hatched) but who did not have to raise young (eggs were removed) survived better than those that raised young – a negative correlation. What is relevant to my study is what occurred among the population of unmanipulated individuals, which was clearly a positive correlation between survival and reproduction (Golet et al. 2004). I do not know if my estimates of var(λ) from these 62 populations are biased. However, given the greater preponderance of positive than negative correlations among vital rates, if there is bias then it seems more likely I have underestimated the effects of process variance in survival on population growth. Thus, my results could be viewed as conservative.

Now, I reconsider the low r^2 values and substantive residual variation evident in Figs. 2–4. It seems possible that covariates such as taxonomic order, landscape (e.g., marine, forested, alpine), or clutch size may account for some of this variation (Gaillard et al. 2005) and the dispersion in data points in Figs. 2 and 3. Such patterning would still be consistent with life-history theory, but use of such covariates seems inadequate to explain the high values in variance of survival, which then manifest as high variance in population growth seen in Fig. 4. Unless large process variances in survival are counter-weighted by negative covariances in reproduction (discussed above), then large values of var(λ) are predicted to have substantive negative effects on populations and would be selectively disfavored relative to the many populations with small var(λ). I suggest that large values of var(λ) may be indicative of these species encountering environments different from that in which their current life history evolved. This could occur through a variety of anthropogenic effects as well as through substantive changes in climate. An obvious example is greater snow geese (Calvert and Gauthier 2005), where substantial variation in adult survival during this study was clearly a function of the initiation of spring hunting, which has been widely viewed as an additive (rather than compensatory) form of mortality for geese (Johnson and Owen 1992). One would expect that if hunting were compensatory, it would have only a small effect on population dynamics, and thus high values of var(λ) for hunted species may potentially be a signal of the additive effects of harvest. Of the ducks studied by Lake et al. (2006), only canvasbacks had high var(λ). In their review of additive and compensatory mortality in ducks, Nichols et al. (1984) found little evidence for additive mortality, except in canvasbacks (Nichols and Hines 1983).

Other types of anthropogenic effects may also impact vital rate variability, albeit less directly. For instance, cerulean warblers have declined in abundance, enough so that the United States Fish and Wildlife Service considered listing them as a threatened species (Link and Sauer 2002). The causes of population change are uncertain but potentially linked to alteration to their wintering habitats in montane evergreen forests of South America, which have recently undergone substantial change due to agriculture, principally coffee and coca production (Stotz et al. 1996; Jones et al. 2004). Many species face significant habitat degradation or loss from human activities, including western bluebirds (logging, grazing, and fire suppression; Guinan et al. 2000) and great bittern (filling of breeding wetlands; Puglisi and Bretagnolle 2005), which are two species with high var(λ) in Fig. 4. The response to and consequence of habitat loss can be conceptualized in light of the ideal free distribution elucidated by Fretwell and Lucas (1970). As the abundance of primary habitats declines, birds distribute among remaining primary habitats and less optimal secondary habitats, with an overall decline in average fitness. In the context of this paper, fitness decline equates to a reduction in mean survival rate. Additionally, because of the increased heterogeneity of habitats exploited, the variability in survival rates among all birds is expected to increase. Thus, changes to the mean *and variance* of vital rates can be a demographic consequence of habitat loss.

Greater environmental variability due to changing climate is a concern for many species. Rainfall patterns appear to impact survival of American redstarts, the species with the largest var(λ) (Fig. 4). Global climate models predict changes in the variability of precipitation, through changes in both the frequency and magnitude of precipitation events (Alley et al. 2007). There is much geographic heterogeneity in global climate models, so the predicted effects on particular species groups must be contemplated in the appropriate regional context of model outputs. Notably, Saltz et al. (2006) used regional outputs of a global climate model to demonstrate how changes in the variance of precipitation (without a change in mean) in Israel is expected to have a negative affect on dynamics of Asiatic wild ass (*Equus hemionus*).

Female barn swallows exhibited the second largest var(λ), although males in this population exhibited much less variation. This species exhibits strong sexual selection and the magnitude of that sexual selection is being heightened by the advanced timing of spring associated with climate change (Spottiswoode et al. 2006; Møller 2007a). Because such climate change is creating more but variable opportunities to breed within a season for this species (Møller 2007b), this may then cause more variability in adult survival as females incur the survival costs of variable and increased reproduction (Golet et al. 2004).

For virtually all species with high var(λ) in Fig. 4, there appears to be recent human-caused perturbations to their habitats or changes in the climate patterns they experience. I think it is useful to view these species as deviations from life-history expectation because current conditions have likely deviated from their evolutionary past, when their particular life history patterns evolved. Of the 62 populations considered, more than 40% were long-lived marine birds and none of these were among the seven populations with high var(λ). This result suggests that perhaps more *r* selected species are more vulnerable to heightened stochasticity. Some degree of environmental variability can be accommodated by species through phenotypic plasticity, which likely explains, for instance, the different balance of survival and fecundity rates among black-legged kittiwakes in Atlantic versus Pacific ocean habitats (Coulson 2002; Frederiksen et al. 2005). Additionally, some adaptation (genetic change through natural selection) has been noted in correspondence to changing season lengths for temperate breeding species (Nussey et al. 2005). However, the magnitude of environmental change is likely too large and rapid for species to adapt fast enough (Bradshaw and Holzapfel 2006; Holland et al. 2006; Alley et al. 2007). The predictions are for an increasingly stochastic world, which reduces population viability, even in the absence of changes in mean vital rates (Boyce et al. 2006). Such environmental change creates a significant conservation challenge.

Acknowledgments M. J. Lawonn and S. E. McCloskey helped obtain data sets for this analysis, and S. Fleischman discussed variance decomposition with me. I thank M. Frederiksen, M. Lindberg and the EURING organizing committee for the invitation to attend this meeting. J. J. Rotella served as a valuable sounding board for my ideas while we sampled the local brew. M. Frederiksen, Lindberg M, and M. Conroy provided comments on the manuscript. I thank the authors of the 62 cited studies whose data I used. Meta-analyses are completely dependent on high quality efforts by others that leads to a body of work substantive enough to be worthy of meta-analysis. Any errors in my use or interpretation of their data are my fault alone.

Appendix: Populations Used for Meta-Analysis of Adult Survival Rates

I located data-sets by searching specific keywords at www.scopus.com, limiting the search to articles printed since 2000, and requiring at least 5 years of annual survival estimates of adult birds, along with annual estimates of the square root of sampling variance. Entries are sorted by mean survival rate.

Latin name	Common name	Region	Sex	Breeding age	Mean survival	Process SD	Years in time series	Figure/ Table	Reference	Estimation model/ Estimation software/ Goodness of fit software
Phylloscopus collybita	Chiffchaff	EU	C	1	0.3032	0.0686	12	Figure 1	Julliard (2004)	CJS / SURGE / NA
Hirundo rustica	Barn Swallow	EU	M	1	0.3260	0.0826	14	Table 2	Møller and Szep (2002)	CJS* / MARK / RELEASE
Hirundo rustica	Barn Swallow	EU	F	1	0.3314	0.1670	14	Table 2	Møller and Szep (2002)	CJS* / MARK / RELEASE
Acrocephalus scirpaceus	Reed Warbler	EU	C	1	0.3826	0.0772	12	Figure 1	Julliard (2004)	CJS / SURGE / NA
Sylvia atricapilla	Blackcap	EU	C	1	0.4139	0.0575	12	Figure 1	Julliard (2004)	CJS / SURGE / NA
Turdus merula	Blackbird	EU	C	1	0.4960	0.0685	12	Figure 1	Julliard (2004)	CJS / SURGE / NA
Dendroica cerulea	Cerulean Warbler	NA	C	1	0.4977	0.1247	6	Figure 1	Jones et al. (2004)	CJS / MARK / RELEASE
Dendroica petechia	Yellow warbler	NA	F	1	0.5090	0.0578	8	Table 2	Mazerolle et al. (2005)	CJS* / MARK / RELEASE
Dendroica caerulescens	Black-throated Blue Warbler	NA	M	1	0.5178	0.0000	14	Figure 2	Sillett and Holmes (2002)	CJS / MARK / RELEASE
Sialia mexicana	Western Blue-bird	NA	F	1	0.5288	0.1376	6	Table 3	Keyser et al. (2004)	Burnham / MARK / Bootstrap
Dendroica petechia	Yellow Warbler	NA	M	1	0.5368	0.0000	8	Table 2	Mazerolle et al. (2005)	CJS* / MARK / RELEASE
Mniotilta varia	Black and White Warbler	NA	C	1	0.5560	0.0000	12	Figure 2	Dugger et al. (2004)	CJS / MARK / RELEASE

(continued)

(continued)

Latin name	Common name	Region	Sex	Breeding age	Mean survival	Process SD	Years in time series	Figure/ Table	Reference	Estimation model/ Estimation software/ Goodness of fit software
Parus caeruleus	Mediterranean Blue Tit	EU	C	1	0.5686	0.0639	14	Figure 6	Grosbois et al. (2006)	CJS / MARK / U-CARE
Seiurus aurocapilla	Ovenbird	NA	C	1	0.5822	0.0000	12	Figure 2	Dugger et al. (2004)	CJS / MARK / RELEASE
Forpus passerinus	Neotropical Parrotlet	TR	M	1	0.5873	0.0000	8	Figure 2	Sandercock et al. (2000)	CJS / SURGE / RELEASE
Parus montanus	Willow Tit	EU	C	1	0.5926	0.0370	10	Table 3	Lampila et al. (2006)	CJS / MARK / Bootstrap
Forpus passerinus	Neotropical Parrotlet	TR	F	1	0.5959	0.0893	8	Figure 2	Sandercock et al. (2000)	CJS / MARK / Bootstrap
Parus atricapillus	Black-capped Chick-adee	NA	C	1	0.6040	0.0632	34	Table 2	Gould and Nichols (1998)	CJS* / SURVIV / SURVIV
Setophaga ruticilla	American Redstart	NA	C	1	0.6150	0.2130	12	Figure 3	Dugger et al. (2004)	CJS / MARK / RELEASE
Anas americana	American Wigeon	NA	M	1	0.6391	0.0332	7	Table 3	Lake et al. (2006)	Brownie / MARK / ESTIMATE
Anser anser	Greylag Goose	EU	C	3	0.6948	0.0841	8	Figure 5a	Frederiksen et al. (2004a)	Burnham / MARK / Bootstrap
Botaurus stellaris	Great Bittern	EU	M	2	0.7107	0.1733	9	Table 3	Gilbert et all. (2002)	CJS / SURGE / RELEASE
Anas acuta	Northern Pintail	NA	M	1	0.7383	0.0628	11	Table 3	Lake et al. (2006)	Brownie / MARK / ESTIMATE

(continued)

(continued)

Latin name	Common name	Region	Sex	Breeding age	Mean survival	Process SD	Years in time series	Figure/ Table	Reference	Estimation model/ Estimation software/ Goodness of fit software
Sterna dougallii	Roseate Tern	NA	C	4	0.7480	0.0876	10	Table 1	Gould and Nichols (1998)	CJS* / SURVIV / SURVIV
Anas acuta	Northern Pintail	NA	F	1	0.7583	0.0000	11	Table 3	Lake et al. (2006)	Brownie / MARK / ESTIMATE
Chen canagica	Emperor Goose	NA	F	4	0.7618	0.1130	7		Schmutz (unpublished)	CJS / MARK/ Bootstrap
Pygoscelis adeliae	Adelie Penguin	SO	C	5	0.7668	0.0502	8	Figure 6	Jenouvrier et al. (2006)	CJS* / MARK / U-CARE
Aythya valisineria	Canvasback	NA	M	2	0.7717	0.1308	7	Table 3	Lake et al. (2006)	Brownie / MARK / ESTIMATE
Anser caerulescens atlantica	Greater Snow Goose	NA	C	3	0.7866	0.1735	12	Figure 3	Calvert and Gauthier (2005)	Brownie / MARK / ESTIMATE
Strix occidentalis occidentalis	Northern Spotted Owl	NA	C	2	0.7948	0.0000	11	Figure 1	LaHaye et al. (2004)	CJS / MARK / RELEASE
Aptenodytes forsteri	Emperor Penguin	SO	M	5	0.8089	0.0998	20	Figure 2	Barbraud and Weimerskirch (2001)	CJS / MARK / RELEASE
Aythya affinis	Lesser Scaup	NA	M	1	0.8122	0.0679	7	Table 3	Lake et al. (2006)	Brownie / MARK / ESTIMATE
Aethia cristatella	Crested Auklet	NA	C	3	0.8127	0.0900	9	Table 47	Drummond (2006)	CJS / MARK / RELEASE
Branta berincla nigrican	Black Brant	NA	C	3	0.8137	0.0566	13	Figure 6	Sedinger et al. (2006)	CJS RD / MARK / RDSURVIV

(continued)

(continued)

Latin name	Common name	Region	Sex	Breeding age	Mean survival	Process SD	Years in time series	Figure/ Table	Reference	Estimation model/ Estimation software/ Goodness of fit software
Branta canadensis	Canada Goose	NA	C	3	0.8167	0.0582	8	Table 11	Schaeffer et al. (2004)	Brownie / MARK / ESTIMATE
Rissa tridactyla	Black-legged Kittiwake	EU	C	5	0.8224	0.1266	11	Figure 1	Oro and Furness (2004)	CJS / SURGE / RELEASE
Anser brachy-rhynchus	Pink-footed Goose	EU	C	3	0.8236	0.0247	14	Figure 7	Frederiksen et al. (2004a)	Burnham / MARK / Bootstrap
Anser brachy-rhynchus	Pink-footed Goose	EU	C	3	0.8237	0.0477	13	Figure 2	Kery et al. (2006)	CJS* / M-SURGE / U-CARE
Sterna dougallii	Roseate Tern	NA	M	4	0.8255	0.0000	8	Table 2	Nichols et al. (2003)	CJS [#] / SURVIV[#] / SURVIV[#]
Aethia pusilla	Least Auklet	NA	C	3	0.8399	0.0364	8	Table 46	Drummond (2006)	CJS / MARK / RELEASE
Aptenodytes forsteri	Emperor Penguin	SO	F	5	0.8484	0.0000	20	Figure 2	Barbraud and Weimerskirch (2001)	CJS* / MARK / RELEASE
Acrocephalus sechel-lensis	Seychelles Warbler	TR	C	2	0.8486	0.0671	18	Figure 1	Brouwer et al. (2006)	CJS / MARK / Bootstrap
Halobaena caerulea	Blue Petrel	SO	C	6	0.8604	0.0972	7	Figure 2	Barbraud and Weimerskirch (2005)	CJS[$]/ M-SURGE / U-CARE
Ciconia ciconia	White Stork	EU	C	4	0.8619	0.0520	17	Figure 2	Schaub et al. (2005)	Burnham / MARK / U-CARE

(continued)

(continued)

Latin name	Common name	Region	Sex	Breeding age	Mean survival	Process SD	Years in time series	Figure/ Table	Reference	Estimation model/ Estimation software/ Goodness of fit software
Phalacrocorax carbo sinensis	Great Cormorant	EU	C	2	0.8650	0.0774	17	Figure 9	Frederiksen and Bregnballe (2000)	CJS* / SURGE / RELEASE
Diomedea chlororhynchos	Yellow-nosed Albatross	SO	C	7	0.8701	0.0774	7	Table 1	Weimerskirch et al. (2001)	CJS / SURGE / RELEASE
Chen rossii	Ross Goose	NA	C	3	0.8809	0.0580	12	Figure 6	Alisauskas et al. (2006)	Brownie / MARK / Bootstrap
Catharacta skua	Great Skua	EU	C	4	0.8850	0.0315	10	Table 2	Ratcliffe et al. (2002)	CJS / MARK / RELEASE
Morus bassanus	Northern Gannet	EU	C	5	0.8921	0.0516	42	Figure 3	Wanless et al. (2006)	Brownie / MARK / Bootstrap
Sula leucogaster	Brown Booby	TR	C	4	0.8959	0.0224	15	Figure 2	Beadell et al. (2003)	CJS / MARK / Bootstrap
Phoebastria irrorata	Waved Albatross	TR	C	7	0.9150	0.0295	5	Figure 2	Awkerman et al. (2006)	CJS/ MARK / Median C-hat
Rissa tridactyla	Black-legged Kittiwake	EU	C	5	0.9182	0.0277	15	Figure 4	Frederiksen et al. (2004b)	CJS* / MARK / U-CARE
Sterna dougallii	Roseate Tern	NA	F	4	0.9030	0.0239	8	Table 2	Nichols et al. (2003)	CJS# /SURVIV / SURVIV
Phaethon rubricauda	Red-tailed Tropicbird	TR	C	4	0.9052	0.0396	15	Table 3	Doherty et al. (2004)	CJS*$/ SURVIV / SURVIV

(continued)

(continued)

Latin name	Common name	Region	Sex	Breeding age	Mean survival	Process SD	Years in time series	Figure/ Table	Reference	Estimation model/ Estimation software/ Goodness of fit software
Aptenodytes patagonicus	King Penguin	SO	F	6	0.9063	0.1278	5	Figure 1	Olsson and van de Jeugd (2002)	CJS / SAS / RELEASE
Aptenodytes patagonicus	King Penguin	SO	M	6	0.9222	0.0621	5	Figure 1	Olsson and van de Jeugd (2002)	CJS / SURGE / RELEASE
Uria aalge	Common Guillemot	EU	C	5	0.9270	0.0000	10	Table 2	Harris et al. (2000)	CJS* / SURGE / RELEASE
Fulmarus glacialoides	Southern fulmar	SO	C	10	0.9276	0.0583	33	Figure 3	Jenouvrier et al. (2003)	CJS* / MARK / RELEASE
Rissa brevirostris	Red-legged Kittiwake	NA	C	5	0.9386	0.0000	11	Table 5	Schmutz and Byrd (2004)	CJS / MARK / Bootstrap
Pagodroma nivea	Snow petrel	SO	C	10	0.9425	0.0400	21	Table 1	Chastel et al. (1993)	CJS / SURGE / NA
Uria aalge	Common Guillemot	EU	C	5	0.9490	0.0206	13	Table 2	Harris et al. (2000)	CJS* / SURGE / RELEASE
Rissa tridactyla	Black-legged Kittiwake	NA	C	5	0.9502	0.0237	8	Table 2	Golet et al. (2004)	CJS$^{\$}$ / MARK / MSSURVIV

Region – North America (NA), Europe (EU), Tropics (TR), Southern Ocean (SO).
Sex – Combined (C), Male (M), Female (F).
Estimation models – Brownie = Band recovery models with Brownie or Seber parameterizations of recovery parameters; Burnham = combined model of band recovery and mark-recapture data; CJS = Cormack-Jolly-Seber models, where * indicates a modification for trap effects or transient individuals, $ indicates an extension to a multi-state approach, # indicates an extension for accommodating uncertainty in sex determination, and RD indicates a CJS model augmented with a Robust Design approach.
Estimation and goodness-of-fit software – ESTIMATE, MARK, MSSURVIV, RELEASE, SAS, SURGE, SURVIV are all recognized names of analytical software, with ESTIMATE and RELEASE available within program MARK. Bootstrap and Median C-hat refer to additional goodness-of-fit options within MARK. A few studies did not report goodness-of-fit testing (NA), but all these used SURGE, which has goodness-of-fit capacity.

References

Alisauskas RT, Drake KL, Slattery SM, Kellett DK (2006) Neckbands, harvest, and survival of Ross's geese from Canada's central Arctic. Journal of Wildlife Management 70:89–100.

Alley R, and 33 others (2007) Climate change 2007: the physical science basis. Contribution of working group I to the fourth assessment report of the Intergovernmental Panel on Climate Change. www.ipcc.ch .

Awkerman JA, Huyvaert KP, Mangel J, Alfaro Shiguetoc J, Anderson DJ (2006) Incidental and intentional catch threatens Galapagos waved albatross. Biological Conservation 133:483–489.

Barbraud C, Weimerskirch H (2001) Emperor penguins and climate change. Nature 411:183–185.

Barbraud C, Weimerskirch H (2005) Environmental conditions and breeding experience affect costs of reproduction in blue petrels. Ecology 86:682–692.

Beadell JS, Schreiber EA, Schreiber RW, Schenk GA, Doherty PF (2003) Survival of brown boobies (*Sula leucogaster*) at Johnston Atoll: a long-term study. Auk 120: 811–817.

Boyce MS, Haridas CV, Lee CT, NCEAS Stochastic Demography Working Group (2006) Demography in an increasingly variable world. Trends in Ecology and Evolution 21:141–148.

Bradshaw WE, Holzapfel CM (2006) Evolutionary response to rapid climate change. Science 312:1477–1478.

Brouwer L, Richardson DS, Eikenaar C, Komdeur J (2006) The role of group size and environmental factors on survival in a cooperatively breeding tropical passerine. Journal of Animal Ecology 75:1321–1329.

Burnham KP, Anderson DR (2002) Model selection and multimodel inference: a practical information-theoretic approach. Second edition. Springer, New York, USA.

Burnham KP, Anderson DR, White GC, Brownie C, Pollock KH (1987) Design and analysis methods for fish survival experiments based on release-recapture. American Fisheries Society Monograph 5.

Burnham KP, White GC (2002) Evaluation of some random effects methodology applicable to bird ringing data. Journal of Applied Statistics 29:245–264.

Calvert AM, Gauthier G (2005) Effects of exceptional conservation measures on survival and seasonal hunting mortality in greater snow geese. Journal of Applied Ecology 42:442–452.

Caswell H (2001) Matrix population models: construction, analysis, and interpretation. Second edition. Sinauer Associates, Inc., Sunderland, Massachusetts, USA.

Chastel O, Weimerskirch H, Jouventin P (1993) High annual variability in reproductive success and survival of an Antarctic seabird, the snow petrel *Pagodroma nivea*. Oecologia 94:278–285.

Clutton-Brock TH (1988) Reproductive success. Pages 472–486 *in* Reproductive success: studies of individual variation in contrasting breeding systems. Chicago University Press, Chicago, Illinois, USA.

Coulson JC (2002) Why do adult kittiwakes survive so long but breed so poorly in the Pacific. Journal of Avian Biology 33:111–112.

Coulson TJ, Gaillard J-M, Festa-Bianchet M. (2005) Decomposing the variation in population growth into contributions from multiple demographic rates. Journal of Animal Ecology 74:789–801.

DeSante DF, Burton KM, Saracco JF, Walker BL (1995) Productivity indices and survival rate estimates from MAPS, a continent-wide programme of constant-effort mist-netting in North America. Journal of Applied Statistics 22:935–947.

Doak DF, Morris WF, Pfister C, Kendall BE, Bruna EM (2005) Correctly estimating how environmental stochasticity influences fitness and population growth. American Naturalist 166: E14–E21.

Doherty PF, Schreiber EA, Nichols JD, Hines JE, Link WA, Schenk GA, Schreiber RW (2004) Testing life history predictions in a long-lived seabird: a population matrix approach with improved parameter estimation. Oikos 105:606–618.

Drummond BA (2006) Biological monitoring in the central Aleutian Islands, Alaska in 2006: summary appendices. United States Fish and Wildlife Service Report, AMNWR 06/06. Adak, Alaska.

Dugger KM, Faaborg J, Arendt WJ, Hobson KA (2004) Understanding survival and abundance of overwintering warblers: does rainfall matter? Condor 106:744–760.

Easterling DR, Meehl GA, Parmesan C, Changnon SA, Karl TR, Mearns LO (2000) Climate extremes: observations, modeling, and impacts. Science 289:2068–2074.

Emanuel K (2005) Increasing destructiveness of tropical cyclones over the past 30 years. Nature 436:686–688.

Frederiksen M, Bregnballe T (2000) Evidence for density-dependent survival in adult cormorants from a combined analysis of recoveries and resightings. Journal of Animal Ecology 69: 737–752.

Frederiksen M, Harris MP, Wanless S (2005) Inter-population variation in demographic parameters: a neglected subject? Oikos 111:209–214.

Frederiksen M, Hearn RD, Mitchell C, Sigfusson A, Swann RL, Fox AD (2004a) The dynamics of hunted Icelandic goose populations: a reassessment of the evidence. Journal of Applied Ecology 41:315–334.

Frederiksen M, Wanless S, Harris MP, Rothery P, Wilson LJ (2004b) The role of industrial fisheries and oceanographic change in the decline of North Sea black-legged kittiwakes. Journal of Applied Ecology 41:1129–1139.

Fretwell SD, Lucas HL (1970) On territorial behavior and other factors influencing habitat distribution in birds. I. Theoretical development. Acta Biotheoretica 19:16–36.

Gilbert G, Tyler GA, Smith KW (2002) Local annual survival of booming male great bittern *Botaurus stellaris* in Britain, in the period 1990–1999. Ibis 144:51–61.

Gaillard J-M, Yoccoz NG, Lebreton J-D, Bonenfant C, Devillard S, Loison A, Pontier D, Allaine D (2005) Generation time: a reliable metric to measure life-history variation among mammalian populations. American Naturalist 166:119–123.

Gaillard J-M, Yoccoz NG (2003) Temporal variation in survival of mammals: a case of environmental canalization? Ecology 84:3294–3306.

Gillespie J (1974) Natural selection for within-generation variance in offspring number. Genetics 76:601–606.

Golet GH, Schmutz JA, Irons DB, Estes JA (2004) Determinants of reproductive costs in the long-lived black-legged kittiwake: a multi-year experiment. Ecological Monographs 74:353–372.

Gould WR, Nichols JD (1998) Estimation of temporal variability of survival in animal populations. Ecology 75:2531–2538.

Grosbois V, Henry P, Blondel J, Perret P, Lebreton J, Thomas DW, Lambrechts MM (2006) Climate impacts on Mediterranean blue tit survival: an investigation across seasons and spatial scales. Global Change Biology 12:2235–2249.

Guinan JA, Gowaty PA, Eltzroth EK (2000) Western bluebird (*Sialia mexicana*). In Birds of North America, no. 510 *in* Poole A, Gill F, eds. Academy of Natural Sciences, Philadelphia, Pennsylvania, USA.

Haridas CV, Tuljapurkar S (2005) Elasticities in variable environments: properties and implications. American Naturalist 166:481–495.

Harris MP, Wanless S, Rothery P, Swann RL, Jardine D (2000) Survival of adult common guillemots *Uria aalge* at three Scottish colonies. Bird Study 47:1–7.

Holland MM, Bitz CM, Tremblay B (2006) Future abrupt reductions in the summer Arctic sea ice. Geophysical Research Letters 33:L23503.

Jenouvrier S, Barbraud C, Weimerskirch H (2003) Effects of climate variability on the temporal population dynamics of southern fulmars. Journal of Animal Ecology 72:576–587.

Jenouvrier S, Barbraud C, Weimerskirch H (2006) Sea ice affects the population dynamics of Adelie penguins in Terre Adelie. Polar Biology 29:413–423.

Johnson DH, Owen M (1992) World waterfowl populations: status and dynamics. Pages 635–652 *in* McCullough DR, Barrett RH, eds. Wildlife 2001: populations. Elsevier Applied Science, New York, USA.

Jones J, Barg JJ, Sillett TS, Veit ML, Robertson RJ (2004) Minimum estimates of survival and population growth for cerulean warblers (*Dendroica cerulea*) breeding in Ontario, Canada. Auk 121:15–22.

Julliard R (2004) Estimating the contribution of survival and recruitment to large scale population dynamics. Animal Biodiversity and Conservation 27:417–426.

Kery M, Madsen J, Lebreton J (2006) Survival of Svalbard pink-footed geese *Anser brachyrhynchus* in relation to winter climate, density and land-use. Journal of Animal Ecology 75:1172–1181.

Keyser AJ, Keyser MT, Promislow BEL (2004) Life-history variation and demography in western bluebirds (*Sialia mexicana*) in Oregon. Auk 121:118–133.

Lahaye WS, Zimmerman GS, Gutierrez RJ (2004) Temporal variation in the vital rates of an insular population of spotted owls (*Strix occidentalis occidentalis*): contrasting effects of weather. Auk 121:1056–1069.

Lake BC, Walker J, Lindberg MS (2006) Survival of ducks banded in the boreal forest of Alaska. Journal of Wildlife Management 70:443–449.

Lampila S, Orell M, Belda E, Koivula K (2006) Importance of adult survival, local recruitment and immigration in a declining boreal forest passerine, the willow tit *Parus montanus*. Oecologia 148:405–413.

Lande R, Engen S, Sæther B-E (2003) Stochastic population dynamics in ecology and conservation. Oxford University Press, Oxford.

Lewontin RC, Cohen D (1969) On population growth in a randomly varying environment. Proceedings of the National Academy of Sciences of the USA 62:1056–1060.

Link WA, Doherty PF (2002) Scaling in sensitivity analysis. Ecology 83:3299–3305.

Link WA, Sauer JR (2002) A hierarchical analysis of population change with application to Cerulean warblers. Ecology 83:2832–2840.

Mazerolle DF, Dufour KW, Hobson KA, den Haan HE (2005) Effects of large-scale climatic fluctuations on survival and production of young in a neotropical migrant songbird, the yellow warbler *Dendroica petechia*. Journal of Avian Biology 36:155–163.

Møller AP (2007a) Tardy females, impatient males: protandry and divergent selection on arrival date in the two sexes of the barn swallow. Behavioral Ecology and Sociobiology 61: 1311–1319.

Møller AP (2007b) Interval between clutches, fitness, and climate change. Behavioral Ecology 18:62–70.

Møller AP, Szep T (2002) Survival rate of adult barn swallows *Hirundo rustica* in relation to sexual selection and reproduction. Ecology 8:2220–2228.

Morris WF, Doak DF (2004) Buffering of life histories against environmental stochasticity: accounting for a spurious correlation between the variabilities of vital rates and their contributions to fitness. American Naturalist 163:579–590.

Nichols JD, Conroy MJ, Anderson DR, Burnham KP (1984) Compensatory mortality in waterfowl populations: a review of evidence and implications for research and management. Transactions of the North American Wildlife and Natural Resources Conference 49:535–554.

Nichols JD, Hines JE (1983) The relationship between harvest and survival rates of mallards: a straightforward approach with partitioned data sets. Journal of Wildlife Management 47: 334–348.

Nichols JD, Kendall WL, Hines JE, Spendelow JA (2003) Estimation of sex-specific survival from capture–recapture data when sex is not always known. Ecology 85:3192–3200.

Nussey DH, Postma E, Gienapp P, Visser ME (2005) Selection on heritable phenotypic plasticity in a wild bird population. Science 310:304–306.

Olsson O, van der Jeugd HP (2002) Survival in king penguins *Aptenodytes patagonicus*: temporal and sex-specific effects of environmental variability. Oecologia 132:509–516.

Oro D, Furness RW (2004) Influences of food availability and predation on the survival of kittiwakes. Ecology 83:2516–2528.

Pfister CA (1998) Patterns of variance in stage-structured populations: evolutionary predictions and ecological implications. Proceedings National Academy Science 95:213–218.

Puglisi L, Bretagnolle V (2005) Breeding biology of the great bittern. Waterbirds 28:392–398.

Ratcliffe N, Catry P, Hamer KC, Klomp NI, Furness RW (2002) The effect of age and year on the survival of breeding adult great skuas *Catharacta skua* in Shetland. Ibis 144:384–392.

Rich TD, Beardmore CJ, Berlanga H, Blancher PJ, Bradstreet MSW, Butcher GS, Demarest DW, Dunn EH, Hunter WC, Inigo-Elias EE, Kennedy JA, Martell AM, Punjabi AO, Pashley DN, Rosenberg KV, Rustay CM, Wendt JS, Will TC (2004) Partners in flight North American landbird conservation plan. Cornell Laboratory of Ornithology, Ithaca, New York, USA.

Roff DA (2002) Life history evolution. Sinauer Associates, Inc., Sunderland, Massachusetts, USA.

Sæther B-E, Bakke, Ø (2000) Avian life history variation and the contribution of demographic traits to the population growth rate. Ecology 81:642–653.

Saltz D, Rubenstein DI, White GC (2006) The impact of increased environmental stochasticity due to climate change on the dynamics of Asiatic wild ass. Conservation Biology 20:1402–1409.

Sandercock BK, Beissinger SR, Stoleson SH, Melland RR, Hughes CR (2000) Survival rates of a neotropical parrot: implications for latitudinal comparisons of avian demography. Ecology 81:1351–1370.

Schaeffer SE, Rusch DH, Humburg DD, Lawrence JS, Zenner GG, Gillespie MM, Caswell FD, Wilds S, Yaich SC (2004) Survival, movements, and harvest of eastern prairie population Canada geese. Wildlife Monographs 156.

Schaub M, Kania W, Koppen U (2005) Variation of primary production during winter induces synchrony in survival rates in migratory white storks *Ciconia ciconia*. Journal of Animal Ecology 74:656–666.

Schmutz JA, Byrd GV (2004) Modeling the influence of productivity, survival, and metapopulation dynamics on trends of kittiwakes and murres at the Pribilof Islands, Alaska. http://www.absc.usgs.gov/research/seabird˙foragefish/products/index.html .

Sedinger JS, Ward DH, Schamber JL, Butler WI, Eldridge WD, Conant B, Voelzer JF, Chelgren ND, Herzog MP (2006) Effects of El Nino on distribution and reproductive performance of black brant. Ecology 87:151–159.

Sillett TS, Holmes RT (2002) Variation in survivorship of a migratory songbird throughout its annual cycle. Journal of Animal Ecology 71:296–308.

Spottiswoode CN, Tøttrup AP, Coppack T (2006) Sexual selection predicts advancement of avian spring migration in response to climate change. Proceedings of the Royal Society B Biological Sciences 273:3023–3029.

Stotz DF, Fitzpatric JW, Parker TA, Moskovits DK (1996) Neotropical birds: ecology and conservation. University of Chicago Press, Chicago, Illinois, USA.

Tuljapurkar S (1982) Population dynamics in variable environments. III. Evolutionary dynamics of r-selection. Theoretical Population Biology 21:141–165.

Wanless S, Frederiksen M, Harris MP, Freeman SN (2006) Survival of gannets *Morus bassanus* in Britain and Ireland, 1959–2002. Bird Study 53:79–85.

Weimerskirch H, Zimmermann L, Prince PA (2001) Influence of environmental variability on breeding effort in a long lived seabird, the yellow-nosed albatross. Behavioral Ecology 12: 22–30.

White GC (2000) Population viability analysis: data requirements and essential analyses. Pages 288–331 *in* Boitani L, Fuller TK, eds. Research techniques in animal ecology: controversies and consequences. Columbia University Press, New York, USA.

White GC, Burnham KP, Barker R (2008) Evaluation of some Bayesian MCMC random effects inference methodology applicable to bird ringing data. In: Thomson DL, Cooch EG, Conroy MJ (eds.) Modeling Demographic Processes in Marked Populations. Environmental and Ecological Statistics, Springer, New York.

Williams BK, Nichols JD, Conroy MJ (2002) Analysis and management of animal populations: modeling, estimation, and decision making. Academic Press, San Diego, California, USA.

Filling a Void: Abundance Estimation of North American Populations of Arctic Geese Using Hunter Recoveries

Ray T. Alisauskas, Kiel L. Drake, and James D. Nichols

Abstract We consider use of recoveries of marked birds harvested by hunters, in conjunction with continental harvest estimates, for drawing inferences about continental abundance of a select number of goose species. We review assumptions of this method, a version of the Lincoln–Petersen approach, and consider its utility as a tool for making decisions about harvest management in comparison to current sources of information. Finally, we compare such estimates with existing count data, photographic estimates, or other abundance estimates. In most cases, Lincoln estimates are far higher than abundances assumed or perhaps accepted by many waterfowl biologists and managers. Nevertheless, depending on the geographic scope of inference, we suggest that this approach for abundance estimation of arctic geese may have usefulness for retrospective purposes or to assist with harvest management decisions for some species. Lincoln's estimates may be as close or closer to truth than count, index, or photo data, and can be used with marking efforts currently in place for estimation of survival and harvest rates. Although there are bias issues associated with estimates of both harvest and harvest rate, some of the latter can be addressed with proper allocation of marks to spatially structured populations if subpopulations show heterogeneity in harvest rates.

1 Introduction

While estimation of population growth rate provides a useful metric of population health, abundance estimation remains of fundamental importance to animal conservation, particularly for exploited populations, or for those which face risk from other factors (e.g., habitat loss) unrelated to direct exploitation. For example, low abundance remains relevant to the status of a population, even if it shows a

R.T. Alisauskas (✉)
Environment Canada, Prairie and Northern Wildlife Research Centre, 115 Perimeter Road, Saskatoon, Saskatchewan S7N 0X4, Canada
e-mail: ray.alisauskas@ec.gc.ca

D.L. Thomson et al. (eds.), *Modeling Demographic Processes in Marked Populations*, Environmental and Ecological Statistics 3, DOI 10.1007/978-0-387-78151-8_20,

high rate of population growth. In some cases, high densities of certain animal populations could pose conservation problems for other reasons, regardless of population trajectory. In the case of animal populations exploited directly by humans, such as through hunting, informed decisions about harvest management should be underpinned minimally by knowledge about abundance of animals and hunters, population trajectory, survival probability and harvest, and how such direct exploitation might influence survival and thus population growth.

There may be few if any reliable annual estimates of continental populations of geese that nest in arctic North America. Most information about annual variation in abundance is based on counts conducted by observers in aircraft that visually estimate the number of geese within flocks. In some cases, flocks contain hundreds of thousands of geese; moreover, counts are conducted in a few seconds without photographic or video aids, without a sampling design or without notions about what fraction of the population is counted (or sampled when sampling designs exist). Such visual estimates have long been known typically to underestimate true numbers, and the bias is more severe with larger groups of birds (Spinner 1949; Ely et al. 1993; Boyd 2000). Direct counts based on such visual estimates are also the method used for annual midwinter counts of waterfowl, often referred to as "the midwinter index". As well, estimates of some species are made irregularly in the arctic using aerial photography of breeding colonies (Kerbes et al. 2006), when colonial species are relatively clumped and sedentary. However, assumptions about complete enumeration (i.e., detection) of colonies, the number of unsampled nonbreeders, and detection of geese from photographic images have not been rigorously tested. False assumptions about these issues lead to incorrect inferences, probably with continental abundance estimates biased low. The extent of this bias for a continental population is related to the magnitude of nondetection, which remains unknown with methods normally used.

We consider the usefulness of Lincoln's (1930) approach, as originally proposed for population estimation of North American ducks, for annual estimation of population size and possibly derived metrics of population change for 4 recognized populations of arctic-nesting geese. We focus on (1) greater snow geese (*Chen caerulescens atlanticus*), (2) midcontinent lesser snow geese (*C. c. caerulescens*), (3) Ross's geese (*C. rossii*), and (4) midcontinent white-fronted geese (*Anser albifrons frontalis*). While all four breed in remote regions, two of these (2 and 4) have relatively widespread breeding distributions, both characteristics which impede enumeration. Populations 1, 2, and 3 are colonial so that breeding concentrations are clumped, and two of these (1 and 3) have relatively restricted breeding ranges. Broad geographic breeding distributions and dispersed nesting, characteristic of white-fronted geese, provide a virtually insurmountable challenge to complete annual coverage using, e.g., appropriate aerial survey methods either with helicopter or fixed-winged aircraft and a design using either strip or line transects.

We review existing sources of abundance information to illustrate the void, and then review assumptions behind Lincoln's estimator for population size. Finally, we apply information from band recovery data and harvest estimates for estimation of continental abundance for these four example populations, in comparison to existing sources of abundance information.

2 Background

2.1 Pertinent Aspects of Life History of Arctic-Nesting Geese

Most species of geese are highly social, and remain in gregarious concentrations throughout the annual cycle (Owen 1980). During most of the year, geese are also highly mobile, especially during spring and fall migration, but also throughout winter. However, after arrival on their breeding ranges in May or June, arctic-nesting geese become relatively sedentary near nest territories, although they are still capable of flight. Although precise timing depends on latitude, geese undergo an annual molt of flight feathers resulting in flightlessness from July until some time in August. Geese that do not attempt to nest, or that experience nest failure, undergo wing molt before those that successfully hatch nests. Portions of some populations undertake a molt migration to areas outside of the normal breeding range (Salomonson 1968; Abraham et al. 1999), whereas successfully breeding adults are confined to brood-rearing areas limited to distances that flightless goslings can either swim or walk. Nevertheless, during this period of flightlessness after nesting, large numbers of arctic-nesting geese (both breeders and nonbreeders) are generally found in habitats that offer abundant graminoid vegetation, in association with relatively large (> 1 ha) water bodies that provide escape from terrestrial predators. Although flightless, most species of arctic-nesting geese remain highly gregarious. These qualities render the flightless period during the arctic summer as the most efficient time of year to capture and mark large numbers of arctic-nesting geese. Although some situations permit researchers to capture reasonably large numbers of flightless geese on foot (e.g. Sedinger et al. 1995), use of helicopters (Timm and Bromley 1976) increases accessibility to large areas permitting mass capture while allowing far greater flexibility in stratification and allocation of marking effort over larger geographic areas.

Consequently, many arctic-nesting geese have been marked in the arctic most recently (Fig. 1). Unlike prairie-nesting ducks, most recoveries of arctic-nesting geese occur far to the south of where they are marked. Whereas most marking of these geese occurs before mid-August, they are generally not harvested by hunters until mid-September in Canada, and most are killed later and farther south in the U.S. (e.g. Alisauskas et al. 2006).

2.2 Lincoln's (1930) Estimator

The estimation approach dates back to Lincoln's (1930:2) reasoning behind the use of marked ducks for estimation of continental population size. He postulated

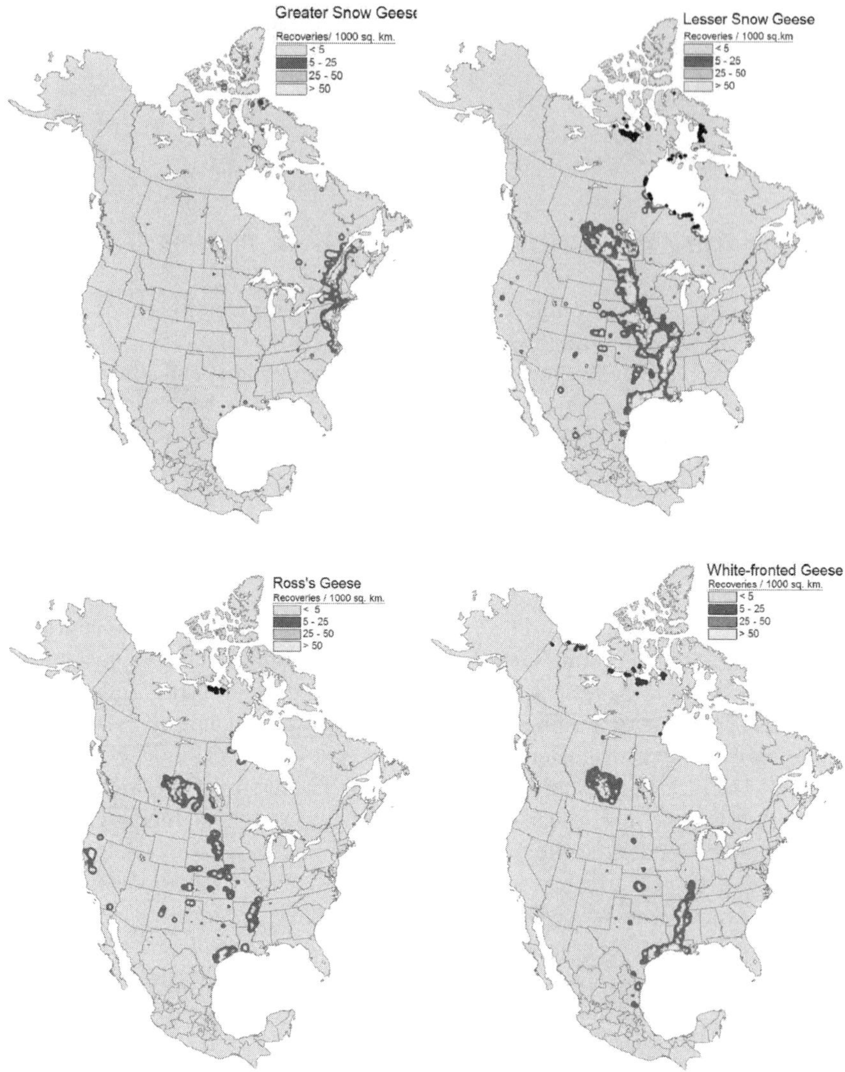

Fig. 1 Locations where adults from arctic goose populations of interest were marked during June to August (1989 – 2004) in Canada's central and eastern arctic (*filled circles*), in relation to distributions of recoveries by hunters. Recovery distribution is represented as density of recoveries per 1000 km^2. Note geographic separation of first sample (during marking) from second samples (during hunting seasons). Only recoveries from Canada or U.S. are used for subsequent Lincoln estimates of population size

> Given a fairly accurate statement showing the number of wild ducks killed in North America in any one season, then the total number of ducks present on the continent for that season may be estimated by a percentage computation, based upon the relation that the total number of banded ducks killed during their first season as band carriers bears to the total number banded.

The estimator of population size, *N*, is simply

$$\hat{N}_i = \frac{\hat{H}_i}{\hat{h}_i} \tag{1}$$

where $\hat{H}_i$ is an estimate of harvest for year *i*, and $\hat{h}_i$ is the estimated probability that a bird alive at the time of banding in year *i* is harvested in year *i*. It may be convenient to think of $\hat{H}_i$ as analogous to a count statistic, with $\hat{h}_i$ functioning as a kind of detection probability, as structured in a general canonical estimator of abundance (Williams et al. 2002:244). For example, the Lincoln–Petersen capture–recapture estimator of abundance for closed populations can be written as:

$$\hat{N} = \frac{n_2}{\hat{p}_2} \tag{2}$$

where n_2 is the number of animals captured in the second of two sampling occasions, and $\hat{p}_2 = m_2/n_1$ or the ratio of recaptures to animals marked in the first sample.

One difference between the harvest and capture–recapture contexts is that the number of animals captured in the second sample is known in the capture–recapture context, whereas in the harvest context, the total number of birds harvested must be estimated. Harvest of each waterfowl species and its associated variance are estimated annually for each province in Canada (Sen 1971). These estimates are reported at the online source of the Canadian Wildlife Service (CWS) National Harvest Database (http://www.cws-scf.ec.gc.ca/harvest/default_e.cfm) for Canadian harvest. U.S. harvest estimates, as desribed by Sen (1971) and Geissler (1990) are also available (Kruse and Sharp 2002; Kruse 2006). Associated sampling variance estimates are not readily available for U.S. harvest data, although methods exist for its computation (Geissler 1990). Otis (2006) assumed $CV_{\hat{H}_i} = 0.05$ for annual harvest of mollards (*Anas platyrhynchos*), as estimated by Geissler (1990).

A second difference between the harvest and capture–recapture contexts is that both the initial number of animals marked, n_1, and the number of recaptures, m_2, are known in the capture–recapture context. In the harvest context, the number of birds banded is known, but the number of banded birds harvested must be estimated. Incomplete reporting of recovered bands has long been recognized, leading to interest in band reporting rate, ρ_i, defined as the probability that a banded bird that is harvested (shot and retrieved) in year *i* is reported (band number conveyed to the Canadian or U.S. Bird Banding Laboratories). In the absence of any special incentives or solicitation to induce complete reporting of recovered bands, we are left with recovery rate, f_i, defined as the *probability that a banded bird alive when a given cohort is banded will be shot and its band reported during the next hunting season* (Brownie et al. 1985), where

$$\hat{f}_i = \hat{h}_i \hat{\rho}_i. \tag{3}$$

Note also that incomplete reporting probability of marked animals during the exploitation process means that there is a detection probability associated with estimation of $\hat{h}$, itself a detection probability applied to $\hat{H}_i$ (Eq. 1). Otis (2006) noted that there are technical concerns about the statistical performance of ratio estimators, in general, that are related to imprecision associated with the estimated denominator. Thus, there are several characteristics inherent in Lincoln's estimator that may render it ill-behaved statistically.

2.3 Previous Application of Lincoln's Estimator to Waterfowl Populations

There are numerous examples of the use of banding data in conjunction with annual harvest statistics for exploited waterfowl populations specifically, but there has been no integration of the method into annual decisions about harvest management, to our knowledge. For example, Bowers and Martin (1975) used this approach with wood ducks (*Aix sponsa*), which are difficult to enumerate during breeding because they are widely distributed at low densities, and are difficult to detect because of their habitat preference for forested wetlands.

Boyd (1976) used Lincoln's index for estimating numbers of arctic-nesting geese. His focus was what he referred to as the Hudson Bay population of lesser snow geese, also known as the midcontinent population of lesser snow geese in reference to their winter quarters. Most recently, Otis (2006) compared the similarity of Lincoln estimates of mallard abundance to independent estimates of breeding population (BPOP) size. BPOP is obtained from the Waterfowl Breeding Population and Habitat Survey conducted annually by the U.S. Fish and Wildlife Service and the Canadian Wildlife Service in cooperation with other state and provincial resource management agencies. Estimation of BPOP is based on an extensive aerial survey with associated ground-based work designed to estimate the fraction of ducks detected from the air. The outcome of Otis's evaluation was a remarkably good consistency between the two methods of estimation that resulted in similar time-series profiles, although the Lincoln estimates were consistently higher than BPOP estimates, sometimes by a factor of ~2. Otis also estimated the coefficient of variation of $\hat{N}_i$, $\widehat{CV}(\hat{N}_i)$, as well as the average instantaneous rate of population growth, $\hat{r}$, and its estimated variance, $\widehat{\text{var}}(\hat{r})$, using annual Lincoln estimates and associated measures of variation. Otis (2006) further advocated derivation of metrics of population change from estimation of mourning dove population size using Lincoln's method for making decisions about harvest regulations.

2.4 Assumptions Behind the Method

Although Eq. 1 is simple and straightforward, a number of assumptions must be satisfied before $\hat{N}$ can be viewed as a suitable estimator for N (e.g., Seber 1982:59; Williams et al. 2002:291). Here, we consider assumptions in the context of the geographic range of the species of arctic geese that we are interested in:

(a) *The population is closed so that N is constant.* The Lincoln estimator has been shown to be robust to some forms of deviation from the closure assumption (e.g., Seber 1982; Williams et al. 2002). In particular, we expect mortality to occur between the time of banding and the time of harvest, so that the population is open to losses. Robson (1969) showed that if mortality occurs between the first and second samples in a 2-sample capture–recapture study, and if this mortality applies equally to marked and unmarked animals, then the Lincoln estimator provides an approximately unbiased estimate of abundance at the time of the first sample. Thus, there can be no mortality resulting as a direct consequence either of mark presence, or of handling. Although there is ample evidence for increased mortality in geese each marked with a neckband and a legband, compared to those only with a legband (Alisauskas and Lindberg 2002; Alisauskas et al. 2006; but see Gauthier et al. 2001), the generally accepted assumption is that survival probability of birds marked with only a legband is identical to that of unmarked birds. Another way in which this assumption can be violated involves nonrepresentative banding in which animals that are marked experience higher or lower mortality between first and second samples than animals marked elsewhere (see below), resulting from a handling bias (Sedinger et al. 1997). Gains to the population between the first and second samples are not expected to occur because of the discrete seasonality of arctic goose breeding and the timing of the banding period. We view Lincoln estimates of abundance as pertaining to the time of banding in late summer.

(b) *Probabilities of animals appearing in a sample may differ between the first and second sampling periods, but should be equal for animals within each period.* This assumption is very likely to be violated in harvest applications for arctic-nesting geese and is the one on which we focus. Banded samples do not represent randomly selected birds from throughout the breeding range, thus providing the potential for different probabilities of harvest for banded and unbanded birds. Heterogeneous capture probabilities with Lincoln estimates are known to produce negatively biased abundance estimates (e.g., Seber 1982; Williams et al. 2002). However, this inference only applies to a specific kind of heterogeneity in which probabilities of a bird appearing in the first and second sample are correlated within individuals (e.g., birds with low probability of being banded also have a low probability of being harvested). In fact, heterogeneity among individuals in the case of independence between the first and second samples yields approximately unbiased estimates (e.g., Robson 1969; Seber 1982). The sampling reality of our application is that we may be banding subpopulations of breeding geese in a manner that is not proportional to their abundance. Thus we get variation across space in probabilities of birds appearing in the first sample. If these birds from different subpopulations then show different probabilities of appearing in the second sample (i.e., different harvest rates), then we expect biased estimates of abundance with the bias depending on the relationship between the probabilities of appearing in the first and second samples.

(c) *Animals do not lose their marks in the time between the two samples and all marks are reported on recovery in the second sample.* Legband loss between summer banding and the subsequent hunting season is likely to be negligible, so we do not consider this as a potential problem. It has long been known that hunters do not report all birds that they harvest. As noted in Eq. 3, band recovery rate, $\hat{f}$, estimated directly from banding data, must be adjusted by band reporting rate, $\hat{\rho}$, to compute $\hat{h}$. Band reporting rate can be estimated from reward band studies whereby hunters are induced to report bands from harvested birds with high probability (hopefully approaching 1) by rewarding them with monetary compensation (Henny and Burnham 1976; Nichols et al. 1991; Nichols et al. 1995; Royle and Garrettson 2005). Such studies have been designed to estimate $\hat{\rho}$ for bands that carry no reward. Until recently (2003–2005), there have been no reward band studies to estimate reporting rates for arctic-nesting geese, as most attention has instead focused on ducks, and specifically on mallards. Thus, we considered using reporting rate estimates for mallard ducks to adjust arctic goose recovery rates. Geographic variation in band reporting rates (Nichols et al. 1995) is a function of many unknown aspects of human dimensions driving hunter behaviour and can be problematic for application to continental estimates of harvest rate when derived in an ad hoc fashion using $\hat{f}/\hat{\rho}$. However, most recoveries of arctic-nesting geese of interest to us in this paper occur in areas of the midcontinent (i.e., prairie Canada, and the Mississippi and Central Flyways, Fig. 1) where hunters also recover most mallards. It is possible that differences exist in the probability that duck and goose hunters report bands, but we reasoned that such differences are minimized when such hunters are from the same geographic area. In addition, many hunters harvest both ducks and geese, so we suspect that they report bands from both. In the absence of direct estimates of harvest rate for geese, estimates of band reporting rate from mallard studies should be reasonable for our purposes. Specific hypotheses about differences in band reporting rates between mallards and different goose species are currently being tested with reward bands that have been applied on different species and populations of arctic-nesting geese during 2003–2005 (Zimmerman et al., USFWS, pers. comm.). Future estimates of goose abundance based on the Lincoln estimator can benefit from use of these population-specific reporting rate estimates.

Historically, point estimates of $\hat{\rho}$ have differed over time and across species (see review in Conroy and Blandin (1984) and subsequent work of Nichols et al. (1995) and Royle and Garrettson (2005)). However, when considered with respect to standard errors associated with estimation, there is little evidence of substantial variation in reporting rate before 1995. In 1995, a new band type was instituted with a toll-free telephone number stamped on the band in addition to the mailing address. The toll-free number has increased band reporting rate substantially (James Dubovsky, USFWS, pers. comm.; Royle and Garrettson 2005), requiring separate treatment of bands with and without the toll-free stamp.

(d) *Geographic area of band recoveries corresponds to geographic area of harvest estimates.* Although the inference about population size pertains to the number alive at the time of marking, the geographic range of recoveries used for estimating $\hat{h}_i$ must match that for which $\hat{H}_i$ is estimated. Ideally, this should capture the full geographic range of distribution during the second sampling period (harvest) for the population of interest. If so, then annual direct recovery rate is maximized leading to improved precision in estimation of $\hat{h}_i$. For example, estimation of $\hat{N}_i$ can be done with recoveries from Canada to estimate Canadian $\hat{h}_i$, so long as only Canadian $\hat{H}_i$ is used. However, even if all other assumptions about absence of heterogeneity in $\hat{h}_i$ are satisfied to reduce bias, precision would be reduced.

2.5 Heterogeneity as a Source of Bias in the Lincoln Estimator

We considered the sampling process as it applies to arctic-nesting geese (i.e., initial captures during summer, and dead recoveries during the subsequent hunting season) and evaluated the potential for sampling heterogeneity to cause bias, $E(\hat{N} - N)$, in Lincoln estimates, $\hat{N}$, of population size, N. Specifically, we address variation in harvest rate among subpopulations, for cases where constituent subpopulations are sampled disproportionately to their abundance. We imagined a simple system where a superpopulation of size N is composed of only two subpopulations, N_1 and N_2, with respective harvest probabilities, h_1 and h_2. We define the binomial mixing parameter, π_N, as the probability that a goose from a superpopulation of size N is a member of subpopulation 1, such that $E(N_1) = N\pi_N$ and $E(N_2) = N(1 - \pi_N)$. Finally, from a marked sample of size b, a proportion π_b is from N_1 (such that expected number marked from subpopulation 1 is $b\pi_b$) and the complement from N_2 [expected number marked from subpopulation 2 is $b(1 - \pi_b)$].

We investigated behavior of Lincoln's $\hat{N}$ using the following expectations for estimators of harvest and harvest rate:

$$E(\hat{H}) = N[\pi_N h_1 + (1 - \pi_N)h_2] \tag{4}$$

$$E(\hat{h}) \approx \frac{b[\pi_b h_1 + (1 - \pi_b)h_2]}{b} = \pi_b h_1 + (1 - \pi_b)h_2. \tag{5}$$

Thus, we approximated the expectation of the abundance estimator as:

$$E(\hat{N}) = E\left[\frac{\hat{H}}{\hat{h}}\right] \sim \frac{N[\pi_N h_1 + (1 - \pi_N)h_2]}{\pi_b h_1 + (1 - \pi_b)h_2}. \tag{6}$$

Equation 6 shows that if members from the two subpopulations are marked in a representative fashion, i.e., in proportion to their abundance in the population such that $\pi_b = \pi_N$, then $\hat{N}$ should be approximately unbiased, regardless of heterogeneity in h. As well, if marking is nonrepresentative of subpopulation composition, then $\hat{N}$ still remains unbiased, so long as there is no heterogeneity in h. However, if subpopulations are marked nonrepresentatively, and respective harvest rates differ, then abundance estimates can be biased; magnitude and direction of bias depends on differences between π_b and π_N, and between h_1 and h_2.

We examined bias further as a function of these differences. We first specified N and π_N, $\pi_N \geq 0.5$. In each exercise, we held h_2 constant at 0.05, and then varied

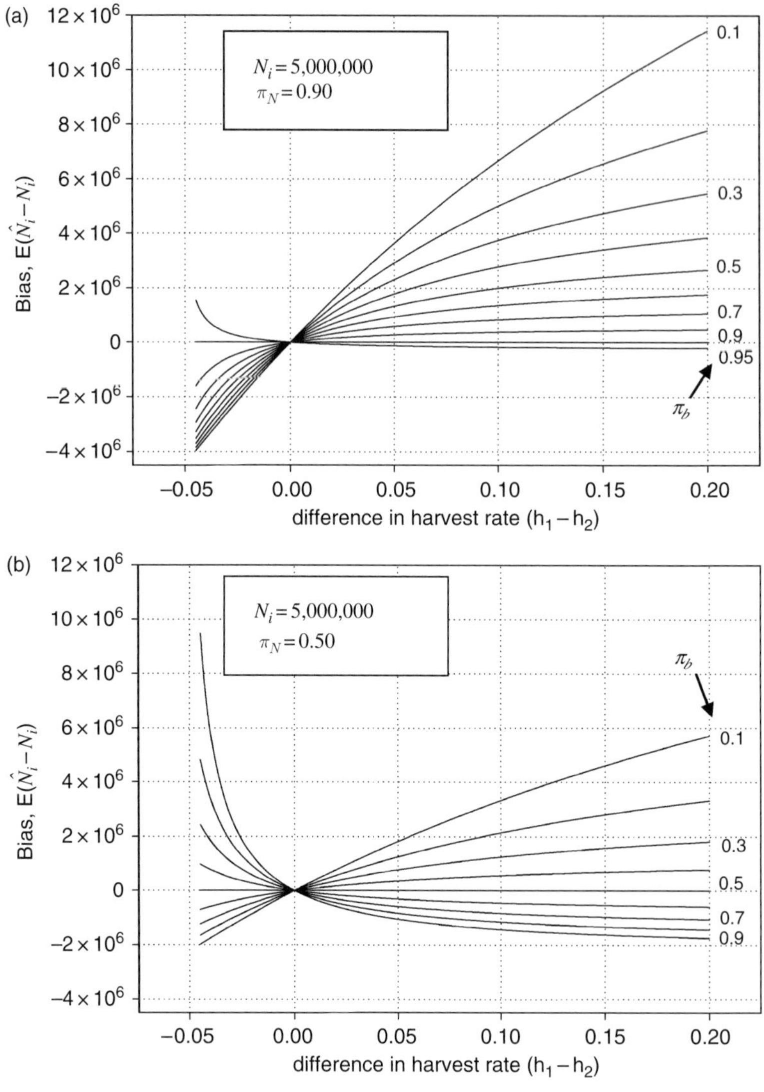

Fig. 2 Bias, E $((\hat{N} - N))$, in Lincoln estimates, $\hat{N}_i$ (as derived from estimates of harvest, $\hat{H}_i$, and harvest rate, $\hat{h}_i$) for a hypothetical population of size $N = 5$ million in which relative abundance of subpopulations, N_1: N_2, was 9:1 in Fig. 2a, and 1:1 in Fig. 2b. Bias is in relation to ratio of marks, π_b (labels beside curves), applied to respective subpopulations, and degree of heterogeneity in harvest rates between subpopulations ($h_{1i} = 0.005$ to 0.25, and $h_{2i} = 0.05$, respectively). Magnitude and direction of bias depend on differences between proportion of all geese in subpopulation 1, π_N, and proportion of all marks in subpopulation 1, π_b

h_1 between 0.005–0.25. We fixed the total number in the banded sample, and then specified a range of π_b. Thus, we explored the bias as a function of differences between π_b and π_N and between h_1 and h_2.

In the situation where the true subpopulation ratio is 9:1 ($\pi_N = 0.9$), if the smaller subpopulation (N_2) is over-represented in the banded sample ($\pi_b < 0.9$), and N_2 has a lower harvest rate (i.e., $h_2 < h_1$), then $\hat{N}$ is biased high. If N_2 is over-represented in the banded sample, and N_2 has a greater harvest rate (i.e., $h_2 > h_1$), then $\hat{N}$ is biased low. Alternatively, if the larger subpopulation (N_1) is over-represented in the banded sample, and $h_1 > h_2$, then $\hat{N}$ is biased low. If N_1 is over-represented in the banded sample, and $h_1 < h_2$, then $\hat{N}$ is biased high (Fig. 2a).

In the situation where the true subpopulation ratio is 1:1 ($\pi_N = 0.5$), then $\hat{N}$ is biased low if the banded sample over-represents the subpopulation with a greater harvest rate. On the other hand, if the banded sample under-represents the subpopulation with the greater harvest rate, then $\hat{N}$ is biased high (Fig. 2b). In summary, if the harvest rate estimate is expected to be smaller than true harvest rate, then $\hat{N}$ should be positively biased. If the harvest rate estimate is expected to be larger than true harvest rate, then $\hat{N}$ should be negatively biased.

It is possible to deal with bias arising from unrepresentative marking, if there is knowledge about the true proportional abundances of the subpopulations, $\hat{\pi}_N$. In an example with 2 subpopulations, each with different, $\hat{h}_i$, an appropriately weighted estimate of the superpopulation $\hat{h}_i^{corr}$ is given by

$$\hat{h}_i^{corr} = \hat{\pi}_N \hat{h}_{1,i} + (1 - \hat{\pi}_N)\hat{h}_{2,i}. \tag{7}$$

This approach (Eq. 7) corresponds to the practice of “weighting” either band recovery data, or estimates computed from them, by relative population size for the purpose of estimating age ratio, and both distribution and derivation of the harvest (e.g., Munro and Kimball 1982; Nichols and Tomlinson 1993).

3 Methods

3.1 Harvest Rate

We first computed numbers of lesser and greater snow geese marked annually with standard USFWS/CWS metal legbands, by selecting only instances that occurred east of 110° W longitude and north of 53° N latitude during June, July or August from 1989 to 2004. We restricted the geographic range of Ross’s goose marking to the Queen Maud Gulf region, where most Ross’s geese breed (Ryder and Alisauskas 1995) although winter and breeding range of this species is expanding (Alisauskas et al. 2006). Also, we used information about adult white-fronted geese with legbands only marked north of 60° N latitude in Canada. Thus, all geese with neckbands, reward bands, or control bands used in conjunction with reward bands were excluded from analysis to avoid potential bias. For example, Alisauskas et al.

(2006) found that recovery probability of adult Ross's geese marked with neckbands was 1.75 times that of adults marked only with standard metal legbands. Most geese were marked with such standard metal legbands in August, or secondarily in July. We then computed numbers of direct recoveries (i.e., only during the hunting season subsequent to marking) from this sample of marked geese and estimated probability of direct recovery with associated binomial variance (Fig. 3).

We adjusted annual recovery rates, $\hat{f}_i$, by band reporting rate, $\hat{\rho}_i$, to estimate harvest rate, $\hat{h}_i$ (Fig. 3). We used estimates of band reporting rate from reward bands applied to mallards during 1989–2002. These estimates were 0.38 ± 0.02 SE for 1989–1993 (Nichols et al. 1995), 0.514 ± 0.077 for 1994, 0.498 ± 0.094 for 1995, 0.491 ± 0.069 for 1996, 0.620 ± 0.089 for toll-free bands for 1997, 0.805 ± 0.033 for 1998–2001 (James Dubovsky, USFWS, pers. comm.), and 0.719 ± 0.034 for 2002 (Royle and Garrettson 2005). In the preceding, a year, i, corresponds to the hunting season beginning in fall of year i and ending in winter of calendar year $i + 1$. As well, reward bands were applied to various species of arctic-nesting geese (greater and lesser snow geese, Ross's geese, but not white-fronted geese) that are the focus of this paper from 2003 to 2005. Thus for 2003 and 2004, we used

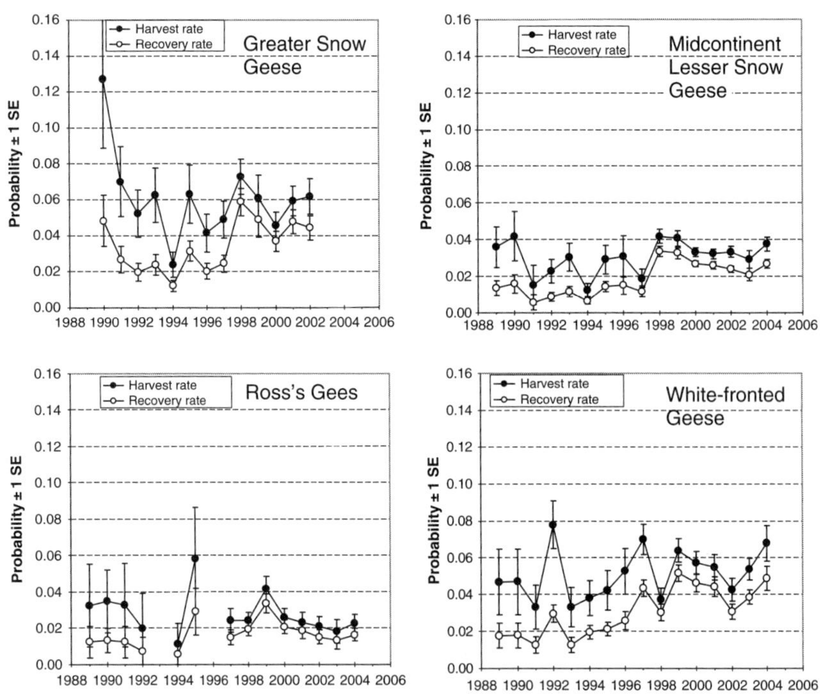

Fig. 3 Time series of recovery probability, $\hat{f}_i$, of geese marked only with legbands and reported as retrieved by hunters during the subsequent hunting season (i.e., direct recovery rate open circles); also shown is probability of harvest, $\hat{h}_i$ (*closed circles*), estimated as the ratio of annual recovery probability to band reporting rate (see text). Only recoveries from Canada and the U.S. are used to correspond with availability of harvest estimates, $\hat{H}_i$, from those countries

$\hat{\rho}_i = 0.730 \pm 0.028$ as a preliminary estimate for all species (Zimmerman et al. pers. comm.). This estimate seems reasonable for use with white-fronted geese because, as with lesser snow geese, most harvest of midcontinent white-fronted geese (Fig. 1) is largely sympatric with mallard harvest. Variance of $\hat{h}_i$ was estimated for year i using the delta method as:

$$V\hat{a}r\left(\hat{h}_i\right) \approx \frac{\hat{\mathrm{var}}\left(\hat{f}_i\right)}{\hat{\rho}_i^2} + \frac{\hat{f}_i^2.\hat{\mathrm{var}}\left(\hat{\rho}_i\right)}{\hat{\rho}_i^4}. \tag{8}$$

3.2 Harvest

We used estimates of harvest in Canada during autumn (i.e. during regular seasons, $\hat{R}_{ijk}$, where i denotes year, j denotes population, and k is jurisdiction) available from the CWS National Harvest Database; specifically, we selected (a) greater snow geese from Quebec, (b) lesser snow geese from Saskatchewan and Manitoba, (c) Ross's geese from Alberta, Saskatchewan and Manitoba, and (d) white-fronted geese from Alberta and Saskatchewan. These correspond largely to the geographic range of band recoveries in Canada for each species (Fig. 1). Harvest estimates of respective populations during autumn and winter regular seasons in pertinent regions of the U.S. were taken from Kruse and Sharp (2002) for 1989–1998 and from Kruse (2006) for 1999–2004. Annual estimates of U.S. harvest included (a) [greater] snow geese from the Atlantic Flyway (b) [lesser] snow geese from the Central and Mississippi Flyways, (c) Ross's geese from the Pacific, Central and Mississippi Flyways, and (d) white-fronted geese from the Central and Mississippi Flyways.

To illustrate the approach, we focus on estimation of adult population size, although we note that population size of juvenile geese could similarly be estimated, and simply added to adult estimates for a total continental population size at the time of marking in August. The proportion of adults, g_{ijk}, harvested in year i was calculated for each population j and jurisdiction k, noted above, as:

$$\hat{g}_{ijk} = 1 - \left[\frac{\hat{a}_{ijk}}{1 + \hat{a}_{ijk}}\right] \tag{9}$$

where $\hat{a}_{ijk}$ is the ratio of juveniles (<1 year old) to adults (>1 year old) harvested (http://www.cws-scf.ec.gc.ca/harvest/default_e.cfm, Kruse and Sharp (2002) and Kruse (2006)). Thus, adult harvest during regular seasons is $\hat{R}_{ijk}^{adult} = \hat{g_{ijk}} \cdot \hat{R}_{ijk}$ and, summing over z jurisdictions, results in regular-season harvest of adults for year i from population j, or

$$\hat{R}_{ij}^{adult} = \sum_{k=1}^{z} \hat{R}_{ijk} \tag{10}$$

There have been large and unprecedented changes in the duration of hunting seasons for greater and lesser snow geese, and Ross's geese (e.g. Moser 2001; Reed and Calvert 2006), in attempts to reduce population size (Batt 1997). Beginning in spring 1999, regulations in Canada and the U.S. allowed harvest during regular seasons, R, to be supplemented with additional harvest, C known as spring harvest in parts of Canada or conservation order harvest in the U.S. Thus, annual harvest, H_{ijk} in year i of species j in jurisdiction k should be estimated from

$$\hat{H}_{ijk} = \hat{R}_{ijk} + \hat{C}_{ijk} \tag{11}$$

Harvest of Ross's geese during spring is not permitted in Canada (Alisauskas et al. 2006), but estimates of spring harvest of lesser and greater snow geese separately, $\hat{C}_{ijk}$, are made by CWS (Michel Gendron, CWS Ottawa). However, species-specific estimates of conservation order harvest, $\hat{C}_{ijk}$, are not available from the U.S. because their survey design does not distinguish between Ross's and lesser snow geese, nor between age classes (Paul Padding, USFWS). Instead, Alisauskas et al. (2006) estimated $\hat{C}_i$ for Ross's Geese by weighting regular reason harvest by the ratio of band recoveries from conservation order seasons, c_i, to those recovered during regular seasons, r_i. Namely, we expect that

$$\frac{C_i}{R_i} = \frac{c_i \rho_{c_i}^{-1}}{r_i \rho_{r_i}^{-1}}. \tag{12}$$

If band reporting rates during regular and conservation order seasons are not different, then Eq. 12 can be rearranged so that

$$\hat{C}_i = \frac{c_i}{r_i} \hat{R}_i. \tag{13}$$

Consequently, we estimated adult harvest in years 1999–2004 for each species of Ross's and snow geese as:

$$\hat{H}_{ij}^{adult} = \hat{R}_{ij}^{adult} + \hat{C}_{ij}^{adult}. \tag{14}$$

Because recovery rate data are used to estimate $\hat{H}$ in this case, the estimates of $\hat{H}$ and $\hat{h}$ are no longer independent and will have some covariance; we assumed that this was negligible for the purposes of this paper.

3.3 Population Size

We used Eq. 1 to estimate $\hat{N}_i$. We took two approaches toward estimation of $\hat{N}_i$: the first was to use $\hat{H}_{ij}$ and $\hat{h}_{ij}$ from the geographic area in both Canada and the U.S. in which each species is normally harvested (i.e., Mexico excluded because of absence of harvest estimates); the second approach was to restrict estimation to use of only Canadian $\hat{H}_{ij}$ and $\hat{h}_{ij}$ because of consistency in design and approach to harvest estimation during 1989–2004, compared to harvest estimation in the U.S. which changed sampling frames in 1999. It was expected that the cost of restriction to Canadian $\hat{h}_{ij}$ would be reduced precision because of the generally low proportion of total continental recoveries that occur in Canada for each of these species (Alisauskas et al. 2006).

The delta method can be used to estimate variance of $\hat{N}_i$ as:

$$\widehat{\text{var}}\left(\hat{N}_i\right) \approx \frac{\widehat{\text{var}}\left(\hat{H}_i\right)}{\hat{h}_i^2} + \frac{\hat{H}_i^2\ \widehat{\text{var}}\left(\hat{h}_i\right)}{\hat{h}_i^4} \tag{15}$$

where $\widehat{\text{var}}\left(\hat{H}_i\right)$ is based on an assumed $\widehat{CV}_{\hat{H}_i} = 0.05$ (Geissler 1990; Otis 2006).

Similarly, Otis (2006) used the delta method to estimate the coefficient of variation $\widehat{CV}_{\hat{N}_i}$ of $\hat{N}_i$

$$\widehat{CV}_{\hat{N}_i} = \sqrt{\widehat{CV}^2_{\hat{H}_i} + \widehat{CV}^2_{\hat{h}_i}}. \tag{16}$$

4 Results

4.1 Harvest Probability

Figure 3 shows annual harvest rate for each species based on recoveries of adult geese, marked only with standard metal legbands, from Canada and the U.S. There were 365 direct recoveries of 11,209 greater snow geese marked from 1990 to 2002, from a marking effort mostly localized near Bylot Island (Fig. 1). Lesser snow geese were marked over a much wider geographic range (Fig. 1) reflecting occurrence of large known breeding concentrations; from 1989 to 2004, there were 1,178 direct recoveries from a sample of 49,867 adults. We used 10,213 adult Ross's geese marked with legbands near the Queen Maud Gulf only from 1989 to 2004 (Fig. 1) of which 191 were direct recoveries. Finally, of 21,997 adult white-fronted geese marked with only legbands in Canada north of 60 degrees N latitude, there were 767 direct recoveries from Canada and the U.S. (Table 1).

In general, precision of estimates improved more recently with larger numbers of adult geese of each species marked annually with only legbands. The exception was that there were insufficient adult greater snow geese marked with only legbands in 2003 and 2004 ($n = 12$) to permit estimation of annual recovery probabilities. As well, there were only 36 and 3 Ross's geese so marked in 1993 and 1996, with no direct recoveries reported. Over the period of study, greater snow geese showed the highest average ($\pm$ 1 SE) annual harvest probabilities (0.061 $\pm$ 0.024), followed by white-fronted geese (0.051 $\pm$ 0.014), lesser snow geese (0.030 $\pm$ 0.009), and Ross's geese (0.028 $\pm$ 0.012).

We further checked to see if there existed heterogeneity in $\hat{h}_i$ for lesser snow geese. From 1998 to 2004, most lesser snow geese marked with legbands only were captured either at Queen Maud Gulf in Canada's central arctic (representing a northern breeding area, b =10,189) or from three areas south of Hudson Bay (La Pérouse Bay, Cape Henrietta Maria, or Akimiski Island representing southern breeding areas, b =28,411). The proportion of banded snow geese that were from northern breeding areas, π_b, each year ranged from 0.13 to 0.55 (Table 2). There was annual variation in $\hat{h}_i$ for each of the two subpopulations, and harvest probability

Table 1 Numbers of adult geese marked annually with legbands only (b_i), direct recoveries from Canada and the U.S. (d_i), and estimated harvest ($\hat{H}_i$) from corresponding geographic range of direct recoveries (see Fig. 1) for 4 populations of arctic-nesting geese marked in Canada

	Greater snow geese			Lesser snow geese			Ross's geese			White-fronted geese		
	b_i	d_i	$\hat{H}_i$	b_i	d_i	$\hat{H}_i$	b_i	d_i	$\hat{H}_i$	b_i	d_i	$\hat{H}_i$
1989	0	–	24,269	806	11	259,321	162	2	7,783	394	7	127,912
1990	228	11	35,250	628	10	242,159	304	4	8,063	393	7	129,543
1991	488	13	18,277	347	2	231,026	161	1	7,737	636	8	133,256
1992	807	16	30,188	1,488	13	229,844	134	1	5,766	1352	40	109,681
1993	716	17	24,149	1,471	17	201,491	36	0	5,922	790	10	105,549
1994	1,309	16	37,692	2,529	16	268,649	173	1	14,045	1,327	26	137,399
1995	986	31	26,860	1,726	25	389,172	172	5	18,343	1,518	32	129,140
1996	1,132	23	39,477	593	9	366,176	3	0	20,304	1,006	26	183,770
1997	1,021	25	26,225	1,191	14	421,513	932	14	20,799	1,985	86	189,823
1998	954	56	104,009	4,005	135	622,842	1,488	29	33,800	1,461	44	179,643
1999	469	23	133,050	3,477	114	650,125	1,165	39	42,410	2,328	120	186,282
2000	1,086	40	93,497	10,680	287	669,954	1,442	30	54,951	2,063	95	294,853
2001	1,111	53	104,350	7,450	194	722,115	965	18	54,600	1,564	69	221,359
2002	902	40	87,940	7,164	172	568,059	1,056	16	54,892	1,729	53	175,427
2003	3	0	62,904	1,900	40	494,663	609	8	35,437	2,381	92	160,327
2004	9	1	78,120	4,412	119	734,008	1,411	23	83,220	1,067	52	153,116

Table 2 Effect of non-representative marking of lesser snow geese from either north (near Queen Maud Gulf) or south of Hudson Bay (La Pérouse Bay, Akimiski Island, or Cape Henrietta Maria) on estimation of population size using Lincoln's (1930) method. Note proportional allocation of marks to northern snow geese, $\hat{\pi}_b$, which varies annually, in relation to proportional contribution of northern snow geese to superpopulation, $\hat{\pi}_N = 0.90$ (Kerbes et al 2006). Also shown are harvest probability, $\hat{h}_i$, for each subpopulation, and for the superpopulation, as well as biased and corrected estimates of superpopulation size

	Number marked, b_i			Subpopulation $\hat{h}_i$		Superpopulation harvest probability		Superpopulation estimate		
Year	North	South	Sampling parameter, $\hat{\pi}_b$	North	South	Naïve $\hat{h}_i$	Corrected $\tilde{h}_i$	Naïve (biased) $\hat{N}_i$	Bias-corrected $\tilde{N}_i$	Bias[a]
1998	2,181	1,777	0.551	0.031	0.057	0.042	0.033	14,874,512	18,679,514	−3,805,002
1999	1,450	1,945	0.427	0.041	0.040	0.041	0.041	15,962,188	15,843,655	118,533
2000	1,378	9,239	0.130	0.031	0.034	0.033	0.031	20,069,204	21,620,524	−1,551,320
2001	1,255	5,992	0.173	0.025	0.034	0.032	0.026	22,323,229	28,129,460	−5,806,231
2002	1,673	5,426	0.236	0.025	0.036	0.033	0.026	17,011,755	21,776,530	−4,764,774
2003	896	979	0.478	0.017	0.040	0.029	0.019	16,893,968	25,570,362	−8,676,395
2004	1,356	3,053	0.308	0.033	0.039	0.038	0.033	19,566,731	21,938,616	−2,371,884

[a] $(\hat{N}_i - \tilde{N}_i)$.

tended to be higher for lesser snow geese that were marked in areas south of Hudson Bay, compared to those marked north of Hudson Bay.

4.2 Harvest

Harvest from all 4 populations showed increases starting during the early 1990s, followed in most cases by declines in the 2000s (Table 1). As well, harvest of midcontinent lesser snow geese, greater snow geese and Ross's Geese all appeared to increase further after initiation of conservation order or spring seasons in 1998. Approximate 95% CL of slope estimates from linear time models of annual harvest suggested an average annual increase of 5,560 ± 2,818 greater snow geese/year, 6,005 ± 10,213 lesser snow geese/year, 4,379 ± 1,121 Ross's geese/year, and 5,844 ± 4,269 white-fronted geese/year, over the period considered.

4.3 Lincoln Estimates

In general, Lincoln's estimates were considerably and consistently higher than other count-based estimates such as spring counts of adult greater snow geese in Quebec, and fall counts of adult and young white-fronted geese in Saskatchewan (Fig. 4). Such counts are often treated as complete inventories of species. In these two examples, there was a discrepancy that may have arisen from (1) bias in the Lincoln estimates, (2) biased inferences about abundances from the counts, or (3) biased inferences from both approaches.

Despite large differences between the Lincoln estimates and other estimates or counts, there was rather good parallelism in time series. Correlations were very high for greater snow geese and Ross's geese, which also have the most restricted breeding ranges of the four examples (Fig. 5). There are no annual indices of abundance other than the midwinter count for lesser snow geese, which was weakly related to Lincoln's estimates ($r = 0.398$). Lincoln estimates of adult white-fronted geese, when based on both Canadian and U.S. recoveries and harvest estimates, were also weakly correlated with fall counts of this species ($r = 0.448$), which include both adults and young of the year. However, when estimation of adult white-fronted geese was based on information restricted to band recoveries and harvest estimates from Canada only, the relationship was much more certain ($r = 0.780$), although Lincoln estimates were even greater for unknown reasons.

Based on this approach, estimates of greater snow geese and Ross's geese had low precision because of comparatively low numbers marked (Table 1), and because of low harvest rates in the case of Ross's geese (Fig. 3). Greater precision for estimates of lesser snow geese was a result of greater numbers marked, especially in more recent years, and for white-fronted geese a function of high harvest rates. Sample size requirements are provided in Appendix as related to the level of precision required under specific harvest rates and precision of harvest estimates. Species differences in precision led to variation in inference strength. For example,

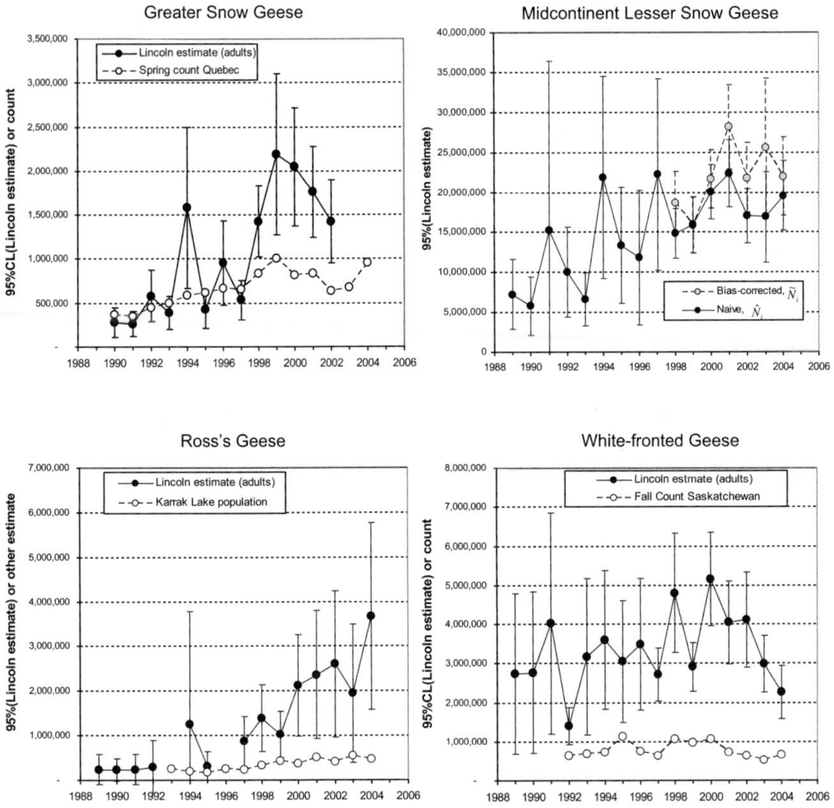

Fig. 4 Time series of Lincoln estimates of population size at time of marking, i.e., August, computed from annual estimates of harvest and harvest probability. For lesser snow geese, shown are (i) naïve Lincoln estimates computed from harvest probabilities based on direct recoveries unstratified by geographic location of marking, $\hat{N}_{1989-2004}$, and (ii) bias-corrected Lincoln estimates based on harvest rates stratified by marking origin, $\tilde{N}_{1998-2004}$; see Table 2

although recent declines in greater snow geese and continued increases in Ross's geese corresponded to patterns in time series from other counts or estimates, low precision precluded confidence about temporal changes in the Lincoln estimates. In contrast, an absence of declines in lesser snow goose numbers, but declines in white-fronted goose numbers (also consistent with the fall count of white-fronted geese in Saskatchewan) can be inferred with greater confidence (Fig. 4). With relatively long time series such as considered herein, inference about population trajectory might still be reasonable if a small number of years with imprecise estimates are excluded from consideration. For example, precision in $\hat{N}_i$ of lesser snow geese was very low in the years 1991, 1994 and 1997 (Fig. 4), and their exclusion from the time series still results in 13 annual estimates of rather good precision from which to draw inference. Log-linear regression of $\log_e \hat{N}_i$ over such a time series could be used for estimation of intrinsic rate of increase, $\hat{\bar{r}}$, (Eberhardt and Simmons 1992), perhaps

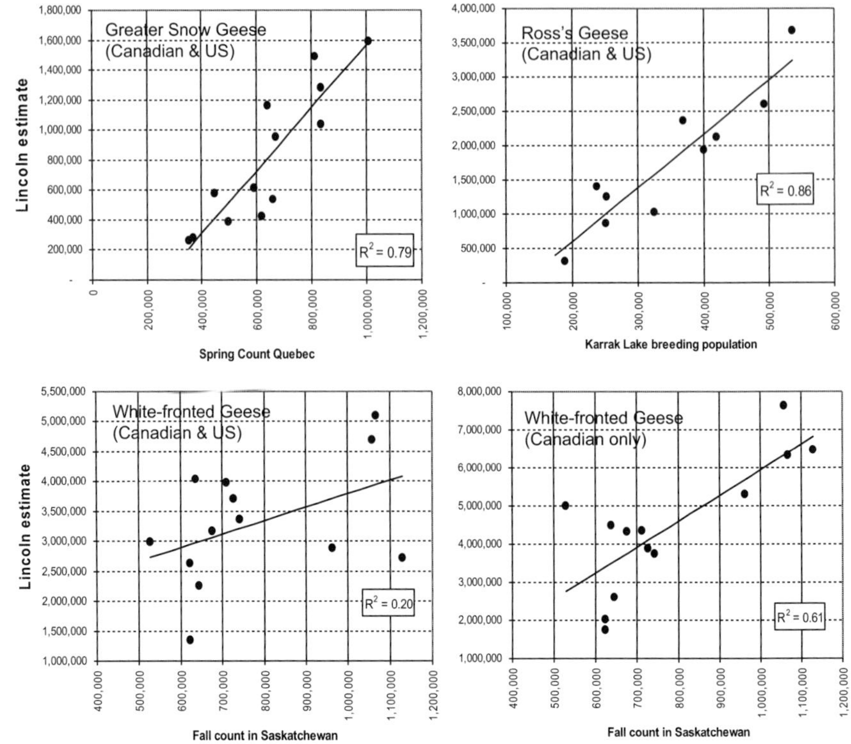

Fig. 5 Relationships of Lincoln estimates of population size, $\hat{N}$, to various count or other estimates available for different species of arctic-nesting geese. Shown are Lincoln's estimates based on recoveries from both Canada and the U.S. for greater snow geese, Ross's geese and white-fronted geese, and also based only on Canadian harvest and Canadian recoveries of white-fronted geese

weighted by $\frac{1}{\hat{\text{var}}(\log_e \hat{N}_i)}$; then, mean annual rate of increase, $\hat{\hat{\lambda}}$, over the time series may be calculated as $\hat{\hat{\lambda}} = e^{\hat{\hat{r}}}$, or alternatively as $\hat{\hat{\lambda}} = 1 + \hat{\hat{r}}$ as recommended by Eberhardt and Simmons (1992) for "birth pulse" populations such as arctic-nesting geese.

4.4 Heterogeneity in Harvest Probability

We estimated harvest probabilities for lesser snow geese marked north vs. south of Hudson Bay for 1998–2004 (Table 2). Kerbes et al. (2006) reported that contribution of snow geese from north of Hudson Bay to the superpopulation was $\pi_N = 0.90$. However, allocation of marks to northern birds was disproportionately low with π_b ranging between 0.13 and 0.55. As well, harvest probabilities of southern birds was 1.1–2.3 times higher than northern birds in 6 of 7 years. Such over-representation of southern snow geese in the marked sample, even though they represented only

$(1-\pi_N) = 0.10$ of the superpopulation, resulted in negatively biased estimates of continental population size because of the pattern of heterogeneity in harvest probability of geese from the two geographic strata (Table 2). Thus, Lincoln estimates of abundance for midcontinent snow geese were likely biased low in all years with exception of 1999 when $\hat{h}_i$ was greater for northern birds resulting in slight overestimation.

5 Discussion

Estimates of geese using Lincoln's (1930) method from these four examples were invariably higher than other independently conducted counts or other methods of estimation. This was consistent with Otis' (2006) findings when he applied the method to midcontinent mallards. Previous presentation of such estimates from our goose examples to professional biologists and wildlife managers familiar with the biology of these species or the process of setting harvest regulations most often resulted in disbelief and reluctance to accept the Lincoln estimates (Alisauskas, pers. obs.). More weight tended to be placed on various counts (e.g., Nieman et al. 2005), or in the case of midcontinent snow geese, on photographic surveys of breeding birds at known colonies (Kerbes 1975).

Although we found correlations between Lincoln estimates and other sources of information, wide disparities remained and it is unknown which estimates exhibit greater bias. In the case of white-fronted goose counts during September in Saskatchewan and Alberta, it is assumed that the superpopulation of migrating geese is completely within the sampled area when counts are made, all flocks are detected within the area, and counts of geese in each flock are without appreciable error. Thus, the discrepancy between counts (all white-fronted geese) in Saskatchewan, and the Lincoln estimates (of adults only) could be due to incomplete detection of geese during counts, incomplete presence of geese during counts, large errors in counts of up to hundreds of thousands of birds/flock, bias in Lincoln estimates, or some combination thereof.

Bias in Lincoln estimates could only emanate, in turn, from biases in $\hat{h}_i$ or $\hat{H}_i$. If $\hat{H}_i$ is unbiased, then a major assumption underpinning Lincoln's method is that there is sufficient mixing of geese from different source populations before harvest occurs. If there is much spatial structuring of populations when marked during summer, then without subsequent mixing before harvest, there is a good likelihood for heterogeneity in $\hat{h}_i$. For example, differences in routes or schedules of migration by different source populations may result in heterogeneity due to spatial variation in harvest rates, or to different duration of harvest. We expect that the potential for biased $\hat{N}_i$ stemming from heterogeneity in $\hat{h}_i$ is greater for our two examples with broad breeding ranges (white-fronted and lesser snow geese) than for those with more restricted ranges (Ross's and greater snow geese).

Such heterogeneity in $\hat{h}_i$ was evident in lesser snow goose, which have a breeding distribution spanning the widest range of latitude among our four examples, yet appear to be composed of two large subpopulations that differ in their migration behavior as well as harvest and survival probabilities (Alisauskas et al. in prep). The distribution of these subpopulations is such that birds from the northern portion of their range (between 65 and 68°N) are thought to represent about 90% , i.e., $\hat{\pi}_N = 0.9$, of the breeding population compared to 10% from south of Hudson Bay (south of 60°N) based on photographic estimates of breeding birds in 1997–1998 (Kerbes et al. 2006). Photographic enumeration likely provides an underestimate of breeding numbers (see below) because, among other things, there likely are undetected colonies (Alisauskas and Boyd 1994). For our example, we assumed a constant probability of colony detection, but this probability may vary over snow goose breeding range. For example, the estimate of $\hat{\pi}_N$ from photo surveys may be biased low if detection of colonies in the northern, more remote portion of midcontinent snow goose breeding range has been lower than in the south. In fact, most newly documented colonies in the last two decades (Alisauskas and Boyd 1994; Kerbes et al. 2006) have been located in this northern stratum, where researchers have been focusing more attention more recently outside of southern Hudson Bay. Thus if $\hat{\pi}_N$ is biased, we suspect that it may be an underestimate.

Snow geese marked on the southern portion of their breeding range (including LaPérouse Bay, Akimiski Island, and Cape Henrietta Maria) tend to be recovered at higher rates compared to those from northern portions including West Hudson Bay, Southampton Island, Baffin Island and Queen Maud Gulf (Table 2). In addition, variation in harvest dates of midcontinent lesser snow geese was most sensitive to the latitude at which geese were marked before migration when compared to age of geese and neckband presence, while controlling for latitude of recovery; Canadian harvest of northern snow geese occurred about 10 October, which was a full two weeks later than the average date of 25 September for southern snow geese (Alisauskas et al. in prep). Indeed, this difference in schedules of harvest persisted into the northern U.S. Thus, it appears as though the vanguard of snow goose migration during fall is composed of the relatively small southern Hudson Bay subpopulation ($1 - \pi_N \cong 0.10$) which is harvested first possibly by a relatively high ratio of hunters to geese. About 2 weeks later, ~9 times as many snow geese from northern breeding areas arrive onto hunting areas, essentially swamping southern birds and likely increasing the ratio of geese to hunters. We suspect that this is the cause for lower $\hat{h}_i$ for snow geese that originate from northern areas of Hudson Bay, compared to areas south of Hudson Bay.

Midcontinent white-fronted geese winter in areas largely sympatric with midcontinent snow geese, but breeding range spans a greater range of longitude from Alaska to Hudson Bay (Fig. 1). However, current knowledge of their density distribution over this summer range is very incomplete. As well, our analysis was restricted to birds marked in Canada because that represented an unbroken time series since 1989. As such, we could not more fully assess the extent to which differences in harvest regimes among source populations (including Alaska) may have biased our abundance estimates for white-fronted geese.

In addition to heterogeneity of $\hat{h}_i$ related to geographic origin, we envision additional sources of heterogeneity related to social or breeding status of geese from the same geographic area. For example, schedules of annual wing molt and ability to regain flight are earlier for nonbreeders and failed breeders than for adults that have produced young capable of flight. Earlier departure from the arctic by unsuccessful adults may predispose them to higher h_i than faced by adults with young, in the same way that snow geese from south Hudson Bay are harvested at a higher rate than individuals from the more numerous northern stratum. On the other hand, adults with young may suffer higher vulnerability from harvest in relation to nonbreeders because of family cohesiveness and the predisposition of juvenile geese to show greater vulnerability to harvest. Indeed, Drake (2006) found that successfully nesting Ross's geese had lower survival probability than those that experienced experimentally induced nest failure, but that failed breeders were more susceptible, or available, for harvest. In any event, harvest heterogeneity between productive and nonproductive adults may vary annually in response to severity of arctic weather; breeding failure due to delays in arctic snow melt is the most important governor of breeding success and recruitment on a continental scale, at least for mid-continent lesser snow geese (Boyd et al. 1982; Alisauskas 2002), and probably most other arctic-nesting geese. This hypothesis of differential migration with accompanying differences in probabilities and schedules of harvest between breeders and nonbreeders requires testing. If important, then we suggest that breeding probability could be a suitable population-level metric, and a special case of the binomial mixing parameter, π_N, (substitute in Eqs. 6 and 7), so long as something is known about the number of bands applied to breeders and nonbreeders.

The disparity between our estimates of midcontinent lesser snow geese and other methods of enumeration were the greatest of all species considered. Our 1998 estimate of 14.9 ± 3.1 million (95% CL) snow geese coincided with 1997–1998 count from periodic photographic surveys done in arctic nesting colonies (Kerbes et al. 2006) of 3.8 million breeding snow geese. Adjustment for relative bias increased Lincoln's estimate for this year to 18.7 million. Part of the discrepancy may be attributable to negative bias associated with the photosurvey method; only breeders within known colonies are sampled, and so the method does not account for (1) nondetection of unknown colonies, mentioned above, and (2) nondetection of non breeders. As well, there likely is incomplete detection of goose images during photo interpretation that varies among observers and among different nesting habitats of geese, and probably between color morphs. It has been assumed, but not confirmed, that flying geese captured on photographs, as well as those on water, or those in groups of >3 are nonbreeding, and so are not counted as part of the photo enumeration. A final assumption is that all breeding geese remain flightless when photo aircraft fly over. The extent of negative bias associated with the photosurvey method remains unknown without further testing and evaluation, but the cumulative effect of each of these undoubtedly causes progressively worse negative bias, and accounts for some of the disparity between photosurvey estimates and those using the Lincoln estimator. In fact, disproportionate marking of southern snow geese likely caused negative bias in Lincoln estimates in most years because of higher harvest of those birds. An important point here is that bias adjustments for this problem are possible.

If Lincoln estimates of snow goose population size are biased high, then bias is likely unrelated to heterogeneity in $\hat{h}_i$ in our example, as such heterogeneity in combination with non-representative marking of birds should result in negative bias in $\hat{N}_i$. Instead, positive bias could result from overestimation of $\hat{H}$. We are unsure of the extent to which biases in $\hat{H}$ influence our estimate of $\hat{N}_i$ Padding (pers. comm.) and Otis (2006) point to potential biases in $\hat{H}$ stemming from sampling frame problems associated with estimation of U.S. harvest. Additional bias in $\hat{H}$ may arise from inaccurate reporting of harvest by hunters due to prestige, memory, or nonresponse bias, although there are efforts to make adjustment for this in both Canada and the U.S. (e.g. Sen 1971; Filion 1975). It is thought that $\hat{H}$ may be overestimated by ~30% but there are currently efforts underway to confirm and rectify this (Paul I. Padding USFWS, pers. comm.)

In summary, the Lincoln method for estimation of goose population size appeared to be useful for retrospective purposes in some of our examples. General relationships between these estimates and other sources of information suggested that inferences about longer term population trajectory were probably reasonable. However, there was a wide range of estimated precision between species and for annual estimates within species. Boyd et al. (1982:8) found that "calculated confidence intervals are so wide as to be of little assistance in discerning what may be happening, though they show that little weight should be give to apparent changes from one year to the next". Increased precision can be achieved with larger numbers of birds marked (Appendix), but precision is also an inverse function of $\hat{h}$ (Eq. 12), so as to be a problem for estimation of species with very low $\hat{h}$, such as Ross's geese. In contrast, precision of estimates for lesser snow geese and white-fronted geese were much better, because of greater numbers marked (particularly in more recent years, Table 1), and comparatively higher harvest probabilities (Fig. 2).

Although heterogeneity of $\hat{h}_i$ as a source of bias in $\hat{N}_i$ is largely beyond the researcher's control, bias associated with $\hat{h}_i$ can, at least in principle, be controlled for with (1) proper allocation of marks consistent with proportional contributions of source populations to the superpopulation, or (2) bias adjustments (Eq. 7) after estimation of $\hat{h}_i$ for each population stratum. Knowledge about spatial distributions appear to be better for colonial species (lesser snow geese, Ross's geese, greater snow geese) than for more dispersed nesters (white-fronted and cackling geese *Branta hutchinsii*), but remoteness of all species and vastness of breeding distributions, especially for dispersed nesters, poses a considerable challenge. Such knowledge may be improved only with further basic survey work that involves systematic sampling and fuller coverage leading to density estimation completely within each species' range. Where there is high likelihood of heterogeneity in harvest probability for different breeding subpopulations, future marking should either focus on proportional representation of subpopulation abundance, or solely on the largest (and thus most representative) strata for marking, if population estimation is the goal. Finally, fundamental to use of Lincoln's method is the estimation and resolution of any bias in $\hat{H}_i$ or $\hat{h}_i$ for each species.

Appendix

Coefficient of variation of population estimate, CV($\hat{N}_i$), using the Lincoln (1930) estimator in relation to number of geese marked with legbands, b, in increments of 1,000, and in relation to harvest probability, h, for different levels of precision in harvest estimates, CV($\hat{H}$) = 0.05, 0.075, 0.10, or 0.15.

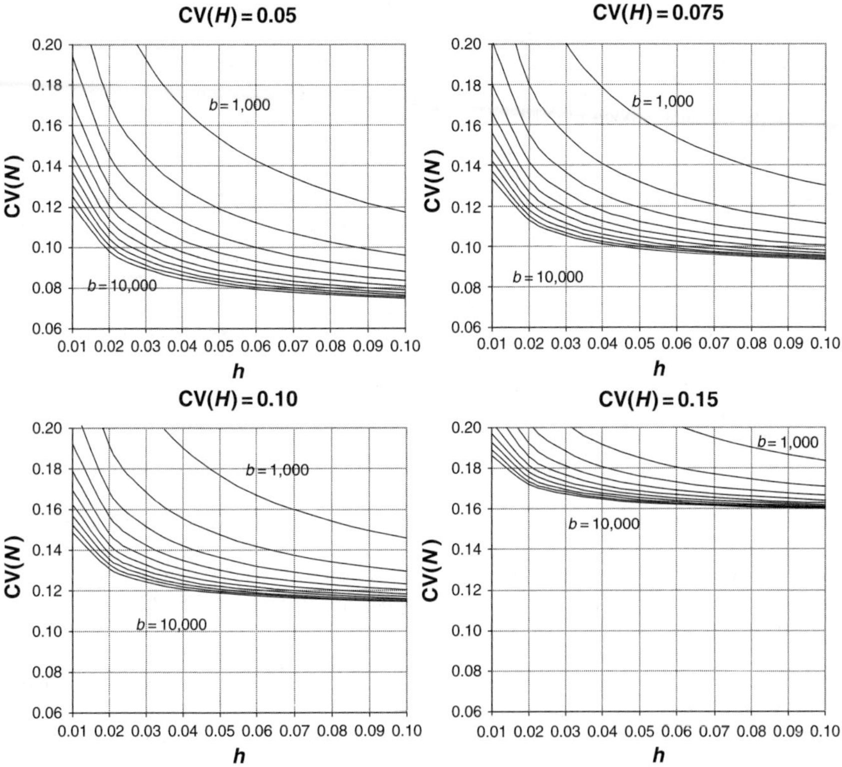

References

Abraham KF, Leafloor JO, Rusch DH (1999) Molt migrant Canada Geese in northern Ontario and western James Bay. Journal of Wildlife Management 63:649–655.

Alisauskas RT (2002) Arctic climate, spring nutrition, and recruitment in midcontinent Lesser Snow Geese. Journal of Wildlife Management 66:181–193.

Alisauskas RT, Boyd H (1994) Previously unrecorded colonies of Ross' and lesser snow geese in the Queen Maud Gulf Bird Sanctuary. Arctic 47:69–73.

Alisauskas RT, Lindberg MS (2002) Effects of neckbands on survival and fidelity of white-fronted and Canada geese captured as non-breeding adults. Journal of Applied Statistics 29: 521–537.

Alisauskas RT, Drake KL, Slattery SM, Kellett DK (2006) Neckbands, harvest and survival of Ross's geese from Canada's central arctic. Journal of Wildlife Management 70:89–100.

Alisauskas RT, Rockwell RF, Cooch EG, Drake KL, Dufour KW, Slattery SM, Caswell FD, Leafloor JO, Reed ET, Moser T (in prep) Effect of population reduction efforts on survival of midcontinent Lesser Snow geese.

Batt BDJ (ed.). (1997) Arctic ecosystems in peril: report of the Arctic Goose Habitat Working Group. Arctic Goose Joint Venture Special Publication. U.S. Fish and Wildlife Service, Washington, DC and Canadian Wildlife Service, Ottawa, ON, 120 pp.

Bowers EF, Martin FW (1975) Managing wood ducks by population units. Transactions North American Wildlife and Natural Resources Conference 40:300–323.

Boyd H (1976) Estimates of total numbers in the Hudson Bay population of Lesser Snow geese, 1964–1973. Canadian Wildlife Service Progress Notes No. 63.

Boyd S (2000) A comparison of photo counts versus visual estimates for determining the size of snow goose flocks. Journal of Field Ornithology 71:686–690.

Boyd H, Smith GEJ, Cooch FG (1982) The Lesser Snow goose of the easternCanadian Arctic: their status during 1964–1979 and their management from 1981 to 1990. Canadian Wildlife Service Occasional Paper 46, 25 pp.

Brownie C, Anderson DR, Burnham KP, Robson DS (1985) Statistical inference from band recovery data – a handbook. 2nd ed. U.S. Fish and Wildlife Service, Resource Publication 156, Washington, DC.

Conroy MJ, Blandin WW (1984) Geographic and temporal differences in band reporting rates for American black ducks. Journal of Wildlife Management 48:23–36.

Drake KL (2006) The role of dispersal in population dynamics of breeding Ross's geese. Ph.D. Thesis, University of Saskatchewan, Saskatoon, p. 115

Eberhardt LL, Simmons MA (1992) Assessing rates of increase from trend data. Journal of Wildlife Management 56:603–610.

Ely CR, Takekawa JY, Wege ML (1993) Distribution, abundance and age ratios of Wrangel Island Lesser Snow geese *Anser caerulescens* during migration on the Yukon-Kuskokwim Delta, Alaska. Wildfowl 44:24–32.

Filion FL (1975) Estimating bias due to nonresponse in mail surveys. The Public Opinion Quarterly 39:482–492.

Gauthier G, Pradel R, Menu S, Lebreton J-D (2001) Seasonal survival of Greater Snow geese and effect of hunting under dependence in sighting probability. Ecology 82:3105–3119.

Geissler PH (1990) Estimation of confidence intervals for federal waterfowl harvest surveys. Journal of Wildlife Management 54:201–205.

Henny CJ, Burnham KP (1976) A reward band study of Mallards to estimate band reporting rates. Journal of Wildlife Management 40:1–14.

Kerbes RH (1975) The nesting population of Lesser Snow geese in the eastern Canadian Arctic: a photographic inventory of June 1973. Canadian Wildlife Service Report Series No. 35.

Kerbes RH, Meeres KM, Alisauskas RT, Caswell FD, Abraham KF, Ross RK. (2006) Surveys of nesting mid-continent Lesser Snow geese and Ross's geese in Eastern and Central Arctic Canada. Canadian Wildlife Service Technical Report Series Number 447.

Kruse KL (2006) Central Flyway harvest and population survey data book. U.S. Fish and Wildlife Service, Denver, CO.

Kruse KL, Sharp DE (2002) Central Flyway harvest and population survey data book. U.S. Fish and Wildlife Service, Denver, CO.

Lincoln FC (1930) Calculating waterfowl abundance on the basis of banding returns. Circular 118, U.S. Department of Agriculture, Washington, DC.

Moser TJ (2001) The status of Ross's geese. Arctic Goose Joint Venture Special Publication. U.S. Fish and Wildlife Service, Washington, DC, and Canadian Wildlife Service, Ottawa, ON.

Munro RE, Kimball CF (1982) Population ecology of the mallard. VII. Distribution and derivation of the harvest. U.S. Fish and Wildlife Service Resource Publication 147.

Nichols JD, Tomlinson RE (1993) Analyses of banding data, p. 269–280. In Baskett TS, Sayre MW, Tomlinson RE, Mirarchi RE (eds.) Ecology and management of the mourning dove. Wildlife Management Institute, Stackpole Books, Harrisburg, PA.

Nichols JD, Blohm RJ, Reynolds RE, Trost RE, Hines JE, Bladen JP (1991) Band reporting rates for mallards with reward bands of different dollar values. Journal of Wildlife Management 55:119–126.

Nichols JD, Blohm RJ, Reynolds RE, Trost RE, Hines JE, Bladen JP (1995) Geographic variation in band reporting rates for mallards based on reward banding. Journal of Wildlife Management 59:697–708.

Nieman D, Warner K, Smith J, Solberg J, Roetker F, Lobpries D, Lyman N, Walters R, Durham S (2005) Fall inventory of midcontinent white-fronted geese, 2005. Canadian Wildlife Service Unpublication Report, Saskatoon, Saskatchewan.

Otis D (2006) Mourning dove hunting regulation strategy based on annual harvest statistics and banding data. Journal of Wildlife Management 70:1302–1307.

Owen M (1980) Wild geese of the world. BT Batsford Ltd., London, UK, 236 pp.

Reed ET, Calvert AM (2006) An evaluation of the special conservation measures for greater snow geese: report of the Greater Snow Goose Working Group. Arctic Goose Joint Venture Special Publication. U.S. Fish and Wildlife Service, Washington, DC, and Canadian Wildlife Service, Ottawa, ON.

Robson DS (1969) Mark-recapture methods of population estimation, In: NL Johnson and H Smith Jr. (eds.) New developments in survey sampling. Wiley Interscience, New York, pp. 120–140.

Royle JA, Garrettson PR (2005) The effect of reward band value on mid-continent mallard band reporting rates. Journal of Wildlife Management 69:800–804.

Ryder JP, Alisauskas RT (1995) Ross' Goose (*Chen rossii*). In: Poole A, F. Gill (Eds.) The birds of North America, No. 162. Philadelphia: The Academy of Natural sciences and Washington DC: The American Ornithologists' Union.

Salomonson F (1968) The moult migration. Wildfowl 19 5–24.

Seber GAF (1982) The estimation of animal abundance and related parameters. Charles Griffin & Co Ltd, London.

Sedinger JS, Flint PL, Lindberg MS (1995) Environmental influence on life-history traits: growth, survival, and fecundity in Black Brant. Ecology 76:2404–2414.

Sedinger JS, Lindberg MS, Rexstad EA, Chelgren ND, Ward DH (1997) Testing for handling bias in survival estimation for Black Brant. Journal of Wildlife Management 61:782–791.

Sen AR (1971) Some recent developments in waterfowl sample survey techniques. Applied Statistics 20:139–147.

Spinner GP (1949) Observations on Greater Snow geese in the Delaware Bay area. Auk 66: 197–198.

Timm DE, Bromley RG (1976) Driving Canada Geese by helicopter. Wildlife Society Bulletin 4:180–181.

Williams BK, Nichols JD, Conroy MJ (2002) Analysis and management of animal populations. Academic Press, San Diego.

Integration of Demographic Analyses and Decision Modeling in Support of Management of Invasive Monk Parakeets, an Urban and Agricultural Pest

Michael J. Conroy and Juan Carlos Senar

Abstract We investigated from 2003 to 2006 the population dynamics of Monk Parakeets (*Myiopsitta monachus*), an invasive, exotic, pest species inhabiting northeastern Spain. Our study focused on several colonies of parakeets in Barcelona. Starting in 2003, we trapped and marked birds at the main Barcelona colony in Ciutadella Park during 2 annual periods: winter (pre-nesting) and late summer (post-nesting), respectively. We marked 459 individuals, and subsequently reencountered marked birds at the colony via recapture, and additionally obtained resightings of parakeets throughout Barcelona (n = 381 recaptures and 570 resightings). We used a variation of the Robust Design in conjunction with reverse-time CR modelling to estimate survival and recruitment rates, and to determine the relative contribution of survival and recruitment to population growth rate. Due to high dispersal, apparent survival rates were low, so we used the combined recapture-resighting data to provide more realistic estimates of demographic survival. We then combined the projections with estimates of survival and recruitment elasticity from our statistical models in a decision model, in order to investigate alternative management scenarios for reducing damage from continued parakeet expansion. Given the logistical and social constraints under which managers operate, it appears that the most effective management strategy would be removal by trapping (in urban areas) or shooting (in rural areas) of birds during summer-winter period.

1 Introduction

Exotic species are now recognized as one of the leading global threats to native biodiversity and ecosystem function (Temple 1992; Kolar and Lodge 2001; Stockwell et al. 2003). They also cause significant economic losses (Pimentel et al. 2000), but their control and removal is normally difficult, and to succeed, requires previous appraisal and study (Feare 1991; Myers et al. 2006).

M.J. Conroy (✉)
USGS, Georgia Cooperative Fish and Wildlife Research Unit, University of Georgia, Athens, GA 30602, USA
e-mail: mconroy@uga.edu

D.L. Thomson et al. (eds.), *Modeling Demographic Processes in Marked Populations*, Environmental and Ecological Statistics 3, DOI 10.1007/978-0-387-78151-8_21,

The Monk Parakeet (*Myiopsitta monachus*) is a typical invasive, exotic, pest bird species. The increase in parrot trade in recent years has facilitated its spread from South America into North America and Western Europe (Hyman and Pruett-Jones 1995; Van Bael and Pruett-Jones 1996; Cassey et al. 2004). The species is considered as a pest in most parts of its range, causing agricultural damage, injuring ornamental trees by picking up small branches for nest-building, damaging electric lines and other human made structures during nesting, acoustic contamination, and problems with falling nests (Bucher and Bedano 1976; Bucher and Martin 1983; Bucher 1984; Temple 1992; Bucher 1992).

The species has been especially successful in Barcelona, Spain, with the city containing one of the largest populations in western Europe (Domènech 1997; Domènech et al. 2003). Monk parakeets became established in Barcelona city in the early seventies, and by 2001 had reached 1441 (±265) individuals, with 313 nests, growing at an average 8% annual rate over 1994–2001 (Domènech et al. 2003).

Economic losses associated with the increase in parakeets have been partially quantified by Senar and Domènech (2001), who conducted surveys of damage in the agricultural area of Baix Llobregat, specifically in the municipalities of Prat de Llobregat, Sant Boi de Llobregat, Viladecans and Gavà. Senar and Domènech (2001) focused on damage to tomatoes, which is the main crop in the area, and prone to damage by parakeets. The birds also damage fruit trees and maize, but these fields are of much lower economic importance in the area. Senar and Domènech (2001) estimated that parakeets damaged 93,679 tomatoes (71,000–136,000). Assuming an average price of 0.50 €/kg and that 1 kg represents about 6 tomatoes (5–9), this translates to an estimated loss of 7,800 € (6,000–11,400 €) during 2001.

Damage by parakeets also occurs within Barcelona city but is more difficult to quantify. Birds harvest large number of branches during nest construction, and damage to trees (especially *Platanus* sp.) can be locally significant (Senar and Domènech 2001). Nests are built high (8–15 m) in trees and present a risk of human injury and property damage when they are dislodged during storms (Senar and Domènech 2001). Finally, large colonies of birds are very noisy and create disturbance around human dwellings; to mitigate complaints and minimize damage from falling nests, nests are periodically removed by the City Council. The costs of damage due to birds and nests, and of efforts to reduce these damages, are, however, poorly quantified.

The aim of this paper is to construct a decision support model that incorporates existing demographic data on Monk Parakeets, in order to evaluate alternative means of controlling the increase of parakeet populations.

2 Methods

2.1 Study Area

Monk Parakeets were studied in Barcelona city and the Baix Llobregat area, north-eastern Spain. The Baix Llobregat area is a traditional agricultural area in the

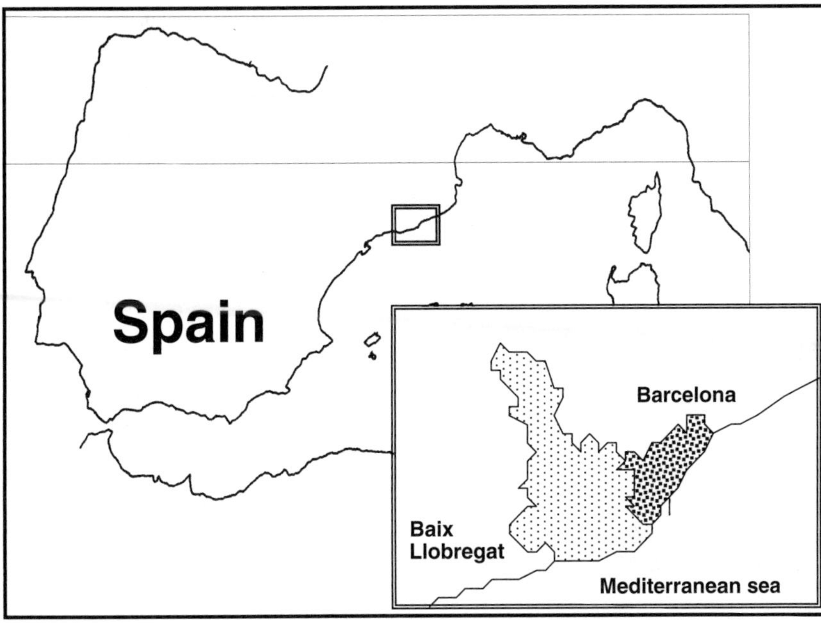

Fig. 1 Location of study areas for Monk Parakeets in Spain, 2003–2006

Llobregat river delta, located 10 km south of Barcelona (Fig. 1). The land is mostly flat and with intensive irrigated agriculture, comprised of many different vegetables. Damage by parakeets, is however concentrated on tomatoes, maize and fruit trees.

2.2 Capture–Recapture and Resighting

We used a variation on the Robust Design (Williams et al. 2002) to allow for estimation of key demographic parameters over 26-month periods, the first (winter–summer; Table 1) of which includes the principal breeding period of parakeets. We obtained capture–recapture data over 6 primary (assumed to be open to survival, recruitment, and movement) and 23 secondary occasions (assumed closed to these factors) from November 2003 to August 2006. We captured and recaptured parakeets with a Yunick Platform Trap (2 × 1× 1 m; Yunick 1971) located at the Natural History Museum of Barcelona, within the Ciutadella Park in Barcelona city, which holds the larger Monk Parakeet colony within the Barcelona city (Domènech et al. 2003) (Fig. 2). Birds were marked with numbered aluminium rings and with special numbered medals, which could be read without having to trap the bird (Ingram 1977). We marked a total of 459 individuals, obtaining 381 recaptures between November 2003 and October 2006. We also obtained 570 resightings of individually marked birds during November 2003–March 2006. During the study period, we obtained resightings via reports from birdwatchers throughout Barcelona, and

Table 1 Encounter periods and survival/emigration intervals for capture-recapture, resighting, and recovery analysis of Monk Parakeets in Barcelona, Spain, 2003–2006

Month	Year	Primary period	Secondary period	Interval length (d)[a]
Nov–Jan[b]	2003–2004	1	1	16
			2	16
			3	15
			4	222[c]
Aug–Sep[d]	2004	2	1	16
			2	15
			3	105[e]
Dec–Feb[b]	2004–2005	3	1	16
			2	15
			3	179[c]
Jul–Sep[c]	2005	4	1	16
			2	15
			3	15
			4	15
			5	124[e]
Jan–Mar[b]	2006	5	1	16
			2	15
			3	15
			4	156[c]
Aug–Oct[c]	2006	6	1	15
			2	15
			3	26
			4	

[a] Interval between end of sampling period and beginning of next primary or secondary period.
[b] Winter capture period.
[c] Winter–summer interval.
[d] Late summer trapping period.
[e] Summer–winter interval.

observations made during the course of other activities, such as censusing parakeet nests. These observations were augmented by a citywide survey over 235 h in the summer of 2004 via a series of transects throughout Barcelona (Senar and Carrillo-Ortiz 2005); this latter effort resulting in 216 of the 570 total resightings. The resightings were then grouped, along with captures and recaptures, according to primary (open) and secondary (closed) occasions for subsequent analysis (Table1).

2.3 Statistical Analyses

2.3.1 Components of Population Growth

We initially used the Robust Design and Pradel's temporal symmetry model (Williams et al. 2002) with the Huggins full heterogeneity models to estimate apparent survival (ϕ) and recruitment (f) between each of the primary periods, as implemented in program MARK (White and Burnham 1999). This model provides estimates of abundance (N) at each primary period as derived parameters. Estimates of population growth rate (λ) and seniority (γ) can be computed either by reparameterizing the Pradel model, or as derived parameters via

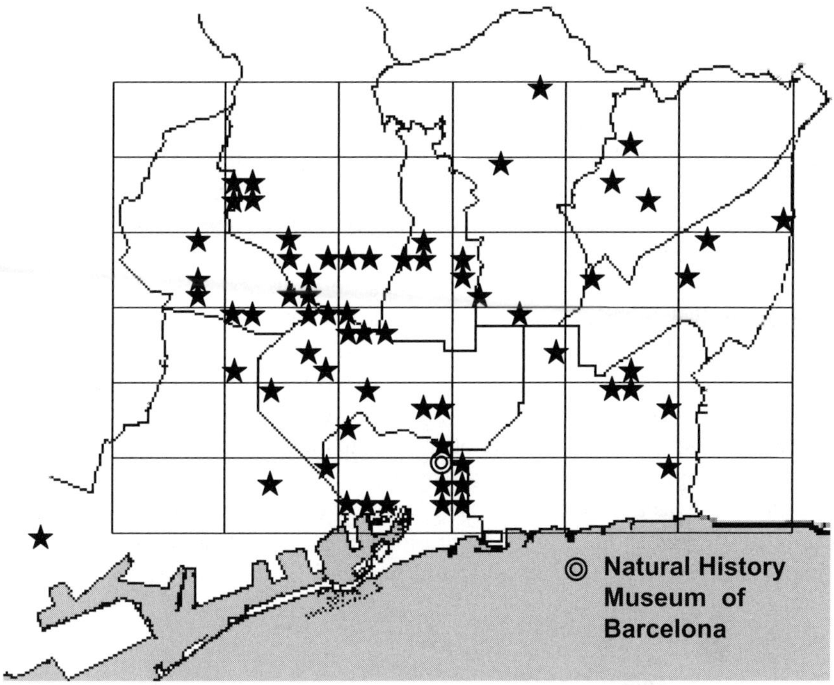

Fig. 2 Locations of Monk Parakeet colonies in Barcelona, Spain, 2003–2006

$$\hat{\lambda}_t = \hat{f}_t + \hat{\phi}_t \tag{1}$$

and

$$\hat{\gamma}_{t+1} = \frac{\hat{\phi}_t}{\hat{\phi}_t + \hat{f}_t}, \tag{2}$$

with estimated variances computed via the delta method (Williams et al. 2002: Appendix F). Expression (2) estimates the proportional contribution of survival to population growth at t+1, important for our later consideration of strategies to optimally control parakeet populations. We preferred the ϕ, f parameterization over the ϕ, λ or ϕ, γ parameterizations available in MARK (White and Burnham 1999) because constraints such as $\phi_t = \phi$ or $f_t = f$ can be specified independently, whereas constraints such as $\lambda_t = \lambda$ force a dependency with the remaining parameter (in this case ϕ). We evaluated goodness of fit under the Pradel model first via program RELEASE (from the MARK interface), which provides a general test of fit for CJS-type models. We also conducted 500 bootstrap simulations under the most general (time-specific survival and recapture) CJS model corresponding to these data, and compared the average deviance value from these simulations to the deviance obtained under this model for our data, finally taking the ratio of the latter to the former as an estimate of c. We then applied this factor to our estimates and AIC values under the Pradel model, to adjust for overdispersion.

We note that our data and the Pradel model do not allow for complete identification of parameters of biological interest. First, the parameter ϕ is denoted apparent survival, because it is the product of the probability of demographic survival (S) and fidelity (F); thus, to the extent that emigration from the study area occurs, ϕ will underestimate S. Second, our estimates of the parameter f cannot separate between in-situ recruitment, and immigration from other sites. We used the joint recapture-resighting data and the Barker (1997) model to address the former problem and to obtain estimates of S and F. We used the data from the Robust Design (Table 1) to estimate survival, fidelity, return, recapture and resighting, with survival and fidelity rates constrained to one and return rates constrained to zero over the secondary (closed) periods. Because no dead recoveries occurred during the study period, we constrained recovery parameters (r_t) to be zero. As with the Pradel analyses, we evaluated goodness-of-fit via bootstrap simulation, adjusting subsequent model comparisons and confidence intervals by the variance inflation estimate as necessary. Our study design, and inability to accurately determine ages of captured birds, precluded us from addressing the latter problem, with implications for the application of our estimates to the decision problem, considered further below.

2.3.2 Decision Model

A decision model requires 3 components: (1) a quantitative statement of the objective, (2) delineation of decision alternatives, and (3) a model relating the decision alternatives or controls to the objective. The task then is to determine the combination of decisions that best meet the resource objective, taking into account biological, economic, or other constraints.

In our case, the biological objective is to reduce population numbers or growth of parakeets, so as to reduce economic and other damages. For either urban or rural situations, economic loss occurs, and it is thus desirable to reduce the number of parakeets. Because we cannot necessarily quantify these losses (particularly for urban birds), we will henceforth assume that the goal is to reduce the losses by reducing number of birds, and use our population and catch-effort models to provide guidance as to the most efficient ways to do so.

The decisions at our disposal include removal of adult and flighted juvenile birds via trapping or shooting, and removal of nests. These controls are assumed to relate to specific parameters of survival (S) and recruitment (f). The exact form of our decision model depends on a number of considerations, including the availability of appropriate information for costs and other constraints. First, we specified controls and model outcomes in terms of per-capita rates, because our most reliable data was in these terms. Second, we considered objective functions of 2 forms: (1) meeting a biological objective (reduced growth), subject to economic and other constraints; and (2) meeting a cost objective, subject to biological constraints. The first form can be generically represented as:

$$\begin{gathered} \min\,[\lambda(x, y)] \\ \text{subject to} \\ C(x, y) < C^* \\ \lambda = \lambda_0 \end{gathered} \tag{3}$$

where $\lambda(x, y)$ is a predicted relation of removal of birds (x) and of nests (y) to population growth, $C(x, y)$ is a function describing the per-unit costs of x, y; C^* is a cost constraint that cannot be exceeded; and λ_0 is an initial (pre-control) value for population growth (currently, >1).

Alternatively, the problem can be expressed in terms of a cost or effort objective, which is constrained to meet the biological objectives. That is,

$$\begin{gathered} \min\,[C(x, y)] \\ \text{subject to} \\ \lambda(x, y) \leq \lambda^* \end{gathered} \tag{4}$$

Somewhat counter-intuitively, this approach may be effective when costs are difficult to quantify but nevertheless it is desirable to keep effort as low as possible. Depending on the situation, we will consider both forms.

2.4 Control Model

Our expression of the relationship of the controls to population growth rate requires further elaboration. This relationship is closely related to our seasonal model of population growth, as parameterized by the reverse-time capture–recapture analysis. Recall that λ has 2 components, survival (ϕ) and recruitment (f), and that the relative contribution of these to λ is captured by the parameter γ (throughout this development we assume that $\phi = S$, i.e. that permanent emigration is negligible; we relax this assumption later by invoking the results from the joint recapture-resighting analysis). Furthermore, γ can be used to model the proportional change in λ that would occur due to a proportional change in either ϕ or f (Williams et al. 2002). In our case, we have these relationships estimated over 2 separate, approximately 6-month periods of the year, one (winter–summer) in which reproduction is important, and the other (summer–winter) in which it is negligible. If we define λ over a single 6-month period, the relationship between our controls x and y (which are proportional decreases in ϕ and f, respectively), is

$$\lambda(x, y) = \lambda_0\,[1 - x\gamma - y(1 - \gamma)]$$

where

$$x = \frac{\Delta\phi}{\phi_0}, y = \frac{\Delta f}{f_0} \tag{5}$$

and ϕ_0, f_0 are the current values and proposed reductions, respectively in ϕ and f.

In our model, demographic parameters are specific to each 6-month period, which in turn has relevance for decision making. We allow "recruitment" to occur during both periods, but assume that it is principally due to reproduction during the first period (and thus subject to control via nest removal) but due to immigration during the second period (and thus not subject to control via removal). Over the first (winter–summer) period when recruitment is significant we have

$$\lambda_1(x_1, y) = \lambda_{01}\,[1 - x_1\gamma - y(1 - \gamma_1)]\,.$$

whereas over the second (summer–winter) period, when reproduction is negligible, we have

$$\lambda_2(x_2) = \lambda_{02}\,[1 - x_2\gamma_2]$$

Finally, annual population growth is obtained as the product:

$$\begin{aligned}\lambda_1(x_1, y) \times \lambda_2(x_2) &= \lambda_{01}\,[1 - x_1\gamma_1 - y(1 - \gamma_1)]\,\lambda_{02}\,[1 - x_2\gamma_2] \\ &= \lambda_0\,[1 - x_1\gamma_1 - y(1 - \gamma_1)]\,[1 - x_2\gamma_2] \qquad (6)\end{aligned}$$

where γ_1, γ_2 are the proportional contributions of survival to population growth during the first (winter–summer) and second (summer–winter) periods, respectively.

2.5 Cost Functions

We could quantify costs for some of our controls, but others were problematic. First, we have estimates of the time and other costs needed to trap a given number of birds, and have related this to our estimates of trapping success. Our records of trapping efforts in 2005 and 2006 indicated that approximately 54.6 h of effort were required to achieve the current capture rate of approximately 0.15; these hourly costs are in addition to fixed costs of trapping (traps, feed, and so forth) and that we need a trap for each main colony. We modelled the relationship of capture (removal) rate to increasing effort via a simple exponential model (Williams et al. 2002: 320) as

$$\Delta\phi = p = 1 - \exp(-kf) \qquad (7)$$

where p is capture rate, f is effort (in this case, hours), and k is a coefficient relating effort to success. Equation (7) is easy to solve for k given specified p and f, providing for an estimate of $k = 0.00297$ given $p = 0.15$ and $f = 54.6$. These relationships can provide an idea of the effort needed to obtain specific reductions (Fig. 3). For instance, 50, 75, 90, and 99% would be predicted to require 232, 465, 773, and 1547 h, respectively. More or less time-efficient procedures would of course result in different values for k and these predictions, but we would expect a similar

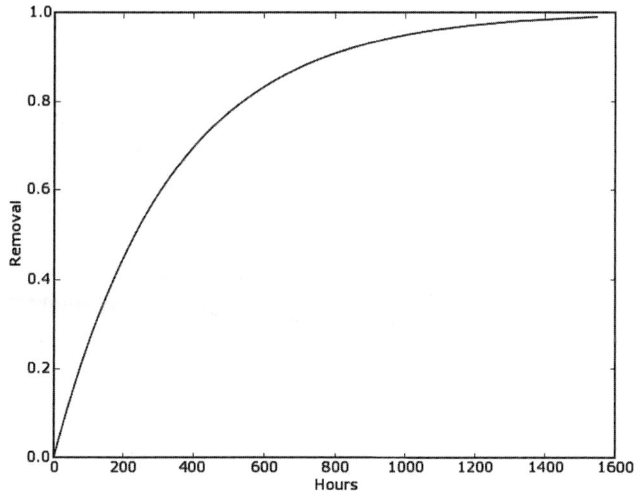

Fig. 3 Capture-effort relationship: hours of removal effort in relation to removal rate for Monk Parakeets in Barcelona, Spain, 2003–2006

relationship of cost to effort over a wide range of abundance and removal efforts. We recognize that equation (7) requires the assumption of a constant relationship (k) between a unit of effort (f) and capture rates (p); however, we have no information on which to base a more realistic model, for instance allowing for a saturation effect of removal. In the Section 4 we describe how this assumption may affect the generality of our results, and offer recommendations for further work.

2.6 Optimization Solution

We used a combination of graphical examination, nonlinear optimization, and simulation-optimization to find control strategies that met our objectives. Because we anticipated that control via nest removal, although potentially effective, would be costly and socially unacceptable (Neidermeyer and Hikey 1977; Van Bael and Pruett-Jones 1996), we first sought to explore the proportional reduction in recruitment or survival required to maintain λ at a specific level λ^*, where λ^* is presumed to be a reduction from the initial (pre-reduction) growth rate of λ_0. Under this special case of equation (2), the relationship between x, y and λ is completely specified by equation (6). For this analysis, we made the additional assumption $x_1 = x_2 = x$; that is, proportional reduction in survival is the same during both seasons. With this proviso, the constraint $\lambda = \lambda^*$ provides the solution

$$y = \left[1 - x\gamma_1 - \frac{\lambda^*}{\lambda_0(1 - x\gamma_2)}\right](1 - \gamma_1)^{-1} \tag{8}$$

for specified values of y. By plotting y on x for selected values of λ^* (e.g., 1.00, 0.95, and 0.90), we can visually determine combinations of x and y that achieve the objective, and, for instance, rule out those that are clearly impracticable.

The above is a useful first step in our decision analysis, but does not consider the relative costs of means of reducing population growth. As suggested above, and according to previous experience in the U.S.A., removal of nests – the principal means of reducing f – may be an unpopular means of reducing growth rates, causing serious political problems (Temple 1992; Van Bael and Pruett-Jones 1996). Our second decision analysis instead assumes that reduction is by means of removal, principally by trapping, and considers how removal effort should be allocated between seasons. We recast equation (2) so that the decision variables were achieved reductions in survival during each season ($\Delta\phi_1, \Delta\phi_2$), and changed the inequality constraint to an equality constraint. The former is required for application of our effort-removal relationship (equation 7), and the latter greatly simplifies the search for optimal solutions. The recast decision problem is

$$\begin{aligned} &\min\left[C(\Delta\phi_1, \phi_2)\right] \\ &\qquad \text{subject to} \\ &\qquad \lambda(\Delta\phi_1, \Delta\phi_2, \Delta f_0; f_0, \phi_0) = \lambda^* \end{aligned} \tag{9}$$

where

$$C(\Delta\phi_1, \Delta\phi_2) = \sum_{i=1}^{2}\left[-\log(1 - \Delta\phi_i)/k\right]$$

and the constraint is satisfied by

$$x_1 = \left[1 - y_0(1 - \gamma_1) - \frac{\lambda^*}{\lambda_0(1 - \gamma_2 x_2)}\right]\gamma_1^{-1} \tag{10}$$

where

$$x_i = \frac{\Delta\phi_i}{\phi_{0i}}, \; y_0 = \frac{\Delta f_0}{f_0}$$

and Δf_0 is a specified, constant reduction in recruitment, which for the purposes of exploring allocation of removal effort can be set to zero. This problem could be solved by nonlinear optimization, for example using Lagrangian multipliers (Taha 1976; Williams et al. 2002). However, because there is a single constraint and 2 control variables, it is easy to solve by substituting the constraint for one of the control variables. Thus, we replaced x_1 by equation (10) and from equation (9) obtained the value of $\Delta\phi_2$ that minimized $C(\Delta\phi_1, \Delta\phi_2)$ via direct search; this value satisfies the constraint, and therefore is optimal. However, the above presupposes that $\lambda_0 > \lambda^*$; if instead $\lambda_0 \leq \lambda^*$ then the optimal decision by definition is no action ($\Delta\phi_1 = \Delta\phi_2 = 0$). Although this condition appears to be pathological, it does arise in the course of stochastic simulations under parameter uncertainty (below). Finally, there are situations where no solution is feasible; that is, no combination of removal actions is sufficient to achieve λ^*. These could occur, for instance, if recruitment alone produces $\lambda > \lambda^*$, and again can arise in the course of stochastic simulations.

Although the above approach leads to a rational way to explore candidate decisions about control, it is fundamentally dependent on assumed values for the underlying parameters- namely, seasonal survival (ϕ), reproduction ($\underline{f}$), and the derived parameters $\underline{\lambda}$ and $\underline{\gamma}$. However, these parameters are not known, but rather are estimated from field data and stochastic models, with varying degrees of statistical reliability. Furthermore, the parameters themselves can and likely do vary over time. Optimization should therefore take into account this stochastic uncertainty, because it potentially will influence decision making. Formal approaches exist in which uncertainty can essentially be added as another constraint to the decision model (Taha 1976); however, these can be extremely complicated to solve. We instead use a simulation-optimization approach, which contains the following steps:

- The objective function and constraints are defined as for the deterministic problem.
- Uncertainty in model parameters is characterized by probability distributions, and a random draw is made of a vector of parameter values.
- An optimal decision and objective value are obtained based on these parameter values.
- The process is repeated a large number (e.g., 10,000) of times and the distributions of the decision variables and objective values examined.

Specifically, we computed standard errors for each parameter θ that incorporated statistical and temporal variation, by

$$SE(\theta) = \sqrt{\sum_i [(\hat{\theta}_i - \bar{\theta})^2 + SE(\hat{\theta}_i)^2]} \qquad (11)$$

where θ is the parameter of interested (seasonal ϕ or f), $\hat{\theta}_i$ is the estimated parameter for the ith study year, $SE(\hat{\theta}_i)$ is the estimated sampling variance for $\hat{\theta}_i$, and $\bar{\theta}$ is the average of the estimates over the k study years (k = 2 or 3, depending on the parameter). These values were used to draw normally distributed random variables $x_i = g_i(\theta)$, where $g_1(\phi) = \log\left[\frac{\phi}{1-\phi}\right]$, $g_2(f) = \log(f)$, with var(x) obtained by delta approximations under the appropriate transformation. The random variates x_i were then back-transformed via $g_i^{-1}(\theta_i)$ to obtain random values for θ and f, which in turn were used to calculate λ and γ. This approach provides optimal decision solutions for each selection of random variates, but is not a formal, stochastic optimization, which would require more computationally intensive methods (Williams et al. 2002). We performed all decision model computations using Enthought Numeric Python version 2.4 (http://python.org).

3 Results

3.1 Survival, Recruitment, and Population Growth

We initially attempted to fit several Pradel models to the full Robust CR data structure, using Huggins heterogeneity estimators in MARK. Because of numerical instability, particularly for capture and abundance estimates within primary periods, we modified our approach to a Pradel symmetry model with survival and recruitment parameters constrained to be closed across secondary periods. This eliminated abundance estimates (which were not of primary interest in this analysis) but retains estimates of $\underline{\phi}$ and $\underline{f}$. This model fit as evaluated by program RELEASE; however, our bootstrap simulations revealed mild overdispersion, and we used our estimate of $\hat{c} = 1.10$ to adjust variances and AIC values (Burnham and Anderson 2002). Based on *QAICc*, we selected model $\phi(t)f(t)p(t)$ (Table 2); the resulting point estimates of $\underline{\phi}$ and f were similar to those from the full Robust analysis, but confidence intervals were narrower (Table 3). However, apparent survival rates seemed to us to be unrealistically low, particularly during the winter–summer periods; we suspect that this was due to high rates of emigration. This was confirmed by the joint recapture-resighting analysis (Tables 4–5), with survival during both periods variable among years, but generally exceeding 0.90; the global model fit these data adequately, with indication of mild extra-binomial variation ($\hat{c} = 1.12$). We used these estimates of S in conjunction with the Pradel estimates of f (Table 3) to obtain estimates of

Table 2 Model selection for reverse-time capture–recapture analysis of Monk Parakeets in Barcelona, Spain, 2003–2006

Model[a]	ΔQAICc[b]	QAICc Weight[c]	Number of parameters
$\phi(t)f(t)p(t)$	0.000	0.796	30
$\phi(t)f(s)p(t)$	4.004	0.107	28
$\phi(s)f(t)p(t)$	4.578	0.081	28
$\phi(p)f(s)p(t)$	7.784	0.016	25
$\phi(.)f(.)p(t)$	32.445	0.000	23
$\phi(.)f(.)p(.)$	655.084	0.000	3

[a]Model subscripts: t denotes variation in parameter over seasons and years; p over seasons only (constant across years), s by period (winter–summer vs. summer–winter), and · denotes no seasonal or annual variation.
[b]Akaike Information Criterion adjusted for quasi-likelihood factor of 1.10 and small-sample correction.
[c]Model weight based on QAIC (Burnham and Anderson 2002).

Table 3 Model-averaged estimates of apparent survival and recruitment from reverse-time capture–recapture analysis of Monk Parakeets in Barcelona, Spain, 2003–2006

Year	Period	$\hat{\phi}$	SE	$\hat{f}$	SE
2003–2004	Winter–Summer	0.300	0.078	1.413	0.567
	Summer–Winter	0.633	0.083	0.118	0.173
2004–2005	Winter–Summer	0.391	0.070	0.894	0.259
	Summer–Winter	0.958	0.100	0.154	0.201
2005–2006	Winter–Summer	0.222	0.057	0.484	0.140

Table 4 Model selection for joint recapture-resighting analysis of Monk Parakeets in Barcelona, Spain, 2003–2006

Model[a]	Delta QAICc[b]	QAICc Weights[c]	Number of parameters
S(t) F(t) F'(t)	0	0.948	39
S(.) F(t) F'(t)	5.825	0.052	36
S(t) F(.) F'(.)	17.389	0.000	34
S(.) F(.) F'(.)	18.151	0.000	31

[a]All models have time-specific recapture and resighting probabilities, and recovery probabilities constrained to zero.

[b]Akaike Information Criterion adjusted for quasi-likelihood factor of 1.12 and small-sample correction.

[c]Model weight based on QAIC (Burnham and Anderson 2002).

Table 5 Model-averaged estimates of survival (*S*), fidelity (*F*) and return (F') from joint recapture-resighting analysis of Monk Parakeets in Barcelona, Spain, 2003–2006

Year	Period	$\hat{S}$	SE	$\hat{F}$	SE	$\hat{F}'$	SE
2003–2004	Winter–Summer	0.971	0.044	0.163	0.056	0.124	0.041
	Summer–Winter	0.853	0.039	1.000	0.002	0.000	0.000
2004–2005	Winter–Summer	0.776	0.055	1.000	0.002	0.600	0.328
	Summer–Winter	0.928	0.037	0.069	0.055	0.245	0.087
2005–2006	Winter–Summer	0.994	0.026	0.184	0.038	0.142	0.036

the derived parameters, λ and γ (Table 6). The estimates reveal both seasonal and annual variation in growth rates and components of growth, with, as expected, a generally a higher proportion $(1 - \gamma)$ due to recruitment occurring during the winter-summer period, which encompasses the breeding season. Annualized growth rate over 2003–2005 was $\bar{\lambda}$=2.04, substantially higher than crude, apparent growth as reflected by population surveys ($\bar{\lambda}$=1.08), possibly due to recruitment from immigration, but also to the fact that surveys are based on the counts of nests, which are routinely removed by the city council when pruning the trees, potentially resulting in underestimates of abundance. We acknowledge that our estimate of annualised growth rate likely overestimates actual population growth rate, and suggest that the true value is likely lower. Nevertheless, we suggest that our estimates of γ and λ reasonably represent the relative contributions of survival and recruitment, and thus proceeded to use the averages of the seasonal estimates, and standard errors incorporating temporal variability and statistical uncertainty, in our decision model (Table 7).

Table 6 Estimated population growth and components of population growth of Monk Parakeets in Barcelona, Spain, 2003–2006

Year	Period	$\hat{\lambda}$	SE[a]	$\hat{\gamma}$	SE
2003–2004	Winter–Summer	2.383	0.568	0.407	0.097
	Summer–Winter	0.971	0.177	0.879	0.156
2004–2005	Winter–Summer	1.670	0.265	0.465	0.074
	Summer–Winter	1.081	0.204	0.858	0.160
2005–2006	Winter–Summer	1.478	0.142	0.672	0.064
	$\bar{\lambda}^b$	2.04			

[a]Estimated via delta method from separate estimates of $SE(\phi)$ and $SE(f)$.

[b]Square root of product of estimated seasonal growth rates through summer–winter 2004–2005.

Table 7 Parameter values used in simulations of Monk Parakeets in Barcelona, Spain, 2003–2006

Parameter	Period	$\bar{x}$	SE[a]
ϕ	Winter–Summer	0.914	0.185
	Summer–Winter	0.891	0.075
f	Winter–Summer	0.931	0.917
	Summer–Winter	0.136	0.266
λ^b	Winter–Summer	1.844	–
	Summer–Winter	1.026	–
γ^b	Winter–Summer	0.515	–
	Summer–Winter	0.868	–

[a]Incorporates both temporal variability and statistical uncertainty in parameter value.
[b]Estimate based on mean ϕ and f; realizations for stochastic simulations use random draws from distributions of ϕ and f.

3.2 Decision Model

Our graphical analysis of combinations of proportional reduction in recruitment and survival resulting in specified values of population growth (Fig. 4) suggests that proportionally less effort is required in lowering ϕ than in lowering f to achieve the same λ^*. Thus, based on purely biological considerations, the most rapid decrease in population growth could theoretically be achieved by concentrating on decreasing survival rates. Further, nest removal, the main technique for lowering f may be difficult because of the high synchronicity in breeding phenology of the species and the need to apply control within the last two weeks of the breeding period, to optimize the effort and to reduce time left to the Parakeets for re-breeding. It may also be socially unacceptable in urban areas, in particular because of perceptions

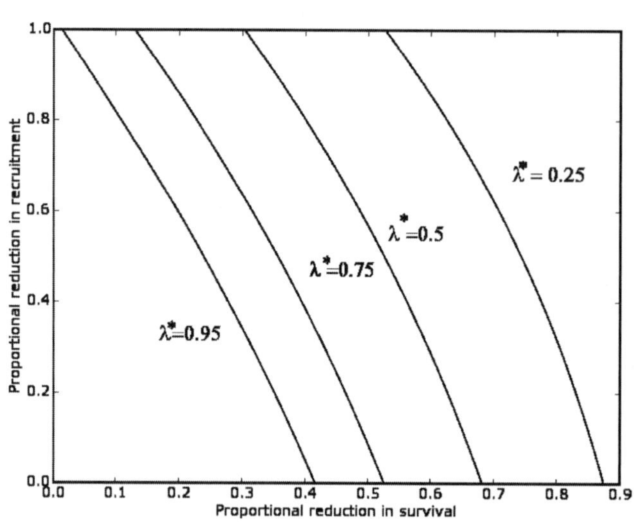

Fig. 4 Combinations of proportional reduction in recruitment and survival resulting in specified values of population growth (λ^*) for Monk Parakeets in Barcelona, Spain, 2003–2006, using mean initial recruitment and survival estimates from Table 7

Table 8 Optimal allocation of removal to seasonal periods for Monk Parakeets in Barcelona, Spain, 2003–2006

	Removal allocation[a]			
λ^*	p_1	p_2	Effort (hours)	Years to achieve 90% reduction
0.95	0.0019	0.510	240	44.9
0.75	0.004	0.618	324	8.0
0.50	0.003	0.754	472	3.3
0.25	0.012	0.889	742	1.7

[a] $p_i = \Delta\phi$, i=1 (winter–summer), 2(summer–winter)

about abandonment and mortality of pulli during nest removal. By contrast, removal of adults can be accomplished in urban areas by trapping and translocation or euthanasia, and can be done in a way that it less disturbing to the public. We thus shifted focus from allocation of effort between Δf vs. $\Delta\phi$, to allocation of removal effort to winter–summer ($\Delta\phi_1$), during which reproduction occurs, and summer–winter ($\Delta\phi_2$), during which reproduction is minimal. We used the constrained optimization approach described earlier to find values of $\Delta\phi_1$, $\Delta\phi_2$ that achieve λ^* = 0.25, 0.5, 0.75, and 0.95, while minimizing cost. The results indicate that the optimal allocation of removal effort is to allocate the bulk of removal to the summer–winter (Table 8). Costs depend on the desired λ^*, with costs obviously increasing as λ^* decreases, and with the assumed value of k, with the costs increasing at lower k (requiring proportionally more removal effort to achieve the same $\Delta\phi$). Assuming that the specified levels of λ^*can be maintained (e.g., there are no density feedbacks), parakeets could be nearly eliminated (reduced by 90%) within <10 years if λ^*<0.75 (Table 8).

The general pattern of these results is confirmed by the stochastic simulations, with median objective values clearly indicating that removal effort allocated to the summer–winter period is optimal (Table 9). Under stochastic assumptions, however, a substantial number of cases arise in which no solution is feasible; i.e., given the parameter values randomly drawn, it is not possible to achieve λ^*via removal. These cases occur, not surprisingly, more frequently as the objective becomes more aggressive, so that at λ^*= 0.25, >25% of simulations had no feasible solution. In a smaller number of cases, simulations resulted in "no action" scenarios; i.e., given the parameter values randomly selected, λ^* had already been achieved without removal; these occurred most frequently under less aggressive reduction objectives (e.g., $\lambda^* = 0.95$).

4 Discussion

In the early 1970's, the United States Fish and Wildlife Service (USFWS) reduced USA Monk Parakeet population size by half in three years, by an extensive program of removal of nests and shooting of adult birds (Neidermeyer and Hikey 1977; Van Bael and Pruett-Jones 1996). However, the relative success of the different control methods was not evaluated, particularly important when dealing with large population sizes, as it is the case with many current Monk Parakeet populations (Butler 2005). Our demographic analyses and optimizations/simulations suggest

Table 9 Influence of parameter uncertainty on optimal decisions and objective values for Monk Parakeets in Barcelona, Spain, 2003–2006; $n = 1000$ simulations, $k = 0.002977$[a]

				Decision variables and objective value				
Objective		n		$\bar{x}$	SE	5% quantile	Median	95% quantile
$\lambda^* = 0.95$	Feasible solutions	803	p_1	0.158	0.010	0.0002	0.003	0.865
			p_2	0.431	0.010	0.000	0.426	0.890
			Effort	361.450	14.281	29.88	241.18	1162.190
	No action to meet objective	69						
	No solution feasible	128						
$\lambda^* = 0.75$	Feasible solutions	822	p_1	0.153	0.010	0.0003	0.005	0.832
			p_2	0.508	0.004	0.000	0.529	0.913
			Effort	415.13	14.483	56.47	295.8	1179.58
	No action to meet objective	27						
	No solution feasible	151						
$\lambda^* = 0.5$	Feasible solutions	825	p_1	0.169	0.010	0.003	0.007	0.869
			p_2	0.613	0.009	0.000	0.655	0.958
			Effort	535.2	16.045	135.563	400.3	1371.2
	No action to meet objective	9						
	No solution feasible	166						
$\lambda^* = 0.25$	Feasible solutions	732	p_1	0.217	0.012	0.001	0.013	0.882
			p_2	0.669	0.011	0.00	0.775	0.948
			Effort	644.8	15.44	219.9	553.9	1378.9
	No action to meet objective	2						
	No solution feasible	266						

[a] k = capture coefficient for catch-effort model, λ^* = desired annual population growth rate, p_i, $i = 1, 2$ is removal rate allocated to winter–summer (1) or summer–winter (2) period.

that the most effective method for reducing parakeet population growth would be removal via trapping or shooting during the summer–winter period, when reproductive recruitment is minimal. Besides being less efficient at reduction, nest removal operations may also be difficult to apply for social and political reasons (Temple 1992; Van Bael and Pruett-Jones 1996). Nest removal operations must also be concentrated in a short period of time of about two weeks, because of the highly synchronous reproduction of the species in our area (Senar and Carrillo-Ortiz 2005) and the need to avoid re-laying. Thus, for both biological as well as social constraints, removal of individuals rather than nests, with efforts concentrated during the summer–winter period, are proportionally more effective at reducing populations. However, we must be clear that these analyses are fundamentally dependent on critical assumptions, some of which cannot be confirmed with existing data.

First, our estimates of demographic parameters (ϕ, f and the derived parameters λ and γ) are, strictly speaking, applicable only to the population of parakeets subject to capture in vicinity of the Museum. Although we think that these rates are reasonably general, and projections based on them agree with the observed population trajectory, other portions of the population in Barcelona and environs may exhibit higher or lower rates. Our examination of resightings of birds marked with visible tags suggests substantial interconnection between the Museum colony and other colonies in Barcelona (Senar and Carrillo-Ortiz 2005). Our analysis of survival and fidelity based on joint recaptures and resightings confirms that emigration from the Museum population to other colonies is high, and that the Museum population also receives substantial immigration from other colonies. However, there is little evidence of regular exchange with populations beyond Barcelona. Future work could focus on CR of birds at multiple colonies, thus providing both colony-specific demographic rates, as well as potentially estimates of movement via multi-state models (Williams et al. 2002).

Second, recruitment and survival are only partially controllable with nest control and removal, respectively. A portion of f is, by definition, survival of juveniles to 1 year of age; therefore, removal of flight-capable birds at least partially can contribute to Δf. Also, as earlier noted, a portion of f is contributed by immigration from outside the local population, and therefore, would not be expected to be influenced by removal of nests, unless control is conducted over the full distribution of the species; for Barcelona birds, this would imply city-wide control.

Third, many of our parameter estimates, particularly seasonal growth rates (λ) but also γ, were estimated with poor precision. More seriously, our data did not allow us to fully evaluate key assumptions underlying our statistical models, so that estimates may not accurately represent the parameter of interest. Specifically, we are reasonably confident in our estimates of demographic survival (S) based on the combined recapture-resighting data. However, we suspect that we have overestimated recruitment (f), in part because of our inability to disentangle in-situ reproduction from immigration; this may have also led to unrealistically high estimates of seasonal population growth rate (λ), and potentially could have biased our estimates of γ. This, in turn, would have caused us not only to overestimate the amount of effort required to achieve specified λ values, but potentially could have misled us in our examination of the relative efficiencies of reductions in recruitment versus survival. The decision model also relies on a very simplistic relationship between effort and trapping success, forced on us by the lack of more informative data on this relationship. If the relationship between removal effort and survival reduction is variable (e.g., a function of density), then our estimates of both absolute and relative cost would be biased. Full exploration of the sensitivity of decisions to these uncertainties would be important to future applications of this approach.

Finally, our models do not incorporate density-dependent feedback or other compensatory mechanisms, and the absence of these components may affect the generality of our conclusions. If, for instance, there is a compensatory response to removal, via density-dependent decreases in other mortality sources, we would expect to see no response of the population to $\Delta\phi$, up to a critical value that is determined by a survival in the absence of removal (Anderson and Burnham 1976). Likewise, depression of breeding populations to low levels could trigger a density-dependent increase in recruitment rates, again at least partially nullifying the effects of management. Managers wishing to apply these models to control decisions might be prudent to explore optimal decisions under alternative assumptions, and, if possible, reduce this type of structural uncertainty via adaptive management (Williams et al. 2002).

Our analyses assume that control could be accomplished over a relatively short time span, by means of a sufficiently aggressive removal program. In the absence of such programs the population is predicted to increase rapidly, and our cost equations do not account for the increased difficulties associated with larger densities of parakeets. If "waiting for the future" to control parakeets is a decision alternative, then managers should examine what these future costs might be. Finally, parakeet populations clearly have an intrinsically great capacity for population growth (Muñoz and Real 2006), and it seems likely that populations, if not eliminated, could rapidly expand following cessation of control. Full consideration of the problem thus may require a more formal, dynamic analysis that takes into account uncertainty in dynamics, financial discounting, and other factors in allocating control decisions through time (Taha 1976; Williams et al. 2002).

4.1 Social Considerations

The Monk Parakeet is attractive to many people, so that many efforts to control the species have become so contentious that they have been abandoned (Temple 1992; Hyman and Pruett-Jones 1995). In a survey based on 1,800 people in Barcelona city, 80% of people were opposed to the control of the species (Senar and Domènech 2001). However, given the rate of increase and spread of the species in newly established populations (Muñoz and Real 2006), the potential damage that this can cause (Bucher and Bedano 1976; Bucher and Martin 1983; Bucher 1984; Temple 1992; Bucher 1992; Senar and Domènech 2001), and the general consideration of the Monk Parakeet as an exotic invasive species, social considerations should not prevent the different relevant governmental organizations from controlling of the species. Nevertheless, social considerations complicate the control problem. For this reason, the use of traps, which are more discrete than shooting, may be more advisable in urban environments. Additionally, shooting in urban areas is not allowed by law and may be highly dangerous and contentious.

In rural environments, however, removal by shooting may be a practicable means of reducing ϕ, and may be socially accepted, especially by the farmers who suffer most of the loss. For instance, in 2000–2001, within the agricultural municipality of Gavà, 76 Monk Parakeets were shot in 24 days, because of attacks on maize fields (Departament de Medi Ambient pers. comm.). The costs of removal by shooting will obviously be different from those of trapping. We have no estimates on these costs for parakeet shooting, but note that costs would only affect k in equation (5), and would not affect either the relative effect of shooting vs. nest removal, or the optimal allocation of shooting between seasons (which should be identical to the optimal allocation of trapping). Since absolute costs would be affected, however, it would be important to quantify these, and ensure that the costs incurred in any removal program were justified by the economic damages avoided. We have mainly focused on the relative costs and benefits of various means of control, on assumption that such control would be economically warranted, because of difficulties in quantifying economic losses from parakeet damage. Future analyses should obtain these cost estimates, and better quantify the total costs of proposed control measures, thus allowing fuller consideration of the economic and social benefits of control.

Acknowledgments The project has been supported by Departament de Medi Ambient de la Generalitat de Catalunya and the Institut de Salut Pública of Ajuntament de Barcelona. We are especially grateful to the late Josep Ballús, who always showed a special sensitivity and insight for the environmental problems, and to Victor Peracho for his constant support. We thank Lluïsa Arroyo, José Carrillo, Jordi Domènech, Javier Quesada, Iker Ruiz, Esther del Val, Laura Cardador, Antoni Tallarba, Raül Aymí, Lluïs Colom, Carme Mate, David Camps and Arcadi Pomer for their help during the work, and Susana Mangar, Encarna Gomez Campos, Sisco Monjo-Luis, Juan Manuel Marín, Blanca Figuerola, Beatriz Arroyo, Esther Sanchez, Francesc Sardà, Jone Aguinagalde, Marta Soler, Ana Ruiz, Natalia Diaz, Josep Piqué, Cristina Cabrera, Manel Duró, Agnès Canals, Xavier Tomas, Raül Aymí, for information on the location of marked parakeets. We also thank collaboration of Institut Municipal de Parcs i Jardins of the Barcelona City Council. Finally, we thank Jim Peterson, Chris Fonnesbeck, Mark Lindberg, Morten Frederiksen, and an anonymous referee, for comments on earlier drafts. The Georgia Cooperative Fish and Wildlife Research Unit is jointly sponsored by USGS, the University of Georgia, Georgia Department of Natural Resources, US Fish and Wildlife Service, and the Wildlife Management Institute.

References

Anderson DR, Burnham KP (1976) Population ecology of the mallard. VI. The effect of exploitation on survival. US Fish and Wildlife Service Resource Publ. 128, Washington, DC, USA

Barker RJ (1997) Joint modelling of live-recapture, tag-resight, and tag-recovery data. Biometrics 53:666–677.

Bucher EH (1984) Las aves como plaga en la Argentina. [9], 3-17. Cordoba, Argentina, Centro de Zoologia Aplicada, Universidad Nacional de Córdoba.

Bucher EH (1992) Neotropical parrots as agricultural pests. Pages 201–219 in New world parrots in crisis.Solutions from conservation biology (Beissinger SR, Snyder NFR Eds.) Smithsonian Intitution Press, New York and London.

Bucher EH, Bedano PE (1976) Bird damage problems in Argentina. Pages 3–16 in International studies on Sparrows (Working group on granivorous birds-intecol, Eds.) PWN- Polish Scientific Publishers, Warsaw, Poland.

Bucher EH, Martin LF (1983) Los nidos de cotorras (Myiopsitta monachus) como causa de problemas en lineas de transmision electrica. Vida Silvestre Neotropical 50–51.

Burnham KP, Anderson DR (2002) Model selection and inference: a practical information theoretic approach. Springer-Verlag, New York, USA.

Butler, CJ (2005) Feral parrots in the Continental United States and United Kingdom: Past, present, and future. Journal of Avian Medicine and Surgery 19:142–149.

Cassey P, Blackburn TM, Russell GJ, Jones KE, Lockwood JL (2004) Influences on the transport and establishment of exotic bird species: an analysis of the parrots (Psittaciformes) of the world. Global Change Biology 10:417–426.

Domènech J (1997) Cotorra gris de la argentina (*Myiopsitta monachus*). Pages 253 Animales de nuestras ciudades (Omedes A, Senar JC, Uribe F Eds.) Planeta, Barcelona.

Domènech J, Carrillo-Ortiz J, Senar JC (2003) Population size of the Monk Parakeet Myiopsitta monachus in Catalonia. Revista Catalana d'Ornitologia 20:1–9.

Feare CJ (1991) Control of bird pest populations. Pages 463–478 in Bird population studies (Perrins CM, Lebreton J-D, Hirons GJ Eds.) Oxford University Press, Oxford.

Hyman J, Pruett-Jones S (1995) Natural history of the Monk Parakeet in Hyde Park, Chicago. Wilson Bulletin 107:510–517.

Ingram JC (1977) Animal marking, recognition marking of animals research.Symposium, London,

Kolar CS, Lodge DM (2001) Predicting invaders. Trends in Ecology and Evolution 16:546

Myers JH, Simberloff D, Kuris AM, Carey JR (2006) Eradication revisited: dealing with exotic species. Trends in Ecology and Evolution 15:316–320.

Muñoz AR, Real R (2006) Assessing the potential range expansion of the exotic monk parakeet in Spain. Diversity and Distributions 12:656–665.

Neidermeyer WJ, Hikey JJ (1977) The monk parakeet in the United States 1970–75. American Birds 31:413–424.

Pimentel D, Lach L, Zuniga R, Morrison AD (2000) Environmental and economic costs of nonindigenous species in the United States. BioScience 50:53–65.

Senar JC, Carrillo-Ortiz J (2005) Dinàmica de poblacions de la Cotorra de pit gris (*Myiopsitta monachus*) i recomanacions per al seu control. Museu Ciències Naturals, Barcelona.

Senar JC, Domènech J (2001) Valoració dels danys per Cotorra de pit gris al Baix Llobregat i a la ciutat de Barcelona. Museu de Ciències Naturals, Barcelona.

Stockwell CA, Hendry AP, Kinnison MT (2003) Contemporary evolution meets conservation biology. Trends in Ecology and Evolution 18:94–101.

Taha H (1976) Operations research: an introduction. MacMillan, New York.

Temple SA (1992) Exotic Birds: a growing problem with no easy solution. The Auk 109:395–397.

Van Bael S, Pruett-Jones S (1996) Exponential population growth of Monk Prakeets in the United States. Wilson Bulletin 108:584–588.

White GC, Burnham KP (1999) Program MARK: survival estimation from populations of marked animals. Bird Study 46:S120–S139.

Williams BK, Nichols JD, Conroy MJ (2002) Analysis and management of animal populations. Academic Press, New York.

Yunick RP (1971) A platform trap. EBBA News 34:122–125.

Section VI
Combining Sources of Information – Kalman Filters, Matrix Methods and Joint Likelihoods

Olivier Gimenez and Hal Caswell

Completing the Ecological Jigsaw

Panagiotis Besbeas, Rachel S. Borysiewicz, and Bryon J.T. Morgan

Abstract A challenge for integrated population methods is to examine the extent to which different surveys that measure different demographic features for a given species are compatible. Do the different pieces of the jigsaw fit together? One convenient way of proceeding is to generate a likelihood for census data using the Kalman filter, which is then suitably combined with other likelihoods that might arise from independent studies of mortality, fecundity, and so forth. The combined likelihood may then be used for inference. Typically the underlying model for the census data is a state-space model, and capture–recapture methods of various kinds are used to construct the additional likelihoods. In this paper we provide a brief review of the approach; we present a new way to start the Kalman filter, designed specifically for ecological processes; we investigate the effect of break-down of the independence assumption; we show how the Kalman filter may be used to incorporate density-dependence, and we consider the effect of introducing heterogeneity in the state-space model.

Keywords Abundance data · Diffuse initialization · Exact initial Kalman filter · Grey heron · Grey seals · Heterogeneity · Initialisation of the Kalman filter · Integrated analysis · Joint likelihood · Lack of independence · Mark-recapture-recovery data · Maximum likelihood · Stable age distribution · State-space model

1 Introduction and Background

State-space models are now widely used in ecology; see for example de Valpine (2002; 2003; 2004); de Valpine and Hastings (2003); de Valpine and Hilborn (2005) and Dennis et al. (2006). They also provide a framework for combining in a single

---

Byron J.T. Morgan (✉)
Institute of Mathematics, Statistics and Actuarial Science, University of Kent, Canterbury, Kent CT2 7NF, England
e-mail: B.J.T.Morgan@kent.ac.uk

D.L. Thomson et al. (eds.), *Modeling Demographic Processes in Marked Populations*, Environmental and Ecological Statistics 3, DOI 10.1007/978-0-387-78151-8_22,

analysis information from several different surveys of wild animals. How this might be done using methods of classical inference is described in Besbeas et al. (2002, 2003), with further illustrations in Besbeas et al. (2005) and Gauthier et al. (2007). One of the examples in Besbeas et al. (2002) concerns the grey heron, *Ardea cinerea*, on which we have available national census data and ring-recovery data, and we use a subset of these data to illustrate this paper. The ring-recovery data provide information on annual survival probabilities, while the census data provide information on both survival and productivity. Under the assumption of independence of the two surveys a combined likelihood is formed from the product of the two component likelihoods, and inference then follows from maximisation of this combined likelihood. The approach extends flexibly to include additional likelihood components, for instance possibly referring to productivity or movement. It is sometimes convenient to approximate certain component likelihoods, as explained in Besbeas et al. (2003). The central likelihood that links the component likelihoods together is that for abundance data, which in classical analysis may be obtained from a Kalman filter approach applied to an underlying state-space model; see for instance Webster and Heuvelink (2006). A major appeal of this type of analysis is its computational tractability and objectivity. While this paper employs classical methods of statistical inference, Bayesian methods can be found in, for example, Brooks et al. (2004); Meyer and Millar (1999) and Millar and Meyer (2000a,b).

However this approach to joint analysis depends on making a number of assumptions and adopting certain procedures. For example, it is assumed that the different surveys are independent; it is assumed that we can approximate discrete distributions by normal distributions; it is assumed that we can suitably start the Kalman filter iterations, and it is assumed that the state-space model adopted correctly partitions variation between its transition and measurement processes, to be described later. The robustness of the normality assumptions has been demonstrated by Brooks et al. (2004). In this paper we explore the other distinct issues described above, and the layout of the paper is as follows. In Section 2 we propose a new method of Kalman filter initiation that is designed specifically for ecological investigations, and demonstrate its good performance in a simulation study. In Section 3 we investigate the effect of relaxing the assumption of independence between different surveys. In Section 4 we show, by means of an example, that it can be possible to include non-linearity in a standard Kalman filter approach. In Section 5 we investigate how one might include heterogeneity in a state-space model, and also combine it with heterogeneity in an associated model for mark-recapture-recovery data. The paper ends with a Discussion section which revisits each issue in turn, and outlines further avenues for research.

2 Starting the Filter Using a Stable Age-Distribution

The Kalman filter is an efficient computational algorithm for fitting state-space models to time-series data; see for example Meinhold and Singpurwalla (1983). In ecological applications, state-space models, which encompass matrix population models (Caswell 2001), can be used to provide a straightforward analysis of census data when fitted by the Kalman filter. Consider the usual linear state-space model

$$\begin{aligned} \boldsymbol{\alpha}_{t+1} &= \boldsymbol{T}_t\boldsymbol{\alpha}_t + \boldsymbol{\epsilon}_t \\ \boldsymbol{y}_t &= \boldsymbol{Z}_t\boldsymbol{\alpha}_t + \boldsymbol{\eta}_t, \qquad t = 1, \ldots, n, \end{aligned} \tag{1}$$

where $\{\boldsymbol{\alpha}_t\}$ is a sequence of $m \times 1$ vectors called state vectors, with an initial state vector $\boldsymbol{\alpha}_1$, $\{\boldsymbol{y}_t\}$ is a sequence of $q \times 1$ observation vectors, there are two sequences of random variables, $\{\boldsymbol{\epsilon}_t\}$ and $\{\boldsymbol{\eta}_t\}$, with zero expectation, and $\{\boldsymbol{T}_t\}$ and $\{\boldsymbol{Z}_t\}$ are two sequences of $m \times m$ and $q \times m$ matrices called transition and measurement matrices respectively. For univariate observations, where $q = 1, \boldsymbol{y}_t$ and $\boldsymbol{\eta}_t$ are scalars. The two equations above are termed, respectively, the transition equation and the observation equation. The system matrices $\boldsymbol{T}_t$ and $\boldsymbol{Z}_t$ typically depend on unknown parameters, and the aim of a state-space analysis is to estimate the unknown parameters and state vectors. In ecological applications both system matrices might be functions of time. Under the two conditions that the initial state vector is Gaussian with mean $E(\boldsymbol{\alpha}_1)$ and covariance $\boldsymbol{P}_1 = Cov(\boldsymbol{\alpha}_1)$, both being given, and also that the two sequences $\{\boldsymbol{\epsilon}_t\}$ and $\{\boldsymbol{\eta}_t\}$ are mutually independent Gaussian, and are mutually independent of $\boldsymbol{\alpha}_1$, with covariances $Cov(\boldsymbol{\epsilon}_t) = \boldsymbol{Q}_t$ and $Cov(\boldsymbol{\eta}_t) = \boldsymbol{H}_t$, respectively, then the model may be fitted by the Kalman filter. Full descriptions of the state-space modelling framework and the Kalman filter are provided by, for example, Harvey (1989) and Durbin and Koopman (2001).

Common to all applications of the Kalman filter is the choice of the mean and covariance matrix of the initial state vector, $\boldsymbol{\alpha}_1$; see for example Gomez and Maravall (1993). The initialisation problem of the Kalman filter is an issue requiring attention in general, in areas such as economics and engineering, but the problem may be more important in population ecology, where typically there are small samples, state-space models may involve a large number of states and unknown parameters, and models are usually non-stationary. For this reason we do not in this paper consider the method of unconditional initialisation (Harvey 1989, p. 121), because this method depends critically upon the assumption of stationarity. We shall now describe three alternative methods of initialisation that may be used, before outlining a new approach. The work of Sections 2.1–2.3 draws upon Besbeas and Morgan (2006).

2.1 Diffuse Initialisation

2.1.1 Approximate Diffuse Initialisation

For non-stationary state-space models the unconditional distribution of the state vector is not defined. In the absence of any prior information, the initial distribution of $\boldsymbol{\alpha}_1$ may be specified in terms of a diffuse prior. That is, we assume that

$$\boldsymbol{\alpha}_1 \sim N(\boldsymbol{0}, \kappa \boldsymbol{I}), \tag{2}$$

where $\boldsymbol{I}$ is the identity matrix, for very large κ. For univariate observations, this results in a prior distribution that is updated from the first m observations, which are not included in the likelihood function (Harvey 1989, pp. 122, 127). In our work we have set $\kappa = 10^6$. The value assigned to κ is crucial to the success of the approximation (Harvey and Phillips 1979); if κ is too small the effect of using Eq. 2 does not diminish quickly enough and the resulting calculations are inaccurate, while if κ is too large there is a loss of precision through rounding errors. Furthermore it is not always clear how to decide on the value of m, for instance in the case of multivariate observations. Unlike in many other areas of application of the Kalman filter, ecological time-series are relatively short, and diffuse initialisation can reduce the length of the series appreciably. For instance, in the illustration of Section 4, we have 21 years of data and $m = 7$.

2.1.2 Exact Diffuse Initialisation

Exact methods of diffuse initialisation also exist; see for example de Jong (1991). The exact diffuse approach considered in this paper is described by Koopman and Durbin (2003) and provides a simple objective algorithm for initialising the Kalman filter. This approach extends the standard Kalman filter recursions, and in principle the method determines m automatically.

2.2 Maximum-Likelihood Initialisation

As an alternative, we can assume that $\boldsymbol{\alpha}_1$ is a vector of fixed constants, so that $\boldsymbol{\alpha}_1 = \boldsymbol{d}$ and $\boldsymbol{P}_1 = \boldsymbol{0}$. Estimating the vector $\boldsymbol{d}$ can then be done by including $\boldsymbol{d}$ in the parameter vector $\boldsymbol{\theta}$ for estimation as part of the maximum-likelihood (ML) procedure. In practice, however, the elements of $\boldsymbol{d}$ may not all be identifiable, and the likelihood may have multiple optima.

2.3 A New Method for Ecological Applications

In population ecology, the state vector $\boldsymbol{\alpha}_t$ is typically a vector denoting the numbers of individuals in the population in a number of classes at time t. Typical elements of $\boldsymbol{T}_t$ include age or stage-specific survival and productivity parameters and rates at which individuals in one state make the transition to another state, for example through immigration or emigration. The matrix $\boldsymbol{T}_t$ is referred to as a Leslie or Lefkovitch matrix, depending on whether the population is age- or stage-classified, respectively.

The Perron–Frobenius theorem states that, for appropriate *constant* transition matrices $\boldsymbol{T}$, there exists a real positive eigenvalue λ that is greater in absolute value (or in modulus, if some of the other eigenvalues are complex) than all other eigenvalues. The implications of this theorem are that the dominant eigenvalue λ represents the asymptotic growth rate of the population, and the normalized right eigenvector associated with λ represents the asymptotic proportion of every age or stage class in the total population. Note that we may obtain $\lambda < 1$. We call the eigenvalue λ *the asymptotic growth rate* and its corresponding right eigenvector, $\boldsymbol{v}$, is called the *stable age (or stage) distribution.*

We propose starting the Kalman filter by taking the initial mean vector $\boldsymbol{a}_1$ to be proportional to the stable age (stage) distribution of a Leslie (Lefkovitch) matrix $\boldsymbol{T}$, with the proportions scaled by the total size of the first observation, $\boldsymbol{y}_1$. In order to choose $\boldsymbol{P}_1$ we take the conservative approach of requiring the lower end of a $100(1-\alpha)\%$ confidence interval for each element of $\boldsymbol{\alpha}_1$ to be non-negative, and that elements are independent. Thus for example for univariate observations, with $\alpha = 0.05$, we take

$$\boldsymbol{a}_1 = \boldsymbol{v}\boldsymbol{y}_1/(\boldsymbol{Z}_1\boldsymbol{v}), \qquad \boldsymbol{P}_1 = \text{diag}((\boldsymbol{a}_1/1.96)^2)$$

where $\boldsymbol{y}_1$ and $\boldsymbol{Z}_1$ are the first observation and measurement vector respectively. In practice, in order to derive the stable age distribution we need to know $\boldsymbol{T}$, which may contain unknown parameters. We select the values for the parameters in $\boldsymbol{T}$ that are in common with the demographic analyses by using their maximum-likelihood estimates obtained from analysing the demographic data alone. Any remaining parameter(s) can be obtained by iteration; for details see Besbeas and Morgan (2006). When the matrix $\boldsymbol{T}_t$ is time-dependent, then we obtain $\boldsymbol{T}$ by an appropriate time-average of the $\boldsymbol{T}_t$.

We now compare these four methods by means of a simulation study based on an analysis of grey heron data.

2.4 A Grey Heron Application

Taken from Besbeas et al. (2002), the example consists of ring-recovery and abundance data. The demographic data are from birds ringed as young in the UK between 1955 and 1997, and the abundance data arise from a census estimating the total numbers of breeding pairs in England and Wales between 1928 and 1998 inclusive. For simplicity here, we do not incorporate demographic data on productivity. However in any case, for this species, it is difficult to obtain such information. The recovery data are modelled using multinomial distributions involving annual survival probabilities, ϕ_1, ϕ_2 and ϕ_a, of first-year, second-year and older animals, respectively, and a recovery probability λ, which is the probability of recovery and reporting of dead marked birds. The survival probabilities are regressed on a measure of winter severity, w, using logistic regression. As before, time-variation is denoted by a t subscript. We take w to measure the number of days in a year when the temperature drops below freezing at a particular Central England location. Thus we have $\text{logit}(\phi_{1t}) = \beta_0 + \beta_1 w_t$, $\text{logit}(\phi_{2t}) = \gamma_0 + \gamma_1 w_t$ and $\text{logit}(\phi_{at}) = \delta_0 + \delta_1 w_t$, where w_t denotes the weather severity in year t. In addition, as reporting rates for dead wild birds in the UK are generally found to decline over time (Baillie and Green 1987), we set $\text{logit}(\lambda_t) = \xi_0 + \xi_1 t$, where t measures years.

The census data are described by means of a linear Gaussian state-space model based on a Leslie matrix, involving a constant productivity measure f and measurement error variance $\boldsymbol{H}_t = \sigma^2$, in addition to the survival probabilities. The model used by Besbeas et al. (2002) is given as

$$\begin{pmatrix} N_1 \\ N_2 \\ N_a \end{pmatrix}_{t+1} = \begin{pmatrix} 0 & f\phi_1 & f\phi_1 \\ \phi_2 & 0 & 0 \\ 0 & \phi_a & \phi_a \end{pmatrix}_t \begin{pmatrix} N_1 \\ N_2 \\ N_a \end{pmatrix}_t + \boldsymbol{\epsilon}_t$$

$$y_t = \begin{pmatrix} 0 & 1 & 1 \end{pmatrix} \begin{pmatrix} N_1 \\ N_2 \\ N_a \end{pmatrix}_t + \eta_t, \tag{3}$$

where $N_{1,t}$, $N_{2,t}$ and $N_{a,t}$ denote, respectively, the numbers of female herons aged 1-year, 2-years and greater than 2-years at time t. Thus we assume that the census estimates $N_{2,t} + N_{a,t}$. Appropriate binomial and Poisson expressions for the variances of the sequences $\{\epsilon_{1,t}\}$, $\{\epsilon_{2,t}\}$, $\{\epsilon_{a,t}\}$, which in obvious notation are the components of $\boldsymbol{\epsilon}_t$, are explained in Besbeas et al. (2002) and Sullivan (1992). In Section 5 of the paper we consider more general models by introducing heterogeneity. Maximum likelihood parameter estimates describing both data sets result from maximizing the joint likelihood

$$L_j(\boldsymbol{\theta}) = L_r(\boldsymbol{\phi}, \lambda) L_c(\boldsymbol{\psi}, \boldsymbol{\phi}), \tag{4}$$

where $\boldsymbol{\theta}$ represents all the model parameters, $\boldsymbol{\phi}$ represents all the survival probability parameters, the census component L_c is formed by the Kalman filter, and $\boldsymbol{\Psi}$ denotes additional parameters f, σ and also $\boldsymbol{d}$, if and only if initialisation by maximum likelihood is to be used, and the ring-recovery component L_r is a product of multinomial distributions, each corresponding to a separate cohort of ringed birds, which is given in Eq. 5.

2.5 Simulation Comparison of Methods

The simulation study is tailored to the real data application of the last section. For the simulation study, recovery data with 10, 15 and 20 years of recoveries were generated using the observed ringing totals and winter severity, w_t, in 1988–1997, 1983–1997 and 1978–1997, respectively. These were combined for joint analysis with independent sets of abundance data from 1984–1998 ($n = 15$), 1979–1998 ($n = 20$), 1974–1998 ($n = 25$) and 1969–1998 ($n = 30$). The sample sizes were selected to reflect values typically encountered in population ecology. The abundance data were generated by simulating a population from 1928 using the model of Eq. 3. The parameter values used to generate the data were obtained from a previous joint analysis of all the observed data, and are shown in Table 1. For each set of simulated ring-recovery and abundance data, the parameters were

Table 1 Parameter values used to generate the simulated data which are used to compare the performance of four different methods for starting the Kalman filter

Parameter	Value
β_0	−0.191
β	−0.022
γ_0	0.379
γ	−0.017
δ_0	1.182
δ	−0.014
ξ_0	−2.030
ξ	−0.833
$\log f$	−0.040
$\log \sigma^2$	12.256
$\log N_{1,1}$	7.278
$\log N_{2,1}$	6.704
$\log N_{a,1}$	7.852

estimated by maximum likelihood from Eq. 4, with the Kalman filter initialised by four different methods. In the stable age (SA) case, the productivity parameter f does not have an estimate from demographic information. In the simulation, for simplicity, rather than employ an iteration method on the parameter f, we instead choose f at random, uniformly over the range [0.5, 1.5]; this means that the SA method results could be improved, had an iterative procedure been used.

The SA method was based on the mean transition matrix over the relevant period to account for annual environmental stochasticity. Thus for each replication we obtained estimates for survival from the ring-recovery data. These estimates were combined with a single simulated value for f, generated uniformly on [0.5, 1.5], to form $n-1$ transition matrices for times $t = 1, \ldots, n$. The mean of these matrices yielded a stable age distribution, $\boldsymbol{v}$, which we used to initialise the Kalman filter in the manner of Section 2.3.

The ML method was implemented with $\boldsymbol{d} = (d_1, d_1, d_2)'$, corresponding to $N_1 = N_2$ in the initial state vector, as only a linear combination of $N_{1,1}$ and $N_{2,1}$ was identifiable in this model. Nonetheless, the method occasionally terminated with a boundary (zero) estimate for d_1, which makes little biological sense, and these replicates are omitted. The results are given in Table 2. We can see relatively small differences between the methods, particularly for the cases with 15 and 20 years of recovery, but across all parameters the SA method performs well in comparison with the alternatives, in terms of root mean square error (RMSE). Differences between the methods are more pronounced for intercept rather than slope parameters, and f. Note that the productivity parameter f appears in the Kalman filter likelihood alone, and for this parameter the SA method had the smallest RMSE in 11 out of 12 cases.

Clearly RMSE and relative performance will vary with the parameter values, and dimensions of the study but the results matched to one typical study are encouraging for use of the SA method. An alternative approach to simply scaling up proportions by the total size of the first observation is to use simple multinomial sampling, based on $\boldsymbol{y}_1$, and this is done in the work of Section 4.2.

Table 2 Comparison of four methods for initialising the Kalman filter, for a joint analysis of ring-recovery and abundance data. The values given are bias and RMSE, and are based on a maximum of 250 independent replications. We use *AD* to denote the approximate diffuse method, *ED* to denote the exact diffuse method, *ML* to denote the maximum-likelihood method, and *SA* to denote the method based on the stable age distribution. For each estimated parameter we show in bold the smallest RMSE. Ties are treated equally. (a) Simulations with 10 years of recovery; (b) Simulations with 15 years of recovery; (c) Simulations with 20 years of recovery. For (b) and (c), only the values for 15 and 30 census years are presented, for brevity

(a)	15 census years								20 census years							
	Bias				RMSE				Bias				RMSE			
Parameter	AD	ED	ML	SA	AD	ED	ML	SA	AD	ED	ML	SA	AD	ED	ML	SA
β_0	0.064	0.001	0.024	0.014	0.242	0.197	0.212	**0.191**	0.032	0.008	0.029	0.006	0.191	0.173	0.188	**0.157**
β	0.004	0.000	−0.000	−0.000	0.018	**0.017**	**0.017**	**0.017**	0.000	0.001	0.001	0.001	0.016	**0.015**	**0.015**	**0.015**
γ_0	0.059	−0.057	0.017	−0.007	0.355	0.326	0.329	**0.286**	0.057	−0.003	0.043	0.005	0.323	0.314	0.327	**0.288**
γ	−0.001	−0.010	−0.003	−0.006	**0.030**	0.035	0.032	0.032	−0.002	−0.004	−0.003	−0.003	**0.035**	**0.035**	0.036	0.036
δ_0	0.075	−0.043	0.009	0.004	0.474	0.391	0.425	**0.364**	0.044	−0.030	0.023	−0.025	0.409	0.383	0.413	**0.341**
δ	0.004	0.005	0.002	0.003	0.025	0.025	**0.023**	**0.023**	0.003	0.002	0.001	0.002	**0.022**	**0.022**	**0.022**	**0.022**
ξ_0	0.008	−0.044	−0.021	−0.029	0.438	0.416	0.419	**0.409**	0.053	0.026	0.047	0.026	0.395	0.383	0.391	**0.377**
ξ	0.016	0.050	0.037	0.039	0.482	0.480	0.480	**0.477**	−0.047	−0.030	−0.042	−0.032	0.455	0.449	0.452	**0.448**
$\log f$	−0.071	0.057	−0.015	−0.000	0.516	0.442	0.473	**0.405**	−0.049	0.032	−0.034	0.025	0.451	0.422	0.453	**0.371**

	25 census years								30 census years							
	Bias				RMSE				Bias				RMSE			
Parameter	AD	ED	ML	SA	AD	ED	ML	SA	AD	ED	ML	SA	AD	ED	ML	SA
β_0	0.047	0.017	0.040	0.029	0.249	0.221	0.250	**0.209**	0.017	−0.010	0.013	0.002	0.182	0.174	0.187	**0.172**
β	−0.002	−0.001	−0.001	−0.001	**0.014**	**0.014**	**0.014**	**0.014**	−0.002	−0.001	−0.002	−0.001	0.014	**0.013**	0.014	**0.013**
γ_0	0.070	0.004	0.043	0.033	0.347	0.320	0.354	**0.297**	0.043	−0.041	0.029	0.005	0.311	0.316	0.322	**0.296**
γ	−0.003	−0.001	−0.004	−0.002	0.032	0.032	0.032	**0.031**	−0.003	−0.007	−0.002	−0.003	**0.032**	0.035	0.033	0.033
δ_0	0.085	−0.017	0.043	0.038	0.470	0.433	0.478	**0.395**	0.049	−0.036	0.025	0.003	0.440	0.424	0.459	**0.418**
δ	0.001	0.001	0.001	0.001	0.020	0.020	0.021	**0.019**	0.001	0.006	0.002	0.003	**0.019**	0.022	0.020	0.020
ξ_0	0.015	−0.021	0.004	−0.006	0.442	0.434	0.446	**0.426**	0.076	0.046	0.069	0.059	0.417	**0.408**	0.414	0.409
ξ	0.016	0.037	0.024	0.026	0.488	0.489	0.490	**0.485**	−0.067	−0.051	−0.063	−0.059	0.489	**0.485**	0.487	0.486
$\log f$	−0.093	0.014	−0.052	−0.043	0.519	0.478	0.529	**0.436**	−0.048	0.055	−0.027	0.003	0.460	0.448	0.479	**0.434**

Table 2 (continued)

(b)	15 census years								30 census years							
	Bias				RMSE				Bias				RMSE			
Parameter	AD	ED	ML	SA	AD	ED	ML	SA	AD	ED	ML	SA	AD	ED	ML	SA
β_0	0.004	−0.000	0.002	0.003	0.092	**0.090**	0.091	**0.090**	−0.009	−0.013	−0.010	−0.010	0.095	0.095	0.095	**0.094**
β	0.002	0.000	0.000	0.000	**0.008**	**0.008**	**0.008**	**0.008**	0.000	0.000	0.000	0.000	**0.007**	**0.007**	**0.007**	**0.007**
γ_0	0.009	−0.007	0.003	0.004	**0.171**	0.172	**0.171**	0.172	0.017	0.002	0.013	0.012	0.149	0.149	0.150	**0.147**
γ	−0.001	−0.003	−0.002	−0.002	**0.013**	0.014	**0.013**	**0.013**	−0.001	−0.001	−0.001	−0.001	**0.013**	**0.013**	**0.013**	**0.013**
δ_0	0.016	−0.016	−0.007	−0.001	0.197	**0.191**	0.193	**0.191**	−0.004	−0.015	−0.005	−0.006	**0.187**	0.188	0.190	**0.187**
δ	0.003	−0.001	−0.001	−0.000	0.014	0.013	0.013	**0.012**	−0.001	−0.000	−0.000	−0.000	**0.009**	**0.009**	**0.009**	**0.009**
ξ_0	0.010	−0.001	0.002	0.003	0.135	**0.134**	**0.134**	**0.134**	0.008	0.005	0.008	0.008	**0.147**	**0.147**	**0.147**	**0.147**
ξ	−0.021	−0.007	−0.010	−0.010	0.202	**0.201**	**0.201**	**0.201**	−0.012	−0.011	−0.013	−0.013	**0.220**	**0.220**	**0.220**	**0.220**
log f	0.008	0.015	0.005	0.001	0.210	**0.204**	0.206	0.207	0.007	0.019	0.006	0.007	0.192	0.193	0.195	**0.190**

(c)	15 census years								30 census years							
	Bias				RMSE				Bias				RMSE			
Parameter	AD	ED	ML	SA	AD	ED	ML	SA	AD	ED	ML	SA	AD	ED	ML	SA
β_0	0.007	0.006	0.007	0.006	0.070	0.070	0.070	**0.069**	−0.002	−0.003	−0.002	−0.002	**0.069**	0.070	0.070	**0.069**
β	0.002	0.000	0.000	0.000	**0.007**	**0.007**	**0.007**	**0.007**	−0.000	−0.000	−0.000	−0.000	**0.006**	**0.006**	**0.006**	**0.006**
γ_0	0.004	−0.005	0.001	−0.002	0.115	0.115	**0.114**	**0.114**	0.000	−0.009	−0.002	−0.003	**0.116**	0.117	0.117	**0.116**
γ	0.001	−0.000	0.000	−0.000	**0.010**	**0.010**	**0.010**	**0.010**	−0.001	−0.001	−0.000	−0.000	**0.001**	**0.001**	**0.001**	**0.001**
δ_0	0.012	0.000	0.004	0.002	0.145	0.143	0.144	**0.140**	0.017	0.010	0.015	0.014	0.144	0.144	0.145	**0.142**
δ	0.002	−0.000	−0.000	0.000	0.009	**0.008**	**0.008**	**0.008**	−0.001	0.000	−0.000	−0.000	**0.007**	**0.007**	**0.007**	**0.007**
ξ_0	0.014	0.008	0.001	0.009	0.084	0.083	0.083	**0.082**	0.002	0.001	0.003	0.002	**0.086**	**0.086**	**0.086**	**0.086**
ξ	−0.018	−0.001	−0.011	−0.010	0.143	**0.142**	0.143	**0.142**	−0.001	−0.001	−0.002	−0.002	**0.140**	**0.140**	**0.140**	**0.140**
log f	0.009	0.000	−0.004	−0.002	0.147	0.145	0.146	**0.144**	−0.005	0.001	−0.006	−0.004	0.143	0.144	0.145	**0.141**

3 Dependent Data: Animals in Common Between Ring-Recovery and Census Data Sets

It is sometimes the case that census data and demographic data are not independent. This can occur, for instance, in studies involving mammals living on small islands, an illustration being provided by Tavecchia et al. (2007). It is therefore important to investigate how such a lack of independence can affect a method of combined analysis in which component likelihoods are multiplied together. Here we consider the effect of having dependence between a ring-recovery data set and the corresponding census data. We do this for a model with two age-classes for survival and with constant parameters, ϕ_1, ϕ_a, and λ; we shall return to this model in Section 5.2.1. Life histories were constructed for a 20 year period, with $\phi_1 = 0.5$, $\phi_a = 0.7$. Separate recapture and recovery tables were derived from the life histories, each of 8 years duration, for a range of values of the probability of reporting of a dead animal, λ, and probability of recapture of a live animal, p. In addition, observation error was added to the life histories, to produce census data. Thus in this illustration there is strong dependence between the various data sets. In the study described by Tavecchia et al. (2007) the dependence was far less extreme, with demographic information gathered on a study site which was a relatively small part of an island where the census took place. The simulations were run 250 times, and mean square errors calculated for each of the model parameters. Here we only give results relating to using the recovery information since, with the simulation configuration described above, the recapture information dominated the census information, so that it was then far less important whether or not the census information was dependent or independent. The RMSE results shown in Table 3 are clear. In this example, when $\lambda = 0.4$, when independent census data are combined with recovery data, then for the three parameters ϕ_1, ϕ_a and λ there is a reduction in RMSE from combining the two data sets. Additionally, we are able, when the data sets are combined, to estimate the productivity parameter f and the standard deviation σ for the observation equation. However, when we combine the recovery information with dependent census data, then RMSEs are mostly increased compared to the case of combining with independent census data. However for the three parameters, ϕ_1, ϕ_a and λ the

Table 3 A simulation study of the effect of combining recovery data with independent and dependent census data. Shown are the RMSEs of each parameter, resulting from 250 simulations

λ	Parameter	Recovery	Recovery + Independent census	Recovery + Dependent census
0.4	ϕ_1	0.3474	0.3131	0.3439
	ϕ_a	0.5505	0.4402	0.5150
	λ	0.2587	0.2184	0.2449
	f		0.5160	0.5933
	σ		0.2027	0.1995
0.7	ϕ_1	0.2245	0.2114	0.2335
	ϕ_a	0.3980	0.3669	0.3855
	λ	0.3961	0.3736	0.5309
	f		0.4074	0.4368
	σ		0.2086	0.2202

RMSEs are slightly reduced, compared with using the recovery data alone to estimate those parameters. When $\lambda = 0.7$, the effect of the dependency is more serious, and although the effect of data combination is again that we can now estimate f and σ, as well as RMSEs increasing relative to the case of combining independent data, the RMSEs for two out of the three parameters ϕ_1, ϕ_a and λ are increased relative to the values that result when only the recovery data are used. While this is only a single study, the message is that one should take care conducting combined analyses for dependent data sets. We note finally that the paper by Barker and Kavalieris (2001) is relevant to the study of this section, as it considers the information gained from combining data sets with common parameters. Where our work is different is in considering small sample results for mean square errors – Barker and Kavalieris obtained large sample results, with no consideration of bias.

4 Non-Linearity: An Illustration Involving Grey Seals

For the case where the state-space model is non-linear, variations on the Kalman filter, such as the extended Kalman filter, have been developed; see, e.g., Chen (1993). However in certain important cases, non-linear models can be regarded as *conditionally Gaussian*, when a standard Kalman filter analysis may be carried out – see Harvey (1989, p. 156). Here system matrices, $\boldsymbol{T}_t$, may depend upon previous observations, up to and including $\boldsymbol{y}_t$, in the notation of Eq. 1. We illustrate this approach by means of an application to populations of grey seals, *Halichoerus grypus*. The study of grey seals on the British coast dates from 1960, when the animal was protected; this is the earliest example of a protected British animal species. In the work that follows we use census data from 1984 to 2004 from one colony, Faray, as previous data from that colony are unreliable. The ecology and scientific study of the grey seals are complicated by the distribution of animals over many different physical locations, and by their movement between these. The research encompasses separate investigations of pup and adult survival, of fecundity, and of annual pup censuses. A novelty of the grey seal application is that in this instance, the annual censuses are aerial, and focus on the newly born pups in each year. This is because the pups are readily counted on the resulting photographs, as pups have white fur.

4.1 Model for Grey Seals

In this work we employ approximating likelihoods for first year and adult seal survival probabilities, obtained respectively from Hall et al. (2001) and Pomeroy (pers. comm.). The capture–recapture work of Pomeroy derives from data collected on the island of North Rona. The estimates from the Hall et al. (2001) paper, and from the Pomeroy capture–recapture study were: from Hall et al. (2001): $\hat{\phi}_1^f = 0.8950(0.0339)$, $\hat{\phi}_1^m = 0.7360(0.0520)$, where we distinguish between male and female first-year survival, and from Pomeroy: $\hat{\phi}_a = 0.9496(0.0100)$. With the exception of pups, when both sexes are included, in the modelling work that follows we just consider female seals, and use ϕ_1 to denote ϕ_1^f. As explained in Besbeas et al. (2003), the appropriate likelihoods for these parameters were approximated using normal distributions based on these values.

The state-space model takes the form shown below, where N_0 denotes the number of pups, both male and female, and N_i denotes the number of females of age i, $i = 1, \ldots, 6+$. Here we assume that female seals do not have pups until they are of age 5, and thereafter productivity is denoted by f. The parameter ν_t is a probability of site-fidelity.

$$\begin{pmatrix} N_0 \\ N_1 \\ N_2 \\ N_3 \\ N_4 \\ N_5 \\ N_{6+} \end{pmatrix}_{t+1} = \begin{pmatrix} 0 & 0 & 0 & 0 & 0 & \phi_a \nu f & \phi_a \nu f \\ 0.5\phi_1 & 0 & 0 & 0 & 0 & 0 & 0 \\ 0 & \phi_a \nu & 0 & 0 & 0 & 0 & 0 \\ 0 & 0 & \phi_a \nu & 0 & 0 & 0 & 0 \\ 0 & 0 & 0 & \phi_a \nu & 0 & 0 & 0 \\ 0 & 0 & 0 & 0 & \phi_a \nu & 0 & 0 \\ 0 & 0 & 0 & 0 & 0 & \phi_a \nu & \phi_a \nu \end{pmatrix}_t \begin{pmatrix} N_0 \\ N_1 \\ N_2 \\ N_3 \\ N_4 \\ N_5 \\ N_{6+} \end{pmatrix}_t + \boldsymbol{\epsilon}_t$$

We make the usual Poisson/binomial variances for the components of $\boldsymbol{\epsilon}_t$, as was done previously. The observation equation is now simply given by

$$y_t = N_{0,t} + \eta_t,$$

ie., $\boldsymbol{Z}_t = (1, 0, 0, 0, 0, 0, 0)$, and we assume that $var(\eta_t) = \sigma^2$.

We incorporate all the available information from setting

$$L_j(\phi_1, \phi_a, \beta, f, \sigma) = \mathcal{N}(\phi_1; \hat{\phi}_1, \widehat{\sigma}_1) \times \mathcal{N}(\phi_a; \hat{\phi}_a, \widehat{\sigma}_a) \times L_c(\phi_1, \phi_a, \beta, f, \sigma)$$

where β is a carrying-capacity related parameter, included to slow down growth, we use $\mathcal{N}$ to denote a univariate normal probability density function, and $\hat{\sigma}_1$ and $\hat{\sigma}_a$ are respectively the appropriate estimated standard errors given above for the two survival probabilities. When we consider the pup census from the colony on Faray, we find that the size of the pup population levels off, corresponding, it is thought, to the operation of some form of density dependence. We considered three alternative ways to deal with this density dependence, as described in the next section.

4.2 Including Density Dependence

We experimented with 3 different models to account for density dependence, and they are described below; in each case, all other model parameters are held constant.

(i) Density-dependent pup survival, where for suitable τ,

$$\phi_{1,t} = \begin{cases} \phi_p & t = 1984, \ldots, \tau \\ \frac{\phi_p}{1+\beta(N_{0,t} - N_{0,\tau})} & t = \tau + 1, \ldots, 2004. \end{cases}$$

(ii) Density-dependent permanent emigration (or, equivalently, adult survival)

$$\nu_t = \begin{cases} 1 & t = 1984, \ldots, \tau \\ \frac{1}{1+\beta(N_{0,t}-N_{0,\tau})} & t = \tau+1, \ldots, 2004. \end{cases}$$

(iii) Broken-stick fecundity. The idea here is that there is a change-point in fecundity at a certain time. How to devise such a model is explained in Besbeas et al. (2005). This results in a linear model and so the details are not given here.

We illustrate performance of these models in Fig. 1 with $\tau = 1990$, replacing $N_{0,t}$ by y_t, as appropriate. The first and second of the above models are non-linear but may be treated as conditionally Gaussian, and can be fitted in the usual way by the Kalman filter. Conditionally Gaussian models arise when the non-linearity involves

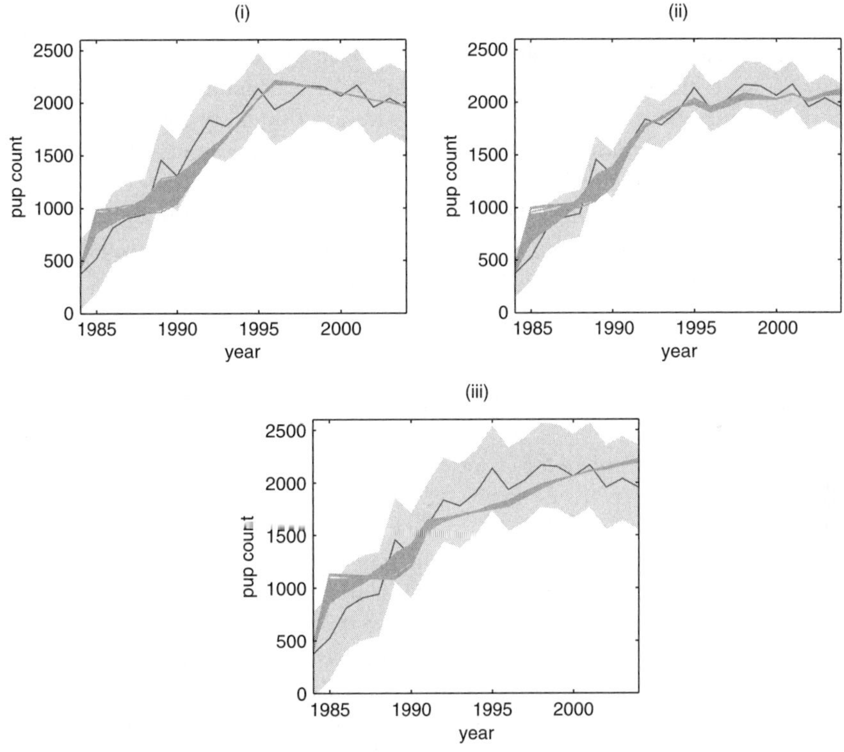

Fig. 1 Pup census data from Faray, showing the relative fits of three alternative models to account for apparent density dependence. The separate solid line denotes the census data. The models are: (i) density dependent pup survival, (ii) density-dependent emigration (or death) and (iii) broken-stick fecundity. The text explains the 100 replicate model fits in each case, which result from smoothing the Kalman filter estimates (Harvey 1989, p. 154). Also shown as the limits of grey shaded areas are the values of $y_t \pm 1.96\hat{\sigma}_m$, where $\hat{\sigma}_m$ is the median of the 100 estimates of σ

elements of the state-space vector from the previous observation time, estimated by the corresponding observations at that time. In this application we used a stable age-distribution start to the Kalman filter, as described in Section 2.3. Instead of obtaining a single set of fitted values, we sampled from the normal distributions for ϕ_1 and ϕ_a, obtained as explained above. In addition we used multinomial sampling based on y_1, as described in Section 2.5. This was then repeated 100 times, to give the several fitted curves for each of the three models, in Fig. 1, each corresponding to a different stable age distribution start to the Kalman filter. The fitted values result from smoothed Kalman filter estimates (Harvey 1989, p. 154). In this application, we averaged the $\boldsymbol{T}_t$ matrices for only the start of the time-period, which also avoided having to obtain an estimate of β when forming the $\boldsymbol{T}_t$ matrices. Had we needed an estimate of β to form a $\boldsymbol{T}_t$ matrix, then we could have done so by sampling from an appropriate prior distribution for β. It appears from Fig. 1 that the model with density-dependent emigration/survival provides the best description of the data, and this is also thought to be the most sensible model from an ecological perspective: as colonies fill up, then animals emigrate.

Seals produce a single pup each year, and the effect of perinatal mortality is that we should expect to obtain $\hat{f} < 1$. In fact, we obtain the boundary estimate, $\hat{f} = 1$, in all of the replications. This incorrect finding is probably due to the model including density-dependent emigration (mortality), but not including corresponding immigration. Work in progress also allows for immigration, and good model fits are then obtained in agreement with a value of $f < 1$.

The fitted curves exhibit little of the irregular variation seen in the count data, and a better-fitting model might arise from the logistic regression of the survival probabilities on a suitable weather covariate, as in Section 2.5.

5 Heterogeneity in Ring-Recovery and State-Space Models

In a state-space model there are two types of variation, in the transition and observation equations, respectively. In this section of the paper we present a preliminary study of how models effectively divide up the total variation between the two equations, in the important case when the transition equation includes parameter heterogeneity (cf., for example Clark et al. 2005). The heterogeneity modelled is between-individual heterogeneity. We shall use the grey heron data to illustrate the work of this section, and we shall start by considering in some detail how to incorporate heterogeneity in a model for ring-recovery data alone. Morgan and Freeman (1989) and Freeman and Morgan (1992) describe more general models involving additional age-dependence in survival, and/or time dependence in all parameters. In order to write down the likelihood function for the data, we introduce the following new notation. Let the number ringed in year $l \leq i \leq r$ be R_i, the number recovered in year $l \leq j \leq c$, having been ringed in year i, be m_{ij}, and let the number unrecovered from the year i cohort be $u_i = R_i - \sum_{j=i}^{c} m_{ij}$. Let the model probability corresponding to m_{ij} be p_{ij} and let $q_i = 1 - \sum_{j=i}^{c} p_{ij}$ be the probability of non-recovery from the ith cohort. A particular model for the data consists of a specification of the probabilities $p_{ij} \equiv p_{ij}(\phi, \lambda)$ in terms of the model parameters. Then, provided the birds suffer independent fates, the likelihood is product-multinomial, with log-likelihood given by

$$\log L_r(\phi, \lambda) = \sum_{i=1}^{r} \sum_{j=i}^{c} m_{ij} \log p_{ij} + \sum_{i=1}^{r} u_i \log q_i, \tag{5}$$

where terms not depending on the parameters have been omitted. This is the ring-recovery log likelihood, as encountered earlier in Eq. 4. In order to display appropriate multinomial cell probabilities below, we shall take a ring-recovery study with birds ringed as nestlings for $r = 3$ successive years, and recoveries recorded for the $c = 4$ years following the initial ringing. In the simplest case, parameters are constant, and the recovery probabilities are given in Table 4 for the case of just two age-classes for survival, with corresponding probabilities, ϕ_1 and ϕ_a. For each cohort, the probabilities of non-recovery are (1− the corresponding row totals).

5.1 Allowing for Heterogeneity in Recovery Data Alone

This section describes how models for ring-recovery data that allow for heterogeneity in their parameters can be specified and fitted in a straightforward manner using maximum likelihood. The new models include all of the models discussed by Freeman and Morgan (1992) as special cases, including those incorporating covariates. For illustration we concentrate on heterogeneity in adult survival, but the approach is general and not restricted to introducing overdispersion in one parameter alone. As we shall see, it is important to consider heterogeneity in tandem with model-structure, and so we provide a detailed study of heterogeneity for ring-recovery data before we consider how also to include heterogeneity in a state-space model, and in the combination of the same with a ring-recovery model.

We can allow for variability in ϕ_a by giving it a beta distribution, with density function

$$f(\phi) = \frac{\Gamma(\alpha)\Gamma(\beta)}{\Gamma(\alpha + \beta)} \phi^{\alpha-1}(1 - \phi)^{\beta-1}, \quad 0 \le \phi \le 1,$$

Table 4 Cell probabilities $\{p_{ij}\}$ for a simple model for recovery data, with no over-dispersion. In this illustrative example there are three years of ringing and four years of recovery

Year of ringing	Year of recovery			
	1	2	3	4
1	$(1 - \phi_1)\lambda$	$\phi_1(1 - \phi_a)\lambda$	$\phi_1\phi_a(1 - \phi_a)\lambda$	$\phi_1\phi_a^2(1 - \phi_a)\lambda$
2		$(1 - \phi_1)\lambda$	$\phi_1(1 - \phi_a)\lambda$	$\phi_1\phi_a(1 - \phi_a)\lambda$
3			$(1 - \phi_1)\lambda$	$\phi_1(1 - \phi_a)\lambda$

where $\alpha, \beta > 0$ are the parameters of the distribution. For related work, see Barry et al. (2003); Pollock and Raveling (1982) and Burnham and Rexstad (1993). It is straightforward to show that

$$E\left[X^g(1-X)^h\right] = \frac{\Gamma(\alpha+\beta)}{\Gamma(\alpha)\Gamma(\beta)} \frac{\Gamma(\alpha+g)\Gamma(\beta+h)}{\Gamma(\alpha+\beta+g+h)} \tag{6}$$

$$var(X) = \frac{\alpha\beta}{(\alpha+\beta)^2(\alpha+\beta+1)} \tag{7}$$

when the random variable X has the beta probability density function $f(\phi)$.

The recovery probabilities in Table 4 are no longer appropriate in the new framework because ϕ_a is now random. In order to account for variation in ϕ_a we replace the random multinomial probabilities p_{ij} by their expected values, p^*_{ij}. Thus this approach involves calculating the expectations $E(p_{ij})$ with respect to the distribution of ϕ_a, which, from Eq. 6, are straightforward to obtain. For example, the expected recovery probabilities corresponding to the probabilities in Table 4 are given in Table 5. The log-likelihood for the data may then be calculated using Eq. 5, with p_{ij} replaced by p^*_{ij}.

In order to incorporate environmental covariates, as in Section 2.4 for example, it is convenient to use an alternative parameterisation, in which the beta distribution is parameterised in terms of its mean, $\mu = \alpha/(\alpha+\beta)$, and a second parameter $\theta = 1/(\alpha+\beta)$. We can now easily introduce logistic regressions on covariates through the parameter μ; see Crowder (1978). The second parameter, θ, is proportional to the variance of the beta distribution, and here we assume that θ is constant, although a logarithmic regression of θ on covariates is just as easily introduced. When $\theta = 0$ the model reduces to the form illustrated in Table 4. The corresponding expressions for the recovery probabilities p^*_{ij}, in terms of μ and θ, are shown in Table 6, where we also indicate time dependence of all parameters except for θ, which is held constant. We now illustrate this approach by application to the grey heron data.

Table 5 Cell probabilities $\{p^*_{ij}\}$ for a model for recovery data when the adult annual survival probabilities of Table 4 have a beta distribution, with parameters α, β

	Year of recovery			
	1	2	3	4
1	$(1-\phi_1)\lambda$	$\phi_1 \frac{\beta}{\alpha+\beta}\lambda$	$\phi_1 \frac{\alpha}{\alpha+\beta} \frac{\beta}{\alpha+\beta+1}\lambda$	$\phi_1 \frac{\alpha}{\alpha+\beta} \frac{\alpha+1}{\alpha+\beta+1} \frac{\beta}{\alpha+\beta+2}\lambda$
2		$(1-\phi_1)\lambda$	$\phi_1 \frac{\beta}{\alpha+\beta}\lambda$	$\phi_1 \frac{\beta}{\alpha+\beta} \frac{\beta}{\alpha+\beta+1}\lambda$
3			$(1-\phi_1)\lambda$	$\phi_1 \frac{\beta}{\alpha+\beta}\lambda$

Table 6 As for Table 5, but now with the (μ, θ) parameterisation for the beta distribution, and with general time-variation for all parameters except θ

	Year of recovery			
	1	2	3	4
1	$(1-\phi_{11})\lambda_1$	$\phi_{11}(1-\mu_1)\lambda_2$	$\phi_{11}\frac{\mu_1}{1+\theta}(1-\mu_2)\lambda_3$	$\phi_{11}\frac{\mu_1}{1+\theta}\frac{\mu_2+\theta}{1+2\theta}(1-\mu_3)\lambda_4$
2		$(1-\phi_{12})\lambda_2$	$\phi_{12}(1-\mu_2)\lambda_3$	$\phi_{12}\frac{\mu_2}{1+\theta}(1-\mu_3)\lambda_4$
3			$(1-\phi_{13})\lambda_3$	$\phi_{13}(1-\mu_3)\lambda_4$

5.2 Application to Grey Heron Data

Previous modelling of British heron survival (North and Morgan 1979; Besbeas et al. 2002) included separate survival probabilities for first-year and second-year animals, and a common probability for older animals, as was done previously in Section 2.4. In this section we shall consider this survival age structure as well as alternative survival models. In particular, we experiment with models where the survival of birds was modelled as being age-dependent to age 1, 2 or 3, and common for older ages thereafter; it would not seem necessary to introduce further age classes for survival for these data. For each of the three survival age structures, we have fitted models involving (i) constant parameters, (ii) time-dependent parameters but constant adult survival, and (iii) time-dependent parameters, including adult survival. As in Section 2.4, when we have time-dependent survival then this is modelled by logistic regression on a weather covariate, while time-dependent reporting probability corresponds to logistic regression on time. For survival probabilities we shall use the same single environmental covariate as was used in Section 2.4. We are interested in the effect of assumptions made with regard to heterogeneity on what we conclude with regard to model structure.

5.2.1 Models with Two Survival Probabilities

In this case the model is that considered in Section 3. The maximum-likelihood estimates and corresponding maximum likelihood values from models involving two age classes for survival are given in Table 7. We can see that the use of covariates greatly improves the maximised likelihood values in both approaches. We can also see from Table 7, panels (i)–(ii), that the (α, β) and (μ, θ) parameterisations have identical likelihoods, as expected. If we also look at panel (iii), we can see that the model with heterogeneity provides a better description of the data, in terms of likelihood/AIC value. There are interesting differences between the maximum-likelihood parameter estimates from the two approaches. Table 7 (iii) also shows differences between estimates of precision for the common parameters, with the new approach providing more conservative estimates of standard error, as a result of the variability introduced into ϕ_a. The estimated beta distribution for ϕ_a from the new approach is shown in Fig. 2 (solid line).

Table 7 Maximum-likelihood estimates of parameters for a range of models for ring-recovery data alone when there are two survival probabilities. Here and later, L_r denotes the maximised likelihood value. The case of constant parameters is when there is no logistic regression of any of the model parameters; the case of constant adult survival is when there is logistic regression of ϕ_1 on the weather covariate w and of λ on time. The case of time-dependent parameters is when, additionally, adult survival is logistically regressed on w. Note that there are two heterogeneity columns for panels (i) and (ii), the first corresponding to the (α, β) parameterisation, and the second corresponding to the (μ, θ) parameterisation. For brevity we only give estimated standard errors in panel (iii)

	(i) Constant parameters			(ii) Constant adult survival			(iii) Time-dependent parameters	
	Heterogeneity			Heterogeneity			Heterogeneity	
Parameter	Without	(α, β)	(μ, θ)	Without	(α, β)	(μ, θ)	Without	(μ, θ)
ϕ_1 intercept	−0.2677	−0.2159	−0.2159	−0.2123	−0.1471	−0.1471	−0.2105(0.0477)	−0.1473(0.0556)
ϕ_1 slope (β_1)				−0.0308	−0.0302	−0.0302	−0.0308(0.0054)	−0.0301(0.0054)
ϕ_a intercept	0.7597			0.8158			0.8220(0.0512)	
ϕ_a slope (δ_1)							−0.0120(0.0040)	
α		1.9094			1.8983			
β		1.2659			1.1514			
μ intercept			0.4110			0.5000		0.5055(0.0747)
μ slope								−0.0124(0.0047)
θ			0.3149			0.3279		0.3222(0.0751)
λ intercept	−2.1854	−2.1600	−2.1600	−2.0357	−2.0023	−2.0023	−2.0351(0.0253)	−2.0026(0.0294)
λ slope				−0.8356	−0.8361	−0.8361	−0.8316(0.0461)	−0.8314(0.0465)
$-\log L_r$	8704.02	8664.60	8664.60	8522.45	8484.52	8484.52	8518.03	8480.91

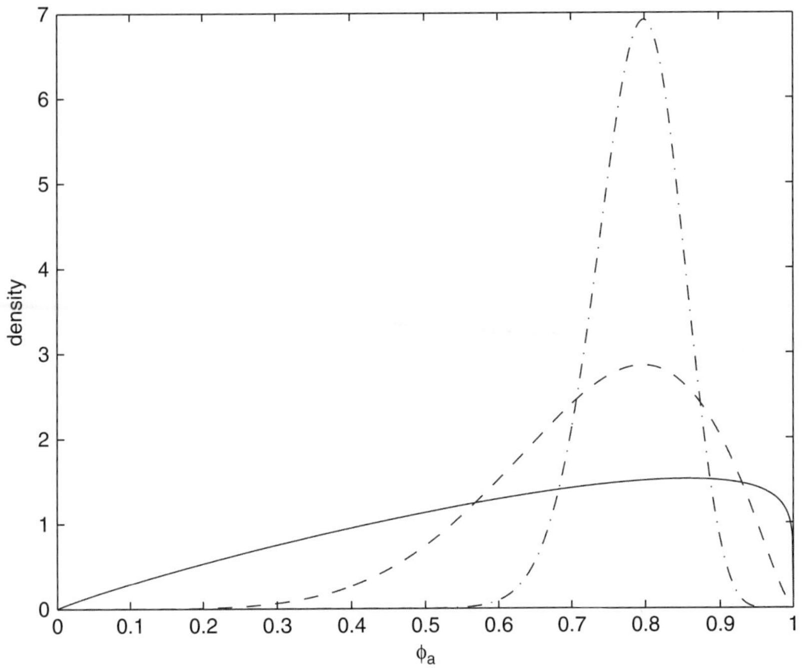

Fig. 2 An illustration of how increasing the age-structure for annual survival probabilities in a model for ring-recovery data can reduce the amount of heterogeneity needed in the model. Shown are the estimated beta distributions assumed for the adult survival probability ϕ_a. *Solid line*: when there are two survival probabilities in the model; *dashed line*: when there are three survival probabilities in the model; *dash-dotted line*: when there are four survival probabilities in the model

5.2.2 Models with Three Survival Probabilities

We also considered models distinguishing a separate probability of survival, ϕ_2, for birds in their second year of life, as in Section 2.4. Thus, in these models, ϕ_1 denotes the probability of first-year survival, as before, but the probability of adult survival, ϕ_a, applies to birds aged 2 years or older. The maximum-likelihood estimates and likelihood values from these models are shown in Table 8.

The use of three age classes for survival is seen to improve the fit of the models, but now less so for the models with heterogeneity. There are smaller differences between the maximum-likelihood estimates from the models with and without heterogeneity. However, the former models still provide a better description of the data. It is interesting to note that the regression of adult survival on weather severity is only significant in the model without heterogeneity. In fact there is no evidence that the mean of the beta distribution changes with time, as a model having a separate value for μ for each year, μ_t, has a maximised log-likelihood of -8456.25. Note that the estimate of θ, which determines the spread of the beta distribution, shrinks as the model becomes more sophisticated. The fitted beta distribution for

Table 8 Maximum-likelihood estimates of parameters for a range of models for ring-recovery data alone when there are three survival probabilities. The convention for column labeling adopted here is the same as that used in Table 7

	(i) Constant parameters			(ii) Constant adult survival			(iii) Time-dependent parameters	
	Heterogeneity			Heterogeneity			Heterogeneity	
Parameter	Without	(α, β)	(μ, θ)	Without	(α, β)	(μ, θ)	Without	(μ, θ)
ϕ_1 intercept	−0.2617	−0.2430	−0.2430	−0.2039	−0.1829	−0.1829	−0.2024(0.0481)	−0.1840(0.0506)
ϕ_1 slope (β_1)				−0.0310	−0.0308	−0.0308	−0.0309(0.0054)	−0.0306(0.0054)
ϕ_2 intercept	0.2985	0.3309	0.3309	0.3721	0.4072	0.4072	0.3745(0.0735)	0.4052(0.0777)
ϕ_2 slope (γ_1)				−0.0155	−0.0153	−0.0153	−0.0155(0.0068)	−0.0153(0.0067)
ϕ_a intercept	1.0926			1.1580			1.1655(0.0707)	
ϕ_a slope (δ_1)							−0.0110(0.0052)	
α		6.2044			6.7211			
β		2.4749			2.4581			
μ intercept			0.9191			1.0059		1.0208(0.0911)
μ slope								−0.0101(0.0054)
θ			0.1152			0.1089		0.0983(0.0502)
λ intercept	−2.1825	−2.1734	−2.1734	−2.0315	−2.0208	−2.0208	−2.0309(0.0254)	−2.0216(0.0268)
λ slope				−0.8332	−0.8330	−0.8330	−0.8302(0.0463)	−0.8298(0.0463)
$-\log L_r$	8668.55	8662.79	8662.79	8484.72	8479.82	8479.82	8482.51	8478.10

ϕ_a is plotted in Fig. 2 (dashed line). The distribution has become more focused, as second-year survival has been decomposed into its constituent rate, ϕ_2, and is no longer incorporated into ϕ_a.

5.2.3 Models with Four Survival Probabilities

Figure 2 and Table 8 (iii) demonstrate that there is still an amount of heterogeneity in adult survival. For example, an approximate Wald-test statistic of $\theta = 0$ equals 1.95, which is nearly significant at the 5% significance level. We therefore considered models also distinguishing a separate probability of survival, ϕ_3, for birds in their third year of life. The results from fitting these models are shown in Table 9 and Fig. 2 (dash-dotted line). Note that the probability of adult survival, ϕ_a, now applies to birds aged 3 years or older

The use of four age classes for survival is seen to improve further the fit of the models, particularly for the models without heterogeneity. It is interesting, therefore, that previous modelling of grey heron survival has concentrated on three age classes. There are now only minor differences between the maximum-likelihood estimates from the models with and without heterogeneity. Estimates of precision are generally larger for the models with heterogeneity, as previously. However there is no evidence of heterogeneity in adult survival, and refitting the models with zero θ only changes the likelihood by < 0.5. There is also no evidence, in Table 9 (iii), from simple Wald tests, that the regressions of third-year and adult survival on winter severity are significant. This is further illustrated in Table 10, where it is seen to hold in both presence and absence of heterogeneity, in contrast to the case of three age classes for survival (cf. Table 8(iii)).

The conclusion from this study of heterogeneity in adult survival when one is modelling ring-recovery data alone is that heterogeneity appears to depend on the model structure. If one has enough survival age-classes in the model then one does not need a heterogeneity parameter to account for age-variation in survival. We now consider how this conclusion may change when one is performing a joint analysis

5.3 Accounting for Heterogeneity in State-Space Models and in Combined Modelling

In state-space modelling it has been natural to use binomial error variances, for when a number n of individuals survive or die in any particular year, and Poisson error variances for when one considers recruitment, as from reproduction. In both of these cases, we can allow for heterogeneity, and in this section we explain how this might be done by simply focussing upon the binomial distribution for adult animals. We replace the binomial variance of $n\phi_a(1-\phi_a)$ by the corresponding beta-binomial variance which is given by

Table 9 Maximum-likelihood estimates of parameters for a range of models for ring-recovery data alone when there are four survival probabilities. The convention for column labeling adopted here is the same as that used in Table 7

	(i) Constant parameters			(ii) Constant adult survival			(iii) Time-dependent parameters	
	Heterogeneity			Heterogeneity			Heterogeneity	
Parameter	Without	(α, β)	(μ, θ)	Without	(α, β)	(μ, θ)	Without	(μ, θ)
ϕ_1 intercept	−0.2582	−0.2539	−0.2539	−0.1982	−0.1948	−0.1948	−0.1978(0.0484)	−0.1946(0.0488)
ϕ_1 slope (β_1)				−0.0309	−0.0308	−0.0308	−0.0309(0.0054)	−0.0308(0.0054)
ϕ_2 intercept	0.3046	0.3121	0.3121	0.3814	0.3872	0.3872	0.3823(0.0742)	0.3875(0.0750)
ϕ_2 slope (γ_1)				−0.0156	−0.0155	−0.0155	−0.0155(0.0068)	−0.0155(0.0068)
ϕ_3 intercept	0.7805	0.7915	0.7915	0.8833	0.8914	0.8914	0.8845(0.1043)	0.8919(0.1066)
ϕ_3 slope				−0.0154	−0.0154	−0.0154	−0.0155(0.0092)	−0.0155(0.0092)
ϕ_a intercept	1.2736			1.3438			1.3432(0.0903)	
ϕ_a slope (δ_1)							−0.0070(0.0064)	
α		26.1016			38.6872			
β		7.7336			10.4819			
μ intercept			1.2164			1.3059		1.3073(0.1057)
μ slope								−0.0069(0.0064)
θ			0.0296			0.0203		0.0188(0.0320)
λ intercept	−2.1808	−2.1787	−2.1787	−2.0287	−2.0270	−2.0270	−2.0285(0.0256)	−2.0270(0.0259)
λ slope				−0.8319	−0.8318	−0.8318	−0.8303(0.0463)	−0.8302(0.0463)
$-\log L_r$	8661.18	8660.68	8660.68	8476.13	8475.89	8475.89	8475.53	8475.32

Table 10 A comparison of minus the maximised log-likelihood values when models are fitted to ring-recovery data alone, there is regression of certain survival probabilities on a measure of winter severity, w, regression of the reporting probability on time and the heterogeneity parameter θ is excluded (case (i)) or included (case (ii))

(i)	Without overdispersion	$-\log L_r$	No. of parameters
	$\phi_1(w), \phi_2(w), \phi_3(w), \phi_a(w), \lambda(t)$	8475.53	10
	$\phi_1(w), \phi_2(w), \phi_3(w), \phi_a, \lambda(t)$	8476.13	9
	$\phi_1(w), \phi_2(w), \phi_3, \phi_a, \lambda(t)$	8477.53	8
(ii)	With overdispersion	$-\log L_r$	No. of parameters
	$\phi_1(w), \phi_2(w), \phi_3(w), \mu(w), \theta, \lambda(t)$	8475.32	11
	$\phi_1(w), \phi_2(w), \phi_3(w), \mu, \theta, \lambda(t)$	8475.89	10
	$\phi_1(w), \phi_2(w), \phi_3, \mu, \theta, \lambda(t)$	8477.28	9

$$n\mu(1-\mu)\left\{1+\frac{\theta(n-1)}{1+\theta}\right\}.$$

Here μ is the mean of the beta distribution assumed for ϕ_a, and θ is proportional to the variance of that distribution, as in the ring-recovery work described above. When this is done, we can now combine the ring-recovery and census data likelihoods, with the beta-distribution parameters μ and θ in common. Of course, in practice it is possible that one or other of the two data sets might experience more over-dispersion than the other, and that possibility is easily accommodated.

We provide an example of joint analysis of ring-recovery and census data with heterogeneity in both cases by returning to the heron example. For illustration, we consider the model with parameters $\phi_1(w), \phi_2(w), \phi_3, \mu, \theta, \lambda(t), \sigma$ and f. The profile log-likelihood for the two dispersion parameters, σ, corresponding to the state-space observation equation, and θ, corresponding to the beta-distribution assumed for the annual survival probability ϕ_a of adult birds is shown in Fig. 3. The change in the maximised log-likelihood values, between fixing $\theta = 0$ and estimating θ is 23, and the estimated value of θ is $\hat{\theta} = 0.0635(0.0129)$. Thus we have strong evidence against $\theta = 0$, in contrast to the analysis of the ring-recovery data alone. We can see here how the introduction of θ reduces σ, and at the same time increases θ. The estimators of these two dispersion parameters are negatively correlated, as one might expect. The heron census includes the winter of 1962, which was extremely severe, resulting in a very large decline in the census count for the following year. In a model without the dispersion parameter θ, this variation had to be described by the observation dispersion parameter σ alone. The inclusion of θ allows the variation to be shared between the two parts of the model.

In this example, although we have seen large, but explainable, changes in the parameter σ, the other model parameters are remarkably stable (results not shown here), and it is gratifying that they and their standard errors change little when overdispersion is added to the model. However, this one example shows that dealing with heterogeneity in modelling ring-recovery and census data needs very careful thought. Furthermore, model-selection in this area may need to be informed by

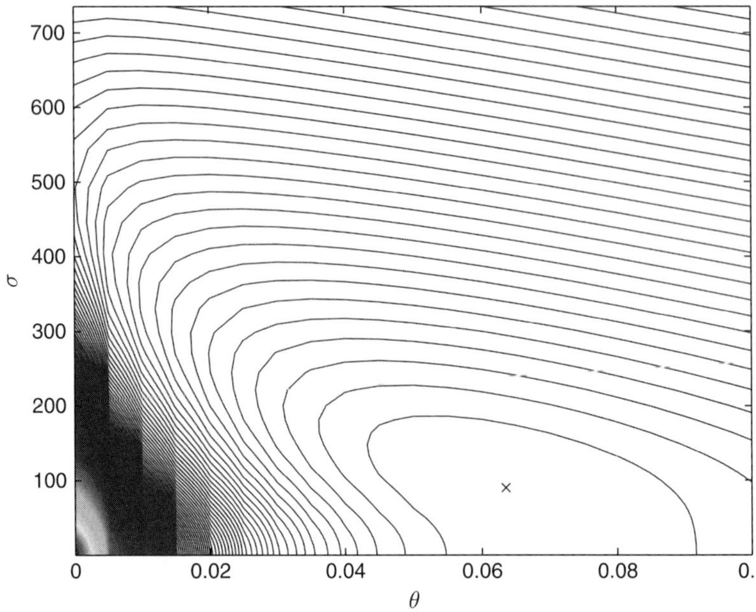

Fig. 3 Combined modelling of ring-recovery and census data for the grey heron, with a common over-dispersion parameter θ. Profile log-likelihood for the two dispersion parameters σ and θ. The model fitted here has the parameters, $\phi_1(w)$, $\phi_2(w)$, ϕ_3, μ, θ, $\lambda(t)$, σ and f. The location of the maximum is denoted by $\times$

additional/biological considerations, which might for instance provide guidance on the appropriate magnitude of measurement error.

6 Discussion and Further Research

We have provided preliminary investigations of a number of aspects of using the Kalman filter to fit state-space models to ecological time-series in conjunction with modelling demographic data. In Section 2 we have presented a way of initiating the Kalman filter which is particularly suited to ecological applications, has good performance relative to alternatives, and is very simple to use, especially for the case of multiple time-series, and we have also shown how the method may be adapted, in the work of Section 4. It is especially attractive to be able to use a method in ecology which depends on an ecological construct such as a stable age distribution. As emphasised by Besbeas and Morgan (2006), not only is this initialisation method easy to use, but it is also superior to competitors when one is dealing with multiple time-series. In Section 3 we provided a brief illustration of the potential dangers of combining dependent data sets, emphasising that this feature of joint analysis might be important in any particular application. In Section 4 we showed how non-linearities may be easily accounted for if a resulting model is treated as conditionally

Gaussian. In Section 5 we showed how to include heterogeneity in models for both census and ring-recovery data, and demonstrated the effect of doing this on grey heron data. From a detailed analysis of the ring-recovery data alone, it was emphasised that it is vital to use a correct model-structure, as otherwise a heterogeneity parameter will describe lack of fit, and this is especially important when combining information.

There is potential for much more research, for instance with regard to heterogeneity in other parts of the state-space model, and also with regard to checking for parameter stability as the dispersion features of the model are changed, and we are continuing work in all of the areas covered in the paper. More generally, fruitful additional areas for research in integrated population modelling relate to goodness-of-fit, model-selection, relative amounts of information described, with regard to how one might model the observation variance in terms of the current population sizes, and in determining the parameter redundancy status of state-space models, and how that might change as models are combined in a single analysis.

Acknowledgments Many people have contributed to the work of this paper, over several years. In particular, we thank: John Harwood, Paddy Pomeroy and Len Thomas for their input to the seal modeling and analysis, particularly Len, for insisting that one cannot estimate $f = 1$, Jean-Dominique Lebreton and Len Thomas for their discussion of Kalman filter initialisation, and Gilles Gauthier and Jean-Dominique Lebreton for their collaboration in joint modelling of data on Greater snow geese, which contributed to our understanding of how to initiate the Kalman filter. We thank the British Trust for Ornithology for permission to use the grey heron data. The paper referees, Perry de Valpine and Jean-Dominique Lebreton made many helpful suggestions which improved the paper.

References

Baillie SR, Green RE (1987) The importance of variation in recovery rates when estimating survival rates from ringing recoveries. Acta Ornithol. 23:41–60.

Barker RJ, Kavalieris L (2001) Efficiency gain from auxiliary data requiring additional nuisance parameters. Biometrics 57:563–566.

Barry SC, Brooks SP, Catchpole EA, Morgan BJT (2003) The analysis of ring-recovery data using random effects. Biometrics 59:54–65.

Besbeas P, Freeman SN, Morgan BJT (2005) The potential of integrated population modelling. Aust. N. Z. J. Stat. 47:35–48.

Besbeas P, Freeman SN, Morgan BJT, Catchpole EA (2002) Integrating mark-recapture-recovery and census data to estimate animal abundance and demographic parameters. Biometrics 58:540–547.

Besbeas P, Lebreton J-D, Morgan BJT (2003) The efficient integration of abundance and demographic data. Appl. Stat. 52:95–102.

Besbeas P, Morgan BJT (2006) Kalman filter initialization for modelling population dynamics. Submitted for publication.

Brooks SP, King R, Morgan BJT (2004) A Bayesian approach to combining animal abundance and demographic data. Anim, Biodivers. Conserv. 27:515–529.

Burnham KP, Rexstad EA (1993) Modeling heterogeneity in survival rates of banded waterfowl. Biometrics 49:1194–1208.

Caswell H (2001) Matrix population models: construction, analysis, and interpretation. 2nd edition. Sinauer Association, Sunderland, MA.

Chen G (Ed.) (1993) Approximate Kalman filtering. World Scientific Publishers, Singapore.

Clark JS, Ferraz G, Oguge N, Hays H, Di Costanzo J (2005) Hierarchical Bayes for structured, variable populations: from recapture data to life-history prediction. Ecology 86:2232–2244.

Crowder MJ (1978) Beta-binomial ANOVA for proportions. Appl. Stat. 27:34–37.

de Jong P (1991) The diffuse Kalman filter. Ann. Statist. 19:1073–1083.

Dennis B, Ponciano JM, Lele SR, Taper ML, Staples DF (2006) Estimating density dependence, process noise, and observation error. Ecol. Monogr. 76:323–341.

de Valpine P (2002) Review of methods for fitting time-series models with process and observation error and likelihood calculations for nonlinear, non-Gaussian state-space models. Bull. Mar. Sci. 70:455–471.

de Valpine P (2003) Better inferences from population-dynamics experiments using Monte Carlo state-space likelihood methods. Ecology 84:3064–3077.

de Valpine P (2004) Monte Carlo state-space likelihoods by weighted posterior kernel density estimation. J. Am. Stat. Assoc. 99:523–534.

de Valpine P, Hastings A (2003) Fitting population models incorporating process noise and observation error. Ecol. Monogr. 72:57–76.

de Valpine P, Hilborn R (2005) State-space likelihoods for nonlinear fisheries time-series. Can. J. Fish. Aquat. Sci. 62:1937–1952.

Durbin J, Koopman SJ (2001) Time series analysis by state space methods. Oxford University Press, Oxford.

Freeman SN, Morgan BJT (1992) A modelling strategy for recovery data from birds ringed as nestlings. Biometrics 48:217–236.

Gauthier G, Besbeas P, Lebreton J-D, Morgan BJT (2007) Population growth in Greater Snow Geese: a modelling approach integrating demographic and population survey information. Ecology 88(6):1420–1429.

Gomez V, Maravall A (1993) Initializing the Kalman filter with incompletely specified initial conditions. pp 39–63 In Approximate Kalman filtering. Chen G. (Ed.). World Scientific Publishers, Singapore.

Hall AJ, McConnell BJ, Barker RJ (2001) Factors affecting first-year survival in grey seals and their implications for life history strategy. J. Anim. Ecol. 70:138–149.

Harvey AC (1989) Forecasting, structural time series models and the Kalman filter. Cambridge University Press, Cambridge.

Harvey AC, Phillips GDA (1979) Maximum-likelihood estimation of regression models with autoregressive-moving average disturbances. Biometrika 66:49–58.

Koopman SJ, Durbin J (2003) Filtering and smoothing of state vector for diffuse state-space models. J. Time Ser. Anal. 24:85–98.

Meinhold RJ, Singpurwalla ND (1983) Understanding the Kalman filter. Am Statist. 37:123–127.

Meyer R, Millar RB (1999) BUGS in Bayesian stock assessments. Can. J. Fish Aquat. Sci. 56:1078–1086.

Millar RB, Meyer R (2000a) Bayesian state-space modeling of age-structured data: fitting a model is just the beginning. Can J. Fish. Aquat. Sci. 57:43–50.

Millar RB, Meyer R (2000b) Non-linear state space modelling of fisheries biomass dynamics by using Metropolis-Hastings within-Gibb sampling. Appl. Stat. 49:327–342.

Morgan BJT, Freeman SN (1989) A model with first-year variation for ring-recovery data. Biometrics 45:1087–1102.

North PM, Morgan BJT (1979) Modelling heron survival using weather data. Biometrics 35: 667–682.

Pollock KH, Raveling DG (1982) Assumptions of modern band-recovery models, with emphasis on heterogeneous survival rates. J. Wild. Man. 46:88–98.

Sullivan PJ (1992) A Kalman filter approach to catch-at-length analysis. Biometrics 48:237–258.

Tavecchia G, Besbeas P, Coulson T, Morgan BJT, Clutton-Brock TH (2007) Estimating population size and hidden demographic parameters with state-space modelling. Submitted for publication.

Webster R, Heuvelink GBM, (2006) The Kalman filter for the pedologist's tool kit. Eur. J. Soil Sci. 57:758–773.

Using a State-Space Model of the British Song Thrush *Turdus philomelos* Population to Diagnose the Causes of a Population Decline

Stephen R. Baillie, Stephen P. Brooks, Ruth King, and Len Thomas

Abstract We investigated the utility of state-space models for determining the demographic causes of population declines, using the Song Thrush as an example. A series of integrated state-space models were fitted to census and ring-recovery data from the United Kingdom for the period 1968–2000. The models were fitted using Bayesian MCMC techniques with uniform priors and were ranked using the Deviance Information Criterion (DIC). Ring-reporting rates were modelled as a declining logit-linear function of year, with separate slopes for first-year birds and adults. The system process involved three demographic parameters, first-year survival, adult survival and productivity. Survival rates were modelled as year-specific, as specific to blocks with uniform population growth rates, or as logit-linear functions of weather or year. Productivity rates were modelled as random annual effects, as block-specific or as log-linear functions of year. We fitted 17 such models chosen on the basis of our prior knowledge of this system, given that it was not practical to fit all potential models. Six models within 10 points of the smallest DIC value were selected for inference. The posterior distributions from these preferred models suggest that population growth rates are best correlated with first year survival and that and that there is also a pattern of consistent but weaker correlations between population growth rate and adult survival. Correlations between population growth rates and productivity were more variable, and may have been influenced by errors in other parts of the model, as productivity is essentially measured by difference. Thus in this analysis the evidence for productivity having a substantial influence of population changes is equivocal. The interpretation of these results and the potential value of integrated state-space models for research into the population dynamics of declining populations are discussed.

1 Introduction

Many wild bird populations have declined in recent years as a result of large-scale environmental changes (Baillie et al. 2006). Some populations need to be managed

S.R. Baillie (✉)
British Trust for Ornithology, The Nunnery, Thetford, Norfolk IP24 2PU, UK
e-mail: stephen.baillie@bto.org

D.L. Thomson et al. (eds.), *Modeling Demographic Processes in Marked Populations*, Environmental and Ecological Statistics 3, DOI 10.1007/978-0-387-78151-8_23,

to reduce their economic impacts or to protect other wildlife, while others need to be managed in ways that will produce a sustainable harvest. To address any of these management objectives it is important to understand the demographic and ecological processes that bring about population changes (Baillie 2001; Caughley 1994; Green 1995, 1999).

The British Trust for Ornithology operates an Integrated Population Monitoring programme, which monitors the abundance and demographic rates of a large number of widespread bird species (Baillie 1990). To-date most applied analyses of these data have involved estimating relative abundance and individual demographic rates from appropriate statistical models, and then exploring the consequences of these estimates using simple demographic models (Baillie and Peach 1992; Siriwardena et al. 1999; Thomson et al. 1997; Freeman and Crick 2003). Freeman et al. (2007) present a partially integrated approach where a population model is fitted directly to the site-specific census data but productivity and survival are still incorporated as "known constants" derived from separate statistical models. Fully integrated approaches allow parameter estimation and demographic modelling to be combined within a state-space model (Besbeas et al. 2002; Buckland et al. 2004; Thomas et al. 2005; Morgan 2008). Lapwing (*Vanellus vanellus*) data from the UK have been used to explore the use of this approach for birds. Models can be fitted using the Kalman filter under certain normality assumptions (Besbeas et al. 2002, 2005; Besbeas and Freeman 2006) or within a more general Bayesian framework based on MCMC methods (Brooks et al 2004; King et al. 2006).

Here we explore the use of state-space models to investigate the demographic causes of population declines using the Song Thrush (*Turdus philomelos*) population of the United Kingdom as an example. Unlike the Lapwing, for which the recovery analysis must be based on birds ringed as nestlings, we were able to model recoveries of Song Thrushes ringed as juveniles and adults. The population of this species declined by 51% between 1967 and 2003 (Baillie et al. 2006) and it is currently red listed. However, Song Thrushes are reasonably abundant and widely distributed in the UK, with an estimated population size of 1.1 million pairs (Gibbons et al. 1993; Baker et al. 2006). They have therefore continued to be reasonably well represented in demographic monitoring schemes throughout the period of the decline. The demographic causes of the decline of this species have been the subject of several previous studies (Baillie 1990; Thomson et al. 1997; Robinson et al. 2004). This earlier work provides a good basis for evaluating the potential role of state-space modelling.

2 Methods

2.1 Field Methods and Sources of Data

This analysis is based on a long-term series population data for Song Thrushes from Britain. Relative abundance was measured using territory counts from the Common

Birds Census and survival estimates were derived from ring recovery data. The data set analysed here was exactly the same as that used by Robinson et al. (2004) except that we have omitted recoveries of birds ringed as nestlings. Details of the data gathering methods have therefore not been repeated here. Methods of analysis are set out below.

2.2 Census Data

The census data were taken from the Common Birds Census using the years 1968–2000. This provides counts of the number of territories present on each plot in each year that it was censused, based on the mapping method. Many plot years are missing from the data set, due mainly to turnover of plots within the scheme. A population index and its variance were calculated using the method of King et al. (2006, Appendix A). Annual indices were calculated initially using a log-linear Poisson regression model fitted as a generalized linear model (GLM) with categorical site and year effects. Thus:

$$e_t = \left(\sum_{k=1}^{K} \exp(\alpha_k + \beta_t) \right)$$

where e_t is the index in year t, α_k is the site effect for year k and β_t is the year effect for year t, both from the GLM. The GLM also gives asymptotic standard errors for the site effects (s_k) and year effects (τ_t).

We used an estimate of the natural logarithm of the total number of territories present on all plots that had been included in the CBC during the study period (1968–2000) as an index of population size. We take logs here due to the skewness of the underlying distribution. Our new index was calculated by Monte Carlo integration using:

$$\gamma_t = \ln\left(\sum_{k=1}^{K} \exp(a_k + b_t) \right)$$

where γ_t is the index of abundance in year t, a_k is the site effect for year k and b_t is the year effect for year t. We treat a_k and b_t as random variables where:

$$a_k \sim N(\alpha_k, s_k^2) \text{ and } b_t \sim N(\beta_t, \tau_t^2)$$

and the two distributions are assumed independent. The γ_t are then calculated by drawing values of a_k and b_t from their respective distributions to obtain a series of values for γ_t and the sample mean taken as an estimate of the population index. Similarly the variance of the γ_t provides an estimate of the variance of the population index. This procedure gives an index on a log-normal scale, with variances that take full account of the errors in the site and year effects.

The census data were also used to identify periods with similar population growth rates (Fewster et al. 2000; Robinson et al. 2004). A generalized additive model with site effects and a smoothed trend function was fitted to the data. Significant turning points ($p<0.05$) were identified from the second derivatives of the smoothed population trend by bootstrapping on sites. We based our analysis on logarithms of the smoothed series so that turning points would represent changes in the relative rate of population change. Within the time series from 1968 to 2000 turning points were identified in 1975, 1979, 1983, 1987, 1992 and 1998. These turning points were used to define blocks of years within which particular demographic parameters were assumed to be constant within some of our models (Table 1).

2.3 Ring Recoveries

The ring recovery models were based on a two age-class model where birds were ringed as juveniles and adults (Brownie et al. 1985). The analysis uses recoveries of Song Thrushes ringed as adults or as free flying juveniles between April and September. Details of exactly how recovery years were defined are given in Fig. 1 of Robinson et al. (2004). The numbers ringed in the first cohort and the expected frequencies of recoveries of birds ringed as adults (first row) and as juveniles (second row) in the first year of ringing and the first three years of recoveries are as follows:

$$\mathbf{N_{a1}} \quad \mathbf{N_{a1}(1 - S_{a1})\,\lambda_{a1}} \quad \mathbf{N_{a1}S_{a1}(1 - S_{a2})\lambda_{a2}} \quad \mathbf{N_{a1}S_{a1}S_{a2}(1 - S_{a3})\lambda_{a3}}$$

$$\mathbf{N_{f1}} \quad \mathbf{N_{f1}(1 - S_{f1})\lambda_{f1}} \quad ,\mathbf{N_{f1}S_{f1}(1 - S_{a2})\lambda_{a2}} \quad \mathbf{N_{f1}S_{f1}S_{a2}(1 - S_{a3})\lambda_{a3}}$$

Table 1 Seventeen combined models ranked using the deviance information criterion. The terminology used to define these models is given in the text

Model	First-year survival	Adult survival	Productivity	Number of parameters	Δ DIC
AWR	Year-specific	Drought	Random effects	73	0.0
BWR	Blocks	Drought	Random effects	48	0.1
AWB	Year-specific	Drought	Blocks	78	1.5
WWR	Frost	Drought	Random effects	43	3.2
AWY	Year-specific	Drought	Year trend	73	5.1
BWB	Blocks	Drought	Blocks	53	7.2
BAR	Blocks	Year-specific	Random effects	78	11.8
AAR	Year-specific	Year-specific	Random effects	103	13.2
BBR	Blocks	Blocks	Random effects	53	15.4
ABR	Year-specific	Blocks	Random effects	78	15.5
WAR	Frost	Year-specific	Random effects	73	16.2
WBR	Frost	Blocks	Random effects	48	17.9
AAY	Year-specific	Year-specific	Year trend	103	23.1
AAB	Year-specific	Year-specific	Blocks	108	24.0
BBB	Blocks	Blocks	Blocks	58	24.2
WAY	Frost	Year-specific	Year trend	73	38.3
YWB	Year trend	Drought	Blocks	48	44.5

The above example shows the full model where adult survival ($\mathbf{S_a}$), first year survival ($\mathbf{S_f}$), adult reporting rate ($\lambda_\mathbf{a}$) and first year reporting rate ($\boldsymbol{\lambda}_\mathbf{f}$) all have annual parameters. $\mathbf{N_{a1}}$ represents the number of adults ringed in year 1 and $\mathbf{N_{f1}}$ the number of juveniles ringed in year 1. In all of our analyses we used reduced parameter models for some or all of these parameters as detailed below.

2.4 Modelling Ring Reporting Rates

A large number of ring reporting rate models for this data set were examined by Robinson et al. (2004). This included a model in which reporting rates were both age and year-specific, and various simpler models with additive age effects and with either year-specific or logit-linear effects of time. There was a significant decline in reporting rates with time, a pattern that has been found in several other datasets (Baillie and Green 1987; Freeman et al. 2007). There was strong support for a model that showed a logit-linear decline in reporting rate, with separate slopes and intercepts for the first year and adult age classes. This model was the most strongly supported for all nine parameterizations of survival rate considered by Robinson et al. (2004). Thus:

$$\text{logit}(\lambda_f) = a_f \text{ year} + b_f$$

$$\text{logit}(\lambda_a) = a_a \text{ year} + b_a$$

This model for reporting rates has therefore been used for all of the analyses presented in this paper.

2.5 Modelling Survival

The survival rates of Song Thrushes have been shown to be dependent on weather during both winter and summer (Thomson et al. 1997; Robinson et al. 2004). Their main food is soil invertebrates, particularly earthworms, which may become inaccessible during periods of severe winter weather. Similarly, drought may severely restrict access to soil invertebrates during the summer.

We extracted weather data for three weather stations that are broadly representative of Lowland England where most Song Thrushes are ringed. For each station we calculated the length of the longest consecutive period of freezing weather (mean air temperature < 0) between October and March and the length of the longest period of consecutive days with negligible rainfall (total daily rainfall less than 1 mm) between June and mid-August. These values were then averaged across the three weather stations and are referred to as frost days and drought days respectively (Robinson et al. 2004).

We therefore considered models where survival is a logit-linear function of weather. These were:

$$\text{logit}(S_f) = a_f \text{ frost} + b_f$$

$$\text{logit}(S_a) = a_a \text{ drought} + b_a$$

We restrict our modelling of relationships between weather and survival to these two equations following the earlier analysis of these data by Robinson et al. (2004). They found that modelling either first year or adult survival in relation to both frost and drought did not increase the amount of variation explained significantly. We also considered models in which survival was year-specific, block-specific or a logit-linear function of time. Block models assumed that survival rates of the age class concerned (first year or adult) were constant within periods of similar population growth rates, determined from turning points in the abundance trajectory as described above.

2.6 Modelling Productivity

Our data set did not include any explicit estimates of reproductive rates. Productivity was therefore calculated by difference using the overall measures of population change and the first year and adult survival data. The productivity measure used here therefore represents the number of young produced per breeding female up to the mean date when birds are ringed as juveniles. It therefore incorporates variation due to number of breeding attempts, number of fledglings produced per attempt and survival over the first 63 days after fledging (Robinson et al. 2004).

We initially considered a model with year-specific estimates of productivity. However, such a model is likely to result in over fitting where productivity is calculated by difference, because the estimated population growth rates from the demographic parameters will always be identical to those derived directly from the census data. We therefore fitted the following random effects productivity model:

$$P_t \sim N(P, \sigma^2)$$

where P_t represents the annual productivity effects, P is the mean productivity and σ^2 is the variance of the annual effects. This was incorporated in our integrated model as a Bayesian hierarchical model. P was assigned a normal prior with mean 1.0 and variance 10. σ^2 was assigned an inverse Gamma prior with parameters 0.001 and 0.001. We undertook additional simulations to check that our results were not sensitive to the exact priors chosen. Productivity was also modelled as, as constant within blocks of uniform population growth rates (block-specific) or as a log-linear function of year.

2.7 State-Space Model

To model the overall dynamics of the population we use a state-space model (Buckland et al. 2004) that comprises observation and system process components. The observation process involves a model of the annual CBC indices and their variances (but not their co-variances) derived using the method described above. We assume that the log of the population index is Normally distributed about the log of the true population size. Thus:

$$\ln y_t \sim N(\ln N_t, \sigma_t^2)$$

where y_t represents the observed annual population indices, N_t represents the true underlying numbers of females in the breeding population and $\sigma_t{}^2$ represents the year-specific variances of population size.

The system process is based on a simple Leslie matrix model. Song Thrushes breed when they are one year old. Survival rates of first year birds differ from those of older individuals but otherwise survival is assumed to be independent of age. Following earlier integrated models of the dynamics of bird populations (Besbeas et al. 2002; Brooks et al. 2004; King et al. 2006) we use a Poisson model to describe the number of offspring produced each year and a Binomial model to describe survival from one year to the next. Our model for the number of females in the population is based on the deterministic relationship:

$$N_t = N_{t-1} P_{t-1} S_{f,t-1} + N_{t-1} S_{a,t-1}$$

Thus we model the number of females in the population as:

$$N_t = N_{a,t} + N_{1,t}$$

where

$$N_{a,t} \sim \text{Bin}(N_{t-1}, S_{a,t-1})$$

and

$$N_{1,t} \sim \text{Po}(N_{t-1} P_{t-1} S_{f,t-1})$$

2.8 Fitting the Joint Model

In the state-space model we consider the population indices y as a function of true population size, productivity, first year survival, adult survival and the variances of the population indices. Thus we have:

$$f(y|N, P, S_f, S_a, \sigma_t^2) = f(y|N, \sigma_t^2)\ f(N|P, S_f, S_a)$$

where $f(y|N, \sigma_t^2)$ is the density corresponding to the observation process and $f(N|P, S_f, S_a)$ is the density corresponding to the system process.

The recovery data m are modelled as the product of multinomial distributions with parameters first year reporting rate, adult reporting rate, first year survival and adult survival. Thus we have:

$$f(m|\,\lambda_f, \lambda_a, S_f, S_a)$$

Under the assumption of independence between the census and recovery data we can obtain a joint probability distribution for the combined data as follows:

$$f(y, m|\ N, P, S_f, S_a, \sigma_t^2, \lambda_f, \lambda_a)$$

$$= f(y|N, P, S_f, S_a, \sigma_t^2)\ f(m|\,\lambda_f, \lambda_a, S_f, S_a)$$

The models were fitted using Bayesian methods. We define priors on the parameters and then draw samples from the joint posterior distribution using MCMC methods.

The models were fitted using a purpose written FORTRAN program to perform sequential Metropolis-Hastings updates on each parameter in turn. We performed 1,000,000 iterations for each set of simulations and discarded the first 100,000 as burn in. The results were thinned by a factor of 10. Our various models took between 2 hours 24 min and 18 hours 42 min to run on various PCs (typically 2.8 GHz processors) running under Fedora Linux. Good convergence was achieved and model runs with different starting values gave similar results.

We did not have prior information that was independent of this study. Therefore survival rates were assigned uniform priors between 0 and 1.0 while productivity rates were assigned uniform priors between 0 and 2.0. Priors for regression coefficients and intercepts were normally distributed with means of 0 and variances of 100. The system equation for the state-space model is recursive, requiring the provision of a starting value for population size (N). The population index for this starting year (1968 in the present analysis) was treated as a normal prior with mean and variance obtained from the analysis of index values described above. Previous studies using similar state space models have found that the results are not sensitive to the exact choice of prior starting value (Brooks et al. 2004; King et al. 2006). This is to be expected, as this is effectively a model of relative abundance.

2.9 Model Selection

Our models included three demographic parameters (productivity, first year survival and adult survival) each of which could vary in four different ways (log or logit-linear function of weather, year-specific/random annual effects, block-specific or log or logit-linear function of time). This gave a set of 64 potential models. It was not possible to fit all of these models using the MCMC methods described above due to constraints of computer time. We therefore selected 17 models that we felt were most likely to describe this system, based on previous work (Table 1). The fit of these models was compared using the Deviance Information Criterion (DIC). Differences in DIC values of more than 10 should definitely rule out the model with the higher DIC value, while differences between 5 and 10 may be regarded as substantial (Spiegelhalter et al. 2002).

2.10 Classical Analysis

We compare some of our results with those from a classical analysis of the same data set which did not incorporate any formal joint modelling of the census and recovery data. For these analyses standard recovery models were fitted using program MARK (White and Burnham 1999). Productivity was calculated as that required to give the observed population change given the survival estimates obtained from the MARK analyses.

3 Results

The new population index generated by simulation is very similar to the original index (Fig. 1). The small but systematic difference between the two indices is because the original index values represent the mode of a skewed distribution while the new index values represent the mean (King et al. 2006, Appendix A).The population declined from the start of the time series in 1968 until 1987, with the steepest decline from 1975 onwards. There were marked fluctuations in abundance between 1987 and 2000 but there was no clear long-term trend during this period.

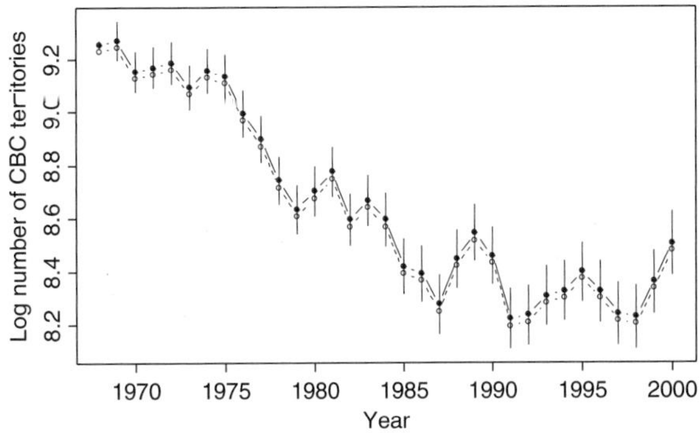

Fig. 1 Abundance of Song Thrushes on CBC plots between 1968 and 2000 calculated from the site and year effects by simulation (*filled circles* joined by *solid line*). Error bars are 95 percentiles from the simulation results. The open circles joined by a broken line show an index based only on the year effects from the generalized linear model. For further details of methods see text

Our integrated models aim to determine the demographic processes that brought about this pattern of population changes.

The numbers of parameters and delta DIC values for our 17 models are shown in Table1. Six models had Δ DIC values of less that 10, including four with Δ DIC values of less than 5. In all six models adult survival was a logit-linear function of drought. First year survival was year-specific in three models, block-specific in two models and a function of frost on one model. Productivity had random annual effects in thee models, block-specific effects in two models and showed a log-linear trend with year in one model. Thus there is good evidence of some form of year-specific variation for both first-year survival and productivity, while variation in adult survival appears to be well modelled as a function of drought. The only difference between the two best models (Δ DIC 0.1) is that one has fully year-specific first year survival, while the other has block-specific first year survival.

Ring reporting rates of both first year birds and adults showed a logit-linear decline with time. The details given here are based on model AWR (posterior means with 95% HPDIs for regression parameters: first year slope –0.424, –0.566 to –0.282; first year intercept –3.616, –3.782 to –3.437; adult slope –0.579, –0.657 to –0.504; adult intercept –3.748, –3.828 to –3.664) but results from the other five preferred models were similar. Reporting rates of first year birds were slightly higher than those of adults (Fig. 2a), a feature found in all six models. First year reporting rates from the MARK analysis were slightly higher than those from the integrated analysis in this model (Fig. 2b) but the differences were less than this in the other five models. In all six analyses the adult reporting rate estimates from MARK and from our integrated analysis were almost identical. Results from the best model (AWR) are plotted in Fig. 3. Broken lines show results from a similar model based on a conventional analysis, where N is taken from the CBC index, survival rates are modelled using Mark with the same parameterization as in model AWR, and annual productivity is calculated by difference. In model AWR the estimates of true population size from the state-space model closely match the original CBC index (Fig. 3a). The system error in the model is represented by a Poisson process and the pattern of system errors therefore closely follows that of true population size (Fig. 3b). Population growth rates from the integrated and classical analyses are similar (Fig. 3c) but the differences between the two approaches can be seen more clearly than in Fig. 3a.

Annual estimates of productivity (Fig. 3d) and first year survival (Fig. 3e) from the integrated model have relatively poor precision, even though the productivity values are shrinkage estimates from a random effects model. Nevertheless there is reasonably good agreement between the integrated analysis and the classical analysis in both cases. The poor precision of the first-year survival estimates was found in both the integrated and MARK analyses, and is likely to be a function of the number of recoveries available for this species. The logit-linear relationship between adult survival and drought days was very similar in both the classical and integrated analyses (posterior means and HPDIs from integrated analysis: intercept 0.509, 0.427 to 0.592; slope –0.135, –0.191 to –0.076). As a result of this the patterns

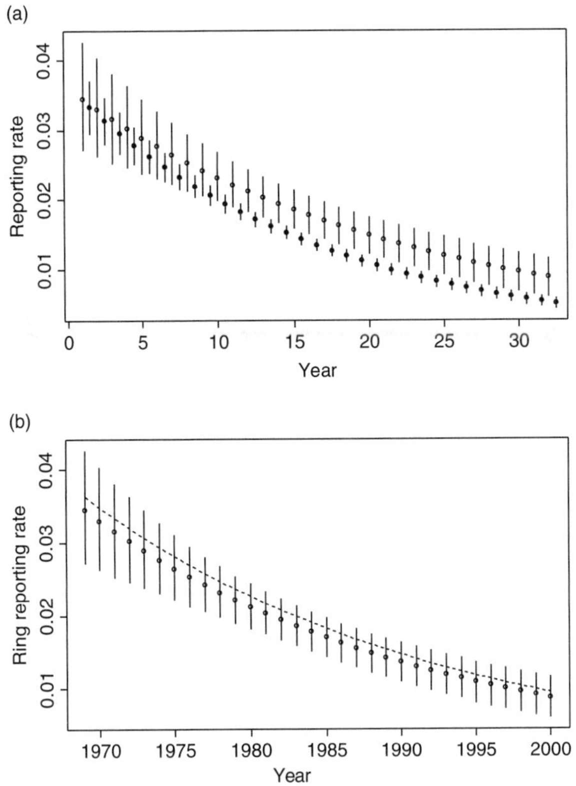

Fig. 2 Temporal trends in ring reporting rates of Song Thrushes, modelled as a logit-linear function of year. Error bars show 95% highest probability density intervals from MCMC analysis. (**a**) First year (*open circles*) and adult (*filled circles*) reporting rates from model AWR. (**b**) First-year reporting rates (*open circles* with *error bars*) from model AWR. Broken line shows the results of a stand-alone mark-recovery analysis of the same data conducted using program MARK

of annual variation in adult survival shown by the two analyses are almost identical (Fig. 3f).

The six best models (Table 2) all showed very similar patterns of change in the population index over time. Patterns of variation in population growth rates were very similar for the top five models (r = 0.879–0.992) but posterior mean population growth rates from model BWB showed a weaker relationship with those from the other models (r = 0.749–0.823). Posterior mean values for annual productivity were less strongly correlated between the different models. The three models that included random annual effects for productivity (AWR, BWR and WWR) showed moderate correlations between posterior mean productivity values (r = 0.725–0.812) as did the three models with block effects or a year trend (AWB, BWB and AWY, r = 0.612–0.892). Posterior mean first year survival rates were well correlated between the three models with year-specific first year survival (AWR, AWB and

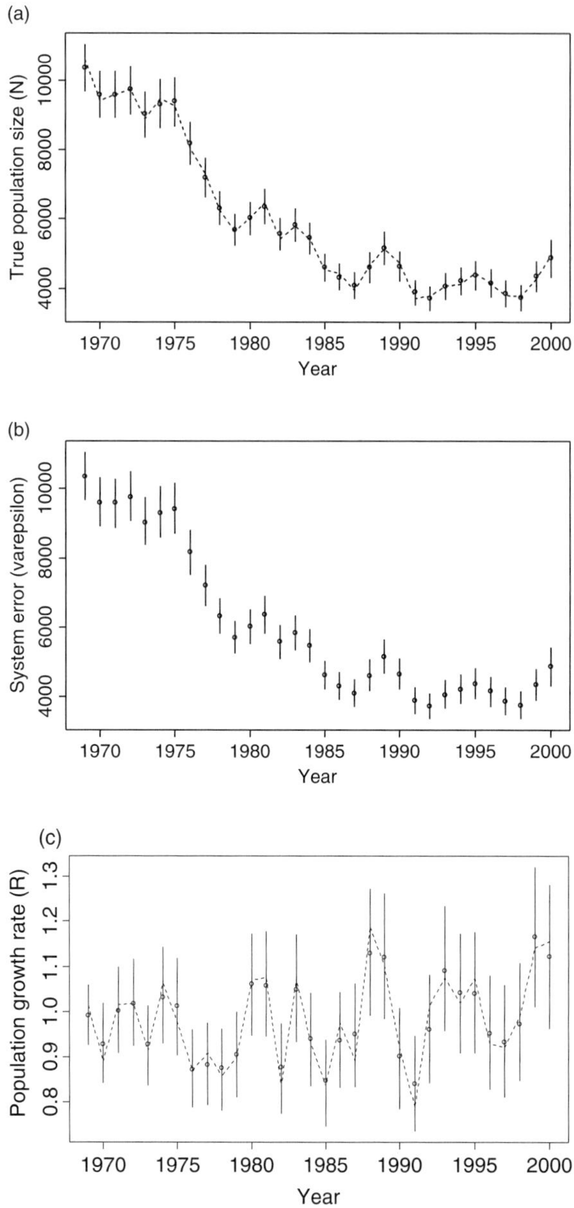

Fig. 3 Time-series of parameter estimates obtained by fitting joint model AWR with year-specific first year survival, adult survival modelled as a logit-linear function of drought days and productivity modelled as annual random effects (*open circles* with *error bars*). Values are posterior means with 95% highest probability density intervals. Results based on separate analyses of the census and ringing data are also shown (*broken lines*). For further details of methods see text. (**a**) Population size (N); (**b**) System error; (**c**) Population growth rate; (**d**) Productivity; (**e**) First-year survival; (**f**) Adult survival

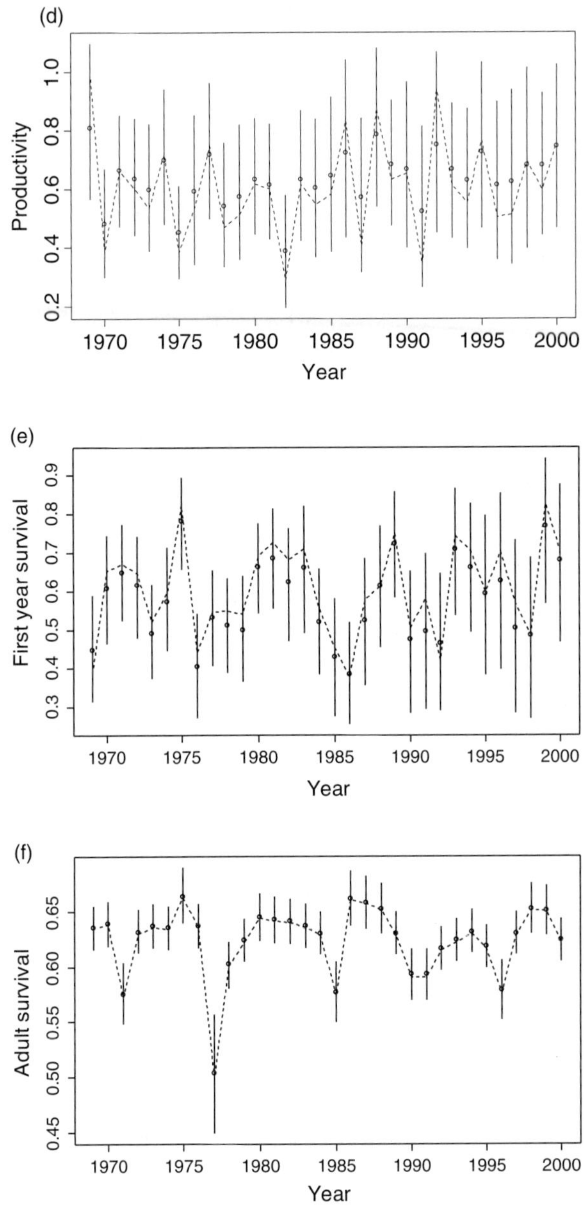

Fig. 3 (continued)

Table 2 Posterior means and 95 percentiles for correlations between population growth rate and demographic parameters. Results are presented for the six best model as ranked by DIC

		Correlation between growth rate and population								
		Productivity			First-year Survival			Adult Survival		
Model	Δ DIC	2.5 percentile	Mean	97.5 percentile	2.5 percentile	Mean	97.5 percentile	2.5 percentile	Mean	97.5 percentile
AWR	0.0	0.206	0.493	0.720	0.331	0.586	0.797	0.170	0.336	0.499
BWR	0.1	0.451	0.697	0.862	0.338	0.525	0.677	0.205	0.380	0.547
AWB	1.5	−0.053	0.323	0.607	0.406	0.725	0.900	0.296	0.437	0.571
WWR	3.2	0.583	0.798	0.926	0.405	0.550	0.678	0.200	0.385	0.559
AWY	5.1	−0.282	0.228	0.342	0.712	0.874	0.956	0.315	0.464	0.600
BWB	7.2	−0.506	0.114	0.660	−0.069	0.530	0.820	0.500	0.660	0.796

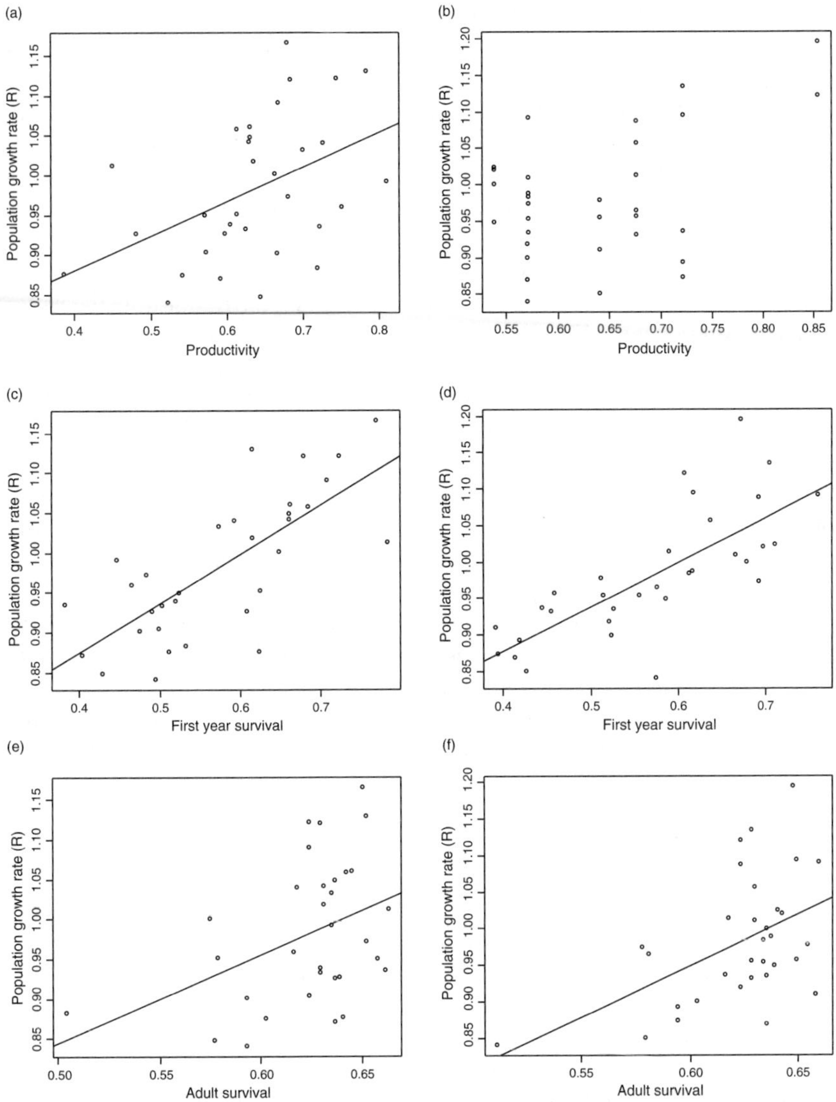

Fig. 4 Relationships between population growth rates and individual demographic parameters from two of our preferred models, AWR and AWB. Both models have year-specific first year survival rates and adult survival modelled as a logit-linear function of drought days. Productivity is modelled as either random annual effects (model AWR) or as constant within periods with uniform population growth rates (model AWB). Points are posterior means from the MCMC analysis. Lines show linear regressions where the 2.5 percentiles of the posterior correlation coefficients are greater than zero. (**a**) Productivity, model AWR; (**b**) Productivity, model AWB; (**c**) First-year survival, model AWR; (**d**) First-year survival model AWB; (**e**) Adult survival, model AWR; (**f**) Adult survival, model AWB

AWY, r = 0.915–0.927). Similarly the first year survival estimates between the two models with block-specific first year survival were well correlated (BWR vs BWB, r = 0.949). In contrast posterior mean first-year survival values from the model where first year survival was logit-linear function of frost days (WWR) were only moderately correlated with those from the other models (r = 0.501–0.606). The six best models all had adult survival modelled as a logit-linear function of drought, and posterior mean adult survival rates from all of these models were almost identical (r > 0.999 in all cases).

A key aim of this type of analysis is to quantify the contributions of changes in different demographic rates to overall population changes. To illustrate the patterns involved posterior means for annual population growth rates were plotted against posterior means for annual productivity, first year survival and adult survival . Sample plots for models AWR and AWB are presented in Fig. 4. The MCMC approach makes it straightforward to carry out formal tests of these relationships. Correlation coefficients for the above relationships were calculated separately using the results from each MCMC iteration. The distributions of these posterior correlation coefficients were then used to asses the strength of these relationships within each of our six preferred models (Table 2). Posterior mean correlations between population growth rate and productivity varied between 0.114 and 0.798, with the 2.5 percentile for three of the six models overlapping zero. In contrast posterior mean correlations between population growth rates and first year survival varied between 0.525 and 0.875, with only that from the model with the greatest delta DIC having a 2.5 percentile less that zero. Posterior mean correlations between population growth rate and adult survival were less that the equivalent correlations for first year survival in five out of six cases, and in all models the 2.5 percentile was greater than zero.

4 Discussion

4.1 Demographic Causes of the Song Thrush Decline

Analyses of the BTO's Song Thrush population data up to the 1980s indicated that there were no trends in nesting success or adult survival that could explain the decline (Baillie 1990). Indeed nesting success had actually increased, a pattern that has since been found in some other declining species and may be a density-dependent response to reduced abundance (Siriwardena et al. 2000). A subsequent analysis of first year and adult survival rates based on constant reporting rate models showed that the observed variation in first year survival was sufficient to explain the observed population decline (Thomson et al. 1997). Further more detailed analyses extended this work by including temporal variation in reporting rates and the estimation of post-fledging survival (Robinson et al. 2004). This block-specific analysis showed that population growth rates were well correlated with first year survival and more weakly correlated with post-fledging survival. There was no correlation with either productivity or adult survival.

The results from the present study are broadly consisted with these earlier analyses of this data set, with strong evidence that population growth rates are well correlated with first year survival. Adult survival is also correlated with population growth rates, although the correlation is weaker than for first year survival. Evidence for a correlation between productivity and population growth rate is equivocal, being sensitive to the choice of model within the set of models selected using DIC (Table 2). The three models where productivity is modelled using random annual effects all show a correlation between productivity and population growth rate but the other models do not. Productivity is essentially calculated by difference and it may therefore incorporate errors in the measurement of the other demographic parameters. Thus a correlation between population growth rate and productivity may partly reflect errors in other parameters or parameters that have not been measured adequately. While our use of a random effects model for productivity may have reduced problems of over-fitting, it will not solve this problem completely. Incorporation of direct measures of productivity from nest record cards or from juvenile to adult ratios in standardized catches could potentially improve the robustness of these results. It should be noted that our measure of productivity includes post-fledging survival (between ringing as a nestling and 63 days later), for which there is some previous evidence for a correlation with population growth rate. The addition of explicit estimates of post-fledging survival to our model would help to clarify this. This study suggests that all three demographic rates included in our model have some influence on population growth rates, as is to be expected. The main aim of such studies is to identify the relative contributions of these demographic parameters, as part of the process of constructing and testing hypotheses about the ecological causes of population changes.

To further explore the differences between the results of this study and those of Robinson et al. (2004) we examined the posterior correlations between population growth rates and demographic parameters using model BBB, with block-specific estimates similar to those from the earlier study. This model had poor support, being ranked 15 out of 17 on the basis of DIC (Table 1). Posterior mean correlation coefficients indicate that under this model population growth rate is most strongly correlated with first year survival ($r = 0.632$) and is only weakly associated with adult survival ($r = 0.227$) and productivity ($r = 0.127$), results that correspond well with the results from our preferred models, and also with those of Robinson et al. (2004). However, the HPDIs for all of these correlations substantially overlap zero.

The key environmental factors responsible for changes in the first year survival of Song Thrushes remain to be identified. Evidence for a range of candidate hypotheses that could explain this change has been reviewed elsewhere (Peach et al. 2004b; Robinson et al. 2004). Song Thrushes feed mainly on soil invertebrates, particularly earthworms, and reductions in the amount and quality of suitable feeding habitat are likely to have contributed to their decline. Important habitat changes are likely to include the loss of permanent pasture to arable and increased under-field drainage leading to more rapid drying out of the soil. There is also good evidence that Song Thrushes are affected by summer moisture levels in surface soils, which in turn affect the availability of their earthworm prey (Peach et al. 2004a). Furthermore, Peach et al. (2004b) demonstrated a strong negative relationship between the percentage of land subjected to under-field drainage and regional changes in Song Thrush populations. This evidence fits well with the relationship between summer droughts and adult survival, which in turn influences population changes. Similar factors may also have affected first year survival, although there is no direct correlation between first year survival and summer drought.

Variation in summer food supply may also affect components of our productivity measure, particularly post-fledging survival. Intensive field studies have demonstrated that in a rapidly declining farmland population Song Thrushes were making too few nesting attempts to sustain the population (Peach et al. 2004b) However, these results are based on comparisons between two local populations after the main national decline had taken place. In contrast, national nest record card data show that success per attempt has increased over the period of the decline (Baillie 1990; Baillie et al. 2006). These two results are not necessarily contradictory but further work is needed to clarify the possible role of productivbity changes in the decline of this species.

4.2 Utility and Development of Integrated Population Modelling

Integrated population modelling of the type presented here is a potentially powerful method for identifying the demographic mechanisms underlying population declines. The need to develop purpose written code currently restricts the speed with which these methods can be applied, and the ease of evaluating new models that may be proposed during the analysis. However, new general software for the these types of methods is developing rapidly, and with increased computing power it should become much easier to perform such integrated analyses in the future. The code that we have developed for the analyses presented here will make it much easier to conduct similar analyses of other BTO population data.

We plan to extend the current study in a number of directions. Obvious first steps are to add explicit estimation of post-fledging survival to the model and to extend our use of random effects, particularly within the co-variate relationships (Hoyle et al. in prep.; Fonnesbeck et al. 2008). We also plan to incorporate more structural relationships within the models, particularly density-dependence (Lebreton 2008), through the use of a hierarchical modelling approach. Only being able to fit a small number of models was a limitation of the present study, although we did examine more models than in most other applications of these techniques to-date. The best way forward here is likely to be the application of reversible-jump MCMC to explore large model spaces. The use of this technique within an integrated population modelling framework has already been demonstrated by King et al. (2006). However, a substantial amount of development work will be needed to apply this technique to the range of models envisaged here. In the short-term one alternative approach may be to use techniques based on the Kalman Filter (Besbeas et al. 2002), which would allow certain classes of models to be fitted more rapidly. Another may be to apply the approach used by Maunder

(2004) for analyses of fisheries data. He developed models that use random effects (or approximations using penalized likelihood) rather than the state space framework (although these are arguably the same) because population sizes were very large and demographic uncertainty was minor.

In addition to ringing and census data the BTO operates a Nest Record Scheme that provides detailed data on individual nesting attempts (Crick et al. 2003). The statistical models used to estimate various components of nesting success can potentially be incorporated in an integrated population model in a way that is analogous to the ring recovery analyses used in this paper (Besbeas unpublished). The analysis presented here is based on only summary statistics derived from the census data. In principle it is possible to fit an integrated population model to the full sites by years matrix from the census data, thus incorporating the full variance–covariance structure of these data within the analyses (Besbeas and Freeman 2006). We do not currently know how much useful information would be obtained by increasing the complexity of the modelling process in this way but it seems likely that this might differ between datasets. Maunder (2001) found that integrated fisheries models showed little difference in point estimates but considerable reductions in estimates of uncertainly. This is probably because the integrated approach takes the covariance among years into consideration. This could also be achieved if the covariance of the years from the census data were included using a multivariate normal likelihood (Besbeas et al. 2003). Finally there is the potential to extend these methods to different types of ornithological monitoring data, such as the Constant Effort Sites Scheme. This scheme provides data on relative abundance, juvenile to adult ratios and mark-recapture survival data all measured at the same sites. Work to develop integrated Bayesian population models for these types of data is currently in progress (Cave et al. 2008).

Acknowledgments This work was undertaken while Stephen Baillie was on sabbatical at the Universities of Cambridge and St Andrews. He is grateful for the Statistical Laboratory, University of Cambridge and the Centre for Ecological and Environmental Modelling, University of St Andrews, for the provision of facilities and to the BTO for supporting this work. Rob Robinson provided data and results from his earlier analyses of these data and Chiara Mazetta gave advice on the analysis of the CBC data. Steve Freeman, Rob Robinson, Takis Besbeas and Mark Maunder provided helpful comments on an earlier version of this manuscript. The Common Birds Census and the Ringing Scheme are undertaken through a partnership between the BTO and the Joint Nature Conservation Committee (on behalf of Natural England, Scottish Natural Heritage, the Countryside Council for Wales and the Department of the Environment for Northern Ireland). This work is only possible due to many thousands of hours of field-work by volunteer census workers and ringers. We are very grateful to all of the above for their help and support.

References

Baillie SR (1990) Integrated population monitoring of breeding birds in Britain and Ireland. Ibis 132:151–166.

Baillie SR (2001) The contribution of ringing to the conservation and management of bird populations: a review. Ardea 89 (special issue):167–184.

Baillie SR, Green RE (1987) The importance of variation in recovery rates when estimating survival rates from ringing recoveries. In North PM (ed.), Ringing Recovery Analytical Methods. Acta Ornithologica 23:41–60.

Baillie SR, Marchant JH, Crick HQP, Noble DG, Balmer DE, Coombes RH, Downie IS, Freeman SN, Joys AC, Leech DI, Raven MJ, Robinson RA, Thewlis RM (2006) Breeding Birds in the Wider Countryside: their conservation status 2005. BTO Research Report No. 435. BTO, Thetford. (http://www.bto.org/birdtrends)

Baillie SR, Peach WJ (1992) Population limitation in Palaearctic-African migrant passerines. Ibis 134(Suppl. 1):120–132.

Baker H, Stroud DA, Aebischer NJ, Cranswick PA, Gregory RD, McSorley CA, Noble DG, Rehfisch MM (2006) Population estimates of birds in Great Britain and the United Kingdom. British Birds 99:25–44.

Besbeas P, Freeman SN (2006) Methods for joint inference from panel survey and demographic data. Ecology 87:1138–1145.

Besbeas P, Freeman SN, Morgan BJT, Catchpole EA (2002) Integrating mark-recapture-recovery and census data to estimate animal abundance and demographic parameters. Biometrics 58:540–547.

Besbeas P, Freeman SN, Morgan BJT (2005) The potential of integrated population modelling. Australian and New Zealand Journal of Statistics 47:35–48.

Besbeas P, Lebreton J-D, Morgan BJT (2003) The efficient integration of abundance and demographic data. Applied Statistics 52:95–102.

Brooks SP, King R, Morgan BJT (2004) A Bayesian approach to combining animal abundance and demographic data. Animal Biodiversity and Conservation 27:515–529.

Brownie C, Anderson DR, Burnham KP, Robson DS (1985) Statistical inference from band recovery data – A handbook, second edition. United States Department of the Interior Fish and Wildlife Service, Resource Publication 156, Washington, DC.

Buckland ST, Newman KB, Thomas L, Kösters N (2004) State-space models for the dynamics of wild animal populations. Ecological Modelling 171:157–175.

Caughley G (1994) Directions in conservation biology. Journal of Animal Ecology 63:215–244.

Cave VM, Freeman SN, Brooks SP, Balmer DE (2008) On adjusting for missing visits in the indexing of abundance from 'Constant Effort' ringing. In: Thomson DL, Cooch EG, Conroy MJ (eds.) Modeling Demographic Processes in Marked Populations. Environmental and Ecological Statistics, Springer, New York, 3:949–964

Crick HQP, Baillie SR, Leech DI (2003) The UK Nest Record Scheme: its value for science and conservation. Bird Study 50:254–270.

Fewster RM, Buckland ST, Siriwardena GM, Baillie SR, Wilson JD (2000) Analysis of population trends for farmland birds using generalized additive models. Ecology 81:1970–1984.

Fonnesbeck CJ, Edwards HH, Reynolds III JE (2008) Hierarchical covariate models for detection and availability of Florida manatees at a warm water aggregation site. In: Thomson DL, Cooch EG, Conroy MJ (eds.) Modeling Demographic Processes in Marked Populations. Environmental and Ecological Statistics, Springer, New York, 3:563–578

Freeman SN, Crick HQP (2003) The decline of the Spotted Flycatcher *Muscicapa striata* in the UK: an integrated population model. Ibis 145:400–412.

Freeman SN, Robinson RA, Clark JA, Griffin BM, Adams SY (2007) Changing demography and population decline in the Common Starling *Sturnus vulgaris*: a multisite approach to integrated population monitoring. Ibis 149:587–596.

Gibbons DW, Reid JB, Chapman RA (1993) The New Atlas of Breeding Birds in Britaiin and Ireland: 1988–1991. T & A D Poyser, London.

Green RE (1995) Diagnosing causes of bird population declines. Ibis 137(Suppl.):S47–S55.

Green RE (1999) Applications of large-scale studies of demographic rates to bird conservation. Bird Study 46(Suppl.):S279–S288.

Hoyle SD, Maunder MN, Veran S, Lebreton J-D (in prep.) A Bayesian integrated population dynamics model to evaluate the impact of fisheries on black-footed albatross using all available information.

King R, Brooks SP, Mazzetta C, Freeman SN (2006) Identifying and diagnosing population declines: a Bayesian assessment of Lapwings in the UK. Technical Report, University of St Andrews.

Lebreton JD (2008) Assessing density-dependence: where are we left? In: Thomson DL, Cooch EG, Conroy MJ (eds.) Modeling Demographic Processes in Marked Populations. Environmental and Ecological Statistics, Springer, New York, 3:19–42

Maunder MN (2001) A general framework for integrating the standardization of catch-per-unit-of-effort into stock assessment models. Canadian Journal of Fisheries and Aquatic Science 58:795–803.

Maunder MN (2004) Population viability analysis, based on combining integrated, Bayesian, and hierarchical analyses. Acta Oecologica 26:85–94.

Morgan BJT (2008) Completing the ecological jigsaw. In: Thomson DL, Cooch EG, Conroy MJ (eds.) Modeling Demographic Processes in Marked Populations. Environmental and Ecological Statistics, Springer, New York, 3:513–540

Peach WJ, Denny M, Cotton PA, Hill IF, Gruar D, Barritt D, Impey A, Mallord J (2004a) Habitat selection by song thrushes in stable and declining farmland populations. Journal of Applied Ecology 41:275–293.

Peach WJ, Robinson RA, Murray KA (2004b) Demographic and environmental causes of the decline of rural song thrushes *Turdus philomelos* in lowland Britain. Ibis 146(Suppl. 2):50–59.

Robinson RA, Green RE, Baillie SR, Peach WJ, Thomson DL (2004) Demographic mechanisms of the population decline of the song thrush *Turdus philomelos* in Britain. Journal of Animal Ecology 73:670–682.

Siriwardena GM, Baillie SR, Wilson JD (1999) Temporal variation in the annual survival rates of six granivorous birds with contrasting population trends. Ibis 41:621–636.

Siriwardena GM, Baillie SR, Crick HQP, Wilson JD (2000) The importance of variation in the nesting success of seed-eating birds in determining their population trends on farmland. Journal of Applied Ecology 37:128–148.

Spiegelhalter DJ, Best NG, Carlin BP, Van der Linde A (2002) Bayesian measures of model complexity and fit (with discussion). Journal of the Royal Statistical Society, Series B 2002 64:583–616.

Thomas L, Buckland ST, Newman KB, Harwood J (2005) A unified framework for modelling wildlife population dynamics. Australian and New Zealand Journal of Statistics 47:19–34.

Thomson DL, Baillie SR, Peach WJ (1997) The demography and age-specific annual survival of song thrushes during periods of population stability and decline. Journal of Animal Ecology 66:414–424.

White GC, Burnham KP (1999) Program MARK: survival estimation from populations of marked animals. Bird Study 46:S120–S139.

A Hierarchical Covariate Model for Detection, Availability and Abundance of Florida Manatees at a Warm Water Aggregation Site

Christopher J. Fonnesbeck, Holly H. Edwards, and John E. Reynolds III

Abstract We constructed a Bayesian hierarchical model for estimating the population size and associated probabilities of availability and conditional detection for Florida manatees aggregating during winter, based on a series of monitoring flights over 3 years, 2001–2003. Building upon the findings of Edwards et al. (2007), our approach combines four sources of monitoring data in a single integrated modeling framework to estimate all model parameters simultaneously. Population size was modeled as a function of availability and detection, which in turn were estimated with covariate models consisting of environmental predictor variables. Previous work estimating manatee abundance from aerial surveys have either serially combined parameters estimated in separate models (Edwards et al. 2007), modeled availability and detection jointly (Craig and Reynolds 2004) or ignored detection bias altogether. Time-specific estimates of availability were high, with some variation among flight series, while estimates of conditional detection were extremely variable from one survey to the next. We obtained improved precision in our estimates of population size relative to Edwards et al. (2007). Our results emphasize the consequences of ignoring detection bias when interpreting survey counts. We hopc that this research will be influential in the design of a new state-wide aerial survey monitoring program for Florida manatees.

1 Introduction

For more then 40 years aerial surveys have been used to assess the distribution and obtain counts of the Florida manatee (*Trichechus manatus latirostris*) population. However, until recently, the utility of aerial surveys for assessing the status of the manatee population was limited because an unknown number of animals went undetected by the observers, resulting in population underestimation (Hartman 1974; Caughley 1977; Eberhardt 1982; Packard et al. 1989; Pollock and Kendall 1987;

C.J. Fonnesbeck (✉)
Department of Mathematics and Statistics, PO Box 56, University of Otago, Dunedin, New Zealand
e-mail: Chris.Fonnesbeck@MyFWC.com

D.L. Thomson et al. (eds.), *Modeling Demographic Processes in Marked Populations*, Environmental and Ecological Statistics 3, DOI 10.1007/978-0-387-78151-8_24,

Lefebvre et al. 1995). The ability of an observer to detect and count manatees during an aerial survey depends on: (1) animals being near the water's surface where they can be seen and counted by the observer, given that they are present (otherwise inducing availability bias); and (2) animals being seen, given that they are available to be detected by the observer (otherwise inducing perception bias; Marsh and Sinclair 1989; Lefebvre et al. 1995; Pollock et al. 2004, 2006). Most aerial survey methods for manatees do not account for either of these biases. However, recent work by Edwards et al. (2007) and Craig and Reynolds (2004) has attempted to estimate components of detectability to improve manatee abundance estimates at winter aggregation sites. Edwards et al. (2007) partitioned and estimated the different components of manatee detection probability (presence at the aggregation site, availability of an individual animal, and then detection conditional on presence and availability of an individual animal) at one warm-water refuge. Craig and Reynolds (2004) analyzed repeated winter counts of manatees at power plants over 20 years using a Bayesian hierarchical model that allowed for uncertain detection that varied by region and by water temperature.

In winter, Florida manatees aggregate at warm-water outfalls of power plants and natural springs to avoid the cold ambient temperatures of surrounding waters. Unlike other marine mammals, sirenians (Order Sirenia, i.e. dugongs and manatees) are poorly adapted to tolerate cold. Because of their low metabolic rate and high thermal conductivity, manatees seek warmer water when water temperatures drop below approximately 20°C, thereby avoiding cold stress injury or death. Since the middle of the last century, Florida manatees have become dependent on the warm-water effluents from power plants and other artificial warm-water sources for survival in winter (Irvine 1982). In 2006, for example, 82% of manatees counted during the statewide synoptic survey were seen at power plants (FWC, unpublished data). Aerial surveys that intensively cover these winter aggregation sites (Packard et al. 1985) have been useful in obtaining minimum population counts (Shane 1984; Packard et al. 1989; Garrott et al. 1994, 1995; Reynolds III and Wilcox 1994; Ackerman 1995). However, they have not been useful for obtaining reliable population estimates or indices because counts obtained from surveys are not adjusted to account for imperfect detection of manatees by the observers. The resulting estimates are therefore not appropriate for either assessing population status or trend analysis (Anderson 2001, 2003).

Availability and detection of manatees during aerial surveys are influenced by several factors, most related to environmental conditions. Results of previous studies at the Tampa Electric Big Bend power plant (TECO) in Tampa Bay (e.g. Edwards et al. 2007) have illustrated the influence that environmental conditions, mainly weather, have on manatee behavior. Temperature, cloud cover and wind can influence a manatee's dive interval, dive depth and basking behavior at the plant. Some of these covariates also influence the timing of migration to the power plant each winter (Deutsch et al. 2003). Since manatees are sensitive to cold, their behavior in winter is almost entirely dominated by their need to thermoregulate. (Deutsch et al. (2003) found that during very cold weather (water temperature ~13–14°C) manatees forego feeding and spend 88% of their time inside the warm-water discharge of the

TECO Big Bend power plant. Tampa Bay aerial survey studies found that counts are almost always higher in the afternoon when sun angle and air temperatures are highest (Edwards et al. 2007; Wright et al. 2002); highest counts were obtained on warm, sunny days, following a cold front, when manatees were basking at the water's surface. On cold, windy and cloudy days, manatees bottom-rested for up to 20 min, making it difficult or impossible to fully enumerate them. Even when animals are at the surface, detection is not guaranteed. Factors that can affect the ability of an observer to see manatees include: environmental conditions (e.g. water turbidity, sea state), air speed and altitude of the aircraft, sun angle, observer fatigue and observer experience (Lefebvre et al. 1995).

Recent research by Edwards et al. (2007) has attempted to account for biases in detectability by calibrating winter survey counts at the TECO Big Bend power plant using estimates of availability and detection based on marked animals. Extending this work, we seek to improve estimates of abundance by (i) developing general covariate models for each of the components of detection and (ii) integrating information from multiple data sources to improve inference. These goals are realized with a single, unified modeling framework.

2 Methods

2.1 Study Area

We conducted an aerial survey study of manatees at the coal-fired TECO Big Bend power plant at Apollo Beach, Florida (Fig. 1), during three winters (December 2000 to March 2003, see Edwards et al. 2007, for description). The Big Bend plant is one of 5 sites along the west coast of Florida that regularly attracts more than 100 manatees during winter (Fig. 2; U.S. Fish and Wildlife Service, 2001). Of these sites, aerial surveys are most difficult at this location; the location of the power plant relative to the discharge canal makes it difficult to survey efficiently, and the water depth and turbidity in the discharge canal where the animals aggregate greatly reduce the observers' ability to see submerged or partly-submerged manatees. The canal dimensions were approximately 1006 m long × 96 m wide × 8 m deep.

2.2 Field Methods

2.2.1 Marking Manatees

Manatees were opportunistically captured and marked in December of 2000, 2001, and 2002 in or near the Big Bend plants thermal discharge canal (see Edwards et al. 2007, for details). The animals were beached for a short time so that their health could be assessed and tracking and marking gear could be attached (Deutsch et al. 1998; Weigle et al. 2001). Fifteen manatees were marked in 2000, 15 in 2001, and 16 in 2002. A 56 × 33 cm colored vinyl flag (green, yellow, white, or

Fig. 1 Aerial photograph of TECO power plant, showing discharge canal on the left

Fig. 2 Aerial view of manatees at the east end of the TECO Big Bend power plant discharge canal

red and white) inscripted with a large, unique black symbol ("X", diagonal or vertical line, square, or solid background with no symbol) was attached to the tail of the animals to mark individuals. When the animals were at or near the surface of the water, the flags were visible to both ground and aerial observers. In addition, time-depth-temperature archival data loggers (hereafter, TDR; model LTD-100 Archival Data Logger, Lotek, Inc., St. Johns, Newfoundland, Canada) were deployed on 5 marked manatees in each year (n=15). The archival tags recorded pressure (PSI) and temperature (C) every 30 s (see Edwards et al. 2007, for details). Pressure readings were converted to depth (m) by $d = p/1.458$.

2.2.2 Aerial and Ground Surveys

Several series of aerial surveys (flight series) were conducted, generally on consecutive days before and after the passage of strong cold fronts (2000–2001, n = 29 surveys, 3 flight series; 2001–2002, n = 15 surveys, 1 flight series; 2002–2003, n = 29 surveys, 2 flight series). Surveys were flown once in the morning and once in the afternoon (10 passes over the canal during each flight at approximately 1000 and 1400 h) to assess the temporal variance of survey counts. Observers had approximately 1 min to count manatees during each pass using intensive search methods (Packard et al. 1985). Each 10-pass survey lasted about 30 min. To avoid double counting, each group was counted separately over a few of the passes, using hand counters for large groups. Large groups were circled until the observer was satisfied that all animals in that group had been counted. Once a group was counted, the observer moved to another group and proceeded down the canal until all groups were counted. Biases introduced by mixing of individuals among groups are avoided if counting of the groups is done with short intervals between counts (Edwards et al. 2007).

All surveys were flown by one of five experienced aerial observers (>150 h of manatee aerial survey experience and >50 counts at power plants) in a high-winged Cessna 172 aircraft at an altitude of about 500 ft (152 m) and a ground speed of 60–70 knots (111–130 km/hr). Each set of surveys began 1–2 days prior to a strong cold front (weather permitting) and ended 3–15 days later depending on the length of the cold period. A second observer in the back seat, seated on the same side of the plane as the primary observer recorded the sightings of individually-marked (flagged) manatees. These sightings were recorded and mapped separately for each pass and combined for the 10 consecutive passes. Herd structure and social organization are not well-defined in manatees (Hartman 1979; Bengtson 1981); individuals join and leave groups frequently, therefore we believe that detection of one individual is independent of the detection of another (Edwards et al. 2007). However, synchronous diving and surfacing may occur, particularly in response to disturbance.

During the surveys, two land observers were positioned on an elevated platform (approximately 3.5 m above the water) overhanging the east end of the discharge canal. These observers documented the presence of flagged animals during the surveys to compare sightings of manatees in a defined aggregation area with those of the aerial observers. The land observers were present at the power plant throughout the day and recorded the time and position of all flagged manatees seen, in addition to those within their view (within a predetermined area) during each aircraft pass over the canal during the survey. On survey days the ground observers were present at the plant observing the manatees from about 0830 to 1630 h. This was adequate time for the observers to move around the discharge canal and document all marked animals present in the aggregation area. Any flagged manatees that were known (from telemetry or visual sightings) to be away from the power plant, or for which the flag gear was known to have been lost, were subtracted from the total number present at the plant on that day. All marked animals present at the plant on any given day were assumed to have been sighted by either the on-ground or the aerial observers (i.e., perfect detection). In a portion of the survey area near the elevated platform, it was possible for both aerial and on-ground observers to simultaneously count all flagged animals seen from their respective vantage points as the plane passed over pre-specified start and stop points. Air-to-ground radios were used to coordinate the counting between all observers. During each of 10 passes over the discharge canal, both aerial and ground observers recorded sightings as described above. Three separate counts were obtained from each pass by cross-referencing and comparing flagged animals mapped by the aerial and ground observers: (i) animals seen from both the air and ground, (ii) animals seen from the air only, and (iii) animals seen from the ground only (Edwards et al. 2007).

2.2.3 Environmental Data

We recorded a variety of environmental parameters to help assess the effects of weather and other environmental factors on our aerial survey counts. Air temperature (°C), wind speed (kts) and wind direction were recorded before take-off of every flight as reported by the Federal Aviation Administration Automated Surface Observation System. Cloud cover (estimated as the percentage of the sky obscured by clouds) was recorded from the airplane just prior to each survey; for use in the model this variable was discretized to a binary value representing the presence or absence of cloud cover. Environmental readings were recorded at 1000 and 1400 h to best reflect the conditions during the surveys. Optic StowAway digital temperature probes (Onset Computer Corp., N. Falmouth, MA, USA) collected water temperatures at three sites around the Big Bend power plant. Two probes were placed within the heated discharge canal, one at the platform at the east end of the canal where manatees aggregate, and the other mid-way down the canal along a shallow area near a sand flat where manatees frequently rested. A third probe was placed at the intake of the plant to record ambient water temperatures from Tampa Bay.

2.3 Modeling and Estimation

The number of manatees using the TECO Big Bend power plant on a given day in winter can be estimated by modeling a relationship between the number of animals C_i counted during each survey i and the true, latent abundance N_i. Recognizing that detection is imperfect, this relationship is mediated by the probability p_i of detecting manatees at the site during survey i:

$$C_i = N_i p_i. \tag{1}$$

During aerial surveys of manatees, some animals present at the site are submerged, thus not available for detection by observers (i.e. $p_i = 0$). It is therefore useful to express p as the product of two quantities: (i) the probability of an animal being at or near the surface (availability) $p_i^{(a)}$ and (ii) the conditional probability of detection, given availability $p_i^{(d|a)}$:

$$p_i = p_i^{(a)} p_i^{(d|a)} \tag{2}$$

Edwards et al. (2007) estimated availability by assuming that probabilities of a manatee being available for detection during each of 10 passes were identical and independent. This yields an estimate of availability that is calculated as the complement of the probability of non-detection during all 10 passes:

$$p_i^{(a)} = 1 - (1 - \theta_i)^{10} \tag{3}$$

where θ_i is the probability of availability for a single pass during survey i. However, the assumption of independence for each pass may not be appropriate. Examination of dive profiles reveals a pattern of autocorrelation among TDR readings; manatees at the surface at any point during the time series tend to remain at the surface in subsequent readings, while those diving tend also to be submerged in one or more subsequent readings.

To account for observed autocorrelation, we instead modeled dive behavior as a stochastic process. First, the proportion of animals available at the beginning of the survey θ_i was estimated by the proportion of TDR readings within 2 m of the surface during survey hours, according to a binomial distribution:

$$x_i^{(\text{up})} \sim \text{Bin}(x_i^{(\text{up})} + x_i^{(\text{down})}, \theta_i) \tag{4}$$

where $x_i^{(\text{up})}$ and $x_i^{(\text{down})}$ are the numbers of TDR readings at and below the surface, respectively. The threshold value of 2 m was chosen because it is an approximation of the deepest plausible depth at which the TDR (attached above the tail of an adult animal) could be submerged, with the head or torso still visible near the water's surface. Smaller values resulted in continuous dives of unrealistic length (>50 min), suggesting that the animal was actually at the surface even when the TDR was below the threshold. Moreover, due to lack of water clarity, animals are rarely visible in the canal when submerged deeper than 2 m. Second, to estimate availability during the entire survey, we estimated the probability of a submerged animal resurfacing at least once during the survey. This probability was derived by modeling surfacing events for submerged animals as a Poisson process, with mean and variance λ; this corresponds to an exponential model of waiting time with parameter λ (Ross 1996). From this, it is straightforward to model the probability of no surfacing events u during the period of the survey t:

$$\begin{aligned} Pr(u \mid \lambda, t) &= \frac{e^{-\lambda_i t}(\lambda t)^u}{u!} \\ Pr(u = 0 \mid \lambda, t) &= e^{-\lambda t} \end{aligned} \tag{5}$$

The fraction of unavailable animals during survey i is the product of the probability of being submerged at the beginning of the survey and the probability of not surfacing at any point during the survey. Thus, the probability of availability $p_i^{(a)}$ is simply the complement of this product:

$$p_i^{(a)} = 1 - (1 - \theta_i)e^{-\lambda_i t} \tag{6}$$

We estimated that the mean time for conducting a 10-pass survey was about 30 min. Estimates for λ_i were derived based on 2-h periods of TDR monitoring, corresponding approximately to the periods during which surveys were conducted (09:00–11:00 and 13:00–15:00), therefore $t = 0.25$. This model of surfacing behavior assumes that there are no "instantaneous" surfacing events between passes that cannot be detected by observers. While we concede that this assumption may be violated, we contend that (a) this model is more biologically appropriate than the independent surfacing probabilities model in (3) (Edwards et al. 2007), based on inspection of manatee dive profiles and (b) such brief surfacing events, should they occur, may not fairly be considered periods of availability, given the limited opportunity to view such events. This implicitly defines availability as prolonged periods of surface activity.

In an effort to account for extra-binomial variation and to aid prediction, covariate models were developed for both λ_i and θ_i. In both cases, linear models were constructed to predict availability parameters based on environmental covariates collected during the study. The proportion of time spent at the surface θ_i was modeled as a function of air temperature (air), wind speed (wnd) and a fixed effect for the flight series s of which the current survey s_i was a member, to account for unmeasured temporal variation:

$$\text{logit}[\theta_i] = \gamma_0 + \gamma_1 \text{wnd}_i + \gamma_2 \text{air}_i + \gamma_{3i} I(s_i) \tag{7}$$

Here, $I(s_i)$ is an indicator function that is equal to 1 when survey i is in flight series s_i, and zero otherwise. Canal and discharge temperatures were standardized and the logit link function was used to guarantee probabilities on the unit interval. Similarly, the Poisson parameter λ from (6) was estimated as a linear function of the same covariate set:

$$\log[\lambda_i] = \alpha_0 + \alpha_1 \text{wnd}_i + \alpha_2 \text{air}_i + \alpha_{3i} I(s_i) \tag{8}$$

with a log link function applied to ensure non-negative values of λ_i. Because θ_i and λ_i are modeled from the same set of covariates, it is possible that the resulting estimates may be correlated. Coping with such correlations is an additional advantage conferred by a joint estimation framework.

The unconditional (on availability) probability of detection p_i was estimated from the number of flagged manatees known to be present at the site y_i during survey i, based on counts from land observers, and those that were actually observed from the air during survey flights x_i. These were related in a binomial likelihood:

$$x_i \sim \text{Bin}(y_i, p_i) \tag{9}$$

where the unconditional detection rate p_i is calculated from (2). The conditional probability of detection $p_i^{(d|a)}$ was calculated as a linear function of factors thought to be related to an observer's ability to see manatees, which included wind speed, cloud cover (cld) and n_s-1 flight series fixed effects:

$$\text{logit}[p_i^{(d|a)}] = \beta_0 + \beta_1 \text{wnd}_i + \beta_2 I(\text{cld}_i) + \beta_{3i} I(s_i) \tag{10}$$

With estimates of detection, we are able to correct the total count of manatees during each flight to yield an estimate of abundance at the power plant for each survey, following (1). Finally, we estimated the mean abundance μ_s for each flight series by modeling the individual survey abundances N_i, $i = 1, \ldots, n_s$ as negative binomial random variables:

$$f(N_i) = \frac{\Gamma(\omega_s + N_i)}{\Gamma(\omega_s) N_i!} \left[\frac{\mu_s}{\mu_s + \omega_s}\right]^{N_i} \left[\frac{1}{1 + \mu_s/\omega_s}\right]^{\omega_s} \tag{11}$$

with positive-valued parameter ω_s.

Estimation of model parameters was via Markov chain Monte Carlo methods (MCMC, Gilks et al. 1996; Gamerman 1997). The model was implemented in PyMC (http://pymc.googlecode.com), a MCMC module for the Python programming language (http://python.org) that implements a random-walk Metropolis–Hastings sampler. All model parameters were assigned uniform (non-informative) prior distributions; for covariate model parameters these were sparse normal priors (precision $\tau = 0.001$), while non-negative-valued parameters were given uniform priors over [0, 1000]. The model was run for 300,000 iterations, with the first 100,000 discarded as a burn-in interval during which proposal distribution variances were tuned to achieve optimal mixing. A time series analysis was conducted to detect evidence of non-convergence, following Geweke (1992); no evidence of lack of convergence was discovered for any model parameters. Finally, goodness-of-fit was assessed by comparing the deviance of the data used to fit the model to values simulated from the model, based on the estimated parameters (Gelman et al. 1996). Results of this test suggested adequate fit for the model.

3 Results

3.1 Covariate Models

The overall probability θ_i of an individual being at the surface was estimated to be negatively correlated with wind speed γ_1 and positively with air temperature γ_2 (Table 1), both with 95% Bayesian credible intervals (BCI) that did not include zero. Conversely, the Poisson mean frequency λ_i of surfacing events was positively correlated with wind speed α_1 and negatively with air temperature α_2. However, the latter 95% BCI included zero, and more than 25% of the posterior distribution of values were positive (Table 1).

Wind speed and cloud cover were modeled as covariates to the conditional probability of detection. Estimates suggest lower rates of detection at higher wind speeds, and higher rates during cloudy conditions.

3.2 Availability and Detection

The probability of availability $p_i^{(a)}$ for each flight, as calculated in (6) ranged from a low of 0.826 (0.786, 0.866) on 28 January 2003 to a high of 0.950 (0.920, 0.977)

Table 1 Estimates of linear mixed model parameters of covariate models for at-surface probability (top), surfacing frequency (middle), and conditional detection probability (bottom) within the MCMC estimation framework, with standard error and 95% Bayesian credible interval. All estimates are on the logit scale

				95% BCI	
Model	Parameter	Estimate	SE	Lower	Upper
Surface Proportion (θ)	intercept (γ_0)	–0.934	0.032	–0.998	–0.871
	wind (γ_1)	–0.095	0.014	–0.124	–0.068
	air temperature (γ_2)	0.504	0.018	0.468	0.539
	series 2 (γ_{30})	–1.689	0.077	–1.843	–1.541
	series 3 (γ_{31})	–0.389	0.042	–0.474	–0.308
	series 4 (γ_{32})	0.720	0.054	0.616	0.826
	series 5 (γ_{33})	0.155	0.043	0.072	0.242
	series 6 (γ_{34})	–0.452	0.042	–0.534	–0.367
Surfacing Frequency (λ)	intercept (α_0)	2.145	0.069	2.011	2.278
	wind (α_1)	0.065	0.032	0.003	0.127
	air temperature (α_2)	–0.014	0.038	–0.090	0.059
	series 2 (α_{30})	–0.026	0.123	–0.266	0.213
	series 3 (α_{31})	0.060	0.087	–0.106	0.231
	series 4 (α_{32})	–0.067	0.132	–0.324	0.187
	series 5 (α_{33})	–0.251	0.100	–0.444	–0.052
	series 6 (α_{34})	–0.253	0.092	–0.433	–0.073
Detection ($p^{(d\|a)}$)	intercept (β_0)	0.628	0.251	0.155	1.126
	wind (β_1)	–0.610	0.103	–0.818	–0.415
	cloud cover (β_2)	0.694	0.215	0.260	1.108
	series 2 (β_{30})	–0.748	0.513	–1.723	0.283
	series 3 (β_{31})	0.460	0.356	–0.256	1.148
	series 4 (β_{32})	1.631	0.625	0.491	2.806
	series 5 (β_{33})	0.464	0.436	–0.345	1.270
	series 6 (β_{34})	0.094	0.329	–0.550	0.742

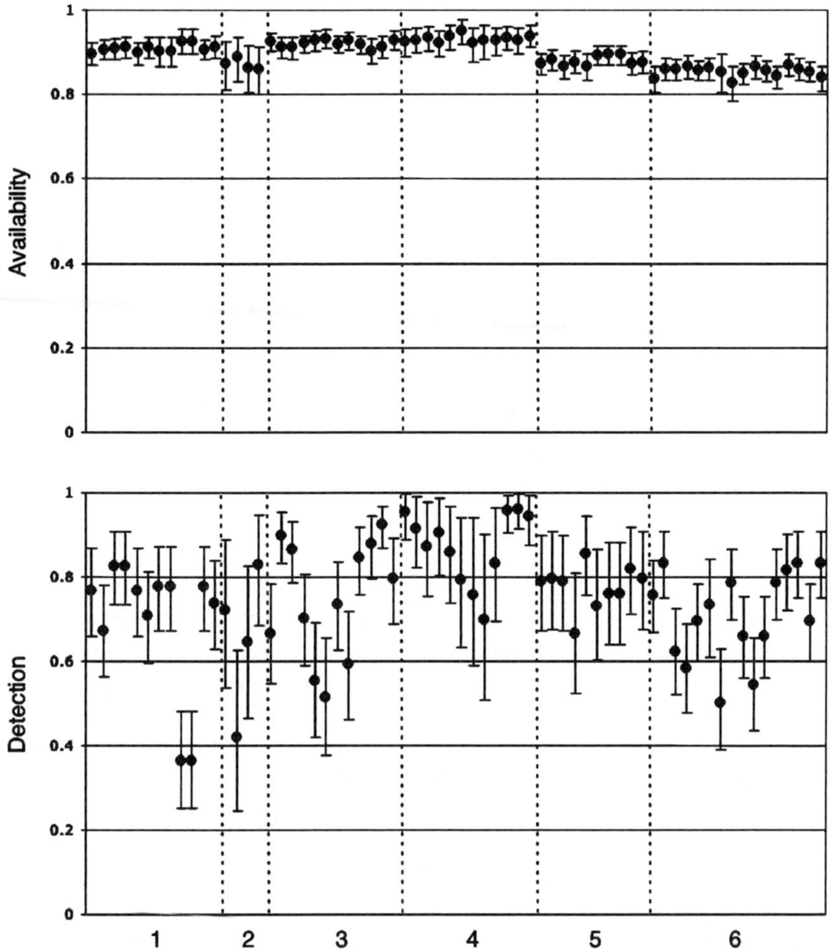

Fig. 3 Estimated probabilities of availability (*top*) and detection given availability (*bottom*) for each flight in sequence, 2001–2003. Estimated mean values are shown with 95% BCI. Dotted vertical lines group individual flights into 6 series

on 1 March 2002. Availability was consistently high, with greater variation among flight series than within (Fig. 3). Estimates of the probability of detection $p_i^{(d|a)}$, given availability, ranged from 0.364 (0.248, 0.481) on 9 January 2001 to 0.962 (0.922, 0.996) on 6 March 2002. There was strong variation in conditional detection among flights, independent of flight series (Fig. 3).

3.3 Abundance

The estimated power plant manatee abundances during each flight series, estimated as the expected value of a log-normal distribution, are shown in Table 2. The lowest abundance was during the lone 2002 flight series, 117.64 (92, 144), while the highest

Table 2 Estimates of manatee abundance at the TECO power plant for each flight series from this research and from (Edwards et al. (2007), with associated standard errors

		This Study		Edwards et al. (2007)	
Series	Year	Estimate	SE	Estimate	SE
1	2001	224.73	29.81	239.30	40.64
2	2001	340.02	76.06	360.78	107.23
3	2001	234.81	26.18	238.53	37.03
4	2002	117.64	13.56	135.27	17.81
5	2003	202.94	16.98	288.03	35.65
6	2003	329.43	32.22	352.12	35.26

was during the second January 2001 series, 340.02 (201, 495). Note that this series was also the least-precisely estimated, yielding a standard error more than 3 × larger than for any other estimate.

4 Discussion

This work brings four sources of information to bear on the problem of estimating manatee abundance at the primary warm-water aggregation site in Tampa Bay: (1) time-depth recorder readings, (2) sightings of known numbers of flagged manatees, (3) total counts of manatees using the TECO Big Bend discharge canal and (4) a suite of environmental covariates recorded during each survey. By implementing a hierarchical model, we were able to integrate these datasets into a single estimation framework. Previous analyses by Edwards et al. (2007) estimated each of the components of detection separately, then combined the estimates to correct survey counts for an estimate of abundance. By making more efficient use of information with the hierarchical model, we were able to derive more precise estimates and develop covariate relationships that may be useful in future survey analysis efforts.

Relative to those of Edwards et al. (2007), our point estimates of abundance during each of the six flight series were lower (Table 2). However, in all but one series (5) the 95% BCI included the estimated value from the previous work. In all series, the standard error associated with each estimate of abundance was smaller for this study. Our lower estimates of abundance (with the exception of series 2) are likely attributable to generally higher estimates of availability in each of the flight series. Recall that Edwards et al. (2007) assumed independent probabilities of availability during each pass within a survey, while we allowed for dependence in availability among passes, modeling the probability of resurfacing as a stochastic process. The validity of our availability model rests both on the assumptions regarding surfacing behavior mentioned in the Methods section, and on the representativeness of the sample of animals fitted with time-depth recorders. Though there are a large total number of readings, these were derived from a small number of subjects, leaving open the possibility of sampling bias.

As this is the same dataset employed by Edwards et al. (2007), sampling error should not be responsible for differences in estimated availability between the studies.

As we expected, the linear covariate model for detection revealed a negative association of conditional detection with wind and higher detection in cloudy conditions (Table 1). Windy conditions cause disturbances on the water's surface, making sighting difficult, particularly for animals that are partly submerged. Manatees also are likely to remain submerged on windy days to avoid its chilling effect, so we expect the proportion of time on the surface to be reduced. Though counter-intuitive to the views of some observers (sunlight is thought to increase visibility into the water column), overcast or partly-cloudy skies can reduce glare, thereby increasing visibility. The resulting estimates of detection are highly variable among flights, with little consistency within flight series (Fig. 3). This variability underlines the importance of estimating detection for manatee surveys, and reinforces the importance of survey conditions. Without such estimates, researchers are forced to make unreasonable assumptions about the magnitude and temporal variability of these biases.

The extreme variability in detection speaks against pooling data within flight series, as done by Edwards et al. (2007). This highlights an advantage of the hierarchical modeling approach for such analyses. Data are often pooled when they are sparse, in spite of the level of heterogeneity among the data. The hierarchical covariate model avoids this trade-off by modeling a common model intercept among flights, but accounting for heterogeneity via covariate relationships. Thus, each detectability estimate borrows strength from the entire dataset without unduly sacrificing precision. This is a primary advantage of using Bayesian estimation via MCMC rather than separately fitting each model component individually.

In comparison, availability proved to be far less variable, with more variation among flight series than within (Fig. 3). The environmental covariates, wind speed and air temperature, influenced at-surface proportion θ and surfacing frequency λ each in opposite directions (Table 1). This may have contributed to the relatively consistent estimates of availability, although the covariates more strongly influenced at-surface proportion. However, observing that the strongest pattern of variation was among flight series suggests that unmeasured covariates varying over a coarser time scale may be more appropriate predictors of availability than wind speed and air temperature.

Further refinement of our model might be achieved by applying model selection e.g. using DIC, Spiegelhalter et al. 2002) to a suite of alternative model parameterizations. Though most of the key covariate parameters, save some of the flight series fixed effects, had 95% credible intervals that excluded zero, a more formal model selection approach may identify variables that do not contribute to better inference. This is not of immediate concern, since the covariates employed in the model were relatively straightforward and inexpensive to obtain, but we concede that model selection could yield an improved model with additional precision. This could include a comparison of alternative functional forms for the distribution of survey population estimates $\{N_i\}$.

The current draft of the Florida manatee management plan (FWC, in prep.) identifies population estimation based on aerial surveys as a research priority, to serve as a monitoring tool for the recovery of species. This research demonstrates the potential for estimating abundance at manatee aggregation sites based on counts that are subject to availability and perception biases. Ultimately, we hope that these results will provide guidance in the design of a new state-wide survey for obtaining reliable estimates of population status. While some aspects of this work are specific to the TECO Big Bend aggregation site, we believe some results to be more general, and therefore, applicable at other similar manatee aggregation sites. The covariate relationships for detection and availability, for example, likely hold true independent of location since wind, air temperature and cloud cover ought to affect these probabilities similarly across space. This notion should be tested with smaller-scale validation studies at other locations. Moreover, it is encouraging that informative models of variation in detection and availability can be constructed using a set of covariates that is cheaply and easily obtained. However, we recognize that correcting for incomplete detection and availability is just one of many issues that impede the use of aerial surveys as a state-wide population monitoring tool (Lefebvre et al. 1995). These problems must be fully addressed in a robust sampling design that accounts for manatees both at and away from power plants during the winter. It may be that a suite of methods is needed to survey across a range of expected manatee densities in a variety of habitats; the methods described here may not be appropriate for very sparse populations away from aggregation sites, for example. Plans for such a redesigned survey are currently in preparation, and we are optimistic that aerial surveys can play a more prominent role in the conservation of the endangered Florida manatee.

Acknowledgments We thank K. Fick, K. Higgs, M. Martz, S. McDonald, S. Tarr, the late E. Argo, and many volunteers for their assistance with ground counts at the TECO Big Bend plant. We are grateful to B. Ackerman, J. Powell, L. Tipton for flying surveys and K. Brill, K. Frisch, K. Higgs, P. Haase, J. Koelsch, A. Smith, and many others for their help with flag-counting during aerial surveys. We thank the staff and pilots at Albert Whitted Flying Club and Bay Air Flight Services in St. Petersburg, FL, for their assistance with flights. Financial support was provided by the FWC Save the Manatee Trust Fund and Mote Marine Laboratory. This work was conducted under USFWS permit #MA773494.

References

Ackerman B (1995) Aerial surveys of manatees: A summary and progress report. In: O'Shea T, Ackerman B, Percival H, (eds) Population Biology of the Florida Manatee, pages 13–33. National Biological Service, Washington, DC.

Anderson D (2001) The need to get the basics right in wildlife field studies. Wildlife Society Bulletin 29(4):1294–1297.

Anderson D (2003) Response to engeman: Index values rarely constitute reliable information. Wildlife Society Bulletin 31(1):288–291.

Bengtson J (1981) Ecology of Manatees (Trichechus manatus) in the St. Johns River, Florida. PhD thesis, University of Minnesota, Minneapolis, MN.

Caughley G (1977) Analysis of Vertebrate Populations. John Wiley and Sons, New York.

Craig B, Reynolds J (2004) Determination of manatee population trends along the atlantic coast of florida using a bayesian approach with temperature-adjusted aerial survey data. Marine Mammal Science 20(3):386–400.

Deutsch C, Bonde R, Reid J (1998) Radio-tracking manatees from land and space: Tag design, implementation, and lessons learned from long-term study. Marine Technology Society Journal 32:18–29.

Deutsch C, Reid J, Bonde R, Easton D, Kochman H, O'Shea T (2003) Seasonal movements, migratory behavior, and site fidelity of west indian manatees along the atlantic coast of the united states. Wildlife Monographs 151:1–77.

Eberhardt L (1982) Censusing Manatees. Technical Report 8-1, Florida Cooperative Fish and Wildlife Research Unit, University of Florida, Gainesville, FL.

Edwards H, Pollock K, Ackerman B, Reynolds J, Powell J (2007) Components of detection probability in manatee aerial surveys during winter. Journal of Wildlife Management 71(6): 2052–2060.

Gamerman D (1997) *Markov* Chain Monte Carlo: Statistical Simulation for Bayesian Inference. Chapman and Hall, London, first edition.

Garrott R, Ackerman B, Cary J, Heisey D, Reynolds III J, Wilcox J (1995) Assessment of trends in sizes of manatee populations at several florida aggregation sites. In: O'Shea T, Ackerman B, Percival H, editors, Population Biology of the Florida Manatee pages 34–35. National Biological Service.

Garrott RA, Ackerman B, Cary J, Heisey D, Reynolds III J, Rose P, Wilcox J (1994) Trends in counts of manatees at winter aggregation sites. Journal of Wildlife Management 58:642–654.

Gelman A, Meng X, Stern H (1996) Posterior predicitive assessment of model fitness via realized discrepencies with discussion. Statistica Sinica 6:733–807.

Geweke J (1992) Evaluating the accuracy of sampling-based approaches to calculating posterior moments. In Bernardo J, Berger J, Dawid A, Smith J, editors, Bayesian Statistics 4, pages 169–193. Oxford University Press, Oxford.

Gilks W, Richardson S, Spiegelhalter D (1996) Introducing Markov chain Monte Carlo, chapter 1, pages 1–19. Interdisciplinary Statistics Series. Chapman and Hall, London.

Hartman D (1974) Distribution, Status and Conservation of the Manatee in the United States. Document PB 81-140725, National Technical Information Service, Springfield, VA.

Hartman D (1979) Ecology and behavior of the manatee (*Trichechus Manatus*) in florida. American Society of Mammalogy Special Publications 5.

Irvine A (1982) West indian manatee. In Davis D, editor, CRC Handbook of Census Methods for Terrestrial Vertebrates, pages 241–242. CRC Press, Inc., Boca Raton, FL.

Lefebvre L, Ackerman B, Portier K, Pollock K (1995) Aerial survey as a technique for estimating trends in manatee population size – problems and prospects. In O'Shea T, Ackerman B, Percival H, editors, Population Biology of the Florida Manatee, pages 63–74. National Biological Service Information and Technology Report 1. U.S. Department of the Interior.

Marsh H, Sinclair D (1989) Correcting for visibility bias in strip transect aerial surveys of aquatic fauna. Journal of Wildlife Management 53:1017–1024.

Packard J, Frohlich R, Reynolds III J, Wilcox J (1989) Manatee response to interruption of a thermal effluent. Journal of Wildlife Management 53:692–700.

Packard J, Summers R, Barnes L (1985) Variation of visibility bias during aerial surveys of manatees. Journal of Wildlife Management 49:347–351.

Pollock K, Kendall W (1987) Visibility bias in aerial surveys: A review of estimation procedures. Journal of Wildlife Management 51:502–510.

Pollock K, Marsh H, Bailey L, Farnsworth G, Simons T, Alldredge M (2004) Separating components of detection probability in abundance estimation: An overview with diverse examples. In: Thompson W, editor, Sampling Rare or Elusive Species: Concepts, Designs, and Techniques for Estimating Population Parameters, pages 43–58. Island Press, New York.

Pollock K, Marsh H, Lawler I, Alldredge M (2006) Estimating animal abundance in heterogeneous environments: An application to aerial surveys of dugong. Journal of Wildlife Management 70:255–262.

Reynolds III J, Wilcox J (1994) Observations of florida manatees (*Trichechus Manatus Latirostris*) around selected power plants in winter. Marine Mammal Science 10:163–177.

Ross S (1996) Stochastic Processes. Wiley and Sons, New York, second edition.

Shane S (1984) Manatee use of power plant effluents in brevard county, florida. Florida Scientist 47:180–187.

Spiegelhalter D, Best N, Carlin B, van der Linde A (2002) Bayesian measures of model complexity and fit. Journal of the Royal Statistical Society, Series B 64:583–639.

U.S. Fish and Wildlife Service (2001) Florida Manatee Recovery Plan, (*Trichechus manatus latirostris*), third revision. Technical report, U.S. Fish and Wildlife Service, Atlanta, GA.

Weigle B, Wright I, Ross M, Flamm R (2001) Movements of Radio-tagged Manatees in Tampa Bay and Along Florida's West Coast, from 1991 to 1996. Technical Report TR-7, Florida Fish and Wildlife Conservation Commission, St. Petersburg, FL.

Wright I, Reynolds III J, Ackerman B, Ward L, Weigle B, Szelistowski W (2002) Trends in manatee (Trichechus manatus latirostris) counts and habitat use in tampa bay, 1987-1994: Implications for conservation. Marine Mammal Science 18:259–274.

An Integrated Analysis of Multisite Recruitment, Mark-Recapture-Recovery and Multisite Census Data

R.S. Borysiewicz, B.J.T. Morgan, V. Hénaux, T. Bregnballe, J.-D. Lebreton, and O. Gimenez

Abstract The statistical analysis of mark-recapture-recovery (MRR) data dates back to the 1960s, when the foundation was laid for stochastic models, fitted to data by the method of maximum likelihood. There have been a number of developments which have proved to be extremely influential. Two of these are: the extension of MRR data and modelling to multi-site inference, and the integrated modelling of single-site MRR and census data. The aim of this study is to unite these two independent research programs, in order to enable effective integrated analysis of multi-site MRR data and multi-site census data. Census data can be described by a state-space model, and the likelihood is formed using the Kalman filter. By making use of movement information provided by MRR data, it is possible to avoid flat likelihood surfaces, thus allowing estimation of site-dependent parameters. This increases the precision of dispersal parameters and allows estimation of parameters inestimable from MRR studies alone.

This paper extends research within the area of integrated population analysis by developing methods for analysing multi-site census data coupled with multi-site capture recapture data. The methodology is explored using a simulated data set, the structure of which is motivated by a dataset of Great cormorants (*Phalacrocorax carbo sinensis*).

Keywords Integrated Analysis · Kalman Filter · Mark-Recapture-Recovery Data· Multistate Models · *Phalacrocorax carbo sinensis* · Recruitment · State-Space Models

1 Introduction

1.1 Mark-Recapture-Recovery Models

The development of models for mark-recapture-recovery (MRR) data began in the 1960s with the Cormack-Jolly-Seber (CJS) model (Cormack 1964; Jolly 1965;

R.S. Borysiewicz (✉)
IMSAS, University of Kent, Canterbury, UK.
e-mail: R.S. Borysiewicz@kent.ac.uk

D.L. Thomson et al. (eds.), *Modeling Demographic Processes in Marked Populations*, Environmental and Ecological Statistics 3, DOI 10.1007/978-0-387-78151-8_25,

Seber 1965), which estimates survival and capture rates from recapture data of an animal population collected at a single site. This model initiated a surge of interest into constructing models for this type of data and numerous extensions have since been developed. These include incorporation of categorical variables for characterising individuals and analysis of multiple data sets using group effects (Lebreton et al. 1992). Also, models that allow the integrated analysis of both recaptures of live animals and recoveries of dead animals have been developed (Burnham 1993; Lebreton et al. 1995; Barker 1997; Catchpole et al. 1998).

A further development which has enhanced the potential of MRR data analysis has been the extension of the CJS model to the multi-site framework. First established by Arnason (1972, 1973) and later developed by Schwarz et al. (1993) and Brownie et al. (1993), this extension allows estimation of survival and transition probabilities as well as recapture parameters and parameter redundancy of these multi-site models has been assessed in Gimenez et al. (2003). The Arnason-Schwarz model has been further generalised to multi-state models (Lebreton et al. 1999) and multi-event models (Pradel 2005). A number of computer software packages for fitting such models are available, including program MARK (White and Burnham 1999) and M-Surge (Choquet et al. 2004).

1.2 Integrated Population Analysis

Integrated population analysis combines data from a variety of sources. Integrating MRR and census data was first proposed by Besbeas et al. (2002) as a method for estimating productivity, otherwise not estimable from either type of data alone. It is possible to form the likelihood for the MRR data as outlined in Section 1.1. A state space model of the census data comprises two parts – the observation equation and the underlying state equation. Gaussian assumptions provide the key to the accessibility of these models and makes it possible to form the likelihood, using a recursive procedure known as the Kalman filter (Harvey 1989; Durbin and Koopman 2001). Assuming independence, it is possible to combine the MRR and census likelihoods to provide one global likelihood which can be optimised to provide maximum likelihood estimates of all parameters.

A multivariate normal approximation to the exact MRR likelihood (Besbeas et al. 2003) more efficiently integrates both data sets. Further, Brooks et al. (2004) introduced a Bayesian approach, while Besbeas et al. (2005) provides a discussion on further possible advances. Besbeas et al. (2008) gives details of the Kalman filter methodology, including initialisation procedures in ecological applications and a discussion of the break-down of independence assumptions and the effects of introducing overdispersion in the state space model.

The integrated analysis performed in this paper is on a simulated multi-site data set which contains both breeding and non-breeding individuals. Recruitment, defined as the progression from a non-breeding state to a breeding state, is a parameter of biological interest, and it is this type of transition, along with dispersals between study areas which will be estimated from the contribution of MRR multi-site/state and multi-site census simulated data.

2 Motivation for the Study

This investigative simulation study was motivated by data on cormorants, *Phalacrocorax carbo sinensis*, collected by the National Environmental Research Institute in Denmark. *P. carbo* is the most widely distributed of all cormorants – known to breed in North America, Europe, Asia, Africa and Australasia. The studied cormorants belong to the Eurasian subspecies *P. carbo sinensis*. This subspecies is smaller than the North Atlantic subspecies *P. carbo carbo*, and often breeds and winters inland (Hatch et al. 2000).

The data were collected as part of a larger ringing programme started in Denmark in 1977, which continues to the current day. The data correspond to a period of population expansion (1981–1993) in 6 colonies located 32–234 km apart. Recapture and recovery data from this period were analysed in detail by Henaux et al. (2007) who estimated dispersal and recruitment.

The oldest of the six colonies, Vorsø, was established in 1944 and along with Ormø (OR, est. 1972) and Brændegård Sø (BR, est. 1973) comprised the only colonies present in Denmark at the start of the ringing study. Colonies Toft Sø (TO), Dyrefod (DY) and Mågeøerne (MA) established during the study in 1982, 1984 and 1985 respectively.

14,018 cormorant chicks were marked between 1981 and 1991 with a standard metal ring on one leg and a coloured plastic band, engraved with a unique combination of 3 alphanumeric characters, on the other leg. Resightings of the ringed cormorants took place from 1983 to 1993. Resightings were of breeding cormorants only and these breeders were identified using strict biological criteria identified in Henaux et al. (2007). Recoveries spanned a large geographical area, ranging from northern United Kingdom to Southern Algeria and from western Spain to eastern Romania. Recoveries of birds for which only one ring was found were excluded to avoid negative bias due to ring loss.

Each of the six colonies were censused in early May. Data consisted of a count of all occupied nests. The location of nests varies between colonies: in MA nests were built on the ground, in BR nests were found in trees and on the ground while in the other colonies all nests were in trees.

Based on the parameters of this real-life investigation, this simulation study demonstrates the statistical gains of performing integrated population modelling on multi-site data and also the ease with which even complex models, such as the recruitment structure, can be incorporated into an integrated population modelling framework.

3 Methods

3.1 Formation of the Mark-Recapture-Recovery Likelihood

The closed-form likelihood for Arnason-Schwarz models is derived in King and Brooks (2003). Suppose captures or recaptures occur for animals age $j \in \mathcal{J} =$

$\{0, \ldots, J\}$ and the study site is split into $\mathcal{R}$ regions. The set of model parameters includes:

- $\phi_j(r)$ is the probability that an animal in location r at age j survives until age $j+1$;
- $\lambda_j(r)$ is the probability that an animal in location r at age j dies and is recovered dead before age $j+1$;
- $p_{j+1}(r)$ is the probability that an animal in location r at age $j+1$ is recaptured; and
- $\psi_j(r, s)$ is the probability that an animal in location r at age j moves to location s by age $j+1$ given that it is alive at age $j+1$.

The encounter history of each animal can be broken down into three partial histories. These are: last live encounters and beyond, consecutive live sightings and dead recoveries. The likelihood can similarly be deconstructed into these elements and full details of the likelihood construction can be found in the Appendix.

3.2 *Formation of the Census Likelihood*

3.2.1 The State Space Model and the Kalman Filter

The Kalman filter is a recursive procedure for computing the optimal estimator of a state vector at time t, based on the information available at time t (Harvey 1989). By imposing Gaussian assumptions it is possible to calculate the maximum likelihood estimates of unknown model parameters. The general linear Gaussian state space model is:

$$y_t = Z_t\alpha_t + \epsilon_t \tag{1}$$

$$\alpha_{t+1} = T_t\alpha_t + \eta_t \tag{2}$$

with $\epsilon_t \sim N(0, H_t)$ and $\eta_t \sim N(0, Q_t)$.

Equation (1) is the observation equation and (2) is the state space equation. The state vector, α_t, is unobserved and y_t is a vector of observations. The matrices Z_t, T_t, H_t and Q_t are assumed to be serially independent and independent of each other at all times.

The initial state vector α_1 is assumed to be $N(a_1, P_1)$ independently of $\epsilon_1, \ldots, \epsilon_n$ and $\eta_1, \ldots, \eta_n$. In practice, some or all of the matrices will depend on elements of an unknown model parameter vector.

The aim is to obtain a conditional distribution of α_{t+1} given $Y_t = \{y_1, \ldots, y_t\}$. Since all distributions are normal, conditional distributions of subsets of variables given other subsets are also normal; the required distribution is therefore determined by a knowledge of $a_{t+1} = E(\alpha_{t+1}|Y_t)$ and $P_{t+1} = Var(\alpha_{t+1}|Y_t)$.

The Kalman Filtering derives the filtering equations and may also compute the smoothed estimates of the error vectors ϵ_t and η_t, given all the observations $y_1, \ldots, y_n$. Denoting the parameter vector by θ and using our previous assumptions of normality, the likelihood is

$$L(\theta \mid y) = p(y_1, \ldots, y_n \mid \theta) = p(y_1 \mid \theta) \prod_{t=2}^{n} p(y_t | Y_{t-1}, \theta)$$

and the log-likelihood is given by

$$\log L(\theta \mid y) = \sum_{t=1}^{n} \log p(y_t | Y_{t-1}, \theta)$$

where $p(y_1 | Y_0, \theta) = p(y_1)$. Following substitution of appropriate parameters we obtain

$$\log L(\theta \mid y) = -\frac{np}{2} \log 2\pi - \frac{1}{2} \sum_{t=1}^{n} \log(|F_t| + v_t' F_t^{-1} v_t) \tag{3}$$

This likelihood is known as the prediction error decomposition form of the likelihood since v_t can be interpreted as a vector of prediction errors, $y_t - E(y_t)$. F_t is the covariance matrix of the conditional distribution of the observations and both vectors v_t and matrix F_t are calculated directly from the Kalman filter.

3.2.2 Ecological Application of the Kalman Filter

Define $N_{i_t}^{x}$ to be the number of animals in state $x = \{N, B\}$ (where N denotes non-breeder and B denotes breeder) in site $i = 1, 2, 3$ at time t. State vector α_t is

$$\alpha_t = \left(N_1^N \; N_2^N \; N_3^N \; N_1^B \; N_2^B \; N_3^B\right)_t^T ,$$

then following the notation of equation (1), since only breeders are observed

$$Z_t = \begin{pmatrix} 0\,0\,0\,1\,0\,0 \\ 0\,0\,0\,0\,1\,0 \\ 0\,0\,0\,0\,0\,1 \end{pmatrix}, \forall t.$$

Vector y_t is the observed census data and the observation error, σ, is assumed to be constant over time and site; we discuss relaxing this assumption later. The underlying state space equation is governed by a Leslie matrix (Caswell 2000), and the explicit form of the Leslie matrix for this application is given in Section 3.4.

The Kalman filter is initialised by specifying values for a_1 and P_1, the mean and covariance matrix of the initial state vector, respectively. Following Besbeas and Morgan (2008) we initialised the filter using the stable age distribution of the population. Other initialisation procedures were implemented on the same simulated data set and the model structure was robust to the assumption of the stable age distribution start.

3.3 Simulation of Multisite Recruitment Data

150 simulated non-breeding animals were marked at each site and MRR data were simulated on this cohort for seven years. All ringing was carried out on non-breeders and recaptures were of breeders only. Recapture and survival probabilities were site- and state-dependent and the recovery probability was independent of recovery site. Three types of transition were considered, which reflect the model structure used by Henaux et al. (2007):

- **natal dispersal**: the movement of a non-breeder from one geographical location to another whilst remaining a non-breeder.
- **recruitment**: the accession from non-breeder to breeder.
- **breeding dispersal**: the movement of a breeder from one geographical location to another.

Cormorants can only start breeding from age two and remain capable of breeding in every subsequent year (Henaux et al. 2007), thus, within this simulation study we assume that animals start recuiting at age 2 and that all animals have recruited by age 5 and once a breeder, they remain a breeder until death. Only breeding birds are ever recaptured and hence constraints applied to natal dispersal and recruitment need to be enforced in order to ensure identifiability of all transitions, and so it is assumed that natal dispersal occurs only in the first year of life and subsequently recruitment then occurs at a single site and no further dispersals occur between sites until the animal has reached the breeding state.

The assumptions, defined above, are in addition to the traditional Arnason-Schwarz model assumptions, and in order to proceed with integrated analysis the MRR data and census data must be independent. Violation of this assumption can lead to biased estimates, as shown in Besbeas et al. (2008).

The simulated census data were site-specific with structure governed by a Leslie matrix. Twenty years of site-dependent census information were simulated which, like the cormorant data set, was of breeding birds only. Parameter values were chosen to reflect biologically reasonable values. Fecundity (defined to be the number of offspring multiplied by the probability of reproduction in a particular year) was fixed at a constant value of 1.2. Fecundity however could be adapted to allow for time, site or density dependence. Transition, survival, capture and recovery probabilities were set to values which were biologically reasonable, with non-breeder survival (0.6) assumed lower than breeder survival (0.8) and recovery rates (0.4) lower than recapture rates (0.7). Dispersal rates varied between 0.05 and 0.3 between sites.

3.4 Specification of the Leslie Matrix

Suppose we denote the three non-breeding states as 4, 5 and 6 and the three breeding states as 1, 2 and 3 for the respective three sites in each case. The parameters involved in the MRR likelihood, following the notation of Section 3.1, are:

Survival Probability: $\phi_j(r) = \phi(r)$ for all j, for $r \in \{1,2,3,4,5,6\}$, where $r \in \{1,2,3\}$ denotes breeder survival and $r \in \{4,5,6\}$ denotes non-breeder survival;
Recovery Probability: $\lambda_j(r) = \lambda$ for all j and r;
Recapture Probabilities: $p_{j+1}(r) = p(r)$ for all j and $r \in \{1,2,3\}$;
Natal Dispersal: $\psi_1(r,s)$ for $r,s \in \{4,5,6\}$
Breeding Dispersal: $\psi(r,s)$ for $r,s \in \{1,2,3\}$
Recruitment: $\psi_{2+}(r, r-3)$ for $r \in \{4,5,6\}$
Non-Maturation: $\psi_{2+}(r,r)$ for $r \in \{4,5,6\}$

The structure of the Leslie matrix for the state space model, using the notation above is then:

$$\begin{pmatrix} \phi(4)\psi_{2+}(4,4) & 0 & 0 & f\phi(4)\psi_1(4,4) & f\phi(5)\psi_1(5,4) & f\phi(6)\psi_1(6,4) \\ 0 & \phi(5)\psi_{2+}(5,5) & 0 & f\phi(4)\psi_1(4,5) & f\phi(5)\psi_1(5,5) & f\phi(6)\psi_1(6,5) \\ 0 & 0 & \phi(6)\psi_{2+}(6,6) & f\phi(4)\psi_1(4,6) & f\phi(5)\psi_1(5,6) & f\phi(6)\psi_1(6,6) \\ \phi(4)\psi_{2+}(4,1) & 0 & 0 & \phi(1)\psi(1,1) & \phi(2)\psi(2,1) & \phi(3)\psi(3,1) \\ 0 & \phi(5)\psi_{2+}(5,2) & 0 & \phi(1)\psi(1,2) & \phi(2)\psi(2,2) & \phi(3)\psi(3,2) \\ 0 & 0 & \phi(6)\psi_{2+}(6,3) & \phi(1)\psi(1,3) & \phi(2)\psi(2,3) & \phi(3)\psi(3,3) \end{pmatrix}$$

where f is the fecundity parameter.

3.5 Computational Implementation

MATLAB was used to code the MRR likelihood and the Kalman filter which constructed the census likelihood. The global likelihood, formed by multipliying the two likelihoods, assuming independence, was then optimised using a built-in optimisation method within the MATLAB software. The logistic link was used to constrain recapture, recovery and survival probabilities between 0 and 1, whilst the generalised logit link, which is an extended logit function was used to ensure that as well as transition probabilities being between 0 and 1, appropriate combinations of the transitions added to 1. Further details of the use of the generalised logit link can be found in Choquet et al. (2005). Figure 1 demonstrates the formation of the appropriate likelihoods and also the common parameters used to model each type of data.

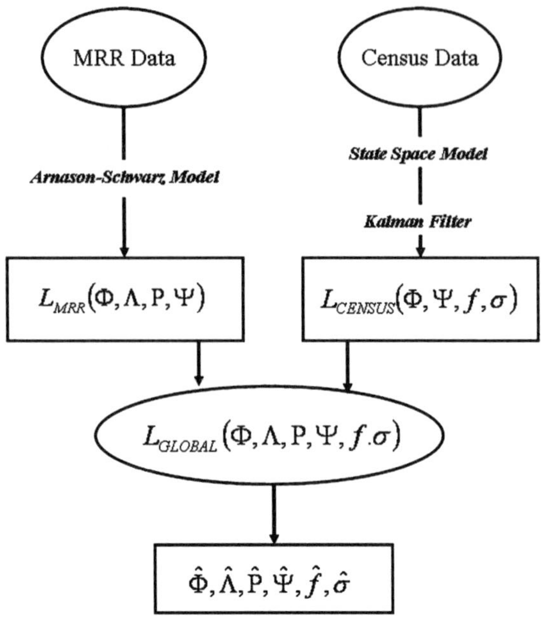

Fig. 1 Directional acyclic graph showing how two types of data (MRR data and census data) are combined to obtain global estimates of parameters: Φ – survival parameters; Λ – recovery parameters; P – recapture parameters; Ψ – transition parameters, f – fecundity, σ – observation error. The Arnason-Schwarz model is used to form the MRR likelihood whilst the Kalman filter is applied to a state space model to form the census likelihood

4 Results

The maximum likelihood estimates (MLEs) are shown in Table 1. There were no issues of either intrinsic or extrinsic parameter redundancy in the integrated model structure, so the MRR data analysed alone is able to estimate all survival, transition and capture/recovery probabilities. Once the MRR data were combined with census information we were also able to estimate the fecundity of the population. The addition of census data improves the precision of capture and recovery probabilities slightly (Table 2), however the change is small as the only information added by the census data is through the correlation structure of these parameters. The largest improvement in precision is for the breeding dispersal parameters, with no improvement in the natal dispersal parameter. This is most likely due to the fact that natal dispersal only appears in the Leslie matrix as a product with the unknown parameter fecundity. Thus, precision is not improved without compromising the precision of the fecundity parameter. Attempting to estimate parameters from the census data alone results in parameter estimates with low precision due to the census data producing a flat likelihood (the fecundity estimate from census data alone is 1.77 with standard error 14.854), however once the census data are combined with MRR data,

Table 1 The maximum likelihood estimates and associated standard errors calculated by a finite difference method, of the MRR data used alone and the integrated MRR and census data

Parameter	True Value	MRR Estimate	MRR SE	Integrated Estimate	Integrated SE
$\phi(1, 2, 3)$	0.80	0.826	0.0312	0.820	0.0279
$\phi(4, 5, 6)$	0.60	0.587	0.0211	0.586	0.0211
p	0.70	0.714	0.0396	0.712	0.0379
$\psi(4, 5) = \psi(4, 6)$	0.10	0.078	0.0395	0.073	0.0350
$\psi(5, 4) = \psi(5, 6)$	0.20	0.189	0.0441	0.193	0.0418
$\psi(6, 4) = \psi(6, 5)$	0.05	0.024	0.0227	0.025	0.0236
$\psi(1, 2) = \psi(1, 3)$	0.20	0.248	0.0367	0.209	0.0213
$\psi(2, 1) = \psi(2, 3)$	0.10	0.083	0.0283	0.106	0.0194
$\psi(3, 1) = \psi(3, 2)$	0.30	0.332	0.0351	0.325	0.0279
$\psi(4, 1) = \psi(5, 2) = \psi(6, 3)$	0.60	0.593	0.0593	0.601	0.0559
λ	0.40	0.406	0.0258	0.405	0.0258
f	1.2	-	-	1.167	0.1420

Table 2 The mean square errors of the parameters for the MRR data used alone and the integrated MRR and census data

Parameter	MSE (MRR Data Alone)	MSE (Integrated Data)
$\phi(1, 2, 3)$	0.0752	0.0525
$\phi(4, 5, 6)$	0.0106	0.0110
p	0.0422	0.0375
$\psi(4, 5) = \psi(4, 6)$	0.4436	0.4621
$\psi(5, 4) = \psi(5, 6)$	0.1504	0.1273
$\psi(6, 4) = \psi(6, 5)$	1.6180	1.5683
$\psi(1, 2) = \psi(1, 3)$	0.2402	0.0362
$\psi(2, 1) = \psi(2, 3)$	0.2184	0.0594
$\psi(3, 1) = \psi(3, 2)$	0.1733	0.1056
$\psi(4, 1) = \psi(5, 2) = \psi(6, 3)$	0.0612	0.0544
λ	0.0120	0.0119
f	–	0.0156
σ	–	0.0232

the fecundity parameter estimate was precise (MSE = 0.0156). Adult survival gains some precision whilst juvenile survival does not.

The generalised variances of the common parameters of the MRR analysis and the integrated analysis are 3.64×10^{-13} and 2.80×10^{-17} respectively. Thus, combining the additional census data has considerably improved the overall precision of common parameters and allowed fecundity to be estimated with good precision.

The combined analysis has also accurately estimated the observation error of the census data. Though it was assumed constant for this simulation, it is interesting to allow observation error to vary proportionally to population size, e.g. $\epsilon_t \sim N(0, y_{t-1}\sigma^2)$ (Tavecchia et al. 2006). Similarly, complex fecundity structures, such as density dependence could also be considered.

5 Conclusions

These procedures are extensions of the single-site integrated population analysis. MRR is now a widely used type of data, with even advanced MRR models in common use. Frequently, however, census information is collected concurrently with MRR data. Until now, census data have been analysed separate from the MRR data, thereby sacrificing their shared information. The methodologies described in this paper are simple to implement and can be completed 'post-hoc' by using a simple multivariate normal approximation to the MRR likelihood; if only MLEs and associated standard errors are available, they can be used to construct an approximate diagonal variance-covariance matrix to facilitate analysis and complicated population transitions can be incorporated into the state space model.

The potential of single-site integrated population analysis has been assessed in numerous studies as a method for estimating previously inestimable parameters. This simulation study has shown that not only does multi-site/state integrated population analysis estimate these parameters, but also greatly improves the precision of some parameters. This is obviously desirable in complex models with a large number of parameters that are frequently estimated with low precision.

Single site integrated population modeling with one dimensional census information allows the estimation of additional parameters, however little change is made to the precision of common parameters. By incorporating multi-dimensional census data, in terms of site or state-specific census data, we have shown that this extra information has greatly improved the precision of parameters within the model.

Acknowledgments We wish to thank Jens Gregersen and Lars Abrahamsen for their great effort in ringing and recording breeding attempts of colour-ringed cormorants.

Appendix

The Construction of the Closed-Form Arnason-Schwarz Likelihood

This appendix gives an outline of the construction of the explicit MRR likelihood first derived in King and Brooks (2003). Recall that the encounter history of an animal can be broken down into three partial histories: last live encounters and beyond, consecutive live sightings, dead recoveries. The likelihood can similarly be deconstructed into these elements. The following derived probabilities facilitate the likelihood construction.

Let $\chi_{(j,k)}(r)$ denote the probability that an animal is seen for the last time at age $j \in \mathcal{J}$ in location $r \in \mathcal{R}$, and would be age k at the end of the study, with $j \leq k \leq \mathcal{J}$. Then,

$$\chi_{(j,k)}(r) = \begin{cases} 1 & (j = k) \\ 1 - \phi_{(j)}(r)\left[1 - \sum_{s \in \mathcal{R}} \psi_j(r,s)\{1 - p_{j+1}(s)\}\chi_{(j+1,k)}(s)\right] & \\ -\{1 - \phi_j(r)\}\lambda_j(r) & (j < k) \end{cases}$$

$O_{(k,j)}(r,s)$ denotes the probability that an animal in location $r \in \mathcal{R}$ at age $k \in \mathcal{J}$ remains unobserved until it is subsequently resighted in location $s \in \mathcal{R}$ at age $j+1$, for $0 \le k \le j \le J-1$. Then,

$$O_{(k,j)}(r,s) = p_{j+1}(s)Q_{(k,j)}(r,s),$$

where $Q_{(k,j)}(r,s)$ denotes the probability that an animal migrates from region $r \in \mathcal{R}$ at age $k \in \mathcal{J}$ to location $s \in \mathcal{R}$ at age $j+1$, for $0 \le k \le j \le J-1$, and is unobserved between these ages, and is given by

$$Q_{(k,j)}(r,s) = \begin{cases} \phi_k(r)\psi_k(r,s) & (k=j) \\ \phi_k(r)\sum_{l=1}^{R}\{1-p_{k+1}(l)\}\psi_k(r,l)Q_{k+1,j}(l,s) & (k<j) \end{cases}$$

$D_{(k,j)}(r)$ denotes the probability that an animal is recovered dead in the interval $(j, j+1)$ given that it is last seen at age $k \le j$ in $r \in \mathcal{R}$ and is given by

$$D_{kj}(r) = \begin{cases} \{1-\phi_j(r)\}\lambda_k(r) & (k=j) \\ \sum_{l=1}^{R}\{1-\phi_j(l)\}\lambda_j(l)\{1-p_j(l)\}Q_{k,j-1}(r,l) & (k<j) \end{cases}$$

The following sufficient statistics which are obtained from the encounter history data are then formed:

$v_{(j,k)}(r)$ denotes the number of animals that are recaptured for the last time in location $r \in \mathcal{R}$ aged j and would be aged $j \le k \le J$ at the end of the study;
$n_{(k,j)}(r,s)$ denotes the number of animals that are observed in location $r \in \mathcal{R}$ at age k and next observed alive in location $s \in \mathcal{R}$ at age $j+1$; and
$d_{(k,j)}(r)$ denotes the number of animals recovered dead between ages j and $j+1$ that are last observed alive at age $k \le j$ in location $r \in \mathcal{R}$.

The likelihood function has the form given below:

$$L(\theta|v,n,d) = \prod_{r\in\mathcal{R}}\left[\prod_{j=0}^{J}\prod_{k=j}^{J}\{\chi_{(j,k)}(r)\}^{v_{(j,k)}(r)}\prod_{k=0}^{J-1}\prod_{j=k}^{J-1}\{D_{(k,j)}\}^{d_{(k,j)}(r)}\right.$$
$$\left.\prod_{k=0}^{J-1}\prod_{j=k}^{J-1}\prod_{s\in\mathcal{R}}\{O_{(k,j)}(r,s)\}^{n_{(k,j)}(r,s)}\right]$$

where θ comprises the model parameters $\{\Phi, \Lambda, P, \Psi\}$ and we denote our MRR likelihood by $L_{MRR}(\Phi, \Lambda, P, \Psi)$.

References

Arnason AN (1972) Parameters estimates from mark-recapture-recovery experiments on two populations subject to migration and death. Researches on Population Ecology 13:97–113.

Arnason AN (1973) The estimation of population size, migration rates, and survival in a stratified population. Researches on Population Ecology 15:1–8.

Barker RJ (1997) Joint modeling of live-recapture, tag-resight, and tag-recovery data Biometrics 53(2):666–677.

Besbeas P, Morgan BJT (2008) Kalman filter initialization for modelling population dynamics. Biometrics Invited Revision.

Besbeas P, Freeman SN, Morgan BJT, Catchpole EA (2002) Integrating mark-recapture-recovery and census data to estimate animal abundance and demographic parameters. Biometrics 58: 540–547.

Besbeas P, Lebreton J-D, Morgan BJT (2003) The efficient integration of abundance and demographic data. Applied Statistics 52:95–102.

Besbeas P, Freeman SN, Morgan BJT (2005) The potential of integrated population modelling. Australian & New Zealand Journal of Statistics 47(1):35–48.

Besbeas P, Borysiewicz RS, Morgan BJT (2008) Completing the ecological jigsaw. In press.

Brooks SP, King R, Morgan BJT (2004) A Bayesian approach to combining animal abundance and demographic data. Animal Biodiversity and Conservation 27(1):515–529.

Brownie C, Hines JE, Nichols JD, Pollock KH, Hestbeck JB (1993) Capture-recapture studies for multiple strata including non-markovian transitions. Biometrics 49:1173–1187.

Burnham KP (1993) A theory for combined analysis of ring-recovery and recapture data. pages 199–213. Basel: Birkhauser Verlag.

Caswell H (2000) Matrix population models: Construction, analysis and interpretation. Sinauer Associates Incorporated, Sunderland, MA.

Catchpole EA, Freeman SN, Morgan BJT, Harris MP (1998) Integrated recovery/recapture data analysis. Biometrics 54:33–46.

Choquet R, Reboulet A-M, Pradel R, Gimenez O, Lebreton J-D (2004) M-Surge: New software specifically designed for multistate capture-recapture models. Animal Biodiversity and Conservation 27(1):207–215.

Choquet R, Reboulet A-M, Pradel R, Gimenez O, Lebreton J-D (2005) M-Surge 1–7 user's manual. CEFE/CNRS, Montpellier, FRANCE, http://ftp.cefe.cnrs.fr/ biom/Soft_CR.

Cormack RM (1964) Estimates of survival from sightings of marked animals. Biometrika 51: 429–438.

Durbin J, Koopman SJ (2001) Time series analysis by state space methods. Oxford University Press, Oxford, U.S.

Gimenez O, Choquet R, Lebreton J-D (2003) Parameter redundancy in multistate capture-recapture models. Biometrical Journal 6:704–722.

Harvey AC (1989) Forecasting, structural time series models and the Kalman filter. Cambridge University Press, Cambridge, U.K.

Hatch JJ, Brown KM, Hogan GG, Morris RD (2000) Great cormorant (*Phalacrocorax carbo*). In: The birds of North America, Eds. A. Poole and F.Gill, (553).

Henaux V, Bregnballe T, Lebreton J-D (2007) Dispersal and recruitment during population growth in a colonial bird, the great cormorant. Journal of Avian Biology 38(1):44–57.

Jolly GM (1965) Explicit estimates from capture-recapture data with both death and immigration-stochastic models. Biometrika 52:225–247.

King R, Brooks SP (2003) Closed-form likelihoods for Arnason-Schwarz models. Biometrika 90(2):435–444.

Lebreton J-D, Burnham KP, Clobert J, Anderson DR (1992) Modeling survival and testing biological hypotheses using marked animals: A unified approach with case studies. Ecological Monohraphs 62:67–118.

Lebreton J-D, Morgan BJT, Pradel R, Freeman SN (1995) A simultaneous analysis of dead recovery and live recapture data. Biometrics 51(4):1418–1428.

Lebreton J-D, Almeras T, Pradel R (1999) Competing events, mixtures of information and multistratum recapture models. Bird Study, 46(Suppl.):S39–46.

Pradel R (2005) Multievent: An extension of multistate capture-recapture models to uncertain states. Biometrics 61.

Schwarz CG, Schweigert JF, Arnason AN (1993) Estimating migration rates using tag-recovery data. Biometrics 59:291–318.

Seber GAF (1965) A note on the multiple-recapture census. Biometrika 52:249–259.

Tavecchia G, Besbeas PT, Coulson T, Clutton-Brock TH, Morgan BJT (2006) Estimating population size and hidden demographic parameters with state-space modelling. Technical Report UKC/IMS/06/15.

White GC, Burnham KP (1999) Program Mark: Survival estimation from populations of marked animals. Bird Study 46:S120–S138.

Section VII
Bayesian Applications – Advances, Random Effects and Hierarchical Models

Ken Burnham and Bill Link

Bayes Factors and Multimodel Inference

William A. Link and Richard J. Barker

Abstract Multimodel inference has two main themes: model selection, and model averaging. Model averaging is a means of making inference conditional on a model set, rather than on a selected model, allowing formal recognition of the uncertainty associated with model choice. The Bayesian paradigm provides a natural framework for model averaging, and provides a context for evaluation of the commonly used AIC weights. We review Bayesian multimodel inference, noting the importance of Bayes factors. Noting the sensitivity of Bayes factors to the choice of priors on parameters, we define and propose nonpreferential priors as offering a reasonable standard for objective multimodel inference.

Keywords Bayes factors · Bayesian inference · Multimodel inference · Nonpreferential priors

1 Introduction

Science is about uncertainty. In the best circumstances, phenomena studied produce data which can be regarded as realizations of a random variable X having probability distribution $f(x;\theta)$, completely specified except for some or all components of a vector θ. We refer to the family of distributions $f(x;\theta)$ as a model; scientific inference focuses on estimation or prediction of the unknown quantities in θ.

Our statement that "data can be regarded" as described is intended to be neutral, as also our use of the term "model." The representation may be exact, or it may be merely a useful approximation. Unfortunately, the term "model" is charged with meaning, frequently carrying the sense of approximation. Used thus, the phrase "true model" is an oxymoron. It is easy when discussing models to become entangled in all sorts of philosophical musings about Truth and the limitations

---

W.A. Link (✉)
U.S. Geological Survey, Patuxent Wildlife Research Center, 12100 Beech Forest Road, MD 20708, USA
e-mail: wlink@usgs.gov

D.L. Thomson et al. (eds.), *Modeling Demographic Processes in Marked Populations*, Environmental and Ecological Statistics 3, DOI 10.1007/978-0-387-78151-8_26,

of Knowledge, musings which pursued with sufficient assiduity leave us with Descartes, questioning our own existence.

When we say that data "can be regarded" as observations of a random variable we mean simply that the family of distributions is taken as granted, and that our inference is conditioned on this assumption. Single model inference operates as though the posited model were an exact depiction of the stochastic processes generating observations.

Our inference is limited by model uncertainty: we are limited by our inability to confidently assert that $f(x;\theta)$ is an exact depiction of the stochastic processes giving rise to the data. Goodness of fit tests may tell us that we ought not to have confidence in a model, but they cannot finally decide the issue of model adequacy. Nor does the evaluation of a single model point the way forward; if the model is deemed inadequate, what then? What is needed is a set of alternative models $f_i(x;\theta^{(i)})$, $i = 1,2,\ldots$, and a formal mechanism for evaluating the relative confidence we place in each of them.

Bayesian multimodel inference provides such a mechanism.

Two aspects of multimodel inference are distinguished: model selection and model averaging. Model selection is the choice of a single model; subsequent inference is conditioned on this choice. The dangers of using the same data for model choice and subsequent inference are well-known (Chatfield 1995; Draper 1995; Hoeting et al. 1999). For instance, significance tests for modeled effects cannot be interpreted as though the model were fixed in advance: effects included in selected models are almost certain to be statistically significant in those models.

Model averaging is the process of making inference conditional on a model set, rather than on a particular model in the set. The resulting inference is based on a weighted average of the models, the weights relating to the confidence we have in the various models.

In this paper, we point out that if model averaging is conducted in accordance with basic rules of probability, it is essentially an application of Bayes theorem. This observation applies regardless of whether one chooses to conduct a Bayesian analysis: one need not be Bayesian to use Bayes theorem. Seen thus, model weights are posterior probabilities for models, conditional on the model set. There exists a corresponding set of prior model probabilities, even if it is not explicitly specified. The prior on models has a substantial role in the resulting inference.

Wildlife and ecological statistics have been substantially and positively influenced by Burnham and Anderson's promotion of Akaike's information criterion (AIC) as a tool for model selection, and as a basis for model averaging (Burnham and Anderson 1998, 2002, 2004). We encourage a careful evaluation of the prior on models associated with AIC weights. This prior will be seen to depend on sample size, and in such a way as to substantially favor more highly parameterized models. We describe alternative weighting schemes similar to AIC weighting, which can be implemented at no additional computational cost while avoiding this tendency.

We believe that despite computational challenges, the best approach to multimodel inference is provided by the Bayesian paradigm. We begin, therefore, with a review of Bayesian multimodel inference (BMI), noting the centrality of a quantity known as the Bayes factor. Bayes factors are determined by the data, the probability distribution $f(x;\theta)$, and the priors on parameters θ, but do not depend on the prior model weights. Indeed, Bayes factors can be thought of as the machinery by which prior model weights are converted to posterior model weights; the mechanism exists independent of the specified prior model weights.

Bayes factors, and consequently BMI, are sensitive to the choice of priors on parameters. The effect of this choice is much greater than its effect in single model applications, such as Bayesian estimation. BMI being a fairly new and somewhat unfamiliar topic in wildlife and ecological applications, we feel that problems of prior sensitivity may not yet have received adequate attention among analysts. We briefly review the issue, describing approaches that have been proposed for dealing with prior sensitivity, and proposing principles that may serve as guidelines for prior selection.

2 Bayesian Multimodel Inference for Fully Specified Models

Bayes theorem, in its simplest form, follows directly from the law of total probability and the definition of conditional probability. Given a mutually exclusive and exhaustive set $\boldsymbol{M} = \{M_1, M_2, ..., M_R\}$ of outcomes, and an event X, the conditional probability of M_j given X can be calculated as

$$\begin{aligned}\Pr(M_j|X) &= \frac{\Pr(M_j \; and \; X)}{\Pr(X)} = \frac{\Pr(X \; and \; M_j)}{\sum_{i=1}^{R}\Pr(X \; and \; M_i)} \\ &= \frac{\Pr(X \;|M_j)\Pr(M_j)}{\sum_{i=1}^{R}\Pr(X \mid M_i)\Pr(M_i)}\end{aligned} \tag{1}$$

In Bayesian multimodel inference, the set $\boldsymbol{M} = \{M_1, M_2, ..., M_R\}$ consists of models.

Let Q be a quantity about which we wish to make a prediction based on the observation X. In Bayesian inference, that prediction will take the form of a probability statement based on a posterior distribution; if we were only considering model j, the prediction would be based on $\Pr(Q|X, M_j)$. If we are interested in combining predictions over the entire model set $\boldsymbol{M}$, we will use

$$\Pr(Q|X) = \sum_{j=1}^{R}\Pr(Q|X, M_j)\Pr(M_j|X); \tag{2}$$

Note in equation (2) that the posterior model probabilities $\Pr(M_j|X)$ serve as weights for the model-specific inferences.

We have referred to $\boldsymbol{M}$ as a "mutually exclusive and exhaustive set." More realistically, we might suppose that $\boldsymbol{M}$ is a subset of some larger collection of models. But there is no difference in the resulting inference. When conducting inference with a single model, we acknowledge that inference is conditioned on that model; in multi-model inference, we acknowledge that inference is conditioned on the model set. Recognizing that $\boldsymbol{M}$ is a subset of some larger collection of models does not change our inference though it might change our notation: it would perhaps be wise for us to include $\boldsymbol{M}$ in the conditional probability notation, as for instance $\Pr(M_j|\boldsymbol{M})$ instead merely of $\Pr(M_j)$. For ease of notation this conditioning is usually not explicitly indicated, just as it is not typically indicated in single model inference. One way or the other, all probability statements are conditioned on this set.

If we are only interested in the relative support for two models, say M_j and M_k, we might consider the ratio of their posterior probabilities, the odds $\Pr(M_j|X)/\Pr(M_k|X)$. From equation (1) it is seen that the posterior odds can be written as

$$\frac{\Pr(M_j|X)}{\Pr(M_k|X)} = \frac{\Pr(X|M_j)}{\Pr(X|M_k)} \frac{\Pr(M_j)}{\Pr(M_k)}. \tag{3}$$

The Bayes factor for comparing models j and k is defined as the ratio $BF_{j,k} = \Pr(X|M_j)/\Pr(X|M_k)$. Thus equation (3) is

$$\frac{\Pr(M_j|X)}{\Pr(M_k|X)} = BF_{j,k} \times \frac{\Pr(M_j)}{\Pr(M_k)}, \tag{4}$$

Note that the Bayes factor is a likelihood ratio for the two models, and that it is the multiplicative factor by which prior model odds are converted to posterior model odds. Expressed another way, the Bayes factor is the ratio of posterior odds to prior odds on models. $BF_{j,k} > 1$ means that the data provide greater support for model j than for model k. It is worth noting that the Bayes factor exists independently of the model set $\boldsymbol{M}$; it is an absolute measure of relative support.

It will be useful to re-write equation (1) in terms of Bayes factors. Substituting $\pi_i = \Pr(M_i)$ for the prior model weights and $w_i = \Pr(M_i|X)$ for posterior model weights, and dividing numerator and denominator by $\Pr(X|M_1)$, we have

$$w_j = \frac{\Pr(X|M_j)/\Pr(X|M_1)\,\pi_j}{\sum_{i=1}^{R} \Pr(X|M_i)/\Pr(X|M_1)\,\pi_i},$$

i.e.,

$$w_j = \frac{BF_{j,1}\,\pi_j}{\sum_{i=1}^{R} BF_{i,1}\,\pi_i}. \tag{5}$$

The choice of expressing model weights in terms of Bayes factors relative to model 1 is of no consequence; one could replace $BF_{i,1}$ and $BF_{j,1}$ in equation (5) with $BF_{i,k}$ and $BF_{j,k}$ since the latter values are simply obtained by multiplying the former by $BF_{1,k}$.

2.1 Example 1

A concrete example may be helpful at this point. Suppose that in writing a computer program, you have the uncomfortable feeling that instead of typing RNDU (generating a variate uniformly distributed on the interval [0, 1]) you may have typed RNDN (generating a standard normal variate). Having compiled the code, there is no easy way to check, except to look at the output X. Let M_1 denote the uniform model and M_2 denote the normal model. Then the Bayes factor based on a single observation $X = x$ is the ratio of density functions

$$BF_{1,2} = \frac{\mathrm{I}(0 \leq x \leq 1)}{\exp(-x^2/2)/\sqrt{2\pi}},$$

portrayed in Fig. 1. Note that observations x falling in the range [0,1] shift support toward the uniform model, but that observations outside of this range yield $BF_{1,2} = 0$; these provide incontrovertible evidence against M_1 in favor of M_2. Observations $X < 0$ or $X > 1$ result in posterior probability of zero for model 1. An observation $X = 0.60$ leads to a Bayes factor of 3 : prior odds of 1–1 are changed to posterior odds of 3–1, prior odds of 1–9 are changed to posterior odds of 3–9. In the first case prior probability of 0.50 for model 1 becomes posterior probability of 0.75; in the latter case, prior probability of 0.10 is increased to posterior probability of 0.25 (3/12).

Three features of the Bayes factor are worth emphasizing: first, that is an evaluation of the evidence in favor of one model versus another; this stands in remarkable

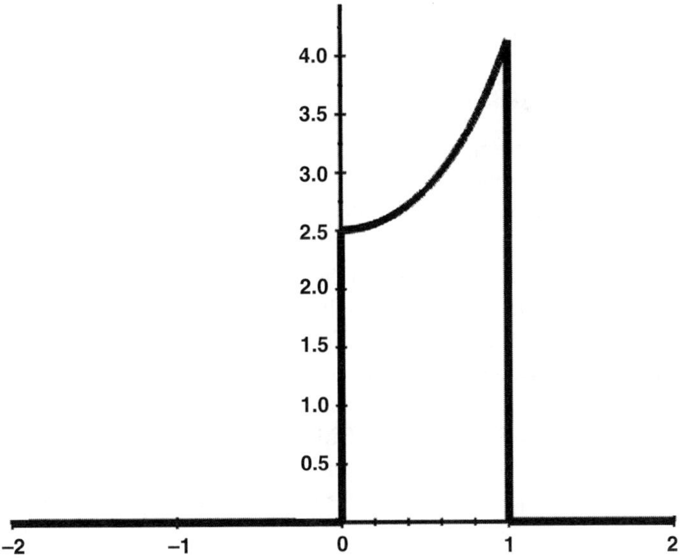

Fig. 1 Bayes factor $BF_{1,2}$ for Example 1, plotted as function of observation x

contrast to test statistics, which only provide evidence against a single model. The second is that Bayes factors operate independently of the prior model weights, indicating only a shift in support: prior probabilities of 0.10 and 0.90 on M_1 and M_2 are only changed to posterior probabilities of 0.25 and 0.75 when $BF_{1,2} = 3$. This feature is neither "good" nor "bad" – it is simply a logical consequence of applying the rules of probability theory to the problem of multimodel inference.

The third feature is that Bayes factors for subsequent independent observations are combined multiplicatively; if you observe $X_i = x_i, i = 1, 2, \ldots, n$ in n runs of the computer program, the Bayes factor for the entire data set is

$$BF_{1,2}(x_1, x_2, ..., x_n) = \prod_{i=1}^{n} BF_{1,2}(x_i).$$

This appealing feature is an example of Bayesian coherency (i.e., adherence to the basic rules of probability). Bayes factors from independent studies are combined by multiplication, yielding the same results as if analyzed simultaneously. In hypothesis testing, analyses of individual data sets need not lead to the same conclusion as analysis of a combined data set; a p-value for a combined data set might not have any direct relation to the p-values for individual data sets.

Kass and Raftery (1995) describe weights of evidence in favor of one model over another as Positive ($3 < \text{BF} \leq 20$), Strong ($20 < \text{BF} \leq 150$), and Very Strong (BF > 150). Not surprisingly, a single observation of X, of necessity producing a Bayes factor $BF_{1,2} < 4.2$, cannot yield strong evidence in favor of the uniform model, though it can provide conclusive evidence against it ($BF_{1,2} = 0$).

2.2 Comments Thus Far

2.2.1 Truth in the Model Set

In the preceding description, we have identified models with a mutually exclusive and exhaustive set of events $\boldsymbol{M} = \{M_1, M_2, ..., M_R\}$. Thus the mathematical structure underlying Bayesian multimodel inference is equivalent to a single draw from a multinomial distribution with cell probabilities $\pi_i = \Pr(M_i), i = 1, 2, \ldots, R$ followed by the generation of data X according to the selected model. Perhaps unfortunately, the selected model is often referred to as the "true model."

This structure has been dismissed as unreasonable, on the grounds that it requires "Truth in the model set"; from thence, one slides easily into the aforementioned philosophical morass about Truth and Models. But " truth in the model set" is merely a model; one might as well criticize the BMI structure for not specifying who made the draw. Berger and Pericchi (1996) refer to "truth in the model set" as "standard Bayesian language" noting that "one does not strictly have to assume that one of the models is true."

In single model inference, we always condition on the model being "true", whether we are certain that to be the case or not Box (1976). We use a t-distribution because under normality the sample mean and variance are independent normal and chi-squared, even though we really might confess uncertainty about these assumptions if pressed hard enough. There is not much of a leap from that modeling abstraction to the multimodel abstraction: "truth in the model set" is just a model.

It is crucial to see multimodel inference as conditional on the model set, just as inference under a single model is conditional on that model. In both cases we acknowledge our inferences as conditional. Posterior probabilities of "truth" are conditional on "truth in the model set." Thus they can be regarded as relative degrees of support, within the model set. If we are radically skeptical of the model set, there is probably no point in model averaging over it; in this case, the use of Bayes factors for comparing models is nevertheless appropriate. Bayes factors stand independent not only of prior model weights, but also of the model set: they can be used simply for pairwise comparisons of models.

So then rather than saying that Bayesian model averaging requires "truth in the model set" we should say that Bayesian model averaging "operates as though truth were in the model set."

2.2.2 Model Weights as Posterior Model Probabilities

Equation (2) can be thought of as a special case of a general formula for a variety of weighted predictions, namely

$$\Pr(Q|X, \boldsymbol{M}) = \sum_{j=1}^{R} w_j \Pr(Q|X, M_j); \tag{6}$$

where w_j summing to one are model weights; here we make explicit that all of the probabilities under consideration are conditional on a model $\boldsymbol{M}$. Equation (6) indicates a recipe for weighted predictions: one obtains a prediction under each of a set $\boldsymbol{M}$ of candidate models, then weights these by the w_j to obtain a single prediction for the entire model set.

Under the Bayesian paradigm, equation (2) indicates that $w_j = \Pr(M_j|X, \boldsymbol{M})$ is a solution of equation (6). If it is reasonable to assign probabilities to models, so that $\Pr(M_j|X, \boldsymbol{M})$ is defined, it can be shown under weak conditions that $w_j = \Pr(M_j|X, \boldsymbol{M})$ is the *only* solution of equation (6); proof is included in Appendix 1. The point of this is that using model weights in predictions of the form equation (6) is essentially treating them as posterior model probabilities, conditional on the model set.

3 Bayesian Multimodel Inference with Unknown Parameters

Equations (1)–(5) were presented as applicable to fully specified models, without unknown parameters. Nothing is changed by supposing that corresponding to model j is an unknown parameter $\theta^{(j)}$, except that we must calculate $\Pr(X|M_j)$ by integrating the conditional distribution $\Pr(X|\theta^{(j)}, M_j)$ against a conditional prior distribution for $\theta^{(j)}$. In the interest of simplicity, we will commence using bracket notation for distributions, with $[A]$ denoting the marginal distribution of A, and $[A|B]$ denoting the conditional distribution of A given B. Thus equations (1)–(5) are generally applicable in BMI, but using

$$\Pr(X|M_j) = \int \left[X|\theta^{(j)}, M_j\right]\left[\theta^{(j)}|M_j\right] d\theta^{(j)}, \tag{7}$$

the marginal distribution of X under model j.

Note that $\left[X|\theta^{(j)}, M_j\right]$ can be regarded as a joint likelihood for parameters and model; indeed, if $\left[\theta^{(j)}|M_j\right]$ were replaced by a point mass prior on the maximum likelihood estimator (MLE), the Bayes factor would reduce to the Neyman–Pearson likelihood ratio. Instead, the marginal distribution is used as a likelihood for the model, averaging the joint likelihood against the parameter uncertainty indicated by the prior.

3.1 Example 2

Let model 1 be that observation X is from a binomial distribution with unknown index N and unknown success parameter p. Assigning a discrete uniform prior for values of N less than or equal to T, and a uniform prior on [0,1] to p, we have

$$\begin{aligned}\Pr(X = x\ |M_1) &= \frac{1}{T+1}\left\{\sum_{N=x}^{T}\binom{N}{x}\int_0^1 p^x(1-p)^{(N-x)}dp\right\} \\ &= \frac{1}{T+1}\sum_{N=x}^{T}\frac{1}{N+1}, 0 \le x \le T.\end{aligned} \tag{8}$$

Let model 2 be that observation X is from a Poisson distribution with unknown parameter λ. Given that λ has a $\Gamma(\alpha, \beta)$ prior, the marginal distribution of X has distribution

$$\Pr(X = x\ |M_2) = \int_0^\infty \frac{\lambda^x e^{-\lambda}}{x!}\frac{\beta^\alpha \lambda^{(\alpha-1)}e^{-\beta\lambda}}{\Gamma(\alpha)}d\lambda$$

$$= \frac{\beta^\alpha \Gamma(\alpha + x)}{(\beta+1)^{\alpha+x}x!\Gamma(\alpha)}, x/, \ge 0 \tag{9}$$

a negative binomial distribution. If $T = 25$ in equation (8), then setting $\alpha = 75/53$ and $\beta = 12/53$ in equation (9) equates mean and variance of the marginal distributions $\Pr(X = x|M)$. The marginal distributions of X under models 1 and 2 are shown in Fig. 2; their ratio, the Bayes factor is given in Fig. 3. Note that since the

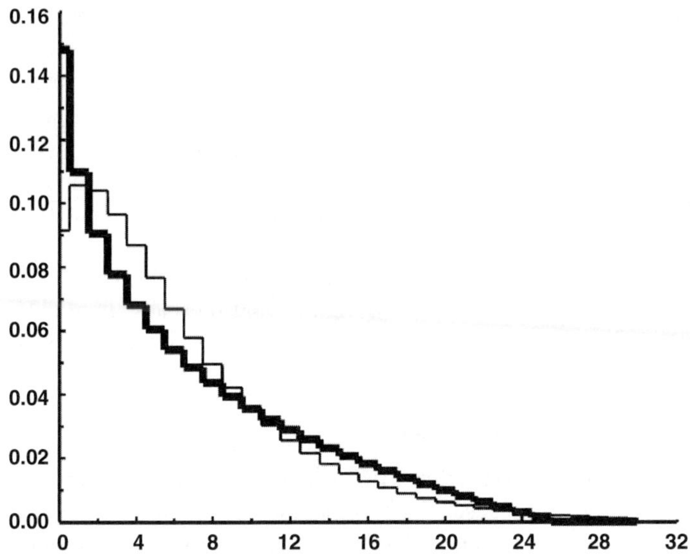

Fig. 2 Marginal distributions for Example 2, under model 1 (*thick line*) and model 2 (*thin line*). Here, $T = 25$, $\alpha = 75/53$, and $\beta = 12/53$

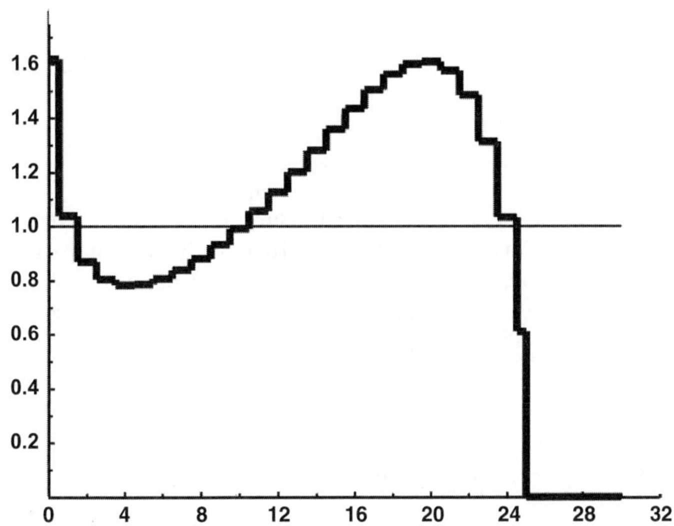

Fig. 3 $BF_{1,2}$ for Example 2, plotted as a function of observation x. Here, $T = 25$, $\alpha = 75/53$, and $\beta = 12/53$

prior on N under model 1 rules out values of $X > 25$, such values are conclusive against model 1.

For now we make no further comment on choice of priors for parameters, a topic of substantial importance to which we return subsequently.

3.2 Approximations for Bayesian Multimodel Inference with Unknown Parameters

Calculation of Bayes factors can be challenging, and various asymptotic approximations have been proposed. As a general rule we do not advocate these, except in preliminary analysis. Given the expenses typically associated with study design and data collection, it seems appropriate to face up to the costs of computation, and give statistical analysis its due.

We mention one such approximation, the Bayesian information criterion, not only because of its usefulness in its own right, but also because it provides a means for evaluating the widely used AIC weights and, perhaps, improving on them.

3.2.1 Bayesian Information Criterion

The Bayesian information criterion (BIC) is unfortunately named, the name having been chosen doubtless because of its computational similarity to the AIC, and not because of any explicit relation to Kullback–Liebler (KL) information. Both are penalized forms of the log-likelihood,

$$\text{AIC} = -2\log\left(f(x,\hat{\theta})\right) + 2k, \tag{10}$$

and

$$\text{BIC} = -2\log\left(f(x,\hat{\theta})\right) + k\log(n), \tag{11}$$

where $\hat{\theta}$ is MLE of θ, k is the number of estimable parameters in the model, and n is the sample size. AIC is an estimator of KL information, but BIC arises in an entirely different fashion, in producing an asymptotic approximation to the Bayes factor in a fully Bayesian analysis, with a specific vague prior on θ. Using subscripts i and j for models, $BF_{i,j}$ is approximated by $\exp\left(-(\text{BIC}_i - \text{BIC}_j)/2\right)$ (Kass and Raftery 1995). Substituting these approximations in equation (5) and simplifying, we have the approximation

$$\hat{w}_j = \frac{\exp(-\text{BIC}_j/2)\,\pi_j}{\sum_{i=1}^{R}\exp(-\text{BIC}_i/2)\,\pi_i}. \tag{12}$$

The quantities typically referred to as BIC weights are obtained using this approximate Bayes factor, and assuming equal prior model weights. There is however nothing requiring that the prior model weights be equal. We will speak of generalized BIC weights as those arising from the BIC approximation to the Bayes factor, but with alternative sets of prior model weights. These can be computed at no additional computational cost relative to AIC weights, allowing (indeed, forcing) the analyst to specify prior model probabilities, rather than to accept default values.

3.2.2 AIC Weights as Generalized BIC Weights

AIC weights are of the same form as equation (12), but with AIC replacing BIC, and equal prior model weights. Thus

$$w_j^{\mathrm{AIC}} = \frac{\exp(-\mathrm{AIC}_j/2)}{\sum_{i=1}^{R} \exp(-\mathrm{AIC}_i/2)},$$

which from the definitions (10) and (11) can be expressed as

$$w_j^{AIC} = \frac{\exp(-\mathrm{BIC}_j/2)\ \ \exp\left((k_j/2)\log n - k_j\right)}{\sum_{i=1}^{R} \exp(-\mathrm{BIC}_i/2)\ \ \exp\left((k_i/2)\log n - k_i\right)}.$$

Thus AIC weights can be considered generalized BIC weights (i.e., approximate posterior model weights) with prior weights

$$\pi_j \propto \exp\left((k_j/2)\log n - k_j\right), \tag{13}$$

as noted by Burnham and Anderson (2004). These priors have the unusual feature of relying on sample size and substantially favor more highly parameterized models when n is large. Thus, since priors are supposed to represent knowledge before collection of data, the more data one plans on collecting, the more credence is placed a priori on the more complex models. Kadane and Lazar (2004) note

> One justification for the AIC is Bayesian (Akaike 1983), namely that asymptotically, comparisons based on Bayes factors and on AIC are equivalent if the precision of the prior is comparable to the precision of the likelihood. This requirement that the prior change with the sample size is unusual asymptotics and, furthermore, is usually not the case. Rather, the data tend to provide more information than the prior.

We recently compared a set of 5 models, having 1, 2, 2, 3, and 4 parameters, respectively. The data set we considered had $n = 1961$; we were taken aback to realize that if we used AIC weights, the corresponding prior model weights form (13) were 0.02, 0. 35, 0. 35, 5.74, and 93.53%, respectively (Link and Barker 2006). Users of AIC weights ought to be aware of equation (13) and its implications; similar results are easily obtained for AIC*c* weights. One of the advantages of AIC weights is their ease of computation. However, we note that generalized BIC weights (equation (12)) can be computed with essentially no additional computational expense, but allowing, indeed forcing the analyst to supply a reasonable set of prior model weights.

4 Effect of Priors on Parameters on Bayes Factors

In Example 2, keeping T= 25 in model 1, but changing the prior under model 2 from having $\{\alpha = 75/53, \beta = 12/53\}$ to $\{\alpha = 75/20, \beta = 12/20\}$ maintains the marginal mean under model 2, but decreases the variance, so that the marginal distributions are given as in Fig. 4. The result is that the Bayes factors are now as in Fig. 5.

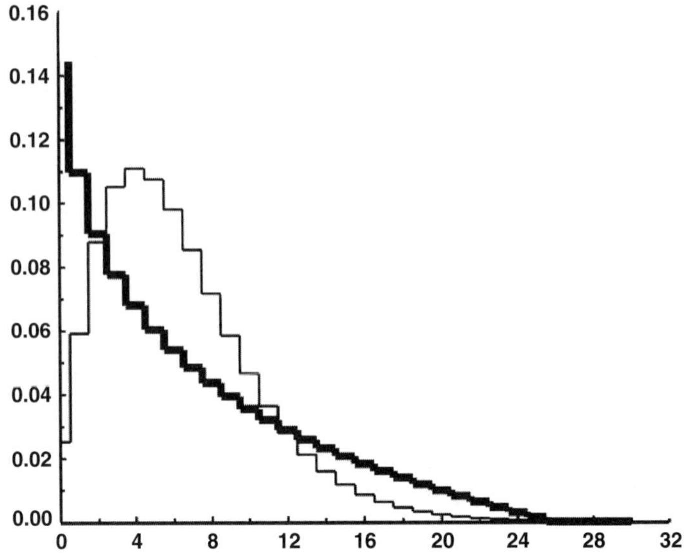

Fig. 4 Marginal distributions for Example 2, under model 1 (*thick line*) and model 2 (*thin line*). Here, T =25, α = 75/5, and β = 12/5

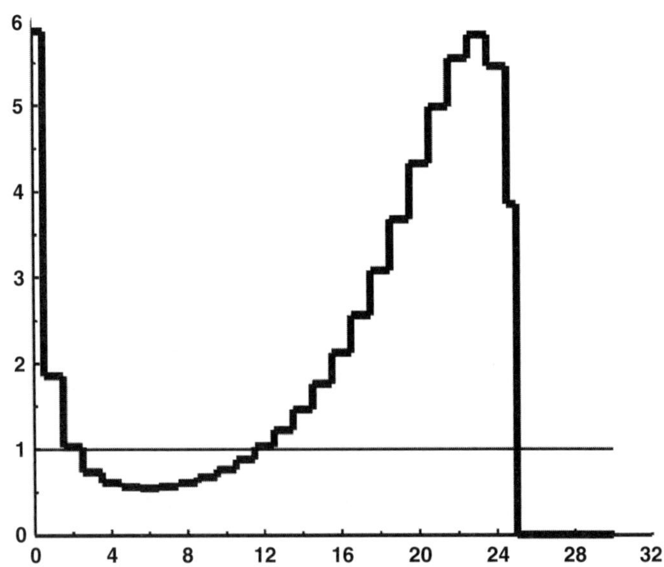

Fig. 5 $BF_{1,2}$ for Example 2, plotted as a function of observation x. Here, T = 25, α = 75/5, and β = 12/5

Thus the choice of priors on unknown parameters plays an important role in determining the Bayes factor for comparing the two models. It is an inevitable and possibly disconcerting feature of BMI that priors on parameters have a substantial influence on Bayes factors. This stands in contrast with single-model Bayesian inference, in which priors on parameters can often be "overwhelmed" by data, in the sense that the influence of priors on posterior distributions diminishes as sample size increases.

The reason for priors on parameters having substantial influence on the model selection problem is that Bayes factors do not merely test, in this case, for Binomial versus Poisson, but for one fully-specified marginal distribution (8) vs. another (9). We do not regard this as a failing of the inferential system, but as a feature that needs to be taken into account in implementation. There is no problem in so doing if informative priors are available, but selecting priors representing limited prior knowledge can be challenging, as indicated by the next example.

4.1 Example 3: Problems Arising from Improper Priors

Bayes factors provide a sensible means for comparing two fully specified models, including comparisons frequently handled by means of hypothesis tests. Suppose for instance that we wish to decide whether an observation X comes from a standard normal distribution (Model 1), or from a normal distribution with mean $\mu = 2$, and standard deviation of one (Model 2). Using the self-explanatory notation $X \sim N(\mu, \sigma^2)$, we have $M_1 : X \sim N(0, 1)$, and $M_2 : X \sim N(2, 1)$. Then $BF_{1,2} = \exp(-2x + 2)$, a strictly decreasing function of x, equal to one when $x = 1$, when the data supports the two models equally (Fig. 6).

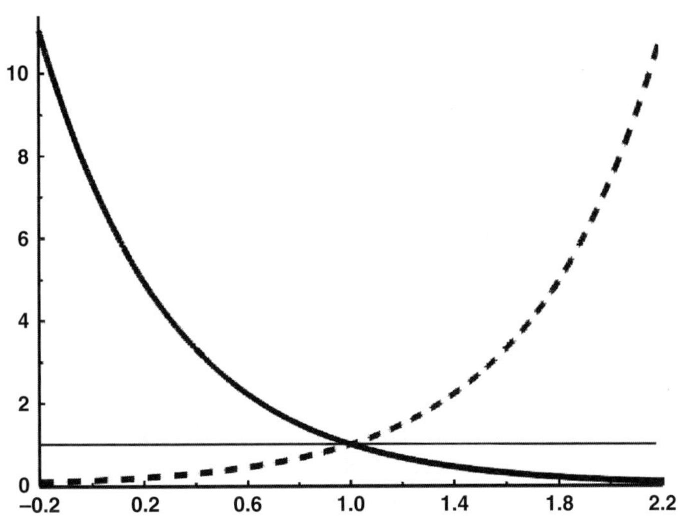

Fig. 6 Bayes factors for Example 3, $BF_{1,2}$ (*solid line*) and $BF_{2,1}$ (*dashed line*), plotted as functions of observation x

Suppose however that in model 2 the mean is unknown; we might suppose that the value "2" still is our best guess, but that our uncertainty about μ is conveniently modeled as that μ is a normal random variable with mean 2 and specified variance σ^2. Then model 2 becomes $M_2 : X|\mu \sim N(\mu, 1), \mu \sim N(2, \sigma^2)$. The marginal distribution for X is the same as if we had specified $M_2 : X \sim N(2, 1 + \sigma^2)$; hence, the Bayes factor is

$$BF_{1,2} = \sqrt{1+\sigma^2}\exp\left(-\frac{1}{2}\;\frac{x^2\sigma^2+4x-4}{1+\sigma^2}\right), \tag{14}$$

The Bayes factor (14) has some disturbing features when either σ or $|x|$ is large.

First, $BF_{1,2}$ is no longer a monotone function of x if $\sigma > 0$. Indeed, for fixed $\sigma > 0$, the limit as x approaches $\pm\,\infty$ is zero. Large negative values of x thus provide more support for model 2 than for model 1, perhaps contrary to intuition. This conundrum is resolved by noting that model 2 is consistent with greater variation in X, noting that Bayes factors provide only a measure of relative support for models, and do not measure model adequacy. With $\sigma = 0.9$, observations $X < -6$ somewhat surprisingly favor model 2 over model 1.

Perhaps more alarming is the observation that for fixed x, the limit as σ approaches ∞ is infinite; a fact that at first note seems to contradict the previous observation (but which involved fixing σ, and varying x). In Bayesian estimation, complete absence of knowledge of an unknown parameter is often represented by use of an improper prior, with infinite variance. Restricting attention to model 2, this approach is completely satisfactory for estimation: the posterior distribution for μ can be shown to be normal with mean X and variance 1, yielding reasonable results, numerically similar to typical frequentist analyses. However, the example considered here shows that the use of improper priors might not be satisfactory in multimodel inference.

Indeed, Berger and Pericchi (1998) note that these problems extend to the use of vague proper priors. The problem is that models having more parameters tend to allow greater prior uncertainty in the range of the data to be produced; this is reflected in typically lower values for the marginal distribution function of the data, hence a tendency for the Bayes factor to be large in comparing a simple model to a more complex model. The greater the uncertainty in the collection of priors, the more serious the problem becomes.

4.2 Choosing Priors on Parameters for BMI

Clearly, implementation of BMI requires careful specification of priors on parameters. We turn our attention once again to the calculation of the marginal distribution,

$$\Pr(X|M) = \int [X|\theta, M]\,[\theta|M]\,d\theta \tag{15}$$

taking heed to the role of $[\theta|M]$. A substantial literature on methods for choosing $[\theta|M]$ exists; see Berger and Pericchi (1996) for a review.

Some examples: Smith and Spiegelhalter (1980) and Spiegelhalter and Smith (1982) proposed using an imaginary minimal training sample that most favors the simpler model as though it were previous data, to inform the prior for the parameter. In the case of Example 3, this would mean supposing a previous observation $X^* = 0$ under model 2, with arbitrarily large prior variance on μ. The resulting posterior would be $[\mu|M_2] = N(0, 1)$, so that the marginal distribution under model 2 would be $N(0, 2)$, and $BF_{1,2} = \sqrt{2}\exp(-x^2/4 - x + 1)$.

Aitkin (1991) suggests using the actual data as a training sample. Thus (for Example 3) the posterior distribution for μ under model 2, having begun with prior having arbitrarily large variance, is $[\mu|M_2] = N(X, 1)$. The resulting Bayes factor is $BF_{1,2} = \sqrt{2\pi}\exp(-x^2/2)$. This procedure has been criticized as contrary to standard Bayesian practice, in using the data twice (Aitkin 1991, Discussion; Berger and Pericchi 1996), but it may perhaps be justified in terms of posterior predictive distributions (see Aitkin 1991, and Aitkin 1991 Response to Comments).

Other authors have suggested splitting data sets into training and analysis portions; using the former to generate informative priors for parameters, and the latter for multimodel inference. This expedient side-steps the criticism of using the data twice, but is itself liable to criticism on various grounds: for one, the results will depend on which subset of the data is selected for training; for another, some loss of information is incurred through not using the training data for model selection. In addition, and more importantly from our perspective, this approach may not deal with the fundamental difficulty that alternative models, incorporating varying amounts of prior uncertainty, will have marginal distributions which prejudice the analysis a priori. This vice appears to be taken as a virtue by some authors, who suggest that Bayes factors "automatically" penalize model complexity. Our perspective is that this phenomenon is inconsistent across applications and hence undesirable; penalties for model complexity, if desired, should be incorporated in priors on models, and not in the Bayes factors.

We thus suggest that priors on parameters be chosen with the marginal distributions in mind, attempting to achieve some level of parity in the effects of priors on parameters across models in the model set.

We will suppose that the observation is of the form $X = (X_1, X_2, ..., X_n)$, where X_i are conditionally independent, given θ and M. The effect of parameter uncertainty on X_i under model M is expressed in the marginal distribution $[X_i|M]$. The guiding principle for selecting priors will be that the marginal distribution of individual observations is as nearly identical as possible, and that the evidence for selecting among models is in the joint marginal distribution. We will say that a set of priors $[\theta^{(j)}|M_j]$ is *nonpreferential over* $\boldsymbol{M}$ if for all i,

$$[X_i|M] = \int [X_i|\theta^{(j)}, M_j]\ [\theta^{(j)}|M_j]\, d\theta^{(j)} \tag{16}$$

is constant for $M_j \in \boldsymbol{M}$.

4.3 Example 4: Random Effects vs. Fixed Effects

Suppose $[X_i|\theta_i, M]$ is specified, and that if $M = 1$, $\theta_i \sim F$, with F known; but under $M = 2$, $\theta_i \equiv \mu$, an unknown. Then any prior satisfying $[\mu|M = 2] = F$ is nonpreferential over the pair of models.

For example, suppose we have n independent observations of Binomial random variables with success rates p_i. Model 1 is that the p_i's are sampled from a specified beta distribution, say β(2,2). Model 2 is that $p_i \equiv \mu$, for some unknown value $0 < \mu < 1$.

A nonpreferential prior is obtained by supposing that $\mu \mid M = 2 \sim \beta(2,2)$. Regardless of whether model 1 or model 2 is correct, the marginal distribution of a single observation is the same beta-binomial. It should be noted, however, that the joint distribution $[X|M]$ depends on M: observations X_i will tend to be more clustered under model 2.

4.4 Example 5: Random Effects vs. Fixed Effects, Continued

A slight generalization of Example 4: Suppose $[X_i|\theta_i, M]$ is specified, and that if $M = 1$, $\theta_i \sim F_\alpha$, with the family F known, but α unknown. As in example 4, under $M = 2$, $\theta_i \equiv \mu$, unknown.

We need to specify prior distributions $[\alpha|M = 1]$ and $[\mu|M = 2]$. If $[\alpha|M = 1] = g$, then suppose that $[\mu|M = 2] = F_\beta$, with $\beta \sim g$. It follows that the marginal distribution $[\theta_i|M]$ is the same mixture of F and g, hence independent of M.

For example, suppose we have n independent observations of Binomial random variables with success rates p_i. Model 1 is that the p_i 's are sampled from an unspecified β(*a*,*b*) distribution, Model 2 is that $p_i \equiv \mu$, for some unknown value $0 < \mu < 1$. Given a specification for priors on a and b under Model 1, we implicitly define a marginal prior $g(p) = [p_i|M = 1]$; a nonpreferential prior is constructed by supposing that given $M = 2$, $\mu \sim g$.

4.5 Example 6: Finite Mixtures of Binomials

Suppose that $X_i|p_i$ are independent binomial random variable with known index N and success parameters p_i. Model 1 is that there is a single unknown value of p_i, say π; model 2 is that p_i is drawn from a two point mixture: $p_i = \pi_L$ with probability w, and $p_i = \pi_H$ with probability $1-w$. Thus

$$p_i = \mathbf{I}(M = 1)\pi + \mathbf{I}(M = 2)(\gamma_i \pi_L + (1 - \gamma_i)\pi_H)$$

where $I(\cdot)$ is the indicator function and γ_i are independent Bernoulli trials with parameter w. A nonpreferential set of priors is obtained as the distributions induced by setting $\pi_L = \min(U_1, U_2)$, $\pi_H = \max(U_1, U_2)$, $w = U_3$, and $\pi = U_4$, where U_i are independent uniformly distributed random variables on [0,1]. Then $[p_i|M]$ is a uniform distribution under either model.

This example can be extended to include k-point mixtures, with mass points $\{\pi_1, \pi_2, ..., \pi_k\}$ having prior distribution equivalent to the order statistics of a sample of size k from a uniform distribution, and weights sampled from a uniform Dirichlet distribution.

4.6 Example 7: Regression Models

Suppose that $[X_i|\theta_i, M]$ is a distribution indexed by parameter θ_i, and that $\{\xi_{i1}, \xi_{i2}, ..., \xi_{ip}\}$ is an associated set of covariates determining θ_i as satisfying, $g(\theta_i) = \sum_{j=1}^{P} \beta_j \xi_{ij}$, for a specified link function g. The problem of choosing covariates is a multimodel inference, with alternative models obtained by setting some of the β_j's equal to zero.

While it may not be possible to construct prior distributions for the unknown β_j's exactly satisfying equation (16), the goal of nonpreferential priors may be approximated by matching typical values of moments of the linear predictor across models.

For example, Link and Barker (2006) considered a set of logistic regression models. Having standardized covariates, they specified mean-zero normal priors for the β_j's, with variances proportional to the number of parameters in the model. The total variance V of the linear predictor was treated as a draw from an inverse-gamma distribution, i.e., so that $1/V$ has mean p/λ = 3.2890 / 7.8014 = 0.4216, and variance/mean ratio of $1/\lambda$ = 1 / 7.8014 = 0.1282. This choice was motivated by the observation that, if logit(p) is a mean-zero normal random variable with variance V, then marginally p has a distribution that is approximately uniform on the unit interval.

5 Discussion

Computational barriers having fallen, the benefits of the Bayesian paradigm are being realized in hierarchical models of increasing complexity. Our capacity as analysts to fit models may be outstripping our efforts at model criticism and evaluation.

There are still computational barriers for BMI. For small model sets, program WinBUGS (Spiegelhalter et al. 2000) can be used, with models sampled according to a multinomial distribution. This approach requires considerable tuning and tweaking. For example, priors for parameters $\theta^{(j)}$ need to be specified not only conditional on their model, but also conditional on models $M^{(k)}$ for $k \neq j$. If these "pseudo-priors" $[\theta^{(j)}|M^{(k)}]$ are vague, mixing across the model space will be slow. These quantities having no bearing on inference, it is wise to choose these as approximating $[\theta^{(j)}|X, M^{(j)}]$, on the basis of preliminary single model analyses.

It is also useful (whether using WinBUGS or not) to take advantage of the fact that BF's are invariant to prior model weights: hence, one may fix a preliminary set of prior model weights, calculate posterior model weights and obtain a working estimate of the BF's. This working estimate of the BF's can then be used to choose new prior model weights which lead to approximately equal posterior model weights, enhancing Markov mixing over model space. The process can be iterated until reliable estimates of the BF's are obtained; these, then can be used with whatever set of prior weights one desires for final model averaging. A rough estimate of the necessary computation time for obtaining a fixed level of precision in estimating Bayes factors can be obtained using methods described in Appendix 2.

A sample set of WinBUGS code for multimodel inference is available online at *Ecological Archives E087-159-S1*, http://esapubs.org/archive/ecol/E087/159/suppl-1.htm, an online appendix to Link and Barker (2006).

Reversible jump Markov chain Monte Carlo (RjMcMC, Green 1995) provides substantial computational advantages over the multinomial model approach. The analysis reported in Link and Barker (2006) conducted using RjMcMC coded in GAUSS, ran in about one tenth the time of the corresponding WinBUGS code. There is, however, a rather steep learning curve associated with RjMcMC.

Regardless of how BMI is implemented, Bayes factors are a lurking presence. Care needs to be taken in choosing priors for parameters; one cannot simply adjust the priors as though they were tuning parameters for the performance of Markov chain simulations.

Appendix 1: Model Weights as Model Probabilities

Theorem *Let Q be an event, let X denote observed data, and suppose that*

$$\Pr(Q|X, M_i) = \Pr(Q|X, M_j)$$

implies $i = j$.

Suppose that posterior model probabilities $\Pr(M_i|X)$ *exist, for models* M_i, $i = 1, 2, \ldots$. *Let* A_i, $i = 1, 2, \ldots$ *be positive constants, corresponding to the models.*

Let $\boldsymbol{M}^{(k)}$ *denote a collection of k models; without loss of generality we will assume* $\boldsymbol{M}^{(k)}$ *consists of the models 1 through k, assuming only that at least one of the models in the set has* $\Pr(M_i|X) > 0$.

Suppose that model weights for the collection $\boldsymbol{M}^{(k)}$ *are defined by*

$$w_{i,k} = A_i \Big/ \sum_{j=1}^{k} A_j.$$

Finally, suppose that for every k,

$$\Pr(Q|X, \boldsymbol{M}^{(k)}) = \sum_{i=1}^{k} w_{i,k} \Pr(Q|X, M_i), \tag{17}$$

Then $w_{i,k} = \Pr(M_i|X, \boldsymbol{M}^{(k)})$.

Proof Let $a_{i,k} = \Pr(M_i|X, \boldsymbol{M}^{(k)}) = \Pr(M_i|X)/\Pr(\boldsymbol{M}^{(k)}|X)$ and $b_i = \Pr(Q|X, M_i)$. Then $\Pr(Q|X, \boldsymbol{M}^{(k)}) = \sum_{i=1}^{k} a_{i,k}\, b_i$, so equation (17) can be written as

$$\sum_{i=1}^{k} a_{i,k}\, b_i = \sum_{i=1}^{k} w_{i,k}\, b_i. \tag{18}$$

Consider the case $k = 2$, and set $w = A_1 \big/ (A_1 + A_2)$. Then equation (18) becomes

$$a_{1,2}\, b_1 + a_{2,2}\, b_2 = w\, b_1 + (1 - w)\, b_2,$$

implying that

$$w = \frac{a_{1,2}\, b_1 + (a_{2,2} - 1)\, b_2}{b_1 - b_2},$$

provided that $b_1 \neq b_2$, which follows from the assumptions. Thus there exists *a unique solution* to equation (18), for $k = 2$. Note that

$$w_{i,k} = a_{i,k} \Big/ \sum_{j=1}^{k} a_{j,k} = \Pr(M_i|X) \Big/ \sum_{j=1}^{k} \Pr(M_j|X) \tag{19}$$

is a solution to equation (18), for all k. Thus equation (19) is the unique solution to equation (18), when $k = 2$, so

$$\frac{A_1}{A_1 + A_2} = \frac{\Pr(M_1|X)}{\Pr(M_1|X) + \Pr(M_2|X)},$$

from which it follows that

$$\Pr(M_1|X)\, A_2 = \Pr(M_2|X)\, A_1. \tag{20}$$

Let $C = A_1/\Pr(M_1|X)$. Then $A_1 = C\ \Pr(M_1|X)$, and from equation (20), $A_2 = C\ \Pr(M_2|X)$. Furthermore, since the reasoning holds under any permutation of the integers i, $A_j = C \Pr(M_j|X)$ for all j. Consequently

$$w_{i,k} = A_i \Big/ \sum\nolimits_{j=1}^{k} A_j = \Pr(M_i|X) \Big/ \sum_{j=1}^{k} \Pr(M_j|X) = \Pr(M_i|X, \boldsymbol{M}^{(k)}). \tag{21}$$

Appendix 2: Precision of Estimated Bayes Factors Using MCMC

The Bayes factor for comparing models i and j is the ratio of posterior odds to prior odds,

$$BF_{i,j} = \frac{{}^{w_i}/_{w_j}}{{}^{\pi_i}/_{\pi_j}}.$$

For specified prior weights, we use MCMC (Markov chain Monte Carlo) to obtain estimated posterior weights $\hat{w}_i$, which are used to obtain estimated Bayes factors, which we distinguish here as $EBF_{i,j}$. A bit of rearrangement yields

$$EBF_{i,j} = BF_{i,j} \left(\frac{{}^{\hat{w}_i}/_{\hat{w}_j}}{{}^{w_i}/_{w_j}} \right).$$

Now supposing that instead of a Markov chain, we had independent draws from the posterior for models, the $\hat{w}_i$ would be cell frequencies for a multinomial random variable, and the delta method approximation would yield

$$SD\left(EBF_{i,j}\right) = BF_{i,j} \sqrt{\frac{(w_i + w_j)}{M\ w_i\ w_j}}.$$

where M is the chain length. Thus in a run of 5.0E6, suppose that the autocorrelation at lag 200 was less than 0.003. We could conservatively treat the full chain of 5.0E6 as equivalent to 2.5E4 independent values. If (through tweaking of the prior model weights) we had approximately equal posterior model weights, and there were 5 models in our model set, we could obtain a rough estimate of $SD\left(EBF_{i,j}\right)$ as

$$SD\left(EBF_{i,j}\right) \approx BF_{i,j} \sqrt{\frac{(\frac{1}{5} + \frac{1}{5})}{2.5\text{E4}\ \frac{1}{5}\ \frac{1}{5}}} = BF_{i,j} \times (0.02).$$

Thus the relative error of the Bayes factors would probably be no more than 2%.

References

Aitkin M (1991) Posterior Bayes factors. Journal of the Royal Statistical Society, Series B, 53: 111–142.

Akaike H (1983) Information measures and model selection. Bulletin of the International Statistical Institute 50:277–290.

Berger JO, Pericchi LR (1996) The intrinsic Bayes factor for model selection and prediction. Journal of the American Statistical Association 91:109–122.

Berger JO Pericchi LR (1998) On criticisms and comparisons of default Bayes factors for model selection and hypothesis testing. ISDS Discussion Paper 97-43. Duke University, Durham, North Carolina, USA.

Box GEP (1976) Science and statistics. Journal of the American Statistical Association 71: 791–799.

Burnham KP, Anderson DR (1998) Model selection and inference: a practical information-theoretic approach. Springer-Verlag, New York, USA.

Burnham KP, Anderson DR (2002) Model selection and multimodel inference: a practical information-theoretic approach. Second edition. Springer-Verlag, New York, USA.

Burnham KP, Anderson DR (2004) Multimodel inference: understanding AIC and BIC in model selection. Sociological Methods and Research 33:261–304.

Chatfield C (1995) Model uncertainty, data mining and statistical inference (with discussion). Journal of the Royal Statistical Society (London), Series A 158:419–466.

Draper D (1995) Assessment and propagation of model uncertainty (with discussion). Journal of the Royal Statistical Society (London), Series B, 57:45–97.

Green PJ (1995) Reversible jump Markov chain Monte Carlo computation and Bayesian model determination. Biometrika 82:711–732.

Hoeting JA, Madigan D, Rafterty AE, Volinsky CT (1999) Bayesian model averaging: a tutorial. Statistical Science 14:382–417.

Kadane JB, Lazar NA (2004) Methods and criteria for model selection. Journal of the American Statistical Association 99:279–290.

Kass RE, Raftery AE (1995) Bayes factors. Journal of the American Statistical Association 90:773–795.

Link WA, Barker RJ (2006) Model weights and the foundations of multimodel inference. Ecology 87:2626–2635.

Smith A, Spiegelhalter D (1980) Bayes factors and choice criteria for linear models. Journal of the Royal Statistical Society, series B, 42:213–220.

Spiegelhalter D, Smith A (1982) Bayes factors for linear and loglinear models with vague prior information. Journal of the Royal Statistical Society, series B, 44:377–387.

Spiegelhalter D, Thomas A, Best NG (2000) WinBUGS, version 1.3 user manual. MRC Biostatistics Unit, Cambridge, UK.

Estimating Demographic Parameters from Complex Data Sets: A Comparison of Bayesian Hierarchical and Maximum-Likelihood Methods for Estimating Survival Probabilities of Tawny Owls, *Strix aluco* in Finland

Charles M. Francis and Pertti Saurola

Abstract We compared a method of moments approach using estimates from a maximum likelihood framework, ultrastructural models within a maximum likelihood framework, and hierarchical models estimated using Markov chain Monte Carlo within a Bayesian framework for estimating survival and recapture probabilities and their variance components for a large, complex 20 year data set consisting of both live recaptures and recoveries. Estimates of mean age-specific survival and recapture probabilities for four age classes (young, second-year, third-year and adult) were similar with all approaches, but the maximum likelihood approach with year-specific parameters estimated some recovery and recapture probabilities on boundaries, leading to overestimates of some individual adult survival probabilities and hence overestimates of adult variance components. All approaches estimated similar coefficients for the relationships between winter temperature and survival probabilities, but the maximum likelihood approaches appeared to exaggerate variation in relation to prey abundance. Annual estimates from the Bayesian hierarchical models were sensitive to the choice of the hierarchical structure; modelling the difference between second-year, third-year and adults in survival and recapture probabilities as random effects better estimated the patterns of annual variation than treating all age classes as independent. Our comparisons suggest that Bayesian hierarchical models may be more likely to produce reliable estimates than maximum likelihood methods, even for large data sets, especially if there are many parameters and considerable annual variation in sample sizes.

Keywords Bayesian hierarchical models · Markov chain Monte Carlo · Mark recapture · Maximum likelihood · Recovery · Survival estimation · *Strix aluco* · Tawny Owl

C.M. Francis (✉)
Canadian Wildlife Service, National Wildlife Research Centre, Environment Canada, ON KIA 0H3, Canada
e-mail: charles.francis@ec.gc.ca

D.L. Thomson et al. (eds.), *Modeling Demographic Processes in Marked Populations*, Environmental and Ecological Statistics 3, DOI 10.1007/978-0-387-78151-8_26,

1 Introduction

Survival probabilities of birds may vary among years due to factors such as annual variation in average weather, individual severe weather events, changes in food supply, population density or other stochastic factors. Understanding the magnitude of this variation and the relative importance of the factors influencing variation is important for modeling population dynamics (e.g., Caswell 2001) and identifying factors that may limit a population (e.g., Barker et al. 2002).

One approach for estimating variance components of annual survival probabilities from mark-recapture data is to use maximum likelihood methods to fit a model with year-specific parameters for all survival probabilities and then estimate relationships based on analysis of these point estimates taking into account their estimation error (Link 1999). Burnham and White (2002) describe such a method of moments approach which has been implemented in program MARK (White and Burnham 1999) to estimate the true variance in a collection of parameters such as survival probabilities, after adjusting for the sampling variances and covariances of the parameters. They showed, through simulations, that this approach can perform well with a standard mark-recapture model. External covariates can also be incorporated into the model to estimate their importance in explaining observed variation (e.g., Barker et al. 2002; Francis and Saurola 2004). However, this approach is dependent on obtaining reasonable estimates not only for the parameters but also their sampling variance–covariance matrix. Especially in large, complex models, it is often necessary to use a logit link function or similar transform to ensure convergence of the estimation procedure and to ensure that estimates remain within bounds (White and Burnham 1999). In this case, if any parameters are estimated near the boundaries, the variance–covariance matrix is not reliably estimated, and the method may not perform well (Burnham and White 2002). This may be a concern not only with sparse data sets, but also with relatively large data sets with substantial annual variation in sample size, such that some cells may have small sample sizes. An additional limitation is that this method only allows modelling random effects in one set of parameters at a time (e.g., adult survival probabilities), whereas in a complex data set it may be appropriate to consider simultaneously several groups of parameters as random effects (e.g., survival, recapture and recovery probabilities of multiple age classes).

An alternative approach for modelling variation in parameters in relation to covariates is to build a fixed effects ultrastructural model that considers selected parameters (e.g., adult survival probabilities) as constant or as a linear function (usually on a logit scale) of external covariates, and then directly estimates the coefficients of the extra parameters (Lebreton et al. 1992). Such models are readily implemented in the software package MARK using a design matrix to specify the constraints (White and Burnham 1999). Some possible options include modelling a set of parameters as constant over time (thus effectively estimating mean values of the parameters), modelling a logit-linear trend over time, or modelling relationships with external covariates, such as weather or food availability. This approach can be used to fit models even with data sets that are too sparse to obtain precise estimates for individual years, but has a number of limitations. The precision of the estimated mean values or regression coefficients may be overstated, as they are based on the unrealistic assumption that the data exactly meet the model without any additional sources of variation. If there is among-year variation in sample size, estimates may be weighted towards years with larger sample sizes. Furthermore, constraining one group of parameters to fit exactly in a deterministic relationship may lead to bias in estimates of coefficients of other parameters if the true model includes extra variation, because of correlations among parameter estimates. For example, modelling recapture probabilities as a linear function of a covariate may bias estimates of survival probabilities if there is additional, unmodeled variation in recapture probabilities that does not fit the linear function. Finally, this approach does not allow estimation of variation around the expected values, because this is assumed to be zero. Estimates of variance components are necessary for any form of stochastic population modelling; realistically, we do not expect demographic parameters to vary in a deterministic way with any given set of external covariates, but to be influenced to at least some degree by additional, unknown sources of variation. This can only be captured with a random effects, rather than a fixed effects model.

Link and Barker (2004) recommended an alternative approach to estimate random effects and relationships with covariates, using Bayesian hierarchical models that directly incorporate all of the potential sources of variance, including fixed effects, random effects, and any desired covariates into the model structure. They suggested that such a conceptual framework has many advantages, including the fact that all desired parameters are estimated simultaneously rather than after the fact in a somewhat ad hoc fashion. The models can be readily fitted using Markov chain Monte Carlo (MCMC) methods within a hierarchical Bayes framework using the MCMC features recently implemented in MARK (White et al. 2008). Several recent papers have promoted the advantages of a Bayesian framework for analysis of population parameters (e.g., Brooks et al. 2004; Jamieson and Brooks 2004; Link and Barker 2004). However, we are not aware of any published papers that have applied the MCMC feature in MARK to a large, complex real mark-recapture data set, and compared the results to the more traditional maximum likelihood (ML) approaches.

We note that Lele et al. (2007) proposed an approach which they called data cloning, to obtain ML estimates for hierarchical models using MCMC methods, that is independent of the choice of priors. However, several aspects of hierarchical models, and in particular random effects models, are inherently Bayesian. For example, a random effects model that assumes particular groups of parameters (e.g., first year or adult survival probabilities) are drawn from a random distribution (e.g., normal) is effectively imposing an empirical Bayes prior on the data that allows information from other years to inform estimates for years with limited data. This is, in some ways, intermediate between a model assuming that parameters remain constant over time, and a complete fixed effects model estimating each parameter separately (Burnham and White 2002). Because we view this as one of the potential advantages of hierarchical models in a mark-recapture context, we consider only the

Bayesian implementation of hierarchical models in this paper, but note that other approaches do exist.

In this paper, we use an empirical approach to compare each of the three methods described above for estimating survival probabilities and their correlates for a large, relatively complex data set for Finnish Tawny Owls (*Strix aluco*), involving combined recapture and recovery data over a 20-year time period. We have previously shown, based on estimates from ML models (Francis and Saurola 2004), that survival probabilities of this species vary in relation to both winter weather (especially mean winter temperature) and the stage of the multi-year abundance cycle of voles and other microtine rodents. Recapture probabilities (which are related to breeding propensity) are also strongly affected by the vole cycle.

We compare the three approaches for estimating several types of parameters: mean values of age-specific survival, recapture, recovery and fidelity parameters; relationships between survival and capture probabilities and both categorical covariates (the stage of the vole cycle) and continuous covariates (measures of winter weather); and variance components of survival probabilities with and without covariates. We also compare two approaches, within the hierarchical model framework, for estimating age-specific variation in survival and recapture probabilities beyond the age classes (young and adults) that were specifically identified at the time of marking.

Unlike a simulation, our approach does not allow a determination of the absolute validity of each approach (because the true values are unknown). Instead we determine whether parameter estimates, and biological conclusions, differ among the approaches and identify any practical advantages or disadvantages of each approach based on a real data set.

2 Methods

2.1 Field Data

Tawny Owls (*Strix aluco*) breed throughout much of Europe. In Finland, the Tawny Owl is a relative newcomer, with the first record in 1875 and the distribution still largely restricted to the south (Saurola 1995). Finnish bird ringers, both amateurs and professionals, have concentrated on ringing owls for many years, with the result that relatively large sample sizes are available for several species. Ringing of Tawny Owls has taken place throughout much of the species breeding range in southern Finland since the 1960s (Saurola 1997; Francis and Saurola 2002, 2004; Saurola and Francis 2004). Many individuals now breed in nest boxes, mostly installed by ringers and inspected regularly during the breeding season. At successful nests, the nestlings are ringed when they are of sufficient age. In addition, most ringers also attempt to trap and ring the adult female and many also trap the male at each nest. We selected data from the 20 year period from 1980 to 1999 for analysis. This period was selected to match the analyses in Francis and Saurola (2004), and also because the 3-year cycle in abundance of voles was very regular during this period, facilitating estimation of the impacts of the vole cycle on demography. The sample sizes during this period (including only breeding season captures) were 18,166 nestlings ringed, leading to 1,737 recaptures and 1,655 recoveries, along with 1,742 adults ringed which generated 1,278 recaptures and 190 recoveries.

Small mammal abundance varied dramatically among years, with variation in abundance of voles between high and low points in the population cycle up to two orders of magnitude (Brommer et al. 2002; Hanski et al. 1991). Previous analyses have shown that patterns were generally well correlated at broad spatial scales with a 3-year cycle in most of southern Finland (Sundell et al. 2004). However, quantitative measures of vole abundance were only available from selected areas of Finland and not for the whole of the time period, so it was necessary to combine these data with qualitative, casual observations in the field to classify years for our study period into one of three categories, here labelled based on the potential impact for survival of owls over the coming year:

1. Poor – voles are at their peak at the beginning of the year (which is good for breeding), but crash later in the year, sometimes starting in the summer, such that owls have very few voles available for food during the year, especially in the following winter;
2. Medium – voles are low at the start of the year, but gradually increase over the course of the year;
3. Good – voles are moderately abundant at the beginning of the year and increase to a peak over the summer and following winter.

As an index of winter severity, we used the mean of the mean monthly temperatures between December and March taken as the mean of data from five weather stations of the Finnish Meteorological Institute in southern Finland: 1201 Jokioinen (60° 49'N, 23° 30'E) ; 1303 Hattula Leteensuo (61° 04'N, 24° 14'E); 1304 Hattula Lepaa (61° 08'N, 24° 20'E); 1306 Pälkäne Myttäälä (61° 20'N, 24° 13'E); 1403 Lammi, Biological station (61° 03'N, 25° 03'E). Mean temperature over the winter varied from a low of –11°C to a high of –2°C. For simplicity of comparison, we did not consider the second measure winter severity, snow depth, examined by Francis and Saurola (2004) which, in any case, had a much weaker relationship.

2.2 *Statistical Analyses*

Survival, recapture and recovery probabilities were modeled using the joint recapture and recovery model of Burnham (1993). Previous analyses have shown that these combined models allow estimation of age-specific survival probabilities with little apparent bias due to emigration from the study areas (Francis and Saurola 2002). These models estimate four classes of parameters: survival (Φ) – the probability that an animal alive at the beginning of the year (here defined as 1 June) will be alive the following year; recapture (p) – the probability that a marked individual alive and present in the study population will be captured in a particular year; recovery (r) – the probability that an individual that dies in a particular year will be found and its ring number reported to the ringing office; and "fidelity" (F) – defined as the probability that a marked surviving individual that was in the local population the previous year is still in the population available for recapture. The "fidelity" parameter is difficult to interpret biologically because both recaptures and recoveries occurred over a large geographical area (Francis and Saurola 2004); hence it does not actually provide an estimate of fidelity to a particular breeding location. Rather, it estimates the probability that a bird will return to breed in an area where a ringer is working.

We used models in which survival and recapture probabilities were allowed to differ with age over 4 age classes, while recovery probabilities and fidelity parameters were allowed to differ between two age classes (many parameters were not estimable if additional age classes were added for recovery or fidelity because too few birds were recaptured as known second or third years and then recovered). Birds ringed as "adults" were treated as if they were all in the highest age class. In practice, these would have included some birds in their second or third year. Although methods are now known for distinguishing these younger age classes based on plumage, they generally were not identified during the early years of this study. However, these were not a large percentage of birds captured (because of the lower capture probabilities of younger birds), and are thus unlikely to cause much bias in the adult estimates.

All model selection and parameter estimation using both ML and MCMC models was performed using program MARK, with all models fitted using a logit-link function (White and Burnham 1999).

For the ML approach, the most general model we fitted allowed all age-specific parameters to vary independently among years. Constrained models were also fitted in which 2nd and 3rd year survival and recapture probabilities were constrained using a design matrix to vary in parallel (on a logit scale) with those of adults and/or fidelity parameters were constrained to be constant over time (Table 1). For comparison of fit among these models, QAICc was calculated using $\hat{c} = 1.20$ as determined previously by Francis and Saurola (2004). MARK underestimated the number of parameters for some models, in which one or more parameter estimates was on a boundary, and it was thus necessary to adjust the parameter counts to determine the correct QAICc.

For the method of moments approach, year-specific parameters were estimated for each age class from the ML models, and then the mean values and process variance were estimated using the variance components procedure implemented in MARK. We also fitted models incorporating stage of the vole cycle, winter temperature or both as covariates, and estimated the change in the residual variance. This procedure fits a regression model to a group of parameters taking into account the estimated variance–covariance matrix (Burnham and White 2002). For each age class, we excluded the final year estimates which were generally not identifiable (e.g., the final survival parameter is confounded with the final recapture or recovery parameter). In a few cases, one or more parameters were estimated at boundaries, in which case we calculated means with and without those boundary estimates.

Table 1 Model selection results for a series of models based on maximum likelihood estimation using MARK, showing models with year-specific survival parameters in increasing order of generality and decreasing parsimony (models 1–5), followed by models with survival constrained by various covariates (models 6–9)

	Model parameters[1]				Model fit[2]		
#	S	P	R	F	ΔQAICc	# Parm.	QDeviance
1	Y,+,+,Y	Y,+,+,Y	Y,Y	C,C	0.0	122	3272.2
2	Y,+,+,Y	Y,Y,Y,Y	Y,Y	C,C	35.4	154	3237.0
3	Y,Y,Y,Y	Y,+,+,Y	Y,Y	C,C	41.9	156	3239.9
4	Y,Y,Y,Y	Y,Y,Y,Y	Y,Y	C,C	76.2	188	3203.2
5	Y,Y,Y,Y	Y,Y,Y,Y	Y,Y	Y,Y	129.4	222	3183.9
6	WV,+,+,WV	Y,+,+,Y	Y,Y	C,C	11.3	92	3358.5
7	W,+,+,W	Y,+,+,Y	Y,Y	C,C	25.6	88	3385.3
8	V,+,+,V	Y,+,+,Y	Y,Y	C,C	109.3	90	3480.9
9	C,C,C,C	Y,+,+,Y	Y,Y	C,C	126.2	86	3510.9

[1] Parameters represent probabilities for survival (S), recapture (P), recovery (R) and "fidelity" (F) – see text for definitions. In each column, letters refer to variation modeled in first-year, second-year, third-year and adult age classes respectively (or first-year and older age classes for R and F): Y = year-specific, C = constant, + = parallel (on a logit scale) to the adult age class, V = varies only with stage of the vole cycle, W = logit-linear relationship with winter temperature, WV = varies with both weather and voles.

[2] $\hat{c}$ was estimated at 1.2 from a parametric bootstrap procedure.

We estimated mean values of all parameters on the real scale (i.e., after back-transforming from the logit scale). However, variance components were estimated on the logit-transformed parameters to allow comparison with estimates from the MCMC approach.

For the ultra-structural, fixed effects ML approach, we fitted models in which first year and adult survival probabilities were further constrained, using a design matrix, to be either constant, to vary only with the stage of the vole cycle, to vary only with weather, or to vary with both the stage of the vole cycle and with weather (Table 1). For this approach, all models were fitted with second and third year survival and recapture parameters constrained to fluctuate in parallel with those of adults. Covariates were modelled only on survival parameters, while first year and adult recapture parameters were allowed to vary freely among years. These models provide estimates of mean values and regression parameters, but not variance components around the model (as the latter are assumed to be zero in this approach).

For the hierarchical models approach, we used the MCMC estimation procedure in MARK based on one of three underlying structural models, as defined through the design matrix. The first was the most general model described above which allowed all age-specific parameters to vary independently. The second modeled second and third year survival and recapture parameters to be equal to those of adults plus a difference parameter (on a logit scale), but these difference parameters were allowed to vary among years. Within an ML framework, this has the same number of parameters as the most general model (and produces identical results to the most general model), but in a hierarchical framework it differs in that prior "hyperdistributions" can be set on the differences between age classes, rather than on the values for each age class separately. The third model constrained the differences between second-year, third-year and adult age classes to be constant over time for both survival and recapture, and further constrained fidelity parameters as being constant over time, thus matching the most parsimonious model fitted through the ML approach (model 1 in Table 1).

Imposing a hierarchical structure to these data in MARK was done by defining the hyperdistribution using a "hyperdesign" matrix to set up expected values for each group of age-specific parameters (White et al. 2008). Each parameter was modeled as a linear function (on a logit scale) of various covariates, with an error term ϵ_t assumed to be normally distributed with mean 0 and variance σ^2 where σ^2 was modeled separately for each group of parameters. The following four linear functions for particular time-specific parameters ϕ_t were considered (b_i represent regression coefficients):

(1) Mean (M): $\text{Logit}(\phi_t) = b_1 + \varepsilon_t$

(2) Voles (V): $\text{Logit}(\phi_t) = b_1X_1 + b_2X_2 + b_3X_3 + \varepsilon_t$
where $X_1 = 1$ if the voles are at stage 1 in year *t*, otherwise $X_1 = 0$; $X_2 = 1$ if voles are at stage 2 in year *t*, otherwise $X_2 = 0$, and $X_3 = 1$ if voles are at stage 3 in year *t*, otherwise $X_3 = 0$.

(3) Weather (W): $\text{Logit}(\phi_t) = b_1 + b_2T_t + \varepsilon_t$
where T_t = the difference in year *t* from the mean temperature over the 20-year time period, standardized to variance 1.

(4) Voles and Weather (VW): $\text{Logit}(\phi_t) = b_1X_1 + b_2X_2 + b_3X_3 + b_4T_t + \varepsilon_t$
where X_i, and T_t are defined as in Equations (2) and (3).

In the most general model, functions were fitted with separate means and variances for 12 groups of parameters (4 age classes each for survival and recapture, and 2 each for recovery and fidelity). For comparisons among the three different structural models used for MCMC, all age-specific groups of parameters were modeled with no covariates (Equation 1). For estimation of variance components as well as changes in expected values with covariates, we considered only models in which second and third year survival and recapture probabilities were parameterized using time-varying differences from adult parameters (models 11–15 in Table 2). For these models, hyperdistributions for first-year and adult survival parameters were modeled with each of Equations 1–4 while those for first-year and adult recapture parameters were modeled with either Equations 1 or 2, as listed in Table 2. Other parameters, including the difference parameters for second and third year survival and recapture probabilities, as well as recovery and fidelity parameters were only modeled with Equation 1 (i.e., no relationships with external covariates were considered).

To complete the hierarchical model, standard non-informative priors were placed on the coefficients of each of the design parameters (b_i) and each of the variance parameters (σ_i^2) for the hyperdistributions, as well as any parameters constrained to be constant over time, as described by White et al. (2008).

Because of the complexity involved with editing the large Parameter Index Matrices, the Design Matrix and the Hyperdesign Matrix, the input files for the analyses were created using custom-written SAS programs (SAS Institute 2003),

Table 2 Model structure and Deviance Information Criteria (DIC) for all hierarchical models considered in this paper and fitted using MCMC

#	Model parameters[1] S	P	R	F	DIC
10	Y,Y,Y,Y	Y,Y,Y,Y	Y,Y	Y,Y	4171
11	Y,+y,+y,Y	Y,+y,+y,Y	Y,Y	Y,Y	4146
12	Y,+y,+y,Y	V,+y,+y,V	Y,Y	Y,Y	4140
13	V,+y,+y,V	V,+y,+y,V	Y,Y	Y,Y	4140
14	W,+y,+y,W	V,+y,+y,V	Y,Y	Y,Y	4135
15	WV,+y,+y,WV	V,+y,+y,V	Y,Y	Y,Y	4136
16	Y,+c,+c,Y	Y,+c,+c,Y	Y,Y	C,C	4153
17	V,+c,+c,V	V,+c,+c,V	Y,Y	C,C	4153
18	W,+c,+c,W	V,+c,+c,V	Y,Y	C,C	4145
19	WV,+c,+c,WV	V,+c,+c,V	Y,Y	C,C	4143

[1] Parameters represent probabilities for survival (S), recapture (P), recovery (R) and "fidelity" (F) – see text for definitions. In each column, letters refer to variation modeled in first, second, third and adult age classes respectively (all models on a logit scale): Y = random effects with no covariates, V = random effects with mean that differs for each stage of the vole cycle, W = random effects around logit-linear relationship with winter temperature, WV = random effects that varies both with winter temperature and voles, +y = difference between second or third year and adults modelled as random effects with no covariates, +c = difference between second or third year and adults constrained to be constant over time, C = constant over time.

and run using the MARK batch facility. For the ML approach, it was sometimes necessary to rerun the models a number of times using starting parameters derived from simpler models, as the ML estimation procedure sometimes converged on local maxima.

All MCMC models were run with the default options in MARK, using a random starting point, 1,000 burn-in samples, 4,000 tuning samples, and 10,000 iterations in each chain. A minimum of six independent chains were run for each model. In all cases, for the models presented here, the Gelman (1996) convergence diagnostics provided by MARK were less than 1.02 (usually less than 1.001), indicating no marked difference between the chains; parameter estimates matched to 2 or 3 significant digits, much less than the standard error estimates. Estimates presented here were derived by combining data from 6 to 10 chains (to increase precision), and determining the 2.5, 50, and 97.5 percentiles (representing the median and 95% posterior credible intervals), as well as the mean and standard error for all parameters and hyperparameters of interest. MCMC models were compared based on the Deviance Information Criterion (DIC) as calculated within MARK; we present the mean DIC of all estimation chains run for a particular model.

3 Results

Model selection in the ML framework, among models in which first-year and adult survival and recovery probabilities were allowed to vary with time (models 1–5 in Table 1) indicated that the most parsimonious approach involved constraining fidelity parameters to be constant over time and second and third year survival and capture probabilities to vary in parallel with those of adults.

ML models incorporating either voles, winter weather or both in an ultrastructural framework, were a much better fit than models assuming survival was constant over time (models 6–9; Table 1), but not nearly as good a fit as the model allowing survival to vary freely among years (model 1; Table 1) indicating that, although some variation in survival probabilities could be explained by vole abundance and/or winter weather, these covariates were clearly not sufficient to explain all of the annual variation in survival.

With the MCMC hierarchical approach, models in which second- and third-year survival and recapture probabilities were treated as differing from those of adults based on time-varying difference parameters (models 11–15; Table 2) generally had a much lower DIC than equivalent models in which these probabilities were modeled independently from those of adults (model 10). They were also somewhat better than equivalent models in which the difference parameters were treated as constant over time (models 16–19; Table 2).

Despite differences in fit of the models, estimated means of age-specific parameters were generally similar regardless of the estimation procedure used (Table 3). The only exceptions were related to problems in the ML model approach with 1 or more estimates at the upper boundary for recovery probabilities in both the general and parsimonious models, and for 1 estimate of recapture probability under the most general model. In these cases, because the standard error was estimated at zero, the method of moments estimator did not perform well, and overestimated the mean.

Table 3 Estimates of mean age-specific survival probabilities with their standard errors derived from selected maximum likelihood and hierarchical models listed in Tables 1 or 2. Means from the ML process were calculated using the method of moments procedure in MARK from the year-specific ML estimates with their estimated variance–covariance matrix. Means from the hierarchical models were estimated directly using MCMC with all year-specific groups of parameters modelled as random effects drawn from a normal distribution with the specified mean and no covariates (see Methods for details)

	Maximum likelihood		MCMC hierarchical model		
Parameter	Model 5	Model 1	Model 10	Model 11	Model 16
$S_{year\ 1}$	32.8 ± 3.0	33.1 ± 3.1	35.0 ± 3.6	34.7 ± 3.4	34.8 ± 3.5
$S_{year\ 2}$	63.9 ± 4.1	64.3 ± 3.4	63.3 ± 2.2	63.7 ± 3.1	64.0 ± 3.1
$S_{year\ 3}$	76.0 ± 3.2	72.0 ± 3.3	72.3 ± 2.2	71.1 ± 3.0	70.9 ± 3.0
S_{adult}	76.4 ± 2.7	74.7 ± 2.5	76.3 ± 1.5	76.0 ± 2.2	75.6 ± 2.3
$P_{year\ 1}$	18.5 ± 3.0	18.6 ± 3.0	16.7 ± 3.2	16.6 ± 3.1	16.5 ± 3.2
$P_{year\ 2}$	37.6 ± 5.0[a] (32.0 ± 2.9)	32.7 ± 2.8	34.4 ± 3.4	32.6 ± 3.2	32.5 ± 3.1
$P_{year\ 3}$	37.6 ± 4.0	38.1 ± 3.2	39.5 ± 4.0	37.7 ± 3.8	38.0 ± 3.5
P_{adult}	42.0 ± 2.6	41.9 ± 2.8	42.3 ± 2.8	42.3 ± 3.0	42.4 ± 3.1
$R_{year\ 1}$	8.0 ± 0.6	8.5 ± 0.7	8.2 ± 0.8	8.1 ± 0.8	8.2 ± 0.8
R_{adult}	23.0 ± 6.5[b] (10.9 ± 0.7)	19.2 ± 4.7[a] (12.1 ± 0.8)	13.6 ± 1.3	14.0 ± 1.1	14.0 ± 1.2
$F_{year\ 1}$	34.5 ± 2.6	37.4 ± 2.7	37.1 ± 2.8	37.8 ± 2.5	37.5 ± 2.7
F_{adult}	90.6 ± 2.8	89.5 ± 1.4	92.9 ± 3.0	91.2 ± 2.4	90.2 ± 1.4

[a, b] includes 1[(a)] or 2[(b)] values estimated at upper boundary with sampling variance estimated at zero. Values in parentheses show estimated means excluding these values.

However, omitting these boundary estimates underestimated means relative to other approaches (Table 3).

Precision of the estimates was also generally similar with all methods with the exception of the adult "fidelity" parameter, which was estimated much more precisely in models that assumed it was constant over time (models 5 and 16). Of course, this greater precision may be misleading if the true parameter varied over time.

In contrast to mean values, estimates of the pattern of annual variation in survival, recapture and recovery probabilities differed among estimation methods (Figs. 1–3). For the ML models, the precision of individual survival estimates for all age classes (as reflected in much narrower 95% confidence limits) was greatly improved by using the more parsimonious models (compare Fig. 1b with Fig. 1a). Both ML models suggest one or more years with exceptionally high adult survival probabilities, but the years are not the same with the two models (1984 and 1987 for the most general model, but 1988 for the parsimonious model). These estimates were related to boundary estimates for the corresponding recovery parameters (Fig. 3a,b). Similar problems with boundary estimates of recapture parameters were apparent in the most general model (Fig. 2a) but all recapture estimates were relatively precise and well behaved based on the more parsimonious model in which 2nd and 3rd year capture probabilities were constrained to vary in parallel with those of adults (Fig. 2b).

The MCMC models did not have problems with boundary estimates in any parameters (Figs. 1–3), but differed with the choice of hierarchical structure: models in which all age-specific survival and recapture parameters were modelled as independent random effects suggested very little annual variation in survival of older age classes (Fig. 1c). In contrast, the model treating differences between second-year, third-year and adult age classes for both survival and recapture as random effects not only had a lower DIC (models 11–15, Table 2), but also indicated much greater annual variation in survival probabilities of older age classes (Fig. 1d). Estimated first-year survival probabilities were similar with either model. Estimated first-year and adult recapture probabilities were also similar with either model, but second and third year recapture probabilities were estimated much more precisely with the difference model (Fig. 2d). Estimated recovery probabilities from the difference model showed much less annual variation (Fig. 3d), suggesting that the extra variation in recovery probabilities in Fig. 3c may have been due to inadequately estimated variation in adult survival probabilities.

Variance components analysis based on the method of moments ML estimator and the hierarchical models MCMC approach gave similar estimates of the variance in first-year survival probabilities (Table 4). This is consistent with the similarity in the pattern of annual variation for first-year survival from both approaches (Fig. 1b,d). Both approaches indicated that the residual variance could be reduced by considering voles and/or winter temperature as covariates, though the ML approach suggested these explained a higher percentage of the total variance compared with the MCMC approach (Table 4).

For adult survival, the estimated process variance based on the ML approach was nearly double that estimated by the MCMC approach (Table 4). Both approaches

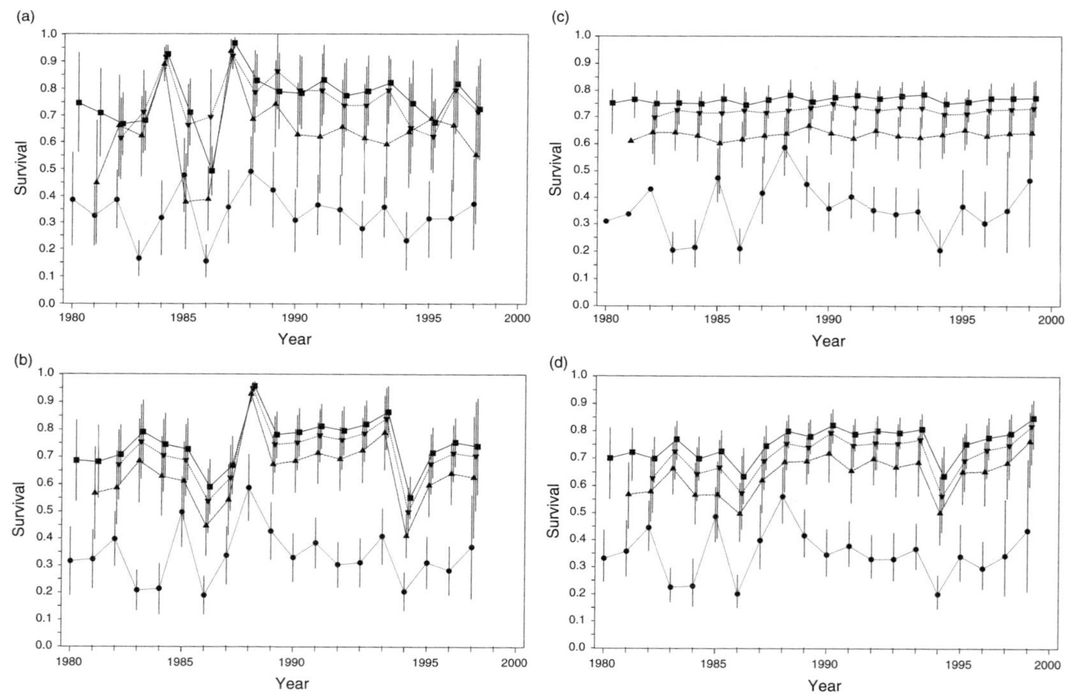

Fig. 1 Estimated annual survival probabilities with 95% confidence/credible intervals for Tawny Owls in their first year of life (*lowest line*, ●), second year of life (*lower middle line*, ▲), third year of life (*upper middle line*, ▼) and as adults (*top line*, ■) from 1980 to 1997 derived from four different models (see Tables 1 and 2 for descriptions of each model number): (**a**) ML model #5; (**b**) ML model #1; (**c**) MCMC model #10; (**d**) MCMC model #11. For ML models, estimates shown are shrinkage estimates based on applying the method of moments estimator in MARK separately to parameters for each age class. No covariates were included in any of the models

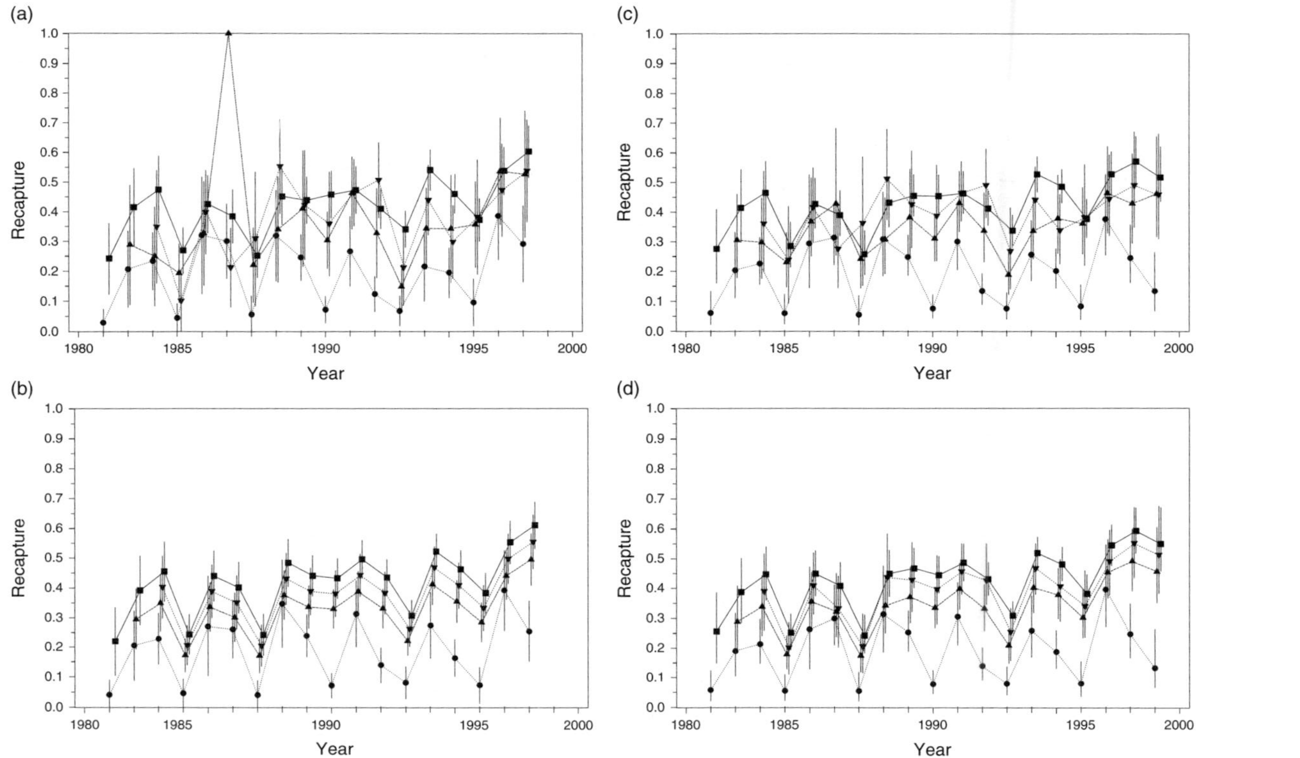

Fig. 2 Estimated annual recapture probabilities with 95% confidence/credible intervals for Tawny Owls one year old (*lowest line*, ●), two years old (*lower middle line*, ▲), three years old (*upper middle line*, ▼) and as adults (*top line*, ■) from 1980 to 1997 derived using the same approaches and same models as in Fig. 1

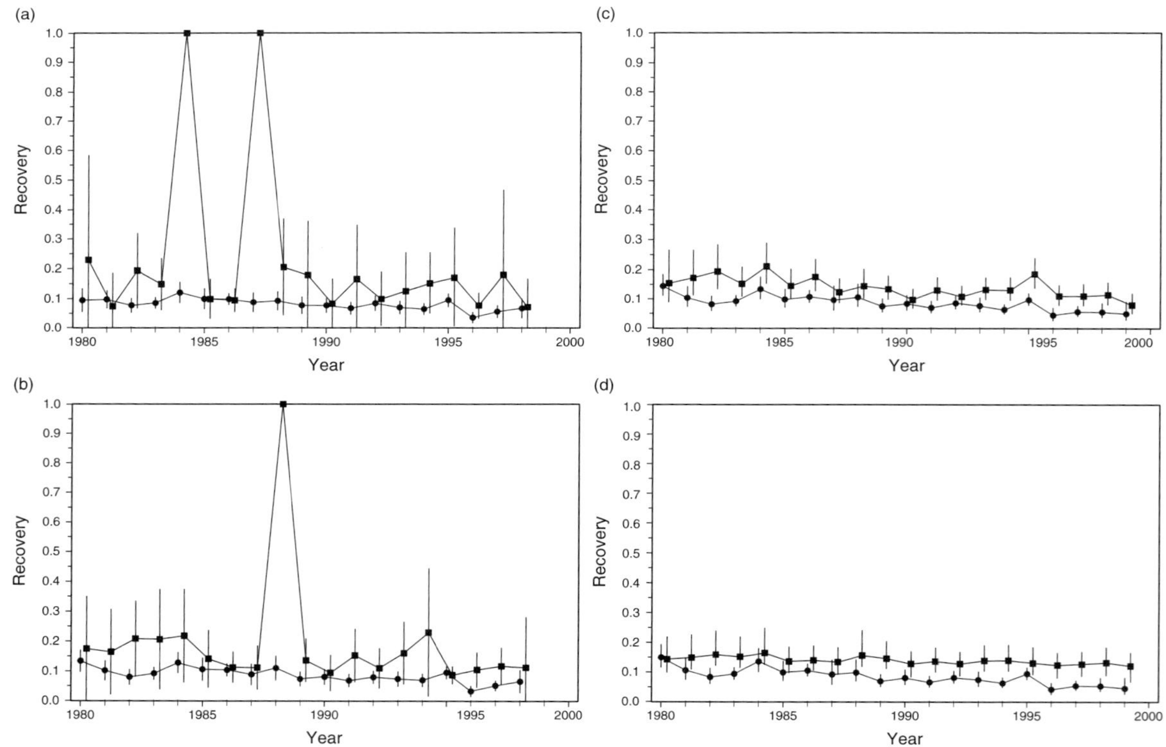

Fig. 3 Estimated annual recovery probabilities with 95% confidence/credible intervals for Tawny Owls in their first year of life (*lowest line*, ●), and in subsequent years (*top line*, ■) from 1980 to 1997 derived using the same approaches and same models as in Fig. 1

Table 4 Estimated process variance (σ^2) on logit scale (with 95% confidence or credible intervals) in first-year and adult survival probabilities of Tawny Owls derived from the maximum likelihood (ML) and hierarchical (MCMC) models. For the ML approach, variance was estimated from the method of moments estimator in MARK using parameter estimates and their estimated sampling-covariance matrix from Model 1 (Table 1) and incorporating appropriate covariates (intercept only, voles, winter temperature or both) using a design matrix. For the MCMC approach, covariates were incorporated directly into the hierarchical structure of the models prior to estimating parameters (models 11–15 in Table 2)

	First-year		Adults	
Model	ML	MCMC	ML	MCMC
Mean only[1]	0.21 (0.09 – 0.57)	0.23 (0.14 – 0.38)	0.38 (0.19 – 0.90)	0.14 (0.05 – 0.26)
Mean only[2]	n/a	0.26 (0.15 – 0.44)	n/a	0.16 (0.05 – 0.29)
Voles[2]	0.13 (0.04 – 0.40)	0.19 (0.09 – 0.40)	0.31 (0.15 – 0.80)	0.17 (0.06 – 0.32)
Weather[2]	0.12 (0.03 – 0.38)	0.16 (0.08 – 0.30)	0.20 (0.09 – 0.54)	0.01 (0.00 – 0.10)
Weather & Voles[2]	0.09 (0.01 – 0.32)	0.14 (0.06 – 0.30)	0.19 (0.08 – 0.56)	0.01 (0.00 – 0.10)

[1] For MCMC, prior distribution of recapture probabilities modeled with no covariates.
[2] For MCMC, prior distribution of recapture probabilities modeled with mean that varies with vole cycle.

indicated a reduction in variance after taking into account winter weather, but the ML approach suggested some effect of prey abundance and a combined effect explaining only about 50% of the variance. In contrast, the MCMC approach indicated that winter weather alone could explain nearly 100% of the annual variation in adult survival.

Incorporating different combinations of covariates into the hierarchical structure of the MCMC models tended to reduce the DIC (Table 2; Models 12–16) suggesting an improvement in the predictive ability of the models. The pattern of individual survival estimates for both age classes was generally similar with any combination of covariates (Fig. 4), though they tended to approach more closely the expected values for models incorporating weather as a covariate. Comparison of individual survival estimates with their expected values suggest that much, but not all, of the variation in first-year survival matches variation in both voles and winter temperature (Fig. 4d), while adult survival can be almost perfectly explained by winter temperature alone (Fig. 4c). Recapture probabilities (which are proportional to breeding propensity) varied strongly with the vole cycle for one-year old owls, but there was additional variation in adult capture probabilities (Fig. 5b). We have previously shown (Francis and Saurola 2004) this can be modeled with a temporal trend of increasing capture probability over time, but we did not fit such models in this study.

Both the ML and MCMC approaches were consistent in indicating that first-year and adult survival was, on average, substantially lower at the bottom of the vole cycle than the top (Table 5). Ultrastructural models that constrain survival probabilities to fit the regression model exactly (ML-1; Table 5) produced qualitatively similar results to the method of moments ML estimates (ML-2) or MCMC estimates, but tended to exaggerate the variation with the vole cycle. This was particularly apparent for adult survival rates, for which the ultrastructural models not only suggest a very large, 23% difference in mean survival probabilities through the vole

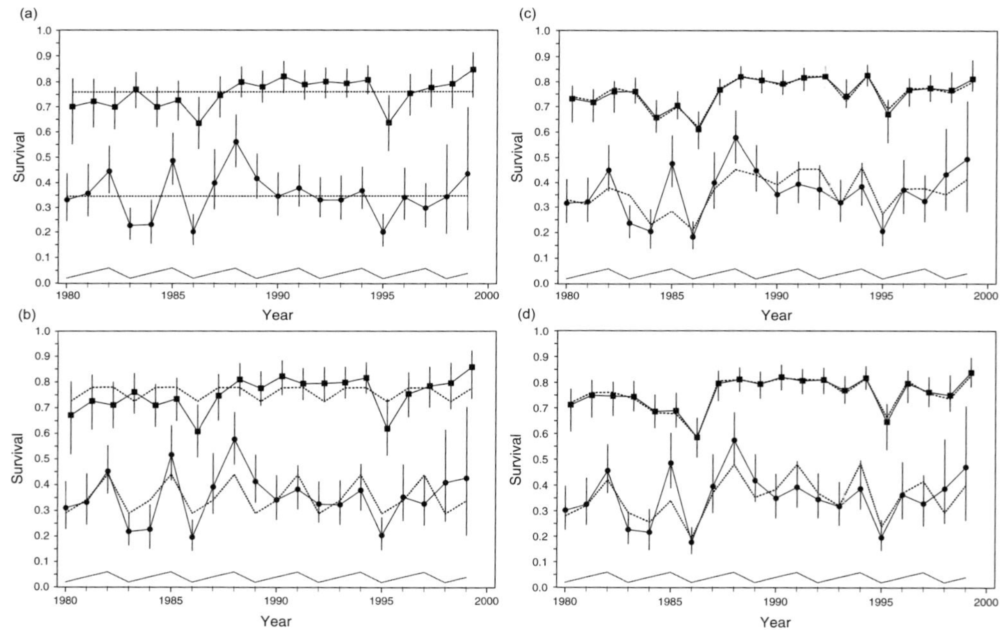

Fig. 4 Estimated annual survival probabilities with 95% credible intervals for Tawny Owls in their first year of life (*lowest line*, ●) and as adults (*top line*, ■) from 1980 to 1997 derived from MCMC models with hyperdistributions in which survival was modeled with: (**a**) no covariates (model 11); (**b**) variation in relation to voles (model 13); (**c**) variation in relation to winter temperature (model 14; see Table 2 for model details); (**d**) variation in relation to both winter temperature and voles (model 15). Expected values under each model are shown with dotted lines. Results for model 12 were essentially the same as model 11 and are not shown. Stage of the vole cycle is shown diagrammatically by a zig-zag line at the bottom of the graphs with the lowest value representing stage 1 and the highest value stage 3

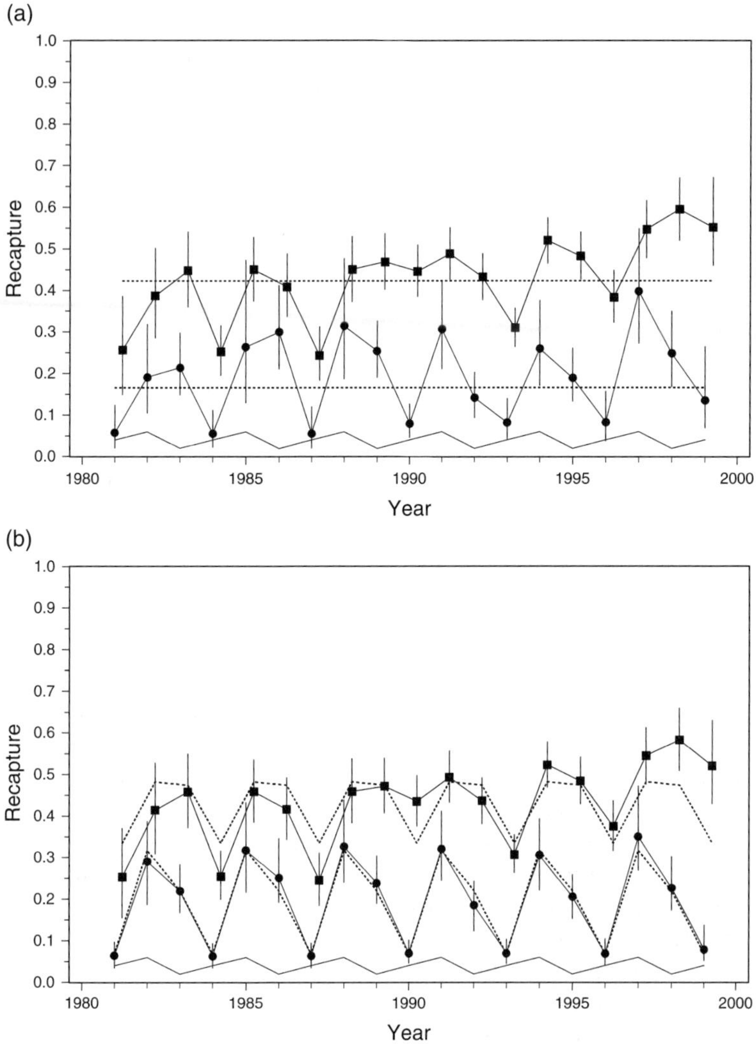

Fig. 5 Estimated annual recapture probabilities with 95% credible intervals for Tawny Owls in their first year of life (*lowest line*, •) and as adults (*top line*, ■) from 1980 to 1997 derived from MCMC models with different hyperdistributions in which survival was modeled with: (**a**) no covariates (model 11); (**b**) variation in relation to voles (model 13; see Table 2 for model details). Stage of the vole cycle is shown as in Fig. 4

cycle, but are also misleadingly precise. In contrast, the MCMC approach suggests relatively little variation in adult survival with the vole cycle. The estimated relationships between survival and winter temperature were very similar with all three approaches (Table 5), possibly reflecting the fact that the MCMC approach suggests nearly all of the variation in adult survival can be modelled with this covariate.

Table 5 Estimated first-year and adult survival probabilities (mean ± SE) of Tawny Owls in relation to each stage of the vole cycle (means at each stage), and in relation to mean winter weather (intercept and slope in relation to standardized temperature expressed on a logit scale) derived using maximum likelihood (ML) or hierarchical models (MCMC). ML-1 represents models with the covariates incorporated as fixed effects into ultrastructural models 7 (for voles) or model 8 (for temperature) in Table 1. ML-2 represents models in which coefficients were estimated using the method of moments approach as described in Table 4. MCMC estimates were derived from model 13 (voles) or model 14 (temperature) in Table 2

	First-year			Adults		
Model	ML – 1	ML – 2	MCMC	ML – 1	ML – 2	MCMC
Vole Stage						
1 – Poor	27.7 ± 2.3	26.2 ± 3.9	29.1 ± 4.5	68.6 ± 2.8	68.1 ± 4.5	72.7 ± 3.9
2 – Medium	28.7 ± 2.4	29.6 ± 4.2	34.1 ± 5.3	73.5 ± 2.8	73.8 ± 4.5	78.0 ± 3.5
3 – Good	48.6 ± 3.6	42.7 ± 4.6	44.0 ± 5.5	91.8 ± 1.5	82.2 ± 4.1	78.0 ± 3.6
Temperature (°C)						
Intercept	–0.60 ± 0.10	–0.73 ± 0.13	–0.59 ± 0.13	1.16 ± 0.08	1.15 ± 0.13	1.14 ± 0.07
Slope	0.33 ± 0.04	0.28 ± 0.09	0.31 ± 0.10	0.37 ± 0.04	0.41 ± 0.11	0.28 ± 0.06

4 Discussion

Our analyses found that, in many cases, estimates of survival, recapture, and recovery probabilities, especially their mean values (Table 3), were similar regardless of the analysis approach used. Estimated relationships with external covariates were also generally similar among approaches, although the ultrastructural ML models tended to exaggerate differences and suggest excessive precision, presumably because they did not allow for additional, unmodelled variation (Table 5).

However, discrepancies between the ML approach, even using the method of moments estimates (Burnham and White 2002), and the hierarchical MCMC models arose in some cases due to estimation of one or more recapture, recovery or survival parameters near boundaries with the ML approach. Because of the strong sampling covariances among parameter estimates, recovery probabilities estimated at boundaries lead to over-estimated survival probabilities. Unfortunately, due in part to the logit transform required to fit such models, the variances and covariances for such parameters in the ML approach were incorrectly estimated as being very small or zero, biasing the method of moments estimators (Burnham and White 2002). These boundary estimates thus affected not only individual survival estimates, but also the variance components estimates, as shown by the much higher ML estimate for variance of adult survival (Table 4).

In contrast, the hierarchical models did not have problems with boundary estimates, because the prior distributions for families of parameters provided additional information, in an empirical Bayes framework for estimating parameters with few data. We agree with Link and Barker (2004) that hierarchical models provide a logical framework for incorporating prior information about the likely structure of the system. For example, in the absence of information to the contrary, it seems logical to assume that recovery probabilities will be similar among years. However, unlike an ultrastructural model in which recovery probabilities can be modelled as constant over time, the Bayesian framework allows individual probabilities to vary over time when sufficient data are available to obtain reliable individual estimates.

Without knowing the true values of the parameters, it is not possible to be certain which approach provided the most reliable estimates. However, two factors suggest the estimates from the hierarchical models are likely to be more reliable. The first is the problems with boundary estimates with the ML approach. The second is the relatively tight fit in the hierarchical models between covariates expected to affect survival or recapture and their estimated values in the MCMC framework (Figs. 4 and 5), as also indicated by the improved fit of these covariate models relative to models without covariates (Table 2).

Estimation of variance components necessarily requires fitting models with annual variation in parameters such as survival, recapture or recovery probabilities. In a large data set, this can lead to a proliferation of parameters; our most general 4-age class model had 222 identifiable parameters, and even the most parsimonious model still had 122 parameters (Table 1). Despite the fact that we have a very large data set involving nearly 20,000 birds and 5,000 subsequent encounters, the data available to estimate some individual parameters was sparse, due in part to the high annual variation in demographic parameters (the estimation of which is one of the objectives of analysing the data). The major advantage of the hierarchical models was that they allowed information from year-specific parameters with better data to inform estimates for parameters with sparser data, in an empirical Bayes fashion. We suspect that this may be an advantage for many different data sets, ranging from small, sparse data sets to which only simple models can be fitted, to large, long-term data sets to which complex models with many parameters can be fitted.

Although not considered in this paper, an additional advantage of hierarchical models is that they can also be used to model and estimate covariances among parameters (e.g., Jamieson and Brooks 2004). For this particular data set, this could potentially be useful for estimating life history tradeoffs, such as tradeoffs between breeding at a young age and subsequent breeding or survival probabilities.

We did find, however, that the hierarchical models were somewhat sensitive to the selection of hyperdistributions, with substantial differences in the observed pattern of annual variation depending on the model selected. Furthermore, the structure imposed on one group of parameters (e.g., survival probabilities) also affected the pattern for other parameters such as recovery probabilities (e.g., Fig. 3) owing to the sampling covariances among estimates. In our models, the sample sizes available to estimate second and third-year survival and capture probabilities were generally small and variable, relative to those of young or adults, due to the relatively small number of birds captured at one-year of age, especially in years of low vole abundance (Fig. 2). Thus, the survival estimates for these age classes were relatively strongly affected by the selection of the prior distribution (Fig. 3). Because of correlations among parameter estimates, these also affected estimates of adult survival (Fig. 3). A priori, we might expect that survival and capture probabilities of 2nd and 3rd year birds would vary in similar ways to those of older age classes. Thus, in the absence of information to the contrary, it seems reasonable to model individual survival or capture probabilities as being similar to those of adults, with a difference parameter, rather than varying completely independently. This is, in fact, analogous to the ultrastructural model used to constrain survival parameters for this age class in the ML framework (model 1, Table 1). However, DIC model selection suggests that imposing these constraints through a prior distribution that still allows for some annual variation if the data warrant, in a standard Bayesian fashion, provides a better fit than constraining the differences to be constant (comparison of models 11–15 with models 16–19 in Table 2). This parameterization had the advantage that data from all three older age classes contributed to improving estimates of annual variation in survival. The improved fit, as measured by DIC, of this model relative to a model in which independent priors were imposed on all age-specific parameters (model 11 vs model 10 in Table 2), as well as the very close match between the resultant estimates of adult survival probabilities and external covariates (Fig. 4c,d) suggest this was a much more appropriate prior distribution.

An important practical consideration in any modeling procedure, especially from the perspective of a biologist, is the availability of computer software to implement the models. We found the MCMC implementation of hierarchical models within MARK (White et al. 2008) to be straightforward and easy to use. Hierarchical models were readily set up with similar design matrices to those used for constraining model parameters in an ultrastructural framework. Because of the size and complexity of our models, and to reduce the risk of errors, we created the input data sets using custom-written SAS programs, but an interactive interface is also available for setting up the hyperdistributions within MARK, similar to that used for setting up parameter index matrices or other design matrices. Although model fitting took much more computing time than the standard ML approach in MARK, a single chain in the MCMC procedure, even for this relatively large complex data set typically completed in a few hours, and we encountered no problems with convergence provided the model was properly specified. In contrast, the ML models sometimes required repeated runs with different starting values to ensure that they had not converged on local maxima. We thus encourage others interested in modeling annual variation in survival or other population parameters to consider using hierarchical models for mark-recapture analyses.

Acknowledgments We are particularly grateful to all of the bird-ringers in Finland for their outstanding work in gathering the data used in this analysis. We also thank the members of the EURING community for continuing to develop new and improved statistical models and computer software, especially Gary White who not only developed MARK but also answered many of our questions about the best way to implement these models in MARK. Two anonymous reviewers and Ken Burnham provided comments that helped us to improve this manuscript.

References

Barker R, Fletcher D, Scofield P (2002) Measuring density dependence in survival from mark-recapture data. Journal of Applied Statistics 29:305–315.

Brommer JE, Pietiäinen H, Kolunen H (2002) Reproduction and survival in a variable environment: Ural Owls (Strix uralensis) and the three-year vole cycle. The Auk 119:544–550.

Brooks SP, King R, Morgan BJT (2004) A Bayesian approach to combining animal abundance and demographic data. Animal Biodiversity and Conservation 27.1:515–529.

Burnham KP (1993) A theory for joint analysis of ring recovery and recapture data. In: Marked individuals in the study of bird population: 199–213 (Lebreton J-D, North PM Eds.) Birkhäuser, Basel.

Burnham KP, White GC (2002) Evaluation of some random effects methodology applicable to bird ringing data. Journal of Applied Statistics 29:245–264.

Caswell H (2001) Matrix population models: construction, analysis and interpretation, 2nd Edition. Sinauer Associates, Sunderland, MA.

Francis CM, Saurola P (2002) Estimating age-specific survival rates of tawny owls—recaptures versus recoveries. Journal of Applied Statistics 29:637–647.

Francis CM, Saurola P (2004) Estimating components of variance in demographic parameters of Tawny Owls, Strix aluco. Animal Biodiversity and Conservation 27.1:489–502.

Gelman A (1996) Inference and monitoring convergence. Markov Chain Monto Carlo in Practice (Gilks WR, Richardson S, and Spiegelhalter DJ, Eds.): 131–143. Chapman and Hall/CRC, Boca Raton, FL.

Hanski I, Hansson L, Henttonen H (1991) Specialist predators, generalist predators, and the microtine rodent cycle. Journal of Animal Ecology 60:353–367.

Jamieson LE, Brooks SP (2004) Density dependence in North American ducks. Animal Biodiversity and Conservation 27.1:113–128.

Lebreton JD, Burnham KP, Clobert J, Anderson DR (1992) Modeling survival and testing biological hypotheses using marked animals: a unified approach with case studies. Ecological Monographs 62:67–118.

Lele SR, Dennis B, Lutscher F (2007) Data cloning: easy maximum likelihood estimation for complex ecological models using Bayesian Markov Chain Monte Carlo methods. Ecology Letters 10:551–563.

Link W (1999) Modeling pattern in collections of parameters. Journal of Wildlife Management 63:1017–1027.

Link W, Barker R (2004) Hierarchical mark-recapture models: a framework for inference about demographic processes. Animal Biodiversity and Conservation 27.1:441–449.

SAS Institute (2003) The SAS System for Windows, Release 9.1. SAS Institute Inc., Cary, NC.

Saurola P (1995) Suomen pöllöt, Owls of Finland. Kirjayhtymä, Helsinki. 271 pp. (in Finnish with English summary).

Saurola P (1997) Monitoring Finnish Owls 1982–1996: methods and results. In: Biology and conservation of owls of the northern hemisphere. 2nd International Symposium, February 5–9, 1997, Winnipeg, Manitoba, Canada: 363–380 (Duncan JR, Johnson DH, Nichols TH, Eds.). USDA Forest Service Gen. Tech. Rep. NC..

Saurola P, Francis CM (2004) Estimating population parameters and dispersal distances of owls from nationally coordinated ringing data in Finland. Animal Biodiversity and Conservation 27.1:403–415.

Sundell J, Huitu O, Henttonen H, Kaikusalo A, Korpimäki E, Pietiäinen H, Saurola P, Hanski I (2004) Large-scale spatial dynamics of vole populations in Finland revealed by the breeding success of vole-eating predators. Journal of Animal Ecology 73:167–178.

White GC, Burnham KP (1999) Program MARK: survival estimation from populations of marked animals. Bird Study 46:S120–S139.

White GC, Burnham KP, Barker RJ (2008) Evaluation of a Bayesian MCMC random effects inference methodology for capture-mark-recapture data. In: Thomson DL, Cooch EG, Conroy MJ (eds.) Modeling Demographic Processes in Marked Populations. Environmental and Ecological Statistics, Springer, New York, 3:1119–1128.

Inference About Species Richness and Community Structure Using Species-Specific Occupancy Models in the National Swiss Breeding Bird Survey MHB

Marc Kéry and J. Andrew Royle

Abstract Species richness is the most widely used biodiversity measure. Virtually always, it cannot be observed but needs to be estimated because some species may be present but remain undetected. This fact is commonly ignored in ecology and management, although it will bias estimates of species richness and related parameters such as occupancy, turnover or extinction rates. We describe a species community modeling strategy based on species-specific models of occurrence, from which estimates of important summaries of community structure, e.g., species richness, occupancy, or measures of similarity among species or sites, are derived by aggregating indicators of occurrence for all species observed in the sample, and for the estimated complement of unobserved species. We use data augmentation for an efficient Bayesian approach to estimation and prediction under this model based on MCMC in WinBUGS. For illustration, we use the Swiss breeding bird survey (MHB) that conducts 2–3 territory-mapping surveys in a systematic sample of 267 1 km^2 units on quadrat-specific routes averaging 5.1 km to obtain species-specific estimates of occupancy, and estimates of species richness of all diurnal species free of distorting effects of imperfect detectability. We introduce into our model species-specific covariates relevant to occupancy (elevation, forest cover, route length) and sampling (season, effort). From 1995 to 2004, 185 diurnal breeding bird species were known in Switzerland, and an additional 13 bred 1–3 times since 1900. 134 species were observed during MHB surveys in 254 quadrats surveyed in 2001, and our estimate of 169.9 (95% CI 151–195) therefore appeared sensible. The observed number of species ranged from 4 to 58 (mean 32.8), but with an estimated 0.7–11.2 (mean 2.6) further, unobserved species, the estimated proportion of detected species was 0.48–0.98 (mean 0.91). As is well known, species richness declined at higher elevation and fell above the timberline, and most species showed some preferred elevation. Route length had clear effects on occupancy, suggesting it is a proxy for the size of the effectively sampled area. Detection probability of most species showed clear seasonal patterns and increased with greater survey effort; these are important results for the planning of focused surveys. The main benefit of our model,

M. Kéry (✉)
Swiss Ornithological Institute, 6204 Sempach, Switzerland
e-mail: marc.kery@vogelwarte.ch

D.L. Thomson et al. (eds.), *Modeling Demographic Processes in Marked Populations*, Environmental and Ecological Statistics 3, DOI 10.1007/978-0-387-78151-8_28,

and its implementation in WinBUGS for which we provide code, is its conceptual simplicity. Species richness is naturally expressed as the sum of occurrences of individual species. Information about species is combined across sites, which yields greater efficiency or may even enable estimation for sites with very few observed species in the first place. At the same time, species detections are clearly segregated into a true state process (occupancy) and an observation process (detection, given occupancy), and covariates can be readily introduced, which provides for efficient introduction of such additional information as well as sharp testing of such relationships.

Keywords Biodiversity · BBS · Breeding bird survey · Community · Data augmentation · MCMC · Monitoring · Metacommunity · Species richness · WinBUGS

1 Introduction

Species richness is the number of species present at a time and place. It is arguably the simplest and therefore also the most widely used measure of biodiversity in ecology and management (Gotelli and Colwell 2001; Purvis and Hector 2000. Furthermore, species richness is comparatively easy to measure in practice, in stark contrast to other, conceptually less well defined surrogates of biodiversity such as ecosystem health, or more costly measures, such as genetic diversity.

Measurement of species richness is not without problems though. In most situations, some species will be overlooked (Schmidt 2005). This will bias observed species richness and its dynamic components, species turnover, extinction and colonisation rates, as well as measures of similarity among sites or species with respect to the true values of these parameters in the community. Although this situation is commonly ignored in ecology and management, earlier efforts to correct for such non-detection biases include extrapolation of the observed number of species using accumulation curves (Gotelli and Colwell 2001) as well as community analogues of closed population capture–recapture models for abundance estimation where species take the place of individuals (Boulinier et al. 1998a, b; Nichols et al. 1998a, b; Boulinier et al. 2001; Cam et al. 2002; Kéry and Schmid 2006). Both approaches have their drawbacks. Accumulation curves are purely phenomenological and do not have a mechanistic basis in a sampling process. Capture–recapture-types of models are inefficient for the frequent case when species richness estimates are desired for multiple surveyed sites that represent samples of the same community, and they may be impossible when the number of detected species is very low (Kéry and Royle 2008). Accounting for additional information such as covariates is difficult for both approaches.

Here, we present a statistical model for the estimation of species richness and related community measures that is applicable in the frequent case when replicate sites (=communities) representing a metacommunity are sampled repeatedly over a short time (Dorazio and Royle 2005; Royle et al. 2007a). Our model consists of component models for site-occupancy of individual species tied together by random effects assumptions about some of the species-specific parameters and is a multi-species generalisation of single-species site-occupancy models (MacKenzie et al. 2002, 2006; Royle and Kéry 2007). We apply the model to the 2001 data of the Swiss national breeding bird survey "Monitoring Häufige Brutvögel" (MHB; Schmid et al. 2004) to estimate the fraction of species present and detected at sampled sites. Importantly, our model enables us to introduce covariate information for both occurrence and detection of species.

2 Material and Methods

2.1 The Swiss Breeding Bird Survey MHB

The national breeding bird survey "Monitoring Häufige Brutvögel" (MHB) has been run by the Swiss Ornithological Institute since 1999 (Schmid et al. 2004). In a systematic sample of 267 1 km^2 quadrats, volunteers conduct three territory-mapping surveys (two in high-elevation quadrats above the timberline) during every breeding season (15 April–15 July) along a quadrat-specific, irregular transect route averaging 5.1 km (range 1.2–9.4) that aims to cover as large part of each quadrat as possible. Route length declines significantly from 5.4 km on average at 210 m to 4.7 km on average at 2710 m elevation ($F_{1,262} = 14.65$, $p < 0.001$), but this relationship was weak ($R^2 = 5\%$), and there was no relationship with forest cover ($F_{1,262} = 1.64$, $p = 0.20$). Mean duration of a single survey in 2001 was 228 min (range 75–410), mean survey effort (time per unit transect length) 48 min km^{-1} (range 14–157) and mean survey dates were 10 May, 29 May, and 9 June, respectively. Although MHB yields abundance information for about 150 detected species (see, e.g., Kéry et al. 2005; Royle et al. 2005; Royle et al. 2007b), we only used survey-specific detection/nondetection records for each of the 134 diurnal species observed anywhere in the 254 surveyed quadrats in 2001. Hence, we are modelling a three dimensional data matrix X where element $x_k(i, j)$ denotes detection (1) or nondetection (0) of species i ($i = 1, \ldots, N$) at quadrat j ($j = 1, \ldots, 254$) during survey k ($k = 1,2,3$).

2.2 The Model

Here we adopt the framework of developing community models based on species-specific models of occurrence (Dorazio and Royle 2005). The basic idea is to formulate the model in terms of a latent binary process $z(i, j)$, where $z(i, j) = 1$ if species i occurs at quadrat j. To account for imperfect observation of this state process, we require data $x_k(i, j)$ for $k = 1,2, \ldots, K$ which are Bernoulli trials if $z(i, j) = 1$, otherwise they are structural zeros. In the context of single species, this is a state-space formulation of the model described by MacKenzie et al. (2002). Like these authors,

Dorazio and Royle (2005) proposed a procedure for estimation based on integrated likelihood, in which the latent $z(i, j)$ variables were removed from the likelihood. Further, Dorazio and Royle (2005) adopted a conditional formulation of the likelihood in which the size of the community, say N, is also removed from the likelihood. They develop a Bayesian analysis based on Markov chain Monte Carlo, in which species-specific occurrence probability parameters are estimated. They used these to develop predictions of missing $z(i, j)$ variables, and functions of those variables such as community similarity and richness. Here we provide a fully Bayesian analysis of the model applied to the MHB data, using the data augmentation parameterization described by Royle et al. (2007a). This formulation was also used in the analysis presented in Dorazio et al. (2006).

The species occurrence model is specified by $z(i, j) \sim$ Bernoulli $(\psi(i, j))$ and the observation model is specified by $x_k\ (i, j) \sim$ Bernoulli $(p_k\ (i, j)^*z(i, j)$. Thus, if $z(i, j) = 0$, the resulting observations are fixed zeros. In the most general formulation, we suppose that $\text{logit}(\psi(i, j)) = \mu_i + \alpha_j$ and $\text{logit}(p(i, j)) = \nu_i + \beta_j$. Here, $\psi(i, j)$ is the probability of occurrence (occupancy) for species i at quadrat j, $p(i, j)$ is the probability of detection for species i at quadrat j, μ_i and ν_i are species effects, and α_j and β_j are quadrat effects. Hence, the logit transforms of occupancy and detection probability are both assumed to be a sum of species and quadrat effects.

The general modelling framework described so far can easily be adapted to the specific data set and objectives of a particular study. For instance, in all three previous applications (Dorazio and Royle 2005; Dorazio et al. 2006; Royle et al. 2007a), it was assumed that quadrat effects α_j and β_j are constant and hence, $\text{logit}(\psi(i)) = \mu_i + \alpha$ and $\text{logit}(p(i)) = \nu_i + \beta$. Since μ_i and ν_i were assumed to be draws from zero-mean normal distributions, α and β were the mean logit-scale parameters of occupancy and detection probability.

Furthermore, these authors modelled species-specific differences in occupancy and detection probability as being correlated by assuming a parametric form for the joint distribution of μ_i and ν_i. That is, they assumed $\left[\mu_i, \nu_i|\mathbf{\Sigma}\right] \sim \text{Normal}(\mathbf{0}, \mathbf{\Sigma})$ with the 2×2 matrix $\mathbf{\Sigma}$ specified by two variances $(\sigma_\mu^2, \sigma_\nu^2)$ and a covariance $(\sigma_{\mu\nu})$. Modelling this correlation can easily be justified on grounds of the expected relationships between quadrat-specific abundance and occupancy on the one hand and abundance-induced heterogeneity in quadrat-specific detection probability on the other (Royle and Nichols 2003; Dorazio and Royle 2005).

Kéry and Royle (2008) assumed a minimal version of the model; $\text{logit}(\psi(i)) = \mu_i$ and $\text{logit}(p(i)) = \nu_i$, with μ_i and ν_i normally distributed, species-specific random effects such that $\mu_i \sim \text{Normal}(\mu_\mu, \sigma_\mu^2)$ and $\nu_i \sim \text{Normal}(\mu_\nu, \sigma_\nu^2)$. This model is analogous to model M_h in a site occupancy context (Otis et al. 1978), since it assumes that the only source of variation in detection and occurrence probabilities is species identity.

In this paper, we consider an obvious but important extension of the model, by allowing for effects of quadrat- and survey-specific covariates in detection probability, and also spatial covariate effects in occupancy. While such model structure would seem necessary in most applications, existing conventional methods of estimating species richness in the presence of imperfect detection do not easily accommodate such structure. We feel that this is one of the main benefits of developing the model based on species-specific models of occurrence. The quadrat effects can be expressed as linear functions of quadrat-specific covariates that are thought to influence the probability that species i occurs at quadrat j. In our present study, we adopted a model where the logit transform of occupancy is a linear function of forest cover, quadrat elevation (linear and squared) and route length:

$$\text{logit}(\psi(i, j)) = \mu_i + \beta_{1i} * \text{forest}_j + \beta_{2i} * \text{elev1}_j + \beta_{3i} * \text{elev2}_j + \beta_{4i} * \text{length}_j$$

Here, μ_i is again a species effect and serves as an intercept, while coefficients β_{1i} through β_{4i} are subscripted i, showing them to be different for each species i. We add route length as a predictor of occupancy rather than detection probability since we envision it as a surrogate of the effectively sampled area but note that a point could be made for adding it into the linear predictor of detection probability instead. Given enough temporal replicates, it could be included in both.

Similarly, we expressed the logit transform of detection probability $p_k\ (i, j)$ for species i, quadrat j and survey k as a sum of three components.

$$\text{logit}(p_k(i, j)) = \nu_i + \beta_{5i} * \text{date1}_{jk} + \beta_{6i} * \text{date2}_{jk} + \beta_{7i} * \text{effort}_{jk}$$

Hence, we assumed that each species had its own logit-scale baseline probability of detection (ν_i) that was modified by the date (linear and quadratic) at which survey k was conducted at quadrat j as well as by survey effort at quadrat j during survey k. We expressed effort by the ratio of survey duration and route length. As before, coefficients β_{5i} through β_{7i} are species-specific. For numerical reasons, all covariates except route length were normalised.

As described so far, this model contains far too many parameters to be useful. Moreover, many of the parameters are liable to be poorly identified because many species will be observed a small number of times (or even once). In addition, there are presumably a number of species in the community that did not appear in the sample at all, and so the information about parameters for those species must derive from the prescribed model structure. Hence, we add to the model one hierarchical layer and assume that β_{1i} through β_{7i} are independent normal random effects (see above for the multivariate normal distribution assumed for μ_i and ν_i,) and we estimate the hyperparameters, i.e. means and (co)variances of these distributions.

2.3 Data Augmentation and Bayesian Analysis

One of the fundamental issues in the development of models of community structure and composition is that a given sample is unlikely to contain all species within the community. Historically, this has been dealt with by formulating models for estimating community structure as capture–recapture models, in which the community size parameter is analogous to population size. The main disadvantage of that approach in the present problem is that calculation of the likelihood would require tedious integration to remove all of the random effects from the likelihood (i.e., worse than Dorazio and Royle 2005).

This computational burden, and the search for a convenient Bayesian treatment of the problem, motivated the data augmentation approach described by Royle et al. (2007a), which we adopt for our analysis of the MHB data. With data augmentation, the observed data comprised of encounter histories on n individuals is physically augmented with a large number of "all zero" encounter histories, say $M - n$, yielding a fixed total number of available "species" of M. This comprises n observed species and $M - n$ pseudo-species represented by all zeros. The resulting model for the augmented data set is a zero-inflated version of the "known-N" model (provided that M is sufficiently large). The nice thing about this formulation is that zero-inflation models yield to a simple Bayesian implementation due to their conditional structure. They also can be implemented directly in WinBUGS (Spiegelhalter et al. 2003). Specifically, we introduce another latent indicator variable w_i for $i = 1, 2, \ldots, n, n + 1, n + 2, \ldots, N, N + 1, N + 2, \ldots, M$ and then assume $z(i, j)$ are Bernoulli $(w_i * \psi(i, j))$ and $w_i \sim \text{Bernoulli}(\Omega)$. The problem of estimating N is thus translated into the equivalent task of estimating that Bernoulli parameter Ω.

The interpretation of M is that of a super-population of species from which the actual sampled community (the species exposed to sampling) was drawn by some hypothetical random sampling mechanism; see Royle et al. (2007a) for additional details and discussion of the data augmentation idea. The main consideration in setting a fixed M is that it must be large enough so as to avoid truncation of the posterior distribution. This is easily diagnosed by looking at a few small trial runs of the MCMC algorithm.

On the other hand, N is not a fixed constant, but rather one of the important structural parameters of the model. However, its interpretation may not be self-evident. In the context of a simple model with no spatial covariates, so that all quadrats are essentially regarded as i.i.d. replicate samples from a homogeneous landscape, N is the asymptote of a species-accumulation curve (Dorazio et al. 2006), i.e., it is the number of species that occur in some area from which the samples were drawn, *as that area tends to infinity*. To obtain an estimate of the number of species that actually occur on any location, or at any set of sampled locations, one must obtain the small-area estimate of the number of species, say $N(j)$ (Dorazio and Royle 2005). This can be obtained by predicting each unobserved $z(i, j)$ for $i = 1,2, \ldots, N$ and $j = 1,2, \ldots, J$, and then summing them up at each quadrat, and aggregating spatially. The interpretation of N is somewhat complicated when covariates on $z(i, j)$ are included in the model. In this case, N still seems to represent a maximum community size that could exist as the size of a geographic region tends to infinity. However, as that hypothetical area increases, it is not clear how the structure of that hypothetical landscape is defined in the context of the covariate model in which the covariates are fixed. One could clarify the interpretation of N when there are covariates in the model, if those covariates were themselves assumed to be random variables. Regardless of this conceptual issue in defining N, we can predict each of the $z(i, j)$'s including for the "unobserved species" and then aggregate those over space and time given data on the covariates at quadrats for which predictions are desired.

The implementation in program WinBUGS (Spiegelhalter et al. 2003) is straightforward and requires only a few lines of code (see Appendix 1). To complete the formulation of our model, we assumed vague prior distributions for all parameters, see Appendix 1; in addition, as priors for the hyperdistributions of the β's above we chose $N(0,10)$ for the means and Uniform(0,10) for the standard deviation.

We have found that in some cases WinBUGS does not appear to update the random effects properly in a logit-normal model. We don't know why this is, but it can be diagnosed by complete non-mixing of the variance components when multiple chains are run from random starting values (see left panels in Appendix 2). We have an ad hoc fix for this problem, which is to truncate the random effects distributions by multiplying with the indicator function $I(-B,B)$ for B fairly large relative to the variance in the random effect, e.g., $B = 21$ (see right panels in Appendix 2). We think that this tricks WinBUGS into choosing a different algorithm, perhaps more general, but one that seems to work, albeit at the same time it slows down the algorithm considerably. While the model can be implemented fairly easily in other packages such as R, and despite the WinBUGS updating problem, we feel that the extensibility and simplicity of the WinBUGS implementation is a benefit to its use in such problems.

For purposes of evaluating convergence, we ran three chains using random starting values, of length 2000 each after a 1000 burnin. Resulting chains were thinned by 2 resulting in 3000 iterations for inference about posterior distributions. Convergence was assessed using the Brooks–Gelman–Rubin diagnostic (Gelman and Rubin 1992), using the facilities provided in the R add-on library BOA (Smith 2005). Results indicated convergence, with scale reduction factors near 1 (0.9997–1.089) for all parameters except for sigma(route length) with 1.647 indicating apparent nonconvergence. (Interestingly, the mean parameter for length appeared to converge). We therefore caution that results associated with that variance component be viewed with caution.

3 Results

3.1 Community Analysis MHB

Among 254 MHB 1 km quadrats, 134 diurnal bird species were observed during 2001 while the estimate under our model was 169.9 (95% CI 151–195; Fig. 1a). Figure 1a indicates that data augmentation by 100 was entirely sufficient. The observed number of species ranged from 4 to 58 per quadrat (mean 32.8, sd 11.3). Under our model, there were between 0.7 and 11.2 further, unobserved species (mean 2.6, sd 1.1) in each quadrat. This translates into a proportion of species actually detected ranging from 0.48 to 0.98 (mean 0.91, sd 0.08). Hence, the estimated species richness in 1 km quadrats in Switzerland in 2001 ranged from 8.1 to 59.3 (mean 35.4, sd 10.8). Figures 1b–d show posterior distributions of species richness for three sample quadrats with virtually no unobserved species (b) and with a

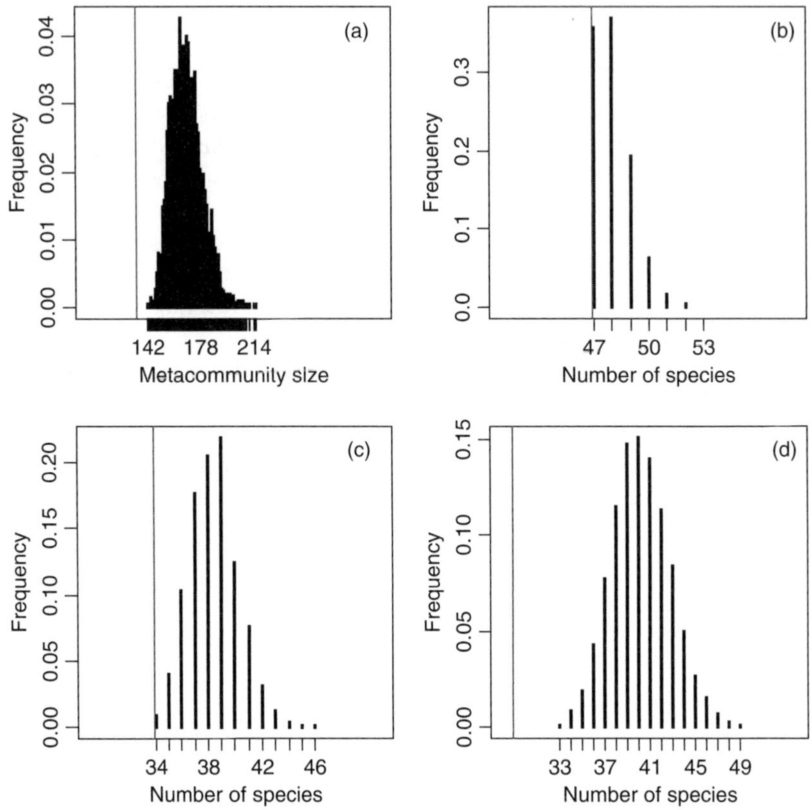

Fig. 1 Examples of posterior distributions of species richness of diurnal breeding birds in Switzerland in 2001. (**a**) Metacommunity size, and community size in (**b**) quadrat 73, (**c**) quadrat 105 and (**d**) quadrat 168. Thin vertical lines indicate the observed number of species

moderate (c) or a large number of unobserved species estimated to be present (d). As expected on theoretical grounds (Royle and Nichols 2003) occupancy and detection probability were clearly correlated with that correlation estimated at 0.62 (95% CI 0.43–0.76).

Avian species richness in Switzerland showed the well-known decline with increasing elevation (Fig. 2a) and was distinctly higher in forested than unforested quadrats (Fig. 2). The estimated number of unobserved species per quadrat under the model remained more or less constant over the elevation and the forest cover gradients (Fig. 2–d), which translated into a decline in the estimated proportion species detected per quadrat at higher elevations and very low forest cover (Fig. 2e–f).

Species responses of occupancy probability to elevation were highly variable, but not surprisingly, most species showed some distinct elevational preference (Fig. 3a). Route length had an important effect on occupancy probability in most species (Fig. 3b), suggesting a larger effectively sampled area was more likely to

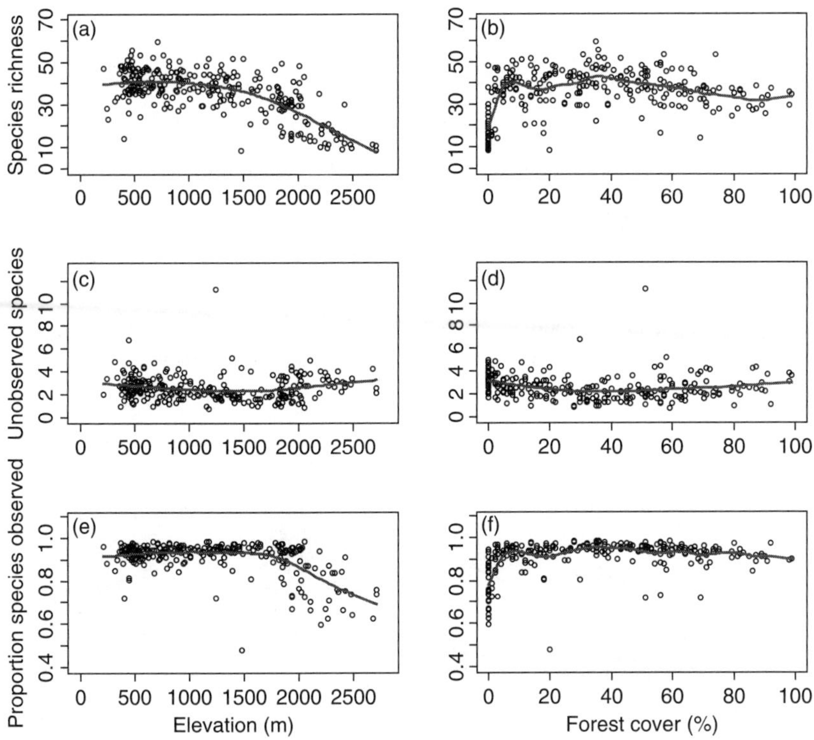

Fig. 2 Relationships between elevation and percentage forest cover, respectively, and (**a** and **b**) Swiss avian species richness (estimated number of diurnal bird species per 1 km MHB quadrat), (**c** and **d**) estimated number of unobserved diurnal species and (**e** and **f**) the estimated proportion of species detected. Smoothing splines are added

contain at least one territory of a species than a smaller sampling area. Species differences in detection probability were huge, and species responses of detection probability to date and survey effort were also variable. Many species showed a maximum detectability in the middle of the survey season, but some also had a negative hyperbolic, negative or positive relationship, or none at all (Fig. 3c). Response of detection probability to effort was clearly positive for most species (Fig. 3d).

4 Discussion

4.1 Modelling Framework

Species richness is of great importance in many branches of ecology as well as in management (Purvis and Hector 2000; Connolly 2005; Orme et al. 2005). Although most often possible complications due to imperfect species detectability are simply ignored in these studies, there exists a large array of methods to

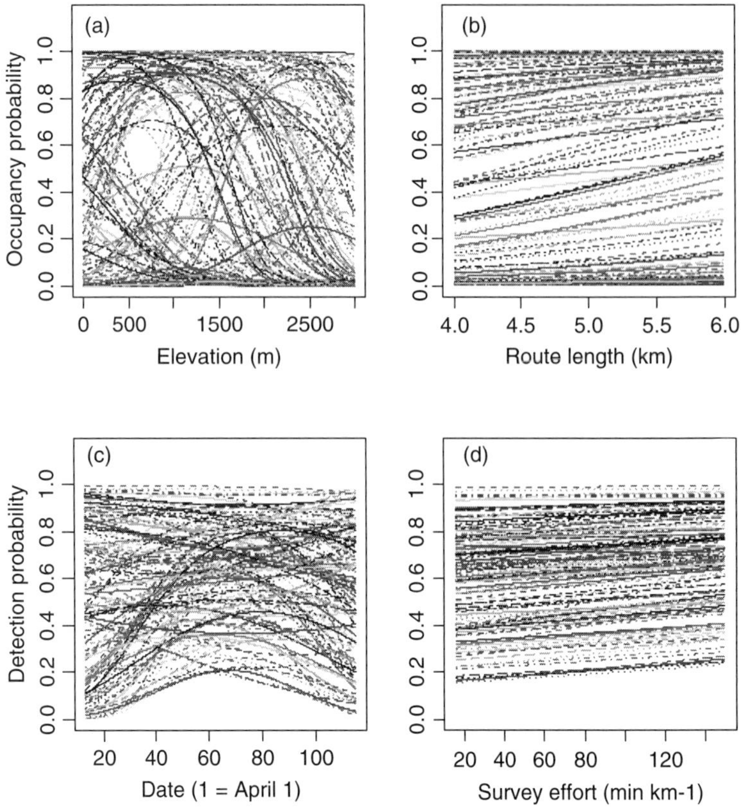

Fig. 3 Species-specific reponses of occupancy probability to (**a**) elevation and (**b**) route length, and of detection probability to (**c**) date and (**d**) survey rate. Note the caveat about non-convergence of the Markov chains with respect to the variance of species-specific responses of occupancy to route length

estimate species richness while correcting for imperfect detectability, e.g. extrapolation of species accumulation curves, parametric modelling of apparent species abundance distributions, nonparametric modelling based on sampling theory and capture–recapture type of models. So what are the benefits of our model?

First and foremost, perhaps, are its conceptual simplicity and elegance. Species detections are clearly separated into a true, but imperfectly observed occupancy state of each species in the metacommunity $z(i, j)$, i.e. the latent structure Z, and into an observation process conditional upon that occupancy state. Consequently, factors affecting the observations can be clearly segregated into those that affect the system state (occupancy), here, forest cover, elevation and route length, and those affecting only the observations of that system (i.e., the detection process), here, survey date and effort. This not only sharpens our thinking about the system but arguably also uses the available information more efficiently.

Second, the model effectively integrates all available information by retaining species identity across quadrats. Arguably the largest amount of variance in detectability p_k (i, j) will be due to species identity i, and our model borrows strength across quadrats j in the estimation of p_k (i, j) and hence, quadrat-specific $N(j)$. Not surprisingly, compared to a quadrat-by-quadrat analysis of species richness, integrated modelling of the metacommunity yields large benefits in terms of precision of the estimates, and in some quadrats with very small numbers of observed species enables an estimate in the first place (Kéry and Royle 2008).

Further, the core of the model is the Z matrix (i.e. the set of indicators $z(i, j)$), a centerpiece of many applications of ecology, management and conservation and the true presence-absence matrix (McCoy and Heck 1987). The Z matrix is a complete description of a metacommunity in terms of the occurrence of individual species. An estimate of Z enables rich inference about a metacommunity; for instance, summing across species yields quadrat-specific species richness and summing across quadrats essentially yields occupancy for each species, another central quantity in ecology and its applications (MacKenzie et al. 2006). Furthermore, species similarity can be assessed in terms of the number of quadrats they co-occur, and quadrat similarity by the pattern of species occurrences (Dorazio et al. 2006). Finally, species richness can be estimated not only for each community (=quadrat) but also for the entire metacommunity or subsets thereof.

Data augmentation greatly simplifies the fitting of the model in a Bayesian mode of analysis by allowing analysis of a zero-inflated version of the model in which N is known. The simple conditional structure of this model is amenable to Bayesian analysis, and the zero-inflation parameter takes the place of N. That is, while there are a number of missing species in the sample, we know the data for all such species to be composed entirely of zeros. Thus, we create a larger data set by augmenting the observed data with a large number of such all-zero encounter histories. The model for the augmented data is acknowledged to have excessive zeros by the inclusion of a single additional zero-inflation parameter. The hierarchical formulation of this zero-inflated version of the "known-N" model is trivial to implement, especially in WinBUGS.

The requirement of spatially and temporally replicated data may be seen as a cost of our model. However, spatial replication is a common feature of a many studies on species richness, and its degree can be far less than in our current study. For instance, we have previously applied the model to a much smaller subset of MHB data consisting of 26 quadrats only (Kéry and Royle 2008). Hence, our model could probably be applied to many data sets provided that a minimum of two temporal replicates are available.

4.2 The Application

Switzerland is a country with a high population density of capable ornithologists and despite its mountains, has hardly any larger, really inaccessible areas. We can thus safely assume that the true metacommunity size is rather well known. From 1995 to

2004, 185 diurnal breeding bird species were known, of which 163 bred regularly and 22 irregularly. In addition, for 13 species 1–3 nesting records are known for the period 1900–2004 (Volet 2006). 134 diurnal species were observed during MHB surveys in 2001. The estimate under our model of 169.9 (95% CI 151–195) therefore appears sensible, even though perhaps slightly low. However, this example shows how elusive a concept such as species richness really is; does it refer to the regular, or to the regular and irregular, or all breeding species ? It is likely that it is the rare species which introduce most uncertainty (and complications !) into such estimates (Mao and Colwell 2005). Therefore, we would not want to put too much emphasis on this estimate of metacommunity size but rather feel comfortable with an apparent consistency between the estimate and "known truth".

At a quadrat level, in 97% of cases the estimated number of overlooked species was only 1–4, which lead to a high estimated proportion of detected species (91%). The relationships of avian species richness with two important landscape predictors, elevation and forest cover, was as expected; there was a strong decline at higher elevations and with very low or no forest cover. Of course the two are confounded, since the fall in species richness above the timberline around 2000 m can be seen in both. Interestingly, the timberline is also evident in the estimated proportion of species detected. Probably this is because only two surveys are conducted at most high-elevation quadrats. This must lead to a smaller proportion of species detected, as can be seen from the the relationship $P^* = 1 - (1 - p)^n$, where P^* is the proportion of species detected, p the per-survey mean species detection probability and n the number of surveys. In addition, the slightly shorter route lengths at higher elevations may have induced a smaller effective sample area within the sampling quadrats and therefore a smaller number of species actually present in that area.

Variation in abundance arguably leads to variation in detection probability (Royle and Nichols 2003), and positive abundance–occupancy relationships belong to the most general patterns in community ecology (He and Gaston 2003; Freckleton et al. 2006). As explained by Dorazio and Royle (2005; also see Royle and Nichols 2003), a positive correlation between occupancy and detection probability was therefore expected.

The estimated occupancy probability of most species showed some preferred elevation which was very much expected on biological grounds. The strong relationship between estimated occupancy and route length is a strong indication that in most quadrats the effective sampling area is smaller than 1 km^2. Under a Poisson assumption for the number of territories, a longer route will clearly expose to detection at least one territory of a species with greater likelihood than a shorter route. We note that our parameterisation of this effect could be improved by assuming an asymptotic rather than a logit-linear relationship between occupancy and route length as was done for models of abundance in Royle et al. (2007b). We are investigating parameterizations that accommodate incomplete quadrat coverage.

Detection probability varied enormously among species and for most species also drastically over the season. There was less variation but still a clear increase of detection probability related to increased sampling effort as quantified by the time spent per unit route length. These are important results because they illustrate very clearly the danger of comparing observed, raw occupancy across species with different detection probability, or observed species richness between surveys conducted at different times in the season or at different sampling intensities. Figure 3c and d can also have great importance for the planning of a survey of an individual species in that the optimal values of season and effort can be easily determined for each species (Kéry 2002).

Our present results regarding the proportion of species detected (91%) concur well with a previous estimate of 89% based on Burnham's Jackknife for a larger number of years (2001–2003, and thus including the data in this study) as they do with those of mixed modelling of the resultant mean species detectability estimates in the earlier study (Kéry and Schmid 2006). We confirmed a decline in the proportion of species detected with elevation and in quadrats surveyed only twice instead of three times, as well as a lack of effects of landscape structure (forest cover) apart from a fall at very low forest cover, which apparently is accounted for by the reduced number of surveys above the treeline. We also confirmed a positive effect on detection probability of route length, although we modelled this into occupancy here, and of survey effort.

In contrast to earlier work, we introduced rich covariate structure into our model here, which is equivalent to introducing additional information. In particular, in the MHB there are four important components that affect detection of a species in a quadrat; route length, duration of survey, season of survey and observer identity. All were previously shown to have considerable effects on counts of species or individuals in avian surveys such as the North American BBS or the Swiss MHB (Sauer et al. 1994; Link and Sauer 1998; Kéry et al. 2005; Royle et al. 2005; Kéry and Schmid 2006; Royle et al. 2007b). In our analysis, we account for all of them except the last one; we note that it would be entirely possible to include a random observer effect if the same observer surveys different quadrats. The ease with which such covariates can be introduced in the analysis is one of the strengths of our modeling framework.

Five of the key assumptions of our model are these: (1) Occupancy probability $\psi(i, j)$ must be conditionally (on covariates such as elevation) independent across species and quadrats. This is obviously violated to the degree that there exists structure in the community caused by predation, competition or facilitation, i.e. by direct interactions among the species. However, it is likely that at the scale of our analysis such effects might be hard to distinguish from a random occurrence of species, which, interestingly, resembles arguments in favour of neutral community models (Gotelli and McGill 2006). Furthermore, it might be possible to formally test hypotheses about such species interactions within our modeling framework by adding hypothesized structure to Z. (2) Detections are also conditionally (on covariates) independent across species and quadrats, i.e., detection probability of a species should not depend on which other species was detected at a quadrat, nor on which other quadrat it had been detected before. In our study, the former is unlikely to be violated, furthermore, virtually all quadrats were surveyed by different observers, hence, the latter is no problem either. (3) Detections are also independent across temporal samples at the same quadrat. This assumption may be violated to some degree for some noteworthy species at least. For instance, rare or attractive species may be more likely to be detected once an observer has recorded them at a quadrat. Alternatively, shy species might become more elusive after first detection. Again, at the scale of our analysis we think it unlikely that such "behavioural response" effects (Otis et al. 1978) would be a problem, and if they were deemed to be, they could be included in the model. (4) The distribution of the logits of occurrence and detection probability is adequately modelled. We think that this assumption is reasonable, but note that parametric forms other than the normal distribution could be chosen if necessary. (5) The metacommunity is closed, i.e., the matrix Z is constant across all temporal samples. In spite of staggered arrival of some migratory species there seems to be little noticeable violation of the traditional "closure assumption" in our MHB data (Kéry and Schmid 2006). Furthermore, known patterns of closure violation, as in the case of migrant arrival, could also be accommodated in the model. Finally, a convenience rather than an assumption is that the surveyed quadrats are a representative sample of some larger area, because then the estimated metacommunity size has a useful interpretation.

5 Conclusions

We have illustrated a promising new modelling framework applicable to spatially and temporally replicated samples that provides an estimate of the imperfectly observed species-by-site occurrence matrix. This enables rich inference about species richness of both the entire metacommunity as well as each community individually, as well as of occupancy of each species and of measures of similarity of species and sites. The clear segregation of the observations into occupancy state and observation process permits an efficient integration of additional covariate information. Possible extensions of the model consist of relaxing the closure assumption partially or entirely. This would allow application of our modelling framework to communities that are dynamic relative to the time frame of a survey (e.g. insects) as well as direct estimation and modelling of parameters of community dynamics such as extinction, colonisation and turnover rates.

Acknowledgments We thank the large number of volunteers that conduct the field work in the Swiss breeding bird survey MHB and H. Schmid for valuable collaboration in MHB analysis projects. We thank R.J. Barker, W.A. Link and H. Steen for valuable comments on a previous manuscript.

Appendix 1

WinBUGS code for a model without covariates for the sake of improved readability. However, we indicate places where covariates would be introduced.

```
model {

#Prior distributions and parameter transformations
omega ~ dunif(0,1)
p0 ~ dunif(0,1)
psi0 ~ dunif(0,1)
sigmap ~ dunif(0,10)
sigmapsi ~ dunif(0,10)
rho ~ dunif(-1,1)

taup <- (1/(sigmap*sigmap))
taupsi <- (1/(sigmapsi*sigmapsi))
mup <- log(p0/(1-p0))
mupsi <- log(psi0/(1-psi0))
var.eta <- taup/(1.-pow(rho,2))

# Likelihood
for(i in 1:(nspec+nzeroes)){                          # Process model
        w[i] ~ dbin(omega,1)
        lpsi[i] ~ dnorm(mupsi,taupsi) I(-21,21)        # Note truncation trick
        mu.lp[i] <- mup +(rho*sigmap/sigmapsi)*(lpsi[i]-mupsi)
        lp[i] ~ dnorm(mu.lp[i],var.eta) I(-21,21)      # Note truncation trick
        for(j in 1:nquadrat){
                logit(psi[j,i]) <- lpsi[i]             # Add covariates here
                mu.psi[j,i] <- psi[j,i]*w[i]
                z[j,i] ~ dbern(mu.psi[j,i])
        }
}

for(k in 1:(nspec+nzeroes)){# Observation model
        for (i in 1:nquadrat) {
                for(j in 1:T){
                        logit(mu[i,j,k]) <- lp[k]      # Add covariates here
                        mu2[i,j,k] <- z[i,k]*mu[i,j,k]
                        x[i,j,k] ~ dbern(mu2[i,j,k])
                }
        }
}

# Species-richness N as a derived parameter
for(i in 1:nquadrat){
        Nquadrat[i] <- sum(z[i,1:(nspec+nzeroes)])
}
n0<-sum(w[(nspec+1):(nspec+nzeroes)])
Ntotal<-nspec+n0
}
```

Appendix 2

WinBUGS time-series plot for some random effects and random effects distribution hyperparameters. *Left*: Plots for model without restrictions; *right*: plots for model where random effects truncated by multiplication with indicator function $I(-21,21)$.

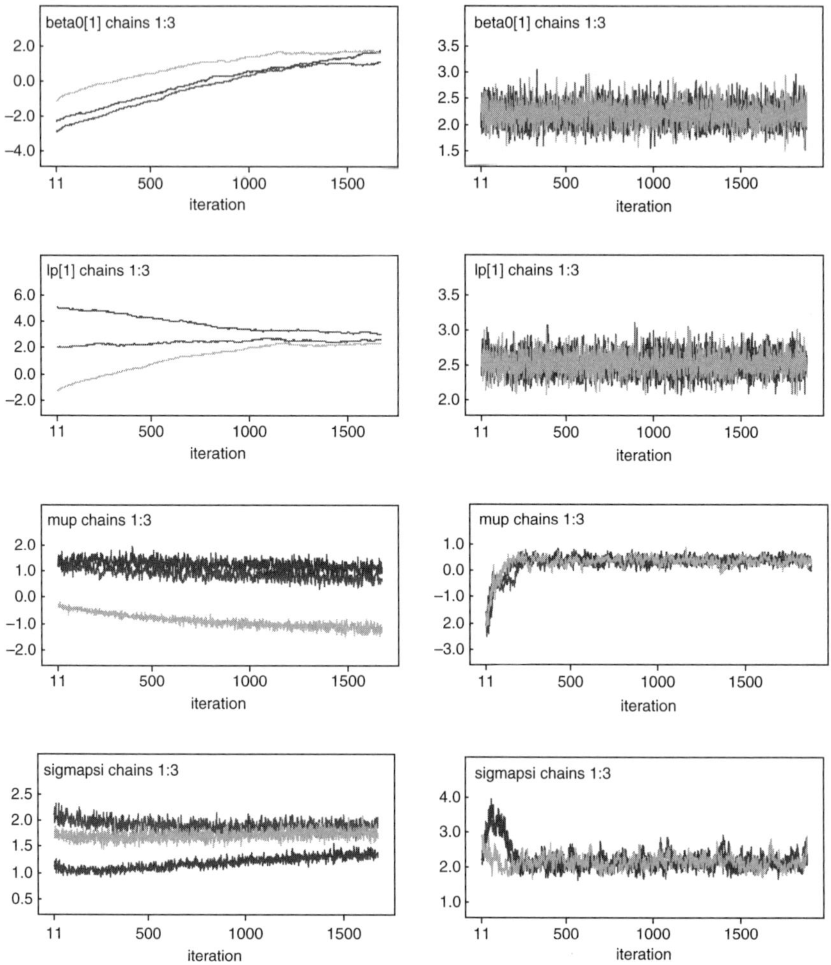

References

Boulinier T, Nichols JD, Sauer JR, Hines JE, Pollock KH (1998a) Estimating species richness: The importance of heterogeneity in species detectability. Ecology 79:1018–1028.

Boulinier T, Nichols JD, Hines JE, Sauer JR, Flather CH, Pollock KH (1998b) Higher temporal variability of forest breeding bird communities in fragmented landscapes. Proceedings of the National Academy of Sciences 95:7497–7501.

Boulinier T, Nichols JD, Hines JE, Sauer JR, Flather CH, Pollock KH (2001) Forest fragmentation and bird community dynamics: Inference at regional scales. Ecology 82: 1159–1169.

Cam E, Nichols JD, Sauer JR, Hines JE (2002) On the estimation of species richness based on the accumulation of previously unrecorded species. Ecography 25:102–108.

Connolly SR (2005) Process-based models of species distributions and the mid-domain effect. American Naturalist 166:1–11.

Dorazio RM, Royle JA (2005) Estimating size and composition of biological communities by modeling the occurrence of species. Journal of the American Statistical Association 100: 389–398.

Dorazio RM, Royle JA, Söderström B, Glimskär A (2006) Estimating species richness and accumulation by modeling species occurrence and detectability. Ecology 87:842–854.

Freckleton RP, Noble D, Webb TJ (2006) Distribution of habitat suitability and the abundance–occupancy relationship. American Naturalist 167:260–275.

Gelman A, Rubin DB (1992) Inference from iterative simulation using multiple sequences. Statistical Science 7:457–511.

Gotelli NJ, Colwell RK (2001) Quantifying biodiversity: Procedures and pitfalls in the measurement and comparison of species richness. Ecology Letters 4:379–391.

Gotelli NJ, McGill BJ (2006) Null versus neutral models: What's the difference? Ecography 29:793–800.

He F, Gaston KJ (2003) Occupancy, spatial variance, and the abundance of species. American Naturalist 162:366–375.

Kéry M (2002) Inferring the absence of a species – A case study of snakes. Journal of Wildlife Management 66:330–338.

Kéry M, Royle JA (2008) Hierarchical Bayes estimation of species richness and occupancy in spatially replicated surveys. Submitted to Journal of Applied Ecology, in press.

Kéry M, Royle JA, Schmid H (2005) Modeling avian abundance from replicated counts using binomial mixture models. Ecological Applications 15:1450–1461.

Kéry M, Schmid H (2006) Estimating species richness: Calibrating a large-scale avian monitoring program. Journal of Applied Ecology 43:101–110.

Link WA, Sauer JR (1998) Estimating population change from count data: Application to the North American breeding bird survey. Ecological Applications 8:258–268.

MacKenzie DI, Nichols JD, Lachman GB, Droege S, Royle JA, Langtimm CA (2002) Estimating site occupancy when detection probabilities are less than one. Ecology 83:2248–2255.

MacKenzie DI, Nichols JD, Royle JA, Pollock KH, Hines JE, Bailey LL (2006) Occupancy estimation and modeling: inferring patterns and dynamics of species occurrence, Academic Press, San Diego, USA.

Mao CX, Colwell RK (2005) Estimation of species richness: Mixture models, the role of rare species, and inferential challenges. Ecology 86:1143–1153.

McCoy ED, Heck Jr. KL, (1987) Some observations on the use of taxonomic similarity in large-scale biogeography. Journal of Biogeography 14:79–87.

Nichols JD, Boulinier T, Hines JE, Pollock KH, Sauer JR (1998a) Inference methods for spatial variation in species richness and community composition when not all species are detected. Conservation Biology 12:1390–1398.

Nichols JD, Boulinier T, Hines JE, Pollock KH, Sauer JR (1998b) Estimating rates of local species extinction, colonization, and turnover in animal communities. Ecological Applications 8:1213–1225.

Orme CDL, Davies RG, Burgess M, Eigenbrod F, Pickup N, Olson VA, Webster AJ, Ding TS, Rasmussen PC, Ridgely RS, Stattersfield AJ, Bennett PM, Blackburn TM, Gaston KJ, and Owens, I.P.F. (2005) Global hotspots of species richness are not congruent with endemism or threat. Nature 436:1016–1019.

Otis DL, Burnham KP, White GC, Anderson DR (1978) Statistical inference for capture data on closed animal populations. Wildlife Monographs 62:1–135.

Purvis A, Hector A (2000) Getting the measure of biodiversity. Nature 405:212–219.

Royle JA, Nichols JD (2003) Estimating abundance from repeated presence-absence data or point counts. Ecology 84:777–790.

Royle JA, Nichols JD, Kéry M (2005) Modeling occurrence and abundance of species with imperfect detection. Oikos 110:353–359.

Royle JA, Dorazio RM, Link WA (2007a) Analysis of multinomial models with unknown index using data augmentation. Journal of Computational and Graphical Statistics 16:67–85.

Royle JA, Kéry M (2007) A Bayesian state-space formulation of dynamic occupancy models. Ecology 88:1813–1823.

Royle JA, Kéry M, Gautier R, Schmid H (2007b) Hierarchical spatial models of abundance and occurrence from imperfect survey data. Ecological Monographs 77:465–481.

Sauer JR, Peterjohn BG, Link WA (1994) Observer differences in the North American Breeding bird survey. Auk 111:50–62.

Schmid H, Zbinden N, Keller V (2004) Überwachung der Bestandsentwicklung häufiger Brutvögel in der Schweiz, Swiss Ornithological Institute, Sempach, Switzerland.

Schmidt BR (2005) Monitoring the distribution of pond-breeding amphibians when species are detected imperfectly. Aquatic Conservation: Marine and Freshwater Ecosystems 15: 681–692.

Smith BJ (2005) Bayesian Output Analysis Program (BOA), Version 1.1.5. The University of Iowa. http://www.public-health.uiowa.edu/boa (accessed March 23, 2005).

Spiegelhalter D, Thomas A, Best NG (2003) WinBUGS User Manual, Version 1.4., MCR Biostatistics Unit, Cambridge.

Volet B (2006) Liste der Vogelarten der Schweiz (Checklist of the birds of Switzerland). Der Ornithologische Beobachter 103:271–294.

Time-Varying Covariates and Semi-Parametric Regression in Capture–Recapture: An Adaptive Spline Approach

Simon J. Bonner, David L. Thomson, and Carl J. Schwarz

Abstract Advances in capture–recapture methodology have allowed the inclusion of continuous, time-dependent individual-covariates as predictors of survival and capture probabilities. The problem posed by these covariates is that they are only observed for an individual when that individual is captured. One solution is to assume a model of the covariate which defines the distribution of unobserved values, conditional on the observed values, and apply Bayesian methods to compute parameter estimates and to test the covariate's effect.

Previous applications of this approach have modeled the survival probability as a linear function of the covariate on some scale (e.g. identity or logistic). In some applications a linear function may not adequately describe the true relationship. Here we incorporate semi-parametric regression to allow for more flexibility in the relationship between the covariate and the survival probabilities of the Cormack–Jolly–Seber model. A fully Bayesian, adaptive algorithm is used to model the relationship with splines, in which the complexity of the relationship is governed by the number and location of the knots in the spline. A reversible jump Markov chain Monte Carlo algorithm is implemented to explore splines with different knot configurations, and model averaging is used to compute the final estimates of the survival probabilities.

The method is applied to a simulated data set and to data collected through the Dutch Constant Effort Sites ringing project to study the survival of reed warblers (*Acrocephalus scirpaceus*) as a function of condition.

Keywords Adaptive splines · Bayesian inference · Capture–recapture · Free-knot splines · Reversible-jump Markov chain Monte Carlo · Semi-parametric regression

---

S.J. Bonner (✉)
Department of Statistics and Actuarial Science, Simon Fraser University, Burnaby, BC V5A 1S6, Canada
e-mail: sbonner@stat.sfu.ca

D.L. Thomson et al. (eds.), *Modeling Demographic Processes in Marked Populations*, Environmental and Ecological Statistics 3, DOI 10.1007/978-0-387-78151-8_29,

1 Introduction

Continuous, individual variables (e.g. body mass) may often be of interest as predictors of survival or catchability of animals in capture–recapture (CR) studies. The difficulty posed by variables that are unique to each animal and change over time is that they can only be observed when an individual is captured. Because of the missing values, these variables cannot be used in standard CR models (see e.g. Lebreton et al. (1992) who employ environmental or fixed individual covariates as predictors of survival). Bonner and Schwarz (2006) introduce one method for including continuous covariates in CR studies by developing a hierarchical model of the unobserved covariate values and then using Bayesian analysis via Markov chain Monte Carlo (MCMC) simulation to obtain parameter estimates. The method is applied to study the effect of body mass on the survival of the meadow vole (*Microtus pennsylvanicus*).

The objective of this paper is to allow more flexibility in the relationship between the covariate and the survival probability. The model of Bonner and Schwarz (2006) assumes that the link between the probability that individual i survives from occasion t to $t+1$ and the covariate is a linear function on the logistic scale:

$$\eta_{it} = \text{logit}\,(P(\text{ind. } i \text{ is alive at time } t\text{+1}|\text{ind. } i \text{ is alive at time } t)) = \beta_0 + \beta_1 z_{it}$$

where z_{it} denotes the value of the covariate for individual i at time t. In this paper we model η_{it} using a large class of non-linear functions. In particular we model:

$$\eta_{it} = s(z_{it})$$

where $s(\cdot)$ is a smoothing spline fit through an adaptive spline approach.

Smoothing splines are flexible functions formed from polynomial segments which connect at selected points called knots. Unlike polynomials which form a global fit to data, splines are locally adaptive such that the coefficients of the polynomial segments may vary over the range of the data (Eubank 1999). The amount of local change possible in a spline is determined by two factors: the number and location of the knots, and the amount by which the spline may change at each knot. This dichotomy leads to two methods for fitting splines. Penalized spline methods use a large number of fixed knots over the range of the data, but introduce a smoothing parameter which limits how the coefficients of the polynomial segments may differ on either side of a knot. Adaptive (or free knot) spline methods include estimation of the number and location of knots as part of the fitting procedure.

The adaptive method can be framed as a problem of fitting a hierarchical model in which splines with different numbers of knots form different submodels. This problem is ideally suited to Bayesian inference incorporating reversible jump Markov chain Monte Carlo (RJMCMC) to explore splines with different numbers of knots and hence different numbers of parameters (Green 1995). Final estimates are computed by model averaging the functions sampled on each RJMCMC iteration.

Section 2 of this paper provides an introduction to adaptive splines and describes the method for fitting survival probabilities as a function of a continuous covariate. Section 3 examines a simulation study in which the survival probability is known to have local dependence on the covariate. Fit of the spline model is compared with the fit of a simpler cubic polynomial model. In Section 4, we apply the method to study the relationship between the survival and condition of reed warblers (*Acrocephalus scirpaceus*) captured as part of the Dutch Constant Effort Sites (CES) ringing program. The final section discusses advantages and disadvantages of the method and provides some suggestions for its use in future CR studies.

2 Methods

2.1 Notation

Observed Data & Latent Variables:

T = number of capture occasions (indexed by t). *Observed.*
a_i = occasion on which individual i is first captured and marked. *Observed.*
z_{it} = value of time-varying covariate for individual i at time t. *Partially observed.*

Parameters:

ϕ_{it} = probability that individual i is available for capture at occasion $t+1$ given that it was available for capture at occasion t
p_t = probability that any individual available for capture at occasion t is captured
$s(z)$ = spline function modelling dependence of ϕ_{it} on z_{it}
κ = number of potential knot locations
K = number of knots in a single realization of the adaptive spline
ξ_k = location of the kth knot
ξ_l, ξ_u = lower and upper boundary knots
$\boldsymbol{\beta}, \boldsymbol{\gamma}$ = vectors of coefficients of $s(z)$ (truncated polynomial basis)
$\boldsymbol{b}$ = vector of coefficients of $s(z)$ (B-spline basis)
μ_t = population mean change in covariate between t and $t+1$
σ^2 = population variance of change in covariate between adjacent capture occasions

2.2 Cormack–Jolly–Seber Model with Individual Covariates

The purpose of our method is to allow flexibility in modelling the relationship between some process in a capture–recapture experiment (e.g. survival or capture) and an individual covariate. Here we illustrate the method by incorporating a time-varying condition measure as a predictor of survival in the Cormack–Jolly–Seber (CJS) model. We assume that animals are captured and marked on T capture occasions and released back into the population where they can be recaptured on subsequent occasions. The CJS models the probability of recapturing individuals in terms of two sets of parameters: the apparent survival probabilities and the capture probabilities. Standard assumptions for the CJS model can be found in many sources (see e.g. Seber (1982), Williams et al. (2002)).

Our model modifies the assumptions by allowing the survival probability to depend on a single time-varying covariate, denoted z_{it}. Bonner and Schwarz (2006) model the relationship between ϕ_{it} and z_{it} using a simple logistic function:

$$\eta_{it} = \text{logit}(\phi_{it}) = \beta_0 + \beta_1 z_{it}.$$

The primary difficulty with fitting this model is the large number of missing values for the covariate that may result when animals are not captured. Bonner and Schwarz (2006) solve this by defining a distribution for the unobserved values of the covariate conditional on the observed values. In particular, the model assumes that changes in the covariate between subsequent capture occasions are normally distributed according to:

$$z_{i,t+1}|z_{it} \sim N(z_{it} + \mu_t, \sigma^2) \tag{1}$$

where μ_t is allowed to vary with t and σ^2 is constant. Estimates of β_0 and β_1 are then generated from Bayesian inference via Markov chain Monte Carlo.

Here we extend the model by allowing η_{it} to be a non-linear function of z_{it}. In particular, we model $\eta_{it} = s(z_{it})$ where $s(z)$ is a spline. For simplicity, the capture probabilities are assumed to depend only on the capture occasion.

2.3 Splines

The class of splines is a set of functions that is commonly used for smoothing – finding a flexible function, $y = f(x)$, to describe coordinate data $(x_1, y_1), \ldots,$ (x_n, y_n). Part of their appeal is that splines can be formulated as an extension of polynomial regression, and much of the methods and theory used to fit simple regression models can be applied to fitting splines. The disadvantage of a straight polynomial regression model for scatterplot smoothing is that its fit is global. The value of the polynomial over any (small) interval determines its value over the entire range of data (Schumaker 1993, p. 103). If the relationship between y and x is complicated then the polynomial model will not fit well or will require many terms to achieve an adequate fit.

Spline models remedy this by introducing extra predictors that allow local changes in the fitted curve. A polynomial of order q is a function formed as a linear combination of the functions $\{1, x, \ldots, x^{q-1}\}$. A spline of order q on the interval $[\xi_l, \xi_u]$ is a function formed as a linear combination of the basis functions

$\{1, \ldots, x^{q-1}, (x-\xi_1)_+^{q-1}, \ldots, (x-\xi_K)_+^{q-1}\}$, for some points $\xi_l < \xi_1 < \ldots < \xi_K < \xi_u$. That is:

$$f(x) = \beta_1 + \beta_2 x + \ldots + \beta_q x^{q-1} + \sum_{k=1}^{K} \gamma_k (x-\xi_k)_+^{q-1}.$$

The function $(x-\xi_k)_+^{q-1}$ is a truncated polynomial of order q defined to be equal to 0 for $x < \xi_k$ and $(x-\xi_k)^{q-1}$ for $x \geq \xi_k$. The points ξ_l and ξ_u are the boundary knots of the spline and the points ξ_k, which partition $[\xi_l, \xi_u]$ into $k+1$ disjoint intervals, are the internal knots of the spline.

Over any interval between adjacent knots, $[\xi_k, \xi_{k+1}]$, the spline is equal to a polynomial of order q. At each knot the form of this polynomial is allowed to change, but only in a constrained manner which ensures that the entire curve over $[\xi_l, \xi_u]$ will be smooth. The curve must be continuous at each knot ($\lim_{x \to \xi_k^-} f(x) = f(\xi_k) = \lim_{x \to \xi_k^+} f(x)$), it will have $q-2$ continuous derivatives over the entire range, and the $(q-1)$th derivative will be continuous except for jumps of size γ_k at each knot. In contrast, a polynomial of order q has $q-1$ non-zero derivatives all of which are continuous. Our method considers cubic splines ($q = 4$) so that the spline is equivalent to a cubic polynomial between adjacent knots and has 2 continuous derivatives (Schumaker 1993, p. 108).

The flexibility of a spline comes from the jumps in the $(q-1)$st derivative allowed at each knot. For a cubic spline, the jumps occur in the 3rd derivative which means that the second derivative is continuous but may change very sharply at each knot. With q fixed, more flexibility can be introduced in the spline in one of two ways: (1) increasing the number of knots or (2) allowing for larger jumps in the derivative. This dichotomy leads to two competing approaches for fitting splines. Penalized spline methods use a large number of knots with fixed positions to allow flexibility anywhere in the range, and maintain smoothness by constraining the size of the jumps in the $(q-1)$st derivative allowed at each knot (Ruppert, Wand, and Carroll 2003, p. 65). This requires specification or estimation of a smoothing parameter which constrains the γ_k. Adaptive or free-knot spline methods allow γ_k to change more freely and include estimation of the number and location of the knots within the fitting procedure so that knots can be placed where needed. This is the approach taken in our method.

2.4 B-Splines

In practice, the truncated polynomial basis $\{1, \ldots, x^{q-1}, (x-\xi_1)_+^{q-1}, \ldots, (x-\xi_K)_+^{q-1}\}$ can often lead to numerical problems. First, the columns of the design matrix can be highly correlated if ξ_k and ξ_{k+1} are close; and second, the entries of the design matrix can be very large so that matrix operations become unstable. These problems can be avoided by using an equivalent basis of functions: i.e. a set of $q+K$ functions such that every spline can be written as a linear combination of the new set of functions and vice versa. A common choice is the set of B-spline basis functions (Ruppert et al. 2003; Schumaker 1993). This basis has several computational advantages including: (1) the basis functions are positive and sum to 1 at any single point, (2) as a result, the values of the design matrix will always be between 0 and 1, and (3) all of the functions in the basis are local in that each is positive only over a sub-interval of $[\xi_l, \xi_u]$. In the new basis, a spline of degree q with K knots can be written as:

$$s(z) = \sum_{j=1}^{q+K} b_j B_j^{(q)}(z)$$

where $B_j^{(q)}(z)$ is the jth B-spline basis function of order q and b_j is its coefficient.

One important consideration in computing the B-spline basis is the choice of the values ξ_l and ξ_u which define the interval over which the spline will be computed. These points are often called the boundary knots, and are required in computing the values of the B-spline basis functions. While the basis functions can be computed even for points which lie outside of this interval, the choice of the boundary knots is crucial for numerical reasons. At any point which lies inside $[\xi_l, \xi_u]$ the value of the B-spline basis functions will all lie in $[0, 1]$ and the sum over all of the basis functions will be exactly 1. This makes computing with the basis very stable. At points outside of this interval the value of the B-spline basis functions may be negative, and may become very large which makes the algorithms prone to numerical errors. This issue is addressed further below. Further discussion and algorithms for computing B-splines are provided in (Schumaker 1993).

2.5 Bayesian Adaptive Splines

To estimate the function $s(z)$ we follow the adaptive Bayesian method of Biller (2000) for fitting B-splines to generalized linear models. First, we select a large number, κ, of potential knot locations prior to the analysis. Here we space these locations evenly across the observed range of the data. The prior distribution is then constructed so that it places all of its mass on functions in the set of splines with between 1 and κ knots at these locations. This prior distribution is then updated with information from the data, in the form of the likelihood, to generate the posterior distribution over the same space of splines. In practice, even though the number of different knot configurations is finite the likelihood cannot be computed for each model and so the posterior cannot be constructed analytically. Instead, Biller provides a Markov chain Monte Carlo algorithm for generating a sample of realizations from the posterior distribution from which inference can be made.

A single spline in the restricted space is identified by three components: K, the number of knots $(0 \leq K \leq \kappa)$; $\boldsymbol{\xi}$, the vector of knot locations; and $\boldsymbol{b}$, the coefficients of the spline. To define a distribution over the space we need to define a joint density for these parameters. Following Biller (2000) again we write the joint prior density as the product of densities:

$$\pi(\boldsymbol{b}, \boldsymbol{\xi}, K) = \pi(\boldsymbol{b}|\boldsymbol{\xi}, K)\pi(\boldsymbol{\xi}|K)\pi(K)$$

where $\pi(\boldsymbol{b}|\boldsymbol{\xi}, K)$ is the conditional density of the coefficients given the number and locations of the knots, $\pi(\boldsymbol{\xi}|K)$ the density of the knot locations given the number of knots, and $\pi(K)$ the marginal density on the number of knots.

First, we assign $\pi(K)$ a truncated Poisson(λ) distribution so that:

$$\pi(K) \propto e^{-\lambda}\lambda^K (K!)^{-1}.$$

Then given K knots we assume that all configurations of the knots are equally likely. As there are $\binom{\kappa}{K}$ possible splines with K knots out of the κ potential locations, this leads to the conditional density:

$$\pi(\boldsymbol{\xi}|K) = \binom{\kappa}{K}^{-1}.$$

Finally, given $\boldsymbol{\xi}$ and K we assign the elements of $\boldsymbol{b}$ independent, diffuse normal priors with mean 0 and variance τ^2.

The likelihood for the model is an extension of the CJS likelihood with time-varying covariates and is exactly as given in Bonner and Schwarz (2006) with $\text{logit}(\phi(z)) = s(z)$. Combined with the prior distribution above, this defines a posterior over the restricted space of splines. Because of the large number of models in the space it is not possible to compute summaries of the posterior distribution analytically. Instead, Biller (2000) provides a MCMC algorithm that samples different realizations from the space. Inference is then made by computing summary statistics from this sample of functions.

The major challenge in the algorithm is that moving between splines with different numbers of knots changes the dimension of the model by increasing or decreasing the length of $\boldsymbol{b}$. Moves between models of different dimension cannot be accommodated in standard MCMC algorithms, like Metropolis-Hastings (MH), and instead, adding or removing knots is performed through RJMCMC. Like the MH algorithm, each iteration of RJMCMC involves proposing a new state for the parameters conditional on the current parameter values. This new function is then accepted with a probability computed from the prior distributions of the current and proposed states, their likelihoods given the observed data, and their densities under the proposal distribution. If accepted, the proposed state becomes the current state, and the chain continues. Otherwise the current state is retained and a new proposal is generated. As in the MH algorithm, the acceptance step in RJMCMC ensures that the posterior distribution will be a stationary distribution of the Markov chain. The RJMCMC acceptance probability was first derived by Green (1995); a simplified derivation is given in Waagepetersen and Sorensen (2001) and more details are available in recent books on Bayesian analysis or statistical computation (see e.g. Chen et al. (2000)).

The specific algorithm of Biller (2000) generates new proposals by adding or deleting one knot from the current spline. On each iteration, a random choice is made to add or delete one knot. When adding a knot, the location of the new knot is selected from all of the currently unoccupied locations with equal probability. When deleting a knot, the knot to delete is randomly chosen from the currently occupied locations. Each iteration also includes a step for moving one chosen knot to a nearby unoccupied location, which helps the chain to move across the function space.

Implementation of this algorithm for our model poses some difficulties because of the unobserved covariate values. In particular, Biller (2000) recommends choosing the boundary knots, ξ_l and ξ_u, equal to the minimum and maximum values of the covariate and spacing the κ potential knot locations between. This is not possible in our application because the minimum and maximum values are not actually observed. Setting ξ_l and ξ_u equal to the observed minimum and maximum, say $z_{\min}^{\text{obs}}$ and $z_{\max}^{\text{obs}}$, it is likely that some of the unobserved covariate values will lie beyond the boundary knots. Instead, we recommend choosing ξ_l and ξ_u to enclose a wide range about the observed data (e.g. $\xi_l = z_{\min}^{\text{obs}} - (z_{\max}^{\text{obs}} - z_{\min}^{\text{obs}})$ and $\xi_u = z_{\max}^{\text{obs}} + (z_{\max}^{\text{obs}} - z_{\min}^{\text{obs}})$) but still spacing the potential knots equally between $z_{\min}^{\text{obs}}$ and $z_{\max}^{\text{obs}}$. This arrangement allows for covariate values to lie outside of $[z_{\min}^{\text{obs}}, z_{\max}^{\text{obs}}]$, but constrains the curve to be a cubic polynomial on the intervals $[\xi_l, z_{\min}^{\text{obs}}]$ and $[z_{\max}^{\text{obs}}, \xi_u]$ where there is no observed data. Sampling from the posterior distribution also requires steps to impute the unobserved covariate values and to update the remaining model parameters.

As in Bonner and Schwarz (2006) steps must be included in the algorithm for updating the remaining parameters of the model and simulating the missing data values. The final algorithm for generating a sample from the posterior distribution of all random variables has the following structure:

Initialization:

(1) Define the boundary knots, ξ_l and ξ_u, and select κ potential knot locations between $z_{\min}^{\text{obs}}$ and $z_{\max}^{\text{obs}}$.
(2) Select initial values for all parameters, hyperparameters, and the missing data values for each individual.

MCMC Iteration:

(1) Latent data:
Simulate the unobserved covariates, z_{it}.
(2) Parameters of covariate distribution:
Update the parameters $\mu_1, \ldots, \mu_{T-1}$, and σ^2.
(3) Spline fit:

  (a) Propose change in dimension (i.e. addition of a knot at an empty location or deletion of a randomly selected knot).
  (b) Propose movement of a single, randomly chosen knot to a vacant location in the same neighbourhood.
  (c) Update the spline coefficients, $\boldsymbol{b}$.

(4) Remaining CJS parameters:
Update the capture probabilities $p_2, \ldots, p_T$.

MCMC iterations are then repeated until the chain converges and a large sample of realizations is generated. Full details of the different MCMC steps are described in a technical report (Bonner 2007).

Because there is no single set of knots, an estimate of $s(z)$ cannot be generated by plugging estimates of the coefficients into the B-spline equation. Instead, $s(z)$ is estimated by the posterior mean over the entire model space which is approximated by averaging over the functions sampled on a large number of iterations from the tail of the Markov chain. Precision of $s(z)$ is assessed with pointwise 95% highest posterior density (HPD) credible intervals. That is, for each value of z in $[\xi_l, \xi_u]$ we compute the shortest interval which covers 95% of the sampled values of $s(z)$.

3 Simulation Study

In our simulation study, capture histories for 500 individuals were generated from a CJS model with 3 capture occasions. Covariate values for each individual were simulated from the diffusion model in (1) with the initial distribution $z_{a_i} \sim N(0, 1)$ and parameters $\mu_1 = \mu_2 = 0.00$ and $\sigma^2 = 1.00$. Survival probabilities for each interval were computed from a bimodal function of the covariate with modes at $z = -1$ and $z = 1$. This function is plotted in Fig. 1. Capture probabilities were $p_2 = p_3 = 0.85$. Marking times were assigned so that half of the individuals were first captured at occasion 1 and half at occasion 2.

Three different models of the survival probability were fit to the simulated data to study the method's performance. The first model fit a cubic polynomial to the logit of the survival probability. The second and third fit the adaptive spline model described above using two different prior distributions for the number of active knots: Poisson with mean 25 and Poisson with mean 75. Markov chains for all three

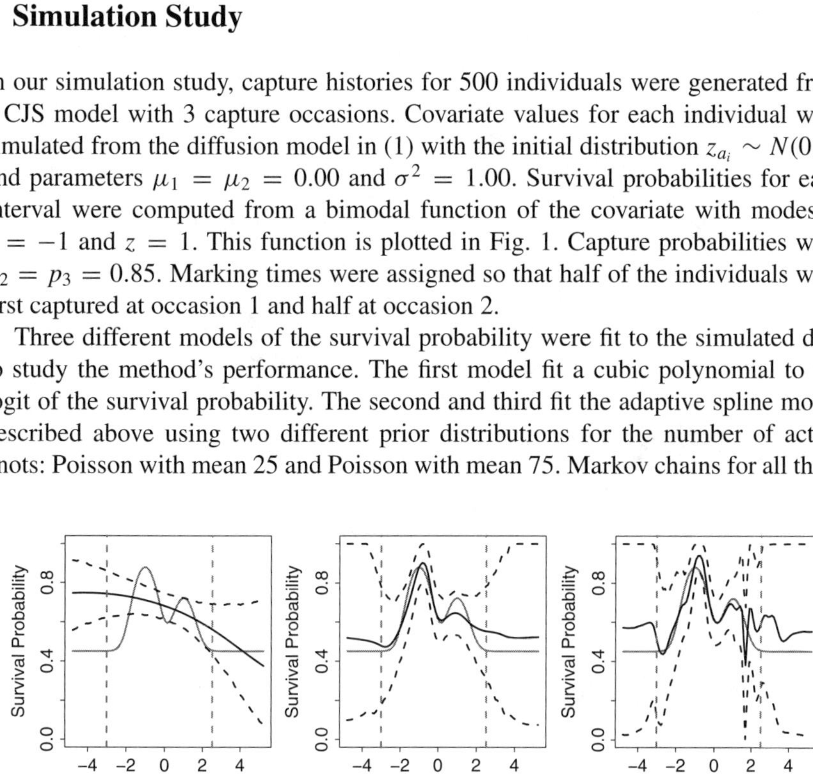

Fig. 1 Estimated survival probability as a function of the covariate for the simulated data. The left plot illustrates the estimated function assuming a cubic fit, the centre plot using the adaptive spline method with a Poisson(25) prior on the number of knots, and the right plot using the adaptive spline method with a Poisson(75) prior on the number of knots. In each plot the solid grey line indicates the true function, solid black line the pointwise posterior mean fit, and dotted black lines the bounds of the pointwise posterior 95% credible interval. The vertical dotted grey lines indicate the 2.5 and 97.5 percentiles of the simulated covariate values

analyses were run for 100,000 iterations. The initial 10,000 were discarded as burn-in, and every 10th of the remaining 90,000 iterations were retained for inference.

Estimates of the survival probability as a function of the covariate for all 3 models are plotted along with their pointwise 95% HPD credible intervals in Fig. 1. The cubic function clearly is too rigid to adjust to the local changes in the survival probability. Instead, the fitted curve decreases throughout the range of the covariate, and the 95% credible intervals fail to cover the true survival probability for much of the range.

In comparison, the spline fit using the Poisson(25) prior easily captures the bimodality of the survival probability. The pointwise 95% credible intervals completely cover the true function, but are between 2 and 3 times wider than the credible intervals for the cubic fit. A trace plot and histogram of the number of active knots in the spline for each MCMC iteration and a plot indicating the average number of times each knot location is occupied are shown in Fig. 2. The number of knots appears to converge very quickly to a stable distribution which places 95% of the posterior probability on models with between 5 and 14 knots. The posterior median is 10 knots. Knot locations that are most often occupied are centred near the

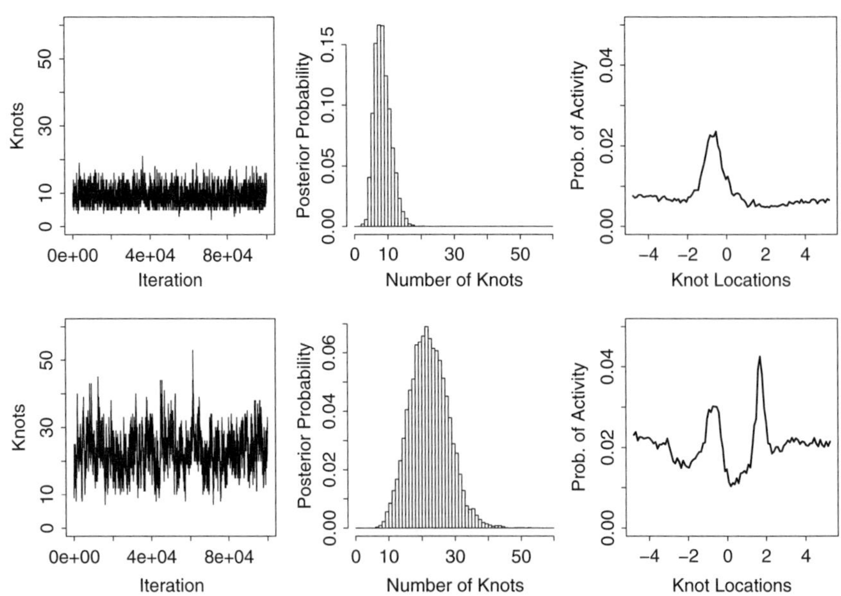

Fig. 2 Number and locations of knots in the spline fits of the survival probability for the simulated data. The upper row shows results using the Poisson(25) prior distribution on the number of knots and the lower using the Poisson(75) prior. The plots illustrate, from left to right, the number of knots on each MCMC iteration, the proportion of knot locations occupied on each iteration, and the proportion of iterations for which each potential knot location is occupied

Table 1 Estimates of capture probabilities and parameters of the covariate distribution for the simulated data. Results on the left are from the model assuming a cubic relationship between survival and the covariate; on the right are from the spline models. Estimates are given by the posterior mean with 95% HPD credible interval

Parameter	True value	Cubic polynomial	Poisson(25) prior	Poisson(75) prior
p_1	0.85	0.85(0.77,0.91)	0.84(0.76,0.90)	0.84(0.76,0.90)
p_2	0.85	0.84(0.73,0.96)	0.82(0.72,0.93)	0.80(0.70,0.90)
μ_1	0.00	−0.06(−0.23,0.10)	−0.06(−0.23,0.10)	−0.06(−0.23,0.10)
μ_2	0.00	0.04(−0.09,0.17)	0.04(−0.09,0.17)	0.04(−0.09,0.17)
σ	1.00	1.03(0.96,1.10)	1.03(0.96,1.10)	1.03(0.96,1.10)

largest mode, although knots throughout the range have a minimum probability of approximately 0.01 of being occupied.

The posterior distribution of the spline fit using the Poisson(75) prior assigns probability to more complex models with many more knots. Ninety-five percent of the posterior probability is assigned to models with between 14 and 35 knots, with a posterior median of 23 knots. The minimum posterior probability of activity is doubled to approximately 0.02.

The result of using the Poisson(75) prior is that more knots are included in the spline on each MCMC iteration and so the estimated survival probability as a function of the covariate is less smooth. In fact, the estimate now contains considerable noise and the most frequently occupied knot locations are associated with an anomalous local change in the survival probability at $z = 1.8$. This spike in the survival probability is caused by a chance grouping of individuals in the data all of which are captured on one occasion with covariate values near 1.8 and fail to survive until the next capture occasion. Examination of the true survival status (available from the simulation) versus the covariate shows exactly the same result. Note that even though the estimated survival probability is far from the true value, the 95% credible intervals still cover the truth at all points.

Results for the remaining parameters are given in Table 1. Changing the model of the survival probabilities has negligible affect on their estimates, and the means and 95% credible intervals produced by all three models are remarkably similar. The reason for this is the combination of high capture and high survival probabilities. Of the 500 individuals, 300 are captured on at least 2 occasions. With this many recaptures good estimates of the capture probabilities can be obtained from direct comparison of the capture histories, and the observed covariate values allow accurate estimation of μ_1, μ_2, and σ^2, without any knowledge of the survival probability.

4 Example

Data for the study of reed warblers (*Acrocephalus scirpaceus*) was obtained from the Dutch Constant Effort Sites (CES) banding project (Speek 2006). In the CES project, volunteer ringers capture birds on 12 day-long visits between April and

August to each of 38 sites in Holland. Ringers optionally record demographic and biometric characteristics of captured birds including age, sex, body mass, wing length, and tarsus length. The program was initiated in 1994 and data was available for 10 years up to 2003.

This analysis applies our extended CJS model to the final 5 years of data using a measure of the birds' condition as a predictor of survival. Each year of the study was considered as a single capture occasion. Multiple captures of the same bird in 1 year were combined by collapsing records into a single capture indicator and averaging the biometrics measurements. The condition measure for a single bird in a single year was defined as the ratio of its average observed body mass to its average observed wing length. After cleaning the data, measurements of the body mass ranged from 9.9 to 17.5 g and wing length from 62 to 71 mm. The range of the condition measure was 0.15 to 0.24 g/mm.

The CES database contains records of approximately 300,000 captures of 25,000 reed warblers captured between 1999 and 2003. The majority of these birds were observed only once, which was taken as evidence of large numbers of transients in the population. To avoid heterogeneity in the survival probability resulting from emigration, we used an ad hoc method restricting our analysis to resident individuals, including only birds captured 2 or more times–even if the 2 captures occurred in the same year. Capture histories were then conditioned on the birds' second release. Observations of juvenile birds were removed from the data because we believed that the probability of survival, and its relation to condition, was likely to differ between juveniles and adults. It was also necessary to remove many individuals for whom the condition measure could not be computed in any year of the study because of missing data on body mass or wing length. The final data set for our analysis contained capture histories of 592 birds, with 111 captured on 2 or more years.

As in the simulation study, three models were fit to the relationship between the survival probability: a cubic model on the logit scale, and two adaptive spline models with different priors on the number of knots. For the spline models, $\kappa = 100$ potential knots were equally spaced between the minimum and maximum observed condition values and the boundary knots were located at 0.05 and 0.34 g/mm. The prior distributions on the number of knots were Poisson with mean 25 in the first run and Poisson with mean 75 in the second. Separate intercept terms were included in all three models to allow the survival probabilities to change over time and capture probabilities were also modelled separately for each capture occasion. Note that $\phi_4(z)$ and p_5 are not completely confounded because of the dependence on z, but are very weakly identifiable separately and so only their product was estimated. The covariate was again modelled according to (1).

Figure 3 compares the survival probabilities estimated from the three different models. Here the fit of the curves is similar for all three models. Some local effects do appear in the spline estimates, but these occur at extreme values of condition where few birds are observed and come at the cost of much lower precision. Indeed, the apparent peak in survival at 0.22 g/mm arises from two birds captured with covariate values near this point. The peak disappears from both spline models when these birds are removed (results not shown) and the credible intervals at this point range from approximately 0.10 to 0.90 indicating extreme uncertainty. As in the

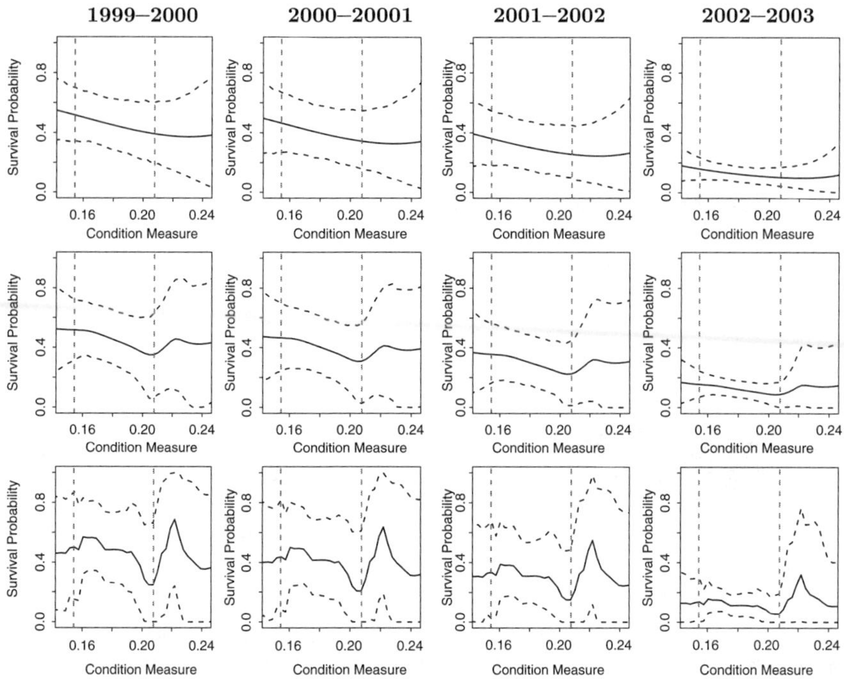

Fig. 3 Estimated survival probabilities as a function of condition for the reed warbler data. Estimates from the cubic model are shown in the top row, from the spline model with Poisson(25) prior on the number of knots in the middle, and from the spline model with Poisson(75) prior on the bottom. Solid black lines indicate the pointwise posterior mean and dashed lines the bounds of the posterior pointwise 95% credible intervals. In each plot, the vertical grey dashed lines indicate the 2.5-th and 97.5-th percentiles of the observed condition values. The plots for 2002–2003 actually estimate the product $\phi_4(z)p_5$ because of the weak identifiability of these parameters separately

simulation, the Poisson(75) prior generates a posterior that places higher probability on models with more knots, which decreases the smoothness and precision of the estimated survival probability (see Figure 4). The 95% credible intervals for all three models overlap at all points.

Estimates of the remaining parameters and a point estimate of survival at the median value of condition, 0.17 g/mm, are provided in Table 2. Also included are the results of fitting a Bayesian implementation of the CJS model with no effect of the covariate. The results are very similar for all four models. In all cases, the estimates suggest no significant change in the capture probability over time, though there is a slight decrease in the survival probability. Credible intervals of $\phi_t(z)$ are wider for the spline models than for the cubic model, but this does not affect the remaining parameters. The estimates of μ_t are all close to 0 indicating that there is no distinct increase or decrease in the birds' condition over any period. The estimate of σ is 0.012 g/mm for all models. For a bird with fixed wing length between 62 and 71 mm, this translates to an estimated standard deviation in mass between 0.74 and 0.85 g.

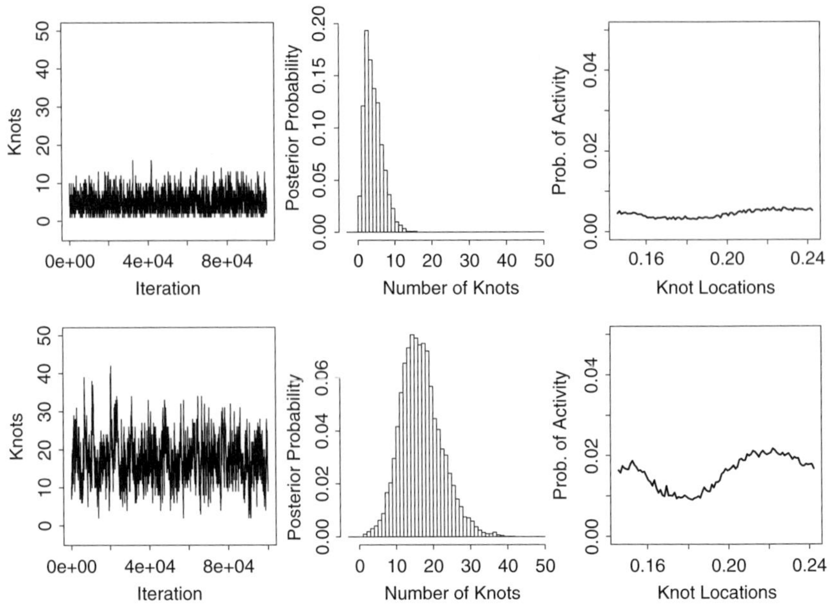

Fig. 4 Number and locations of knots in the spline fits of the survival probabilities for the reed warbler data. The upper row shows results using the Poisson(25) prior distribution on the number of knots and the lower using the Poisson(75) prior. The plots illustrate, from left to right, the number of knots on each MCMC iteration, the proportion of knot locations occupied on each iteration, and the proportion of iterations for which each potential knot location is occupied

5 Discussion

To our knowledge, only one other author has suggested the use of splines to allow more flexibility in modelling a covariate's effect in capture recapture methods. Gimenez, Covas, Brown, and Anderson (2006a) incorporate splines to model the effect of the southern oscillation index (SOI) on the survival of Snow petrels living in Terre Adèlie, and Gimenez, Crainiceanu, Barbraud, Jenouvrier, and Morgan (2006b) to model the effect of body mass on the survival of sociable weavers in South Africa. Both applications employ Bayesian methods but differ from our model in two respects. First, there is no missing data in the covariate: in the first application, (SOI) is an environmental covariate that can be observed regardless of the capture of individual birds, and in the second, a single, static covariate is computed by averaging all observations of body mass for each bird. The second difference is that both applications make use of penalized splines with fixed number of knots and fixed knot locations, rather than adaptive (free-knot) splines.

Another application of spline methods in a capture–recapture study is given by Fewster (2008). Their objective is to estimate the distribution of residence times for southern right whales' on their breeding grounds in the Auckland Islands from multiple sighting data. Penalized cubic splines are fit to the density function of the

Table 2 Estimates of capture probabilities and parameters of the covariate distribution for the reed warbler data. Results on the left are from the CJS model assuming no effect of the covariate on survival, then from a model assuming a cubic relationship between survival and the covariate, from the spline model with Poisson(25) prior on the number of knots, and on the right are from the spline model with Poisson(75) prior on the number of knots. Posterior means with 95% credible intervals are provided for each parameter. Estimated survival probabilities are computed at the median observed value of condition, 0.17 g/mm. Note that p_5 is almost confounded with $\phi_4(z)$ and their product was estimated as a single parameter

Quantity	CJS model	Cubic polynomial	Poisson(25) prior	Poisson(75) prior
$\phi_1(0.17)$	0.48(0.33,0.67)	0.47(0.32,0.64)	0.48(0.32,0.68)	0.52(0.32,0.78)
$\phi_2(0.17)$	0.41(0.27,0.63)	0.42(0.25,0.60)	0.43(0.26,0.65)	0.45(0.23,0.71)
$\phi_3(0.17)$	0.33(0.19,0.55)	0.32(0.17,0.49)	0.32(0.17,0.50)	0.34(0.14,0.58)
$\phi_4(0.17)p_5$	0.13(0.09,0.19)	0.13(0.08,0.19)	0.14(0.08,0.19)	0.13(0.06,0.20)
p_2	0.45(0.29,0.62)	0.46(0.29,0.64)	0.45(0.29,0.63)	0.42(0.26,0.60)
p_3	0.46(0.28,0.66)	0.46(0.28,0.67)	0.45(0.27,0.66)	0.43(0.24,0.65)
p_4	0.46(0.25,0.71)	0.48(0.26,0.73)	0.48(0.26,0.72)	0.44(0.23,0.70)
μ_1	NA	0.000(−0.004,0.004)	0.000(−0.004,0.004)	0.000(−0.004,0.004)
μ_2	NA	0.000(−0.005,0.004)	−0.001(−0.006,0.004)	−0.001(−0.006,0.004)
μ_3	NA	−0.002(−0.007,0.004)	−0.002(−0.007,0.003)	−0.002(−0.007,0.002)
μ_4	NA	−0.002(−0.010,0.007)	−0.001(−0.010,0.007)	−0.003(−0.011,0.005)
σ	NA	0.012(0.010,0.014)	0.012(0.010,0.014)	0.012(0.010,0.014)

residence times to ensure a smooth result. In contrast to our method, and the work above, inference in this application is based on classical likelihood methods.

The advantage of an adaptive spline over a penalized spline is that the number and location of the knots are estimated as part of fitting the spline, so that knots can be located where more flexibility is needed. One consequence is that the degree of smoothing can vary across the spline by clustering knots where the curve changes most rapidly and placing no knots where the curve is most smooth. The posterior distribution computed from the simulated data set with the Poisson(25) prior assigns low probability of being occupied to most knot locations, except for those in the area of the largest mode which are 2–3 times more likely to be occupied. Denison, Mallick, and Smith (1998) provide some examples of adaptive spline fits to even more rapidly changing functions with jump discontinuities. In contrast, a penalized spline fit has an upper limit to the flexibility which is determined by the spacing of the fixed knots. Of course, the added flexibility also increases the potential for overfitting as seen in both the simulation and the analysis of the reed warbler data.

The Bayesian approach to adaptive spline fitting also provides a natural way to incorporate uncertainty concerning the location of the knots. Selecting a single set of knots for the model, based on the posterior mode or any other criteria, ignores the fact that the knot locations aren't known and that different knot configurations provide different fitted curves. The Bayesian methodology provides posterior model probabilities that naturally rank the different models. Model averaging using these probabilities to weight the different models then provides a clear way to aggregate the different fitted curves into a single estimate. The posterior mean over all knot locations will favour the features of the most probable models, but also includes some features of the less probable models weighted according to their posterior probabilities. Credible intervals computed from the entire set of models also allow for uncertainty across different sets of knots and will generally be wider than those computed from a single set of knots (Hoeting, Madigan, Raftery, and Volinsky 1999).

The primary difficulty with Bayesian adaptive spline methods is selecting the joint prior distribution on the set of models indexed by $\boldsymbol{b}$, $\boldsymbol{\xi}$, and K. As is evident from the examples in this paper, the choice of prior is a very important determinant of the smoothness of the final curve. A prior distribution that places too much mass on simple models risks ignoring important aspects of the data, while a prior that favours complex models risks overfitting. Apart from the prior distribution there is no penalty for the model complexity, and overfitting is a serious concern. In simple smoothing of a scatterplot it is possible to choose the priors subjectively and then plot the fitted curve over-top of the raw data to assess the fit. The difficulty in our application is that neither the covariate nor the response are completely observed, so the fit of the curve cannot be visualized directly.

Our recommendation is that several prior distributions be selected and the resulting curves compared to see how the fit changes and whether the changes are biologically plausible. For the reed warbler data, the obvious difference between the cubic fit to the survival probability and the spline fits with Poisson(25) and Poisson(75) prior is the peak at 0.22 g/mm. The size of the peak in the last set of curves in Fig. 3 is striking, but the point where this occurs is well beyond the 97.5%-ile of the observed covariate values, where there is little data, and the 95% credible intervals at this point are very wide. Further analysis reveals that the peak is the effect of 2 birds and disappears once these birds are removed. Despite the size of the peak, it seems clear that there is no evidence of a jump in the survival probability at 0.22 g/mm.

A second subjective decision that must be made in applying our method is the choice of boundary knots $[\xi_l, \xi_u]$. In theory, as long as $[\xi_l, \xi_u]$ encloses the observed data this choice should have little effect on the fitted curve. The challenge is that if $[\xi_l, \xi_u]$ is very wide then the distance from the lower boundary to the first internal knot or from the last knot to the upper boundary will be large. This will lead to small values in the design matrix and the numerical algorithms may become unstable. Conversely, if $[\xi_l, \xi_u]$ is too narrow then it is possible for imputed values of the covariate to lie far outside this range and similar problems can occur. To assess the impact of this choice, we repeated the analysis of the simulated data with several values for of ξ_l and ξ_u and found no effect on the fitted survival probability. We also encountered no numerical problems using our default choice for the boundary knots.

One source of confusion with the adaptive-spline method might be the apparent discrepancy between the prior and posterior distribution on the number of knots. In both the simulation and example, the posterior distribution concentrates its mass on much simpler models than the prior distribution. This seems contradictory, but exactly the same behaviour appears in the original examples of Biller (2000). In his discussion of another paper on Bayesian adaptive splines, Holmes (2002) explains that the apparent discrepancy is a result of the Bartlett–Lindley paradox. Ignoring the prior distribution on $\boldsymbol{\xi}$ and K, the vague multivariate-normal prior on the spline coefficients, $\pi(\boldsymbol{b}|\boldsymbol{\xi}, K)$, induces a prior for the data whose variance increases with the dimension of $\boldsymbol{b}$. As a result any observed data has lower prior probability under more complex models and the distribution actually places less and less mass on models of higher and higher complexity.

The key in the adaptive spline model is that the prior on $\boldsymbol{b}|\boldsymbol{\xi}, K$ which favours simple models is partially offset by the Poisson prior on K which assigns very little mass to these models–when λ is large enough. The resulting prior distribution is a balance that assigns its mass to models simpler than those favoured by $\pi(K)$ alone. In essence, there is no discrepancy; rather, the prior on the set of models has to be interpreted through the full joint distribution, $\pi(\boldsymbol{b}, \boldsymbol{\xi}, K)$, and not simply the marginal prior $\pi(K)$.

Another caution with our approach, and with methods incorporating time-varying covariates in general, is the amount of data needed to provide adequate estimates. The final data set for the reed warblers contained 592 animals, but only 111 were captured on two or more occasions. Further, the condition measure was recorded two or more times for only 77 birds. This provides very little information regarding how the covariate changes over time and how differences in condition might affect the birds' survival. Fitting a Bayesian implementation of the standard CJS model to the data (assuming time-dependent survival and capture probabilities and ignoring any effect of the covariate) yields credible intervals for the survival probabilities that are between 0.34 and 0.36 wide. In light of this uncertainty when ignoring the covariate, it seems unlikely that any model will be able to detect an effect of condition on survival.

Based on our experience, including time-dependent covariates in the CJS model requires more data (i.e. capture of more individuals at given capture and survival rates) than models assuming homogeneity of individuals, or using environmental or static individual predictors. Using splines to model the survival probability as a function of the covariate will require even more data. Whereas a parametric curve borrows information from all values of the covariate to estimate the survival probability at any given value of the covariate, the spline only uses information from a local neighbourhood of covariate values. The result is that if few birds are observed with values in a given range of the covariate then in that range the estimate from the spline model will be highly variable though the estimate from the parametric model may still be precise.

As a final note we address the removal of transient birds from the CES reed warbler data. The majority of individuals captured are never recaptured and it is likely that many of these individuals are passing through the sites while migrating to other locations. The apparent survival probability of these individuals is 0. Pradel, Hines, Lebreton, and Nichols (1997) developed a model to account for transient individuals in the standard CJS model and compared it to the ad hoc method of conditioning on second release. They found that the ad hoc method produced unbiased estimates of the survival probabilities and was almost as efficient as the more complicated model when capture probabilities are high and the model cannot be simplified. With capture probabilities of 0.4 they found relative efficiency greater than 0.8.

An added effect in our application may be filtering of the values of the covariate. Resident individuals captured with values that equate to low survival probabilities will have less chance of being recaptured and more chance to be removed from the analysis. This should not bias the estimates of survival probability, but will decrease the precision of the estimates where these individuals are removed. The analysis is not intended to be an exhaustive examination of the reed warbler data, and how to deal with transients properly remains an open question.

Acknowledgments Simon Bonner was partially funded by a CGS-D scholarship from the National Science and Engineering Research Council of Canada (NSERC). We are indebted to Dr. Rachel Fewster, Dr. Ken Burnham, and an anonymous reviewer for their constructive comments on earlier drafts.

References

Biller C (2000) Adaptive Bayesian regression splines in semiparametric generalized linear models. Journal of Computational and Graphical Statistics 9(1):122–140.

Bonner SJ (2007) Time-Varying Covariates and Semi-Parametric Regression in Capture–Recapture: An Adaptive Spline Approach. Technical report, Simon Fraser University, Burnaby, BC.

Bonner SJ, Schwarz CJ (2006) An extension of the Cormack–Jolly–Seber model for continuous covariates with application to *Microtus pennsylvanicus*. Biometrics 62(1):142–149.

Chen M-H, Shao Q-M, Ibrahim JG (2000) Monte Carlo Methods in Bayesian Computation. Springer, New York.

Denison D, Mallick B, Smith A (1998) Automatic Bayesian curve fitting. Journal of the Royal Statistical Society, Series B 60:333–350.

Eubank RL (1999) Nonparametric Regression and Spline Smoothing. Marcell Dekker, Inc., New York, 2nd edition.

Fewster R (2008) Cubic splines for estimating the distribution of residence times using indvidual resighting data. In: Thomson DL, Cooch EG, Conroy MJ (eds.) Modeling Demographic Processes in Marked Populations. Environmental and Ecological Statistics, Springer, New York 3:393–416.

Gimenez O, Covas R, Brown CR, Anderson MD (2006a) Nonparametric estimation of natural selection on a quantitative trait using mark-recapture data. Evolution 60(3):460–466.

Gimenez O, Crainiceanu C, Barbraud C, Jenouvrier S, Morgan B (2006b) Semiparametric regression in capture–recapture modelling. Biometrics 62:691–698.

Green PJ (1995) Reversible jump Markov chain Monte Carlo computation and Bayesian model determination. Biometrika 82(4):711–7332.

Hoeting JA, Madigan D, Raftery AE, Volinsky CT (1999) Bayesian model averaging: a tutorial. Statistical Science 14:382–417.

Holmes C (2002) Discussion of: spline adaptation in extended linear models. Statistical Science 17(1):22–24.

Lebreton J-D, Burnham KP, Clobert J, Anderson DR (1992) Modelling survival and testing biological hypotheses using marked animals: a unified approach with case studies. Ecological Monographs 62(1):67–118.

Pradel R, Hines JE, Lebreton J-D, Nichols JD (1997) Capture–recapture survival models taking account of transients. Biometrics 53:60–72.

Ruppert D, Wand M, Carroll R (2003) Semiparametric Regression. Cambridge University Press, Cambridge, UK.

Schumaker LL (1993) Spline Functions: Basic Theory. John Wiley & Sons, Inc., New York, 2nd edition.

Seber GAF (1982) The Estimation of Animal Abundance: And Related Parameters. Macmillan, New York, 2nd edition.

Speek HG. Ces Handleiding. Technical report, Institute of Ecology Centre for Terrestrial Ecology, Heteren, Netherlands, December 2006. URL http://www.vogeltrekstation.nl/ces_handleiding.htm.

Waagepetersen R, Sorensen D (2001) A tutorial on reversible jump MCMC with a view toward applications in QTL-mapping. International Statistical Review 69:49–61.

Williams BK, Nichols JD, Conroy MJ (2002) Analysis and Management of Animal Populations. Academic Press, San Diego, CA.

A Further Step Toward the Mother-of-All-Models: Flexibility and Functionality in the Modeling of Capture–Recapture Data

Matthew R. Schofield and Richard J. Barker

Abstract The idea behind the mother-of-all-models is to have the likelihoods for commonly used capture–recapture models factorized into conditional likelihoods that can be called and combined on request to give a user specified model. Barker and White (2004) mapped out a conceptual plan for the mother-of-all-models that included the robust design model and joint recapture, live re-sighting models. However they were unable to obtain a factorization that could easily include the multi-state model. Including any missing data directly into the model using data augmentation allows us to write the model in terms of the complete data likelihood (CDL). The CDL is a more natural representation of the model that factors into separate components that can be combined to give many different capture–recapture models, including the multi-state model. Overcoming the obstacles in the factorization brings the mother-of-all-models one step closer with the development of software the next step.

1 Introduction

Modern software for capture–recapture models, such as MARK (White and Burnham 1999) and M-SURGE (Choquet et al. 2004) allows ecologists to consider a wide range of capture–recapture models. However, these programs suffer from two inherent problems: (i) there is often a significant lag time between models being published and becoming available to users in the software, and (ii) matching data to the correct likelihood requires a thorough understanding of the capture–recapture literature. For example, at least six types of model are available for open populations including birth in MARK. There are three different parameterizations of Pradel (1996), POPAN type models (Schwarz and Arnason 1996) as well as the parameterizations of Burnham (1991) and Link and Barker (2005).

M.R. Schofield (✉)
University of Otago, Department of Mathematics and Statistics, P.O. Box 56, Dunedin, New Zealand
e-mail: mschofield@maths.otago.ac.nz

D.L. Thomson et al. (eds.), *Modeling Demographic Processes in Marked Populations*, Environmental and Ecological Statistics 3, DOI 10.1007/978-0-387-78151-8_30,

These problems, together with the recent proliferation of capture–recapture models (Barker and White 2004; Schofield and Barker 2007b), gave rise to the "mother-of-all-models" (MoM) concept. The idea behind the MoM is to have conditional likelihood components that are selected by the user to give the appropriate capture-recapture model (Barker and White 2004). The MoM overcomes the problems in more traditional software because (i) new and complex models are simpler to incorporate in the program as only the additional conditional likelihood components are required, and (ii) users only need to understand the features of their data in order to select the components they need to specify an appropriate model. If someone wanted to include a per-capita birth rate index in their analysis they would simply need to include a birth component and select the per-capita birth rate index parameterization without needing to know the appropriate reference (Pradel 1996). While the MoM concept is good in theory, it will only work if a factorization of the likelihood is found that can easily be broken into conditional components that distinguish between different capture and recapture models.

The observed data from a capture–recapture study is denoted by the $u. \times k$ matrix $\boldsymbol{X}^{obs}$, which consists of the capture histories of the $u.$ observed individuals over the k sampling occasions. The value $X_{ij}^{obs} = 1$ means that individual i was caught in sample j and $X_{ij}^{obs} = 0$ otherwise.

Barker and White (2004) proposed writing the capture–recapture likelihood in terms of a Cormack–Jolly–Seber (CJS) core (Cormack 1964; Jolly 1965; Seber 1965). The CJS can be expressed as

$$[\boldsymbol{X}^{obs}|\boldsymbol{u}],$$

where $\boldsymbol{u} = (u_1, \ldots, u_k)$ and u_j is the number of unmarked individuals caught in sample j. The notation $[y]$ denotes the probability density function for continuous y or probability mass function for categorical y. This representation assumes that there is no loss on capture, but this can easily be relaxed (Barker and White 2004). The only demographic parameter in the CJS model are survival probabilities. Other demographic parameters such as birth rates can only be included when the first captures $\boldsymbol{u}$ are modeled. Barker and White (2004) showed how the Jolly–Seber model (Jolly 1965; Seber 1965) and the Crosbie–Manly–Arnason–Schwarz model (Crosbie and Manly 1985; Schwarz and Arnason 1996) could be obtained by multiplying additional likelihood components to the CJS core. For example, the Jolly–Seber (JS) model is obtained by adding a term that models $\boldsymbol{u}$,

$$[\boldsymbol{X}^{obs} \mid \boldsymbol{u}][\boldsymbol{u} \mid \boldsymbol{U}],$$

where $\boldsymbol{U} = (U_1, \ldots, U_k)$ and U_j is the total number of unmarked individuals available for capture at the time of sample j. Though the concept was promising, a difficulty identified by Barker and White (2004) is the lack of convenient factorization for the multi-state model. The multi-state model does not easily factorize into a CJS core with additional components.

Barker and White (2004) factorized the model in terms of the observed data likelihood (ODL). The ODL is the likelihood of the data after any missing or unknown information has been removed through integration (or summation). For the CJS model, the missing data are the interval censored times of death for every individual ever observed. An alternative to the ODL is to include the missing data directly into the likelihood using data augmentation (Tanner and Wong 1987). Instead of explicitly integrating out the unknown information, it is removed by choosing an appropriate computational algorithm to fit the model. Commonly used algorithms for data augmentation are the EM algorithm and Markov chain Monte Carlo (MCMC) algorithms (Dempster et al. 1977; van Dyk and Meng 2001). Including the missing data means that we are able to model in terms of the complete data likelihood (CDL). We show how the CDL can be factored so that many of the popular capture–recapture models are obtained as products of the conditional likelihood components.

2 Development of the Mother-of-All-Models Framework

2.1 *Closed Population Modeling*

The first step in completing the data is to introduce N, the total number of individuals ever available for capture throughout the study (Crosbie and Manly 1985; Schwarz and Arnason 1996) as a parameter. Conditional on N, the capture histories for every individual ever available for capture are known and are denoted by $\boldsymbol{X}$, which comprises $\boldsymbol{X}^{obs}$ and a $(N - u.) \times k$ matrix of zeros.

The CDL for closed population modeling is

$$[\boldsymbol{X}|\boldsymbol{p}, N] = \frac{N!}{u.!(N-u.)!}\prod_{i=1}^{N}\prod_{j=1}^{k} p_{ij}^{X_{ij}}(1-p_{ij})^{1-X_{ij}},$$

where $\boldsymbol{p}$ is a $N \times k$ matrix of capture probabilities with p_{ij} being the probability of capture for individual i in sample j. Different models for the capture process, such as models M_0, M_t, M_b, M_h, M_{tb}, M_{th}, M_{bh} and M_{tbh} (Otis et al. 1978) are obtained through placing constraints on the parameter p_{ij}. For example, model M_t assumes the probability of capture varies between capture periods, but is constant between individuals in each capture period,

$$p_{ij} = p_j, \forall i, j.$$

Note that for closed populations the CDL is identical to the ODL.

A useful step, particularly for generalizing the closed population model into open populations is splitting the capture matrix $\boldsymbol{X}$ in two parts. The first component $\boldsymbol{X_1}$ is a $N \times k$ matrix containing the information up to and including the first capture for every individual (if caught). The value X_{1ij} takes the value 1 if individual i is first caught in sampling occasion j and is $X_{1ij} = 0$ otherwise. The second component

$\boldsymbol{X}_2$ is a $u. \times k$ matrix containing all information after the first capture of each individual. The value $X_{2ij} = 1$ if individual i was recaptured in sampling occasion j and $X_{2ij} = 0$ otherwise.

The CDL $[\boldsymbol{X}|\boldsymbol{p}, N]$ can be factored as

$$[\boldsymbol{X}|\boldsymbol{p}, N] = [\boldsymbol{X}_1|\boldsymbol{p}, N][\boldsymbol{X}_2|\boldsymbol{p}, \boldsymbol{X}_1].$$

These components are specified as

$$[\boldsymbol{X}_1|\boldsymbol{p}, N] = \frac{N!}{u.!(N-u.)!} \prod_{i=1}^{N} \prod_{j=1}^{f_i} p_{ij}^{X_{ij}} (1-p_{ij})^{1-X_{ij}},$$

$$[\boldsymbol{X}_2|\boldsymbol{p}, \boldsymbol{X}_1] = \prod_{i=1}^{u.} \prod_{j=f_i+1}^{k} p_{ij}^{X_{ij}} (1-p_{ij})^{1-X_{ij}},$$

where f_i is the sample of first capture for individual i. If i was never caught, then $f_i \equiv k$.

2.2 Cormack–Jolly–Seber Model

The Cormack–Jolly–Seber (CJS) model relaxes the closed population assumption by allowing death to occur during the study. Mortality is expressed through the $N \times k$ matrix $\boldsymbol{d}$, with $d_{ij} = 1$ meaning that individual i died between the times of sample j and $j+1$, and $d_{ij} = 0$ otherwise, with the constraint $\sum_{j=1}^{k} d_{ij} = 1$ (note that $d_{ik} = 1$ means the individual was alive at the end of the study). As individuals can only die after their last capture, $d_{ij} = 0$ for all j before the last capture of i with the values of d_{ij} missing after the time of last capture. As with the capture histories we split $\boldsymbol{d}$ into two components: the $N \times k$ matrix $\boldsymbol{d}_1$ contains the information in $\boldsymbol{d}$ up to and including the first capture of each individual and the $u. \times k$ matrix $\boldsymbol{d}_2$ contains the information in $\boldsymbol{d}$ after first capture.

The CDL for the CJS model is obtained by combining the capture component for $\boldsymbol{X}_2$ from the closed population model, with a component that models mortality,

$$[\boldsymbol{X}_2|\boldsymbol{d}_2, \boldsymbol{p}, \boldsymbol{X}_1][\boldsymbol{d}_2|\boldsymbol{S}, \boldsymbol{X}_1],$$

where $\boldsymbol{p} = (p_2 \ldots, p_k)$ is the vector of time specific capture probabilities and $\boldsymbol{S} = (S_1, \ldots, S_{k-1})$ is the vector of time specific survival probabilities with S_j being the probability of surviving from the time of sample j until the time of sample $j+1$.

The mortality component for each individual is specified to be the outcome a single multinomial draw,

$$[\boldsymbol{d}_2|\boldsymbol{S}, \boldsymbol{X}_1] = \prod_{i=1}^{u.} MN(1, \zeta_i).$$

The vector ζ_i is

$$\zeta_i = \left((1 - S_{f_i}), S_{f_i}(1 - S_{f_i+1}), \ldots, \prod_{j=f_i}^{k-2} S_j(1 - S_{k-1}), \prod_{j=f_i}^{k-1} S_j \right).$$

The likelihood component for $\boldsymbol{X}_2$ is the same as for the closed population models except that the individuals are only available for capture until the censored period of death,

$$[\boldsymbol{X}_2|\boldsymbol{d}_2, \boldsymbol{p}, \boldsymbol{X}_1] = \prod_{i=1}^{u.} \prod_{j=f_i+1}^{t_{i2}} p_j^{X_{ij}}(1 - p_j)^{1-X_{ij}},$$

where t_{i2} is the period of death ($d_{it_{i2}} = 1$).

The defining feature of the CJS model is that we do not include the likelihood component for $\boldsymbol{X}_1$ because we are unwilling to make any assumptions about the birth process. If we included the conditional likelihood component for $\boldsymbol{X}_1$ with no component for birth, we would assume all individuals were available for capture at the beginning of the study. Including another conditional likelihood component to allow for birth yields the Crosbie–Manly–Arnason–Schwarz model in Section 2.4.

2.3 Jolly–Seber model

Even though most of our focus will be on the Crosbie–Manly–Arnason–Schwarz (CMAS) formulation of the model (see Section 2.4), we present the Jolly–Seber (JS) model for completeness.

The CDL for the JS model is obtained by combining the CDL for the CJS model and a conditional likelihood component that models the first captures of individuals in terms of the number of unmarked individuals in the population at the time of each sample,

$$[\boldsymbol{X}_2|\boldsymbol{d}_2, \boldsymbol{p}, \boldsymbol{X}_1][\boldsymbol{d}_2|\boldsymbol{S}, \boldsymbol{X}_1][\boldsymbol{u}|\boldsymbol{p}, \boldsymbol{U}].$$

Note that the vector $\boldsymbol{p}$ now also includes p_1. The components $[\boldsymbol{X}_2|\boldsymbol{d}_2, \boldsymbol{p}, \boldsymbol{X}_1]$ and $[\boldsymbol{d}_2|\boldsymbol{S}, \boldsymbol{X}_1]$ are the same as in the CJS with

$$[\boldsymbol{u}|\boldsymbol{p}, \boldsymbol{U}] = \prod_{j=1}^{k} \frac{U_j!}{u_j!(U_j - u_j)!} p_j^{u_j}(1 - p_j)^{U_j - u_j}.$$

This formulation completes the data by including the random variables $\boldsymbol{U}$ as parameters, instead of N.

2.4 Crosbie–Manly–Arnason–Schwarz Model

The Crosbie–Manly–Arnason–Schwarz (CMAS) model (Crosbie and Manly 1985; Schwarz and Arnason 1996) relaxes the closed population assumption by allowing birth and death. Strictly, by "birth" we mean recruitment of individuals into the population with members not distinguished by age. We express mortality in the same way as for the CJS model. Birth is expressed through the $N \times k$ matrix $\boldsymbol{b}$, with $b_{ij} = 1$ meaning that individual i was born between the times of sample j and $j+1$, with $b_{ij} = 0$ otherwise, with the constraint $\sum_{j=0}^{k-1} b_{ij} = 1$ (note that $b_{i0} = 1$ means the individual was born before the study began). As individuals must have been born before their first capture, $b_{ij} = 0$ for all j after first capture with the values of b_{ij} missing before first capture.

The CDL for the CMAS model is obtained by taking the CDL for the closed population studies and including the birth and death,

$$[\boldsymbol{X}_1|\boldsymbol{b}, \boldsymbol{p}, N][\boldsymbol{X}_2|\boldsymbol{d}_2, \boldsymbol{p}, \boldsymbol{X}_1][\boldsymbol{b}|\beta, N][\boldsymbol{d}_1|\boldsymbol{b}, \boldsymbol{S}, N][\boldsymbol{d}_2|\boldsymbol{S}, \boldsymbol{X}_1].$$

The components $[\boldsymbol{X}_2|\boldsymbol{d}_2, \boldsymbol{p}, \boldsymbol{X}_1]$ and $[\boldsymbol{d}_2|\boldsymbol{S}, \boldsymbol{X}_1]$ are same as for the CJS model. The likelihood component for $\boldsymbol{X}_1$ is the same as for closed population modeling except that the individual is only available for capture after it has been born,

$$[\boldsymbol{X}_1|\boldsymbol{b}, \boldsymbol{p}, N] = \frac{N!}{u.!(N-u.)!} \prod_{i=1}^{N} \prod_{j=t_{i1}}^{f_i} p_j^{X_{ij}} (1-p_j)^{1-X_{ij}},$$

where t_{i1} is the first sample after birth for individual i. The likelihood component for $\boldsymbol{b}$ is

$$[\boldsymbol{b}|\beta, N] = \prod_{i=1}^{N} MN\left(1, (\beta_0, \beta_1, \ldots, \beta_{k-1})\right),$$

where β_j is the probability of being born between time of sample j and $j+1$ conditional on being alive and available for capture at some time during the study and has the constraint that $\sum_j \beta_j = 1$ (Schwarz and Arnason 1996). The likelihood component for $\boldsymbol{d}_1$ is

$$[\boldsymbol{d}_1|\boldsymbol{b}, \boldsymbol{S}, N] = \prod_{i=1}^{u.} \prod_{j=t_{i1}}^{f_i-1} S_j \prod_{i=u.+1}^{N} MN(1, \zeta_i),$$

$$\zeta_i = \left((1-S_{t_{i1}}), S_{t_{i1}}(1-S_{t_{i1}+1}), \ldots, \prod_{j=t_{i1}}^{k-2} S_j(1-S_{k-1}), \prod_{j=t_{i1}}^{k-1} S_j \right).$$

The likelihood is split into two components because all $u.$ individuals caught are known to have not died before first capture.

An advantage of this factorization is that different parameterizations can easily be included for any component. For example, the birth models of Burnham (1991), Pradel (1996), Link and Barker (2005) or any other parameterization can be specified instead of β_j (Schofield and Barker 2007b). Switching between these different models does not involve re-writing the complete likelihood of the model, but re-parameterizing the birth component of the model.

There are many possible factorizations for the CMAS model. The factorization in equation (1) was chosen due to its simplicity. However, to include more complex models, such as models that allow density-dependence, a more complex factorization is required (Schofield and Barker 2007b). If we factorize the birth and death components according to the natural ordering of time events, we can write the CDL of the CMAS model as

$$[\boldsymbol{X}_1|\boldsymbol{b}, \boldsymbol{p}, N][\boldsymbol{X}_2|\boldsymbol{d}_2, \boldsymbol{p}, \boldsymbol{X}_1]\times$$
$$\prod_{i=1}^{N}\left\{[b_{i0}|\beta_0]\prod_{j=1}^{k-1}[b_{ij}|\boldsymbol{b}_{:(0:j-1)}, \boldsymbol{d}_{:(1:j-1)}, \boldsymbol{\beta}_{0:j-1}][d_{ij}|\boldsymbol{b}_{:(0:j)}, \boldsymbol{d}_{:(0:j-1)}, S_j]\right\},$$

where $\boldsymbol{d}_{:(1:j-1)}$ is a matrix containing columns 1 through $j-1$ of the matrix $\boldsymbol{d}$. This factorization is general with the factorization specified earlier a special case.

2.5 Covariates

2.5.1 Multi-State Model

The multi-state model is obtained when we include a partially-observed individual-specific categorical covariate that provides information about various parameters in the model. The values of the covariate are known for the sampling occasions when the individual was caught, but are missing when the individual was not caught. We follow Dupuis (1995) and include the missing covariate values using data augmentation to obtain the CDL. An example of the multi-state model is when tagging occurs in three different locations, A, B or C, and the probability of capture and survival rates are thought to differ by location, with the location unknown when the individual is not caught. Denoting the covariate by the matrix z, we split the covariate into z_1, the covariate values up to and including the first capture and z_2, the covariate values after first capture. Conditioning on first capture, the CDL for the multi-state model is the CDL for the CJS model combined with a component for the covariate,

$$[\boldsymbol{X}_2|\boldsymbol{d}_2, \boldsymbol{p}, z, \boldsymbol{X}_1][\boldsymbol{d}_2|\boldsymbol{S}, z, \boldsymbol{X}_1][z_2|z_1, \psi;],$$

where $\boldsymbol{\psi}$ is a matrix/vector of parameters used to model the covariate. Note that both the likelihood components for $\boldsymbol{X}_2$ and $\boldsymbol{d}_2$ condition on z as the capture and mortality components can depend on the covariate. The model for the covariate

$[z_2|z_1, \psi]$ can be specified to give various models, such as the first order Markovian model (Schwarz et al. 1993) or the second order Markovian model (Brownie et al 1993). For a first order Markovian model,

$$
\begin{aligned}
[z_2|z_1, \psi] &= \prod_{i=1}^{N} \prod_{j=f_i+1}^{t_{i2}} [z_{ij}|z_{ij-1}, \psi] \\
\left[z_{ij} = A|z_{ij-1} = A, \psi\right] &= \psi_{11} \\
\left[z_{ij} = B|z_{ij-1} = A, \psi\right] &= \psi_{12} \\
\left[z_{ij} = C|z_{ij-1} = A, \psi\right] &= 1 - \psi_{11} - \psi_{12} \\
\left[z_{ij} = A|z_{ij-1} = B, \psi\right] &= \psi_{21} \\
\left[z_{ij} = B|z_{ij-1} = B, \psi\right] &= \psi_{22} \\
\left[z_{ij} = C|z_{ij-1} = B, \psi\right] &= 1 - \psi_{21} - \psi_{22} \\
&\vdots \\
\left[z_{ij} = C|z_{ij-1} = C, \psi\right] &= 1 - \psi_{31} - \psi_{32}.
\end{aligned}
$$

Availability for capture is another potential categorical covariate, where $z_{ij} = 1$ means that individual i was available for capture in sample j and $z_{ij} = 0$ otherwise. Various movement assumptions such as permanent, random or first order Markovian emigration can be specified through the likelihood component $[z_2|z_1, \psi]$. For example, permanent emigration is,

$$
\begin{aligned}
\left[z_{ij} = 1|z_{ij-1} = 0, \psi\right] &= 0 \\
\left[z_{ij} = 0|z_{ij-1} = 0, \psi\right] &= 1 \\
\left[z_{ij} = 1|z_{ij-1} = 1, \psi\right] &= F_j \\
\left[z_{ij} = 0|z_{ij-1} = 1, \psi\right] &= 1 - F_j
\end{aligned}
$$

Permanent emigration causes confounding between S_j and F_j. The most convenient way to include this movement structure into the model is to not include the likelihood component for z and consider the parameter $\phi_j = S_j F_j$ instead of S_j. Likewise, for random emigration, we can consider the parameter $p'_j = p_j F_j$ instead of p_j.

2.5.2 Multi-Event Models

Multi-event models arise when covariates are observed with uncertainty. The true underlying value of the covariate is referred to as the state and denoted by the matrix z with the covariate values observed with error referred to as the events and denoted by z' (Pradel 2005). The only extension to the CDL of the multi-state model above is the introduction of a conditional likelihood component that models how the true covariate values z were corrupted to give z',

$$[\boldsymbol{X}_2|\boldsymbol{d}_2, \boldsymbol{p}, z, \boldsymbol{X}_1][\boldsymbol{d}_2|\boldsymbol{S}, z, \boldsymbol{X}_1][z'|z, \boldsymbol{\theta}][z_2|z_1, \psi],$$

where $\boldsymbol{\theta}$ models the corruption process. In many cases the covariate model $[z_2|z_1, \psi]$ and the corruption model $[z'|z, \boldsymbol{\theta}]$ can be specified to give a hidden Markov model (Pradel 2005; Cappé et al. 2005).

2.6 Continuous Covariates

A model with time-varying individual-specific continuous covariates has the same factorization as a model with time-varying individual-specific categorical covariates. Bonner and Schwarz (2006) included the individual body weight as a covariate and modeled the covariate as

$$[z_2|z_1, \psi] = \prod_{i=1}^{N} \prod_{j=f_i+1}^{t_{i2}} [z_{ij}|z_{ij-1}, \psi]$$

$$\left[z_{ij}|z_{ij-1}, \psi\right] = N(z_{ij-1} + \psi_{1j-1}, \psi_2),$$

where ψ_{1j} is the mean increase in weight for the population between the time of sample j and $j+1$.

2.7 Joint Re-sighting Models

The joint models of Burnham (1993) and Barker (1997) are used when dead recoveries and/or re-sightings of individuals occur between sampling occasions. The recoveries and re-sightings give us additional information about $\boldsymbol{d}$. If an individual was recovered dead we have complete knowledge of the ith row of $\boldsymbol{d}$. Furthermore, we also obtain additional information about $\boldsymbol{d}$ if an individual was re-sighted after the sample of last capture.

The information from dead recoveries is expressed through the $u. \times k'$ matrix $\boldsymbol{Y}_1$, where $Y_{1ij} = 1$ means that individual i was recovered dead between samples j and $j+1$ and $Y_{1ij} = 0$ otherwise. The value k' is the number of periods re-sighting and recovery data was collected. The information from live re-sightings is expressed through the $u. \times k'$ matrix $\boldsymbol{Y}_2$, where $Y_{2ij} = 1$ if individual i was re-sighted alive between sample j and $j+1$ and $Y_{2ij} = 0$ otherwise. Note that if $k' \geq k$ then the matrix $\boldsymbol{d}$ becomes a $N \times (k'+1)$ matrix to include the additional information on death from the re-sightings and recoveries.

To incorporate a model with only dead recoveries, we include a likelihood component for $\boldsymbol{Y}_1$. Conditioning on first capture, the CDL becomes,

$$[\boldsymbol{X}_2|\boldsymbol{d}_2, \boldsymbol{p}, \boldsymbol{X}_1][\boldsymbol{d}_2|\boldsymbol{S}, \boldsymbol{X}_1][\boldsymbol{Y}_1|\boldsymbol{r}, \boldsymbol{d}],$$

where $\boldsymbol{r} = (r_1, \ldots, r_{k'})$ and r_j is the probability of being recovered dead between time of sample j and $j+1$ given the individual died between time of sample j and $j+1$. The likelihood component for the dead recoveries is

$$[\boldsymbol{Y}_1|\boldsymbol{r}, \boldsymbol{d}] = \prod_{i=1}^{N} r_{t_{i2}}^{Y_{1it_{i2}}}(1 - r_{t_{i2}})^{1-Y_{1it_{i2}}}.$$

The CDL for the joint re-sighting and recovery model includes an additional likelihood component for $\boldsymbol{Y}_2$,

$$[\boldsymbol{X}_2|\boldsymbol{d}_2, \boldsymbol{p}, \boldsymbol{X}_1][\boldsymbol{d}_2|\boldsymbol{S}, \boldsymbol{X}_1][\boldsymbol{Y}_1|\boldsymbol{r}, \boldsymbol{d}][\boldsymbol{Y}_2|\boldsymbol{R}, \boldsymbol{R}', \boldsymbol{d}, \boldsymbol{Y}_1, \boldsymbol{X}_1],$$

where $\boldsymbol{R} = (R_1, \ldots, R_{k'})$, $\boldsymbol{R}' = (R'_1, \ldots, R'_{k'})$, R_j is the probability of being re-sighted alive between time of sample j and $j+1$ conditional on surviving between time of sample j and $j+1$ and R'_j is the probability of re-sighting an individual alive between time of sample j and $j+1$ conditional on it not surviving between time of sample j and $j+1$. We follow the model of Barker et al. (2004) and specify the likelihood component for the live re-sightings as

$$[\boldsymbol{Y}_2|\boldsymbol{R}, \boldsymbol{R}', \boldsymbol{d}, \boldsymbol{Y}_1, \boldsymbol{X}_1 = \prod_{i=1}^{N} \Bigg\{ \left({R'_{t_{i2}}}^{Y_{2it_{i2}}}(1 - R'_{t_{i2}})^{1-Y_{2it_{i2}}} \right)^{1-Y_{1it_{i2}}} \times \prod_{j=f_i}^{t_{i2}-1} R_j^{Y_{2ij}}(1 - R_j)^{1-Y_{2ij}} \Bigg\}.$$

2.8 *Known Fate Models*

In known fate models the capture probabilities p_j are specified to be 1. This means that the interval censored times of death are known and are no longer included using data augmentation.

2.9 *Robust Design Models*

Robust design models include closed population sampling periods within open population sampling periods. To implement a robust design model, we constrain the survival probabilities to 1 and (if included) the birth probabilities to 0 during the closed population periods. In the usual way capture probabilities can be specified to reflect belief about the closed population capture process.

3 Discussion

We have mapped out a factorization of the CDL that allows for conditional likelihood components to be combined to form user specified models. The conditional likelihood components relate to features of the capture–recapture experiment, such as mortality and birth, with many of the well used capture–recapture models being able to be specified. The choice of matrices to express mortality and birth is arbitrary. We choose $\boldsymbol{b}$ and $\boldsymbol{d}$ because this offers a distinction between the birth and death processes, however, there are numerous possibilities, including using the matrix $\boldsymbol{a}$, where $a_{ij} = 1$ if individual i is alive in sample j and $a_{ij} = 0$ otherwise.

The CDL is a natural representation of the likelihood that allows us to concentrate on modeling the demographic features of interest instead of focusing on the complexities of sampling. For example, the most meaningful birth parameter for the dataset can be obtained by switching between the parameterizations of Burnham (1991), Schwarz and Arnason (1996), Pradel (1996), Link and Barker (2005) or any other birth rate parameter, without having to alter the rest of the model. We do not need four separate models to include different birth parameters, just four different parameterizations of the conditional birth component.

Data augmentation can also be used to specify hierarchical models that offer the ability to include relationships between parameters (Link and Barker 2005), and between parameters and random variables. An example is density dependent relationships, where both survival and per-capita birth rates are related to population size (Schofield and Barker 2007b).

To fully utilize the MoM concept, we require user friendly software with efficient algorithms that integrate over the missing data. Unfortunately, no such software currently exists. Making inference from the CDL is difficult in a frequentist setting because algorithms such as the EM algorithm can be difficult to implement in practice, especially with many missing values. The Bayesian framework appears more promising, with the development of computational methods such as MCMC making it feasible to use the CDL in practice. The leading Bayesian computation program is WinBUGS (Spiegelhalter et al. 1999). A virtue of WinBUGS is that it was developed to combine conditional likelihood components as outline in the models above. Even though Schofield and Barker (2007a) showed how WinBUGS can be used to fit many capture–recapture models using the CDL, there are still problems:

1. WinBUGS is not as fast as user-written complied code, particularly when the data set is large or the model complex.
2. In the CMAS model, the size of the $\boldsymbol{b}$, $\boldsymbol{d}$ and $\boldsymbol{X}$ all depend on N, a parameter in the model. One possible approach is to specify the matrices $\boldsymbol{b}$, $\boldsymbol{d}$ and $\boldsymbol{X}$ as $M \times k$ matrices, where M is an upper size limit for the parameter N (Durban and Elston 2005; Royle et al. 2007). The problem with this approach is that in practice M may need to be very large, dramatically slowing the program.

One possible solution to the difficulties is to modify the source code for WinBUGS. Open source code is available and specialized toolboxes could be included to solve any problems in implementation. This would allow use of the good features of the WinBUGS software, such as smart algorithm choices and the ability to easily specify hierarchical relationships between parameters, while overcoming many of the deficiencies.

Acknowledgments The first author would like to thank the Tertiary Education Commission of New Zealand for the Ph.D. scholarship that funded this research. We also thank Dr. Len Thomas and an anonymous referee for the helpful comments on the paper.

References

Barker RJ (1997) Joint modeling of live-recapture, tag-resight, and tag-recovery data. Biometrics 53:666–677.

Barker RJ, Burnham KP, White GC (2004) Encounter history modeling of joint mark-recapture, tag-resighting and tag-recovery data under temporary emigration. Statistica Sinica 14: 1037–1055.

Barker RJ, White GC (2004) Towards the mother-of-all-models: customised construction of the mark-recapture likelihood function. Animal Biodiversity and Conservation 27:177–185.

Bonner SJ, Schwarz CJ (2006) An extension of the Cormack–Jolly–Seber model for continuous covariates with application to *Microtus pennsylvanicus*. Biometrics 62:142–149.

Brownie C, Hines JE, Nichols JD, Pollock KH, Hestbeck JB (1993) Capture–recapture studies for multiple strata including non-Markovian transition probabilities. Biometrics 49:1173–1187.

Burnham KP (1991) On a unified theory for release-resampling of animal populations. In: Chao MT, Cheng PE (eds.) Proceedings 1990 Taipei Simposium in Statistics. Institute of Statistical Science, Academia Sinica, Taipei, pp. 11–35.

Burnham KP (1993) A theory for combined analysis of ring recovery and recapture data. In: Lebreton J-D, North P (eds.) Marked Individuals in Bird Population Studies. Birkhauser Verlag, Basel, pp. 199–213.

Capp'e O, Moulines E, and Ryd'en T (2005), Inference in Hidden Markov Models, Springer, New York.

Choquet R, Reboulet AM, Pradel R, Gimenez O, Lebreton JD (2004) M-SURGE: new software specifically designed for multistate capture–recapture models. Animal Biodiversity and Conservation 27:1–9.

Cormack RM (1964) Estimates of survival from the sighting of marked animals. Biometrika 51:429–438.

Crosbie SF, Manly BFJ (1985) Parsimonious modelling of capture-mark-recapture studies. Biometrics 41:385–398.

Dempster AP, Laird NM, Rubin DB (1977) Maximum likelihood from incomplete data via the EM algorithm. Journal of the Royal Statistical Society B 39:1–38.

Dupuis JA (1995) Bayesian estimation of movement and survival probabilities from capture-recapture data. Biometrika 82:761–772.

Durban JW, Elston DA (2005) Mark-recapture with occasion and individual effects: abundance estimation through Bayesian model selection in a fixed dimensional parameter space. Journal of Agricultural, Biological, and Environmental Statistics 10:291–305.

Jolly GM (1965) Explicit estimates from capture-recapture data with both death and immigration-stochastic model. Biometrika 52:225–247.

Link WA, Barker RJ (2005) Modeling association among demographic parameters in analysis of open population capture–recapture data. Biometrics 61:46–54.

Otis DL, Burnham KP, White GC, Anderson DR (1978) Statistical inference from capture data on closed animal populations. Wildlife Monographs 62:1–135.

Pradel R (1996) Utilization of capture-mark-recapture for the study of recruitment and population growth rate. Biometrics 52:703–709.

Pradel R (2005) Multievent: an extension of multistate capture–recapture models to uncertain states. Biometrics 61:442–447.

Royle J, Dorazio R, Link W (2007) Analysis of multinomial models with unknown index using data augmentation. Journal of Computational and Graphical Statistics 16:67–85.

Schofield MR, Barker RJ (2007a) Flexible hierarchical mark-recapture modeling for open populations using WinBUGS. Submitted to Ecological and Environmental Statistics.

Schofield MR, Barker RJ (2007b) A unified capture–recapture model. Journal of Agricultural, Biological and Environmental Statistics, In Press.

Schwarz CJ, Arnason AN (1996) A general methodology for the analysis of capture–recapture experiments in open populations. Biometrics 52:860–873.

Schwarz CJ, Schweigert JF, Arnason AN (1993) Estimating migration rates using tag-recovery data. Biometrics 49:177–193.

Seber GAF, (1965) A note on the multiple-recapture census. Biometrika 52:249–259.

Spiegelhalter DJ, Thomas A., G., B. N., Lunn D (2003) WinBUGS Version 1.4 User Manual, MRC Biostatistics Unit.

Tanner MA, Wong WH (1987) The calculation of posterior distributions by data augmentation (with discussion). Journal of the American Statistical Society 82:529–550.

van Dyk DA, Meng XL (2001) The art of data augmentation. Journal of Computational and Graphical Statistics 10:1–50.

White GC, Burnham KP (1999) Program MARK: survival estimation from populations of marked animals. Bird Study 46S:120–139.

Section VIII
The Robust Design – Sampling, Applications and Advances

Bill Kendall and Jim Sedinger

Exploring Extensions to Multi-State Models with Multiple Unobservable States

Larissa L. Bailey, William L. Kendall, and Don R. Church

Abstract Many biological systems include a portion of the target population that is unobservable during certain life history stages. Transition to and from an unobservable state may be of primary interest in many ecological studies and such movements are easily incorporated into multi-state models. Several authors have investigated properties of open-population multi-state mark-recapture models with unobservable states, and determined the scope and constraints under which parameters are identifiable (or, conversely, are redundant), but only in the context of a single observable and a single unobservable state (Schmidt et al. 2002; Kendall and Nichols 2002; Schaub et al. 2004; Kendall 2004). Some of these constraints can be relaxed if data are collected under a version of the robust design (Kendall and Bjorkland 2001; Kendall and Nichols 2002; Kendall 2004; Bailey et al. 2004), which entails >1 capture period per primary period of interest (e.g., 2 sampling periods within a breeding season). The critical assumption shared by all versions of the robust design is that the state of the individual (e.g. observable or unobservable) remains static for the duration of the primary period (Kendall 2004). In this paper, we extend previous work by relaxing this assumption to allow movement among observable states within primary periods while maintaining static observable or unobservable states. Stated otherwise, both demographic and geographic closure assumptions are relaxed, but all individuals are either observable or unobservable within primary periods. Within these primary periods transitions are possible among multiple observable states, but transitions are not allowed among the corresponding unobservable states.

Our motivation for this work is exploring potential differences in population parameters for pond-breeding amphibians, where the quality of habitat surrounding the pond is not spatially uniform. The scenario is an example of a more general case where individuals move between habitats both during the breeding season (within primary periods; transitions among observable states only) and during the non-breeding season (between primary periods; transitions between observable and unobservable states). Presumably, habitat quality affects demographic parameters

L.L. Bailey (✉)
U.S. Geological Survey, Patuxent Wildlife Research Center, 12100 Beech Forest Road, Laurel, MD, 20708-4017, USA

D.L. Thomson et al. (eds.), *Modeling Demographic Processes in Marked Populations*, Environmental and Ecological Statistics 3, DOI 10.1007/978-0-387-78151-8_31,

(e.g. survival and breeding probabilities). Using this model we are able to test this prediction for amphibians and determine if individuals move to more favorable habitats to increase survival and breeding probabilities.

Keywords Capture-mark-recapture · Clearcut · Identifiability · Multistate mark-recapture models · Parameter redundancy · Pond-breeding amphibians · Robust design · *Ambystoma* · salamanders · Unobservable states

1 Introduction

In many biological systems a portion of the target population may be unobservable because not all geographic areas are sampled or because some individuals in the population may not make themselves available for detection. Examples involving marine species include seabirds (Hunter and Caswell this issue; Converse et al. this issue; Lebreton et al. 2003), marine mammals (Fujiwara and Caswell 2002), sea turtles (Kendall and Bjorkland 2001; Dutton et al. 2005; Rivalan et al. 2005), and anadromous or spawning fish (Sulak and Clugston 1999; Fox et al. 2000). Terrestrial systems also have species that have been modeled with multi-state models with unobservable states, including waterfowl (Barker et al. 2005), dormant plants (Shefferson et al. 2001; Kéry and Gregg 2004) and amphibians (Schmidt et al. 2002; Bailey et al. 2004; Church et al. 2007).

Failing to account for these unobservable states may cause severe bias in demographic parameters obtained from mark-recapture models. The severity and direction of the bias depends on the proportion of the population that is unobservable and whether movement to and from observable states is completely random or Markovian (Burnham 1993; Kendall et al. 1997; Kendall 1999). Utilizing multi-state mark-recapture (MSMR) models with unobservable state(s) (Arnason 1972, 1973; Lebreton and Pradel 2002), with or without the robust design (Pollock 1982; Kendall et al. 1997; Kendall 2004), is a way of reducing or eliminating the bias caused by unobservable states. Furthermore, these models allow for estimation of transitions between unobservable and observable states, which may, in some instances, be of primary biological interest (e.g. estimation of breeding probabilities).

One consideration before applying any mark-recapture model is to determine whether the model is 'identifiable' (i.e. that it does not contain redundant parameters). A model is defined to be identifiable if no two values of the parameters yield the same maximized likelihood (Gimenez et al. 2004). A classic example of a model that does not meet this criterion is the fully time-specific Cormack–Jolly–Seber (CJS) model. It is well known that this model is not 'full-rank' (i.e. contains redundant parameters) because the last survival and recapture probabilities cannot be estimated separately, only their product is estimable (Lebreton et al. 1992; Gimenez et al. 2004). Determining the number and identity of uniquely identifiable parameters is crucial for valid biological inference and proper interpretation of parameters and model selection (Burnham and Anderson 2002; Gimenez et al. 2003, 2004). To date, parameter redundancy in MSMR models has been thoroughly explored for models lacking unobservable states (Gimenez et al. 2003), and for scenarios with a single observable and single unobservable state (Kendall and Nichols 2002; Fujiwara and Caswell 2002; Schaub et al. 2004) and in situations where extra information is available (e.g. robust design (Kendall and Nichols 2002; Schaub et al. 2004; Bailey et al. 2004) or recoveries (Barker et al. 2005)). In cases without recovery information investigators have explored both open population and robust design MSMR models, and determined the scope and constraints under which models are identifiable. In general, investigators found that robust designs yield more estimable parameters with better precision and less restrictive assumptions (Kendall and Nichols 2002; Schaub et al. 2004). Among robust design models a common, critical assumption is that the state of the individual remains static for all sampling occasions within the 'primary' period (Kendall 2004). The sampling takes place at 2 temporal scales: multiple samples are collected over a relatively short time period where the state of the individual is maintained (over secondary occasions within primary periods), then transitions between states are possible between these primary periods. Kendall (2004) outlined three existing versions of the robust design: (1) the classic Pollock's robust design (Pollock 1982; Kendall et al. 1997) where within primary periods both geographic and demographic closure is maintained; (2) the open-population robust design (Schwarz and Stobo 1997; Kendall and Bjorkland 2001) where geographic closure is relaxed, allowing individuals to enter and exit the sampling area once between secondary surveys, but demographic closure is maintained over the primary period; and (3) the gateway robust design (Bailey et al. 2004) where geographic closure is maintained but demographic closure is relaxed to allow mortality within the primary period. In each case the state of the individual (observable or unobservable) is maintained within the primary period. Here, we extend this previous work by further relaxing the gateway robust design by allowing movement among observable states within primary periods while maintaining a static observable or unobservable state. Stated otherwise, both demographic and geographic closure assumptions are relaxed, but the observable state is maintained within primary periods (Fig. 1).

Our motivation for this work was to explore potential differences in population parameters for pond-breeding amphibians, where the quality of habitat surrounding the breeding pond was not uniform. Specifically, a portion of the forest surrounding the pond(s) had been clearcut. Several studies have documented reduced counts of pond-breeding and terrestrial salamanders in clearcut areas (reviewed in deMaynaider and Hunter 1995, 1998), and this was also true for the collection of ponds studied by Church (2004) in Virginia, USA. This observation naturally leads one to pose the question: 'why are the counts reduced in clearcut areas?' To address this question, we offer 3 possible explanations: (1) apparent survival is higher in the forested area, (2) breeding probability is higher in the forested area, or (3) individuals are selecting or moving to the forested habitat, perhaps but not necessarily, because the first 2 explanations hold true. In other words, individuals may be selecting the forested areas because it represents 'better' habitat, leading to higher survival and breeding probabilities, or they may

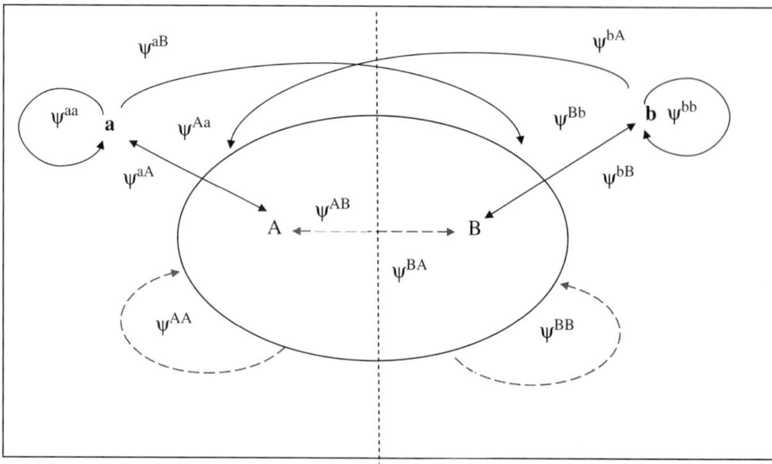

Fig. 1 Diagram of 4 states with 2 observable (A, B), and 2 unobservable (a, b). Transitions denoted with dotted lines may happen between any sampling occasions, while the remaining transitions only occur between primary periods. In our amphibian example, A and a represent breeders and non-breeders, respectively, in the forested habitat; B and b represent the same breeding states in the clearcut habitat. Transition probabilities, denoted with ψ, include both movement and breeding probabilities. For example, ψ^{AB} represents the probability a breeder in the forested habitat at time t, remains a breeder and moves to the clearcut habitat before the next sampling occasion $t+1$. ψ^{Aa} represents the probability a breeder in the forested habitat at time t, remains in same habitat but skips a breeding opportunity (i.e. transitions into an unobservable, nonbreeding state). Notice that $(1-\psi^{Aa})$ represents the probability a breeder in the forested habitat at time t, remains a breeder between t and $t+1$. Individuals are assigns to different habitat states based on which habitat they are captured in entering and exiting the pond

be selecting forested habitat in the absence of any apparent demographic advantage (Schlaepfer et al. 2002).

In this paper, we explore identifiably for the case of a single pond with 2 observable and 2 unobservable states, representing the corresponding breeding and non-breeding state in both clearcut and forested habitats (Fig. 1). We apply our findings to data from a single pond and species to test our 3 a priori hypotheses about the species response to a 20-year-old clearcut.

2 Investigating Intrinsic Parameter Redundancy

2.1 Methods

Gimenez et al. (2004) provides an outstanding review of tools used to investigate parameter redundancy in mark-recapture models. We applied three of the methods they detailed (The Hessian, simulation or analytic-numeric, and the formal derivative matrix) to our MSMR models to investigate intrinsic redundancy of model parameters. By exploring intrinsic redundancy, we sought to detect redundancy problems in the structure of the model, independent of any specific data set. We utilized programs MARK and M-SURGE and an analytic-numeric method (Burnham et al. 1987, pp. 214–217, 292–295, referred to as the simulation method in Gimenez et al. 2004) with several sets of parameter values to explore approximate bias and precision of parameter estimates and determine the rank of the resulting Hessian matrix (a matrix of second partial derivative of the multivariate likelihood function). We also tried to utilize the formal derivative matrix (Catchpole and Morgan 1997; Catchpole et al. 2002; Gimenez et al. 2003) and a numerical version of this method, termed the numerical Catchpole Morgan Freeman (numerical CMF), included in program M-SURGE (Choquet et al. 2005). The fully general formal method is based on the analytic computation of a symbolic matrix of derivatives of the multinomial cell probabilities with respect to the model parameters. It has been recommended because it is more reliable in determining if a model is fully identifiable, yielding less ambiguous results compared to the rank of the Hessian matrix or analytic-numeric methods. In addition, the formal method can determine the estimable functions of redundant parameters in cases where the models are not fully identifiable (Catchpole et. al. 2002; Gimenez et al. 2003, 2004). While the formal method, using the symbolic math features of software Maple, has been successfully used to examine parameter redundancy with a single unobservable state (Schaub et al. 2004), we found that the problem was too large for the software with more complex MSMR models involving multiple unobservable states (also see Hunter and Caswell this issue).

The analytic-numeric method involves using a known, 'true' model with realistic parameter values, θ_{true}, to generate an artificial data set consisting of either the expected value of each detection history or expected sufficient statistics (Nichols et al. 1981; Burnham et al. 1987; Gimenez et al. 2004). The generated data are analyzed as if they were real data, under any model of interest. The method is strictly numerical, but 'analytical', not Monte Carlo-based (Burnham et al. 1987, p. 215). Using large numbers of released individuals, this method can be used to approximate estimator bias and precision, and using criteria determined a priori, a model can be deemed 'identifiable'. For example, Kendall and Nichols declared a model identifiable if all parameter estimates were unbiased to the 5th decimal place and had coefficients of variation $<100\%$ (Kendall and Nichols 2002). The analytic-numeric method is only valid for the fixed parameter values chosen to generate the expected values, thus several sets of different, realistic values should be explored to insure that models are not conditionally full rank (Gimenez et al. 2004).

We employed the analytic-numeric method and determined the rank of the Hessian matrix using both programs MARK (White and Burnham 1999) and M-SURGE (Choquet et al. 2005). For each scenario, we considered 4 primary periods, each with 2 secondary surveys (8 total survey occasions). The number of newly released individuals, R_{ij} was set at 1000 for the first time period ($R_{11} = 1000$), with 100 newly marked individuals released for every occasion thereafter. For the exploration in program MARK, we generated expected capture histories for 2 scenarios (Tables 1 and 2) and analyzed these data using the default selection (2nd Part) to invert the Hessian matrix and calculate variances (source code for generating expected value data is available from J. Hines, jhines@usgs.gov). Often initial values had to be provided to achieve convergence: we explored at least 2 sets of

Table 1 Parameter values used to generate expected value datasets for one scenario used to evaluate intrinsic redundancy of various model. For transition probabilities, the pair of values corresponds to within primary periods (within the breeding season) and between primary periods (between breeding seasons) respectively. Letters represent different states (e.g. habitats): capital letters represent observable states and corresponding unobservable states are denoted by lower lowercase letters

$S^A = 0.65$	$S^a = 0.65$	$S^B = 0.90$	$S^b = 0.90$
$p^A = 0.60$	$p^a = 0$	$p^B = 0.50$	$p^b = 0$
$\psi^{AA} = 0.90,\ 0.50^\dagger$	$\psi^{aA} = 0.00, 0.30$	$\psi^{BA} = 0.20, 0.20$	$\psi^{bA} = 0.00, 0.40$
$\psi^{Aa} = 0.00, 0.40$	$\psi^{aa} = 1.00,\ 0.20^\dagger$	$\psi^{Ba} = 0, 0$	$\psi^{ba} = 0, 0$
$\psi^{AB} = 0.10, 0.10$	$\psi^{aB} = 0.00, 0.50$	$\psi^{BB} = 0.80,\ 0.40^\dagger$	$\psi^{bB} = 0.00, 0.40$
$\psi^{Ab} = 0, 0$	$\psi^{ab} = 0, 0$	$\psi^{Bb} = 0.00, 0.40$	$\psi^{bb} = 1.00,\ 0.20^\dagger$

†Transitions calculated by subtraction.

Table 2 Parameter values used to generate expected value datasets for a second scenario. For transition probabilities, the pair of values corresponds to within primary periods (within the breeding season) and between primary periods (between breeding seasons) respectively. Letters represent different states (e.g. habitats): capital letters represent observable states and corresponding unobservable states are denoted by lower lowercase letters

$S^A = 0.80$	$S^a = 0.80$	$S^B = 0.70$	$S^b = 0.70$
$p^A = 0.60$	$p^a = 0$	$p^B = 0.50$	$p^b = 0$
$\psi^{AA} = 0.70,\ 0.50^\dagger$	$\psi^{aA} = 0.00, 0.30$	$\psi^{BA} = 0.40, 0.40$	$\psi^{bA} = 0.00, 0.40$
$\psi^{Aa} = 0.00, 0.20$	$\psi^{aa} = 1.00,\ 0.50^\dagger$	$\psi^{Ba} = 0, 0$	$\psi^{ba} = 0, 0$
$\psi^{AB} = 0.30, 0.30$	$\psi^{aB} = 0.00, 0.20$	$\psi^{BB} = 0.60,\ 0.30^\dagger$	$\psi^{bB} = 0.00, 0.30$
$\psi^{Ab} = 0, 0$	$\psi^{ab} = 0, 0$	$\psi^{Bb} = 0.00, 0.30$	$\psi^{bb} = 1.00,\ 0.30^\dagger$

†Transitions calculated by subtraction.

initial values for each scenario. Initial values were within $\pm$ 0.20 (θ_{true}) but were never set equal to the true values. For the exploration in program M-SURGE, we used these same sets of parameters values and one other (Tables 1–3) to generate expected sufficient statistics (i.e. m-array, Choquet et al. 2005). To reduce the possibility of the likelihood search settling at a local minima, we used multiple (10) random initial values to initialize each model and parameter set combination.

Table 3 Set of parameter values used to generate the third set of expected value data. This analysis was performed in M-SURGE only. Parameter values were chosen to mimic estimates obtained from an analysis of female salamander data from a single breeding pond. In this scenario survival probability varies between primary periods and secondary surveys, corresponding to the non-breeding and breeding seasons, respectively. For survival and transition probabilities, the pair of values corresponds to within primary periods (within the breeding season) and between primary periods (between breeding seasons) respectively. States A and a correspond to breeders and nonbreeders in the forest habitat, respectively. States B and b correspond to those same 2 breeding states in the clearcut habitat

$S^A = 0.97, 0.63$	$S^a = 0.97, 0.63$	$S^B = 0.97, 0.63$	$S^b = 0.97, 0.63$
$p^A = 0.82$	$p^a = 0$	$p^B = 0.82$	$p^b = 0$
$\psi^{AA} = 0.98,\ 0.45^\dagger$	$\psi^{aA} = 0.00, 0.75$	$\psi^{BA} = 0.50, 0.10$	$\psi^{bA} = 0.00, 0.15$
$\psi^{Aa} = 0.00, 0.53$	$\psi^{aa} = 1.00,\ 0.20^\dagger$	$\psi^{Ba} = 0, 0$	$\psi^{ba} = 0, 0$
$\psi^{AB} = 0.02, 0.02$	$\psi^{aB}\psi^{aB} = 0.00, 0.05$	$\psi^{BB} = 0.50,\ 0.20^\dagger$	$\psi^{bb} = 0.00, 0.40$
$\psi^{Ab} = 0, 0$	$\psi^{ab} = 0, 0$	$\psi^{Bb} = 0.00, 0.70$	$\psi^{bb} = 1.00,\ 0.70^\dagger$

†Transitions calculated by subtraction.

M-SURGE computes the rank of the Hessian matrix by a finite difference scheme using an analytic gradient method which improves precision in inverting the Hessian (Gimenez et al. 2004) over methods which determine the first derivatives numerically. Furthermore, M-SURGE contains a numerical version of the formal method (the numerical CMF) that has proven quite reliable (Choquet et al. 2005). The rank is estimated as the number of non-zero singular values and the redundant parameters are given (Choquet et al. 2005).

We evaluated the performance of each model by examining the resulting rank of the Hessian matrix, parameter estimates and their associated standard errors, and the maximum rank of the derivatives matrix calculated at the MLE and several neighbors (i.e. rank determined by numerical CMF method in M-SURGE). Kendall and Nichols (2002) conservatively identified models without redundancy problems as those where 'all estimators were unbiased to the 5th decimal place and had coefficients of variation $<100\%$'. As this is an admittedly arbitrary criterion, we report results for this as well as a more liberal criterion that estimates be unbiased to 3 decimals with CV $<100\%$. Although we admit that requiring CV$<100\%$ might exclude parameters that are structurally estimable, an estimator with a larger CV would provide little biological information. We did not consider fully parameterized models, rather we maintained constraints that were determined necessary in previous explorations of MSMR models with a single unobservable state (Kendall and Nichols 2002; Schaub et al. 2004; Bailey et al. 2004). These constraints include: survival probabilities for unobservable and observable animals are set equal ($S_{ij}^{U} = S_{ij}^{O}$), no transitions are allowed between unobservable states ($\psi^{u_1u_2} = \psi^{u_2u_1} = 0$), and for models with time-dependent transition probabilities, the last nonzero transition probability was set equal to the previous nonzero transition probability ($\psi_t = \psi_{t-1}$) and the first nonzero transition probability involving movement from an unobservable state was similarly constrained to the next like parameter (e.g., $\psi_1^{uO} = \psi_2^{uO}$ in scenarios described next).

In our biological setting, individuals are captured entering and leaving a common location (e.g. breeding area) each season, yielding $j = 2$ secondary samples within each of 4 primary periods, $i = 1,2,3,4$. It is possible for individuals to transition between breeding areas during the primary period, i.e., between the 2 observable states, but they maintain their breeding/nonbreeding status: only movement between observable states is allowed (i.e., no movement is allowed between the unobservable states; Fig. 1). We use capital letters to denote observable states and lower case letters to denote the corresponding unobservable states. Under this setting we explored parameter redundancy using the previously described analytic-numeric and numerical CMF method with 3 sets of parameters (Tables 1–3). In addition, we explore models where capture probability is known from another source of data or by employing the gateway robust design.

2.2 Results

All examined models were found to be non-redundant even without extra information on capture probability, using the numerical CMF method in

Table 4 Results of analytic-numeric analysis with expected values for parameters defined in Tables 1 and 2. K is the number of parameters in each model and the rank of the Hessian was determined by both programs MARK and M-SURGE. Additionally, we report whether the model would be defined as 'estimable' under the conservative criteria used by Kendall and Nichols (2002): unbiased to 5 decimal places and coefficient of variation <100% for all parameters (Level 1) or whether parameter estimates are unbiased to only 3 decimals and CV<100% (Level 2). Occasionally models did not converge utilizing the analytic-numeric method in program MARK, with any set of initial values (denoted DNC). The model is deemed full rank if the numerical approximation of the formal model, implemented in M-SURGE, showed no parameter redundancy

Survival	Transition[a]	Detection	K	Analytic-numeric method Level 1	Level 2	Rank of Hessian	Numeric CMF: full rank
Parameter values: $S^{A=a} = 0.65$, $S^{B=b} = 0.90$, $p^A = 0.60$, $p^B = 0.50$, ψ given in Table 1							
$S_{ij}^{A=a}, S_{ij}^{B=b}$	$\psi.^{All}$	Known	22	Y	Y	Full	Y
		$p.^A, p.^B$	24	Y	Y	Full	Y
		p_{ij}^A, p_{ij}^B	34¶	N	Y	Full	Y
	ψ_i^{All}	Known	34	Y	Y	Full	Y
		$p.^A, p.^B$	36	N	Y	Full	Y
		p_{ij}^A, p_{ij}^B	48	DNC	DNC	Not Full†	N
$S.^{A=a}, S.^{B=b}$	$\psi.^{All}$	Known	10	Y	Y	Full	Y'
		$p.^A, p.^B$	12	Y	Y	Full	Y
		p_{ij}^A, p_{ij}^B	24	DNC	DNC	Full†	Y
	ψ_i^{All}	Known	22	DNC	DNC	Full†	Y
		$p.^A, p.^B$	24	Y	Y	Full	Y
		p_{ij}^A, p_{ij}^B	36	DNC	DNC	Full†	Y
Parameter values: $S.^{A=a} = 0.80$, $S.^{B=b} = 0.70$, $p.^A = 0.60$, $p.^B = 0.50$, ψ given in Table 2							
$S_{ij}^{A=a}, S_{ij}^{B=b}$	$\psi.^{All}$	Known	22	Y	Y	Full	Y
		$p.^A, p.^B$	24	Y	Y	Full	Y
		p_{ij}^A, p_{ij}^B	34¶	N	Y	Full	Y'
	ψ_i^{All}	Known	34	N	Y	Full	Y
		$p.^A, p.^B$	36	N	Y	Full	Y
		p_{ij}^A, p_{ij}^B	48	DNC	DNC	Not Full†	N
$S.^{A=a}, S.^{B=b}$	$\psi.^{All}$	Known	10	Y	Y	Full	Y
		$p.^A, p.^B$	12	DNC	DNC	Full†	Y
		p_{ij}^A, p_{ij}^B	24	N	Y	Full	Y
	ψ_i^{All}	Known	22	Y	Y	Full	Y
		$p.^A, p.^B$	24	N	Y	Full	Y
		p_{ij}^A, p_{ij}^B	36	Y	Y	Full	Y

[a]Transitions include $\psi^{AB}, \psi^{Aa}, \psi^{BA}, \psi^{Bb}, \psi^{aA}, \psi^{aB}, \psi^{bA}$, *and* ψ^{bB}. Four additional transitions are calculated by subtraction, $\psi^{AA}, \psi^{BB}, \psi^{aa}$, *and* ψ^{bb}. For models with time dependence in transitions, the transitions in the last time period were set equal to the previous time step, $\psi_i^{All} = \psi_{i-1}^{All}$ and transitions out of the unobservable state in the first time period were set equal to the second $\psi_1^{uO} = \psi_2^{uO}$.

¶ Last S_i and p_{ij} are cleanly confounded.

†Rank of Hessian obtained from M-SURGE only.

M-SURGE, except the most general model with time-specificity in all parameters, $S_{ij}^{A=a}$, $S_{ij}^{B=b}$, ψ_i^{All}, p_{ij}^A, p_{ij}^B (Table 4). For the first 2 parameter sets, some models would not have met the conservative criteria employed by Kendall and Nichols (2002) as they contained parameter estimates that were not unbiased to 5 decimals, but all these estimators were unbiased to at least 3 decimals. Results from parameter set 3 are identical to the findings in Table 4 except all estimates in non-redundant models were unbiased to 5 decimal places. All combinations of models and parameter values produced estimators with CV <100%. In cases where identifiably would be questionable under the analytic-numeric method only (i.e. some parameters were not unbiased to 5+ decimal places), we found that survival probabilities were generally the least biased estimators and generally met the criteria, but there was no consistency in which of the remaining parameters were most biased, failing to meet the a priori criteria. Furthermore, the parameters that failed to meet the criteria often changed when different initial values were used with the same model and generating scenario. Movements from unobservable to observable states were always the least precise estimators, with CV often >50%, which is consistent with previous studies and was expected given that relatively little information is available to estimate these parameters (Kendall and Nichols 2002; Schaub et al. 2004; Bailey et al. 2004).

3 Example

3.1 Amphibian Study System

The scenario described above represents a general case where only breeding individuals are observable, but breeders may move between habitats of variable quality within the breeding season (within primary periods). Both geographic movement and breeding state transitions are possible between sampling seasons (i.e. between primary periods). We apply our findings above to data from a single sinkhole pond within the George Washington National Forest in Virginia, USA. The forest surrounding 1/4 of the pond had been clearcut in 1980 and partially replanted with pine species. The remaining 3/4 of the pond is embedded in 100+ year-old, primarily deciduous forest. A drift fence, approximately 290 m in circumference, was monitored between September 1999 to August 2004, with pitfall traps spaced on each side of the fence at 10 m intervals. Pitfall traps were opened and checked daily during the active migratory and breeding seasons for a suite of salamander species (mostly ambystomatids). In this analysis, we focus on a single species, the marbled salamander (*Ambystoma opacum*). Adult *A. opacum* migrate to breeding ponds beginning in September to court and lay eggs in the dry pond basin or along the dry margins of the pond. Males generally leave the pond basin after breeding, but females will remain at the nest until late fall or winter rains fill the pond and flood the nest.

During migration to and from the pond, individuals are captured by the drift fence and collected in pitfall traps. Each captured individual was digitally photographed on the dorsal side and a computer assisted pattern matching software developed specifically for this species was used to match individuals and construct capture histories (Lex Hiby, Conservation Research Ltd, Cambridge, UK). Males are easily distinguished by swollen cloacae, and females entering the pond have enlarged abdomens. During the 4 breeding seasons, 4742 females and 6476 male captures were recorded on the forested side of the pond, compared to only 630 female and 1229 male captures on the clearcut side. We formulated 3 hypotheses to explain why captures are six times higher in forested compared to clearcut habitat, despite only a 3:1 ratio of forest:clearcut pitfall sampling effort. First, demographic rates (apparent survival and breeding probabilities) may be lower in the clearcut habitat, perhaps due to limited shelter or food resources. Alternatively, individuals may simply be moving or selecting forested habitat, perhaps (but not necessarily) because it represents superior habitat leading to higher survival or breeding probabilities. Finally, if detection probabilities are approximately equal among the 2 habitats, low counts may simply be a consequence of historic effects of the clearcut, and populations of salamanders have not yet rebounded from acute mortality, reproductive failure, or mass migration caused by clearcutting activities.

3.2 Methods and Models

To test these hypotheses we used data from a single pond consisting of 4 states (Fig. 1) with 2 groups (males and females), during 4 breeding seasons (4 primary periods each with 2 secondary surveys, 8 total capture opportunities). We formulated a candidate model set consisting of 72 models, representing a combination of 2 capture probability structures, 9 survival probability structures and 4 transition probability structures determined a priori. We modeled capture probability as either completely time, sex and habitat specific, or a reduced model that is sex and habitat specific, but with only 3 different time periods. During one occasion in the first and third years, conditions were such that it was impossible or potentially dangerous to the animals to process all individuals in the pitfall traps, thus a subset of individuals were photographed and the rest escaped or were released without photographing. Finally, to avoid any confounding in models with time-specific survival probabilities, we set capture probabilities in the last primary period equal, $p_{4,1} = p_{4,2}$.

Survival probabilities were always assumed to be equal between observable and unobservable individuals (e.g. $S_{ij}^{A} = S_{ij}^{a}$). In our most time-constrained survival model, we assumed survival was different during the breeding and non-breeding season (i.e., survival probabilities within the pond were always modeled differently than survival probabilities outside the pond). Generalizing from this time-constrained survival structure, we considered models with: no year effects, each year different, or years grouped into wet (i = 1,2) and dry (i = 3,4) years determined by the pond conditions during the fall breeding season. In addition to these time-specific structures, we modeled survival probability as sex-specific, habitat-specific, or sex and habitat-specific, yielding 9 total combinations of survival structures. Finally, we modeled transition probabilities (movement and breeding probabilities) as either sex and habitat-specific or just sex-specific, with 2 time structures: constant among years or with years grouped into wet and dry seasons.

The major emphasis of the modeling was to account for known or suspected time and sex differences in capture, survival and transition probabilities, in order to focus on the main objective of exploring whether habitat state (forest, clearcut: denoted with different letters in Fig. 1) was an important factor in estimating survival, breeding, and movement parameters. Because several of the biological models in the candidate set were not clearly constrained versions of more general, non-redundant models tested above, we verified that all biological models were identifiable using the 3 parameter sets and methods described above. We used M-SURGE to analyze the real data, but estimated overdispersal using the median c-hat method with the global model in program MARK (Cooch and White 2006).

3.3 Amphibian Results

All biological models in the candidate set were found to be non redundant. A single model with sex and time-specific survival, habitat- and sex-specific transition probabilities and habitat, sex and time-specific capture probabilities fit the data better (more parsimoniously) than any other model in the candidate set (Table 5). Consistent with our a priori expectations, capture probabilities clearly dropped during the occasions where not all individuals were photographed, but there was time variation even among occasions where we didn't expect it based on field observations (Fig. 2, Table 5). Apparent survival probabilities were slightly lower for females, especially during the non-breeding season, but there was no indication of differences in survival probabilities among individuals inhabiting forest or clearcut habitats (Fig. 3, Table 5). Transition probabilities were constant among wet and dry years, but varied among sex and forest and clearcut habitats. Breeders moved to the forested side of the pond at a higher rate than would be expected

Table 5 Summary of the fit and selection statistics for the top 5 models for marbled salamander (*Ambystoma opacum*) data. Model selection was based on Akaike Information Criteria adjusted for overdispersal (QAIC); the model with the lowest QAIC value is considered 'best'. ΔQAIC is the difference in QAIC values between each model and the low-QAIC model; *K* is the number of estimated parameters; *w* is the Akaike model weight. Parameters include survival (S), transition (ψ), and capture probability (p); these parameters were permitted to vary according to time (t), habitat (hab), and sex (sx). The model in bold is the most parameterized (global) model in the parameter set. The median c-hat estimate based on the global model was $\hat{c} = 1.74$ (SE = 0.03)

Model	Deviance	*K*	QAIC	ΔQAIC	*w*
S(sx,t) ψ(hab,sx) p (hab,sx,t)	24089.26	58	13960.40	0.00	1.0
S(sx,t) ψ(hab,sx) p (hab,sx,t*)	24169.84	43	13976.71	16.31	0.0
S(hab,sx,t) ψ(hab,sx) p (hab,sx,t*)	24136.21	57	13985.39	24.98	0.0
S(hab,t†) ψ(hab,sx,t†) p (hab t,sx,t)	24179.53	52	14000.28	39.88	0.0
S(hab,sx)ψ(hab,sx,t†)p (hab,sx,t)	**24114.85**	**72**	**14003.11**	**42.71**	**0.0**

* Denotes capture probability with 3 time parameters: p_{12}, p_{31}, p_{other}.

† Denotes probabilities that differ for wet (i=1,2) and dry years (i=3,4).

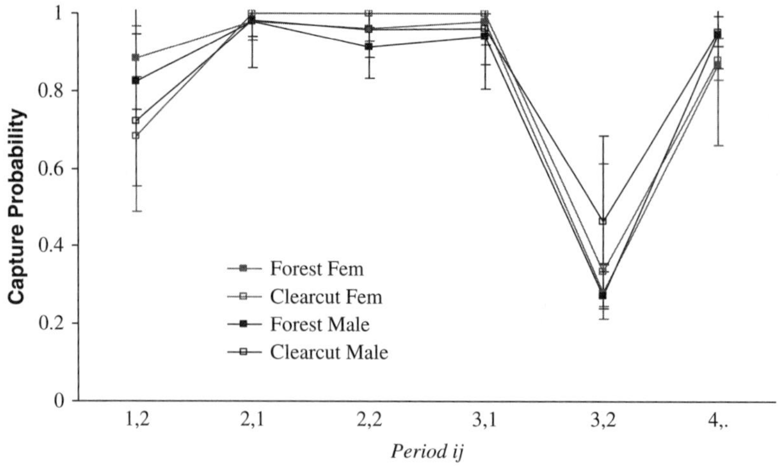

Fig. 2 Estimates and 95% confidence intervals for time, habitat and sex-specific capture probabilities using the top model in the candidate model set (*S*(sx,t) ψ(hab,sx) *p* (hab,sx,t))

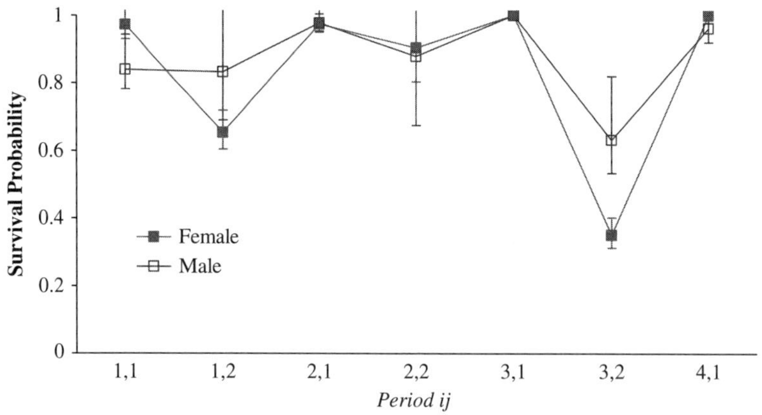

Fig. 3 Estimates and 95% confidence intervals for time and sex-specific apparent survival probabilities using the top model in the candidate model set (*S*(sx,t) ψ(hab,sx) *p* (hab,sx,t))

based on the 3:1 ratio of forest:clearcut area. Within the pond, males were slightly more philopatric with movement $\psi^{BA} = 0.38$ (SE = 0.03, 95% CI = (0.33,0.43)) and $\psi^{AB} = 0.03$ (SE = 0.004, 95% CI = (0.02,0.04)) compared to females, $\psi^{BA} = 0.48$ (SE = 0.04, 95% CI = (0.41,0.55)) and $\psi^{AB} = 0.03$ (SE = 0.005, 95% CI = (0.02,0.04)). Females were more likely to skip breeding opportunities than males, but the probability was higher for individuals that exited the clearcut side of the pond: $\psi^{Bb} = 0.67$ (SE = 0.05, 95% CI = (0.55,0.77)) vs. $\psi^{Aa} = 0.53$ (SE = 0.02, 95% CI = (0.48,0.57)) for females and $\psi^{Bb} = 0.40$ (SE = 0.06, 95% CI = (0.29,0.51)) vs. $\psi^{Aa} = 0.36$ (SE = 0.05, 95% CI = (0.27,0.45)) for males. Notice that because we had no a priori reason for expecting habitat to differentially affect one sex, but not the other, we did not include such model structures in the candidate model set. Successive breeders were highly philopatric, during the non-breeding season (between primary periods), and their movement probabilities did not suggested preferential migration to the forest habitat given the 3:1 availability of forest:clearcut area: $\psi^{AB} = 0.05$ (SE = 0.007, 95% CI = (0.04,0.07)) vs. $\psi^{BA} =$ 0.10 (SE = 0.02, 95% CI = (0.06,0.15)) for males, and $\psi^{AB} = 0.03$ (SE $\psi^{BA} =$ 0.005, 95% CI = (0.02,0.04)) vs. $\psi^{BA} = 0.10$ (SE = 0.03, 95% CI = (0.05,0.18)) for females. The most striking differences among habitats were seen in those individuals that skipped breeding opportunities. Those individuals were more likely to return to breed the following year if they exited into the forested area. For females, these probabilities were high: $\psi^{aA} = 0.74$ (SE = 0.07, 95% CI = (0.59,0.85)) and $\psi^{aB} = 0.04$ (SE = 0.01, 95% CI = (0.02,0.08)) vs. $\psi^{bB} =$ 0.46 (SE = 0.13, 95% CI = (0.24,0.70)) and $\psi^{bA} = 0.13$ (SE = 0.07, 95% CI = (0.04,0.33)). Notice, $\psi^{aA} + \psi^{aB} = \psi^{aO} = 0.78$ compared to $\psi^{bO} = 0.59$, where O represents transitioning to either observable breeding state (A or B). Males that skipped breeding opportunities took longer to return to a breeding state, but the probability of returning appeared higher for those individuals that exited into the forested area: $\psi^{aA} = 0.41$ (SE = 0.12, 95% CI = (0.21,0.66)) and $\psi^{aB} = 0.02$ (SE = 0.011, 95% CI = (0.01,0.06)) (or $\psi^{aO} = 0.43$) vs. $\psi^{bB} = 0.22$ (SE = 0.08, 95% CI = (0.10,0.42)) and $\psi^{bA} = 0.07$ (SE = 0.04, 95% CI = (0.02,0.21)) (or $\psi^{bO} = 0.29$).

Overall, movement probabilities between habitats were highest for breeders within the pond basin and this movement preferentially favored forested habitat for both sexes. Probabilities that correspond to movements that occur outside the pond basin did not seem to differ from expectations based on ratios of available clearcut and forested habitats. While individuals in the forested habitat did not have substantially higher apparent survival probabilities, they did have higher estimates of breeding probabilities, which together with movement probabilities, help explain why salamander counts are higher in forested areas.

4 Discussion

Here we explored a common biological scenario where only breeding individuals are observable, but these breeders may move between habitats of variable quality within the breeding season (within primary periods). Both geographic and breeding state transitions are possible between sampling seasons (i.e. between primary periods). While previous work had explored identifiably issues for systems with a single observable and unobservable state, we needed to expand the state space to address our biological questions. We explored 3 different methods to investigate intrinsic identfiability (non redundancy) and found that it was possible to relax the geographic closure assumption of the previously described robust design models, to allow movement among observable states within primary periods. Both the analytic-numeric (also termed simulation method in Gimenez et al. 2004) and the numeric approximation of the formal method implemented in M-SURGE gave similar results, but the numerical CMF method helped verify model identfiability in situations where results from the analytic-numeric method were ambiguous. Criteria for declaring a model intrinsically identifiable under the analytic-numeric method are somewhat arbitrary; thus determining whether a model meets a priori defined criteria is not simply a function of the model structure, but also depends on the parameter values used to generate the expected data set(s) and the values used to initiate the optimization of the likelihood.

For models that converged, the rank of the Hessian obtained via programs MARK and M-SURGE were nearly identical, but we were unable to obtain convergence for some combination of models and generated data sets using program MARK (Tables 1 and 2). Our results indicated that the Hessian matrix was full rank for all but our most general model (i.e. no identified nonzero eigenvalues or all singular values were considered > 0). Still, we acknowledge that evaluating only the rank of the Hessian may be flawed by numerical issues, such as deciding which eigenvalues are truly non-zero. A perfect tuning of a zero threshold value is probably not possible (Viallefont et al. 1998). The formal derivative method and the numerical CMF method implemented in M-SURGE do not require numerical approximations of the Hessian and thus avoid this complication (Choquet et al. 2005). Also see Hunter and Caswell (this issue) for an alternative method involving automatic differentiation and exploration of a more general multi-state problem involving multiple unobservable states. We found that using the combination of analytic-numeric and the numerical CMF methods was useful for evaluating whether models were intrinsically identifiable as well as anticipating expected precision of various parameters.

Utilizing our results from the intrinsic identfiability investigation, we were able to address a series of biological hypotheses about the current effects of a 20-year-old clearcut on a single amphibian species. Our results suggested that forested habitats provide resources that allow individuals to breed more often compared to clearcut habitats, and perhaps as a result, salamanders tend to move into forested habitats despite apparently higher densities. Interestingly, most of this movement occurs within, or across, the pond basin with relatively little movement occurring between breeding seasons. One possible explanation for this finding is that salamanders may use different orientation cues when exiting the pond basin (favoring the forest habitat) compared to those they use to locate and migrate to breeding habitats.

Female *A. opacum* have lower apparent survival probabilities than males, especially during the non-breeding season, contrary to previous findings for two other ambystomid species that breed in this pond (*A. tigrinum,* Bailey et al. 2004; Church et al. 2007, and *A. maculatum* unpublished results). If females incur a cost of reproduction associated with brooding eggs, the cost seems to be incurred after females have exited the pond basin. The habitat that the females exit into does not seem to affect their apparent survival probability, although it does appear to influence their subsequent breeding probability.

The biological scenario presented here is common to most pond-breeding amphibian systems, but one could conceive of numerous biological systems involving unobservable states. We concur with other authors that both intrinsic and extrinsic parameter redundancy needs to be explored in complex MSMR scenarios (e.g. Lebreton and Pradel 2002; Gimenez et al. 2004), especially in cases with multiple unobservable states. Multiple tools are available within programs MARK and M-SURGE to allow investigators to tailor the investigation to their own biological systems. An added advantage of program M-SURGE is the numerical CMF method for non redundancy, and the ability to quickly run the optimization for a single model with multiple, randomly chosen, initial values to help insure a global minimum is found.

Acknowledgments We are indebted to R. Choquet, O. Gimenez, and J. D. Lebreton for their thorough explanation and teaching of the methods incorporated in program M-SURGE. This manuscript has greatly benefited from useful discussions with these researchers and S. J. Converse, C. M. Hunter, C. A. Langtimm, J. D. Nichols, H. Caswell and one anonymous reviewer. This work was funded in part by the National Science Foundation (DEB 0414118 H. M. Wilbur) and USGS Amphibian Research and Monitoring Initiative.

References

Arnason AN (1972) Parameter estimates from mark-recapture experiments on two populations subject to migration and death. Researches on Population Ecology 13:97–113

Arnason AN (1973) The estimation of population size, migration rates, and survival in a stratified population. Researches on Population Ecology 15:1–8

Bailey LL, Kendall WL, Church DR, Wilbur HM (2004) Estimating survival and breeding probabilities for pond-breeding amphibians: a modified robust design. Ecology 85:2456–2466

Barker RJ, White GC, McDougall, M (2005) Movement of Paradise Shelduck between molt sites: a joint multistate-dead recovery mark-recapture model. Journal of Wildlife Management 69:1194–1201

Burnham KP (1993) A theory for combined analysis of ring recovery and recapture data. In: Marked Individuals in the Study of Bird Populations, Lebreton JD, North PM (eds) Birkhauser-Verlag, Basel, pp. 199–214

Burnham KP, Anderson DR (2002) Model Selection and Multimodel Inference. Springer-Verlag, New York

Burnham KP, Anderson DR, White GC, Brownie C, Pollock KH (1987) Design and analysis of methods for fish survival experiments based on release-recapture. American Fisheries Society Monograph 5:1–437

Catchpole EA, Morgan BJT (1997) Detecting parameter-redundant models. Biometrika 88: 187–196

Catchpole EA, Morgan BJT, Viallefont A (2002) Solving problems in parameter redundancy using computer algebra. Journal of Applied Statistics 29:625–636

Choquet R, Reboulet M, Pradel R, Gimenez O, Lebreton JD (2005) M-SURGE 1-8 User's Manual. CEFE, Montpellier, France (http://ftp.cefe.cnrs.fr/biom/soft-cr/)

Church DR (2004) Population Ecology of *Ambystoma tigrinum* and Occupancy Dynamics in an Appalachian Pond-Breeding Amphibian Assemblage. Dissertation. University of Virginia, Charlottesville, VA, USA

Church DR, Bailey LL, Wilbur HM, Kendall WL, Hines JE (2007) Iteroparity in the variable environment of the salamander *Ambystoma tigrinum*. Ecology 88:891–903.

Converse SJ, Kendall WL, Doherty PF, Naughton, N (2008) A traditional and less-invasive robust design: choices in optimizing effort allocation for seabird studies. In: Thomson DL, Cooch EG, Conroy MJ (eds.) Modeling Demographic Processes in Marked Populations. Environmental and Ecological Statistics, Springer, New York.

Cooch, EG,, White GC (2006) Program MARK: A Gentle Introduction. http://www.phidot.org/software/mark/docs/book/

deMaynaider PG, Hunter ML (1995) The relationship between forest management and amphibian ecology: a review of the North American literature. Environmental Reviews 3:230–261

deMaynaider PG, Hunter ML (1998) Effects of silvercultural edges on the distribution and abundance of amphibians in Maine. Conservation Biology 12:340–352

Dutton DL, Dutton PH, Chaloupka M, Boulon RH (2005) Increase of a Caribbean leatherback turte *Dermochelys coriacea* nesting population linked to long-term nest protection. Biological Conservation 126:186–194

Fox DA, Hightower JE, Paruka FM (2000) Gulf sturgeon spawning migration and habitat in the Choctawhatchee River system, Alabama–Florida. Transactions of the American Fisheries Society 129:811–826

Fujiwara M, Caswell H (2002) A general approach to temporary emigration in mark-recapture analysis. Ecology 83:3266–3275

Gimenez O, Choquet R, Lebreton JD (2003) Parameter redundancy in multi-state capture–recapture models. Biometrical Journal 45:704–722

Gimenez O, Viallefont A, Catchpole EA, Choquet R, Morgan BJT (2004) Methods for investigating parameter redundancy. Animal Biodiversity and Conservation 27:561–572

Hunter CM, Caswell H (2008) Rank and redundancy of multistate mark-recapture models for seabird populations with unobservable states. In: Thomson DL, Cooch EG, Conroy MJ (eds.) Modeling Demographic Processes in Marked Populations. Environmental and Ecological Statistics, Springer, New York.

Kendall WL (1999) Robustness of closed capture–recapture methods to violations of the closure assumption. Ecology 80:2517–2525

Kendall WL (2004) Coping with unobservable and mis-classified states in capture–recapture studies. Animal Biodiversity and Conservation 27:97–107

Kendall WL, Bjorkland R (2001) Using open robust design models to estimate temporary emigration from capture–recapture data. Biometrics 57:1113–1122

Kendall WL, Nichols JD (2002) Estimating state-transition probabilities for unobservable states using capture–recapture/resighting data. Ecology 83:3276–3284

Kendall WL, Nichols JD, Sauer JS, Hines JE (1997) Estimating temporary emigration using capture–recapture data with Pollock's robust design. Ecology 78:563–578

Kéry M, Gregg KB (2004) Demographic estimation methods for plants with dormancy. Animal Biodiversity and Conservation 27.1: 129–131

Lebreton JD, Pradel R (2002) Multi-state recapture models: modeling incomplete individual histories. Journal of Applied Statistics 29:353–369

Lebreton JD, Burnham KP, Clobert J, Anderson DR (1992) Modeling survival and testing biological hypotheses using marked animals: a unified approach with case studies. Ecological Monographs 62:67–118

Lebreton JD, Hines JE, Pradel R, Nichols JD, Spendelow JA (2003) Estimation by capture–recapture of recruitment and dispersal over several sites. OIKOS 101:253–264

Nichols JD, Noon BR, Stokes SL, Hines JE (1981) Remarks on the use of mark-recapture methodology in estimating avian population size. In: Estimating the Numbers of Terrestrial Birds Ralph CJ, Scott MJ (eds), Study of Avian Biology, Cooper Ornithological Society, Allen Press, Inc. Lawrence, Kansas, USA, vol 6, 121–136

Pollock KH (1982) A capture–recapture design robust to unequal probability of capture. Journal of Wildlife Management 46:757–760

Rivalan P, Prevot-Julliard AC, Choquet R, Pradel R, Jacquemin B, Girondot M (2005) Trade-off between current reproductive effort and delay to next reproduction in the leatherback sea turtle. Oecologia 145:564–574

Schaub M, Gimenez O, Schmidt BR, Pradel R (2004) Estimating survival and temporary emigration in the multistate capture-recapture framework. Ecology 85: 2107–2113

Schlaepfer MA, Runge MC, Sherman, PW (2002) Ecological and evolutionary traps. Trends in Ecology and Evolution 17: 474–480

Schmidt BR, Schaub M, Anholt BR (2002) Why you should use capture–recapture methods when estimating survival and breeding probabilities: on bias, temporary emigration, overdispersion, and common toads. Amphibia-Reptilia 23:375–388

Schwarz CJ, Stobo WT (1997) Estimation temporary migration using the robust design. Biometrics 53:178–194

Shefferson RP, Sandercock BK, Proper J, Beissinger SR (2001) Estimating dormancy and survival of a rare herbaceous perennial using mark-recapture models. Ecology 82:145–156

Sulak KJ, Clugston JP (1999) Recent advances in life history of Gulf of Mexico sturgeon, *Acipenser oxyrinchus desotoi*, in the Suwannee River, Florida, USA: a synopsis. Journal of Applied Ichthyology-Zeitschrift für Angewandte Ichthyologie 15:116–128

Viallefont A, Lebreton JD, Reboulet AM, Gory G (1998) Parameter identifiability and model selection in capture–recapture models: a numerical approach. Biometrical Journal 40:1–13

White GC, Burnham KP (1999) Program MARK: survival estimation from populations of marked animals. Bird Study 46:120–139

Extending the Robust Design for DNA-Based Capture–Recapture Data Incorporating Genotyping Error and Laboratory Data

Paul M. Lukacs, Kenneth P. Burnham, Brian P. Dreher, Kim T. Scribner, and Scott R. Winterstein

Abstract For many species, non-invasive sampling of feathers, hair, feces or other tissue has the potential to be very useful and in some cases is already widely used to answer ecological questions. These samples are genotyped and the genotypes are used to identify individuals. There is some level of uncertainty when identifying individuals from genotyping results. We present an extension to the robust design capture–recapture model that allows for the estimation of genotyping error rate and properly estimates population size, survival, temporary emigration, and capture probability in the face of genotyping error. The model uses information contained in the secondary encounter occasions to estimate genotyping error which would otherwise be impossible for an open-population model with a robust design component. We further extend the model to allow estimation of the probability of correctly genotyping a sample from laboratory data. We demonstrate that with an additional data source for genotyping error, parameters are more precisely estimated by allowing some genotyping error and a larger sample size than by culling samples to eliminate the potential for errors in genotypes and reducing model complexity. We use noninvasive and hunter collected data from black bears in Michigan as an example.

Keywords Abundance · Capture–recapture · Microsatellites · Non-invasive sampling · Tag misread

1 Introduction

Non-invasively sampled feathers, hairs and scats are being widely collected to answer ecological questions. DNA can be extracted from cells found in these samples. One use for these data is to obtain microsatellite genotype of the animal that left the sample in order to identify individuals. This can be a very powerful

---

P.M. Lukacs (✉)
Colorado Division of Wildlife, 317 W. Prospect Rd., Fort Collins, CO 80526, USA
e-mail: Paul.Lukacs@state.co.us

D.L. Thomson et al. (eds.), *Modeling Demographic Processes in Marked Populations*, Environmental and Ecological Statistics 3, DOI 10.1007/978-0-387-78151-8_32,

method to study species which are expensive and dangerous to handle or are very difficult to capture. Thus far, DNA-based capture–recapture studies have been largely used for estimating abundance (Woods et al. 1999; Boulanger et al. 2003), but capture–recapture theory has been developed to estimate a wide range of ecologically interesting parameters such as survival, emigration rates and population growth rates. Therefore, it is important to broaden the scope at which these data are being applied.

Standard capture–recapture theory assumes that individual animals are correctly identified each time they are captured. For DNA-based studies, not all individuals are necessarily identified correctly. Errors may arise in genotyping which are caused by a number of sources including degradation of the sample, mutation within the PCR reaction and other sources. When errors are present in genotyping, it will appear as if there are more individuals in the population than actually exist. It will also appear as if survival rates are lower than they actually are because errant genotypes are not likely to be observed again. Some studies have been able to minimize error to a great degree (Paetkau 2003). Other studies showed error rates that are much larger (Creel et al. 2003). Given that error in genotyping is a distinct possibility and if it occurs it will bias parameter estimates, it is important to be able to explicitly model the error rate in a capture–recapture model.

An important issue facing the use of genotypes as marks is that the researcher never knows the list of marks in the population (Lukacs and Burnham 2005a). With conventional capture–recapture studies, no animals in the population are marked prior to the study. Then, each animal captured is given a unique mark which is subsequently recorded by the researcher. Therefore, the list of marks in the population is known. When genotypes are used as marks, every animal has a mark prior to the beginning of the study. Unfortunately, the researcher does not know what any of the marks are or even how many marks exist. This causes uncertainty in interpreting the mark of an individual because there is often no way to verify that the mark is read correctly. In a conventional capture–recapture study, if a researcher recorded a ring number that does not match a ring on the list of known marks, then the ring would either be read again or the capture would be ignored. Researchers involved in DNA-based capture–recapture studies do not have this luxury.

Lukacs and Burnham (2005b) developed a method to estimate genotyping error rate in demographically and geographically closed population capture-recapture models. The method uses three basic assumptions to develop a likelihood function to estimate genotyping error rate, population size and capture probability. First, the model assumes that a molecular marker set is used that has sufficient power to resolve individuals with a high degree of certainty. Typically, these marker sets are microsatellites, but they can be single nucleotide polymorphisms or any other molecular marker with sufficient power to uniquely resolve individuals. Such marker sets exist for a large number of species and more are being developed. Second, it is assumed that a genotyping error will not result in a genotype that exists in the population. Given the typically small population sizes relative to the huge number of possible genotypes, it is unlikely for an error to result in the genotype of an individual that already exists in the population. This assumption may not hold for populations with little genetic variation, but is useful in many situations. Moreover, if this assumption is violated, the result only changes the sufficient statistics slightly and in some cases does not change the sufficient statistics of a closed capture–recapture model and therefore will not change the estimates much. Third, it is assumed that the same error will not be generated twice. A large number of rare events would have to occur for two errors to result in identical incorrect genotypes. This assumption may be violated if, for example, one allele at a locus tends to have a much higher dropout rate than the other alleles. These assumptions are practical and can be realistically met in DNA-based capture–recapture studies. In addition to their use in capture–recapture, the assumptions have been asserted as useful in the examination of error in molecular techniques (Paetkau 2003).

The model in Lukacs and Burnham (2005b) uses some simplifying assumptions to make the problem more tractable. Notably, the model assumes that all encounter histories are independent, while in fact correct and incorrect observations of genotypes from the same individual are dependent. This assumption allows for a useful model and is often reasonably well approximated because animals frequently leave more than one sample within an occasion and therefore both the correct and incorrect genotype can be observed in the same occasion. In addition, capture probability and genotyping error are assumed to be equal across individuals. Variation in capture probability is common in trapping studies and is just as likely in noninvasive sampling. Therefore, the potential for bias due to parameter heterogeneity exists in this model just as is does in standard closed-population capture–recapture models.

We extend the results of Lukacs and Burnham (2005b) to a demographically and geographically open population model and sampling scheme often referred to as Pollock's robust design (Pollock 1982; Kendall et al. 1997). The robust design is composed of two types of sampling periods. Primary sampling periods are separated by a relatively long length of time during which the population is assumed to be demographically and geographically open. Within each primary period are secondary periods which are very close together in time and the population is assumed to be demographically and geographically closed. An extension to the robust design relaxing the closure assumption has been developed (Schwarz and Stobo 1997), but that model will not be addressed in this paper. The robust design allows the estimation of population size at each primary period, survival probability between primary periods, temporary emigration between primary periods, and capture probability during a secondary period. In addition, our extension allows for the estimation of genotyping error rate at each primary period.

We further extend the robust design with genotyping error model to incorporate data from laboratory studies of genotyping error. The ability to combine multiple sources of data helps estimate parameters more precisely than could be done with any one data source analyzed alone. There is also a potential trade-off between eliminating genotyping error from noninvasive samples at a considerable expense of effort versus using auxiliary data to more precisely estimate the error rate while allowing some errors to exist. Moreover, these data are often already being collected; therefore it would be most efficient to use the data in a single analysis.

2 Statistical Model and Notation

Our notation largely follows Kendall et al. (1997) and Lukacs and Burnham (2005b). The robust design is a sampling scheme with t primary sampling periods. Each $i = (1, 2, \ldots, t)$ of the primary periods contains l_i secondary sampling periods. The l_i secondary periods are not required to be equal in number. The population is assumed to be open between primary periods and closed between secondary periods within a primary period. An example of a design for a relatively long lived species could have $t = 5$ primary periods each separated by 1 year and $l_1, \ldots, l_4 = 4$ secondary periods per primary period separated by 1 day (Fig. 1).

2.1 Parameters

- φ_i Probability of an animal surviving from primary period i to $i + 1$ and remaining faithful to the population given it is alive at i
- γ'_i Probability of being off the sampling area at time i given the animal was off the area at $i - 1$
- γ''_i Probability of being off the sampling area at time i given the animal was on the area at $i - 1$
- p_{ij} Probability of initially observing a genotype at secondary sample j of primary period i
- c_{ij} Probability of subsequently observing a genotype at secondary sample j of primary period i. The probability of subsequent observation is different than p because p includes both correct and incorrect genotypes while c only includes correct genotypes and because there may be a behavioral response to the sampling device.

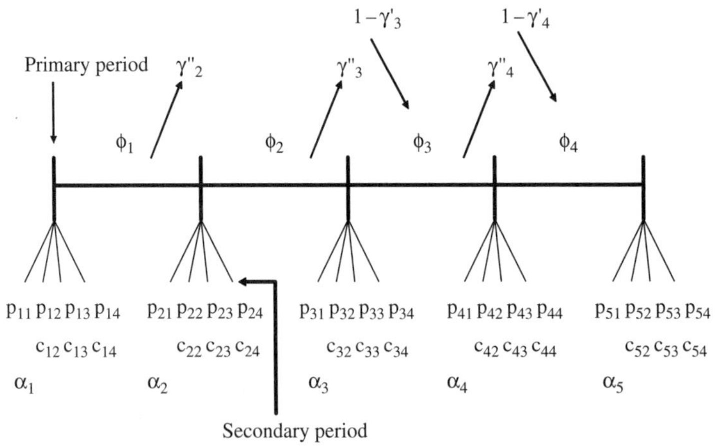

Fig. 1 An example robust design with $t = 5$ primary periods and $l_1, \ldots, l_5 = 4$ secondary periods

p_{i*} Probability of being captured at least once during primary period i.

$$p_i^* = 1 - \prod_{j=1}^{l_i} (1 - p_{ij})$$

α_i Probability that the first observation of a genotype in primary period i is a correct genotype

f_{i0} the number of genotypes in the population which are never observed at primary period i

N_i Population size at primary period i

2.2 Statistics

L total number of secondary capture occasions. $L = \sum_{i=1}^{t} l_i$

$\mathbf{h} = \{h_1, h_2, \ldots, h_L\}$ encounter history vector h_i=1 if the genotype is observed, 0 otherwise

$\mathbf{h}_i$ encounter history of the secondary samples within primary period i

$\mathbf{h}_\mathrm{p}$ encounter history of the primary periods. $h_{\mathrm{p}i}$ equals the sum of the encounter history for the secondary occasions within primary period i. This subset of **h** is needed to help delineate whether a genotype is detected in a given primary period.

x_{ij} an indicator variable that equals 1 if the genotype is observed at secondary occasion i and 0 otherwise

$M_{i,t+1}$ Number of unique genotypes observed during primary period i

We used the same assumptions here as were used in Lukacs and Burnham (2005b) and discussed in the introduction to estimate genotyping error rate within a primary period. From those assumptions cell probabilities were computed for each possible encounter history and the likelihood was constructed. The robust design likelihood was composed of two pieces, the likelihood for the open population primary periods, $\mathcal{L}_1$, and the likelihood for the secondary periods within each primary period, $\mathcal{L}_2$.

The likelihood for the secondary periods within each primary period was developed from the cell probabilities of each possible encounter history. For genotypes first observed at secondary occasion k and subsequently observed again within primary period i, the probability of the encounter history is

$$\Pr[\mathbf{h}_i] = \left[\prod_{j=1}^{k-1} (1 - p_{ij})\right] [p_{ik}\alpha_i] \left[\prod_{g=k+1}^{t} c_{ig}^{x_{ig}} (1 - c_{ig})^{1-x_{ig}}\right]$$

For a genotype which is only observed at secondary occasion k and never observed again, the probability of the encounter history is

$$\Pr[\mathbf{h}_i] = \left[\prod_{j=1}^{k-1}(1-p_{ij})\right]\left[p_{ik}\alpha_i\left(\prod_{g=k+1}^{t}(1-c_{ig})\right)+p_{ik}(1-\alpha_i)\right]$$

For a genotype that is never observed, the probability of the encounter history is

$$\Pr[h_i] = \prod_{i=1}^{t}(1-p_{ij}).$$

An example of the cell probabilities and their expected values is in Appendix 1 for a five-occasion case. The multinomial likelihood for the secondary periods is the product of the likelihood of each secondary period within a primary period

$$\mathcal{L}_2 = \prod_{i=1}^{t}\left[\frac{(f_{i0}+M_{i,t+1})!}{f_{i0}!}\prod_{h_i}\Pr[\mathbf{h}_i]^{n_{\mathbf{h}_i}}\right]$$

The likelihood for the primary periods is also based on a multinomial distribution. For a K= 4 primary occasion case the cell probabilities for the first encounter after release are given in Appendix 2. From these cell probabilities, it is possible to construct an expression for an encounter history. Let **A** be an upper triangular matrix of cell probabilities similar to those represented in Appendix 2 and let a_{ij} represent the elements of **A**. In addition, let b be the set of occasions on which the genotype was observed and subsequently seen again. Let r be the occasion of the last observation of the genotype. For example, for a genotype observed at times 1, 3 and 4, $b\subset\{(1,3),(3,4)\}$ and $r = 4$. For a genotype that was observed more than once, otherwise stated the sum of the elements of $\mathbf{h}_\mathrm{p} > 1$, the probability of the encounter history is

$$\Pr[\mathbf{h}_\mathrm{p}] = \alpha_t\left[\prod_{(i,j)\in b}a_{ij}\right]\left(1-\sum_{j=r+1}^{t}a_{rj}\right)$$

For a genotype that is only observed once (sum of $\mathbf{h}_\mathrm{p} = 1$), the probability of the encounter history across primary periods is

$$\Pr[\mathbf{h}_\mathrm{p}] = \alpha_t\left[1-\sum_{j=r+1}^{t}a_{rj}\right]+(1-\alpha_t)$$

The multinomial likelihood across primary periods is

$$\mathcal{L}_1 \propto \prod_{\mathbf{h}_\mathrm{p}}\Pr[\mathbf{h}_\mathrm{p}]^{n_{\mathbf{h}_\mathrm{p}}}$$

The full likelihood for the robust design incorporating genotyping error is given by the product of $\mathcal{L}_1$ and $\mathcal{L}_2$. Parameter estimates can be found by numerical optimization of the log-likelihood. We used a quasi-Newton optimization in SAS PROC IML (SAS Institute Inc. 2002). The variance–covariance matrix of the parameters was computed by inverting the Hessian matrix and taking its negative. This model is now available in Program MARK (White and Burnham 1999).

Up to this point, N has not appeared in the likelihood and it does not enter into the likelihood. N must be estimated as a derived parameter for this likelihood. It is estimated as

$$\hat{N}_i = \hat{\alpha}_i(\hat{f}_{i0} + M_{i,t+1})$$

The variance of $\hat{N}$ is estimated as

$$\hat{\text{v}}\text{ar}[\hat{N}] = \hat{\alpha}_i^2\, \hat{\text{v}}\text{ar}[\hat{f}_{i0}] + \left(\hat{f}_{i0} + M_{it+1}\right)^2\, \hat{\text{v}}\text{ar}[\hat{\alpha}_i] + 2\hat{\alpha}_i\, M_{it+1}\, \hat{\text{c}}\text{ov}[\hat{f}_{i0},\, \hat{\alpha}_i]$$

The estimates of N across primary periods may covary depending on the model structure of the capture probabilities and genotyping error rates.

Model parameters may be functions of group covariates as is commonly used in general linear models (McCullagh and Nelder 1989). The α parameter will typically be modeled on a sine link because it is near the boundary of 1.0 in many studies. If the likelihood of the secondary periods within a primary period is replaced with the conditional likelihood given in Lukacs and Burnham (2005b), parameters, except for N, may be modeled as functions of individual covariates as well as group covariates.

When considering the likelihood presented, it is important to note a simplifying assumption that was made in the model development. Information about genotyping error across primary periods is not completely used to help estimate genotyping error. This assumption was made to reduce the parameter set. If all information is used across primary periods, cohort specific p and c parameters must be added to the model and abundance must be estimated for every cohort. This quickly leads to a model with too many parameters to estimate.

The robust design likelihood can be supplemented with additional sources of data to better estimate the probability of correctly assigning a genotype and survival probability. Laboratory results can be used to help estimate the probability of correctly assigning a genotype. Genetic data are likely to exist from known individuals in the population. Genotypes from tissue or blood samples can be compared to those from feather, hair or scat samples. The tissue or blood samples can be genotyped with virtually no error and therefore serve as a reference sample. The count of correctly and incorrectly matched noninvasive samples should follow a binomial distribution. Therefore, the likelihood for estimating the probability of correctly assigning a genotype is

$$\mathcal{L}_3 = \prod_{i=1}^{K} \binom{n_i}{x_i} \alpha_i^{x_i}\, (1 - \alpha_i)^{n_i - x_i}$$

where n_i is the number of reference samples from primary period i, x_i is the number of correctly matched samples from primary period i and α_i is the probability of correctly assigning the genotype as defined above. As with the robust design, α_i can be modeled as a function of covariates through a link function.

Now the full likelihood for the robust design with multiple sources of data is the product of the three likelihood functions

$$L = \mathcal{L}_1\mathcal{L}_2\mathcal{L}_3$$

If no laboratory data are available, $\mathcal{L}_3 = 1$. Without laboratory data, there must be at least three secondary sampling occasions per primary occasion to estimate α because α is not identifiable with only two secondary occasions. With laboratory data available only one secondary sample is needed if abundance is not of interest. Despite that, it is always beneficial to have multiple secondary occasions to better estimate all model parameters.

3 Model Testing

We used simulated data to test the properties of the robust design model incorporating genotyping error. First, we concentrated model evaluation on only the open population portion of the model. Capture–recapture data were simulated in a factorial design with four levels of the probability of correctly genotyping a sample (0.96–0.99), five levels of capture probability (0.1–0.5) and two levels of survival probability (0.6, 0.9). Initial population size was 1,000 for all simulations. Each design point was replicated 100 times. All simulations used four primary occasions and five secondary occasions within each primary occasion. Population size, capture probability and recapture probability were estimated separately for each primary occasion and survival probability and the probability of correctly genotyping a sample were estimated as equal across primary occasions. The probability of temporary emigration was fixed to zero for all simulations because it is known to be very difficult to estimate precisely and to allow clear inferences about the estimates of abundance and survival. The data were simulated under a model that is more complicated than the estimating model, so exactly unbiased results are not expected.

We further evaluated the utility of laboratory data to better estimate α and consequently N. To do this we used black bear (*Ursus americanus*) genetic capture-recapture data from Dreher et al. (2007). These data consist of five occasions of noninvasive hair sampling and one occasion consisting of a black bear hunt where hunters were required to submit a hair and tissue sample from their harvested bear. For complete study design details see Dreher et al. (2007). These data represent a single closed-population capture–recapture session or one primary sampling period of the robust design. The genetic analysis of the noninvasive hair samples consisted of three stages with additional error checking and quality control at each stage. The genetic analyses followed the protocol of Paetkau (2003). At the first stage

Table 1 Sample sizes and genotyping error rates for the eight data sets used to examine the effectiveness of auxiliary laboratory data

Stage	Loci amplified	Sample size	Error rate
Initial run	≥3	508	0.11
	≥4	501	0.13
	5	350	0.06
Rerun	≥3	564	0.20
	≥4	561	0.20
	5	508	0.26
Mismatch at 1 or 2 loci	≥4	546	0.01
	5	504	0.01

all samples were genotyped and samples with three or more amplifying loci were retained (n =508). At the second stage, samples failing to amplify at all loci in the first stage were reanalyzed and those with three or more amplifying loci were retained ($n = 561$). Finally, at the third stage, samples that differed at only one or two loci were reanalyzed to check for potential errors. Only those samples considered accurate at four or five loci were retained (n = 504). This provides data sets with varying accuracy of identification and sample size (Table 1). In addition, 96 paired hair and tissue samples were collected from the hunter harvested bears to directly test the error rate of the hair sample analysis using genotypes from tissue samples as a known genotype.

We developed encounter histories for each of the eight data sets by matching common genotypes across sampling occasions. We then selected 500 bootstrap samples with replacement from each of the data sets. We fit a standard closed capture-recapture model (Otis et al. 1978), a model incorporating genotyping error (Lukacs and Burnham 2005b) and the model presented here using the laboratory data. We then compared the average standard error size for each model to determine which method would be expected to provide the most precise results.

4 Results

The models provided good parameter estimates in the simulated examples. Abundance at the first primary occasion was well estimated across varying levels of the probability of correctly genotyping a sample (Table 2). As would be expected, the estimates of abundance improved as the rate of genotyping error decreased. Survival was well estimated across the range of probabilities of correctly genotyping a sample (Table 3). Beyond the first primary period, abundance is a function of survival and therefore a fixed true value of abundance is not available to compare the estimated abundance to for future sampling periods.

Capture probability 0.2 or more provided reliable estimates of abundance (Table 4). With a capture probability of 0.1, estimates of abundance were biased high (>12% bias, SE = 1.2), were imprecise and failed to converge in 29 of 900 replicates. As capture probability decreases below 0.1 abundance estimates will further degrade unless the number of secondary occasions is increased. Abundance

Table 2 Average abundance estimates for the first primary occasion of a robust design model including genotyping error at four levels. The probability of correctly genotyping a sample is denoted α and abundance is N and SE is standard error of the mean of N. Mean N is averaged over five levels of capture probability from 0.1 to 0.5. Mean N is based on 1,000 simulation replicates per level of α. True $N = 1{,}000$ and $\alpha = 0.95$

α	Mean N	SE
0.96	1098	3.3
0.97	1073	2.9
0.98	1054	3.0
0.99	1037	3.2

Table 3 Average estimated survival between primary occasions for a robust design model including genotyping error at four levels. The probability of correctly genotyping a sample is denoted α and survival is φ. Means and standard errors (SE) are computed over 500 simulation replicates at each level and averaged over varying levels of capture probability

	Mean estimated φ			
α	True $\varphi = 0.6$	SE	True $\varphi = 0.9$	SE
0.96	0.595	0.001	0.911	0.002
0.97	0.596	0.001	0.915	0.002
0.98	0.591	0.001	0.899	0.002
0.99	0.601	0.001	0.909	0.002

Table 4 Average abundance estimates by capture probability for the first primary occasion of a robust design model including genotyping error at four levels. The capture probability is denoted p and abundance is N and SE is standard error of the mean of N. Mean N is averaged over five levels of the probability of correctly genotyping a sample from 0.95 to 0.99. Mean N is based on 1,000 simulation replicates per level of p

p	Mean N	SE
0.1	1120	12.17
0.2	1041	3.11
0.3	1050	1.50
0.4	1059	1.19
0.5	1067	1.19

was best estimated when capture probability was in the range of 0.2–0.3. Bias in estimated abundance increased slightly at higher capture probabilities. Survival was generally well estimated across the entire range of capture probabilities (Table 5).

Without additional laboratory information about misidentification, the standard error was always larger for the misidentification models than for a standard capture–recapture model (Fig. 2). When the additional data are added, the misidentification models estimate abundance more precisely. In the case where nearly all genotyping error has been removed including laboratory data does not improve precision. In addition, the standard error for model using the laboratory data and a higher misidentification rate is less than that of the standard capture-recapture model using the nearly error free data, yet the nearly error free data are more expensive to achieve.

Table 5 Average survival estimates by capture probability for a robust design model including genotyping error at four levels. The capture probability is denoted p and survival is φ and SE is standard error of the mean of N. Mean φ is averaged over five levels of the probability of correctly genotyping a sample from 0.95 to 0.99. Mean φ is based on 1,000 simulation replicates per level of p

	Mean estimated φ			
p	True $\varphi = 0.6$	SE	True $\varphi = 0.9$	SE
0.1	0.620	0.001	0.927	0.002
0.2	0.609	0.001	0.960	0.002
0.3	0.596	0.001	0.905	0.001
0.4	0.583	0.001	0.890	0.001
0.5	0.574	0.001	0.876	0.001

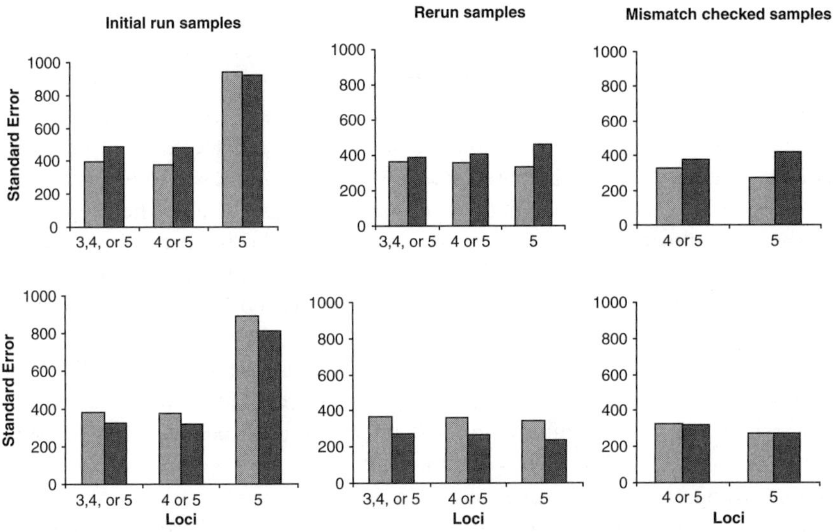

Fig. 2 Mean standard errors of estimated abundance for models including genotyping error (*dark shaded*) and not including genotyping error (*light shaded*) with (*bottom panels*) and without (*top panels*) the inclusion of auxiliary laboratory derived genotyping error rate data based on real data from black bears in Michigan

5 Discussion

Noninvasive DNA sampling techniques open many opportunitiesfor capture–recapture studies where they were not previously feasible. This will allow ecologists to answer important questions regarding species that are difficult to detect or handle. Thus far, genetic capture–recapture data has been used only for estimating abundance. One necessary step toward answering broader ecological questions is to expand DNA-based capture–recapture studies beyond estimates of abundance at a single point in time. The robust design allows ecologists to explore more interesting ecological questions involving survival, abundance and emigration rates.

The robust design is a natural extension from abundance estimation to explorations of survival and temporary emigration rates. The expanded robust design model presented here allows abundance, survival and emigration to be estimated properly in the face of genotyping error. For studies currently estimating abundance annually in a closed population capture–recapture framework, the robust design offers survival and emigration rates with little or no additional effort. Moreover, these parameters can be modeled as functions of covariates in a general linear models framework to answer a wider array of ecological questions.

The simulation results presented suggest the estimator performs well across varied capture probability and genotyping error levels. It is not surprising that the estimator performed poorly when a capture probability of 0.1 is used because in those cases an animal has only a 0.08 probability of being caught two or more times within a primary occasion containing five secondary sampling occasions. Therefore, relatively few individuals are recaptured. Recaptures are essential in order to estimate genotyping error rate because it is estimated from the over abundance of genotypes only detected once relative to those detected more than once. Without recaptures or an alternate data source, there is no way to estimate genotyping error. If capture probability is low and there is no direct field method to increase it, estimates of abundance and survival can be improved by increasing the number of secondary sampling occasions. This increases p^* that in turn increases the total number of unique animals caught and improves the parameter estimates.

A useful feature of this robust design model is that it shares the same likelihood as a robust design model not including genotyping error. Therefore, model comparisons can be made with information – theoretic criteria such as AIC_c (Burnham and Anderson 2002). This allows one to examine the information loss in assuming no genotyping error when some may be present. If genotyping error rate is relatively small, it may be advantageous in terms of mean squared error to assume zero error. Conversely, when genotyping error rate is large the added bias would suggest assuming zero error is a poor choice. AIC_c can quantitatively assess such tradeoffs. Given the debate in the literature over the effect of genotyping error on capture–recapture studies (McKelvey and Schwartz 2004a; Paetkau 2004; McKelvey and Schwartz 2004b) and evidence supporting both no genotyping error (Paetkau 2003) and considerable genotyping error (Creel et al. 2003), it is important to have a data driven way to resolve the issue.

Combining data sources is an important way to gain a better understanding of an ecological system. Here we present a model that incorporates noninvasive genetic data and laboratory testing of the accuracy of the genotyping methods. For species such as bears (*Ursus* spp.) these data are routinely available through hunter harvested bears. It is not surprising that when misidentification rate is estimated, the abundance estimate is less precise given identical data are used to estimate abundance because an additional parameter is estimated. The addition of the laboratory data allows abundance to be more precisely estimated while reducing the cost of reanalysis of samples to eliminate genotyping errors from the data set. This is the real power of this method because genotyping error rates are likely to change over time and thus will likely be estimated as a time varying parameter. Without the additional lab data, the error rate parameter would be very poorly estimated.

The bootstrap analysis also lends insight into the trade-off between spending a lot of money and effort to eliminate genotyping error versus allowing some error and estimating its rate. If no auxiliary data are available on genotyping error rate, the most precise estimate of abundance can be obtained by eliminating genotyping error (Fig. 2). If auxiliary data are available on genotyping error rate, a more precise estimate of abundance can be obtained by allowing some error and a larger sample size and combining the two sources of data (Fig. 2). Given that the potential exists for individual variation to exist in the quality of noninvasive samples, using more low quality samples and allowing some genotyping error may reduce amount of individual variation in detection probability.

Without accounting for genotyping error when using DNA-based capture–recapture methods, the risk exists to over-estimate abundance and under-estimate survival. DNA-based sampling is often used for rare and exploited species. Therefore, accurate assessment of demographic parameters is a key aspect for species management. The robust design model presented here allows for appropriate estimates of abundance and survival in the face of genotyping error.

Acknowledgments The Colorado Division of Wildlife, Great Outdoors Colorado and Dr. Tanya Shenk provided support for this research. Dr. William Kendall and three anonymous referees provided helpful comments on the manuscript.

Appendix 1

Cell probabilities and expected values for the secondary capture occasions within a primary occasion for a 5-occasion case with p =0.3, $c = 0.3$, $\alpha = 0.95$ and $N =$ 1,000. A population with 1,000 individuals and a 0.95 probability of the genotype being correct is expected to have 1,052 different genotypes available for detection. Note that a constraint must be placed on p_5 for all of the model parameters to be identifiable (Table 6).

Appendix 2

Cell probabilities for the primary sessions of a robust design can be constructed through the use of matrix algebra. To do so, a few matrices must be defined.

Table 6 Cell probabilities and expected values for the secondary capture occasions within a primary occasion for a 5-occasion case with $p = 0.3$, $c = 0.3$, $\alpha = 0.95$ and $N = 1,000$. A population with 1,000 individuals and a 0.95 probability of the genotype being correct is expected to have 1,052 different genotypes avilable for detection. Note that a constraint must be placed on p_5 for all of the model parameters to be identifiable

History	Cell probability	Probability	E[n]
00001	$(1-p_1)(1-p_2)(1-p_3)(1-p_4)[p_5\alpha+p_5(1-\alpha)]$	0.0720	75.8211
00010	$(1-p_1)(1-p_2)(1-p_3)[p_4\alpha(1-c_5)+p_4(1-\alpha)]$	0.0736	77.4458
00011	$(1-p_1)(1-p_2)(1-p_3)p_4\alpha c_5$	0.0293	30.8700
00100	$(1-p_1)(1-p_2)[p_3\alpha(1-c_4)(1-c_5)+p_3(1-\alpha)]$	0.0758	79.7668
00101	$(1-p_1)(1-p_2)p_3\alpha(1-c_4)c_5$	0.0293	30.8700
00110	$(1-p_1)(1-p_2)p_3\alpha c_4(1-c_5)$	0.0293	30.8700
00111	$(1-p_1)(1-p_2)p_3\alpha c_4 c_5$	0.0126	13.2300
01000	$(1-p_1)[p_2\alpha(1-c_3)(1-c_4)(1-c_5)+p_2(1-\alpha)]$	0.0789	83.0826
01001	$(1-p_1)p_2\alpha(1-c_3)(1-c_4)c_5$	0.0293	30.8700
01010	$(1-p_1)p_2\alpha(1-c_3)c_4(1-c_5)$	0.0293	30.8700
01011	$(1-p_1)p_2\alpha(1-c_3)c_4 c_5$	0.0126	13.2300
01100	$(1-p_1)p_2\alpha c_3(1-c_4)(1-c_5)$	0.0293	30.8700
01101	$(1-p_1)p_2\alpha c_3(1-c_4)c_5$	0.0126	13.2300
01110	$(1-p_1)p_2\alpha c_3 c_4(1-c_5)$	0.0126	13.2300
01111	$(1-p_1)p_2\alpha c_3 c_4 c_5$	0.0054	5.6700
10000	$p_1\alpha(1-c_2)(1-c_3)(1-c_4)(1-c_5)+p_1(1-\alpha)$	0.0834	87.8195
10001	$p_1\alpha(1-c_2)(1-c_3)(1-c_4)c_5$	0.0293	30.8700
10010	$p_1\alpha(1-c_2)(1-c_3)c_4(1-c_5)$	0.0293	30.8700
10011	$p_1\alpha(1-c_2)(1-c_3)c_4 c_5$	0.0126	13.2300
10100	$p_1\alpha(1-c_2)c_3(1-c_4)(1-c_5)$	0.0293	30.8700
10101	$p_1\alpha(1-c_2)c_3(1-c_4)c_5$	0.0126	13.2300
10110	$p_1\alpha(1-c_2)c_3 c_4(1-c_5)$	0.0126	13.2300
10111	$p_1\alpha(1-c_2)c_3 c_4 c_5$	0.0054	5.6700
11000	$p_1\alpha c_2(1-c_3)(1-c_4)(1-c_5)$	0.0293	30.8700
11001	$p_1\alpha c_2(1-c_3)(1-c_4)c_5$	0.0126	13.2300
11010	$p_1\alpha c_2(1-c_3)c_4(1-c_5)$	0.0126	13.2300
11011	$p_1\alpha c_2(1-c_3)c_4 c_5$	0.0054	5.6700
11100	$p_1\alpha c_2 c_3(1-c_4)(1-c_5)$	0.0126	13.2300
11101	$p_1\alpha c_2 c_3(1-c_4)c_5$	0.0054	5.6700
11110	$p_1\alpha c_2 c_3 c_4(1-c_5)$	0.0054	5.6700
11111	$p_1\alpha c_2 c_3 c_4 c_5$	0.0023	2.4300
00000	$(1-p_1)(1-p_2)(1-p_3)(1-p_4)(1-p_5)$	0.1681	176.9158
Total		1.0000	1052.6316
Derived abundance estimate			1000

Matrices

$$\mathbf{f}_i + 1 = \left[\gamma''_{i+1}\ (1-\gamma''_{i+1})(1-p^*_{i+1})\right]$$

$$\mathbf{G}_i = \begin{bmatrix} \gamma'_i & (1-\gamma'_i)(1-p^*_i) \\ \gamma''_i & (1-\gamma''_i)(1-p^*_i) \end{bmatrix}$$

$$\mathbf{d}_i = \begin{bmatrix} (1-\gamma'_i) \\ (1-\gamma''_i) \end{bmatrix}$$

Table 7 Multinomial cell probabilities for primary periods of a 4-period robust design. Cell probabilities are conditional on a genotype that is released at the occasion listed in row label and next detected at occasion listed in the column label. The values represent the elements of matrix **A**

	Primary recapture occasion		
Primary release occasion	2	3	4
1	$\varphi_1(1-\gamma_2'')p_2^*$	$\varphi_1\mathbf{f}_2\,\varphi_2\mathbf{d}_3 p_3^*$	$\varphi_1\mathbf{f}_2\varphi_2\mathbf{G}_3\varphi_3\mathbf{d}_4 p_4^*$
2		$\varphi_2(1-\gamma_3'')p_3^*$	$\varphi_2\mathbf{f}_3\,\varphi_3\mathbf{d}_4\,p_4^*$
3			$\varphi_3(1-\gamma_4'')p_4^*$

$$p_i^* = 1 - \prod_{j=1}^{l_i}(1 - p_{ij})$$

Given these matrices and the other model parameters, cell probabilities can be constructed as is shown in Table 7.

References

Boulanger J, White GC, McLellen BN, Woods J, Proctor M, Himmer S (2003) A meta-analysis of grizzly bear DNA mark-recapture projects in British Columbia, Canada. Ursus 13:137–152.

Burnham KP, Anderson DR (2002) Model selection and multimodel inference, second edition. Springer-Verlag, New York.

Creel S, Spong G, Sands JL, Rotella J, Zeigle J, Joe L, Murphy KM, Smith D (2003) Population size estimation in Yellowstone wolves with error-prone noninvasive microsatellite genotypes. Molecular Ecology 12:2003–2009.

Dreher BP, Winterstein SR, Scribner KT, Lukacs PM, Etter DR, Rosa GJM, Lopez VA, Libants S, Filcek KB (2007) Non-invasive estimation of black bear abundance in Michigan incorporating genotyping errors and harvested bears. Journal of Wildlife Management 71: 2689–2693.

Kendall WL, Nichols JD, Hines JE (1997) Estimating temporary emigration using capture–recapture data with Pollock's robust design. Ecology 78:563–578.

Lukacs PM, Burnham KP (2005a) Review of capture–recapture methods applicable to noninvasive genetic sampling. Molecular Ecology 14:3909–3919.

Lukacs PM, Burnham KP (2005b) Estimating population size from DNA based capture–recapture data incorporating genotyping error. Journal of Wildlife Management 69:396–403.

McCullagh P, Nelder JA (1989) Generalized linear models, second edition. Champman and Hall/CRC, London.

McKelvey KS, Schwartz MK (2004a) Genetic errors associated with population estimation using non-invasive molecular tagging: problems and new solutions. Journal of Wildlife Management 68:439–448.

McKelvey KS, Schwartz MK (2004b) Providing reliable and accurate genetic capture–mark–recapture estimates in a cost-effective way. Journal of Wildlife Management 68:453–456.

Otis DL, Burnham KP, White GC, Anderson DR (1978) Statistical inference from capture data on closed animal populations. Wildlife Monographs 62.

Paetkau D (2003) An empirical exploration of data quality in DNA-based population inventories. Molecular Ecology 12:1375–1387.

Paetkau D (2004) The optimal number of markers in genetic capture–mark–recapture studies. Journal of Wildlife Management 68:449–452.

Pollock KH (1982) A capture–recapture design robust to unequal probability of capture. Journal of Wildlife Management 46:752–757.

Schwarz CJ, Stobo WT (1997) Estimating temporary migration using the robust design. Biometrics 53:178–194.

SAS Institute Inc. (2002) SAS version 9.0. Cary, North Carolina, USA.

White GC, Burnham KP (1999) Program MARK: survival estimation from populations of marked animals. Bird Study 46:S120–S139.

Woods JG, Paetkau D, Lewis D, McLellan BN, Proctor M, Strobeck C (1999) Genetic tagging of free-ranging black and brown bears. Wildlife Society Bulletin 27:616–627.

A Traditional and a Less-Invasive Robust Design: Choices in Optimizing Effort Allocation for Seabird Population Studies

Sarah J. Converse, William L. Kendall, Paul F. Doherty Jr., Maura B. Naughton, and James E. Hines

Abstract For many animal populations, one or more life stages are not accessible to sampling, and therefore an unobservable state is created. For colonially-breeding populations, this unobservable state could represent the subset of adult breeders that have foregone breeding in a given year. This situation applies to many seabird populations, notably albatrosses, where skipped breeders are either absent from the colony, or are present but difficult to capture or correctly assign to breeding state. Kendall et al. (in press) have proposed design strategies for investigations of seabird demography where such temporary emigration occurs, suggesting the use of the robust design to permit the estimation of time-dependent parameters and to increase the precision of estimates from multi-state models. A traditional robust design, where animals are subject to capture multiple times in a sampling season, is feasible in many cases. However, due to concerns that multiple captures per season could cause undue disturbance to animals, Kendall et al. (in press) developed a less-invasive robust design (LIRD), where initial captures are followed by an assessment of the ratio of marked-to-unmarked birds in the population or sampled plot. This approach has recently been applied in the Northwestern Hawaiian Islands to populations of Laysan (*Phoebastria immutabilis*) and black-footed (*P. nigripes*) albatrosses. In this paper, we outline the LIRD and its application to seabird population studies. We then describe an approach to determining optimal allocation of sampling effort in which we consider a non-robust design option (nRD), and variations of both the traditional robust design (RD), and the LIRD. Variations we considered included the number of secondary sampling occasions for the RD and the amount of total effort allocated to the marked-to-unmarked ratio assessment for the LIRD. We used simulations, informed by early data from the Hawaiian study, to address optimal study design for our example cases. We found that the LIRD performed as well or nearly as well as certain variations of the RD in terms of root mean square error, especially when relatively little of the total effort was allocated to the assessment of the marked-to-unmarked ratio versus to initial captures. For the RD, we found no

S.J. Converse (✉)
Colorado Cooperative Fish and Wildlife Research Unit, Patuxent Wildlife Research Center, Laurel, MD, USA
e-mail: sconverse@usgs.gov

D.L. Thomson et al. (eds.), *Modeling Demographic Processes in Marked Populations*, Environmental and Ecological Statistics 3, DOI 10.1007/978-0-387-78151-8_33,

clear benefit of using 2, 4, or 6 secondary sampling occasions per year, though this result will depend on the relative effort costs of captures versus recaptures and on the length of the study. We also found that field-readable bands, which may be affixed to birds in addition to standard metal bands, will be beneficial in longer-term studies of albatrosses in the Northwestern Hawaiian Islands. Field-readable bands reduce the effort cost of recapturing individuals, and in the long-term this cost reduction can offset the additional effort expended in affixing the bands. Finally, our approach to determining optimal study design can be generally applied by researchers, with little seed data, to design their studies at the outset.

1 Introduction

For many animal populations, one or more life stages are not accessible to sampling. Traditional open-population capture–recapture models that do not account for this are likely to produce biased estimates of survival (Kendall et al. 1997). Several authors have suggested the use of multi-state capture–recapture models including one or more unobservable states to account for temporary inaccessibility to sampling (Fujiwara and Caswell 2002; Kendall and Nichols 2002; Schaub et al. 2004). Such models allow for estimation of the probabilities of transition between states, in addition to survival and detection probabilities. However, multi-state capture-recapture models integrating an unobservable state generally suffer from a lack of identifiability of parameters in the model, especially when time-specific forms are used (Kendall and Nichols 2002; Schaub et al. 2004). In the face of unobservable states, Kendall and Nichols (2002) showed that use of Pollock's (1982) robust design (RD) increases parameter identifiability as well as the precision of estimates of state transitions.

A familiar example of an unobservable life stage occurs with colonially-breeding bird populations where each year some adults do not breed and non-breeding individuals are largely unobservable because they either remain at sea or visit the colony occasionally but are not easily captured or assigned to breeding state. This situation applies to many seabird populations, and notably albatrosses (Diomedeidae). Kendall et al. (in press) have proposed design strategies for investigations of albatross demography where such temporary emigration occurs, suggesting the use of the robust design to permit the estimation of time-dependent parameters and to increase the precision of estimates from multi-state models. These design strategies arose to meet the needs of US Fish and Wildlife Service biologists studying the demography of Laysan albatross (*Phoebastria immutabilis*) and black-footed albatross (*P. nigripes*) breeding in the Northwestern Hawaiian Islands, particularly on 2 islands of Midway Atoll. Monitoring objectives in this ongoing study include estimation of survival and breeding probabilities. Due to concern by investigators that multiple captures per season could cause undue disturbance to animals, Kendall et al. (in press) developed a less-invasive robust design (LIRD), where initial captures of individuals each year are followed by a tally of birds used to calculate the ratio of the sampled population previously captured in that year.

Therefore, multiple options exist for capture-recapture study design in the face of unobservable states, including a non-robust design option (nRD), in which animals are subject to capture only once in a sampling season, and variations of both the RD (where animals are subject to capture multiple times in a sampling season, during which the population is assumed to be closed to demographic changes) and the LIRD (where animals are subject to capture once and are subject to a tally once in a sampling season). Simulation is an excellent tool for determining optimal study design for a particular case, and given particular study objectives. As an example, we developed a simulation study to determine optimal sampling strategies for albatross studies. Here, we present the results of our simulation study of RD and LIRD sampling scenarios, and sampling scenarios not including the robust design (nRD). With our simulations, we addressed the following questions: (1) What is the performance of the LIRD versus the RD and nRD in terms of root mean square error and confidence interval coverage of survival and state transition parameters used to estimate breeding probabilities; (2) For the LIRD, what is the optimal allocation of effort to initial captures versus tally effort; and (3) For the RD, what is the optimal number of secondary sampling occasions? We asked these questions for studies of different lengths and under different sampling contexts. We used the 2 albatross species resident on Midway Atoll in the Northwestern Hawaiian Islands as focal species in designing our simulations.

1.1 The Less-Invasive Robust Design

Before introducing our simulation methods we present further details on the LIRD. We use sampling in an albatross colony as an example to illustrate the LIRD, which is described in greater detail in Kendall et al. (in press). Here we imagine applying this design to a colony in which no birds are banded at the outset.

Each year just after the egg-laying period, investigators move through the colony or chosen plots within the colony, randomly choosing nesting individuals and fitting them with a standard metal (e.g., U.S. Geological Survey Bird Banding Laboratory) band bearing a unique numeric code (hereafter, metal band). In some designs, an auxiliary field-readable band may also be attached to the other leg, and for these individuals, subsequent "recaptures" will consist of reading the field-readable band rather than the metal band, i.e., the animal need not be physically handled after a field-readable band is applied. A field-readable band is made of hard, colored plastic and has a numeric code in a contrasting color. This band can be read from a short distance (typically 3–4 m) while the bird is either standing or sitting on a nest (in the latter case, this may require approaching the bird and moving feathers away from the leg manually).

After the first year, a portion of the individuals will have already been banded from previous years, and these individuals will be subject to recapture, while previously unmarked birds will be banded as in previous years. To indicate that an individual was captured in the current year, a temporary mark (one that will last for the duration of the field season but will disappear before the next) is applied.

Examples of temporary marks include a streak of acrylic nail polish or felt-tipped marker on the individual's head. Individuals with temporary marks are avoided for the remainder of the first sampling period, and on subsequent sampling trips (e.g., days) during the first sampling period, only birds without temporary marks are captured. For albatrosses, the first sample should be conducted over a long enough period to have access to both members of a breeding pair – perhaps 2 weeks or more – as members of a pair switch between incubating and foraging at sea. When applying the LIRD to other taxa, the first sample may be conducted over a shorter period of time as appropriate.

Shortly after the first sample is concluded, an investigator walks through the study area choosing active nests at random, then checking and recording whether or not the individual on the nest has a temporary mark, resulting in a tally of temporarily marked and unmarked individuals (the second sample). It is critical that there be no bias in selecting marked versus unmarked individuals for inclusion in the tally, i.e., the tally sample must be a random sample of the population subject to marking. After the tally is completed (a fixed number of nests to tally might be determined beforehand), the proportion of temporarily marked individuals in the sample should be an unbiased estimate of the proportion of the population physically captured in the first sample: $\hat{p}_t = m_{t2}/n_{t2}$ where n_{t2} is the total number of nesting individuals examined and m_{t2} is the number of those individuals with the temporary mark in the second (tally) sample.

For the statistical model, consider a multiyear study of a population where there are two life history states of interest, e.g., an observable breeder (O) and an unobservable skipped breeder (U) in the case of albatrosses. Let S_t^O, S_t^U be the probability an animal in state O or U, respectively, in year t survives to year $t + 1$ and remains faithful to the population. Furthermore, let Ψ_t^{OO}, Ψ_t^{UO} be the probability an animal in state O or U, respectively, in year t breeds (i.e., becomes observable) in year $t + 1$, given that it survives to year $t + 1$. Finally, let p_t^O, p_t^U be the probability an individual is captured in year t, given that it is in state O or U, respectively, and where we assume that $p_t^U = 0$. This leads to the Arnason–Schwarz multistate capture–recapture model (Schwarz et al. 1993) with an unobservable state.

The parameters of this model cannot be estimated without further restrictive assumptions, such as time-constancy in parameters (Fujiwara and Caswell 2002; Kendall and Nichols 2002; Schaub et al. 2004) or an additional source of data (Kendall 2004). The additional source of data can come from the LIRD described above, where the multi-state multinomial likelihood is supplemented by a product of simple binomial models for each year:

$$\prod_{t=1}^{\#years} \frac{n_{t2}^O!}{m_{t2}^O!(n_{t2}^O - m_{t2}^O)!}(p_t^O)^{m_{t2}^O}(1 - p_t^O)^{(n_{t2}^O - m_{t2}^O)}.$$

This second source of information is sufficient to estimate p_t^O from only the within-year data, and therefore permits estimation of time-specific survival and state transition probabilities (Kendall et al. 1997).

2 Simulation Methods

2.1 *Study Designs*

We considered 3 different basic study designs in our simulation study, including designs without robust design information (nRD), the traditional robust design (RD), and the less-invasive robust design (LIRD). Each of these designs was examined both with metal bands only and also with field-readable bands. Field-readable bands may be useful in albatross capture–recapture studies because both effort and disturbance to birds can be reduced in recaptures if birds need not be handled. In 2005, the US Fish and Wildlife Service began widespread application of field-readable bands on adult albatrosses in several study colonies of the Northwestern Hawaiian Islands.

For the 2 robust design types, we considered 3 variations of each. For the RD, we varied the number of secondary sampling occasions (2 occasions, 4 occasions, or 6 occasions) within each year. For the LIRD, we varied the proportion of the population that is included in the tally (5%, 25%, or 45%). For example, in a sampled population of 5000 birds, a tally of 5% of the birds would indicate that, on average, 250 birds would be tallied to calculate the ratio of temporarily marked-to-unmarked birds.

For each of these designs, we considered studies of either 5 or 10 years. We also considered 2 species: both Laysan and black-footed albatrosses. Because black-footed albatross tend to be less docile than Laysan albatross, most types of sampling activities, such as banding a previously unbanded bird, take longer when working with black-footed albatross; this has implications for optimal effort allocation.

In all of the simulation sets, we assumed a population of 10,000 individuals and the same between-year population dynamics: a 2-state multi-state model, with one observable and one unobservable state, an annual survival probability of 95% (equal for the 2 states and across years), annual probability of movement from the observable to the unobservable state of 80%, and annual probability of movement from the unobservable to the observable state of 80% (both also equal across years). On average, half of the individuals in the population were observable in any given year.

In the first set of simulations, we considered a time-constant estimation model on the survival and transition parameters. Only detection probabilities varied across simulation sets and over time within simulation sets, as described below. Under the time-constant model, model parameters were identifiable both with and without robust-design information.

In the second set of simulations, we considered a time-dependent estimation model on the survival and transition parameters. In the time-dependent estimation model the last of each of the transition parameters was set equal to the penultimate parameter, and the first transition out of the unobservable state was set equal to the second (Kendall and Nichols 2002). This set of simulations was designed to compare design types when identifiability of parameters was compromised under the nRD (Kendall and Nichols 2002; Schaub et al. 2004).

We considered 56 simulations in the first set, including 1 variation of the nRD, 3 variations of the LIRD, and 3 variations of the RD for 2 marking types (metal or field-readable), 2 study lengths (5 or 10 years), and 2 species (Laysan albatross or black-footed albatross). In the second set, we considered a smaller set of 6 simulations, including 1 variation each of the nRD, RD (2 occasions), and LIRD (5% of effort in tally) for 2 marking types (metal or field-readable), 1 study length (10 years), and 1 species (Laysan albatross).

2.2 Simulations – Seed Data

A combination of the RD and the LIRD was applied to Laysan and black-footed albatrosses in 2005 on a pilot basis (Kendall et al. in prep) providing seed data for our simulations. Birds banded at the U.S. Fish and Wildlife study at Midway Atoll are generally not physically captured; rather, a bander reaches under the bird and bands the leg(s) while another worker shields the bander from the bird's bill with a hard plastic tray. Occasionally birds are physically captured for banding if they attempt to leave the nest. Banding birds under these conditions may take more time, but is thought to reduce the risk that birds will temporarily abandon eggs, leaving eggs at risk for thermal stress or predation, or, more seriously, will permanently abandon eggs.

We used a sample of the pilot data from Sand Island, Midway Atoll, Northwestern Hawaiian Islands, to estimate average times necessary to complete various components of the sampling process under each of the different designs; these times were used as seed data in our simulations. We broke the sampling process down as follows: Sampling under the different designs consisted of various combinations of (1) marking previously unmarked birds; (2) "recapturing" birds last captured in a previous year (all designs) or in the present year (RD only), which may involve simply reading and writing down a field-readable band number; (3) tallying birds, and (4) traveling between birds. Banding unbanded birds involved some combination of affixing a metal band, affixing an additional field-readable band, and adding a temporary mark. Recapturing birds involved some combination of reading a metal band, reading a field-readable band, and adding a temporary mark. For example, under the LIRD with field-readable bands, 2 investigators would spend on average approximately 5 min banding a previously unbanded Laysan albatross, including 3 min affixing a metal band, 1 min affixing a field-readable band, 0.5 min adding a temporary mark, and 0.5 min traveling between birds. More effort was needed for the less docile black-footed albatross. The average numbers of minutes spent by 2 investigators per individual bird for different types of sampling activities are recorded in Table 1.

We then used these numbers to develop our simulation parameters. We considered a fixed amount of effort (total minutes spent by 2 investigators in a sampling season) of 60 h, or 3600 min, per year. A fixed effort of 60 h per year allowed us to simulate constant effort throughout the length of the study; if we had used a greater fixed effort we would not have been able to keep it constant over the length

Table 1 Mean times necessary, in minutes, for 2 investigators to complete different types of sampling activities on an individual bird under the different study designs considered, for both Laysan and black-footed albatrosses

		Minutes to complete activity	
Design[a]	Capture type[b]	Laysan albatross	Black-footed albatross
nRD m	New	3.5	5
	Recapture	2	2.5
nRD fr	New	4.5	6.5
	Recapture	0.5	0.5
LIRD m	New	4	5.6
	Recapture	2.5	3.1
	Tally	0.325	0.4
LIRD fr	New	5	7.1
	Recapture	0.75	0.8
	Tally	0.325	0.4
RD m	New	3.5	5
	Recapture	2	2.5
RD fr	New	4.5	6.5
	Recapture	0.5	0.5

[a]Designs include no robust design without (nRD m) or with (nRD fr) a field-readable band, the less-invasive robust design without (LIRD m) or with (LIRD fr) a field-readable band, and the traditional robust design without (RD m) or with (RD fr) a field-readable band.
[b]A new capture is a bird that was previously unmarked. A recapture under the LIRD is a bird captured in a previous year and recaptured in the present year. A recapture under the RD includes both recaptures from a previous year and recaptures within the present year.

of the simulated study because the total possible effort expended in later years would have been lower than the total effort allotted. Next, we determined expected detection probabilities given this effort, assuming an observable population of 5000 birds (i.e., half the total simulated population of 10,000 birds). For example, for the first variation (5% of population in tally) of the LIRD design with metal bands for Laysan albatross, in the first year, the detection probability (p_1) can be determined by solving the equation

$$\mathrm{E}_T = N \times p_1 \times \mathrm{E}_n + N \times p_{tal} \times \mathrm{E}_{tal}$$

for p_1, where E_T is the total available effort (3600 min), N is the observable population size (5000), p_{tal} is the proportion of the total population in the tally (0.05), E_n is the effort required to process a new capture of a single bird under this design (4 min), and E_{tal} is the effort required to tally a single bird under this design (0.325 min). Solving for p_1, we find that we would expect a detection probability of approximately 0.176 in the first year. In subsequent years, detection probability increases as some birds that are encountered will already have metal bands and captures of these individuals will require less effort (i.e., an additional component will be added to the equation, the previously marked population multiplied by the detection probability p_1 multiplied by the effort to recapture previously marked individuals), so a greater proportion of the population may be captured for the same effort. In this way, detection probabilities were calculated for each simulation set in each year.

2.3 Simulation Methods

We simulated data sets in a SAS data step (SAS 9.1, SAS Institute, Cary, North Carolina, USA) and used Program MARK (White and Burnham 1999) to obtain parameter estimates. We assumed a constant population size of 10,000 individuals (i.e., releases of 10,000 in year 1, releases of 500 in subsequent years to make up for a mortality probability of 5%). We used a Monte Carlo step to assign approximately half of releases to the observable state, and additional Monte Carlo steps for survival and transitions between states each year.

A Monte Carlo step was also used to simulate the capture of individuals, and thus to compile encounter histories consisting of a series of "1"s and "0"s to denote capture or non-capture, respectively, in a particular year or secondary occasion within a year. In the nRD simulations, there were an equal number of capture occasions as years. These data were analyzed under the multi-state data type in Program MARK. In the RD designs, there were 2, 4, or 6 capture occasions per year, each with the same capture probability within a year, and these data were analyzed under the multi-state robust design data type in Program MARK. Finally, for the LIRD, we devised a strategy for simulating data so that it could also be analyzed in Program MARK (code can be obtained by contacting the first author).

We completed 1000 simulations of each simulation set. We calculated the average bias and variance in estimators of survival probability (S) and the 2 transition parameters between states (Ψ_t^{OU}, Ψ_t^{UO}) and used these to calculate average root mean square error $\left(\text{MSE} = \sqrt{\text{bias}^2 + \text{variance}}\right)$ on each of these 3 parameters. For the time-constant estimation model, comparative results between the different designs were generally similar for the 3 different demographic parameters (S, Ψ^{OU}, and Ψ^{UO}), so we present graphical results for the survival parameter only.

For the time-dependent models, we calculated the average bias and variance in the estimators across the first 4 estimates (corresponding to the first 4 years) and across all of the identifiable estimates (i.e., 8, except for transitions out of the unobservable state, where there were 7). We also computed average confidence interval coverage (95% nominal) on these parameters. Confidence intervals computed in Program MARK are computed based on the standard error on the transformed scale (in this case, with a sine transformation) and then back-transformed to the real scale to ensure that they fall on the interval 0–1.

3 Results

With the time-constant model, MSE on S was between 3 and 5% for all variations of the RD and LIRD and the nRD based on Laysan albatross over 5 years (top panel, Fig. 1) and between 4 and 8% for all designs based on black-footed albatross over 5 years (top panel, Fig. 2). Over 10 years, these values were much lower: between 0.3 and 1% for Laysan albatross (bottom panel, Fig. 1) and between 1 and 2% for black-footed albatross (bottom panel, Fig. 2).

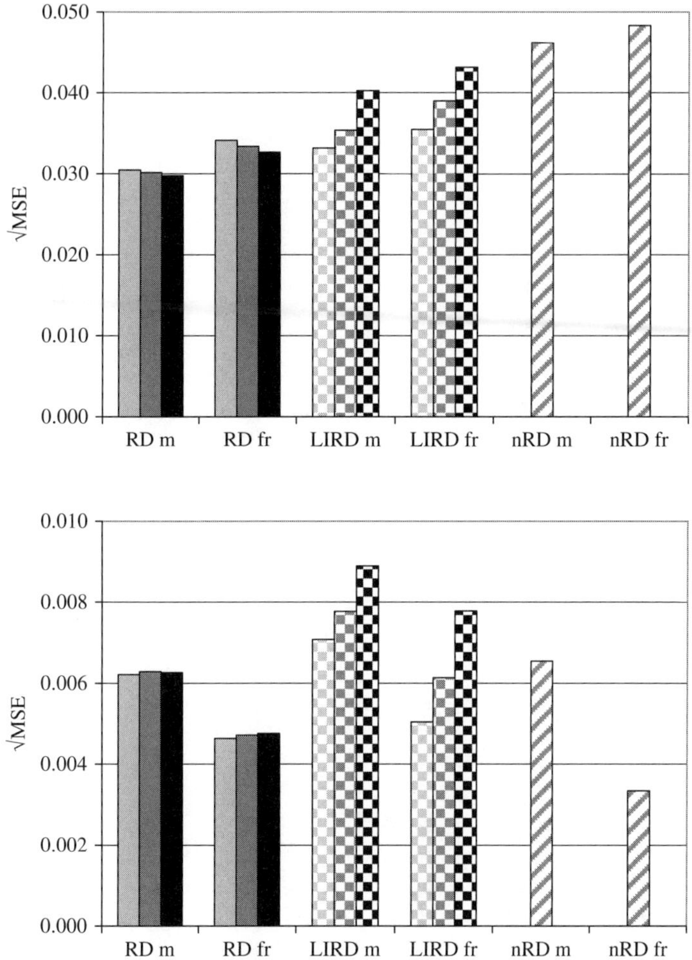

Fig. 1 Root mean square error (MSE) of survival (*S*) for study designs of Laysan albatross over 5 (*top panel*) and 10 (*bottom panel*) years, based on constant survival and transitions over years. Designs include the robust design (*solid*) using metal (RD m) and field-readable bands (RD fr), the less-invasive robust design (*checked*) using metal (LIRD m) and field-readable bands (LIRD fr), and non-robust design (*striped*) using metal (nRD m) and field-readable bands (nRD fr). Within the RD, light bars are 2, medium are 4, and dark are 6 secondary occasions per year. Within the LIRD, light bars are 5%, medium are 25%, and dark are 45% of the population in the tally. Note the difference in scale between the panels

For transition Ψ^{OU} with the time-constant model, MSE was between 2 and 4% for Laysan albatross and between 4 and 7% for black-footed albatross over 5 years. Over 10 years, these values were between 1 and 2% for Laysan albatross and between 2 and 3% for black-footed albatross. For transition Ψ^{UO}, MSE was between 8 and 13% for Laysan albatross and between 11 and 21% for black-footed

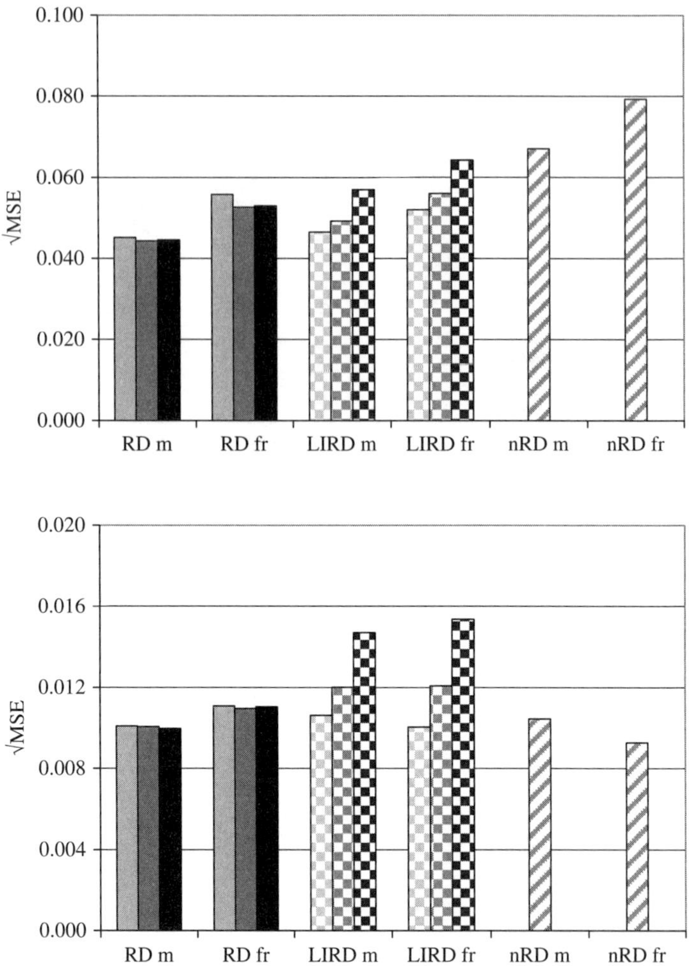

Fig. 2 Root mean square error (MSE) of survival (S) for study designs of black-footed albatross over 5 (*top panel*) and 10 (*bottom panel*) years, based on constant survival and transitions over years. Designs include the robust design (*solid*) using metal (RD m) and field-readable bands (RD fr), the less-invasive robust design (*checked*) using metal (LIRD m) and field-readable bands (LIRD fr), and non-robust design (*striped*) using metal (nRD m) and field-readable bands (nRD fr). Within the RD, light bars are 2, medium are 4, and dark are 6 secondary occasions per year. Within the LIRD, light bars are 5%, medium are 25%, and dark are 45% of the population in the tally. Note the difference in scale between the panels

albatross over 5 years. Over 10 years, these values were between 2 and 5% for Laysan albatross and 5 and 7% for black-footed albatross.

Confidence interval coverage was generally quite close to the nominal value of 95% (i.e., between 93 and 97%) for the time-constant model, with a few exceptions. For S over 5 years, coverage was much below nominal for both Laysan and black-footed albatrosses (top panels, Figs. 3 and 4). Poor coverage in these cases was

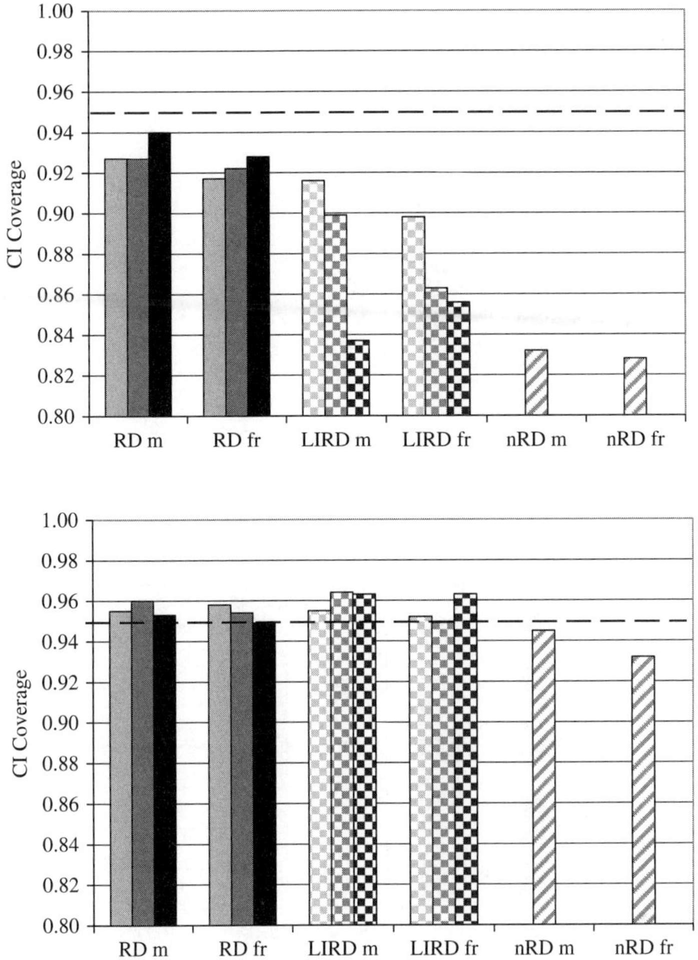

Fig. 3 Confidence interval (CI) coverage (95% nominal – *dashed line*) of survival (*S*) for study designs of Laysan albatross over 5 (*top panel*) and 10 (*bottom panel*) years, based on constant survival and transitions over years. Designs include the robust design (*solid*) using metal (RD m) and field-readable bands (RD fr), the less-invasive robust design (*checked*) using metal (LIRD m) and field-readable bands (LIRD fr), and non-robust design (*striped*) using metal (nRD m) and field-readable bands (nRD fr). Within the RD, light bars are 2, medium are 4, and dark are 6 secondary occasions per year. Within the LIRD, light bars are 5%, medium are 25%, and dark are 45% of the population in the tally

likely related to the fact that the true parameter value was near the maximum value of 1.0. Also, for transition Ψ^{UO} over 5 years for black-footed albatross, coverage was also somewhat below nominal.

In all cases, the LIRD performed best, in terms of MSE, when the proportion of the population in the tally was lowest (i.e., 5% of the population sampled in the tally). This was true for simulation sets of both 5 and 10 years, but in simulations of 10 years, designs with a greater proportion of the population in the tally (i.e.,

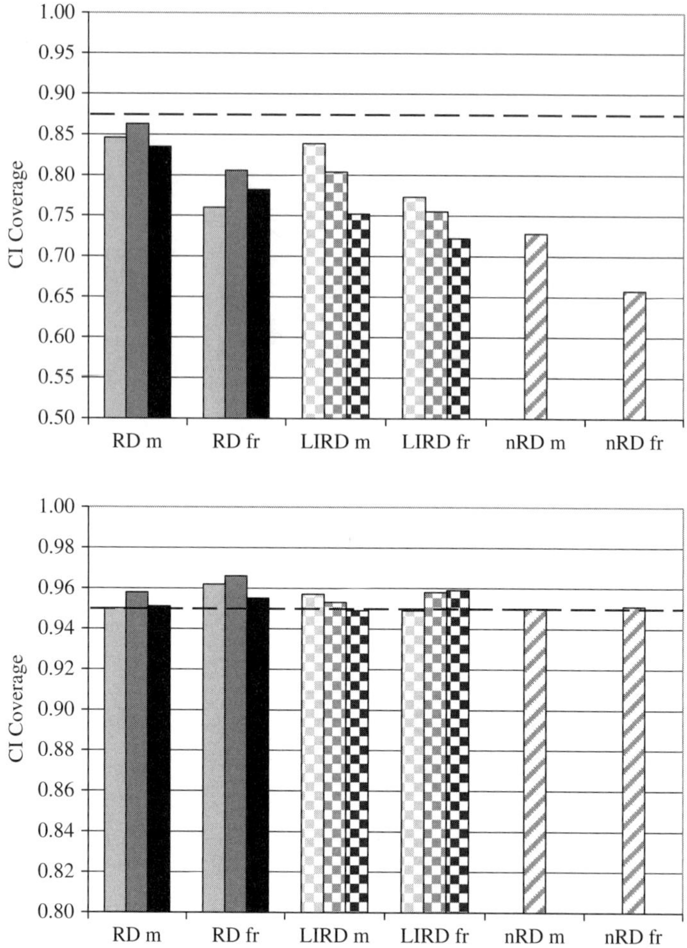

Fig. 4 Confidence interval (CI) coverage (95% nominal – *dashed line*) of survival (*S*) for study designs of black-footed albatross over 5 (*top panel*) and 10 (*bottom panel*) years, based on constant survival and transitions over years. Designs include the robust design (*solid*) using metal (RD m) and field-readable bands (RD fr), the less-invasive robust design (*checked*) using metal (LIRD m) and field-readable bands (LIRD fr), and non-robust design (*striped*) using metal (nRD m) and field-readable bands (nRD fr). Within the RD, light bars are 2, medium are 4, and dark are 6 secondary occasions per year. Within the LIRD, light bars are 5%, medium are 25%, and dark are 45% of the population in the tally. Note the difference in scale between the panels

45%) faired even worse, with a greater difference between the performance of the strongest and weakest of the LIRD variations. Based on additional *a posteriori* simulations of the LIRD for Laysan albatross (with metal bands only, for 10 years) we found that the lowest average MSE over the 3 demographic parameters in the time-constant estimation model (S, Ψ^{OU}, Ψ^{UO}) occurred with 9% of the population included in the tally. In this case, the tally accounted for only 4.1% of the total sampling effort.

Within the RD, we found little difference between 2, 4, and 6 sampling occasions in terms of MSE, and no consistent pattern. While overall detection probability (p^*, the probability that an individual is captured at least once during a sampling year) is greater with 2 sampling occasions (because less effort is allocated to recapturing individuals that were previously captured in that year) overall sample size of captures is greater with 6 sampling occasions. For example, given RD with field-readable bands for Laysan albatross, the average p^* over 10 years for 2 secondary occasions was 0.286, the total number of individuals in the data set was 9067, and the total number of detections was 15,780. For 4 secondary occasions, these numbers were 0.280, 8951, and 16,147, respectively; and for 6 secondary occasions these numbers were 0.278, 8914, and 16,272, respectively.

For the 5-year simulation studies of both species, the MSE on all 3 parameters of interest was lowest for the RD variations with metal bands. Within the Laysan albatross simulations, the next lowest mean square errors on the 3 parameters occurred with the RD variations with field-readable bands or with the LIRD with 5% of the population in the tally, with metal bands. However, with the black-footed albatross, the next lowest mean square errors always occurred with the LIRD with 5 and 25% of the population in the tally, with metal bands. In all cases, the nRD simulations fared worst in terms of MSE.

In the 10-year simulation studies, the simulations including field-readable bands performed better in terms of MSE than they did in the 5-year studies. For the Laysan albatross, the nRD with field-readable bands performed best for all 3 parameters, followed by the RD with field-readable bands, and the LIRD variation 1 (5% in tally) with field-readable bands, while the LIRD variation 3 (45% in the tally) with metal bands performed worst. For the black-footed albatross simulations, designs integrating metal bands continued to perform well. For survival, the nRD with field-readable bands performed best followed by the RD with metal bands. For the 2 transition parameters, the smallest MSE was found for the RD with metal bands along with the LIRD variation 1 (5% in the tally) with both metal and field-readable bands. For black-footed albatross, the LIRD variation 3 (45% in the tally) with field-readable bands performed worst in the 10-year simulations in terms of MSE.

In the time-dependent model, MSE on S, Ψ^{OU}, and Ψ^{UO} was much higher with the nRD designs than with the RD and LIRD designs. For example, for survival, MSE was between 0.045 and 0.049 for the RD and LIRD, but was between 0.216 and 0.233 for the nRD, when averaged over the first 4 estimable parameters (Fig. 5). If all 8 estimable parameters were considered, the averages were between 0.048 and 0.062 for the RD and LIRD, but were between 0.872 and 0.973 for the nRD. This difference in performance was even greater for transition parameters (Figs. 6 and 7).

In the time-dependent estimation model, confidence interval coverage was poor (i.e., coverage was always less than 75%) for the nRD designs. For the RD and LIRD designs, confidence interval coverage was near nominal for the transition parameters, but was poor for survival (Fig. 8).

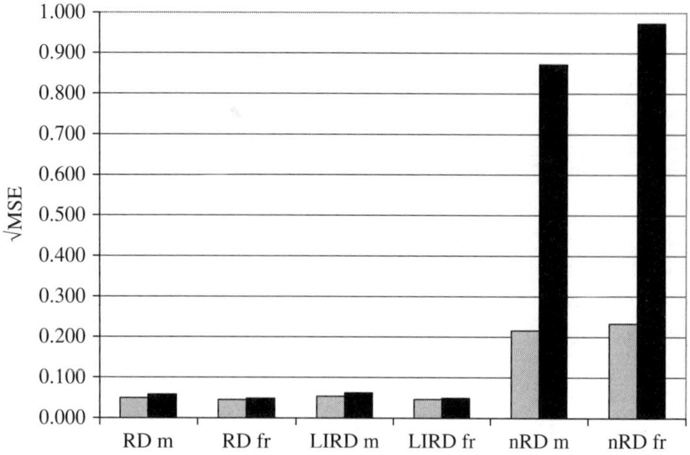

Fig. 5 Root mean square error (MSE) of survival (*S*) for study designs of Laysan albatross over 10 years, based on time-dependent survival and transitions over years. Designs include the robust design using metal (RD m) and field-readable bands (RD fr), the less-invasive robust design using metal (LIRD m) and field-readable bands (LIRD fr), and non-robust design using metal (nRD m) and field-readable bands (nRD fr). The gray bars represent the average over the first 4 estimable parameters, and the black bars over all 8 estimable parameters

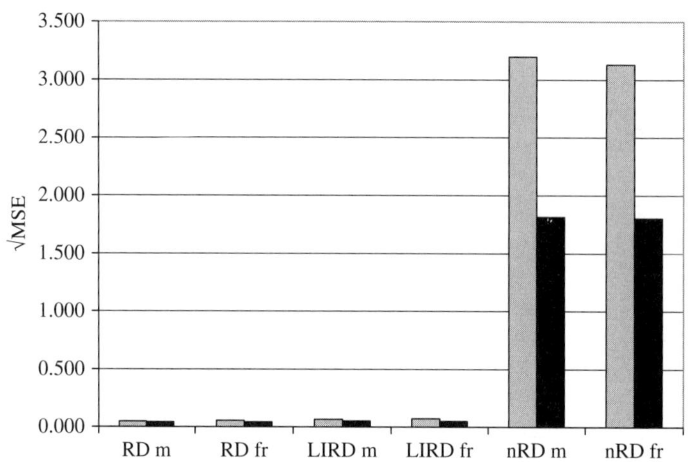

Fig. 6 Root mean square error (MSE) of transition Ψ^{OU} for study designs of Laysan albatross over 10 years, based on time-dependent survival and transitions over years. Designs include the robust design using metal (RD m) and field-readable bands (RD fr), the less-invasive robust design using metal (LIRD m) and field-readable bands (LIRD fr), and non-robust design using metal (nRD m) and field-readable bands (nRD fr). The gray bars represent the average over the first 4 estimable parameters, and the black bars over all 8 estimable parameters

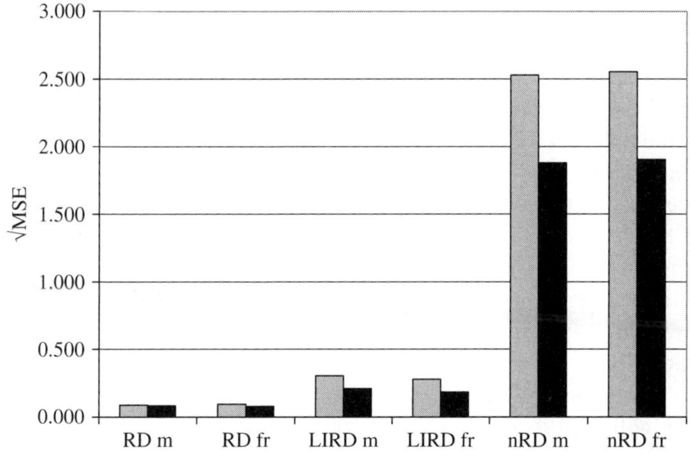

Fig. 7 Root mean square error (MSE) of transition Ψ^{UO} for study designs of Laysan albatross over 10 years, based on time-dependent survival and transitions over years. Designs include the robust design using metal (RD m) and field-readable bands (RD fr), the less-invasive robust design using metal (LIRD m) and field-readable bands (LIRD fr), and non-robust design using metal (nRD m) and field-readable bands (nRD fr). The gray bars represent the average over the first 4 estimable parameters, and the black bars over all 7 estimable parameters

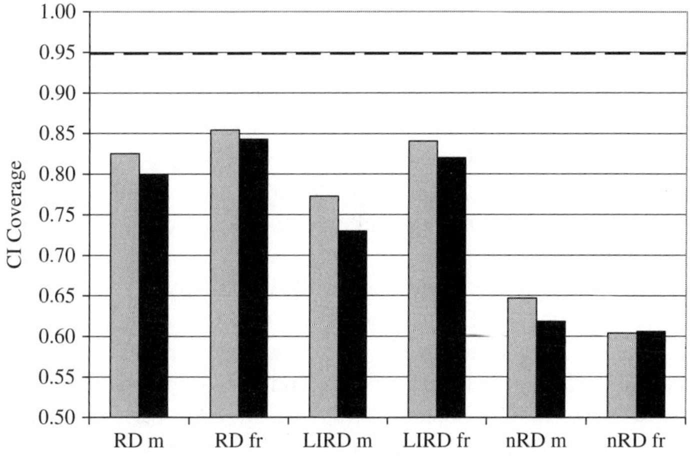

Fig. 8 Confidence interval (CI) coverage (95% nominal – *dashed line*) of survival (*S*) for study designs of Laysan albatross over 10 years, based on time-dependent survival and transitions over years. Designs include the robust design using metal (RD m) and field-readable bands (RD fr), the less-invasive robust design using metal (LIRD m) and field-readable bands (LIRD fr), and non-robust design using metal (nRD m) and field-readable bands (nRD fr). The gray bars represent the average over the first 4 estimable parameters and the black bars over all 8 estimable parameters

4 Discussion

Overall, we found that the LIRD could, in some cases, perform as well as the traditional RD. Designs where greater effort was placed on the first sampling occasion, versus on the tally, showed improved performance for both species in terms of MSE on the estimates for both 5- and 10-year studies. Such designs also had improved confidence interval coverage, especially noticeable in the 5-year simulations.

These results indicate that the LIRD can be a viable study design for albatrosses, most likely if the large majority of the effort is placed on new within-year captures, rather than on tallying the marked-to-unmarked count. The premium on new within-year captures was even greater in longer-term studies (i.e, 10-year studies) likely because more new within-year captures translate to a greater number of newly marked individuals over time, and a greater sample size of marked individuals will tend to increase estimator performance, especially for long-lived animals.

We developed a method for fitting the LIRD in Program MARK which was, while possible, not straightforward. The capability to analyze data under the LIRD has also been added to program MSSRVRD (http://www.mbr-pwrc.usgs.gov/software), which was originally developed to analyze data under the multistate RD.

With the RD, there was not a clear benefit of few versus many secondary occasions. While the total number of individuals in the data set decreases with an increase in the number of secondary occasions, the total number of detections increases. This is because recaptures are less costly in our simulations, in terms of effort, than first captures. With a greater number of secondary occasions, more effort is allocated to less costly recaptures, thus allowing a greater total number of detections. The tradeoff between the total number of individuals and total number of detections will depend on the relative costs of first captures versus recaptures and the length of the study. In our case, we found that, for 5-year studies, 4 or 6 occasions were almost always optimal, while for 10-year studies, 2 occasions were more frequently optimal. Again, this reflects the increased benefit of first captures in longer-term studies.

In the RD, we assumed a model without heterogeneity in capture probabilities across individuals, which can be fit in an unbiased fashion with 2, 4, or 6 occasions. However, if there is heterogeneity, it is necessary to have more secondary occasions to fit it well. In addition, if there are more complicated movement factors in the model, such as, in colonially-breeding birds, late arrival or early departure from the nesting area, the open robust design model is a more appropriate choice, and this model benefits from more time periods (Kendall and Bjorkland 2001).

We found that field-readable bands were also more valuable in longer-term studies, where the increased effort placed on new captures early in the study has greater opportunity to pay off in lower effort placed on future recaptures. This was especially true for the simulations based on Laysan albatross, which require less effort to band. Field-readable bands used with black-footed albatross would presumably show a greater benefit in even longer-term studies. The LIRD will probably be more attractive to investigators who do not have the option of using field-readable bands. Recapture effort, and disturbance to individuals, in albatross studies with field-readable bands may be so much reduced that the benefits of a true robust design or an open robust design; (Kendall and Bjorkland 2001) are more easily afforded. However, as Kendall et al. (in prep) note, there are some situations where field-readable bands are not a viable choice (e.g., burrow-nesting species), and the LIRD could also have many benefits in these cases. It is important to note that while in some cases the RD with metal bands performed best in terms of MSE, this design is not an acceptable option for Hawaiian albatrosses because of the concern about excessive disturbance.

Because we restricted our effort to be equal over time, by necessity detection probabilities in our simulations started out fairly low and increased over time, as a larger proportion of the population was marked, and so a larger proportion could be captured for the same effort. However, in the Northwestern Hawaiian Islands, investigators have focused on adding field-readable bands to a large majority of individuals early in the study. In future years, this will pay off with increased precision for a lower effort.

In the time-constant models, over longer time periods (i.e., 10 years), the designs without robust design information sometimes fared best (e.g., nRD with field-readable bands performed best for all 3 parameters estimated for Laysan albatross, in terms of MSE). This result likely reflects the increased value in longer-term studies of putting all the effort into marking new birds, rather than recapturing birds. However, for the time-dependent model, there was a substantially larger MSE for the nRD designs. This reflects the increase in bias and variance that occurs in the absence of robust design information with a time-dependent model, due to the fact that parameter identifiability and precision are compromised in the time-dependent model without the robust design (Kendall and Nichols 2002). Strong time constancy, especially in state transition parameters, may be the rule with albatrosses. So we recommend that collection of robust design information always be seriously considered. Robust design sampling can be integrated with other elements of sampling in seabird colonies. For example, nest success monitoring can be combined with robust design sampling of adults. With field-readable bands, this can be more easily accomplished because the presence of a particular individual on a nest can be ascertained by reading the leg band, and the nest status can be determined at the same time.

The approach we took to optimizing study design can be usefully applied in a variety of circumstances. Results will vary for different species, where different amounts of effort would be required to accomplish sampling tasks. Furthermore, as we found in our results, planned study length will also strongly influence the optimal study design. Seed data similar to what we used here for constructing simulations can be gathered in a relatively short amount of time, and can be highly valuable in determining the optimal use of study resources. Simulation code for conducting the simulations described herein, and for analyzing LIRD data in Program MARK, can be obtained by contacting the first author.

Acknowledgments We would like to thank the following individuals who contributed to data collection efforts in the pilot field season at Midway Atoll: Mike Johnson, John Klavitter, Leona Laniawe, Sarah Money, Jill Morgyn, Edith Nonner, Eldridge Park, Thelma Park, Greg Schubert, LeeAnn Woodward, and Rebecca Woodward; and at French Frigate Shoals: Angela Anders, Morgan Gilmour, Brett Hartl, Stefan Kropidlowski, Jennifer Lavers, Frank Mayer, and Paula Shannon. We also thank Gary White for advice about simulation design and Program MARK. Finally, Mike Erwin, Paul Lukacs, Jeff Spendelow, and one anonymous reviewer provided helpful suggestions on the draft manuscript. Financial support for SJC was provided by USFWS Regions 1 and 7 and USGS Patuxent Wildlife Research Center, through the Colorado Cooperative Fish and Wildlife Research Unit (Cooperators: Colorado Division of Wildlife, Colorado State University, USGS Biological Resources Discipline, and the Wildlife Management Institute).

References

Fujiwara M, Caswell H (2002) A general approach to temporary emigration in mark-recapture analysis. Ecology 83:3266–3275.

Kendall WL (2004) Coping with unobservable and mis-classified states in capture–recapture studies. Animal Biodiversity and Conservation 27.1:97–107.

Kendall WL, Bjorkland R (2001) Using open robust design models to estimate temporary emigration from capture–recapture data. Biometrics 57:1113–1122.

Kendall WL, Converse SJ, Doherty PF, Jr, Naughton MB, Anders A, Hines JE, Flint E (in press) Design considerations in demographic studies of animal populations: a case of colonial seabirds. Ecological Applications: in press.

Kendall WL, Nichols JD (2002) Estimating state-transition probabilities for unobservable states using capture–recapture/resighting data. Ecology 83:3276–3284.

Kendall WL, Nichols JD, Hines JE (1997) Estimating temporary emigration using capture–recapture data with Pollock's robust design. Ecology 78:563–578.

Pollock KH (1982) A capture–recapture design robust to unequal probability of capture. Journal of Wildlife Management 46:757–760.

Schaub M, Gimenez O, Schmidt BR, Pradel R (2004) Estimating survival and temporary emigration in the multistate capture–recapture framework. Ecology 85:2107–2113.

Schwarz CJ, Schweigert JF, Arnason AN (1993) Estimating migration s using tag-recovery data. Biometrics 49:177–193.

White GC, Burnham KP (1999) Program MARK: survival estimation from populations of marked animals. Bird Study Supplement 46:120–138.

Non-random Temporary Emigration and the Robust Design: Conditions for Bias at the End of a Time Series

Catherine A. Langtimm

Abstract Deviations from model assumptions in the application of capture–recapture models to real life situations can introduce unknown bias. Understanding the type and magnitude of bias under these conditions is important to interpreting model results. In a robust design analysis of long-term photo-documented sighting histories of the endangered Florida manatee, I found high survival rates, high rates of non-random temporary emigration, significant time-dependence, and a diversity of factors affecting temporary emigration that made it difficult to model emigration in any meaningful fashion. Examination of the time-dependent survival estimates indicated a suspicious drop in survival rates near the end of the time series that persisted when the original capture histories were truncated and reanalyzed under a shorter time frame. Given the wide swings in manatee emigration estimates from year to year, a likely source of bias in survival was the convention to resolve confounding of the last survival probability in a time-dependent model with the last emigration probabilities by setting the last unmeasurable emigration probability equal to the previous year's probability when the equality was actually false. Results of a series of simulations demonstrated that if the unmeasurable temporary emigration probabilities in the last time period were not accurately modeled, an estimation model with significant annual variation in survival probabilities and emigration probabilities produced bias in survival estimates at the end of the study or time series being explored. Furthermore, the bias propagated back in time beyond the last two time periods and the number of years affected varied positively with survival and emigration probabilities. Truncating the data to a shorter time frame and reanalyzing demonstrated that with additional years of data surviving temporary emigrants eventually return and are detected, thus in subsequent analysis unbiased estimates are eventually realized.

Knowing the extent and magnitude of the potential bias can help in making decisions as to what time frame provides the best estimates or the most reliable opportunity to model and test hypotheses about factors affecting survival probability.

C.A. Langtimm (✉)
U.S. Geological Survey, Florida Integrated Science Center, 2201 NW 40th Terrace, Gainesville, FL 32605, USA
e-mail: clangtimm@usgs.gov

D.L. Thomson et al. (eds.), *Modeling Demographic Processes in Marked Populations*, Environmental and Ecological Statistics 3, DOI 10.1007/978-0-387-78151-8_34,

To assess bias, truncating the capture histories to shorter time frames and reanalyzing the data to compare time-specific estimates may help identify spurious effects. Running simulations that mimic the parameter values and movement conditions in the real situation can provide estimates of standardized bias that can be used to identify those annual estimates that are biased to the point where the 95% confidence intervals are inadequate in describing the uncertainty of the estimates.

1 Introduction

Capture–recapture estimates of survival probabilities with the Cormack–Jolly–Seber (CJS) model can be biased in the presence of non-random temporary emigration (Kendall et al. 1997). The extension of the robust design (Pollock 1982) capture–recapture framework to model temporary emigration (Kendall et al. 1997) has improved survival estimation, as well as advanced our ability to answer questions and test hypotheses about factors affecting movement and survival. Nonetheless, deviations from the model assumptions in the application of models to real life situations can introduce unknown bias. Understanding the type and magnitude of the bias under these conditions is important to interpreting model results. Robust estimates of changes in annual rates of survival and movement are critical to both biological and legal assessments of population status.

Here I present the results of a robust design analysis of photo-documented sighting histories of the Florida manatee (*Trichechus manatus latirostris*). Estimates of annual adult survival rates and how they vary temporally are key components in population models used to assess status and recovery of this federally-listed endangered species (Runge et al. 2004). Previous analysis of manatee data for one subpopulation under the standard (CJS) model identified significant annual variation in non-random temporary emigration rates and possible bias in survival estimates, which threw suspicion on a highly ranked model of declining trend in survival (Langtimm et al. 2004). However, annual survival probabilities estimated in this study under a robust design framework that allows for estimation of temporary emigration also showed similar, but less severe, suspect low estimates near the end of the time series, which again either could be due to a real decline or an artifact from biased estimates. To determine the nature of the low estimates, I implemented a series of Monte Carlo simulations of robust design data to mimic the manatee system and identified a source of bias. With additional simulations I explored factors that affect the magnitude and proliferation of the bias back through time and examined approaches to reduce bias.

The Florida manatee's biology and our current monitoring program present what may be an unusual case of high and extremely variable non-random temporary emigration. Delineating bias for special cases, however, can alert others to potential bias in their own systems, and lead to caution and improvements in current applications. Most importantly, understanding the performance of models with regard to annual estimation during long time series is crucial, as long-term studies designed specifically for analysis with capture–recapture models are becoming more common. Furthermore, many long-term monitoring studies focus on large, long-lived endangered or threatened species with life history strategies of relatively constant, high adult survival, which tend to be the most sensitive/influential component of population growth rate.

2 Methods

2.1 Monitoring and Data Acquisition

Annual adult manatee survival rates are estimated from a long-term photo-identification monitoring program where individuals are identified by unique features, most often healed scars acquired from boat strikes (Langtimm et al. 2004). Sighting records, images, and life history information of individuals are entered into the Manatee Individual Photo-identification System (MIPS), a multi-partner database maintained by the United States Geological Survey, the Fish and Wildlife Research Institute of the Florida Fish and Wildlife Conservation Commission, and Mote Marine Laboratory.

Photographs are taken annually at several primary winter aggregation sites. In summer manatees are dispersed along the coast of Florida and the southeastern United States where they forage on seagrass and freshwater aquatic vegetation (Hartman 1979). In winter the manatee's home range contracts to areas near a limited number of natural and industrial warm-water discharge sites, which manatees primarily use when water temperatures drop below 20°C. Manatees are tropical, and at their northern limits in Florida are subject to death and debilitating effects from cold stress during extended periods of cold weather (Buergelt et al. 1984; Bossart et al. 2002). These seasonal movements and mass attraction to warm-water facilitate photo-identification and monitoring.

Casting the biological and sampling system in terms of robust design capture–recapture models, the study areas consist of very small areas in and around the warm-water sites, where photographers have access to manatees through an underwater or telephoto lens. Manatees are not resident to these sites for the entire winter, but move in and out of the sites according to their physiological needs for warm refuge, forage, and fresh drinking water. Manatees beyond the limits of camera range are not in the study area and are not available for capture. With regard to defining what constitutes a temporary emigrant, the manatee situation is analogous to the example presented by Kendall et al. (1997) on white-footed mice and torpor. With the mice, some individuals could be considered "temporary emigrants" during cold periods if they remain underground (potentially in torpor) during trapping occasions. In the manatee case, some individuals could be considered "temporary emigrants" if during warmer winters they do not frequent the sampled high quality, primary refuges and instead rely on unsampled lower quality secondary refuges nearer to seagrass beds, freshwater, or other resources. Other factors could produce temporary emigrants as well, such as changes in warm-water discharges at power plants due to electricity demand and market forces or down-time for maintenance. It has also been proposed that hurricane strikes could displace individual manatees into poorly monitored areas due to storm-generated currents or storm-related loss of habitat, and may account for some of the correlation of lower adult apparent survival rates with years of extreme storms (Langtimm and Beck 2003).

Although variation in temporary emigration is most likely non-random among years (open population among primary periods), the movement and detection of individual manatees using the warm-water sites within a season is most likely a random process (closed population within the primary period). Manatees prone to use warm-water sites in a given year should continually enter those sites during the coldest periods, where field personnel randomly photograph individuals within view. Thus the manatee system should meet the assumptions for analysis with a closed robust design framework.

2.2 Robust Design Analysis

I constructed capture histories for a first analysis employing the CJS model for two regional subpopulations in Florida: Northwest Gulf Coast (NW, 22 years of data) and Atlantic Coast (AC, 18 years of data). These subpopulations are spatially distinct and rely on different types of warm-water refuges. In the NW, large warm-water artesian springs dominate; while on the AC artificial warm-water effluents from power plants are primary refuges. These two regions also differ in population attributes, habitat, and threats to manatees (O'Shea and Langtimm 1995); thus comparative analysis can be useful in discerning if temporary emigration and effects on bias in survival estimates is a potential problem for analysis of all Florida manatee subpopulations. Data selection criteria and protocols are described in Langtimm et al. (1998, 2004). Multiple sightings of individuals during a winter sample are routine, and for this analysis I collapsed the data to represent one sighting per sampling occasion.

I used the goodness-of-fit (GOF) tests of U-Care 2.2 (Choquet et al. 2005) to assess the fit of the data to the most general CJS model. Schaub et al. (2004) demonstrated that subtest 2.Ct has good power to detect Markovian temporary emigration (probability of being an emigrant depends on whether or not an individual was an emigrant on the previous occasion). The significant lack of fit suggested non-random emigration within these 2 subpopulations and led me to use the robust design to model and estimate manatee temporary emigration and survival.

I rearranged the photo-identification data to conform to a robust design by partitioning the multiple-sightings during each winter (primary period) into two secondary surveys of equal length. I applied the multi-state robust design temporary emigration model, with Huggins closed captures, to the new capture histories using Program MARK (White and Burnham 1999; Kendall 2001).

The model is composed of capture probabilities (p), movement probabilities (Ψ) and survival probabilities (S). I used a step-down approach to evaluate each of these parameters (Lebreton et al. 1992), first keeping temporary emigration and survival general, I determined the top model among candidate capture probability models and then using that capture probability structure, I compared each of the candidate models of temporary emigration and finally of survival probabilities. For within period capture probabilities, I allowed p to vary between the two surveys and set $p_t = c_t$ (where p is the probability of initial capture and c is the probability of recapture). I then evaluated two models of variation in capture probabilities among the primary periods; one constant ($p.*$) and the other time specific (p_t*).

Temporary emigration in this model is further specified as two parameters conditional on the state of a given animal at time t: Ψ_t^{OU} = the probability that an animal observable in the study area in period t emigrates out of the study area and is unobservable for period $t + 1$; Ψ_t^{UU} = the probability that an emigrant out of the study area in period t and unobservable remains away from the study area and is unobservable in period $t + 1$. Temporary emigration is random when $\Psi^{OU} = \Psi^{UU}$ (Kendall et al. 1997); the larger the difference between Ψ^{OU} and Ψ^{UU}, the greater the Markovian effect. I evaluated the following temporary emigration models: (1) no temporary emigration ($\Psi^{OU} = \Psi^{UU} = 0$), (2) time-invariant random temporary emigration ($\Psi^{OU} = \Psi^{UU}$), (3) time-specific random temporary emigration ($\Psi_t^{OU} = \Psi^{UU}$), (4) time-invariant Markovian emigration (Ψ^{OU}, Ψ^{UU}), and (5–7) three variants of time-specific Markovian emigration (Ψ_t^{OU}, Ψ_t^{UU}), (Ψ^{OU}, Ψ_t^{UU}), (Ψ_t^{OU}, Ψ^{UU}). When Markovian temporary emigration is time-specific, the last estimate of Ψ^{OU} and Ψ^{UU} are confounded with survival in models with time-specific survival probabilities. Following Kendall et al. (1997), I set the final two emigration probabilities as equal ($\Psi_T^{OU} = \Psi_{T-1}^{OU}$ and $\Psi_T^{UU} = \Psi_{T-1}^{UU}$) to resolve this confounding. Alternatively, Kendall (2006) recommends constraining the last emigration estimates to be functions of one or more covariates that would better predict what the unmeasurable temporary emigration probabilities were between the last 2 primary periods. Although some hypotheses about factors affecting manatee temporary emigration have been proposed, in this analysis I elected to use the time-specific model results for an exploratory examination of patterns and to identify possible covariates for use in future analyses.

I evaluated two models of survival (S) – one time-specific (S_t) and the other time-invariant ($S.$). The time-specific model is important because it is the basis for estimates of temporal variance (Langtimm et al. 2004) necessary for the construction of manatee population models used in assessments of population dynamics and status (Runge et al. 2004). Under the multi-state robust design, survival probabilities for unobservable and observable animals were set equal ($S_{ij}^U = S_{ij}^O$, Kendall and Nichols 2002).

I used Akaike's Information Criterion adjusted for small sample size (AICc,) and normalized Akaike weights to evaluate the evidence for a given model relative to all the other candidate models (Burnham and Anderson 1998). Maximum-likelihood was used for estimation of parameters and their standard errors.

2.3 Truncation of Data to Test for Bias

Examination of the annual survival point estimates indicated a suspicious drop in survival near the end of the 20 year study under the time dependent survival model for both the CJS and robust design approach. To test for a spurious effect, I right truncated the capture histories to span fewer years of data, reran the best models under the robust design model, and generated new survival estimates for comparison to estimates in the first analysis. By truncating the capture histories, estimates for years in the middle of the original time series, now occurred at the end. If the model was accurately estimating annual survival probabilities, the point estimate for any given year in either the full or truncated datasets should be similar regardless of its location at the end or middle of the time series. A large change in a point estimate when moved to the end of the series could indicate some source of bias affecting the analysis.

2.4 Monte Carlo Simulations

To examine expected bias and precision of parameter estimates of the robust design models I conducted Monte Carlo simulations, using the simulation option in Program MARK (Cooch and White 2006). I used time-invariant robust design models to generate capture histories to study the conditions I had identified in the manatee system. By constraining the underlying values to be constant, it made it easier to examine bias under time-dependent estimation models, which I found to be important in previous analyses of manatee data (Langtimm et al. 1998, 2004). The models and values to generate data (Tables 1–2) included high annual survival rate and significant annual variation in Markovian temporary emigration. For all true models I considered 12 primary periods with 2 secondary surveys with 500 newly released individuals at each capture occasion (a total of 24 capture occasions). I held survival probabilities constant ($S.$) and capture probabilities constant ($p^* = 0.5$) and modeled temporary emigration as Markovian with $\Psi^{OU} < \Psi^{UU}$. In one scenario I held emigration probabilities constant over time (Ψ^{OU},Ψ^{UU}); in another set of scenarios I mimicked significant annual variation in migration probabilities by alternating between two sets of values for Ψ^{OU} and for Ψ^{UU}. All generated capture histories from the true models were evaluated with the estimation model S_t, Ψ_t^{OU}, Ψ_t^{UU}, $p^*..$, with survival probabilities for unobservable and observable animals set equal ($S_{ij}^U = S_{ij}^O$) Kendall and Nichols (2002). Because the last emigration probabilities were confounded with the last survival probability under the time-dependent model, following Kendall et al. (1997) I constrained the final two emigration probabilities as equal ($\Psi_T^{OU} = \Psi_{T-1}^{OU}$ and $\Psi_T^{UU} = \Psi_{T-1}^{UU}$). I also evaluated additional estimation models, in which no or alternative constraints were placed on the last annual emigration probabilities.

I calculated absolute bias (difference of parameter estimate from the true value) and coefficient of variation (CV) for the annual survival estimates. I also calculated the standardized bias (the ratio of absolute bias/standard error) to evaluate the expected coverage of the estimated 95% confidence interval under the estimated bias. If the standardized bias is ≤ 0.5, the effect on coverage of the 95% confidence intervals is negligible (Cochran 1963:14, Burnham et al. 1987:284).

Table 1 Bias and coefficient of variation (CV) in annual survival probabilities for the robust design model of time-dependent survival and time-dependent Markovian temporary emigration, S_t, Ψ_t^{OU}, Ψ_t^{UU}, $c_t = p_t$, where $\Psi_T^{OU} = \Psi_{T-1}^{OU}$ and $\Psi_T^{UU} = \Psi_{T-1}^{UU}$. The model is evaluated for data generated with Monte Carlo simulations of constant capture probability of 0.5 and under different conditions of constant survival and time-dependent Markovian temporary emigration. In scenarios with time-dependent temporary emigration temporary, Ψ^{OU} and Ψ^{UU} were alternated over time between the two values listed, with the final time period ending with the higher value. Absolute bias = mean estimate –true value, standardized bias = absolute bias/standard error (Burnham et al. 1987:215), %CV = mean SE/mean estimate X 100. Values in **bold** highlight standardized bias greater than 0.50 where expected coverage of the 95% confidence interval would be reduced to below 93%

	Scenario 1: constant temporary emigration S= 0.95 $\Psi^{OU} = 0.10$ $\Psi^{UU} = 0.60$ replicates = 500			Scenario 2: time dependent temporary emigration $S = 0.95$ $\Psi^{OU} = 0.30, 0.50$ $\Psi^{UU} = 0.50, 0.75$ replicates = 500			Scenario 3: time dependent temporary emigration $S = 0.70$ $\Psi^{OU} = 0.30, 0.50$ $\Psi^{UU} = 0.50, 0.75$ replicates = 1000			Scenario 4: time dependent temporary emigration $S = 0.50$ $\Psi^{OU} = 0.30, 0.50$ $\Psi^{UU} = 0.50, 0.75$ replicates = 1000		
S year	Absolute bias	Standardized bias	%CV	Absolute bias	Standardized bias	%CV	Absolute bias	Standardized bias	%CV	Absolute bias	Standardized bias	%CV
1	0.000	0.000	1.8	0.002	0.111	1.9	0.001	0.021	6.7	0.006	0.086	13.8
2	0.001	0.077	1.4	0.000	0.000	1.4	0.001	0.030	4.7	0.001	0.023	8.8
3	0.000	0.000	1.3	0.000	0.000	1.3	0.000	0.000	4.9	–0.001	–0.019	10.6
4	0.000	0.000	1.2	0.000	0.000	1.1	–0.001	–0.036	4.0	0.001	0.024	8.2
5	0.000	0.000	1.1	–0.002	–0.182	1.2	0.000	0.000	4.4	0.001	0.020	10.2
6	0.000	0.000	1.2	–0.001	–0.100	1.1	0.001	0.037	3.9	0.001	0.024	8.2
7	0.000	0.000	1.3	–0.012	**–1.000**	1.3	–0.008	–0.258	4.5	0.003	0.051	11.7
8	0.000	0.000	1.7	–0.008	**–0.667**	1.3	–0.004	–0.143	4.0	0.000	0.000	9.8
9	0.002	0.080	2.6	–0.057	**–2.850**	2.2	–0.032	**–0.653**	7.3	0.018	0.133	26.1
10	0.001	0.032	3.3	–0.046	**–2.556**	2.0	–0.031	**–0.939**	4.9	–0.009	–0.158	11.6
11	0.000	0.000	3.2	–0.248	**–13.053**	2.7	–0.180	**–6.429**	5.4	–0.119	**–2.644**	11.8

Table 2 Additional comparisons of bias and coefficient of variation (CV) in annual survival probabilities for the robust design model of time-dependent survival and time-dependent Markovian temporary emigration, S_t, Ψ_t^{OU}, Ψ_t^{UU}, $c_t = p_t$, where $\Psi_T^{OU} = \Psi_{T-1}^{OU}$ and $\Psi_T^{UU} = \Psi_{T-1}^{UU}$. The model is evaluated for data generated with Monte Carlo simulations of constant capture probability of 0.5, constant survival of 0.95, and additional conditions of Markovian temporary emigration. In scenarios with time-dependent temporary emigration Ψ^{OU} and Ψ^{UU} were alternated over time between the two values listed. In Scenario 5, the magnitude of the Markovian effect is reduced compared to Scenario 2 (Table 1) by lowering the difference between the values of Ψ^{OU} and Ψ^{UU}. In Scenario 6, Markovian temporary emigration alternates over time between two values, but the final time period ends with the lower rather than higher values. Absolute bias = mean estimate –true value, standardized bias=absolute bias/mean SE of the estimate (Burnham et al. 1987:215), %CV = mean SE/mean estimate X 100. Values in **bold** highlight standardized bias greater than 0.50 where confidence interval coverage should be adversely affected

	Scenario 5: lesser Markovian effect $\Psi^{OU} = 0.20, 0.30$ $\Psi^{UU} = 0.30, 0.40$ replicates = 500			Scenario 6: time dependent temporary emigration with low emigration at T $\Psi^{OU} = 0.50, 0.30$ $\Psi^{UU} = 0.75, 0.50$ replicates = 500			Scenario 7: probability of becoming a new emigration constant, probability of remaining an emigrant time dependent $\Psi^{OU} = 0.10$ $\Psi^{UU} = 0.50, 0.75$ replicates = 500		
S year	Absolute bias	Standardized bias	%CV	Absolute bias	Standardized bias	%CV	Absolute bias	Standardized bias	%CV
1	–0.002	–0.118	1.8	–0.001	–0.063	1.7	0.000	0.000	1.6
2	0.001	0.083	1.3	–0.001	–0.071	1.5	–0.001	–0.091	1.2
3	0.001	0.091	1.2	–0.003	–0.273	1.2	0.000	0.000	0.9
4	0.000	0.000	1.1	–0.003	–0.273	1.2	–0.001	–0.111	0.9
5	0.001	0.111	0.9	–0.002	–0.200	1.1	–0.002	–0.250	0.8
6	–0.001	–0.111	0.9	–0.005	–0.455	1.2	–0.004	–0.500	0.8
7	0.000	0.000	0.9	–0.005	–0.500	1.1	–0.005	**–0.625**	0.8
8	0.000	0.000	0.9	–0.023	**–1.769**	1.4	–0.011	**–1.100**	1.1
9	0.000	0.000	1.6	–0.011	**–0.733**	1.6	–0.014	**–1.167**	1.3
10	–0.004	–0.222	1.9	–0.086	**–5.375**	1.9	–0.029	**–1.933**	1.6
11	–0.120	**–6.316**	2.3	0.048	–[a]	0.0	–0.021	**–1.500**	1.5

[a]SE = 0 as an artifact of S_{11} estimated at 1.0, therefore standard bias cannot be estimated.

3 Results

3.1 Robust Design Model Outcomes

The global GOF test for the CJS model was significant for both regional subpopulations (NW: $\chi^2_{98} = 143.066$, P = 0.002, $\hat{c} = 1.6$; AC: $\chi^2_{136} = 342.032$, P < 0.0001, $\hat{c} = 2.5$). Subtest 2.Ct was significant for both the subpopulations (NW: $\chi^2_{20} = 46.516$, P = 0.0007; AC: $\chi^2_{17} = 144.391$, P < 0.0001), indicating non-random temporary emigration as a probable cause for the lack of fit. The variance inflation factor calculated from the test 2.Ct results showed a larger effect of temporary emigration on GOF for the AC ($\hat{c} = 8.5$) as compared to the NW ($\hat{c} = 2.3$).

Non-random temporary emigration was substantiated with the model selection results in the robust-design analysis for both regions, suggesting a pattern of emigration common to the Florida manatee. For both regions, the best models included Markovian movement (temporary emigration). There was no support for random movement (Akaike weights <0.00001) or permanent emigration (i.e. models with no temporary emigration, Akaike weights <0.00001). For the AC, the best Markovian model included significant annual variation in probabilities of being out of the study area for both new emigrants (Ψ^{OU}) and emigrants of the previous year (Ψ^{UU}) (Akaike weight = 0.735). For the NW however, variation was negligible for new emigrants, but significant for emigrants of the previous year (Akaike weight = 0.999). Examination of the annual point estimates of Ψ^{OU} and Ψ^{UU} (Figs. 1 and 2) shows several interesting patterns. In both regions, generally the differences in

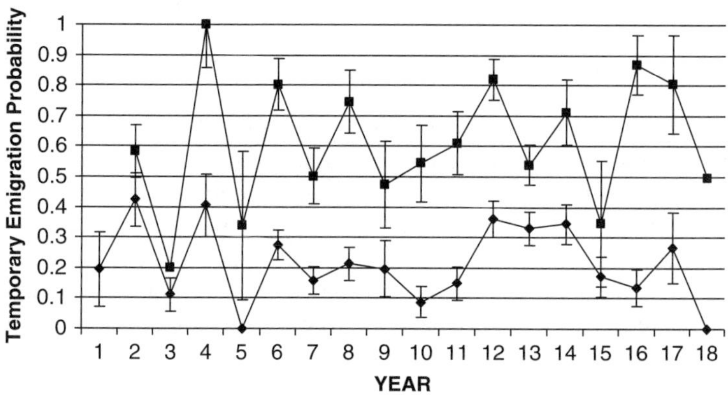

Fig. 1 Annual probabilities of Ψ_t^{OU} (♦) and Ψ_t^{UU} (■) and SE for Florida manatees on the Atlantic Coast of Florida estimated under the robust design model S_t, Ψ_t^{OU}, Ψ_t^{UU}, $c_t = p_t$, where $\Psi_T^{OU} = \Psi_{T-1}^{OU}$ and $\Psi_T^{UU} = \Psi_{T-1}^{UU}$ (T = 19 years, Akaike weight = 0.735). Ψ_t^{OU} = the probability that an animal observable in the study area in period *t* emigrates out of the study area and is unobservable for period *t* + 1. Ψ_t^{UU} = the probability that an animal out of the study area in period *t* and unobservable remains out of the study area and is unobservable in period *t* + 1. Temporary emigration is random when $\Psi_t^{OU} = \Psi_t^{UU}$; the larger the difference between Ψ_t^{OU} and Ψ_t^{UU}, the greater the Markovian effect

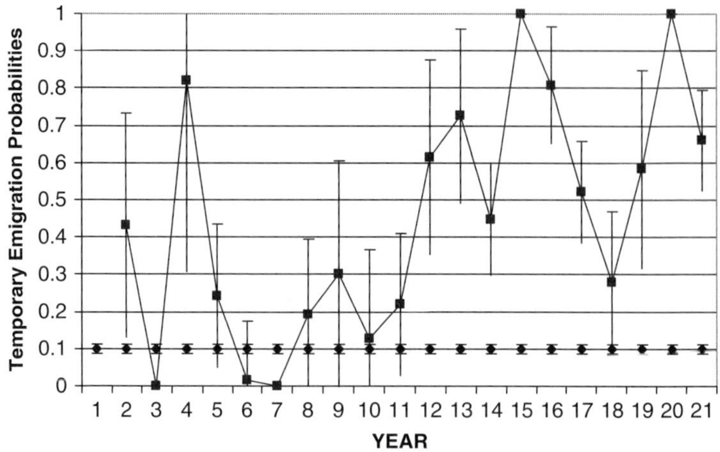

Fig. 2 Annual probabilities of Ψ_t^{OU} (υ) and Ψ_t^{UU} (■) and SE for Florida manatees in Northwest Florida estimated under the robust design model $S_.$, $\Psi_.^{OU}$, Ψ_t^{UU} $c_t = p_t$, (T = 22 years, Akaike weight = 0.999). Ψ^{OU} = the probability that an animal observable in the study area in period t emigrates out of the study area and is unobservable for period $t+1$. Ψ_t^{UU} = the probability that an animal out of the study area in period t and unobservable remains out of the study area and is unobservable in period $t+1$. Temporary emigration is random when $\Psi_t^{OU} = \Psi_t^{UU}$; the larger the difference between Ψ_t^{OU} and Ψ_t^{UU}, the greater the Markovian effect

probabilities are quite pronounced between Ψ^{OU} and Ψ^{UU}, and in most cases $\Psi^{OU} < \Psi^{UU}$, i.e. there is a much higher probability for manatees to remain away from the study area, than to leave if they were present the previous year. Although the SE is large for some of the annual estimates, it is apparent that emigration probabilities can vary broadly from year to year, particularly for individuals absent the previous year.

Based on the differences in manatee aggregation behavior and variable monitoring effort between warm and cold winters, it was expected that the best models of capture probability included significant annual variation, and indeed, there was no support for constant capture probabilities (Akaike weights < 0.0001). The best survival model under time-specific capture probabilities and time-specific Markovian emigration (S_t, Ψ_t^{OU}, Ψ_t^{UU}, $c_t = p_t$) contained significant annual variation for the AC (Akaike weight = 0.735), but time-invariant survival probabilities ($S_.$, $\Psi_.^{OU}$, Ψ_t^{UU}, $c_t = p_t$) for the NW (Akaike weight = 0.999).

3.2 Annual Survival Estimates After Truncation of Capture Histories

Examination of the annual survival probabilities for the Atlantic Coast estimated under the time-specific model indicated high survival rates, but with a large drop at the end of the 20-year time series. After truncating the capture histories to shorter

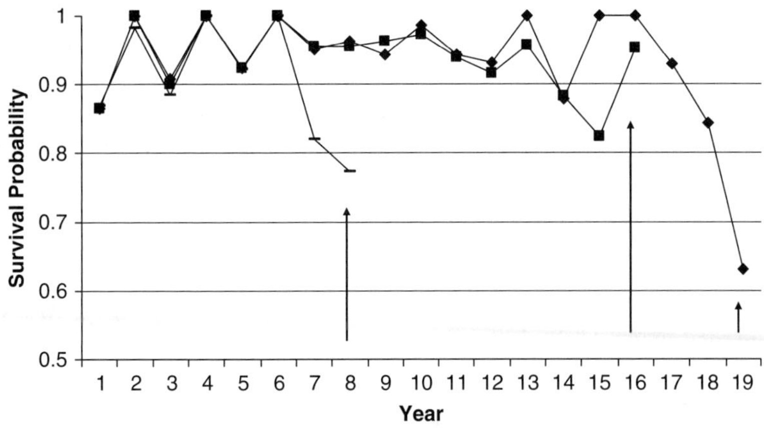

Fig. 3 Annual survival probabilities for Florida manatees on the Atlantic coast estimated under the robust design model S_t, Ψ_t^{OU}, Ψ_t^{UU}, $c_t = p_t$, where $\Psi_T^{OU} = \Psi_{T-1}^{OU}$ and $\Psi_T^{UU} = \Psi_{T-1}^{UU}$ before and after the capture histories for the 20 year study (♦) were truncated to 17 (■) and 9 (–) years. Arrows identify the final survival estimate for each set of time series

time frames of 17 and 9 years, lower survival probabilities near the end of these shorter series remained (Fig. 3).

3.3 Bias and Coefficients of Variation

Because the last few annual survival estimates appeared to change with additional years of data, a possible source of bias was the constraint I placed on the last two emigration parameter estimates, namely setting $\Psi_T^{OU} = \Psi_{T-1}^{OU}$ and $\Psi_T^{OU} = \Psi_{T-1}^{UU}$. Given the wide swings in manatee emigration estimates from year to year (Figs. 1 and 2) it was likely that the equality constraint was not realistic and the bias introduced to those final two emigration estimates also biased the survival estimates. The series of simulation analyses supported this assumption. Annual survival estimates were unbiased under the time-dependent estimation model, if Markovian temporary emigration in truth was constant over time (Scenario 1, Table 1). However if emigration varied between consecutive years, the estimation model produced biased estimates at the end of the series that propagated back through more than the last two survival estimates. Bias was evident, regardless of the value of the survival rate; however, both the magnitude of the bias and the number of biased estimates increased as true survival probability increased (Scenarios 2–4, Table 1). Standardized bias was greater than 0.50 (with reduced coverage in the 95% confidence intervals) for the last 5 years at a survival rate of 0.95; for the last 3 years at a survival rate of 0.70; but only for the last year at a survival rate of 0.50.

Note that Scenarios 2–4 (Table 1: time-dependence by alternating two values for Ψ^{OU} and two for Ψ^{UU})were run with the order of the true values for the last estimates of Ψ^{UU} set to 0.50 and 0.75 and Ψ^{OU} set to 0.3 and 0.5. By constraining the last two probabilities as equal in the estimation model, emigration probability in the last time period should be biased lower, while emigration in the previous period should be biased high. Survival estimates for these time periods under this situation were negatively biased (Table 1), resulting in the spurious appearance of a downward trend. Reversing the order of the values of the last two Ψ^{UU} to 0.75 and 0.50, and Ψ^{OU} to 0.50 and 0.30 would reverse the direction of known bias in emigration probabilities for these time periods; consequently survival estimates under this situation showed a positive bias at the last time period, but negative bias in the years before (Scenario 6, Table 2). Applying the estimation model to datasets describing other situations with a known inequality between the last two emigration probabilities resulted in biased survival estimates at the end of the time series; however the magnitude of bias differed. Lowering the emigration probabilities and reducing the difference in true values between the last two time periods for Ψ^{OU} (0.2 and 0.3) and Ψ^{UU} (0.3 and 0.4) reduced the bias to only the last year (Scenario 5, Table 2). If Ψ^{OU} was held constant and only Ψ^{UU} was time-specific (Scenario 7, Table 2), or the reverse (Ψ^{OU} time-specific and Ψ^{UU} constant), survival estimates were still biased at the end of the series. However, if emigration probability for the last time

Table 3 Comparisons of bias and coefficient of variation (CV) in annual survival probabilities for the robust design model of time-dependent survival and time-dependent Markovian temporary emigration, S_t, Ψ_t^{OU}, Ψ_t^{UU}, $c_t = p_t$, with different constraints for the final Ψ^{OU} and Ψ^{UU}. The estimation models are evaluated for data generated with Monte Carlo simulations of constant capture probability of 0.5, constant survival of 0.95, and with time-dependent temporary emigration, Ψ^{OU} and Ψ^{UU}, alternated over time between the two values listed, with the final time period ending with the higher value. In Scenario 8, no constraints are placed on the estimation model. In Scenario 9, the last emigration estimates are constrained to the true value at *T*–2. Absolute bias = mean estimate –true value, standardized bias = absolute bias/mean SE of the estimate (Burnham et al. 1987:215), %CV = mean SE/mean estimate X 100. Values in **bold** highlight CVs with unusually high values indicative of high SE that make the survival estimates meaningless

	Scenario 8: Estimation Model: No Constraints $\Psi^{OU} = 0.30, 0.50$ $\Psi^{UU} = 0.50, 0.75$ replicates = 500			Scenario 9: Estimation Model: Last emigration rates constrained to a true equality $\Psi^{OU} = 0.30, 0.50$ $\Psi^{UU} = 0.50, 0.75$ replicates = 500		
S year	Absolute bias	Standardized bias	%CV	Absolute bias	Standardized bias	%CV
1	0.000	0.000	1.9	–0.004	–0.222	1.9
2	–0.001	–0.077	1.4	–0.003	–0.231	1.4
3	0.000	0.000	1.4	–0.004	–0.333	1.3
4	0.000	0.000	1.2	–0.004	–0.400	1.1
5	–0.002	–0.065	3.3	–0.003	–0.273	1.2
6	–0.001	–0.037	2.8	–0.004	–0.400	1.1
7	–0.012	–0.075	**17.0**	–0.003	–0.214	1.5
8	–0.010	–0.079	**13.5**	–0.003	–0.214	1.5
9	–0.060	–0.077	**88.1**	–0.005	–0.167	3.2
10	–0.048	–0.082	**65.0**	–0.002	–0.057	3.7
11	–0.165	–0.071	**295.4**	–0.005	–0.143	3.7

period was constrained to a year where the equality was indeed true, annual survival estimates were unbiased (Scenario 9, Table 3). If no constraints were placed on the last estimates, the standard errors were too broad yielding the last survival estimates meaningless (Scenario 8, Table 3).

4 Discussion

The analysis of the real manatee datasets identified and described a biological system with strong annual variation in non-random temporary emigration. Our unusual monitoring system, where cold weather is the only inducement to draw manatees into our extremely small "study areas" at spatially dispersed warm-water sites, undoubtedly contributes to the magnitude of the observed Markovian variation. In both regions, there is a much higher probability for manatees to remain away from the study area, than to leave if they were present the previous year. This makes sense in that individuals have different home ranges, which will vary in availability and quality of secondary passive thermal refuges that can be used in lieu of the monitored primary refuges. Use of these secondary sites will vary with overall winter severity. Telemetry studies (Deutsch et al. 2003) indicate that manatees have high fidelity to their seasonal home ranges. If we were able to sample individuals in a larger study area that included their foraging areas, and not just their winter refuges, undoubtedly more individuals in the global population would be observable compared to those of the present design. Potentially this would reduce the magnitude of the difference between Ψ^{OU} and Ψ^{UU}, and reduce bias in survival rates as seen in simulation Scenario 5 (Table 2). Logistical constraints however limit our ability to effectively sample outside the refuges.

Although, our sample design may be unusual compared to other wildlife species, the combination of high survival rates, high rates of non-random temporary emigration, significant time-dependence, and a complexity of potential factors affecting emigration making it difficult to model temporary emigration in any meaningful fashion may be common to other species. The simulations presented here to examine bias under those conditions are not all inclusive of the range of possible scenarios, but this case study presents an approach and conclusions that can be useful to a broader suite of studies.

Results of the Monte Carlo simulations demonstrated that if the unmeasurable temporary-emigration probabilities in the last time period are not appropriately constrained, an estimation model with annual variation in survival can produce bias in some of the survival estimates at the end of the study. Examination of the sampling variance–covariance matrix for each of the original analyses and of select simulations showed decreasing covariance of the last survival estimate to each preceding survival estimate. This attenuating covariance illuminates why there is attenuation in bias in survival estimates with decreasing time. Comparing covariance of the last Ψ^{OU} and Ψ^{UU} showed the highest dependence on survival parameters, with decreasing dependence moving back through time in the annual estimates. The pattern of covariance is easily discerned from examination of the

Table 4 Cell probabilities underlying the primary period capture–recapture data for Markovian temporary emigration, summarized in m_{ij}–array format under the robust design (taken from Kendall et al. 1997). ϕ_i = probability of an animal in the superpopulation in period i surviving to, and remaining in the superpopulation at period $i + 1$; Ψ_i^{OU} = probability that a nonemigrant at time i is a temporary emigrant at $i + 1$; p^*_i = probability that an animal is caught at least once in primary occasion i, given that the animal is in the population during that sampling period; f_{h+1} = a 1 X 2 vector of probabilities of not being captured during the first primary period after release (period $h + 1$), given that an animal survives from primary period h to $h + 1$ and is in the superpopulation at $h + 1$; G_i = a 2 X 2 transition matrix of probabilities that an animal is outside the study area but not captured in primary period $i = h + 2, h + 3, \ldots, K{-}1$, given that it is outside or inside the study area in primary period $i - 1$, survives to period i, and is in the superpopulation in both periods; d_j = a 2 X 1 vector of probabilities of an animal being in the study area in primary period j, given that it is either outside or inside the study area in primary period $j - 1$ and survives to period j. When $\Psi_T^{OU} = \Psi_{T-1}^{OU}$ and $\Psi_T^{UU} = \Psi_{T-1}^{UU}$ is assumed but false in the time-dependent estimation model, $(1 - \Psi_i^{OU})$ and d_j in primary periods 4 and 5 are biased and affect the estimation of all ϕ_i in the cell probabilities for those periods

	Primary period of next recapture			
Primary period of release	2	3	4	5
1	$\phi_1(1 - \Psi_1^{OU})p_2^*$	$\phi_1 f_2\phi_2 d_3 p_3^*$	$\phi_1 f_2\phi_2 G_3\phi_3 d_4 p_4^*$	$\phi_1 f_2\phi_2 G_3 G_4\phi_4 d_5 p_5^*$
2		$\phi_2(1 - \Psi_2^{OU})p_3^*$	$\phi_2 f_3\phi_3 d_4 p_4^*$	$\phi_2 f_3\phi_3 G_4\phi_4 d_5 p_5^*$
3			$\phi_3(1 - \Psi_3^{OU})p_4^*$	$\phi_3 f_4\phi_4 d_5 p_5^*$
4				$\phi_4(1 - \Psi_4^{OU})p_5^*$

cell probabilities underlying the primary period capture–recapture data of the robust design, as summarized in the m_{ij} -array format (Table 4). The cell probabilities for individuals returning in the last period 5, when originally released at periods 1, 2, or 3 incorporate estimates of the unmeasurable emigration probabilities in the last period. Estimation of S_1 through S_4 in those cells is dependent on the estimate of Ψ_T^{OU} and Ψ_T^{UU}. If setting the last two estimates as equal is not valid, those parameters are biased and will affect the estimation of the survival parameters in the cell probabilities at time T (period 5 in Table 4) as well as at time T–1 (period 4 in Table 4).

The simulation results demonstrated that the number of biased annual survival estimates depended on both the true survival rate and the true emigration rates. Maintaining high emigration rates, but decreasing the survival rates resulted in fewer affected years (Scenarios 2–4, Table 1). Maintaining a high survival rate, but decreasing the Markovian effect also resulted in fewer affected years (Scenario 2, Table 1 compared to Scenario 5, Table 2). The results suggest the interaction of these parameters influence how far the bias propagates back through the annual survival estimates. Examination of the cell probabilities in Table 4 shows why this is true. If released individuals have high survival and a high probability of emigrating and staying away, eventually they will return to the study area. If the number of returned emigrants $(1{-}\Psi^{UU})$ in each of the last two cells is large, the bias of parameters derived from the improper constraint of setting the last two estimates of Ψ^{OU} and

Ψ^{UU} will cause bias in the estimates of annual survival probabilities near the end of the study. If survival rates are lower or temporary emigration rates are lower, the number of returning emigrants is smaller, particularly for individuals released earlier in the study, and the contribution of that cell to the overall estimation of S is smaller or even negligible.

The convention to constrain the last two emigration estimates as equal (Kendall et al. 1997) may not be the best option in the presence of significant annual variation in temporary emigration. As a preferred approach, Kendall (2006) suggested modeling the last estimate as a function of predictive covariates. Nonetheless, if the covariates do not accurately describe the variation, or some event affecting emigration in the last period is not immediately identified and adequately modeled (for example displacement from an extreme storm), the time-specific survival rates still could be biased. Another option would be to use radio-telemetry or some other approach to monitor and estimate temporary emigration during the last period and construct a joint likelihood incorporating both approaches. Detection rates are close to one with telemetry and there is greater certainty on the location of individuals inside and out of the study area. However, sample size will probably be an issue due to the time, effort and expense of a tracking program. USGS manatee research includes a successful satellite telemetry program (Deutsch et al. 2003), but it is not a realistic option for the AC or NW areas. The number of individuals we could simultaneously monitor is small and the analysis would require monitoring individuals at each of the aggregation sites where we conduct photo-documentation. This option, however, may be appropriate for studies of other species. A final option would be to use a random effects model, rather than a time-specific model for survival probabilities. The random effects model accounts for the correlation that exists between successive estimates and for variation in parameters across individuals. Use of such a model would lessen the magnitude of bias due to the unmeasureable final parameters and provide better estimates of trend. This approach would provide valid estimates of temporal variance necessary for population models, but does not allow one to test for a fixed effect from a proposed mortality source that just recently occurred. With increasing environmental variability and uncertainty with climate change and human development, testing for fixed effects is becoming increasingly important.

If no other options are available to reduce bias, then constraining emigration probabilities for the last two time periods may be the best approach. The propagation of bias back through the time-specific survival rates, however, could be problematic. The degree of bias one is willing to accept will depend on the objectives of the analysis for research and management. The good news, however, is that with additional years of data, unbiased estimates eventually should be realized. Nonetheless, regardless of bias or statistical fixes, inherently there will be less information about survival at the end of the time series compared to the beginning, especially for long-lived species. Later estimates will always show greater uncertainty.

Knowing the extent and magnitude of the potential bias can help in making decisions as to what time frame provides the best estimates to meet the objective, or the most reliable opportunity to model and test hypotheses about factors affecting survival. To assess bias, I recommend the two approaches I took in this study. First, truncating the capture histories to shorter time frames and reanalyzing the data to compare time-specific estimates may help identify spurious effects. Second, running simulations that mimic the parameter values and movement conditions evident from the initial analysis provide estimates of standardized bias that can be used to identify those annual estimates that are biased to the point where the 95% confidence intervals are inadequate in describing the uncertainty of the estimates. The simulation option in Program MARK is easy to use and can greatly facilitate this type of assessment.

Reducing the magnitude of the Markovian movement should help reduce bias, given the results of Scenario 5 (Table 2). The biology and behavior of the study animals determine a great deal of this and are beyond the control of the investigator. However, sampling design and monitoring protocols may exacerbate the effect, and this can be reviewed and possibly improved. With regard to our manatee research, we hope to move to a more systematic monitoring schedule. With greater effort during the warmer winters, we may be able to identify more individuals that visit the aggregation sites. This should not only reduce the gaps in the sighting histories, but also provide better representation for analysis of the movement behavior.

Acknowledgments This analysis and paper benefited greatly from discussions with Bill Kendall. Comments and suggestions from two anonymous peer reviewers greatly improving the manuscript. Discussions with Mike Runge lead to my identification of the bias problem. I thank Cathy Beck and the USGS MIPS team – Amy Teague, Gaia Meigs-Friend, Ron Osborn, and Howard Kochman – as well as the many MIPS photographers for the opportunity to work with such a great database.

References

Bossart GD, Meisner RA, Rommel SA, Ghim S-J, Jenson AB (2002) Pathological features of the Florida manatee cold stress syndrome. Aquatic Mammals 29:9–17.

Buergelt CD, Bonde RK, Beck CA, O'Shea TJ (1984) Pathologic findings in manatees in Florida. Journal of the American Veterinary Medical Association 185:1331–1334.

Burnham KP, Anderson DR (1998) Model selection and inference: a practical information-theoretic approach, 2nd edn. Springer-Verlag, New York, USA.

Burnham KP, Anderson DR, White GC, Brownell C, Pollock KH (1987) Design and analysis methods for fish survival experiments based on release-recapture. Monograph 5. American Fisheries Society, Bethesda, MD, USA.

Cooch E, White G (2006) Program MARK- a gentle introduction, 5th edn. (http://www.phidot.org/software/mark/docs/book/)

Cochran WG (1963) Sampling techniques, 2nd edn. John Wiley & Sons, New York.

Choquet R, Reboulet AM, Lebreton JD, Gimenez O, Pradel R (2005) U-Care 2.2 User's Manual. CEFE, Montpellier, France. (http://ftp.cefe.cnrs.fr/biom/Soft-CR/).

Deutsch CJ, Reid JP, Bonde RK, Easton DE, Kochman HI, O'Shea TJ (2003) Seasonal movements, migratory behavior, and site fidelity of West Indian manatees along the Atlantic Coast of the United States. Wildlife Monographs 151:1–77.

Hartman DS (1979) Ecology and behavior of the manatee (*Trichechus manatus*) in Florida. American Society of Mammalogists Special Publication 5:1–153.

Kendall WL (2001) The robust design for capture–recapture studies: analysis using Program MARK. In: Field R, Warren RJ, Okarma H, Seivert PR (eds) Wildlife, land, and people:

priorities for the 21st century. Proceedings of the Second International Wildlife Management Congress. The Wildlife Society, Bethesda, MD, pp 357–360.

Kendall WL (2006) The "robust design". In: Cooch E, White G (eds) Program MARK- a gentle introduction, 5th edn., pp 16.1–16.35.

Kendall WL, Nichols JD (2002) Estimating state-transition probabilities for unobservable states using capture–recapture/resighting data. Ecology 83:3276–3284.

Kendall WL, Nichols JD, Hines JE (1997) Estimating temporary emigration using capture–recapture data with Pollock's robust design. Ecology 78:563–578.

Langtimm CA, Beck CA (2003) Lower survival probabilities for adult Florida manatees in years with intense coastal storms. Ecological Applications 13:257–268.

Langtimm CA, Beck CA, Edwards HH, Fick-Child KJ, Ackerman BB, Barton SL, Hartley WC (2004) Survival estimates for Florida manatees from the photo-identification of individuals. Marine Mammal Science 20:438–463.

Langtimm CA, O'Shea TJ, Pradel R, Beck CA (1998) Estimates of annual survival probabilities for adult Florida manatees (*Trichechus manatus latirostris*). Ecology 79:981–997.

Lebreton J-D, Burnham KP, Clobert J, Anderson DR (1992) Modeling survival and testing biological hypotheses using marked animals: a unified approach with case studies. Ecological Monographs 62:67–118.

O'Shea TJ, Langtimm CA (1995) Estimation of survival of adult Florida manatees in the Crystal River, at Blue Spring, and on the Atlantic coast. In: O'Shea TJ, Ackerman BB, Percival HF (eds) Population biology of the Florida manatee. US Department of the Interior, National Biological Service, Information and Technology Report 1. pp 194–222.

Pollock KH (1982) A capture–recapture design robust to unequal probability of capture. Journal of Wildlife Management 46:757–760.

Runge MC, Langtimm CA, Kendall WL (2004) A stage-based model of manatee population dynamics. Marine Mammal Science 20:361–385.

Schaub M, Gimenez O, Schmidt BR, Pradel R (2004) Estimating survival and temporary emigration in the multistate capture–recapture framework. Ecology 85:2107–2113.

White GC, Burnham KP (1999) Program MARK survival estimation from populations of marked animals. Bird Study 46:120–139.

Section IX
State Uncertainty – Assignment Error and Unobservable States

Michael Schaub and Jean-Dominique Lebreton

One Size Does Not Fit All: Adapting Mark-Recapture and Occupancy Models for State Uncertainty

William L. Kendall

Abstract Multistate capture–recapture models continue to be employed with greater frequency to test hypotheses about metapopulation dynamics and life history, and more recently disease dynamics. In recent years efforts have begun to adjust these models for cases where there is uncertainty about an animal's state upon capture. These efforts can be categorized into models that permit misclassification between two states to occur in either direction or one direction, where state is certain for a subset of individuals or is always uncertain, and where estimation is based on one sampling occasion per period of interest or multiple sampling occasions per period. State uncertainty also arises in modeling patch occupancy dynamics. I consider several case studies involving bird and marine mammal studies that illustrate how misclassified states can arise, and outline model structures for properly utilizing the data that are produced. In each case misclassification occurs in only one direction (thus there is a subset of individuals or patches where state is known with certainty), and there are multiple sampling occasions per period of interest. For the cases involving capture–recapture data I allude to a general model structure that could include each example as a special case. However, this collection of cases also illustrates how difficult it is to develop a model structure that can be directly useful for answering every ecological question of interest and account for every type of data from the field.

Keywords Capture–recapture · Disease model · Multistate models · Occupancy · State uncertainty

1 Introduction

Multistate mark-recapture models (MSMR), in which an animal is captured and marked, then tracked over time through multiple locations or life history states, were first developed in the early 1970s (Arnason 1972, 1973). They were more fully

---

W.L. Kendall (✉)
USGS Patuxent Wildlife Research Center, Laurel, MD, USA
e-mail: wkendall@usgs.gov

D.L. Thomson et al. (eds.), *Modeling Demographic Processes in Marked Populations*, Environmental and Ecological Statistics 3, DOI 10.1007/978-0-387-78151-8_35,

developed in the early 1990s (Schwarz et al. 1993; Brownie et al. 1993). Beginning with a metapopulation study of geese (Hestbeck et al. 1991), they have been applied in numerous studies to model survival, transitions between states, or state-specific abundance. Lebreton et al. (1999) and Lebreton and Pradel (2002) conceptualized death as an absorbing life history state, and therefore under this approach most capture-mark-recapture models (CMR) become MSMR models.

Especially in taking the approach of Lebreton and Pradel (2002), several sources of uncertainty about an animal's state arise. Many of these revolve around the occasions when an individual is not captured. Where an animal is missed at a given occasion but it is captured before and thereafter, was it present in one of the states subjected to capture effort but simply missed? Alternatively, at that time did it occupy an unobservable state, where no capture effort was applied? For time periods before an animal is first captured, was it present but not captured, or had it not yet recruited into one of the states that were being monitored? Conversely, for occasions after its last capture, was the animal available but not captured, had it temporarily moved to an unobservable state, permanently dispersed outside of the set of observable states, had its ring or other mark fallen off, or was it dead? These cases all represent state uncertainty, and have been treated by several authors (see review by Williams et al. 2002).

Other cases of state uncertainty occur on occasions where an animal is captured or sighted. The animal is detected but there is uncertainty about which state it occupies. For geographic states this is mostly not a problem, but can still arise. For example, for breeding ground studies there could be animals captured in the southern part of the breeding range where it is not clear whether they have reached their breeding grounds or are transients on their way to sites further north. Similarly, a migratory bird might be captured on the wintering grounds, but it is not clear to which breeding population it belongs (see Kendall et al. 2006).

More commonly, uncertainty arises about a detected animal's life history state. For example, in breeding colonies it is not always clear if a sighted animal is a breeder that year, a pre-breeder, or an adult who has skipped breeding that year. Observing an individual feeding a chick on a nest might indicate a breeder with certainty. Seeing an individual that is known to have nested in previous years eliminates the possibility that it is a pre-breeder, but does not by itself determine its exact state. In addition, for disease studies it is not always clear upon inspection whether or not an animal has a disease. A similar problem can arise for cases where an animal's state does not change. For sexually monomorphic species, many observations of an animal can be made without knowing its sex with certainty (Nichols et al. 2004; Pradel 2005).

Ignoring state misclassification can cause substantial bias in survival and transition probabilities. Mistakenly assigning individuals to the wrong state will lower the power to detect differences between survival probabilities for those states. Ignoring an uncertain state will bias survival estimates for all states involved. For example, through simulation Nichols et al. (2004) found that if the probability of assigning a bird to the proper sex is 0.3 each time it was encountered, thus producing a cohort of birds that are never sexed, survival estimates for males and females were positively biased by 8 and 12%, respectively (500 and 400% increase in mean square error, respectively). Transition probabilities are also affected. In a study of manatees Kendall et al. (2004) found that under a MSMR model conditional breeding probability was estimated to be 0.31, whereas when corrected for misclassification the estimate was 0.43 (100% increase in mean square error).

In this paper I will focus on the case where state is not certain at the time of detection. Pradel (2005, 2008) presented a general structure for this type of model, where each period of interest consists of a single detection occasion. Under this framework "events" (observed states) are accounted for in the process of modeling vital rates and transitions among true phenotypic states. For perspective I will briefly present an example of this type of model where an animal that occupies one of two live dynamic states can be misclassified into the opposite state at any time. Although the list of cases of misclassification that can be modeled under this general framework is long, the list is not exhaustive. I present five case studies that do not fall directly under this model structure. Three involve the modeling of captures of individuals under a robust design approach. These three cases point to the value of developing a robust design generalization of the Pradel (2005) model. The last two cases involve occupancy modeling of disease, viewing them as misclassification problems.

2 Dynamic States, One Capture Occasion per Period

Models correcting for misclassification of dynamic states can be viewed as adjusting a MSMR model. Therefore, for perspective I first review the structure of an Arnason–Schwarz MSMR model without misclassification. For a K-period study let S_t^r = probability an animal in state r at time t lives to time $t+1$ and remains faithful to the population or metapopulation; ψ_t^{rs} = probability that an animal in state r at time t transitions to state s at time $t+1$, given that it lives to time $t+1$; and p_t^s = probability that an animal in state s at time t is detected (e.g., captured). Below I provide a simple example of a detection history over two periods, and its probability structure under the Arnason–Schwarz model, for an animal that can occupy either of two states: diseased (D) and not diseased (N).

$$\text{DD} \quad S_1^D \psi_1^{DD} p_2^D \tag{1}$$

I have purposefully provided a simple example, with a reminder that even with two states, these probabilities become very complex quickly as the number of time periods or states grows (Schwarz et al. 1993; Brownie et al. 1993).

Now assume that at each point in time, it is possible to mistake a diseased animal for one without disease (e.g., it shows no clinical sign), or vice versa (an animal that has recovered fully from a disease in terms of mortality risk still shows vestigial clinical signs). To adjust the Arnason–Schwarz model for these possibilities we need additional notation. Let $\pi_{t_1}^r$ = probability that an animal is in state r at time t_1, given that the investigator has captured it at that time; and $\delta_{t_2}^s$ = probability that an investigator assigns an animal to state s at time t_2, given that the animal occupies that state and is captured. In this case the probability structure for the history above, conditional on release at time 1 in an apparently diseased state, becomes:

$$\begin{aligned}\text{DD}\quad & \pi_1^D[S_1^D\psi_1^{DD}p_2^D\delta_2^D + S_1^D\psi_1^{DN}p_2^N(1-\delta_2^N)] \\ & + (1-\pi_1^D)[S_1^N\psi_1^{ND}p_2^D\delta_2^D + S_1^N\psi_1^{NN}p_2^N(1-\delta_2^N)] \qquad (2)\end{aligned}$$

The first and second terms describe the cases where the true state of the individual at time 1 is diseased or not diseased, respectively. Within each set of brackets the first and second terms refer to the cases where the state assignment at time 2 is correct and incorrect, respectively. So $\pi_1^D S_1^D \psi_1^{DN} p_2^N (1 - \delta_2^N)$ represents the case where the animal is diseased at time 1 (π_1^D), survives to time 2 (S_1^D), recovers from the disease (ψ_1^{DN}), is detected at time 2 (p_2^N), but is misclassified as still being diseased ($1 - \delta_2^N$).

This model is a simple case of the very general model of Pradel (2005). He found that without further information, if the state of the animal does not change over time (e.g., sex), under maximum likelihood exactly two combinations of parameters produce the minimum deviance. The biologically realistic set of values is chosen by the investigator. When sex was known with certainty for a subset of animals in the study, the problem of bimodal deviance disappeared (see also Nichols et al. 2004). For another static state problem, Runge et al. (2007) corrected for misclassification of species under a similar model where state does not change: classification to species of vole (*Microtus spp.*). They had direct supplemental information on π_t^r and δ_t^s from a subset of study animals that died in the trap and could be examined for species indicators internally. For dynamic states involving misclassification, if true state were not known for a subset of individuals one would expect the multimode problem to be exacerbated.

3 Dynamic States, Multiple Capture Occasions per Period

The following case studies have three things in common. First, each is conducted under a robust design (Pollock 1982; Schwarz and Stobo 1997; Kendall and Bjorkland 2001), consisting of multiple sampling occasions for each period of interest. Relatedly, the state of an individual is assumed to remain static for the duration of a primary period (Kendall 2004). Second, in each case misclassification occurs in only one direction. For example, state 1 can be mistaken for state 2, but state 2 is never mistaken for state 1. Therefore, there is a subset of animals at each time period for which the true state is known (e.g., those assigned to state 1). However, in each case we assume there is no set of animals in the opposite state whose state is unambiguous through observation. Therefore assignment to state 2 is always ambiguous. In each case, an animal is assumed to occupy this ambiguous state until a given definitive behavior is displayed. The correction for uncertainty is based on the detection probability for that behavior, conditional on observing the individual across multiple sampling periods within a season. Third, in each case the state structure of the population (or at least for those states accessible to sampling) is estimated as a set of parameters in the model. In case three this state structure among juveniles, not dynamics over time, is the focus of the analysis.

3.1 Breeding State of Florida Manatees

Kendall et al. (2003, 2004) considered a MSMR for adult female Florida manatees (*Trichechus manatus latirostris*) that occupied one of two states: breeder (B) and skipped breeder (b). Individual adult females were identified from photographs taken by observers in the water or from a vantage point out of the water, based on scar patterns created by boat propeller strikes. If a first year calf was in close attendance when the photo was taken, then the adult was classified as a breeder. If no young calf was seen, a calf was present but was considered >1 year old, or if it was not clear to which adult a calf belonged, then the marked adult was assigned to state b. By being conservative in assigning a calf to an adult female, one-way misclassification was maintained. An assignment to the breeder state was unambiguous, whereas an assignment to non-breeder was ambiguous. In the terminology of Pradel (2005, 2008) the "event" of a clearly attendant first-year calf maps directly into the breeder state, whereas the events of an adult female apparently alone or with an apparently older calf maps into either the breeder or skipped-breeder state.

The model in Kendall et al. (2004) is less restrictive than the model in Kendall et al. (2003), and therefore is outlined here. The sampling effort for manatees is intense during winter aggregations at warm-water sites. Observers make multiple circuits of manatee winter range, taking photos. Therefore sampling effort can be characterized as consisting of multiple sighting occasions within the season (i.e., a robust design). Kendall et al. (2004) partitioned sighting data within each year into two sampling periods. The breeding state of a female was assumed to remain constant within the season (i.e., no calves were born and a calf was assumed not to wean during that time), thus calling for a closed population model within a season. Therefore the Kendall et al. (2004) model consists of a multistate closed robust design model, adjusted for one-way misclassification. I summarize this model below.

Beginning with within-season modeling, for simplicity I present the probability structure for two of the nine possible sighting histories within a year:

$$\begin{aligned} &\text{BB} \quad \omega_t^B p_{t1}^B \delta_{t1}^B p_{t2}^B \delta_{t2}^B / denom \\ &\text{bb} \quad [\omega_t^B p_{t1}^B (1-\delta_{t1}^B) p_{t2}^B (1-\delta_{t2}^B) + (1-\omega_t^B) p_{t1}^b p_{t2}^b] / denom, \end{aligned} \tag{3}$$

where p and δ are now defined on a secondary sampling period basis, and capture history is conditioned on being captured at least once [$denom = \omega_t^B(p_t^{B\delta} + p_t^{B(1-\delta)}) + (1-\omega_t^B)p_t^b$]; ω_t^B = the probability that an adult female manatee is a breeder in year t, given that she is alive and part of the population; $p_t^{B\delta} = 1 -$

$\prod_{j=1}^{2}(1-p_{tj}^{B}\delta_{tj}^{B})$, the probability a breeder is seen with her calf at least once in the season; $p_t^{B(1-\delta)} = \prod_{j=1}^{2}(1-p_{tj}^{B}\delta_{tj}^{B}) - \prod_{j=1}^{2}(1-p_{tj}^{B})$, the probability a breeder is seen at least once in a season but her calf is never seen; and $p_t^b = 1-\prod_{j=1}^{2}(1-p_{tj}^{b})$, the probability a skipped breeder is seen at least once in a season.

The parameters from within-season modeling are used to correct the MSMR model between primary periods for misclassification. Simple examples of sighting histories for two years are given below. In this case within-season histories are pooled into one. If an adult were seen at least once with her first-year calf, then her history would be a B (e.g., a history within season of Bb is treated as a B). Otherwise she would receive a b or 0, depending on whether or not she were seen during the season:

$$\begin{aligned}
&\text{BB} \quad S_1^B \psi_1^{BB} p_2^{B\delta} \\
&\text{Bb} \quad S_1^B \left[\psi_1^{BB} p_2^{B(1-\delta)} + \psi_1^{Bb} p_2^{b}\right] \\
&\text{bB} \quad \left[\pi_1^b S_1^b \psi_1^{bB} + \left(1-\pi_1^b\right) S_1^B \psi_1^{BB}\right] p_2^{B\delta},
\end{aligned} \tag{4}$$

where $\pi_t^b = (1-\omega_t^B)p_t^b/[\omega_t^B p_t^{B(1-\delta)} + (1-\omega_t^B)p_t^b]$, the probability that an adult female detected in season t is not a breeder. Therefore all parameters in the likelihood are defined at the secondary sampling level except the S_t^r and ψ_t^{rs}.

Note the relationship between the π_t^b and ω_t^B in the expression above, keeping in mind Equation (2). ω_t^B represents the probability that an adult female manatee is a breeder, given that she is an adult and *alive in* season t. Conversely, π_t^b represents the probability an adult female manatee is a skipped breeder, given she is *captured* in season t. Therefore, ω_t^B represents stage structure (Caswell 2001), whereas π_t^b is simply a nuisance mixture parameter. Use of the robust design permits the direct estimation of stage structure in this case.

3.2 Squatting Behavior in Kittiwakes

A misclassification problem similar to the manatee case arises in the study of colonial nesting seabirds. Cam et al. (2002) characterized the life history of kittiwakes (*Ryssa tridactyla*) as consisting of four life history stages after the chick stage: pre-breeder, squatter, breeder, skipped breeder. A squatter is an advanced pre-breeder that displays some nesting behavior when the owners of a nest are away from it. Cam et al. (2002) found that individual squatters tended to be more successful future breeders, thus distinguishing themselves from other pre-breeders in terms of fitness. However, some squatters are misclassified as pre-breeders, because investigators do not always see the squatting behavior.

In the case of the kittiwake in Cam et al. (2002), the field season can be partitioned into multiple sampling occasions. Therefore, statistically the kittiwake case is very similar to the manatee case, with two life history states involved with one-way misclassification (squatters can be mistaken for pre-breeders but not vice versa). The kittiwake example is different only in that there are additional states in the model that are not involved with misclassification (breeders and skipped breeders).

3.3 Weaning of Sea Lions

Estimating the proportion of Steller sea lion (*Eumetopias jubatus*) juveniles, by age, that have been weaned in a given year (Pendleton et al. unpublished manuscript) presents an estimation problem similar to the manatee and kittiwake problems. Repeated surveys are done within a month, and then across multiple months within a given winter, at the major aggregation site for the population. Observers search for juveniles branded with unique codes. In some instances the juvenile is seen alone or its association is not clear, and therefore it is tentatively classified as weaned. On other occasions an observer might see the same juvenile suckling from its mother. In this case the individual has clearly not weaned. Therefore a risk for misclassifying animals exists. Although the state (W = weaned, N = not weaned) of the juvenile might be considered static for the entire season (i.e., they do not wean during the season), the collection of individuals at the aggregation site is not. The arrival time of individuals, as well as their departure times, is staggered over time. Therefore an open robust design (Schwarz and Stobo 1997; Kendall and Bjorkland 2001) is a more appropriate starting point for modeling than a closed robust design.

I present below a simple example of an appropriate probability structure for estimating the proportion weaned under the scenario described above. Consider a winter where surveys of the sea lion population are conducted four times. The first two occasions are spaced closely in time (perhaps a couple of days apart). The next two occasions are also spaced close to one another, but occur in the following month. Therefore enough time elapses between the pairs of surveys that new individuals could arrive and some that were there during the first pair of occasions could have departed. Under this scenario I provide two example sighting histories and their probability structures, where event S denotes observed suckling, and event A denotes observed alone:

$$
\begin{aligned}
&\text{00 SS} \quad && \omega^N[\beta_o^N(1-p_{11}^N)(1-p_{12}^N)\phi_1^N+\beta_1^N]p_{21}^N\delta_{21}^N p_{22}^N\delta_{22}^N/denom \\
&\text{AA 0A} \quad && [(1-\omega^N)\beta_o^W p_{11}^W p_{12}^W\phi_1^W(1-p_{21}^W)p_{22}^W + \\
& && \omega^N\beta_o^N p_{11}^N(1-\delta_{11}^N)p_{12}^N(1-\delta_{12}^N)\phi_1^N(1-p_{21}^N)p_{22}^N(1-\delta_{22}^N)]/denom, \qquad (5)
\end{aligned}
$$

where ω^N = probability a juvenile has not yet weaned, given that it is present in the population; β_i^N, β_i^W = probability an individual that is not weaned (N) and weaned (W), respectively, arrives at the aggregation site before the first survey of month i is conducted, given that it arrives at all; ϕ_i^N, ϕ_i^W = probability that a juvenile sea lion that is not weaned and weaned, respectively, present during the two surveys in month i is still present during the surveys in month $i+1$; δ_{ij}^N = probability a juvenile that is not weaned is seen suckling, given that the juvenile is sighted during survey j of month i; and *denom* = the probability a sea lion is sighted $\geq$1 time during the season:

$$\omega^N[\beta_0^N\{1-(1-p_{11}^N)(1-p_{12}^N)\}+\{\beta_0^N(1-p_{11}^N)(1-p_{12}^N)\phi_1^N+\beta_1^N\}$$
$$\{1-(1-p_{21}^N)(1-p_{22}^N)\}]+(1-\omega^N)[\beta_0^W\{1-(1-p_{11}^W)(1-p_{12}^W)\}$$
$$+\{\beta_0^W(1-p_{11}^W)(1-p_{12}^W)\phi_1^W+\beta_1^W\}\{1-(1-p_{21}^W)(1-p_{22}^W)\}].$$

This simple example could be extended to more surveys within a season in a straightforward manner. If transitions from unweaned to weaned across years were of interest, a between-years component could be developed as well. It would look like the manatee corrected MSMR model above, except that the annual probabilities of detection for each state would be composed of a different function of within-season parameters.

4 One-Way Misclassification in Occupancy Modeling

Occupancy modeling involves the detection or nondetection of a group of interest on a collection of well defined spatial units. This group is often a species or other taxonomic group, but could be anything, including a disease. In recent years occupancy modeling has included accounting for the possibility that the group of interest was present at a spatial unit but was not detected (see MacKenzie et al. 2006). A simple example of the most basic occupancy model of this type is illustrated by the following example. To estimate occupancy, each spatial unit of interest is surveyed for the group of interest multiple times (e.g., twice) over a short period of time (so that the unit is occupied or not for both surveys).

Consider two possible detection histories that could result from two samples, where a 1 indicates the group was detected and a 0 indicates it was not detected:

$$\begin{aligned} &10 \quad \psi_t p_{t1}(1-p_{t2}) \\ &00 \quad \psi_t(1-p_{t1})(1-p_{t2})+(1-\psi_t). \end{aligned} \tag{6}$$

In this case I define ψ_t = the probability that a given unit is occupied at time t (different from the definition of ψ provided in Section 2 but consistent with the notation of Mackenzie et al. 2006), and p_{tj} = probability the group of interest is detected during survey j of time period t, given that the unit is occupied. For the first example detection history in (6) the unit is clearly occupied because it was detected at least once (during survey 1 but not survey 2). In the second example history the group was not detected, so the first term of the probability structure is the probability the group was there but not detected in either survey. The second term describes the probability the unit was not occupied. Mackenzie et al. (2006) describe more complicated models, including a robust design approach where local extinction and recolonization are modeled over time.

One can see the similarity between model structures 3 and 6, indicating a similarity between occupancy modeling and correcting for misclassification in MSMR models. Whereas δ in (3), the detection probability of a behavior, serves to correct for misclassification in MSMR models, p in (6) serves to correct for a misclassification of a spatial unit as occupied or not. Each also requires a mixture parameter to account for an ambiguity about the state of an individual or the status of a patch. However, even within occupancy modeling there is opportunity to refine the classification of an occupied spatial unit. I present two cases below. In the first case occupancy is estimated and then conditional on a patch being occupied, it is further classified into one of two states. In the second example occupancy is modeled at one spatial scale, and then conditional on occupancy at that scale occupancy, at smaller scales within that patch is estimated.

4.1 State Assignment in Disease Occupancy Modeling

In some cases there could be interest not only in the proportion of spatial units occupied by a species, but in the prevalence of a disease among those areas where the species exists. This case arose in considering the spread of West Nile Virus (WNV) through northern spotted owl (*Strix occidentalis caurina*) habitat, via a vector such as mosquitoes (Franklin, personal communication). The spatial units of interest are owl territories. In each territory mosquito traps are set out over, for example, two sampling occasions close together in time. If mosquitoes are captured in the trap they are analyzed for the presence of the pathogen that causes WNV. Below I present two example detection histories for these two samples. In this case V indicates the vector (mosquito) was found in a territory, but the WNV pathogen was not detected in these mosquitoes; D indicates the vector was found and the lab analysis indicated the presence of the WNV pathogen; and 0 indicates the vector was not found (and hence neither was the pathogen). Note that the pathogen cannot be found if the vector is not. In this case the probability structure for three example detection histories would be:

$$\begin{aligned} &\text{DV} \quad \psi_t^V \psi_t^{D|V} p_{t1}^V \delta_{t1}^D p_{t2}^V \left(1 - \delta_{t2}^D\right) \\ &\text{VV} \quad \psi_t^V \psi_t^{D|V} p_{t1}^V \left(1 - \delta_{t1}^D\right) p_{t2}^V \left(1 - \delta_{t2}^D\right) + \psi_t^V \left(1 - \psi_t^{D|V}\right) p_{t1}^V p_{t2}^V \\ &\text{00} \quad \psi_t^V \left(1 - p_{t1}^V\right) \left(1 - p_{t2}^V\right) + \left(1 - \psi_t^V\right), \end{aligned} \tag{7}$$

where ψ_t^V = probability the vector is present in the owl territory at time period t; $\psi_t^{D|V}$ = probability the WNV pathogen is present in the territory at time period t, given the vector is present; p_{tj}^V = probability the vector is detected in sampling period j of time period t, given it is present; and δ_{tj}^D = probability the pathogen is detected in sampling period j of time period t, given it is present in the vector. In the first history both the vector and the pathogen are present, so there is no uncertainty.

In the second history only the vector is detected. Therefore the first term in the probability structure accounts for the pathogen being present and missed, and the second term accounts for the pathogen being absent. In the third history the vector is not detected (and therefore neither is the pathogen), so the two terms in the probability structure account for the vector being present and missed and for the vector being absent, respectively.

There are many applications in disease studies where this type of model could be applied. However, the use of this misclassification model is not limited to disease applications. This same model was independently developed by Nichols et al. (2007), ironically for another population of spotted owls, where the parameters of interest were the probability an owl territory was occupied by a mated pair (equivalent to ψ_t^V above), and conditional on that occupancy, the probability that pair successfully breeds (equivalent to $\psi_t^{D|V}$). These models can be considered special cases of the species co-occurrence models in MacKenzie et al. (2006).

4.2 Multilevel Sampling in Disease Occupancy Modeling

This case was motivated by the problem of designing a surveillance strategy for the presence of a highly pathogenic avian influenza pathogen (HPAI) among waterfowl in the continental United States. This is essentially an occupancy problem, where waterfowl feces are collected and sent to a lab for analysis. There are several issues to contend with in designing such a strategy (Farnsworth et al., unpublished report), but I will focus on estimation issues at various levels. Consider for this example that the primary measure of interest is the prevalence of HPAI across all 10-min blocks in the continental USA. A sample of 10-min blocks from this frame is chosen for assessment. Within each of those 10-min blocks multiple waterfowl refuges could exist and be sampled. Within each refuge multiple wetlands could be sampled, and within each of those wetlands multiple samples of feces could be collected. Viewing it this way 10-min blocks, refuges, wetlands, and samples represent primary, secondary, tertiary, and quaternary levels of sampling, respectively (Fig. 1).

In Fig. 1 I provide an example of possible detection histories for HPAI at the various levels of sampling. At the top there is a 1, indicating that for 10-min block i, HPAI was found. Looking below at the level of refuge, HPAI was detected at refuge 2 but not refuge 1. Two wetlands were sampled within refuge 2, and HPAI was found at both wetlands. Looking below again at the level of samples within these two wetlands, for wetland 1 HPAI was detected in both samples but for wetland 2 HPAI was detected in sample 1 but not sample 2. As in CMR problems, at a given level detection in one sample but not in another implies that detection probability for HPAI is <1.0.

To develop an occupancy model for this problem we begin at the level of refuge within 10-min block i. At this level the detection history and probability structure can be expressed as

$$01 \quad \psi_i(1-p_{i1})p_{i2}$$

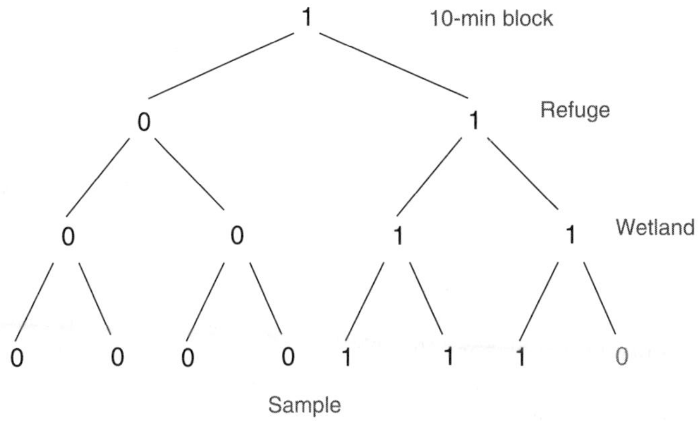

Fig. 1 Example of occupancy data for an avian influenza pathogen in waterfowl, collected at multiple nested spatial levels: 10-min block, refuge, wetland, and sample. A "1" indicates detection of the pathogen at the specified spatial scale, and a "0" indicates nondetection

where parameters are defined as in the basic occupancy model with two samples (6). However, here samples are collected across space instead of over time (MacKenzie et al. 2006). At this stage we are potentially ignoring a source of misclassification. Presence of HPAI is determined by laboratory analysis of the fecal samples. Given that HPAI is present in the sample, the probability it is detected is potentially <1.0. Here we ignore the potential of a false positive and define $\delta^{HPAI} = \delta_{AI}\delta_{subtype}\delta_{HP}$ as the probability that laboratory tests detect HPAI, given it is present in a sample. As can be seen, this probability is a product of three other detection probabilities associated with three steps in the lab procedure: δ_{AI} = probability a PCR test detects that an avian influenza pathogen is present; $\delta_{subtype}$ = probability that a subsequent test for subtype correctly identifies it as H5 or H7 subtype; and δ_{HP} = probability that a third test correctly determines that it is highly pathogenic. Lab experiments can determine these detection probabilities and their variances. I will not address that aspect here, but the statistical model used for the lab tests could be incorporated into a joint likelihood with the occupancy model I describe here. The overall error rate can then be incorporated into the occupancy model:

$$01 \quad \psi_i(1 - p_{i1}\delta^{HPAI})p_{i2}\delta^{HPAI}, \tag{8}$$

so that p_{ij} reflects the probability that HPAI is actually present in a given sample, rather than including the probability it is properly detected. However sampling variability within a refuge is still included in p_{ij}.

Extra information can be utilized by considering tertiary sampling within refuge, across multiple flocks or wetlands. At this level the detection history consisted of two samples for each refuge (Fig. 1):

$$00\ 11 \quad \psi_i[\psi_{i1}(1-p_{i11}\delta^{HPAI})(1-p_{i12}\delta^{HPAI})+(1-\psi_{i1})]\psi_{i2}p_{i21}\delta^{HPAI}p_{i22}\delta^{HPAI} \tag{9}$$

where ψ_{ij} = conditional prevalence, the probability that HPAI is present at refuge j of 10-min block i, given that it is present in 10-min block i; and detection probability is modified to p_{ijk} = probability HPAI is in the sample taken from wetland k at refuge j in 10-min block i. Because multiple samples were taken within a refuge, we have the potential to estimate how prevalent the disease is within an infected geographic unit. The first two 0's are ambiguous (it was present at refuge 1 and missed or it was absent), which is reflected in the two terms within the brackets in (9).

Because multiple samples are taken from each wetland, we can take one step further down this cascade, to consider quaternary sampling. At this level the sample detection history in Fig. 1 consists of four pairs: two samples per wetland, two wetlands per refuge, two refuges within the 10-min block. The history and its associated probability structure is

$$\begin{aligned} 00\ 00\ 11\ 10 \quad & \psi_i[\psi_{i1}\left\{1-\psi_{i11}\left(1-\left(1-p_{i111}\delta^{HPAI}\right)\left(1-p_{i112}\delta^{HPAI}\right)\right)\right\} \\ & \left\{1-\psi_{i12}\left(1-\left(1-p_{i121}\delta^{HPAI}\right)\left(1-p_{i122}\delta^{HPAI}\right)\right)\right\}+(1-\psi_{i1})] \\ & \left[\psi_{i2}\psi_{i21}p_{i211}\delta^{HPAI}p_{i212}\delta^{HPAI}\psi_{i22}p_{i221}\delta^{HPAI}\left(1-p_{i222}\delta^{HPAI}\right)\right] \end{aligned} \tag{10}$$

where ψ_{ijk} = prevalence within a refuge, the probability that HPAI is present at wetland k in refuge j of 10-min block i, given that it is present at refuge j; and detection probability is modified to p_{ijkl} = probability HPAI is in the sample l taken from wetland k at refuge j in 10-min block i. The last line of (10) describes the last two pairs in the detection history. It is detected at both wetlands in refuge 2, so the refuge and both sampled wetlands within the refuge are clearly infected. Therefore this part of the probability structure is straightforward. The first two lines of (10) describe the ambiguous nature of all 0's in the first two pairs within the detection history. Refuge 1 might or might not be infected, represented by the first and second terms within the bracket, and if the refuge is infected, the two sampled wetlands within that refuge might or might not be infected. This model permits the possibility that neither sampled wetland within a refuge is infected, but the refuge is infected. If every wetland within every refuge were to be sampled (i.e., exhaustive sampling within a refuge), then occupancy of a refuge could be written in terms of occupancy of its wetlands [in this case $\psi_{ij} = 1-(1-\psi_{ij1})(1-\psi_{ij2})$]. The same would be true for refuges within 10-min block if sampling were exhaustive at that level.

What is the value of developing this cascade of modeling levels, rather than focusing on the simple corrected occupancy model presented in (8)? There are two potential reasons: one from the epidemiological perspective and the other from the statistical perspective. First, this cascade of sampling levels permits the development of hierarchical epidemiological models. Especially given that HPAI could be "dropped on" a 10-min block due to the migratory nature of waterfowl, knowing how widespread the disease might be within that spatial unit could be of interest. If the occupancy model I outlined here were extended to modeling occupancy dynamics, then the spread of the pathogen across and within 10-min blocks could be modeled. Second, different factors could be driving detection or occupancy at various levels of sampling. If so, different predictors might be used to model these values at different levels. If these parameters are pooled (e.g., expression 8), this might induce unspecified heterogeneity that is more difficult to model or account for, thus inducing bias in estimates of occupancy at the level of the 10-min block. As with any consideration of adding additional levels of modeling, its value will depend on how much data are being collected at each level. How much effort would be needed at each level is an open question.

5 Discussion

I began with a general expression (1) for describing misclassification between two states that could go in either direction. A general framework for this is presented in Pradel (2005, 2008) and associated software E-SURGE (Choquet et al. 2008), assuming there is one sampling occasion per time period of interest. Knowing the true state of at least a subset of individuals reduces or eliminates the problem of multimodality in the deviance. This is achieved by the investigator selecting a subset of individuals for confirming state (e.g., bleeding a bird to determine its sex), or by modeling the probability that an unambiguous event occurs (e.g., a marked sea lion pup demonstrates it is not weaned by suckling from an adult female).

In each of the recapture cases I presented misclassification was in one direction, which simplifies the likelihood. More importantly for analyzing the data, each primary period of interest consists of multiple sampling occasions. In each case the state was considered static within a primary period. Whereas with the manatee and kittiwake examples demographic and geographic closure was assumed, in the sea lion case the geographic closure assumption was relaxed. A next step would be to develop a likelihood structure for which these three cases would represent special cases. This would constitute an open robust design counterpart to the Pradel (2005) model (Fig. 2). This would represent a very complex structure, with the following types of parameters: For each primary period, the probability an individual occupies a given state, then for each state the probability of surviving to the next period and transitioning to a given state. Within each primary period, for each secondary sampling occasion and for each state occupied there would be a probability of arrival to the study area just prior, the probability of detection, the probability of assigning the individual to each possible state (based on an event), and the probability an individual remains in the study area until the next sampling occasion. So the inclusion of time-dependent state structure and state transition probabilities in the same model permits modeling of the relationship between the two.

The West Nile Virus example illustrates that misclassification also arises in occupancy modeling. Nevertheless the approach to adjusting for this phenomenon was similar to the case of recapture studies. In this case also misclassification occurred in only one direction. I used it in a disease modeling context, and Nichols et al. (2007) developed and applied the same model in a case of classifying

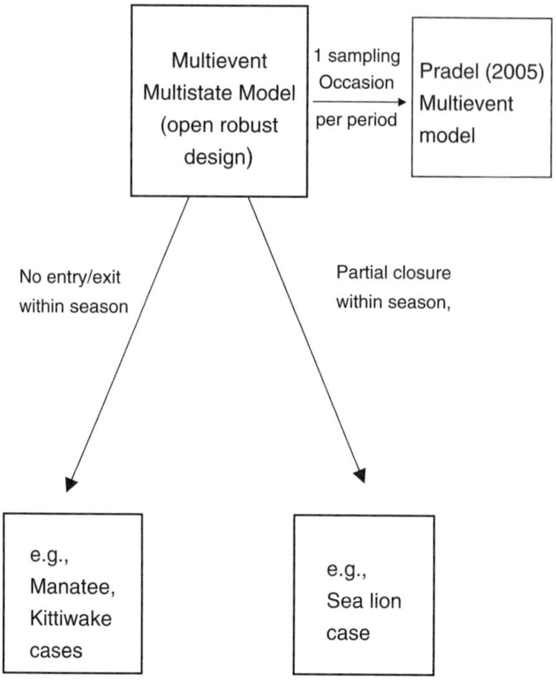

Fig. 2 An outline of the relationship between the Pradel (2005) multievent model, the manatee, kittiwake, and sea lion case studies presented here, and an open robust design multievent model that could be developed for an arbitrarty number of states and events

breeding status. Given the recent surge in occupancy modeling, more misclassification issues can be expected to arise. The avian influenza case demonstrated how misclassification can be brought into the modeling of occupancy at various spatial scales, while carrying adjustments for misclassification through each level, including calibration in the lab. This sampling at multiple spatial or temporal levels provides data for parameterizing multi-level or hierarchical models of system dynamics.

Accounting for state uncertainty in capture–recapture models is a relatively recent area of research. Therefore there is plenty of opportunity to investigate general and application-specific questions. The Pradel (2005) model provides a very nice general framework for analyzing many state uncertainty problems, and E-SURGE (Choquet et al. 2008) will prove a very useful tool to use in that pursuit. Nevertheless, the cases I have presented here reminds us that new ecological field projects and new study designs often require new or customized statistical models. The general multievent open robust design model alluded to in Fig. 2 could provide a general tool for three of the cases described in this paper, as well as many others, but this model and the Pradel (2005) model would still not likely be exhaustive in their combined coverage of cases.

Besides model development, there are plenty of open questions related to this issue in terms of study design and data selection, which could be evaluated using simulation or expected value methods. In previous work the use of robust design models have been shown to increase precision in estimators, in comparison with methods that ignore this structure and pool across secondary samples. The same can be expected with state uncertainty models, but the extent of improvement, or how it performs with small sample sizes, should be investigated. Precision should also be investigated as a function of the proportion of encounters that result in unambiguous assignment to state.

When misclassification can be viewed as occurring in one direction, the problem becomes simpler conceptually and some parameters are eliminated. For example, in the manatee study when Kendall et al. (2004) were assembling data for the analysis, in some cases female that were noted to possibly have a calf were assigned to the skipped breeder state to maintain the one-way classification. Are there cases (e.g., where very few animals are unambiguously in a state) where forcing this one-way misclassification is not advisable?

Finally, the assessment of goodness of fit is also an open question with these models, both with and without the robust design. One could envisage an extension of the method of Pradel et al. (2003) to assess fit. There is also the possibility of using the median c-hat method found in program MARK (White and Burnham 1999). A Pearson Chi-square test could also be conducted.

In conclusion, I have presented several disparate ecological studies where the issue of misclassification arises. My purpose was to illustrate that the problem can be fairly common, that with various types of extra information it can be dealt with statistically, and that the approaches needed to address the problem can vary by case study. Therefore generalized software like E-SURGE (Choquet et al. 2008) can be used to address the problem in some cases, and other software like MSSURIVmis (www.mbr-pwrc.usgs.gov/software.html) might be needed for other cases. As with detection probability, in many cases ignoring misclassification can produce substantially biased estimates of demographic parameters of interest, including survival and state transition probabilities. More work is needed on eliciting such problems from field ecologists, and developing statistical models and software to adjust for them. This interaction and basic modeling will be the same, regardless of whether ultimately frequentist or Bayesian statistical models are used.

Acknowledgments I thank Catherine Langtimm, Cathy Beck, and Bob Bonde for discussions about manatees; Emmanuelle Cam for discussion about kittiwakes; Grey Pendleton for discussion about sea lions; Alan Franklin, Matt Farnsworth, Ryan Miller, Gary White, Paul Doherty, and Ken Burnham for discussions of disease modeling, and Jim Nichols and Jim Hines for discussions of misclassification in general. Roger Pradel and an anonymous referee provided helpful comments on an earlier draft of this paper

References

Arnason AN (1972) Parameter estimates from mark-recapture experiments on two populations subject to migration and death. Researches on Population Ecology 13:97–113.

Arnason AN (1973) The estimation of population size, migration rates, and survival in a stratified population. Researches in Population Ecology 15:1–8.

Brownie C, Hines JE, Nichols JD, Pollock KH, Hestbeck JB (1993) Capture–recapture studies for multiple strata including non-Markovian transitions. Biometrics 49:1173–1187.

Cam E, Cadiou B, Hines JE, Monnat JY (2002) Influence of behavioural tactics on recruitment and reproductive trajectory in the kittiwake. Pages 163–185 *in* B. J. T. Morgan and Thomson DL, editors. 'Statistical Analysis of Data from Marked Bird Populations', Journal of Applied Statistics 29, nos 1–4. ISSN 0266-4763.

Caswell H (2001) Matrix Population Models: Construction, Analysis, and Interpretation. Sinauer Associates, Sunderland, MA.

Choquet R, Lauriane R, Pradel R (2008) Program E-SURGE: a software for fitting multievent models. In: Thomson DL, Cooch EG, Conroy MJ (eds.) Modeling Demographic Processes in Marked Populations. Environmental and Ecological Statistics, Springer, New York, 3: 845–866.

Hestbeck JB, Nichols JD, Malecki RA (1991) Estimates of movement and site fidelity using mark-resight data of wintering Canada Geese. Ecology 72:523–533.

Kendall WL (2004) Coping with unobservable and mis-classified states in capture–recapture studies. Animal Biodiversity and Conservation 27:97–107.

Kendall WL, Bjorkland R (2001) Using open robust design models to estimate temporary emigration from capture–recapture data. Biometrics 57:1113–1122.

Kendall WL, Conn PC, Hines JE (2006) Combining multistate capture–recapture data with tag recoveries to estimate demographic parameters. Ecology 87:169–177.

Kendall WL, Hines JE, Nichols JD (2003) Adjusting multistate capture–recapture models for misclassification bias: manatee breeding proportions. Ecology 84:1058–1066.

Kendall WL, Langtimm CA, Beck CA, Runge MC (2004) Capture–recapture analysis for estimating manatee reproductive rates. Marine Mammal Science 20:424–437.

Lebreton J-D, Almeras T, Pradel R (1999) Competing events, mixtures of information and multistatum recapture models. Bird Study 46:S39–S46.

Lebreton J-D, Pradel, R (2002) Multistate recapture models: modelling incomplete individual histories. Pages 353–370 *in* B. J. T. Morgan and Thomson DL, editors. 'Statistical Analysis of Data from Marked Bird Populations', Journal of Applied Statistics 29, nos 1–4. ISSN 0266-4763.

MacKenzie DI, Nichols JD, Royle JA, Pollock KH, Bailey LL, Hines JE (2006) Occupancy estimation and modeling. Academic Press, San Diego, CA.

Nichols JD, Hines JE, MacKenzie DI, Seamans ME, Gutierrez RJ (2007) Occupancy estimation and modeling with multiple states and state uncertainty. Ecology 88:1395–1400.

Nichols JD, Kendall WL, Hines JE, Spendelow JA (2004) Estimation of sex-specific survival from capture–recapture data when sex is not always known. Ecology 85:3192–3201.

Pollock KH (1982) A capture–recapture design robust to unequal probability of capture. Journal of Wildlife Management 46:757–760.

Pradel R (2005) Multievent: an extension of multistate capture–recapture models to state uncertainty. Biometrics 61:442–447.

Pradel R (2008) The stakes of models with state uncertainty. In: Thomson DL, Cooch EG, Conroy MJ (eds.) Modeling Demographic Processes in Marked Populations. Environmental and Ecological Statistics, Springer, New York, 3:781–796.

Pradel R, Wintrebert C, Gimenez O (2003) A proposal for a goodness-of-fit test to the Arnason–Schwarz multisite capture–recapture model. Biometrics 59:43–53.

Runge JP, Hines JE, Nichols JD (2007) Estimating species-specific survival and movement when species identification is uncertain. Ecology 88:282–288.

Schwarz CJ, Schweigert JF, Arnason AN (1993) Estimating migration rates using tag-recovery data. Biometrics 49:177–193.

Schwarz CJ, Stobo WT (1997) Estimating temporary migration using the robust design. Biometrics 53:178–194.

White GC, Burnham KP (1999) Program MARK: survival estimation from populations of marked animals. Bird Study 46:120–139.

Williams BK, Nichols JD, Conroy MJ (2002) Analysis and Management of Animal Populations. Academic Press, San Diego, CA.

The Stakes of Capture–Recapture Models with State Uncertainty

Roger Pradel

Abstract The development of the use of CR multistate models is a major feature of the last 5 years. However, concerns have rightfully appeared about uncertainty in state assignment. I examine situations where uncertainties seem to be intrinsic such as with breeding status. But I also argue that uncertainty is not just a liability, it can be an opportunity – for instance, to exploit more fully the data at hand and limit disturbance. Then I examine the methodological answers that have been proposed. They mainly concern the models conditional on first release and are of a more or less general applicability. I advocate a general approach that can be adapted to each particular case and be used to expand extant specialized approaches. I will also consider how uncertainty could be incorporated into non-conditional models such as models of stopover duration. I conclude that, with the advent of genetic sampling, the new challenges for CR models will be uncertainty in individual identity and dependence among individuals.

Keywords Heterogeneity of capture · Hidden Markov model · Label switching · Mixture models · Multievent · Sex uncertainty · Stopover duration

1 Introduction

Capture–recapture (CR) studies have spread rapidly in the ecological literature since the beginning of the 1990s (Fig. 1). At the same time, their focus has changed. Once almost exclusively used to address questions of population dynamics, its pertinence

---

R. Pradel (✉)
CEFE, 1919 rte de Mende, 34293 Montpellier cedex 5, France
e-mail: roger.pradel@cefe.cnrs.fr

R. Pradel is devoted to developping sound quantitative methods for population biology and has specialized in capture–recapture models. He has developped goodness-of-fit tests and has shown that capture–recapture models are a particular type of heterogeneous hidden Markov models, opening the way to a number of generalizations. The author or coauthor of several computer programs, he organizes workshops for biologists on a regular basis and has contributed to many conservational and evolutionary studies. He is currently the leader of the Biometry and Population Biology team in CEFE, Montpellier, France.

D.L. Thomson et al. (eds.), *Modeling Demographic Processes in Marked Populations*, Environmental and Ecological Statistics 3, DOI 10.1007/978-0-387-78151-8_36,

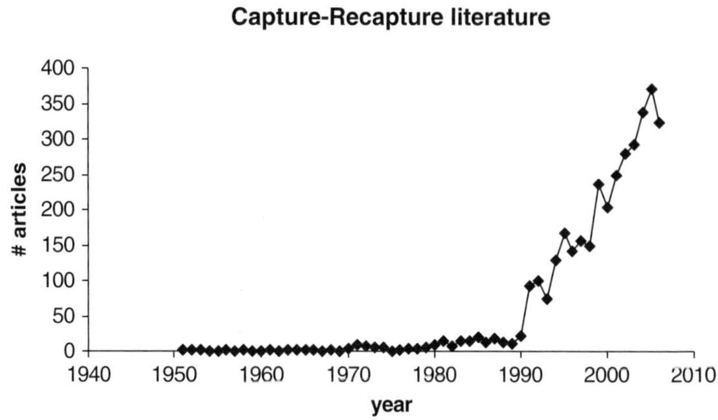

Fig. 1 The spread of capture–recapture studies: research in Web of Science of articles with the keywords 'capture–recapture' or 'mark-recapture' or 'sight–resight' or 'band recovery' or 'marked animals'

in addressing evolutionary questions and studying life history strategies is now recognized. Indeed, CR studies are an excellent means of assessing an individual's performance against what it did at other times (strategy questions) but also according to its distinctive features as compared to other individuals (individual variability). The models for the analysis of CR data have had to evolve to meet the new demands. In the early 90s, models that account for the state of an individual (Arnason 1973) have been rediscovered and developped and have seen an ever increasing use since 1998 and particularly for the last five years. Today they represent 5% of all CR publications and this proportion is likely to increase in the foreseeable future (Fig. 2). The first multistate papers to appear were theoretical papers dealing

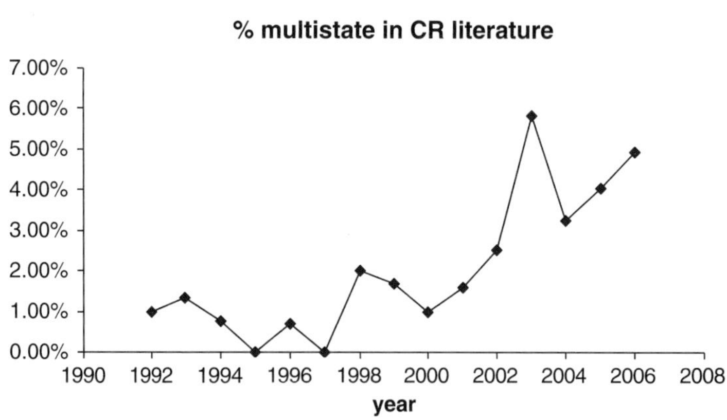

Fig. 2 Percentage of multistate papers in the capture–recapture literature: research in Web of Science among the articles of the research in Fig. 1 with the keywords 'multistate' or 'multisite' or 'multistrata'

Table 1 Multistate papers by category. The original use of multistate models was the study of geographical movements (column 2), the study of transitions between physiological or developmental states followed (column 3 and 4) with an emphasis on reproduction (column 3). Technical uses where the states refer to a methodological status (presence or absence of the tag, live or dead encounter...) have developped recently (column 5). The papers dealing with the statistical properties of the multistate models are now less important in proportion than they used to be (column 6)

Year	Geographical movement	Breeding status + recruitment	Stages (physiological, developmental ...)	Technical uses (mixture of information, tag loss ...)	Purely statistical papers
2006	5	5	2	2	1
2005	3	5	1	2	1
2004	4	0	1	5	1
2003	9	4	2	0	4
2002	4	0	1	0	2
2001	2	1	0	0	1
2000	1	1	0	0	0
1999	2	0	0	0	2
1998	1	1	1	0	0
1997	0	0	0	0	0
1996	0	0	0	0	1
1995	0	0	0	0	0
1994	0	1	0	0	1
1993	0	0	0	0	1
1992	0	0	1	0	1
All years	31	18	9	9	16

with the statistical aspects of these models. Then came papers describing from a theoretical point of view the adaptation of the general formulation to particular uses. Eventually, actual uses have become dominant, which is a sign of maturity (see Table 1). Most of the early uses were about movement but now we see a greater variety of applications where the states considered are physiological or developmental with a dominance of the breeding status (several studies of trade-off and recruitment). However, along with this new type of studies, a growing concern about uncertainty has been rightfully voiced (Fig. 3). Back in the 60s, the concern was about the uncertainty of detection of the individuals in the field – obviously it is very difficult to meet with certainty a free-ranging individual and the CR models have actually been designed to deal with this problem – but today the concern is about state assessment. Uncertainty in state assessment may arise for instance with the breeding status of an animal seen without young during the breeding season, or the health status in an epidemiological study where only outer symptoms can be observed. Uncertainty is also intrinsically present with any hidden condition, even static, like the sex in monomorphic species or more generally the membership in a particular class of a heterogeneous population (Pledger et al. 2003). In a more technical ground, tag loss can be treated as a problem of uncertain state where the

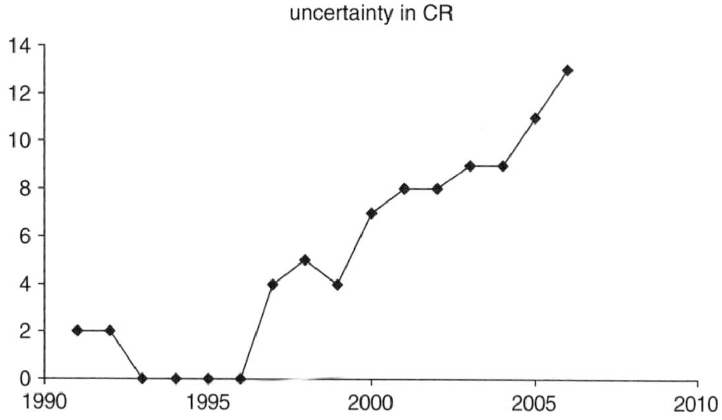

Fig. 3 Concern about uncertainty in the CR literature: number of papers with the keyword 'uncertainty' among those from the research of Fig. 1

state is the presence or absence of the mark. All in all, just like for uncertainty of detection, state uncertainty seems to be unavoidable in many situations.

But I believe that state uncertainty has also positive aspects that should be stressed. Being able to tackle state uncertainty means making use of all the information that is currently unused for fear of being mistaken and this is often the bulk of the information collected. For instance, the behavioral clues are rarely 100% sure and yet they are often the basis for determining the breeding status or the sex of an individual. Biometrical measures are another example of 'imperfect' information that is nonetheless commonly used to e.g. sex animals. By becoming able to incorporate such information, we can not only retain a greater number of individuals in our data set but also change the way we carry the field work. In devoting less time to each individual, we achieve two desirable aims: we limit disturbance and can monitor a larger sample of the population with the same amount of time and effort. In limiting disturbance, we show more consideration for animal welfare and at the same time observe a more natural system.

Yet, some colleagues have voiced concern about the risk of being unable to estimate the parameters with such data and consider that the status of some individuals must be known with certainty to avoid redundancy problems. It is too early to answer unequivocally. However, I am more optimistic and do not believe that we should shy away from the general treatment of uncertain states. The applications I have considered so far do not behave differently than the mutistate models with respect to redundancy. I will say more on this point in the discussion but let us already note that (1) redundancy problems can now be detected automatically so that the use of models with state uncertainty is not a blind move (Choquet et al. this issue), (2) it is often possible to incorporate some hardcore data with no uncertainty gathered with more laborious but more reliable techniques such as genetic sexing in the case of sex uncertainty, and (3) if needed, a Bayesian approach can be adopted to incorporate prior information (see Gimenez et al. this issue).

2 Conditional CR Models with Uncertainty

Up to now, a number of specific responses have been given to the cases perceived as most important (breeding status: Kendall et al. 2004; sex membership: Fujiwara and Caswell 2002, Nichols et al. 2004; hidden heterogeneity: Pledger et al. 2003). Yet, another approach consists in bringing a common response to all cases at once by treating the question of uncertainty in general terms (multievent models, Pradel 2005). A practical disadvantage of the broad approach is that any specific use will suppose some tuning of the general model; on the other hand, it provides maximum flexibility, which is all the more desirable that all situations are unlikely to be anticipated and covered otherwise. Also, providing a unified frame of thought usually proves fruitful in stimulating new ideas. The multievent approach I advocate seeks in fact to achieve a balance between the inconvenience of a general formulation and its advantage in terms of flexibility by introducing simple common concepts that underlie all specific situations. It is on this basis that a unique computer program has been developped (Choquet et al. this issue) and can be used for particular applications.

The main idea is to decouple the observation from the state assessment. In a multistate approach dealing with the breeding status, a typical capture history can go like this: (breeder, unseen, breeder). This is what is coded and therefore analyzed, but the untold truth is perhaps (seen on the colony, not seen, seen feeding nearby the colony). In effect, the biologist has made the decision that the first and last observations were tantamount to breeding status. I argue that this decision is in fact better left to the model. To this aim, we must code in the capture history not the states, which we leave open, but the particular type of observation that was made when the animal was encountered or the fact that the animal was not observed : as a general term, I speak of 'events'. Thus, the central idea is to shift the emphasis from the states to the events (Fig. 4). The multievent model then describes two processes: the process of transition among states, and the process of generation of events given the underlying state. It has three fundamental types of parameters (Fig. 4):

(1) The 'initial state probabilities', which describe the probability that an individual is in one or another state when it is first encountered

$$\mathbf{\Pi}_t = (\pi_1 \pi_2 \dots \pi_s) .$$

This is an entirely new type of parameters from multistate models. Dependent on the kind of study, it can be related to the sex-ratio, the prevalence of a disease or the percentage of breeders in the population.

(2) The 'transition probabilities', which are exactly as in the multistate models except that the state 'dead' is explicit; thus, the corresponding matrix below with departure state in rows is row-stochastic:

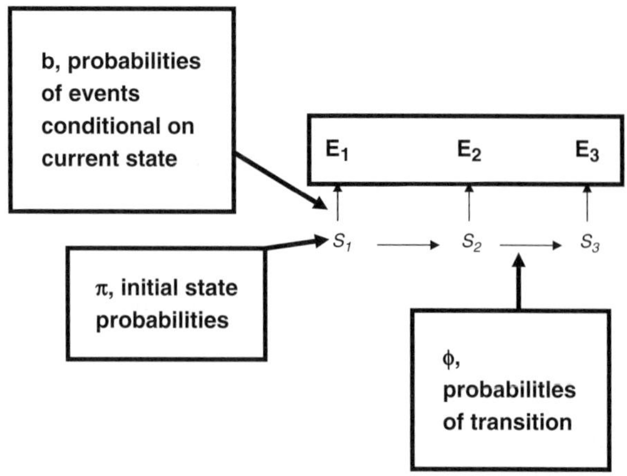

Fig. 4 The rationale of multievent models. The encounter history of a marked individual (boxed above) is made of the events E_i, not the states. The states S_i are related to the observed encounter history through the probabilities of generation of events given the states, b

$$\mathbf{\Phi}_t = \begin{pmatrix} \phi_{11} & \phi_{12} & \dots & \phi_{1s} \\ \phi_{21} & & & \\ \vdots & & \ddots & \\ \phi_{s1} & & & \phi_{ss} \end{pmatrix}.$$

(3) The 'conditional event probabilities', which are the probabilities of the events conditional on the underlying state. They generalize the capture probabilities of multistate models. For instance, with the event 'seen' and the state 'breeder', the corresponding conditional event probability is the probability that a breeder is seen, i.e. the traditional capture probability of breeders, p_{breeder}. With the states in rows and the events in columns, the corresponding matrix is row-stochastic.

$$\mathbf{B}_t = \begin{pmatrix} b_{11} & b_{12} & \dots & b_{1l} \\ b_{21} & & & \\ \vdots & & \ddots & \\ b_{s1} & & & b_{sl} \end{pmatrix}.$$

To see how the probability of a capture history can be written with these parameters, let us consider again the capture history (seen on the colony, not seen, seen feeding nearby the colony). As we never know what the real state is, we have to consider in turn all $2^3 = 8$ possible trajectories between the breeding and non-breeding states over the three years. For each trajectory, we calculate the associated probability and then add them all together to get the probability of the capture history itself. For instance, when considering the possibility that the animal was breeding in all three years, we will write: P(breeder)P(seen on the colony|breeder)P(breeder in year 2|breeder in year 1)P(not seen| breeder) P(breeder in year 3|breeder in year 2) P(seen feeding nearby the colony | breeder). P(breeder) is an initial state probability. Then, the rest of the formula alternates between event probabilities and transition probabilities. Here the event probabilities are P(seen on the colony|breeder), P(not seen| breeder) and P(seen feeding nearby the colony | breeder). The transition probabilities are P(breeder in year 2|breeder in year 1) and P(breeder in year 3|breeder in year 2). The probability of this capture history is found as the sum over all 8 possible paths. It can be written in compact matrix form: $\mathbf{\Pi} D(\mathbf{B}_1(.,\text{seen on the colony}))\mathbf{\Phi}_1 D(\mathbf{B}_2(.,\text{not seen}))\mathbf{\Phi}_2 D(\mathbf{B}_3(.,\text{seen feeding near by the colony}))$ where $D\,(\mathbf{B}_t\,(.,\text{ event } i))$ is the diagonal matrix with diagonal elements taken from the column of $\mathbf{B}_t$ corresponding to event i (Pradel 2005).

Starting from the general layout of multievent models that we have just seen, one has to specify the exact form taken by the matrices of parameters for each particular application. A module in program E-SURGE (Choquet et al. this issue) facilitates this step. This program then maximizes the likelihood through a quasi-Newton algorithm. Following the steps of Gimenez et al. (2007), it is also possible to express the multievent model as a state-space model and, on this basis, to use MCMC or filtering algorithms. I illustrate now how the general model can be particularized with the example of two previously treated situations. I then show how the new formulation leads easily and naturally in each case to the treatment of a broader biological context.

2.1 Models with Unknown Sex Individuals

Nichols et al. (2004) have studied the situation of a monomorphic species where the sex of the individuals cannot always be ascertained but, when it can, this is done without error. To rewrite this model as a multievent model, we consider 3 states: live female (F), live male (M) and dead individual (†), and 3 exclusive events: ‘not encountered’, ‘sex not ascertained’ and ‘sex ascertained’. Because the first time an individual is encountered it is necessarily alive, the initial state probabilities involve only the proportion of females among unmarked, π, as an independent parameter. (Note that, because of sampling considerations, π is not simply related to the sex ratio. I will come back to this point in the discussion.)

$$\mathbf{\Pi}_t = \begin{pmatrix} \pi & 1-\pi & 0 \end{pmatrix}.$$

The transition probabilities are very simple. A female (resp. male) survives with probability ϕ^F (resp. ϕ^M) and remains a live female (resp. male) or dies and joins the state ‘dead’. A dead individual remains dead forever. Hence,

$$\mathbf{\Phi}_t = \begin{pmatrix} \phi^F & 0 & 1-\phi^F \\ 0 & \phi^M & 1-\phi^M \\ 0 & 0 & 1 \end{pmatrix}.$$

The conditional event probabilities involve the sex-dependent probabilities of encounter and the probability to ascertain the sex of an encountered individual: δ^F

(resp. δ^M) for a female (resp. male) (Nichols et al. notation). For instance, the probability that a female is encountered and its sex is ascertained is p^F, the probability that a female is encountered, times δ^F, the probability that its sex is then ascertained.

$$\mathbf{B}_t = \begin{pmatrix} 1-p^F & p^F\left(1-\delta^F\right) & p^F\delta^F \\ 1-p^M & p^M\left(1-\delta^M\right) & p^M\delta^M \\ 1 & 0 & 0 \end{pmatrix} \quad (1)$$

(columns: not encountered; sex not ascertained; sex ascertained)

(This corresponds to sampling situation A of Nichols et al.)

One interest of putting this model into the multievent framework is that it is now fairly easy to expand to more complex situations. Often, the clues used to determine the sex are not perfect. They nevertheless can be used if a probability of error is introduced. Pradel et al. (2008) present a thorough treatment of this situation. I give here a summary. Let $x^F<1$ and $x^M<1$ be the probabilities to sex resp. females and males correctly. The states remain the same as above but we have now 4 possible outcomes or events to consider: 'not encountered', 'sex not determined', 'named sex female', 'named sex male'. The conditional event probability matrix then becomes:

$$B_t = \begin{pmatrix} 1-p^F & p^F\left(1-\delta^F\right) & p^F\delta^F x^F & p^F\delta^F\left(1-x^F\right) \\ 1-p^M & p^M\left(1-\delta^M\right) & p^M\delta^M\left(1-x^M\right) & p^M\delta^M x^M \\ 1 & 0 & 0 & 0 \end{pmatrix}$$

Contrasting this form of $\mathbf{B}_t$ from that of formula (1), we see that the last event has been duplicated to incorporate the possibility that the named sex is not correct. The $\mathbf{\Phi}_t$ and $\mathbf{\Pi}_t$ matrices remain unchanged.

I have applied this enlarged model to the treatment of sex specificity in parameters of an Audouin's gull population where the sex clues are behavioural displays and relative body size. Four criteria are used: 'copulation', 'begging food', 'courtship feeding' and 'relative body size'. The determination of sex is based on the role played by an individual in a pair. The previous model was further enlarged to allow for different error rates with different criteria. There are then 9 events to consider (see Table 2). Eight are obtained by crossing the 4 criteria with the 2 named sexes; the last corresponds to when the sex is not determined. The output from this model (Table 3) shows that relative body size is the less reliable criterion but position during copulation is not so good a criterion as was originally thought. In

Table 2 Events considered and codes used in the encounter histories of Audouin's gulls found at the Ebro delta (Spain)

Criterion used	Named male	Named female	No sex named
Copulation	1	5	9
Begging food	2	6	
Courtship feeding	3	7	
Body size	4	8	

Table 3 Results from the model with uncertain sex assessment for Audouin's gulls at the Ebro Delta, Spain (excerpt from Pradel et al. 2008, Table 5)

Parameter	Estimate (SE)
Proportion of females	0.47 (0.03)
Female survival	0.91 (0.01)
Male survival	0.86 (0.01)
Error copulation	0.06 (0.04)
Error begging food	0.05 (0.03)
Error courtship feeding	0.00 (0.16)
Error body size	0.11 (0.06)

designing future field work, the quality of the information brought by the different criteria could be balanced with the difficulty of gathering this particular piece of information.

2.2 Heterogeneity Models

There is almost always some sort of heterogeneity among individuals in a population. In particular, heterogeneity of catchability can be high. Pledger et al. (2003) have brought a general answer to these questions by considering the existence of several hidden classes of individuals each with their own parameters. This model in all its generality has a straightforward formulation in the multievent framework. The states are the classes plus the state 'dead'; the events are simply 'encountered' and 'not encountered'. With c classes, we have:

1. initial state probabilities (remember that the last state is 'dead' and is not observable as an initial state)

$$\mathbf{\Pi}_t = \begin{pmatrix}\pi_1 \; \pi_2 \; \ldots \; \pi_c \; 0\end{pmatrix}$$

   with $\pi_1 + \pi_2 + \ldots + \pi_c = 1$.
2. transition probabilities

$$\mathbf{\Phi}_t = \begin{pmatrix} \phi_1 & 0 & \ldots & 1-\phi_1 \\ 0 & \phi_2 & & 1-\phi_2 \\ \vdots & & \ddots & \\ & & & 1 \end{pmatrix}$$

3. conditional event probabilities (first column is 'not encountered')

$$B_t = \begin{pmatrix} 1-p_1 & p_1 \\ 1-p_2 & p_2 \\ \ldots & \ldots \\ 1 & 0 \end{pmatrix}$$

This model can now easily be generalized to allow transitions among the different classes. This may be relevant if the membership in one class correspond for instance to a social status. Here I consider a different situation. In the course of a long-term study of black-headed gulls in central France, a breeding colony inhabiting a pond is screened each year from a floating hide. This colony has open water and clusters of vegetation. The gulls nesting on the edge of the vegetation clusters are easy to monitor unlike those deep inside. Because the gulls tend to come back from year to year to the same place in the colony (Prévot-Julliard et al. 1998b), this introduces a type of heterogeneity of catchability among individuals that can be reduced in a first approximation to two classes of catchability (Prévot-Julliard et al. 1998a). However, over the years an individual bird may move and change area. The model can be modified to account for this last fact by introducing transitions between classes conditional on survival, ψ_{12} and ψ_{21}. In the specific model used for this data set, I assumed two classes of catchability ($c = 2$), no heterogeneity of survival ($\phi_1 = \phi_2$) and for the sake of simplicity that the parameters were constant over time. I also assumed that no attempt was made at identifying the class membership from the position in the colony and the activity at the time of an observation. The states, events, $\mathbf{\Pi}$ and $\mathbf{B}$ matrices are thus essentially unchanged from the general multievent formalization of the Pledger, Pollock and Norris model above. Only the transition matrix changes structure:

$$\mathbf{\Pi}_t = \begin{pmatrix}\pi_1 & \pi_2 & 0\end{pmatrix}$$

$$\mathbf{\Phi}_t = \begin{pmatrix} \phi(1-\psi_{12}) & \phi\psi_{12} & 1-\phi \\ \phi\psi_{21} & \phi(1-\psi_{21}) & 1-\phi \\ 0 & 0 & 1 \end{pmatrix}$$

$$\mathbf{B}_t = \begin{pmatrix} 1-p_1 & p_1 \\ 1-p_2 & p_2 \\ 1 & 0 \end{pmatrix}$$

Table 4 Results from models of heterogeneity of capture for Black-headed gulls inhabiting the La Ronze pond, central France. The data set has 176 individuals monitored over 19 years. Parameters with s.e. between parentheses are: ϕ survival, π proportion of low-catchability individuals, p_1 low capture probability, p_2 high capture probability, ψ_{12} transition from the low to the high catchability class, ψ_{21} transition from the high to the low catchability class. In the first model 'w/o movement' individuals are not allowed to move from one catchability class to the other; they can do so in the second model 'w/ movement'

Model	ϕ	π	p_1	p_2	ψ_{12}	ψ_{21}	Dev
w/o movement	0.82	0.84	0.08	0.32	–	–	2880.25
	(0.08)	(0.02)	(0.02)	(0.06)			
w/ movement	0.83	0.95	0.09	0.48	0.02	0.09	2873.62
	(0.02)	(0.06)	(0.02)	(0.10)	(0.01)	(0.11)	

For the black-headed gull data set, the model allowing transitions between the two classes fits better than the one with frozen states even though the estimated movements are relatively low as expected. They are more frequent toward the zone of low catchability. This is in agreement with the relative number of nests with low and high catchability: those on the edge of the vegetation clusters and those in the middle (Table 4).

3 Uncertainty in Unconditional Models

So far, I have considered only models conditional on the time of first encounter. But there are several areas where it is important to look at the part of the capture history that precedes the first capture and uncertainty can actually arise from the ignorance of when an animal entered the population. A clear example of this is provided by the question of stopover duration: how long does a bird stay on a stopover site before flying on? Its probability of departure likely depends on how long it has already been present on the site because the more time it has been around, the more likely it is to have refuelled enough to start its next migration leg (Pradel et al. 2005; for an early model of departure dependent upon arrival time, see Crosbie and Manly 1985). The uncertainty of the time elapsed since arrival can be handled by considering hidden states: 'not yet arrived' (–), 'just arrived' (pr0), 'arrived one day earlier' (pr1), 'arrived two days earlier' (pr2)..., 'departed' (†). The events are just two: 'encountered', 'not encountered'. Eventually, we need an additional type of parameter for the transitions from the state 'not yet arrived', the probability of arrival on day t for a bird 'not yet arrived', a_t. As for ϕ_i the probability of remaining on the site for one more day, it will be dependent on age i. With this model, the description of each encounter history starts at date 1 so that the initial state probabilities are only needed at this time. Theoretically, we could keep open the possibility that a bird present on day 1 has been here for several numbers of days. However, if the study is started sufficiently early during the migration period, we can assume that all the birds present on day 1 have just arrived. This is the simplifying assumption I am making here. Thus there are just two initial states possible on day 1: the birds that have not yet arrived and those which are arriving.

$$\mathbf{\Pi}_1 = \left(1 - a_1 \; a_1 \; 0 \ldots 0\right)$$

After its arrival, a bird may depart any time. If it does not, the transition to the next state will be automatic:

$$\mathbf{\Phi}_t = \begin{pmatrix} 1-a_t & a_t & 0 & \cdots & 0 \\ & & \phi_1 & & 1-\phi_1 \\ & & & \ddots & 1-\phi_2 \\ & & & & \vdots \\ & & & & 1 \end{pmatrix} \quad \begin{matrix} - \\ \text{pr1} \\ \text{pr2} \\ \vdots \\ \text{departed} \end{matrix}$$

Table 5 Estimated probabilities of departure of reed buntings (*Emberiza schoeniclus*) from a stopover site Bolle di Magadino when probability of departure depends on the time since arrival. The data set comes from a previous study (Schaub et al. 2001). It has 1712 individuals. The data are pooled in 5-day periods. The calculated average stopover duration is 7.03 days. It was 7.30 days in the original study with time-dependent departure rates

Time since arrival	Just arrived <5 days	5–9 days	10–14 days
Probability of departure before 5 days	0.19	0.65	1

I also assume for the sake of simplicity that the encounter probability is constant over time and does not depend on how long a bird has been present:

$$\mathbf{B} = \begin{pmatrix} 1 & 0 \\ 1-p & p \\ \vdots & \vdots \\ 1-p & p \\ 1 & 0 \end{pmatrix}$$

As an example, let us examine the capture history 010. This is a bird that may or may not have been present the first day and may or may not have departed before the third day. In combining presence or absence in the first and last days, we have to consider 4 different possibilities. If the bird was present all three days, it was present the first day with probability a_1 in state pr0, it remained one day more reaching state pr1 (transition probability ϕ_1), it was spotted on day 2 (probability p), then it remained for another day moving to state pr2 (probability ϕ_2) but was missed on day 3 (probability $1-p$) . These five probabilities put together make for the first term of the complete expression for the probability of capture history 010, which is:

$$\begin{aligned} \mathrm{P}(010) = & a_1\phi_1 p\phi_2(1-p) + a_1\phi_1 p(1-\phi_2) + (1-a_1)a_2 p\phi_1(1-p) \\ & + (1-a_1)a_2 p(1-\phi_1). \end{aligned}$$

Actually, this model is not entirely unconditional. To appear in the data set, a bird must have been contacted at least once. Hence, we must condition on this fact: the corresponding probability appears as a denominator in the likelihood. Because of this last conditioning, this model is not a multievent model of the type defined in Pradel (2005) and it cannot currently be fitted with E-SURGE. I have written a small MATLAB program to implement it. Applied to one example of the original stopover paper (Schaub et al. 2001), the model yields probabilities of departure that as expected increase strongly with the time spent on the site (Table 5).

4 Discussion

The advantage of a general treatment of state uncertainty in CR studies is to provide a unifying framework and a level of abstraction favorable to clear thinking and generalizations. In a way, it forces to better comprehend the structure of the study and hence to clarify the question investigated. This advantage largely offsets in my mind the inconvenience of having to identify the specifics of the study. This is all the more true that the idea of decoupling events and states is simple enough to be easily grasped and applied to different situations. A practical advantage is that a common tool can then be used to analyse the different types of data; a computer program with a model description language has been developed to this effect (Choquet et al. this issue). We have also seen that the simple idea of decoupling events and states can be carried out to non-conditional models. *All in all, the current problem of state uncertainty is no different from the old problem of uncertainty of detection* to which CR models have been the answer. The risks are similar: if the probability of misclassification depends on the state, estimates would inevitably be biased in the absence of correction. In our field of work today nobody thinks of ignoring the necessary correction for $p < 1$; in the future, probably nobody will think of not correcting for state uncertainty.

A possible concern with the new models is that they may be unstable and especially that their parameters might often be non-identifiable. The risk is real but it does not seem to be as acute as one might have feared. With my still limited experience with these models, I have acquired the feeling that the risk of redundancy is of the same order of magnitude as in multistate models, the only difference being that, with time-dependent models, redundancy occurs at the start as well as at the end of the experiment. Today, there is a variety of tools for studying redundancy (see for instance Hunter et al. this issue). However, the really good news is that the identification of redundant parameters has been automated (Choquet et al. this issue) so that there is no need to study the formal structure of the models in advance.

The occurrence of label switching (Redner and Walker 1984) is another, entirely different problem, new to the CR area, and inherited from the hidden Markov model structure of the multievent models (Pradel 2005). Here, several discrete statistically equivalent solutions exist. For instance, in a study of sex assignment in monomorphic species, a second solution is easily obtained from the correct one by renaming males the females and females the males (see Pradel et al. in press, for a proof). However, the question is moot most of the time because one of the solutions is totally unrealistic; we would probably have to accept unrealistically high rates of error in sex assignment with the 'wrong' solution. Of course, this is a very simple situation. In general, there will be more than two states and they will not be static. However, if the events are carefully chosen, it should always be easy to distinguish the correct solution from another one obtained by arbitrarily shuffling the states. Suppose that we know beforehand that a particular event has a high probability of realisation under state 1 but not under the others. The examination of the conditional probabilities of this event will rule out any statistically valid solution where state 1 is misplaced. Ultimately, the label switching problem seems to be only a problem if the events are not discriminatory.

In fact, if an ambiguity exists – be it linked to redundancy or to label switching – the right attitude is probably to try to augment the information. This can be done by obtaining additional information from new data or by incorporating knowledge from previous studies. To carry on with the example of sex assignment in monomorphic species, additional information can consist in sexing genetically a few individuals. The initial state probabilities for these individuals can then be fixed to the appropriate value to specify what their sex is. Little additional information seems to be needed to eliminate the wrong solution (Pradel et al. in press). The other option is to use an a priori distribution for the parameters, i.e. doing a Bayesian analysis. The initial state probabilities are good candidate for this: again with the sex assignment example, the proportion of each sex in the population is usually fairly well known in advance. However, the initial state probabilities relate to the new unmarked individuals encountered, not to the population as a whole. If the sampling scheme overselects one state, then the proportion known to be valid for the population must be corrected. If μ is the proportion of males in the population and p_M and p_F are the encounter probabilities of males and females respectively, the initial probability that a newly encountered individual is a male is $(p_M\mu)/(p_M\mu + p_F(1-\mu))$. This is strictly valid for the initial sample only; in subsequent samples, the initial probability has to be further corrected for the proportion of males remaining among the unmarked (see however Oro et al. 2004 for a way around this problem). The initial state probabilities are thus not directly population parameter and must be rewritten carefully if a Bayesian approach is to be implemented.

To conclude, I would like to mention what seems to me the natural continuation of the current topic. So far, we have seen how to deal with state uncertainty. But there is another type of uncertainty that has not yet been really addressed in CR (but see Schwarz and Stobo 1999) and yet that should become more and more important with the likely and desirable merging of population genetics and population dynamics: uncertainty in the individual. Being able to deal with this would allow using incomplete or ambiguous mark reading. This is exactly what is happening with genetic sampling when there are genotyping errors (Lukacs and Burnham 2005); true twins are another cause of uncertainty with genetic sampling. I believe that this is the new challenge we will have to face. Uncertainty in the individual and genetic sampling are also linked to another aspect of the population dynamics which has long been ignored in CR, the dependence among individuals. With the advent of genetic sampling, we will inevitably have to deal with information about pedigree and more generally kinship and can no longer ignore the relationships among individuals.

Acknowledgments I want to thank Bill Kendall and an anonymous referee for their stimulating comments. Valuable comments have also been provided by R. Choquet and O. Gimenez.

References

Arnason AN (1973) The estimation of population size, migration rates and survival in a stratified population. Researches on Population Ecology 15:1–8

Choquet R, Rouan L, Pradel R (this issue) Program E-SURGE: a software application for fitting Multievent models.

Crosbie SF, Manly BFJ (1985) Parsimonious modelling of capture-mark-recapture studies. Biometrics 41:385–398

Fujiwara M, Caswell H (2002) Estimating population projection matrices from multi-stage mark-recapture data. Ecology 83:3257–3265

Gimenez O, Morgan, BJT, Brooks SP (this issue) Weak identifiability in models for mark-recapture-recovery data.

Gimenez O, Rossi V, Choquet R, Dehais C, Doris B, Varella H, Vila J-P, Pradel R (2007) State-space modelling of data on marked individuals. Ecological Modelling 206:431–438

Hunter CM, Caswell H, Forcada J, Croxall J (this issue) Survival and breeding of South Georgia albatrosses: accounting for unobservable states

Kendall WL, Langtimm CA, Beck CA, Runge MC (2004) Capture–recapture analysis for estimating manatee reproductive rates. Marine Mammal Science 20:424–437

Lukacs PM, Burnham KP (2005) Review of capture-recapture methods applicable to noninvasive genetic sampling. Molecular Ecology 14:3909–3919

Nichols JD, Kendall WL, Hines JE, Spendelow JA (2004) Estimation of sex-specific survival from capture-recapture data when sex is not always known. Ecology 85:3192–3201

Oro D, Cam E, Pradel R, Martínez-Abraín A (2004) Influence of food availability on demography and local population dynamics in a long-lived seabird. Proceedings of the Royal Society of London Series B-Biological Sciences 271:387–396

Pledger S, Pollock KH, Norris JL (2003) Open capture–recapture models with heterogeneity: I. Cormack–Jolly–Seber model. Biometrics 59:786–794

Pradel R (2005) Multievent: an extension of multistate capture–recapture models to uncertain states. Biometrics 61:442–447

Pradel R, Maurin-Bernier L, Gimenez O, Genovart M, Choquet R, Oro D (2008) Estimation of sex-specific survival with uncertainty in sex assessment. Canadian Journal of Statistics 36(1): 29–42

Pradel R, Schaub M, Jenni L, Lebreton J-D (2005) Migrating birds stop over longer than usually thought: reply. Ecology 86:3418–3419

Prévot-Julliard A-C, Lebreton J-D, Pradel R (1998a) Re-evaluation of adult survival of Black-headed Gull (Larus ridibundus) in presence of recapture heterogeneity. The Auk 115:85–95

Prévot-Julliard A-C, Pradel R, Lebreton J-D, Cézilly F (1998b) Evidence for birth-site tenacity in breeding Black-headed Gulls, Larus ridibundus. Canadian Journal of Zoology 76:2295–2298

Redner RA, Walker HF (1984) Mixture densities, maximum-likelihood and the EM algorithm. SIAM Review 26:195–237

Schaub M, Pradel R, Jenni L, Lebreton J-D (2001) Migrating birds stop over longer than usually thought: an improved capture-recapture analysis. Ecology 82:852–859

Schwarz CJ, Stobo WT (1999) Estimation and effects of tag-misread rates in capture–recapture studies. Canadian Journal of Fisheries and Aquatic Sciences 56:551–559

Rank and Redundancy of Multistate Mark-Recapture Models for Seabird Populations with Unobservable States

Christine M. Hunter and Hal Caswell

Abstract Unobservable stages are common in many life cycles. Estimates of the vital rates, such as survival and breeding probabilities, of these stages are essential for demographic analysis but difficult to obtain. Explicit modeling of these states in multi-state mark-recapture methods can provide such estimates. However, models can be rank-deficient, meaning that not all parameters can be estimated. Determining whether a model is full rank is essential for interpretation of model selection and estimation results. Full rank models can be obtained by imposing biologically reasonable constraints on parameters. Developing such models requires an efficient way to assess model rank and determine which parameters, if any, are redundant. We introduce the use of automatic differentiation (AD) for this purpose. It generates the Jacobian matrix of the likelihood function in a way that is numerically stable, can accommodate large complicated models, and produces rank estimates accurate to machine precision. It reveals whether a model is full rank or rank-deficient (either intrinsically or for a particular data set), how many parameters or parameter combinations can be estimated, and which parameters are confounded. We use the method to explore three examples relevant to seabirds: a model with multiple breeding sites, a model distinguishing successful and failed breeders, and a model for pre-breeder survival and recruitment. We find a surprisingly large number of time-invariant and time-varying models to be of full rank, thus allowing estimation of all parameters, despite the unobservable states. We present a biological example for the Wandering Albatross (*Diomedea exulans*). Reliable assessment of model rank for multi-state mark-recapture models with unobservable stages will make it possible to use these methods in demographic applications.

C.M. Hunter (✉)
Department of Biology and Wildlife, Institute of Arctic Biology, PO Box 757000, University of Alaska Fairbanks, AK 99775, USA
e-mail: christine.hunter@uaf.edu

D.L. Thomson et al. (eds.), *Modeling Demographic Processes in Marked Populations*, Environmental and Ecological Statistics 3, DOI 10.1007/978-0-387-78151-8_37,

1 Introduction

Unobservable states, in which individuals have zero capture or resighting probabilities, often appear in multi-state capture-mark-recapture analyses. Unobservable states can result from factors such as the behavior, physiological state, or location of individuals in those states. This includes, but is not limited to, temporary emigration, the special case where individuals leave and then return to a sampling area over multiple sampling occasions. As multi-state models become more widely applied to demographic analysis (Caswell and Fujiwara 2004), unobservable states will be encountered more frequently.

If unobservable states are ignored, heterogeneity in capture probabilities can bias survival and transition estimates. Explicitly modeling unobservable states can solve this problem (introduced by Lebreton et al. (1999) and independently by Fujiwara and Caswell (2002a); see also Kendall and Nichols (2002)) and provides greater flexibility in modeling biologically important processes. However, the lack of information on individuals in unobservable states typically renders models parameter redundant, meaning that fewer parameters or parameter combinations can be estimated than the original number of parameters in the model (Catchpole et al. 1996; Gimenez et al. 2003). For simplicity, we hereafter refer to parameters or parameter combinations that can be estimated as estimable parameters and to parameter redundant models as rank-deficient.

Determining rank deficiency is necessary for correctly interpreting model comparison measures such as Akaikie's Information Criterion (AIC) and parameter estimates. One solution to parameter redundancy is to impose certain constraints on parameters in the model (Fujiwara and Caswell 2002a; Kendall and Nichols 2002). Parameter constraints can represent a priori information about species biology or biological hypotheses to be tested by model selection. To the extent that the constraints are biologically reasonable, this procedure can be a valuable approach to constructing and testing biological hypotheses.[1]

Determining rank deficiency and the number of separably estimable parameters is a challenging problem. Until recently, there was no general and reliable way to do so. Here we present a new method that complements currently available techniques. Our method provides estimates of both redundancy resulting from the structure of the model and irrespective of the data, i.e. intrinsic model redundancy, and redundancy resulting from the limitations of a particular data set, i.e. extrinsic model redundancy (Gimenez et al. 2003). It can be applied to a wide class of model structures, parameter constraints, temporal constraints and additional information. Our approach shows which parameters are separately estimable and which are confounded in some way, although it does not reveal the form of confounded parameter functions.

[1] In sufficiently simple models, including extra information on capture probabilities obtained using Pollock's robust design can sometimes resolve parameter redundancy (Kendall et al. 1997; Kendall and Nichols 2002). However, when the unobservable states are more richly structured, merely knowing capture probabilities in the observable states will generally not solve the problem.

We describe our method in the context of parameter estimation for long-lived colonial seabirds, the demographic analysis of which (e.g., Hunter et al. 2000; Hunter and Caswell 2005) motivated this work. We explore three multi-state models. The first describes transitions among 2 observable and one unobservable breeding sites. The second describes the transition of adults among observable breeding stages and unobservable non-breeding stages. The third describes the recruitment of fledglings to the breeding population; where fledglings and breeders are observable but intervening pre-breeding stages are not. For each of these models we evaluate parameter redundancy under a variety of biologically interesting constraints, both with and without time variation. As an example, we apply the method to the Wandering Albatross (*Diomedea exulans*). Our results show that a surprising number of complex models with unobservable stages are estimable from mark-recapture data.

2 Model Rank and Identifiability

The theory for determining estimability of mark-recapture models was laid out in a series of papers by Catchpole and Morgan (1997, 2001) and Catchpole et al. (1996, 1998, 2001). Estimation is carried out by maximizing the likelihood $L(\boldsymbol{\theta})$ where $\boldsymbol{\theta}$ is a $q \times 1$ vector of parameters. Let $f_i(\boldsymbol{\theta})$, $i = 1, \ldots, n$, be a set of quantities jointly sufficient for the calculation of $L(\boldsymbol{\theta})$. In this paper, we use the set of probabilities corresponding to the entries of the multistate m-array (Brownie et al. 1993). The same approach can be applied to the probabilities of the individual capture histories (Nichols et al. 1992; Fujiwara and Caswell 2002b; Caswell and Fujiwara 2004). Estimability is determined by the Jacobian matrix of the likelihood,

$$\mathbf{J} = \left(\frac{\partial f_i}{\partial \theta_j} \right) \tag{1}$$

The rank of this matrix of derivatives gives the number of parameters or parameter combinations that can be separately estimated. The *rank deficiency* of the model is the difference between the number q of parameters and the rank of $\mathbf{J}$. An orthonormal basis for the null space of $\mathbf{J}$ provides information on which parameters are confounded and which are separately estimable.

Evaluating model rank requires: (i) defining the model structure, (ii) calculating the likelihood for the model, (iii) computing the Jacobian matrix of the likelihood, and (iv) finding the rank of the Jacobian. We discuss each of these steps below. Then we present three seabird life cycle models, describing the model structure, sets of constraints and model estimability.

2.1 Defining Model Structure

We define the overall model structure with a life cycle graph, a directed graph with nodes corresponding to the stages (or states; we use the terms interchangeably) of interest. Stages are numbered from 1 to s. An arc in the graph from stage j to stage i

implies that it is possible for an individual to move from stage j to stage i during one time step. This life cycle graph differs from those used in demography (e.g., Caswell 2001) in that it need not include the entire life cycle and describes only transitions of living individuals, not production of new individuals by reproduction.

2.2 Calculating the Likelihood

To calculate the likelihood function we define a transition matrix $\boldsymbol{\Phi}_t$, of dimension $s \times s$, corresponding to the life cycle graph. The element $\phi_{ij}(t)$ of $\boldsymbol{\Phi}_t$ is the probability of transition from stage j at time t to stage i at time $t+1$, for $t = 1, \ldots, N-1$, where N is the number of sampling occasions. Note the column-to-row orientation of these matrices, which corresponds to their use in projecting probabilities of state occupancy (Fujiwara and Caswell 2002b; Caswell and Fujiwara 2004) and to their use as components of demographic models. We also define a diagonal capture matrix $\mathbf{P}_t$, with $p_{ii}(t)$ the probability of recapture of an individual in stage i at time t, for $t = 2, \ldots, N$.

Sufficient statistics for the calculation of the likelihood are contained in the multistate m-array, which is an array of matrices $\mathbf{M}^{(r,c)}$, where r denotes the release time ($r = 1, \ldots, N-1$) and c denotes the time of first recapture ($c = r+1, \ldots, N$), and $m_{ij}^{(r,c)}$ is the number of individuals released in stage i at occasion r that are next captured in stage j at occasion c, with $c > r$. Note the row-to-column orientation of these matrices, which is customary in mark-recapture analysis, but opposite to that in population modeling.

The probabilities of these release-recapture combinations are calculated from the matrices $\boldsymbol{\Phi}_t$ and $\mathbf{P}_t$ in the form of an array we call the p-array. The p-array is an array of matrices $\boldsymbol{\Psi}^{(r,c)}$, where $\psi_{ij}^{(r,c)}$ is the probability that an individual released in stage i at time r is next captured in stage j at time c. The matrices composing the p-array are calculated from $\boldsymbol{\Phi}$ and $\mathbf{P}$ as

$$\boldsymbol{\Psi}^{(r,c)} = \begin{cases} (\mathbf{P}_{r+1}\boldsymbol{\Phi}_r)^\top & c = r+1 \\ (\mathbf{P}_c\boldsymbol{\Phi}_{c-1}(\mathbf{I}-\mathbf{P}_{c-1})\boldsymbol{\Phi}_{c-2}\cdots(\mathbf{I}-\mathbf{P}_{r+1})\boldsymbol{\Phi}_r)^\top & c > r+1 \end{cases} \tag{2}$$

Note the transposition required to achieve the row-to-column orientation.

In most (but perhaps not all) cases, individuals cannot be marked and released in unobservable stages. Thus some of the conditional probabilities in $\boldsymbol{\Psi}^{(r,c)}$ are conditioned on an event of probability zero, and therefore undefined. Thus we set

$$\psi_{ij}^{(r,c)} = 0 \qquad \forall i \in \{\text{unobservable states}\} \tag{3}$$

Some individuals are released but never recaptured. However, given the numbers released, this component is not independent of the fates of the individuals that are recaptured, so it makes no contribution to the model rank and is ignored here.

The log likelihood from the recaptured individuals is

$$\log L = \sum_{r=1}^{N-1} \sum_{c=r+1}^{N} \sum_{i=1}^{s} \sum_{j=1}^{s} m_{ij}^{(r,c)} \log \Psi_{ij}^{(r,c)} \tag{4}$$

$$= \sum_{h=1}^{\frac{1}{2}N(N-1)s^2} f_h(\boldsymbol{\theta}) \tag{5}$$

where a generic component of the log likelihood has been denoted by $f_h(\boldsymbol{\theta})$.

2.3 Computing the Jacobian and its Rank

The Jacobian matrix (1) has dimension $\frac{1}{2}N(N-1)s^2 \times q$. The rank of the model is the rank of $\mathbf{J}$. Rank deficiency may be intrinsic to the model or due to imperfect data (e.g., Gimenez et al. 2003). The intrinsic rank is the rank that would result from perfect data, i.e., those in which the m-array is directly proportional to the p-array,

$$m_{ij}^{(r,c)} \propto \psi_{ij}^{(r,c)} \tag{6}$$

for all i, j, r, and c. These are the data that would arise in an infinite sample, and it is in this sense that we refer to them as "perfect." The (h, k) element of $\mathbf{J}$ is

$$\frac{\partial f_h}{\partial \theta_k} = \frac{\partial}{\partial \theta_k} m_{ij}^{(r,c)} \log \psi_{ij}^{(r,c)} \tag{7}$$

$$= \frac{m_{ij}^{(r,c)}}{\psi_{ij}^{(r,c)}} \frac{\partial \psi_{ij}^{(r,c)}}{\partial \theta_k} \tag{8}$$

Thus, if the data are perfect, the elements of $\mathbf{J}$ satisfy

$$\frac{\partial}{\partial \theta_k} \left(m_{ij}^{(r,c)} \log \psi_{ij}^{(r,c)} \right) \propto \frac{\partial \psi_{ij}^{(r,c)}}{\partial \theta_k} \tag{9}$$

and $\mathbf{J}$ can be calculated directly from the p-array without any data. We will denote this Jacobian by $\mathbf{J}_{\text{intrinsic}}$. The Jacobian matrix for a model with a specific (and perhaps less than perfect) data set is easily obtained from $\mathbf{J}_{\text{intrinsic}}$ as

$$\mathbf{J}_{\text{data}} = \text{diag}\left(\frac{m_{ij}^{(r,c)}}{\psi_{ij}^{(r,c)}} \right) \mathbf{J}_{\text{intrinsic}} \tag{10}$$

where the elements in the diagonal matrix are arranged in the same order as those in $\mathbf{J}_{\text{intrinsic}}$.

2.4 Differentiation as a Challenge

The entries of the p-array are complicated functions of the parameters. The derivatives of those entries are even more complicated. Thus determining the derivatives that make up $\mathbf{J}$ is the key to determining the rank deficiency of the model. Computing these derivatives is a challenging problem, which has been approached in several ways. We introduce a new approach (automatic differentiation) here.

The accuracy of the derivatives has profound consequences for the rank of $\mathbf{J}$, because the rank of a matrix depends on the exact independence, or lack thereof, of the columns of that matrix. Arbitrarily small perturbations, caused by errors in $\mathbf{J}$, can render dependent columns independent. The rank of a matrix is determined operationally as the number of non-zero singular values, so the ability to detect rank deficiency depends on the sensitivity of the singular values to errors in the matrix entries. This sensitivity is measured by the condition number (roughly, the magnitude of the error in the result relative to the error in the input). The condition number for singular values is 1 (Stewart 1991), so errors of order ϵ in $\mathbf{J}$ can lead to errors of order ϵ in the singular values. This is good news for computing the largest singular values, but bad news for determining rank (Stewart 1992), which depends on the smallest singular values. Small errors in the derivatives in $\mathbf{J}$ can make it difficult or impossible to tell which singular values should be considered zero (Table 1).

Before presenting our solution to computing these derivatives, we briefly review other approaches.

Table 1 Comparison of the singular values of a Jacobian calculated by numerical differentiation and automatic differentiation. The underlying model, a time-invariant version of Fig. 2, has two observable and two unobservable states and 14 parameters. The correct model rank, verified by symbolic calculation, is 12

	Numeric differentiation	Automatic differentiation
Apparent rank	14	12
Singular values	0.570	0.569
	0.507	0.506
	0.370	0.229
	0.326	0.154
	0.275	0.061
	0.248	0.041
	0.219	0.033
	0.153	0.024
	0.064	0.017
	0.054	0.010
	0.045	0.005
	0.025	0.002
	0.017	1.825e-017
	0.006	1.100e-017

2.4.1 Symbolic Differentiation

The most general way to calculate the Jacobian matrix $\mathbf{J}$ is to derive a symbolic formula for each of the partial derivatives in $\mathbf{J}$ and then to evaluate the rank of $\mathbf{J}$ symbolically. This calculation is too complicated for solution by hand, but for simple models can be carried out using symbolic mathematics packages such as Maple or Mathematica (Gimenez et al. 2003). This method involves forming a symbolic vector of the log-probabilities $\log(f_i(\boldsymbol{\theta}))$, differentiating each of the f_i with respect to $\boldsymbol{\theta}$, and determining the rank of the resulting symbolic matrix $\mathbf{J}(\boldsymbol{\theta})$. If a model is rank-deficient it is possible to determine the separably estimable components of $\boldsymbol{\theta}$ and the form of the estimable functions of the remaining parameters from the orthogonal basis vectors of the null space (Gimenez et al. 2003).

Symbolic differentiation produces accurate results, is not dependent on specific values of $\boldsymbol{\theta}$, and has the advantage of providing information on the estimable functions of the parameters. Unfortunately, even relatively simple problems can quickly exceed the capabilities of symbolic math packages. So, at least at present, the application of this method is limited in scope.

2.4.2 Analytic Differentiation

The likelihood function (4) is computed as a series of matrix products. Using matrix calculus (e.g., Magnus and Neudecker 1988), it is possible to derive analytical formulae for the derivatives of each component of the likelihood. These formulae can then be evaluated for specified values of the parameters and the rank of the resulting Jacobian determined numerically. This approach of combining the analytical calculation of the Jacobian with a numerical calculation of the rank is the basis for the rank calculations in the software M-Surge and E-Surge (R. Choquet personal communication, Rouan et al. unpublished). It produces accurate estimates of the rank, and the analytical formulae also provide the gradient of the likelihood, which is of great use in the optimization process. However, analytic formulae are difficult to derive and implement and their use requires considerable programming skills.

2.4.3 Numerical Differentiation

A tempting solution that avoids the need to derive and evaluate the analytical formulae, is the use of numerical differentiation. Unfortunately, numerical differentiation is a notoriously unstable problem. It is plagued with two kinds of errors. Discretization error arises from using a finite difference to approximate a continuous derivative. It goes to zero as the size of the finite difference goes to zero. Conditioning error, on the other hand, arises from roundoff errors caused by taking the difference between two nearly equal quantities. It becomes infinite as the size of the finite difference goes to zero. Any particular numerical difference scheme will have its own balance of these two errors, depending on the function and the parameter values. It is difficult, or impossible, to determine how accurate a particular numerical approximation to a derivative is.

The combination of discretization error and conditioning error in a numerical estimate of the Jacobian makes it impossible to decide how many singular values are "effectively zero" without imposing an arbitrary threshold. Based on extensive exploration of this approach, we strongly recommend against using numerical differentiation to determine the rank of mark-recapture models.

2.4.4 Automatic Differentiation

After quickly exceeding the capacity of symbolic differentiation, obtaining unreliable results from numerical differentiation and struggling to implement analytic formulae for the derivatives our search for an alternative led us to a recently developed approach called *automatic differentiation* (or sometimes *algorithmic differentiation*) abbreviated as AD (Griewank 2000, 2003; Shampine et al. 2005). AD calculates derivatives numerically, but does not use finite differences. It returns results with the same accuracy as analytical differentiation; i.e., to machine accuracy, and is more efficient than symbolic computation.

The idea behind AD is to compute the derivative of a quantity simultaneously with the calculation of the quantity itself. This is accomplished by replacing each variable with an ordered pair of objects, the first of which is the value and the second of which is the derivative. Mathematical operations applied to the variables are redefined to also provide the derivatives of the variable at each step. At the end of a calculation, the derivatives are automatically (hence the name) returned.

For example, a variable y would be replaced in AD by an ordered pair

$$y \mapsto (y, dy). \tag{11}$$

The mathematical operations applied to y are redefined to operate on this new object. Thus addition becomes

$$(y, dy) + (z, dz) = [\, y + z, dy + dz] \tag{12}$$

Multiplication and division follow the product rule for derivatives

$$(y, dy)(z, dz) = [\, yz,\ \ y(dz) + (dy)z] \tag{13}$$

$$\frac{(y, dy)}{(z, dz)} = \left[\frac{y}{z}, \frac{z(dy) - y(dz)}{z^2}\right] \tag{14}$$

Applied to the likelihood function, AD carries the differentiation forward as the f_i are constructed, in contrast to the usual differentiation methods that work backwards using differentiation rules to decompose the f_i.

Redefining mathematical operations so that derivatives are properly mapped in AD has required a major effort in software engineering over the past decade (see www.autodiff.org). One such routine is the MATLAB Automatic Differentiation

(MAD) package available from TOMLAB Optimization, Inc. (Forth and Edvall 2006; see http://tomopt.com).

Because the derivatives provided by AD are accurate to machine precision, the smallest singular values of the Jacobian are (again to within machine precision) zero, rather than ambiguously small, if the model is rank-deficient. The rank of Jacobian matrices calculated by AD can be determined directly, e.g. using the MATLAB rank command. For example, Table 1 compares the results of applying numerical differentiation and AD to a model known, from symbolic calculation, to have a rank of 12. The two smallest singular values calculated by AD are zero to machine precision ($\approx 10^{-17}$), but when calculated from the numerical derivatives, there is no clear distinction between the two smallest (0.017 and 0.006) and the next largest (0.025) singular values.

3 Models for Seabird Life Cycles

We present three life cycle examples that are relevant in the context of long-lived seabirds. Many seabirds can be marked, released, and recaptured only on their breeding colonies. Thus, they are unobservable between fledging and recruitment and between breeding attempts. The life cycles we consider describe these unobservable pre-breeding and inter-breeding situations (see, e.g., Hunter et al. 2000 for an example of how these would fit into the entire life cycle). For each of these models we describe the life cycle structure and the family of constraints that we consider.

3.1 Multiple Breeding Sites

Figure 1 shows a model with two breeding locations (stages 1 and 2) and an unobservable non-breeding state (stage 3). We parameterize the model in terms of a

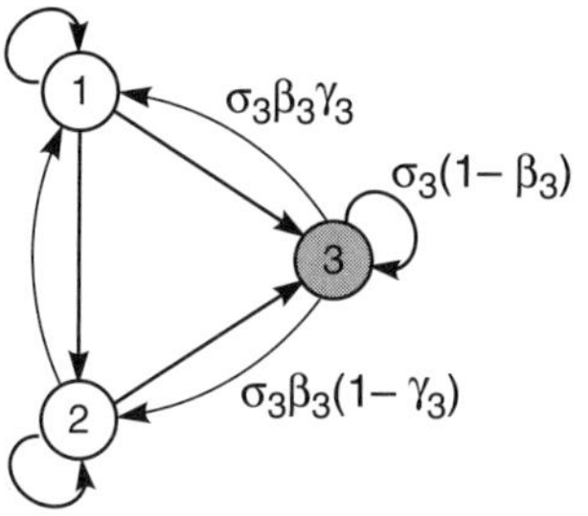

Fig. 1 The three-stage breeding site model. Stages 1 and 2 represent breeders at two observable locations; the unobservable stage 3 (*shaded*) can be interpreted as non-breeding or breeding in an unobservable location. Transition probabilities are shown from stage 3; other stages are similar (σ_i = survival, β_i = breeding at an observable site given survival, γ_i = probability of breeding at site 1, given survival and breeding)

sequence of conditional probabilities. The probability of survival of a bird in stage i is σ_i. The probability of breeding in an observable site, given survival, is β_i. The probability of breeding in site 1 given a bird is in stage i, conditional on surviving and breeding, is γ_i. By definition the probability of breeding in site 2 given survival and breeding is thus $1-\gamma_i$. The capture probability of stage i is p_i, with the additional constraint that $p_3 = 0$ because stage 3 is unobservable.

3.1.1 Constraints

Many biological hypotheses might be used to specify parameter constraints in this model. We consider five possible equality constraint models for each parameter type, i.e. for the σ_i, β_i, and γ_i. For example, possible constraint models for survival are:

Model	Constraint
1	$\sigma_1 = \sigma_2 = \sigma_3$
2	$\sigma_1 = \sigma_3, \sigma_2$
3	$\sigma_1 = \sigma_2, \sigma_3$
4	$\sigma_1, \sigma_2, = \sigma_3$
5	$\sigma_1, \sigma_2, \sigma_3,$

Constraint Model 1 estimates a common survival probability for all stages. At the other extreme, Model 5 estimates a separate survival for each stage. Between these extremes, Models 2–4 estimate two parameters among the three stages.

These constraints can be treated in two equally valid ways: as hypotheses to be evaluated or as models imposed a priori based on the goals of the study. For example, if stage 3 represents non-breeding birds, a study of the effects of breeding on survival might use Model 3, which estimates survival separately for breeding and non-breeding stages, because that estimates the effect of interest. On the other hand, a study whose goal was a single survival estimate for all stages would obtain the best such estimate using Model 1.

Applying these five constraint Models to σ_i, β_i, and γ_i allows a total of 125 possible constraint model combinations.

3.2 *Breeding Success and Failure*

In species with significant parental investment, the difference between successful and failed breeding attempts can have important consequences. For the large albatrosses, which are biennial breeders, failed breeders may breed the following year but successful breeders cannot. In general, the success or failure of breeding in year t may affect the probability of breeding and of breeding success in year $t+1$.

Figure 2 shows a life cycle graph (currently being used for analyses of albatross demography) that distinguishes successful and failed breeders, and non-breeders

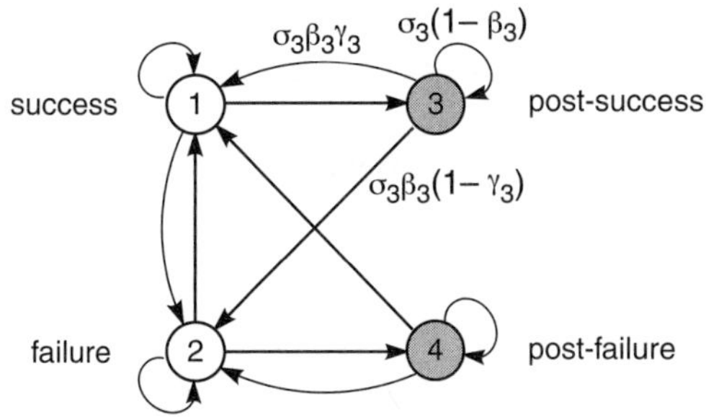

Fig. 2 The four-stage model for breeding success and failure. Stage 1 is successful breeders, stage 2 is unsuccessful breeders, stage 3 is non-breeders whose previous breeding attempt was successful, and stage 4 is non-breeders whose previous breeding attempt failed. Unobservable stages are *shaded*. Transition probabilities are shown for stage 3; other stages are similar (σ_i = survival, β_i = breeding given survival, γ_i = success given breeding)

whose previous breeding attempt was a success or a failure. Breeding stages (1 and 2) are observable; non-breeding stages (3 and 4) are not. The model is parameterized in terms of the probability of survival, σ_i, the probability of breeding given survival, β_i, and the probability of success given breeding, γ_i. The capture probabilities are denoted by p_i and by assumption, $p_3 = p_4 = 0$. For biennial species $\beta_1 = 0$ and γ_1 is undefined.

3.2.1 Constraints

For the breeding success model we consider four constraint models for each parameter type; e.g., for survival, we have:

Model	Constraint
1	$\sigma_1 = \sigma_2 = \sigma_3 = \sigma_4$
2	$\sigma_1 = \sigma_3, \sigma_2 = \sigma_4$
3	$\sigma_1 = \sigma_2, \sigma_3 = \sigma_4$
4	$\sigma_1, \sigma_2, \sigma_3, \sigma_4$

Constraint Model 1 estimates a common probability for all stages. Model 4 estimates a separate probability for each stage. Between these extremes, Model 3 estimates one probability for breeding individuals and another for non-breeding individuals. It corresponds to the biological hypothesis that the energetic and physiological costs of breeding are the most important influence on the vital rates. Model 2 estimates one probability for breeders that are (currently or previously) successful, and another for failed breeders. As a hypothesis, it emphasizes the distinction between successful and unsuccessful birds. The four constraint models for σ_i, β_i, and γ_i give a total of 64 possible constraint combinations.

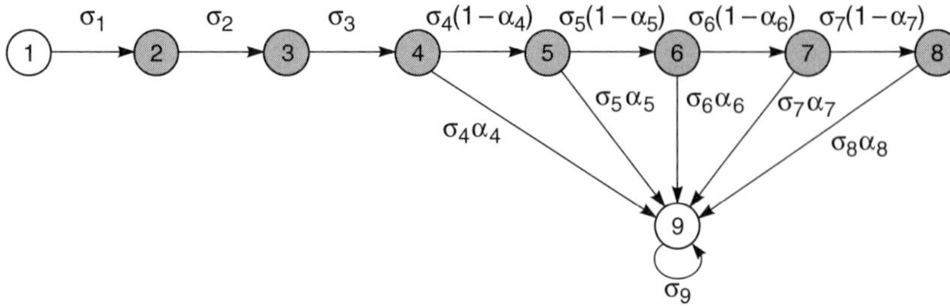

Fig. 3 Prebreeder recruitment life cycle. Stages are (1) first year, (2) second year, ..., (8) eighth year, (9) breeders. Stages (1) and (9) are observable, all other stages are unobservable (*shaded*). Probability of survival of a bird in stage i is σ_i, probability of first breeding at age i is α_i

3.3 Pre-breeding Survival and Recruitment

Most seabirds with delayed maturity are unobservable between fledging and recruitment. Figure 3 shows an example, parameterized in terms of age-specific survival probability σ_i and the age-specific conditional probability of recruitment α_i, given survival. See Clobert et al. (1994) for an analysis in terms of the Cormack–Jolly–Seber model. The minimum age at recruitment, k, and the number of recruiting age classes, y, are specified as part of the model structure. Stages 1 and $k + y$ are observable; all other stages are unobservable. In Fig. 3, for example, $k = 4$ and $y = 5$.

3.3.1 Constraints

One goal of this model is to untangle, as much as possible, the age dependence of survival and recruitment during the pre-breeding period. Complete age-dependence is impossible to estimate (Clobert et al. 1994). Many constraint models for survival and recruitment are possible. As examples we explore six constraint models for survival and two for recruitment. To specify these, we use the MATLAB notation $[a : b] = [a, a + 1, \ldots, b]$.

Model	Survival constraint
1	$\sigma_1 = \cdots = \sigma_{k+y}$
2	$\sigma_1, \sigma_2 = \cdots = \sigma_{k+y}$
3	$\sigma_1 = \cdots = \sigma_{k-1}, \sigma_k = \cdots \sigma_{k+y}$
4	$\sigma_1 = \cdots = \sigma_{k-1}, \sigma_k = \cdots \sigma_{k+y-1}, \sigma_{k+y}$
5	$\sigma_1 = \cdots = \sigma_{k-1}$, logistic on $[k : k + y]$
6	$\sigma_1 = \cdots = \sigma_{k-1}$, logistic on $[k : k + y - 1], \sigma_{k+y}$

Model	Recruitment constraint
1	$\alpha_k = \cdots \alpha_{k+y-1}$
2	logistic on $[k : k + y - 1]$

Survival constraint Model 1 sets all survival probabilities equal. Model 2 estimates a survival probability for fledglings and a common survival probability for all older age classes. Model 3 estimates one survival probability for all ages less than k and another for all ages greater than k. Model 4 estimates one survival probability for all ages less than k, a second for all ages between k and $k + y - 1$ (all those that can potentially recruit), and a third for breeding adults. Model 5 estimates a single survival probability for all ages less than k and a logistic increase in survival from age k to the breeding adult stage. Model 6 estimates a single survival probability for all ages less than k, a logistic increase in survival from age k to $k + y - 1$, and a separate survival for breeding adults. The logistic models permit exploration of hypotheses about the improvement of survival with age. They require two parameters; for example for survival Model 5:

$$\text{logit}\,(\sigma_i) = a + bi \qquad i = k, \ldots, k + y \tag{15}$$

Recruitment Model 1 estimates a single recruitment probability for all ages beyond the age of first recruitment. Model 2 estimates a logistic increase with age in recruitment probability beyond the age of first recruitment. We emphasize that these constraint models are only a sampling of those that might be biologically interesting.

3.4 Time-Varying Models

We investigated model rank for both time-invariant and time-varying versions of the models. We present results for three classes of temporal variation; free time-variation, additive time variation, and a temporal trend.

1. Free time-variation yields a separate value at each time for a given parameter. For example, a freely time-varying model for survival in the breeding site life cycle (Fig. 1) with constraint Model 4 ($\sigma_2 = \sigma_3$) would estimate

$$\begin{aligned} &\sigma_1(t) \quad t = 1, \ldots, N - 1 \\ &\sigma_2(t) \quad t = 1, \ldots, N - 1 \end{aligned} \tag{16}$$

   for a total of $2(N - 1)$ estimated parameters.
2. Additive time-variation requires parameters to vary in parallel on the logit scale. For example, an additive model for survival in the breeding site life cycle, again with constraint Model 4, would estimate

$$\begin{aligned} &\sigma_1(t) && t = 1, \ldots, N - 1 \\ &\text{logit}\,[\sigma_2(t)] = \text{logit}\,[\sigma_1(t)] + c_2 && \end{aligned} \tag{17}$$

   with $\sigma_3 = \sigma_2$ because of the constraints, for a total of N estimated parameters.
3. Temporal trends in a parameter are described by a logistic dependence of the parameter on time; e.g.,

$$\text{logit}\,[\sigma_i(t)] = a_i + b_i t \tag{18}$$

The same structure can describe dependence on an environmental covariate $x(t)$, in which case

$$\text{logit}\,[\sigma_i(t)] = a_i + b_i x(t) \tag{19}$$

In this perspective, time is simply a covariate with a particularly simple time dependent structure.

For simplicity, we chose to examine only models that impose the same type of time variation on all parameters of the same type (i.e. all survival parameters, or all breeding probabilities, etc.). For example, we consider a model where the σ_i are freely time-varying but the β_i are not, but we exclude models where σ_1 is time varying but σ_2 is not.

3.4.1 Constraints on Time Variation

In time-varying mark-recapture models it is common for parameters at the last time period (e.g., survival and capture) to be confounded. In addition, time-varying models with unobservable states also experience confounding at the beginning of the capture sequence. Consider Fig. 2 and suppose that individuals can be released only in the observable states 1 and 2. From $t = 1$ to $t = 2$, no marked individuals can make any of the transitions originating in stages 3 or 4. From $t = 2$ to $t = 3$, however, some individuals originating in states 1 or 2 at $t = 1$ can make those transitions and arrive at observable states. Thus time variation in the parameters defining the transitions from stages 3 and 4 cannot be resolved until two sampling occasions have passed. Thus, it seems that identifiability requires that freely time-varying parameters be constrained to be equal at $t = 1$ and $t = 2$.

In more complicated life cycles (e.g., Fig. 3), we conjecture that a sufficient condition is to constrain parameters long enough to assure a positive probability of individuals released in the observable states traversing all pathways in the life cycle graph. For Figs. 1 and 2 this requires two time steps.

To address these issues, we constrained all freely time-varying parameters to be equal at $t = 1$ and $t = 2$, and at $t = N - 2$ and $t = N - 1$.

4 Implementation using Automatic Differentiation

To implement these calculations, we wrote a series of MATLAB programs (www.mathworks.com) to accept a set of parameter values, generate the matrices $\boldsymbol{\Phi}_t$ and $\mathbf{P}_t$, and compute the p-array. We used the MATLAB Automatic Differentiation (MAD) package produced by TOMLAB (Forth and Edvall 2006) to generate the Jacobian matrix. We evaluated the rank of the Jacobian using the MATLAB `rank` command, which counts the number of singular values greater than a threshold. We examined the singular values and found no ambiguity between zero and the non-zero singular values.

For rank-deficient models we used the MATLAB command `null` to obtain an orthonormal basis for the null space of the Jacobian. Parameters whose entries are zero in all of these basis vectors are separately estimable. All other parameters are confounded in some way.

As an example, consider the time-invariant 3-state breeding site model with $N = 5$ sampling occasions (Fig. 1). We specify values for the eleven model parameters:

γ_1	γ_2	γ_3	β_1	β_2	β_3	σ_1	σ_2	σ_3	p_1	p_2
0.60	0.65	0.62	0.90	0.85	0.45	0.95	0.94	0.92	0.90	0.80

The Jacobian $\mathbf{J}$, which is of dimension 90×11 in this case, has singular values

$$\begin{array}{c} 5.69\text{e} - 001 \\ 5.05\text{e} - 001 \\ 2.31\text{e} - 001 \\ 1.54\text{e} - 001 \\ 7.02\text{e} - 002 \\ 4.36\text{e} - 002 \\ 2.95\text{e} - 002 \\ 1.72\text{e} - 002 \\ 7.65\text{e} - 003 \\ 2.69\text{e} - 003 \\ 1.24\text{e} - 017 \end{array}$$

The smallest singular value is zero to MATLAB precision, and the rank of $\mathbf{J}$ as calculated by MATLAB is 10, implying a rank deficiency of 1. The one-dimensional null space is spanned by the vector

$$\begin{pmatrix} 0 \\ 0.0000 \\ 0.0000 \\ 0.1493 \\ 0.1493 \\ 0.1493 \\ -0.2985 \\ -0.3731 \\ 0.8395 \\ 0.0000 \\ 0.0000 \end{pmatrix} \tag{20}$$

The zero entries of this vector correspond to the parameters that can be separately estimated; in this case only γ_1, γ_2, γ_3, p_1, and p_2. All the survival and breeding probabilities are confounded in some combination.

4.1 Rank, Generic Rank, and Parameter Dependence

Our results are obtained for a particular parameter vector $\boldsymbol{\theta}$, raising the question of whether a model found to be full rank for one set of parameters might be rank-deficient for another. However, the theory for generic matrix properties suggests in a very powerful and general sense that this is unlikely to happen.

Consider a matrix, such as **J**, whose entries are functions of a set of parameters $\boldsymbol{\theta} \in R^q$. A property of such a matrix is said to be *generic* if it holds at all points in a dense open set in parameter space (Hirsch and Smale 1974, p. 154), or if it holds at all points $\boldsymbol{\theta}$ except for those in a subset of R^q composed of the intersection of the zeros of a set of polynomials in $\boldsymbol{\theta}$ (e.g., Wonham 1985, p. 28). Such an intersection, called a proper algebraic variety, has measure zero in R^q. Thus, by either definition, a generic property is true of "almost all" matrices, or holds with probability 1 if we think of the entries of $\boldsymbol{\theta}$ as being selected "at random".

Within the set of all matrices of size $m \times n$ the subset of full-rank $m \times n$ matrices is known to be open and dense if the entries of the matrix are independently varying quantities (Golub and van Loan 1996, p. 73). Thus, under this condition, having full rank is a generic property of a matrix. However, we are interested in matrices whose entries are functions of parameters and therefore not generally independent. The generic-rank of a matrix $\mathbf{A}[\boldsymbol{\theta}]$ whose entries are functions of a set of q parameters is the rank obtained for all values of $\boldsymbol{\theta}$ outside of a proper algebraic variety in R^q (Murota 2003, p. 38).

The critical conclusion relevant here is that the generic rank is equal to the maximum, over all parameter values, of the rank obtained as a function of $\boldsymbol{\theta}$,

$$\text{generic-rank}(\mathbf{A}) = \max_{\boldsymbol{\theta} \in R^q} \text{rank}(\mathbf{A}[\boldsymbol{\theta}]) \tag{21}$$

(Murota 2003).[2] Thus if an arbitrary set of parameters yields a Jacobian of full rank, the Jacobian is of full generic rank and the corresponding model will be rank-deficient only for very particular combinations of parameters (i.e., from a set of measure zero). If an arbitrary set of parameters yields a rank-deficient Jacobian, there are two possibilities. Either the obtained rank is the generic rank, and the model is rank-deficient everywhere (since its rank can only go down from the generic rank). Or the parameters happened to fall exactly on one of the measure-zero proper algebraic varieties in the parameter space, in which case the generic-rank will be greater, perhaps even full rank.[3]

---

[2]Murota states this result for matrices whose entries are rational functions of the parameters, so it would apply directly to Jacobians calculated in terms of the probabilities (the identity link) rather than the logit transform of the probabilities. When calculated using the logit link, the entries of **J** are analytic, but not rational, functions of $\boldsymbol{\theta}$. However, analytic functions are given everywhere by their Taylor series, so the result applies equally to the logit link case (M. Golubitsky personal communication).

[3]Note that this same argument is used independently by Rouan et al. (unpublished).

When investigating a particular model, the chance of the second outcome can be reduced by avoiding parameter vectors with obvious relations among the entries. This includes parameters that are equal to each other. Thus, for example, while a time-varying model with $\sigma_i(t) = 0.5$ for all t is a possible time-varying model, it may give different rank results from the same model in which the $\sigma_i(t)$ vary, however small the variation.

5 Results: Model Rank for Seabird Life Cycles

Even though these life cycles contain few stages, the combinations of constraints and time variation for the different parameters (e.g. survival, breeding, capture) yield many thousands of models. It is surprising how many of these models are full rank. To reveal patterns we present examples of rank deficiency as a function of parameter constraints and time variation.

5.1 Multiple Breeding Sites

5.1.1 Time-Invariant Models

Our explorations suggest that constraints on σ_i and β_i have a greater effect on rank deficiency than constraints on γ_i or p_i. With no constraints on γ_i or p_i, there is a set of 25 different constraint models for β_i and σ_i (Table 2). Most of these are full rank. It appears that some constraints on survival are necessary for models to be full rank. All models with three separate survival probabilities (constraint Model 5; σ_1, σ_2, and σ_3) were rank-deficient. With only two exceptions, all models with some constraint on survival were fully estimable. The two exceptions were models where survival of the unobservable state was unconstrained.

The combination of 5 constraint models for each of the σ_i, β_i, and γ_i yields 125 models (imposing no constraints on the p_i). Constraints on γ_i had little effect on model rank. An additional four models were rank deficient when two γ_i were set equal (γ_i constraint models 2–4) but all 25 models were rank deficient if all states

Table 2 Rank deficiency (and total number of parameters) for the time-invariant 3-stage breeding site model (Fig. 1) with constraints on σ_i and β_i. In all cases, γ_i and p_i are estimated separately for each stage

Breeding constraint model	Survival constraint model				
	$\sigma_1 = \sigma_2 = \sigma_3$	$\sigma_1 = \sigma_3,\ \sigma_2$	$\sigma_1 = \sigma_2,\ \sigma_3$	$\sigma_1,\ \sigma_2 = \sigma_3$	$\sigma_1,\ \sigma_2,\ \sigma_3$
$\beta_1 = \beta_2 = \beta_3$	0 (7)	0 (8)	1 (8)	0 (8)	1 (9)
$\beta_1 = \beta_3,\ \beta_2$	0 (8)	0 (9)	0 (9)	0 (9)	1 (10)
$\beta_1 = \beta_2,\ \beta_3$	0 (8)	0 (9)	1 (9)	0 (9)	1 (10)
$\beta_1,\ \beta_2 = \beta_3$	0 (8)	0 (9)	0 (9)	0 (9)	1 (10)
$\beta_1,\ \beta_2,\ \beta_3$	0 (9)	0 (10)	0 (10)	0 (10)	1 (11)

had an equal probability of breeding in site 1 (γ_i constraint model 1). Thus, 60 of the 125 time-invariant models for this life cycle are fully estimable.

5.1.2 Time-Varying Models

Adding time variation, even with the restriction that all parameters of a particular type are treated the same, creates a very large number of possible models. More than 27,000 combinations of constraint models and time variation in σ_i, β_i, γ_i, and p_i are possible. Table 3 shows an example set for all constraint and time variation combinations of σ_i and β_i with no constraints and free time variation for γ_i and p_i. Comparing Tables 2 and 3, it appears that if the time-invariant model is rank deficient, the time-varying model is also rank deficient when survival varies freely with time and breeding probability is either constant, additive or freely time-varying. The remainder of the 361 models in Table 3, i.e. 93% are fully estimable. This suggests the primary influence on parameter redundancy is the specification of survival probabilities.

Other choices of constraints can, of course, change model rank. Although when all γ_i are separate, time variation in γ_i or p does not affect rank deficiency; when all γ_i are equal an additional 116 of the 361 models become rank deficient. This emphasizes the fact that reducing the number of parameters does not always make the model easier to estimate, because of interaction among the parameters in determining the likelihood.

We calculated the rank of the 27,436 models defined by all choices of constraints and time-variation. Of these, fully 88 were full rank. It is interesting that some combinations of constraints that are rank-deficient for time-invariant models become full rank with the addition of time variation in σ_i and/or β_i.

If the p_i are known independently, constraints on γ_i and time variation in γ_i or p_i had no effect on rank deficiency. A known value for the p_i does not allow any additional combinations of constraint or time variation models in Table 3 to be fully estimable. However, a few additional models with constraints on γ_i become fully estimable when the p_i are known.

The results in Tables 2 and 3 were calculated for six sampling occasions. Gimenez et al. (2003) showed that rank deficiency does not change with the number of capture occasions, assuming the number of occasions is greater than or equal to the minimum number that allows for all possible transitions.

5.2 Breeding Success and Failure

5.2.1 Time-Invariant Models

Table 4 shows the rank of the 16 models defined by 4 constraints on the σ_i and β_i, with breeding success γ_i estimated separately for each stage and p_i estimated separately for each observable stage. All of the models that estimate survival separately for the four stages (constraint model 4) were rank-deficient, as were

Table 3 Rank deficiency for time-varying 3-stage breeding site models with constraints on σ_i and β_i and $N = 6$ capture occasions. In all cases, γ_i and p_i are estimated separately for each stage and are freely time-varying. Codes used for time variation are: c = constant, x = trend or covariate, a = additive, and f = free time variation

Breeding constraint and time variation	Survival constraint and time-variation																		
	$\sigma_1 = \sigma_2 = \sigma_3$			$\sigma_1 = \sigma_3,\ \sigma_2$				$\sigma_1 = \sigma_2,\ \sigma_3$				$\sigma_2 = \sigma_3,\ \sigma_1$				$\sigma_1,\ \sigma_2,\ \sigma_3$			
	c	x	f	c	x	a	f	c	x	a	f	c	x	a	f	c	x	a	f
$\beta_1 = \beta_2 = \beta_3$																			
c	0	0	0	0	0	0	0	1	0	0	1	0	0	0	0	1	0	0	1
x	0	0	0	0	0	0	0	0	0	0	0	0	0	0	0	0	0	0	0
f	0	0	0	0	0	0	0	0	0	0	2	0	0	0	0	0	0	0	2
$\beta_1 = \beta_3,\ \beta_2$																			
c	0	0	0	0	0	0	0	0	0	0	0	0	0	0	0	1	0	0	1
x	0	0	0	0	0	0	0	0	0	0	0	0	0	0	0	0	0	0	0
a	0	0	0	0	0	0	0	0	0	0	0	0	0	0	0	0	0	0	2
f	0	0	0	0	0	0	0	0	0	0	0	0	0	0	0	0	0	0	2
$\beta_1 = \beta_2,\ \beta_3$																			
c	0	0	0	0	0	0	0	1	0	0	1	0	0	0	0	1	0	0	1
x	0	0	0	0	0	0	0	0	0	0	0	0	0	0	0	0	0	0	0
a	0	0	0	0	0	0	0	0	0	0	2	0	0	0	0	0	0	0	2
f	0	0	0	0	0	0	0	0	0	0	2	0	0	0	0	0	0	0	2
$\beta_2 = \beta_3,\ \beta_1$																			
c	0	0	0	0	0	0	0	0	0	0	0	0	0	0	0	1	0	0	1
x	0	0	0	0	0	0	0	0	0	0	0	0	0	0	0	0	0	0	0
a	0	0	0	0	0	0	0	0	0	0	0	0	0	0	0	0	0	0	2
f	0	0	0	0	0	0	0	0	0	0	0	0	0	0	0	0	0	0	2
$\beta_1,\ \beta_2\ \beta_3$																			
c	0	0	0	0	0	0	0	0	0	0	0	0	0	0	0	1	0	0	1
x	0	0	0	0	0	0	0	0	0	0	0	0	0	0	0	0	0	0	0
a	0	0	0	0	0	0	0	0	0	0	0	0	0	0	0	0	0	0	2
f	0	0	0	0	0	0	0	0	0	0	0	0	0	0	0	0	0	0	2

Table 4 Rank deficiency (and total number of parameters) for the time-invariant 4-stage breeding success model (Fig. 2) with constraints on σ_i and β_i. In all cases, γ_i and p_i are estimated separately for each stage

Breeding constraint model	Survival constraint model			
	$\sigma_1=\sigma_2=\sigma_3=\sigma_4$	$\sigma_1=\sigma_3,\ \sigma_2=\sigma_4$	$\sigma_1=\sigma_2,\ \sigma_3=\sigma_4$	$\sigma_1,\ \sigma_2,\ \sigma_3,\ \sigma_4$
$\beta_1=\beta_2=\beta_3=\beta_4$	0 (8)	0 (9)	1 (9)	1 (11)
$\beta_1=\beta_3,\ \beta_2=\beta_4$	0 (9)	0 (10)	0 (10)	2 (12)
$\beta_1=\beta_2,\ \beta_3=\beta_4$	0 (9)	0 (10)	1 (10)	1 (12)
$\beta_1,\ \beta_2,\ \beta_3,\ \beta_4$	0 (11)	0 (12)	0 (12)	2 (14)

combinations of survival constraint model 3 with breeding probability constraint models 1 or 3. Constraints on γ_i had little effect on parameter redundancy. An additional two models become rank deficient for γ_i constraint models 1 and 3. As in the 3-stage model, survival constraints had a greater effect on parameter redundancy than conditional breeding parameters.

5.2.2 Time-Varying Models

Of the many possible time-varying models, we show results in Table 5 for 225 models defined by combinations of constraints and time variation for the σ_i and β_i. No constraints are imposed on the γ_i or p_i, both of which were allowed to vary freely with time. Of these models, 203 were full rank. For these cases, where the probability of breeding success is independent for each stage, rank-deficiency occurs only in constraint models for which time-invariant models are also rank-deficient.

Constraints on γ_i cause some additional models in Table 5 to become rank deficient, especially when all probabilities of breeding success are equal. The combination of 4 constraint models for each of the σ_i, β_i, and γ_i, combined with 4 types of time-variation yields more than 14,000 models.[4] Of these, more than 86% were fully estimable. An additional 2 of models become fully estimable if the p_i are known independently rather than estimated.

5.3 *Pre-breeding Survival and Recruitment*

As an example of the pre-breeding model, we compute rank deficiency for a model with a minimum age of first breeding of 4 and a maximum age of first breeding of 8 (as shown in Fig. 3).

[4]The number is slightly less than a simple combinatorial calculation would suggest because some possibilities (e.g., all σ_i equal, with additive time variation) are not possible.

Table 5 Rank deficiency for the time-varying 4-stage breeding success models with constraints on the σ_i and β_i with $N = 8$ capture occasions. In all cases γ_i and p_i are estimated separately for all stages and are freely time-varying. Codes used for time variation are: c = constant, x = trend or covariate, a = additive, and f = free time variation

Breeding constraint and time variation	Survival constraint and time variation														
	$\sigma_1 = \sigma_2 = \sigma_3 = \sigma_4$			$\sigma_1 = \sigma_3,\ \sigma_2 = \sigma_4$				$\sigma_1 = \sigma_2,\ \sigma_3 = \sigma_4$				$\sigma_1,\ \sigma_2,\ \sigma_3,\ \sigma_4$			
	c	x	f	c	x	a	f	c	x	a	f	c	x	a	f
$\beta_1 = \beta_2 = \beta_3 = \beta_4$															
c	0	0	0	0	0	0	0	1	0	0	1	1	0	0	1
x	0	0	0	0	0	0	0	0	0	0	0	0	0	0	0
f	0	0	0	0	0	0	0	0	0	0	4	0	0	0	4
$\beta_1 = \beta_3,\ \beta_2 = \beta_4$															
c	0	0	0	0	0	0	0	0	0	0	0	2	0	0	2
x	0	0	0	0	0	0	0	0	0	0	0	0	0	0	0
a	0	0	0	0	0	0	0	0	0	0	0	0	0	0	5
f	0	0	0	0	0	0	0	0	0	0	0	0	0	0	8
$\beta_1 = \beta_2,\ \beta_3 = \beta_4$															
c	0	0	0	0	0	0	0	1	0	0	1	1	0	0	1
x	0	0	0	0	0	0	0	0	0	0	0	0	0	0	0
a	0	0	0	0	0	0	0	0	0	0	4	0	0	0	4
f	0	0	0	0	0	0	0	0	0	0	4	0	0	0	4
$\beta_1,\ \beta_2,\ \beta_3,\ \beta_4$															
c	0	0	0	0	0	0	0	0	0	0	0	2	0	0	2
x	0	0	0	0	0	0	0	0	0	0	0	0	0	0	0
a	0	0	0	0	0	0	0	0	0	0	0	0	0	0	5
f	0	0	0	0	0	0	0	0	0	0	0	0	0	0	8

5.3.1 Time-Invariant Models

All the time-invariant models we examined were fully estimable, regardless of the survival and recruitment constraint combination. These time-invariant models are shown as special cases within Table 6, which includes results for the time-varying models.

5.3.2 Time-Varying Models

To account for confounding of parameters at the initial and final time periods as discussed in Section 3.4, for all models with freely time-varying survival

Table 6 Rank deficiency for time-varying pre-breeder recruitment models with constraints on σ_i and α_i for $N = 15$ capture occasions. The probability of survival in stage i is σ_i and the probability of first breeding at age i is α_i. Codes used for time variation are: c = constant, x = trend or covariate, a = additive, and f = free time variation

	Recruitment constraint					
	$\alpha_4 = \ldots = \alpha_8$			logistic $[\alpha_4 : \alpha_8]$		
Survival constraint	c	x	f	c	x	f
$\sigma_1 = \ldots = \sigma_9$						
c	0	0	0	0	0	0
x	0	0	0	0	0	0
f	0	0	0	0	0	0
σ_1						
$\sigma_2 = \ldots = \sigma_9$						
c	0	0	0	0	0	0
x	0	0	0	0	0	0
a	0	0	0	0	0	0
f	0	0	0	0	0	0
$\sigma_1 = \ldots = \sigma_3$						
$\sigma_4 = \ldots = \sigma_9$						
c	0	0	0	0	0	0
x	0	0	0	0	0	0
a	0	0	0	0	0	0
f	0	0	0	0	0	0
$\sigma_1 = \ldots = \sigma_3$						
$\sigma_4 = \ldots = \sigma_8$						
σ_9						
c	0	0	0	0	0	0
x	0	0	0	0	0	0
a	0	0	0	0	0	0
f	1	0	14	0	0	0
$\sigma_1 = \ldots = \sigma_3$						
logistic $[\sigma_4 : \sigma_9]$						
c	0	0	0	0	0	0
x	0	0	0	0	0	0
a	0	0	0	0	0	0
f	0	0	0	0	0	0
$\sigma_1 = \ldots = \sigma_3$						
logistic $[\sigma_4 : \sigma_8]$						
σ_9						
c	0	0	0	0	0	0
x	0	0	0	0	0	0
a	0	0	0	0	0	0
f	0	0	0	0	0	0

probabilities, the probabilities of survival at the first 4 time periods were constrained to be equal and the probabilities of survival at the last 5 time periods were constrained to be equal. Similarly, for freely time-varying recruitment, the probabilities of recruitment were constrained to be equal for the first 4 time periods.

All but two combinations of parameter constraints and time-variation models were full rank (Table 6). The only rank-deficient models had free time-variation in three survival probabilities ($\sigma_1 = \sigma_2 = \sigma_3$, $\sigma_4 = \ldots = \sigma_8$, and σ_9) and a common recruitment probability across ages 4–8 that was either constant or freely time-varying.

5.4 Applying Constraint Models to the Wandering Albatross

Evaluation of model rank is a useful first step in parameter estimation to aid in model selection and estimation results. As an example of the use of parameter constraints, we show some preliminary results from an ongoing study of the Wandering Albatross (*Diomedea exulans*) at Bird Island, South Georgia. The Wandering Albatross is classified as vulnerable by the IUCN due to observed population declines (about 28% from 1961 to 1996 at Bird Island, South Georgia) and the impact of longline fisheries. Albatross are long-lived, reaching maturity between 5 and 15 years, and have low fecundity, often skipping years between breeding attempts and laying only one egg each breeding attempt.

We examine parameter redundancy for the 4-stage breeding success model (Fig. 2) for Wandering Albatross data collected by the British Antarctic Survey during more than 40 years of banding and monitoring on Bird Island (Croxall et al. 1998). We use a data set for known-age birds from 1986 to 2005, during this period the recapture effort has been relatively constant. The data set includes 4884 individual capture histories.

We are particularly interested in whether the survival of successful and failed breeders differs, suggesting differences in individual quality, or whether survival of breeding and non-breeding birds differs, suggesting a trade-off between reproduction and survival. To address these questions we fit time-invariant versions of the 4-stage breeding success model (Fig. 2).

Before fitting the models using MATLAB and program E-Surge (Choquet et al. 2004), we checked extrinsic model redundancy for the data set. In only two cases did we find $\mathbf{J}_{\text{intrinsic}}$ of full rank while $\mathbf{J}_{\text{data}}$ was rank-deficient. These were for survival constraint Model 4 and breeding constraint Models 2 or 4 and both had free time variation in survival and a covariate for breeding probability. We have no intuitive explanation for why these two models should be rank-deficient.

We used AIC to compare support for the models. Table 7 contains the ΔAIC values (i.e., the AIC for the model relative to the minimum AIC for the model set). For this set of time-invariant models, the choice of survival constraint had little effect on AIC or on parameter estimates (not shown here). However, models providing separate estimates of β_2, β_3 and β_4 performed much better than models with constrained breeding probabilities. This suggests that there is little difference in survival probability among the four stages but that important differences exist

Table 7 ΔAIC values for time-invariant breeding success models (Fig. 2) for the Wandering Albatross on Bird Island, South Georgia. In all cases, γ_i and p_i are estimated separately for each stage

Breeding constraint model	Survival constraint model			
	$\sigma_1 = \sigma_2 = \sigma_3 = \sigma_4$	$\sigma_1 = \sigma_3,\ \sigma_2 = \sigma_4$	$\sigma_1 = \sigma_2,\ \sigma_3 = \sigma_4$	$\sigma_1,\ \sigma_2,\ \sigma_3,\ \sigma_4$
$\beta_1 = \beta_2 = \beta_3 = \beta_4$	360	333	307	304
$\beta_1 = \beta_2,\ \beta_3 = \beta_4$	292	211	211	211
$\beta_1 = \beta_3,\ \beta_2 = \beta_4$	242	219	219	221
$\beta_1,\ \beta_2,\ \beta_3,\ \beta_4$	1.13	0.00	0.00	0.0001

among the breeding probabilities. The model in row 4 column 4 is rank-deficient, and hence would be of no use for parameter estimation. We would therefore exclude it from model averaging and further analysis.

The estimation procedure has difficulty distinguishing between the models in columns 2 and 3. Analyses of simulated data (not shown here) confirm that this is a function of the similarity among the σ_i, not a feature of the model structure.

6 Discussion

Unobservable stages are common in the life cycles of many organisms. They are common in long-lived marine species where individuals can usually only be individually identified on the breeding grounds. This means that pre-breeding individuals are unobservable between leaving the natal area and recruiting to the breeding population, as are mature individuals that do not breed in a given year. This problem has been recognized in birds (e.g., Clobert et al. 1994; Kendall and Nichols 1995; Cam et al. 1998; Pradel and Lebreton 1999; Schwarz and Arnason 2000; Weimerskirch et al. 1987; Lindberg et al. 2001; Spendelow et al. 2002; Lebreton et al. 1992; Crespin et al. 2006), marine mammals (Schwarz and Stobo 1997; Fujiwara and Caswell 2002a) and sea turtles (Kendall et al. 1997; Kendall and Bjorkland 2001). The phenomenon is not limited to marine species (Kendall et al. 1997; Kery et al. 2005; Bailey et al. 2004).

The rank-deficiency created by unobservable stages can be removed by imposing constraints on the parameters. In the models studied here, the number of models we found to be fully estimable (with the exception of the usual confounding of parameters at the last time period and equivalent confounding at the first time period) is a pleasant surprise. This suggests it is possible to estimate demographic models that encompass interesting biological features even when the life cycle includes unobservable states.

In some cases, adding parameters can make a rank-deficient model estimable. For example, the time-invariant 4-stage breeding success model (Fig. 2) is rank-deficient when the four survival probabilities are separate. Yet, in a time-varying model with survival probabilities modeled as functions of a covariate or a temporal trend, the model becomes fully estimable even though it has an additional $N - 4$ parameters. This phenomenon has been observed previously.

A Jolly–Seber model with fully age-dependent time-invariant survival probabilities and a time-invariant recovery probability, is rank-deficient. Adding extra parameters by allowing first-year survival to be time-dependent makes the model non-redundant (Morgan and Freeman 1989; Catchpole et al. 1996).

Some discussions of temporary emigration have emphasized the use of robust design models to provide independent information on capture probabilities. Our results suggest this helps primarily for simple or more highly constrained models. For example, we found no differences in rank deficiency between models with p_i known versus p_i estimated when all γ_i were modeled separately. Knowing p has also been shown to improve estimability of a 2-stage (one observable and one unobservable state) first-order Markov model (Kendall and Nichols 2002). Our method suggests more of the models in Table 2 of Kendall and Nichols (2002) are full rank, particularly those with fewer time constancy or parameter constraints. This difference is partly due to our use of both initial and final constraints. Kendall and Nichols (2002) considered models with the usual confounding of parameters at the last time period to be estimable, but did not allow for not being able to release individuals in the unobservable state at the first time step.

It is not always obvious from the confounded parameters whether imposing final and initial constraints will make a model estimable. For example, in the models considered by Kendall and Nichols, with separate time-varying transition probabilities, separate but constant survival probabilities, and time-varying capture probability, all parameters except one (capture probability at $t = 2$) are confounded, but when initial and final constraints are imposed this model becomes fully estimable.

Although many of the models presented here are intrinsically estimable, they may not be simple to estimate. Long-term and relatively complete data sets will be required to fit multi-state models with unobservable states. It is known that multi-state models tend to produce multiple maxima because they involve sums of products, rather than only the products that occur in single-state models (Lebreton and Pradel 2002). Unobservable states are likely to add to this problem and make estimation under these models even more difficult.

The method we present here extends and complements previous work on the rank of multi-state models (Catchpole and Morgan 2001; Fujiwara and Caswell 2002a; Kendall and Nichols 2002; Gimenez et al. 2003, 2004). It has the advantage of being relatively easy to implement, given the matrix formulation of the likelihood, the availability of the TOMLAB implementation of automatic differentiation, and a working knowledge of MATLAB. It can be used to determine both intrinsic and extrinsic parameter redundancy. Only the formal symbolic analytic method provides full information on how parameters are confounded in redundant models, but the method we present does provide information on which parameters are separably estimable and which are confounded in some way. Our method is completely general in terms of model structure (including use of conditional transitions), parameter constraints, temporal constraints, and availability of additional information, limited only by biological reality (and the investigator's imagination).

Estimates of survival and transition rates are essential to a meaningful investigation of population dynamics, especially in colonial seabirds. Multi-state

Table 8 Rank deficiency for time-varying biennial 4-stage breeding success models with constraints on σ_i and β_i with $N = 8$ capture occasions. All γ_i and p_i are estimated separately and freely time-varying. Codes used for time variation are: c = constant, x = trend or covariate, a = additive, and f = free time variation

Breeding constraint	Survival constraint and time variation														
	$\sigma_1 = \sigma_2 = \sigma_3 = \sigma_4$			$\sigma_1 = \sigma_3,\ \sigma_2 = \sigma_4$				$\sigma_1 = \sigma_2,\ \sigma_3 = \sigma_4$				$\sigma_1,\ \sigma_2,\ \sigma_3,\ \sigma_4$			
and time variation	c	x	f	c	x	a	f	c	x	a	f	c	x	a	f
$\beta_1 = \beta_2 = \beta_3 = \beta_4$															
c	0	0	0	0	0	0	0	0	0	0	0	1	0	0	1
x	0	0	0	0	0	0	0	0	0	0	0	0	0	0	0
f	0	0	0	0	0	0	0	0	0	0	0	0	0	0	4
$\beta_1 = \beta_3,\ \beta_2 = \beta_4$															
c	0	0	0	0	0	0	0	0	0	0	0	2	0	0	2
x	0	0	0	0	0	0	0	0	0	0	0	0	0	0	1
a	0	0	0	0	0	0	0	0	0	0	0	0	0	0	5
f	0	0	0	0	0	0	0	0	0	0	0	0	0	0	8
$\beta_1 = \beta_2,\ \beta_3 = \beta_4$															
c	0	0	0	0	0	0	0	0	0	0	0	1	0	0	1
x	0	0	0	0	0	0	0	0	0	0	0	0	0	0	0
a	0	0	0	0	0	0	0	0	0	0	0	0	0	0	4
f	0	0	0	0	0	0	0	0	0	0	0	0	0	0	4
$\beta_1,\ \beta_2,\ \beta_3,\ \beta_4$															
c	0	0	0	0	0	0	0	0	0	0	0	2	0	0	2
x	0	0	0	0	0	0	0	0	0	0	0	0	0	0	1
a	0	0	0	0	0	0	0	0	0	0	0	0	0	0	5
f	0	0	0	0	0	0	0	0	0	0	0	0	0	0	8

mark-recapture analysis provides a powerful tool for obtaining such estimates. We hope the methods we present here will improve the utility and efficiency of this analysis, providing a means to determine parameter redundancy of a wider variety of models than is currently possible.

Acknowledgments We acknowledge support and funding from NSF grant DEB-0343820, NOAA grant NA03-NMF4720491, EPA Grant R-82906901-4, and the Ocean Life Institute of the Woods Hole Oceanographic Institution. The methods described here were developed as part of the efforts of the WHOI Albatross Demography Working Group. Discussions with Remi Choquet, John Croxall, Jaume Forcada, Martin Golubitsky, Stephanie Jenouvrier, and Jean-Dominique Lebreton were particularly valuable. We thank Marcus Edvall of Tomlab for helpful advice on automatic differentiation. We also thank Stephen Freeman and an anonymous reviewer for helpful comments. The Wandering Albatross example is from an ongoing collaboration with the British Antarctic Survey, full results of the analysis will appear elsewhere.

Appendix

Section 5.2 presents rank deficiency results for the 4-stage breeding model for a species capable of breeding annually. This appendix gives rank deficiency results for biennial versions of that model as used in the Wandering Albatross example (Table 8). For such species, $\beta_1 = 0$, so the transition from successful breeder to post-success non-breeder (Fig. 2) becomes deterministic. This makes additional models with survival constraint Model 3 fully estimable. Most models with four separate survivals and full time variation in survival are still parameter redundant.

References

Bailey LL, Kendall WL, Church DR, Wilbur HM (2004) Estimating survival and breeding probability for pond-breeding amphibians: a modified robust design. Ecology 85:2456–2466

Brownie C, Hines JE, Nichols JD, Pollock K, Hestbeck JB (1993) Capture–recapture studies for multiple strata including non-Markovian transitions. Biometrics 49:1173–1187

Cam E, Hines JE, Monnat J-Y, Nichols JD, Danchin E (1998) Are adult nonbreeders prudent parents? The kittiwake model. Ecology 79:2917–2930

Caswell H (2001) Matrix population models. Second edition. Sinauer, Sunderland, MA

Caswell H, Fujiwara M (2004) Beyond survival estimation: mark-recapture, matrix population models, and population dynamics. Animal Biodiversity and Conservation 27:471–488

Catchpole EA, Morgan BJT (1997) Detecting parameter redundancy. Biometrika 84:187–196

Catchpole EA, Morgan BJT (2001) Deficiency of parameter-redundant models. Biometrika 88:593–598

Catchpole EA, Freeman SN, Morgan BJT (1996) Steps to parameter redundancy in age-dependent recovery models. Journal of the Royal Statistical Society Series B- Methodological 58: 763–774

Catchpole EA, Morgan BJT, Freeman SN (1998) Estimation in parameter-redundant models. Biometrika 85:462–468

Catchpole EA, Kgosi PM, Morgan BJT (2001) On the near-singularity of models for animal recovery data. Biometrics 57:720–726

Catchpole EA, Morgan BJT, Viallefont A (2002) Solving problems in parameter redundancy using computer algebra. Journal of Applied Statistics 29:625–636

Choquet R, Reboulet AM, Pradel R, Gimenez O, Lebreton JD (2004) M-SURGE: new software specifically designed for multistate capture-recapture models. Animal Biodiversity and Conservation 27:207–215

Clobert J, Lebreton J-D, Allaine D, Gaillard JM (1994) The estimation of age-specific breeding probabilities from recaptures or resightings in vertebrate populations: II. Longitudinal models. Biometrics 50:375–387

Crespin L, Harris MP, Lebreton J-D, Frederikesen M, Wanless S (2006) Recruitment to a seabird population depends on environmental factors and on population size. Journal of Animal Ecology 75:228–238

Croxall J, Prince P, Rothery P, Wood AG (1998) Population changes in albatrosses at South Georgia. In: Robertson G, Gales R (eds) Albatross biology and conservation. Surrey Beatty & Sons Pty Ltd., Chipping Norton, NSW, pp 69–83

Forth SA, Edvall MM (2006) User guide for MAD – a Matlab automatic differentiation toolbox. TOMLAB/MAD. Version 1.4. The forward mode. Tomlab Optimization Inc., 855 Beech St #121, San Diego, CA, USA

Fujiwara M, Caswell H (2002a) A general approach to temporary emigration in mark-recapture analysis. Ecology 83:3266–3275

Fujiwara M, Caswell H (2002b) Estimating population projection matrices from multi-stage mark-recapture data. Ecology 83:3257–3265

Gimenez O, Choquet R, Lebreton J-D (2003) Parameter redundancy in multi-state capture-recapture models. Biometrical Journal 45:704–722

Gimenez O, Viallefont A, Catchpole EA, Choquet R, Morgan BJT (2004) Methods for investigating parameter redundancy. Animal Biodiversity and Conservation 27:561–572

Golub GH, Van Loan CF (1996) Matrix computations. Third edition. Johns Hopkins University Press, Baltimore, MD

Griewank A (2000) Evaluating derivatives: applications of algorithmic differentiation. SIAM, Philadelphia, PA

Griewank A (2003) A Mathematical view of automatic differentiation. Acta Numerica 12:321–398

Hirsch MW, Smale S (1974) Differential equations, dynamical systems, and linear algebra. Academic Press, New York

Hunter CM, Caswell H (2005) Selective harvest of sooty shearwater chicks: effects on population dynamics and sustainability. Journal of Animal Ecology 74:589–600

Hunter CM, Moller H, Fletcher D (2000) Parameter uncertainty and elasticity analyses of a population model: setting research priorities for shearwaters. Ecological Modelling 134:299–323

Kendall WL, Nichols JD (1995) On the use of secondary capture–recapture samples to estimate temporary emigration and breeding proportions. Journal of Applied Statistics 22:751–762

Kendall WL, Bjorkland R (2001) Using open robust design models to estimate temporary emigration from capture–recapture data. Biometrics 57:1113–1122

Kendall WL, Nichols JD (2002) Estimating state-transition probabilities for unobservable states using capture–recapture/resighting data. Ecology 83:3276–3284

Kendall WL, Nichols JD, Hines JE (1997) Estimating temporary emigration using capture–recapture data with Pollock's robust design. Ecology 78:563–578

Kery M, Gregg KB, Schaub M (2005) Demographic estimation methods for plants with unobservable life-states. Oikos 108:307–320

Lebreton J-D, Pradel R (2002) Multistate recapture models: modelling incomplete individual histories. Journal of Applied Statistics 29:353–369

Lebreton J-D, Burnham KP, Clobert J, Anderson DR (1992) Modeling survival and testing biological hypotheses using marked animals: a unified approach with case studies. Ecological Monographs 62:67–118

Lebreton J-D, Almeras T, Pradel R (1999) Competing events, mixtures of information, and multistratum recapture models. Bird Study 46:S39–S46

Lindberg MS, Kendall WL, Hines JE, Anderson MG (2001) Combining band recovery data and Pollock's robust design to model temporary and permanent emigration. Biometrics 57: 273–282

Magnus JR, Neudecker H (1988) Matrix differential calculus with applications in statistics and econometrics. John Wiley & Sons, New York

Morgan BJT, Freeman SN (1989) A model with first-year variation for ring-recovery data. Biometrics 45:1087–1101

Murota K (2003) Matrices and matroids for ststems analysis. Springe-Verlag, New York

Nichols JD, Sauer JR, Pollock KH, Hestbeck JB (1992) Estimating transition probabilities for stage-based population projection matrices using capture–recapture data. Ecology 73:306–312

Pradel R, Lebreton J-D (1999) Comparison of different approaches to the study of local recruitment of breeders. Bird Study 46(Supplement):74–81

Rouan L, Choquet R, Pradel R (unpublished) A general framework for modeling memory in capture–recapture data. Submitted for publication.

Schwarz CJ, Arnason AN (2000) Estimation of age-specific breeding probabilities from capture-recapture data. Biometrics 56:59–64

Schwarz CJ, Stobo WT (1997) Estimating temporary migration using the robust design. Biometrics 53:178–194

Shampine LF, Ketzscher R, Forth SA (2005) Using AD to solve BVPs in MATLAB. ACM Transactions on Mathematical Software 31:79–94

Spendelow JA, Nichols JD, Hines JE, Lebreton J-D, Pradel R (2002) Modeling post-fledging survival and age-specific breeding probabilities in species with delayed maturity: a case study of Roseate terns at Faulkner Island, Connecticut. Journal of Applied Statistics 29:385–405

Stewart GW (1991) Perturbation theory for the singular value decomposition. in Vaccaro RJ (ed) SVD and signal processing II: algorithms, analysis, and implementation. Elsevier, Amsterdam

Stewart GW (1992) Determining rank in the presence of error. In: Moonen MS, Golub GH, de Moor BL (eds) Linear algebra for large scale and real-time applications. Kluwer, Dordrecth

Weimerskirch H, Clobert J, Jouventin P (1987) Survival in five southern albatrosses and its relationship with their life history. Journal of Animal Ecology 56:1043–1055

Wonham WM (1985) Linear multivariate control: a geometric approach. Third edition. Springer, New York

Mark-Recapture Jolly-Seber Abundance Estimation with Classification Uncertainty

Wendell O. Challenger and Carl J. Schwarz

Abstract Wildlife managers and ecologists are often interested in estimating abundance of animals belonging to a certain fixed group (e.g. sex), but in some cases group membership cannot always be ascertained. Group assignment uncertainties can occur either through the inability to assign group membership because of a lack of group-specific characteristics (e.g. males and females look alike), lack of training (e.g. volunteers), or through errors in assignment. Recently, methodological advances in closed population capture-recapture models have allowed for the inclusion of classification uncertainties in parameter estimates. We build on this work by addressing identification uncertainty in abundance estimation (open population models), providing a general method for dealing with multiple groups/states when the true underlying group/state can be considered fixed for the duration of the experiment. We then apply this methodology to estimate the sex-specific abundances of walleyes (*Stizostedion vitreum*) in Mille Lacs, Minnesota.

Keywords Capture–recapture · Unknown sex · Jolly–Seber · Maximum likelihood · Group uncertainty · Sexing errors

1 Introduction

The capture–recapture (CR) experimental protocol provides a flexible approach for making inferences on animal populations that may be hard to observe otherwise. Inferences can range from individual level processes, such as survivorship, to population level processes such as abundance and recruitment. The Jolly-Seber (JS) model is used for open populations, where both births (or immigration) and deaths (or emigration) are explicitly modelled. Current applications of the JS abundance model include estimating overall and group specific abundances.

W.O. Challenger (✉)
Department of Statistics, and Actuarial Sciences, Simon Fraser University, Burnaby, BC V5A 1S6, Canada
e-mail: wchallen@stat.sfu.ca

D.L. Thomson et al. (eds.), *Modeling Demographic Processes in Marked Populations*, Environmental and Ecological Statistics 3, DOI 10.1007/978-0-387-78151-8_38,

The JS class of models make key assumptions on the marked and unmarked animals in the population of interest. Specifically, both segments of the population are assumed to be homogeneous in catchability and demographic parameters such as survivorship. This assumption of homogeneity is not only required at the population level but also at the individual level, where parameters are assumed to be the same among all animals. However, populations often have underlying structures that can violate the latter assumption of animal homogeneity. For example, animals with differing age and/or sex may exhibit differences in catchabilities and/or different apparent survivorship.

If left unmodeled the heterogeneity may result in over-dispersion, model misfit and potentially biased estimates. Heterogeneity can be handled by modeling affected parameters in terms of individual covariates or, at a courser level, by modeling the heterogeneity through discrete partitions. When over the course of the experiment membership to a partition is fixed (e.g. sex), the term "group" is often used, while the term "state" often implies the potential for membership can change. For systems where membership to a partition is fixed (i.e. groups) classically parameter grouping methods (e.g. Jolly 1965; Lebreton et al. 1992) are often used, while systems where membership can change (non-fixed states) are handled by multi-state (e.g. Lebreton and Pradel 2002) and multi-event models (Pradel 2005). These approaches not only assume that heterogeneity may be effectively partitioned, but that the state/group can also be accurately assessed on capture. Therefore, the ability to assign state/group represents an additional source of variability, which up until recently has not been included in many models.

State/group assignment uncertainty can be divided into two distinct types, identification uncertainty and mis-assignment uncertainty. Identification uncertainty represents a probability of identifying a state/group when observed, while mis-assignment is the probability of assigning the incorrect state/group. Depending on the system, one or both of these types of error may occur separately or in synchrony. Models have been developed to deal with uncertainties in identification (e.g. Nichols et al. 2004), state assignment (e.g. Fujiwara and Caswell 2002, Kendall et al. 2003; Pradel 2005), as well as both types simultaneously (Pradel 2005). However, these advances are restricted to the Cormack-Jolly-Seber (CJS) framework and as such do not allow for abundance estimates. Since abundance estimates are often of interest to wildlife managers, there is a need to extend these approaches to the JS framework.

In extending the JS framework to deal with identification uncertainty, we deal with characteristics that are fixed over the duration of the experiment. Fixed characteristics, such as sex, are often of interest to wildlife managers and may also partition heterogeneity in the population. However, due to factors, such as lack-of-training, identification of these characteristics on captures is not always known. To this end we consider the situation where sex can only be accurately assigned on some of the capture occasions or not at all.

When dealing with group uncertainty for fixed characteristics, it can be appropriate to look at identification uncertainty rather than mis-assignment (e.g. see Nichols et al. 2004). Primarily, each sampling occasion gives an independent opportunity to determine the true underlying group. As such, group assignments are not required on each sampling occasion, since future and previous occasions provide useful information. In this manner positive group assignments may be restricted to situations where the group is known with certainty, with all remaining scenarios handled by assignment to the unknown group. This is in contrast to non-fixed characteristics (e.g. breeding state) that can transition between sampling occasions and where there is an added emphasis on making a state assignment each every capture occasion (but see Pradel 2005).

In the present paper we extend the approach used by Nichols et al. (2004) to deal with sex uncertainty, to the JS framework by modifying the super-population formulation of Schwarz and Arnason (1996). Positive assignments were restricted to situations where sex was known with certainty and all other situations were dealt with as unknown designations. This model was then applied to walleyes (*Stizostedion vitreum*) data from Mille Lacs, Minnesota, where sexing designations were not always possible due to staffing restrictions. The model was then used to determine sex-specific abundances and sex-specific recruitment over multiple years.

2 Survey Protocol

The survey protocol follows the standard JS protocol with groups except that on each capture occasion, 3 possible events may occur: unobserved, group unknown, or a positive group identification. We consider the case of 2 groups, males and females. We assume that any positive assignment is definitive and as a consequence strictly forbid the assignment of more than one sex to a tag history. For situations where group uncertainty exists the *unknown* designation should be used.

This is intended for situations where there are a mixture of experts and non-experts conducting the survey and each type of observer has different capabilities to assign group membership (i.e. sex) on capture. Both types of observers are expected to record tag numbers correctly, while only expert observers are allowed to assign group membership. Upon capture experts may choose to assign group membership (i.e. male or female), or designate group membership as *unknown*, while captures by non-experts are given the *unknown* group membership. In doing so we give expert observers the option of forgoing group membership assignment, should the need arise, and allow non-expert observations to be incorporated.

This treatment of groups differs slightly from the more classical grouping methods (i.e. Jolly 1965; Lebreton et al. 1992; Schwarz and Arnason 1996) that assume all animals can be definitively assigned to their respective groups at some point during the experiment. We relaxed this assumptions by modeling the probability of encountering individuals with unresolved sex as being a composite of both sexes. In this manner we take an approach that is similar both Nichols et al. (2004) and the post-stratification method developed by Conroy et al. (1999), which used covariates to assign sex post capture to juvenile animals. However, both approaches were developed for the CJS framework (excluding abundance estimation) and unlike the post-stratification we make no requirements on the availability of covariates that can be used to predict sex. Our approach also differs from ad hoc techniques that use a unique category with its own demographic parameters to deal with the unknown sex designations. If handled in such a manner group specific demographic parameters, such as apparent survivorship, will be positively biased due to the expected increase in encounters for histories where sex has been resolved (Nichols et al. 2004).

Finally, in the most general case of our model, it is assumed group assignments will be attempted on each and every sampling occasion. While useful for estimating assignment probabilities, it is not an absolute requirement. If proper model restrictions are employed, experimental protocols can be designed where group assignment is only attempted on certain capture occasions (see the example section).

3 Notation

Classical mark-recapture experiments use a capture history vector of "1" or "0" to represent whether an animal was caught (1) or not caught (0) during k sampling occasions. In our case we distinguish between the unobserved state (0) and three possible observed "groups": unknown (1), female (2) and male (3). Here the *female* and *male* represent a positive identification, while *unknown* represents uncertainty in assignment. As such the capture history will still be a vector of length k, but will consisting of 0's, 1's, 2's and 3's instead of the traditional 0's and 1's. For example, the history {313} would be a male that was caught on all three sampling occasions but was positively identified as male on the 1st and 3rd occasions only.

Lastly, since positive identifications are considered definitive, a single tag history cannot contain both male and female identifications. As a result histories such as {312} are not allowed and possible observed tag histories will not be a permutation of all states, but instead a restricted subset.

3.1 Statistics and Indicator Variables

k – number of sampling occasions in the experiment.
m – number of uniquely observable tag histories. Note that this does not include the unobserved tag history ($\boldsymbol{\omega}_0$).
i – the index for tag histories, where $i = 0, 1, 2, \ldots, m$; $i = 0$ is used to denote the unobserved tag history.
j – the index for sampling occasion, where $j = 1, 2, \ldots, k$.
n_j – total number of animals caught at sampling occasion j.
v_j – total number of animals lost on capture at sampling occasion j.
ω_i – capture and identification history vector,
where $\omega_i = [\omega_{i1}, \omega_{i2}, \ldots, \omega_{ik}]'$ and

$$\omega_{ij} = \begin{cases} 3 \text{ for animals captured at sampling occasion } j \text{ and identified as male} \\ 2 \text{ for animals captured at sampling occasion } j \text{ and identified as female} \\ 1 \text{ for animals captured at sampling occasion } j \text{ and identified as unknown sex} \\ 0 \text{ for animals not captured at sampling occasion } j \end{cases}$$

ζ_i – the sex indicator for history $\boldsymbol{\omega}_i$, where

$$\zeta_i = \begin{cases} \{M\} & \text{for } \max(\omega_{i1},\omega_{i1},...,\omega_{ik}) = 3 \\ \{M\} & \text{for } \max(\omega_{i1},\omega_{i1},...,\omega_{ik}) = 2 \\ \{M, F\} & \text{for } \max(\omega_{i1},\omega_{i1},...,\omega_{ik}) \leq 1 \end{cases}.$$

Animals with an unknown sex designation are modeled as a composite of both sexes.

$\boldsymbol{\omega}_i^c$ – a capture history vector, where $\boldsymbol{\omega}_i^c = [\omega_{i1}^c, \omega_{i2}^c, \ldots, \omega_{ik}^c]'$ and

$$\omega_{ij}^c = \begin{cases} 1 \text{ if } \omega_{\text{ij}} \geq 1 - \text{a captured animal} \\ 0 \text{ otherwise} \end{cases}$$

$\boldsymbol{\omega}_i^s$ – sex identification vector, where $\omega_i^s = [\omega_{i1}^s, \omega_{i2}^s, \ldots, \omega_{ik}^s]'$ and

$$\omega_{ij}^s = \begin{cases} 1 \ \text{ if } \omega_{\text{ij}} > 1 - \text{a sexed animal} \\ 0 \ \text{ otherwise} \end{cases}$$

n_{ω_i} – number of animals with tag history ω_i.
f_i – first occasion when animals with tag history ω_i were captured.
l_i – last occasion when animals with tag history ω_i were captured.
κ_i – loss-on-capture indicator for tag history ω_i, where

$$\kappa_i = \begin{cases} 1 \ \text{ lost on sampling occasion } \mathrm{l_i} \\ 0 \ \text{ not lost on any sampling occasions} \end{cases}$$

3.2 Fundamental Model Parameters

N – the *super-population* is the total of animals that were present at the start of the study or entered the system between any pair of sampling occasion and survived to the next sampling occasion.

β_j – the expected fraction of the super-population that enters the population at sampling occasion j and survived to sampling occasion $j+1$ for $\{j : j = 0, 1, \ldots, k-1\}$. β_0 is the expected fraction of animals alive just prior to the first sampling occasion and $\sum_{j=0}^{k-1} \beta_j = 1$.

π_j^s – the probability that an animal entering the population between j and $j+1$ is of sex s, for $s \in \{M, F\}$ and $\{j : j = 0, 1, \ldots, k-1\}$. The parameter

π_0^s is the proportion of animals of sex s alive in the population prior to the first sampling occasion and $\sum_s \pi_j^s = 1$, for all j.

p_j^s – the sex-specific probability an animal will be caught at occasion j, given that the animal is alive at sampling occasion j, for $s \in \{M, F\}$ and $\{j : j = 1, 2, \ldots, k\}$.

δ_j^s – the probability, given an animal has been caught, that the sex will be positively identified at sampling occasion j, for $s \in \{M, F\}$ and $\{j : j = 1, 2, \ldots, k\}$. The probability the animal's sex cannot be identified is $(1 - \delta_j^s)$. Positive identifications are considered to be definitive.

ϕ_j^s – The sex-specific probability that an animal survives and remains in the population from sampling occasion j to sampling occasion $j + 1$, given it was alive and in the population at sampling occasion j, for $s \in \{M, F\}$ and $\{j : j = 1, 2, \ldots, k\}$.

ν_j – the probability of losing an animal when it is captured at sampling occasion j. Equal probability is assumed for both sexes. A loss is any event whereby the animal is not released into the catchable population after capture.

3.3 Functions of Parameters

β_j^* – the expected fraction of the super-population remaining to enter the population between sample occasion j and $j+1$, for $\{j : j = 0, 1, \ldots, k-1\}$. Note that $\beta_0^* = \beta_0$, $\beta_j^* = \frac{\beta_j}{\sum_{i=j}^{k-1} \beta_i}$, and $\beta_{k-1}^* = 1$. The β_j^*'s were used when fitting the likelihood, because they are unconstrained, taking on any value in $[0, 1]$, yet maintain the $\sum_{j=0}^{k-1} \beta_j = 1$ constraint.

χ_j^s – The probability an animal of sex s is not observed again when last captured in period j.

$$\chi_j^s = \begin{cases} 1 - \phi_j^s + \phi_j^s(1 - p_{j+1}^s)\chi_{j+1}^s & j < k \\ 1 & j = k \end{cases} \quad \text{for } s \in \{M, F\}.$$

ψ_j^s – The probability of an animal of sex s entering the population and remaining unobserved prior to sampling occasion j.

$$\psi_j^s = \begin{cases} \beta_0 \pi_0^s & j + 1 = 1 \\ \psi_j^s(1 - p_j^s)\phi_j^s + \beta_j \pi_j^s & j + 1 > 1 \end{cases} \quad \text{for } s \in \{M, F\},$$

where π_0^s is the proportion of animals of sex s alive just prior to the first sampling period, while π_j^s, for $j > 0$, is the sex-specific proportion of the entrants.

B_j^s – The total number of animals of sex s that enter the system between sampling occasion j and $j + 1$, for $\{j : j = 0, 1, \ldots, k - 1\}$ and $s \in \{M, F\}$. B_0^s represents the number of animals alive (male or female) just prior to the first sampling occasion. The remaining B_j^s's are referred to as the net births and represent the process of live births, immigration and/or recruitment into the population of interest. Also note $E\left[B_j^s\right] = N\beta_j\pi_j^s$ and $N = \sum_s \sum_{j=0}^{k-1} B_j^s$.

N_j^s – The sex-specific population size at sampling occasion j.

$$E[N_{j+1}^s] = \begin{cases} N\beta_0\pi_0^s & j+1=1 \\ E[N_j^s]\phi_j^s + N\beta_j\pi_j^s & j+1>1 \end{cases} \text{ for } s \in \{M, F\}.$$

When losses on capture occur, a different formulation is required (see Appendix A).

4 Model Development

4.1 Assumptions

The standard mark-recapture assumptions are made. As well, we assume that both marked and unmarked animals of the same sex exhibit the same sex-specific catchability $\{p_i^s\}$ and survivorship $\{\phi_i^s\}$ (homogeneity). Also important is the assumption that newly captured unmarked animals are a random sample of all unmarked animals in the population. Additional assumptions specific to this model include equal probability of being sexed for caught males and females and homogeneity of demographic parameters for animals that were sexed and animals that were not sexed. This last assumption could be violated for example if juveniles are harder to sex and exhibit different survivorship. It is also assumed that the underlying characteristic (sex) is fixed throughout the experiment and the underlying population can be dichotomously divided into either males or females. Finally, as with most capture-recapture experiments it is assumed that tags are not lost, not misread and are unique to each animal, that sampling is instantaneous, and that the study area is constant throughout the experiment.

4.2 Likelihood

The model we propose is a direct extension of the super-population model proposed by Schwarz and Arnason (1996). While Schwarz and Arnason did allow for group specific demographic parameters (i.e. $\{\beta_{jg}\}, \{p_{jg}\}, \{\phi_{jg}\}$), we go a step further by modeling the probability of a positive group assignment (the identification probability), as well as modeling tag histories in which assignment did not occur.

With the JS capture–recapture models, it is common practice to break the likelihood into distinct portions. Following the notation of Schwarz and Arnason (1996) the likelihood of the super-population model can be written as

$$L = L_1^A(N, \{\beta_i\}, \{p_i\}, \{\phi_i\}) \times L_1^B(\{\beta_i\}, \{p_i\}, \{\phi_i\}) \times L_2(\{\nu_i\}) \times L_3(\{p_i\}, \{\phi_i\}).$$

These components make up $L_1^A \times L_1^B = P(\text{first capture}|\{\beta_i\}, \{p_i\}, \{\phi_i\})$, $L_2 = P(\text{loss-on-capture}|\nu_i)$ and $L_3 = P(\text{recapture}|\{p_i\}, \phi_i)$ respectively. Schwarz and Arnason divided the probability of first capture into two components (L_1^A and L_1^B), by first conditioning the complete likelihood on the total number of unmarked animals observed in the experiment (we use n_{obs} in place of $u.$). This formulation was used to develop a conditional estimate of $\hat{N}$ by first maximizing $L_1^B \times L_2 \times L_3$ and then using the estimates to derive the conditional MLE of $\hat{N}$. Asymptotically, the final result will be equivalent to maximizing the entire likelihood (Sanathanan 1972; Schwarz and Arnason 1996).

We formulate our model in a similar manner except we combine the L_1^B, L_2 and L_3 terms ($L_1^{B*} = L_1^B \times L_2 \times L_3$) so that the probability expression for each observed tag history can be modelled directly. Also similar to the group specific model proposed by Schwarz and Arnason (1996), we have male and female specific demographic parameters. However, we also model the probability an entrant will be of a particular sex (π_j^s) and the probability the sex can be identified, conditional on capture (δ_j^s).

We start by modeling the total number of observed tag histories ($n_{\text{obs}} = \sum_{i=1}^{m} n_{\omega_i}$ and $n_{\omega_0} = N - n_{\text{obs}}$), where $L_1^A = [n_{obs}|N] \sim$ Binomial(N, 1 − P($\boldsymbol{\omega}_0$)). Here $\boldsymbol{\omega}_0$ is the unobserved capture history ($\boldsymbol{\omega}_0 = [0, 0, \ldots, 0]'$) and $P(\boldsymbol{\omega}_0)$ is the probability of a male or female entering the population at some time prior to, or during, the study and remaining unobserved for the remainder of the sampling occasions.

$$P(\boldsymbol{\omega}_0) = \sum_{s \in \{M,F\}} \sum_{j=0}^{k-1} \beta_j \pi_j^s (1 - p_{j+1}^s) \chi_{j+1}^s \tag{1}$$

Next, we model the distribution conditional upon being seen at least once, $L_1^{B^*} = \left[\{n_{\omega_i}\}|n_{\text{obs}}\right] \sim \text{Multinomial}\left(n_{\text{obs}}, \{\lambda_{\omega_i}\}\right)$, where $\lambda_{\omega_i} = \frac{P(\boldsymbol{\omega}_i)}{(1-P(\boldsymbol{\omega}_0))}$. The probability of the observed tag history $P(\boldsymbol{\omega}_i)$, is

$$\begin{aligned} P(\boldsymbol{\omega}_i) = \sum_{s \in \zeta_i} \psi_{f_i}^s & \left\{ \prod_{j=f_i}^{l_i} (p_j^s)^{\omega_{ij}^c} (1 - p_j^s)^{(1-\omega_{ij}^c)} \right\} \\ \times & \left\{ \prod_{j=f_i}^{l_i} (\delta_j^s)^{\omega_{ij}^s \times \omega_{ij}^c} (1 - \delta_j^s)^{(1-\omega_{ij}^s) \times \omega_{ij}^c} \right\} \\ \times & \left\{ \prod_{j=f_i}^{l_i - 1} \phi_j^s \right\} (\chi_{l_i}^s)^{(1-\kappa_i)} \\ \times & \left\{ \prod_{j=f_i}^{l_{i-1}} (1 - \nu_j)^{\omega_{ij}^c} \right\} \times (1 - \nu_{l_i})^{(1-\kappa_i)} (\nu_{l_i})^{\kappa_i} \end{aligned} \tag{2}$$

Note that for the case where an animal is seen only once ($f_i = l_i$) the terms $\prod_{j=f_i}^{l_i-1} \phi_j^s$ and $\prod_{j=f_i}^{l_i-1} \nu_j$ are both treated as evaluating to 1.

The probability of identifying the correct sex is represented by the δ_j^s parameter, which is conditional upon capture (i.e. $\omega_{ij}^s \times \omega_{ij}^c$). For animals where the sex identification has been positively determined at some point during the experiment $\zeta_i \in \{M\}$ or $\zeta_i \in \{F\}$, otherwise $\zeta_i \in \{M, F\}$. As a result, for tag histories where sex has been confirmed, only demographic parameters associated with that sex (or group) are used. For example the tag history 0133 (or 0UMM) is a male that was captured on the second, third and fourth sampling occasions. On the first capture, its sex could not be identified, but on subsequent occasions a definitive sex assignment was made. The probability of this history (excluding the loss-on-capture component) will be

$$P(0133) = \psi_2^M p_2^M (1 - \delta_2^M) \phi_2^M p_3^M \delta_3^M \phi_3^M p_4^M \delta_4^M .$$

However, for tag histories where a definitive group identification is not possible, the histories are modelled as belonging to both groups, but with an unidentified status (i.e. $\omega_{ij}^s = 0$ for all j where $\omega_{ij}^c = 1$). As a result the $(1 - \delta_j^s)$ term appears in association with each capture occasion. For example the tag history 0111 (or 0UUU) has the same capture history as the previous example, but without a definitive sex identification. In this case the probability of this tag history will be

$$\begin{aligned} P(0111) = {} & \psi_2^M p_2^M (1 - \delta_2^M) \phi_2^M p_3^M (1 - \delta_3^M) \phi_3^M p_4^M (1 - \delta_4^M) \\ & + \psi_2^F p_2^F (1 - \delta_2^F) \phi_2^F p_3^F (1 - \delta_3^F) \phi_3^F p_4^F (1 - \delta_4^F). \end{aligned}$$

This form of grouping differs from more classical grouping methods (i.e. Jolly 1965, Lebreton et al. 1992, Schwarz and Arnason 1996) that assume all animals can be definitively assigned to their respective groups during at least one of the encounters. In the classical approach the fully paramaterized model (no model constraints, either group or time) will be equivalent to running multiple independent capture recapture experiments in parallel on each of the possible groups. In contrast, for the approach proposed here the *unknown* designation will always be a composite of the possible positive assignments, even in the most parameterized model. As a result, the complete likelihood for this model will be

$$\begin{aligned} L_{\text{complete}} &= L_1^A \times L_1^{B*} \\ &= \left\{ \binom{N}{n_{obs}} (1 - P(\boldsymbol{\omega}_0))^{n_{obs}} P(\boldsymbol{\omega}_0)^{N - n_{obs}} \right\} \\ &\quad \times \left\{ \frac{n_{obs}!}{n_{\omega_1}! n_{\omega 2}! \ldots n_{\omega_m}!} \prod_{i=1}^{m} \left(\frac{P(\boldsymbol{\omega}_i)}{(1 - P(\boldsymbol{\omega}_0))} \right)^{n_{\omega_i}} \right\}. \end{aligned} \tag{3}$$

4.3 Model Constraints, Link Functions and Covariates

We used parameter index matices (PIM) as implemented in MARK (White and Burnham 1999) to provide a flexible modeling environment, as suggested by

Lebreton et al. (1992). This allows a general way to specify parameter restrictions on the fully time dependent model. Four separate PIM's were used for the $\{\pi_j^M\}$, $\{p_j^s\}$, $\{\delta_j^s\}$ and $\{\phi_j^s\}$ model parameters respectively. Covariates may be included through the use of design matrices. Similar to MARK, parameters may also be fixed at known values. Fixing parameter values can be useful for creating testable hypothesis such as a 50:50 incoming sex ratio (i.e. $\pi_j^M = 0.5$ for all $j > 0$).

A fifth, limited PIM, was implemented for $\{\beta_j\}$ parameters, where only basic constraints may be imposed. Some examples include death only ($\beta_0 = 1, \beta_{1:k-1} = 0$), no recruitment for certain periods (e.g. $\beta_1 = 0$), and constrained entrance (e.g. $\beta_1 = \beta_2$ or $\beta_1 = \beta_2 = 0.1$). Covariates cannot however be supplied for the $\{\beta_j\}$ parameters (see Schwarz and Arnason 1996, for an explanation).

Finally, common link functions were used to restrict parameter estimates to remain between 0 and 1 (see Lebreton et al. 1992). Possible link functions follow the MARK implementation and include the sin, logit, log and identity links.

4.4 Parameter Redundancy

As mark-recapture models become more complex, the large parameter sets needed to describe the modelled processes may lead to parameter redundancy (Catchpole and Morgan 1997).

Recently, Catchpole and Morgan (1997), Catchpole et al. (1998) and Catchpole and Morgan (2001) developed a technique for assessing parameter redundancy using existing computer algebra packages that are capable of performing symbolic math. The advantage over numeric techniques is that numerical criteria are not needed to identify uniquely estimable parameters and in some cases estimable parameter combinations can also be determined.

Following the implementation by Gimenez et al. (2003), we used the software package Maple (Version 10) to determine the model deficiency as well as the uniquely identifiable parameters. In our case the most general, time dependent, model was found to have a parameter deficiency of 4. The full set of the time dependent classification parameters $\{\delta_j^s\}$ were identifiable. As in the simple JS super-population model, $\{\phi_j^s : j = 1, 2, \ldots, k-2\}$ and $\{p_j^s : j = 2, 3, \ldots, k-1\}$ were identifiable. Finally, as the sampling occasions increase the middle proportion of entrants $\{\beta_j : j = 2, 3, \ldots, k-2\}$ and the probability of a male entrant $\{\pi_j : j = 2, 3, \ldots, k-2\}$ also become uniquely estimable. The remaining parameters formed complex estimable combinations. It should be noted that sparse data sets may further increase the parameter deficiency, as is the case with the example study.

4.5 Parameter Estimation

Parameter estimates were derived in a manner similar to the procedure described by Schwarz and Arnason (1996). Estimates of $\{\hat{\upsilon}_j\}$ were found by maximizing the L_2 loss-on-capture component. The L_1^{B*} was then maximized with respect to the remaining parameters given $\{\hat{\upsilon}_j\}$. Finally, the estimates $\{\hat{\beta}_j\}$, $\{\hat{\pi}_j\}$, $\{\hat{p}_j^s\}$, $\{\hat{\delta}_j^s\}$ and $\{\hat{\phi}_j^s\}$ were used to derive the estimate $\hat{N}$ using L_1^A, where $\hat{N} = \left[n_{obs}/(1-\widehat{P(\omega_0)})\right] \left(\text{greatest integer} \le n_{obs}/(1-\widehat{P(\omega_0)})\right)$ as per Sanathanan (1972).

At this point it should be noted that in this formulation of the JS model, beyond the assumption on unmarked animals (which is untestable), there is no information about N or β_0 in the capture histories (see Link and Barker 2005). That said, the capture histories do contain information on the remaining β_j parameters and given reasonable survival estimates, the estimates of N and β_0 may still be fair.

Finally, the L_1^{B*} was maximized numerically using a quasi-Newton (variable metric) procedure in the free statistical software package R. The Hessian matrix was numerically determined using the full likelihood (3) and point estimates from the conditional fit. The delta method was used to determine the variance-covariance matrix for all derived parameters, such as $\{\hat{B}_i^s\}$ and $\{\hat{N}_i^s\}$. It should be noted that the variance calculations for $\{\hat{N}_i^s\}$ excluded the additional variance associated with estimating the sex ratio of the loss-on-captures that were not successfully sexed (see Appendix A). This represents a rather rare scenario as well as a small source of variation in most applications.

4.6 Goodness of Fit

The goodness-of-fit tests for a JS model involves the L_1^{B*} component, where the likelihood has been conditioned on the observed histories. Both model deviance and the Pearson chi-squared statistic were used. The model deviance is calculated by comparing the likelihood of the fitted model to the saturated model, where every ω_i history has the probability $\frac{n_{\omega_i}}{n_{obs}}$. The difference between the saturated and fitted model (4) will be $D \sim \chi^2_{(m-p-1)}$, where the number of degrees of freedom is equal to the difference in the number of parameters between the two models ($m-p-1$). The number of parameters in the saturated model is the number of unique tag histories m minus 1, while p is the number of parameters estimated in the L_1^{B*} component.

$$D = 2\left[\sum_{i=1}^{m} n_{\omega_i}\left(\log\left(\frac{n_{\omega_i}}{n_{obs}}\right) - \log\left(\lambda_{\omega_i}\right)\right)\right] \tag{4}$$

In addition to deviance, the Pearson chi-squared goodness-of-fit (GOF) statistic was also used

$$\chi^2 = \sum_{i=1}^{m} \frac{(o_i - e_i)^2}{e_i} \tag{5}$$

where $o_i = n_{\omega_i}$ is the observed tag history frequency and $e_i = N \times P(\omega_i) = n_{obs} \times \frac{P(\omega_i)}{1-P(\omega_0)}$ is the expected frequency. Under the hypothesis that the model is correct $\chi^2 \sim \chi^2_{(m-p-1)}$. Asymptotically, both tests should be equivalent, however differences can occur depending on the adequacy of the $\chi^2_{(m-p-1)}$ approximation. Specifically, for small frequencies there is some evidence to suggest that χ^2 may perform better that D, since D can be unduly influenced by very small frequencies (see Cressie and Read 1989).

4.7 Model Selection

By using the PIM's and design matrices, many different models can be specified following the notation of Lebreton et al. (1992). Of particular interest will be time-varying parameterization (e.g. $\{\pi_j\}$), time independent (e.g. $\pi.$) and group independent parameterization (e.g. $p_j^M = p_j^F$ for $j = 1, 2, \ldots, k$). Also of interest may be entrance restrictions such as a death only model (e.g. $\beta_0 = 1; \beta_j = 0$ for $j = 1, 2, \ldots, k-1$), or death only for specific time periods (e.g. $\beta_2 = 0$). Finally, also of interest may be situations where sex assignment does not occur in every sampling occasion (i.e. $\delta_j^s = 0$ for some j). These models can be specified using the respective parameter PIMs and specifying constants for the fixed parameters.

A variety of techniques can be used to choose between competing sets of models and model averaging for models that have equal support (see Burnham and Anderson 2002). Techniques used for model selection included the Akaike information criterion (AIC$_c$) and QAIC$_c$. Both balance overall model fit to the data with number of parameters needed, but QAIC$_c$ also corrects for lack of fit and effective sample size. It should be noted the lack of fit used for the QAIC$_c$ calculations was based on the Pearson GOF statistic (5). Finally, a candidate model set was chosen *a prior* and the best fit was selected from competing models.

5 Example

Our example is concerned with estimating the number of walleye in Mille Lacs Lake, Minnesota, for which the study design is outlined in Schwarz (2004). Briefly, a three-year mark-recapture study was initiated in 2002. In all years tags were applied in two phases, first via trap netting on the spawning grounds and second during the angling season from launch-boats. Recoveries came from angling, trap netting, tribal harvests and a gillnet assessment at the end of the season. Nearly complete sexing occurred during the spawning ground releases and gillnet assessment at the beginning and end of each season respectively. However, due to manpower restrictions, only partial sexing was possible during the angling portion of the season, resulting in capture histories with unknown sex designations.

Table 1 Summary of Mille Lacs study design

	2002		2003		2004	
	$j = 1$	$j = 2$	$j = 3$	$j = 4$	$j = 5$	$j = 6$
Survey Type	Spring Tagging + Tribal Harvest	Launch Boats + Angling	Spring Tagging + Tribal Harvest	Launch Boats + Angling	Spring Tagging + Tribal Harvest	Launch Boats + Angling

For the purpose of the example all 3 years are considered (2002 – 2004), but only two sampling occasions per year are included, the spawning ground releases and harvest at the start of the season and the angling season that occurs over the summer, but prior to the end of the season gill net assessment, which was not included. This results in a total of six sampling occasions, with two sampling occasions occurring every year (Table 1). The data set is available upon request.

5.1 Model Constraints

The first sampling event within each year (sampling occasions $j = 1, 3, 5$), was limited to fish greater than 14 in. in length. For the second event within each year (sampling occasions $j = 2, 4, 6$), the minimum size requirement was increased 16 in. so as to preclude recruitment. Therefore, in the model recruitment was not allowed between occasions 1 and 2, occasions 3 and 4, and occasions 5 and 6 (i.e. the following values were fixed: $\beta_1, \beta_3, \beta_5 = 0$). New recruits were however allowed between occasions 2 and 3, and between occasions 5 and 6 (i.e. $\beta_2, \beta_4 \geq 0$). These restrictions reflect allowable catch restrictions and as a result the $\{\beta_j\}$ parameters represent recruitment into the size restricted fishery, rather than births.

The walleye data set also suffered from issues of tag loss, mainly between 2002 and 2003 (see Cowen and Schwarz 2006). As a result, the estimates of ϕ_2^s (apparent survivorship between periods 2 and 3) are expected to be lower than the estimates of ϕ_4^s as such these values were never restricted to be equal.

5.2 Model Selection and Best Model Estimates

Due to the known issue of tag loss in this population (see Cowen and Schwarz 2006) model misfit was expected. Since the example is for illustration purposes, the best model presented was judged by a combination of model goodness of fit criteria (model deviance and the over-dispersion estimate $\hat{c}$), derived parameter estimates (population sizes), and AIC model ranking. The fundamental parameter estimates from the best fitting model are displayed in Table 2, with derived parameters, population size (N_j^s) and recruitment (Bj^s), displayed in Table 3.

Table 2 Point estimates (SE) from the best model fit. Values have not been adjusted for over-dispersion ($\hat{c} = 128$). Fixed parameters are shown without a SE

	2002		2003		2004	
	$j = 1$	$j = 2$	$j = 3$	$j = 4$	$j = 5$	$j = 6$
β_{j-1}	0.846 (0.029)	0	0.052 (0.026)	0	0.102 (0.014)	0
π^M_{j-1}	0.339 (0.024)	0	0.489 (0.050)	0	0.489 (0.050)	0
p^M_j	0.131 (0.013)	0.013 (0.002)	0.283 (0.008)	0.010 (0.001)	0.191 (0.017)	0.005 (0.001)
p^F_j	0.011 (0.001)	0.025 (0.002)	0.024 (0.002)	0.019 (0.002)	0.029 (0.005)	0.008 (0.002)
δ^M_j	0.981 (0.001)	0.183 (0.011)	0.981 (0.001)	0.183 (0.011)	0.981 (0.001)	0.183 (0.011)
δ^F_j	0.968 (0.005)	0.072 (0.002)	0.968 (0.005)	0.072 (0.002)	0.968 (0.005)	0.072 (0.002)
ϕ^M_j	0.750 (0.067)	0.624 (0.055)	0.829 (0.071)	0.903 (0.068)	0.829 (0.071)	–
ϕ^F_j	0.903 (0.068)	0.650 (0.054)	0.795 (0.093)	0.829 (0.151)	0.795 (0.093)	–

Table 3 Estimated male and female specific birth, abundances and sex ratio by sampling occasion. Note that population size (N^s_j) are in 1000's. Also note that B^s_0 are not shown as they are the same as N^s_1 values. SE have not been adjusted for over-dispersion. The final column represents estimates from a simple JS model pooled over sex

	Births (B^s_{j-1})		Population sizes (N^s_j)			Population total (N_j)	
Year	Females	Males	Females	Males	F/M Ratio	Sex-specific	Simple JS
2002 ($j = 1$)	–	–	490 (34)	252 (25)	1.9	743 (46)	695 (10)
($j = 2$)	0 (–)	0 (–)	441 (27)	176 (31)	2.5	617 (45)	677 (10)
2003 ($j = 3$)	22 (11)	23 (11)	302 (23)	125 (3)	2.4	427 (23)	342 (5)
($j = 4$)	0 (–)	0 (–)	238 (27)	92 (10)	2.6	330 (28)	327 (5)
2004 ($j = 5$)	45 (10)	44 (5)	231 (39)	118 (13)	2.0	351 (42)	326 (5)
($j = 6$)	0 (–)	0 (–)	182 (41)	89 (20)	2.0	271 (45)	312 (5)

The model fit is poor ($\hat{c} = 128$) for two reasons. First, the very large sample sizes imply that small discrepancies in fit can be detected quite easily. Second, as noted in Cowen and Schwarz (2006), tag loss, particularly on the first winter, is a serious problem in this study. The latter would require extensive modeling to incorporate properly. Nevertheless, despite the apparent lack-of-fit, the estimates are reasonable and match quite well to those from other work (Schwarz 2004; Cowen and Schwarz 2006).

In particular, the model picks up the known higher catchability of male fish versus female fish on the spawning ground (occasions $j = 1, 3, 5$; p^M_j and p^F_j estimates, Table 2) and the lower catchability during the angling season (occasions $j = 2, 4, 6$; p^M_j and p^F_j estimates, Table 2). When compared to estimates from an equivalent simple JS, that does not include sex, the naive catchability estimates occurred somewhere between the sex-specific estimates (Table 4).

The initial male population estimate was low ($\hat{N}^M_1 \approx$ 252,000 versus $\hat{N}^F_1 \approx$ 490,000, Table 3), but is not unexpected. Tribal harvest (during the spring) consists mostly of male fish. Over the years, this will tend to lower the sex ratio of the standing population, as seen in the female/male ratio column (Table 3). We are puzzled though, by the apparent high male proportion of new recruits prior to 2003. We tried several different initial values, but the resulting estimates were always high.

Table 4 Recapture probabilities estimates (SE) from the sex-specific JS and a simple JS model. SE have not been adjusted for over-dispersion

	Sex-specific		Simple JS
Year	p_j^F	p_j^M	p_j
2002 ($j = 1$)	0.011 (0.001)	0.142 (0.013)	0.055
($j = 2$)	0.024 (0.001)	0.014 (0.002)	0.020
2003 ($j = 3$)	0.025 (0.002)	0.282 (0.008)	0.131
($j = 4$)	0.020 (0.002)	0.010 (0.001)	0.017
2004 ($j = 5$)	0.028 (0.005)	0.193 (0.018)	0.101
($j = 6$)	0.008 (0.002)	0.005 (0.001)	0.007

Yearly survival rates for 2002 are likely biased low because of the excessive tag loss observed between 2002 and 2003. The estimates appear more reasonable for later periods when tag loss was much less prevalent. Arnason and Mills (1981) showed that homogeneous tag loss resulted in unbiased estimates of population size, but biased estimates of recruitment.

Finally, our estimates of population size are comparable to those in Cowen and Schwarz (2006) which accounted for tag loss, but not a lack of sex identification. The estimate of total population size are also comparable to those from a simple JS model pooled over both sexes (Table 3).

6 Discussion

Abundances, both general and sex or group-specific, are of often of interest to biologists and managers alike, but obtaining such estimates can be difficult in cases where the group status is not always measurable. If the population sex ratio is known and constant then such assignments are not needed (Zhang et al. 2005). However, situations where the sex ratio is known and stable is rare. The method presented here provides a general approach to deal with unknown designations that avoids biased demographic estimates that can arise if classification uncertainty is otherwise dealt with in an ad hoc fashion.

The extension is an intermediate between the classical grouping approach, where the state is fixed and always measurable and the multi-state extension of the JS model (Schwarz and Dupuis 2007). If sex is always known then the experiment may be analyzed as a simple stratification, where male and female abundances are estimated separately. However, biologically relevant restrictions, such as a 50:50 sex ratio of incoming entrants, cannot be imposed with stratification. One of the nice features of the proposed model is that the probability of an entrant's sex is directly modelled in the likelihood. This makes it a fairly trivial task to impose constraints on the sex ratio of entrants, which in turn could be useful for testing a variety of biological hypotheses in a multi-model framework (see Burnham and Anderson 2002). The model presented is also flexible in regards to identification probabilities. If definitive sex assignments can always be made (over either a portion or the entire experiment), then this formulation can still be used by fixing the the appropriate δ_j^s parameter to 1.

While formulation presented here dealt explicitly with sex as the underlying group of interest, the model notation can also handle multiple fixed groups. In this case the superscript s would be used to represent the various groups, rather than sex. Also an additional restriction is needed so that probability of entrant belonging to a specific group sums to one for all sampling occasions (i.e. $\sum_s \pi_j^s = 1$ for all j). Lastly, for the unknown group tag history i the assigned group ζ_i would be the set of all possible fixed groups in the system.

This extension is not however without limitations. In dealing with classification uncertainty, explicit assumptions were made on how group assignments occurred. Specifically, any positive assignments were considered to be definitive. This was intended for sampling situations where there is a mixture of observers (e.g. experts and non-experts) and only a subset is capable of making positive assignments. This does not however protect against mis-assignments. If the proportion of mis-assignment is large, then group-specific estimates may become biased. While it is possible to adapt the modeling framework to deal with mis-assignment in addition to group identification uncertainty, mis-assignment can be largely controlled if field practitioners ensure that any uncertainty in an animal's group assignment is treated as an inability to assign a group (i.e. unknown).

Finally, it is also assumed that there is no unmodelled heterogeneity in the identification probabilities. This may also not be the case if animals that have less physical development or are diseased have identification probabilities that differ from the rest of the population. Such conditions will not only affect the identification probabilities, but such animals may also exhibit different demographic parameters which would violate homogeneity assumptions. In this case further groupings or the addition of covariates may be needed to model the potential heterogeneity in identification probabilities and demographic parameters.

Acknowledgments Funding for this work was provided by the Natural Sciences and Engineering Research Council (NSERC) of Canada and the Minnesota Department of Natural Resources.

Appendix A: Calculating $N_i{}^s$ when Removals Are Present

If removals or losses occur, then the $\{N_j^s\}$ estimates need to be adjusted using

$$E\left[N_{j+1}^s\right] = \begin{cases} N\beta_0\pi_0^s & j=0 \\ E\left[N_j^s\right]\phi_j^s + N\beta_j\pi_j^s - L_j^{s*}\phi_j^s & j>0 \end{cases} \quad \text{for } s \in \{M, F\} \tag{6}$$

where L_j^{s*} are the sex-specific removals/losses at time j and the term $-L_j^{s*}\phi_j^s$ represent the number of losses expected to have have survived to time $j+1$ if they were not removed from the population at time j. The value of L_j^{s*} may be known if all losses can be sexed, otherwise the sex ratio of removals and/or losses must be estimated.

Let α_j^s be the expected proportion of unknown captures that are of sex s, for $s \in \{M, F\}$. An estimate of α_j^s can be found as

$$E\left[\alpha_i^s\right] \approx \frac{E\left[N_j^s\right] p_j^s\left(1-\delta_j^s\right)}{E\left[N_j^s\right] p_j^s\left(1-\delta_j^s\right)+E\left[N_j^{s^c}\right] p_j^{s^c}\left(1-\delta_i^{s^c}\right)} \quad \text{for } s \in \{M, F\} \tag{7}$$

where s^c is the complement sex and $N_j^s p_j^s(1-\delta_j^s)$ represents the expected number of animals, at time j, of sex s, caught, but whose sex remains unidentifiable. The expected number of sex-specific losses for time j can then be determined as

$$E\left[L_j^{s*}\right] = L_j^s + L_i^u E\left[\alpha_j^s\right] \text{ for } s \in \{M, F\} \tag{8}$$

where L_j^u is the number of losses with unknown sex and L_j^s is the number of losses where sex is known. The estimated value L_j^{s*} is then used in (6).

References

Arnason AN, Mills KH (1981) Bias and loss of precision due to tag loss in Jolly-Seber estimates for mark-recapture experiments. Canadian Journal of Fisheries and Aquatic Sciences 38: 1077–1095.

Burnham KP, Anderson DR (2002) Model selection and multi-model inference: a practical information-theoretic approach, 2nd Edition. Springer-Verlag, New York.

Catchpole EA, Morgan BJT (1997) Detecting parameter redundancy. Biometrika 84: 187–196.

Catchpole EA, Morgan BJT (2001) Deficiency of parameter-redundant models. Biometrika 88:593–598.

Catchpole EA, Morgan BJT, Freeman SN (1998) Estimation in parameter-redundant models. Biometrika 85:462–468.

Conroy MJ, Senar JC, Hines JE, Domenech J (1999) Development and application of a mark-recapture model incorporating predicted sex and transitory behaviour. Bird Study 46 Suppl.:62–73.

Cowen L, Schwarz CJ (2006) The Jolly-Seber model with tag loss. Biometrics 62:699–705.

Cressie N, Read TRC (1989) Pearson's χ^2 statistic and the loglikelihood ratio statistic G^2: a comparative review. International Statistical Review 57:19–43.

Fujiwara M, Caswell H (2002) Estimating population projection matrices from multi-stage mark-recapture data. Ecology 83:3257–3265.

Gimenez O, Choquet R, Lebreton JD (2003) Parameter redundancy in multistate capture–recapture models. Biometrical Journal 45:704–722.

Jolly GM (1965) Explicit estimates from capture-recapture data with both death and immigration-stochastic model. Biometrika 52:225–247.

Kendall WL, Hines JE, Nichols JD (2003) Adjusting multistate capture–recapture models for misclassification bias: manatee breeding proportions. Ecology 84:1058–1066.

Lebreton JD, Burnham KP, Clobert J, Anderson DR (1992) Modeling survival and testing biological hypotheses using marked animals – a unified approach with case-studies. Ecological Monographs 62:67–118.

Lebreton JD, Pradel R (2002) Multistate recapture models: modelling incomplete individual histories. Journal of Applied Statistics 29:353–369.

Link WA, Barker RJ (2005) Modeling association among demographic parameters in analysis of open population capture–recapture data. Biometrics 61:46–54.

Nichols JD, Kendall WL, Hines JE, Spendelow JA (2004) Estimation of sex-specific survival from capture–recapture data when sex is not always known. Ecology 85:3192–3201.

Pradel R (2005) Multievent: An extension of multistate capture–recapture models to uncertain states. Biometrics 61:442–447.

Sanathanan L (1972) Estimating size of a multinomial population. Annals of Mathematical Statistics 43:142–152.

Schwarz CJ (2004) Estimating the number of walleye in Mille Lacs, Minnesota – Final Report. Tech. rep., Minnesota Department of Natural Resources.

Schwarz CJ, Arnason AN (1996) A general methodology for the analysis of capture–recapture experiments in open populations. Biometrics 52:860–873.

Schwarz CJ, Dupuis JA (2007) A Bayesian approach to the multistate Jolly–Seber capture–recapture model. Biometrics. Online: doi:10.1111/j.1541-0420.2007.00815.x

White GC, Burnham KP (1999) Program MARK: survival estimation from populations of marked animals. Bird Study 46:120–139.

Zhang LK, Liu LP, You N (2005) Estimating population size in logistic capture–recapture models with a known sex ratio. Communications in Statistics -Theory and Methods 34:37–44.

Program E-SURGE: A Software Application for Fitting Multievent Models

Rémi Choquet, Lauriane Rouan, and Roger Pradel

Abstract Multievent models (Pradel 2005, 2008) handle state uncertainty, and they therefore cover a range of situations like hidden capture heterogeneity and sex determination from behaviour which cannot be treated in the multistate paradigm. We introduce a new software application called E-SURGE, built upon the concepts developed in program M-SURGE (Choquet et al. 2004) to encompass this new class of capture–recapture models. It also improves on M-SURGE by allowing the decomposition of transitions into several steps. We present the new concepts involved, notably the event and the multistep process, and how they are implemented in E-SURGE. We then illustrate the use of E-SURGE with three examples. One example deals with breeding propensity where the breeding state cannot always be ascertained; a further deals with emigration which is considered as a two-step process (Grosbois and Tavecchia 2003) and the last one with a version of a memory model where survival can be handled directly.

Keywords Capture–Recapture · Hidden Markov Chain

1 Introduction

Capture–recapture (CR) data, i.e. the histories of encounters of individually recognizable animals, have long been a main source of information on the dynamics of animal populations. Their use is regularly enlarged to address new questions like movements, trade-offs between reproduction and survival, transitions among life stages or the spread of a disease. The concept of state plays a key role in these new developments. It is flexible and can indifferently represent the breeding status, the developmental stage or the spatial location. However, it is not always possible to ascertain the state of an individual when it is encountered. For instance, in an epidemiological study, a diseased animal may not present any outer symptoms;

R. Choquet (✉)
Centre d'Ecologie Fonctionnelle et Evolutive, UMR 5175, CNRS, 1919 Route de Mende, F34293 Montpellier cedex 5, France
e-mail: choquet@cefe.cnrs.fr

D.L. Thomson et al. (eds.), *Modeling Demographic Processes in Marked Populations*, Environmental and Ecological Statistics 3, DOI 10.1007/978-0-387-78151-8_39,

conversely, a medical test may be falsely positive. Yet, the multistate models currently used to analyze these data do not allow for uncertainty.

Recently, Pradel (2005) has developped a new approach to handle state uncertainty. This approach introduces a new concept, that of event, which is what is actually known. In our epidemiological study for instance, the event would be the test result or the observed presence or absence of the symptom. There is no one–one correspondance between event and state and a particular event may arise under different states. The new approach is not reducible to the multistate approach. It has logically been named multievent. This paper presents the first program for fitting multievent models to CR data, called E-SURGE (which stands for Multi-**E**vent **SUR**vival **G**eneralized **E**stimation). E-SURGE extends M-SURGE, a previous program for fitting mutistate models developed by the same team.

In addition to handling the different structure required for multievent analysis, E-SURGE also extends M-SURGE in another respect: it allows the specification of transitions between states as a multistep process. For instance, in a study of movement between geographical locations, the transition toward a particular site can be decomposed as the probability to emigrate times the probability to settle on the arrival site given departure. With these two extensions, the new program allows to fit a much greater variety of models and to address in the same unified framework several problems which had been tackled individually:

- heterogeneity of capture (Pledger et al. 2003; Pradel 2005),
- sex uncertainty (Nichols et al. 2004),
- memory in movements (Hestbeck et al. 1991; Brownie et al. 1993; Rouan et al. 2008),
- mixture of live and dead encounters over several sites (Lebreton et al. 1999; Véran et al. 2007).

The motivation for developing program E-SURGE was to bring to biologists the multievent framework in a powerful and user-friendly environment. Several programs exist for CR analysis (Hines 1994; White and Burnham 1999; Choquet et al. 2005) but E-SURGE is the first general program for multievent models. It benefits from the experience gained in developing M-SURGE. M-SURGE introduced a powerful language (Choquet 2008) for describing the set of multistate CR models and used reduced statistics and advanced numerical algorithms to produce faster and more reliable estimates. E-SURGE has similar capabilities for the maximum likelihood optimization of complex age and time-dependent models with linear constraints among parameters in a generalized linear model (GLM) fashion. Its features include:

- A module called GEPAT where the steps making up the transitions (see above for an example) are specified. Similarly, in this same module, events can be declared to be linked to states through a multistep process. For instance, a female animal may be wrongly judged to be a male through three successive steps: it is encountered, its sex is assessed, an error occurs. This detailed decomposition of

the event generation conditional on the state allows for the incorporation in the same analysis of encounters where no judgement is made (Pradel et al. 2008). Multievent models also include initial state probabilities (see below) which can equally be decomposed at this stage.
- A module called GEMACO, inherited from M-SURGE, where the mode of variation of the parameters defined in GEPAT is specified by means of a powerful model description language.
- Advanced convergence options. Convergence is a very sensitive issue in multievent as well as in multistate models. In E-SURGE, the user gains a greater control over convergence through a choice of starting options including the results of previous models, random initial values and multiple random initial values.
- Acute rank estimation. E-SURGE calculates, with an algorithm described in Rouan et al. (2008) very precise estimate of the rank of a model. This is one of the key steps for correct model selection using the AIC (Akaike 1987).
- The detection of redundant parameters. E-SURGE analyzes the likelihood in the neighborhood of the point of convergence and lists the parameters that are apparently redundant. Redundancy can then be double-checked by drawing profile likelihood curves.

In the next section, we present briefly the theory of multievent models; then in Sections 3 and 4, the way these models are implemented in E-SURGE in theory and in practice. Three examples are detailed in the application section. The first one concerns breeding status and shows an example of uncertainty in state assessment. The second is the classical Arnason–Schwarz multistate model but the movement between sites is decomposed into two steps. The last is an original version of the memory model where survival probabilities can be handled directly thanks to an appropriate decomposition of the movement.

2 Short Recalls on Multievent Models

Multievent models assume that individuals move independently among a finite set $\mathbb{E}$ of states over a finite number K of sampling occasions, and that the transitions among the successive states obey a Markov chain. The successive states occupied by an individual are not observed directly. Rather, at each occasion k, one member of a finite set Ω of events is observed. The event observed at occasion k is assumed to depend only on the unobserved underlying state of the individual at that occasion (see Fig. 1). Thus, these models belong to the class of Hidden Markov Models (HMM), see for example MacDonald and Zucchini (2000) and Cappé et al. (2005).

Multievent models are defined in terms of three kinds of parameters: initial state probabilities π, transition probabilities ϕ, and event probabilities b. Our presentation of multievent models will use the following general notation which generally follows Pradel (2005):

...	V_{k-1}		V_k		V_{k+1}	...	observed events
	$\uparrow$		$B^{(k)} \uparrow$		$\uparrow$		
...	E_{k-1}	$\xrightarrow[\Phi^{(k-1)}]{}$	E_k	$\rightarrow$	E_{k+1}	...	hidden states
occasions	$k-1$		k		$k+1$		

Fig. 1 Graph summarising the relations of dependence between successive states (transition matrix Φ) and observed events (event generation matrix B) in the multievent time-dependent model. V_k is the random variable of the event observed at occasion k, E_k that of the underlying state of the animal at the same occasion

N	the number of states.
U	the number of events.
K	the number of occasions.
K	the number of occasions.
A	the maximum age class.
NG	the number of groups.
LI, LT, LB	the number of step processes for initial state, transition and event.
$i = 1, \ldots, N$	the index of the previous (or departure) state.
$j = 1, \ldots, N$	the index of the current (or arrival) state.
$u = 1, \ldots, U$	the index of the current event.
$k = 1, \ldots, K$	the occasion index.
$a = 1, \ldots, A$	the index of current age classes.
$ng = 1, \ldots,$ NG	the index of the current group.
$\ell = 1, \ldots,$ LI, LT, or LB	the step process index
$\mathbb{E} = \{e_1, ..., e_N\}$	the set of states, where e_N = '†' for the death.
$\Omega = \{v_1, ..., v_U\}$	the set of events, where the v_1 = 'not seen'

Unlike traditional practice in CR but similar to Fujiwara and Caswell (2002), Pradel (2005) and consistent to Markov Chain practice MacDonald and Zucchini (2000), the dead (†) is explicity included in $\mathbb{E}$. All transition matrices are written with i as row index and j as column index, thus following the Markov chain convention in which transitions are from rows to columns. Event matrices use j (denoting the state) as row index and u (denoting an event) as column index. Transition and event matrices are row-stochastic.

For the time-dependent model, the elementary parameters are

- $\pi_i^{(k)}$ the probability of being in state e_i when first encountered at occasion k,
- $\phi_{ij}^{(k)}$ the probability of being in state e_j at time $k+1$ if in state e_i at occasion k,
- $b_{ju}^{(k,1)}$ the probability of event v_u for an animal in state e_j encountered for the first time at occasion k,
- $b_{ju}^{(k,2)}$ the probability of event v_u for an animal reencountered in state e_j at occasion k,
- $\mathbf{\Pi} = (\pi_i)$ denotes the $(1 \times N)$ vector of initial state probabilities,

- $\boldsymbol{\Phi} = \left(\phi_{ij}\right)$ denotes the $(N \times N)$ matrix of unconditional transition probabilities, i.e. the matrix of probabilities that an individual moves from one state to another state over a time interval.
- $\boldsymbol{B} = \left(b_{ju}\right)$ denotes the $(N \times U)$ matrix of event probabilities. There are potentially two such matrices, one for first encounters $(b_{ju}^{(k,1)})$ and one for reencounters $(b_{ju}^{(k,2)})$. They may as well be considered as instances of the same matrix corresponding to two different age-classes. This is what we will do thereafter.

Together, $\boldsymbol{\Pi}$, $\boldsymbol{\Phi}$ and $\boldsymbol{B}$ define the multievent model. A particular multievent model will be obtained by specifying the mode of variation (by time, age, group or state...) of each type of parameter.

The likelihood of a model is proportional to the probability of the data given that model. The basic unit of data in E-SURGE is the capture history. Let $h = (o_1, \ldots, o_K)$ be a capture history with first encounter at occasion e, event o_k, $k = e, \ldots, K$, has any value between 1 and U and β is the vector of all parameters. Then

$$P(h|\beta) = \boldsymbol{\Pi}^{(e)} D\left(\boldsymbol{B}^{(e,1)}(\cdot, o_e)\right) \left(\prod_{k=e+1}^{K} \boldsymbol{\Phi}^{(k-1)} D\left(\boldsymbol{B}^{(k,2)}(\cdot, o_k)\right) \right) 1_N \qquad (1)$$

where $\boldsymbol{B}^{(k,2)}(\cdot, o_k)$ is the o_kth column of the re-encounter matrix at occasion k, i.e. the instance of $\boldsymbol{B}$ corresponding to the second age-class at occasion k. $D(x)$ denotes the matrix with the elements of vector $\mathbf{x}$ on the diagonal and zeros elsewhere, and $\mathbf{1}_N$ is an N vector of ones.

Assuming that individuals are independent, the likelihood for the entire set of capture histories is obtained as the product of the likelihoods for each history,

$$L(\beta) = \prod_{h} P(h|\beta)^{n_h}$$

where n_h is the number of copies of capture history h in the data set.

3 Concepts in E-SURGE

3.1 Decomposition in Elementary Steps

In E-SURGE, all types of parameters (transition probabilities, initial state probabilities and event probabilities) may be decomposed in an indefinite number of steps as arising from a sequence of 'life processes'. The first task is to specify the actual structure that will be used. The familiar decomposition of the transition matrix into survival and transition conditional on survival (implemented in M-SURGE) is the classical example in CR, but in some cases more steps may be involved. For instance, if we want to model the movement between geographical locations with probabilities of emigration and settlement on the new site, the transition probabilities between locations has to be decomposed into two steps. In the second step a migrant will not be allowed to settle on its departure site. Thus, in addition to specifying that there are three steps in total (one for survival and two for movement), we will also have to forbid certain transitions. These specifications are mainly done in module GEPAT.

In E-SURGE, the matrices $\mathbf{\Pi}$, $\mathbf{\Phi}$, $\boldsymbol{B}$ are called the full initial state vector, the full transition matrix, and the full event matrix, respectively. In an approach similar to that used for periodic matrix population models (Caswell 2001, Chapter 14) but between two states, the full matrices are written as products of elementary matrices.

$$\begin{aligned} \mathbf{\Pi} &= \prod_{\ell=1}^{\mathrm{LI}} \mathbf{\Pi}^{(\ell)} \\ \mathbf{\Phi} &= \prod_{\ell=1}^{\mathrm{LT}} \mathbf{\Phi}^{(\ell)} \\ \boldsymbol{B} &= \prod_{\ell=1}^{\mathrm{LB}} \boldsymbol{B}^{(\ell)} \end{aligned} \tag{2}$$

For Equation (2), we will say that we need a DES(LI,LT,LB) (Decomposition in Elementary Steps), where LI, LT and LB define the number of steps for the matrices $\mathbf{\Pi}$, $\mathbf{\Phi}$, $\boldsymbol{B}$, respectively.

3.2 Umbrella Model

After having specified the elementary steps, we have defined what we call the umbrella model (UM). The parameters of the UM are further constrained to compare different biological hypotheses of interest. In other words, the UM is the most general model that can be fitted, and the one within which all other models examined are nested.

There are six potential sources of variation in the parameters:

1. previous state,
2. current state,
3. current event,
4. time, i.e. occasions elapsed since the beginning of the capture session,
5. age, i.e., occasions elapsed since first capture,
6. groups, i.e. permanent categories of individuals, such as sexes or species, or discrete unconnected study sites.

In the UM, parameters are allowed to vary freely over time and among groups. Only the number of states and number of age classes can be set to specific values.

Survival, transitions, and encounter probabilities may depend on age (i.e., time since first capture, not necessarily true chronological age). The user specifies an oldest relevant age class; all animals with this age or older are combined into a single age-class. Constraining the range of ages restricts the range of models that can be fitted but it can greatly save memory and reduce computation time. For transitions or survival, common choices for the maximum relevant age are 1 (A = 1), which implies no age effect, and 2 (A = 2), which creates a model in which the first age class is contrasted to animals older than one year (this is particularly useful when animals are marked as young). Setting A = 2 can also be used to model transience (Pradel et al. 1997).

Specifying age-dependence in encounters is slightly more complicated. In multistate (as opposed to multievent) applications, all calculations are conditional on the first encounter and hence the probability of first encounter is not estimated. In multievent applications, E-SURGE has the option of modelling the probability of the initial event $b_{ju}^{(k,1)}$. Therefore, E-SURGE always considers at least two age classes for encounters, allowing the first event probability (first age class) to be modelled or not (probability to be first captured fixed to 1). Thus if one chooses a maximum age A = 1, which implies no age effect, E-SURGE creates nonetheless 2 age classes for events. If one sets A = 2, E-SURGE creates 3 age classes for events (see Table 1).

3.3 Constrained Models

Model-building in E-SURGE (as in M-SURGE) proceeds by imposing linear constraints on the parameters of the umbrella model in the spirit of GLM (Lebreton et al. 1992). The vector $\boldsymbol{\theta}$ of 'biological parameters' (parameters of direct interest to the biologist e.g., $\boldsymbol{\theta} = (\boldsymbol{\Pi}, \boldsymbol{\Phi}, \boldsymbol{B})$, organized as a vector) is expressed as a linear transformation of a vector β of 'mathematical parameters'. To keep the biological parameters, which are probabilities, in their permissible range (0,1), a link function f is generally applied:

$$f(\boldsymbol{\theta}) = \boldsymbol{X}\beta$$

The matrix $\boldsymbol{X}$ is a 'matrix of constraints' or genuine design matrix in the case of a designed experiment. In general, it expresses hypotheses about the dependence of the parameters on state (of departure or arrival), age, time, group, covariates, and

Table 1 Variations considered in the parameters of the umbrella models of E-SURGE. The type of variation is represented by upper indices for time(k), age(a), group(g) and step(l)

$A \neq 1$	$\boldsymbol{\Pi}^{(k,g,(l))}$ $\boldsymbol{\Phi}^{(k,a\leq A,g,(l))}$ $\boldsymbol{B}^{(k,a\leq A+1,g,(l))}$
$A = 1$	$\boldsymbol{\Pi}^{(k,g,(l))}$ $\boldsymbol{\Phi}^{(k,a=1,g,(l))}$ $\boldsymbol{B}^{(k,a\leq 2,g,(l))}$

so on. The design matrix is built by the module GEMACO (**GE**nerator of **MA**trices of **CO**nstraints), using the model definition language described in Choquet et al. (2004) and Choquet (2008). Often $\boldsymbol{X}$ will contain both discrete indicator 0/1 variables (for equality constraints) and continuous covariates (e.g., effort or weather covariates). An overview of linear constraints in CR models with a single state is given by Lebreton et al. (1992), linear constraints in multistate models are considered in Choquet (2008).

An important difference in the application of GEMACO in E-SURGE, as compared to M-SURGE, is that the GEMACO keywords ('from', 'to', etc.) refer to the elementary matrices in E-SURGE. The rows (from) and columns (to) in these matrices do not necessarily correspond to the states in the model (e.g., in the encounter matrix, the columns refer to events, not states), whereas in M-SURGE they always correspond to states.

3.4 Maximum Likelihood Estimation

Reduced-form data descriptions like the *m*-array are not available for multievent models. Thus the likelihood calculation depends on the application of the transition and event probabilities to individual capture histories. The maximum likelihood algorithm is as follows

1. Calculate the number of identical histories.
2. Select initial values for the vector β of mathematical parameters.
3. Calculate the vector of biological parameters $\boldsymbol{\theta} = f^{-1}(\boldsymbol{X}\beta)$, with f the generalized (or multinomial) logit link or the identity link.
4. Calculate the elementary matrices, and (as the product of the elementary matrices) each of the full matrices $\boldsymbol{\Pi}$, $\boldsymbol{\Phi}$, and $\boldsymbol{B}$ according to Equation (2).
5. Use the full matrices $\boldsymbol{\Pi}$, $\boldsymbol{\Phi}$, and $\boldsymbol{B}$ to calculate the individual probability $P(h|\beta)$ of each capture history according to equation (1).
6. Calculate the deviance

$$\mathrm{Dev}(\beta) = -2\log L(\beta) = -2\sum_h n_h \log P(h|\beta)$$

7. Iterate steps 3–6 in a Quasi-Newton minimization method updating the vector of mathematical parameters to decrease the deviance, until convergence.
8. Obtain the MLE's and the deviance and various by-products of Maximum Likelihood estimation like AIC (Akaike 1987) and derivate quantities (QAIC, QAICc).

Note 1 The choice of initial values is critical in multievent models as the problem of local minima is worse than for multistate (Lebreton and Pradel 2002). For that purpose, advanced features were developed in E-SURGE. Users can choose between several options for initial values, among them are random initial values, starting values estimated from a previous model and starting values given in files. In the context of Markov chains, the well known Expectation Maximization (EM) algorithm has also poor properties regarding local minima, see Celeux et al. (1995).

Note 2 AIC or derivate quantities (Burnham and Anderson 2002) permit model selection inside an umbrella model. A likelihood ratio test between two models under two different umbrella models is not currently available and would require bootstrapping. We emphasize the difficulty to do bootstrapping in the presence of local minima. For a more complete discussion about model selection, see Burnham and Anderson (2002) and Pledger et al. (2003).

4 E-SURGE in Practice

In this section we present three examples illustrating the potential of E-SURGE. The first example with breeding propensity is a good example to demonstrate its application when uncertain states occur; the second one considers the decomposition of the survival-transition probabilities into several steps and the last one combines the two previous issues by showing the implementation of a particular case of memory model.

4.1 First Example: A Simple Model for Studying Breeding Propensity

The study of transitions between the states breeder (Br) and non-breeder (NBr) is a topic of major interest to biologists. Nevertheless, the assessment of the breeding status can be difficult; indeed animals are not always observed in breeding activity and their status can remain unknown which motivates the use of multievent model. Let us define formally the underlying multievent model.

To be consistent with CR practice, events are originally coded in the encounter histories as:

0 for $v_1 =$ 'not seen';
1 for $v_2 =$ 'seen breeding';
2 for $v_3 =$ 'seen but status unknown'.

Inside E-SURGE, each event code is increment by 1 to be consistent with Equation (1) but this is transparent for the user.

According to the available information, the 'hidden states' chosen to define the model are:

Br for 'breeder';
NBr for 'non breeder';
† for 'dead'.

We have first to specify the parameterization of the model (i.e $\mathbf{\Pi}$, $\mathbf{\Phi}$, and $\boldsymbol{B}$):

- The initial state probabilities: at the first encounter, the individuals can be 'breeder' or 'non breeder' but cannot be 'dead'.

$$\mathbf{\Pi} = \begin{pmatrix} \text{Br} & \text{NBr} & \dagger \\ \pi_B & 1-\pi_B & 0 \end{pmatrix},$$

- The survival-transition probabilities: all the transitions between states Br and NBr are possible. Transitions between Br and NBr to † are constraint to sum to one, and transitions between † and Br and between † and NBr are not possible:

$$\mathbf{\Phi} = \begin{matrix} \\ \text{Br} \\ \text{NBr} \\ \dagger \end{matrix} \begin{pmatrix} \text{Br} & \text{NBr} & \dagger \\ \phi_{\text{Br,Br}} & \phi_{\text{Br,NBr}} & 1-\phi_{\text{Br,Br}}-\phi_{\text{Br,NBr}} \\ \phi_{\text{NBr,Br}} & \phi_{\text{NBr,NBr}} & 1-\phi_{\text{NBr,Br}}-\phi_{\text{NBr,NBr}} \\ 0 & 0 & 1 \end{pmatrix}$$

- The event probabilities: in this particular context, a 'breeder' can be 'not seen', 'seen breeding' or 'seen but status unknown'; a 'non breeder' is either 'not seen' or 'seen but status unknown' and a 'dead' individual can only be 'not seen'.

$$\boldsymbol{B}^{(a=2)} = \begin{matrix} \\ \text{Br} \\ \text{NBr} \\ \dagger \end{matrix} \begin{pmatrix} 0 & 1 & 2 \\ 1-p_{1|\text{Br}}-p_{2|\text{Br}} & p_{1|\text{Br}} & p_{2|\text{Br}} \\ 1-p_{\text{NBr}} & 0 & p_{\text{NBr}} \\ 1 & 0 & 0 \end{pmatrix} \quad (3)$$

Since the first encounter of the individuals is modelled explicitly here, one needs to pay attention to the initial-event probabilities. At first encounter, all individuals alive are obligatorily observed which leads to fix to 0 the probabilities of the event 'not seen'.

$$\boldsymbol{B}^{(a=1)} = \begin{matrix} \\ \text{Br} \\ \text{NBr} \\ \dagger \end{matrix} \begin{pmatrix} 0 & 1 & 2 \\ 0 & p^0_{1|\text{Br}} & 1-p^0_{1|\text{Br}} \\ 0 & 0 & 1 \\ 1 & 0 & 0 \end{pmatrix}$$

To fit this model in E-SURGE we have to execute five main stages. In the first stage, we implement the structure of the model (see Fig. 2) with:

1. the definition of the number of events and states;
2. the definition of the general structure of the matrices of parameters using the GEPAT interface. The different matrices are represented symbolically using

(a)

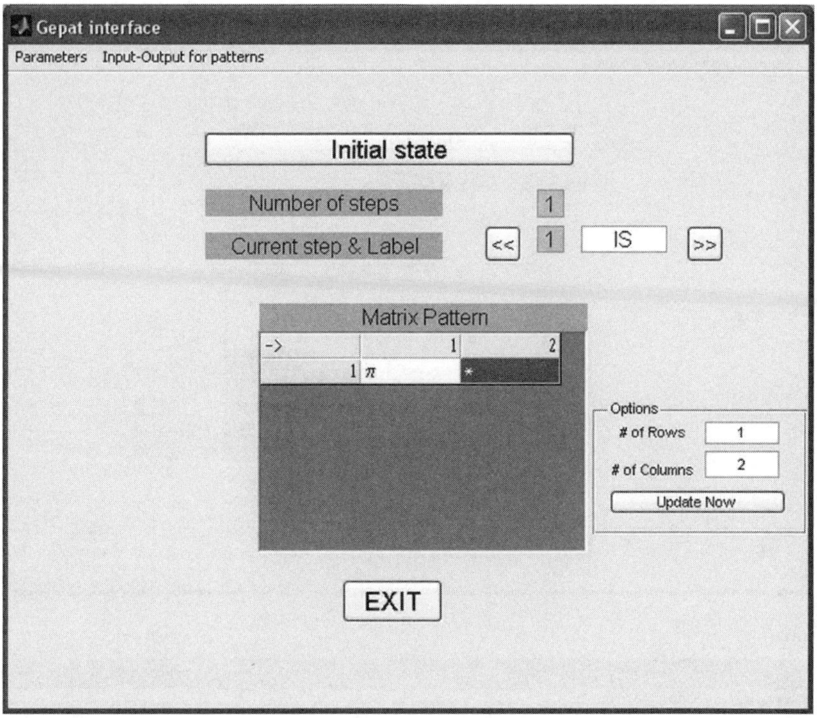

Fig. 2 In the window (1), we specified the number of states (N = 3), events (U = 3), age classes (A = 1) and groups (G = 1). The GEPAT interface is then opened. Patterns of $\mathbf{\Pi}$, $\mathbf{\Phi}$ and $\boldsymbol{B}$ are defined in windows 2a, 2b, 2c, respectively using the following rules: the parameters of interest are indicated using any greek letter, the parameters constraint to zero are indicated using '-' and the complementary parameters are indicated using '*'

an 'excel-like' interface; this step consists in filling different cells using the following rules:

- a greek letter indicates a parameter of interest i.e. one that will be estimated or fixed;
- '*' indicates the complementary parameter (there is one and only one '*' by row due to the row-stochasticity of the matrices);
- '–' indicates parameters constrained to zero.

Because the pattern must be the same for the two kinds of event matrices $\boldsymbol{B}^{(a=1)}$ and $\boldsymbol{B}^{(a=2)}$, which are considered by the program as two instances of the same matrix corresponding to two different age-classes. A particular attention must be payed for this example, to the matrices of the event probabilities. For $\boldsymbol{B}^{(a=1)}$, we need to fix to 0 the probabilities of the event 'not seen' conditioning respectively on the states Br and NBr. This constraint leads us to consider this cell as a parameter of interest rather than a complementary parameter. As a consequence,

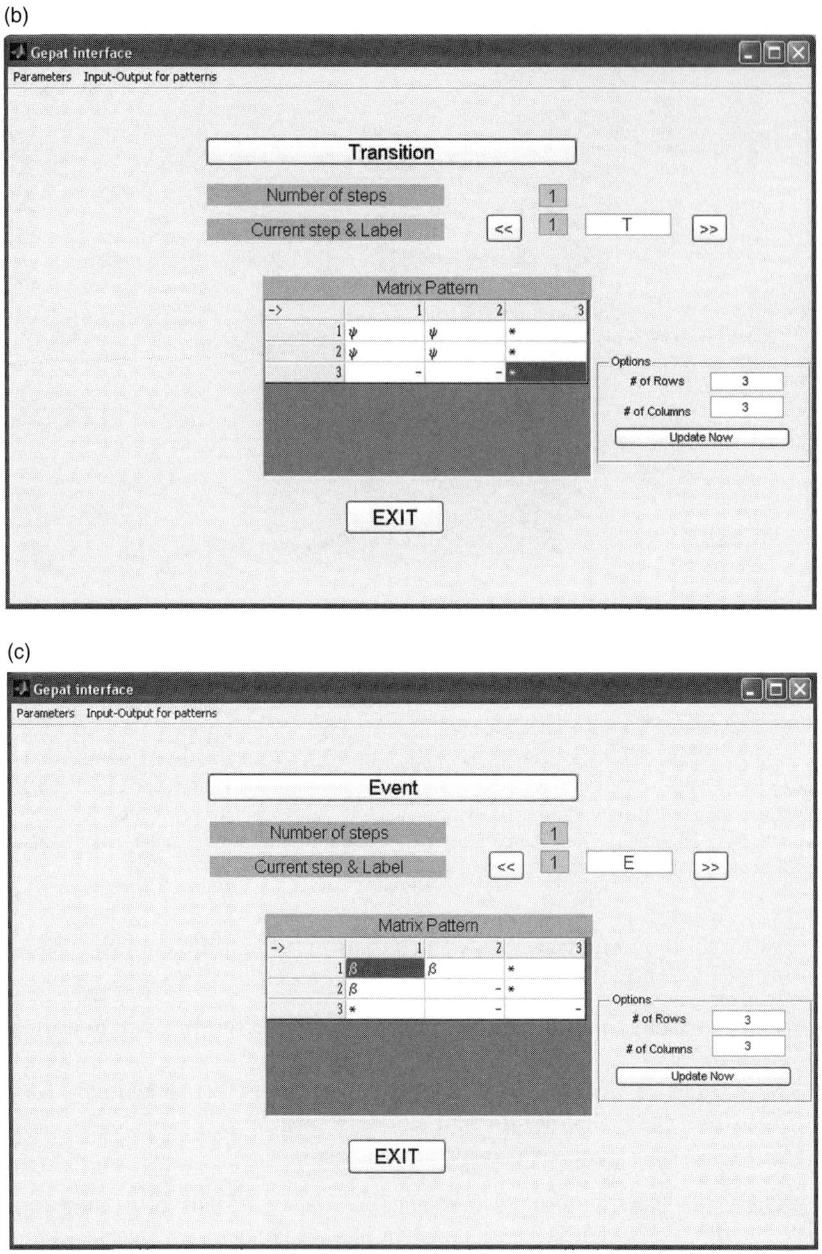

Fig. 2 (continued)

the pattern used in E-SURGE is not a direct transcription of the previous matrix $\boldsymbol{B}^{(a=2)}$ as it was the case for the vector $\mathbf{\Pi}$ or the matrix $\mathbf{\Phi}$.

In the second stage, we define the constraints of the model using the GEMACO interface (Choquet 2008) as shown in Fig. 3. Once again, particular attention must be payed to constraints applied to the event probabilities: the first encounter corresponds to the first age-class of b (denoted 'a(1)' or 'firste' in the GEMACO syntax)

(a)

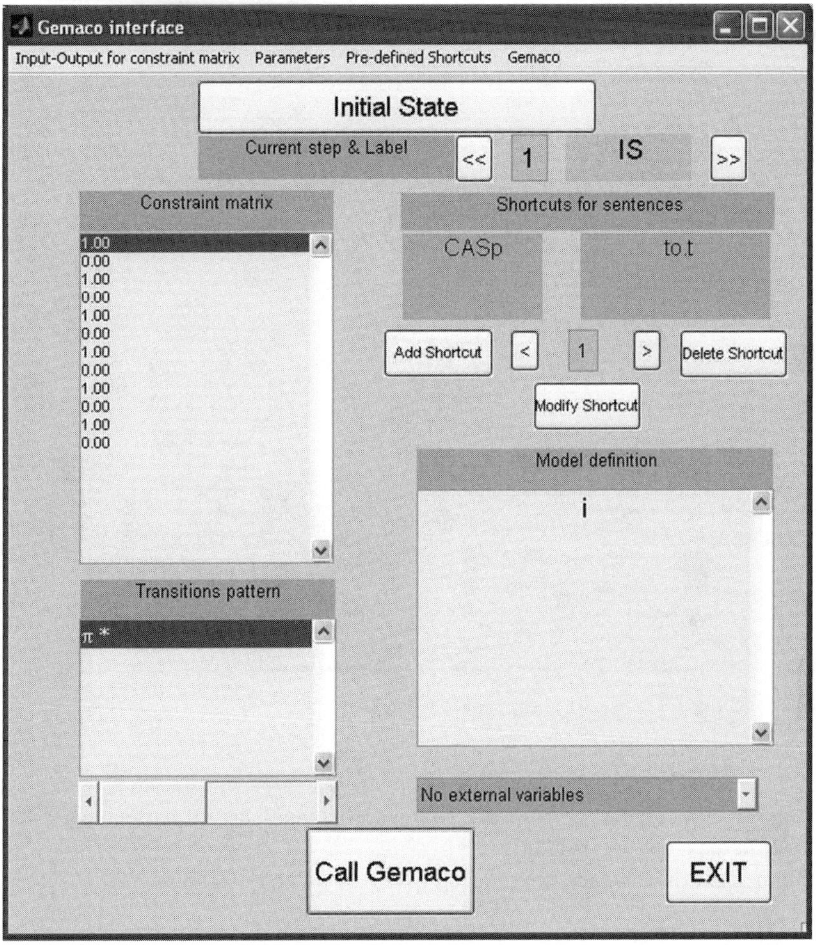

Fig. 3 The GEMACO interface defines constraints for the UM. In window (**a**), the user has entered the phrase 'i' to define a constant initial state probability. In window (**b**), the user has entered the phrase 'from.to.t' to define a state and time dependent effect for survival-transition. In window (**c**), the user has entered the phrase 'firste+a(2).from.to'. The first mathematical parameter corresponding to 'firste' is fixed to 0 in the next stage. The parameters $p_{1|\mathrm{Br}}$, $p_{2|\mathrm{Br}}$, p_{NB} in equation (3) are set to be different and constant across time with the phrase 'a(2).from.to'

(b)

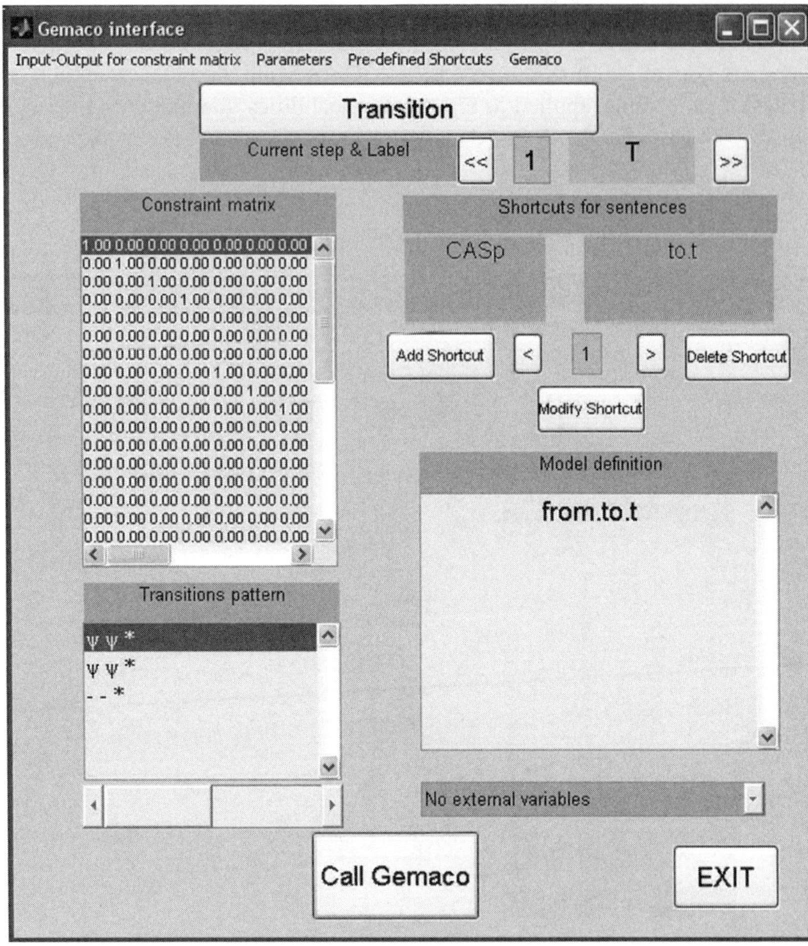

Fig. 3 (continued)

and the next encounters corresponds to the second age-class of *b* (denoted 'a(2)' or 'nexte' in the GEMACO syntax).

In the next three stages, we proceed like in M-SURGE (Choquet et al. 2004):

- we fix the parameter of ***B*** corresponding to the probability to be 'not seen' when first encountered to 0 (in the probabilities scale) and change the initial values if needed using the IVFV interface;
- we run the model;
- we examine and interpret the results.

(c)

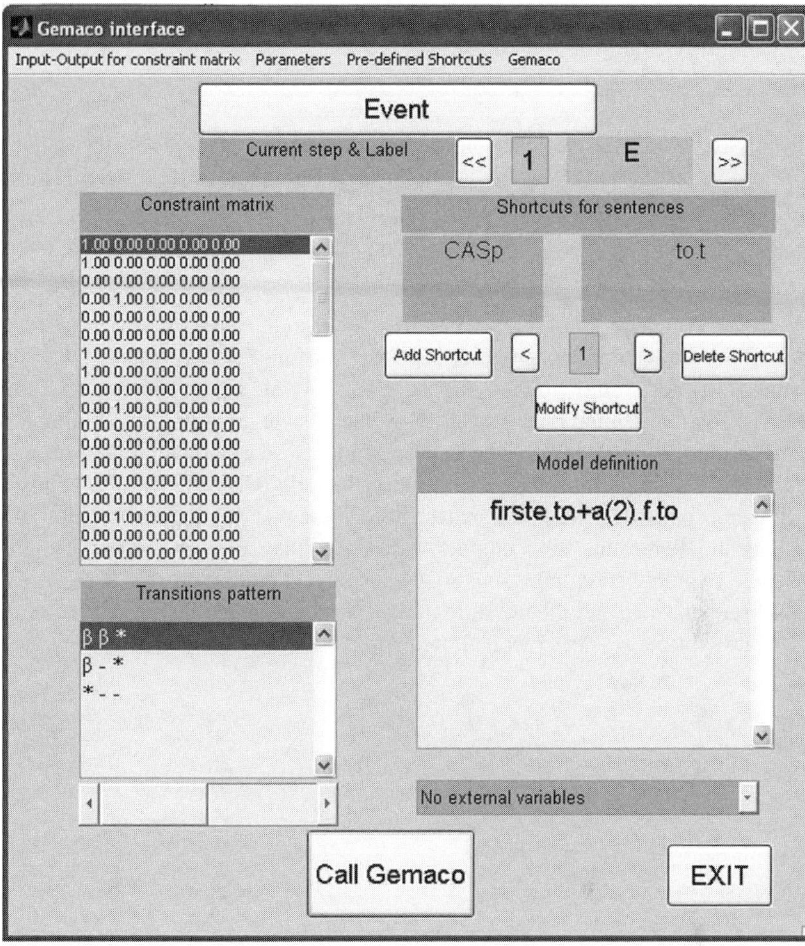

Fig. 3 (continued)

4.2 *Second Example: A Version of the Arnason–Schwarz Model Incorporating Site Fidelity*

Now we consider a version of the Arnason–Schwarz model in which the probability of transition conditional on survival is further subdivided into a probability of leaving the site (the complement of site fidelity), and a probability of moving to each other site conditional on leaving (Grosbois and Tavecchia 2003). For this example, we need a DES(1,3,1). With 3 geographical sites and assuming that if an animal is seen, its site is known without error, the set of events (i.e., the results of observations) is

$$\Omega = \{\text{'not seen', 'seen at 1', 'seen at 2', 'seen at 3'}\}$$

In this case, we will see that the intermediate states involved in the sequence of life processes are not the same as the basic set of states in the model. Thus the elementary matrices are not necessarily square.

4.2.1 Defining the Sets of Intermediate States

In the classical separate formulation of the Arnason–Schwarz model, the set of possible states for an individual is the same for both elementary matrices (survival and transition conditional on survival). However, as we are going to see, there may be a different set of states at each of the elementary steps. It can be helpful to use the graphical formulation of periodic matrix models (Caswell 2001) to construct the elementary matrices and their states sets. Figure 4 shows the formulation of the Grosbois and Tavecchia (2003) model, where the states of each elementary matrix are denoted as numbered nodes on a row. Each step in the life process is represented by a subsequent row, and the possible transitions are denoted by arrows. The initial set of states at time $k + 1$ is reconstituted at the bottom of the graph.

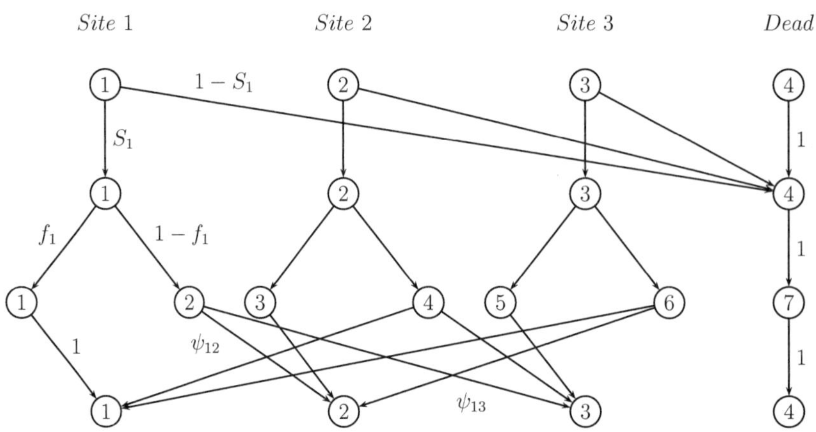

Fig. 4 Graphical representation of the Grosbois and Tavecchia (2003) model. Shown are the transitions for survival, fidelity given survival, and destination given movement. The elementary transition probabilities are shown on the pathways originating from site 1 and dead (at the top); the transition probabilities from the other sites 2 and 3 follow the same pattern. The first row-stochastic matrix $\mathbf{\Phi}^{(1)}$ maps row 1 onto row 2 of the diagram (survival step S). The matrix $\mathbf{\Phi}^{(2)}$ maps row 2 onto row 3 (fidelity given survival, f), and the matrix $\mathbf{\Phi}^{(3)}$ maps row 3 onto row 4 (destination given movement, ψ). In row 3, for practical reason, the numbers 1–7 do not refer to the original states. For instance, the intermediate state 2 corresponds to an individual that has moved from site 1 and the intermediate state 3 corresponds to an individual that has remained in site 2

In the Grosbois and Tavecchia (2003) site fidelity model, the sets of states are

$$\begin{aligned}
\mathbb{E}^{(0)} &= \{\text{'site 1', 'site 2', 'site 3', '†'}\}\\
\mathbb{E}^{(1)} &= \{\text{'site 1', 'site 2', 'site 3', '†'}\}\\
\mathbb{E}^{(2)} &= \{\text{'staying in 1', 'leaving 1', 'staying in 2', 'leaving 2',}\\
&\qquad \text{'staying in 3', 'leaving 3', '†'}\}
\end{aligned}$$

4.2.2 Matrices

The initial state matrix is

$$\mathbf{\Pi}^{(k)} = \left(\pi_1^{(k)}\ \pi_2^{(k)}\ 1 - \pi_1^{(k)} - \pi_2^{(k)}\ 0\right)$$

The first elementary transition matrix (for survival) maps $\mathbb{E}^{(0)}$ onto $\mathbb{E}^{(1)}$, and hence is a 4×4 matrix:

$$\mathbf{\Phi}^{(k,(1))} = \begin{pmatrix} s_1^{(k)} & 0 & 0 & 1 - s_1^{(k)} \\ 0 & s_2^{(k)} & 0 & 1 - s_2^{(k)} \\ 0 & 0 & s_3^{(k)} & 1 - s_3^{(k)} \\ 0 & 0 & 0 & 1 \end{pmatrix}$$

The second elementary matrix (for site fidelity given survival) maps from $\mathbb{E}^{(1)}$ to $\mathbb{E}^{(2)}$, and hence is a 4×7 matrix. Letting f_i be the probability of remaining in site i given survival, we have

$$\mathbf{\Phi}^{(k,(2))} = \begin{pmatrix} f_1^{(k)} & 1 - f_1^{(k)} & 0 & 0 & 0 & 0 & 0 \\ 0 & 0 & f_2^{(k)} & 1 - f_2^{(k)} & 0 & 0 & 0 \\ 0 & 0 & 0 & 0 & f_3^{(k)} & 1 - f_3^{(k)} & 0 \\ 0 & 0 & 0 & 0 & 0 & 0 & 1 \end{pmatrix}$$

The third elementary matrix (for movement conditional on emigration, ψ) maps from $\mathbb{E}^{(2)}$ back to $\mathbb{E}^{(0)}$, and so is a 7×4 matrix:

$$\mathbf{\Phi}^{(k,(3))} = \begin{pmatrix} 1 & 0 & 0 & 0 \\ 0 & \psi_{12}^{(k)} & 1 - \psi_{12}^{(k)} & 0 \\ 0 & 1 & 0 & 0 \\ \psi_{21}^{(k)} & 0 & 1 - \psi_{21}^{(k)} & 0 \\ 0 & 0 & 1 & 0 \\ \psi_{31}^{(k)} & 1 - \psi_{31}^{(k)} & 0 & 0 \\ 0 & 0 & 0 & 1 \end{pmatrix}$$

The event matrices $\boldsymbol{B}$ map from the set $\mathbb{E}^{(0)}$ of states to the set Ω of events, and thus are of dimension 4×4.

$$\boldsymbol{B}^{(a=1)} = \begin{pmatrix} 0 & 1 & 0 & 0 \\ 0 & 0 & 1 & 0 \\ 0 & 0 & 0 & 1 \\ 1 & 0 & 0 & 0 \end{pmatrix}$$

$$\boldsymbol{B}^{(k,a=2)} = \begin{pmatrix} 1-p_1^{(k)} & p_1^{(k)} & 0 & 0 \\ 1-p_2^{(k)} & 0 & p_2^{(k)} & 0 \\ 1-p_3^{(k)} & 0 & 0 & p_3^{(k)} \\ 1 & 0 & 0 & 0 \end{pmatrix}$$

Note The choice of states for the intermediate transitions is not always unique. There may be more than one way to group individuals, and at present the only advice we can give is to determine from the structure of the model what information needs to be kept at any one step in order to define the probability of subsequent transitions.

4.3 Third Example: The Separate Formulation of the Memory Model

Finally, we consider a version of the memory model given in Pradel (2005) in which the survival-transition probability is further subdivided into the probability of survival and the probability of movement from each site to the others conditional on surviving. We need a DES(1,2,1). This formulation is the analogue of the separate formulation of the conditional Arnason–Schwarz model.

With the 2 geographical sites 1 and 2, $\mathbb{E}$ consists of 5 states {'11', '12', '21', '22', '†'}. State '11' denotes presence at site 1 at time $k-1$ and k, state '12' denotes presence at site 1 at time $k-1$ and presence at site 2 at time k, and so on. The set of states remains constant across the two life processes steps

$$\mathbb{E}^{(0)} = \mathbb{E}^{(1)} = \{\text{'11', '12', '21', '22', '†'}\}$$

Assuming that if an animal is seen and its current site is known without error, the set of events (i.e., the results of observations) is

$$\Omega = \{\text{'not seen', 'seen at 1', 'seen at 2'}\}$$

The initial state matrix is

$$\boldsymbol{\Pi}^{(k)} = \begin{pmatrix} \pi_{11}^{(k)} & \pi_{12}^{(k)} & \pi_{21}^{(k)} & \pi_{22}^{(k)} & 0 \end{pmatrix}$$

There are now two elementary transition matrices, one corresponding to survival(noted s) and one to movements (noted ψ) conditional on survival:

$$\mathbf{\Phi}^{(k,(1))} = \begin{pmatrix} s_{11}^{(k)} & 0 & 0 & 0 & 1-s_{11}^{(k)} \\ 0 & s_{12}^{(k)} & 0 & 0 & 1-s_{12}^{(k)} \\ 0 & 0 & s_{21}^{(k)} & 0 & 1-s_{21}^{(k)} \\ 0 & 0 & 0 & s_{22}^{(k)} & 1-s_{22}^{(k)} \\ 0 & 0 & 0 & 0 & 1 \end{pmatrix}$$

$$\mathbf{\Phi}^{(k,(2))} = \begin{pmatrix} \psi_{111}^{(k)} & 1-\psi_{111}^{(k)} & 0 & 0 & 0 \\ 0 & 0 & 1-\psi_{122}^{(k)} & \psi_{122}^{(k)} & 0 \\ \psi_{211}^{(k)} & 1-\psi_{211}^{(k)} & 0 & 0 & 0 \\ 0 & 0 & 1-\psi_{222}^{(k)} & \psi_{222}^{(k)} & 0 \\ 0 & 0 & 0 & 0 & 1 \end{pmatrix}$$

There are two elementary event matrices. The first one is fixed and the second is varying:

$$\boldsymbol{B}^{(a=1)} = \begin{pmatrix} 0 & 1 & 0 \\ 0 & 0 & 1 \\ 0 & 1 & 0 \\ 0 & 0 & 1 \\ 1 & 0 & 0 \end{pmatrix}$$

$$\boldsymbol{B}^{(k,a=2)} = \begin{pmatrix} 1-p_1^{(k)} & p_1^{(k)} & 0 \\ 1-p_2^{(k)} & 0 & p_2^{(k)} \\ 1-p_1^{(k)} & p_1^{(k)} & 0 \\ 1-p_2^{(k)} & 0 & p_2^{(k)} \\ 1 & 0 & 0 \end{pmatrix}$$

This model is an extension of the model presented in Pradel et al. (2008), but the model of Brownie et al. (1993) can also be decomposed in the same way.

5 Discussion and Perspectives

In this paper, we have presented E-SURGE, the first program to fit non-homogeneous hidden Markov chains models in CR, called multievent models. This program has an interface (named GEPAT) to build various parameterizations and goes far beyond the classical combined and separate parameterizations. We can consider any product of elementary transition matrices, any product of elementary encounter matrices and any product of elementary initial matrices, which allows great flexibility to fit different kind of models. This approach permits incorporation of multievents models in a GLM framework and avoids dealing with non-linear constraints, which are difficult to manage. GEPAT coupled with GEMACO for model constraints gives an easy, flexible and powerfull way to construct models, that are useful to address biological questions.

Advanced features of E-SURGE are not described in this paper, yet we enumerate some of them. Essential but hidden are elaborate algorithms used to fit a model, compute its rank and detect redundant parameters. More visible are the results stored in a spreadsheet file, which allows an easy access to the estimated parameters of all fitted models for further calculations or for drawing graphs. Lastly, the post allocation of classes proposed in Pledger et al. (2003) is also available.

We are continuously exploring new features with working versions. We hope to have soon implemented the profile deviance (and more generally all the tools already implemented in M-SURGE), estimation of the Lifetime Reproductive Success (Rouan et al. 2008), the possibility to model individual covariates (with fixed or random effect) as a promising link with genetics (Gimenez et al. 2006) and the Stochastic EM algorithm (Celeux et al. 1995) to deal with local minima in a more efficient way.

6 Program Availability

Program E-SURGE can be downloaded freely from the WWW at
http://www.cefe.cnrs.fr/BIOM/logiciels.htm

Instructions for installation and a user guide are provided.

Acknowledgments Especial thanks are due to Michael Schaub for proof reading and suggestions for this article. We wish also to thank Evan Cooch and Jim Hines for suggestions on the paper.

References

Akaike H (1987) Factor analysis and AIC. Psychometrika 52:317–332.

Brownie C, Hines J, Nichols J, Pollock K, Hestbeck J (1993) Capture–recapture studies for multiple strata including non-markovian transitions. Biometrics 49:1173–1187.

Burnham K, Anderson D (2002) Model selection and inference: A practical information-theoretic approach, 2nd edition. Springer-Verlag, New York.

Cappé O, Moulines E, Rydn T (2005) Inference in Hidden Markov models (Springer series in Statistics). Springer, New York.

Caswell H (2001) Matrix population models 2nd edition., Sinauer Associates, Sunderland, MA.

Celeux G, Chauveau D, Diebolt J (1995) On stochastic versions of the EM algorithm. Technical Report 2514, INRIA Rhone-Alpes.

Choquet R (2008) Automatic generation of multistate capture–recapture models. The Canadian Journal of Statistics 36(1):43–57.

Choquet R, Reboulet AM, Pradel R, Gimenez O, Lebreton JD (2004) M-SURGE: new software specifically designed for multistate capture–recapture models. Animal Biodiversity and Conservation 27(1):207–215.

Choquet R, Reboulet A, Pradel R, Gimenez O, Lebreton JD (2005) M-SURGE: 1.8 user's manual. Technical report, CEFE UMR 5175, Mondpellier.

Fujiwara M, Caswell H (2002) Estimating population projection matrices from multi-stage mark-recapture data. Ecology 83(12):3257–3265.

Gimenez O, Covas R, Brown C, Anderson M, Brown M, Lenormand T (2006) Nonparametric estimation of natural selection on a quantitative trait using mark-recapture data. Evolution 60(3):460–466.

Grosbois V, Tavecchia G (2003) Modeling dispersal with capture–recapture data: Disentangling decisions of leaving and settlement. Ecology 84(5):1225–1236.

Hestbeck J, Nichols J, Malecki R (1991) Estimates of movement and site fidelity using mark-resight data of wintering Canada Geese. Ecology 72:523–533.

Hines JE (1994) Mssurviv user's manual, National Biological Service, Patuxent Wildlife Research Center, Laurel, MD 20708–4017.

Lebreton JD, Almeras T, Pradel R (1999) Competing events, mixtures of information and multi-stratum recapture models. Bird Study 46:39–46.

Lebreton J-D, Burnham K, Clobert J, Anderson D (1992) Modeling survival and testing biological hypotheses using marked animals: A unified approach with case studies. Ecological Monographs 62:67–118.

Lebreton JD, Pradel R (2002) Multistate recapture models: modelling incomplete individual histories. Journal of Applied Statistics 29(1–4):353–369.

MacDonald IL, Zucchini W (2000) Hidden Markov and other models for discrete-valued time series. Monographs on Statistics and applied probability. Chapman and Hall, London.

Nichols JD, Kendall WL, Hines JE, Spendelow JA (2004) Estimation of sex-specific survival from capture-recapture data when sex is not always known. Ecology 85(12):3192–3201.

Pledger S, Pollock KH, Norris J (2003) Open capture–recapture models with heterogenety: I. Cormack–Jolly–Seber model. Biometrics 59:786–794.

Pradel R (2005) Multievent: An extension of multistate capture recapture models to uncertain states. Biometrics 61:442–447.

Pradel R (2008) The stakes of models with state uncertainty. In: Thomson DL, Cooch EG, Conroy MJ (eds.) Modeling Demographic Processes in Marked Populations. Environmental and Ecological Statistics, Springer, New York, 3:781–796.

Pradel R, Johnson AR, Viallefont A, Nager RG, Cezilly F (1997) Local recruitment in the Greater Flamingo: A new approach using capture-mark-recapture data. Ecology 78:1431–1445.

Pradel R, Maurin-Bernier L, Gimenez O, Genovart M, Choquet R, Oro D (2008) Estimation of sex-specific survival with uncertainty in sex assessment. The Canadian Journal of Statistics 36(1):29–42.

Rouan L, Choquet R, Pradel R (2008) A general framework for modeling memory in capture–recapture data. Jabes, In Press.

Rouan L, Gaillard JM, Guédon Y, Pradel R (2008) Estimation of lifetime reproductive success when reproductive status cannot always be assessed. In: Thomson DL, Cooch EG, Conroy MJ (eds.) Modeling Demographic Processes in Marked Populations. Environmental and Ecological Statistics, Springer, New York, 3:867–880.

Véran S, Gimenez O, Flint E, Kendall B, Doherty P, Lebreton JD (2007) Quantifying the impact of longline fisheries on adult survival in the black-footed Albatross. Journal of Applied Ecology 44(5):942–952

White G, Burnham K (1999) Program MARK: Survival estimation from populations of marked animals. Bird Study 46(suppl.):120–139.

Estimation of Lifetime Reproductive Success When Reproductive Status Cannot Always Be Assessed

Laurine Rouan, Jean-Michel Gaillard, Yann Guédon, and Roger Pradel

Abstract The Lifetime Reproductive Success (LRS) of an individual i.e. the number of young raised during its lifespan is an indicator of its contribution to future generations and thus a measure of fitness. Nevertheless, the LRS is hard to estimate because of the difficulty to keep track of the outcome of each breeding attempt (successful or failed and, if successful, number of young raised). We propose two new methods to estimating the LRS that takes into account the uncertainty about the reproductive status when the individuals are not detected or when the reproductive status cannot be assessed. We illustrate these two methods using roe deer reproductive histories and discuss their advantages and disadvantages.

Keywords Capture-Recapture · Counting Algorithm · Fitness · Generalized Viterbi algorithm · Hidden Markov Models

1 Introduction

Lifetime Reproductive Success (LRS) is a commonly used estimate of individual fitness (Clutton-Brock 1988; Newton 1989). It can be defined as the total number of offspring an individual produces over its entire lifespan after some critical stage has been successfully passed (e.g. number of weaned young in mammals or number of fledglings in birds, see Clutton-Brock (1988) for case studies). As LRS is a measure of the lifetime reproductive performance of an individual, its actual calculation supposes the knowledge of the individual's entire reproductive history. Nevertheless, in wild populations, the exhaustive monitoring of a large number of individuals over a long time period is difficult; there are not only problems of detection but also problems in assessing the reproductive status. For instance, an individual seen during the breeding season could have produced and/or raised young or not and if it has,

L. Rouan (✉)
Centre d'Écologie Fonctionnelle et Évolutive, Montpellier, France
e-mail: lauriane.rouan@cefe.cnrs.fr

D.L. Thomson et al. (eds.), *Modeling Demographic Processes in Marked Populations*, Environmental and Ecological Statistics 3, DOI 10.1007/978-0-387-78151-8_40,

the number of its offspring could be difficult to determine. So, the capture-recapture (CR) data are inevitably incomplete. In the absence of established statistical tools, some 'ad hoc' methods are used and consist of counting only the reproductive events observed leading thus to a 'minimum LRS' or of completing the reproductive histories by assuming that an individual not observed in a given year during the reproductive season, but captured (or observed) a previous year and captured (or observed) in a later year has some chance to be an effective breeder (Jensen et al. 2004). Pradel (2005) proposes a new approach, called multievent models, to deal with the uncertainty in the breeding states assessment. The specificity of these models, belonging to the hidden Markov models class (Rabiner 1989; Ephraim and Merhav 2002), lies in the distinction between the (hidden) states (in our case, the reproductive status) and the 'events' observed and recorded in the reproductive histories (for example: 'not seen', 'seen with one offspring', 'seen during the breeding period', ...). Thus, contrary to the hypotheses of the now well established multistate CR models (Arnason 1973), the underlying state of a captured individual is not necessarily known for certain i.e. each observation is potentially linked to several underlying states. Thus, a particular encounter history can be associated with different sequences of 'hidden states' (see Fig. 1); but which of these has actually generated the sequence of observations (encounter history)? Answering this question would enable to calculate the LRS but it is, unfortunately, impossible. So, we suggest in this paper to estimate the LRS by taking into account all these potential sequences. We propose two different methods. The first method (Section 3.1) consists in the actual determination of the potential hidden state sequences in order of probability of occurrences ('associated probability'). It uses the Generalized Viterbi Algorithm (Foreman 1993). To each sequence corresponds a number of young produced and/or raised and the LRS is estimated as the mean of these numbers weighted by the

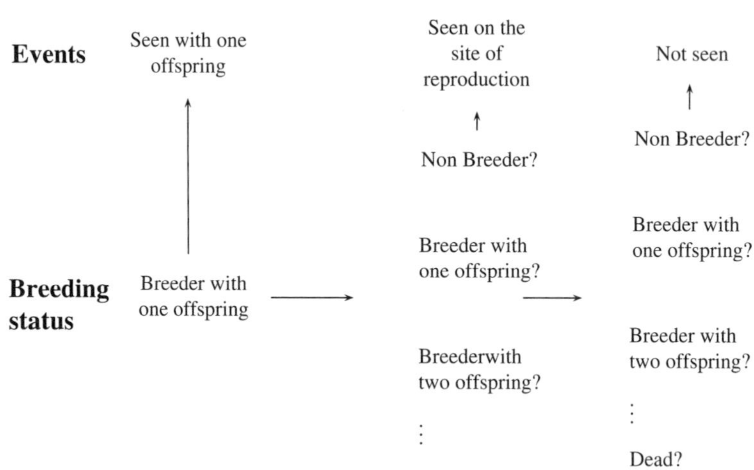

Fig. 1 The events recorded at each capture occasion are potentially linked to several states. In other words different state sequences can generate the same encounter history

associated probabilities. But the calculation of the LRS does not require the knowledge of the timing of reproduction, only the number of reproductive episodes is required. So, we present a second method, less expensive in terms of computation time (Section 3.2). It is an adaptation to the CR field of a counting algorithm initially designed for the general framework of the hidden semi-Markov chains (Guédon 1999). It does not enable the access to the underlying states sequences but provides the distribution of the number of occurences of any underlying state during the animal lifespan and enables us to estimate the LRS.

Both approaches make use of a common notation and are illustrated with a common example presented in the following section.

2 Illustrating the Two Methods Using the Roe Deer *Capreolus capreolus* Life History

The 'multievent approach' is based on the differentiation of the observations made at each capture session and the actual reproductive status of the females. In other words, as illustrated in Fig. 1 we consider two separate stochastic processes:

- the state process denoted $(S_t)_{t>0}$ that describes the succession of the reproductive statuses;
- the observation process denoted $(O_t)_{t>0}$ that describes the different 'events' observed over the study period.

These two processes are linked using what we call, hereafter, the 'observation probabilities' or 'conditional event probabilities'. Based on the roe deer life history, we can distinguish the following different 'events' (see e.g. Gaillard et al. 1998):

0: the animal is not seen;
1: the animal is seen without any fawn;
2: the animal is seen with one fawn;
3: the animal is seen with two fawns;
4: the animal is seen with three fawns.

The associated underlying states are:

Non Breeder (NB);
Breeder with one fawn (B1);
Breeder with two fawns (B2);
Breeder with three fawns (B3);
Dead (†).

This defines the event and state sets: $\Omega = \{0, 1, 2, 3, 4\}$ and $E = \{NB, B1, B2, B3, \dagger\}$. It remains to specify the parameters that define the multievent model and appear in the different algorithms that we propose to estimate the LRS. There are three kinds of parameters:

1. the transition probabilities:
   - $\phi_{ij}^t = P(S_{t+1} = j | S_t = i)$, probability of being in state j at $t + 1$ for an animal in state i at t;
2. the initial state probabilities
   - $\pi_i^t = P(S_t = i)$, probability of being in state i when first captured at t. In the general case, females are initially marked during early life to be considered as known-aged individuals. For instance, roe deer are often marked in their first winter at about 8 months of age and can only be in state 'non breeder' since the first reproduction is only possible at 2-year-old (Gaillard et al. 1992). Thus $\pi_{NB}^t = 1$.
3. the conditional event probabilities:
   - $b_s^t(o) = P(O_t = o | S_t = s)$, probability of event o conditional on current state s.

   These last parameters can actually be expressed as functions of the familiar encounter probabilities of the multistate models:

   - p_j^t, probability to be encountered in site j at time t for an individual alive and in site j at that time.

For roe deer, we can assume that the uncertainty in assessing the reproductive status is only due to problems of detection i.e. if a female is seen, its reproductive status corresponds exactly to the observation made in the right period (September–December) because fawns closely follow their mother during that period (see Gaillard et al. 2000). So, most of the $b_s^t(o)$ are equal to zero. The only not null parameters are: $b_{NB}^t(1) = p_{NB}^t$, $b_{B1}^t(2) = p_{B1}^t$, $b_{B2}^t(3) = p_{B2}^t$, $b_{B3}^t(4) = p_{B3}^t$, $b_j^t(0) = 1 - p_j^t$ for $j \in \{NB, B1, B2, B3\}$ and $b_{\dagger}^t(0) = 1$. This assumption has been made in agreement with our knowlegde about roe deer life history in order to simplify the statistical model but is not necessary generally. In the general case, two kinds of uncertainty can occur: the lack of detection at some reproductive attempts and an uncertain determination for animals detected. Both kinds can be easily accounted for in the two methods detailed below.

The estimation of these different parameters is the first step, common to our two approaches, to estimate the LRS.

3 Estimation of the Lifetime Reproductive Success

To calculate the LRS, one needs to know the number of offspring an individual has produced at each reproductive occasion. The first, intuitive, idea is to reconstruct the reproductive life of the individual.

3.1 Method 1: Estimating LRS Using the Generalized Viterbi Algorithm

3.1.1 Method

Given any observation sequence $O = o_1 o_2 \cdots o_T$, the generalized Viterbi algorithm (Foreman 1993) seeks to find the L $(L \geq 1)$ most probable underlying state sequences $\{S^i = s_1^i\, s_2^i \cdots s_T^i / i = 1, 2, \cdots, L\}$ maximizing the associated probability $P(S|O)$. This algorithm has been originally designed for the area of automatic speech recognition and is adapted to the treatment of homogeneous Markov chains and observations sequences starting at time $t = 1$. Its use for the CR data needs some specific adaptations to:

1. allow for time varying parameters;
2. handle histories starting after date 1;
3. use a stopping criteria based on the cumulative probability of occurrence i.e. our 'adapted' generalized Viterbi algorithm doesn't generate a fixed number of state sequences but more precisely generates the L_α state sequences such that the cumulative probability $\sum_{i=1}^{L_\alpha} P\left(S^i|O\right)$ reaches a fixed threshold α. In this way, to obtain all the possible state sequences, α must be fixed to 1.

3.1.2 LRS Estimation

The parameters of the retained model, estimated using program E-SURGE (Choquet et al. 2007), are used to calculate the different quantities needed to define the Viterbi algorithm. The second step of this method is to apply the algorithm to a particular encounter history O thus generating the L_α state sequences and their associated probability. Finally the LRS is estimated as the weighted mean of the number of offspring given by the different state sequences:

$$LRS = \frac{\sum_{i=1}^{L_\alpha} n^i_{offspring} P\left(S^i|O\right)}{\sum_{i=1}^{L_\alpha} P\left(S^i|O\right)}$$

3.1.3 Examples

We illustrate this approach using two encounter histories taken from the data set

1. $O^1 =$ 0 0 0 0 0 0 0 0 0 0 0 0 0 0 0 0 0 0 0 0 0 1 0 4 0 2 2 0 2 2 2 0
2. $O^2 =$ 0 0 0 0 0 0 0 0 0 0 0 1 3 2 3 2 0 3 2 2 2 1 0 0 0 0 0 0 0 0

the second one being 'less complete' than the first in the sense that there are fewer detection issues (number of 0 after the individual first capture) in O^1 than in O^2.

To illustrate this first method, we have chosen to fix α to one.

The application of our 'adapted' generalized Viterbi algorithm to O^1 provides the nine following state sequences (in descending order of associated probability):

- $S^1 =$ NB NB B3 B1 B1 B1 B1 B1 B1 B1 B1 with $P\left(S^1|O^1\right) \approx 0.2228$;
- $S^2 =$ NB B1 B3 B1 B1 B1 B1 B1 B1 B1 B1 with $P\left(S^2|O^1\right) \approx 0.1827$;

- S^3 = NB B3 B3 B1 B1 B1 B1 B1 B1 B1 B1 with $P\left(S^3|O^1\right) \approx 0.1421$;
- S^4 = NB NB B3 B1 B1 B1 B1 B1 B1 B1 † with $P\left(S^4|O^1\right) \approx 0.1359$;
- S^5 = NB B1 B3 B1 B1 B1 B1 B1 B1 B1 † with $P\left(S^5|O^1\right) \approx 0.1114$;
- S^6 = NB B3 B3 B1 B1 B1 B1 B1 B1 B1 † with $P\left(S^6|O^1\right) \approx 0.0866$;
- S^7 = NB NB B3 B1 B1 B1 B1 B1 B1 B1 B2 with $P\left(S^7|O^1\right) \approx 0.0482$;
- S^8 = NB B1 B3 B1 B1 B1 B1 B1 B1 B1 B2 with $P\left(S^8|O^1\right) \approx 0.0395$;
- S^9 = NB B3 B3 B1 B1 B1 B1 B1 B1 B1 B2 with $P\left(S^9|O^1\right) \approx 0.0307$.

Thus, the LRS is estimated as:

$$
\begin{aligned}
LRS_1 &= 11 \times 0.2228 + 12 \times 0.1827 + 14 \times 0.1421 + 10 \times 0.1359 + 11 \\
&\quad \times 0.1114 + 13 \times 0.0866 + 12 \times 0.0482 + 13 \times 0.0395 + 15 \times 0.0307 \\
&\approx 11.8965
\end{aligned}
$$

The same method applied to the encounter history O^2 generates 492 state sequences! The 'best' and 'worst' sequences with their associated probability are:

- S^1 = NB B2 B1 B2 B1 B1 B2 B1 B1 B1 NB with $P\left(S^1|O^2\right) \approx 0.6991$.
- S^{492} = NB B2 B1 B2 B1 B1 B2 B1 B1 B1 NB B2 B1 B2 B1 B2 B2 B1 B2 with $P\left(S^{492}|O^2\right) \approx 6.1806e-007$.

The estimated LRS is, for this second encounter history: 12.7639.

The generalized Viterbi algorithm is thus a reasonable approach to estimate the LRS but presents some limits in terms of computation time. The number of reconstructed state sequences is variable depending on the associated encounter history; the computation time can vary between few seconds like in case of O^1 and tens of minutes for O^2. However, as the mere calculation of the LRS does not require the knowledge of the exact sequence of reproductive events but only the total number of offspring produced and/or raised, we propose another approach for estimating the LRS, which yields the same estimate as the generalized Viterbi algorithm with a threshold $\alpha = 1$ but is much less expensive in computation time.

3.2 Method 2: Estimating LRS Using the Counting Algorithm

3.2.1 Method

Our general idea to calculate the LRS is to count the occurrences of the breeding hidden states. More precisely, this counting algorithm determines, giving an encounter history $O = o_e o_{e+1} \cdots o_T$, the probability $P\left(N_k(T) = n | o_e o_{e+1} \cdots o_T\right)$ that an underlying state k has occurred n times ($n \in \{0,\ 1,\ 2, \cdots,\ T-e+2\}$, e is the date of first capture and T the last capture occasion) by the end of the study.

Let us so introduce the auxiliary quantities

$$\beta_t(i,n) = P\left(S_t = i, N_k(t) = n, O_e = o_e, \cdots, O_t = o_t\right) \tag{1}$$

The quantity $\beta_t(i, n)$ is the unconditional probability that the underlying state k has occurred n times until the date t and that the underlying state at this date is i.

The probability $P\left(N_k(T) = n|o_e o_{e+1} \cdots o_T\right)$ that the state k occurs n times can be expressed using these quantities as follows:

$$P\left(N_k(T) = n|o_e o_{e+1} \cdots o_T\right) = \frac{\sum_{i=1}^{N} \beta_T(i, n)}{\sum_{n=0}^{T} \sum_{i=1}^{N} \beta_T(i, n)} \tag{2}$$

The denominator corresponds to the conditioning on the encounter history O. The counting algorithm corresponds then to the recursive calculation of $\beta_t(i, n)$ for all t in $\{0, 1, 2, \cdots, T\}$ and n in $\{0, 1, 2, \cdots, t\}$ (see Appendix 1).

3.2.2 LRS Estimation

As in the Viterbi algorithm case, the first step is to introduce the estimated parameters of the retained model in the algorithm. In a second step, applying it to an encounter history $O = o_e o_{e+1} \cdots o_T$, provides for each underlying state k the probabilities: $P\left(N_k(T) = n|O\right)$ for n in $\{0, 1, 2, \cdots, T\}$ i.e. the distribution of the number of occurrences of each state k.

For each state k, the conditional expectancy $E\left[N_k(T)|O\right]$ can be interpreted as the average number of occasions in state k. We can easily show that $\sum_{all\ states\ k} E\left[N_k(T)|O\right] = T - e + 1$ is the number of capture occasions since the first capture of the individual (see Appendix 2). Given these quantities, the LRS is estimated in this way:

$LRS = \sum_{k=1}^{K} k \times E\left[N_{Bk}(T)|O\right]$ where K is the maximum number of offspring an individual can produce at each breeding season and Bk corresponds to the status 'breeder with k offspring'.

3.2.3 Examples

In our examples, only the states B1, B2 and B3 are interesting, so we will give only the distributions relative to these particular states.

For the first encounter history O^1 of Section 3.1.3:

n	0	1	2	3	4	5	6	7
$P\left(N_{B1}(T) = n|O^1\right)$	0	0	0	0	0	0	0	0.3015
$P\left(N_{B2}(T) = n|O^1\right)$	0.8815	0.1185	0	0	0	0	0	0
$P\left(N_{B3}(T) = n|O^1\right)$	0	0.7405	0.2595	0	0	0	0	0

n	8	9	10	11
$P\left(N_{B1}(T) = n|O^1\right)$	0.5158	0.1827	0	0
$P\left(N_{B2}(T) = n|O^1\right)$	0	0	0	0
$P\left(N_{B3}(T) = n|O^1\right)$	0	0	0	0

The estimated LRS is so:

$$\begin{aligned} LRS_1 &= 1 \times E\left[N_{B1}(T)|O^1\right] + 2 \times E\left[N_{B2}(T)|O^1\right] + 3 \times E\left[N_{B3}(T)|O^1\right] \\ &= 1 \times 7.8811 + 2 \times 0.1185 + 3 \times 1.2595 \\ &\approx 11.8966 \end{aligned}$$

For the second encounter history O^2:

n	0	1	2	3	4	5	6
$P\left(N_{B1}(T)=n\|O^2\right)$	0	0	0	0	0	0	0.7237
$P\left(N_{B2}(T)=n\|O^2\right)$	0	0	0	0.9048	0.0729	0.0151	0.0049
$P\left(N_{B3}(T)=n\|O^2\right)$	0.9971	0.0029	0	0	0	0	0

n	7	8	9	10	11	12
$P\left(N_{B1}(T)=n\|O^2\right)$	0.1197	0.1095	0.0329	0.0099	0.0030	0.0009
$P\left(N_{B2}(T)=n\|O^2\right)$	0.0016	0.0005	0.0002	0.0001	0	0
$P\left(N_{B3}(T)=n\|O^2\right)$	0	0	0	0	0	0

n	13	14	15	16	17	18	19
$P\left(N_{B1}(T)=n\|O^2\right)$	0.0002	0	0	0	0	0	0
$P\left(N_{B2}(T)=n\|O^2\right)$	0	0	0	0	0	0	0
$P\left(N_{B3}(T)=n\|O^2\right)$	0	0	0	0	0	0	0

The estimated LRS is so:

$$\begin{aligned} LRS_2 &= 1 \times 6.4998 + 2 \times 3.1282 + 3 \times 0.0029 \\ &\approx 12.7649 \end{aligned}$$

In the two cases, the counting algorithm provides the same results as the generalized Viterbi algorithm but with a computation time approaching few seconds.

4 Discussion

This article has presented new general methods to estimate Lifetime Reproductive Success with missing values for some reproductive events (generated either by a lack of detection or by an uncertain reproductive status). The two methods provided identical LRS estimates. The main difference between the two methods lies in the reconstruction or not of the possible reproductive sequences associated with a given encounter history. The 'Viterbi approach' is the more detailed in that it reveals the state sequences associated with the encounter history and their probability of occurrence. The LRS is then simply calculated as the weighted average of the numbers of offspring provided by these state sequences. However, this algorithm can be very expensive in terms of computation time depending on the level of uncertainty of the encounter history. The counting algorithm only provides the expected distribution of the number of offspring produced and/or raised over the lifetime. The exact sequence of reproductive events along the lifetime is not informed using this latter method but is much faster than the Viterbi algorithm. The counting algorithm seems thus the most appropriate method to estimate the LRS. However, LRS is just one measure of single-generation individual fitness, and other single-generation measurements have been proposed to be better in some situations. For instance, the 'individual λ' method (McGraw and Caswell 1996) accounts for the timing of reproductive events, which LRS does not (Käär and Jokela 1998). Although a recent comparative analysis of fitness measures showed that LRS generally performs better than 'individual λ' (Brommer et al. 2004), the latter measure is likely to be preferred in markedly increasing or decreasing populations when the timing of reproductive events has a major impact on fitness. By using the Viterbi algorithm, biologists could not only obtain an estimate of LRS, but also get estimates of reproductive output at each breeding attempt. Therefore, the calculation of individual λ, or of the promising measure of individual contributions (Coulson et al. 2006) would be straightforward by using the Viterbi algorithm. On the other hand, only an estimate of LRS can be obtained using the counting method.

In conclusion, the counting method we proposed here should allow biologists to get a way of estimating quickly LRS in a large range of field conditions. By using the Viterbi method, biologists could even obtain different time-sensitive and time-insensitive fitness measures. These two methods are currently implemented in MATLAB and will be soon available in program E-SURGE.

Appendix 1: The Counting Algorithm

As mentioned above for the adaptation of the generalized Viterbi algorithm, in capture-recapture studies, probabilities are usually calculated conditionally on the first capture of the individuals; so the recursion will begin with $t = e$.

1. Initialization
   $t = e$,
   $i \in E$,
   $n \in \{0, 1\}$,

   If $i = k$,

$$\begin{aligned}
\beta_e(k, 0) &= 0 \\
\beta_e(k, 1) &= P\left(S_e = k, N_k(e) = 1, O_e = o_e\right) \\
&= P\left(S_e = k, O_e = o_e\right) \\
&= \pi_k^e b_k^e(o_e)
\end{aligned}$$

   Si $i \neq k$,

$$\begin{aligned}\beta_e(i,0) &= P\left(S_e = i, N_k(e) = 0, O_e = o_e\right)\\ &= P\left(S_e = i, O_e = o_e\right)\\ &= \pi_i^e b_i^e(o_e)\\ \beta_e(k,1) &= 0\end{aligned}$$

2. Recursion

$t \in \{e+1, e+2, \ldots, T\}$,
$i \in E$,
$n \in \{0, 1, 2, \cdots, t\}$,

If $i = k$,

$$\begin{aligned}\beta_t(i,n) &= P\left(S_t = k, N_k(t) = n, O_e = o_e, \cdots, O_t = o_t\right)\\ &= \sum_{j=1}^{N} P\left(S_t = k, S_{t-1} = j, N_k(t-1) = n-1, O_e = o_e, \cdots, O_{t-1}\right.\\ &\left.= o_{t-1}, O_t = o_t\right)\end{aligned}$$

If $i \neq k$,

$$\begin{aligned}\beta_t(i,n) = &\ P\left(S_t = i, N_k(t) = n, O_e = o_e, \cdots, O_t = o_t\right)\\ = &\sum_{j=1}^{N} P\left(S_t = i, S_{t-1} = j, N_k(t-1) = n, O_e = o_e, \cdots, O_{t-1} = o_{t-1},\right.\\ &\left.O_t = o_t\right)\end{aligned}$$

But, for all time t, all state λ and all number m:

$$\begin{aligned}P\left(S_t\right. &= \lambda, S_{t-1} = j, N_k(t-1) = m, O_e = o_e, \cdots, O_{t-1} = o_{t-1}, \left.O_t = o_t\right)\\ &= P\left(O_t = o_t | S_t = \lambda, S_{t-1} = j, N_k(t-1) = m, O_e = o_e, \cdots, O_{t-1}\right.\\ &\quad \left.= o_{t-1}\right) \times P\left(S_t = \lambda, S_{t-1} = j, N_k(t-1) = m, O_e = o_e, \cdots, O_{t-1}\right.\\ &\quad \left.= o_{t-1}\right)\\ &= P\left(O_t = o_t | S_t = \lambda\right) \times P\left(S_t = \lambda | S_{t-1} = j, N_k(t-1) = m, O_e = o_e,\right.\\ &\quad \left.\cdots, O_{t-1} = o_{t-1}\right) \times P\left(S_{t-1} = j, N_k(t-1) = m, O_e = o_e, \cdots, O_{t-1}\right.\\ &\quad \left.= o_{t-1}\right)\\ &= P\left(O_t = o_t | S_t = \lambda\right) \times P\left(S_t = \lambda | S_{t-1} = j\right)\\ &\quad \times P\left(S_{t-1} = j, N_k(t-1) = m, O_e = o_e, \cdots, O_{t-1} = o_{t-1}\right)\end{aligned}$$

So:
If $i = k$,

$$\begin{aligned}\beta_t(i,n) &= \sum_{j=1}^{N} b_k^t(o_t)\phi_{jk}^{t-1}\beta_t(j,n-1)\mathbf{1}_{[n>0]} \\ &= b_k^t(o_t)\sum_{j=1}^{N}\beta_{t-1}(j,n-1)\phi_{jk}^{t-1}\mathbf{1}_{[n>0]}\end{aligned}$$

If $i \neq k$,

$$\begin{aligned}\beta_t(i,n) &= \sum_{j=1}^{N} b_i^t(o_t)\phi_{ji}^{t-1}\beta_t(j,n) \\ &= b_i^t(o_t)\sum_{j=1}^{N}\beta_{t-1}(j,n)\phi_{ji}^{t-1}\end{aligned}$$

where

$$\begin{cases}\mathbf{1}_{[n>0]} = 0, \text{ if } n \leq 0 \text{ (here, if } n = 0) ; \\ \mathbf{1}_{[n>0]} = 1, \text{ if } n > 0.\end{cases}$$

Appendix 2: $\sum E_{all\ states\ k}\left[N_k(T)|O\right]$ =Number of Occasions

For any state k and encounter history $O = o_e o_{e+1} o_{e+2} \cdots o_T$,

$$\begin{aligned}E\left[N_k(T)|O\right] &= E\left[\sum_{t=e}^{T}\mathbf{1}_{[S_t=k]}|O\right] \\ &= \sum_{t=e}^{T} E\left[\mathbf{1}_{[S_t=k]}|O\right] \\ &= \sum_{t=e}^{T} P\left(S_t = k|O\right)\end{aligned}$$

Thus,

$$\sum_{all\ states\ k} E\left[N_k(T)|O\right] = \sum_{all\ states\ k} \sum_{t=e}^{T} P\left(S_t = k|O\right)$$
$$= \sum_{t=e}^{T} \sum_{all\ states\ k} P\left(S_t = k|O\right)$$
$$= \sum_{t=e}^{T} 1$$
$$= T - e + 1$$

References

Arnason AN (1973) The estimation of population size, migration rates and survival in a stratified population. Researches on Population Ecology 15(1):1–8.

Brommer JE, Gustafson L, Pietiainen H, Mérila J (2004) Single-generation estimates of individual fitness as proxies for long-term genetic contribution. American Naturalist 163: 505–517.

Choquet R, Rouan L, Pradel R (2008) Program E-SURGE: a software application for fitting multi-event models. In: Thomson DL, Cooch EG, Conroy MJ (eds.) Modeling Demographic Processes in Marked Populations. Environmental and Ecological Statistics, Springer, New York.

Clutton-Brock TH (1988) Reproductive Success. Studies of Individual Variation in Contrasting Breeding Systems. University of Chicago Press, Chicago.

Coulson T, Benton TG, Lundberg P, Dall SRX, Kendall BE, Gaillard JM (2006) Estimating individual contributions to population growth: evolutionary fitness in ecological time. Proceedings of the Royal Society of London, Series B 273:547–555.

Ephraim Y, Merhav N (2002) Hidden markov processes. IEEE Transactions on Information Theory 48(6):1518–1569.

Foreman LA (1993) Generalization of the viterbi algorithm. Journal of Mathematics Applied in Business and Industry 4:351–367.

Gaillard JM, Andersen R, Delorme D, Linnell J (1998) Family effects on growth and survival of juvenile roe deer. Ecology 79:2878–2889.

Gaillard JM, Festa-Bianchet M, Delorme D, Jorgenson J (2000) Body mass and individual fitness in female ungulates: bigger is not always better. Proceedings of the Royal Society of London, Series B 267:471–477.

Gaillard JM, Sempéré AJ, Boutin JM, Van Laere G, Boisaubert B (1992) Effects of age and body weight on the proportion of females breeding in a population of roe deer (Capreolus capreolus). Canadian Journal of Zoology 70:1541–1545.

Guédon Y (1999) Computational methods for discrete hidden semi-markov chains. Applied Stochastics Models in Business and Industry 15:195–224.

Jensen H, Saether BE, Ringsby TH, Tufto J, Griffith SC, Ellegren H (2004) Lifetime reproductive success in relation to morphology in the house sparrow passer domesticus. Journal of Animal Ecology 73:599–611.

Käär P, Jokela J (1998) Natural selection on age-specific fertilities in human females: comparison of individual-level fitness measures. Proceedings of the Royal Society of London, Series B 265:2415–2420.

McGraw JB, Caswell H (1996) Estimation of individual fitness from life-history data. American Naturalist 147:47–64.

Newton I (1989) Lifetime Reproduction in Birds. Academic Press, London.

Pradel R (2005) Multievent: an extension of multistate capture-recapture models to uncertain states. Biometrics 61(2):442–447.

Rabiner LR (1989) A tutorial on hidden markov models and selected applications in speech recognition. Proceedings of the IEEE 77:257–286.

Section X
New Software Developments for Modeling Demographic Processes

Jim Hines

WinBUGS for Population Ecologists: Bayesian Modeling Using Markov Chain Monte Carlo Methods

Olivier Gimenez, Simon J. Bonner, Ruth King, Richard A. Parker, Stephen P. Brooks, Lara E. Jamieson, Vladimir Grosbois, Byron J.T. Morgan and Len Thomas

Abstract The computer package WinBUGS is introduced. We first give a brief introduction to Bayesian theory and its implementation using Markov chain Monte Carlo (MCMC) algorithms. We then present three case studies showing how WinBUGS can be used when classical theory is difficult to implement. The first example uses data on white storks from Baden Württemberg, Germany, to demonstrate the use of mark-recapture models to estimate survival, and also how to cope with unexplained variance through random effects. Recent advances in methodology and also the WinBUGS software allow us to introduce (i) a flexible way of incorporating covariates using spline smoothing and (ii) a method to deal with missing values in covariates. The second example shows how to estimate population density while accounting for detectability, using distance sampling methods applied to a test dataset collected on a known population of wooden stakes. Finally, the third case study involves the use of state-space models of wildlife population dynamics to make inferences about density dependence in a North American duck species. Reversible Jump MCMC is used to calculate the probability of various candidate models. For all examples, data and WinBUGS code are provided.

Keywords Bayesian statistics · Density dependence · Distance sampling · External covariates · Hierarchical modeling · Line transect · Mark-recapture · Random effects · Reversible jump MCMC · Spline smoothing · State-space model · Survival estimation

1 Introduction

The Bayesian approach dates back to the Reverend Thomas Bayes and the 18th century. However, due to practical problems of implementing the Bayesian approach, little advance was made for over two centuries. The development of new

O. Gimenez (✉)
CEFE/CNRS, UMR 5175, 1919, Route de Mende, 34293 Montpellier Cedex 5, France
e-mail: olivier.gimenez@cefe.cnrs.fr

D.L. Thomson et al. (eds.), *Modeling Demographic Processes in Marked Populations*, Environmental and Ecological Statistics 3, DOI 10.1007/978-0-387-78151-8_41,

methodology coupled with recent advances in computational power and the availability of flexible and reliable software have led to a great increase in the application of Bayesian methods within the last three decades, population ecology being no exception (Clark 2005; Ellison 2004; McCarthy 2007). Indeed, the application of the Bayesian theory in population ecology has been greatly facilitated by the implementation of algorithms known as Markov chain Monte Carlo (MCMC) methods (Gilks et al. 1996 and Link et al. 2002 for an introduction for ecologists) in flexible and reliable software. For example, MARK (White and Burnham 1999), one of the most popular computer programs in population ecology, now includes an MCMC option which implements a simple MCMC algorithm (White and Burnham this volume). AD Model Builder (ADMB; Fournier 2001) is a general modeling environment for fitting complex models to data, that has been used mainly in fisheries stock assessment (Maunder et al. submitted), and has an MCMC option to implement Bayesian analysis (see Maunder et al. this volume). Here we focus on the program WinBUGS (Bayesian inference Using Gibbs Sampling; Spiegelhalter et al. 2003), which implements up-to-date and powerful MCMC algorithms that are suited to a wide range of target distributions for analyzing complex models.

The paper is organized as follows. We first review the Bayesian framework and show how it can be fruitfully implemented using MCMC algorithms and program WinBUGS. We then focus on three case studies to illustrate how WinBUGS can be used to apply Bayesian methods using MCMC algorithms in population ecology. The first example deals with mark-recapture models to estimate survival probabilities and shows how to incorporate covariates with maximum flexibility. The second example shows how to estimate population density while accounting for detectability by using distance sampling methods. Finally, the third case study involves modeling count data using state-space models. We conclude with a short discussion of various possible extensions to both the methods and software that we have illustrated.

When presenting the examples, we include short illustrations in WinBUGS (`code is indicated using this typeface`). For all three examples, the relevant data and full WinBUGS code are given at http://eprints.st-andrews.ac.uk/archive/00000450/.

2 The Bayesian Method Using MCMC Algorithms: Practical Implementation in WinBUGS

Typical statistical problems involve estimating a vector of parameters, θ, using the available data. The classical approach assumes that the parameters are fixed, but have unknown values to be estimated. Classical maximum likelihood estimates generally provide a point estimate of the parameter of interest. The Bayesian approach assumes that the parameters themselves have some unknown distribution. The approach is based upon the idea that the experimenter begins with some prior beliefs about the system, and then updates these beliefs on the basis of observed data. Using Bayes' Theorem, the posterior distribution of the parameters given the data $\pi(\theta|data)$ has density proportional to the product of the likelihood of the data given the parameters $L(data|\theta)$ and the prior distribution of the parameters $\pi(\theta)$: $\pi(\theta|data) \propto L(data|\theta) \times \pi(\theta)$. The prior distribution represents the expert's belief, before observing any data. If there is no strong prior information on the parameters, vague priors are typically specified on the parameters which represent very weak opinion concerning the model parameters. Unfortunately, in most realistic applications the posterior distribution is generally of such high dimension that little useful inference can be obtained directly. As a consequence, while the joint posterior distribution (or the corresponding marginal distributions) provide the best summaries of the parameters, point estimates and uncertainty intervals are often more interpretable. It is the process of summarizing the posterior that is the source of the computational complexity of the Bayesian approach. Estimating the summary statistics of interest (for a vector of parameters θ) requires elimination of the other parameters. The Bayesian approach does this through integration using the MCMC algorithm. The high-dimensional integral associated with the posterior density is actually estimated using appropriate Monte Carlo integration, which consists of constructing a Markov chain with stationary distribution equal to the posterior distribution of interest. Then, once the chain has converged, realizations can be regarded as a dependent sample from this distribution. WinBUGS implements powerful ways of constructing these chains, adapting to a wide range of target (posterior) distributions and therefore allowing a large number of possible models to be fitted. Further details on Bayesian modeling using MCMC algorithms can be found in Gilks et al. (1996) and Congdon (2003, 2006). The WinBUGS software is currently freely available at http://www.mrc-bsu.cam.ac.uk/bugs/.

A typical WinBUGS session proceeds as follows: the user specifies the model to run in the form of the likelihood and prior distributions for all parameters to be estimated. Data and initial values must also be provided. Following the validation of the user specification, MCMC simulations are generated such that the stationary distribution of the Markov chain is the posterior distribution of interest. Thus, this algorithm provides a sample from the posterior distribution of interest from which, it is possible to produce estimates of the posterior distributions using kernel density estimates, and summary statistics of interest such as posterior medians and credible intervals. Convergence diagnostics are also available either directly in WinBUGS or using the R packages CODA (Plummer et al. 2004) or BOA (Smith 2004). Note that we will not discuss this crucial issue here, but recommendations can be found in Kass et al. (1998). An important feature of WinBUGS is that it comes with a tutorial designed to provide new users with a step-by-step guide to running an analysis in WinBUGS. There are also a wide range of varied and detailed examples, including, for instance: logistic regression with random effects, analyses of variance with repeated measurements, meta-analyses and survival analyses with frailties. It is often useful to call WinBUGS from other programs in order to input complex sets of data and initial values, avoid specifying the parameters to be monitored in each run, post-process the results in other software, display complex graphics or perform Monte Carlo studies running WinBUGS iteratively in a loop. Together with data and WinBUGS codes, we give an illustration of the use of the R (Ihaka and Gentleman 1996; R Development Core Team 2007) package R2WinBUGS (Sturtz et al. 2005), as well as an illustration of how to call WinBUGS from MATLAB using the package MATBUGS (http://www.cs.ubc.ca/~murphyk/Software/MATBUGS/matbugs.html) at http://eprints.st-andrews.ac.uk/archive/00000450/. Other programs that can be used to interface to WinBUGS are listed on the WinBUGS web page given above. General and complementary introductions to WinBUGS are given in Congdon (2006) and McCarthy (2007). We now turn to the analysis of real case studies to illustrate the use of WinBUGS. Note that likelihoods and priors are implemented by defining their probability distribution based on the model parameters using the tilde ($\sim$) symbol. This notation will be used throughout the paper.

3 Estimating Survival Using Mark-Recapture Data

As an illustration, we use data on the white stork *Ciconia ciconia* population in Baden Württemberg (Germany), consisting of 321 capture histories of individuals ringed as chicks between 1956 and 1971. From the 60 s to the 90 s, all Western European stork populations were declining (Bairlein 1991). This trend is thought to be the result of reduced food availability (Schaub et al. 2005) caused by severe droughts observed in the wintering ground of storks in the Sahel region of Africa. This hypothesis has been examined in several studies (Kanyamibwa et al. 1990; Barbraud et al. 1999; Grosbois et al. in revision). In this section, we use WinBUGS and several of its features to further explore the relationship between rainfall in the Sahel and survival probabilities of the Baden Württemberg white stork population.

3.1 Simple Models

The standard Cormack–Jolly–Seber model (CJS, Cormack 1964; Jolly 1965; Seber 1965; Lebreton et al. 1992) considers time-dependence for the probability ϕ_i that an individual survives to occasion $i + 1$ given that it is alive at time i, and for the probability p_j that an individual is recaptured at time j. The data consist of encounter histories for each individual made of 1's corresponding to recapture or resighting and 0's otherwise. These data can be efficiently condensed in the so-called reduced m-array (e.g. Lebreton et al. 1992) which summarizes the data in the form of the number of individuals released per occasion i, denoted R_i, and the number of first recaptures given release at occasion i at the succeeding occasions j, denoted m_{ij}. The m-array for the white stork data is provided in Table 1.

Conditioning on the numbers released and assuming independence among cohorts, the CJS model likelihood can be written as a product of multinomial probability distributions corresponding to each row of the m-array. The probabilities corresponding to the m-array cells are complex nonlinear functions of the survival and detection probabilities. For example, the probability of the number of individuals released at occasion 3 and recaptured for the first time at occasion 5, given the number of released individuals at occasion 3 is:

Table 1 The m-array for the White stork data set. The number of individuals released at occasion i (R_i) and the number of first recaptures at occasion j, given release at occasion i (m_{ij}) are provided. For example, 38 birds were released in 1969 among which, 22 were first recaptured in 1970, and 16 (= 38–22) were never observed again

Year of release (19-)	Number released	Year of first recapture (19-)															
		57	58	59	60	61	62	63	64	65	66	67	68	69	70	71	72
56	26	19	2	0	0	0	0	0	0	0	0	0	0	0	0	0	0
57	50	0	33	3	0	0	0	0	0	0	0	0	0	0	0	0	0
58	53	0	0	35	4	0	0	0	0	0	0	0	0	0	0	0	0
59	69	0	0	0	42	1	0	0	0	0	0	0	0	0	0	0	0
60	73	0	0	0	0	42	1	0	0	0	0	0	0	0	0	0	0
61	71	0	0	0	0	0	32	2	1	0	0	0	0	0	0	0	0
62	64	0	0	0	0	0	0	46	2	0	0	0	0	0	0	0	0
63	64	0	0	0	0	0	0	0	33	3	0	0	0	0	0	0	0
64	66	0	0	0	0	0	0	0	0	44	2	0	0	0	0	0	0
65	55	0	0	0	0	0	0	0	0	0	43	1	0	0	1	0	0
66	60	0	0	0	0	0	0	0	0	0	0	34	1	0	0	0	0
67	53	0	0	0	0	0	0	0	0	0	0	0	36	1	0	0	0
68	51	0	0	0	0	0	0	0	0	0	0	0	0	27	2	0	0
69	38	0	0	0	0	0	0	0	0	0	0	0	0	0	22	0	0
70	33	0	0	0	0	0	0	0	0	0	0	0	0	0	0	15	1
71	23	0	0	0	0	0	0	0	0	0	0	0	0	0	0	0	15

$$\phi_3(1 - p_4)\phi_4\, p_5. \tag{1}$$

For further details of fitting the CJS model in a Bayesian framework, see Brooks et al. (2000). We start with a simple mark-recapture model, a simplification of the CJS model where, based on the conclusions of previous studies (Kanyamibwa et al. 1990; Grosbois et al. in revision), the recapture probabilities are considered constant over time.

3.1.1 Defining Priors

We define priors for the survival probabilities and the recapture probability as Beta distributions with parameters 1 and 1 (equivalently uniform distributions between 0 and 1). Within WinBUGS, this is specified as:

```
for(i in1:ni){phi[i] ~ dbeta(1,1)}
p ~ dbeta(1,1)
```

where `ni` is the number of occasions of release in the study.

3.1.2 Constructing the Likelihood

The likelihood is defined as a product of multinomial distributions using the function `dmulti`:

```
for(i in1:ni){m[i,1:(nj+1)] ~ dmulti(q[i,],r[i])}
```

where the `m` object is the m-array matrix of data (augmented by the number of individuals never seen again after release in the last column), `nj` is the number of recapture occasions within the study, `r` is the vector of released individuals and `q` is a matrix of the m-array cells probabilities. The `q` matrix and `r` vector are calculated in the WinBUGS code.

3.1.3 Results

The posterior medians of the survival probabilities are displayed in Fig. 1a, along with their posterior 95% credible intervals.

To check that the temporal variations in the survival are worth considering, we also consider a compromise approach in which survival is taken as constant over time. Starting from the code of the previous model, one way to proceed would be to consider one scalar parameter for the survival, specify the prior distribution as for the detection probability and modify the likelihood accordingly. A neat trick which avoids modifying the likelihood part of the code, is to define a single dummy variable with a Beta prior and then set all survival probabilities equal to that variable:

```
#U(0,1)prior distribution for dummy variable
constant.phi~dbeta(1,1)
#All survival probabilities equal to dummy variable
for(i in1:ni){phi[i] <- constant.phi}
```

3.1.4 DIC for Model Selection

As a preliminary model selection technique, we use the Deviance Information Criterion (DIC; Spiegelhalter et al. 2002). One interpretation of the DIC is as a Bayesian counterpart to the AIC for model selection. Essentially, the DIC is a diagnostic that balances the requirements of model fit and low complexity. Typically, as models get more complex by the addition of extra parameters, their fit improves. The DIC diagnostic therefore penalizes additional parameters so that a parsimonious model is chosen, and the smaller the DIC value, the better the compromise is. One advantage is that the DIC can be calculated directly in WinBUGS from the chains produced by an MCMC run. However, the DIC statistic is in its infancy and is controversial (see the discussion papers following Spiegelhalter et al. 2002 and Celeux et al. 2006). Here we consider the DIC as a preliminary tool for comparing competing models, and we will discuss a more rigorous approach later, in the form of posterior model probabilities.

Examining the DIC values in Table 2, we see that the time-dependent model appears to outperform the constant model, and hence is better supported by the data. This suggests that dependence upon time is needed to explain variations in the survival probabilities. To better understand these findings, we will consider in the next section environmental covariates as possibly explaining time variation in the survival probabilities.

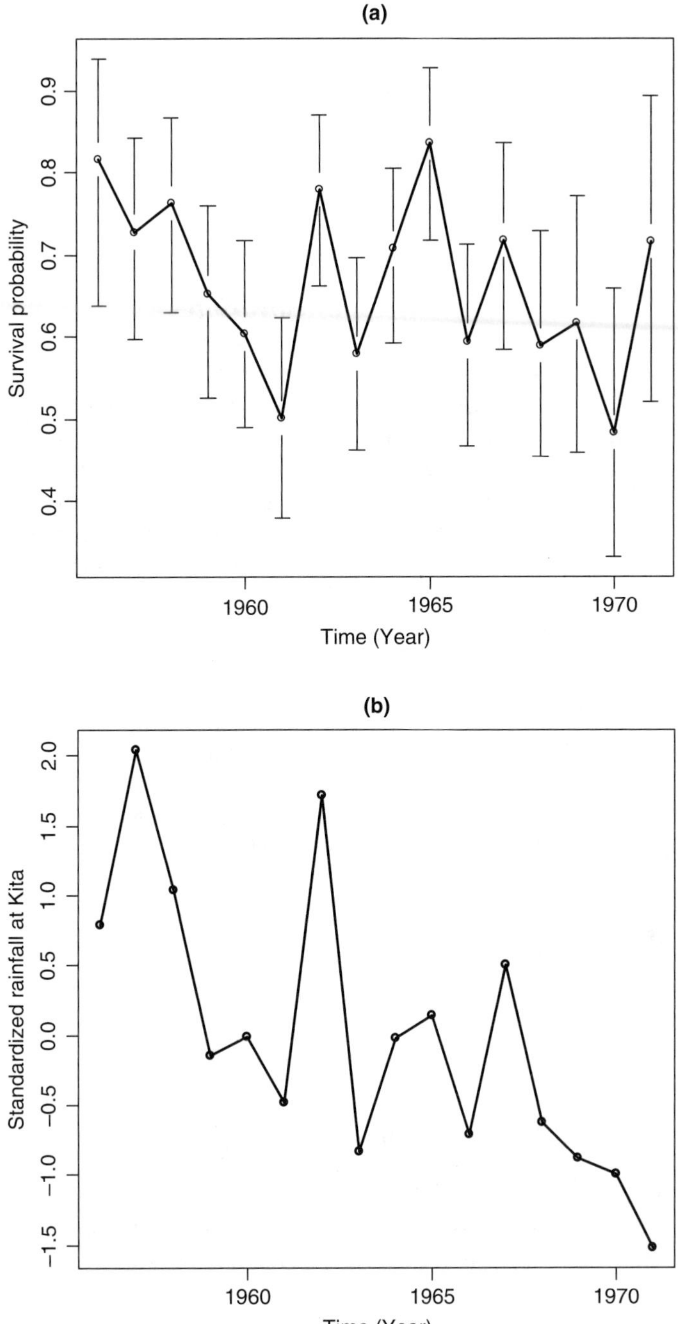

Fig. 1 (**a**) White stork survival estimates from model with time-dependent survival probabilities and constant detection probabilities; vertical bars represent 95% pointwise credible intervals; (**b**) rainfall time series at meteorological station Kita in the Sahel

Table 2 Models fitted to the white stork data. DIC is the deviance information criterion and pD is the number of effective parameters. See text for details

Model	DIC	pD
Constant survival probabilities	174.3	1.9
Time-dependent survival probabilities	166.0	16.3
Covariate-dependent survival probabilities	159.4	3.1
Covariate-dependent as well as random-effect survival probabilities	161.0	10.6
Nonparametric survival probabilities	158.1	7.4

3.2 Incorporating Linear Effects of Covariates

We now turn to the incorporation of covariates in the CJS model (North and Morgan 1979; Pollock et al. 1984; Clobert and Lebreton 1985; Lebreton et al. 1992; see Pollock 2002 for a review). As we mentioned earlier, the variation in white storks survival is likely to be related to rainfall variations. As expected, it can be seen that the variations in the survival estimates (Fig. 1a) are correlated to Sahel rainfall variations (Fig. 1b). According to Williams et al. (2002, p. 373), we therefore consider a model including a linear effect of the rainfall covariate on the logit scale:

$$\text{logit}(\phi_i) = \log\left(\frac{\phi_i}{1-\phi_i}\right) = \beta_1 + \beta_2 x_i, \qquad (2)$$

where x_i is the value of the covariate between occasions i and $i+1$, and the β 's are regression parameters to be estimated. We use normal distributions with mean 0 and large variance (10^6) as vague prior distributions for those parameters. The rainfall measurements are standardized to improve mixing within the Markov chain. Note that the standardization can be implemented in WinBUGS:

```
for(i in1:ni){cov[i] <- (cov[i] -mean(cov[]))/sd(cov[])}
```

where `cov[i]` denotes the covariate value in year i.

The code provided in the previous section is amended as follows:

```
for(i in1:ni){logit(phi[i]) <-beta[1] +beta[2] * cov[i]}

for(j in1:2){beta[j] ~dnorm(0,1.0E-6)}
```

Note that in WinBUGS, normal distributions are described in terms of a mean and precision, where precision = 1/variance. As a consequence, a variance of 1,000,000 corresponds to a precision of 0.000001. In addition, we note that this model makes the strong assumption that variation in the survival probabilities is explained by the covariate. This can be relaxed by the inclusion of additional random effects.

3.3 Incorporating Random Effects

We consider two models with random effects in this section, both addressing two different questions. Note that incorporating random effects is also a way to share information among parameters, particularly improving estimates for years where there is little information in the data (e.g. Harley et al. 2004).

First, specifying constant survival probabilities can be too restrictive to capture sources of temporal variability, while estimating as many parameters as time intervals may be too costly to assess specific time trends (Burnham and White 2002; Royle and Link 2002). We consider a compromise model where time is treated as a random effect, ε, with a normal distribution with mean 0 and variance σ^2. We therefore estimate the mean logit survival probability, say μ, and the temporal process variance in survival probability σ^2 (Gould and Nichols 1998; Burnham and White 2002):

$$\text{logit}(\phi_i) = \mu + \varepsilon_i. \tag{3}$$

Considering random effects raises the problem of calculating the likelihood, which is obtained by integrating over the random effect ε. This is, indeed, a problem involving a high-dimensional integral that could be handled by using approximations (Chavez-Demoulin 1999), circumvented by resorting to asymptotic arguments (Gould and Nichols 1998; Burnham and White 2002), or numerical integration (e.g. importance sampling: Skaug and Fournier 2006 or Gaussian quadrature: Wintrebert et al. 2005). By contrast, the Bayesian approach provides an exact solution to this problem (Brooks et al. 2000, 2002, note that both references contain WinBUGS code) and WinBUGS offers a powerful and flexible alternative to standard software such as MARK (White and Burnham 1999) or M-SURGE (Choquet et al. 2005).

The specification of the model for the survival probabilities was as follows:

```
for(i in 1:ni){
logit(phi[i]) <- logitphi[i]
logitphi[i] ~ dnorm(mu,taueps)
}
```

We consider an inverse-gamma distribution with parameters 0.01 and 0.01 and a normal distribution with mean 0 and large variance (100) as vague prior distributions for taueps and respectively mu:

```
taueps ~ dgamma(0.01,0.01)
mu ~ dnorm(0,0.01)
```

Note that a gamma distribution for the precision is equivalent to an inverse-gamma distribution for the variance. In this case, these are typical specifications of vague priors (see also Lambert et al. 2005; van Dongen 2006; Gelman 2006). The posterior distribution of the variance can easily be obtained by monitoring the quantity `sigma2eps` defined as:

```
sigma2eps <- 1/taueps
```

Second, the inclusion of random effects allows there to be additional variability within the survival rates that can be attributed to natural variability, or temporal variability not explained by the covariates within the study. This is a simple extension of the above covariate model. In particular, we specify an additional random effect term denoted by ε, which has a normal distribution with mean 0 and variance σ^2. In particular we model the survival rate to be of the form:

$$\text{logit}(\phi_i) = \beta_1 + \beta_2 x_i + \varepsilon_i. \tag{4}$$

Then, the parameters to be estimated are the regression coefficients (β's) and the random effect variance parameter σ^2. In a particular application, Barry et al. (2003) noticed that omitting the random effect can lead to overestimation of the significance of the covariate on survival. To include these additional random effects, the code is modified as follows:

```
for(i in1:ni){
   logit(phi[i]) <- logitphi[i]
   logitphi[i]~dnorm(f[i],taueps)
   f[i] <- beta[1] + beta[2] * cov[i]
}
taueps~dgamma(0.01,0.01)
```

In our model, slope estimates produced using Eqs. (2) and (4) to model survival are very close to each other: posterior medians for the slope β_2 were 0.36 in both cases with 95% credible intervals [0.14; 0.58] and [0.20; 0.55] (see Fig. 2). This may indicate that the random effect was not needed in the model, as the estimates tend to confirm (the distribution of σ^2 places all of its mass near 0 with posterior median 0.04 and 95% credible interval [0.01; 0.22]), and indicated by the preliminary DIC analysis (see Table 2).

A formal way of testing the null hypothesis $\sigma^2 = 0$ will be discussed later. In both cases, the effect of rainfall is positive, indicating that the more it rained in the Sahel zone, the better storks survived.

3.4 Nonparametric Modeling

There is another strong assumption made in Eq. (2), namely that the effect of the covariate on the survival probability is linear on the logit scale. However, nonlinear relationships involving the impact of environmental factors on population dynamics may occur (Mysterud et al. 2001). More flexible models for the survival probability are therefore needed. Gimenez et al. (2006a; see also Gimenez and Barbraud this volume and Gimenez et al. 2006b for a similar approach applied to individual covariates) have recently proposed a method in which the shape of the relationship is determined by the data without making any prior assumption regarding its form, by using penalized splines (P-splines; Ruppert et al. 2003). Here,

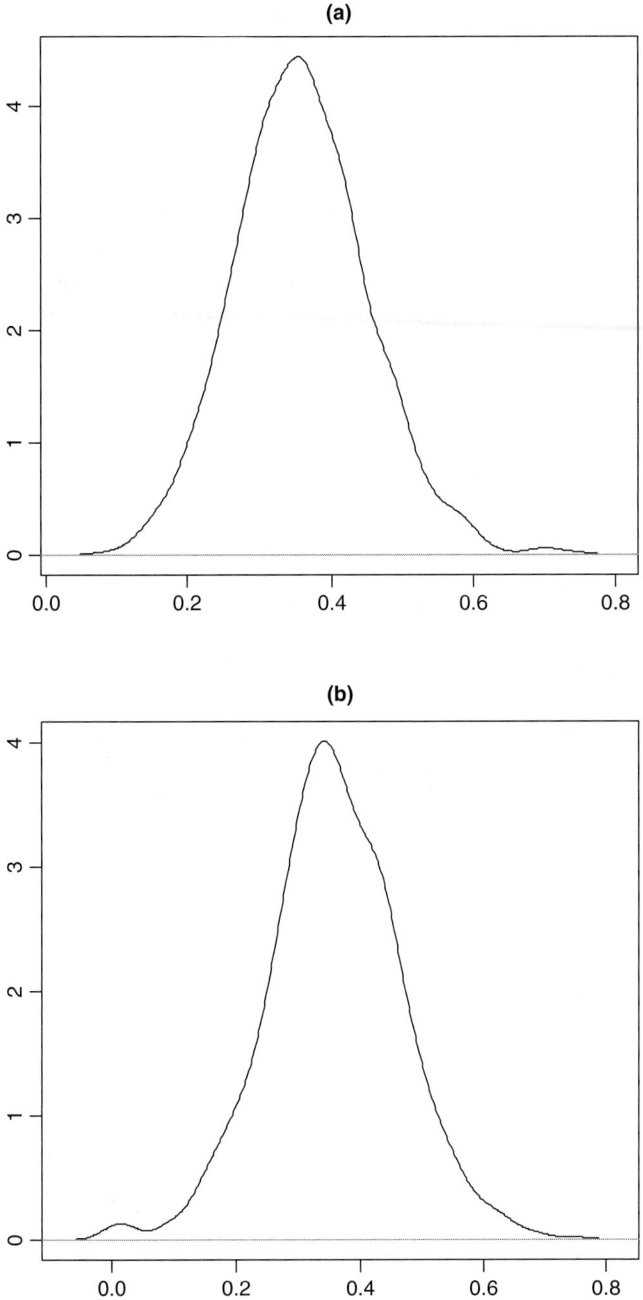

Fig. 2 Posterior distributions of the regression parameter β_2 corresponding to the rainfall effect on annual white stork survival (**a**) without and (**b**) with a random effect

we give details of how to implement their approach in WinBUGS. We consider the following regression model for the survival probability ϕ_i:

$$\text{logit}(\phi_i) = f(x_i) + \varepsilon_i, \tag{5}$$

where x_i is the value of the covariate between occasions i and $i+1$, f is a smooth function and ε_i are i.i.d. random effects $N\left(0, \sigma_\varepsilon^2\right)$. The function f specifies a nonparametric flexible relationship between the survival probability and the covariate that allows nonlinear environmental trends to be detected. Following Gimenez et al. (2006a), we use a truncated polynomial basis to describe f:

$$f(x) = \beta_0 + \beta_1 x + \ldots + \beta_p x^p + \sum_{k=1}^{K} b_k (x - \kappa_k)_+^p, \tag{6}$$

where x is the covariate, and $\beta_0, \beta_1, \ldots, \beta_p, b_1, \ldots, b_K$ are regression coefficients to be estimated, $P \geq 1$ is the degree of the spline, $(u)_+^p = u^p$ if $u \geq 0$ and 0 otherwise, and $\kappa_1 < \kappa_2 < \ldots < \kappa_K$ are fixed knots. We use $K = \min\left(\frac{1}{4}I, 35\right)$ knots to ensure the desired flexibility, and let k_k be the sample quantile of x's corresponding to probability $\frac{k}{K+1}$. Those quantities are calculated outside WinBUGS in program R. In particular, we model the relationships using a linear ($P = 1$) P-spline with $K = 4$ knots implemented through the WinBUGS constants `degree and nknots`. To avoid overfitting, we penalize the b's by assuming that the coefficients of $(x - \kappa_k)_+^P$ are normally distributed random variables with mean 0 and variance σ_b^2 to be estimated. This is the reason why this approach is referred to as penalized splines (Ruppert et al. 2003). Note that an alternative to P-splines called adaptive splines (Biller 2000) is considered in the mark-recapture context by Bonner et al. (this volume). The penalization is achieved by specifying:

```
for(k in 1:nknots){b[k] ~ dnorm(0,taub)}
```

where the variance parameter is given an inverse-gamma distribution (i.e. the precision has a gamma distribution):

```
taub ~ dgamma(0.001, 0.001)
```

A by-product of this approach is that the amount of smoothing is automatically calculated as $\sigma_b^2/\sigma_\varepsilon^2$. To implement the P-splines model in WinBUGS, it is convenient to express it as a Generalized Linear Mixed Model (GLMM), as shown by Crainiceanu et al. (2005). If X is the matrix with ith row $X_i = \left(1, x_i, \ldots, x_i^P\right)^T$ and Z the matrix with the ith row $Z_i = \left((x_i - \kappa_1)_+^P, \ldots, (x_i - \kappa_K)_+^P\right)^T$, then an equivalent model representation of Eqs. (5) and (6) in the form of a GLMM is given by Gimenez et al. (2006a):

$$\text{logit}(\phi) = X\beta + Zb + \varepsilon, \text{cov}\begin{pmatrix} b \\ \varepsilon \end{pmatrix} = \begin{pmatrix} \sigma_b^2 I & 0 \\ 0 & \sigma_\varepsilon^2 I \end{pmatrix} \tag{7}$$

We are now able to implement the P-splines model in WinBUGS. To code Eq. (7), we used:

```
for(i in 1:n){
   logit(phi[i]) <- logitphi[i]
   logitphi[i] ~ dnorm(f[i],taueps)
   f[i] <- inprod(beta[],X[i,]) + inprod(b[],Z[i,])
}
```

The first statement corresponds exactly to Eq. (6), the second implements the random effects distribution and the last one specifies the structure of the mean logit survival, where the function inprod denotes the inner product of two vectors. The first part of the last statement contains the fixed effect of Eq. (7), where `beta[]` is the vector $\beta = (\beta_0, \ \beta_1, \beta_2)$, `X[i,]` is X_i and inprod`(beta[],X[i,])` is the polynomial part. The second part of the last statement contains the random effects, where `b[]` is the vector $b = (b_1, b_2, b_3, b_4)$, `Z[i,]` is Z_i and inprod`(b[],Z[i,])` is the truncated polynomial part of the regression in Eq. (7).

We then obtain matrices X and Z directly in WinBUGS, although this step could be done in program R for example. Matrix X is obtained as:

```
for(i in 1:n){
   for(l in 1:degree+1){
    X[i,l] <- pow(covariate[i],l-1)
   }
}
```

where pow is the power function, and `pow(a,b)` is a^b. Matrix Z is obtained using:

```
for(i in 1:n){
   for(k in 1:nknots){
     Z[i,k] <- pow((covariate[i]-knot[k]) *
      step(covariate[i]-knot[k]),degree)
   }
}
```

where the function step is used to obtain the truncation, where `step(x)` is 1 if x is positive and 0 otherwise, so that `Z[i,k]` is positive only for $x_i > \kappa_k$. For further details see Crainiceanu et al. (2005) and Gimenez et al. (2006a). With the possibility of fitting nonparametric models, one is obviously interested in testing for the presence of nonlinearities in the survival probability regression. We address this question by using the DIC and also using visual comparison for comparing the model with a linear effect of rainfall as well as a random effect (see previous section)

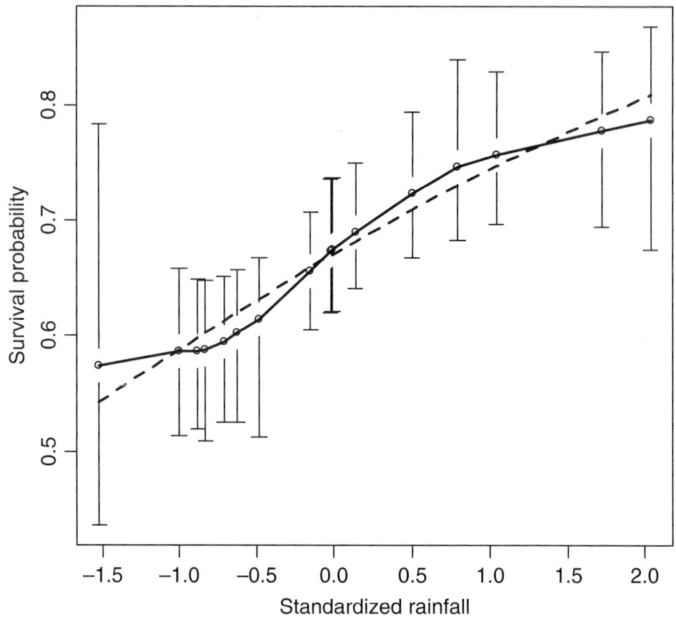

Fig. 3 Annual variations in white stork survival as a function of the standardized rainfall using a nonparametric model. Medians (*solid line*) with 95% pointwise credible intervals (*vertical solid lines*) are shown, along with the estimated linear effect (*dotted line*)

to its nonparametric counterpart. Figure 3 shows that the relationship between rainfall in Sahel and white stork survival can be taken as linear. This is confirmed by DIC values that are similar for these two models (Table 2). Although we have clues for linearity in this example, the issue of formally detecting nonlinearity deserves further investigation.

3.5 Dealing with Missing Data

Bayesian modeling via MCMC also provides a simple method for handling data with missing covariate values. Missing data might occur in capture–recapture studies if the value of an environmental covariate is not recorded on all occasions or if an individual covariate changes over time and can only be observed on the occasions when the specific animal is captured (Bonner and Schwarz 2006). Essentially, a completed data set is generated on each iteration of the MCMC algorithm by specifying an underlying model for the covariate and imputing the missing values of the covariate using the current values of the parameters, and then the completed data set is used to update the parameter values. The result is a sample from the joint posterior distribution of both the parameters and the missing data values, which can be used in Bayesian inference. We illustrate the issue of dealing with missing data by estimating the effect of rainfall in the Sahel on the survival of the white storks in Baden Württemberg after deleting the covariate for several years.

As with the model incorporating random effects, computing the value of the likelihood for a given set of parameter values requires integration with respect to the missing covariates. This can be a complicated numerical problem, especially if several values are missing, and is an obstruction to computing maximum likelihood estimates and their standard errors. From a Bayesian perspective, we view the missing covariates as random variables to which we can assign a probability distribution, just like the model parameters. We define a prior distribution for the missing covariate values and then compute the posterior distribution of both the parameters and the missing values conditional on the observed data. The likelihood function used in the analysis is exactly the same function used when all covariate values are observed, and if MCMC is used to obtain a sample from the posterior distribution then no additional integration is required. Instead, a sample of probable values for the missing covariates is generated by sampling new values on each iteration of the MCMC algorithm in exactly the same way that model parameters are sampled. The prior distribution of the missing covariate can be chosen to capture prior beliefs about the values of the missing covariates and their relation to the rest of the data. A simple, vague prior for the rainfall in year i, x_i, is the normal distribution with mean 0 and large variance $x_i \sim N\left(0, 10^6\right)$. This prior distributes its mass evenly over a very wide range of values and assumes independence of the rainfall across the years of the study. Alternative prior distributions will relate the values of the covariates to each other or to other quantities. Here we use a hierarchical prior that models the change in the covariate over time as $x_i \sim N\left(x_{i-1} + \mu, \sigma_x^2\right)$. This asserts that the change in the covariate between adjacent years is normally distributed with the same mean and variance for all years. Information from the observed covariate values will then be used in determining the posterior mean and variance of the missing values. To complete the prior distribution we must also specify the marginal distribution of the first covariate value, x_1, and the distributions for the hyperparameters, μ and σ_x^2. Here we use the vague prior $x_1 \sim N(0, 10^6)$ for marginal prior of the first covariate, and the standard vague priors for a normal mean and variance: $\mu \sim N(0, 10^6)$ and $\sigma_x^2 \sim \Gamma^{-1}(0.01, 0.01)$. Alternate prior specifications for the covariate values include autoregressive models, regression of the covariate against time, or relation of the covariate to other variables that might have been recorded. Adapting the WinBUGS code to account for the missing covariate values requires two simple changes: (i) adding the prior distribution for the covariates, and (ii) modifying the input data. The WinBUGS code for the hierarchical prior is:

```
mu~dnorm(0,1.0E-6)
taucov~dgamma(.01,.01)
sigma2cov<-1/taucov
cov[1]~dnorm(0,1.0E-6)
for(i in1:(ni-1)){
```

```
    mucov[i] <- cov[i] + mu
    cov[i+1] ~ dnorm(mucov[i], taucov)
}
```

The first three lines of code define the hyperpriors for the hyperparameters (μ is mu and σ_x^2 is sigma2cov). The 4th line defines the marginal prior for x_1 and the for loop defines the distribution of each of the remaining covariate values conditional on the previous value (x_i is cov[i]). Missing values in the input data are specified by replacing the observed value with 'NA'. Suppose that the rainfall is observed in all years except year 15; the input vector for the covariate is:

```
cov = c(.79,2.04,1.04,-.15,-.01,-.48,1.72,-.83,-.02,.14,
        -.71,.50,-.62,-.88,NA,-1.52)
```

Given this data and the model above, WinBUGS will simulate values for the hyperparameters and the missing rainfall observation for year 15 on each MCMC iteration and produce posterior summaries for these quantities, exactly as it does for the other model parameters. Posterior summary statistics for a single run are shown in Table 3.

Estimates of the survival probabilities are almost exactly identical to the estimates produced from the full data; differences in the posterior means and standard

Table 3 Summary statistics for the posterior distributions of the model fitted to the white stork data with survival as a function of rainfall: no missing value, missing value in 1 year (15), missing value in 5 years (5, 6, 11, 12, 13). Reported statistics are the estimated mean, standard deviation (SD), and the 95% credible interval [CI]

Parameter	No missing value Post. mean (SD) [CI]	1 missing value Post. mean (SD) [CI]	5 missing values Post. mean (SD) [CI]
ϕ_1	0.74 (0.04) [0.65;0.83]	0.74 (0.05) [0.66;0.85]	0.74 (0.05) [0.65;0.84]
ϕ_2	0.79 (0.05) [0.69;0.86]	0.79 (0.04) [0.70;0.86]	0.77 (0.05) [0.67;0.86]
ϕ_3	0.75 (0.04) [0.67;0.82]	0.75 (0.04) [0.66;0.82]	0.74 (0.04) [0.66;0.82]
ϕ_4	0.65 (0.04) [0.58;0.73]	0.66 (0.04) [0.58;0.74]	0.66 (0.04) [0.58;0.74]
ϕ_5	0.64 (0.04) [0.56;0.71]	0.64 (0.04) [0.56;0.71]	0.62 (0.05) [0.52;0.72]
ϕ_6	0.57 (0.05) [0.46;0.65]	0.57 (0.05) [0.48;0.65]	0.57 (0.05) [0.46;0.67]
ϕ_7	0.79 (0.04) [0.70;0.86]	0.78 (0.04) [0.70;0.85]	0.77 (0.04) [0.68;0.85]
ϕ_8	0.59 (0.04) [0.51;0.67]	0.59 (0.04) [0.51;0.68]	0.60 (0.05) [0.51;0.69]
ϕ_9	0.69 (0.04) [0.61;0.76]	0.69 (0.04) [0.61;0.77]	0.69 (0.04) [0.62;0.77]
ϕ_{10}	0.73 (0.05) [0.65;0.83]	0.73 (0.05) [0.65;0.83]	0.74 (0.05) [0.66;0.84]
ϕ_{11}	0.61 (0.04) [0.51;0.69]	0.61 (0.04) [0.53;0.70]	0.63 (0.05) [0.52;0.72]
ϕ_{12}	0.71 (0.04) [0.64;0.79]	0.71 (0.04) [0.63;0.79]	0.69 (0.05) [0.58;0.80]
ϕ_{13}	0.61 (0.05) [0.51;0.69]	0.61 (0.05) [0.51;0.70]	0.62 (0.05) [0.51;0.71]
ϕ_{14}	0.60 (0.05) [0.51;0.70]	0.60 (0.05) [0.51;0.70]	0.62 (0.05) [0.53;0.72]
ϕ_{15}	0.56 (0.05) [0.45;0.65]	0.53 (0.07) [0.39;0.65]	0.57 (0.06) [0.44;0.67]
ϕ_{16}	0.57 (0.07) [0.45;0.72]	0.58 (0.06) [0.46;0.71]	0.61 (0.07) [0.49;0.75]
β_1	0.70 (0.09) [0.52;0.89]	0.71 (0.10) [0.53;0.91]	0.74 (0.10) [0.53;0.95]
β_2	0.36 (0.11) [0.14;0.58]	0.35 (0.11) [0.14;0.56]	0.27 (0.13) [-0.03;0.51]
μ	–	−0.15 (0.35) [−0.85;0.56]	−0.16 (0.38) [−0.91;0.64]
$\sigma^2{}_x$	–	1.73 (0.84) [0.80;3.75]	2.03 (1.17) [0.75;5.10]
σ^2	0.06 (0.06) [0.01;0.22]	0.06 (0.06) [0.01;0.21]	0.07 (0.07) [0.01;0.27]
p	0.91 (0.01) [0.88;0.94]	0.91 (0.01) [0.88;0.94]	0.91 (0.01) [0.88;0.94]

deviations are the magnitude as the MCMC error. There is a very slight increase in the posterior variability of the regression coefficients, β_1 and β_2, however the lower bound of the 95% credible interval for β_2 is still well above 0 indicating a clear positive link between rainfall and the storks' survival. The estimated mean change in rainfall is $-.15$ with 95% credible interval $(-.85,.56)$ which suggests that there is no consistent trend over time. The standard deviation of the change in rainfall is relatively large which indicates that there is little association between rainfall in adjacent years. Because of this, the posterior distribution for rainfall in the missing year is uninformative about the true value.

When 5 missing values are generated, there are only minor differences in the posterior distribution of the survival probabilities with small increases in the standard deviation apparent for the years with the covariate deleted (Table 3). This is not surprising because the capture probabilities are very high so that most information about the survival probabilities is derived from a direct comparison of the capture histories rather than the regression on the covariate. There is, however, significant change in the inference for the regression coefficients. The posterior mean of the slope, β_2, is closer to 0 in Table 3, though whether the mean is increased or decreased depends on which years are missing the covariate. More importantly, the posterior standard deviation is increased from 0.11 to 0.13 and the 95% credible interval contains 0 which brings the effect of rainfall on survival into doubt.

To close this section, we note that we have only considered rainfall at a single meteorological station in the Sahel region. However, rainfall measurements at other stations are available, therefore possibly providing a better spatial representation of the white storks' wintering area. The question is then to determine which combination of the stations best explains the variation in survival. If we have 10 stations, we need to perform model selection among a set of 1024 (2^{10}) possible candidates, which would be intractable using classical model selection criteria such as AIC, BIC or DIC. Fortunately, an alternative method can be used that allows model selection among a large set of candidate models. An example is given later in the section dealing with state-space modeling of count data, and we have made available the WinBUGS code to implement this approach on the stork dataset.

4 Estimating Abundance and Population Density Using Line-Transect Data

Line transect surveys are widely used to estimate the density and/or abundance of wildlife populations. The methods, which are a special case of a general approach called distance sampling, are described in detail, from a classical perspective, by Buckland et al. (2001, 2004a). Observers walk along a set of randomly located transect lines recording the perpendicular distance to all detected objects of interest (usually animals) within some detected with some perpendicular truncation distance w. Not all objects within distance w are assumed to be detected; rather a (semi-) parametric model is specified for the probability of detecting an object given it is at perpendicular distance y from the transect line. Under various assumptions (detailed in Buckland et al. 2001), it is then possible to derive the probability density function $f(y)$ of observed distances. This can be fitted to the observed distance data using maximum likelihood methods, and used to correct for the objects missed during the survey. The standard formula for estimating object density, D, is (Buckland et al. 2001):

$$\hat{D} = \frac{n\hat{f}(0)}{2L} \tag{8}$$

where n is the number of objects detected, L is the total length of the transect lines and $\hat{f}(0)$ is the estimated probability density function of observed distances evaluated at zero distance.

As an illustration, we consider a line transect study where a known number of wooden stakes are placed in a sagebrush meadow east of Logan, Utah (Buckland et al. 2001). The true density of stakes is known to be 37.5 stakes/hectare. Eleven different graduate students walked a 1,000 m long transect through the study area independently of one another and recorded perpendicular sighting distances to stakes. One student's data are given in Table 4.

These data consist of 68 observations with a truncation width, w, of 20 m. The same data set is analyzed by Karunamuni and Quinn (1995) who propose a Bayesian approach for line transect sampling. For the sake of simplicity, we make the same assumptions as Karunamuni and Quinn (1995), i.e. we assume that the probability density function $f(y)$ for the detection distances is half-normal and that the data are neither truncated nor grouped into distance intervals (see Buckland et al. 2001 for more on the latter). Thus,

$$f(y) = \sqrt{2/\pi\sigma^2}\exp\left(-y^2/2\sigma^2\right) = c\sqrt{\lambda}\exp\left(-\lambda y^2/2\right), y > 0 \tag{9}$$

where $c = \sqrt{2/\pi}$ and $\lambda = 1/\sigma^2$. Given n detection values, $y_1, \ldots y_n$, the maximum likelihood estimator of $f(0)$ is then given by:

$$\hat{f}(0) = \sqrt{\frac{2n}{\pi\sum y_i^2}} = c\left(\frac{T}{n}\right)^{-\frac{1}{2}} \tag{10}$$

where $T = \sum y_i^2$. The maximum likelihood estimator of the density is given by Eq. (8), above.

Table 4 Sequence of perpendicular distance values for the Stakes line transect example (in meters)

2.02	0.45	10.40	3.61	0.92	1.00	3.40	2.90	8.16	6.47
5.66	2.95	3.96	0.09	11.82	14.23	2.44	1.61	31.31	6.50
8.27	4.85	1.47	18.60	0.41	0.40	0.20	11.59	3.17	7.10
10.71	3.86	6.05	6.42	3.79	15.24	3.47	3.05	7.93	18.15
10.05	4.41	1.27	13.72	6.25	3.59	9.04	7.68	4.89	9.10
3.25	8.49	6.08	0.40	9.33	0.53	1.23	1.67	4.53	3.12
	3.05	6.60	4.40	4.97	3.17	7.67	18.16	4.08	

Adopting a Gamma prior distribution with parameters a and b for λ, Karunamuni and Quinn (1995) show that the posterior distribution of λ is also a Gamma distribution with parameters $a+\frac{n}{2}$ and $\left(\frac{1}{b}+\frac{T}{2}\right)^{-1}$. Although classical Monte Carlo simulations could be used to simulate observations from the posterior distribution of λ, we use WinBUGS to draw random samples using MCMC techniques. This is motivated by generalizations to other probability density functions for the detection distances as well as spatial modeling for which explicit posterior distributions are difficult to obtain. We use the so-called "zeros trick" to implement the half-normal likelihood distribution because it is not included in the list of standard WinBUGS sampling distributions. This method consists of considering an observed data set made of 0's distributed as a Poisson distribution with parameter ϕ so that the associated likelihood is $\exp(-\phi)$. Now, if we set `phi[i]` to $-\log(L(i))$ where the likelihood term $L(i)$ is the contribution of observed perpendicular distance `y[i]`, then the likelihood distribution is clearly found to be $L(i)$. See the WinBUGS manual for further details. The WinBUGS code is as follows:

```
for(i in1:n){
  zeros[i] <- 0
  zeros[i] ~ dpois(phi[i]) # likelihood is exp(-phi[i])
  # -log(likelihood)
  phi[i] <- - (log(2*lambda/3.14)/2 - lambda * pow(y[i],2)/2)
}
```

Karunamuni and Quinn (1995) conduct a sensitivity analysis showing that changing values of the prior distribution has little effect on the posterior results. To allow comparisons with Karunamuni and Quinn's results, we use $a = b = 0.1$ in our analyses. For parameter λ, we therefore specify a gamma distribution with both parameters set equal to 0.1:

```
lambda ~ dgamma(0.1, 0.1)
```

Finally, we calculate an estimate of and the density D (Eqs. (8) and (10)):

```
f0 <- sqrt(2 * lambda / 3.14)

D <- (n * f0)/(2 * L)
```

The results are given in Table 5 and show close agreement with the Bayesian analysis of Karunamuni and Quinn (1995).

Table 5 Results for the Stakes line transect data analysis

	$f(0)$	Standard deviation	D	Standard deviation
True	0.110		0.00375	
Maximum likelihood Estimator	0.098		0.00332	
Karunamuni and Quinn (1995)	0.097	0.008	0.00325* 0.00330**	
This study	0.097	0.009	0.003301	0.000292

*Relative squared error loss, **absolute squared error loss.

5 State-Space Models of Count Data: Assessing Density Dependence

In this section, we describe the use of WinBUGS to fit population models of density dependence that simultaneously account for both process and observation error. The example data we use are annual estimates of the population size of North American duck species on their breeding grounds from 1955 to 2002, derived from the Waterfowl Breeding Population and Habitat Survey (WBPHS, US Fish and Wildlife Service 2003).

Assessing the importance of population size or density in regulating population growth rate is fundamental to population biology, ecology and conservation. However, devising robust tests for this so-called "density dependence" has been controversial (e.g. Lebreton this volume). One problem has been that available data on population sizes or densities are almost always estimates, with some level of observation error, and ignoring this observation error can lead to biased tests (e.g. Shenk et al. 1998).

A potential solution is to use a state-space modeling framework, where one can explicitly specify models for both the underlying population dynamics that change population size over time and the observation process that links true population size to the estimates. Such models describing density dependence were constructed by Jamieson (2004) and Jamieson and Brooks (2004). Here we take as an example their "logistic" model for the population dynamics ("state process model"), which can be written as follows:

$$n_t = n_{t-1}\exp\left(\beta_0 + \sum_{j=1}^{k} \beta_j n_{t-j} + \sigma_p z_{p,t}\right) \tag{11}$$

where n_t is the population size at time t ($t = 1\ldots,T$), β_0 determines the expected rate of population growth when the population size is zero, β_j determines the rate at which growth is changed depending on population size in time period $t-j$, $z_{p,t}$ is a Gaussian N(0,1) random variable that represents un-modeled variation in population growth between time periods ("process error") and σ_p determines the size of these random fluctuations. This is coupled with an "observation process model", which can be written

$$y_t = n_t + s_{o,t} z_{o,t} \tag{12}$$

where y_t is the estimated population size at time t, $z_{o,t}$ is a Gaussian N(0,1) random variable that represents measurement error and $s_{o,t}$, which is assumed known (it is provided as part of the WBPHS data, for example), determines the size of the measurement errors.

The state-space model defined by Eqs. (11) and (12) is non-linear and non-normal (because of Eq. (11)), and therefore is difficult to fit using standard frequentist methods, such as the Kalman filter (although see de Valpine 2002, 2003;

de Valpine and Hastings 2002; Besbeas et al. 2005; Besbeas et al. this volume). Jamieson (2004) and Jamieson and Brooks (2004) describe how the model can be formulated in a Bayesian context, and how the parameters may be estimated, for fixed k, using MCMC. Further, they show how a recent extension of the MCMC algorithm – Reversible Jump MCMC (RJMCMC; Green 1995) – can be used to compute the posterior probability for each of a set of possible values of k, and thereby estimate the probability of the presence of density dependence (i.e., the probability that $k>0$) in a population (although we note that autocorrelated process error can affect such assessments – see Lebreton this volume). For the use of RJMCMC in population ecology, see for example, King and Brooks (2002a, b, 2003, 2008) and King et al. (2006). RJMCMC can also be used to produce model-averaged predictions of future population size. Jamieson and Brooks (2004) apply these methods using custom-written MCMC and RJMCMC samplers, implemented in the computer language C, to data for 10 species of duck from the WBPHS. Three species (Northern Pintail *Anas acuta*, Redhead *Aythya americana* and Canvasback *Aythya valisineria*) appear to show some form of density dependence.

Similar models were fitted to Canvasback and Mallard data from the WBPHS (as well as simulated data) by Viljugrein et al. (2005) using WinBUGS, although code was not included with that paper. An additional covariate, number of breeding ponds, was included and model discrimination was via DIC. In that paper, both species were found to show density dependence.

Our aim is to demonstrate how these models may be fitted using WinBUGS, to investigate the use of the beta version of the RJMCMC plug-in for WinBUGS, and to validate the results by comparing them with the independent sampler and C code written by Jamieson. We present some of this work here; it is described in detail in Parker et al. (in prep.). To save space, we only present results for Canvasback.

5.1 Logistic Model

For computational convenience, we re-parameterized the model presented above so that time periods $t = 1,\ldots, k$ are the times before data are available and $t = k + 1, \ldots, k + T$ are times when data were collected. Note that missing data are easily accommodated in this framework. We also turned Eq. (11) into an additive model by log-transforming:

$$p_t = p_{t-1} + \beta_0 + \sum_{j=1}^{k} \beta_j \exp(p_{n-j}) + \sigma_p z_t \tag{13}$$

where $p_t = \log(n_t)$.

Bayesian methods require specification of prior distributions on all unknown quantities; for the purposes of comparison we used exactly the same distributions as used in Chapter 2 of Jamieson (2004; note these are slightly different from those of Jamieson and Brooks 2004): $\beta_j \sim N(0, 100)$ for $j = 0, \ldots, k, \sigma_p^2 = \Gamma^{-1}(0.001, 0.001)$ and $n_t \sim N(0.540, 0.130)$ for $t = 1, \ldots, k$. Note that numbers of ducks are expressed x10^6 and that the distribution is truncated so that $n_t > 0$ (by setting all sampled values of n_t to the maximum of the value drawn from the above normal distribution and 0.00001). Priors are not required on n_t, $t = k + 1, \ldots, k + T$ due to the Markovian structure of the state process model: priors for these quantities are implicitly specified when priors are set for n_t, $t = 1, \ldots, k$. (See Jamieson 2004 for an in-depth discussion of this; see also de Valpine 2002 and Maunder et al. this volume).

Our WinBUGS program was based on code originally written by Steve Brooks for a workshop on Bayesian methods (Brooks et al. 2005). The key parts are specification of the observation process equation (Eq. (12)) and system process equation (Eq. (13)). The observation process equation code is:

```
for(t in (k+1):T){
  prec[t] <- 1/(s[t]*s[t])
  m[t] ~ dnorm(n[t],prec[t])
}
```

while the system process equation code is:

```
for(t in (k+1):T){
  #mm is used to build up equation 3 - note that b[1] here is
  #beta_0 in equation 1, b[2] is beta_1, etc.
  mm[1,t] <- p[t-1] + b[1]
  for(j in 1:k){mm[j+1,t] <- mm[j,t] + b[j+1]*exp(p[t-j])}
  # Expected value of p[t]
  Ep[t] <- mm[k+1,t]
  # Realized value, with process error - tau is 1/sig_p^2
  p[t] ~ dnorm(Ep[t],tau)
}
```

Predictions of future states, for example up to time $T + 10$, could easily be obtained by replacing the first line of the above loop with

```
for(t in (k+1):(T+10)){
```

Summaries of the posterior parameter estimates for Canvasback for $k = 1,2$, and 5 and runs with burn-in of 50,000 and then 1,000,000 samples are given in Table 6, as are results from the same model reproduced from Jamieson (2004, Table 2.5).

The results are very similar, with differences within the bounds of Monte-Carlo variation. Convergence and mixing were relatively slow; diagnostics are reported in Parker et al. (in prep.).

A naïve way to look for evidence of density dependence is to examine posterior credibility intervals (CI) on the β parameters. For example, in the first-order time lag model ($k = 1$), the 95% posterior CI does not contain 0 throughout, providing support for the notion of first order density dependence in this species.

Table 6 Summary of parameter estimates for the logistic model of Jamieson (2004) applied to Canvasback data for $k = 1, 2, 5$. In each cell, the top line shows results from WinBUGS, while the bottom line shows results reproduced from Jamieson (2005, Table 2.5). Values are posterior means with associated 95% posterior credibility intervals in brackets

k	σ_p^2	β_0	β_1	β_2	β_3	β_4	β_5
1	0.019 (0.008, 0.037)	0.450 (0.135, 0.849)	−0.831 (−1.556, −0.259)				
	0.020 (0.007,0.035)	0.454 (0.113, 0.825)	−0.838 (−1.516, −0.220)				
2	0.018 (0.005, 0.036)	0.428 (0.122, 0.833)	−0.653 (−1.687, 0.593)	−0.140 (−1.167, 0.797)			
	0.018 (0.003, 0.034)	0.435 (0.104, 0.816)	−0.655 (−1.747, 0.520)	−0.149 (−1.149, 0.811)			
5	0.014 (0.003, 0.032)	0.521 (0.133, 1.033)	−0.597 (−1.672, 1.032)	−0.164 (−2.144, 1.304)	0.350 (−1.103, 2.054)	−0.249 (−1.657, 0.919)	−0.312 (−1.231, 0.745)
	0.014 (0.002, 0.029)	0.522 (0.094, 0.993)	−0.586 (−1.725, 0.714)	−0.186 (−2.046, 1.395)	0.347 (−1.205, 1.974)	−0.256 (−1.608, 0.979)	−0.293 (−1.261, 0.730)

5.2 Model Comparison

The above program was extended to allow selection among models using RJMCMC. This algorithm searches over the different models, given the observed data, so that the number of possible models is no longer restrictive. We consider an extension to the standard Bayes Theorem, where we simply consider the model itself to be a (discrete) parameter. The standard formula still applies, but now the posterior distribution is defined over both the parameter and model space. Integrating over the parameters we are able to calculate the marginalized posterior probability for each model. However, this integration is analytically intractable and so we resort to an MCMC-type approach. The standard MCMC algorithm cannot be used in the presence of model uncertainty, and RJMCMC is therefore used to explore simultaneously the parameter and model space within a single Markov chain. We used the Jump extension to WinBUGS (Lunn et al. 2006) to implement RJMCMC. This extension allows the sampler to move between models that include all possible combinations of a set of potential covariates – in our case β_1 to β_V where V is the maximum time lag allowable (set to 5 in our code). k indexes the number of β parameters (excluding β_0, which is in all models) in the model for a particular draw from the chain (i.e., the dimension of the model). In the code, an indicator variable id, indicates which particular model is in a particular draw – for example if id was 10101, that would indicate that the parameters β_1, β_3, and β_5 were in the model for that draw (and therefore that $k = 3$).

In the Jump protocol one specifies a prior on the models by specifying a prior distribution on k. The following gives a prior probability of 0.5 that any β_j $(0 > j \geq V)$ is in the model (Lunn 2006, p. 3):

```
k~dbin(0.5,V)
```

We then specify a design matrix (see Lunn 2006, Eq. (1) with the number of rows equal to the number of time periods and V columns. The elements of each row correspond to the sum in Eq. (13). In the following code, C is the first time period about which we make posterior inferences in states – i.e., C=V+1.

```
for(t inC:T){
  for(j in1:V){
    X[(t-C+1),j]<-exp(p[t-j])
  }
}
```

To set up the reversible jump, we use the two Jump-specific commands `jump.lin.pred` and `jump.model.id`, as follows:

```
#Jumpprocess
psi[1:(T-C+1)]<-jump.lin.pred(X[1:(T-C+1),1:V],k,taub)
id<-jump.model.id(psi[1:(T-C+1)])
```

where `psi` is a vector representing the current values of the linear predictor (Lunn 2006, Eq. (1), and `taub` is the prior precision on the β parameters (in our case

1/100; note that the prior on all β parameters is assumed to be multivariate normal, with mean 0 and the specified precision – this distribution is fixed by the software).

We note that the priors specified on the parameters can influence the corresponding posterior model probabilities. In other words the posterior model probabilities are often sensitive to the prior parameter specification. Thus we recommend that a prior sensitivity analysis should always be performed, and care taken when specifying the priors for the parameters, to represent sensible prior beliefs.

Lastly, we specify the system process equation in terms of the `psi` variable:

```
for(t in C:T){
  #Expected value of p[t]
  Ep[t]<-p[t-1]+psi[(t-C+1)]
  #Realized value, with process error - tau is 1/sig_p^2
  p[t]~dnorm(Ep[t],tau)
}
```

Posterior model probabilities can be calculated from the proportion of time the chain visited each model of interest. This information can be obtained from the `Jump` menu that is added to the WinBUGS interface when the Jump extension is installed, and reports the proportion of time spent in each value of the `id` variable. Note, however, some of the models included in the chain are not of interest – we are only interested in models that for any given k contain parameters $\beta_1, \ldots, \beta_k$: for example with $k = 2$ we are only interested in id 11000, and not 10100, 01100, etc. We therefore select out from the list of id's only those we are interested in, and re-normalize so that the proportion of times in these models of interest sum to 1. These proportions are then estimates of posterior model probability.

Model-averaged estimates of other unknown quantities, such as the n_ts, can also be produced by WinBUGS, but just as with the ids above, these contain both models we are interested in and those we are not. It is necessary to save the value for the variable of interest generated in each sample (the CODA button in the sample monitor tool will do this), as well as the corresponding id values, and then select out only those samples that were generated under id values corresponding to models of interest.

Posterior model probabilities for Canvasback for runs with burn-in of 50,000 and then 1,000,000 samples are given in Table 7, as are results from the equivalent model from a run of the Jamieson C code using burn-in of 20,000 and 100,000 samples

Table 7 Posterior model probabilities for $k = 0, \ldots, 5$ for the logistic model of Jamieson (2004) applied to Canvasback data

k	WinBUGS	Jamieson C code
0	0.279	0.265
1	0.685	0.697
2	0.034	0.036
3	0.002	0.002
4	0.000	0.001
5	0.000	0.000

(a run of the Jamieson code was required because posterior model probabilities were not given for these priors in Jamieson 2004).

While the results are similar, they are not identical. This is likely to be caused by a small difference between the implementations: in the algorithm of Jamieson, the acceptance probabilities for between-model moves do not depend on the priors for the β parameters (Jamieson 2005, Section 3.1.1), while in the WinBUGS algorithm it is not possible to achieve such tuning, and in the default algorithm the priors on the β parameters do affect acceptance rates. Despite these minor differences, the overall conclusions are the same: the best supported model (posterior model probability 0.6–0.7) is the one with first-order density dependence.

6 Discussion

In this paper, we have seen how Bayesian theory can be applied to stochastic models for population ecology using MCMC algorithms as implemented in program WinBUGS.

In a mark-recapture data modeling context, WinBUGS can handle many complex models, without additional effort once the likelihood has been written down. This includes (i) random effects that allow unexplained residual variance to be coped with when dealing with covariates, automatic calculation of the amount of smoothing when splines are to be used but also temporal autocorrelation to be incorporated (Johnson and Hoeting 2003), (ii) missing data in the covariate values to be handled and (iii) variable selection. Note that those advantages may also be applied in distance sampling models in order to incorporate covariates in the modeling of the detection function (Marques and Buckland 2003; Marques et al. 2007). Random effects can also be used to address spatial variation in both families of models, allowing the survival and the detection function to depend on spatial coordinates (e.g. longitude and latitude) using splines in two dimensions (Gimenez and Barbraud this volume) or a combination of various random effects (Grosbois et al. in revision) or alternatively, using the geostatistical tools as available through the GeoBUGS adds-on of WinBUGS and the possibility of interfacing WinBUGS with Geographic Information System (GIS) software (WinBUGS manual; see Wyatt 2003 for an application in fisheries).

In our experience, using R or MATLAB to call WinBUGS makes its use much easier for pre- and post-processing data. Note also that an open-source version of the WinBUGS code has recently been published as OpenBUGS. Among other advances, it can be made to perform block updates (i.e., update multiple unknown quantities simultaneously), which might be of interest for experienced programmers. OpenBUGS also runs under Linux.

Our introduction may make WinBUGS appear like a panacea. However, like all computer programs, WinBUGS is not always the perfect tool for Bayesian methods in population ecology, and developments are taking place to improve it. However, as can be appreciated from the three case-studies, it is capable of producing informative results for sophisticated models. In using WinBUGS, one should be aware of the following potential problems. First, one should be aware that experience is needed to be able to debug WinBUGS programs. Also, the computational burden may be discouraging, and it is sometimes preferable to resort to Fortran or C++ to implement efficient MCMC algorithms for specific problems. Finally, although user-specific functions can be programmed (see the WinBUGS manual), there are no tools for matrix calculus so that, e.g., multistate mark-recapture models are difficult to implement (see however Durban et al. 2005 for closed populations). Interestingly, a state-space modeling approach for data on marked animals proposed by Gimenez et al. (2007) might be a solution to this problem (see also Royle [in press] for a similar state-space formulation allowing modeling individual effects). More generally, in line with Buckland et al. (2004b; see also Newman et al. 2006; Buckland et al. 2007), we believe that state-space modeling can provide a convenient and flexible framework for specifying many stochastic models for the dynamics of wild animal populations. In doing so, WinBUGS may provide an efficient and flexible tool to fit such models, possibly nonlinear and non Gaussian – as has been realized for several years in fisheries (Meyer and Millar 1999; Millar and Meyer 2000; Rivot and Prévost 2002; Lewy and Nielsen 2003; Rivot et al. 2004). We note that other fitting algorithms, such as variations on the Kalman filter, Monte-Carlo particle filter, Laplace approximation, importance sampling may also be applicable (see Buckland et al. 2007 for a review). These ideas open the area to numerous applications including the integration of several sources of information (recovery and recapture data, see Catchpole et al. 1998; count data and demographic data, see Besbeas et al. 2002, 2005, 2008; Brooks et al. 2004; Maunder 2004; Schaub et al. 2007).

We end by providing a non-comprehensive list of applications of Bayesian methods in population ecology. An important advantage of the Bayesian framework is the possibility to incorporate prior information in the analysis. McCarthy and Masters (2005a) show how to use prior information on body mass to improve survival estimates using the CJS models, while Pearce et al. (2001); Yamada et al. (2003); Kuhnert et al. (2005) and Martin et al. (2005a) show how to integrate expert knowledge. Several authors have dealt with important issues regarding the specification of vague priors (Lambert et al. 2005; van Dongen 2006; Gelman 2006), assessment of the sensitivity of the posterior distribution to the specified prior distribution (Millar 2004; Millar and Stewart 2005), parameter identifiability in a Bayesian context (Gimenez et al. this volume) and goodness-of-fit tests (Brooks et al. 2000; Barry et al. 2003; Michielsens and McAllister 2004). Meta-analyses have been successfully carried out to estimate demographic parameters (Tufto et al. 2000) and assess animal movement (Jonsen et al. 2003). Further applications of WinBUGS to analyze animal movement data can be found in Morales et al. (2004) and Jonsen et al. (2005). WinBUGS can be used to address issues associated with binomial and Poisson data such as spatial autocorrelation (Thogmartin et al. 2004; Wintle and Bardos 2006), imperfect detection (Royle and Dorazio 2006), heterogeneity in the detection process (Durban and Elston 2005), excess of zeros (Martin et al. 2005b; Ghosh et al. 2006), observer effects (Thogmartin et al. 2004), detecting trends (Link and Sauer 2002) and missing data (Lens et al. 2002). WinBUGS has allowed a better understanding of the impact in assessing complex effects of density-dependence and predicting the impact of climate change and human exploitation in population dynamics (Bjornstad et al. 1999; Saether et al. 2000; Stenseth et al. 2003; Conroy et al. 2005). Regarding model selection, alternatives to DIC and RJMCMC using WinBUGS are given by Ntzoufras (2002; Gibbs variable selection), Link and Barker (2006; Bayesian information criterion; see also Link and Barker this volume) and Ghosh and Norris (2005; minimum posterior predictive loss). Finally, Link and Barker (2005) considered association among demographic parameters (e.g. recruitment and survival) in analysis of open population mark-recapture data (see also Cam et al. 2002 and Wintrebert et al. 2005 when detectability is equal to one).

In conclusion, we hope this paper will encourage ecologists to explore the potential of the flexible and useful WinBUGS software, and the methods underlying it, for carrying out future applications.

References

Bairlein F (1991) Population studies of white storks ciconia ciconia in Europe, with reference to the western population. In: Perrins C, Lebreton J-D, Hirons G (eds), Bird Population Studies: Relevance to Conservation and Management, pages 207–229. Oxford University Press. Oxford.

Barbraud C, Barbraud JC, Barbraud M (1999) Population dynamics of the White Stork Ciconia ciconia in western France. Ibis 141:469–479.

Barry SC, Brooks SP, Catchpole EA, Morgan BJT (2003) The analysis of ring-recovery data using random effects. Biometrics 58:54–65.

Besbeas P, Freeman SN, Morgan BJT, Catchpole EA (2002) Integrating mark-recapture-recovery and census data to estimate animal abundance and demographic parameters. Biometrics 58:540–547.

Besbeas P, Freeman SN, Morgan BJT (2005) The potential of integrated population modelling. Australian New Zealand Journal of Statistics 47:35–48.

Biller C (2000) Adaptive Bayesian Regression Splines in semiparametric generalized linear models. Journal of Computational and Graphical Statistics 9:122–140.

Bjornstad ON, Fromentin JM, Stenseth NC, Gjosaeter J (1999) Cycles and trends in cod populations. Proceedings of the National Academy of Sciences of the USA 96:5066–5071.

Bonner SJ, Schwarz CJ (2006) An extension of the Cormack–Jolly–Seber model for continuous covariates with applications to Microtus pennsylvanicus. Biometrics 62:142–149.

Brooks SP, Catchpole EA, Morgan BJT (2000) Bayesian animal survival estimation. Statistical Science 15:357–376.

Brooks SP, Catchpole EA, Morgan BJT, Harris MP (2002) Bayesian methods for analysing ringing data. Journal of Applied Statistics 29:187–206.

Brooks SP, King R, Morgan BJT (2004) A Bayesian approach to combining animal abundance and demographic data. Animal Biodiversity and Conservation 27:515–529.

Brooks SP, Gimenez O, King R (2005) Bayesian methods for population ecology – workshop codes. National Centre for Statistical Ecology Software Library. [http://www.ncse.org.uk/software.html].

Buckland ST, Anderson DR, Burnham KP, Laake JL, Borchers DL, Thomas L (2001) Introduction to Distance Sampling. Oxford University Press, Oxford.

Buckland ST, Anderson DR, Burnham KP, Laake JL, Borchers DL, Thomas L Eds. (2004a) Advanced Distance Sampling. Oxford University Press, Oxford.

Buckland ST, Newman KB, Thomas L, Koesters NB (2004b) State-space models for the dynamics of wild animal populations. Ecological Modelling 171:157–175.

Buckland ST, Newman KB, Fernández C, Thomas L, Harwood J (2007) Embedding population dynamics models in inference. Statistical Science 22:44–58.

Burnham KP, White GC (2002) Evaluation of some random effects methodology applicable to bird ringing data. Journal of Applied Statistics 29:245–264.

Cam E, Link WA, Cooch EG, Monnat JY, Danchin E (2002) Individual covariation between life-history traits: seeing the trees despite the forest. The American Naturalist 159:96–105.

Catchpole EA, Freeman SN, Morgan BJT, Harris MJ (1998) Integrated reovery/recapture data analysis. Biometrics 54:33–46.

Celeux G, Forbesy F, Robert CP, Titterington DM (2006) Deviance information criteria for missing data models. Bayesian Analysis 1:651–674.

Chavez-Demoulin V (1999) Bayesian inference for small-sample capture–recapture data. Biometrics 55:727–731.

Choquet R, Reboulet A-M, Pradel R, Gimenez O, Lebreton J-D (2005) M-SURGE: new software specifically designed for multistate capture–recapture models. Animal Biodiversity and Conservation, 27: 207–215. Available from http://www.cefe.cnrs.fr/biom/logiciels.htm/

Clark JS (2005) Why environmental scientists are becoming Bayesians. Ecology Letters 8:2–14.

Clobert J, Lebreton J-D (1985) Dépendance de facteurs de milieu dans les estimations de taux de survie par capture–recapture. Biometrics 412:1031–1037.

Congdon P (2003) Applied Bayesian Modelling, John Wiley, New York.

Congdon P (2006) Bayesian Statistical Modelling, 2 nd edition. John Wiley, New York.

Conroy MJ, Fonnesbeck CJ, Zimpfer NL (2005) Modeling regional waterfowl harvest rates using Markov Chain Monte Carlo. Journal of Wildlife Management 69:77–90.

Cormack RM (1964) Estimates of survival from sighting of marked animals. Biometrika 51:429–438.

Crainiceanu CM, Ruppert D, Wand MP (2005) Bayesian analysis for penalized spline regression using WinBUGS. Journal of Statistical Software 14:1–24.

de Valpine P (2002) Review of methods for fitting time-series models with process and observation error, and likelihood calculations for nonlinear, non-Gaussian state-space models. Bulletin of Marine Science 70:455–471.

de Valpine P (2002) Review of methods for fitting time-series models with process and observation error and likelihood calculations for nonlinear, non-Gaussian state-space models. Bulletin of Marine Science 70:455–471.

de Valpine P (2003) Better inferences from population-dynamics experiments using Monte Carlo state-space likelihood methods. Ecology 84:3064–3077.

de Valpine P, Hastings A (2002) Fitting population models incorporating process noise and observation error. Ecological Monographs 72:57–76.

Durban JW, Elston DA (2005) Mark-recapture with occasion and individual effects: abundance estimation through Bayesian model selection in a fixed dimensional parameter space. Journal of Agricultural, Biological, and Environmental Statistics 10:291–305.

Durban JW, Elston DA, Ellifrit DK, Dickson E, Hammond PS, Thompson P (2005) Multisite mark-recapture for cetaceans: population estimates with Bayesian model averaging. Marine Mammal Science 21:80–92.

Ellison AM (2004) Bayesian inference in ecology. Ecology Letters 7:509–520.

Fournier D (2001) An introduction to AD MODEL BUILDER Version 6.0.2 for use in nonlinear modeling and statistics. Available from http://otter-rsch.com/admodel.htm.

Gelman A (2006) Prior distributions for variance parameters in hierarchical models (Comment on Article by Browne and Draper). Bayesian Analysis 1:515–534.

Ghosh SK, Norris JL (2005) Bayesian capture–recapture analysis and model selection allowing for heterogeneity and behavioral effects. Journal of Agricultural, Biological, and Environmental Statistics 10:35–49.

Ghosh SK, Mukhopadhyay P, Lu J-C (2006) Bayesian analysis of zero-inflated regression models. Journal of Statistical Planning and Inference 136:1360–1375.

Gilks W, Richardson S, Spiegelhalter D (1996) Markov chain Monte Carlo in practice. Chapman & Hall, London.

Gimenez O, Crainiceanu C, Barbraud C, Jenouvrier S, Morgan BJT (2006a) Semiparametric regression in capture–recapture modelling. Biometrics 62:691–698.

Gimenez O, Covas R, Brown CR, Anderson MD, Bomberger Brown M, Lenormand T (2006b) Nonparametric estimation of natural selection on a quantitative trait using mark-recapture data. Evolution 60:460–466.

Gimenez O, Rossi V, Choquet R, Dehais C, Doris B, Varella H, Vila J-P, Pradel R (2007) State-space modeling of data on marked individuals. Ecological Modelling, 206:431–438.

Gimenez O, Barbraud C (2008) The efficient semiparametric regression modelling of capture–recapture data: assessing the impact of climate on survival of two Antarctic seabird species. In: Thomson DL, Cooch EG, Conroy MJ (eds.) Modeling Demographic Processes in Marked Populations. Environmental and Ecological Statistics, Springer, New York.

Gimenez BJT, Morgan O, Brooks SP (2008) Weak identifiability in models for mark-recapture-recovery data. In: Thomson DL, Cooch EG, Conroy MJ (eds.) Modeling Demographic Processes in Marked Populations. Environmental and Ecological Statistics, Springer, New York.

Gould WR, Nichols JD (1998) Estimation of temporal variability of survival in animal populations. Ecology 79:2531–2538.

Green PJ (1995) Reversible jump MCMC computation and Bayesian model determination. Biometrika 82:711–732.

Grosbois V, Gimenez O, Gaillard J-M, Pradel R, Barbraud C, Clobert J, Møller AP, Weimerskirch H (in revision) Assessing the impact of climate variation on survival in vertebrate populations. Biological Reviews.

Grosbois V, Harris MP, Anker-Nilssen T, McCleery RH, Shaw DN, Morgan BJT, Gimenez O (in revision) Spatial modelling of survival using capture-recapture data. Ecology.

Harley SJ, Myers RA, Field CA (2004) Hierarchical models improve abundance estimates: spawning biomass of hoki in Cook Strait, New Zealand. Ecological Applications 14: 1479–1494.

Ihaka SP, Gentleman EA (1996) R: a language for data analysis and graphics. Journal of Computational and Graphical Statistics 5:299–314.

Jamieson LE (2004) Bayesian model discrimination with application to population ecology and epidemiology. PhD Thesis, University of Cambridge.

Jamieson LE, Brooks SP (2004) Density dependence in North American ducks. Animal Biodiversity and Conservation 27:113–128.

Johnson DS, Hoeting JA (2003) Autoregressive models for capture–recapture data: a Bayesian approach. Biometrics 59:341–350.

Jolly GM (1965) Explicit estimates from capture-recapture data with both death and immigration-stochastic model. Biometrika 52:225–247.

Jonsen ID, Myers RA, Flemming JM (2003) Meta-analysis of animal movement using state-space models. Ecology 84:3055–3063.

Jonsen ID, Flemming JM, Myers RA (2005) Robust state-space modeling of animal movement data. Ecology 86:2874–2880.

Kanyamibwa S, Schierer A, Pradel R, Lebreton JD (1990) Changes in adult survival rates in a western European population of the White Stork Ciconia ciconia. Ibis 132:27–35.

Karunamuni RJ, Quinn II TJ (1995) Bayesian estimation of animal abundance for the line transects sampling. Biometrics 51:1325–1337.

Kass RE, Carlin BP, Gelman A, Neal RM (1998) Markov chain Monte Carlo in practice: a round-table discussion. American Statistician 52:93–100.

King R, Brooks SP (2002a) Bayesian model discrimination for multiple strata capture–recapture data. Biometrika 89:785–806.

King R, Brooks SP (2002b) Model selection for integrated recovery/recapture data. Biometrics 58:841–851.

King R, Brooks SP (2003) Survival and spatial fidelity of mouflons: the effect of location, age and sex. Journal of Agricultural, Biological and Environmental Statistics 8:486–513.

King R, Brooks SP, Morgan BJT, Coulson T (2006) Bayesian analysis of factors affecting soay sheep. Biometrics 62:211–220.

King R, Brooks SP (2008) On the Bayesian estimation of a closed population size in the presence of heterogeneity and model uncertainty. Biometrics, In press.

Kuhnert PM, Martin TG, Mengersen K, Possingham HP (2005) Assessing the impacts of grazing levels on bird density in woodland habitat: a Bayesian approach using expert opinion. Environmetrics 16:717–747.

Lambert PC, Sutton AJ, Burton PR, Abrams KR, Jones DR (2005) How vague is vague? A simulation study of the impact of the use of vague prior distributions in MCMC using WinBUGS. Statistics in Medicine 24:2401–2428.

Lebreton J-D, Burnham KP, Clobert J, Anderson DR (1992) Modeling survival and testing biological hypothesis using marked animals: a unified approach with case studies. Ecological Monographs 62:67–118.

Lebreton J-D (2008) Assessing density-dependence: where are we left? In: Thomson DL, Cooch EG, Conroy MJ (eds.) Modeling Demographic Processes in Marked Populations. Environmental and Ecological Statistics, Springer, New York.

Lens L, van Dongen S, Norris K, Githiru M, Matthysen E (2002) Avian persistence in fragmented rainforest. Science 298:1236–1238.

Lewy P, Nielsen A (2003) Modeling stochastic fish stock dynamics using Markov Chain Monte Carlo. ICES Journal of Marine Science 60:743–752.

Link WA, Cam E, Nichols JD, Cooch EG (2002) Of bugs and birds: Markov Chain Monte Carlo for hierarchical modelling in wildlife research. Journal of Wildlife Management 66:277–291.

Link WA, Sauer JR (2002) A hierarchical model of population change with application to cerulean warblers. Ecology 83:2832–2840.

Link WA, Barker RJ (2005) Modeling association among demographic parameters in analysis of open population capture–recapture data. Biometrics 61:46–54.

Link WA, Barker RJ (2006) Model weights and the foundations of multimodel inference. Ecology 87:2626–2635.

Lunn DJ (2006) WinBUGS 'jump' interface: beta-release user manual. Technical report dated February 14th 2006, School of Medicine, Imperial College London, UK.

Lunn DJ, Whittaker JC, Best N (2006) A Bayesian toolkit for genetic association studies. Genetic Epidemiology 30:231–247.

Marques FFC, Buckland ST (2003) Incorporating covariates into standard line transect analyses. Biometrics 59:924–935.

Marques TA, Thomas L, Fancy SG, Buckland ST (2007) Improving estimates of bird density using multiple covariate distance sampling. The Auk 124:1229–1243.

Martin TG, Kuhnert PM, Mengessen K, Possingham HP (2005a) The power of expert opinion in ecological models using Bayesian methods: Impact of grazing on birds. Ecological Applications 15:266–280.

Martin TG, Wintle BA, Rhodes JR, Kuhnert PM, Field SA, Low-Choy SJ, Tyre AJ, Possingham HP (2005b) Zero tolerance ecology: improving ecological inference by modelling the source of zero observations. Ecology Letters 8:1235–1246.

Maunder MN (2004) Population viability analysis based on combining Bayesian, integrated, and hierarchical analyses. Acta Oecologica 26:85–94.

Maunder MN, Schnute JT, Ianelli JN (in press) Computers in fisheries population dynamics. In: Megrey B, Moksness E (eds) Computers in Fisheries Research. Kluwer Academic Publishers, Dordrecht.

Maunder MN, Skaug HJ, Hoyle SD (2008) Comparison of estimators for mark-recapture models: random effects, hierarchical Bayes, and AD Model Builder. In: Thomson DL, Cooch EG, Conroy MJ (eds.) Modeling Demographic Processes in Marked Populations. Environmental and Ecological Statistics, Springer, New York.

McCarthy MA, Masters P (2005) Profiting from prior information in Bayesian analyses of ecological data. Journal of Applied Ecology 42:1012–1019.
McCarthy MA (2007) Bayesian Methods for Ecology. Cambridge University Press, Cambridge.
Meyer R, Millar RB (1999) Bugs in bayesian stock assessments. Canadian Journal of Fisheries Aquatic Sciences 56:1078–1087.
Michielsens CGJ, McAllister MK (2004) A Bayesian hierarchical analysis of stock–recruit data: quantifying structural and parameter uncertainties. Canadian Journal of Fisheries Aquatic Sciences 61:1032–1047.
Millar RB, Meyer R (2000) Bayesian state-space modeling of age-structured data: fitting a model is just the beginning. Canadian Journal of Fisheries Aquatic Sciences 57:43–50.
Millar RB (2004) Sensitivity of Bayes estimators to hyper-parameters with an application to maximum yiel from fisheries. Biometrics 60:536–542.
Millar RB, Stewart WS (2005) Automatic calculation of the sensitivity of Bayesian fisheries models to informative priors. Canadian Journal of Fisheries and Aquatic Sciences 62: 1028–1036.
Morales JM, Haydon DT, Frair J, Holsinger KE, Fryxell JM (2004) Extracting more out of relocation data: building movement models as mixtures of random walks. Ecology 85:2436–2445.
Mysterud A, Stenseth NC, Yoccoz NG, Langvatn R, Steinheim G (2001) Nonlinear effects of large-scale climatic variability on wild and domestic herbivores. Nature 410: 1096–1099.
Newman KB, Buckland ST, Lindley ST, Thomas L, Fernández C (2006) Hidden process models for animal population dynamics. Ecological Applications 16:74–86.
North PM, Morgan BJT (1979) Modeling heron survival using weather data. Biometrics 35: 667–681.
Ntzoufras I (2002) Gibbs variable selection using BUGS. Journal of Statistical Software, 7(7).
Pearce JL, Cherry K, Drielsma M, Ferrier S, Whish G (2001) Incorporating expert opinion and fine-scale vegetation mapping into statistical models of faunal distribution. Journal of Applied Ecology 38:412–424.
Plummer M, Best NG, Cowles K, Vines K (2004) CODA: output analysis and diagnostics for MCMC. R package version 0.9-1, URL http://www-fis.iarc.fr/coda/.
Pollock KH, Hines JE, Nichols JD (1984) The use of auxiliary variables in capture–recapture and removal experiments. Biometrics 40:329–340.
Pollock KH (2002) The use of auxiliary variables in capture–recapture modelling: an overview. Journal of Applied Statistics 29:85–102.
R Development Core Team (2007) R: a language and environment for statistical computing. R Foundation for Statistical Computing, Vienna, Austria. ISBN 3-900051-07-0, URL http://www.R-project.org.
Rivot E, Prévost E (2002) Hierarchical Bayesian analysis of capture-mark-recapture data. Canadian Journal of Fisheries Aquatic Sciences 59:1768–1784.
Rivot E, Prévost E, Parent E, Baglinière JL (2004) A Bayesian state-space modelling framework for fitting a salmon stage-structured population dynamic model to multiple time series of field data. Ecological Modelling 179:463–485.
Royle JA, Link WA (2002) Random effects and shrinkage estimation in capture-recapture models. Journal of Applied Statistics 29:329–351.
Royle JA, Dorazio RM (2006) Hierarchical models of animal abundance and occurrence. Journal of Agricultural, Biological, and Environmental Statistics 11:249–263.
Royle JA (2008) Modeling individual effects in the Cormack–Jolly–Seber model: a state-space formulation. Biometrics, In press.
Ruppert D, Wand MP, and Carroll R (2003) Semiparametric Regression. Cambridge University Press, Cambridge.
Saether BE, Tufto J, Engen S, Jerstad K, Rostad OW, Skatan JE (2000) Population dynamical consequences of climate change for a small temperate songbird. Science 287:854–856.
Schaub M, Kania W, Koppen U (2005) Variation of primary production during winter induces synchrony in survival rates in migratory white storks Ciconia ciconia. Journal of Animal Ecology 74:656–666.

Schaub M, Gimenez O, Sierro S, Arlettaz R (2007) Assessing population dynamics from limited data with integrated modeling: life history of the endangered greater horseshoe bat. Conservation Biology 21:945–955.

Seber GAF, (1965) A note on multiple-recapture census. Biometrika 52:249–259.

Shenk TM, White GC, Burnham KP (1998) Sampling-variance effects on detecting density dependence from temporal trends in natural populations. Ecological Monographs 68:445–463.

Skaug HJ, Fournier DA (2006) Automatic approximation of the marginal likelihood in non-Gaussian hierarchical models. Computational Statistics and Data Analysis 51:699–709.

Smith BJ (2004) BOA: Bayesian Output Analysis Program (BOA) for MCMC. R package version 1.1.2-1, URL http://www.public-health.uiowa.edu/boa.

Spiegelhalter DJ, Best NG, Carlin BR, van der Linde A (2002) Bayesian measures of model complexity and fit. Journal of the Royal Statistical Society Series B-Statistical Methodology 64:583–616.

Spiegelhalter D, Thomas A, Best N, Lunn D (2003) WinBUGS user manual. version 1.4 (http://www.mrc-bsu.cam.ac.uk/bugs.). Technical report, Medical Research Council Biostatistics Unit. Cambridge.

Stenseth NC, Viljugrein H, Saitoh T, Hansen TF, Kittilsen MO, Bølviken E, Glöckner F (2003) Seasonality, density dependence, and population cycles in Hokkaido voles. Proceedings of the National Academy of Sciences of the USA 100:11478–11483.

Sturtz S, Ligges U, Gelman A (2005) R2WinBUGS: a package for running WinBUGS from R. Journal of Statistical Software 12:1–16.

Thogmartin WE, Sauer JR, Knutson MG (2004) A hierarchical spatial model of avian abundance with application to cerulean warblers. Ecological Applications 14:1766–1779

Tufto J, Saether B-E, Engen S, Arcese P, Jerstad K, Rostad OW, Smith JNM (2000) Bayesian meta-analysis of demographic parameters in three small, temperate passerines. Oikos 88:273–281.

US Fish and Wildlife Service (2003) Waterfowl Population Status, 2003, US Department of the Interior, Washington, DC.

van Dongen S (2006) Prior specification in Bayesian statistics: three cautionary tales. Journal of Theoretical Biology 242:90–100.

Viljugrein H, Stenseth NC, Smith GW, Steinbakk GH (2005) Density dependence in North American ducks. Ecology 86:245–254.

White GC, Burnham KP (1999) Program MARK: survival estimation from populations of marked animals. Bird Study 46:120–129. Available from http://www.cnr.colostate.edu/gwhite/software.html.

Williams BK, Nichols JD, Conroy MJ (2002) Analysis and Management of Animal Populations. Academic Press, New York.

Wintle BA, Bardos DC (2006) Modeling species-habitat relationships with spatially autocorrelated observation data. Ecological Applications 16:1945–1958.

Wintrebert C, Zwinderman AH, Cam E, Pradel R, van Houwelingen JC (2005) Joint modeling of breeding and survival of Rissa tridactyla using frailty models. Ecological Modelling 181: 203–213.

Wyatt RJ (2003) Mapping the abundance of riverine fish populations: integrating hierarchical Bayesian models with a geographic information system (GIS). Canadian Journal of Fisheries and Aquatic Sciences 60:997–1006.

Yamada K, Elith J, McCarthy M, Zerger A (2003) Eliciting and integrating expert knowledge for wildlife habitat modelling. Ecological Modelling 165:251–264.

Comparison of Fixed Effect, Random Effect, and Hierarchical Bayes Estimators for Mark Recapture Data Using AD Model Builder

Mark N. Maunder, Hans J. Skaug, David A. Fournier, and Simon D. Hoyle

Abstract Mark-recapture studies are one of the most common methods used to obtain demographic parameters for wildlife populations. Time specific estimates of parameters representing population processes contain both temporal variability in the process (process error) and error in estimating the parameters (observation error). Therefore, to estimate the temporal variation in the population process, it is important to separate these two errors. Traditional random effect models can be used to separate the two errors. However, it is difficult to implement the required simultaneous maximization and integration for dynamic nonlinear non-Gaussian models. An alternative hierarchical Bayesian approach using MCMC integration is easier to apply, but requires priors for all model parameters.

AD Model Builder (ADMB) is a general software environment for fitting parameter rich nonlinear models to data. It uses automatic differentiation to provide a more efficient and stable parameter estimation framework. ADMB has both random effects using Laplace approximation and importance sampling, and MCMC to implement Bayesian analysis.

To demonstrate ADMB and investigate methods to analyze mark-recapture data, we implement fixed effect, random effect, and hierarchical Bayes estimators in ADMB and apply them to three mark-recapture data sets. Our results showed that unrestricted time-effects, random effects, and hierarchical Bayes methods often give similar results, but not in all cases or for all parameters.

1 Introduction

> We need easy ways to embed general and flexible random effects into extant capture–recapture models, without each time deriving estimators. (Burnham and White 2002)

Mark-recapture studies are one of the most common methods used to obtain demographic parameters for wildlife populations (Lebreton et al. 1992). These

---

M.N. Maunder (✉)
Inter-American Tropical Tuna Commission, 8604 La Jolla Shores Drive, La Jolla, CA, 92037-1508, USA
e-mail: mmaunder@iattc.org

D.L. Thomson et al. (eds.), *Modeling Demographic Processes in Marked Populations*, Environmental and Ecological Statistics 3, DOI 10.1007/978-0-387-78151-8_42,

demographic parameters can then be used in population dynamics models to investigate the persistence of the populations or to evaluate management strategies (Williams et al. 2002). In particular, mark-recapture studies can be used to estimate the survival rate, which is subsequently used in population viability analysis (Boyce 1992; White 1999). If the survival rate is in balance with the other population processes, the population will remain stable over time. However, if the survival rate is low the population will decline or if the survival rate is high the population will increase. Density dependence may cause the population to reach equilibrium at a higher or lower population size.

The above deterministic description is an overly simplistic representation of reality because population processes change over time. These temporal changes may be simple random processes, have trends, or be autocorrelated, and they may be important for the evaluation of the population under study. For example, temporal variation or a downward trend in survival may increase the probability of extinction. Therefore, development of methods to estimate temporal variation in population processes is important (White 1999; Maunder 2004).

Time specific estimates of parameters representing population processes contain both temporal variability in the process (process error) and error in estimating the parameters (observation error). Therefore, to estimate the temporal variation in the population process, it is important to separate these two errors (Gould and Nichols 1998; Barry et al. 2003). For example, mark-recapture studies can be used to estimate a survival parameter for each year of the analysis. The survival estimate in a particular year may differ from the mean survival either because the real survival differed in that year or because the randomly collected data caused a different estimate by chance. The variance calculated from the vector of time specific point estimates of survival includes both true temporal variation in survival and the estimation error. More sophisticated approaches that separate the observation and process error are needed to estimate the true temporal variation in survival, particular when data is limited.

State-state modeling approaches (Harvey 1989) have traditionally been used to separate out process and observation error. For a state-space approach used in the mark-recapture context, see Gimenez et al. (2007). The term state-space is used because the procedure evaluates the entire range of possible trajectories (states) through time (de Valpine 2002). Following de Valpine (2002), a simple state-space model can be described by two equations:

Process model

$$X_t = aX_{t-1} + v_{t-1}$$

Observation model

$$Y_t = bX_t + \varepsilon_t$$

where X_t and Y_t represent the states and observations, respectively, and v_t and ε_t represent the process and observation error, respectively, in time t. Although, process variability may be a more appropriate term for v_t (Francis and Shotton 1997).

For fixed parameters, the two random variables (**X**,**Y**) define everything about a specific population trajectory. **Y** being known and **X** being unknown. States are unknown random variables under both frequentist and Bayesian frameworks and therefore they should be integrated out of the likelihood function. This is equivalent to integrating out the process error (**v** in the process model above can replace **X** as the unknown random variable; de Valpine 2002). All possible underlying trajectories (states **X**) can be represented by all the possible process errors, **v**, that could produce those states from the initial conditions. Process error can be modeled as random effects (as described below) and state space models are equivalent to using models with process error represented by random effects. Usually, there will be additional fixed effect parameters in the model (e.g. parameters a and b in the equations above) that also need to be estimated, thus requiring a mixed effects framework. Other terms used to describe similar concepts include variance components, random coefficient models, latent variables, and empirical Bayes.

Frequentist estimation methods applied to mixed effects models require both integration and optimization. Early state-space models involved linear models with Gaussian error implemented using the Kalman filter or extended Kalman filter (e.g. Sullivan 1992). Unfortunately, simultaneous integration and maximization becomes a difficult problem when nonlinear and non-Gaussian models are used, which are typical in mark-recapture and population dynamics models of wildlife populations. It becomes even more complex in dynamic models (Maunder and Deriso 2003; Besbeas et al. 2002). The complexity of models has become even greater as more data types are integrated into a single analysis (Maunder 2003). For this reason, many analysts turn to Bayesian MCMC methods, which are much more amenable to these problems (Punt and Hilborn 1997; Brooks et al. 2004). In a Bayesian context, random or mixed effects models are often called hierarchical models. Many applications, particularly mixed effect (hierarchical) models, can be fitted in a fully Bayesian framework, but seem beyond the current capability of likelihood or empirical Bayes analyses (Gelfand 1996; George 1996; Clark 2005). However, Bayesian analysis requires a completely different inference framework and, particularly in data limited cases, the choice of priors may influence the results (Brooks et al. 2000). Analysts should always investigate the sensitivity of the results to the priors (Brooks et al. 2000).

Inference using non-linear non-Gaussian random effects models requires efficient software. AD Model Builder (ADMB; Fournier 2001) is a software package that provides a general modeling environment for fitting parameter rich nonlinear models to data that has been used extensively in fisheries stock assessment (Maunder 2004; Schnute et al. 2007; Maunder et al. submitted). ADMB is not a statistical method but a customizable tool to efficiently and flexibly implement a wide range of statistical methods. It uses automatic differentiation to provide a more efficient and stable parameter estimation framework. ADMB has been used to fit complex nonlinear models with thousands of parameters simultaneously to multiple types of data (e.g. Maunder and Watters 2003a) and to fit nonlinear models with fewer parameters to hundreds of thousands of data points (e.g. Maunder 2001a; Maunder et al. 2006). In particular, ADMB has Laplace approximation and importance sampling to implement frequentist random effects models (Skaug and Fournier 2006) and MCMC to implement Bayesian analysis.

We start by discussing issues related to statistical inference and then describe implementation of random effects in mark-recapture models. Next we introduce AD Model Builder and apply mark-recapture models developed in AD Model builder to three data sets using both frequentist and Bayesian inference.

2 Nuisance Parameters

Elimination of nuisance parameters is a central problem in virtually all approaches to statistical inference (Berger et al. 1999). Let $\mathbf{y}$ be a vector of observations. Let θ be the parameter of interest and let $\boldsymbol{\varphi}$ be a vector of nuisance parameters all influencing the value of $\mathbf{y}$. Denote the joint density of $\mathbf{y}$, θ, and $\boldsymbol{\varphi}$ by $g(\mathbf{y}, \theta, \boldsymbol{\varphi})$.

The likelihood function for θ is

$$L(\theta) \propto g(\mathbf{y}, \theta, \boldsymbol{\varphi})$$

The goal of statistics is to make inferences on θ given $\mathbf{y}$, but this is complicated by the presence of $\boldsymbol{\varphi}$. There are several approaches to eliminate nuisance parameters from the model. The most common are conditioning, maximizing, and integrating (Royle 1997). The standard frequentist approach, maximum likelihood, maximizes over $\boldsymbol{\varphi}$. This is generalized for confidence interval calculations in the profile likelihood procedure. Bayesian methods automatically integrate over $\boldsymbol{\varphi}$.

Traditionally, the mark-recapture literature refers to the recapture probability parameters as nuisance parameters because survival is the quantity of interest. However, any parameter can be considered and treated as a nuisance parameter if it is not the quantity of interest. The goal is to provide inference about the quantity of interest independent of the nuisance parameters. Generally, random effects are considered nuisance parameters, but this is not always the case.

3 Profile Likelihood

The Profile likelihood method, which produces asymmetrical confidence intervals, often produces a better representation of uncertainty in comparison to the normal approximation method, particularly for nonlinear models. Morgan and Freeman (1989) advocate using the profile likelihood approach (see also Gimenez et al. 2005). In parallel to maximum likelihood, the profile likelihood maximizes out the nuisance parameters. A profile likelihood is created by choosing a series of values for the quantity of interest, and then maximizing the likelihood by estimating the remaining parameters of the model for each value in the series. Confidence intervals are then generated using the χ^2 distribution with the appropriate degrees of freedom (Hilborn and Mangel 1997). The profile likelihood method, which replaces the nuisance parameters with their conditional maximum likelihood estimates, ignores some of the uncertainty in the nuisance parameters and modified profile likelihood methods have been developed to adjust for this loss of uncertainty (Berger et al. 1999).

4 Random Effects and Hierarchical Models

Treating nuisance parameters as random effects and integrating them out of the likelihood is one of the main methods to eliminate nuisance parameters. Random effects models also provide a parsimonious compromise between over simplistic and more realistic complex models (Royle and Link 2002). The random effects models provide a shrinkage estimate of the realizations of the random effects, which are an improvement (in mean squared error) over MLE unrestricted time-effects models (Burnham and White 2002). Let $\boldsymbol{\varphi} = (\mathbf{u}, \mathbf{v})$, where $\mathbf{u}$ is a vector of latent random variables (random effects) and $\mathbf{v}$ is a vector of unknown parameters. In a hierarchical context let $g(\mathbf{y}, \boldsymbol{\theta}, \mathbf{u}, \mathbf{v}) = f(\mathbf{y}, \boldsymbol{\theta}, \mathbf{v}|\mathbf{u})\, h(\mathbf{u})$. The likelihood function for θ must be based on the marginal distribution of $\mathbf{y}$, which is obtained by integrating out $\mathbf{u}$ from the joint distribution $f(\mathbf{y}, \boldsymbol{\theta}, \mathbf{v}|\mathbf{u})\, h(\mathbf{u})$. This yields the marginal likelihood

$$L(\theta) = \int f(\mathbf{y}, \boldsymbol{\theta}, \mathbf{v}|\mathbf{u})\, h(\mathbf{u}) d\mathbf{u}.$$

$h(\mathbf{u})$ is the random effects or hyper distribution, which describes how the realizations of the random effect are distributed. $h(\mathbf{u})$ usually has hyper parameters defining the distribution and these may be nuisance parameters or the parameters of interest and are contained in $\boldsymbol{\theta}$ or $\mathbf{v}$, but are suppressed in the above notation.

In the random effects case, both frequentist and Bayesian approaches integrate out the random effects. However, for inference purposes, there still remains the need to eliminate the nuisance parameters $\mathbf{v}$. In this case, Bayesian analysis integrates out these additional nuisance parameters while frequentist approaches generally maximize out these (fixed effect) parameters.

Random-effects or hierarchical models have, for example, been used to determine individual heterogeneity in breeding and survival rates (Link et al. 2002), temporal variability in survival rates (Burnham and White 2002), population trends (Sauer and Link 2002), individual covariation between life-history traits (Cam et al. 2002), and joint modeling of breeding and survival (Wintrebert et al. 2005).

5 Estimation Methods

Most statistical frameworks accept elimination of random effects by integration (Berger et al. 1999). This is because methods based only on maximization perform poorly due to the likelihood being singular when the variance of the random-effects distribution equals zero and the local maximum is an inconsistent estimator (Berger et al. 1999; Maunder and Deriso 2003). Integration over the random effects distribution, which is equivalent to integrating over the state-space, creates a (true) likelihood that has much better properties. In a frequentist context, mixed effects models, which contain both random effects and fixed effects, require both maximization and integration. Hence the problem with frequentist methods. Applying maximization and integration simultaneously with nonlinear non-Gaussian models is difficult. This is a major reason why Bayesian hierarchical nonlinear models have become so popular (Link et al. 2002; Royle and Link 2002). Integrating across all model parameters using MCMC is much easier than simultaneous maximization and integration with nonlinear non-Gaussian models.

5.1 Frequentist: The Laplace Approximation and Importance Sampling

There arc several approaches that have been developed to maximize and integrate at the same time. Several of these are very efficient and do not take much computational requirements (e.g Kalman filter), but they are generally applied to linear models (e.g. in purely Gaussian models a closed-form expression for the integral can be found). Parameters of nonlinear mixed models can be estimated by maximizing an approximation to the likelihood integrated over the random effects. Different approximations to the integral are available: adaptive Gaussian quadrature, first-order Taylor series approximation, Laplacian approximation (Pinheiro and Bates 1995; Davidian and Giltinan 1993; Beal and Sheiner 1992; Skaug 2002). Simulated likelihood (e.g. Millar 2004a) is a general approach applicable to nonlinear models. Unfortunately, simplistic simulated likelihood approaches are inefficient in some dynamic models. For dynamic models, importance sampling methods appear to work well (Maunder and Deriso 2003). However, effective importance functions need to be generated. Maunder and Deriso (2003) used the variance–covariance matrix estimated when treating the realizations of the random effects as fixed effects to generate the importance function. This method can be improved upon by using the Laplace approximation to generate the importance function (Kuk 1999; Skaug 2002; Skaug and Fournier 2006).

The Laplace approximation, which is based on a second order Taylor expansion, can be used to approximate the integral required to integrate out the random effects. For complex models, calculating the second order derivatives required by the Laplace approximation is difficult. Hand calculation is tedious and error prone and numerical calculation is computationally demanding and inexact. An alternative approach is to apply automatic differentiation (AD) (Griewank 2000). Skaug and Fournier (2006) describe how AD is used to calculate the Hessian matrix and its use in the Laplace approximation. AD produces derivatives that are accurate to machine precision, which increases the stability and reduces computation time compared to methods that use the finite difference approach.

5.2 Bayesian: MCMC

Bayesian inference is based on the posterior distribution, which is proportional to the product of the likelihood and the prior. Prior distributions are required for all model parameters. The random effects distribution is a prior for the realizations of the random effects. However, additional priors (hyperpriors) are needed for the parameters (hyperparameters) of the random effects distribution (hyperdistribution; Gelman et al. 1995). Priors are also needed for the other parameters of the model. Frequently, no additional information is available to develop priors and default or objective priors are used to avoid subjectivity. These priors are not necessarily uninformative for the quantities of interest. For example, the constant of proportionality used to scale a relative index of abundance to the total abundance predicted by a population dynamics model can have different implications depending on what "uninformative" prior is assumed. The two "uninformative" priors that have been historically used: uniform and uniform on the log scale; are just uniform priors on a simple parameter transformation. Cordue and Francis (1994) showed using simulation analysis that the choice between these two priors can have a huge consequence on the estimates of risk in fisheries stock assessments.

Vounatsou and Smith (1995) described how Markov chain Monte Carlo (MCMC) (see Gelfand and Smith (1990) for details on the MCMC algorithm) can be used to implement fully Bayesian methods for analysis of mark-recovery data. They implemented both a Gibbs sampling algorithm and a variant of the Hastings–Metropolis algorithm. However, they used uniform priors and did not use a hierarchical/random effects approach. Link and Barker (2004, 2005) used Bayesian analysis to implement hierarchical extensions to the Cormack–Jolly–Seber model. Barry et al. (2003) use the hierarchical Bayes method to model covariates. The BUGS software has made MCMC procedures accessible to a wide range of practitioners. Brooks et al. (2000) used BUGS to implement a mark-recapture model and Brooks et al. (2004) used WinBUGS to implement a mark-recapture model integrated with a population dynamics model and index of abundance data (see also Gimenez et al. this volume). McCarthy and Masters (2005) used WinBUGS to add prior information on survival in a mark-recapture analysis.

6 Mark-Recapture Models

Mark-recapture or mark-resight studies are often designed to estimate survival rates of a population. The standard approach to analyzing the data is to use a multinomial likelihood to model the outcome of a marked individual. The possible outcomes from a release are first recaptured in periods 1, 2, . . . , T or not recaptured at all. The probabilities of these outcomes are a function of the survival in each period and the probability of being recaptured in each period given that you survived to that period. Once an individual has been recaptured (or resighted), it is usually treated as a new release independent of any previous releases or recaptures. Individuals that were released and recaptured in the same periods can be combined and the data presented in an m-array (see the data files in the Appendix).

Let ϕ_t and p_t represent the probability of surviving time t and the probability of being recaptured in time t, respectively. Let R_j represent the number of individuals released in time j and $m_{j,i}$ represent the number of individuals that were released in time j and recaptured in time i. The probability of an individual marked in time j being recaptured in time i, but not recaptured in a previous time is $\Pr_{j,i} = \phi_i p_i \prod_{t=j}^{i-1} \phi_t (1 - p_t)$ and the natural logarithm of the likelihood, ignoring constants, is

$$\ln\left[L(\phi, p, R, m)\right] = \sum_j \left\{ \left(R_j - \sum_{i:i>j} m_{j,i} \right) \ln\left[1 - \sum_{i:i>j} \Pr_{j,i} \right] + \sum_{i:i>j} m_{j,i} \ln\left[\Pr_{j,i}\right] \right\},$$

where the first term represents the likelihood component attributed to individuals that were never recaptured.

The likelihood function can be maximized to estimate the survival and probability of recapture. Hypothesis tests (e.g. based on AIC) are generally used to test several different models based on whether the survival or probability of recapture are constant over time $\{\phi, p\}$, change over time $\{\phi_t, p_t\}$, or a combination of the two $\{\phi, p_t\}$ or $\{\phi_t ,p\}$. Burnham and White (2002) refer to models that estimate a parameter for each time period "unrestricted time effects" to differentiate them from random effects and time invariant models. In the case of both time varying survival and time varying recapture probability, the parameters for the last time period are confounded and only a single parameter representing the combination of survival and probability of recapture can be estimated (Lebreton et al. 1992).

7 Random Effects in Mark-Recapture Models

An alternative to assuming the survival and probability of recapture either change independently over time or are constant over time, is to treat them as random effects. For example, Royle and Link (2002) used temporal random effects for both survival and reporting rates for band recovery data. Both survival and the probability of recapture lie between 0 and 1, and an appropriate distribution for the random effects that has this property is desirable. A common approach is to use a normal distribution transformed based on a logit link function (Brooks et al. 2002; Royle and Link 2002).

$$p_t = \frac{1}{1 + \exp\left[-\varepsilon_p\right]} \text{ where } \varepsilon_p \sim N\left(\mu_p, \sigma_p\right) \text{ and}$$

$$\phi_t = \frac{1}{1 + \exp\left[-\varepsilon_\phi\right]} \text{ where } \varepsilon_\phi \sim N\left(\mu_\phi, \sigma_\phi\right).$$

In this case, the random effects are the survival and probability of recapture. The fixed effects are the parameters of the random effects distributions $(\mu_\phi, \sigma_\phi, \mu_p, \sigma_p)$.

If random effects distribution for survival is the quantity of interest, then the parameters of the random effects distribution for the probability of recapture are nuisance parameters and need to be eliminated from the analysis.

When survival and the recapture probability are both treated as temporal random effects, the confounding of parameters for the last time period does not cause the model to be over parameterized when random effects are used.

In contrast to Brooks et al. (2002) and Royle and Link (2002), Burnham and White (2002) used a beta distribution for the random effect distribution used in their simulations. Royle and Link (2002) preferred the normal distribution with logit link over a beta because it is less difficult to develop priors and straightforward to include covariates.

8 Bayesian Hierarchical Models in Mark-Recapture Models

The main difference between developing a frequentist random-effects mark-recapture model and a Bayesian hierarchical mark-recapture model is the development of priors. Brooks et al. (2002) and Royle and Link (2002) use a diffuse normal for the mean and an inverse Gamma for the variance (conjugate prior) for hyper priors of a logit transformed normally distributed random effect. It is important to determine the sensitivity of parameter estimates to prior distributions (see Millar (2004b) for sensitivity analysis to priors for hierarchical models).

9 AD Model Builder

We illustrate the use of the AD Model builder software package by first developing a simple maximum likelihood mark-recapture model and applying it to the European dipper data presented in Lebreton et al. (1992). The model developed using ADMB has time specific parameters for both survival and probability of detection to replicate the results of Table 10 in Lebreton et al. (1992). Symmetric confidence intervals are typically calculated using the normal approximation method. However, we also describe the profile likelihood method, which allows the confidence intervals to be asymmetric and may be more appropriate in some circumstances. We then describe how this model can be modified to include random effects and estimation within both frequentist and Bayesian frameworks. We finally apply these methods to data on albatross and yellow-eyed penguin, which have more years of data and are therefore more applicable to random effects modeling. Details of the ADMB code for the mark-recapture models used in this study are presented in the Appendix.

AD Model Builder (ADMB; Fournier 2001) is a software tool for developing parameter rich nonlinear models and has become the dominant software environment for estimating the parameters of complex, highly-parameterized, fisheries stock assessment models. It has several features that make it effective at estimating the parameters of these types of models. The main concept behind ADMB is that supplying machine precision derivatives to the function optimizer greatly reduces run time and improves stability. The derivatives are supplied automatically through precompiled derivative code for common functions (e.g. matrix algebra) and the chain rule. Therefore, any calculations used to derive the objective function from the model parameters and the data, automatically have the derivatives calculated for them. Theory shows that the type of automatic differentiation used by ADMB requires less than 5 times the original calculations to provide the derivative values (Griewank 2000; Skaug and Fournier 2006). The standard approach of using numerical derivatives, which provides less precise derivatives, requires $q + 1$ times the original calculations, where q is the number of model parameters. Inaccuracy of the numerical derivative approximations causes instability in the minimization process and produces unreliable results for ill-conditioned problems.

ADMB also makes use of phased optimization. This is the process of adding additional estimated parameters in a series of steps (phases). At each phase the parameters estimated in the previous phase and the new parameters for the current phase are all estimated, with the values from the previous phase used as starting points. This approach allows the estimation of influential parameters in the early phases and fine tuning parameters in latter phases, which can avoid the estimation procedure getting stuck in unrealistic parameter space. Similarly, the method can also be used to fix parameters that are relatively well known from other sources until latter phases. Parameter bounds are also implemented to avoid unrealistic parameter space.

In addition to function optimization, which can be used for maximum likelihood parameter estimation, ADMB also has a MCMC routine to perform Bayesian integration, a Laplace approximation to implement random effects, and automatic profile likelihood calculations, as well as numerous other functions. The MCMC routine uses the mode of the joint posterior, estimated by the optimization routine, to initiate the MCMC chain and the covariance matrix to develop the jumping rule. The Laplace approximation uses automatic second derivatives and can also be used to develop the sampling distribution for use in an importance sampling method to integrate out the random effect parameters.

10 Applications

Mark-recapture models are applied to three applications: (1) European dipper (Lebreton et al. 1992); (2) yellow-eyed penguins (David Houston personal communication); and (3) black footed albatross (Véran 2006; Véran et al. in press). These applications are used for illustrative purposes only. Three inference approaches are applied to each application (a) maximum likelihood estimation with time varying survival and probability of recapture (an unrestricted time-effects model) (denoted MLE); (b) random effects implemented using Laplace approximation (denoted RE); and (c) hierarchical Bayes (denoted Bayesian). The hierarchical Bayes method used uniform priors on all parameters with the standard deviations of the hyper distributions modeled on the natural logarithmic scale. One million samples were taken using the MCMC algorithm and every thousandth was used to create the posterior distribution. The Bayesian estimates were taken as the mean of the posterior distribution and the uncertainty intervals were calculated using the 2.5 and 97.5 percentiles of the posterior distribution. The survival and probability of recapture can not be separated in the last time period of the MLE method, and the combination of the two has to be estimated as a single parameter (see Gimenez et al. (2004) for a review). We also apply the profile likelihood method to develop confidence intervals for the MLE method. The data is provided in m-array format and the standard algorithm can be used to calculate the multinomial likelihood function from the annual values of survival (ϕ_t) and probability of recapture (p_t). We use the term uncertainty intervals to generally refer to both confidence intervals and Bayesian credibility intervals.

The European dipper data consists of 6 years of releases and 6 years of recaptures (see Appendix). The parameter estimates from the MLE model and their standard errors are identical to those presented by Lebreton et al. (1992) (Table 1). The

Table 1 Estimates of probabilities of survival and recapture, and their uncertainty intervals for the European dipper application. LB = lower bound of uncertainty interval, UB = upper bound of uncertainty interval, other terms defined in text

	MLE					
Year	ϕ	LB	UB	p	LB	UB
1981–1982	0.718	0.407	1.029			
1982–1983	0.435	0.297	0.572	0.696	0.365	0.231
1983–1984	0.478	0.359	0.598	0.923	0.777	0.135
1984–1985	0.626	0.508	0.745	0.913	0.797	0.106
1985–1986	0.599	0.486	0.711	0.901	0.793	0.097
1986–1987				0.932	0.841	0.085
1987–1988						
	RE					
Year	ϕ	LB	UB	p	LB	UB
1981–1982	0.564	0.481	0.647			
1982–1983	0.542	0.454	0.630	0.902	0.845	0.960
1983–1984	0.543	0.453	0.633	0.902	0.845	0.960
1984–1985	0.573	0.485	0.661	0.902	0.845	0.960
1985–1986	0.571	0.483	0.659	0.902	0.845	0.960
1986–1987	0.565	0.479	0.652	0.902	0.845	0.960
1987–1988				0.902	0.845	0.960
	Bayesian					
Year	ϕ	LB	UB	p	LB	UB
1981–1982	0.568	0.483	0.680			
1982–1983	0.540	0.420	0.612	0.899	0.826	0.954
1983–1984	0.542	0.443	0.609	0.902	0.829	0.957
1984–1985	0.575	0.512	0.660	0.904	0.836	0.956
1985–1986	0.571	0.509	0.649	0.904	0.842	0.955
1986–1987	0.566	0.499	0.644	0.905	0.837	0.956
1987–1988				0.904	0.836	0.959
	Profile likelihood					
Year	ϕ	LB	UB			
1981–1982	0.718	0.451	0.979			
1982–1983	0.435	0.292	0.595			
1983–1984	0.478	0.353	0.610			
1984–1985	0.626	0.495	0.752			
1985–1986	0.599	0.480	0.722			

random effects and Bayesian analyses give very similar estimates and uncertainty intervals for the annual estimates of survival and the probability of recapture, but they differ from the MLE estimates. The estimates of both survival and probability of detection are essentially constant over time for the random effects and Bayesian methods. All three methods estimate lower survival in the second and third years, particularly the MLE method, which is consistent with the preferred model for this data set, which assumes a different survival rate in these years due to a flood in 1983 (Lebreton et al. 1992; Brooks et al. 2000). The profile likelihood uncertainty intervals are similar to the normal approximation intervals. As expected, and unlike the normal approximation uncertainty intervals, the profile likelihood uncertainty intervals do not exceed one. The estimates and uncertainty intervals of the means and standard deviations of the random effects distributions are similar for the RE and Bayesian methods, except for the lower bound for the RE uncertainty interval for the standard deviations that is negative (Table 4).

The yellow-eyed penguin data has 20 years of releases and 20 years of recaptures (see Appendix). The Bayesian and RE models give similar estimates of the annual values of survival and probability of recapture, which are less variable than the MLE fixed effects estimates (Table 2). The uncertainty intervals for all three methods are similar. The estimates and uncertainty intervals of the means of the random effects distributions are similar for the RE and Bayesian methods (Table 4). The estimates of the standard deviations of the random effects distributions are similar for the Bayesian method comparedto the RE method (Table 4).

The albatross data has 12 years of releases and 12 years of recaptures (see Appendix). The Bayesian, RE and MLE fixed effects models give similar estimates

Table 2 Estimates of probabilities of survival and recapture, and their uncertainty intervals for the yellow-eyed penguin application. See Table 1 for definitions

	MLE					
Year	ϕ	LB	UB	p	LB	UB
1	0.667	0.395	0.939	1.000	1.000	1.000
2	0.668	0.395	0.940	0.874	0.638	1.110
3	1.000	1.000	1.000	0.817	0.584	1.051
4	0.953	0.825	1.082	0.852	0.658	1.045
5	0.930	0.737	1.122	0.893	0.693	1.094
6	0.531	0.310	0.752	0.494	0.205	0.782
7	1.000	1.000	1.000	0.726	0.493	0.958
8	0.928	0.819	1.037	0.726	0.561	0.891
9	0.927	0.817	1.037	0.896	0.783	1.010
10	0.932	0.850	1.014	0.873	0.767	0.979
11	0.956	0.894	1.017	1.000	1.000	1.000
12	0.875	0.794	0.956	0.941	0.874	1.007
13	0.749	0.656	0.843	0.898	0.819	0.977
14	0.826	0.720	0.933	0.703	0.583	0.823
15	0.812	0.704	0.920	0.855	0.760	0.949
16	0.831	0.740	0.922	0.884	0.802	0.966
17	0.858	0.752	0.963	0.861	0.756	0.966
18	0.654	0.526	0.783	0.729	0.590	0.868
19	0.784	0.639	0.928	0.817	0.683	0.952
20						

Table 2 (continued)

	RE					
Year	ϕ	LB	UB	p	LB	UB
1	0.787	0.598	0.976	0.878	0.747	1.009
2	0.770	0.570	0.970	0.849	0.697	1.001
3	0.895	0.778	1.011	0.837	0.684	0.990
4	0.891	0.775	1.007	0.853	0.714	0.991
5	0.881	0.758	1.004	0.878	0.756	1.000
6	0.678	0.433	0.923	0.670	0.412	0.928
7	0.895	0.779	1.011	0.792	0.617	0.967
8	0.895	0.786	1.005	0.771	0.593	0.948
9	0.899	0.791	1.007	0.877	0.762	0.991
10	0.905	0.803	1.008	0.864	0.747	0.981
11	0.923	0.832	1.014	0.940	0.869	1.012
12	0.872	0.760	0.984	0.913	0.822	1.004
13	0.764	0.599	0.928	0.883	0.779	0.988
14	0.822	0.688	0.956	0.735	0.530	0.939
15	0.825	0.691	0.958	0.853	0.736	0.970
16	0.835	0.711	0.960	0.875	0.767	0.982
17	0.851	0.729	0.973	0.862	0.746	0.977
18	0.682	0.431	0.934	0.760	0.578	0.942
19	0.802	0.646	0.957	0.826	0.689	0.964
20	0.836	0.672	0.999	0.835	0.675	0.995
	Bayesian					
Year	ϕ	LB	UB	p	LB	UB
1	0.783	0.575	0.917	0.867	0.694	0.969
2	0.758	0.546	0.901	0.840	0.682	0.950
3	0.891	0.747	0.978	0.829	0.652	0.938
4	0.889	0.765	0.976	0.849	0.707	0.948
5	0.880	0.753	0.973	0.871	0.750	0.961
6	0.678	0.459	0.852	0.661	0.388	0.848
7	0.890	0.747	0.981	0.787	0.608	0.910
8	0.895	0.793	0.968	0.773	0.625	0.878
9	0.896	0.792	0.967	0.871	0.764	0.949
10	0.906	0.824	0.970	0.861	0.760	0.935
11	0.920	0.847	0.977	0.936	0.852	0.986
12	0.873	0.797	0.939	0.908	0.830	0.966
13	0.764	0.676	0.846	0.879	0.799	0.941
14	0.821	0.724	0.903	0.736	0.619	0.836
15	0.822	0.718	0.909	0.851	0.761	0.921
16	0.834	0.748	0.914	0.870	0.789	0.938
17	0.852	0.757	0.938	0.859	0.773	0.926
18	0.685	0.560	0.802	0.757	0.624	0.859
19	0.799	0.674	0.898	0.822	0.695	0.915
20	0.836	0.682	0.956	0.833	0.678	0.958

(Table 3). The uncertainty intervals for the MLE fixed effects and Bayesian models are similar. The uncertainty intervals for the RE model are similar to the others for the early years, however they are much wider for the latter years. The uncertainty intervals on the probability of recapture for the MLE model are wider than the other methods for the early years years. The estimates of the means and the standard deviations of the random effects distributions differ between the Bayesian and RE methods (Table 4).

Table 3 Estimates of probabilities of survival and recapture, and their uncertainty intervals for the albatross application. See Table 1 for definitions

	MLE					
Year	ϕ	LB	UB	p	LB	UB
1	0.673	0.358	0.988	0.676	0.305	1.047
2	0.877	0.631	1.123	0.871	0.632	1.111
3	0.618	0.347	0.889	0.553	0.221	0.885
4	0.995	0.975	1.015	0.632	0.537	0.726
5	1.000	1.000	1.000	0.803	0.735	0.871
6	0.969	0.927	1.012	0.700	0.624	0.777
7	0.900	0.869	0.930	0.829	0.792	0.865
8	0.927	0.902	0.953	0.831	0.799	0.863
9	0.876	0.848	0.904	0.791	0.759	0.824
10	0.903	0.878	0.927	0.801	0.772	0.831
11	0.941	0.914	0.969	0.777	0.746	0.808
12						
	RE					
Year	ϕ	LB	UB	p	LB	UB
1	0.774	0.412	1.136	0.763	0.655	0.871
2	0.914	0.747	1.082	0.784	0.684	0.884
3	0.699	0.212	1.185	0.746	0.634	0.859
4	0.982	0.942	1.023	0.685	0.531	0.840
5	0.988	0.959	1.018	0.794	0.713	0.874
6	0.962	0.892	1.032	0.725	0.604	0.846
7	0.903	0.768	1.038	0.820	0.744	0.896
8	0.928	0.824	1.031	0.824	0.746	0.901
9	0.877	0.702	1.051	0.789	0.714	0.865
10	0.903	0.769	1.037	0.799	0.725	0.872
11	0.940	0.850	1.031	0.777	0.697	0.856
12	0.884	0.700	1.068	0.759	0.648	0.870
	Bayesian					
Year	ϕ	LB	UB	p	LB	UB
1	0.778	0.488	0.972	0.748	0.612	0.843
2	0.916	0.717	0.999	0.770	0.661	0.868
3	0.671	0.412	0.877	0.725	0.591	0.822
4	0.987	0.961	0.999	0.680	0.593	0.766
5	0.991	0.971	1.000	0.790	0.736	0.849
6	0.968	0.928	0.998	0.718	0.643	0.785
7	0.903	0.872	0.929	0.818	0.785	0.849
8	0.927	0.902	0.951	0.821	0.789	0.850
9	0.876	0.847	0.903	0.789	0.758	0.819
10	0.903	0.880	0.925	0.799	0.772	0.825
11	0.942	0.916	0.968	0.776	0.748	0.803
12	0.920	0.820	0.998	0.732	0.662	0.811

11 Discussion

11.1 Performance of the Estimators

In a previous evaluation of random effects models for mark-recapture and mark-recovery data using moment estimators, Burnham and White (2002) found that their estimator performed well under a wide range of situations. They investigated

Table 4 Estimates of hyper distribution parameters and associated uncertainty intervals. See Table 1 for definitions

		Dipper		Penguin		Albatross	
		RE	Bayesian	RE	Bayesian	RE	Bayesian
μ_ϕ	Estimate	0.241	0.244	1.741	1.772	2.604	3.181
	LB	0.016	−0.005	1.355	1.394	1.858	2.589
	UB	0.465	0.494	2.127	2.210	3.351	4.277
σ_ϕ	Estimate	0.120	0.133	0.643	0.718	1.123	1.582
	LB	−0.246	0.012	0.243	0.304	0.418	0.757
	UB	0.486	0.520	1.043	1.288	1.829	4.259
μ_p	Estimate	2.223	2.284	1.729	1.742	1.224	1.161
	LB	1.584	1.636	1.349	1.358	1.000	0.968
	UB	2.862	2.997	2.110	2.172	1.447	1.344
σ_p	Estimate	0.001	0.079	0.623	0.661	0.284	0.316
	LB	−0.497	0.011	0.243	0.213	0.084	0.109
	UB	0.499	0.430	1.004	1.166	0.484	0.567

different number of recapture occasions, releases, capture probability, mean survival, and survival process variation. Their analysis differed from ours in that they treated the probability of recapture as a constant rather than as a random effect. They also did no make any distributional assumptions about the random effects distribution. They found no or little bias in the estimates of the standard deviation of the random effects distribution for survival, uncertainty interval coverage close to the desired value, and symmetrical coverage of uncertainty intervals. However, the performance did degrade when the true survival was constant over time. Burnham and White (2002) concluded that the random effect shrinkage estimates of an individual year survival are better than MLE unrestricted time-effects estimates. Burnham and White (2002) also found that the uncertainty intervals for individual year survival estimates using the random effects method were consistently narrower than the unrestricted time-effects method. Initial simulation analysis of the methods presented here produced poorer performance compared to that obtained by Burnham and White (2002) and more analyses are needed to provide guidance on which methods should be used for a particular application.

11.2 Integrated Analysis

Temporal variation or pattern among parameters has historically (prior to Lebreton et al. 1992) been modeled as a two-step approach: (1) estimate the parameters and (2) fit a model to the estimates. Link and Barker (2004) term this two-step approach "doing statistics on statistics". Alternatively, in the case of modeling pattern among parameters, a deterministic relationship (ultrastructural model) has been used (Link and Barker 2004). These approaches are not completely satisfactory and can introduce bias in the case of the two step approach, or produce overstated precision or biased hypothesis tests if ultrastuctured models are used (Link 1999; Barry et al. 2003; Maunder and Watters 2003b). When parameters of the model are related to covariates, the most appropriate method is to integrate the covariate into the analysis and use random-effect models to represent the additional process error (Link 1999; Barry et al. 2003; Maunder and Watters 2003b). This reduces bias caused by estimation error, provides additional information to estimate the model parameters, and improves the performance of hypothesis tests.

More generally, integrating all available data into the model can reduce bias and ensure that uncertainty is propagated through the analysis (Fournier and Archibald 1982; Pascual and Kareiva 1996; Maunder 2003; Lebreton 2005). Maunder (2004) details how traditional mark-recapture analysis can be restructured and integrated with population dynamics models. Maunder (1998, 2001b) and Hampton and Fournier (2001) give examples of integrating tag-recapture data into age-structured population dynamics models in a fisheries stock assessment context. Besbeas et al. (2002) and Besbeas and Freeman (2006) provide methods to approximate integrated likelihood approaches in a mark-recapture context. Brooks et al. (2004) provide a Bayesian example of integrating mark-recapture analysis with a population dynamics model and an index of abundance.

Statistical inference involves knowing when and how to summarize and combine data. Integrating all the data into a single analysis may result in a model that is too computationally intensive to analyze or too complicated to understand. There has been a large amount of research on traditional mark-recapture models and they are well understood. There are well developed goodness of fit tests and other diagnostics for these traditional mark-recapture models. These are much less developed for integrated models. Brooks et al. (2004) suggest that goodness-of-fit can be evaluated using Bayesian p-values (Gelman et al. 1996). In complicated cases it might be wise to first apply traditional mark-recapture models before integrating the data into a population dynamics model or to use approximate likelihood (Besbeas et al. 2003).

11.3 When to Use Random Effects

Discussing random effects will ultimately lead to the issue of when should random effects be used and for which parameters. Random effects offer an intermediate between a single fixed effect that represents a constant value for the parameter over time, for example, and a separate fixed effect parameter for each time period. Therefore, random effects can be viewed as a generalization that encompasses the two extremes as the standard deviation of the random effects distribution approaches zero and infinity, respectively. For this reason, random effects eliminate the need to test between the constant and time varying models. However, the issue of testing between the constant model, the random effects model, or some other model (consider the dipper example), and whether the model fits the data may arise. There is also the issue of choosing the form of the random effects distribution.

Pawitan (2003) takes the practical view that random effects can be considered as just a method to deal with large numbers of parameters. In this context, information from data rich time periods is used to inform parameters from data poor time periods. Therefore, candidate parameters are those that are expected to vary over time (or other characteristic), but for which there is a varying degree of information in the data for the different time periods. Another practical method for determining which parameters to treat as random effects in a population dynamics context is to consider which parameters would be considered stochastic in future projections. The random effects distribution represents the process variation and can be used to draw from for future projections.

11.4 Comparison of Inference Frameworks

Bayesian and frequentist inference frameworks differ in the way they treat nuisance parameters and because priors are required for all model parameters in the Bayesian framework. If data are available to develop data-based priors, the frequentist (and Bayesian) framework can either integrate these data into the analysis (Maunder 2003), or use prior likelihoods (Pawitan 2003) or approximate likelihoods (Besbeas et al. 2002). In the case that no data are available to generate data based priors for some model parameters, the Bayesian framework still requires priors for these parameters. If no prior is explicitly used, many methods (e.g. Markov Chain Monte Carlo) implicitly imply a uniform prior. Uniform priors are the basis of inverse probability as used by Bayes and Laplace (Berger 2000), which dominated statistics in the 19th Century. However, the inverse probability method depends on the particular parameterization and uniform priors may not be the least informative priors. This has led to the use of objective or reference priors (Berger et al. 1999). These priors are objective in the context that they are used as a standard approach or to reduce the influence of the prior and are not due to the subjective judgment of the analyst. For example, Jeffrey's priors are commonly used and they are invariant to reparameterization (Gelman et al. 1995). Unfortunately, there are concerns with the behavior of Jeffrey's priors in high dimensions (Berger et al. 1999) and, with the wrong choice, they can be undesirably informative for the quantities of interest. Jeffrey's priors are also very difficult to develop for complex nonlinear models (Millar 2002). Alternatively, the reference prior algorithm can be used (Bernardo 1979; Berger and Bernardo 1992), which has been shown to perform consistently better than the Jeffrey's prior in multivariate examples, and typically has good frequentist properties (Berger et al. 1999). Frequentist-matching (i.e. use simulation testing to determine if the proportion of times that the true value falls inside the credibility interval corresponds to the probability definition of that interval) is a common method to test these objective Bayesian methods (Bayarri and Berger 2004). Frequentist matching suggests the use of confidence distributions (Schweder and Hjort 2002) as objective Bayes posteriors with implied objective priors.

Link and Barker (2004) suggest that a simple solution to the problem of choosing priors is to try several and see whether and how the choice influences posterior inference. Miller (2004b) provides a framework for prior sensitivity analysis in a hierarchical Bayesian framework. Link and Barker (2004) note that the Bernstein-Von

Mises Theorem (also known as the Bayesian Central Limit Theorem) implies that, subject to minor constraints, the influence of the prior diminishes as the sample size increases. They found that for the mark-recapture application they presented, the choice of prior did not have a large influence on the results, and that this is generally the case provided there is adequate data. In general, when the log likelihood is near quadratic and the prior is multivariate normal or vague, the frequentist and Bayesian approaches gives similar results (Schweder 1998).

Our results showed that the frequentist and Bayesian frameworks often give similar results, but not in all cases or for all parameters. Vounatsou and Smith (1995) found that MLE and Bayesian estimates were similar for mark-recovery data for 9 years of mallard data, but standard errors were smaller for the Bayesian analysis. For simulated data they found that MLE and Bayesian approaches gave very similar results for large sample sizes, but for smaller sample sizes the MLE approaches resulted in unrealistically small standard errors for reporting rates. Brooks et al. (2000) found that the sensitivity to priors depended on the complexity of the model. Models with more parameters (time varying survival) were more sensitive to the choice of priors. They argue that parameters that are not supported by the data will be most sensitive to priors and that mark-recovery models will be more prone to prior sensitivity compared to mark-recapture models. Link and Barker (2005) found that the variance and covariance parameters of the random effects distributions were sensitive to priors. Brooks et al. (2000) use sensitivity to priors, in conjunction with parsimony and posterior model probability, as a criterion for judging the appropriate complexity of their model.

Sensitivity to priors can be linked to the comparison of results from Bayesian and frequentist analyses. If confidence distributions are viewed as objective Bayes posteriors with implied objective priors, then comparing results between Bayesian and frequentist analysis is one way of looking at sensitivity to priors. For example, in all but the albatross example, the results from the Bayesian and frequentist analyses were similar indicating lack of sensitivity to priors.

A more comprehensive simulation analysis is needed to compare and contrast the different estimation methods under different situations. The simulation design of Burnham and White (2002) could be used as a starting point.

11.5 Computational Time of Methods

When deciding on a statistical framework or method to use, ease of use and computational requirements are important factors, particularly if the different methods perform similarly. Computational demands are particularly important when using simulation analysis to test the performance of an estimator. For example, Burnham and White (2002) took over 4 months of CPU time to do the 64,000 simulations used to evaluate a random effects estimator for a mark-recapture model. This is nearly 3 min per model run. They also found that Bayesian implementation using MCMC took 100 times as much CPU time. In contrast, the ADMB Laplace approximations used in a similar simulation study took less than 3 s. Royle and

Link (2002) promote the Bayesian framework because of ease of application to complex models. Fixed effect MLE methods are generally computationally efficient compared to the other methods. The Laplace approximation method using automatic differentiation (Skaug and Fournier 2006), as implemented in AD Model builder, is efficient, but the methods high memory requirements causes limitations for complex models. Importance sampling using the Laplace approximation as the importance function is more computational intensive. MCMC implementation of non-linear Bayesian hierarchical models often requires a large number of draws and is therefore computationally intensive. Brooks et al. (2004) illustrate how WinBUGS can be used to implement an integrated mark-recapture and population dynamics model, but explain how inefficient it is compared to custom MCMC code written in Fortran. The profile likelihood method to calculate asymmetrical confidence intervals requires separate profile likelihoods to be carried out for each quantity of interest. In contrast, MCMC and bootstrap only need to be done once to developed uncertainty measures for all quantities of interest. Therefore, when there are many quantities of interest, MCMC and bootstrap methods may be less computationally demanding. The profile likelihood method has not yet been implemented in the random effects version of ADMB.

Acknowledgments Richard Barker provided assistance on the algorithm for the mark-recapture model. John Darby and Dave Houston provided the yellow-eyed penguin data. Jean-Dominique Lebreton and Sophie Véran provided the albatross data. The reviewers, Andy Royle and Olivier Gimenez, provided valuable comments that improved the manuscript.

Appendix: AD Model Builder Code for the Mark-Recapture Model, with Application to the European Dipper Data

ADMB is written in C++ and most of the code used to develop the model is C++ or overloaded C++ functions and operators (for more information on ADMB see admb-project.org). There are also several key words used in ADMBs template system, which was created to simplify model development. The template is comprised of several sections, but only three are necessary for model development: the `DATA_SECTION`, `PARAMETER_ SECTION`, and `PROCEDURE_SECTION`. Most applications also incorporate a `REPORT_SECTION` to produce custom output. The `DATA_SECTION` contains definitions of data to use in the model and variables that are used for intermediate calculations that don't require derivative calculations (e.g. used for data manipulation). The `PARAMETER_SECTION` contains definitions of the parameters of the model to be estimated and variables that are used for intermediate calculations that do require derivative calculations (i.e. are functions of the model parameters). The `PROCEDURE_SECTION` is where the code is written to develop the model (i.e. the calculation of the objective function from the data and the model parameters). The code is written in the *.tpl file (* representing the root of the model name). There are two other files used for developing the model. The first is *.dat file, which is associated with the `DATA_SECTION`, and contains all the data that was defined in the DATA_SECTION. The second is the *.pin file, which is associated with the PARAMETER_SECTION, and provides initial values for all the estimable model parameters defined in the PARAMETER_SECTION.

The code in the DATA_SECTION reads in (the init_ prefix defines a variable that has data read in from the *.dat file) the number of years of data (NRCperiods), a vector of releases for each year (Releases), and the m-array matrix (data). An intermediate variable (temp) is defined and filled with an integer sequence from 1 to the number of years of data so that the variable data can be defined as a ragged array to fit the m-array data format.

The PARAMETER_SECTION defines two vectors of estimated parameters (estS and p), which are bounded between 0 and 1 [the dimensions always come first, followed by the bounds if it is a bounded parameter. The phase always comes last and is optional]. Survival does not have a parameter for the last year as it is subsumed in the parameter for the probability of recapture for the last year for programming convenience. Survival is estimated in the first phase and probability of recapture in the second phase. A intermediate variable (S) is defined to hold all values of survival and two intermediate variables (Scum and Pcum) are defined for the calculation of the multinomial likelihood. Finally, a required variable (f) is defined to hold the objective function value.

The PROCEDURE_SECTION contains all the code to implement the model equations. This section calculates the objective function as a function of the data and the model parameters. The code is written in standard C++ with some exceptions. The ADMB function elem_prod does an element-wise multiplication of two arrays (in this case vectors), sum sums the components of an array. The double slash "//" is used to indicate a comment.

The REPORT_SECTION allows custom output of the model results. The object report is a file with the model's name as the root and "rep" as the extension.

```
DATA_SECTION
  init_int NRCperiods
  init_vector Releases(1,NRCperiods)
  ivector temp(1,NRCperiods)
  !!temp.fill_seqadd(1,1);
  init_matrix data(1,NRCperiods,temp,NRCperiods)
PARAMETER_SECTION
  init_bounded_vector estS(1,NRCperiods-1,0,1,1)
  init_bounded_vector p(1,NRCperiods,0,1,2)
  vector S(1,NRCperiods)
  matrix Scum(1,NRCperiods,1,NRCperiods)
  matrix Pcum(1,NRCperiods,1,NRCperiods)
  objective_function_value f
PROCEDURE_SECTION
  S(1,NRCperiods-1)=estS;
  S(NRCperiods)=1; //so p is combined S and p for last
year
```

```
  for (int i=1;i<=NRCperiods;i++)
  {
      Scum(i,i)=S(i);
      Pcum(i,i)=1;
      for (int j=i+1;j<=NRCperiods;j++)
      {
           Scum(i,j)=Scum(i,j-1)*S(j);
           Pcum(i,j)=Pcum(i,j-1)*(1.0-p(j-1));
      }
      Pcum(i)=elem_prod(Pcum(i),p);
      for (j=i;j<=NRCperiods;j++) f+=-
data(i,j)*log(Scum(i,j)*Pcum(i,j));
      f+=-(Releases(i)-sum(data(i)))*log(1-
sum(elem_prod(Scum(i),Pcum(i))));
  }
REPORT_SECTION
  report <<"S"<<S<<endl;
  report<<"p"<<p<<endl;
```

The *.dat file includes all the data to be read in the DATA_SECTION. The # symbol is used to identify a comment.

```
#init_int NRCperiods
6
#init_vector Releases(1,NRCperiods)
22 60 78 80 88 98
#init_matrix data(1,NRCperiods,temp,NRCperiods)
11  2  0  0  0  0
   22  1  0  0  0
      34  2  0  0
         45  1  2
            51  0
               52
```

The *.pin file includes initial values for all estimable parameters defined in the PARAMETER_SECTION.

```
#init_bounded_vector estS(1,NRCperiods-1,0,1,1)
0.7 0.7 0.7 0.7 0.7
#init_bounded_vector p(1,NRCperiods,0,1,2)
0.7 0.7 0.7 0.7 0.7 0.7
```

The model is compiled using the command

```
admb name
```

where “name” is the root used for the files. This creates an executable file “name.exe”.

The model is run using the command

```
name
```

Random Effects

Several modifications of the code are required to convert the MLE with all fixed effects into a MLE model with random effects for survival and probability of recapture. To avoid some computational issues it is often better to use the following formulation

$$p_t = \frac{\exp\left[\mu_p + \sigma_p \varepsilon_p\right]}{1 + \exp\left[\mu_p + \sigma_p \varepsilon_p\right]} \text{ where } \varepsilon_p \sim N(0, 1)$$

and

$$\phi_t = \frac{\exp\left[\mu_\phi + \sigma_\phi \varepsilon_\phi\right]}{1 + \exp\left[\mu_\phi + \sigma_\phi \varepsilon_\phi\right]} \text{ where } \varepsilon_\phi \sim N(0, 1)$$

First, the (hyper) parameters of the random effects distributions are defined in the PARAMETER_SECTION. The standard deviations are defined on the log scale to ensure they are positive.

```
init_number meanS(1)
init_number ln_sdS(2)
init_number meanp(1)
init_number ln_sdp(2)
```

Next, the realizations of the random effects are defined by converting the time specific parameters into random effects parameters. Random effect parameters have to be defined after all fixed effect parameters. Due to the use of random effects, the survival and probability of recapture can now be estimated in the last year.

```
random_effects_vector Sdev(1,NRCperiods,2)
random_effects_vector pdev(1,NRCperiods,2)
```

The *.pin file is updated with associated values for the new and updated parameters.

Four intermediate variables are defined in the PARAMETER_SECTION to hold values of the standard deviations of the random effects distributions, the survival, and the probability of recapture.

```
number sdS
number sdp
vector S(1,NRCperiods)
vector p(1,NRCperiods)
```

At the start of the PROCEDURE_SECTION, the standard deviations of the random effects distributions are calculated from the logarithmic parameters that are estimated.

```
sdS=mfexp(ln_sdS);
sdp=mfexp(ln_sdp);
```

The survival and probability of recapture are calculated from the random effects parameters using a logistic function in the PROCEDURE_SECTION. The ADMB function elem_div does an element-wise division of two arrays (in this case vectors).

```
S=elem_div(mfexp(meanS+Sdev*sdS),(1+mfexp(meanS+Sdev*sdS)));
p=elem_div(mfexp(meanp+pdev*sdp),(1+mfexp(meanp+pdev*sdp)));
```

The random effects distribution penalty is added to the objective function at the end of the PROCEDURE_SECTION. The ADMB function norm2, which calculates the sum of squares of an array, is used.

```
f+=0.5*norm2(Sdev)+0.5*norm2(pdev);
```

Full program

```
DATA_SECTION
  init_int NRCperiods
  init_vector Releases(1,NRCperiods)
  ivector temp(1,NRCperiods)
  !!temp.fill_seqadd(1,1);
  init_matrix data(1,NRCperiods,temp,NRCperiods)

PARAMETER_SECTION
  init_number meanS(1)
  init_number ln_sdS(2)
  init_number meanp(1)
  init_number ln_sdp(2)

  random_effects_vector Sdev(1,NRCperiods,2)
  random_effects_vector pdev(1,NRCperiods,2)

  number sdS
  number sdp
  vector S(1,NRCperiods)
  vector p(1,NRCperiods)

  matrix Scum(1,NRCperiods,1,NRCperiods)
  matrix Pcum(1,NRCperiods,1,NRCperiods)

  objective_function_value f
```

```
PROCEDURE_SECTION
  sdS=mfexp(ln_sdS);
  sdp=mfexp(ln_sdp);
  S=elem_div(mfexp(meanS+Sdev*sdS),
  (1+mfexp(meanS+Sdev*sdS)));
  p=elem_div(mfexp(meanp+pdev*sdp),
  (1+mfexp(meanp+pdev*sdp)));
  for (int i=1;i<=NRCperiods;i++)
  {
       Scum(i,i)=S(i);
       Pcum(i,i)=1;
       for (int j=i+1;j<=NRCperiods;j++)
       {
            Scum(i,j)=Scum(i,j-1)*S(j);
            Pcum(i,j)=Pcum(i,j-1)*(1.0-p(j-1));
       }
       Pcum(i)=elem_prod(Pcum(i),p);
       for (j=i;j<=NRCperiods;j++) f+=-
data(i,j)*log(Scum(i,j)*Pcum(i,j));
       f+=-(Releases(i)-sum(data(i)))*log(1-
sum(elem_prod(Scum(i),Pcum(i))));}
  }
  f+=0.5*norm2(Sdev)+0.5*norm2(pdev);

REPORT_SECTION
    report<<''S "<<S<<endl;
    report<<''p "<<p<<endl;
```

*.pin file

```
#init_number meanS(1)
1
#init_number ln_sdS(2)
0
#init_number meanp(1)
1
#init_number ln_sdp(2)
0
#init_vector Sdev(1,NRCperiods,1)
0 0 0 0 0 0 0
#init_vector pdev(1,NRCperiods,1)
0 0 0 0 0 0 0
```

The model is compiled using the command line option –re

```
admb -re name
```

For the penguins and albatross applications, an additional command line option is needed when the model is run to ensure there is enough memory -mno 2000

```
name -mno 2000
```

Standard Deviations

To determine the standard deviations of the annual survival and annual probabilities of recapture, which are now derived parameters, the variables for survival and probability of recapture are redefined in the `PRAMETER_SECTION` using the keyword `sdreport_` as a prefix to indicate that standard deviations are calculated for these parameters.

```
sdreport_vector S(1,NRCperiods)
sdreport_vector p(1,NRCperiods)
```

Bayesian

The only change required to convert the random effects model into a Bayesian model that uses MCMC integration is to add some code at the end of the `PROCEDURE_SECTION` that outputs the quantities of interest.

```
if(mceval_phase())
{
ofstream outsamples("samples.out",ios::app);
outsamples.precision(10);
outsamples<<meanS<<" "<<sdS<<" "<<meanp<<"
"<<sdp<<" "<<S<<" "<<p<<endl;
outsamples.close();
}
```

We also added bounds on the standard deviations of the random effects to avoid the MCMC procedure getting stuck at values close to zero. This is done by adding the additional prefix `bounded_` to the parameter definition and then adding bounds in brackets after the parameter name.

```
init_bounded_number ln_sdS(-4.6,2.3,4)
init_bounded_number ln_sdp(-4.6,2.3,5)
```

The model is first run using the following command line options to define the number of samples to take and how often to save the sample to the file.

```
name -mcmc2 1000000 --mcsave 1000
```

Then the model is run with the following command line option to evaluate the model for each of the sets of parameter values saved in the previous command and output the posterior values of the desired quantities

```
name -mceval
```

Profile Likelihood

To automatically calculate a profile likelihood, a variable needs to be defined in the `PARAMETER_SECTION` using the prefix `likeprof_`

```
likeprof_number Sprof1
```

and set equal to the parameter of interest in the `PROCEDURE_SECTION`

```
Sprof1=S(1);
```

Automatic profile likelihood is not implemented in the random effects version of ADMB at the time of writing of this paper. To run the model with automatic profile likelihood the –lprof command line option is used.

```
name -lprof
```

Yellow-Eyed Penguin *.dat File

```
# init_int NRCperiods
20
#init_vector Releases(1,NRCperiods)
12 12 10 14 17 21 9 30 25 41 45 79 97 80 62 80 74 74 46 45
#init_matrix data(1,NRCperiods,temp,NRCperiods)
8 0 0 0 0 0 0 0 0 0 0 0 0 0 0 0 0 0 0 0
  7 1 0 0 0 0 0 0 0 0 0 0 0 0 0 0 0 0 0
    8 0 2 0 0 0 0 0 0 0 0 0 0 0 0 0 0 0
     13 0 0 0 0 0 0 0 0 0 0 0 0 0 0 0 0
       14 0 1 0 0 0 0 0 0 0 0 0 0 0 0 0
          6 4 1 0 0 0 0 0 0 0 0 0 0 0
            6 1 1 0 1 0 0 0 0 0 0 0 0
             21 5 1 0 0 0 0 0 0 0 0 0
               22 0 1 0 0 0 0 0 0 0 0
                 35 3 0 0 0 0 0 0 0 0
                   43 0 0 0 0 0 0 0 0
                     65 2 1 0 0 0 0 0
                       66 2 1 0 2 0 0
                        48 13 1 0 0 0 0
                           44 4 1 0 0 0
                             61 3 0 1 0
                               55 4 1 0
                                 36 7 2
                                   30 4
                                     31
```

*Albatross *.dat File*

```
#init_int NRCperiods
12
#init_vector Releases(1,NRCperiods)
11 12 13 101 101 136 469 606 707 798 938 1106
#init_matrix data(1,NRCperiods,temp,NRCperiods)
5 1 1 0 0 0 0 0 0 0 0 0
 10 0 0 0 0 0 0 0 0 0 0
    4 2 1 1 0 0 0 0 0 0
    64 30 5 1 0 0 0 0 0
      81 11 8 0 1 0 0 0
        94 26 2 3 1 0 0
        350 49 13 2 1 0
          477 52 15 2 2
            496 84 21 3
              583 95 24
                691 127
                    741
```

References

Barry SC, Brooks SP, Catchpole EA, Morgan BJT (2003) The analysis of ring-recovery data using random effects. Biometrics 59:54–65.

Bayarri MJ, Berger J (2004) The interplay between Bayesian and frequentist analysis. Statistical Science 19:58–80.

Beal SL, Sheiner LB (1992) NONMEM User's Guide. NONMEM Project Group, UCSF San Francisco.

Berger J (2000) Bayesian analysis: a look at today and thoughts of tomorrow. Journal of the American Statistical Association 95:1269–1276.

Berger J, Bernardo JM (1992) Ordered group reference priors with applications to a multinomial problem. Biometrika 79:25–37.

Berger J, Liseo B, Wolpert R (1999) Integrated likelihood methods for eliminating nuisance parameters. Statistical Science 14:1–28.

Bernardo JM (1979) Reference posterior distributions for Bayesian inference. Journal of the Royal Statistical Society B 41:113–147.

Besbeas P, Freeman SN (2006) Methods for joint inference from panel survey and demographic data. Ecology 87:1138–1145.

Besbeas P, Freeman SN, Morgan BJT, Catchpole EA (2002) Integrating mark-recapture-recovery and census data to estimate animal abundance and demographic parameters. Biometrics 58:540–547.

Besbeas P, Lebreton J-D, Morgan BJT (2003) The efficient integration of abundance and demographic data. Journal of the Royal Statistical Society Series C – Applied Statistics 52:95–102.

Boyce MS (1992) Population viability analysis. Annual Reviews in Ecology and Systematics 23:481–506.

Brooks SP, Catchpole EA, Morgan BJT (2000) Bayesian animal survival estimation. Statistical Science 15:357–376.

Brooks SP, Catchpole EA, Morgan BJT, Harris MP (2002) Bayesian methods for analysing ringing data. Journal of Applied Statistics 29:187–206.
Brooks SP, King R, Morgan BJT (2004) A Bayesian approach to combining animal abundance and demographic data. Animal Biodiversity and Conservation 21:515–529.
Burnham KP, White GC (2002) Evaluation of some random effects methodology applicable to bird ringing data. Journal of Applied Statistics 29(1–3):245–264.
Cam E, Link WA, Cooch EG, Monnat JY, Danchin E (2002) Individual covariation between life-history traits: seeing the trees despite the forest. The American Naturalist 159:96–105.
Clark JS (2005) Why environmental scientists are becoming Bayesians? Ecology Letters 8:2–14.
Cordue PL, Francis RICC (1994.) Accuracy and choice in risk estimation for fisheries assessment. Canadian Journal of Fisheries and Aquatic Science 51:817–829.
Davidian M, Giltinan DM (1993) Some simple methods for estimating intraindividual variability in nonlinear mixed effects models. Biometrics 49:59–73.
de Valpine P (2002) Review of methods for fitting time-series models with process and observation error and likelihood calculations for nonlinear, non-Gaussian state-space models. Bulletin of Marine Science 70:455–471.
Fournier D (2001) An introduction to AD MODEL BUILDER Version 6.0.2 for use in nonlinear modeling and statistics, available from http://otter-rsch.com/admodel.htm.
Fournier DA, Archibald CP (1982) A general theory for analyzing catch at age data. Canadian Journal of Fisheries and Aquatic Sciences 39:1195–1207.
Francis RICC, Shotton R (1997) "Risk" in fisheries management: a review. Canadian Journal of Fisheries and Aquatic Science 54:1699–1715.
Gelfand AE (1996) Comment: empirical Bayes methods for combining likelihoods. Journal of the American Statistical Association 91:551–552.
Gelfand AE, Smith AFM (1990) Sampling based approaches to calculating marginal densities. Journal of the American Statistical Association 85:398–409.
Gelman A, Carlin JB, Stern HS, Rubin DB (1995) Bayesian Data Analysis. Chapman and Hall, London.
Gelman A, Meng X, Stern H (1996) Posterior predictive assessment of model fitness via realized discrepancies – with discussion. Statistica Sinica 6:733–807.
George EI (1996) Comment: empirical Bayes methods for combining likelihoods. Journal of the American Statistical Association 91:553–555.
Gimenez O, Bonner S, King R, Parker RA, Brooks S, Jamieson LE, Grosbois V, Morgan BJT, Thomas L (in press) WinBUGS for population ecologists: Bayesian modeling using Markov Chain Monte Carlo methods. This volume.
Gimenez O, Choquet R, Lamor L, Scofield P, Fletcher D, Lebreton J-D, Pradel R (2005) Efficient profile-likelihood confidence intervals for capture–recapture models. Journal of Agricultural, Biological, and Environmental Statistics 10(2):184–196.
Gimenez O, Rossi V, Choquet R, Dehais C, Doris B, Varella H, Vila J-, Pradel R (2007) State-space modelling of data on marked individuals. Ecological Modelling 206:431–438.
Gimenez O, Viallefont A, Choquet R, Catchpole EA, Morgan BJT (2004) Methods for investigating parameter redundancy. Animal Biodiversity and Conservation 27(1):561–572.
Gould WR, Nichols JD (1998) Estimation of temporal variability of survival in animal populations. Ecology 79:2531–2538.
Griewank A (2000) Evaluating Derivatives: Principles and Techniques of Algorithmic differentiation. SIAM, Philadelphia, PA.
Hampton J, Fournier DA (2001) A spatially disaggregated, length-based, age-structured population model of yellowfin tuna (*Thunnus albacares*) in the western and central Pacific Ocean. Marine and Freshwater Research 52:937–963.
Harvey AC (1989) Forecasting, structural time series models, and the Kalman filter. Cambridge University Press, Cambridge.
Hilborn R, Mangel M (1997) The Ecological Detective: Confronting Models with Data. Princeton University Press, Princeton, NJ.

Kuk AYC (1999) Laplace importance sampling for generalized linear mixed models. Journal of Statistical Computing and Simulation 63:143–158

Lebreton JD (2005) Dynamical and statistical models for exploited populations. Australian and New Zealand Journal of Statistics 47:49–63.

Lebreton JD, Burnham KP, Clobert J, Anderson DR (1992) Modeling survival and testing biological hypotheses using marked animals: a unified approach with case studies. Ecological monographs 62:67–118.

Link WA (1999) Modeling pattern in collections of parameters. Journal of Wildlife Management 63(3):1017–1027.

Link WA, Barker RJ (2004) Hierarchical mark-recapture models: a framework for inference about demographic processes. Animal Biodiversity and Conservation 27:441–449.

Link WA, Barker RJ (2005) Modeling association among demographic parameters in analysis of open population capture–recapture data. Biometrics 61:46–54.

Link WA, Cam E, Nichols JD, Cooch EG (2002) Of BUGS and birds: Markov chain Monte Carlo for hierarchical modeling in wildlife research. Journal of Wildlife Management 66:277–291.

Link WA, Cooch EG, Cam E (2002) Model-based estimation of individual fitness. Journal of Applied Statistics 29:207–224.

Maunder MN (1998) Integration of tagging and population dynamics models in fisheries stock assessment. Ph.D. dissertation, University of Washington, Seattle, WA.

Maunder MN (2001a) A general framework for integrating the standardization of catch-per-unit-of-effort into stock assessment models. Canadian Journal of Fisheries and Aquatic Science 58:795–803.

Maunder MN (2001b) Integrated Tagging and Catch-at-Age ANalysis (ITCAAN). In: Kruse GH, Bez N, Booth A, Dorn MW, Hills S, Lipcius RN, Pelletier D, Roy C, Smith SJ, Witherell D (eds) Spatial Processes and Management of Fish Populations. Alaska Sea Grant College Program Report No. AK-SG-01-02, University of Alaska Fairbanks, Fairbanks, pp 123–146.

Maunder MN (2003) Paradigm shifts in fisheries stock assessment: from integrated analysis to Bayesian analysis and back again. Natural Resource Modeling 16:465–475.

Maunder MN (2004) Population viability analysis, based on combining integrated, Bayesian, and hierarchical analyses. Acta Oecologica 26:85–94.

Maunder MN, and Deriso RB (2003) Estimation of recruitment in catch-at-age models. Canadian Journal of Fisheries and Aquatic Sciences 60:1204–1216.

Maunder MN, Hinton MG, Bigelow KA, Langley AD (2006) Developing indices of abundance using habitat data in a statistical framework. Bulletin of Marine Science 79(3):545–559.

Maunder MN, Schnute JT, Ianelli JN (in press) Computers in fisheries population dynamics. In: Megrey B, Moksness E (eds) Computers in Fisheries Research. Kluwer Academic Publishers.

Maunder MN, Watters GM (2003a) A-SCALA: an age-structured statistical catch-at-length analysis for assessing tuna stocks in the Eastern Pacific Ocean. Inter-American Tropical Tuna Commission Bulletin 22:433–582.

Maunder MN, Watters GM (2003b) A general framework for integrating environmental time series into stock assessment models: model description, simulation testing, and example. Fisheries Bulletin 101:89–99.

McCarthy MA, Masters P (2005) Profiting from prior information in Bayesian analyses of ecological data. Journal of Applied Ecology 42:1012–1019.

Millar RB (2002) Reference priors for Bayesian fisheries models. Canadian Journal of Fisheries and Aquatic Science 59:1492–1502.

Millar RB (2004a) Simulated maximum likelihood applied to non-Gaussian and nonlinear mixed effects and state-space models. Australian and New Zealand Journal of Statistics 46: 515–526.

Millar RB (2004b) Sensitivity of Bayes estimators to hyper-parameters with an application to maximum yield from fisheries. Biometrics 60:536–542.

Morgan BJT, Freeman SN (1989) A model with first year variation for ring-recovery data. Biometrics 45:1087–1101.

Pascual MA, Kareiva P (1996) Predicting the outcome of competition using experimental data: maximum likelihood and Bayesian approaches. Ecology 77:337–349.

Pawitan Y (2003) In all likelihood: statistical modeling and inference using likelihood. Oxford University Press, Oxford, UK.

Pinheiro JC, Bates DM (1995) Mixed-Effects Models Methods and Classes for S-PLUS. Technical Report No. 89, Department of Biostatistics, University of Wisconsin – Madison.

Punt AE, Hilborn R (1997) Fisheries stock assessment and decision analysis: the Bayesian approach. Reviews in Fish Biology and Fisheries 7:35–63.

Royle RM (1997) Statistical Evidence: A Likelihood Paradigm. International Thomson Publishing, New York.

Royle JA, Link WA (2002) Random effects and shrinkage estimation in capture–recapture models. Journal of Applied Statistics 29:329–351.

Sauer JR, Link WA (2002) Hierarchical modeling of populations stability and species group attributes from survey data. Ecology 86(6):1743–1751.

Schnute JT, Maunder MN, Ianelli JN (2007) Designing tools to evaluate fishery management strategies: can the scientific community deliver? ICES Journal of Marine Science 64:1077–1084.

Schweder T (1998) Fisherian or Bayesian methods of integrating diverse statistical information? Fisheries Research 37:61–75.

Schweder T, Hjort NL (2002) Confidence and likelihood. Scandinavian Journal of Statistics 29:309–332.

Skaug HJ (2002) Automatic differentiation to facilitate maximum likelihood estimation in nonlinear random effects models. Journal of Computational and Graphical Statistics 11: 458–470.

Skaug HJ, Fournier DA (2006) Automatic approximation of h marginal likelihood in non-Gaussian hierarchical models. Computational Statistics and Data Analysis 51:699–709

Sullivan PJ (1992) A Kalman filter approach to catch-at-length analysis. Biometrics 48:237–257.

Véran S (2006) Quantifier l'impact des activités humaines à partir de données incomplètes : effet de la pêche palangrière sur l'albatros à pieds noirs. PhD Thesis, Université Montpellier, Montpellier.

Véran S, Gimenez O, Flint E, Kendall W, Doherty P, Lebreton JD (in press) Quantifying the impact of longline fisheries on adult survival in the Black-footed Albatross. Journal of Applied Ecologyb 44:942–952.

Vounatsou P, Smith AFM (1995) Bayesian analysis of ring-recovery data via Markov chain Monte Carlo simulation. Biometrics 51:687–708.

White GC (1999) Population viability analysis: data requirements and essential analysis. In Boitani L. and Fuller T.K. (eds) Research techniques in animal ecology. Columbia University Press, New York, pp 288–327.

White GC (this volume).

Williams BK, Nichols JD, Conroy MJ (2002) Analysis and management of animal populations. Academic Press, San Diego, California, USA.

Wintrebert C, Zwinderman AH, Cam E, Pradel R, van Houwelingen JC (2005) Joint modeling of breeding and survival of Rissa tridactyla using frailty models. Ecological Modelling 181: 203–213.

Section XI
Open Forum

Charles Francis and Andy Royle

On Adjusting for Missed Visits in the Indexing of Abundance from "Constant Effort" Ringing

Vanessa M. Cave, Stephen N. Freeman, Stephen P. Brooks, Ruth King, and Dawn E. Balmer

Abstract Producing accurate, reliable indices of abundance, enabling the status of breeding bird populations to be monitored is of interest to government, conservation groups and other bodies. Indices for Sedge Warblers *Acrocephalus schoenobaenus* from 1983 to 2002 were produced using catch data from the British Trust for Ornithology's (BTO) Constant Effort Scheme (CES). This is a ringing programme based on standardised mist-netting across up to 12 annual visits to each of a large number of sites. A feature of these data is that some yearly site counts are "censored" due to visits missed within certain years. Peach et al. (1998) developed an intuitive, non-parametric method for correcting for missed visits, prior to model-fitting in the form of a Poisson regression model with an additive offset. In this paper a novel Bayesian approach is introduced, which produces annual indices of abundance whose uncertainty also incorporates a component due to the correction for missed visits. We describe the method in detail, applied to the Sedge Warbler data and to simulated data, and compare the results with those from the current method of Peach et al. (1998).

1 Introduction

Standardisation of effort (in terms of length and position of nets, and duration of their operation) brings a number of advantages to the subsequent analysis of ringing data. Efficient estimation of survival via mark-recapture models (Peach et al. 1990, 1995; Peach 1993; Bonner et al. this volume) arises from the capacity to set probabilities of recapture constant over time, for example. As individual birds caught repeatedly can be identified by their ring number, total numbers of birds caught annually can also be produced, and used to index both

---

V.M. Cave (✉)
CREEM, University of St. Andrews, The Observatory, St. Andrews KY16 9LZ, UK
e-mail: vanessa@mcs.st-and.ac.uk

D.L. Thomson et al. (eds.), *Modeling Demographic Processes in Marked Populations*, Environmental and Ecological Statistics 3, DOI 10.1007/978-0-387-78151-8_43,

abundance (Peach et al. 1998) and productivity (Robinson et al. 2007). The British Trust for Ornithology's (BTO) Constant Effort Scheme (CES) is a long-running, large scale ringing programme based upon such consistency of effort; the breeding season is divided into 12 roughly equal periods, within each of which birds are caught and ringed. The precise timing of ringing, and the position of nets, varies between sites according to local conditions but, importantly, is invariant and replicated at every visit to any *particular* site. Any variability between years in the numbers of birds caught, therefore, is attributable only to annual changes in the population level and to stochastic variation, and not to varying intensity of capture. If all sites in such a scheme are visited over a series of years in accordance with this protocol, then the total numbers of birds caught provide an index of the species' abundance.

In practice, there is considerable deviation from this ideal as sites enter or leave the scheme, thus any two sites might neither start nor cease operation in the same year. This turnover is readily accommodated by fitting a simple linear model to the incomplete series of observations (ter Braak et al. 1994; Peach et al. 1998). The subject of the present paper is the absence of data from isolated individual visits within a site's period of operation, some years receiving fewer than the full twelve visits due to weather, or ringers' personal circumstances. This, therefore, violates the assumption of constant effort between years, yet to omit that year's records wastes valuable data. Peach et al. (1998) proposed a non-parametric approach to address the problem, imputing values for censored counts prior to model fitting. Underlying this approach is the assumption that while there is variation throughout the season in the numbers of birds caught, this variation is consistent between years, otherwise bias is likely to be introduced. Further the approach is conservative in its overall effect, mitigating against extremely high or low imputed values, and ignores uncertainty arising from the correction. In applying this method to seven CES data sets Miles et al. (2007) showed that the estimated ratios of juvenile to adult birds (a measure of productivity) were similar to those based only upon sites without missed visits and hence are unbiased by any such correction. Otherwise, this key step in the process has received little attention, nor has any alternative been proposed. Such an alternative must be unaffected by the sampling imbalance invariably observed – not only the numbers of birds caught but also the likelihood of a visit being made at all vary through the season. In this paper, we consider a Bayesian correction method in which for the first time imputation is simultaneous with model fitting. The model adopts a Poisson distribution for the annual numbers of birds ringed at each site, and a truncated form of this in appropriate years for any site lacking one or more visits. No assumptions about the pattern of inter-annual catches are made. The Bayesian approach also more accurately reflects the uncertainty involved in the estimation of the year and site effects by incorporating the additional uncertainty arising from missed visits within the model fitting algorithm, rather than imputing data for censored counts prior to fitting. In addition, the Bayesian approach will provide an estimate of the uncertainty in the corrected counts if these are of interest in their own right.

2 Data and a Classical Model

2.1 The Constant Effort Scheme

The scheme began in 1983, after a brief pilot period, with around 40 sites and quickly proved popular with volunteer ringers. Over 100 sites now provide data to the scheme annually. Many sources (Peach et al. 1998; Miles et al. 2007; Robinson et al. 2007) describe the history and the field procedure in greater detail, and we restrict ourselves to a brief account here. The breeding season (May–August) is divided into twelve consecutive periods of 10–11 days by the BTO and ringing is carried out at each site on one day selected (by the ringers) within each, with a minimum of 3 days between visits to a site.

Sites are selected by the ringers themselves and cover the entire UK, though they are concentrated in the South-Eastern quadrant where the human population is highest. Sites are largely woodland, scrub or reedbed, habitats at which succession can be controlled (keeping these "Constant" too) by appropriate management and which are amenable to ringing appreciable numbers of birds, largely common passerines. These birds include both young of the year (juveniles, a few weeks old at most) and adult birds, and these can be separated in the hand on the basis of plumage. We concentrate here on the adult birds, numbers of which are assumed to index changes in the breeding population. In isolation, a similar index based on juveniles is less often used, as fluctuations reflect and confound changes both in numbers breeding and in the seasonal productivity per pair, making useful interpretation difficult.

A little more than 20 species are now caught at CES sites in sufficient number for indices to be calculated (Grantham and Robinson 2007). Catches of Sedge Warblers *Acrocephalus schoenobaenus* form the basis of the analyses in this paper. This species is one of the most often encountered, as although its wetland habitats are highly localised in the UK, they are often favoured by ringers. Further, at sites especially suitable very large numbers of Sedge Warblers can be caught, making this a survey of particular value in tracking the fortunes of a species known to be susceptible to environmental change (Peach et al. 1991). Data from 1983 to 2002 provide 1756 annual site counts from a total of 249 sites recording Sedge Warblers at least once. In the following subsection we describe the Peach et al. (1998) method, currently employed by the BTO, for producing annual indices of abundance for Sedge Warblers and other CES monitored species, before we present an alternative Bayesian approach.

2.2 The Model of Peach et al. (1998)

We define the following quantities:

n_{it} denotes the number of different adult birds caught at site i in year t when all 12 visits are made. If less than 12 visits are made n_{it} is unknown but represents the total number of individuals that would have been caught had all 12 visits been made.

N_i denotes the total number of different adults caught at site i for all years of complete coverage (i.e. in which all 12 visits were made).

N'_{it} denotes the number of birds that belong to N_i that are caught at visits corresponding to those made in year t. Thus, for example, if visits 2 and 5 to site i in year t were missed then N'_{it} is equal to N_i minus the number of those birds caught only at visits 2 and/or 5.

l_{it} denotes the observed number of different birds caught at site i in year t. When one or more visits are missed this is a "lower bound" for n_{it}. In years of complete coverage $l_{it} = n_{it}$.

In a year of missed visits, the corrected count E_{it} proposed by Peach et al. (1998) for a site is then:

$$E_{it} = l_{it} \frac{N_i}{N'_{it}}$$

which, it is readily verified, reduces to n_{it} in a year of complete coverage. The number of birds caught varies between visits due to migratory behaviour, thus the ratio of N_i to N'_{it} accounts not only for the number of missed visits but also their precise timing. The model assumes that the l_{it} have a Poisson distribution with mean λ_{it} where:

$$\ln(\lambda_{it}) = \beta_0 + S_i + Y_t + \ln\left(\frac{l_{it}}{l_{it}\left(N_i/N'_{it}\right)}\right) \tag{1}$$

Without missing visits, the expected count at site i, in year t, is simply derived from an intercept β_0, the ith site effect (S_i) and t^{th} year effect (Y_t), additive on the logarithmic scale. Due to the parameter redundancy, the year effect for the first year, and the site effect at an arbitrary site, are constrained to zero. The rightmost term in (1) is an offset, the log-transformed ratio of the observed total count (albeit not necessarily from 12 visits) and the corrected value, used to adjust for the missed visits where appropriate. The model is then readily fitted by maximum likelihood in any Generalized Linear Modelling package.

Sites without any years of complete coverage cannot be accommodated quite this way, as $N_i = N'_{it} = 0$. An analogous correction is thus employed based upon catches at all sites, rather than catches only from the site in question (Peach et al. 1998) – a "global" rather than a "local" correction factor. Counts at a site are however omitted from the analysis altogether if <8 visits were made in a given year.

3 Parameter Estimation and Imputing Censored Counts: A Bayesian Alternative

As the approach above imputes censored counts, and treats them as known before parameters are estimated, no account is taken of their inherent uncertainty. We now consider a new approach, in which the imputation and the model fitting are combined in a single process, and imprecision is determined more accurately. The Poisson distribution is retained, but a truncated form with lower bound l_{it} is adopted for those cases with fewer than twelve visits. A full, uncensored count for these is therefore estimated but this is not required prior to model fitting, as in the previous section. The Metropolis-Hastings algorithm (Gelman et al. 2004) is employed (i) to update the model parameters $\{\beta_0, \mathbf{S}, \mathbf{Y}\}$ and (ii) to update censored counts, in two steps as described in the appendix.

3.1 Implementation

Code for implementing the algorithm described in the appendix was written in C. A total of 100,000 samples from the posterior distributions of the parameters and imputed censored counts were drawn after an initial burn-in of 100,000 updates was discarded, so the method is computationally demanding. Posterior medians, which were essentially identical to the posterior means, were extracted to estimate the model parameters and the imputed counts. The algorithm was run several times and essentially identical results were obtained from each.

4 Results

4.1 Sedge Warbler Data

Abundance indices were calculated using both the Peach et al. (1998) correction and Bayesian method conditioning on the lower bounds for censored data, applied to the annual site-totals from 216 sites over 20 years, 1983–2002. 96 counts from a further 33 sites are omitted as no complete count, based on a full set of visits, has been made in any single year. The Bayesian truncated Poisson model cannot be applied to such data as at least one completed count is required to identify the site effect. Of the counts used 452 (26%) were lacking one or more of the twelve visits.

Estimates of the year effects Y_t from both methods, MLE from the Classical approach of Peach et al. (1998) and posterior medians from the Bayesian approach, are compared in Fig. 1. The trends are similar, though it is notable that the Bayesian estimates are larger (relative to the base year, constrained to zero in each case), especially in later years. Also shown in Fig. 1 are estimates derived only from those counts based on a full set of twelve visits, for which no correction is required. The lack of censored values in such circumstances means that these latter estimates are identical under the two methods. They are seen to bear a greater resemblance to those in which the Peach et al. (1998) correction is used with censored data than they do to those arising from the truncated Poisson.

In years of missed visits predictions of the unknown counts n_{it} via a priori correction (Peach et al. 1998), and those arising from the Bayesian conditional model are highly correlated with each other ($\rho = 0.916$). Both are also highly correlated

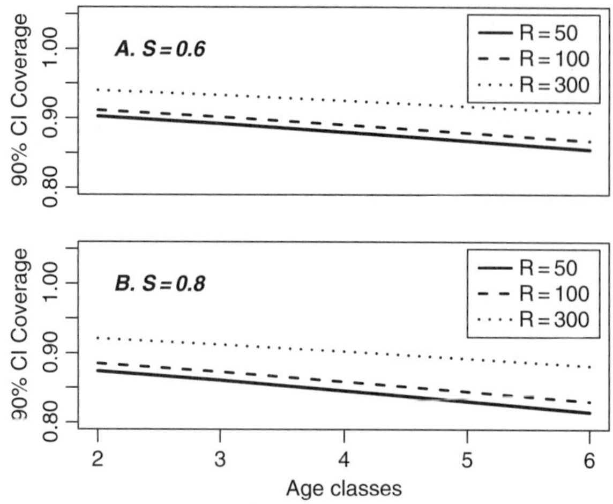

Fig. 1 Estimated year effects Y_t using the Sedge Warbler data and the Bayesian approach (posterior medians) which conditions on the lower bounds of the censored counts (*bold line*), and the standard Classical approach (MLE) which corrects censored counts prior to model fitting using the method of Peach et al. (1998) (*thin line*). The analysis of the reduced subset of the data with censored counts deleted is shown as a *line* with *filled circles*

with their observed lower bounds l_{it}. The Bayesian method however inflates the censored counts by adding an average of 5.49 to these lower bounds, rather than only 1.23 under a priori correction. This explains the Bayesian estimation of abundance in 1983, relative to subsequent years, being reduced in comparison to the Peach et al. (1998) method (Fig. 1): only two (5.3%) of the counts in 1983 are censored (Table 1), and neither method of inflating censored values will have much effect. Estimates are similar in 1984 too, when there are also few censored counts (7.7% of the annual total). In subsequent years, with more frequent censored values, more of the annual totals (and hence the year effects) will become larger relative to those in the base year under the Bayesian method than they do following Peach et al. (1998).

Similar results were obtained in an identical analysis of the numbers of juvenile birds caught (Cave 2006).

Table 1 Number of counts per year, and percentages of these censored (i.e. based upon <12 visits) in adult Sedge Warbler CES data

Year	1983	1984	1985	1986	1987	1988	1989
counts taken	38	39	41	60	71	76	82
% censored	5.3%	7.7%	19.5%	35.0%	23.9%	28.9%	19.5%
Year	1990	1991	1992	1993	1994	1995	1996
counts taken	90	93	97	104	109	108	112
% censored	21.1%	25.8%	25.8%	24.8%	26.6%	25.9%	30.1%
Year	1997	1998	1999	2000	2001	2002	–
counts taken	118	118	122	96	68	78	–
% censored	34.7%	30.5%	28.5%	27.1%	29.4%	25.6%	–

4.2 A Simulation Study

We now describe two analyses of simulated data. In the first we generate artificial data from a Poisson distribution with known parameter values, and truncate certain of these at random. We consider both a realistic level of truncation (based on that in the Sedge Warbler data), and one considerably more extreme than is likely to be seen in practice. Trends are then estimated via the Bayesian method to investigate its performance (as these data are not determined at the level of the individual visits, the a priori correction is not possible). A second set of simulations then introduces an element of lack of fit by selecting series of observations randomly from a set of the Sedge Warbler data. This allows a test of the Bayesian method in circumstances where the Poisson assumption is violated to a degree; furthermore, as complete records from all visits to a site in a given year are randomly selected, or not, the method of Peach et al. (1998) can also be used, and the two methods compared.

To obtain the Poisson data we first fitted the conventional model (1), with counts for incomplete years corrected a priori as above, to Sedge Warbler data from 1986 to 2005. The resulting parameter estimates were then used to simulate data:

$$n_{it} \sim \text{Poisson}(\lambda_{it}), \quad \lambda_{it} = \exp(\hat{\beta}_0 + \hat{Y}_t + \hat{S}_i)$$

A fully balanced data set (20 years of data for each of 178 sites) was thus produced, and a set of estimates for year effects **Y**, unaffected by censoring, was obtained. Each count n_{it} was then either selected for censoring (with probability 0.35) or retained at the existing value. Selected counts were censored by subtracting a censoring amount from $0,1,\ldots,n_{it}$ with probability $p(i) = 1/(2^{i+1})$ for $i \in [0, n_{it} - 1]$ and $p(n_{it}) = p(n_{it} - 1)$. This produces data sets in which the proportion and magnitude of the censored observations roughly approximate that in the Sedge Warbler data (Table 2). The Bayesian approach was then used to produce further estimates of **Y** (posterior medians) using the data set with these censored observations, for comparison with those from the full, uncensored data.

Close agreement is seen between the trends from uncensored and censored data (Fig. 2). Posterior medians of the imputed counts themselves are also in close agreement with their true (pre-censoring) values, though they are inflated, on average, by

Table 2 Predicted degree of censoring in observed adult Sedge Warbler data from 1986 to 2005 and in the simulated data under the two censoring mechanisms. The observed degree of censoring is that obtained by the Peach et al. (1998) corrected count minus the observed count

Degree	**0**	**1**	**2**	**3**	**4**	**5**	**6**	**7**	**8**	**9+**
Observed	.51	.23	.12	.05	.03	.02	.02	.01	.01	.01
Simulated	.57	.24	.1	.05	.02	.01	.002	.002	.004	.0008
Simulated (more severe)	.27	.12	.08	.05	.06	.04	.04	.03	.02	.27

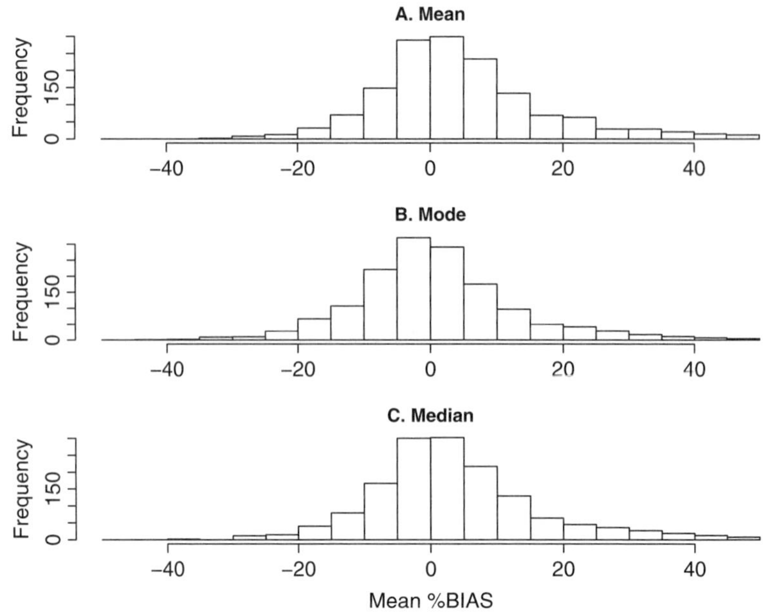

Fig. 2 Posterior medians estimating the year effects from simulated data; from the uncensored (*line* with *filled circles*) and censored (*thin line* under the first censoring mechanism, and bold line under the more severe censoring) data. Censored data are modelled using a Bayesian approach which conditions on the lower bounds of the censored counts

about 1.8. This happens because posterior distributions tend to be skewed (Fig. 3), meaning that medians are likely to be higher.

Given the relatively limited effect of the censoring mechanism above, we repeated these analyses choosing the same censoring amounts with probability $p(i) = 1/(n_{it} + 1)$ for $i \in [0, n_{it}]$ – all possible censorings chosen with equal probability, resulting in more counts censored more severely (Table 2). The censored counts arising were much less correlated ($\rho = 0.807$) with the true observations than were the set of the previous simulations ($\rho = 0.997$). In spite of this, the estimated trend is still little altered (Fig. 2). The posterior medians of the imputed counts again agree well with their true counterparts, with some over-inflation due to the skewed posteriors.

In the second scenario, to simulate data not predetermined to arise from a Poisson distribution, we first removed from the Sedge Warbler data all censored yearly-site counts and all data outside the period 1989–2002 (in each year of which at least 32 sites contributed information). In the remainder, 60 sites had at least four years of observations, a total of 12,844 captures of 7504 birds in 14 years, and 429 annual site totals. Model (1) was then applied to data from these 60 sites to generate estimated year effects for comparison. Censoring was then applied by deleting a randomly-selected 7% of the visits made, this proportion matching that in the genuine data. In

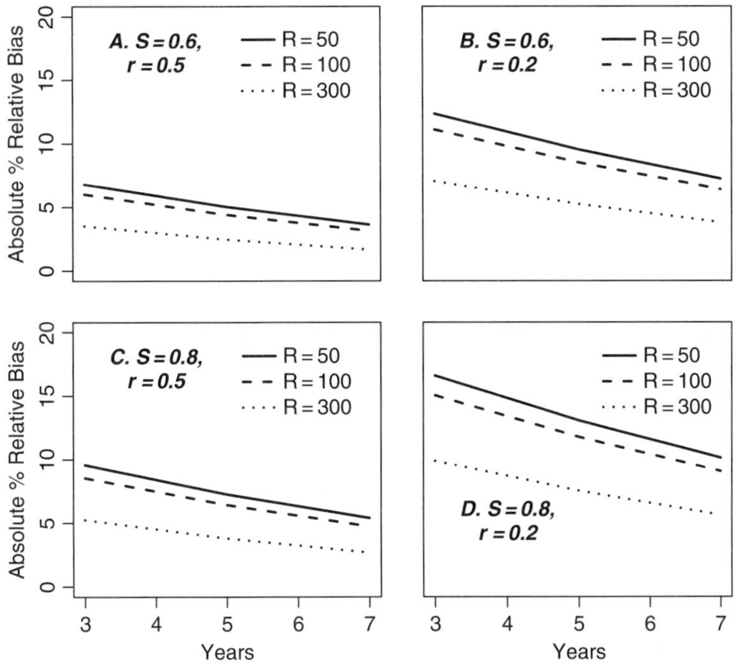

Fig. 3 Four, arbitrarily chosen, posterior distributions for counts estimated under the Bayesian approach. Data were initially simulated from a Poisson distribution and censored to a degree comparable to the Sedge Warbler data

truth, visits are not missed independently (there is, for example, a tendency towards more missed visits later in the season) but the procedure still gives us a useful opportunity to compare methods for accommodating censoring. As these data are produced at the level of individual visits, the method of Peach et al. (1998) can also be used, and compared with the results of the Bayesian method conditioning on the lower bounds. Adjustment in the former required the "global", rather than "local", correction in 23% of the cases.

Figure 4 shows excellent agreement between the annual trend under the Peach et al. (1998) method and that obtained using the data prior to censoring. The trend in the posterior medians from the Bayesian method also matches well except for the years 2000–2001 in which a marked discrepancy is revealed. Confidence limits under the Peach et al. (1998) method are, as expected, narrower than Bayesian credibility intervals as uncertainty in the correction process is ignored (Fig. 5). The correlation between the Bayesian posterior medians and the true counts is high (ρ = 0.935), though that of the Peach et al. (1998) correction is somewhat higher (ρ = 0.993). The Bayesian estimates here are inflated, on average, by about 5.5.

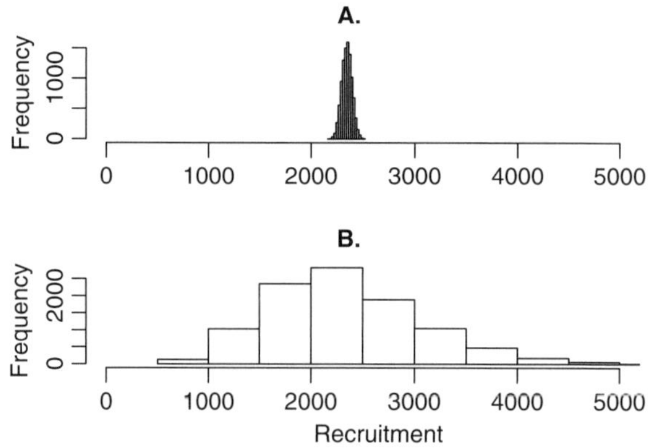

Fig. 4 Estimated year effects from simulated data; from the uncensored (*line* with *filled circles*), and from the censored data using the Peach et al. (1998) correction method (*thin line*) and the posterior medians from the Bayesian approach (*bold line*). Data are simulated by random selection from the sites in the Sedge Warbler data set, hence are overdispersed with respect to a Poisson distribution

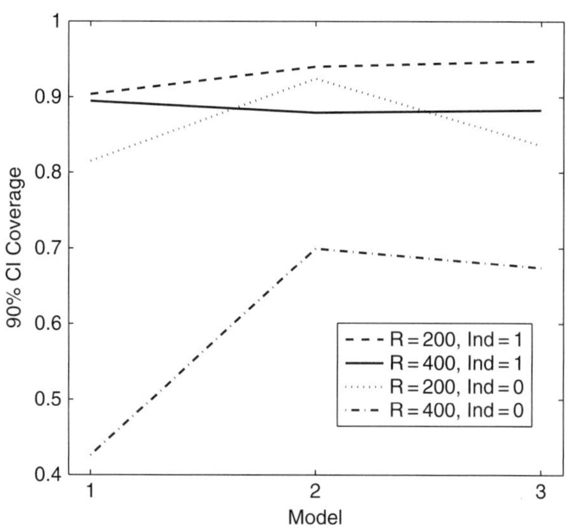

Fig. 5 95% confidence and credible intervals for the estimated year effects of Fig. 4 using the Peach et al. (1998) correction method (*thin bars*) and Bayesian approach (*bold bars*). The *dashed line* represents the year effects derived from the data prior to censoring

5 Discussion and Areas of Future Work

Accurately quantifying the uncertainty in parameter estimates is particularly important when the findings will be used for identifying and monitoring populations of species at risk. Bayesian techniques have recently proved useful in modelling annual territory counts with this aim in mind, permitting for example the calculation of (posterior) probabilities that a species has declined by at least a specified amount (Mazzetta et al. 2007; King et al. 2008). For many species data from constant effort ringing provide an alternative source of information. Modelling techniques however need modification to match the different properties of the data. Rather than being counts of breeding pairs at a site, CES counts are merely numbers of birds caught in mist-nets, assumed proportional to abundance if the ringing effort is unchanged. Failure in this latter, crucial assumption requires explicit consideration in the analysis.

Our first conclusion is that trends from the non-parametric method of Peach et al. (1998) are shown to be robust, even to a degree of model failure. The simulation studies presented here indicate that when the underlying Poisson model is correct, the Bayesian method for incorporating censored counts also works well, even in the presence of extreme censoring. Population trends are unbiased and the approach correctly reports the degree of precision. However this model-based approach may perform worse if the underlying model is incorrectly specified. For a specific data set, namely CES yearly-site counts for adult Sedge Warblers, in which the particularly simple Poisson assumption appears to be violated in the later years especially, the Peach et al. (1998) correction method was shown to perform better. The reasons for the greater differences in the later years will merit further investigation. These findings do not necessarily imply that the non-parametric method is better generally. Such an approach underestimates the standard errors of the year effects. By ignoring the uncertainty arising from the correction a priori of censored counts these parameters are reported with more precision than they merit. By falsely reporting the precision of these abundance indices inference becomes unreliable, and declines might be inferred where a population is in fact stable.

The Bayesian method can readily be adapted to accommodate different, possibly more appropriate models. For example the full (non-censored) adult Sedge Warbler data showed evidence of a lack of fit. The ratio of the deviance to the residual degrees of freedom was 2.44, indicating some overdispersion. Peach et al. (1998) noted that overdispersion was negligible for the majority of adult CES data sets they considered, with only 7 of 28 species having ratios greater than 2.0. In cases where count data are overdispersed the Negative Binomial distribution provides an alternative to the Poisson since it allows the variance to exceed the mean. The Bayesian method described in this paper can be easily rewritten with a Negative Binomial model by simply replacing the Poisson likelihood with the Negative Binomial likelihood and including an extra Metropolis-Hastings updating step for the overdispersion parameter. This is an area of ongoing research.

An additional, viable approach for dealing with missed visits is to model the counts at the visit level. That is, instead of combining counts across all 12 visits and modelling the yearly-site counts, the counts at each visit are modelled instead. An extra factor, the "visit", with twelve levels is added to the model, a visit effect term is added to the RHS of equation (1), and the offset is omitted. As missed visits contribute no information this model can be readily fitted using standard Classical and Bayesian methods. This extension has three potential advantages. Firstly, the problem of missed visits is overcome. Secondly, a greater degree of precision in the estimates may arise simply due to the greater amount of data. Thirdly, the visit effects themselves are of interest as they provide information on the availability of birds for capture over the summer mist-netting season. Further, this latter feature permits investigation of the assumption underlying the Peach et al. (1998) correction that there is no variation in the inter-annual pattern of catches. Preliminary analysis of the adult Sedge Warbler data (Cave 2006) has demonstrated that similar abundance trends were produced, using both Classical and Bayesian methods, whether adjustments were made for missed visits, or data were analysed at the visit, rather than annual, level. Comparison of results with analyses of the data at the visit level will be published separately in due course.

"Constant Effort" schemes are becoming increasingly prevalent. Over a dozen EURING schemes now have similar programmes; for example Finland, France and Spain all have schemes operating since the early nineties (EURING 2006), and the MAPS scheme in the United States adopts a similar protocol (DeSante et al. 1999). We note too that surveys based on bird counts within a specified area, such as that described by Link and Sauer (1999), require consideration of varying effort. A Bayesian approach of the kind described here might also prove useful outside the ringing context. Further investigation into modelling methods is therefore timely as more data sets accrue with the potential to produce long-term indices.

Acknowledgments The authors would like to thank the CES ringers who have devoted a great deal of time and effort in the collection of the CES data over the years. The UK ringing scheme is jointly funded by the BTO, the Joint Nature Conservation Committee (on behalf on Natural England, the Countryside Council for Wales, Scottish Natural Heritage, the Environment & Heritage Service of Northern Ireland) and the ringers themselves. The research in this paper was made possible by funding from the National Centre for Statistical Ecology through an Engineering and Physical Sciences Research Council grant. Additional funding, for Vanessa Cave, was also generously provided by AgResearch (New Zealand). The authors are grateful to Will Peach for discussion, Stephen Baillie and Rob Robinson who commented on the draft of the manuscript, and to Andy Royle and an anonymous reviewer whom made invaluable suggestions for its improvement.

Appendix

A two-step procedure for simultaneous model-fitting and imputation of censored counts. The superscripts z, and z+1, denote the updates at iteration z and (z+1) respectively.

Step One: Updating the Model Parameters

Assuming the n_{it} are Poisson distributed with mean $\lambda_{it} = \exp(\beta_0 + S_i + Y_t)$ the likelihood is given by:

$$f(\mathbf{n}^z; \boldsymbol{\lambda}) = f(\mathbf{n}^z; \beta_0, \mathbf{S}, \mathbf{Y}) = \prod_{it} \frac{\exp(-\exp(\beta_0 + S_i + Y_t))(\exp(\beta_0 + S_i + Y_t))^{n_{it}^z}}{n_{it}^z!}$$

where $\mathbf{n}^z = (\mathbf{n}^{obs}, \mathbf{n}^{cen(z)})$ is a data vector containing both the fully observed uncensored counts $\mathbf{n}^{obs}$ and the current updates of the imputed censored counts $\mathbf{n}^{cen(z)}$ rather than their observed values, thus accommodating the truncated distribution where appropriate. Note that the elements of $\mathbf{n}^{cen(z)}$ are constrained such that $n_{it}^{cen(z)} \geq l_{it}$ for all i, t. See Step 2 for more details.

Vague, independent, Normal priors p with mean 0 and variance 10,000 are given to all model parameters. The posterior distribution π of the model parameters is therefore given by:

$$\pi(\beta_0, \mathbf{S}, \mathbf{Y}; \mathbf{n}^z) \propto f(\mathbf{n}^z; \beta_0, \mathbf{S}, \mathbf{Y}) p(\beta_0) \prod_i p(S_i) \prod_t p(Y_t)$$

The normal proposal is used for updating all parameters, which being symmetric cancels out in the expression for the acceptance probability. The proposal variance was set to 0.5 for all parameters as a priori tuning runs indicated that this would result in acceptance rates, for the majority of the model parameters, near the optimal rate of 0.234 (Roberts et al. 1997).

At iteration $(z + 1)$ the algorithm begins by updating β_0, then one by one the site and year effects. To update β_0, where the current value is denoted by $\beta_0{}^z$, a new v alue, $z\beta_0{}^*$, is proposed from the Normal$(\beta_0{}^z,0.5)$ distribution, which is accepted with probability:

$$\alpha(\beta_0^z, \beta_0^*) = \min\left(1, \frac{f(\mathbf{n}^z; \beta_0^*, \mathbf{S}^z, \mathbf{Y}^z) p(\beta_0^*)}{f(\mathbf{n}^z; \beta_0^z, \mathbf{S}^z, \mathbf{Y}^z) p(\beta_0^z)}\right)$$

If the move is accepted we set $\beta_0{}^{z+1} = \beta_0{}^*$, else $\beta_0{}^{z+1} = \beta_0{}^z$.

Next, an update for S_1 is proposed from the Normal$(S_1{}^z,0.5)$ distribution, which is accepted with probability:

$$\alpha(S_1^z, S_1^*) = \min\left(1, \frac{f(\mathbf{n}^z; \beta_0^{z+1}, S_1^*, \mathbf{S}_{(1)}^z, \mathbf{Y}^z) p(S_1^*)}{f(\mathbf{n}^z; \beta_0^{z+1}, S_1^z, \mathbf{S}_{(1)}^z, \mathbf{Y}^z) p(S_1^z)}\right)$$

Here $\mathbf{S}^z{}_{(i)}$ denotes the vector of site effects, excluding S_i, at iteration z. Likewise the remaining parameters are updated.

Step Two: Imputing Censored Counts

To update an imputed censored count, where the current value is denoted by $n_{it}{}^z$, a new value, $n_{it}{}^*$, is proposed from the discrete Uniform(A,B) distribution (note that for notational convenience the "cen" superscript has been omitted). Here A = $\max(l_{it}, n_{it}{}^z - \varepsilon)$, is used to guaranteed that $n_{it}{}^*$ is at least l_{it}, and B = $n_{it}{}^z + \varepsilon$. The random walk jump, ε, was set at 6 for all imputed censored counts as a priori tuning runs indicated that this would result in acceptance rates, for the majority of the imputed censored counts, again near the optimal rate of 0.234. Under this scheme the proposal distribution is not necessarily symmetric (for example when $n_{it}{}^{z}-\varepsilon < l_{it}$ or $n_{it}{}^{*}-\varepsilon < l_{it}$) and must be incorporated explicitly into the expression for the acceptance probability. The proposal distribution is given by:

$$\text{if } n_{it}^{z} - \varepsilon < l_{it} \text{ then } q(n_{it}^{*}|n_{it}^{z}) = \frac{1}{n_{it}^{z} + \varepsilon + 1 - l_{it}} \text{ else } q(n_{it}^{*}|n_{it}^{z}) = \frac{1}{2\varepsilon + 1}$$

$$\text{if } n_{it}^{*} - \varepsilon < l_{it} \text{ then } q(n_{it}^{z}|n_{it}^{*}) = \frac{1}{n_{it}^{*} + \varepsilon + 1 - l_{it}} \text{ else } q(n_{it}^{z}|n_{it}^{*}) = \frac{1}{2\varepsilon + 1}$$

Uniform priors, with upper bound sufficiently large such that imputed values are extremely unlikely to exceed it, are given to all imputed censored counts. Consequently the prior terms cancel out in the expression for the acceptance probability, as do all likelihood terms in expression (2) aside from the contribution by the n_{it} being updated. The proposed update is therefore accepted with probability:

$$\alpha(n_{it}^{z}, n_{it}^{*}) = \min\left(1, \frac{f(n_{it}^{*}; \beta_0^{z+1}, S_i^{z+1}, Y_t^{z+1})q(n_{it}^{z}|n_{it}^{*})}{f(n_{it}^{z}; \beta_0^{z+1}, S_i^{z+1}, Y_t^{z+1})q(n_{it}^{*}|n_{it}^{z})}\right)$$

References

Bonner S, Thomson DL, Schwarz CJ (2007) Time-varying covariates and semi-parametric regression in capture–recapture. This Volume.

Cave VM (2006) Integrated Modelling of UK Songbird Populations using Data from the British Trust for Ornithology's Constant Effort Sites Scheme. Unpublished manuscript, University of Cambridge.

DeSante DF, O'Grady DR, Pyle P (1999) Measures of productivity and survival derived from standardized mist-netting are consistent with observed population changes. *Bird Study* 46 (Suppl.) S178–S188.

EURING (2006) http://www.euring.org/research/ces_in_europe/methods.htm.

Gelman, A, Carlin JB, Stern HS, Rubin DB (2004) *Bayesian Data Analysis*. 2 nd edn., Chapman and Hall, London.

Grantham M, Robinson R (2007) Keeping it constant to measure the changes. *BTO News* 269: 12–13.

King R, Brooks SP, Mazzetta C, Freeman SN, Morgan BJT (2008) Identifying and diagnosing population declines: a Bayesian assessment of Lapwings in the UK. *Journal of the Royal Statistical Society Series C* In press.

Link WA, Sauer JR (1999) Controlling for varying effort in count surveys – an analysis of Christmas Bird Count data. *Journal of Agricultural, Biological and Environmental Statistics* 4: 116–125.

Mazzetta C, Brooks SP, Freeman SN (2007) On smoothing trends in population index modelling. *Biometrics*, 63(4): 1007–1014.

Miles W, Freeman SN, Harrison NM, Balmer DE (2007) Measuring passerine productivity using Constant Effort Sites: the effect of missed visits. *Ringing & Migration*, 23(14): 231–237.

Peach WJ (1993) Combining mark-recapture data sets for small passerines. In: Lebreton J-D, North PM (eds) *Marked Individuals in the Study of Bird Population*. Birkhauser-Verlag, Basel/Switzerland, pp 107–122.

Peach WJ, Baillie SR, Balmer DE (1998) Long-term changes in the abundance of passerines in Britain and Ireland as measured by constant effort mist-netting. *Bird Study* 45: 257–275.

Peach WJ, Baillie SR, Underhill LG (1991) Survival of British Sedge Warblers *Acrocephalus schoenobaenus* in relation to west African rainfall. *Ibis* 133: 300–305.

Peach WJ, Buckland ST, Baillie SR (1990) Estimating survival rates using mark-recapture data from multiple ringing sites. *The Ring* 13 (1–2): 87–102.

Peach WJ, Crick HQP, Marchant JH (1995) The demography of the decline in the British Willow Warbler population. *Journal of Applied Statistics* 22: 905–922.

Roberts GO, Gelman A, Gilks WR (1997) Weak Convergence and Optimal Scaling of Random Walk Metropolis Algorithms. *The Annals of Applied Probability* 7(1): 110–120.

Robinson RA, Freeman SN, Balmer DE, Grantham MJ (2007) Cetti's Warbler Cettia cetti: analysis of an expanding population. *Bird Study*, 54: 230–235.

ter Braak CJF, Van Strien AJ, Meijer R, Verstrael TJ (1994) Analysis of monitoring data with many missing values: which method? In: Hagemeijer EJM, Verstrael TJ (eds) *Bird Numbers 1992. Distribution, Monitoring and Ecological Aspects*, pp 663–673. Proceedings of the 12th International Conference, International Bird Census Council and European Ornithological Atlas Committee. Beek-Ubbergen, The Netherlands, Sovon.

Simulation Performance of Bayesian Estimators of Abundance Employing Age-at-Harvest and Mark-Recovery Data

Paul B. Conn, Gary C. White, and Jeffrey L. Laake

Abstract The age structure of harvests has long been an important source of information in fisheries stock assessments, especially when augmented with data from catch-effort or research vessel surveys. Age-at-harvest data are also collected for many terrestrial species, a fact which has recently prompted several authors to propose models for analyzing wildlife age-at-harvest data, with the object of estimating abundance, survival, harvest parameters, and recruitment. Since analysis with age-at-harvest data alone often leads to problems with parameter identification, these authors suggested that data from studies of marked animals could be used to inform the estimation of survival and recovery rates. However, little work has been done to examine estimator performance, particularly when model assumptions are violated, as when aging errors occur or when mark-recovery and age-at-harvest data are non-independent. Similarly, we know of no studies that have investigated the efficacy of posterior simulation when Bayesian estimation methods are used for such problems. In this paper, we employ a suite of simulation modules to quantify estimator performance under a number of hypothetical biological scenarios. When all assumptions are satisfied, we show that bias is typically of small magnitude, coefficient of variation is small, and that credible interval coverage is satisfactory. Estimators were robust to errors in age determination but precision had the potential to be severely overestimated when data from marked animals were also included in age-at-harvest summaries. Nevertheless, joint analysis of age-at-harvest and mark-recovery data may represent a viable monitoring strategy for many terrestrial species.

Keywords Abundance · Age-at-harvest · Cohort model · Demography · Mark-recovery · State-space model · Survival

---

P.B. Conn (✉)
Colorado Cooperative Fish & Wildlife Research Unit, Department of Fish, Wildlife, and Conservation Biology, Colorado State University, Fort Collins, CO 80523, USA
e-mail: paul.conn@noaa.gov

D.L. Thomson et al. (eds.), *Modeling Demographic Processes in Marked Populations*, Environmental and Ecological Statistics 3, DOI 10.1007/978-0-387-78151-8_44,

1 Introduction

Accurate estimation of animal population trends is essential for effective wildlife management. However, many of the traditional methods of estimating population parameters may be too expensive to be practical in some circumstances. For instance, closed population mark-recapture estimators often require samples of hundreds or thousands of individuals to achieve adequate precision on abundance (Seber 1982). Thus, a recent focus of research has been to develop methods for jointly analyzing data form disparate sources in hopes of enhancing the precision of abundance estimators without substantially increasing costs.

For exploited populations, one such source of data is the age-structure of hunter harvests. Data of this sort have frequently been used in fisheries stock assessments as a primary source of information for estimating cohort abundances and estimating fishing mortality parameters (Megrey 1989). Methods have ranged from simple deterministic calculations to fully formulated statistical likelihoods capable of incorporating a flexible amount of auxiliary data such as fishing effort and research vessel indices (Fournier and Archibald 1982; Dupont 1983; Deriso et al. 1985). Recently, some fisheries analysts have adopted a Bayesian perspective, allowing them to fit increasingly intricate models for fish population dynamics and to summarize posterior distributions for parameters of interest (McAllister and Ianelli 1997; Virtala et al. 1998; Meyer and Millar 1999; Lewy and Nielson 2003).

While it is possible to formulate various likelihoods relating abundance and mortality to age-at-harvest data, parameters typically cannot be identified without auxiliary information on survival, the harvest process, or abundance (Megrey 1989; Laake 1992; Gove et al. 2002). Studies of marked animals, while impractical for many oceanic fisheries, constitute one such potential source of auxiliary data (Maunder 2001; Gove et al. 2002). Indeed, data from such studies have been successfully used together with wildlife age-at-harvest data to estimate demographic parameters. For instance, Gove et al. (2002) used data from radio-telemetry and reporting rate studies together with age-at-harvest data to jointly estimate survival, abundance, and harvest parameters for elk in Idaho. Similarly, Conn (2007) and Conn et al. (in press) used mark-recovery and age-at-harvest data to estimate demographic parameters for black bear in Pennsylvania.

Since age-at-harvest records are readily collected for many wildlife populations, modeling with such data appears to be a promising avenue for population monitoring. For instance, Conn (2007) and Conn et al. (in press) reported substantially better precision on abundance for a joint age-at-harvest, mark-recovery estimator than for approaches ignoring age structure. Nevertheless, little research has been done to explore estimator performance, particularly when model assumptions are violated. Further, performance of MCMC methods for addressing such problems has not been evaluated.

In this paper, we conduct a suite of simulations in order to explore the performance of abundance estimators when age-at-harvest data are used in the estimation process. We focus on the case where mark-recovery data are available to help model harvest and survival processes, and specifically on the state-space, or "hidden process" (Newman et al. 2006) formulation described by Conn (2007) and Conn et al. (in press). Our goal is not to provide a thorough mathematical derivation of their likelihood. Instead, we start by providing a heuristic description of model structure. Next, we explore the efficacy of posterior simulation. The majority of the article is then devoted to obtaining simulation based measures of model performance. In particular, we include three different simulation modules. In the first (Section 4.1), we explore large sample performance when all model assumptions are met. In the second (Section 4.2), we explore model performance when there are aging errors. Finally, in simulation module III (Section 4.3), we examine consequences when records from marked animals are simultaneously included in age-at-harvest and mark-recovery datasets.

2 Model Description

Conn (2007) and Conn et al. (in press) viewed age-at-harvest data as having arisen from a two-tiered stochastic process. The first part of the process consisted of a population model describing temporal changes in age-specific abundance. The population model involves a survival process, as well as a recruitment process relating the abundance of recruits to previous levels of adult abundance. The recruitment process is summarized by a Poisson distribution, where

$$N_{i+1,1} \sim \text{Poisson}\left(\sum_{j=1}^{A} N_{ij} f_{ij}\right),$$

where definitions of parameters and statistics are given in Table 1. We assume that survival is a binomial process, such that

$$N_{i+1,j+1} \sim \text{Binomial}(N_{ij}, S_{ij}).$$

Table 1 Definitions of parameters, latent variables, and statistics used in age-at-harvest models

S_{ij}	Probability that an age *j* individual survives to time *i*+1 given it was alive at time *i*
h_{ij}	Probability that an age *j* individual is harvested and reported to wildlife personnel in [*i*, *i*+1], given that it was alive at time *i* (i.e., recovery probability)
r_{ij}	Probability that an age *j* individual is reported to wildlife personnel in [*i*, *i*+1], given that it dies in [*i*, *i*+1]
f_{ij}	Per breeder recruitment rate over [*i*, *i*+1], with reference to the number of age *j* unmarked breeders in the population at time *i* and the number of new recruits at time *i*+1
N_{ij}	Number of age *j* individuals in the population in year *i* immediately prior to harvest. The N_{1j} are parameters while the remaining N_{kj} $(k > 1)$ are treated as latent variables
C_{ij}	Number of age *j* unmarked individuals that are harvested and reported to wildlife personnel in year *i*
R_{ij}	Number of of age *j* unmarked individuals that are marked and released in year *i*
A	Age at which an individuals age cannot be reliably distinguished from older age classes
Y	Duration of the study (e.g., years)

Introduction of a self-loop (i.e., an absorbing "+" age class) requires some further development which is omitted here for brevity. The second part of the process consists of a sampling model, whereby age-at-harvest data are drawn from probability distributions dependent on age- and time- specific abundance and harvest parameters. For instance, if the Brownie parameterization (Brownie et al. 1985) is chosen for harvests,

$$C_{ij} \sim \text{Binomial}(N_{ij}, h_{ij}).$$

Reducing the complexity of model structure by assuming time- or age constancy in model parameters is not enough to render them identifiable. Instead, either informative prior distributions or auxiliary data such as from a mark-recovery or radio telemetry study are needed. For the purposes of this paper, we assume that a mark-recovery dataset is available to help model the harvest process. We follow the suggestion of Gove et al. (2002) and base inference on the likelihood $L = L_1 L_2$, where L_1 gives the likelihood for age-at-harvest data, and L_2 gives the likelihood of the auxiliary dataset (in this case from the mark-recovery study). Each piece of the likelihood shares the same survival and harvest parameters, which allows abundance and recruitment to be identified whenever survival (S_{ij}) and recovery (h_{ij}) probabilities can be estimated from the mark-recovery data alone.

3 Posterior Simulation

Estimation via traditional maximum likelihood techniques is difficult in this case, due to the large number of (latent) abundance parameters that need to be integrated out of the likelihood. As Link et al. (2003) pointed out, there may also be considerable interest in retaining these parameters. For instance, calculating annual abundance, $N_{i\cdot} = \sum_j N_{ij}$ requires knowledge of all of the N_{ij} parameters. For these reasons, as well as for ease with which hierarchical extensions to model structure can be implemented, we preferred a Bayesian estimation scheme using Markov Chain Monte Carlo (MCMC).

The full conditional posterior distributions for model parameters do not belong to recognizable families of probability density functions. Therefore, it is not possible to simulate parameter values directly as in traditional Gibbs sampling. One possibility for simulating these parameters is to use a Metropolis-within-Gibbs hybrid update (Brooks 1999; Gelman et al. 2004). While this approach has been shown to work well in Bayesian analysis of mark-recovery and mark-recapture data alone (Brooks 1999), the addition of age-at-harvest data into the likelihood presents additional challenges. Abundance is an integer value, subject to a number of constraints; thus, proposals must also be integers and the Metropolis ratio must be corrected for asymmetries resulting from these constraints. One possibility for generating univariate proposals for abundance, which we adopt throughout, involves the following algorithm:

1. Generate $\theta_i^* \sim \frac{\text{Normal}(\theta_i^{(t-1)}, \sigma_i^2)}{\int_{\theta_i^{\min}-0.5}^{\theta_i^{\max}+0.5} \text{Normal}(\theta_i^{(t-1)}, \sigma_i^2)}$

   for $(\theta_i^{min} - 0.5) \leq \theta_i^* < (\theta_i^{max} + 0.5)$ by rejection sampling.
2. Round θ_i^* to the nearest integer
3. Calculate Metropolis-Hastings ratio as

$$\frac{P(\theta_i^* | \mathbf{Y}, \boldsymbol{\theta}^{(t)}) \int_{\theta_i^{min}-0.5}^{\theta_i^{max}+0.5} \text{Normal}(\theta_i^{(t-1)}, \sigma_i^2)}{P(\theta_i^{(t-1)} | \mathbf{Y}, \boldsymbol{\theta}^{(t)}) \int_{\theta_i^{min}-0.5}^{\theta_i^{max}+0.5} \text{Normal}(\theta_i^*, \sigma_i^2)}.$$

Here, θ_i^* gives the proposed abundance value, $\theta_i^{(t-1)}$ gives the abundance value at the previous iteration of the Markov chain, and θ_i^{min} and θ_i^{max} give the minimum and maximum values that are permissable given the constraints (with $\theta_i^{max} = \infty$ if there is no upper constraint). The first author was unable to improve on the Metropolis-Hastings hybrid update by considering correlated proposals (Conn 2007), so univariate proposals of this type are considered throughout.

Using this approach, mixing of Markov chains was poor. For instance, lag-200 autocorrelation of Markov chains often fell between 0.65 and 0.95 in the simulations reported below. Given this low degree of mixing, one would naturally wish to know how many MCMC iterations would be required to accurately summarize features of the posterior distribution.

While the answer to this question ultimately depends on the population, data, and estimation model, we describe a "best case scenario" approach for answering this sort of question. In particular, we summarize the variation between multiple Markov Chains at different landmarks of simulation time for a number of design points for model $S(a+t)h(a)f(\cdot)$ (using the notation of Lebreton et al. (1992) to describe model structure).

To start, we generated expected value data for 12 design points, rounding harvest numbers to the nearest integer. Possible configurations were

- True abundance, number of releases/year = (5000,150), (5000,300)
- Reporting rate (r) = 0.2 or 0.5
- Number of years (Y), cohorts (A) = (3,3), (5,3), or (5,5).

Here, r is related to the recovery rate h by the formula $h = (1 - S)r$, and is used to conform to analyses later in the manuscript. Study durations were relatively short in all cases to reduce computing time. Note that these values of r are relatively high and indicative of situations in which harvest represents a substantial portion of total mortality and where hunters/fishermen report a large percentage of harvests to wildlife personnel.

All expected value data sets assumed time- and cohort-constant survival and recruitment of 0.6 and 0.4, respectively, and an initial population size of 5,000. Each population was started at it's stable stage distribution by assuming a stable population (λ=1.0) and using standard approaches to calculating the stable stage distribution (Caswell 2001). For each design point and candidate proposal scheme, we employed 10 sets of Markov chains to approximate features of the posterior distribution. We were interested in the similarity of Markov chains at different points of simulation time; that is, how repeatable results were as a function of simulation time. We quantified "similarity" by calculating (a) the average standard deviation of the posterior mean of total abundance (where the standard deviation is calculated across Markov chains for each year of the study and then averaged across years), and (b) the standard deviation of the deviance information criterion (DIC) (Spiegelhalter et al. 2002) when the posterior mean is used to calculate DIC. Each chain was run for 2.1 million iterations, with the first 100,000 iterations discarded as a burn- in

Table 2 Performance of Markov Chains in summarizing marginal posterior distributions for animal abundance and DIC at different landmarks of simulation time for model $S(a+t)h(a)f(\cdot)$. Part (A) gives the average between chain standard deviation of annual abundance as determined by the posterior mean, while part (B) gives the average between chain standard deviation of DIC. *A* gives the number of age classes, *Y* gives the number of years of the study, *R* gives the number of marked releases per year, and *r* gives reporting rate (i.e. the probability a tag is reported given death). Initial abundance was 5000 and model $S(a+t)h(a)f(\cdot)$ was used for estimation in all cases

				MCMC Iterations ($\times 10^5$)				
A	*Y*	*R*	*r*	1	5	10	15	20
A. Between chain SD($\hat{N}$)								
3	3	150	0.2	210.9	71.1	34.2	32.7	44.8
3	3	150	0.5	25.2	12.8	10.7	10.1	9.4
3	3	300	0.2	81.7	28.1	29.3	25.8	18.9
3	3	300	0.5	27.7	11.8	8.8	5.7	4.9
3	5	150	0.2	115.0	69.9	45.2	34.5	27.0
3	5	150	0.5	19.5	12.9	13.6	9.0	7.6
3	5	300	0.2	39.8	13.2	15.9	10.3	8.0
3	5	300	0.5	11.5	7.9	5.1	3.8	3.5
5	5	150	0.2	111.7	39.7	45.3	22.1	17.9
5	5	150	0.5	22.4	14.1	10.9	5.5	2.2
5	5	300	0.2	67.7	20.4	12.9	14.1	9.5
5	5	300	0.5	10.0	5.5	2.9	3.1	2.1
B. Between chain SD(DIC)								
3	3	150	0.2	6.5	2.6	1.7	1.8	1.8
3	3	150	0.5	0.4	0.2	0.1	0.1	0.1
3	3	300	0.2	3.3	1.8	0.9	0.6	0.6
3	3	300	0.5	0.5	0.2	0.1	0.1	0.1
3	5	150	0.2	16.4	4.6	3.8	2.7	1.6
3	5	150	0.5	0.4	0.3	0.2	0.2	0.2
3	5	300	0.2	4.3	1.0	0.7	0.4	0.5
3	5	300	0.5	0.2	0.2	0.1	0.1	0.1
5	5	150	0.2	8.1	5.5	4.4	3.5	3.2
5	5	150	0.5	0.6	0.2	0.2	0.2	0.1
5	5	300	0.2	4.5	1.5	1.3	1.1	1.4
5	5	300	0.5	0.4	0.2	0.2	0.1	0.1

and the chain thinned by recording only 1 in 10 observations to reduce memory requirements. Markov chains were then summarized at 1.0×10^5, 5.0×10^5, 1.0×10^6, 1.5×10^6 and 2.0×10^6 iterations. In all cases, and throughout the remainder of this article, we assumed a set of "diffuse" prior distributions. In particular, we set $\Pr(N_{1j}) \propto c$ (Link et al. 2003), $[\beta_{ij}^S] \sim$ Normal(0, 3), $[\beta_{ij}^h] \sim$ Normal(0, 3), and $[\beta_{ij}^f] \sim$ Normal(0.25, 1). The β_{ij}^θ parameters correspond to fixed effect parameters on the logit scale for $\theta = S$ and $\theta = h$, and the log scale for $\theta = f$.

Several observations may be made based on results of this experiment (Table 2). First, the number of iterations required to summarize the posterior distribution decreases as the quantity of mark-recovery data increases. This could occur either through inclusion of additional years of data, higher recovery probabilities, or higher numbers of initial releases. Number of age classes did not appear to have much of an effect on convergence rates, at least with the range of experimental inputs considered here. A second issue that arose had to do with sparse mark-recovery data. In particular, when numbers of releases and reporting rates were low (e.g., R = 150 and $r = 0.2$), the variance in posterior estimates of abundance from different Markov chains was extremely high even after 2.1 million iterations. This was likely because of near non-estimability of parameters when data were too sparse. Apparently, more data are required to get sensible estimates in these cases, or perhaps stronger priors. As one reviewer noted, it may be inappropriate to use these results to make inferences about requisite simulation times for real life estimation problems. We view these results as a "best case scenario" in the sense that it may be easier to sample the posterior distribution associated with expected value data than would be the case for real life, "messy" data.

4 Estimator Performance

Although estimators of abundance appear to converge if Markov chains are run for long enough, this does not preclude bias or guarantee estimators with good properties. Thus, we used simulation to investigate estimator performance under a variety of hypothetical scenarios. In total, we considered 3 simulation modules to summarize estimator performance and diagnostics. In the first module, we evaluated bias, coefficient of variation (CV) and 90% Bayesian credible interval coverage (BCOV) for a variety of models, parameter combinations, number of age classes, years of data, and number of individuals marked per year. In the second module, we examined the performance of several estimators when there were errors in age determination. Finally, in the third module, we examined consequences of using data from marked animals in both portions (age-at-harvest and mark-recovery) of the likelihood.

4.1 Simulation Module I: Large Sample Performance

In the first simulation module, our goal was to quantify estimator performance when model assumptions were perfectly satisfied and when enough data were available to

Table 3 Combinations of initial population size, number of individuals marked and released per year, and estimation model that are considered in simulation module I

Initial N	Number of releases	Estimation model
1000	50/year	$S(\cdot)h(\cdot)f(\cdot)$
1000	100/year	$S(\cdot)h(\cdot)f(\cdot)$
1000	100/year	$S(a)h(t)f(\cdot)$
2000	100/year	$S(a)h(t)f(\cdot)$
2000	300/year	$S(a)h(t)f(\cdot)$
2000	300/year	$S(a+t)h(a)f(t)$
5000	300/year	$S(a+t)h(a)f(t)$

avoid issues with parameter estimability. We thus assumed combinations of initial population sizes, number of marked individuals, and estimation models such that estimation models included a reasonable level of complexity for the given values of population size and number of marked animals. We further assumed that marked individuals were not part of the population being estimated (that is, they did not contribute information to the age-at-harvest matrix). A complete representation for each combination of initial population size, number of individuals marked per year, and estimation model are presented in Table 3.

We generated data with values for λ of 0.9 and 1.0, values for S of 0.6 or 0.8, and values for r of 0.2 or 0.5. Thus, we considered 8 possible combinations of initial parameter values, with a possibility of 3, 5, or 7 years of data, and 2, 3, or 6 age classes. Studies were relatively short in all cases to reduce computing time.

We used different population models to simulate data depending on the number of age classes. In each case, we specified values of S, λ, and population size and derived the implied value of recruitment and the stable stage distribution. Recruitment was determined by setting det $|\mathbf{A} - \lambda\mathbf{I}| = \mathbf{0}$ and solving for f in terms of λ and S. The stable stage distribution was determined by solving the system of equations $[\mathbf{A} - \lambda\mathbf{I}]\,\mathbf{N}^{\mathrm{T}} = \mathbf{0}$, subject to the constraint $N = N_{11} + N_{12} + \ldots + N_{1A}$, where $N = [N_{11}, N_{12}, \ldots, N_{1A}]$ (Caswell 2001). Possible population models, presented here as matrices (Caswell 2001), included

$$\mathbf{A}_1 = \begin{bmatrix} 0 & 0 & f & f & f & f \\ S & 0 & 0 & 0 & 0 & 0 \\ 0 & S & 0 & 0 & 0 & 0 \\ 0 & 0 & S & 0 & 0 & 0 \\ 0 & 0 & 0 & S & 0 & 0 \\ 0 & 0 & 0 & 0 & S & S \end{bmatrix}, \ \mathbf{A}_2 = \begin{bmatrix} 0 & 0 & f \\ S & 0 & 0 \\ 0 & S & S \end{bmatrix} \text{ and } \mathbf{A}_3 = \begin{bmatrix} 0 & f \\ S & S \end{bmatrix}.$$

Here, the matrix model format is used only to represent structural features of the population models; actual simulations assumed stochasticity in all processes. Model $\mathbf{A}_1$ describes a model with 6 age classes and a pre-breeding census; the last 4 age classes can produce young. Individuals older than five years old cannot be differentiated, a feature incorporated with a self-loop. This scenario corresponds roughly to the population biology and age identification criterion of black bear in Pennsylvania, a population of interest to us. Model $\mathbf{A_2}$ describes a model for a population with 3 age classes, as when juveniles, yearlings, and adults may be differentiated, but finer scale resolution is not possible. In this case, only adults contributes to recruitment in the following year. Note that this model is structurally equivalent to $\mathbf{A_1}$ if one were to pool individuals in age classes 3–6 into a single age class. Finally, model $\mathbf{A_3}$ describes a population where only 2 age classes are recognizable, as with many bird species.

Values for recruitment and stable stage distribution for the 3 models were as follows:

MODEL $\mathbf{A_1}$

$$f = \frac{\lambda^2(\lambda - S)}{S^2} \qquad N_{11} = \frac{1}{a}N \qquad N_{12} = \frac{S}{a\lambda}N \qquad N_{13} = \frac{S^2}{a\lambda}N$$

$$N_{14} = \frac{S^3}{a\lambda^3}N \qquad N_{15} = \frac{S^4}{a\lambda^4}N \qquad N_{1,6+} = \frac{S^5}{a\lambda^4(\lambda - S)}N \qquad a = \frac{\lambda}{\lambda - S}$$

MODEL $\mathbf{A_2}$

$$f = \frac{\lambda^2(\lambda - S)}{S^2} \qquad N_{11} = \frac{1}{a}N \qquad N_{12} = \frac{S}{a\lambda}N \qquad N_{1,3+} = \frac{S^2}{a\lambda(\lambda - S)}N$$

MODEL $\mathbf{A_3}$

$$f = \frac{\lambda(\lambda - S)}{S} \qquad N_{11} = N - S\lambda N \qquad N_{1,2+} = S\lambda N$$

Limiting the number of simulation inputs was not enough to decrease computing time to the level that would allow a "normal" number of Monte Carlo simulations to be run per input configuration (e.g., 1,000–10,000). The number of possible simulation input combinations was 504, and computing time for each combination ranged from around 30 min to several days, depending upon the problem's dimensionality. We thus conceptualized the problem as one of estimating a response surface (Box and Draper 1987), where the number of simulation runs per input configuration was low ($n = 3$), but where strength could be borrowed from the entire ensemble of simulations to summarize estimator performance in different regions of the input parameter space.

Although certain performance measures such as bias, CV, and BCOV are fairly standard, deciding how to summarize estimator performance was difficult. There were a large number of initial parameter combinations and estimation models, and for each estimation model there were a large number of estimated parameters. To condense the number of estimator performance variables to a manageable number, we thus calculated a mean performance value for each simulation. If there were 5 estimates of annual abundance at a given level of simulation inputs, response variables would be calculated for a given simulation as the average value across all 5 estimates. An exception was for BCOV, in which the number of estimates in a given simulation run were taken to be binomial trials, and the number of successes (i.e., the true value of a parameter was within the 90% credible interval) was recorded. For brevity, we only report results for annual abundance; estimator performance for survival, recovery rate, and recruitment rate is described by Conn (2007).

We determined convergence by running 2 Markov chains per simulation and monitoring whether Gelman–Rubin statistics were less than 1.2 immediately after the burn-in period (Gelman et al. 2004). Starting abundance values for the first Markov chain were set equal to the lowest possible values with positive support given the data, and overestimates of survival, recovery rate, and recruitment rate were provided. For the second chain, initial values were automated to produce overestimates of abundance and underestimates of other parameters. Ostensibly, if the ratio of between-chain to within-chain variance (as measured by the Gelman–Rubin statistic) declined sufficiently, this would be evidence that the effects of opposite types of overdispersed starting values had been overcome.

Our approach to summarizing posterior distributions involved running each Markov chain for 1,000,000 iterations. If convergence was determined to have occurred after 500,000 iterations, we combined the second halves of each Markov chain together to produce a sample from the posterior distribution. However, in order to conserve memory, we thinned each chain by only recording every fifth iteration throughout the estimation process, thus yielding a sample of 200,000 from the posterior. Over the course of this study, only 2 simulations did not converge according to the Gelman–Rubin convergence diagnostic.

4.1.1 Bayesian Credible Interval Coverage

Because each simulation produced multiple estimates, we treated the problem of estimating the effects of dependent variables on estimator coverage as one of estimating the regression coefficients of a generalized linear model (glm) with binomial error (McCullagh and Nelder 1989). Under this approach, success probability for the ith simulation, p_i, is determined according to relationship

$$\text{logit}(p_i) = \beta_0 + \beta_1 y_{i1} + \cdots + \beta_k y_{ik},$$

for k dependent variables and regression coefficients. The number of times that a specific type of parameter overlaps it's posterior interval in simulation i, X_i, is then modeled as

$$X_i \sim \text{Binomial}(x|Y_i, p_i),$$

where Y_i is the number of real parameters of a given type estimated in simulation i. For instance, if there are 7 years of data, there are 7 Bayesian credible intervals for abundance (one for each year). Using this approach, we assumed that particular simulations were more or less prone to failure of coverage, depending upon the predictors. One issue with this approach is that there are sampling covariances between abundance estimators derived from the same analysis, which violates the independence assumption needed to employ a binomial model. This problem can be remedied to some degree by estimating an extrabinomial overdispersion parameter (McCullagh and Nelder 1989).

Potential dependent variables affecting BCOV included A, Y, R, N, λ, S, r, and the estimation model ($EstMod$). We fit models with all possible combinations of these main effects, estimating an overdispersion parameter, $\hat{c}$, from the most general model for use with QAIC$_c$ (Burnham and Anderson 2002). Once we had attained model rankings, we selected a model within 2.0 QAIC$_c$ units of the top model for inference. This selection was made somewhat subjectively, with a predilection for simpler models that contained predictors occurring in most of the highly ranked models. Model selection results are suppressed here for brevity but are available elsewhere (Conn 2007).

Using this approach, it appeared that survival, the number of age classes, and the number of newly released marked animals were the most consistent predictors of BCOV for annual abundance. Credible interval coverage increased with R, but decreased with S and A (Fig. 1). Nevertheless, 90% Bayesian interval coverage on annual abundance was close to "nominal" in most instances.

4.1.2 Percent Relative Bias

Disregarding input simulation parameter specifications, average percent relative bias for estimated abundance was 7.6% (SE = 0.9%), 1.8% (SE = 0.4%), and 5.8% (SE = 0.7%) for posterior mean, mode, and median moment estimators, respectively (Fig. 2). As a moment estimator for abundance, the mode thus appeared to have the least bias, which is expected given the manner in which data were simulated.

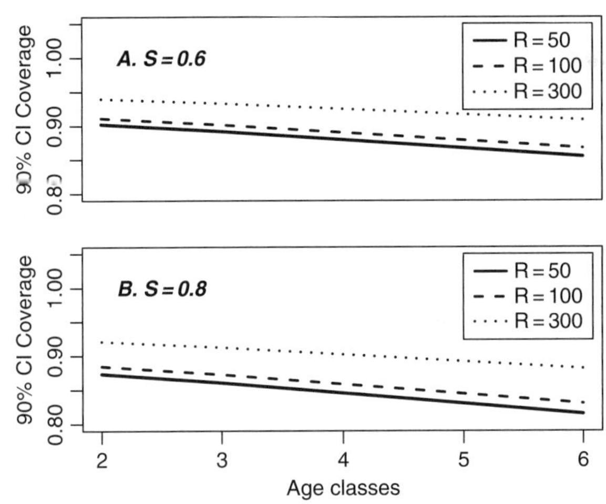

Fig. 1 90% Bayesian credible interval coverage for abundance at different levels of number of released animals per year (R) and number of age classes. Panel (A) gives performance for the case where $S = 0.6$, while panel (B) is for $S = 0.8$

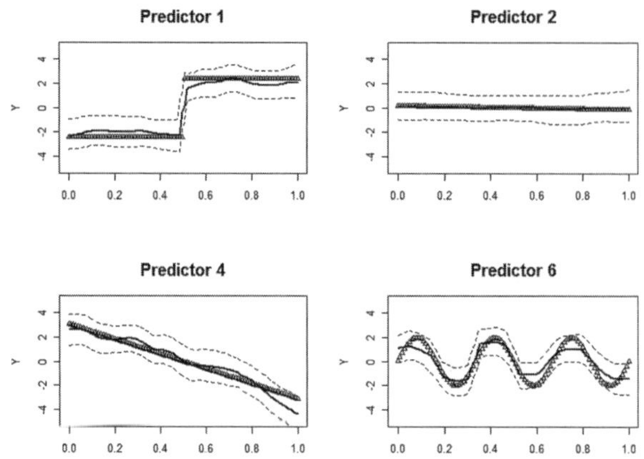

Fig. 2 The distribution of average percent relative bias for abundance over all simulations, as determined by (**A**) the posterior mean, (**B**) the posterior mode, and (**C**) the posterior median

In order to investigate which factors affected the bias of the posterior mode estimator for abundance, we compared the parsimony of alternative linear models that expressed the response variable, |%BIAS|, the absolute value of the average percent relative bias, as a function all possible combinations of predictor variables (A, Y, R, N, λ, S, r, and the $EstMod$). We considered |%BIAS| as a response variable because finding factors that potentially affect the magnitude of bias is perhaps more important than those that affect it's direction. Also, the performance of glms was poor when signed bias was used as a response variable.

Inspection of quantile plots in the statistical language R (R Development Core Team 2008) indicated that residuals from the most general model were not normally distributed, and that there was one major outlier in the data. For analysis, we removed the outlier and systematically considered different power transformations for |%BIAS| until quantile plots indicated that residuals were normally distributed. A power transformation of 0.25 seemed sufficient for this purpose, and a plot of studentized residuals against fitted values further indicated that the residual variance was largely constant under this approach.

Most predictor variables proved important for abundance |%BIAS|, although inspection of parameter estimates and standard errors from top-ranked AIC_c models indicated that N, λ, and the interaction $N \times R$ were not as important. We thus based inferences on the fourth-ranked AIC_c model. In general, |%BIAS| increased with A and S, and decreased with R, Y, and r (Fig. 3). Absolute bias was also negatively associated with estimation model complexity.

4.1.3 Coefficient of Variation

Conducting preliminary explorations of the data, it appeared that there were a number of outliers associated with CV for annual abundance. All such instances were associated with the case of 50 releases per year, 3 years of data, and survival

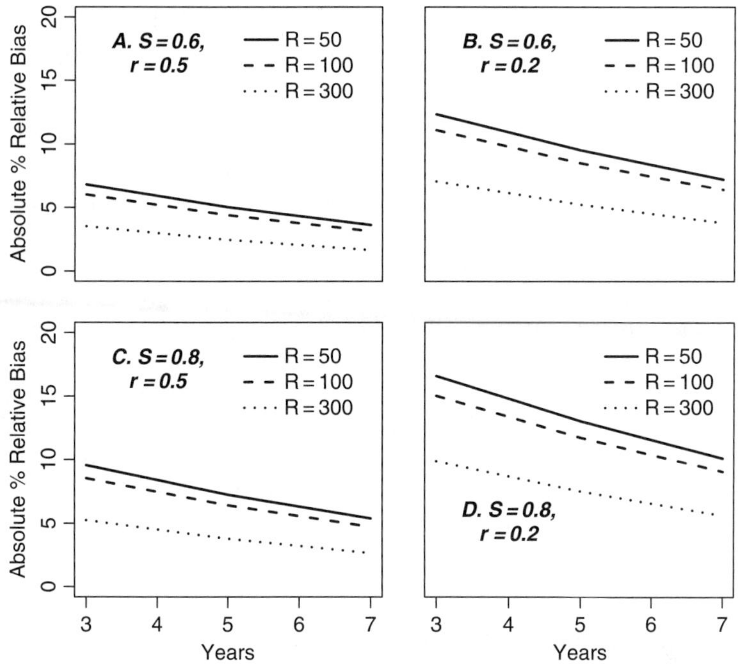

Fig. 3 Estimated relationship between absolute average % relative bias for abundance and R, Y, S, and r. All predictions are given for the case where A and $EstMod$ are set to three and $S(a)h(t)f(\cdot)$ in the linear model, respectively. Panel (**A**) gives results for $S = 0.6$ and $r = 0.5$; panel (**B**) depicts the case when $S = 0.6$ and $r = 0.2$; panel (**C**) is for $S = 0.8$ and $r = 0.5$.; panel (**D**) is for $S = 0.8$ and $r = 0.2$

and reporting rates of 0.8 and 0.2, respectively. In this scenario, the number of expected recoveries of marked animals is quite small (2.0, 3.6, and 4.9 for years 1,2, and 3, respectively) and can lead to unstable estimates. Interestingly, other than one outlier removed from the analysis of bias in abundance, this combination of simulation inputs did not seem to produce unduly extreme outliers for other measures of model performance. Nevertheless, removing these points was essential for analysis of CV.

A power transformation of -0.4 on the response variable proved adequate for meeting linear model assumptions, and the same suite of models were fit to the data. The top AIC_c model in this case was the most general model, with the second closest model > 3 ΔAIC_c units behind. There were too many important predictors in this case to visually portray cumulative effects of all simulation inputs; nevertheless, approximate predictions for mean CV can be made using estimated regression coefficients as

$$\left[\text{CV}(\hat{N})^{-0.4}\right] = 1.31 + 0.000402N + 0.00202R + 0.00324A + 0.111Y$$
$$+ 0.427\lambda - 1.517S + 1.844r - 0.150MOD1 - 0.0234MOD2$$
$$+ 0.00285(R \times MOD1) + 4.93 \times 10^{-8}(N \times R).$$

Because the response variable is taken to a negative power, the interpretation of the sign of regression coefficients is reversed; for instance, a positive coefficient leads to a decrease in CV. As with any regression analysis, it is important not to extrapolate predictions from the linear model past the particular input configurations of the response surface analysis, especially given that the design was not a complete factorial. Over the support of simulation inputs, mean CV($\hat{N}$) was predicted to be lowest (= 0.016) when $N = 5000$, $R = 300$, $A = 6$, $Y = 7$, $\lambda = 1$, $S = 0.6$, and $r = 0.5$. Note that the only estimation model employed for this configuration was $S(a+t)h(a)f(t)$; a lower coefficient of variation could be expected if a simpler estimation model had been used instead. The highest prediction for CV($\hat{N}$) was 0.24, which corresponded to the situation where $N = 1000$, $R = 100$, $A = 2$, $Y = 3$, $\lambda = 0.9$, $S = 0.8$, $r = 0.2$, and $EstMod = S(a)h(t)f(\cdot)$.

4.2 Simulation Module II: Effects of Aging Error

Misclassification of an individual's age might be expected to produce bias in parameter estimates, particularly if error rates vary systematically by age. For instance, small magnitude positive biases are typically observed in age estimates of young black bears, while larger negative biases are observed for older bears (Beck 1991; Costello et al. 2004; Harshyne et al. 1998). Even if there are no systematic age-mediated changes in the direction and magnitude of bias, age misclassification will still tend to result in misrepresentation of "strong" age classes (Fournier and Archibald 1982). In effect, the power to discriminate cohorts with large abundances will be somewhat obscured, and process error in recruitment will be underestimated. In the context of fisheries stock assessment models, it has also been shown that aging error can lead to negative bias in estimates of harvest mortality and measures of abundance, possibly leading to management advice which would favor over-harvesting (Reeves 2003).

In this module, we examined the effect of several magnitudes and types of aging error on estimates of abundance and other parameters under a number of hypothetical scenarios. In each case, we considered a model for aging errors whereby assigned age A' is related to true age A by the relationship

$$A' = \text{round}(A + \epsilon_j,)$$

where ϵ_j denotes a random effect associated with age j. Variation in error type and magnitude are produced by considering different models for ϵ_j. In total, five a priori models for aging error were considered:

- Model 1: No aging error (i.e., $\epsilon_j = 0$)
- Model 2:

$$\epsilon_j \sim \frac{\text{Laplace}(0, \sigma_j)\text{I}_{[0.5-j,\infty)}(\epsilon_j)}{1 - \int_{-\infty}^{0.5-j} \text{Laplace}(0, \sigma_j)} \qquad \sigma_j = 0.1j$$

- Model 3:

$$\epsilon_j \sim \frac{\text{Laplace}(0, \sigma_j)\text{I}_{[0.5-j,\infty)}(\epsilon_j)}{1 - \int_{-\infty}^{0.5-j} \text{Laplace}(0, \sigma_j)} \qquad \sigma_j = 0.2j$$

- Model 4:

$$\epsilon_j \sim \frac{\text{Laplace}(\mu_j, \sigma_j)\text{I}_{[0.5-j,\infty)}(\epsilon_j)}{1 - \int_{-\infty}^{0.5-j} \text{Laplace}(\mu_j, \sigma_j)} \qquad \sigma_j = 0.1j, \qquad \mu_j = 0.4 - 0.2j$$

- Model 5:

$$\epsilon_j \sim \frac{\text{Laplace}(\mu_j, \sigma_j)\text{I}_{[0.5-j,\infty)}(\epsilon_j)}{1 - \int_{-\infty}^{0.5-j} \text{Laplace}(\mu_j, \sigma_j)} \qquad \sigma_j = 0.2j, \qquad \mu_j = 0.4 - 0.2j$$

Here, $\text{I}_{[\Omega]}()$ gives an indicator function for the set Ω, and the Laplace (also known as the double exponential) distribution is the probability density function

$$f_X(x) = \frac{1}{2\sigma}\exp\left(\frac{|x-\mu|}{\sigma}\right).$$

The Laplace distribution has fatter tails than a normal distribution, and may be more useful for describing aging error in natural populations (Conn and Diefenbach 2007). Models 2 and 4 specify a relatively high level of precision on age estimates, while models 3 and 5 are relatively imprecise, particularly at older ages. The magnitude of aging errors often increase with an animal's age (Harshyne et al. 1998), a feature incorporated in models 2–5. Models 2 and 3 assume no bias in age estimation, while models 4 and 5 assume a positive aging bias for yearlings and a negative one for individuals over 2 years of age. In particular, the degree of bias increases with age. For all models, we assume that age 0 individuals are aged definitively (as with cubs in the case of black bears).

For simplicity, we only considered two biological scenarios for which to quantify possible effects of aging error on estimator performance, both of which corresponded roughly to the demography, harvest numbers, and sampling effort associated with female black bears in Pennsylvania. For each scenario, we set initial population size at 5,000, and assumed that 400 new individuals were marked and released per year, the ages of which were in proportion to their relative abundance in the population. We treated the population as if it consisted of 3 demographically relevant age classes: cubs (0–1 year old), yearlings (1–2 years old), and adults (ages 2+), with associated survival rates (S) of 0.4, 0.55, and 0.7, and harvest rates (h) of 0.1, 0.3, and 0.2, respectively. The two different biological scenarios were described by different parameterizations for recruitment (f). Scenario A assumed that recruitment was a Poisson process with a mean of 1.364 female cubs per adult female, a number approximately necessary for a stationary population. Scenario B assumed the same mean, but with a hyperprior on λ, such that $f_i \sim \text{Poisson}(\lambda)$, where $\lambda \sim$ Gamma(10,7.33). Thus the 2 scenarios embodied quite different assumptions about the nature of process error in recruitment (Fig. 4). We considered this potentially relevant since aging error will typically serve to obscure the detection of high abundance cohorts, and thus may lead to underestimates of recruitment process error.

Instead of pooling virtual animals into a "+" category to start with, we allowed them to advance to a possible age of 20, at which time they were automatically removed from the population. In this manner, aging error could be appropriately applied to an animal's real age, and data could be pooled to a pre-specified level for analysis. In this case, we chose to pool back to 7 age classes (0–5 and 6+). Only adults were assumed to contribute to recruitment the following year, and populations were started at a stable stage distribution, as in Section 4.1.

We ran a total of 50 simulations for each combination of the 5 aging error models and 2 biological scenarios. For each simulation and pooling option, we ran 2 chains of length 1,000,000 starting at overdispersed values. If after 500,000 iterations Gelman–Rubin statistics confirmed that the chains had approximately the same within- and between- chain variance, we combined the final 500,000 samples of each chain to arrive at a sample of 1,000,000 from the posterior distribution, which was thinned to 200,000 to save memory. We calculated the same statistics as in Section 4.1 to quantify estimator performance; model $S(a)h(a)f(\cdot)$ was used to estimate parameters for all Scenario A simulations, while model $S(a)h(a)f(t)$ was used for all Scenario B simulations. Here, an a denotes the case where 3 parameters are estimated, corresponding to cubs, yearlings, and adults.

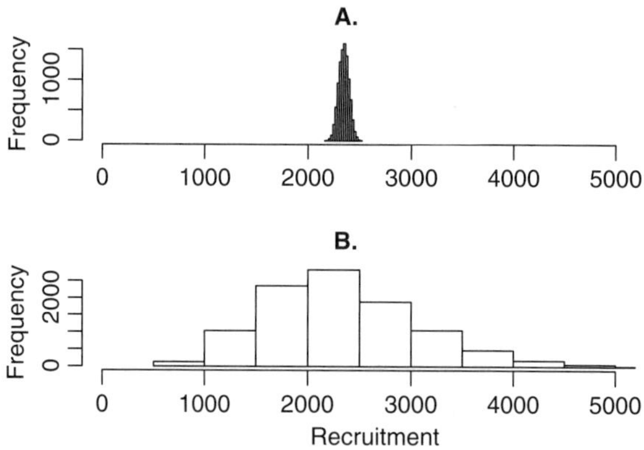

Fig. 4 Histograms representing the relative likelihood of obtaining different values of recruitment under Scenarios **A** and **B** in simulation module II when populations are started at a stable stage distribution and an initial population size of 5,000. In particular, Scenario **A** dictates a relatively fixed number of recruits, while Scenario **B** allows recruitment to differ drastically from year to year

To quantify effects of aging error on estimator performance, we once again compared the relative parsimony of different models where estimator performance was related to all possible subsets of predictor variables. Here, we considered average absolute percent relative bias and 90% credible interval coverage as possible response variables; predictor variables were μ (2 levels), σ (3 levels), and recruitment variance scenario (2 levels). We entertained the possibility of including first level interaction terms whenever the variables comprising the interaction were included in the list of predictor variables. However, one of the interactions between σ and μ was not estimable because simulations did not include the design point where $\mu_j = 0.4 - 0.2j$ and $\sigma_j = 0$.

A total of 36 models were fit for each response variable. When bias was modeled, our strategy was to find a transformation of the response variable which approximately satisfied linear model assumptions for the most general model, as indicated by quantile plots and plots of studentized residuals versus fitted values. In this case, we used AIC_c for model selection. When coverage was of interest, we once again considered a binomial model for the response variable within a generalized linear model framework where success probability was related to predictors. In this case, we used a logit link, estimated an overdispersion parameter, c, and used QAIC_c for model selection.

For bias on estimators of annual abundance, a power transformation of 0.1 seemed adequate for meeting linear model assumptions. An intercept model, in which bias was constant across simulation inputs, was ranked second with $\Delta\text{AIC}_\text{c} = 0.3$. Parameter estimates from other top ranked models actually predicted that bias would decrease with the amount of aging error (for both σ and μ), although confidence intervals overlapped zero. We thus concluded that there was no evidence that aging error increased average percent absolute relative bias in annual abundance estimators.

Models selected for 90% Bayesian credible interval coverage on annual abundance tended to include μ and σ effects on coverage, as well as interactions with these terms and the level of recruitment variance. Investigation of parameter estimates and confidence intervals from top ranked models indicated that coverage decreased slightly when $\sigma > 0$ and when recruitment variance was high. Nevertheless, BCOV was predicted to be close to "nominal" in all cases, indicating that the types and magnitude of aging error considered here had little effect on the accuracy of credible intervals.

4.3 Simulation Module III: Marked Individual Data in Both L_1 and L_2

Up to this point, all evaluations of model performance have assumed independence of L_1 and L_2. In practical applications, records from marked animals will likely be included in both age-at-harvest and auxiliary datasets. Technically, this invalidates the independence assumption needed for a coherent joint likelihood. For instance, sample sizes will be inflated under this approach and thus we might expect an artificially high level of precision. However, the importance of this assumption for obtaining estimators with good properties has not yet been explored.

In this module, we used simulation to compare the performance of estimators with and without data from marked individuals included in the age-at-harvest portion of the likelihood. We surmised that performance would likely be influenced by (i) the number of animals marked each year, and (ii) the complexity of the model fit to the data. For instance, if the number of marked animals is low in comparison to unmarked animals, then most of the data informing inference about population size comes from unmarked animals. Similarly, if fewer parameters are used to describe the survival, harvest, and recruitment processes, there is more information in the age-at-harvest likelihood about them. We thus expected to see a greater degree of bias in coverage and CV when employing simpler models.

We considered a total of 6 scenarios to evaluate estimator performance, which differed by the number of marked animals released each year and by the estimation model considered. The number of marked animals newly released each year was set to either 200 or 400, with the number released in each age category proportional to the number of animals in that age class. Three possible estimation models were considered in order to compare results across different levels of model complexity: $S(\cdot)h(\cdot)f(\cdot)$, $S(a)h(t)f(\cdot)$, and $S(a+t)h(a)f(t)$.

For each scenario, we generated data for cases where marked animals were either part of or not part of the target population. When not part of the target population, they were assumed to have equivalent survival and harvest probabilities to the target population. Fifty replicate data sets were simulated in each case. We used age- and time-constant population parameters, with survival probability set to 0.6, a recovery rate or 0.2, and a recruitment probability of 0.4, with 5 age classes, an initial population size of 5,000, and 5 years of data.

To quantify effects of non-independence on estimator performance, we once again compared the relative parsimony of different models where estimator performance was related to all possible subsets of predictor variables. In this case, we did not expect any changes with respect to bias, but we did expect that precision would be overestimated. Thus, we treated 90% BCOV for abundance as the response variable (see Conn 2007 for analyses involving estimators for survival, recovery rate, and recruitment). Predictor variables were the number of releases each year (200 or 400), the generating model $EstMod$ (3 levels), and an indicator for whether or not data in each part of the likelihood were completely independent (Ind), which equaled 1 if data were independent and 0 otherwise. In addition, whenever main effects terms were included in the model, we considered additional models with all possible combinations of 1-way interactions. If all 1-way interactions were present, we also considered the possibility of a 2-way interaction.

Using the most general model, overdispersion was estimated as $\hat{c} = 2.5$. Nevertheless, model selection using $QAIC_c$ favored highly parameterized models. We chose to base inference on the most general model, which was ranked second $\Delta QAIC_c = 0.6$, because it was the only model that included a highly influential two-way interaction effect. In particular, when data from marked animals were also included in the age-at-harvest matrix, coverage was predicted to be much worse for

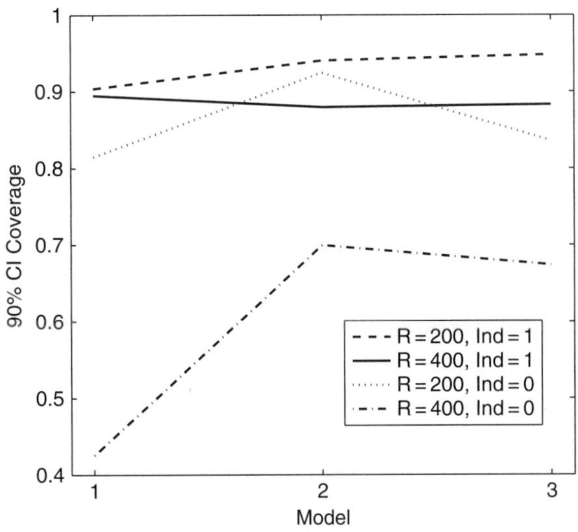

Fig. 5 90% Bayesian credible interval coverage on abundance for different model configurations, number of releases (R), and depending on whether mark-recovery data were independent ($Ind = 1$) from age-at-harvest data or not ($Ind = 0$). Models "1," "2," and "3" were $S(\cdot)h(\cdot)f(\cdot)$, $S(a)h(t)f(\cdot)$, and $S(a+t)h(a)f(t)$, respectively. The $R = 200$ scenario starts with approximately 4% of the population marked and grows to about 9% by the end of the study, while the $R = 400$ scenario starts with 8% of the population marked, growing to around 18% by the end of the five year study. These percentages are roughly in line with the degree of overlap that could be expected between mark-recovery and age-at-harvest datasets at different points in simulation time

the simple estimation model $S(\cdot)h(\cdot)f(\cdot)$ (0.42) than for the other, more complex estimation models (≈ 0.7) (Fig. 5).

5 Discussion

In this paper, we explored whether Bayesian implementation of a state-space formulation for the joint analysis of age-at-harvest and mark-recovery data was likely to result in estimators with good properties. In general, long Markov chains (along the order of 1.0×10^6 iterations) were needed to generate consistent estimators of model parameters and for repeatability in estimation of DIC. However, for the scenarios considered, Bayesian analysis of age-at-harvest and mark-recovery data generally resulted in estimators of population parameters with a low degree of bias and high degree of precision. Bias was positive for abundance and decreased with sample size. Increased sample size could result from longer studies, more individuals marked each year, or higher harvest rates. Further, when all model assumptions were satisfied, credible interval coverage was close to "nominal". Thus, when all assumptions are met, joint analysis of age-at-harvest and mark-recovery data produces reasonable estimates of model parameters.

We also considered what would happen to estimator performance when assumption violations such as aging errors occurred or when individuals contributed information to >1 dataset. Bias and coverage were quite robust to aging errors, at least with the range of models considered here. However, when data from marked animals were included in the age-at-harvest matrix, the potential for understatement of uncertainty was substantial. While use of estimation models with a sufficient degree of complexity helped, coverage was still poor when a large proportion of the population was marked each year. Ostensibly, including marked animals in both L_1 and L_2 artificially inflated sample sizes, resulting in inaccurate inferences about precision. We expect that a similar situation exists for other, previously proposed model formulations for analysis of age-at-harvest data (Gove et al. 2002; Skalski et al. 2005), and for other state-space estimation approaches in which the same animal may contribute to >1 dataset (Besbeas et al. 2002). Due to it's importance, the general issue of non-independence in state-space models is a subject of current research by other research groups (B. J. T. Morgan, Personal Communication).

Although it is hard to come up with a general rule from the limited number of simulations we have performed here, there is some indication that a 10-15% overlap in datasets may be sufficient to produce substantial shortfalls in credible interval coverage (e.g., Fig. 5). Unfortunately, there is no simple solution to this problem, at least for the model structure considered in this paper. For instance, the approach of removing marked animals from the age-at-harvest dataset does not work because it effectively ignores the dynamical process of "removing" individuals from the population by marking, leading to bias in estimators of recruitment and abundance. If the age distribution of marked and unmarked animals are different, this approach could also result in biases in survival and recovery probabilities. An alternative would be to explicitly model the process by which unmarked animals become marked animals (e.g., by including a detection probability parameter in model structure). This would introduce a number of extra parameters into the likelihood and require one to keep track of the latent numbers of marked and unmarked animals in the population at different points in time. This may be a reasonable solution for the joint age-at-harvest model we have evaluated here but would require substantial development. One ad hoc approach that may be useful in practice is to conduct simulations with input values relevant to the study population, artificially increasing the length of credible intervals until coverage is close to nominal. Another suggestion is to consider highly parameterized models in an effort to reduce the amount of "duplicate" information about each parameter.

Extensive assessments of the efficacy of Bayesian analyses of this sort are rarely seen in the literature, in part because requisite computing time can be daunting. We regard response surface simulation designs as a natural way to conduct such simulation experiments in these cases. Strength is borrowed from the entire ensemble of design points, invoking interpolation from a response surface (in our case, a generalized linear model) to increase predictive accuracy anywhere along the surface. In this manner, the number of simulation replicates needed at any one design point decreases dramatically. Use of information-theoretic criterion allowed us to select simulation input values that were important determinants of estimator performance, and thus to provide parsimonious predictions of estimator performance under different scenarios.

For the scenarios considered here, coefficient of variation on abundance was satisfactory (< 0.25) for almost all design points, suggesting that 50 releases of marked animals per year may be sufficient for monitoring purposes, especially when tag return rates are high. Additional simulations would be required to determine the number of marked releases necessary to use this approach for monitoring purposes when tag return rates are low, as with North American waterfowl, passerine ringing programs in Europe, or fisheries applications. Increased precision can be expected if a known fate analysis is employed in addition to mark-recovery analysis (Conn 2007); however, the underlying tag return rate is fundamental for estimating sampling parameters such as recovery probability, and thus for estimating abundance from age-at-harvest data.

References

Beck TDI (1991) Black bears in west-central Colorado, Tech. Report Tech. Publ. 39 DOW-R-T-39-91, Colorado Division of Wildlife, Fort Collins, Colorado.

Besbeas P, Freeman SN, Morgan BJT, Catchpole EA (2002) Integrating mark-recapture-recovery and census data to estimate animal abundance and demographic parameters. Biometrics 58:540–547.

Box GEP, Draper NR (1987) Empirical Model-Bulding and Response Surfaces, John Wiley & Sons, New York, NY, USA.

Brooks SP (1999) Bayesian analysis of animal abundance data via MCMC, Bayesian Statistics 6 (Bernardo JM, Berger JO, Dawid AP, and Smith AFM, eds), Oxford University Press, Oxford, pp. 723–731.

Brownie C, Anderson DR, Burnham KP, Robson DS (1985) Statistical inference from band recovery data: A handbook, US Fish and Wildlife Resource Publication (156), Washington, DC, USA.

Burnham KP, Anderson DR (2002) Model selection and multimodel inference: a practical information-theoretic approach, 2nd Edition, Springer-Verlag, New York, USA.

Caswell H (2001) Matrix Population Models. Sinauer, Sunderland, MA, USA.

Conn PB (2007) Bayesian analysis of age-at-harvest data, with focus on wildlife monitoring programs, Ph.D. thesis, Colorado State University, Fort Collins, CO, USA.

Conn PB, Diefenbach DR (2007) Adjusting age and stage distributions for misclassification errors. Ecology 88:1977–1983.

Conn PB, Laake JL, Diefenbach DR, White GC, Ternent MA (in press) Bayesian demographic analysis of wildlife age-at-harvest data. Biometrics.

Costello CM, Inman KH, Jones DE, Inman RM, Thompson BC, Quigley HB (2004) Reliability of the cementum annuli technique for estimating age of black bears in New Mexico. Wildlife Society Bulletin 32:169–176.

Deriso RB, II Quinn TJ, Neal PR (1985) Catch-age analysis with auxiliary information. Canadian Journal of Fisheries and Aquatic Sciences 42:815–824.

Dupont WD (1983) A stochastic catch-effort method for estimating animal abundance. Biometrics 39:1021–1033.

Fournier D, Archibald CP (1982) A general theory for analyzing catch at age data. Canadian Journal of Fisheries and Aquatic Sciences 39:1195–1207.

Gelman A, Carlin JB, Stern HS, Rubin DB (2004) Bayesian Data Analysis, 2nd Edition, Chapman and Hall, Boca Raton, FL, USA.

Gove NE, Skalski JR, Zager P, Townsend RL (2002) Statistical models for population reconstruction using age-at-harvest data. Journal of Wildlife Management 66:310–320.

Harshyne WA, Diefenbach DR, Alt GL, Matson GM (1998) Analysis of error from cementum-annuli age estimates of known-age Pennsylvania black bears. Journal of Wildlife Management 62:1281–1291.

Laake JL (1992) Catch-effort models and their application to elk in Colorado, PhD thesis, Colorado State University, Fort Collins, CO, USA.

Lebreton JD, Burnham KP, Clobert J, Anderson DR (1992) Modeling survival and testing biological hypotheses using marked animals: A unified approach with case studies. Ecological Monographs 62:67–118.

Lewy P, Nielson A (2003) Modelling stochastic fish stock dynamics using Markov chain Monte Carlo. ICES Journal of Marine Science 60:743–752.

Link WA, Royle JA, and Hatfield JS (2003) Demographic analysis from summaries of an age-structured population. Biometrics 59:778–785.

Maunder MN (2001) Integrated tagging and catch-at-age analysis (ITCAAN), Spatial Processes and the Management of Fish Populations (Kruse GH, Bez N, Booth A, Dorn MW, S. Hills, Lipcius RN, Pelletier D, Roy C, Smith SJ, Witherell D eds), Alaska Sea Grant College Program Report No. AK-SG-01-02, University of Alaska, Fairbanks.

McAllister MK, Ianelli JN (1997) Bayesian stock assessment using catch-age data and the sampling-resampling importance algorithm. Canadian Journal of Fisheries and Aquatic Sciences 54:284–300.

McCullagh P, Nelder JA (1989) Generalized Linear Models, Chapman and Hall, New York, USA.

Megrey BA (1989) Review and comparison of age-structured stock assessment models from theoretical and applied points of view. American Fisheries Science Symposium 6:8–48.

Meyer R, Millar RB (1999) Bayesian stock assessment using a state-space implementation of the delay difference model. Canadian Journal of Fisheries and Aquatic Sciences 56:37–52.

Newman KB, Buckland ST, Lindley ST, Thomas L, Fernàndez C (2006) Hidden process models for animal population dynamics. Ecological Applications 16:74–86.

R Development Core Team (2008) R:A language and environment for statistical computing. R Foundation for statistical Computing, Vienna, Austria. ISBN 3-900051-07-0, http://www.R-project.org.

Reeves SA (2003) A simulation study of the implications of age-reading errors for stock assessment and management advice. ICES Journal of Marine Science 60:314–328.

Seber GAF (1982) The Estimation of Animal Abundance and Related Parameters, 2nd edition, Blackburn Press, Caldwell, NJ, USA.

Skalski JR, Ryding KE, Millspaugh JJ (2005) Wildlife demography: analysis of sex, age, and count data, Elsevier Academic Press, San Diego, CA, USA.

Spiegelhalter DJ, Best NG, Carlin BP, van der Linde A (2002) Bayesian measures of model complexity and fit. Journal of the Royal Statistical Society Series B 64:583–639.

Virtala M, Kuikka S, Arjas E (1998) Stochastic virtual population analysis. ICES Journal of Marine Science 55:892–904.

A Spatial Model for Estimating Mortality Rates, Abundance and Movement Probabilities from Fishery Tag-Recovery Data

J. Paige Eveson, Geoff M. Laslett, and Tom Polacheck

Abstract Spatial heterogeneity in survival and capture probabilities is a critical issue to consider in tagging experiments. If a non-trivial level of spatial heterogeneity exists and is not accounted for, it can lead to unreliable estimates of mortality rates and abundance, and of the uncertainty in these estimates. Here we present a spatial model for analysing multiyear tag-recovery and fishery catch data that allows for mortality rates and abundance to differ among discrete regions and for fish to move among these regions at discrete time intervals. For a given cohort of fish tagged in consecutive years in all regions, this model can provide year- and region-specific estimates of both natural mortality and fishing mortality, region-specific estimates of abundance at the time of initial tagging, as well as year-specific movement probabilities between regions. The precision of parameter estimates can be poor with such a full model, but can be improved with more restricted model parameterizations. Tagging in some regions may be logistically difficult and/or very expensive. We show that if tagging is conducted in all regions in the first year of the experiment, but only in one region thereafter, accurate and precise parameter estimates can sometimes still be achieved. It is not always the regional estimates of mortality rates and abundance that are of primary interest, but rather the population-wide estimates (over all regions). Such population-wide estimates can be obtained by applying a non-spatial model to the data pooled across regions; however, simulation results suggest that there are many situations for which large biases are incurred by using a non-spatial model. Simulations also suggest that there is almost no loss in precision from using the spatial model to obtain population-wide estimates even when the non-spatial model would suffice.

1 Introduction

A fundamental assumption in the use of tagging experiments to estimate mortality rates and/or abundance is that tagged and untagged animals are fully mixed

---

J.P. Eveson (✉)
CSIRO Marine and Atmospheric Research, GPO Box 1538, Hobart, Tasmania 7001, Australia
e-mail: paige.eveson@csiro.au

D.L. Thomson et al. (eds.), *Modeling Demographic Processes in Marked Populations*, Environmental and Ecological Statistics 3, DOI 10.1007/978-0-387-78151-8_45,

throughout the range of the population. This can be difficult to achieve in wild populations, particularly ones distributed over large geographic areas. In cases where complete mixing is not achieved, spatial heterogeneity in survival and capture probabilities of animals will lead to biased estimates of mortality rates and abundance, as well as the uncertainty in these estimates, if it is not accounted for. Thus, spatial heterogeneity is a critical issue to consider in the design and analysis of tagging experiments. In the current paper, we present a spatial model for estimating mortality rates (both natural and human-related), abundance and movement from a multiyear tag-recovery experiment, and use it to investigate the importance of accounting for spatial heterogeneity on the accuracy of the parameter estimates. The model assumes that individuals from a population are tagged and recaptured in a number of discrete regions over multiple years. Population size and mortality rates are allowed to differ among regions, and individuals are allowed to move between regions at discrete time intervals (assumed here to be years). The model was developed for fishery applications in which tags are recovered from dead animals, so that only tag-recovery (i.e., single-recapture) experiments are feasible.

The term multi-state model is often used for the type of model developed here (i.e., with discrete states and transitions between states). We use the term spatial model because the states we are interested in are spatial strata. The states could, however, correspond to physiological conditions, such as breeder or non-breeder (spawner for fish), as long as these states were observable on release and recovery. A number of papers have developed multi-state models in the context of capture–recapture tagging experiments, where multiple recaptures are available for each tagged animal (e.g., Arnason 1973; Hestbeck et al. 1991; Nichols et al. 1992; Brownie et al. 1993). Physiological states are common in applications of multi-state capture–recapture models, but the principles are the same for spatial states. Less work has been done on multi-state models for single–recapture data, exceptions being Joe and Pollock (2002) and Schwarz et al. (1993). The focus of these two studies, as well as most studies in the multiple–recapture setting, has been on estimating movement (transition) probabilities, whereas our focus is on obtaining robust estimates of mortality rates and abundance.

The model presented is a spatial extension of the model developed in Polacheck et al. (2006a). It combines two traditional, but fundamentally different, approaches for analyzing tag-recovery data. The first approach, generally referred to as a Brownie model (Brownie et al. 1985), uses tag-recovery data from multiple years of tagging to provide annual estimates of mortality rates by comparing return rates over time from the releases in consecutive years. Only the numbers of releases and returns by year are required, not the size of the sample examined for tags. The second approach, known as a Petersen model (e.g., Seber 1982), uses data from a single release event to provide an estimate of population size at the time of tagging based on the ratio of the number of tags returned from a sample of the population to the total number of tags in the population. Because our model was developed in a fishery context, commercial catch data constitute the sample from which tags are returned, and thus there is uncertainty in the sample size. However, the approach would be equally valid in a more controlled situation where the exact size of the sample examined for tags was known. We refer to our model as the spatial Brownie–Petersen (BP) model.

The Brownie, or survival, component of the spatial BP model is similar to that of Schwarz et al. (1993), except it is formulated to have separate parameters for survival and movement rates, and separate parameters for tag recovery rates and reporting rates. These separations are important in fisheries for providing advice to management. Also, the survival and recovery rates are parameterized in terms of instantaneous rates of natural mortality and exploitation (i.e., fishing mortality) (Pollock et al. 1991; Hoenig et al. 1998). This parameterization is common in population modeling of fish resources (Beverton and Holt 1957; Hilborn and Walters 1992) and in fishery applications of tagging experiments (e.g. Frusher and Hoenig 2001; Polacheck et al. 2006a).

Although a model with a discrete spatial structure will be artificial for any fish population, it can often provide an adequate approximation. For example, it is common for fish species to have distinct feeding, spawning and/or nursery areas between which they migrate. Furthermore, discrete regions are often defined for fishery management purposes, with regional estimates of exploitation rates and abundance being sought. Thus, applications of discrete-space models to fishery tagging data are quite common (e.g., Cappo et al. 2000; Xiao 2000; Schwarz et al. 1993). Our motivation for developing the spatial BP model was to investigate whether a spatial design and analysis was needed for an ongoing tagging experiment being conducted on juvenile southern bluefin tuna (SBT, *Thunnus maccoyii*). The fishery for juvenile SBT can be divided into distinct components: an inshore purse-seine fishery in the Great Australian Bight that operates during the austral summer, and several offshore longline fisheries spanning much of the southern ocean that operate during the austral winter. While the spatial BP model, as presented, does not fully accommodate the complex movement and fishery dynamics for SBT, it provides the general framework necessary for such investigations. Also, by keeping the model general, it can more easily be modified to suit a large range of situations (e.g., a modified version for SBT is presented in Appendix 12 of Polacheck et al. 2006b).

The current paper begins with a description of the spatial BP model, starting with the assumptions required by the model and followed by the development of the likelihood components. We illustrate that, for a cohort of fish tagged and recovered in all years and regions of the study, the spatial BP model can provide year- and region-specific estimates of both natural mortality and fishing mortality, region-specific estimates of abundance at the time of initial tagging, as well as year-specific movement probabilities between regions. With such a full model, the precision of some parameter estimates can be poor. Thus, we explore how much the estimates can be improved under a range of reduced model parameterizations (e.g., natural mortality only year dependent, fishing mortality only region dependent, movement rates the same across years). Furthermore, for each of these model parameterizations, we investigate parameter identifiability and model performance under two reduced experimental designs: tagging in all regions in the first year then in only one region thereafter, and tagging in only one region in all years. Because tagging in some regions may be logistically very difficult and/or expensive (e.g., in off-shore areas that are not easily accessible), it is of practical interest to know whether it is possible to still achieve reasonable parameter estimates for experimental designs that do not include tagging in all regions and years. Finally, it is often not the regional parameter estimates for mortality and abundance that are of primary interest, but rather the population-wide estimates (i.e., over all regions). This is particularly true in a fishery context, where the assessment of human impacts on the status and sustainability of the resource is the principal focus. Such population-wide estimates can be obtained by applying a non-spatial model to the tag release and recovery data pooled across regions, but are expected to be biased if spatial heterogeneity exists. To investigate the consequences (biases) of not allowing for spatial heterogeneity when it exists, we compared estimates of total fishing mortality and abundance when regional data are analyzed using the spatial BP model versus when the data are pooled across all regions and analyzed using the equivalent non-spatial model. Results for a range of scenarios are presented in order to gain insight into those situations for which spatial tag designs and models are necessary.

2 Methods

Consider a multiyear tagging study in which a cohort of fish is tagged in each of regions 1 to K in years 1 to I (i.e., at age 1 in year 1, age 2 in year 2, up to age I in year I). The tags must be specific to a region and year (generally in fisheries, the tags are specific to an individual via a unique number). Fish from the cohort are subsequently caught in years, or equivalently at ages, 1 to J ($J \geq I$) in a fishery operating in all regions, and tags are returned from a proportion of the fish recovered with tags. Estimates of the numbers of fish from the cohort caught in each region and year are available. The parameters to be estimated from the tag-recovery and catch data are natural mortality and fishing mortality by year and region, cohort abundance in each region at the start of tagging (i.e., at the start of year 1), and movement probabilities between regions in each year. Note that because we are dealing with a single cohort, year and age are interchangeable (e.g., parameters stated to vary with year could instead be stated to vary with age).

The model is developed in terms of a cohort of fish because a cohort is the most natural example of a population closed to births and immigration. This type of closure is required by the Petersen component of the model in order to get abundance estimates with a meaningful interpretation. If one leaves off the Petersen component (i.e., omits the likelihood for the catch data from the overall likelihood – see Model development section), then the requirement for a population without births or immigration is no longer necessary. Nevertheless, if any of the model parameters vary with age, then tagging still needs to be specific to age-classes (i.e., cohorts) in order to get age-specific parameter estimates.

We present the model for a single cohort because this is the minimum required by the model and makes the notation less burdensome. In practice, it is likely that several cohorts (age-classes) would be tagged in each year of tagging. To include multiple cohorts in the model, one simply needs to develop the likelihood for each cohort as described, then multiply them together to form a joint likelihood (provided the data are independent between cohorts, which we discuss later). Note that if all parameters being estimated vary with both year and age, then maximizing the likelihood for each cohort separately is equivalent to maximizing the joint likelihood (i.e., will yield the same parameter estimates). More likely, however, some parameters will be shared; for example, if fishing mortality varies with year but not with age, then all cohorts caught in a given year and region will have a common fishing mortality parameter.

2.1 Model Assumptions

To apply the model to a cohort (or multiple cohorts) of fish requires that the ages of fish at release be known. For some species, release ages can be determined by reading scales collected from the tagged fish. When this is not possible, age can often be inferred from length. The former method has been used in a tagging study on Australian salmon (Cappo et al. 2002), whereas the latter method has been used in tagging experiments on juvenile SBT (Polacheck et al. 2006a). The accuracy of the age estimates will depend on the method, the species, and the specifics of the situation. Here, we assume the release ages are known accurately enough that the error can be ignored.

To allocate the catch in each year and region to cohorts requires knowing the age distribution of the catch in each year and region. Often, the age distribution of the catch will be determined by taking a sample, estimating the ages of fish in the sample (either from lengths or from direct aging of hard parts), and scaling up the estimated age frequencies of the sample to the total catch in numbers (which may itself be estimated from weight). The model directly incorporates uncertainty in the estimated numbers of fish caught from the cohort in each year and region, but it assumes these numbers are unbiased and independent.

Not all tags that are recovered are likely to be reported; this is especially true in commercial fisheries. We assume that estimates of reporting rates are available, and treat them as known in the model.

To separate mortality from movement, it is necessary to model the timing of movement in relation to the mortality processes. We assume that during a given year, fish in a region stay within the region, where they may be caught or die naturally. Exactly at the end of each year, fish move between regions according to a Markov chain model; i.e., a fish's movement at the end of the current year does not depend on its movement at the end of previous years. Furthermore, we assume that a fish's movement is independent of the movement of other fish.

In addition to the above assumptions, the usual assumptions for non-spatial tag-recovery models, as outlined by Brownie et al. (1985) and summarized clearly by Pollock et al. (1991), are required. The most important of these are: the fish tagged within a region are a representative sample of the fish in that region (i.e., that tagged and untagged fish are thoroughly mixed throughout each region where tagging occurs); the fate of each fish is independent of the fate of other fish; all fish in a given region and year have the same survival and capture probabilities; there is no tag shedding or tag-induced mortality.

A section of the Discussion is devoted to discussing the above model assumptions, their implications for parameter estimation if violated, and possible methods for dealing with violations.

2.2 Model Development

The model is developed in a maximum likelihood framework. The overall likelihood is comprised of two components: one for the tag-recovery data and one for the catch data.

First consider the tag-recovery component of the likelihood. Define tagging group (*i,k*) to be the N_i^k fish tagged in year *i* in region *k*. Let $R_{i,j}^{k,k'}$ be the number of tags returned (i.e., recovered and reported) from tagging group (*i, k*) in year *j* and region *k*′, and let $p_{i,j}^{k,k'}$ be the probability that a tag from tagging group (*i, k*) is returned in year *j* and region *k*′.

For tagging group (*i, k*), the numbers of returned tags by region and year, plus the number of unreturned tags, are modeled as multinomial with a log-likelihood (apart from an additive constant) given by

$$\log L_i^k = \left(N_i^k - R_{i,\bullet}^{k,\bullet}\right)\log\left(1 - p_{i,\bullet}^{k,\bullet}\right) + \sum_{j=1}^{J}\sum_{k'=1}^{K} R_{i,j}^{k,k'} \log p_{i,j}^{k,k'}$$

where $R_{i,\bullet}^{k,\bullet} = \sum\limits_{j=1}^{J}\sum\limits_{k'=1}^{K} R_{i,j}^{k,k'}$ and $p_{i,\bullet}^{k,\bullet} = \sum\limits_{j=1}^{J}\sum\limits_{k'=1}^{K} p_{i,j}^{k,k'}$.

The data from each tagging group are independent, so the log-likelihood over all tagging groups is given by

$$\log L_{\text{tag}} = \sum_{i=1}^{I}\sum_{k=1}^{K} \log L_i^k. \tag{1}$$

The return probabilities can be defined in terms of survival, movement, capture and reporting parameters. In particular, $p_{i,j}^{k,k'}$ is given by the $(k,\ k')$ th element of the matrix

$$\mathbf{p}_{i,j} = \begin{cases} \mathrm{D}\left(\mathbf{u}_j\right)\mathrm{D}\left(\boldsymbol{\lambda}_j\right) & j = i \\ \mathrm{D}\left(\mathbf{S}_i\right)\prod_i \ldots \mathrm{D}\left(\mathbf{S}_{j-1}\right)\prod_{j-1}\mathrm{D}\left(\mathbf{u}_j\right)\mathrm{D}\left(\boldsymbol{\lambda}_j\right) & j > i \end{cases} \tag{2}$$

where:

- $\mathbf{S}_j$ is a $K \times\ 1$ vector whose *k* th element, S_j^k, is the probability that a fish alive at the beginning of year *j* in region *k* survives the year;

$\mathbf{u}_j$ is a $K \times 1$ vector whose k th element, u_j^k, is the probability that a fish alive at the beginning of year j is caught during year j in region k;

$\boldsymbol{\lambda}_j$ is a $K \times 1$ vector whose k th element, λ_j^k, is the probability that a tag will be reported from a tagged fish caught during year j in region k;

$\boldsymbol{\Pi}_j$ is a $K \times K$ matrix whose (k, k') th element, $\pi_j^{k,k'}$, is the probability that a fish moves from region k to region k' at the end of year j; and,

$D(\boldsymbol{\theta})$ is an operator that transforms a $K \times 1$ vector $\boldsymbol{\theta}$ into a $K \times K$ diagonal matrix containing the elements of $\boldsymbol{\theta}$ on its diagonal.

We can further express the survival and capture probabilities in terms of natural mortality and fishing mortality, as follows:

$$S_j^k = \exp\left(-\left(F_j^k + M_j^k\right)\right)$$

and

$$u_j^k = \frac{F_j^k}{F_j^k + M_j^k}\left(1 - S_j^k\right)$$

where M_j^k and F_j^k are the instantaneous rates of natural mortality and fishing mortality, respectively, for fish in year j and region k.

Next, consider the catch component of the likelihood. Let P_1^k be the number of fish from the cohort in region k at the start of the tagging experiment (i.e., at the start of year 1), $C_j^{k'}$ be the number of fish from the cohort caught in year j and region k', and $q_j^{k'}$ be the probability of a fish from the cohort being caught in year j and region k'.

If the numbers of fish caught by year and region were known exactly, then they could be modeled as multinomial with probabilities given by $q_j^{k'}$ (similar to the tag return data). However, the catch data will almost always contain uncertainty, sometimes large, due to a number of reasons given in the Model assumptions section. These sources of uncertainty, which we will refer to jointly as sampling error, will generally dominate the multinomial variance (see Polacheck et al. 2006a). Thus, we only include sampling error in the model, and approximate it as Gaussian. Specifically, the number of fish caught in year j and region k' are assumed to have a Gaussian distribution with coefficient of variation (CV) $\upsilon_j^{k'}$.

Assuming that the catches for each region and year are statistically independent, the log-likelihood for the catch data (apart from an additive constant) is given by

$$\log L_{\text{catch}} = -\frac{1}{2}\sum_{j=1}^{J}\sum_{k'=1}^{K}\left(\log \mathrm{V}\left[C_j^{k'}\right] + \frac{\left(C_j^{k'} - \mathrm{E}\left[C_j^{k'}\right]\right)^2}{\mathrm{V}\left[C_j^{k'}\right]}\right) \tag{3}$$

where $\mathrm{E}\left[C_j^{k'}\right] = q_j^{k'} \sum_{k=1}^{K} P_1^k$ and $\mathrm{V}\left[C_j^{k'}\right] = \left(\upsilon_j^{k'}\,\mathrm{E}\left[C_j^{k'}\right]\right)^2$.

Similar to the tag return probabilities, we can express the catch probabilities in terms of survival and movement parameters. Consider a fish that originated in year 1 in region k. The probability, $q_{1,j}^{k,k'}$, of this fish subsequently being caught in year j in region k' is given by the (k, k') th element of the matrix

$$\mathbf{q}_{1,j} = \begin{cases} \mathrm{D}\left(\mathbf{u}_j\right) & j = 1 \\ \mathrm{D}\left(\mathbf{S}_1\right)\prod_1 \ldots \mathrm{D}\left(\mathbf{S}_{j-1}\right)\prod_{j-1}\mathrm{D}\left(\mathbf{u}_j\right) & j > 1 \end{cases},$$

which is analogous to (2) except it does not contain parameters for tag reporting rates. However, for the catch data, we do not know the origins of fish (i.e., their regions in year 1). Thus, the probability, $q_j^{k'}$, of a fish being caught in region k' in year j regardless of its origin is given by

$$q_j^{k'} = \frac{\sum_{k=1}^{K} q_{1,j}^{k,k'} P_1^k}{\sum_{k=1}^{K} P_1^k},$$

which can be substituted into the expected catch formula in (3).

The overall log-likelihood for the tag-recovery and catch data is the sum of the two likelihood components, (1) and (3):

$$\log L_{\text{total}} = \log L_{\text{tag}} + \log L_{\text{catch}}. \tag{4}$$

The parameters that can be estimated by maximizing (4) are $\left\{M_j^k\right\}$, $\left\{F_j^k\right\}$, $\left\{P_1^k\right\}$, and $\left\{\pi_j^{k,k'}\right\}$. Recall that the reporting rates, $\left\{\lambda_j^k\right\}$, are assumed to be known. Also, the catch CVs, $\left\{\upsilon_j^k\right\}$, are not directly estimable in the model, so are assumed to be known (see next paragraph). Furthermore, when a population is tagged in I consecutive years, only $I - 1$ natural mortality rate parameters (per region) can be estimated; this is a well-known feature of non-spatial Brownie models. As such, we impose the constraint that $M_j^k = M_{I-1}^k$ for $I \leq j \leq J$, noting that other constraints are possible. In the applications presented, we only consider an experiment with $I = J$, but there is no problem with having more recapture years than release years as long as natural mortality is appropriately constrained. Finally, we impose the obvious constraint that the movement probabilities for a given year and region sum to one; i.e., $\sum_{k'=1}^{K} \pi_j^{k,k'} = 1$.

Polacheck et al. (2006a) give a detailed explanation of why the catch CVs cannot be estimated reliably in the non-spatial BP model, even when constrained to be constant across years, and a similar argument would apply here. In theory, information about the catch CVs should be available based on the sampling design used to obtain the catch estimates, but such information is often poor or insufficient in practice. Fortunately, Polacheck et al. (2006a) found that, for a constant catch CV across years, the model results were fairly insensitive to the value assumed for the CV so long as it was in the right ballpark (e.g., within ~40% of the true value).

2.3 *Simulations*

2.3.1 Model Performance

We conducted a series of Monte-Carlo simulations to investigate model performance under six different model parameterizations (Table 1). Model 1 allows for all estimated parameters to vary with region and year (i.e., age for a single cohort). In many situations such a full model will not be necessary. Thus, we explored a range of models with more restricted parameterizations to determine the relative improvement that can be achieved in the parameter estimates. We chose parameterizations expected to be encountered in fishery situations. For example, natural mortality will often vary with year (i.e., age for a cohort), but may not vary with region (model 2). In addition, fishing mortality may not vary with year (i.e., age), but will likely vary with region, given that regions are often defined to correspond to distinct fishery components (model 3). Finally, movement probabilities between regions are likely be similar across years for species with annual migration patterns, so we considered time-invariant movement probabilities in combination with each of the above mortality parameterizations (models 4–6).

Throughout our simulations, we considered a tagging experiment involving a single cohort of fish with $I = 3$ release years, $J = 3$ recapture years, and $K = 3$ regions. We assumed 500 fish were tagged in each year and region, for a total of 4500 tags. The parameter values used to generate tag return counts and catch data were: $P_1^k = 100{,}000$; $M_j^k = 0.2$; $F_j^k = 0.3$; $\pi_j^{k,k'} = 0.25$ for $k \neq k'$, $\pi_j^{k,k'} = 0.50$ for $k = k'$; $\lambda_j^k = 1.0$; and $\upsilon_j^k = 0.2$. Note that even though the parameter values used to generate data were the same across years and regions, year- and region-specific parameters were still estimated in accordance with the model being fitted. Parameters were kept constant in order to make comparison of their estimates between models straightforward. The values for the mortality rate and abundance parameters were chosen to be within a plausible range for SBT, but they should be reasonable for a number of fish species.

Table 1 Parameter dependency (year-dependent, region-dependent, or both) for the six models considered. Note that when natural mortality is stated to be year-dependent, it only varies with year up to year $I - 1$, where I is the number of release years (see Model development section)

	Parameter dependency		
Model	Movement probabilities, π	Fishing mortality, F	Natural mortality, M
1	Year and region	Year and region	Year and region
2	Year and region	Year and region	Year
3	Year and region	Region	Year
4	Region	Year and region	Year and region
5	Region	Year and region	Year
6	Region	Region	Year

For each model, we generated 500 tag return and catch data sets. In particular, we generated multinomial return counts by year and region, $\left\{R_{i,j}^{k,k'}\right\}$, for each tagging group (i, k), and Gaussian catch data by year and region, $\left\{C_j^{k'}\right\}$. We then estimated the model parameters by maximizing log-likelihood (4), with the reporting rates and catch CVs fixed at their true values. Data were generated and models were fitted using AD Model Builder (Otter Research Ltd, P.O. Box 2040, Sidney, BC V8L 3S3, Canada). For each parameter estimated, we calculated the mean and standard deviation (SD) of the 500 estimates, and used these to compute the percent relative bias (i.e., (mean – true) / true $\times$ 100 %) and CV (i.e., SD/mean).

2.3.2 Parameter Identifiability Under Reduced Tagging Designs

Tagging in some regions may be difficult or costly, so it is of interest to know whether parameters are still estimable with experimental designs that do not include tagging in all regions in all years. We again considered the same six models as in the previous section, but under two reduced tagging designs: (I) tagging in all three regions in the first year, then only in region 1 thereafter; and (II) tagging only in region 1 in all three years. In both cases, we used a total of 4500 tag releases. For reduced design I, which has five tagging groups, this meant 900 releases per tagging group, and for design II, which has only three tagging groups, this meant 1500 releases per tagging group.

To determine whether the parameters for a given model and experimental design combination were identifiable, we used the analytic-numeric approach described by Burnham et al. (1987) and applied by Kendall and Nichols (2002). Specifically, for each model we generated non-random data (i.e., expected return counts and expected catch numbers) using the same parameter values as in the previous section (except for the tag release numbers), and fitted the model to these data by maximizing log-likelihood (4) using AD Model Builder. We considered a parameter to be identifiable if the parameter value returned was within 0.0001 of the true value and the Hessian-derived CV was less than 100%. We will refer to a model as being identifiable if all its parameters were identifiable.

Alternative approaches for determining parameter identifiability have been developed. For example, Viallefont et al. (1998) present a numerical method for determining parameter identifiability in single-state capture–recapture models based on the rank of the information matrix. Catchpole and Morgan (1997) present a related approach based on symbolic calculation of the rank of a defined derivative matrix (which has the same rank as the information matrix but is algebraically simpler), and apply it to models for capture–recapture and ring(tag)-recovery data. Giminez et al. (2003) generalize the Catchpole and Morgan approach to multi-state capture–recapture models. While these methods have some advantages in terms of exactness (i.e., less ambiguity), the analytic-numeric method described above was simple to apply and adequate for our purposes.

Just because a model is identifiable does not mean it will yield parameter estimates that are useful in practice. For each model and reduced tagging design found to be identifiable we conducted 500 simulations using random data sets. Again, the parameter values used to generate the data were the same as in the Model performance section, except that the distribution of the 4500 releases depended on the reduced tagging design (as described above). The percent relative biases and CVs of the parameter estimates were calculated and compared with the results from the equivalent model in the Model performance section to see how much the estimates were degraded by using a reduced tagging design.

2.3.3 Comparison of Estimates from Spatial and Non-spatial Models

In many situations, the population-wide estimates (i.e., over all regions) of mortality rates and abundance are of as much, if not more, interest than the regional estimates. Although such population-wide estimates can be obtained using a non-spatial model, we would expect them to be biased if spatial heterogeneity exists. To investigate, we compared population-wide estimates of mortality rates and abundance obtained from regional data analyzed using the spatial BP model with those obtained from pooled data (pooled across regions) analyzed using the non-spatial BP model of Polacheck et al. (2006a).

Simulations were carried out using model 1 and the standard tagging design of releases in all regions and years. Model 1 was chosen to maximize the number of parameters that vary by region; otherwise it would not be as important to use a spatial model (in fact it would be unnecessary if none of the parameters varied by region).

We again considered a tagging experiment for a single cohort of fish with I = 3 release years, J = 3 recapture years, and K = 3 regions. Six scenarios were investigated (Table 2). Scenarios 1–5 have different sets of parameter values for the mortality and movement parameters; all have 500 releases in each year and region. Scenario 6 has the same parameter values as scenario 1, but instead of having equal releases in all years and regions, has releases distributed across years and regions in proportion to abundance. In addition to the values specified in Table 2, all scenarios use $P_1^k = k \times 100000$, $\lambda_j^k = 1.0$ and $\upsilon_j^k = 0.2$. Qualitative descriptions of the scenarios are as follows: scenario 1 has low, symmetric movement probabilities between regions; scenario 2 has high, symmetric movement probabilities between regions; scenario 3 has no direct movements between regions 1 and 3 and a net movement to region 3 over time; scenario 4 has mortality rates (fishing and natural) that vary with region but not year; in contrast, scenario 5 has mortality rates that vary with year but not region; finally, scenario 6 has tag releases that are in proportion to regional abundance. Note that symmetric movement probabilities mean that, for a given year, the movement probability from region k to k' is equal to the movement probability from region k' to k.

For each scenario, we generated 500 tag return and catch data sets and fitted the spatial BP model to get estimates of year- and region-specific M's, year- and region-specific F's, region-specific P_1's and year- and region-specific π's. We also pooled the tag-recovery data and the catch data over regions and analyzed the pooled data using the non-spatial BP model to get estimates of year-specific M's, year-specific F's, and total P_1 (i.e., population-wide estimates rather than regional estimates).

Table 2 Parameter values used for the natural mortality rates (*M*), fishing mortality rates (*F*), movement probabilities (π), and numbers of releases (*N*) in the six scenarios comparing population-wide parameter estimates from the spatial and non-spatial Brownie-Petersen models

	Year	Region	Scenario 1	2	3	4	5	6
M	1	1	0.2	0.2	0.2	0.1	0.3	0.2
	1	2	0.3	0.3	0.3	0.2	0.3	0.3
	1	3	0.4	0.4	0.4	0.3	0.3	0.4
	2	1	0.1	0.1	0.1	0.1	0.2	0.1
	2	2	0.2	0.2	0.2	0.2	0.2	0.2
	2	3	0.3	0.3	0.3	0.3	0.2	0.3
F	1	1	0.05	0.05	0.05	0.05	0.1	0.05
	1	2	0.15	0.15	0.15	0.15	0.1	0.15
	1	3	0.25	0.25	0.25	0.25	0.1	0.25
	2	1	0.25	0.25	0.25	0.05	0.2	0.25
	2	2	0.25	0.25	0.25	0.15	0.2	0.25
	2	3	0.25	0.25	0.25	0.25	0.2	0.25
	3	1	0.45	0.45	0.45	0.05	0.3	0.45
	3	2	0.25	0.25	0.25	0.15	0.3	0.25
	3	3	0.05	0.05	0.05	0.25	0.3	0.05
π	1	1→2	0.1	0.3	0.3	0.1	0.1	0.1
	1	1→3	0.1	0.3	0	0.1	0.1	0.1
	1	2→1	0.1	0.3	0.15	0.1	0.1	0.1
	1	2→3	0.1	0.3	0.3	0.1	0.1	0.1
	1	3→1	0.1	0.3	0	0.1	0.1	0.1
	1	3→2	0.1	0.3	0.15	0.1	0.1	0.1
	2	1→2	0.05	0.15	0.3	0.05	0.05	0.05
	2	1→3	0.05	0.15	0	0.05	0.05	0.05
	2	2→1	0.05	0.15	0.15	0.05	0.05	0.05
	2	2→3	0.05	0.15	0.3	0.05	0.05	0.05
	2	3→1	0.05	0.15	0	0.05	0.05	0.05
	2	3→2	0.05	0.15	0.15	0.05	0.05	0.05
N	1	1	500	500	500	500	500	378
	1	2	500	500	500	500	500	756
	1	3	500	500	500	500	500	1134
	2	1	500	500	500	500	500	343
	2	2	500	500	500	500	500	474
	2	3	500	500	500	500	500	551
	3	1	500	500	500	500	500	249
	3	2	500	500	500	500	500	300
	3	3	500	500	500	500	500	314

To compare parameter estimates from the spatial and non-spatial models, it is first necessary to define population-wide parameters for the spatial model. The population-wide year 1 abundance is simply the sum of the regional year 1 abundance parameters, which we denote by $P_1^\bullet = \sum_{k=1}^K P_1^k$. For the population-wide fishing and natural mortality rates, we are seeking average rates across all regions, for which the definitions are not as clear-cut. Suppose we define the average fishing mortality rate in year j, $\bar{F}_j$, and average natural mortality rate in year j, $\bar{M}_j$, as the parameters satisfying

$$P_{j+1}^\bullet = P_j^\bullet \exp\left(-\left(\bar{F}_j + \bar{M}_j\right)\right) \tag{5}$$

where $P_j^\bullet$ denotes the total number of fish at the start of year j over all regions.

If P_j^k denotes the number of fish in region k at the *start* of year j, and P_{j+}^k denotes the number of fish in region k at the *end* of year j, then

$$P_{j+}^k = P_j^k \exp\left(-\left(F_j^k + M_j^k\right)\right). \tag{6}$$

Now, using the fact that $P_{j+1}^\bullet = \sum_{k=1}^{K} P_{j+}^k$, i.e., the total number of fish at the start of year $j+1$ is equal to the total number of fish at the end of year j (fish may have moved between regions, but none have died), and substituting from (5) and (6), gives

$$P_j^\bullet \exp\left(-\left(\bar{F}_j + \bar{M}_j\right)\right) = \sum_{k=1}^{K} P_j^k \exp\left(-\left(F_j^k + M_j^k\right)\right). \tag{7}$$

This type of reasoning is called a *counting argument*.

By applying an analogous counting argument on the number of fish caught in each year, as opposed to abundance of fish, we obtain the equation

$$\begin{aligned} &P_j^\bullet \frac{\bar{F}_j}{\bar{F}_j + \bar{M}_j}\left\{1 - \exp\left(-\left(\bar{F}_j + \bar{M}_j\right)\right)\right\} = \\ &\qquad \sum_{k=1}^{K} P_j^k \frac{F_j^k}{F_j^k + M_j^k}\left\{1 - \exp\left(-\left(F_j^k + M_j^k\right)\right)\right\} \end{aligned} \tag{8}$$

We now have two equations, (7) and (8), which can be solved numerically for $\bar{F}_j$ and $\bar{M}_j$.

For each simulation scenario, we calculated the true values of $\bar{F}_j$ and $\bar{M}_j$ using the true values of F_j^k, M_j^k and P_j^k in (7) and (8). Note that to do so involved first calculating the true P_j^k values iteratively. The same procedure was used to calculate estimated values of $\bar{F}_j$ and $\bar{M}_j$ for each simulation run within a scenario, except using the estimated values of F_j^k, M_j^k and P_j^k in (7) and (8) instead of the true values.

It is worth noting that if we had considered a model for which natural mortality in a given year is constant across regions (i.e., models 2, 3, 5 and 6), then (7) would reduce to

$$\begin{aligned} \bar{F}_j &= -\log\left(\frac{1}{P_j^\bullet} \sum_{k=1}^{K} P_j^k \exp\left(-F_j^k\right)\right) \\ &= -\log\left(\frac{1}{\sum_{k=1}^{K} P_j^k} \sum_{k=1}^{K} P_j^k \exp\left(-F_j^k\right)\right). \end{aligned}$$

This is simply the weighted average of the regional fishing mortality rates, with the weights equal to the regional population sizes. A slightly different value for the average fishing mortality would be obtained using (8). This reflects the fact that the population dynamics have been modeled in terms of exponential and competing natural and fishing mortality rates (i.e., both occur simultaneously and continuously throughout the year).

3 Results

3.1 Model Performance

We first consider the results from the simulations using 500 tag releases per year and region. For all six models, the parameters estimates were essentially unbiased, with the largest percent relative bias for any parameter being less than 4% in magnitude. For model 1, the CVs of the natural mortality estimates were quite high (0.36–0.50); however, the CVs of the other parameter estimates were much lower (Table 3). In particular, the fishing mortality estimates had CVs ~0.10, and the movement probability and abundance estimates had CVs ~0.20. By restricting natural mortality rates to be constant across regions (model 2), the CVs for these parameters were reduced by roughly 30%, from an average of 0.36 in year 1 to 0.26 and from an average of 0.48 in year 2 to 0.31 (Table 3). Further restricting fishing mortality to be constant across years (model 3) reduced the CVs from ~0.10 to 0.06; while this may be a small absolute change, it represents a 40% improvement. Restricting movement probabilities to be the same across regions (model 4) led to ~30% reductions in the CVs for these parameters. Interestingly, restrictions on a given type of parameter (e.g., natural mortality) led to significant improvements in the precision of parameters of that type, but to very small improvements in parameters of other types (the only exception being that restrictions on fishing mortality led to a noticeably smaller CV for natural mortality in year 2). This was true even in models with restrictions on more than one parameter type (models 5 and 6); for example, even when constraints were placed on both natural mortality and movement (model 5), the fishing mortality estimates were only slightly improved and the abundance estimates were unaffected. It is worth noting that the precision of the abundance estimates showed no discernible change in any of the models.

Overall, the CVs of the parameter estimates were better than might have been expected, especially for model 1 with so many parameters. The level of precision that can be achieved depends largely on the number of tag returns, which is determined by the number of releases, the natural mortality and fishing mortality rates, and the tag reporting rates. The values used for these parameters in the above simulations resulted in higher tag returns than may be realistic for many tagging experiments. Thus, we repeated the same set of simulations for a situation with significantly lower tag returns, which we achieved by reducing the number of releases to 100 in each year and region (for a total of 900 tags). As expected, the

Table 3 Coefficient of variation (i.e., standard deviation/mean) of parameter estimates obtained from simulations using models 1 to 6 (500 runs per model). Qualitative descriptions of models are given in Table 1; refer to text for parameter values used to generate data

			Model					
	Year	Region	1	2	3	4	5	6
M	1	1	0.36	0.26	0.23	0.34	0.24	0.22
	2	1	0.49	0.31	0.19	0.45	0.31	0.17
	1	2	0.37	–	–	0.38	–	–
	2	2	0.50	–	–	0.45	–	–
	1	3	0.37	–	–	0.38	–	–
	2	3	0.46	–	–	0.48	–	–
F	1	1	0.09	0.09	0.06	0.10	0.09	0.06
	2	1	0.10	0.09	–	0.09	0.08	–
	3	1	0.12	0.11	–	0.12	0.11	–
	1	2	0.09	0.10	0.06	0.10	0.09	0.06
	2	2	0.09	0.09	–	0.09	0.08	–
	3	2	0.13	0.11	–	0.12	0.11	–
	1	3	0.09	0.09	0.06	0.11	0.10	0.05
	2	3	0.10	0.10	–	0.09	0.08	–
	3	3	0.12	0.11	–	0.12	0.10	–
π	1	1→2	0.20	0.22	0.21	0.14	0.14	0.13
	1	1→3	0.22	0.21	0.20	0.14	0.14	0.14
	1	2→1	0.21	0.21	0.20	0.14	0.14	0.14
	1	2→3	0.22	0.21	0.20	0.15	0.14	0.13
	1	3→1	0.22	0.21	0.19	0.15	0.14	0.13
	1	3→2	0.21	0.20	0.18	0.14	0.14	0.14
	2	1→2	0.21	0.22	0.20	–	–	–
	2	1→3	0.21	0.21	0.20	–	–	–
	2	2→1	0.20	0.20	0.20	–	–	–
	2	2→3	0.20	0.20	0.20	–	–	–
	2	3→1	0.21	0.21	0.20	–	–	–
	2	3→2	0.21	0.21	0.20	–	–	–
P	1	1	0.20	0.19	0.20	0.20	0.21	0.19
	1	2	0.19	0.19	0.20	0.19	0.20	0.19
	1	3	0.20	0.20	0.20	0.20	0.19	0.19

M = natural mortality; F= fishing morality; π = movement probability; P = abundance.

precision of the estimates became much poorer, but more interesting and relevant is that the relative changes were quite consistent. The CVs of the natural mortality, fishing mortality and movement estimates averaged 2.1 times larger over all models, ranging from 1.7 to 2.4 for the natural mortality estimates, from 1.8 to 2.6 for the fishing mortality estimates, and from 2.0 to 2.4 for the movement estimates. Based on the variance formula for multinomial counts, a decrease in sample size of five times increases the CV of the data by $\sqrt{5} = 2.2$ times, which is very close to the observed increase in the parameter CVs. To verify this pattern we ran the simulations again with releases increased by three times (from 500 to 1500 in each year and region), and found, as would be predicted, that the CVs of the natural mortality, fishing mortality and movement estimates averaged 1.7 ($= \sqrt{3}$) times smaller. We note that the CVs of the abundance estimates were not affected to the same degree (they averaged 1.2 times greater with releases reduced by 5 times, and 1.1 times smaller with releases increased by 3 times). This is because changing the number of tag releases (i.e., the number of tag returns) does not have a direct effect on the catch component of the likelihood, from which abundance is estimated; these estimates are much more influenced by the variance of the catch data.

3.2 Parameter Identifiabiltity Under Reduced Tagging Designs

Under reduced design I (tagging only in region 1 each year), none of the models were found to be identifiable. However, under reduced tagging design II (tagging in all regions in the first year, then only in region 1 thereafter), models 3, 5 and 6 were all found to be identifiable. Moreover, for model 2, all of the true parameter values were returned, but the estimates of M_2, F_3^2, F_3^3 and all of the year 2 movement probabilities had CVs over 100% (the maximum being 342% for M_2). In cases where the models were clearly unidentifiable, the CVs for most parameters were huge (over 10,000%) or else not even attainable due to the Hessian matrix not being positive definite. This suggests that model 2 is in fact identifiable, but that the likelihood surface has an almost flat ridge. Thus, we concluded that the only two models not identifiable under reduced tagging design II were those for which natural mortality varied by both year and region (models 1 and 4).

Given the above identifiability results, we ran 500 simulations using random data sets for each of models 2, 3, 5 and 6 under reduced tagging design II. For models 3 and 6, all parameters were estimated with less than 3% relative bias. For models 2 and 5, which differ from models 3 and 6 in that they have year- and region-specific fishing mortality rates (as opposed to just region-specific), the only significant biases occurred for F_3^2 and F_3^3, both being ~18% for model 5 and much higher (~75%) for model 2. Otherwise, all biases in these two models were less than 10% and the majority less than 5%. (Note that F_3^2 and F_3^3 were the two fishing mortality parameters in model 2 that had CVs over 100% in the simulations to determine identifiability.) The CVs of the parameter estimates are presented in Table 4. For model 6, which is the most constrained model, the CVs of the parameter estimates were as good as those obtained with releases in all regions and years (compare Table 4 with Table 3). For models 2, 3 and 5, many parameters had CVs comparable to those obtained with the full tagging design, but there were some parameters with noticeably higher CVs. Generally, the parameters that were estimated poorly were logical when bearing in mind that tagging did not occur in regions 2 and 3 in years 2 and 3. For instance, in the models where fishing mortality varied with year and region (models 2 and 5), the fishing mortality estimates in regions 2 and 3 in years 2 and 3 had high CVs (0.36–0.67). This, in turn, is likely the cause of the high CVs in the natural mortality estimates, since estimates of natural mortality and fishing mortality are known to be highly correlated (e.g., Polacheck et al. 2006a). As another example, in the models with year-dependent movement probabilities (models 2 and 3), the year 2 movement probability estimates had very high CVs (0.59–0.94), and particularly so for movements out of regions 2 and 3 (where no tagging took place in year 2).

It is important to note that the identifiability results are dependent on the number of release and recapture years, the number of regions, and the true parameter values (in particular how much contrast there is in the data). We considered a number of

Table 4 Coefficient of variation (i.e., standard deviation/mean) of parameter estimates obtained from simulations using reduced tagging design II (tagging in all regions in year 1, then only in region 1 thereafter). Shown are the results for all models found to be identifiable (500 runs per model). Qualitative descriptions of models are given in Table 1; refer to text for parameter values used to generate data

			Model			
	Year	Region	2	3	5	6
M	1	1	0.49	0.23	0.40	0.26
	2	1	1.09	0.19	0.59	0.19
	1	2	–	–	–	–
	2	2	–	–	–	–
	1	3	–	–	–	–
	2	3	–	–	–	–
F	1	1	0.08	0.05	0.08	0.04
	2	1	0.13	–	0.08	–
	3	1	0.13	–	0.11	–
	1	2	0.08	0.08	0.08	0.08
	2	2	0.41	–	0.36	–
	3	2	0.67	–	0.54	–
	1	3	0.08	0.07	0.08	0.07
	2	3	0.39	–	0.36	–
	3	3	0.67	–	0.51	–
π	1	1→2	0.28	0.16	0.29	0.11
	1	1→3	0.27	0.16	0.29	0.11
	1	2→1	0.20	0.15	0.19	0.13
	1	2→3	0.30	0.15	0.31	0.14
	1	3→1	0.19	0.15	0.19	0.14
	1	3→2	0.31	0.15	0.31	0.14
	2	1→2	0.62	0.17	–	–
	2	1→3	0.61	0.16	–	–
	2	2→1	0.77	0.61	–	–
	2	2→3	0.94	0.59	–	–
	2	3→1	0.79	0.59	–	–
	2	3→2	0.93	0.59	–	–
P	1	1	0.19	0.20	0.20	0.18
	1	2	0.20	0.19	0.20	0.19
	1	3	0.20	0.20	0.19	0.18

M = natural mortality; F = fishing morality; π = movement probability; P = abundance.

scenarios with these factors varied, and although a more thorough investigation is needed before drawing any conclusions, we could not determine an obvious pattern. A researcher would need to investigate his or her specific situation, in a similar manner as was done here, to determine whether a reduced tagging design would be viable in that situation.

3.3 *Comparison of Population-Wide Estimates from Spatial and Non-spatial Models*

A comparison of the percent relative biases in the population-wide parameter estimates obtained from the spatial versus non-spatial BP model is presented (Fig. 1;

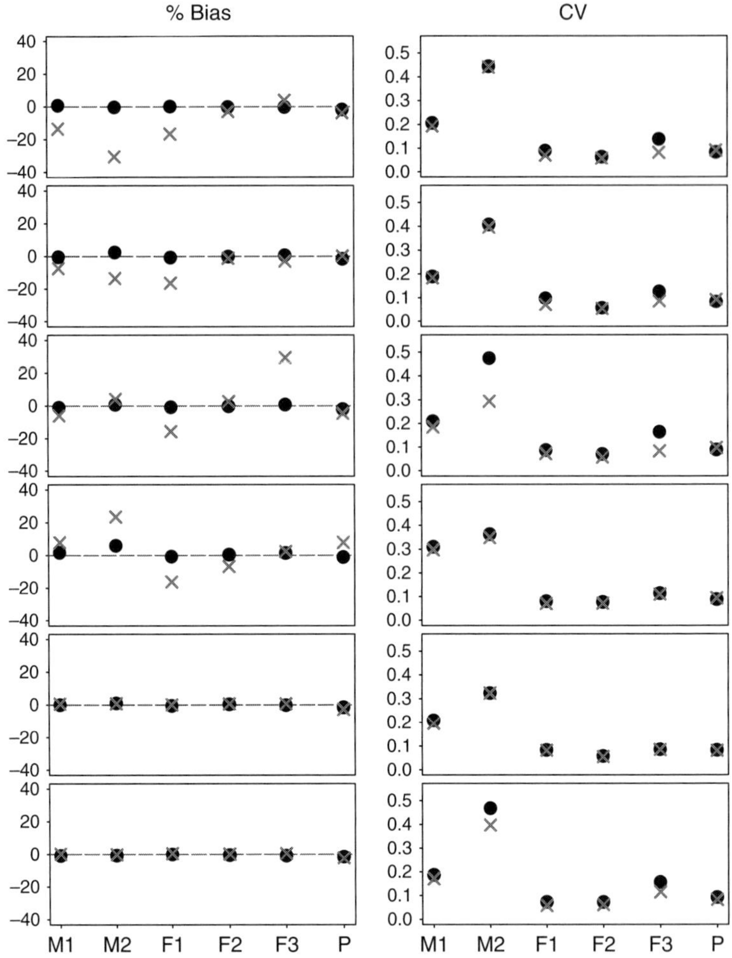

Fig. 1 Comparison of percent relative biases and coefficients of variation of the population-wide parameter estimates obtained using the spatial (*filled circles*) versus non-spatial (*crosses*) Brownie–Petersen model. Results are shown, from top to bottom, for the six scenarios specified in the text and in Table 2. Mj denotes the average (population-wide) natural mortality rate in year *j*; Fj denotes the average fishing mortality rate in year *j*; P denotes total abundance in year 1

left column). The parameter estimates from the spatial model were unbiased in all scenarios. In scenarios 1–4, several of the parameter estimates from the non-spatial model were biased, with the size and direction of the bias depending on the parameter and the scenario. The biases tended to be 10–20% in magnitude, but were higher than 30% in a couple of cases (e.g., for M_2 in scenario 1 and F_3 in scenario 3). The only difference between scenarios 1 and 2 is that scenario 2 has higher movement probabilities and thus more complete mixing of tagged and untagged fish across the population; therefore, it is not surprising that the same parameters were biased in both scenarios, but that these biases were smaller in scenario 2. Scenario 3, which differs from scenario 1 in its movement probabilities (with net movement towards region 3), did not have the same large biases in the natural mortality estimates as scenario 1, but instead had a high bias in F_3. Scenario 4, which only differs from scenario 1 in the values used for the fishing mortality rates (still varying with region in both cases), had positive rather than negative biases in the natural mortality estimates, and was the only scenario to show a small bias in the abundance estimate. From these results, we conclude that high biases can occur from applying a non-spatial model when a spatial one is more appropriate, but that the nature of the biases is not predictable.

For scenarios 5 and 6, the non-spatial model produces unbiased estimates of all parameters. Scenario 5 has mortality rates that are constant across regions, which means no spatial heterogeneity; scenario 6 has tag releases in proportion to regional abundance, which ensures complete mixing. It was therefore expected that these scenarios should give unbiased results, as we have verified. Note that the movement probabilities chosen for these scenarios were the same as scenario 1, because these were seen to produce some large biases. We wanted to ensure that scenarios 5 and 6 would produce unbiased estimates even under "worst-case" conditions.

A comparison of the CVs of the population-wide parameter estimates obtained from the spatial versus non-spatial model is also presented (Fig. 1; right column). The CVs that resulted from both models were very similar, with the exception of year 2 natural mortality in scenario 3. This is of particular interest for scenarios 5 and 6, for which the non-spatial model was capable of producing unbiased estimates, because it suggests that there is almost no loss in precision from using a spatial analysis to obtain population-wide parameter estimates even when a non-spatial analysis would be sufficient. Thus, even though the CVs of the region-specific estimates from the spatial model are larger than the CVs of the population-wide estimates from the non-spatial model (as would be expected for a model with more parameters), when these region-specific estimates are averaged (or summed for abundance) to give population-wide estimates, they are virtually as precise as those derived from the non-spatial model.

4 Discussion

4.1 *Model Assumptions: Plausibility, Effects of Violations and Possible Solutions*

The model requires a number of assumptions that may not be satisfied in a field tagging experiment. The assumption that fish only move between regions at the end of each time period will not be exactly true; however, for many populations, large scale movements occur only seasonally and over relatively short time periods. Furthermore, Hestbeck (1995) investigated how much violations of the assumption of end-of-year movements can bias the parameter estimates in a capture–recapture setting. He found that biases in the movement estimates were variable and could be high in some situations, but that biases in the survival and capture rate estimates were consistently small. Nevertheless, it would be possible to develop the model in terms of an alternative parameterization for the movement process, such as at the start of each year or continuous throughout each year, if this was considered more appropriate (e.g., Beverton and Holt 1957; Joe and Pollock 2002). Alternative assumptions regarding the dependency of movements between years that allow for some degree of memory are also possible (e.g., Brownie et al. 1993; Appendix 11 of Polacheck et al. 2006b).

Tag shedding and tag-induced mortality may not be negligible, in which case additional parameters and additional data (e.g., from double tagging) can be introduced into the model to account for them. Much work has been done on approaches for estimating these quantities (e.g., Beverton and Holt 1957; Kirkwood and Walker 1984; Barrowman and Myers 1996) and incorporating them into non-spatial tag-recovery models (e.g., Seber and Felton 1981; Wetherall 1982; Leigh et al. 2006; Appendix 15 of Polacheck et al. 2006). These approaches could be extended in a straightforward manner to the spatial case.

Although the model, as presented, assumes that tag reporting rates are known, their estimation can be incorporated directly provided data for doing so are available. For example, if reporting rate data are available that are independent of the tag-recovery and catch data (such as data from planted tag experiments), then a likelihood can be developed for these data and simply multiplied to the tag-recovery and catch likelihoods, analogously to the way it was done in Polacheck et al. (2006a) for the non-spatial BP model. If the reporting rate data are not independent of the tag-recovery and catch data, such as when observers are placed on a portion of fishing vessels, then the likelihoods for the tag-recovery and catch data need to be modified; refer to Eveson et al. (2007) for details in the non-spatial case.

When tagging is specific to cohorts, the release ages (and thus cohorts to which fish belong) are assumed to be known without error. This assumption should not bias the parameter estimates as long as the distribution of release ages is not biased. When ages are determined from scales, biases are not expected; however, when ages are determined from length and a given age-length relationship, biases can exist. Investigations with a non-spatial Brownie model suggest that in many situations the effects of aging errors, in terms of biases in the mortality rate estimates, will be minimal (Appendix 8 of Polacheck et al. 2006b). However, even if no biases are incurred from assuming the release ages are known, the standard errors of the parameter estimates derived from the model (i.e., from the Hessian matrix) will be too small. If aging errors are considered a significant problem, then the standard errors could be calculated using an alternative approach, such as bootstrapping.

The assumption that the catch data for a cohort are independent between years and regions should be reasonable for many fisheries, especially those in which catch samples are collected in each year and region. However, if more than one cohort is being modeled, then catch data from multiple ages within the same year will enter the model, and aging errors within a year will be correlated across ages. If the correlation is strong, a more sophisticated error structure for the catch data than independent Gaussians may be required. The appropriate error structure will depend on the sampling design, but possibilities include multinomial within a year or multivariate normal.

Finally, the model makes the common assumption that the tag-recovery data are multinomially distributed. A recognized issue with non-spatial analyses of tag-recovery data is that the data will exhibit more variation than a multinomial model predicts (i.e., will be overdispersed multinomial) due to spatial heterogeneity and incomplete mixing. Although the spatial model is intended to account for much of this, some overdispersion is still likely to exist due to the patchy distribution (e.g., schooling nature) of many fish species. If overdispersion exists, the estimates derived from the multinomial model should still be unbiased, but the model-based standard error estimates will be too small. Again, bootstrapping is one possible method for obtaining more realistic standard errors. Another possibility is to estimate the amount of overdispersion from the model residuals and inflate the model-based standard error estimates accordingly. Unfortunately, the amount by which to inflate the estimates is not simple with such a complicated model; this issue has been investigated in non-spatial versions of the BP model (Polacheck et al. 2006a; Eveson et al. 2007).

4.2 Summary and Conclusions

We have shown how the BP model of Polacheck et al. (2006a) can be expanded to incorporate a discrete spatial structure. Using simulations, we demonstrated that when tagging is conducted in all regions and years, this model is capable of providing estimates of abundance by region, and of natural mortality, fishing mortality and movement probabilities by both year and region. The precision of the estimates will depend on several factors, including the model parameterization, the true parameter values (and their implication for the number of tag returns), and the parameter being considered. However, our simulation results suggest that high precision can be achieved in many situations. In relative terms, we found precision to be consistently highest for the fishing mortality estimates, lowest for the natural mortality estimates, and somewhere in between for the movement probabilities and abundance estimates.

An important and practical finding that emerged from our investigations is that, although tagging in only one region failed to give satisfactory results in any of the models we considered, tagging in only one region after the first year of the experiment is a viable option in many situations. In particular, the only models for which all parameters were not identifiable were ones where natural mortality varied by region as well as year (models 1 and 4); it seems plausible for natural mortality to be similar across regions and/or a given year (i.e., age for a cohort). Not only were the parameters identifiable for many models using this reduced tagging design, but, more importantly, good accuracy and precision could often be achieved. Not needing to release tags in all regions beyond the first year could be highly beneficial in terms of making a tagging experiment viable and more cost-effective. For example, in many instances it may be logistically difficult and/or highly expensive to tag fish in some regions, such as high seas regions of tuna longline fisheries.

We should note that there are specific situations with highly structured movement dynamics in which tagging in only one region can produce reliable results. This was true, for example, in the version of the spatial BP model modified to have movement dynamics representative of juvenile SBT (see Appendix 12 of Polacheck et al. 2006b). Briefly, this model incorporated a seasonal migration pattern, with fish migrating from three offshore regions to a single inshore region at the end of winter, then migrating back to the offshore regions at the end of summer.

Our investigation showed that, if population-wide estimates of mortality rates and abundance are of primary interest, then there are situations in which using a non-spatial analysis would be sufficient. This will obviously be true when neither fishing mortality nor natural mortality varies across the spatial distribution of the population of interest (i.e., when there is no spatial heterogeneity in survival and capture probabilities). Similarly, if movement probabilities are very high, little or no bias will result from a non-spatial analysis. However, the researcher is unlikely to have knowledge about whether these conditions apply prior to conducting the tagging experiment. Also, it seems unlikely for fishing mortality to be spatially constant because fishing vessels would be expected to concentrate their effort in areas of high density. More interesting is that a non-spatial model also provides unbiased estimates when spatial heterogeneity exists among discrete regions if the distribution of tag releases across these regions is in proportion to abundance. In practice, this would be difficult to achieve because the researcher would need to have a priori information on the relative abundances by region, and also because tagging the necessary numbers of animals may not be feasible in some regions. Spreading tagging effort randomly throughout the entire area might be one way to achieve this, but this is likely to be inefficient, with much effort spent attempting to tag fish in areas of low abundance. Also, catchability (i.e., the effectiveness of the sampling gear to catch fish that are present) is likely to differ spatially with differences in the physical and biological environment. Nevertheless, the results do suggest that spreading tagging effort spatially should help to minimize biases.

In summary, there are situations in which a non-spatial model can provide unbiased estimates of population-wide mortality rates and abundance; however, they all require some knowledge about the population and fishery dynamics that is unlikely to be available prior to conducting the experiment. The biases that result from using a non-spatial model when one of these situations does not apply can be large. Our simulation results show there is almost no cost in terms of precision in using a spatial analysis to obtain population-wide estimates even when a non-spatial analysis would be sufficient. Thus, in order to guard against potential biases, it is recommended that a tagging experiment involving a population with a wide geographic distribution be designed and implemented so that the results allow for a spatial analysis to be conducted. Moreover, the region-specific estimates obtained from a spatial analysis can provide insights and improved understanding of the underlying biology of the population being studied. This in turn can have important implications for conservation and management.

Acknowledgments We thank S. Foster, J.A. Royle and an anonymous reviewer for helpful comments and suggestions. The Australian Fisheries Research and Development Corporation (FRDC) provided funding support for this research.

References

Arnason AN (1973) Estimates of population size, migration and survival rates in a stratified population. Researches on Population Ecology 15: 1–8.

Barrowman NJ, Myers RA (1996) Estimating tag-shedding rates for experiments with multiple tag types. Biometrics 52: 1410–1416.

Beverton RJH, Holt SJ (1957) On the dynamics of exploited fish populations. Fishery Investigations, Series 2, 19. Ministry of Agriculture, Fisheries and Food, London, 533pp.

Brownie C, Anderson DR, Burnham KP, Robson DS (1985) Statistical inference from band recovery data: a handbook. 2 nd edition. US Fish and Wildlife Service Resource Publication 156, Washington, DC.

Brownie C, Hines JE, Nichols JD, Pollock KH, Hestbeck JB (1993) Capture–recapture studies for multiple strata including non-Markovian transitions. Biometrics 49: 1173–1187.

Burnham KP, Anderson DR, White GC, Brownie C, Pollock KH (1987) Design and analysis methods for fish survival experiments based on release-recapture. American Fisheries Society Monograph 5.

Cappo M, Walters CJ, Lenanton RC (2000) Estimation of rates of migration, exploitation and survival using tag recovery data for western Australian "salmon" (*Arripis truttaceus*: Arripidae: Percoidei). Fisheries Research 44: 207–217.

Catchpole EA, Morgan BJT (1997) Detecting parameter redundancy. Biometrika 84: 187–196.

Eveson JP, Polacheck T, Laslett GM (2007) An integrated catch-at-age and multiyear tag-recapture model for estimating mortality rates and abundance in a fishery with observers. Fishery Bulletin 105: 493–508.

Frusher SD, Hoenig JM (2001) Estimating natural and fishing mortality and tag reporting rate of southern rock lobster (*Jasus edwardsii*) from a multiyear tagging model. Canadian Journal of Fisheries and Aquatic Sciences 58: 2490–2501.

Giminez O, Choquet R, Lebreton J-D (2003) Parameter redundancy in multistate capture–recapture models. Biometrical Journal 45: 704–722.

Hestbeck JB, Nichols JD, Malecki RA (1991) Estimates of movement and site fidelity using mark resight data of wintering Canada Geese. Ecology 72: 523–533.

Hestbeck JB (1995) Bias in transition-specific survival and movement probabilities estimated using capture–recapture data. Journal of Applied Statistics 22: 737–750.

Hilborn R, Walters CJ (1992) Quantitative fisheries stock assessment: choice, dynamics and uncertainty. Chapman and Hall, London. 570 pp.

Hoenig JM, Barrowman NJ, Hearn WS, Pollock KH (1998) Multiyear tagging studies incorporating fishing effort data. Canadian Journal of Fisheries and Aquatic Sciences 55: 1466–1476.

Joe M, Pollock KH (2002) Separation of survival and movement rates in multi-state tag-return and capture–recapture models. Journal of Applied Statistics 29: 373–384.

Kendall WL, Nichols JD (2002) Estimating state-transition probabilities for unobservable states using capture-recapture/resighting data. Ecology 83: 3276–3284.

Kirkwood GP, Walker MH (1984) A new method for estimating tag shedding rates, with application to data for Australian Salmon, *Arripes trutta esper* Whitley. Australian Journal of Marine and Freshwater Research 35: 601–606.

Leigh GM, Hearn WS, Pollock KH (2006) Time-dependent instantaneous mortality rates from multiple tagging experiments with exact times of release and recovery. Environmental and Ecological Statistics 13: 89–108.

Nichols JD, Sauer JR, Pollock KH, Hestbeck JB (1992) Estimating transition-probabilities for stage-based population projection matrices using capture–recapture data. Ecology 73: 306–312.

Polacheck T, Eveson JP, Laslett GM, Pollock KH, Hearn WS (2006a) Integrating catch-at-age and multiyear tagging data: a combined Brownie and Petersen estimation approach in a fishery context. Canadian Journal of Fisheries and Aquatic Sciences 63: 534–548.

Polacheck T, Eveson JP, Laslett GM (2006b) Estimation of mortality rates from tagging data for pelagic fisheries: analysis and experimental design. Final Report to the Fisheries Research and Development Corporation. FRDC Project 2002/015. ISBN 1 921061 03 0.

Pollock KH, Hoenig JM, Jones CM (1991) Estimation of fishing and natural mortality when a tagging study is combined with a creel survey or port sampling. American Fisheries Society Symposium 12: 423–434.

Schwarz CJ, Schweigert JF, Arnason AN (1993) Estimating migration rates using tag-recovery data. Biometrics 49: 177–193.

Seber GAF, Felton R (1981) Tag loss and the Petersen mark-recapture experiment. Biometrika 68: 211–219.

Seber GAF (1982) The estimation of animal abundance and related parameters. 2 nd edition. Charles Griffin, London. 654 pp.

Viallefont A, Lebreton J-D, Reboulet A-M, Gory G (1998) Parameter identifiability and model selection in capture–recapture models: a numerical approach. Biometrical Journal 40: 1–13.

Wetherall JA (1982) Analysis of double-tagging experiments. Fishery Bulletin 80: 687–701.

Xiao Y (2000) An individual-based approach to evaluating experimental designs for estimating rates of fish movement from tag recoveries. Ecological Modelling 128: 149–163.

Gaussian Semiparametric Analysis Using Hierarchical Predictive Models

Daniel Fink and Wesley Hochachka

Abstract The Hierarchical Predictive Model (HPM) is a semiparametric mixed model where the fixed effects are fit with a user-specified non-parametric component. This approach extends current spline-based semiparametric mixed model formulations, allowing for more flexible nonparametric estimation. Greater adaptability simplifies model specification making it easier to analyze data sets with large numbers of predictors. Greater automation also extends the scope of exploratory analyses that may be performed with mixed models. Using a HPM, the analyst may select the predictive model to best suit their needs, exploiting the strengths of currently available predictive methods. A simulation study is used to demonstrate the advantages of accounting for known hierarchical structure in predictive models and to illustrate the adaptability of current decision-tree based predictive models. A HPM of the relative abundance of the North American House Finch (*Carpodacus mexicanus*) is used to demonstrate exploratory analysis with a real data set.

1 Introduction

Hierarchical models have emerged as the preferred tool for analyzing large complicated data sets. Multifaceted processes can be factored into a series of simpler, conditionally independent sub-processes and a wide variety of parametric models can be incorporated. Bird monitoring data lend themselves to hierarchical treatment because data arise as the result of a stochastic observational process conditional on spatio-temporally varying biological processes. By separating these two processes, researchers have been able to address a number of important complications that arise in the analysis of ecological monitoring data. For example, parametric models have been developed to account for imperfect capture of species (e.g., Jolly 1965; Seber 1965; Amstrup et al. 2006), varying detection during gathering of observational data (MacKenzie et al. 2002; Gelfand et al. 2005), observer-specific effects such as mis-identifications or incorrect counts

D. Fink (✉)
Cornell Laboratory of Ornithology, 159 Sapsucker Woods Rd, Ithaca, NY 14850, USA
e-mail: df36@cornell.edu

D.L. Thomson et al. (eds.), *Modeling Demographic Processes in Marked Populations*, Environmental and Ecological Statistics 3, DOI 10.1007/978-0-387-78151-8_46,

(e.g., Geissler and Sauer 1990; Thogmartin et al. 2004), and spatial correlation (Thogmartin et al. 2004; Wikle 2003; Wikle and Hooten 2006).

However, for many problems there is insufficient a priori knowledge to justifiably specify parametric models at all stages of the hierarchy. Often, it is not known which predictors should be included in a model. Even when important predictors have been identified, an appropriate functional form for their inclusion is unknown. In either case, the ability to specify a fully parametric model would still be desirable under many circumstances. Efficient exploratory tools are needed to discover patterns in data to account for and describe potentially complicated relationships between predictors and response. Exploratory analyses are an important means of hypothesis generation, and, ultimately, enable the specification of better parametric models.

Semiparametric models use as much parametric structure as is warranted by subject-area knowledge while relying on nonparametric techniques to automatically account for additional predictors and processes that are less well understood. This is a hybrid modeling strategy where the nonparametric components function as exploratory tools, automatically detecting and fitting patterns in the data while simultaneously taking into account parametric structure. Several successful semiparametric techniques have been built upon the Linear Mixed Model (LMM). These methods include the spline-based methods of Ruppert et al. (2003), Wood (2006) and Gu (2002). Each of these spline-based models can adapt over sums of smooth low-dimensional predictor functions while inheriting a well developed set of inferential tools from the LMM. Ideally, we would like to extend these methods to deal with large sets of predictors by utilizing nonparametric methods capable of automatically identifying important predictors and interactions, including high order interactions among predictors, and functional forms of relationships.

Within the last decade, data mining and machine learning techniques have emerged as some of the most successful tools for modeling complex, multidimensional data (Hand et al. 2001). These techniques are sophisticated nonparametric tools for data exploration with a focus on producing accurate predictions. Data mining methods include neural networks, decision trees, and support vector machines. Many of these methods have been gaining recognition within the ecological community (De'ath and Fabricius 2000; Elith et al. 2006; Hochachka et al. 2007). These methods are capable of sifting through large number of predictors to identify important ones, their interactions, and functional forms (Hastie et al. 2001). The weakness of these methods are their limited ability to incorporate prior information, especially patterns of correlation. Most current implementations of these tools assume independence among the data.

The purpose of this article is to develop a modeling framework that combines the complementary strengths of the LMM and modern nonparametric predictive models. We call this framework the Hierarchical Predictive Model (HPM). It is a semiparametric mixed model where the fixed effects are fit with a user-specified predictive model. We do this by fitting the HPM as a Bayesian model using a simple Gibbs sampler. The Gibbs sampler allows us to iteratively update fixed and random effects from nonparametric and parametric models, respectively. Taking an Empirical Bayes approach, we estimate the conditional fixed effects with the nonparametric predictive model. Thus, a wide variety of predictive models may be used to explore the fixed effects. Using data mining and machine learning methods, the HPM extends current spline-based semiparametric formulations, allowing for more flexible nonparametric estimation. Greater adaptability simplifies model specification making it easier to analyze data sets with large numbers of predictors. This will be of increasing importance as larger data sets become available to ecologists.

As a hybrid methodology, the HPM draws upon several other modeling frameworks. In Section 2 we review these modeling frameworks. The HPM and its fitting algorithm are developed in Section 3. In Section 4 we present results from a simulation study to demonstrate the advantages of accounting for known hierarchical structure into predictive models and to illustrate the adaptability of current decision-tree based predictive models. In Section 5, a HPM of the relative abundance of the North American House Finch is used to demonstrate an exploratory analysis with a real data set. We conclude with a brief discussion.

2 The Models

In this section we review several modeling frameworks developed for correlated and uncorrelated observations. Emphasis is placed on brief descriptions of the model frameworks noting their scope of application, strengths, and limitations. We distinguish between parametric and nonparametric frameworks and note how their strengths make them well suited for confirmatory and exploratory analysis, respectively.

2.1 Predictive Models

We use the term "predictive model" to refer to any model that extracts information from a set of predictors and independent responses to make future predictions. Let y_i, $i = 1, \ldots, N$ be the responses each associated with p predictors $x_i = \left[x_{1,i}, \ldots, x_{p,i}\right]$. It is assumed that each observation, y_i, arises as an independent realization from some true but unknown function, $F(x_i)$ that maps x_i to y_i. The goal of predictive modeling is to use the data to estimate $F(x)$ while minimizing the expected value of some specified loss function. Predictive models have been developed in various disciplines with their own unique sets of terminology. In statistics, this predictive problem is known as regression. In the machine learning and data mining communities it is known as the supervised learning problem and the term "regression" refers more specifically to supervised learning problems with a continuous response.

In this paper we will restrict our attention to normally distributed observations and write $y = F(X) + \varepsilon$ where y is the $n \times 1$ vector of observations and X is the n

$\times$ p design matrix of predictors. The $n \times 1$ error vector, ε, is assumed arise from an uncorrelated normal distribution with zero mean and variance σ^2.

2.1.1 Parametric Predictive Models

Parametric models have an explicit parametric form where model parameters describe a known or hypothesized process of interest. For example, consider the classic normal linear model in statistics, $y = X\beta + \varepsilon$, where β is a $p \times 1$ vector of parameters. The conditional mean of y is modeled parametrically as a linear combination of predictor effects. Parametric modeling requires enough knowledge about the process being investigated to specify the model. Constructing good parametric models can take considerable time and effort. The strengths of parametric models are the ease of interpretation and the availability of inferential tools. A well developed body of statistical methods, both Frequentist and Bayesian, can be used to make inferences about the parameters of interest and about predicted observations. For this reason, parametric models are most often used to make confirmatory inferences.

2.1.2 Nonparametric Predictive Models

Nonparametric models are often described as models without parameters or without parameters of direct inferential interest. Here, we use the term "nonparametric" to describe predictive models that automatically adapt to patterns in data – this being the essential distinguishing quality. Adaptive models are designed to automatically *discover* patterns. This makes them especially well suited for exploratory analysis. The more adaptive the methodology, the greater the scope of exploration.

Generalized Additive Models (GAMs) (Buja et al. 1989; Hastie and Tibshirani 1990; Wood 2006) are a popular class of nonparametric statistical models for representing a response as the sum of low-dimensional smooth functions of predictors. The simple GAM:

$$y = f_1(x_1) + \varepsilon,$$

can be used to detect and describe nonlinear functional effects of x_1. This GAM can be extended to simultaneously estimate smooth joint effects of x_2 and x_3 by adding an appropriate term, like a tensor plate spline, to yield,

$$y = f_1(x_1) + f_2(x_2, x_3) + \varepsilon.$$

This GAM makes 3 assumptions about the systematic effects of the predictors on the response; (1) the functional effects f_1 and f_2 vary smoothly with the predictor values, (2) predictor x_1 does not interact with x_2 or x_3, and (3) predictors x_2 and x_3 are allowed to interact. This GAM can be used to detect if there is a 2-way interaction among the user specified pair of predictors. Conceptually, one can extend this idea using higher order terms to automatically adapt to more complex multivariate functional forms.

Decision Trees were designed to automatically fit high-dimension multivariate functional forms. Using a strategy of binary recursive partitioning, these models adapt over high dimensional tensor-product predictor spaces to fit models with possibly high-order interactions. Thus, a decision tree model of the form

$$y = f(x_1, \dots, x_p) + \varepsilon$$

can be used to investigate numerous functional relationships. Predictive experiments can be used to extract information for identifying important predictors, describing their effects, and identifying interactions within sets of predictors; see Sections 4 and 5.

Nonparametric models vary widely in the type of adaptation they do and the strategies used to achieve them. Many highly-adaptive nonparametric predictive methods have been developed within the data mining and machine learning communities where problems are characterized by very large data sets, both in terms of the number of responses (*N*) and the number of predictors (*p*). Consequently, these methods are designed to be very efficient, both in terms of analyzing large numbers of responses as well as extracting predictive information from large sets of predictors. These methods include decision trees (e.g., Breiman et al. 1984), neural nets (e.g., Mitchell 1997), Support Vector Machines (SVMs) (e.g., Cristianini and Shawe-Taylor 2000), and ensemble variants of tree-based methods (e.g., bagged and boosted decision trees, random forests; e.g., Breiman 1996; Breiman 2001). Recently, these methods have enjoyed increasing visibility and application within the ecological literature, see De'ath and Fabricius (2000), Elith et al. (2006), and Hochachka et al. (2007).

2.2 Hierarchical Models

With hierarchical models, one can factor complicated, multifaceted processes into a series of simpler conditionally independent parametric sub-processes. When data have obvious hierarchical structure, it is advantageous to model this structure parametrically. The hierarchical model is a formal mechanism for pooling information from correlated responses, potentially making substantial improvements in model efficiency. In disciplines where correlated data are frequently confronted, specialized statistical models have been developed to deal with these correlations. For example, Kriging was developed for geo-statistical analysis where spatial correlation play is very important. Longitudinal analyses explicitly take into account the correlation induced by making several observations on individual experimental units over the duration of an experiment, e.g., patients in clinical trial.

Hierarchical models have also been developed to model a wide variety of processes including spatial data with varying support (Wikle and Berliner 2005;

Banerjee et al. 2004), measurement error models (Berry et al. 2002), dispersion processes (Wikle 2003) and dynamic processes (West and Harrison 1997; Banerjee et al. 2004).

In this paper we will focus on the Linear Mixed Model (LMM), a two-level parametric hierarchical model. The strength of this model is its success as a powerful framework within which to model patterns of correlation. By connecting LMMs together one may assemble more complex hierarchical structures and patterns of correlation, e.g., multilevel models (Goldstein 1995).

2.3 Linear Mixed Models

The Linear Mixed Model extends the linear model by incorporating random effects, which can be regarded as additional error terms, to account for correlations among observations. The general form of the LMM is

$$y = X\beta + Zu + \varepsilon$$

where y is a vector of N observable random variables, β is a vector of p unknown parameters having fixed values (fixed effects), X is the $n \times p$ fixed effect design matrix, and Z is the $n \times q$ random effect design matrix. Both u and ε are unobservable random vectors (random effects) of length q and n, respectively. We will refer to u to as the "random effects" and ε as the "error" term to distinguish them. It is assumed that both the random effects and errors are normally distributed and uncorrelated with each other. Specifically, $u \sim N(0, \Sigma(\varphi))$ where $\Sigma(\varphi)$ is assumed to be a parametric covariance model with variance component(s) φ and $\varepsilon \sim N(0, \sigma^2 I)$ where σ^2 is a positive constant and I is the n-dimensional identity matrix.

The LMM is one of the most useful models in modern statistics, allowing many complications to be handled within the familiar linear model framework. This model has become a standard approach to model genetic effects, longitudinal data, blocked designs, crossed designs, nested designs, varying coefficient models, and numerous problems with temporal and spatial correlation (see Robinson 1991; McCulloch and Searle 2001; Zhao et al. 2006 for good reviews). One of the reasons for the success of the LMM is the ease and efficiency with which correlation structure can be incorporated into the model. Often, a basic understanding of the correlation structure is sufficient knowledge to specify useful covariance models for the random effects. The vast literature on LMM is a testament to this fact.

Like any parametric model, the LMM requires enough *a priori* information to specify the entire model. For each process that is included in the model, the analyst must decide which predictors to include, which predictors interact, and the functional form of all effects. When there is more predictor information that prior knowledge, it may difficult to specify a good fixed effect model, ultimately limiting the amount of covariate information that can be admitted into the model. This becomes a bigger problem as the number of predictors grows and a priori information does not increase proportionally. In practice, this limits the amount of information that may be brought to bear on confirmatory analyses and it is the reason that the LMM is not often used for exploratory analyses.

2.4 Semiparametric Mixed Models

Semiparametric predictive models incorporate flexible nonparametric model components within a parametric framework. This gives the analyst the ability to include as much parametric structure as can be justified by subject-area knowledge while using adaptive nonparametric components to automatically search for additional signal in the data. The use of this hybrid modeling strategy can improve confirmatory analysis by automatically incorporating additional predictor information, with fewer unjustified assumptions, than possible in traditional parametric models. Semiparametric models may also be used to conduct focused exploratory analyses by adaptively searching for patterns after accounting for known parametric structure.

The SemiParametric Mixed Model (SPMM) includes nonparametric model components to automatically incorporate fixed effect predictor information within the LMM framework. Extending the LMM of the previous section, we write the general SPMM as

$$y = f(X) + Zu + \varepsilon$$

where $f(X)$ represents a nonparametric predictive component for fixed predictor effects.

Some of the most effective semiparametric modeling strategies to take advantage of the mixed model framework have been based on penalized splines. These approaches use spline basis-expansions as flexible function effects and then control the complexity of the fit by means of penalization. The key to incorporating penalized splines within the mixed model framework is to recast the penalty as a random effect. Current implementations differ in the types of spline functions and fitting strategies used. Current examples include the penalized regression splines of Ruppert et al. (2003), the generalized additive models of Wood (2006) and the smoothing spline ANOVA models of Gu (2002), though the connection between penalized spline methods and the mixed model has a much longer history (see Wahba 1990).

There are two main advantages to bringing this nonparametric smoothing technique to the mixed model. First, it allows splines to be used with a wide variety of data types and diverse applications where mixed models are already used. Second, it give practitioners access to many of the inferential tools developed for the mixed model. A serious limitation of this strategy is computational. In order to adapt to functional forms in high-dimensional spaces, it is necessary to generate very large spline-basis expansions which in turn require the manipulation of equally large matrices. This is why most current techniques limit the response to be the sum of several low-dimensional smooth functions of predictors.

3 Hierarchical Predictive Models

The HPM, is a SPMM

$$y = f(X) + Zu + \varepsilon,$$

where y is a vector of N observable random variables, X is thc $n \times p$ fixed effect design matrix, Z is the $n \times q$ random effects design matrix, and $f(X)$ is a vector of N predictions. For notational convenience, we will denote the vector of fixed effects as f, suppressing its dependence on the predictors in the fixed effect design matrix. The random effects are normally distributed, $u \sim N(0, \Sigma(\varphi))$ where $\Sigma(\varphi)$ is assumed to be a parametric covariance model with variance component(s) φ. The errors are independent and normally distributed $\varepsilon \sim N(0, \sigma^2 I)$ where σ^2 is a positive constant and I is the n-dimensional identity matrix. The errors and random effects are assumed to be independent of each other.

Although hierarchical models are not inherently Bayesian, complex hierarchical models are most easily fit within the Bayesian framework using simulation-based Markov Chain Monte Carlo (MCMC) techniques. Bayesian inferences are based on the posterior distribution of the unknown model parameters conditioned on all observed, known quantities. The posterior distribution for the HPM is $[f, u, \varphi, \sigma^2 | y, X, Z]$. We denote the distribution of a random vector x by $[x]$ and the conditional distribution of y given x is by $[y|x]$. The conditional dependence of posterior distributions on X and Z will be omitted for notational convenience. The MCMC sampler used to fit the Bayesian HPM is described below.

The Gibbs sampler (Robert and Casella 2004) is used to simulate the posterior by breaking the vector of model parameters into convenient subsets and iteratively sampling from the resulting conditional distributions. The hierarchical structure of the mixed model naturally breaks down into conditional distributions for u, f, and the variance components φ and σ^2,

$$\begin{gathered}[u|f, \varphi, \sigma^2, y] \\ [f|u, \varphi, \sigma^2, y] \\ [\varphi|f, u, \sigma^2, y] \\ [\sigma^2|f, u, \varphi, y]\end{gathered}$$

The Gibbs sampler generates samples from each posterior conditional distribution to sequentially update the parameters. Strategies for updating the parameters vary depending on the form of the conditional distribution.

The conditional distribution of u is proportional to the product of normal distributions,

$$[u|f, \varphi, \sigma^2, y] \propto [u|\Sigma(\varphi)][y|f, u, \sigma^2]$$
$$= \exp\left[-\frac{1}{2}u^T \Sigma^{-1}(\varphi)u\right] \exp\left[-\frac{\sigma^2}{2}(f + Zu)^T (f + Zu)\right].$$

This distribution is conditionally conjugate, meaning that it has an analytically tractable form. In this case, the conjugate posterior is also normal (Lindley and Smith 1972), making it is straightforward to simulate. Most non-normal random effects will give rise to non-standard, analytically intractable full conditionals which require MCMC techniques.

Instead of sampling directly from the conditional distribution of f, our strategy is to use a predictive model to *estimate* the expected conditional fixed effects, $\hat{f} = E[f|u, \varphi, \sigma^2, y]$. Theses estimates are plugged into the Gibbs sampler to update f. Conditioning on the random effects, u, we consider Zu as an initial estimate of the predicted observations. This estimate can be improved by taking into account the systematic effects of the predictors, X. This is where we use the predictive model to estimate the expected responses $\hat{f}$ by regressing the residuals $r = y - Zu$ on predictors X.

The best strategy for sampling from the full conditionals of the variance components depends on the specific covariance model $\Sigma(\varphi)$ and the prior distributions specified for φ and σ^2. For example, when $\Sigma(\varphi) = \varphi I$, as in repeat measures designs or the error term, the inverse gamma distribution is conditionally conjugate. Other prior specifications will require MCMC methods, e.g., reference priors (Zhao and Wells 2005). When the form of $\Sigma(\varphi)$ is more complex, e.g., autoregressive (AR) processes or Matern covariance models, a general purpose algorithm like Metropolis-Hastings can be used to generate samples from the conditionals.

To summarize, the Gibbs sampling algorithm is:

1. Initialize MCMC parameters: $u^{(0)}, \varphi^{(0)}, \sigma^2(0)$
2. For m in *1 to M* do:
3. Predict $f^{(m)}$ from the residuals $r = y - Zu^{(m-1)}$ and covariates X
    4. Sample random effects $u^{(m)} \sim [u|f^{(m)}, \varphi^{(m-1)}, \sigma^{2(m-1)}, y]$
    5. Sample variance component $\varphi^{(m)} \sim [\varphi|f^{(m)}, u^{(m)}, \sigma^{2(m-1)}, y]$
    6. Sample variance component $\sigma^{2(m)} \sim [\sigma^2|f^{(m)}, u^{(m)}, \varphi^{(m)}, y]$
7. end For
8. end Algorithm.

Because we estimate the fixed effects, f, this algorithm is not, strictly speaking, Bayesian. Methods that replace unknown quantities with data-based estimates and then perform Bayesian analysis are known as "empirical Bayes". Empirical Bayesian methods are often used because they allow the analyst to take advantage of prior information in a simplified way without having to specify prior distributions. The resulting empirical Bayes estimators often have good frequentist properties, though theoretical results have been established only for certain estimators (Lehmann and Casella 1999). One disadvantage of estimating parameters with the empirical Bayes approach is that the method does not account for the variability in the estimation step. For this reason, we suggest that all confidence regions based on the HPM posterior be considered only approximate, and most likely biased small. Discussions of the this underestimation in posterior variance, along with remedies, can be found in Carlin and Louis (2000).

The Gibbs sampler can be started with initial values $u^{(0)}$, $\varphi^{(0)}$, $\sigma^{2(0)}$ set equal to estimates from a LMM with the same random effects design and some reasonably simplified fixed effects model. The number of iterations, M, required for convergence to the stationary distribution depends on the complexity of the random effect design and the degree of correlation between fixed and random effects. With relatively simple random effects and little correlation between fixed and random effects, we have found that chains of several thousand iterations are sufficient to achieve convergence.

4 Simulation Study

The following simulation study demonstrates the potential advantages of accounting for known hierarchical structure with predictive models. The predictive performances of several decision-tree based models are compared when used on their own and when embedded within a HPM with known hierarchical structure. With the HPMs predictive power improves and functional structure may be fit and discovered. The posterior distribution of fixed-effect predictions is explored to illustrate the adaptability of decision-tree based predictive models.

4.1 Performance Comparisons with Dependent Data

The data for this simulation were constructed to include several functional features commonly found in ecological data. Dependence among the observations arise from two separate processes; spatial correlation that describes the similarity of neighboring observations and observer effects that describe the similarity among observations made by the same observer. The parametric hierarchical model used to generate the observations is

$$y = f(X) + Z_s u_s + Z_o u_o + \varepsilon,$$

where y is a vector of N observations. Observation errors ε are normally distributed conditionally independent on the process with variance $\sigma^2 = 4$. The fixed effects model is

$$f(X) = -4.5 + 5I(x_1 > 0.5) - 6x_4 + 2\sin(6\pi x_6) + \frac{\sin(6\pi r)}{r},$$

where $I(x_1 > 0.5)$ is the indicator function that takes on the value of 1 when $x_1 > 0.5$ and zero otherwise and $r = \sqrt{(x_9 - 0.5)^2 + (x_{10} - 0.5)^2}$. This model includes a threshold effect (x_1), linear effect (x_4), oscillating effect (x_6), and a complex 2-way interaction between x_9 and x_{10}.

The spatial effects u_s are modeled as a zero-mean, isotropic Gaussian process. Let $u_s \sim N(0, \Sigma(s))$ where the covariance matrix $\Sigma(s)$ describes the covariance between locations, s. The covariance between locations s_i and s_j decays exponentially as a function of the distance between them $\Sigma_s\left(s_i, s_j|\rho, \sigma_s^2\right) = \sigma_s^2 \exp\left(-\left\|s_i - s_j\right\|/\rho\right)$ with range parameter $\rho = 0.05$ and scale parameter $\sigma_s^2 = 16$. For computational convenience we assume that the range parameter is known.

In order to control the size of the spatial effects, and the computations necessary to handle them, we model the spatial correlation u_s as a 50×1 vector of spatial effects at 50 selected "reference locations". The resulting spatial covariance $\Sigma(s)$ is a 50×50 reduced rank correlation matrix (Ruppert et al. 2003, Section 13.4). Reference locations were determined as the centroids of the neighborhoods generated from a k-nearest neighbor analysis of the observation locations, reflecting the spatial density of the observations. The spatial design matrix Z_s is the corresponding $N \times 50$ exponential covariance matrix between the observed locations and the reference locations. The spatial correlation among the N observations is calculated as the product $Z_s u_s$, similar to the Kriging prediction equations.

It is assumed that each observation was made by one of ten individual observers selected at random with equal probability. We further assume that each observer is biased and that the population of these biases or "observer effects", u_o, are independent and normally distributed with variance $\sigma_o^2 = 16$. The observer effect design matrix is an $N \times 10$ indicator matrix with elements $\{Z_o\}_{i,j}$ equal 1 if the i-th observation was made by the j-th observer and 0 otherwise. Thus, the factor $Z_o u_o$ induces correlation among observations made by the same observer.

Each simulated data set consists of $N = 2000$ observed responses from the model specified above. A total of ten fixed effect predictors were generated of which only the 5 indicated above influence the response. Each predictor $x_{i,j}$, $i = 1, \ldots, N$, $j = 1, \ldots, 10$ was generated independently on a uniform random distribution between 0 and 1, $U[0, 1]$, and stored in the $N \times 10$ fixed effect design matrix, X. Locations s_i, $i = 1, \ldots, N$ were generated randomly as independent latitude-longitude pairs on $U[0, 1]$, denoted as predictors x_{11} and x_{12}, respectively. Predictor x_{13} is the $N \times 1$ vector of randomly generated labels for the ten observers. The signal-to-noise ratio is

$$\frac{\mathrm{var}\left(f(X) + Z_s u_s + Z_o u_o\right)}{\sigma_\varepsilon^2 + \sigma_s^2 + \sigma_o^2} \approx 1.24.$$

We compare the performance of four decision-tree methods. Decision trees, as a general class of models, have several features that make them a good choice of predictive model: (1) they are relatively easy to implement and understand, (2) they automatically discover and fit interactions including high-order interactions and (3) most implementations automatically impute missing predictor values. The simplest decision tree approach used here is the "rpart" model (Therneau and

Atkinson 2007) which produces a single Classification and Regression Tree (CART) fit by cost complexity pruning (Breiman et al. 1984). In order to control the highly variable predictions of CART trees, Breiman (1996) suggested averaging predictions from a bootstrap sample of deliberately overfit CART trees. These "Bootstrap AGgregations" are known as "bagged" decision trees and usually outperform single trees. Boosting is another successful method used to average predictions across many simpler decision trees. It is equivalent to fitting an additive expansion in a set of basis functions (Hastie et al. 2001). We use the boosted decision trees implemented in the gbm library in R (Ridgeway 2006). RuleFit is another ensemble method that uses LASSO penalization (Tibshirani 1996) to combine predictions from individual trees (Friedman and Popescu 2005).

Each realization of the data was fit with all four decision-tree models and their corresponding HPMs. In order to make a fair comparison between the decision tree models and the HPMs we gave each of the decision-tree models access to the same predictor information utilized by the HPMs. Thus, each decision tree was fit using *all* 13 predictors including the latitude, longitude, and the vector of observer identifiers.

Model performance is measured as the Mean Squared Error (MSE) between the true and predicted responses. To guard against overfitting, the MSE is computed on an independent test set of data. All test predictions are made at new locations, for new observers so as to avoid any potential overfitting of the random effects, that is, overfitting location-specific or individual observer effects. The LMM BLUP estimator is $\hat{y} = \hat{f}(X) + Z_P \hat{u}_s$, where "hats" denote estimates and Z_P is the covariance between the new locations, s_i $i = 1, \ldots, 1000$ and the reference locations s_j $j = 1, \ldots, 50$, $\{Z_P\}_{i,j} = \Sigma_s\left(s_i, s_j | \rho, \sigma_s^2\right)$. The HPM predictions use the mean marginal posterior estimates for the fixed effects, variance components, and spatial effects. For the decision tree models, we "average out" estimated observer-specific biases by computing the mean predicted response where the mean is taken over the set of observers in the data set used for model training.

Half of each data realization was randomly assigned to training and testing sets. A single training-test set was used, instead of k-fold cross validation to expedite calculations. The simulation study was based on 100 trials. Diffuse Inverse Gamma (IG) priors were used for all the variance components, $\left[\sigma^2\right] = \left[\sigma_s^2\right] = \left[\sigma_o^2\right] = \mathrm{IG}(a = 0.1, b = 0.1)$. MCMC chains were initiated with true values to reduce computations time. Each chain was run for 1000 iterations. All computations in this paper were performed with the R statistical computing language (R Development Core Team 2006).

Boxplots of the test set MSE are shown in Fig. 1. The variation in MSEs is due to the Monte Carlo error, estimate and model uncertainty, and variation from the test-train split. The mean square error is seen to vary among the decision tree models with the largest errors for rpart and smaller errors for each of the ensemble methods. The performance of all decision tree models improves when the methods are embedded in the hierarchical model. The HPM based on RuleFit was the best overall performer. These results suggest the kinds of performance gains possible when covariance patterns exist and are correctly modeled in the hierarchy rather than modeled nonparametrically as fixed effects.

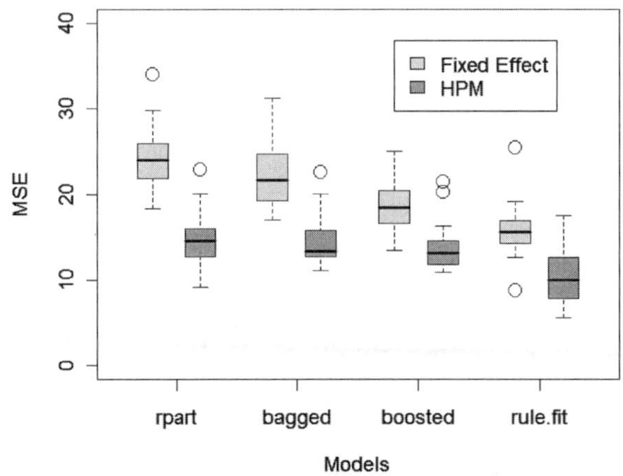

Fig. 1 The test set Mean Squared Error (MSE) between the true responses and the predicted responses are shown for Decision Tree models (Fixed Effect) and HPMs, organized by the type decision tree model. The boxplots show the variation in MSEs due to Monte Carlo error, estimate and model uncertainty, and variation from the test-train split. The performance of all decision tree models improves when the methods are embedded in the hierarchical model

4.2 Partial Dependence Plots for Effect Exploration

Although nonparametric predictive models have good predictive performance, many are essentially "black box" methods, making them difficult to interpret. The same is true of the HPM where all fixed effect information is stored as a high-dimensional joint posterior distribution of predictions. Partial dependence functions (Friedman 2001; Hastie et al. 2001; Hooker 2007; Hochachka et al. 2007) are a simple general purpose tool for visualizing and exploring predictor effects. We use partial dependence plots to explore a fixed-effects posterior distribution and use these plots to illustrate the adaptability of RuleFit within the HPM.

We begin by investigating the effects of each of the 10 individual fixed effect predictors from a single data realization. A natural approach to this investigation is to plot the predictions as a function of a single predictor. Unfortunately, the resulting trend may be simultaneously affected by any number of predictors that affect the response, making it difficult to isolate and describe the effects of any individual predictor. In order to better isolate the effect of each individual predictor we compute the effect of the predictor on the modeled response after accounting for the average effect of all other predictors. This is done by marginalizing over the joint distribution of all other predictors. These are one dimensional partial dependence plots. They best represent the effect of an individual predictor on the predicted response when the predictor's effects are nearly additive. All partial dependence plots are centered at zero.

Univariate partial dependence plots of the posterior conditional means and approximate pointwise 90% Bayesian confidence regions for predictors x_1, x_2, x_4, and x_6 are shown in Fig. 2. The partial dependencies for each mean effect were calculated at 100 equally spaced locations along the x-axis. Linear interpolations were plotted for the effects and confidence bounds. The approximate pointwise

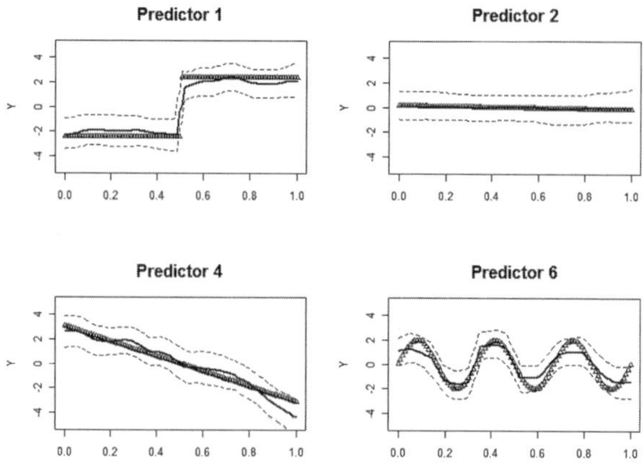

Fig. 2 Univariate partial dependence plots of the posterior conditional means and approximate pointwise 90% Bayesian confidence regions for predictors x_1, x_2, x_4 , and x_6 (solid and dashed, respectively) and true effects (triangles) used in the simulation model described in Section 4.1

confidence regions were estimated with the 95th and 5th quantiles of the posterior partial effects. The RuleFit estimates of the posterior mean effects capture the main features of the true effects for all four predictors. Because of the discrete support of the DT basis, RF is also able estimate the sharp threshold in x_1. RuleFit's penalization strategy may produce smooth effects like the oscillations in x_6. Predictor x_2 is correctly identified as uninformative. The other four predictors were also identified as uninformative.

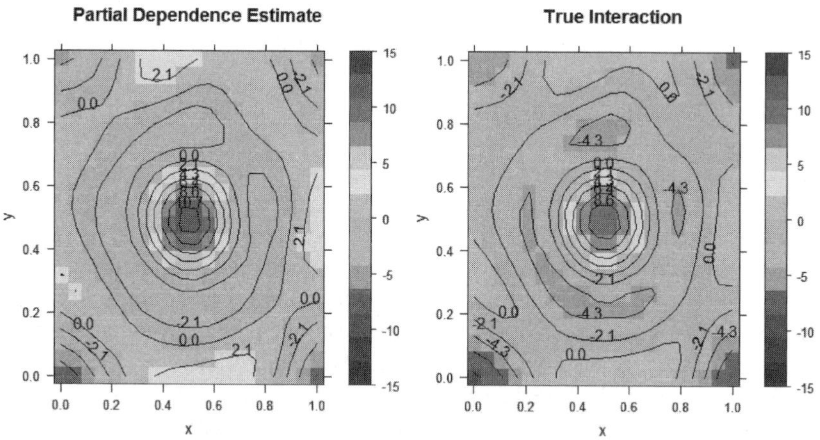

Fig. 3 The two-dimensional partial dependence plots and true interaction surface for $x_9 - x_{10}$ used in the simulation model described in Section 4.1. The interaction between these two predictors and their joint functional form were automatically detected and estimated by the predictive model, RuleFit

We calculated the two-dimensional partial dependence plots to investigate the RuleFit estimate of the $x_9 - x_{10}$ interaction surface, Fig. 3 (left). Partial dependence estimates were made over a 20×20 grid on the $x_9 - x_{10}$ unit square and interpolated using a penalized spline. The true interaction surface, Fig. 3 (right), was evaluated and smoothed on the same grid. Contours and shading are the same for both panels to facilitate comparison. The strong similarity between these plots confirm RuleFit's ability to detect and fit complex interactions within HPM. RuleFit has automatically determined which predictors are additive and which interact. This makes RuleFit a good tool for automatically detecting interactions.

5 HPM Exploration of House Finch Abundance

The goal of this section is to demonstrate how the HPM can be used to explore patterns in real data. The discussion is presented at a conceptual level focusing more on analysis techniques and interpretation than the biological results. For this reason, we have deliberately chosen a model based on a simple hierarchical structure from a well understood species. We use HPM to model the relative abundance of North American House Finch at back yard feeders using data from the citizen-science winter monitoring program, Project FeederWatch (PFW, http://www.birds.cornell.edu/pfw/). The HPM has a random feeder-effect and a large set of previously unused predictors. The exploratory analysis is used to identify important new predictors and estimate some functional effects.

5.1 The Data

PFW is a winter-long "citizen science" monitoring project in which members of the general public throughout the United States and Canada record the maximum number of birds seen together, for each of the bird species that they see at their bird feeders. Observation periods occur over two consecutive days, at weekly or biweekly intervals. The program begins in mid-November and runs till the beginning of April. Participants record the location, date, bird numbers and effort expended during each observation period. They are also asked to provide data describing the weather and the environments around their feeder locations, such as presence or absence of coniferous and deciduous trees, water bodies, and the degree to which landscapes are altered by humans. Information is recorded about factors that may attract or deter nearby birds from being observed at a feeder such as the types of feed available, the number and configuration of the feeders, and the presence of pets and squirrels.

In addition to the information provided by PFW participants, we acquired several other descriptors of sites from the Avian Knowledge Network (see http://www.avianknowledge.net/content/) including descriptions of the general biogeographic region, local habitat, elevation, and human population density. These data were extracted based on the latitudes and longitudes of the PFW feeder sites. The complete data set included a total of 76 predictors, see Table 1.

Table 1 Fixed effect predictors in HPM analysis used these 76 predictors. Seventy-two of the predictors were reported by PFW participants plus each site's Bird Conservation Region (BCR, see http://www.nabci-us.org/map.html), U.S. Census Bureau census block-level human population density estimate from 2000, elevation (2 from different digital elevation data sources and resolutions: USGS National Elevation Dataset, 10 m resolution data http://www.mapmart.com/DEM/DEM.htm; and GTOPO30, 30 arcsec resolution data http://edc.usgs.gov/products/elevation/gtopo30/gtopo30.html), and habitat type from the U.S. National Land Cover Database (NLCD) recorded as (one of 9 separate Anderson level 1 habitat classification categories within the grid block of the count site; U.S. National Landcover Data, 1992 version)

Temporal(2)	Attraction & deterrence to feeders (29)	Local habitat (31)
Season	count_area_size	nlcd
Date	fed_yr_round	yard_type_garden
	fed_in_jan	yard_type_landsca
	fed_in_feb	yard_type_woods
Effort (2)	fed_in_mar	yard_type_desert
	fed_in_apr	yard_type_pavement
	fed_in_may	hab_dcid_woods
	fed_in_jun	hab_evgr_woods
Effort 1	fed_in_jul	hab_mixed_woods
Effort 2	fed_in_aug	hab_orchard
	fed_in_sep	hab_park
	fed_in_oct	hab_water_fresh
	fed_in_nov	hab_water_salt
Human Population	fed_in_dec	hab_residential
Density (1)	numfeeders_suet	hab_industrial
	numfeeders_ground	hab_agricultural
Human.pop.density	numfeeders_hanging	hab_desert_scrub
	numfeeders_platfrm	hab_young_woods
	numfeeders_humming	hab_swamp
	numfeeders_water	hab_marsh
Weather (6)	numfeeders_thistle	hab_other
	numfeeders_fruit	evgr_trees_atleast
	bird_baths_atleast	evgr_shrbs_atleast
	high_feeders	dcid_trees_atleast
	nearby_feeders	dcid_shrbs_atleast
temp_lo	squirrels	fru_trees_atleast
temp_hi	cats	cacti_atleast
snow_coverage	dogs	brsh_piles_atleast
snow_depth	humans	water_srcs_atleast
snow_crusty		evgr_any_atleast
precipitation		dcid_any_atleast
Physiographic (5)		
latitude		
longitude		
elevation (categorical)		
elevation_1		
elevation_2		

Although the PFW data set contains a large number of potentially informative predictors, most of them have never been used to model the distribution or relative abundance of backyard species (Lepage and Francis 2002; Wells et al. 1998; Hochachka and Dhondt 2000; Hochachka and Dhondt 2006). Currently, there is insufficient landscape-level understanding of the necessary ecological processes to specify a parametric model for all available predictors, or even a large subset of them. Such a task would require substantial exploration to determine (1) which predictors to include in the model, as well as (2) the functional form of the predictors, and (3) their interactions.

Missing data are another serious data complication that often affects how data are modeled. For example, of the 11,066 observations for NABCI Bird Conservation Region 13 (Lower Great Lakes/Saint Lawrence Plain; http://www.nabci-us.org/map.html), 9850 were missing at least one of the 72 PFW predictors — 89% of the records were incomplete. The expedient solution to the missing data problem is to throw out responses and/or predictors with missing data, but this reduces the information available for analyses and may introduce bias and increase the variance of results. More rigorous imputation options are often far more difficult to implement and require additional assumptions. For these reasons, many analyses are based on only a subset of the available data.

We analyzed the magnitude of positive group sizes from the 1993–1994 to the 2003–2004 seasons within the eastern North American range of the House Finch (Fig. 4). This species is well understood and has been independently analyzed in

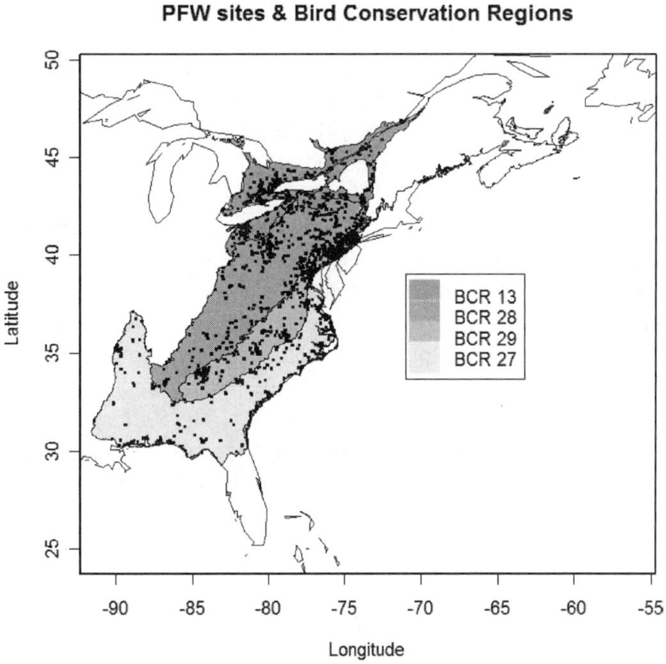

Fig. 4 This map shows the Project Feeder Watch (PFW) feeder sites as black squares within each of the four BCRs analyzed. BCR 13 is the Lower Great Lakes/St. Lawrence Plain region, BCR 28 is the Appalachian Mountains region, BCR 29 is the Piedmont region, and BCR 27 is the Southeast Coastal Plain region

the literature using PFW data (Hochachka and Dhondt 2006) providing a basis for validation. The spatial domain consists of four distinct Bird Conservation Regions. In order to simplify the comparison of regional variation in species' winter distributions, analyses were conducted separately for each of four different BCRs. For each BCR, a random sample of up to 400 "frequently participating" feeder sites were selected for analysis. "Frequently participating" feeders were defined to be sites that contributed at least 15 reports over the 11 season study period. The top 1% of maximum group sizes were trimmed to focus inference on smaller abundances by limiting the influence of the largest observations. Sample sizes were 11066, 5321, 12064, 11701 for BCRs 13, 27, 28, and 29 respectively. Four hundred locations were used for each BCR except BCR 27 which had 232.

5.2 The Model

We use a mixed model with random feeder effects to account for correlation among observations from the same feeder. Let y_i, $i = 1, \ldots, N$ be the natural log of the observed maximum count. We transform to the log scale to model errors as conditionally independent, additive normal noise $y = f(X) + Zu + \varepsilon$, as in Section 3. Random effects u estimate systematic differences among the q feeder locations. They are normally distributed $[u|\sigma_f^2] = \text{Normal}\left(0, \sigma_f^2 I\right)$ where σ_f^2 describes the amount of variation among the sites and I is the q dimensional identity matrix. The feeder effect design matrix Z is a $N \times q$ indicator matrix with elements $\{Z\}_{i,j}$ equal 1 if the i-th observation was made at the j-th feeder and 0 otherwise. We use RuleFit (Section 4.1) to estimate the HPM fixed effect, $f(X)$. The fixed effect design X is the $N \times 76$ matrix of the predictors.

The data from each BCR were fit separately using diffuse inverse gamma (IG) priors for the variance components, $\sigma^2 \sim \text{IG}(a = 0.1, b = 0.1)$ and $\sigma_f^2 \sim \text{IG}(a = 0.1, b = 0.1)$. Initial values for fixed effects were estimated using RuleFit and the residuals from this fit were used to initialize u. Realizations of $(\sigma^2, \sigma_f^2, u, f)$ along with variable importances and partial dependences were collected on each iteration of the Gibbs sampler. Due to the simple structure of the hierarchy, each Gibbs sampler was expected to reach convergence quickly. The Raftery and Lewis (1992) diagnostic estimated that convergence sufficient to estimate the 5th percentile to within 1% accuracy with probability of 0.95 would be achieved with 1825 iterations. We computed 2600 iterations and discarded a burn-in of 100.

5.3 Exploratory Results

A first step towards uncovering the signal detected by the HPM is to rank the relative importance of its predictors. RuleFit computes a measure of relative variable importance designed to identify those variables that are used in its most influential predictive rules (see Section 7, Friedman and Popescu 2005). Relative importances

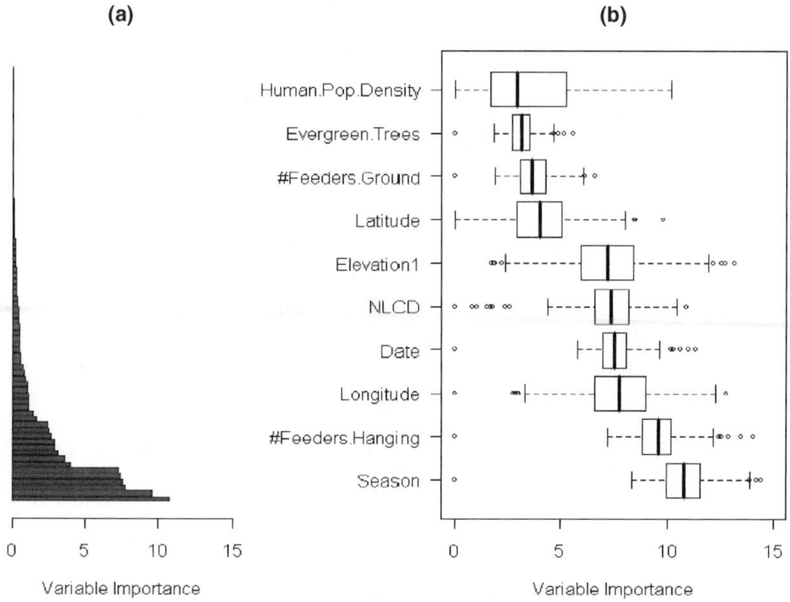

Fig. 5 These plots show the relative variable importances for fixed effect predictors. The barchart in (**a**) shows the ordered marginal posterior median relative variable importance score for all 76 predictors. The boxplots (**b**) of the marginal posterior distributions for the relative variable importances of the 10 most important predictors, ordered according to their medians

are scaled to sum to 100 with larger values representing more important predictors. We collected the vector of relative variable importance at each Gibbs iteration. Figure 5a shows the barchart of the ordered marginal posterior medians for all 76 predictors. The exponential decay in importance is common among data sets with large numbers of predictors. Most information is concentrated among a small set of predictors.

Figure 5b shows Boxplots of the marginal posterior distributions for the 10 most important predictors ordered according to their medians. This group of predictors captures many sources of variation known to be important for PFW data and for this House Finch population. The *Latitude–Longitude* pair describe the location of each feeder. *Season* is a 11-level ordinal predictor denoting the 11 different winter seasons. This is consistent with known changes in population trend over this time period due to the emergence of a novel bacterial pathogen, *Mycoplasma gallisepticum* (e.g., Dhondt et al. 2005). *Date* is a continuous Julian date starting at 1 on November 1st running to 150 on April 1st. This predictor could account for the known partial winter migrations of this House Finch population, as well as seasonal variation in propensity of the birds to visit feeders. RuleFit also identified two factors that attract birds to feeders as important. The *Number of Hanging Feeders* and the *Number of Ground Feeders* were ranked 2nd and 7th respectively. Three local habitat variables were ranked among the top 10 predictors. Landcover classification

(*NLCD*), elevation, and the presence or absence of evergreen trees, were ranked 5th, 6th, and 9th, respectively. *Human.Pop.Density* is human population density at the feeder location. This predictor may describe the association of House Finches with humans in suburban environments. Two observer effort variables describing the total hours of observer effort over each consecutive two day observation period and a 4-level ordinal variable that measures how many "halfday" periods were used for observation were also ranked highly, 13th and 17th, respectively.

A benefit of a Bayesian analysis is the ability to produce estimates of uncertainty. These estimates incorporate the uncertainty from estimation and model uncertainty at the fixed effect level. The model uncertainty comes from the predictive model adaptation over complex model spaces. The boxplots of the marginal posterior variable importances show considerable uncertainty. Note that some of this uncertainty also arises because of predictor multicollinearity.

Partial dependence plots are used to visualize particular predictors effects. Figure 6 shows the partial dependence plots of the posterior conditional means and approximate 90% confidence regions for intra- and inter-seasonal trends in BCR 13. The inter-seasonal trend agrees with our expectation of a decline (Dhondt et al. 2005).

We also used partial dependence plots to explore how NLCD landcover effects vary by region, or BCR. For each BCR, we computed the partial dependence among the 9 NLCD Anderson level-1 land cover classes. Boxplots of the posterior partial dependence effect estimates were plotted for the six landcovers which were represented by at least three of the four BCR's, Fig. 7. The partial dependence among

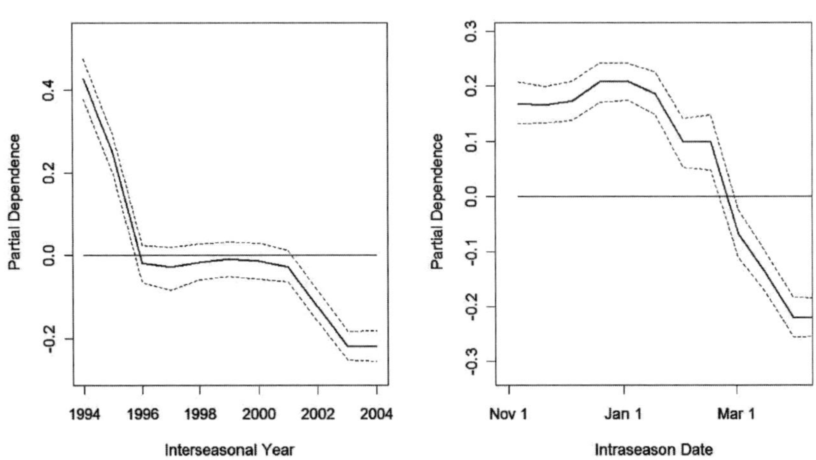

Fig. 6 Partial dependence plots for Intra- and Inter-seasonal trends in BCR 13 with pointwise 90% HPD confidence regions The automatically detected intra-seasonal trends shown here agree with known changes in population trend over this time period due to the emergence of a novel bacterial pathogen, *Mycoplasma gallisepticum* (e.g., Dhondt et al. 2005). The inter-seasonal trend could account for the known partial winter migrations of this House Finch population, as well as seasonal variation in propensity of the birds to visit feeders

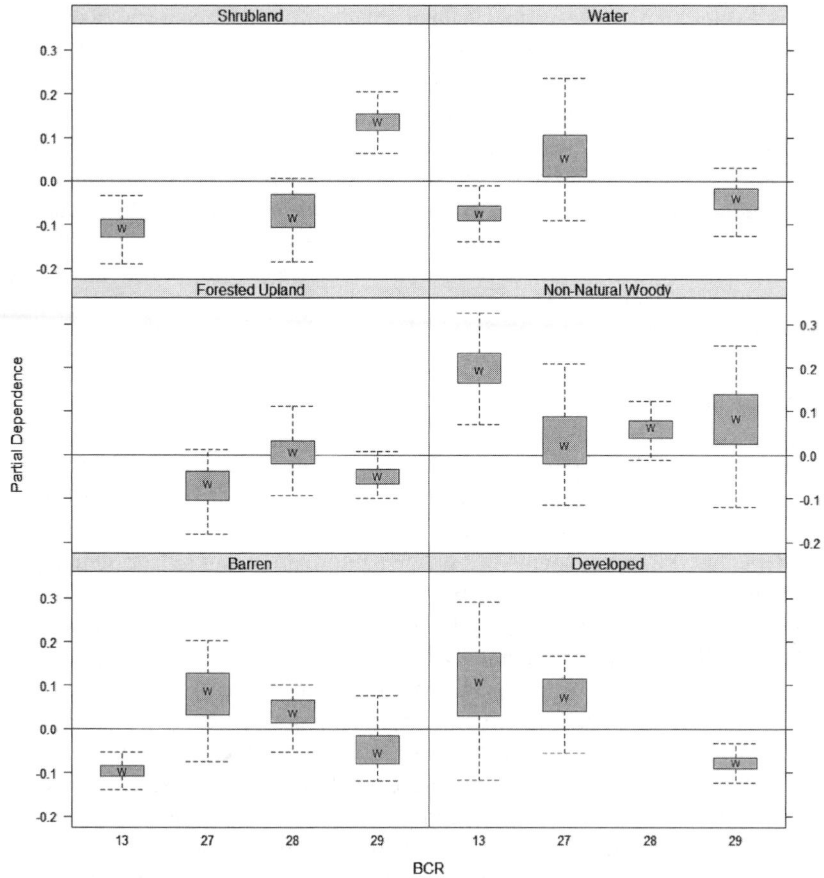

Fig. 7 The partial dependence among the NLCD classes are shown here with boxplots of posterior partial dependence among BCRs plotted by NLCD Anderson level-1 landcover classes. Boxes are centered *within* each BCR, facilitating the comparison of landcover effects within that BCR. However, to compare the *relative* effects of each landcover class across BCRs we grouped these relative effects by landcover. Substantial differences in the relative effects of several landcover classes across different BCR's suggest that local effects of land cover on group size vary regionally. For example, the observed group size of House Finches was smaller on shrublands, compared to other landcover classes, in BCR 13 and 28 while observed group sizes were larger on shrublands in BCR 29. The center "w" represents the median posterior partial dependence. Only landcover classes which were represented by at least three of our four BCR's have been plotted

the NLCD classes are centered *within* each BCR, facilitating the comparison of landcover effects within that BCR. However, to compare the *relative* effects of each landcover class across BCRs we grouped these relative effects by landcover. Substantial differences in the relative effects of several landcover classes across different BCR's suggest that local effects of land cover on group size vary among regions. For example, the observed group size of House Finches was smaller on shrublands, compared to other landcover classes, in BCRs 13 and 28 while observed group sizes were larger on shrublands in BCR 29.

This exploration suggests that habitat effects may operate over two distinct scales – the regional scale represented by BCR and a local scale of the 30 $\times$ 30 m resolution NLCD layers. However, using these data it is not possible to tell if the habitat affects the probability to detection or the ecological process that governs abundance, or both. Interpretations of these effects are also confounded by multicollinearity, especially with other spatial predictors. The use of observational data may introduce sampling bias. For example, a known limitation of PFW data is its spatial footprint, which is concentrated towards anthropogenic habitats. Additional, carefully collected data would be needed to untangle and confirm the causes of the patterns that emerged from our analysis.

6 Discussion

In this article we have developed a highly adaptive semiparametric model by harnessing the complementary strengths of hierarchical and predictive models. From a mixed-model perspective, the strengths of the HPM are its great automation and adaptability. Large amounts of predictor information can be conveniently and quickly included in an analysis and explored.

While HPMs inherit many strengths from their parent models, they also inherit weaknesses. The parametric hierarchy itself must be specified by the analyst and this means that there will be a risk of misspecification. For many important problems patterns of correlation can be specified a priori with confidence, e.g., spatial analyses or repeated measures. In many hierarchies, the risk of misspecification may be mitigated by the adaptability of the hierarchical structure itself. For example, in several common LMM models, when there is insufficient evidence of variation among the random effects, the estimated variance components will take on limiting values that tend to "flatten" the hierarchy, effectively limiting the effect of misspecification.

Like any other model, multicollinearity among predictors makes it difficult to separately identify relationships between the correlated predictors and the response. This is an especially important challenge when exploring large environmental data sets where it is not uncommon to find large sets of multicollinear predictors. Finally, because HPMs require the computation of both the LMM and a predictive model, they can be computationally intensive.

It is important to remember that the data-mining component of an HPM does not carry the negative inferential properties of "data dredging", a term that unfortunately is often viewed as synonymous with "data mining". Chatfield (1995) defined the practice of data dredging as when an "analyst goes to great lengths to obtain a good fit. When a model is formated, fitted, and checked with the same data set in an iterative, interactive way". Within the machine learning community there is a strong insistence on the use of independent data for testing and validation to guard against overfitting and dredging. Indeed, by performing exploration and regression in a single procedure, HPM actually avoids many common "dredging" problems that arise in more traditional multi-step approaches to regression model development and testing.

Because of their relative strengths and weaknesses, we view HPMs as serving three distinct roles in the analysis of observational data:

1. Where accurate predictions are the desired product of an analysis, HPMs are a logical class of models to use, because of their ability to make use of available information from the predictor variables, both when the forms of structural relationships are known and also where these things are unknown.
2. While many "products" from analysis of ecological data can be viewed as hypothesis validations, and there is good reason to conduct hypothesis-driven analysis of data, the specification of realistic and useful hypotheses requires prior knowledge. Where such knowledge does not exist, more exploratory analyses are appropriate, and efficient exploratory analyses will lead to creation of appropriate hypotheses more quickly. HPMs are particularly suited for such exploratory (hypothesis-generating) analyses when there is some amount of prior information available, as this prior knowledge can be incorporated into the exploratory model-building work.
3. Even when an analyst believes that (s)he has sufficient prior knowledge to specify an accurate parametric model, this is still merely a faith-based assertion unless there is some objective way of validating the appropriateness of a parametric model. HPMs can be used to assess the validity of fixed effect components within a hierarchical model by replacing them with more flexible nonparametric components and then comparing the overall predictive performances. Additional information may be gleaned from such a comparison by using partial dependence functions to compare the functional form and interactions of specific predictor effects estimated under both models.

Acknowledgments We thank the Avian Knowledge Network (AKN) team, Marty Wells, Charles Francis, and two anonymous reviewers for many helpful discussions and insightful comments. We would also like to thank the participants of Project FeederWatch and the staff in the Information Sciences unit at the Cornell Laboratory of Ornithology for their work in gathering and maintaining the data used in Section 5. Work on this paper was funded under the NSF Information Technology Research (ITR) for National Priorities program (award #EF-0427914).

References

Amstrup S, MacDonald L, Manly B (2006) Handbook of Rapture-Recapture Analysis. Princeton University Press, Englewood Cliffs, NJ 296 pp.

Banerjee S, Carlin BP, Gelfand AE (2004) Hierarchical Modeling and Analysis for Spatial Data. Chapman & Hall/CRC, London BocaRadon, FL 472 pp.

Berry S, Carroll R, Ruppert D (2002) Bayesian smoothing and regression splines for measurement error problems. Journal of the American Statistical Association 97:160–169.

Breiman L (1996) Bagging predictors. Machine Learning 24:123–140.

Breiman L (2001) Random forests. Machine Learning 45:5–32.

Breiman L, Friedman JH, Olshen RA, Stone JC (1984) Classification and Regression Trees. Chapman & Hall, New York.

Buja A, Hastie T, Tibshirani R (1989) Linear smoothers and the additive model. The Annals of Statistics 17:453–555.

Carlin BP, Louis TA (2000) Bayes and Empirical Bayes Methods for Data Analysis. Chapman & Hall/CRC, Boca Raton, FL.

Chatfield C (1995) Model uncertainty, data mining, and statistical inference. Journal of the Royal Statistical Society, Series A, 158:419–466.

Cristianini N, Shawe-Taylor J (2000) An Introduction to Support Vector Machines and Other Kernel-Based Learning Methods. Cambridge University Press, Cambridge, UK.

De'ath G, Fabricius KE (2000) Classification and regression trees: a powerful yet simple technique for ecological data analysis. Ecology 81:3178–3192.

Dhondt AA, Altizer S, Cooch EG, Davis AK, Dobson A, Driscoll MJL, Hartup BK, Hawley DM, Hochachka WM, Hosseini PR, Jennelle CS, Kollias GV, Ley DH, Swarthout ECH, Sydenstricker KV (2005) Dynamics of a novel pathogen in an avian host: mycoplasmal conjunctivitis in House Finches. Acta Tropica 94(1):77–93.

Elith J, Graham CH, Anderson RP, Dudik M, Ferrier S, Guisan A, Hijmans RJ, Huettmann F, Leathwick JR, Lehmann A, Li J, Lohmann LG, Loiselle BA, Manion G, Moritz G, Nakamura M, Nakazawa Y, Overton JMcC, Peterson AT, Phillips SJ, Richardson K, Scachetti-Pereira R, Schapire RE, Soberón J, Williams S, Wisz MS, Zimmermann NE (2006) Novel methods improve prediction of species' distributions from occurrence data. Ecography 29:129–151.

Friedman JH, Popescu BE (2005) Predictive learning via rule ensembles. Technical Report, Stanford University.

Geissler PH, Sauer JR (1990) Topics in route-regression analysis. In: Sauer JR, Droege S (eds) Survey Designs and Statistical Methods for the Estimation of Avian Population Trends. U.S. Fish and Wildlife Service, Biological Report 90(1):54–57.

Gelfand A, Schmidt AM, Wu S, Silander JA, Latimer A, Rebelo AG (2005) Modelling species diversity through species level hierarchical modeling. Applied Statistics, 54:1–20.

Goldstein H (1995) Multilevel Statistical Models. Halstead Press, New York..

Gu C (2002) Smoothing Spline ANOVA Models. Springer, New York..

Hand DJ, Mannila H, Smyth P (2001) Principles of Data Mining. MIT Press, Cambridge.

Hastie T, Tibshirani R (1990) Generalized Additive Models. Chapman and Hall, London.

Hastie T, Tibshirani R, Friedman J (2001) The Elements of Statistical Learning: Data Mining, Inference, and Prediction. Springer Verlag, New York, 552 pp.

Hochachka WM, Dhondt AA (2000) Density-dependent decline of host abundance resulting from a new infectious disease. Proceedings of the National Academy of Sciences USA 97:5303–5306.

Hochachka WM, Dhondt AA (2006) House finch (*Carpodacus mexicanus*) population- and group-level responses to a bacterial disease. Ornithological Monographs 60:30–43.

Hochachka WM, Caruana R, Fink D, Kelling S, Munson A, Riedewald M, Sorokina D (2007) Data mining for discovery of pattern and process in ecological systems. Journal of Wildlife Management 71(7):2427–2437.

Hooker G (2007) Generalized functional ANOVA diagnostics for high dimensional functions of dependent variables. Journal of Computational and Graphical Statistics 16(3).

Jolly GM (1965) Explicit estimates from capture-recapture data with both death and immigration-stochastic models. Biometrika 52:225–247.

Lepage D, Francis CM (2002) Do feeder counts reliably indicate bird population changes? 21 years of winter bird counts in Ontario, Canada. Condor 104:255–270.

Lehmann E, Casella G (1999) Theory of Point Estimation, 2nd Edition. Springer-Verlag, New York.

Lindley DV, Smith AFM (1972) Bayes estimates for the linear model. Journal of the Royal Statistical Society, Series B, 1–41.

Mackenzie DI, Nichols JD, Lachman GB, Droege S, Royle JA, Langtimm CA (2002) Estimating site occupancy rates when detection probabilities are less than one. Ecology 83(8):2248–2255.

McCulloch CE, Searle SR (2001) Generalized, Linear, and Mixed Models. John Wiley and Sons, New York.

Mitchell T (1997) Machine Learning. McGraw-Hill, New York.

Raftery AE, Lewis SM (1992). One long run with diagnostics: Implementation strategies for Markov chain Monte Carlo. Statistical Science 7:493–497.

R Development Core Team (2006). R: A language and environment for statistical computing. R Foundation for Statistical Computing, Vienna, Austria. ISBN 3-900051-07-0, URL http://www.R-project.org.

Ridgeway G (2006). gbm: Generalized Boosted Regression Models. R package version 1.5-7. http://www.i-pensieri.com/gregr/gbm.shtml.

Robert CP, Casella G (2004) Monte Carlo Statistical Methods, 2nd Edition. Springer, New York.

Robinson GK (1991) That BLUP is a good thing: the estimation of random effects. Statistical Science 8(1):15–51.

Ruppert D, Wand MP, Carroll RJ (2003) Semiparametric Regression. Cambridge University Press, Cambridge, 402 pp.

Seber GAF (1965) A note on the multiple-recapture census. Biometrika 52:249–259

Therneau TM, Atkinson B, R port by Ripley B (2007) rpart: Recursive Partitioning. R package version 3.1-35. http://mayoresearch.mayo.edu/mayo/research/biostat/splusfunctions.cfm

Thogmartin WE, Sauer JR, Knutson MG (2004) A hierarchical spatial model of avian abundance with application to Cerulean Warblers. Ecological Applications 14(6):1766–1779.

Tibshirani R (1996) Regression shrinkage and selection via the lasso. Journal of the Royal Statistical Society B 58(1):267–288.

Wahba G (1990) Spline models for observational data SIAM [Society for Industrial and Applied Mathematics] (Philadelphia).

Wells JV, Rosenberg KV, Dunn EH, Tessaglia-Hymes DL, Dhondt AA (1998) Feeder counts as indicators of spatial and temporal variation in winter abundance of resident birds. Journal of Field Ornithology 69:577–586.

West M, Harrison J (1997) Bayesian Forecasting and Dynamic Models. Springer-Verlag, New York.

Wikle CK (2003) Hierarchical Bayesian models for predicting the spread of ecological processes. Ecology 84:1382–1394.

Wikle CK, Berliner ML (2005) Combining information across spatial scales. Technometrics 47:80–91.

Wikle CK, Hooten MB (2006) Hierarchical bayesian spatio-temporal models for population spread. In: Clark JS and Gelfand A (eds) Applications of Computational Statistics in the Environmental Sciences: Hierarchical Bayes and MCMC Methods. Oxford University Press, Oxford.

Wood SN (2006) Generalized Additive Models: An Introduction with R. Chapman & Hall/CRC, London/Boca Rarm, FL, 416 pp.

Zhao X, Wells MT (2005) Reference priors for linear models with general covariance structures, Cornell Department of Statistical Sciences Technical Report.

Zhao Y, Staudenmayer J, Coull BA, Wand MP (2006) General design Bayesian generalized linear mixed models. Statistical Science 21:35–51.

Effect of Senescence on Estimation of Survival Probability When Age Is Unknown

David Fletcher and Murray G. Efford

Abstract Adult survival probability is a key parameter in any population model for a long-lived species. For many species, information on adult survival comes from a capture–recapture study involving individuals for whom age is unknown. If the species experiences senescence, the estimate of overall adult survival probability will be negatively biased. The purpose of this paper is to assess the extent to which the estimate is biased and the implications for population modelling. We show that the amount of bias depends on the capture probability and the strength of senescence. If the capture probability is greater than 0.5, the expected bias is at most 1%, unless senescence is strong and begins early in adulthood. Individual heterogeneity in capture probability can also lead to negative bias in estimates of survival probability, meaning that moderate effects from senescence and heterogeneity may combine to produce a non-negligible amount of bias. Capture–recapture methods for survival are also used to estimate the time that migrating animals spend at intermediate "stop-over" sites. In this context, an increase in departure probability with time since arrival is analogous to senescence, leading to a negative bias in estimated stop-over duration. This bias will often be large because capture probabilities in such studies are generally very low and departure probability may increase abruptly once animals have rested and re-fueled.

Keywords Bias · Population model · Capture–recapture · Senescence · Stop-over duration · Survival

1 Introduction

In using a population model to help manage a long-lived species, a key demographic parameter is adult survival probability (Caswell 2000). Suppose the species experiences senescence (risk of mortality increases with age) and we plan to use a capture–recapture study to estimate annual adult survival probabilities. If we can determine

D. Fletcher (✉)
Department of Mathematics and Statistics, University of Otago, Dunedin, New Zealand
e-mail: dfletcher@maths.otago.ac.nz

D.L. Thomson et al. (eds.), *Modeling Demographic Processes in Marked Populations*, Environmental and Ecological Statistics 3, DOI 10.1007/978-0-387-78151-8_47,

the age of each individual, we can estimate age-specific survival probabilities and use these in an age-based population model. This will be possible when individuals are marked at birth or when the age of a newly-captured individual can be reliably determined.

For many species, information on adult survival comes from a study involving individuals that are marked as adults and for whom reliable determination of age is not possible. We might then use an estimate of the overall adult survival probability, with all adults in our model being assigned this survival probability. This approach is equivalent to using the (unknown) age-specific survival probabilities in an age-based model, as long as the age distribution of the marked individuals is the same as the stable age distribution predicted by the population model (Yearsley and Fletcher 2002).

The purpose of this paper is to assess the amount of bias in an estimate of overall adult survival probability from a capture–recapture study when age is unknown and there is a senescent decline in survival probability. The only previous work of this kind in the capture–recapture literature is the pioneering simulation study of Manly (1970). He compared different methods for estimating birth probabilities, survival probabilities and population sizes from a capture–recapture study, when the assumptions of the analyses are not valid. Part of his study concerned the bias in estimates of annual survival probability from the fully time-dependent Cormack–Jolly–Seber (CJS) model when survival probability varies with age. Using ten simulations of a capture–recapture study involving five sampling occasions, he found that senescence caused the estimates to be negatively biased, and commented (p. 17) that "...this is to be expected since marked animals will tend to be older than the animals in general and therefore subject to a relatively high mortality rate". The degree to which the age distribution of the marked individuals differs from that of the whole population will be affected by the capture probability for an unmarked individual (the capture probability for a marked individual does not affect this distribution). We therefore expect the amount of bias in survival to be influenced by the capture probability for an unmarked individual.

Our aim, with substantially greater computer power than was available to Manly (1970), is to chart the extent of the bias for parameter values likely to be encountered in mark–recapture studies of long-lived vertebrates and to clarify its relation to other factors. In addition, we consider the implications of the bias for population modelling: it is possible that a seemingly small bias in the overall adult survival probability will lead to a practically important bias in the estimate of population growth rate obtained from a population model.

An increase in the risk of mortality with age is a common, although not universal, feature of mammalian life histories (e.g. Caughley 1966; Promislow 1991; Gaillard et al. 1994; Loison et al. 1999). Relatively constant and high survival in early- to mid-adulthood is typically followed by a steady decline. Senescent decline in survival has also been documented for several species of birds (Dunnet and Ollason 1978; Loery et al. 1987; Bradley et al. 1989; Aebischer and Coulson 1990; Wooller et al. 1992; Pugesek et al. 1995; McDonald et al. 1996; Newton and Rothery 1997; Orell and Belda 2002; Crespin et al. 2006; Sidhu et al. 2007), although the effect is less marked and more variable than in mammals. It is usual for senescence to be more marked in one sex (e.g. Loison et al. 1999). For simplicity, we do not consider species for which mortality decreases with age in early adulthood (e.g. sparrowhawk *Accipiter nisus* Newton and Rothery 1997).

Capture–recapture methods are also used to estimate stop-over durations for migratory bird species (Kaiser 1995; Schaub et al. 2001; Efford 2005; Pledger et al. 2008). This provides another context in which the results of this paper are relevant. The probability of departure from a stop-over site may increase steeply with time-since-arrival (after an initial lag). As the arrival times of individuals are unknown it is usual to ignore this effect in the analyses (e.g. Schaub et al. 2001; but see Pledger et al. 2008). This is exactly equivalent to ignoring a senescent decline in survival probability, and will lead to a negative bias in the estimated stop-over duration.

Our main focus is on the bias in the estimate of overall adult survival probability, and the consequences of this for the population growth rate resulting from use of a population model. An alternative approach would be to consider the impact of senescence on direct estimation of the population growth rate using capture–recapture data (Pradel 1996). We have chosen to focus on estimation of adult survival probability, as there may be a number of reasons one wishes to use a population model other than to obtain an estimate of population growth rate; for all these uses one would want to be aware of the potential amount of bias in the estimate of survival. In addition, a reliable estimate of the adult survival probability will be of use in other settings, such as estimating maximum population growth rate for a bird species (Niel and Lebreton 2005).

2 Methods

We used simulation to estimate the bias in overall survival probability as follows. For each run of the simulation, we generated a reduced m-array corresponding to a capture–recapture study, using the calculations given in the Appendix. We then analysed the data in the m-array by fitting a CJS model in which both survival probability and capture probability were constant, using maximum likelihood. The estimate of overall adult survival probability was then compared with the true overall adult survival probability. In carrying out the simulations, we needed to specify the following variables:

- N Population size, assumed constant over the period of the study
- T Duration of the study (years)
- p Recapture probability, assumed to be time-independent (and equal to first-capture probability)
- $\phi^{(x)}$ Survival probability for an individual of age x, assumed to be independent of year

The severity of senescence determines the values for the age-specific survival probabilities. We chose to use a senescence function in which survival is constant to a specified age, with a linear-logistic decline thereafter. The linear-logistic form is commonly used in tests for senescence (e.g. Nichols et al. 1997; Crespin et al. 2006), and the threshold linear-logistic has been fitted to data by Loison et al. (1999) and Sidhu et al. (2007). Alternatives include the Gompertz function, which is linear on a complementary log-log scale, and therefore very close to the linear-logistic (Gaillard et al. 2004), or a quadratic-logistic function (Loison et al. 1999). We would expect these alternatives to provide similar results to those presented in this paper. The linear-logistic function, combined with a variable onset of senescence, captures the main features of senescence in a relatively simple form.

We assumed that all individuals mature at the same age, or that senescence is determined by the number of years since maturity. The calculations we use can easily be modified to include between-individual variation in age at maturity when senescence is determined by actual age. Throughout the rest of the paper, we therefore use the term "age" to mean the number of years since maturity.

We specified senescence as beginning at ages 0, 5 or 10. We set the survival probability at age 0 to be 0.75, 0.85 or 0.95, corresponding to the likely range of survival probabilities in early adulthood for long-lived species. Finally, we set the decline in survival probability to correspond to linear-logistic slopes of −0.05, −0.1, −0.2 or −0.4. We chose these levels to represent the range of senescence patterns evident in capture–recapture studies of birds and mammals, based on the information available in the studies cited above. Figure 1 illustrates the shape of the function for the case where senescence begins at age 5 and survival at age 0 is 0.85.

In performing the simulations, we assumed that the population had a stable age-distribution, as predicted by a deterministic population model. The model also had all individuals mature at the same age, and both the reproductive rate and adult

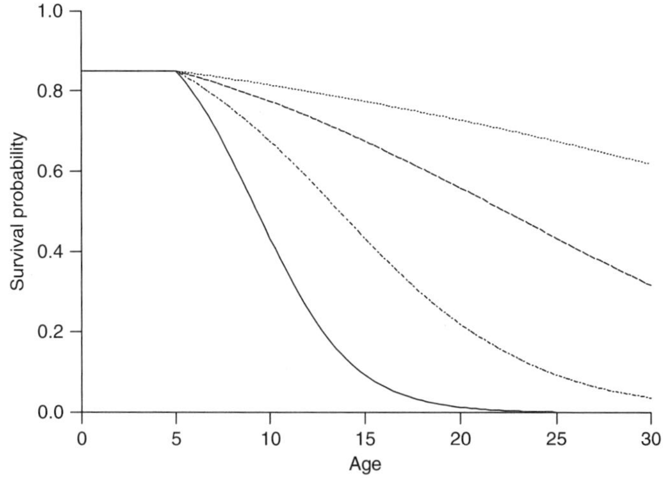

Fig. 1 Age-dependent survival plotted against age (time-since-maturity), where senescence begins at age 5, the survival probability at age 0 is 0.85 and the strength of senescence being determined by a linear-logistic slope of −0.05 (*dotted line*), −0.1 (*dashed line*), −0.2 (*dotted* and *dashed line*) or −0.4 (*solid line*)

survival probability independent of age. The overall survival probability was then the weighted mean of the age-specific survival probabilities, with the weights being the proportion of adults in each age-class, as given by the stable age distribution.

As well as performing the simulations for a range of population sizes, we estimated the large-sample bias using the analytical approach described by Burnham et al. (1987, p. 214), which involved calculating the expected m-array (see Appendix).

We assessed the impact of negative bias in the estimate of overall adult survival probability on an estimate of population growth rate from a population model in which individuals mature at age α, and in which reproductive rate and adult survival probability are the same for all ages. For this model, the sensitivity of population growth rate to adult survival probability is given by

$$s_{\phi} = \frac{\lambda}{\phi + \alpha(\lambda - \phi)}$$

where λ is the population growth rate, α is the age at maturity and ϕ is the overall adult survival probability (Heppell et al. 2000).

Individual heterogeneity in capture probability also leads to negative bias in estimates of survival probability (Carothers 1973; 1979). It is therefore possible that mild senescence and mild capture heterogeneity may act together to produce a non-negligible amount of bias. We therefore considered the combined effect of two-group heterogeneity in capture probability (Burnham et al. 1987, p. 287) and senescence, focussing on the case where the two group sizes are the same.

All calculations were carried out using R (R Development Core Team 2006), and the code is available from the authors.

3 Results

We initially focus on the results for a ten-year study, as this would be considered a minimum study-length for estimation of an overall adult survival probability. Table 1 shows the estimated bias in the estimate of overall adult survival probability, for a stable population of 100 individuals, for each senescence function and for a range of capture probabilities. The bias clearly increases as the capture probability decreases. Likewise, it increases if senescence occurs earlier or if the rate at which senescence occurs is higher. The overall pattern in Table 1 also suggests that the bias increases as the survival probability in early-adulthood decreases.

We ran equivalent simulations for stable populations containing either 1,000 or 10,000 individuals, and found that these were in close agreement with those obtained using the analytical large-sample estimate of bias. We therefore summarise the latter in Table 2. The differences between the results in Tables 1 and 2 are generally small, being at most 0.5% points when the capture probability is greater than 0.1, and at most 0.1% point when the capture probability is greater than 0.4. This suggests that use of the large-sample estimate of bias (Table 2) will be reasonable for

Table 1 Estimates of bias in the estimate of overall survival probability (percentage points) for a 10-year study, for a range of senescence functions and capture probabilities. Each estimate is based on 5,000 simulations. The population size is equal to 100 in each year of the study. For those cases where the bias is greater than one percentage point, the standard error is at most 10% of the estimate; in all other cases, the standard error is at most 0.1 percentage points.

Survival at age 0	Onset at age	Linear-logistic slope	Overall survival	Capture probability 0.1	0.2	0.3	0.4	0.5	0.6	0.7	0.8	0.9
0.75	0	−0.05	0.726	−3.3	−2.2	−1.4	−1.1	−0.8	−0.6	−0.4	−0.3	−0.1
		−0.1	0.707	−5.6	−3.5	−2.6	−2.0	−1.5	−1.1	−0.7	−0.4	−0.2
		−0.2	0.678	−9.1	−6.4	−4.9	−3.7	−2.8	−2.0	−1.4	−0.8	−0.4
		−0.4	0.636	−14.7	−10.8	−8.5	−6.7	−5.1	−3.7	−2.5	−1.6	−0.7
	5	−0.05	0.745	−1.7	−0.8	−0.6	−0.3	−0.3	−0.1	−0.1	0.0	0.0
		−0.1	0.741	−2.1	−1.2	−0.7	−0.5	−0.4	−0.3	−0.2	−0.1	0.0
		−0.2	0.736	−3.4	−1.9	−1.2	−0.9	−0.6	−0.4	−0.3	−0.2	−0.1
		−0.4	0.730	−4.2	−2.6	−1.8	−1.3	−0.9	−0.6	−0.3	−0.1	−0.1
	10	−0.05	0.749	−1.1	−0.4	−0.3	−0.1	0.0	0.0	0.0	0.0	0.0
		−0.1	0.748	−1.1	−0.5	−0.3	−0.1	−0.2	−0.1	−0.1	0.0	0.0
		−0.2	0.747	−1.2	−0.6	−0.3	−0.3	−0.2	−0.1	−0.1	−0.1	0.0
		−0.4	0.746	−1.5	−0.6	−0.5	−0.3	−0.2	−0.1	−0.1	0.0	0.0
0.85	0	−0.05	0.821	−2.4	−1.5	−1.1	−0.8	−0.6	−0.4	−0.3	−0.1	−0.1
		−0.1	0.800	−4.1	−2.8	−2.0	−1.5	−1.1	−0.8	−0.5	−0.3	−0.2
		−0.2	0.767	−7.5	−5.4	−4.1	−3.0	−2.2	−1.6	−1.1	−0.6	−0.3
		−0.4	0.720	−13.3	−9.8	−7.6	−5.7	−4.3	−3.1	−2.1	−1.2	−0.6
	5	−0.05	0.839	−1.1	−0.7	−0.5	−0.4	−0.3	−0.2	−0.1	−0.1	0.0
		−0.1	0.831	−1.8	−1.2	−0.9	−0.6	−0.4	−0.3	−0.2	−0.1	0.0
		−0.2	0.822	−3.2	−1.9	−1.3	−0.9	−0.7	−0.4	−0.3	−0.2	−0.1
		−0.4	0.811	−4.5	−2.9	−2.1	−1.4	−1.0	−0.7	−0.4	−0.2	−0.1
	10	−0.05	0.845	−0.7	−0.3	−0.2	−0.1	−0.1	−0.1	−0.1	0.0	0.0
		−0.1	0.842	−0.9	−0.5	−0.3	−0.2	−0.2	−0.1	−0.1	0.0	0.0
		−0.2	0.839	−1.3	−0.7	−0.5	−0.4	−0.2	−0.2	−0.1	0.0	0.0
		−0.4	0.835	−1.7	−1.0	−0.7	−0.5	−0.3	−0.2	−0.1	−0.1	0.0

Table 1 (continued)

Survival at age 0	Onset at age	Linear-logistic slope	Overall survival	Capture probability 0.1	0.2	0.3	0.4	0.5	0.6	0.7	0.8	0.9
0.95	0	−0.05	0.920	−1.3	−0.6	−0.4	−0.3	−0.2	−0.2	−0.1	−0.1	0.0
		−0.1	0.900	−2.0	−1.4	−1.0	−0.7	−0.5	−0.4	−0.2	−0.2	0.0
		−0.2	0.869	−4.4	−3.0	−2.2	−1.6	−1.2	−0.8	−0.5	−0.3	−0.1
		−0.4	0.823	−9.3	−6.8	−5.0	−3.6	−2.6	−1.8	−1.2	−0.7	−0.3
	5	−0.05	0.930	−1.2	−0.4	−0.3	−0.2	−0.1	−0.1	−0.1	0.0	0.0
		−0.1	0.919	−1.6	−0.8	−0.5	−0.4	−0.3	−0.2	−0.1	−0.1	−0.1
		−0.2	0.904	−2.2	−1.4	−1.0	−0.7	−0.5	−0.3	−0.2	−0.1	−0.1
		−0.4	0.888	−3.5	−2.4	−1.7	−1.2	−0.8	−0.6	−0.3	−0.2	−0.1
	10	−0.05	0.936	−1.0	−0.3	−0.2	−0.2	−0.1	−0.1	0.0	0.0	0.0
		−0.1	0.929	−1.1	−0.5	−0.3	−0.2	−0.2	−0.1	0.0	0.0	0.0
		−0.2	0.921	−1.4	−0.7	−0.5	−0.4	−0.3	−0.2	−0.1	0.0	0.0
		−0.4	0.912	−1.9	−1.1	−0.7	−0.5	−0.4	−0.3	−0.2	−0.1	−0.1

Table 2 Estimates of large-sample bias in the estimate of overall survival probability (percentage points) for a 10-year study, for a range of senescence functions and capture probabilities

Survival at age 0	Onset at age	Linear-logistic slope	Overall survival	Capture probability								
				0.1	0.2	0.3	0.4	0.5	0.6	0.7	0.8	0.9
0.75	0	−0.05	0.726	−2.2	−1.7	−1.3	−1.0	−0.7	−0.5	−0.4	−0.2	−0.1
		−0.1	0.707	−4.2	−3.3	−2.6	−1.9	−1.4	−1.0	−0.7	−0.4	−0.2
		−0.2	0.678	−7.6	−6.0	−4.7	−3.7	−2.7	−2.0	−1.3	−0.8	−0.4
		−0.4	0.636	−12.7	−10.4	−8.3	−6.5	−5.0	−3.7	−2.5	−1.5	−0.7
	5	−0.05	0.745	−0.7	−0.5	−0.4	−0.3	−0.2	−0.1	−0.1	0.0	0.0
		−0.1	0.741	−1.3	−0.9	−0.7	−0.5	−0.3	−0.2	−0.1	−0.1	0.0
		−0.2	0.736	−2.1	−1.6	−1.1	−0.8	−0.5	−0.4	−0.2	−0.1	−0.1
		−0.4	0.730	−3.3	−2.4	−1.7	−1.2	−0.8	−0.5	−0.3	−0.2	−0.1
	10	−0.05	0.749	−0.2	−0.1	−0.1	−0.1	0.0	0.0	0.0	0.0	0.0
		−0.1	0.748	−0.3	−0.2	−0.1	−0.1	−0.1	0.0	0.0	0.0	0.0
		−0.2	0.747	−0.4	−0.3	−0.2	−0.2	−0.1	−0.1	0.0	0.0	0.0
		−0.4	0.746	−0.6	−0.5	−0.3	−0.2	−0.2	−0.1	−0.1	0.0	0.0
0.85	0	−0.05	0.821	−1.7	−1.3	−1.0	−0.7	−0.5	−0.4	−0.3	−0.2	−0.1
		−0.1	0.800	−3.4	−2.6	−2.0	−1.5	−1.1	−0.8	−0.5	−0.3	−0.1
		−0.2	0.767	−6.6	−5.1	−3.9	−3.0	−2.2	−1.5	−1.0	−0.6	−0.3
		−0.4	0.720	−11.9	−9.5	−7.4	−5.7	−4.2	−3.0	−2.0	−1.2	−0.6
	5	−0.05	0.839	−0.8	−0.6	−0.4	−0.3	−0.2	−0.2	−0.1	−0.1	0.0
		−0.1	0.831	−1.4	−1.1	−0.8	−0.6	−0.4	−0.3	−0.2	−0.1	0.0
		−0.2	0.822	−2.4	−1.8	−1.3	−0.9	−0.6	−0.4	−0.3	−0.2	−0.1
		−0.4	0.811	−3.8	−2.8	−2.0	−1.4	−1.0	−0.6	−0.4	−0.2	−0.1
	10	−0.05	0.845	−0.3	−0.2	−0.2	−0.1	−0.1	−0.1	0.0	0.0	0.0
		−0.1	0.842	−0.5	−0.4	−0.3	−0.2	−0.1	−0.1	−0.1	0.0	0.0
		−0.2	0.839	−0.8	−0.6	−0.4	−0.3	−0.2	−0.1	−0.1	−0.1	0.0
		−0.4	0.835	−1.1	−0.8	−0.6	−0.4	−0.3	−0.2	−0.1	−0.1	0.0

Table 2 (continued)

Survival at age 0	Onset at age	Linear-logistic slope	Overall survival	Capture probability								
				0.1	0.2	0.3	0.4	0.5	0.6	0.7	0.8	0.9
0.95	0	−0.05	0.920	−0.8	−0.6	−0.4	−0.3	−0.2	−0.2	−0.1	−0.1	0.0
		−0.1	0.900	−1.7	−1.3	−1.0	−0.7	−0.5	−0.4	−0.2	−0.1	−0.1
		−0.2	0.869	−3.9	−2.9	−2.2	−1.6	−1.1	−0.8	−0.5	−0.3	−0.1
		−0.4	0.823	−8.6	−6.6	−4.9	−3.6	−2.6	−1.8	−1.2	−0.7	−0.3
	5	−0.05	0.930	−0.5	−0.4	−0.3	−0.2	−0.1	−0.1	−0.1	0.0	0.0
		−0.1	0.919	−1.0	−0.7	−0.5	−0.4	−0.3	−0.2	−0.1	−0.1	0.0
		−0.2	0.904	−1.8	−1.3	−1.0	−0.7	−0.5	−0.3	−0.2	−0.1	−0.1
		−0.4	0.888	−3.1	−2.3	−1.7	−1.2	−0.8	−0.5	−0.3	−0.2	−0.1
	10	−0.05	0.936	−0.3	−0.2	−0.2	−0.1	−0.1	−0.1	0.0	0.0	0.0
		−0.1	0.929	−0.6	−0.4	−0.3	−0.2	−0.2	−0.1	−0.1	0.0	0.0
		−0.2	0.921	−0.9	−0.7	−0.5	−0.4	−0.2	−0.2	−0.1	−0.1	0.0
		−0.4	0.912	−1.3	−1.0	−0.7	−0.5	−0.4	−0.2	−0.2	−0.1	0.0

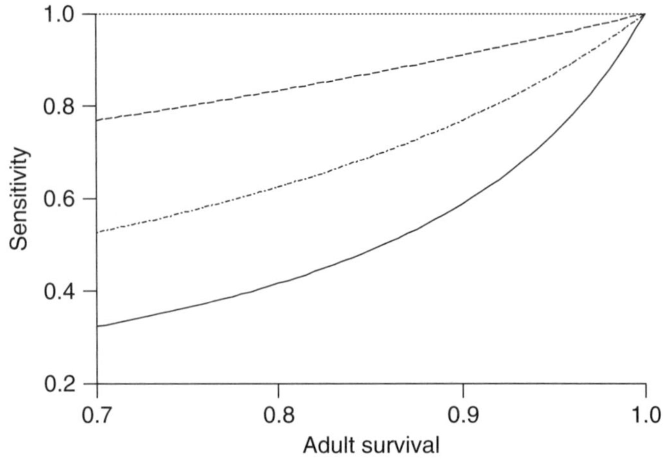

Fig. 2 Sensitivity of population growth rate to adult survival plotted against adult survival, for a deterministic age-based model in which actual age at maturity is 1 year (*dotted line*), 2 years (*dashed line*), 4 years (*dotted* and *dashed line*) or 8 years (*solid line*), and for which the population growth rate is 1

a wide range of studies. Figure 2 shows this sensitivity for a deterministic age-based model in which the actual age at maturity (M) is the same for all individuals, and for which the population growth rate is equal to 1. The sensitivity is plotted against overall adult survival probability for different values of M. Thus if overall adult survival is 0.85 and M is 4 years, the sensitivity is 0.69. This means that a negative bias of 1% point in the estimate of adult survival leads to a negative bias of 0.69% points in the population growth rate. The sensitivity increases as M decreases and as overall adult survival increases.

The results in Tables 1 and 2, and in Fig. 1, can be used to provide guidance as to the potential magnitude of this problem, even when the degree of senescence that the species is likely to experience is not well known. For example, suppose the strength of senescence is likely to be no greater than that represented by a slope of −0.4 on the logistic scale. If the capture probability is at least 0.5, the survival probability in early adulthood is around 0.85, and the onset of senescence is unlikely to occur until at least 5 years after maturity, the bias in the estimate of overall adult survival probability will then be at most 1% point in a 10-year study, regardless of the population size (Tables 1 and 2).

Figure 3 shows how the large-sample estimate of bias changes with study-duration, for a range of capture probabilities, when survival probability in early adulthood is 0.85 and declines from age 5 at a rate of −0.2 on the logistic scale. The bias increases with study duration in all cases, particularly when the capture probability is low.

Figure 4 shows the combined effect of senescence and two-group capture heterogeneity: the large-sample estimate of the bias is plotted against the coefficient of variation (CV) in capture probability, for the case where the two group sizes are

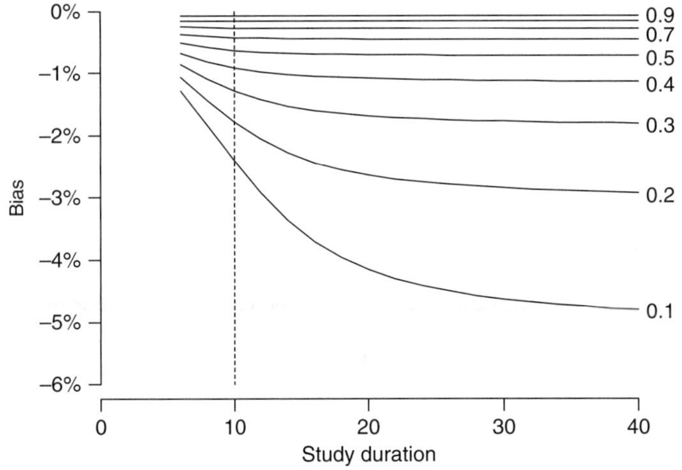

Fig. 3 Effect of study duration on the large-sample estimate of bias in the estimate of overall adult survival, for a range of capture probabilities, when survival in early adulthood is 0.85 and declines from age 5 at a rate of −0.2 on the logistic scale. The vertical line indicates the study-duration used in all other analyses

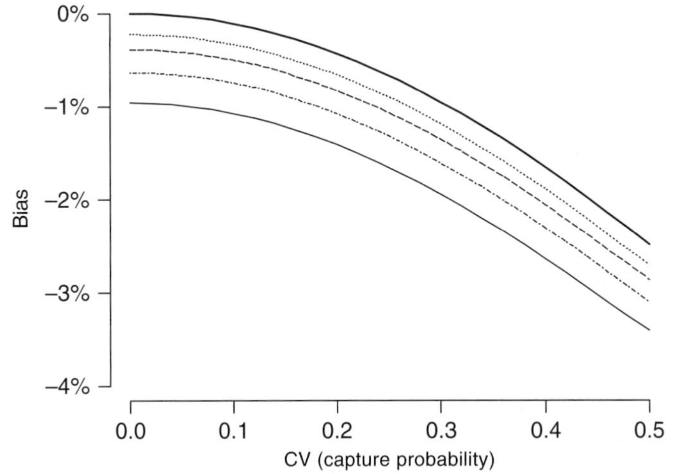

Fig. 4 Bias in the estimate of overall adult survival probability plotted against the coefficient of variation in capture probability, where the group sizes are equal, the mean capture probability is 0.5, and senescence begins at age 5. The survival probability at age 0 is 0.85 and the decline in survival corresponds to a linear-logistic slope of −0.05 (*dotted line*), −0.1 (*dashed line*), −0.2 (*dotted* and *dashed line*) or −0.4 (*solid line*). Bias in the absence of senescence is indicated by the *thick solid line*

the same and the mean capture probability is 0.5. The five curves correspond to no senescence plus the four senescence functions corresponding to survival at age 0 being 0.85 and onset of senescence being at age 5. For example, with a CV of 0.3 in capture probability (corresponding to capture probabilities of 0.35 and 0.65 for the two groups) and no senescence, the bias is −1.0%; if the decline in survival corresponds to a linear-logistic slope of −0.2, this bias increases to −1.8%.

4 Discussion

Our results suggest that if capture probability is high enough, or the degree of senescence is small, the bias in the estimate of overall adult survival probability will be small. It is often assumed that estimates of survival from a CJS-model are reasonably robust to departures from the assumptions (Carothers 1973). Our results suggest that robustness of the estimate of overall survival probability to senescence will depend on the species and the study, and should not be automatically assumed. For example, few data are available for small- to medium-sized mammals, and senescence in this group may be more extreme than we simulated. We have evidence from a long-term study of *Trichosurus vulpecula*, a 2.3 kg marsupial introduced to New Zealand, of a linear-logistic decline in survival of $-0.36\ y^{-1}$ in males, starting at maturity, and a decline of $-0.43\ y^{-1}$ in females, starting in the fifth year after maturity (Efford 2000; J. Taillon and M.G. Efford unpubl.).

Capture–recapture studies of persistence at migratory stopover sites are especially vulnerable to the biases we describe, because the capture probability of an unmarked individual is usually low (<0.25) (e.g. Kaiser 1995; Morris et al. 2005). Departure from migratory stopover sites may also prove to be highly dependent on time since arrival (mimicking extreme "senescence"); data on this relationship are currently lacking (but see Pledger et al. 2008).

Major reviews of the CJS model have generally cited Manly (1970) to the effect that estimates of survival are robust to "moderate" age-dependent mortality, or at least more so than the Fisher-Ford method (Cormack 1979; Seber 1982; Pollock et al. 1990). Williams et al. (2002, p. 435) cited Manly (1970) to the effect that "... use of single-age models in the face of age-specific variation in survival probabilities can result in *positive* bias in survival estimates, although the bias is not large for small to moderate variation in age-specific survival..." (our italics). This appears to refer to simulations with low juvenile survival and high adult survival, and does not reflect findings for senescence.

Pollock et al. (1990, p. 26) suggested that "[age-specific] differences in survival probabilities can be thought of as a special kind of heterogeneity". We agree, but some care is needed. Heterogeneity in the intrinsic survival probabilities of individuals causes *positive* bias in estimates of survival probability from a single marked sample followed through time, due to the depletion of individuals most at risk (e.g. Rexstad and Anderson 1992; Sheil and May 1996; Zens and Peart 2003). A mixed-age population that experiences senescence can also be thought of as heterogeneous with respect to survival probabilities, but leads to *negative* bias. The key point, as noted by Manly (1970), is that the age structure of the marked part of the population will differ from that of the population at large, and that estimates of survival from the marked individuals are therefore not representative of the population. The size of the bias will be determined by this difference in age structure and by the age-specific survival probabilities.

We have found that bias in the estimate of survival increases with study-duration. Zens and Peart (2003) noted a similar effect when there is individual heterogeneity in survival that is not age-dependent. They ignored the issue of detectability of individuals, which also has a large impact, both on the bias itself and on the effect of study-duration on that bias (Fig. 3).

The impacts of senescence and heterogeneous capture probability on the estimate of overall survival probability are approximately additive (Fig. 4). Intrinsic differences between individuals in survival probability are likely to counteract these effects. Likewise, lower survival in early adulthood would be expected to counteract the effects of senescence on the estimate of overall adult survival. An interesting topic for further research would be to assess the relative contribution that the different sources of bias are likely to make in practice.

In carrying out an analysis of capture–recapture data, it is important to assess goodness-of-fit. If survival is age-dependent, we might expect to detect this using Test 3SR (Burnham et al. 1987, p. 74). We could then use a model in which survival is "age"-dependent, where age refers to the number of years since first capture. Crespin et al. (2006) suggest that the power of such a model to detect senescence will be almost the same as when age is known. However, it is difficult to see how such an analysis could be used to improve the estimate of overall adult survival probability.

Pledger (2006; see also Pledger et al. 2008) provides a method of estimation of age-specific survival when age is unknown. The results of this paper allow one to assess the conditions under which an estimate of overall adult survival probability from a CJS-model is likely to be unreliable, thereby indicating when methods such as those outlined in Pledger (2006) will be of particular benefit.

If capture probability is also age-dependent, and this is not allowed for in the analysis, the bias in the estimate of overall adult survival probability will differ from the results presented in this paper. Thus if older adults are easier to capture than younger adults, the bias will be greater than shown here. In the spirit of Crespin et al. (2006), one could fit a "relative-age" model for capture probability as well as survival probability: if this model suggests that capture probability increases with age, one would expect the bias to be larger than indicated by the results in Section 3.

Heppell et al. (2000) suggested use of a correction-factor for adult survival in a population model, in order to allow for senescence. Their formula involves specifying a maximum amount of time spent in the adult stage and effectively leads to a reduction in the estimate of overall adult survival probability. Even if this estimate were not biased in the way described in this paper (e.g. if capture probability were equal to 1), the formula provided by Heppell et al. (2000) would be inappropriate, as the estimate of overall adult survival probability would already include the effect of senescence. Given the negative bias that might be present in this estimate, use of this formula would make the situation worse.

Our estimates of bias involve a number of assumptions about the population. We assume that the population is stable, has a stable age distribution, that both survival and capture probabilities are not time-dependent, and that first-capture probability and recapture probability are identical. It is difficult to give general guidance as to what to expect if the first two assumptions are not met, but the calculations given in the Appendix can easily be modified to allow for a population that is increasing or decreasing, or for an alternative age distribution. If there is time-dependence in the survival and capture probabilities, we would expect the bias to be approximately the same as the estimate obtained from Tables 1 or 2 by setting both the survival probability in early adulthood and the capture probability equal to an estimate of the corresponding mean value. If the first-capture probability differs from the recapture probability, we would expect the bias to be different from the results given here. For example, if we were to have a high first-capture probability in year 1, which reduced to zero thereafter, and the recapture probability remained constant throughout the study, we would expect the bias to be greater than given here, as this would lead to the marked individuals being even older on average than the population.

We have focussed attention on obtaining an estimate of overall adult survival probability using a capture–recapture model in which adult survival probability is constant, with the estimate being used in a deterministic population model. Our results should also provide useful guidance for the case where we are anticipating using a stochastic population model, for which estimates of the mean and between-year variation in adult survival probability would typically be required: we would then be using an estimate of mean adult survival probability from a "random effects" capture–recapture model for adult survival (Burnham and White 2002), and the bias in this estimate is likely to be close to that presented in this paper.

Acknowledgments We are grateful to the Associate Editor and the referees for their many helpful comments.

Appendix

We generate the data used in each run of the simulations in the form of a reduced m-array (Lebreton et al. 1992), using the following calculations. The reduced m-array consists of the following variables:

R_i Number of individuals captured and released in year i $(i = 1, 2, \ldots, T-1)$
m_{ij} Number of individuals that were released in year i and were recaptured for the first time in year j $(i = 1, 2, \ldots, T-1; j = i+1, i+2, \ldots, T)$

We generate these variables as follows:

$$R_i \sim Binomial\,(N, p)\,(i = 1, 2, \ldots, T-1)$$
$$\{m_{i,i+1}, m_{i,i+2}, \ldots, m_{iT}\,|R_i\} \sim Multinomial\,(R_i; \pi_{i,i+1}, \pi_{i,i+2}, \ldots, \pi_{iT})$$
$$(i = 1, 2, \ldots, T-1)$$

where π_{ij} is the probability that an individual released in year i is recaptured for the first time in year j. For computational convenience, we set $\phi^{(x)} = \phi^{(50)}$ for $x \geq 50$, as the proportion of the population aged 50 years or more will be very small

($< 0.1\%$) for the scenarios that we consider. As the capture probability is assumed to be the same for marked and unmarked individuals, each release cohort constitutes a random sample from the population, and we can express π_{ij} as

$$\pi_{ij} = p(1-p)^{j-i-1}\left\{\sum_{x=0}^{49}\left(\rho^{(x)}\prod_{y=x}^{x+j-i-1}\phi^{(y)}\right) + \rho^{(50+)}\left(\phi^{(50)}\right)^{j-i}\right\}$$

$$(i = 1, 2, \ldots, T-1; j = i+1, i+2, \ldots, T)$$

where $\rho^{(x)}$ is the stable age distribution for the adults in the population, given by

$$\rho^{(x)} \propto l(x)\ (x = 0, 1, \ldots, 49) \qquad \rho^{(50+)} \propto \rho^{(49)}\frac{\phi^{(49)}}{1-\phi^{(50)}} \qquad \sum_{x=0}^{50+}\rho^{(x)} = 1$$

where $\rho^{(50+)}$ is the proportion of individuals aged 50 years or more, and $l(x)$ is the probability of surviving to age x, given by

$$l(x) = \begin{cases} \prod_{y=0}^{x-1}\phi^{(y)} & x > 0 \\ 1 & x = 0 \end{cases}$$

For the large-sample case, we generated the expected m-array using

$$R_i = Np\ (i = 1, 2, \ldots, T-1)$$
$$m_{ij} = R_i\pi_{ij}\ (i = 1, 2, \ldots, T-1; j = i+1, i+2, \ldots, T)$$

For the case of two-group heterogeneity in capture probability, where the two group sizes are the same, we generated the expected m-array for each group using the calculations above, and then summed the corresponding elements in the two arrays, i.e. the final reduced m-array had elements given by

$$R_i = R_{1i} + R_{2i} \text{ and } m_{ij} = m_{1ij} + m_{2ij}$$

where R_{ki} and m_{kij} are the elements of the expected m-array for group k ($k = 1, 2; i = 1, 2, \ldots, T-1; j = i+1, i+2, \ldots, T$).

References

Aebischer NJ, Coulson JC (1990) Survival of the kittiwake in relation to sex, year, breeding experience and position in the colony. Journal of Animal Ecology 59:1063–1071.

Bradley JS, Wooller RD, Skira IJ, Serventy DL (1989) Age-dependent survival of breeding short-tailed shearwaters *Puffinus tenuirostris*. Journal of Animal Ecology 58:175–188.

Burnham KP, Anderson DR, White GC, Brownie C, Pollock KH (1987) Design and analysis methods for fish survival experiments based on release-recapture. American Fisheries Society Monograph 5.

Burnham KP, White GC (2002) Evaluation of some random effects methodology applicable to bird ringing data. Journal of Applied Statistics 29:245–264.

Carothers AD (1973) The effects of unequal catchability on Jolly–Seber estimates. Biometrics 29:79–100.

Carothers AD (1979) Quantifying unequal catchability and its effect on survival estimates in an actual population. Journal of Animal Ecology 48:863–869.

Caswell H (2000) Matrix population models: construction, analysis, and interpretation. 2nd edn. Sinauer, Massachusetts.

Caughley G (1966) Mortality patterns in mammals. Ecology 47:906–918.

Cormack RM (1979) Models for capture–recapture. In: Cormack RM, Patil GP, Robson DS (eds) Sampling biological populations. International Co-operative Publishing House, Fairland, Maryland, pp 217–255.

Crespin L, Harris MP, Lebreton J-D, Wanless S (2006) Increased adult mortality and reduced breeding success with age in a population of common guillemot *Uria aalge* using marked birds of unknown age. Journal of Avian Biology 37:273–282.

Dunnet GM, Ollason JC (1978) The estimation of survival rate in the fulmar, *Fulmarus glacialis*. Journal of Animal Ecology 47:507–520.

Efford MG (2000) Possum density, population structure and dynamics. In: Montague T (ed) The brushtail possum biology, impact and management of an introduced marsupial. Manaaki Whenua Press, Christchurch, Chapter 5, pp 47–61.

Efford MG (2005) Migrating birds stop over longer than usually thought: comment. Ecology 86:3415–3418.

Gaillard J.M, Allainé D, Pontier D, Yoccoz NG, Promislow DEL (1994) Senescence in natural populations of mammals: a reanalysis. Evolution 48:509–516.

Gaillard J-M, Viallefont A, Loison A, Festa-Bianchet M (2004) Assessing senescence patterns in populations of large mammals. Animal Biodiversity and Conservation 27:47–58.

Heppell SS, Caswell H, Crowder LB (2000) Life histories and elasticity patterns: perturbation analysis for species with minimal demographic data. Ecology 81:654–665.

Kaiser A (1995) Estimating turnover, movements and capture parameters of resting passerines in standardized capture–recapture studies. Journal of Applied Statistics 22:1039–1047.

Lebreton J-D, Burnham KP, Clobert J, Anderson DR (1992) Modeling survival and testing biological hypotheses using marked animals: a unified approach with case studies. Ecological Monographs 62:67–118.

Loery G, Pollock KH, Nichols JD, Hines JE (1987) Age-specificity of black-capped chickadee survival rates: analysis of capture–recapture data. Ecology 68:1038–1044.

Loison A, Festa-Bianchet M, Gaillard J-M, Jorgenson JT, Jullien JM (1999) Age-specific survival in five populations of ungulates: evidence of senescence. Ecology 80:2539–2554.

McDonald DB, Fitzpatrick JW, Woolfenden GE (1996) Actuarial senescence and demographic heterogeneity in the Florida Scrub Jay. Ecology 77:2373–2381.

Manly BFJ (1970) A simulation study of animal population estimation using the capture–recapture method. Journal of Applied Ecology 7:13–39.

Morris S, Turner EM, Liebner DA, Larracuente AM, Sheets DA (2005) Problems associated with pooling mark-recapture data prior to estimating stopover length for migratory passerines. In: Ralph CJ, Rich TD (eds) Bird conservation, implementation and integration in the Americas: proceedings of the third international Partners in Flight conference 2002. General Technical Report PSW–GTR–191, USDA Forest Service, Pacific Southwest Research Station, Albany, California, pp 673–679.

Newton I, Rothery P (1997) Senescence and reproductive value in sparrowhawks. Ecology 78:1000–1008.

Nichols JD, Hines JE, Blums P (1997) Tests for senescent decline in annual survival probabilities of common pochard, *Aythya ferina*. Ecology 78:1009–1018.

Niel C, Lebreton J-D (2005) Using demographic invariants to detect overharvested bird populations from incomplete data. Conservation Biology 19:826–835.

Orell M, Belda EJ (2002) Delayed cost of reproduction and senescence in the willow tit *Parus montanus*. Journal of Animal Ecology 71:55–64.

Pledger SA (2006) Age-related capture–recapture parameter estimation when true age is unknown. Technical Report, School of Mathematics, Statistics and Computer Science, Victoria University of Wellington, New Zealand.

Pledger S, Efford MG, Pollock K, Collazo J, Lyons J. (2008) Stopover duration analysis with departure probability dependent on unknown time since arrival. In: Thomson DL, Cooch EG, Conroy MJ (eds.) Modeling Demographic Processes in Marked Populations. Environmental and Ecological Statistics, Springer, New York, 3:349–364.

Pollock KH, Nichols JD, Brownie C, Hines JE (1990) Statistical inference for capture–recapture experiments. Wildlife Monographs 107:1–97.

Pradel R (1996) Utilization of capture–mark-recapture for the study of recruitment and population growth rate. Biometrics 52:703–709.

Promislow DEL (1991) Senescence in natural populations of mammals: a comparative study. Evolution 45:1869–1887.

Pugesek BH, Nations C, Diem KL, Pradel R (1995) Mark-resighting analysis of a California gull population. Journal of Applied Statistics 22:625–639.

R Development Core Team (2006) R: a language and environment for statistical computing. R Foundation for Statistical Computing, Vienna, Austria. http://www.R-project.org.

Rexstad EA, Anderson DR (1992) Heterogeneous survival rates of mallards (*Anas platyrhynchos*). Canadian Journal of Zoology 70:1878–1885.

Schaub M, Pradel R, Jenni L, Lebreton J-D (2001) Migrating birds stop over longer than usually thought: an improved capture–recapture analysis. Ecology 82:852–859.

Seber GAF (1982) The estimation of animal abundance and related parameters. 2nd edn. Griffin, London.

Sheil D, May RM (1996) Mortality and recruitment rate evaluations in heterogeneous tropical forests. Journal of Ecology 84:91–100.

Sidhu LA, Catchpole EA, Dann P (2007) Mark-recapture-recovery modeling and age-related survival in Little Penguins (*Eudyptula Minor*) The Auk 124:815–827.

Williams BK, Conroy MJ, Nichols JD (2002) Analysis and management of animal populations. Academic Press, London.

Wooller RD, Bradley JS, Croxall JP (1992) Long-term population studies of seabirds. Trends in Ecology and Evolution 7:111–114.

Yearsley JM, Fletcher D (2002) Equivalence relationships between stage-structured population models. Mathematical Biosciences 179:131–143.

Zens MS, Peart DR (2003) Dealing with death data: individual hazards, mortality and bias. Trends in Ecology and Evolution 18:366–373.

Weak Identifiability in Models for Mark-Recapture-Recovery Data

Olivier Gimenez, Byron J.T. Morgan, and Stephen P. Brooks

Abstract The percentage overlap between prior and posterior distributions is obtained easily from the output of MCMC samplers. A 35% guideline for overlap between univariate marginal prior and posterior distributions has been suggested as an indicator of weak identifiability of a parameter. As long as uniform prior distributions are adopted for all of the model parameters, then the suggested guideline has been found to work well for a range of models of mark-recapture-recovery data, where all the parameters are probabilities. Its use is illustrated on models for ring-recovery data on male mallards, and the Cormack-Jolly-Seber model for capture-recapture data on dippers.

Keywords Bayesian identifiability · Cormack-Jolly-Seber model · Mark-recapture-recovery models · Parameter-redundancy · Prior/posterior overlap · Sensitivity · Survival of wild animals

1 Introduction

1.1 Parameter Redundancy and Identifiability

Models may be devised for mark-recapture-recovery (mrr) data without all the parameters being estimable. A model is said to be *identifiable* if no two values of the parameters give the same maximum likelihood for the data, while *parameter-redundant* models can be re-expressed in terms of fewer than the original number of parameters (Catchpole and Morgan 1997), resulting in that case in likelihood surfaces with completely flat ridges or surfaces. The obvious way to check for parameter redundancy for a particular application is to examine the likelihood surface by computing the rank of the observed Hessian (Viallefont et al. 1998; Formann 2003). Catchpole and Morgan (1997) considered the rank of the model itself, regardless

B.J.T. Morgan (✉)
Institute of Mathematics, Statistics and Actuarial Science, University of Kent, Canterbury, Kent CT2 7NF, UK
e-mail: B.J.T.Morgan@kent.ac.uk

D.L. Thomson et al. (eds.), *Modeling Demographic Processes in Marked Populations,* Environmental and Ecological Statistics 3, DOI 10.1007/978-0-387-78151-8_48, of the data, using symbolic algebra. It is then possible to determine how many and which parameter combinations can be estimated (Catchpole et al. 1998; Catchpole et al. 2001). Applications of this approach can be found in, for example, Gimenez et al. (2003), Gimenez et al. (2005), Schaub et al. (2004), Nasution et al. (2004) and Kéry et al. (2005). In addition, Catchpole et al. (2001) considered *near redundant* models, which, while formally not parameter-redundant, can in some cases produce estimates of certain parameters with poor precision.

1.2 Weak Identifiability

The use of computational Bayesian methods for model-fitting in biology has increased in recent years (Ellison 2004; Clark 2005), including population ecology (Brooks et al. 2000a, 2002). With the increase in computing power, the temptation is to fit more and more complex models using MCMC methods. However, models that are parameter-redundant may be fitted using Bayesian methods, as a result of the information in the prior distribution, and a dramatic illustration of this is provided by Brooks et al. (2000b). Let us write a marginal posterior distribution as $\pi(\theta|Y)$, for data Y, parameter θ and prior distribution $p(\theta)$. Then the parameter θ is said to be *weakly identifiable* when $\pi(\theta|Y) \approx p(\theta)$; see Gelfand and Sahu (1999) and Garrett and Zeger (2000).

Thus in Bayesian analysis, weak identifiability arises when data supply little information about certain parameters. Weak identifiability is the counterpart of near redundancy in a classical analysis, and poses appreciable problems for a Bayesian approach. For example:

- conclusions based on the examination of weakly identified parameters can be misleading (Garrett and Zeger 2000);
- weak identifiability may result in strong correlations between parameters in the posterior distribution, which in turn implies poor mixing in the MCMC samples and very slow convergence (Carlin and Louis 1996; Rannala 2002);
- even with large sample sizes, the likelihood may be unable to overcome the prior (Neath and Samaniego 1997);
- a too-informative prior can drive posterior inference, while a prior too close to improper can yield improper posteriors (Gelfand and Sahu 1999; Bayarri and Berger 2004).

There are various ways to check for weak identifiability:

- one might conduct a classical test for parameter redundancy;
- one might undertake a detailed prior sensitivity analysis;
- one can examine the correlation matrix of the parameter estimators in the posterior distribution;
- as in Garrett and Zeger (2000), one can display the marginal prior/posterior pair plots as a visual aid;
- one can evaluate numerically, and calibrate, the overlap for each marginal prior-posterior pair.

Gimenez et al. (2006) have compared these alternative procedures, and found that the last two of these methods are simple and effective. We outline this approach in Section 2, and then illustrate its use on a data set resulting from marking male mallards, *Anas platyrhynchos*, in Section 3. In Section 4 the method is applied to a data set of capture-recapture data on dippers *Cinclus cinclus*. In Section 5 we examine the alternative approach of sensitivity analysis, while in Section 6 we present correlations between parameter estimates. The paper ends with general discussion in Section 7.

2 Testing for Weak Identifiability

2.1 Theory

In order to check the weak identifiability of any parameter θ, Garrett and Zeger (2000) compared its marginal prior distribution to its marginal posterior distribution by directly evaluating the overlap between the two distributions. This quantity, denoted τ_θ, can be computed as

$$\tau_\theta = \int \min(p(\theta), \pi(\theta|Y))d\theta. \tag{1}$$

Values of τ_θ lie in the interval [0,1], and when τ_θ is above some pre-determined threshold then θ is declared weakly identifiable. The *ad-hoc* threshold of 0.35 has been suggested by Garrett and Zeger (2000), and used in applications of this method.

2.2 Bayesian Inference

The mrr models that we shall consider only involve probabilities, and in this paper we have only presented results arising from taking uniform distributions on the interval [0,1] as priors for all of the model probabilities. Based on preliminary runs, we generated four chains of length 50,000, discarding the first 25,000 as burn-in. Convergence was assessed using the Brooks/Gelman/Rubin statistic (Gelman 1996), and we found that in general the Markov chains exhibited good mixing and moderate autocorrelation.

Simulations were performed using WinBUGS (Gimenez et al. 2008; Lunn et al. 2000), and the R (R Development Core Team 2008) package R2WinBUGS (Sturtz et al. 2005) was used both to call WinBUGS and examine results in R.

2.3 Practical Computation of τ

The computation of τ follows suggestions by Schmid and Schmidt (2006): we estimate the posterior distribution $\pi(\theta|Y)$ by means of a kernel density estimator $\hat{f}$ based on the MCMC generated values x_i, $i = 1, \ldots, n$ (we took $n = 1000$, corresponding to the last 1000 MCMC values obtained), namely $\hat{f}(x) = \frac{1}{n}\sum_{i=1}^{n}\frac{1}{h}K\left(\frac{x - x_i}{h}\right)$ where K is a kernel function centered at the data points x_i and h is the bandwidth. We used a standard Gaussian kernel $K(x) = \frac{1}{\sqrt{2\pi}}\exp\left(-\frac{x^2}{2}\right)$ with its associated optimal bandwidth $h^* = 1.06\,\hat{\sigma}\,n^{-\frac{1}{5}}$, where $\hat{\sigma}$ = min (standard deviation, interquartile range/1.34) (Silverman 1986, page 48).

We then obtained a sample from the distribution of the *min* function in Equation (1) by calculating $y_i = \min(1, \hat{f}(x_i))$ for all i. Finally, a Monte Carlo approximation to τ is given by $\hat{\tau} = \sum_i y_i/n$.

3 Mark-Recoveries: Application to the Freeman-Morgan Model

3.1 The Freeman-Morgan Model

As an example of the approach, we consider a model developed by Freeman and Morgan (1992; FM model hereafter) involving the survival probability $\phi_{1,i}$ of birds in their first year of life, possibly varying with the year i, the probability ϕ_a of survival of adult individuals (i.e. of age $\geq$ 1), taken as constant over time, and constant probabilities of reporting of rings from dead birds in their first year λ_1, or older λ_a.

Taking the FM model as an illustration, Catchpole et al. (2001) showed that probability models that are formally not parameter redundant may behave poorly when fitted to data. The main reason for this near-singularity is that the FM model contains as a sub-model the model with constant first year survival, which is parameter redundant. As a consequence, the smallest eigenvalue of the expected information matrix may be very small rather than zero as is the case in parameter-redundant models.

When Catchpole et al. (2001) applied this model to ring-recovery data obtained from animals marked as young, very poor results were sometimes obtained, with unrealistic estimates of $\phi_{1,i}$ and λ_1 and large associated standard errors.

We fitted the FM model to data on mallards, with 9 years of recovery (Table 1). The data are the result of a ringing study of males ringed as young in the San Luis Valley, Colorado, 1963–1971; Brownie et al. (1985), p48.

Displays of the marginal prior-posterior distribution pairs for the FM model parameters are given in Fig. 1, and the corresponding estimated percentages of overlap are given in Table 2, corresponding to the shaded areas in Fig. 1.

The marginal prior-posterior distribution pairs in Fig. 1 clearly suggest that all parameters except two may exhibit weak identifiability problems, viz., ϕ_a and λ_a which have relatively sharp marginal posterior distributions. Examination of Table 2

Table 1 Recovery data for male mallards Anas platyrhynchos

Year of ringing	Year of recovery (−1962)									Number never seen again
	1	2	3	4	5	6	7	8	9	
1963	83	35	18	16	6	8	5	3	1	787
1964		103	21	13	11	8	6	6	0	534
1965			82	36	26	24	15	18	4	927
1966				153	39	22	21	16	8	942
1967					109	38	31	15	1	1005
1968						113	64	29	22	927
1969							124	45	22	940
1970								95	25	786
1971									38	315

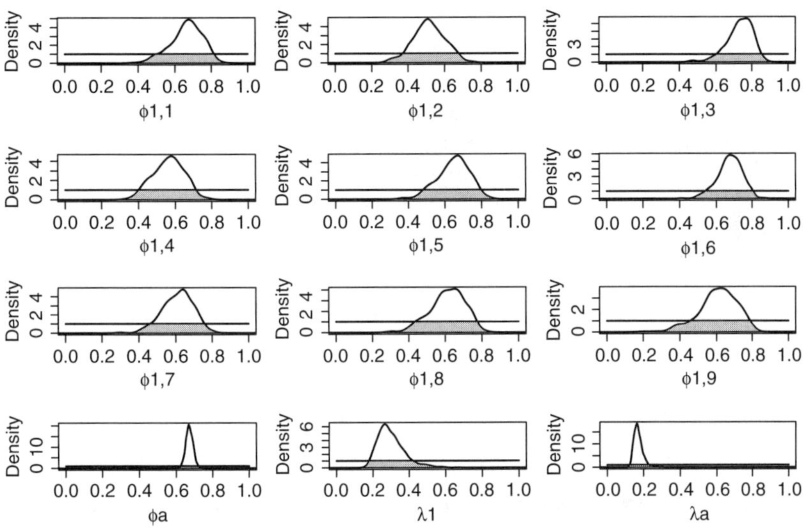

Fig. 1 Display of the prior-posterior distribution pairs for $\phi_{1,i}$, ϕ_a, λ_1 and λ_a in the FM model applied to the mallard data. In order to compute τ, the shaded area of overlap has to be calculated (Equation 1). The priors for all parameters were chosen as $U(0, 1)$ distributions. Note that scales in panels may differ

leads to the same conclusion with all τ values greater than or close to 0.35, except for ϕ_a and λ_a and the young reporting rate λ_1 which was also found to be identifiable using the 35% threshold. This is in agreement with the results obtained by Catchpole et al. (2001), Table 5(a).

3.2 Fitting the Parameter-Redundant Sub-model

It is of interest here to fit the parameter-redundant sub-model, which arises when there is just a single survival probability, ϕ_1, for birds in their first year of life.

Table 2 τ values expressed as percentages for the FM model fitted to the mallard data

Parameter	τ
$\phi_{1,1}$	40.2
$\phi_{1,2}$	41.5
$\phi_{1,3}$	34.6
$\phi_{1,4}$	40.1
$\phi_{1,5}$	41.7
$\phi_{1,6}$	34.7
$\phi_{1,7}$	41.5
$\phi_{1,8}$	43.4
$\phi_{1,9}$	45.3
ϕ_a	10.7
λ_1	30.8
λ_a	13.4

The resulting graph showing the overlaps between priors and posteriors is shown in Fig. 2. It appears from this graph that the only parameter that is not weakly identifiable is ϕ_a. This result is in agreement with the classical methodology of Catchpole et al. (2001), which formally identifies the three estimable parameters in this case as ϕ_a, and the two parameter combinations $\phi_1\lambda_a$, and $\phi_1(1-\lambda_1)$. The values for the

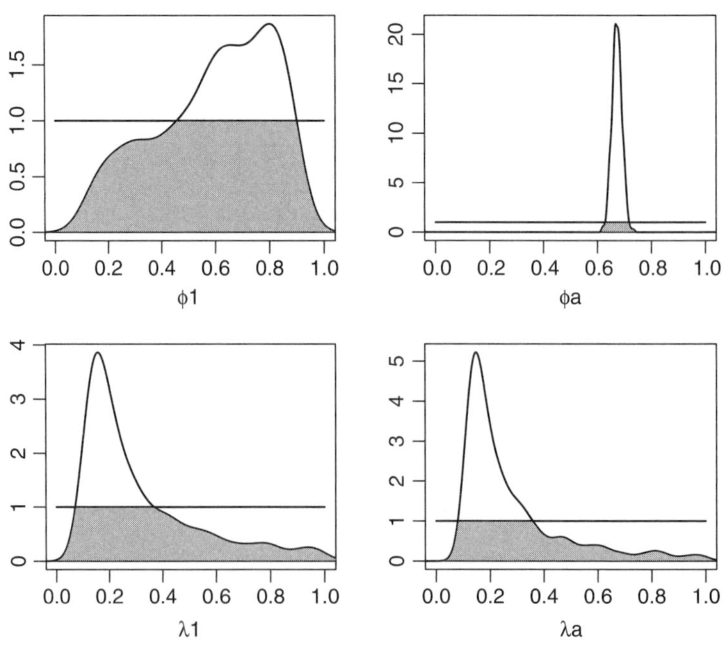

Fig. 2 Display of the prior-posterior distribution pairs for ϕ_1, ϕ_a, λ_1 and λ_a in the parameter redundant sub model of the FM model, applied to the mallard data. The priors for all parameters were chosen as $U(0, 1)$ distributions. Note that scales in panels differ

overlaps are, $\tau_{\phi_1} = 75.93$, $\tau_{\phi_a} = 10.78$, $\tau_{\lambda_1} = 60.66$, $\tau_{\lambda_a} = 51.33$. The difference between Figs. 1 and 2 is one of degree, with the weakly-identified parameters of Fig. 2 exhibiting more overlap with the prior than in the case of Fig. 1.

4 Mark-Recapture: Application to the Cormack-Jolly-Seber Model

We consider now the fully time-dependent capture-mark-recapture Cormack-Jolly-Seber (CJS) model originally developed by Cormack (1964), Jolly (1965) and Seber (1965), for which all parameters are time dependent. We define ϕ_i as the probability that an animal survives to time t_{i+1} given that it is alive at time t_i, and p_j the probability of being recaptured at time t_j. Even though it is parameter-redundant, the model can be useful for analyzing capture-mark-recapture data. We fitted the CJS model to the dipper data set which consists of 7 capture occasions (Table 3). The data are the result of a recapture study of both male and female birds ringed as adults in eastern France, in 1981–1986. In this application, the last survival (ϕ_6) and detection (p_7) probabilities are known to be confounded while all the other parameters are estimable.

Displays of the prior-posterior distribution pairs for survival and detection probabilities are given in Fig. 3.

The overlap percentages are given in Table 4 and correspond to the shaded areas in Fig. 3.

From examining Fig. 3, it appears that almost all CJS model parameters are well identified given the small overlap between the prior and posterior distributions. However, as expected, the last survival and detection probabilities are the exception, their posterior distributions being relatively flat and therefore providing more coverage of the uniform priors. In addition, it should be noted that the first survival and the first detection probabilities are also clearly weakly identifiable, due to the fact that very few individuals were marked at the first sampling occasion (approximately 7% of the full data set). The visual diagnostic of Fig. 3 is confirmed by looking at the numerical values of τ in Table 4. In addition, parameter p_3 seems to be marginally weakly identifiable ($\tau_{p_3} = 35.9$).

Table 3 Recapture data for European dippers *Cinclus cinclus* (data taken from Lebreton et al. 1992)

Year of release	Year of recapture (1981+)						Number never recaptured
	1	2	3	4	5	6	
1981	11	2	0	0	0	0	9
1982		24	1	0	0	0	35
1983			34	2	0	0	42
1984				45	1	2	32
1985					51	0	37
1986						52	46

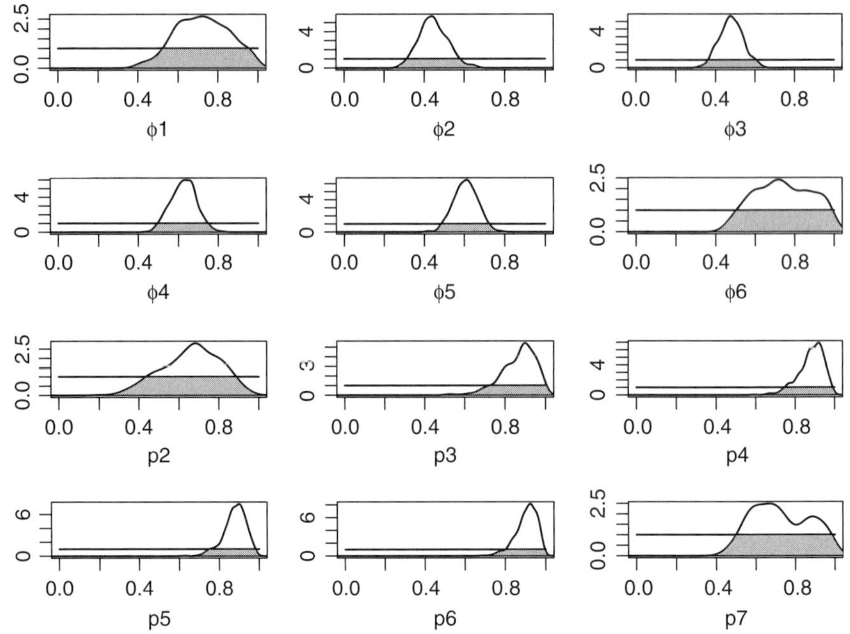

Fig. 3 Display of the prior-posterior distribution pairs for ϕ and p in the CJS model applied to the dipper data. The priors for all parameters were chosen as $U(0, 1)$ distributions. Note that scales in panels differ

Table 4 τ values expressed as percentages for the CJS model applied to the dipper data. For all parameters we use $U(0, 1)$ priors throughout

Parameter	τ
ϕ_1	53.6
ϕ_2	33.4
ϕ_3	29.2
ϕ_4	27.9
ϕ_5	27.4
ϕ_6	57.2
p_2	54.7
p_3	35.9
p_4	28.4
p_5	26.4
p_6	23.4
p_7	56.5

5 Sensitivity Analysis

If parameters are weakly identifiable, then we would expect this to be revealed by a sensitivity analysis, in which we repeat the analysis several times, each time for a different configuration of prior distributions. To illustrate this approach, we consider the CJS model applied to the dipper data set. We take a $U(0, 1)$ prior for all of the

Table 5 Sensitivity analysis for the CJS model applied to the dipper data. In rows we show the parameter for which the prior is changed in turn (from a $U(0, 1)$ to a Beta(1,9)). In columns we give the posterior means for all the parameters. The last column gives the Euclidean distance between each row and the case where all priors are $U(0, 1)$

	ϕ_1	ϕ_2	ϕ_3	ϕ_4	ϕ_5	ϕ_6	p_2	p_3	p_4	p_5	p_6	p_7	eucl. dist.
ϕ_1	0.47	0.46	0.48	0.63	0.60	0.72	0.75	0.87	0.88	0.87	0.90	0.74	0.27
ϕ_2	0.74	0.39	0.49	0.62	0.61	0.73	0.65	0.89	0.88	0.88	0.90	0.73	0.07
ϕ_3	0.72	0.46	0.43	0.63	0.60	0.73	0.67	0.86	0.89	0.88	0.91	0.74	0.05
ϕ_4	0.72	0.45	0.48	0.56	0.61	0.73	0.67	0.87	0.87	0.89	0.90	0.74	0.07
ϕ_5	0.72	0.45	0.48	0.63	0.55	0.74	0.66	0.87	0.88	0.87	0.92	0.74	0.06
ϕ_6	0.72	0.45	0.48	0.63	0.61	0.53	0.67	0.87	0.88	0.88	0.90	0.91	0.28
p_2	0.83	0.43	0.48	0.63	0.60	0.73	0.42	0.87	0.88	0.88	0.91	0.74	0.26
p_3	0.72	0.55	0.45	0.63	0.60	0.73	0.67	0.60	0.88	0.87	0.90	0.74	0.30
p_4	0.72	0.45	0.53	0.60	0.61	0.74	0.67	0.87	0.67	0.87	0.90	0.73	0.21
p_5	0.73	0.45	0.48	0.69	0.57	0.74	0.66	0.87	0.88	0.71	0.91	0.73	0.18
p_6	0.72	0.45	0.48	0.63	0.67	0.71	0.67	0.87	0.88	0.87	0.73	0.71	0.19
p_7	0.72	0.45	0.48	0.63	0.61	0.91	0.67	0.87	0.88	0.87	0.89	0.53	0.27

parameters except for one, which is given a $Beta(1, 9)$ prior, and we then change, in turn, which parameter has the different prior. As we can see from the results displayed in Table 5, the conclusions are not straightforward.

However it appears that there is little sensitivity of parameters to the prior, except for parameters ϕ_1, p_2, ϕ_6, p_3 and p_7. This is in line with the findings from using the overlap measure τ, but there is a difference in order. As with the overlap measure, there is an issue of calibration here. Carrying out the sensitivity analysis by changing the prior for one parameter at a time is time consuming since we need to run the MCMC chains as many times as the number of parameters (note also that to check for convergence of the MCMC, we always run two chains in parallel). Thus this approach could soon become intractable with more complex models, particularly as one would typically use several alternative beta distributions.

Because of the relative complexity of a detailed sensitivity analysis, we now consider a simpler alternative approach, of examining correlations between parameters, obtained from the MCMC output in the usual way.

6 Correlations Between Estimates

6.1 The Dipper Data

The correlation matrix of the parameters for the CJS model applied to the dipper data is given in Table 6.

We have high negative correlation between ϕ_6 and p_7 (-0.89) as expected, moderate negative correlation between ϕ_1 and p_2 (-0.50), and all remaining pairs of parameters give correlations in the range $(-0.32,\ 0.16)$. We note that the value of -0.32 relates to parameter p_3, as well as to parameter ϕ_2. Thus in this example

Table 6 Correlation matrix for the parameters of the CJS model applied to the dipper data

	ϕ_1	ϕ_2	ϕ_3	ϕ_4	ϕ_5	ϕ_6	p_2	p_3	p_4	p_5	p_6	p_7
ϕ_1	1.00	−0.15	0.02	−0.03	−0.03	0.00	−0.50	0.02	0.01	−0.02	−0.05	0.00
ϕ_2	−0.15	1.00	−0.08	0.00	0.00	−0.03	0.09	−0.32	0.05	−0.03	0.00	0.05
ϕ_3	0.02	−0.08	1.00	−0.06	0.02	0.03	−0.03	0.16	−0.18	0.03	0.01	−0.04
ϕ_4	−0.03	0.00	−0.06	1.00	−0.09	−0.02	0.03	−0.01	0.15	−0.32	0.00	−0.02
ϕ_5	−0.03	0.00	0.02	−0.09	1.00	0.01	0.02	−0.02	0.01	0.15	−0.30	−0.06
ϕ_6	0.00	−0.03	0.03	−0.02	0.01	1.00	0.04	0.05	−0.02	−0.07	0.02	−0.89
p_2	−0.50	0.09	−0.03	0.03	0.02	0.04	1.00	0.01	−0.02	0.04	−0.02	−0.03
p_3	0.02	−0.32	0.16	−0.01	−0.02	0.05	0.01	1.00	−0.04	−0.01	0.02	−0.06
p_4	0.01	0.05	−0.18	0.15	0.01	−0.02	−0.02	−0.04	1.00	−0.02	−0.04	0.01
p_5	−0.02	−0.03	0.03	−0.32	0.15	−0.07	0.04	−0.01	−0.02	1.00	−0.04	0.07
p_6	−0.05	0.00	0.01	0.00	−0.30	0.02	−0.02	0.02	−0.04	−0.04	1.00	0.03
p_7	0.00	0.05	−0.04	−0.02	−0.06	−0.89	−0.03	−0.06	0.01	0.07	0.03	1.00

considering the correlation structure between estimates has proved to be useful, and is relatively easy to implement.

6.2 The Mallard Data and the FM Model

The correlation matrix of the parameters for the FM model applied to the Mallard data is given in Table 7.

In this case the correlation matrix is not so easy to interpret. As expected, there are generally low correlations between ϕ_a and all of the other parameters, but that is not true of parameters λ_1 and λ_a, which is therefore out of line with the findings of Section 3.1.

6.3 The Mallard Data and the Parameter-Redundant Sub-model

The correlation matrix of the parameters for the FM submodel applied to the mallard data is given in Table 8.

Table 7 Correlation matrix for the parameters of the FM model applied to the mallard data

	ϕ_a	$\phi_{1,1}$	$\phi_{1,2}$	$\phi_{1,3}$	$\phi_{1,4}$	$\phi_{1,5}$	$\phi_{1,6}$	$\phi_{1,7}$	$\phi_{1,8}$	$\phi_{1,9}$	λ_1	λ_a
ϕ_a	1.00	−0.06	−0.07	−0.05	−0.04	−0.03	−0.02	0.00	−0.01	−0.03	−0.02	0.22
$\phi_{1,1}$	−0.06	1.00	0.85	0.87	0.88	0.88	0.87	0.88	0.86	0.77	0.86	−0.89
$\phi_{1,2}$	−0.07	0.85	1.00	0.82	0.87	0.86	0.84	0.86	0.84	0.73	0.89	−0.83
$\phi_{1,3}$	−0.05	0.87	0.82	1.00	0.88	0.88	0.87	0.88	0.86	0.77	0.85	−0.90
$\phi_{1,4}$	−0.04	0.88	0.87	0.88	1.00	0.90	0.88	0.89	0.88	0.78	0.91	−0.87
$\phi_{1,5}$	−0.03	0.88	0.86	0.88	0.90	1.00	0.88	0.90	0.88	0.78	0.88	−0.89
$\phi_{1,6}$	−0.02	0.87	0.84	0.87	0.88	0.88	1.00	0.88	0.86	0.77	0.88	−0.87
$\phi_{1,7}$	0.00	0.88	0.86	0.88	0.89	0.90	0.88	1.00	0.88	0.77	0.90	−0.87
$\phi_{1,8}$	−0.01	0.86	0.84	0.86	0.88	0.88	0.86	0.88	1.00	0.76	0.87	−0.85
$\phi_{1,9}$	−0.03	0.77	0.73	0.77	0.78	0.78	0.77	0.77	0.76	1.00	0.77	−0.78
λ_1	−0.02	0.86	0.89	0.85	0.91	0.88	0.88	0.90	0.87	0.77	1.00	−0.81
λ_a	0.22	−0.89	−0.83	−0.90	−0.87	−0.89	−0.87	−0.87	−0.85	−0.78	−0.81	1.00

Table 8 Correlation matrix for the parameters of the FM submodel applied to the mallard data

	ϕ_a	ϕ_1	λ_1	λ_a
ϕ_a	1.00	0.00	0.01	0.04
ϕ_1	0.00	1.00	0.86	−0.86
λ_1	0.01	0.86	1.00	−0.59
λ_a	0.04	−0.86	−0.59	1.00

We can see that the only parameter from the original parameter set that is estimable, ϕ_a, is essentially uncorrelated with the other estimates, however other conclusions are elusive, and this is true also if one considers the results of a principal component analysis of the correlation matrix.

7 Discussion

Experience with a range of different models and data sets in the general mrr area consistently suggests that, as proposed by Garrett and Zeger (2000), the prior/posterior overlap threshold of $\tau = 35\%$ works well as a guideline for diagnosing weak identifiability in models for the annual survival of wild animals, when the approach is confined to the case of uniform prior distributions. It has been shown by Gimenez et al. (2006) that if other priors are used then it is difficult to calibrate τ.

A difference between using a classical approach to parameter redundancy based on symbolic algebra and estimating the prior/posterior overlap, as in this paper, is that the latter approach takes account of the effect of both the data and the model. In practice, if possible it is important to understand both the redundancy structure of individual models as well as the influence of data. This has been seen in the analysis of Section 3; the estimates of reporting probabilities for the Mallard data are quite different, but in contrast the estimates of first-year survival are not. What this means is that the fitted FM model for the mallard data is essentially similar to the parameter-redundant sub-model with parameters $(\lambda_1, \lambda_a, \phi_1, \phi_a)$.

While in practice it is important to know the parameter-redundancy status of a model, that is not always possible (Jiang et al. 2007; Pradel et al. 2008; Schaub et al. 2004). As we have seen, weak identifiability may be due to the model and/or the data, and the cause of weak identifiability can be investigated once a model is fitted, by fitting the model again to a larger data set simulated from the fitted model. If that were done for the three examples of this paper, then nothing would change for the parameter-redundant sub-model of the FM model. No amount of additional data can change the fact that of the original four parameters, it is only ϕ_a that can be estimated precisely. For the FM model, increased precision for the estimates of first-year survival should improve overall performance, as in this case the model is not parameter redundant. For the CJS model, analysis of the dipper data has produced estimates of low precision either because of parameter redundancy or because of lack of data. Here increasing cohort sizes will improve the precision of parameters such as ϕ_1, but the basic parameter redundancy due to the confounding of parameters ϕ_6 and p_7 will remain.

We believe that the use of a 35% overlap threshold for τ, combined with uniform priors, is an important guide for interpreting the results of Bayesian analyses of mrr data, and we recommend its use as a simple, general guideline in the area. Of course, it is only a guideline, and needs to be interpreted sensibly.

Because the correlation matrix between parameters is easily obtained from the MCMC output, then we recommend that it is also examined. As any Bayesian analysis will involve some sensitivity analysis, then the results of such an analysis might also be of value. We note finally that when uniform priors are used then overlap with the posterior may be related to the variance of the corresponding parameter, though this will also depend on features such as skew (see for example the results of Fig. 2). A crude alternative to the measure of overlap considered in this paper is simply to use (and calibrate) the posterior inter-quartile range, which provides a measure of spread that is less affected by skew.

The R and WinBUGS programs used in this paper are available on request from the first author.

Acknowledgments The authors would like to thank R. Pradel for very stimulating and helpful discussions. The paper was improved as a result of comments from two referees. O. Gimenez's research was supported by a Marie-Curie Intra-European Fellowship within the Sixth European Community Framework Programme.

References

Bayarri MJ, Berger JO (2004) The interplay of Bayesian and frequentist analysis. Statistical Science 19:58–80.

Brooks SP, Catchpole EA, Morgan BJT (2000a) Bayesian animal survival estimation. Statistical Science 15:357–376.

Brooks SP, Catchpole EA, Morgan BJT, Barry SC (2000b) On the Bayesian analysis of ring-recovery data. Biometrics 56:951–956.

Brooks SP, Catchpole EA, Morgan BJT, Harris MP (2002) Bayesian methods for analysing ringing data. Journal of Applied Statistics 29:187–206.

Brownie C, Anderson D, Burnham KP, Robson DS (1985) Statistical Inference from Band Recovery Data a Handbook. Second Edition. U. S. Fish and Wildlife Service, Resource Publication 156. Washington, D. C., USA. 305pp.

Carlin BP, Louis TA (1996) Bayes and Empirical Bayes Methods for Data Analysis. Chapman & Hall/CRC, London.

Catchpole EA, Kgosi PM, Morgan BJT (2001) On the near-singularity of models for animal recovery data. Biometrics 57:720–726.

Catchpole EA, Morgan BJT (1997) Detecting parameter redundancy. Biometrika 84:187–196.

Catchpole EA, Morgan BJT, Freeman SN (1998) Estimation in parameter-redundant models. Biometrika 85:462–468.

Clark JS (2005) Why environmental scientists are becoming Bayesians. Ecology Letters 8:2–14.

Ellison AM (2004) Bayesian inference in ecology. Ecology Letters 7:509–520.

Formann AK (2003) Latent class model diagnosis from a frequentist point of view. Biometrics 59:189–196.

Garrett ES, Zeger SL (2000) Latent class model diagnosis. Biometrics 56:1055–1067.

Gelfand AE, Sahu K (1999) Identifiability, improper priors, and Gibbs sampling for generalized linear models. Journal of the American Statistical Association 94:247–253.

Gelman A (1996) Inference and monitoring convergence. In Gilks WR, Richardson S, Spiegelhalter DJ, editors, Markov Chain Monte Carlo in Practice., pages 131–143. Chapman & Hall, London.

Gimenez O, Bonner S, King R, Brooks SP, Jamieson LE, Grosbois V, Morgan BJT (2008) WinBUGS for population ecologists: Bayesian modeling using Markov chain Monte Carlo methods. In: Thomson DL, Cooch EG, Conroy MJ (eds.) Modeling Demographic Processes in Marked Populations. Environmental and Ecological Statistics, Springer, New York, 3:883–916.

Gimenez O, Brooks SP, Morgan BJT (2006) Assessing weak identifiability in mark-recapture and mark-recovery models. Technical Report UKC/IMS/06/030, University of Kent.

Gimenez O, Choquet R, Lebreton JD (2003) Parameter redundancy in multistate capture-recapture models. Biometrical Journal 45:704–722.

Gimenez O, Viallefont A, Choquet R, Catchpole EA, Morgan BJT (2005) Methods for investigating parameter redundancy. Animal Biodiversity and Conservation 27:561–572.

Jiang H, Pollock KH, Brownie C, Hightower JE, Hoeing JM, Hearn WS, (2007) Age-dependent tag return models for estimating fishing mortality, natural mortality, and selectivity. Journal of Agricultural, Biological, and Environmental Statistics 12:177–194.

Kéry M, Gregg KB, Schaub M (2005) Demographic estimation methods for plants with unobservable life-states. Oikos 108:307–320.

Lunn DJ, Thomas A, Best N, Spiegelhalter D (2000) WinBUGS – a Bayesian modelling framework: Concepts, structure, and extensibility. Statistics and Computing 10:325–337.

Nasution MD, Brownie C, Pollock KH, Powell RA (2004) The effect on model identifiability of allowing different relocation rates for live and dead animals in the combined analysis of telemetry and recapture data. Journal of Agricultural Biological and Environmental Statistics 9:27–41.

Neath AA, Samaniego FJ (1997) On the efficacy of Bayesian inference for nonidentifiable models. American Statistician 51:225–232.

Pradel R, Maurin-Bernier L, Gimenez O, Genovart M, Choquet R, Oro D (2008) Estimation of sex-specific survival with imperfect clues about sex. Canadian Journal of Statistics 36:29–42.

R Development Core Team (2008) R: A Language and Environment for Statistical Computing. Vienna, Austria. ISBN 3-900051-07-0.

Rannala B (2002) Identifiability of parameters in MCMC Bayesian inference of phylogeny. Systematic Biology 51:754–760.

Schaub M, Gimenez O, Schmidt BR, Pradel R (2004) Estimating survival and temporary emigration in the multistate capture-recapture framework. Ecology 85:2107–2113.

Schmid F, Schmidt A (2006) Nonparametric estimation of the coefficient of overlapping – theory and empirical application. Computational Statistics and Data Analysis 50:1583–1596.

Silverman BW (1986) Density Estimation for Statistics and Data Analysis. Chapman & Hall, London.

Sturtz S, Ligges U, Gelman A (2005) R2WinBUGS: A package for running WinBUGS from R. Journal of Statistical Software 12:1–16.

Viallefont A, Lebreton JD, Reboulet AM, Gory G (1998) Parameter identifiability and model selection in capture-recapture models: a numerical approach. Biometrical Journal 40:313–325.

Estimating N: A Robust Approach to Capture Heterogeneity

Byron J.T. Morgan and Martin S. Ridout

Abstract We evaluate the performance of a new mixture model for heterogeneity in capture probability when estimating the size of a closed population of wild animals. The new model expresses the capture probability as a mixture of a binomial distribution and a beta-binomial distribution. For real data sets, it is shown how the new model can provide a suitable framework for model discrimination. When there is no best model from within the family of models represented by the new mixture, we recommend adopting a conservative approach to estimating population size.

Keywords Binomial Model · Beta-Binomial Model · Closed Population Size · Logistic Normal Distribution · Mixture Model · Non-identifiability

1 The Problem of Model Identification

We fit models to data so that we can use them for activities such as summarising, extrapolation and prediction. Models are only simplifications of reality, and extrapolation can depend crucially upon the particular model chosen. This is a well-known problem; for instance in the area of estimation of virtually safe doses in bioassay, usually a range of models is considered, in order to be able to judge the effect of the model on extrapolation.

The important paper by Link (2003) showed that different models for heterogeneity of capture probability in the estimation of the size of closed populations can fit the observed data equally well, yet differ in the prediction of both population size and its precision. Additional discussion is given in Link (2006).

Various approaches may be adopted in this area; for example, one may

- regard the problem as insurmountable, and refuse to provide any inferences at all, other than use the data to provide a lower bound to the population size.

B.J.T. Morgan (✉)
Institute of Mathematics, Statistics and Actuarial Science, University of Kent, Canterbury, Kent CT2 7NF, UK
e-mail: B.J.T.Morgan@kent.ac.uk

D.L. Thomson et al. (eds.), *Modeling Demographic Processes in Marked Populations*, Environmental and Ecological Statistics 3, DOI 10.1007/978-0-387-78151-8_49,

- ignore the problem, and pretend that it does not exist. Papers are continuing to be written that adopt this strategy.
- attempt to model the heterogeneity, using covariate information on observed individuals, in the hope that the covariates will account for the heterogeneity. This approach may work to some extent, but the association between covariates and capture probability may be poor. Thus this is not a general solution to the problem.
- try to select a model which is thought to correspond to the real-life situation under study. Although this has been proposed, it is typically difficult, and an incorrect choice of model can lead to substantial bias in the estimator of N. It is safer, and more robust, to use the approach of the next item.
- base inferences on an extended model, in order to produce a more robust and conservative approach.

It the last of these approaches that is adopted by Morgan and Ridout (2008), henceforth referenced as MR. In this paper we describe their approach, and summarise its performance on a number of real data sets. We then show how the approach performs for the examples of Link.

2 A Strategy Based on a New Mixture Model

First we give the mathematical formulation of a range of possible models for estimating the size of a closed population with unknown size N. It is assumed that on each of k occasions, a closed population of wild animals is sampled at random and those animals in the sample that have not been marked previously are uniquely marked and returned to the population, while those that have been marked previously are recorded. If we denote the population size by N, and the remaining model parameters by the vector $\boldsymbol{\eta}$, then, omitting constants of proportionality, we obtain the general form of the likelihood as

$$L(N, \boldsymbol{\eta}) \propto \frac{N!}{(N-D)!} \prod_{j=0}^{k} p_j^{f_j}.$$

Here, f_j denotes the number of animals that have been captured j times, for $j \geq 0$ and D is the number of distinct animals caught, given by $D = \sum_{j=1}^{k} f_j$, so that $f_0 = N - D$. The probability p_j denotes the probability that an animal is caught j times out of the k occasions; the $\{p_j\}$ are determined by the model, and in this paper we shall consider the following five possibilities.

- A mixture of A binomial distributions, so that we have

$$p_j \propto \left\{ \sum_{a=1}^{A} \gamma_a \phi_a^j (1-\phi_a)^{k-j} \right\}.$$

In this paper, we consider only the cases of $A = 1$ and $A = 2$. When $A = 1$ we have a single binomial, corresponding to homogeneity of capture probability over animals. We refer to this as the *Bin* model; it corresponds to the model M_0 of Otis et al. (1978). When $A = 2$ we set $\gamma_1 = \gamma$ and $\gamma_2 = 1 - \gamma$; the model abbreviation used in this case is *2 Bins*. Finite mixture models for describing capture heterogeneity were suggested by Norris and Pollock (1996), and have been substantially developed by Pledger (2000, 2005).

- A beta-binomial model. Here

$$p_j \propto \frac{\prod_{r=0}^{j-1}(\mu + r\theta)\prod_{r=0}^{k-j-1}(1 - \mu + r\theta)}{\prod_{r=0}^{k-1}(1 + r\theta)} \equiv p_j^{Be},$$

say.
The model abbreviation used here is *Beta-bin*. This model was investigated by Burnham (1972), and has been considered further in some detail by Dorazio and Royle (2003, 2005).

- A mixture of a binomial and a beta-binomial, so that

$$p_j \propto \left\{\gamma\phi^j(1-\phi)^{k-j} + (1-\gamma)p_j^{Be}\right\},$$

where ϕ is the binomial probability of capture. This is the new mixture model of MR, and it contains the three models above as special cases. In this instance, the model abbreviation used is $Bin + Beta\text{-}bin$.

- The logistic-normal binomial model of Coull and Agresti (1999). Here

$$p_j \propto \phi^j(1-\phi)^{k-j},$$

where $\phi = \int_{-\infty}^{\infty} \frac{1}{(1+e^{-x})} f(x)dx$, and $f(x)$ denotes the normal $N(\mu, \sigma^2)$ density. The model abbreviation used here is *LNB*.

3 Examples

3.1 Real Data

We start by presenting five real data sets, extracted from a larger set of fifteen given by MR. The data are displayed in Table 1; the first four examples involve wild animals, and the last refers to the number of taxicabs in Edinburgh. MR give the source details of all of these examples. The last column of Table 1 indicates the model selected using an appropriate likelihood-ratio test at the 5% level. The only models considered in these comparisons were the sub-models of the $Bin{+}Beta\text{-}bin$ model.

This subset of real data has been selected because the examples provide instances of when each of four models is selected as the best single model for the data set, based on likelihood-ratio tests of nested models, and also one example (pocket mice) where no single model is selected as the best model for the data. This can

Table 1 The $\{f_j\}$ for 5 real data sets. For the Wood mice example we have $f_j = 0$ for $j > 18$. The entries under the "Model" column are explained in the text

Data	j 1	2	3	4	5	6	7	8	9	10	11	12	13	14	15	16	17	18	k	Model
Skinks	56	19	28	18	24	14	9	–	–	–	–	–	–	–	–	–	–	–	7	*Bin + Beta-bin*
House mice	2	64	40	31	16	13	5	1	0	1	–	–	–	–	–	–	–	–	10	2 *Bins*
Pocket mice	16	15	6	5	5	5	3	–	–	–	–	–	–	–	–	–	–	–	7	–
Wood mice	71	59	41	39	20	26	19	12	9	5	8	4	9	2	1	3	3	3	21	*Beta-bin*
Taxicabs	142	81	49	7	3	1	0	0	0	0	–	–	–	–	–	–	–	–	10	*Bin*

Table 2 Values of $-\ell_{max}$ for 5 illustrative data sets. Shown in bold face are the values corresponding to selected models, when a single model is selected for the data

Data	Binomial	Beta-bin	2 Bins	Bin + Beta-bin
Skinks	86.71	22.35	23.04	**18.57**
House mice	44.43	43.55	**39.54**	39.47
Pocket mice	33.15	14.04	12.37	12.33
Wood mice	357.27	**47.45**	87.73	45.34
Taxicabs	**16.95**	16.44	16.34	16.34

be appreciated from the maximised log-likelihood values of Table 2. Note that the binomial model is nested within the beta-binomial model, and that all models are nested within the new mixture model. It is interesting to observe for both the house mice and wood mice examples that there is a clear preference for a particular sub-model of the $Bin + Beta\text{-}bin$ model, and we do not know of such a comparison having been made previously. In order to compare the beta-binomial model to the new mixture model the appropriate chi-square distribution to use has two degrees of freedom. This is because we obtain the beta-binomial model when $\gamma = 0$, at a boundary to the parameter-space. Brooks et al. (1997) and MR investigated the appropriate asymptotic chi-square distribution for this case. When a single model is selected for inference, then we may use that model to estimate N and its precision from an appropriate profile confidence interval. If no model is selected then we recommend the conservative approach of only presenting an interval estimate, taking as the lower end of the interval the smallest lower interval value from the models considered, and taking as the upper end the largest corresponding upper interval values. Typically this interval will be wide, which is appropriate when one cannot discriminate between alternative possible models. An alternative strategy is provided in Section 4.

3.2 Link's Three Examples

Link has presented a number of examples to demonstrate the problem of non-identifiability. We are aware of three such examples, which we believe are all artificial, although that does not detract from their relevance. We shall now examine how our approach performs for these three data sets. The relevant maximised log-likelihood values are given in Table 3.

Table 3 Values of $-\ell_{\max}$ for Link's three data sets. For data set 1 there are not enough observations to allow the Bin + Beta-bin model to be fitted

Data Set	Binomial	Beta-bin	2 Bins	Bin + Beta-bin
1	23.85	10.84	10.48	–
2	52.65	15.00	15.81	14.55
3	878.07	47.01	105.89	45.74

$$data\ set\ 1: \{f_j, j \geq 1\} = [84, 54, 36, 21].$$

This example was communicated directly to us by Link, and like the second example below, was presented as proportions. We have therefore scaled these to provide realistic data sets for analysis. For this example there are not enough observations to allow us to fit the $Bin + Beta\text{-}bin$ model.

Here a lower bound for N is provided by $D = 195$. The binomial model is clearly excluded, but the other two models are indistinguishable. Thus we would produce a 95% interval for N of $(221, \infty)$, from applying the conservative approach above to the confidence intervals of the two indistinguishable models.

Referees have queried whether one can obtain an upper confidence limit $N = \infty$, and a heuristic demonstration of how this can occur is provided in the Appendix.

$$data\ set\ 2: \{f_j, j \geq 1\} = [32, 20, 14, 11, 9, 8, 6].$$

The situation here is identical to that above, and the resulting interval is $(104.7, \infty)$, whereas the lower bound from taking the sum of the f_j is 100. Thus in this instance the modelling exercise barely improves upon the lower bound for N that arises directly from the data.

For both of these data sets, when a *Beta-bin* model is fitted to the data, then the corresponding beta density rises to infinity at zero. It has been shown (Link 2006) that for the second data set, the data may be fitted equally well by the *LNB* model, but that the two models provide different estimates of N. That is not surprising, given that the *LNB* density is zero at the origin. Exactly the same issue arises with the last of the three examples of this section, which we now consider.

The last example is taken from Link (2003), and is substantially larger than one would expect to encounter in practice.

$$data\ set\ 3: \{f_j, j \geq 1\} = [679, 531, 379, 272, 198, 143, 99, 67, 46, 32, 22, 14, 9, 5, 3, 1, 0, 0, 0, 0].$$

In this case we have $D = 2500$, and out of the nested set of models that we consider in this paper, the best model for the data is the beta-binomial model, with $\hat{N} = 3494$ and with 95% profile confidence interval of (3308, 3730). For this example, Link writes that "N ...is for all practical purposes not identifiable," the reason being that the *LNB* model produces an estimate of 3111, with 95% profile confidence interval of (3018, 3218), which does not overlap with the interval given above for the *Beta-bin* model. The discrepancy arises because of the different way in which heterogeneity of capture probability is handled near the origin, as we can see from Fig. 1. The *LNB* model, although having the same number of parameters as the *Beta-bin* model, has a density that is always zero at the origin, and that is a severe restriction. For this data set, the $Bin + Beta\text{-}bin$ model acts as a compromise between the *Beta-bin* and the *LNB* models. The beta-binomial component is selected 98.9% of the time, and the rest of the time a binomial model applies, with

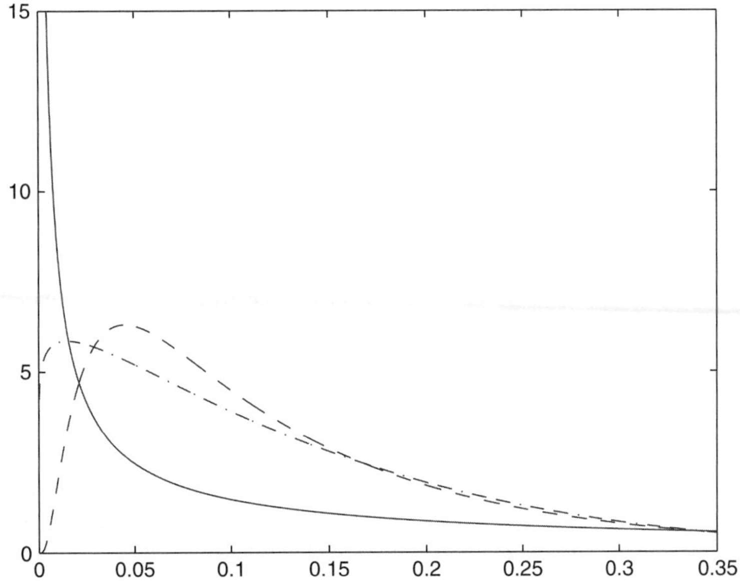

Fig. 1 A comparison of model fits for the third Link data set. Here we show how three models describe the capture probability near zero. Dashed line: *LNB* model; solid line: *Beta-bin* model; dashed-dotted line: the beta component from the $Bin + Beta\text{-}bin$ model

a probability of $p = 0.54$. In terms of likelihood-ratio tests, the $Bin + Beta\text{-}bin$ model is not in this example a significant improvement over the *Beta-bin* model, with a log-likelihood difference of only 1.27. The $Bin + Beta\text{-}bin$ model estimates $N = 3372$, with a 95% confidence interval of $(2966, \infty)$.

4 Discussion

We have presented the new mixture model of MR, and illustrated its behaviour on a range of real and artificial examples. We note that when the *Beta-bin* model is fitted to *all three* of the problem data sets of Link then the fitted beta distribution has an infinite mode at zero. As far as we know, this point has not been made previously.

We speculate that major discrepancies between estimates of N from the *Beta-bin* and *LNB* models will not arise if the beta density corresponding to a fitted beta-binomial distribution also has zero density at the origin.

We have compared nested models using likelihood-ratio tests, and we suggest also that the $Bin + Beta\text{-}bin$ model provides a robust approach if one is uncertain about the distribution of heterogeneity that is appropriate for any particular problem. Alternatively, if one compares models in terms of AICs, as was done by Link (2003) for the last of the three examples in the last section, then the *LNB*, *Beta-bin* and $Bin + Beta\text{-}bin$ models will be indistinguishable. The lack of overlap of the confidence intervals for the *Beta-bin* and *LNB* models for that example can then be placed in the context of the fits of all three models, as the interval for the *Bin* + *Beta-bin* model overlaps with both of the other two intervals, and here too a conservative approach could be adopted.

We see from the illustration of Fig. 1 that both of the two-parameter models illustrated there allow the possibility of radically different behaviour of the probability density function of the capture probability near the origin, with quite different estimates of N resulting, as we have seen. The context of the problem might exclude a particular form of probability density function for the capture probability, though as observed earlier this will be unlikely in general.

An alternative approach would be always to fit the *Bin* + *Beta-bin* model, and base inferences on that model. One would expect that always using a five-parameter model would result in conservative inference, and this is currently under consideration. Further research is also taking place in this area, notably with regard to mixing the *LNB* distribution with a binomial distribution, and considering absolute, rather than just relative, measures of goodness-of-fit.

A referee has observed that producing a confidence interval for population size with an infinite upper bound has no direct relevance to conservation, perhaps when one is surveying an endangered population. Such an interval provides an expression of uncertainty, based on the data available.

The important message of this paper is that informed use of the *Bin* + *Beta-bin* model allows conservative inference to take place in the area of estimating the size of closed populations of wild animals, in the presence of heterogeneity of capture probability.

The computer programs used to fit this model and its sub-models were written in R (R Development Core Team 2008) by MSR, and are available from http://www.kent.ac.uk/ims/personal/msr/estimateN.html. Multiple starts were used to reduce problems with local optima.

Acknowledgments We thank the referees for their comments which improved the paper, and in particular resulted in the Appendix.

Appendix: The Behaviour of the Profile Log-Likelihood for N as $N \to \infty$, for the Beta-Binomial Model

The likelihood can conveniently be written in the factorised form

$$L(N,\eta) \propto \left\{ \frac{N!}{(N-D)!} p_0^{(N-D)} (1-p_0)^D \right\} \times \left\{ \prod_{j=1}^{k} \left(\frac{p_j}{1-p_0} \right)^{f_j} \right\},$$

where the p_j's are the beta-binomial probabilities, dependent on parameters μ and ϕ. The second term is proportional to the conditional likelihood for individuals seen at least once.

As N increases, we expect p_0 to decrease in such a way that

$$N(1 - p_0) \approx D, \tag{1}$$

where D is the number of individuals observed at least once.

Provided $N - D$ is not too small, we have the following approximation, which is based on Stirling's approximation

$$\log\left(\frac{N!}{(N - D)!}\right) \approx N\log(N) - (N - D)\log(N - D) - D.$$

Then some simple algebra shows that the logarithm of the first term of the factorised likelihood, assuming that p_0 is given by equation (1), is approximately independent of N and given by

$$D(\log D - 1).$$

Thus, for large N, and in particular in the limit as $N \to \infty$, the value of the likelihood depends only on the second (conditional) term.

As $N \to \infty$, we expect the conditional likelihood also to approach a constant. Indeed we can anticipate the approximate value of the parameter μ by equating the conditional mean number of times an individual is seen to its expectation. This gives

$$\frac{\sum_{j=1}^{k} j f_j}{D} \approx \frac{k\mu}{1 - p_0} \approx \frac{Nk\mu}{D}$$

or

$$\mu = \frac{1}{Nk}\sum_{j=1}^{k} j f_j. \tag{2}$$

Examples

We now provide two contrasting illustrations from the data sets of Table 1, viz., pocket mice and house mice.

Pocket Mice

Here $D = 55$ and the conditional mean number of times an individual is seen is $32/11 = 2.909$. Table 4 shows, for different values of N, the contributions to the log-likelihood from the first and second terms in the factorised likelihood (denoted $\log L_1$ and $\log L_2$), the overall log-likelihood (denoted by $\log L$) and also the maximum-likelihood estimate of μ, conditional on N and the corresponding value

Table 4 The behaviour of the profile log-likelihood for N: numerical results for pocket mice

N	$\log L_1$	$\log L_2$	$\log L$	$\hat{\mu}$	eqn(2)	$\hat{\theta}$
55	162.9451	−244.1642	−81.2190	0.4226	0.4156	0.7320
62.4764	165.8720	−241.3921	−75.5200	0.3728	0.3659	0.2990
98.2389	165.8110	−239.4072	−73.5962	0.2347	0.2327	0.5939
200	165.5634	−239.4919	−74.0886	0.1134	0.1143	0.8675
2000	165.4173	−239.8120	−74.3947	0.0112	0.0114	1.1048
20000	165.4047	−239.8465	−74.4418	0.0011	0.0011	1.1284

from equation (2). The numerical values were obtained using Maple, with 15 digits of precision.

The second and third rows in the table correspond to the (approximate) lower confidence limit for N and the maximum likelihood estimate of N respectively. All the calculations here suggest that the numerical values are stabilising as $N \to \infty$. In particular, note that the approximation

$$\log L_1 \approx D(\log D - 1) = 165.4033$$

is not only appropriate as $N \to \infty$ but is quite accurate except when N is close to D. Similarly, the approximate value of $\hat{\mu}$ given by equation (2) is reasonably accurate for all values of N considered.

It therefore appears that in this example, the upper 95% profile confidence limit for N is infinite.

House Mice

Here $D = 173$ and the conditional mean number of times an individual is seen is $585/173 = 3.3815$. Table 5 is structured and derived in the same way as Table 4.

The first 3 rows in the table correspond to the (approximate) lower confidence limit for N and the maximum-likelihood estimate of N and the approximate upper confidence limit for N respectively. All the calculations here suggest that the numerical values are stabilising as $N \to \infty$, though compared with Table 4, one requires somewhat larger values of N before $\log L_2$ stabilises. In particular, note that the approximation

Table 5 The behaviour of the profile log-likelihood for N: numerical results for house mice

N	$\log L_1$	$\log L_2$	$\log L$	$\hat{\mu}$	eqn(2)	$\hat{\theta}$
173.1590	718.9907	−1103.394	−384.4040	0.3379	0.3378	0.0069
176.9660	720.3686	−1102.853	−382.4841	0.3307	0.3306	0.0170
183.6701	719.1309	−1103.524	−384.3930	0.3187	0.3185	0.0353
250	715.7774	−1128.042	−409.5225	0.2306	0.2340	0.2073
2500	718.5412	−1166.812	−448.2921	0.0210	0.0234	0.7095
25000	718.5228	−1168.815	−450.2961	0.0021	0.0023	0.7568
250000	718.5197	−1169.006	−450.4866	0.0002	0.0002	0.7614
2500000	718.5195	−1169.025	−450.5056	0.00002	0.00002	0.7619

$$\log L_1 \approx D(\log D - 1) = 718.5914$$

is again not only appropriate as $N \to \infty$ but is very accurate for $N \geq 2500$ and the approximate value of $\hat{\mu}$ given by equation (2) is reasonably accurate for all values of N considered. Thus, in this example too, the profile log-likelihood tends to a limit as $N \to \infty$. However, in this case the limit is such that a finite upper 95% (or 99% or 99.9%) confidence limit exists for N. The relatively narrow confidence interval for N in this example occurs because almost all animals that are seen are seen at least twice, so that the estimated probability that an animal is unseen is necessarily small.

Summary

As is well known, the likelihood for this problem can be factored into two components, one of which is the conditional likelihood for the individuals that have been observed.

In terms of profile likelihoods for N, this conditional likelihood is expected to tend to a constant as $N \to \infty$. The other term is approximately constant, at least when N is large. Thus the overall log-likelihood tends to a constant as $N \to \infty$. Depending on the value of this constant in relation to the log-likelihood at the overall maximum-likelihood estimate, this may lead to an infinite upper confidence limit for N (and it may also result in the lower confidence limit being the observed number of individuals, D).

The basic arguments here are that

i. For large N, one expects the expected number of distinct individuals seen to be similar to the observed number. This implies that the first term in the factorised likelihood is approximately independent of N.
ii. For large N, the conditional log-likelihood is expected to be roughly independent of the choice of N.

These arguments are not specific to the beta-binomial model and one might expect similar conclusions to hold for other models of heterogeneity of capture probability.

References

Brooks SP, Morgan BJT, Ridout MS, Pack SE (1997) Finite mixture models for proportions. Biometrics 53:1097–1115.

Burnham KP (1972) Estimation of population size in multiple capture-recapture studies when capture probabilities vary among animals. PhD thesis, Oregon State University, Corvallis.

Coull BA, Agresti A (1999) The use of mixed logit models to reflect heterogeneity in capture-recapture studies. Biometrics 55:294–301.

Dorazio RM, Royle JA (2003) Mixture models for estimating the size of a closed population when capture rates vary among individuals. Biometrics 59:351–364.

Dorazio RM, Royle JA (2005) Rejoinder to "The performance of mixture models in heterogeneous closed populations". Biometrics 61:874–876.

Link WA (2003) Nonidentifiability of population size from capture-recapture data with heterogeneous detection probabilities. Biometrics 59:1123–1130.

Link WA (2006) Rejoinder to "On identifiability in capture-recapture models" Biometrics 62: 936–939.

Morgan BJT, Ridout MS (2008) A new mixture model for recapture heterogeneity. To appear Applied Statistics.

Norris JL, Pollock KH (1996) Nonparametric mle under two closed capture-recapture models with heterogeneity. Biometrics 52:639–649.

Otis DL, Burnham KP, White GC, Anderson DR (1978) Statistical inference from capture data on closed animal populations. Wildlife Monographs 62:1–135.

Pledger S (2000) Unified maximum likelihood estimates for closed capture-recapture models using mixtures. Biometrics 56:434–442.

Pledger S (2005) The performance of mixture models in heterogeneous closed population capture-recapture. Biometrics 61:868–873.

R Development Core Team (2008) R: A language and environment for statistical computing. R Foundation for Statistical Computing, Vienna, Austria. ISBN 3-900051-07-0, URL http://www.R-project.org

Evaluation of Bias, Precision and Accuracy of Mortality Cause Proportion Estimators from Ring Recovery Data

Michael Schaub

Abstract Knowledge about proportions of specific mortality causes is important for the design of efficient conservation measures or the determination of harvest regulations. Unfortunately, these proportions are difficult to estimate. We (Schaub and Pradel 2004) have recently introduced a multistate capture-recapture model that allows one to estimate proportions of specific mortality causes from recoveries of dead animals with known cause of death. However, parameter estimation was found to be difficult, because the likelihood surface of the model relative to most parameters has a flat ridge, unless the proportions of mortality causes vary with time and the cause-specific recovery rates are constant. These conditions are likely to be violated in most empirical situations. For the application of this model, it is therefore important to study the sensitivity of parameter estimates to violations of these assumptions. I use a Bayesian implementation of the model to evaluate bias, precision and accuracy of parameter estimates under variable means and temporal variation of mortality cause proportions and recovery rates. Survival rate estimates were unbiased in all scenarios. Bias and precision of the proportion of mortality causes and of the cause-specific recovery probabilities decreased with increasing temporal variance of the proportion of mortality causes while their accuracy increased. The bias of these estimates also decreased with decreasing difference between cause-specific recovery probabilities and with decreasing temporal variation of them. Moreover, informative priors affected the posterior distribution of the parameters when temporal variation in the proportion of mortality causes was low. Temporal variance of the proportion of mortality causes could be estimated reliably regardless of bias. This result is important, since it allows one to assess whether accuracy of the estimates of mortality proportions is acceptable for the objectives of a study. The bias of the naïve estimator (quotient of the number of animals reported dying from a particular cause to the total number reported altogether) was usually much larger than the bias of the corresponding estimator from the multistate model. In conclusion, a careful

M. Schaub (✉)
Division of Conservation Biology, Zoological Institute, University of Bern, Baltzerstrasse 6, CH-3012 Bern, Switzerland; Swiss Ornithological Institute, CH-6204 Sempach, Switzerland
e-mail: michael.schaub@nat.unibe.ch

D.L. Thomson et al. (eds.), *Modeling Demographic Processes in Marked Populations*, Environmental and Ecological Statistics 3, DOI 10.1007/978-0-387-78151-8_50,

application of the multistate capture-recapture model can give useful information about the proportion of mortality causes that is otherwise hard to obtain.

Keywords Bayesian · Bias · Multistate Model · Proportion of Mortality Causes · Recovery Probability · Simulation

1 Introduction

Overall mortality rates are typically due to a variety of mortality causes such as predation, disease, accidents or human harvest, and each of these may influence population dynamics differentially. If the frequency of these causes changes, overall mortality is likely to change also and thereby population growth rate. Thus, studying the proportion of different sources of mortality is important for understanding proximate causes of population dynamics.

Unless animals are radio-tagged (e.g. Bro et al. 2001; Buner and Schaub 2008), the importance of a particular mortality cause is difficult to assess. This is because the probability of finding an animal that has died due to a specific mortality cause depends on the mortality cause. For example, animals that die from human-related mortality causes such as hunting or accidents with cars or windows are usually much more likely to be found and reported than animals that die from predation, disease or starvation. Therefore, estimates of the proportion of mortality causes based on the ratio of the number of animals reported dead from a particular cause and the total number reported (Newton et al. 1999; Hüppop and Hüppop 2002) are likely to be seriously biased.

Recently, we (Schaub and Pradel 2004) introduced a multistate capture-recapture model which allows estimation of overall survival rates (S) and the proportion (α) of animals dying from a particular mortality cause under question from recovered dead animal whose cause of death is known. In this model, which I term cause-specific mortality model, the probability of finding and reporting an animal that died from a particular cause of death is estimated, allowing to estimate the parameters of interest (S, α) without bias. The parameters of interest are intrinsically identifiable, if α is time-dependent and the cause-specific recovery rates constant across time (Schaub and Lebreton 2004). However, even in intrinsically identifiable models the model likelihood relative to α and the recovery rates has a flat ridge, such that the estimation of these parameters can be difficult (see discussion in Schaub and Pradel 2004). The problem arises because a nested model with fewer parameters, where α is not variable over time, is intrinsically unidentifiable (Schaub and Lebreton 2004). Therefore, the information matrix of the model is near-singular (Catchpole et al. 2001; Catchpole and Morgan 2001) and the model provides only unbiased estimates when α varies across time. However, it is not clear how large the temporal variation of α needs to be to obtain accurate parameter estimates.

A further important issue is that often additional information is available on some of the parameters in the model. For example it is known that the recovery rate of hunted individuals is larger than the recovery rate of individuals dying from natural causes. In this situation, it may well be that parameter estimation from a Bayesian analysis is less delicate owing to the inclusion of such additional information via the prior distributions or order constraints. I therefore evaluated how much the estimates can be improved through the use of additional information.

Another challenge is that the parameters (α, recovery rates) are not separately estimable when the recovery probabilities are time-dependent. One must therefore fit a model with constant recovery probabilities, yet is it not clear how strongly α and the recovery rates will be biased when there is such variation.

The aim of the paper is to enhance the understanding of the cause-specific mortality model to facilitate its application. I developed a Bayesian implementation of the model and used simulation to explore the conditions when the model provides reasonably accurate parameter estimates. The sensitivity of the parameter estimates under the model is explored along several dimensions, (i) magnitude of temporal variance of α, (ii) magnitude of temporal variance of the recovery rates, (iii) magnitude of cause-specific recovery rates, (iv) sample size (numbers of data years), and (v) use of additional information. I also studied bias and precision of the naïve estimator (quotient of the number of animals reported dying from a particular cause to the total number reported altogether), to evaluate how much the estimates from the cause-specific mortality model improve over the naïve estimates. Based on the results I propose guidelines for the practical application of the model.

2 Material and Methods

2.1 The Cause Specific Mortality Model

The data required for the cause-specific mortality model (Schaub and Pradel 2004) are capture-mark-recovery data where the cause of death of each recovered animal is known without error. Such data is frequently available, for example in the database of the European ringing schemes. The various causes of death observed among recovered individuals are then allocated to the two groups A and B, where group A refers to the mortality cause one is particularly interested in (e.g. hunting) and B refers to all other mortality causes (not A). A multistate capture-recapture history is then constructed for each individual. For example, capture history 010A0 denotes an individual that was marked in the second year and died from cause A and was recovered in year four.

To obtain estimates of α, survival and the recovery rates, I used a multistate capture-recapture model with the three states "alive", "died due to cause A", and "died due to cause B". The model is presented here with a matrix of transition probabilities and a state-specific vector of "sighting" probabilities. Note that states are from top to bottom (states of arrival) and from left to right (states of departure) in the order as indicated above:

$$\begin{bmatrix} S_t & (1-S_t)\alpha_t & (1-S_t)(1-\alpha_t) \\ 0 & 1 & 0 \\ 0 & 0 & 1 \end{bmatrix} \begin{bmatrix} 0 \\ \lambda_{A,t} \\ \lambda_{B,t} \end{bmatrix}.$$

Here, S_t is the probability that an individual survives from time t to time $t + 1$; α_t is the probability that the mortality cause of an individual is A if it dies between t and $t + 1$; and $\lambda_{A,t}$ and $\lambda_{B,t}$ are the probabilities that an individual dying from cause A or B, respectively, between t and $t + 1$ is found and reported. A more detailed description of the model is presented in Schaub and Pradel (2004) and in Schaub and Lebreton (2004).

While we used a frequentist approach in Schaub and Pradel (2004) and Schaub and Lebreton (2004), here I apply a Bayesian analysis. The likelihood is formed from the products of multinomial distributions whose cell probabilities are functions of S, α, λ_A and λ_B (see also Brooks et al. (2000) for an equivalent formulation of a classic ring-recovery model). I used several different priors (see below more details) and Markov chain Monte Carlo (MCMC) methods to sample from the posterior distribution.

2.2 Intrinsic Identifiability of the Model

Evaluation of the model using formal methods (Catchpole and Morgan 1997, Catchpole et al. 2002, Gimenez et al. 2003) has shown that only models in which α is time-dependent and the recovery probabilities are constant across time, i.e. models [S, α_t, λ_A, λ_B] and [S_t, α_t, λ_A, λ_B], are intrinsically identifiable. Other models, where either all parameters are constant across time or where one or both recovery probabilities are time-dependent, are not intrinsically identifiable (Schaub and Lebreton 2004). The fact that model [S, α, λ_A, λ_B] is not fully identifiable (i.e. only the quantities S, $\alpha^*\lambda_A$, and $(1 - \alpha)^*\lambda_B$ are separately estimable) can pose problems for the parameter estimation also under the two identifiable models: it is the time variation of α which renders α, λ_A and λ_B separately estimable. One aim of my study was therefore to understand how large the temporal variation of α needs to be in order to get useful estimates of S, α, λ_A, and λ_B.

2.3 Simulation Methods

To evaluate estimator performance I conducted a simulation study with different scenarios with 500 generated data sets each. Generation of a data set first required selection of those parameters values (θ) that were variable across time. I chose the beta distribution to model temporal variation in a parameter. For time interval t, a θ_t was taken from a beta distribution with mean $\bar{\theta}$ and variance σ_θ^2. The two parameters a and b of the beta distribution were calculated as $a = \bar{\theta}(\bar{\theta}(1-\bar{\theta})/\sigma_\theta^2 - 1)$, and $b = (1-\bar{\theta})(\bar{\theta}(1-\bar{\theta})/\sigma_\theta^2 - 1)$, respectively. Considering a specific number of study years, then I assumed that 1000 individuals were newly marked in each year and created a multistate m-array (Burnham et al. 1987) using multinomial distributions. Parameter estimates were then obtained from analysis of the m-array under the cause-specific mortality model described above.

In a first set of simulations, I studied the effect of different means of α and the two recovery probabilities on the parameter estimates. I generated data from a model with constant survival and recovery probabilities and with time-dependent mortality cause probabilities [S, α_t, λ_A, λ_B]. I considered five different mean values of α ($\bar{\alpha} = \{0.1, 0.35, 0.5, 0.65, 0.9\}$), seven different temporal variances of α ($\sigma_\alpha^2 = \{0.0001, 0.0005, 0.001, 0.005, 0.01, 0.05, 0.1\}$) and two different options of recovery probabilities. In the first option the values of the two recovery probabilities were closer ($\lambda_A = 0.2$, $\lambda_B = 0.1$) than in the second option ($\lambda_A = 0.25$, $\lambda_B = 0.05$). I considered 10 and 30 study years, whereas the mean value for survival ($S = 0.4$) remained the same in all scenarios. These values appeared to cover realistic levels of variation to be expected in practical applications of the model. Non-informative (uniform) $\beta(1,1)$ priors were considered for all parameters and all scenarios. In total I considered 140 different scenarios in this set.

In a second set of simulations, I studied the impact of the inclusion of additional information on the parameter estimates. I generated data under a model with constant survival and recovery probabilities and with time-dependent mortality cause probabilities [S, α_t, λ_A, λ_B]. The mean values of survival ($S = 0.4$), the proportion dying ($\bar{\alpha} = 0.35$) and the number of study years (10) remained the same in all scenarios. I evaluated different magnitudes of temporal variances of α ($\sigma_\alpha^2 = \{0.0001, 0.0005, 0.001, 0.005, 0.01, 0.05, 0.1\}$), two different options of recovery probabilities ($\lambda_A = 0.2$, $\lambda_B = 0.1$, and $\lambda_A = 0.25$, $\lambda_B = 0.05$, respectively) and the inclusion of additional information. For the latter I used four possibilities. First, I assumed that no additional information is available and used non-informative $\beta(1,1)$ priors for all parameters. Second, I used a $\beta(1,1.857)$ prior for α and non-informative $\beta(1,1)$ priors for the other parameters. The $\beta(1,1.857)$ distribution is an almost linearly decreasing distribution with mean 0.35. This is exactly the same as the mean value of $\bar{\alpha}$ used to simulate the data. This prior distribution was chosen in order to explore the potential gain in parameter accuracy if the mean of α were known. Third, I used a uniform prior for α within the interval $\{0, 0.5\}$, and non-informative $\beta(1,1)$ priors for the other parameters. This choice was motivated because in practice it may be possible to have an idea about likely magnitude of α. Fourth, additional structural information was incorporated into the model in form of an order constraint on the recovery probability. Typically, it will be known which mortality cause is associated with a higher recovery probability, which can be translated into an order constraint such as $\lambda_A > \lambda_B$ (specifically I used $\beta(1,1)$ priors for λ_B and Δ, and calculated $\lambda_A = \lambda_B + \Delta$). For example, in hunted species it may be sensible to assume that the recovery probability associated with a mortality cause related to human activity is larger than the recovery probabilities associated to other mortality causes (e.g. "natural" mortality). Non-informative $\beta(1,1)$ priors were used for α and S. In total I considered 56 different scenarios in this set of simulations.

In a third set of simulations, I assessed the impact of the temporal variance of the recovery probabilities on the parameter estimates. Consequently, I constructed the data under model [S, α_t, $\lambda_{A,t}$, $\lambda_{B,t}$]. As before, the mean parameter values (S $= 0.4$, $\bar{\alpha} = 0.35$, $\bar{\lambda}_A = 0.2$, $\bar{\lambda}_B = 0.1$) and the number of study years (10) were constant in all simulations. I considered the effects of the temporal variance of α ($\sigma^2_\alpha = \{0.0001, 0.0005, 0.001, 0.005, 0.01, 0.05, 0.1\}$), two common levels of temporal variance for both recovery probabilities ($\sigma^2_\lambda = 0.001, 0.01$), and two different prior distributions for α (non-informative: $\beta(1,1)$, informative: $\beta(1,1.857)$). In total 28 scenarios were considered for this set.

2.4 Data Analyses

All data sets were analyzed with an intrinsically identifiable model [S, α_t, λ_A, λ_B]. In addition the naïve estimate of the proportion dying due to cause A (η_t: quotient of the number of recoveries associated with cause A in year t and the total number of recoveries in year t) was calculated for each data set. For each parameter θ I calculated the mean bias as,

$$B(\hat{\theta}) = \frac{1}{500} \sum_{sim=1}^{500} \frac{1}{T} \sum_{t=1}^{T} \left(\theta_{sim,t} - \hat{\theta}_{sim,t}\right),$$

i.e., as the mean over the 500 simulated data sets and T years of the difference between the generating parameter and its recovered estimate, where $\theta_{sim,t}$ is the parameter value at time t to construct the simulated data and $\hat{\theta}_{sim,t}$ is the estimated parameter at time t. To evaluate the precision of the estimators, I calculated the coefficient of variation as,

$$CV(\hat{\theta}) = 100 \sqrt{\text{var}\left(\frac{1}{T}\sum_{t=1}^{T}\hat{\theta}_{sim,t}\right)} \Big/ \frac{1}{500}\sum_{sim=1}^{500}\frac{1}{T}\sum_{t=1}^{T}\hat{\theta}_{sim,t}.$$

Finally, to evaluate the accuracy of the estimators I calculated the mean squared error as,

$$MSE(\hat{\theta}) = \text{var}\left(\frac{1}{T}\sum_{t=1}^{T}\hat{\theta}_{sim,t}\right) + B(\hat{\theta})^2.$$

Low MSE indicate high accuracy of the estimator, high MSE indicate low accuracy. I also calculated the mean variance of the estimated α across time in order to evaluate how well this estimated the true underlying variance σ^2_α.

I used R (R Development Core Team 2004) to simulate the data and analyzed them in WinBUGS (Spiegelhalter et al. 2004) from R using the package R2WinBUGS (Sturtz et al. 2005). Initial trials showed that convergence of the

MCMC chains occurred very quickly (after about 50 iterations) as evidenced by the Brooks–Gelman-Rubin diagnostics (Brooks and Gelman 1998). I used 2000 MCMC samples and conservatively discarded the first 1000 to avoid transient (preconvergence) effects in all simulations.

3 Results

In the first set, I evaluated bias, precision and accuracy of the estimators under different mean values of α, λ_A, λ_B and different temporal variation of α. Generally, bias and precision of $\hat{\alpha}$ decreased and accuracy increased with increasing σ_α^2 (Fig. 1). The bias of $\hat{\alpha}$ was mostly positive, but it could become negative when α was high. When the difference between the two cause-specific mortality rates became larger, bias of $\hat{\alpha}$ increased, but the precision did not change. The pattern of bias of the two recovery probabilities was similar to that of $\hat{\alpha}$: when $\hat{\alpha}$ was strongly biased, at least one the recovery probabilities was biased as well (Table 1). Absolute bias of survival rate estimates was low (< 0.002) in all scenarios. The number of study years only had a marginal effect with slightly lower bias of $\hat{\alpha}$ and the two recovery probabilities with more study years (Table 1). The bias of the naïve estimate $\hat{\eta}$ depended on α, but only slightly on σ_α^2, and it increased with increasing difference of the two recovery rates. Moreover, the bias of $\hat{\eta}$ was smaller when $\hat{\alpha}$ was either high or low compared to when $\hat{\alpha}$ was medium. Bias of $\hat{\eta}$ was much larger than bias of $\hat{\alpha}$, while precision of both were similar in almost all conditions (Fig. 1). Consequently, accuracy of $\hat{\eta}$ was lower than accuracy of $\hat{\alpha}$. Exceptions occurred when σ_α^2 was low and the two recovery rates were close.

In the second set of scenarios I assessed the effect of the inclusion of additional information on the estimator bias, precision and accuracy. As before, bias and precision of $\hat{\alpha}$ decreased strongly with increasing σ_α^2, regardless of whether additional information was considered (Fig. 2). The use of informative priors for α or the order constraint for the recovery probabilities had strong impacts, but their impact decreased with increasing σ_α^2. The bias of $\hat{\alpha}$ was reduced and the accuracy increased when an appropriate prior for α was chosen. For example, the use of a $\beta(1,1.857)$ prior distribution for α, which has the same mean as the values used to simulate α, resulted in considerably reduced bias of $\hat{\alpha}$. The uniform prior for α within the interval $\{0, 0.5\}$ also considerably reduced the bias and increased accuracy of $\hat{\alpha}$, when σ_α^2 was low. However, as σ_α^2 increases, bias of $\hat{\alpha}$ increases as well and accuracy declined. The prior distribution which constrained $\hat{\lambda}_A$ to be higher than $\hat{\lambda}_B$ also resulted in slightly reduced bias of $\hat{\alpha}$ and higher accuracy. However, the impact of this choice of prior on parameter accuracy decreased with increasing difference between means of the two recovery probabilities. Absolute bias of survival rate was again negligible (< 0.002) in all scenarios (results not shown). Bias of the recovery probabilities followed the same pattern as that of $\hat{\alpha}$: when bias of $\hat{\alpha}$ was low, bias of both recovery probabilities was low as well (results not shown).

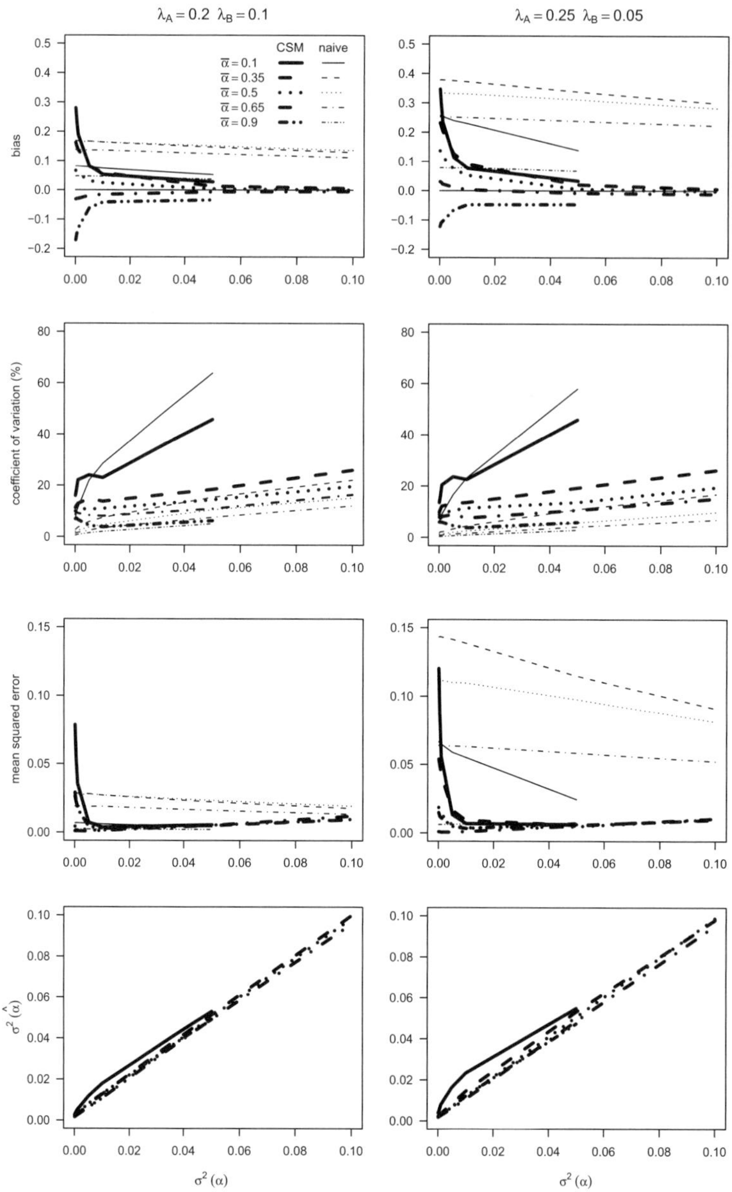

Fig. 1 Bias, coefficient of variation, mean squared error and temporal variance of the proportion of mortality causes estimated from the cause-specific mortality rate model (CSM) and the naïve estimator (naive), respectively, in relation to the simulated temporal variance of α (σ_α^2), the cause-specific recovery probabilities and different levels of $\bar{\alpha}$. Shown are mean values originating from 500 simulations. These simulations were performed assuming a survival rate of 0.4, 10 study years and 1000 newly released individuals in each year. Non-informative β (1,1) priors were used for all parameters and simulations. Some combinations of $\bar{\alpha}$ and $\sigma_\alpha^2 = 0.1$ are not possible, because the beta distribution is not defined

Table 1 Mean bias of survival probability ($\hat{S}$), proportion of mortality cause A ($\hat{\alpha}$), recovery probability associated with cause A ($\hat{\lambda}_A$), recovery probability associated with other causes than A ($\hat{\lambda}_B$) and of the naïve estimate of proportion of mortality cause A ($\hat{\eta}$) from the analysis of 500 generated data sets in relation to different levels of temporal variation of α (σ_α^2), different mean $\bar{\alpha}$, different recovery probabilities (λ_A, λ_B) and number of study years. Non-informative β (1,1) priors were used for all parameters and simulations. Data were generated under model [S, α_t, λ_A, λ_B] and analyzed with the same model

	10 years					30 years				
σ_α^2	$\hat{S}$	$\hat{\alpha}$	$\hat{\lambda}_A$	$\hat{\lambda}_B$	$\hat{\eta}$	$\hat{S}$	$\hat{\alpha}$	$\hat{\lambda}_A$	$\hat{\lambda}_B$	$\hat{\eta}$
$\bar{\alpha} = 0.35,\ \lambda_A = 0.2,\ \lambda_B = 0.1$										
0.0001	0.0007	0.1622	−0.0561	0.0408	0.1689	0.0002	0.1460	−0.0567	0.0308	0.1685
0.005	0.0005	0.0806	−0.0291	0.0175	0.1668	−0.0001	0.0702	−0.0312	0.0128	0.1670
0.1	0.0008	0.0055	−0.0025	0.0014	0.1291	−0.0001	0.0043	−0.0021	0.0007	0.1314
$\bar{\alpha} = 0.35,\ \lambda_A = 0.25,\ \lambda_B = 0.05$										
0.0001	0.0011	0.2320	−0.0936	0.0342	0.3781	0.0002	0.2253	−0.0960	0.0283	0.3789
0.005	−0.0001	0.1267	−0.0583	0.0146	0.3758	−0.0001	0.1172	−0.0608	0.0115	0.3761
0.1	0.0007	0.0058	−0.0035	0.0008	0.2984	0.0002	0.0038	−0.0028	0.0003	0.3067
$\bar{\alpha} = 0.65,\ \lambda_A = 0.2,\ \lambda_B = 0.1$										
0.0001	0.0001	0.0306	−0.0055	0.0117	0.2527	−0.0001	−0.0432	0.0160	−0.0092	0.1381
0.005	−0.0006	−0.0115	0.0004	0.0060	0.2514	0.0002	−0.0221	0.0081	−0.0043	0.1368
0.1	0.0001	−0.0117	0.0047	−0.0014	0.2223	−0.0003	−0.0055	0.0016	−0.0015	0.1162
$\bar{\alpha} = 0.65,\ \lambda_A = 0.25,\ \lambda_B = 0.05$										
0.0001	0.0001	0.0306	−0.0055	0.0111	0.2527	0.0001	0.0286	−0.0091	0.0063	0.2527
0.005	−0.0006	0.0115	−0.0004	0.0060	0.2514	0.0001	0.0077	−0.0017	0.0023	0.2516
0.1	−0.0001	−0.0118	0.0047	−0.0014	0.2222	0.0004	−0.0096	0.0037	−0.0011	0.2282

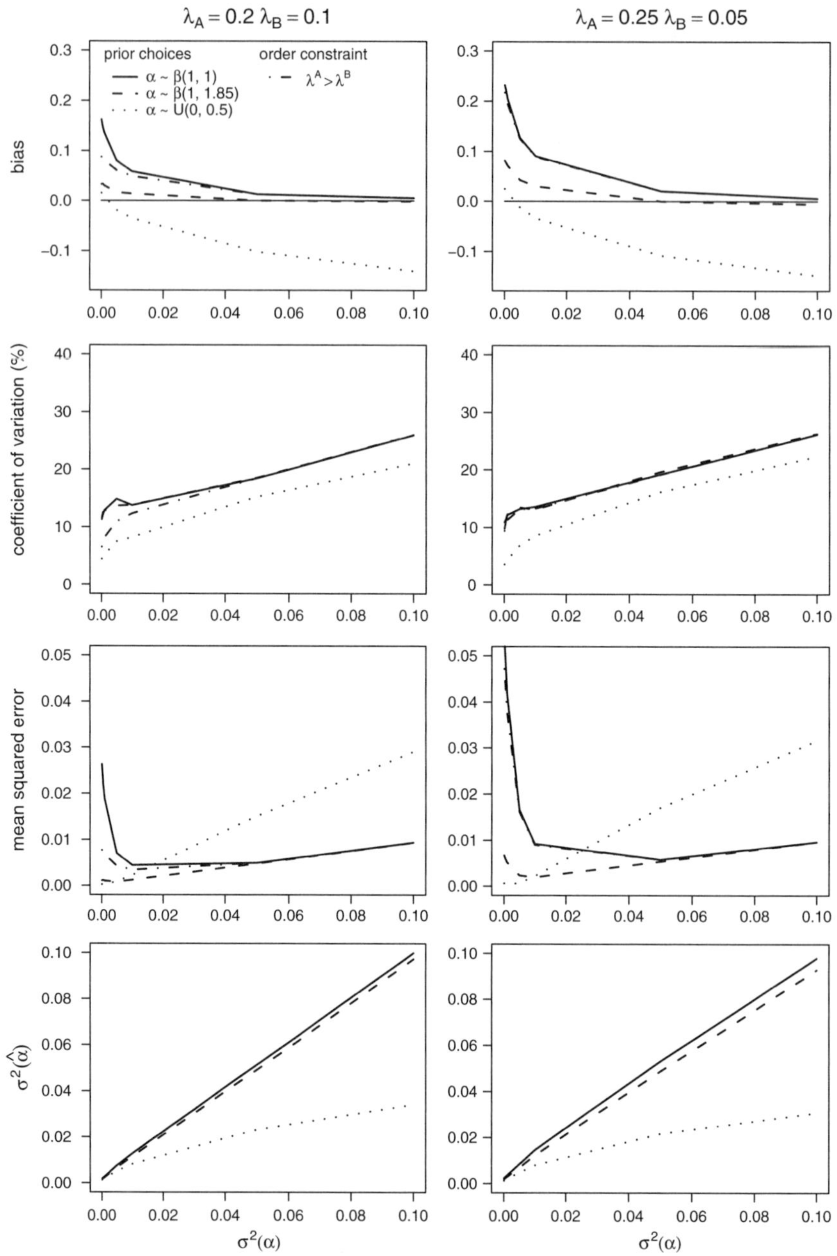

Fig. 2 Bias, coefficient of variation, mean squared error and temporal variance of the proportion of mortality causes estimated from the cause-specific mortality rate model ($\hat{\alpha}$) in relation to the simulated temporal variance of α (σ_{α}^{2}), the cause-specific recovery probabilities and different prior distributions or order constraints. Shown are mean values originating from 500 simulations. These simulations were performed assuming a survival rate of 0.4, $\bar{\alpha} = 0.35$, 10 study years and 1000 newly released individuals in each year

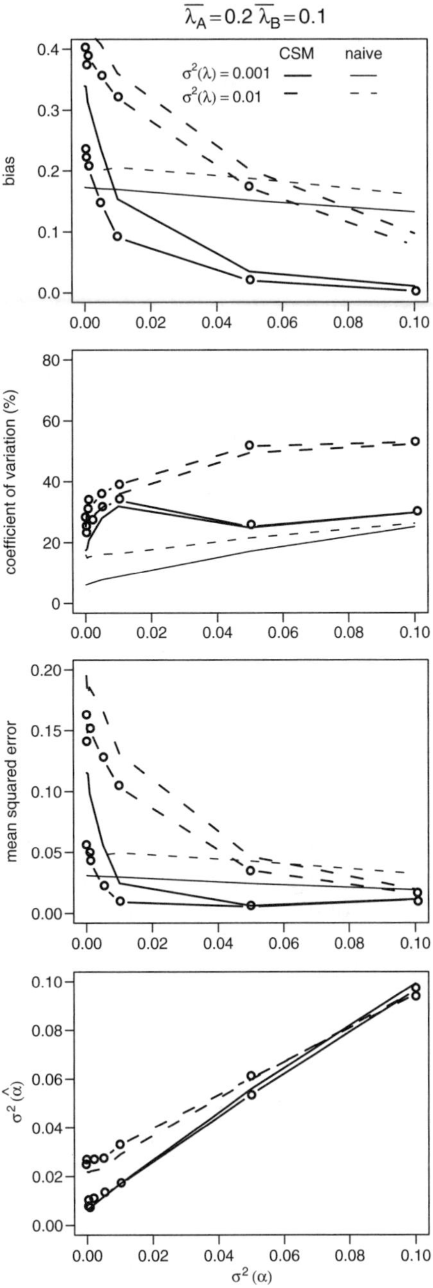

Fig. 3 Bias, coefficient of variation, mean squared error and temporal variance of the proportion of mortality causes estimated from the cause-specific mortality rate model (CSM) and the naïve estimator (naive), respectively, in relation to the simulated temporal variance of α (σ_α^2), temporal variance of cause-specific recovery probabilities and different prior distributions. The lines without dots refer to analyses using a β (1,1)-prior for α, the lines with open dots refer to analyses using a β (1,1.857)-prior for α. Shown are mean values originating from 500 simulations. These simulations were performed assuming a survival rate of 0.4, 10 study years and 1000 newly released individuals in each year

In the third set, I studied the estimator performance when the recovery probabilities were variable across time. The pattern of bias, precision and accuracy of $\hat{\alpha}$ remained the same: bias and precision decreased and accuracy increased with increasing σ_α^2 (Fig. 3). Bias of $\hat{\alpha}$ increased and precision of $\hat{\alpha}$ decreased with increasing σ_λ^2. Use of an informative prior distribution for α reduced the bias and slightly increased accuracy. As in the other sets, bias in survival estimates was always minimal. Bias of recovery probabilities followed again the same pattern as bias of $\hat{\alpha}$ (results not shown). Bias of $\hat{\eta}$ was important in all conditions, and it increased with increasing σ_λ^2 (Fig. 3). Precision of $\hat{\eta}$ decreased with increasing σ_λ^2 and σ_α^2. Bias of $\hat{\eta}$ was larger than bias of $\hat{\alpha}$ when σ_λ^2 was low, otherwise bias of both estimators was important.

Since σ_α^2 plays a major role for bias and accuracy of $\hat{\alpha}$ and the two recovery probabilities, there is clearly an interest to know whether the estimated temporal variance of $\hat{\alpha}$ is a good estimator of the true underlying temporal variation of $\hat{\alpha}$ (σ_α^2); owing to the bias of $\hat{\alpha}$, this was not clear a priori. As shown in Figs. 1, 2 and 3, the temporal variance of $\hat{\alpha}$ was mostly a good indicator of σ_α^2. However, when the uniform prior (U(0, 0.5)) for α was chosen and σ_α^2 was high, the temporal variance of $\hat{\alpha}$ underestimated σ_α^2 (Fig. 2). Moreover, when recovery probabilities were variable across time, σ_α^2 was slightly overestimated (Fig. 3).

4 Discussion

I evaluated bias, precision and accuracy of the estimators of the cause-specific mortality model of Schaub and Pradel (2004). While survival rate estimates were never strongly biased and always very accurate, bias, precision and accuracy of the proportion of the mortality causes ($\hat{\alpha}$) and of the cause-specific recovery probabilities ($\hat{\lambda}_A$, $\hat{\lambda}_B$) strongly depended on the specific situation: generally biases and precision decreased and accuracy increased with increasing temporal variance of $\hat{\alpha}$, with decreasing difference between the two recovery probabilities and with decreasing temporal variation of the two recovery probabilities. Increasing the number of study years only had limited impact on parameter accuracy and bias. The commonly used naïve estimator $\hat{\eta}$ was in almost all cases much more strongly biased and less accurate than $\hat{\alpha}$. These results followed expectations and are in line with our previous assessment of the intrinsic identifiablity of the parameters in the cause-specific mortality model (Schaub and Pradel 2004, Schaub and Lebreton 2004).

The inclusion of additional information sometimes had a strong effect on the parameter estimates, in particular when the temporal variation of $\hat{\alpha}$ was low. According to the evaluation of the intrinsic identifiability this result was expected. When the temporal variation of $\hat{\alpha}$ decreases, the likelihood of the model becomes a more flat ridge, and thus the information in the data to separately estimate all parameters becomes smaller. Consequently, the prior distributions obtain more weight compared to the likelihood until they dominate the posterior distribution (Brooks 1998).

I also found that the estimated temporal variation of the proportion of mortality causes is a fairly good estimator of the true variability of this proportion. Since bias, precision and accuracy of the parameters of interest strongly depend on the temporal variation of the proportion of mortality causes, this is an important result that may help in practical applications of the model. Based on the estimated parameters it can be judged whether accuracy is acceptable.

The naïve estimate $\hat{\eta}$ is only unbiased if the cause-specific recovery rates are identical. As shown here, bias of $\hat{\eta}$ increases the more this condition is violated. In addition, bias of $\hat{\eta}$ was usually much stronger than bias of $\hat{\alpha}$. In contrast to $\hat{\alpha}$, where the magnitude of bias can be assessed by the inspection of the temporal variance of $\hat{\alpha}$, magnitude of possible bias of $\hat{\eta}$ cannot be assessed, as the cause-specific recovery rates remain unknown. This clearly shows that $\hat{\alpha}$ is a superior and more rigor estimator of the proportion of mortality causes than $\hat{\eta}$, even if $\hat{\alpha}$ is biased in some situations.

For the practical application of the model, it would be important to evaluate how often or to what degree the assumptions of high level of variation of mortality proportions across time and constant recovery probabilities are met in practice. Classical analyses of dead recovery data have shown that sometimes recovery probabilities may be fairly constant across years (Thomson et al. 1997); yet, perhaps more often, they appear to vary across time (Piper 1995, Frederiksen and Bregnballe 2000, Schaub et al. 2005, Altwegg et al. 2006). Because these analyses did not distinguish between recovery probabilities and proportions of mortality causes, the estimated recovery probability is a combination of the two. Therefore, it is impossible to know which of these parameters was variable across time or whether both were. In the only empirical analysis so far that separated these parameter types, we (Schaub and Pradel 2004) found little evidence for temporal variation in the recovery probabilities and strong evidence for fairly large temporal variation of the proportion of white storks killed by power lines.

Based on the present study, I here propose some guidelines about how the model will be most fruitfully applied in practice. Such guidelines are important, because parameter estimates may not be adequate in terms of unbiasedness or accuracy in every situation. The first step in the application of the cause-specific mortality model should be the selection of a proper model using AIC or similar methods, although the parameter estimation will later be performed with a model in which the parameters of interest are intrinsically identifiable, i.e. model [S, α_t, λ_A, λ_B] or [S_t, α_t, λ_A, λ_B]. The candidate set should include all combinations of time-specific and constant proportion of the mortality causes and time-specific and constant recovery probabilities. Despite the fact that some models in the candidate set may have parameters that are not separately identifiable, such models can validly be used for model selection. This model selection exercise will help to evaluate the expected accuracy of the parameter estimates. There are three possible outcomes. (1) If it turns out that the recovery probabilities are variable across time, it will be very difficult to obtain useful estimates of $\hat{\alpha}$, even if $\hat{\alpha}$ is highly variable over time. In this situation it may be best to conduct a simulation study that mimics the current situation in order to explore how large the bias in the parameters of interest may possibly be. The estimates under the model could then be reported along with an acknowledgement of the likely magnitude and direction of bias. (2) If recovery probabilities are constant and $\hat{\alpha}$ is variable across time, the parameter estimates will be fairly accurate. For a further assessment of the potential bias, temporal variation in $\hat{\alpha}$ may be considered. This is best combined with a simulation study adapted to the specific situation, which will indicate the likely magnitude and direction of the bias. (3) If model selection will favor a model with constant parameters, α is probably not strongly variable over time and $\hat{\alpha}$ will be biased. To minimize bias a wise use of additional information, if available, may be helpful. Because the true proportion of mortality causes is not likely to be known in practice and because of their strong impact, I would not recommend to use informative priors for α. Constraining the range of the prior for α or forcing one recovery probability to be larger than the other one seems reasonable, because this kind of information is more likely to be available and is quite confident. Constraining the prior range of α (i.e. U(0,0.5)) worked well in cases where the temporal variation of $\hat{\alpha}$ was low. $\hat{\eta}$ may be useful to define the upper limit of this uniform prior. Forcing one recovery probability to be larger than the other, resulted in some bias reduction, in particular when the values of the two recovery probabilities were not far apart.

Model performance might be improved if additional auxiliary data could be added, that must, however, contain information about mortality causes. For example, the inclusion of live recapture data would not help to increase accuracy of $\hat{\alpha}$ and the recovery probabilities, since this kind of data only adds more information about survival rate. Rather, an independent covariate that is correlated with the proportion of mortality causes might lead to progress. Such a covariate could be included using ultrastructural modeling (Link 1999). In the context of hunting, the annual proportion of hunted individuals among all individuals in a population might be useful. Another kind of information would be the mortality causes evaluated with radio tagged individuals. The frequencies obtained from such smaller scale studies could be used as a priori knowledge in the model. In practice we are often interested in long term trends. For example, we may want to know whether a disease over time becomes more or less important as a mortality cause, or whether the impact of harvesting on the overall mortality of a population changes over time. Future research ought therefore to investigate whether such trends can accurately be estimated in the presence of bias in the parameter estimates themselves.

This evaluation of the cause-specific mortality rate model suggests that in only few situations may the required assumptions be sufficiently well met that the model provides completely unbiased parameter estimates. Encouragingly, in many cases resulting biases are not very strong though and it is possible to assess the likely magnitude of bias. Furthermore, it is important to keep in mind that even if the estimated proportion of mortality causes under the model may be biased somewhat, these estimates will be an improvement over the usual naïve estimate based on the actual numbers of animals reported dead from some cause. Therefore, I conclude that a combination of careful application of the cause-specific mortality model along with custom-designed simulations can provide useful inference about the proportion of mortality causes.

Acknowledgments I would like to thank Marc Kéry, Andy Royle and two reviewers for valuable comments on the paper. Marina Savelieva-Praz helped developing the WinBUGS code and Olivier Gimenez with the assessment of the identifiability. Financial support was provided by the Swiss National Foundation (grant No. A0-107539 to Michael Schaub).

Appendix

WinBUGS code used to estimate the parameters under model [S, α_t, λ_A, λ_B]. I also provide a test data set in the format of a multistate m-array that was generated using the following parameter values S $= 0.4$, $\alpha = \{0.4835, 0.0150, 0.5636, 0.0003, 0.9408, 0.0476, 0.3089, 0.8420, 0.6130, 0.0276\}$ (corresponding to $\sigma_\alpha^2 = 0.1$), $\lambda_A = 0.2$, $\lambda_B = 0.1$, and a set of initial values.

The input data is a matrix consisting of the m-array matrix and an additional column (the last one) with the total number of animals from each cohort that were never recovered. The odd columns (except the last) contain the number of recovered animals dying from cause A, the even columns the number of recovered animals dying from other causes than A. The different lines refer to release cohorts.

Code

```
{
# priors
S ~ dbeta(1,1)
for (i in 1:T) {alpha[i] ~ dbeta(1,1)}
lambdaA ~ dbeta(1,1)
lambdaB ~ dbeta(1,1)

# likelihood
for (i in 1:ni) {m[i,1:(2*nj+1)] ~ dmulti(p[i, ],
r[i])}

# calculate the number of birds released each year
for(i in 1:ni) {r[i]<- sum(m[i, ]) }

# cell probabilities of the multistate m-array
# above main diagonal
for (i in 1:(ni-1)) {
   p[i, 2*i+1]<- S*(1-S)*alpha[i]*lambdaA
   p[i, 2*i+2] <- S*(1-S)*(1-alpha[i])*lambdaB}
```

```
# main diagonal
for (i in 1:ni) {
   p[i, 2*i-1] <- (1-S)*alpha[i]*lambdaA
   p[i, 2*i] <- (1-S)*(1-alpha[i])*lambdaB

# further above
for (j in (i+2):nj) {
   p[i, 2*j-1] <- pow(S,(j-i-1))*S*(1-
S)*alpha[i]*lambdaA
   p[i, 2*j] <- pow(S,(j-i-1))*S*(1-S)*(1-
alpha[i])*lambdaB}

# below main diagonal
for (j in 1:(2*i-2)) {p[i,j] <- 0}

# last column: probability of non-recovery
p[i, 2*nj+1] <- 1-sum(p[i, 1:2*nj])
}
}
```

Test Data Set

```
list(ni=10, nj=10, m= structure(.Data= c(58, 32, 17,
11, 7, 5, 3, 1, 1, 1, 0, 0, 0, 0, 0, 0, 0, 0, 0, 0,
864, 0, 0, 0, 59, 0, 18, 0, 4, 0, 3, 0, 2, 0, 2, 0,
0, 0, 0, 0, 0, 912, 0, 0, 0, 0, 71, 22, 28, 9, 12,
4, 7, 1, 1, 0, 1, 1, 0, 0, 1, 0, 842, 0, 0, 0, 0, 0,
0, 0, 61, 0, 19, 0, 9, 0, 5, 0, 1, 0, 0, 0, 0, 905,
0, 0, 0, 0, 0, 0, 0, 0, 100, 5, 46, 1, 17, 1, 9, 0,
3, 0, 1, 0, 817, 0, 0, 0, 0, 0, 0, 0, 0, 0, 0, 4,
63, 2, 29, 0, 5, 0, 3, 0, 0, 894, 0, 0, 0, 0, 0, 0,
0, 0, 0, 0, 0, 0, 49, 48, 19, 17, 3, 7, 2, 3, 852,
0, 0, 0, 0, 0, 0, 0, 0, 0, 0, 0, 0, 0, 0, 88, 10,
37, 9, 11, 3, 842, 0, 0, 0, 0, 0, 0, 0, 0, 0, 0,
0, 0, 0, 0, 0, 73, 16, 34, 9, 868, 0, 0, 0, 0, 0, 0,
0, 0, 0, 0, 0, 0, 0, 0, 0, 0, 0, 4, 54, 942),
.Dim=c(10, 21)), T=10)
```

Initial Values

```
list(alpha=c(0.5, 0.5, 0.5, 0.5, 0.5, 0.5, 0.5, 0.5,
0.5, 0.5), S=0.5, lambdaA=0.5, lambdaB=0.5)
```

References

Altwegg R, Roulin A, Kestenholz M, Jenni L (2006) Demographic effects of extreme winter weather in the barn owl. Oecologia 149:44–51.

Bro E, Reitz F, Clobert J, Migot P, Massot M (2001) Diagnosing the environmental causes of the decline in Grey Partridge *Perdix perdix* survival in France. Ibis 143:120–132.

Brooks SP (1998) Markov chain Monte Carlo method and its application. Journal of The Royal Statistical Society Series D – The Statistician 47:69–100.

Brooks SP, Catchpole EA, Morgan BJT, Barry SC (2000) On the Bayesian analysis of ring-recovery data. Biometrics 56:951–956.

Brooks SP, Gelman A (1998) Alternative methods for monitoring convergence of iterative simulations. Journal of Computational and Graphical Statistics 7:434–455.

Buner F, Schaub M (2008) How do different releasing techniques affect the survival of re-introduced grey partridges *Perdix perdix*? Wildlife Biology 14:26–35.

Burnham KP, Anderson DR, White GC, Brownie C, Pollock KH (1987) Design and analysis methods for fish survival experiments based on release-recapture. American Fisheries Society Monograph Number 5.

Catchpole EA, Kgosi PM, Morgan BJT (2001) On the near-singularity of models for animal recovery data. Biometrics 57:720–726.

Catchpole EA, Morgan BJT (1997) Detecting parameter redundancy. Biometrika 84:187–196.

Catchpole EA, Morgan BJT (2001) Deficiency of parameter-redundant models. Biometrika 88:593–598.

Catchpole EA, Morgan BJT, Viallefont A (2002) Solving problems in parameter redundancy using computer algebra. Journal of Applied Statistics 29:625–636.

Frederiksen M, Bregnballe T (2000) Evidence for density-dependent survival in adult cormorants from a combined analysis of recoveries and resightings. Journal of Animal Ecology 69:737–752.

Gimenez O, Choquet R, Lebreton JD (2003) Parameter redundancy in multistate capture-recapture models. Biometrical Journal 45:704–722.

Hüppop K, Hüppop O (2002) Atlas zur Vogelberingung auf Helgoland Teil 1: Zeitliche und regionale Veränderungen der Wiederfundraten und Todesursachen auf Helgoland beringter Vögel (1909 bis 1998). Die Vogelwarte 41:161–181.

Link WA (1999) Modeling pattern in collections of parameters. Journal of Wildlife Management 63:1017–1027.

Newton I, Wyllie I, Dale L (1999) Trends in the numbers and mortality patterns of sparrowhawk *Accipiter nisus* and kestrels *Falco tinnunculus* in Britain, as revealed by carcass analyses. Journal of Zoology 248:139–147.

Piper SE (1995) A model of the ring-recovery reporting process for the Cape Griffon *Gyps coprotheres*. Journal of Applied Statistics 22:641–659.

R Development Core Team (2004) R: a language and environment for statistical computing. Vienna, R Foundation for Statistical Computing.

Schaub M, Kania W, Köppen U (2005) Variation of primary production during winter induces synchrony in survival rates in migratory white storks *Ciconia ciconia*. Journal of Animal Ecology 74:656–666.

Schaub M, Lebreton JD (2004) Testing the additive versus the compensatory hypothesis of mortality from ring recovery data using a random effects model. Animal Biodiversity and Conservation 27.1:73–85.

Schaub M, Pradel R (2004) Assessing the relative importance of different sources of mortality from recoveries of marked animals. Ecology 85:930–938.

Spiegelhalter D, Thomas A, Best NG (2004) WinBUGS User Manual, Version 1.4. Cambridge, MCR Biostatistics Unit.

Sturtz S, Ligges U, Gelman A (2005) R2WinBUGS: a package for running WinBUGS from R. Journal for Statistical Software 12:1–16.

Thomson DL, Baillie SR, Peach WJ (1997) The demography and age-specific annual survival of song trushes during periods of population stability and decline. Journal of Animal Ecology 66:414–424.

Standardising Terminology and Notation for the Analysis of Demographic Processes in Marked Populations

David L. Thomson, Michael J. Conroy, David R. Anderson, Kenneth P. Burnham, Evan G. Cooch, Charles M. Francis, Jean-Dominique Lebreton, Mark S. Lindberg, Byron J.T. Morgan, David L. Otis, and Gary C. White

Abstract The development of statistical methods for the analysis of demographic processes in marked animal populations has brought with it the challenges of communication between the disciplines of statistics, ecology, evolutionary biology and computer science. In order to aid communication and comprehension, we sought to root out a number of cases of ambiguity, redundancy and inaccuracy in notation and terminology that have developed in the literature. We invited all working in this field to submit topics for resolution and to express their own views. In the ensuing discussion forum it was then possible to establish a series of general principles which were, almost without exception, unanimously accepted. Here we set out the background to the areas of confusion, how these were debated and the conclusions which were reached in each case. We hope that the resulting guidelines will be widely adopted as standard terminology in publications and in software for the analysis of demographic processes in marked animal populations.

Keywords Mark-recapture · Mark-recovery · Terminology · Notation

1 Introduction

Recent decades have seen rapid developments in the analysis of demographic processes in marked animal populations (Senar et al. 2004; Morgan and Thomson 2002; Baillie et al. 1999; North and Nichols 1995; Lebreton and North 1993; North 1987; Morgan and North 1984). This has in large part been achieved through the successful collaboration of biologists, biometricians, statisticians and computer scientists. Thanks to partnerships across these disciplines, we have been able to advance

D.L. Thomson (✉)
Max Planck Institute for Demographic Research, Konrad Zuse Strasse 1, D18057 ROSTOCK, Germany; Netherlands Institute of Ecology, Centre for Terrestrial Ecology, Boterhoeksestraat 48, Postbus 40, 6666 ZG Heteren, Netherlands; School of Biological Sciences, University of Hong Kong, Pokfulam, Hong Kong
e-mail: thomson@demogr.mpg.de

D.L. Thomson et al. (eds.), *Modeling Demographic Processes in Marked Populations*, Environmental and Ecological Statistics 3, DOI 10.1007/978-0-387-78151-8_51,

our understanding through the development of new models and methods, better insights on how to design experiments and collect data (Schwarz 2002), and through the development of sophisticated software packages. These developments have revolutionized the way we conduct demographic analysis and the progress is clear to see, but the interdisciplinary nature of this field and the widespread uptake and implementation of these statistical models by biologists also brings with it challenges of communication between disciplines. This communication is not made easier when ambiguities, inaccuracies and redundancies in terminology and notation appear in the literature. In principle, provided terminology and notation are clearly defined, each author can exercise their right to use whatever notation and terminology are most suited to the issues upon which they are working, but there are many cases where authors have given different names to the same parameters, used the same name for different parameters, or used terminology which is not an accurate descriptor. In an effort to avoid confusion and make communication and comprehension easier, we tried to identify all places where there were problems or potential problems and by open debate and consensus we then tried to establish a series of accepted standards that we hope will be useful and widely followed in publications, software packages, and in all aspects of work in this field, until such time as further revision of terminology becomes desirable.

2 Methods

The EURING conferences constitute the premier forum for discussion and interaction on the subject of modeling demographic processes in marked populations, and it was through this medium that we tried to reach all involved in this field with a view to airing views and reaching consensus. All members of the EURING mailing list were contacted and asked to suggest topics where resolution and standardization would be beneficial, and they were asked to contribute their own views on what they felt would be the best standards to adopt. A particular effort was made to poll the views of those authoring software packages as they have particular influence on the way demographic analyses are approached and the terms and notation which are used. With the resulting agenda, all members of the list, and indeed in principle any other interested parties, were invited to attend a discussion forum at the EURING2003 conference in Radolfzell, Germany. Each of the points was discussed, and as far as was possible we tried to reach consensus on recommended standards. In most cases, it was possible to reach unanimous conclusions.

3 Results and Discussion

A summary of the recommended standards is given in Table 1. In more detail, the issues were debated as follows:

Table 1 Summary of conclusions and recommendations

'Apparent survival', 'local survival', 'true survival', Φ and S
-we should discontinue use of the term 'local survival' and use instead only the term 'apparent survival'
-we should denote 'apparent survival' with the Greek letter Φ (*capital* phi), and 'true survival' with capital S. This means we should not normally use Φ in dead-recovery models.
-these parameters should be denoted by capital letters in all cases
-if these parameters are 'probabilities' then they should be referred to as such and should not be referred to as 'rates'.
-it enhances clarity when we make reference to time periods with terms such as 'annual' or 'monthly' survival probabilities etc.
'Recovery probability', 'Reporting probability', f, λ and r
A fully unanimous recommendation could not be reached on the core issues here, but
-the word 'recovery' should in any case only be used to refer to *dead* re-encounters of marked animals.
-even if standardisation can not be achieved, terms and symbols should be clearly defined in such a way that avoids confusion
As well as these unanimous recommendations, strong arguments were presented for adopting the terminology of Seber (1970, 1971) which in fact defines 'reporting' probability λ as the probability that a marked bird which has died will be found and reported.
The term 'recovery' probability f can then be used sensu Brownie et al. (1985) to refer to the probability that a marked animal alive at the start of the time period will be shot and have its mark reported.
Reporting probabilities and population growth rates are mostly not yet modeled simultaneously, but care is needed to ensure clarity if they are as both are widely denoted with the same symbols.
'Multi-state models', 'Robust design', temporary emigration, resighting probability, Ψ, γ, c
-We should use the term 'multi-state' and not 'multi-strata'
-In robust design models, transitions between the observable state inside the trapping area and the unobservable state outside it should be labeled with the terminology and notation (Ψ) of multi-state models
-the terms 'temporary emigration' γ' and 'temporary immigration' γ'' need not normally be used
-In robust design models, there is no need to introduce a new parameter 'c' or label it 'resighting' probability; instead structure akin to modeling trap-dependence can be introduced into the capture probability whereby a distinction can be made between probabilities of first and subsequent captures within sessions.

3.1 'Apparent Survival', 'Local Survival', 'True Survival', Φ, and S

In many mark-recapture studies where intensive observations are made on small study sites, estimates of survival probabilities are valid under the assumption that animals do not permanently leave the area within which they can be encountered. In recognition of the fact that this assumption is rarely likely to hold, we often use the term 'apparent survival probability', Φ, the probability that an animal will not die and will not permanently leave the study site during the time period. By using the expression 'apparent survival probability', Φ, a clear distinction is made with 'survival probability', S. The survival probability can usually be estimated in mark-recovery models where dead birds can normally be found and reported even if they move some considerable distance from the point of marking. The link between apparent survival probability Φ and probability of survival S is usually through a probability of fidelity (Burnham 1993), and $\Phi = S \times$ fidelity. The probability of permanent emigration is 1-fidelity.

There are two main sources of confusion in this area. Firstly, the term 'local survival' has also been used extensively to describe 'apparent survival', Φ. Secondly, some authors have used Φ to denote 'survival probability', S, in mark-recovery models, and S is sometimes used to denote apparent survival in mark-recapture models.

The forum felt that the term 'apparent survival' made a clearer acknowledgement that the estimated parameter was not a true survival probability, and that the term 'local survival' did not do this and could be interpreted as meaning simply that the survival probability was specific to a local area. We therefore chose unanimously to recommend only the use of the term 'apparent survival' and to discontinue use of the term 'local survival'.

The forum further recommended that apparent survival should always and only be denoted Φ, and that survival probability should always and only be denoted S. It was emphasized that these parameters should be denoted by capital and not lower-case letters. Later it was added that if confusion may be caused by the use of capitals for the matrices used in multi-state models, then the matrices could be denoted with bold-face capitals.

During the discussions, the point was raised that a distinction should be made between 'rates' and 'probabilities' and since these models estimate probabilities they should be referred to as such and not as 'rates'.

In other fields of statistics and demographic analysis, 'survival' often refers to survival from age zero, while mark-recapture and mark-recovery models typically concern survival through a specified time period conditional on being alive at the start of it. The forum agreed that the use of words to specify this time-period (e.g. 'annual' or 'monthly' survival) could help to clarify the meaning where there was potential for confusion.

3.2 'Recovery Probability', 'Reporting Probability', *f*, λ, and *r*

Brownie et al. (1985) used the term 'recovery' probability, f, to denote the probability that a marked animal alive at the start of the time period will be shot and have its mark reported. 'f' can be partitioned further to estimate the probability ('reporting' probability) that a hunter who has shot a marked bird will retrieve the mark and report it. 'f' is an index of hunting pressure and these models are popular for hunted populations in North America. Even when not hunted, marked birds are found dead and reported, and Seber (1970, 1971) used 'Reporting' probability λ to denote the probability that a marked animal that has died will be found and reported. This formulation has been popular in Europe where many non-hunted species are studied and where the probabilities of dead marked birds being found and reported are higher. Others have since referred to Seber's 'reporting' probability as 'Recovery' probability and denoted it 'r' instead of 'λ'. Despite the unfortunate ambiguities and redundancy here, these issues proved very difficult to resolve, and the only unanimous recommendation that could be made was that:

– even if standardisation can not be achieved, terms and symbols should be clearly defined in such a way that avoids confusion

As well as this, strong arguments were presented for adopting the terminology of Seber which had historical precedence. Seber did not in fact use 'Recovery' probability or 'r' in these papers. If Seber's 'Reporting' probability λ is adopted then this avoids the confusion with the 'recovery' probability 'f', but we need to avoid confusion when using 'reporting' probability to refer to the probability that a hunter will report an animal he has shot.

In discussing these issues, two other points were raised and unanimous conclusions were reached. Firstly, the word 'recovery' should only be used to refer to *dead* re-encounters of marked animals. This is distinct from live 'recaptures' and 'resightings'. The collective word for all of these is 're-encounters', and particularly in analyses which combine different types of encounters, it makes sense to refer to 'encounter histories' as opposed to 'capture histories'. The words 'ring recoveries' or 'band recoveries' are often used to describe all forms of re-encounter, but in the context of formal models we urge people not to use the word 'recovery' when referring to live animals.

Secondly, in demography, the symbols 'r' and 'λ' are also both widely used to denote measures of population growth. The forum debated whether the notation we use for reporting probability could lead to confusion in this sense, but concluded that context would normally ensure there was no ambiguity in practice. With the increasingly integrated nature of demographic analyses, it is to be expected that population growth rate and reporting probability will increasingly be handled simultaneously in the same model (Pradel 1996; Besbeas et al. 2002), and care should be taken to avoid confusion when this is the case. In integrated models, the use of 'p' to denote capture probability of live organisms could similarly lead to confusion with p for productivity, though currently this will normally be clear from context.

3.3 Multi-State Models, Robust Design, 'Temporary Emigration', 'Resighting' Probability, and Ψ, γ, and c

In multi-state models, as well as the estimation of survival probabilities, we can estimate the probabilities of transition, Ψ, between states (Brownie et al. 1993; Hestbeck et al. 1991). These states could for example be distinct geographical sites, or they could be behavioural or physiological conditions such as breeding or non-breeding, healthy or diseased.

While conventional open population mark-recapture studies involve single short trapping sessions at regularly spaced time intervals, Robust Design models can be used when each of these conventional trapping sessions are further divided into a short series of closely spaced repeat samples leading to a number of extremely short time periods, 'secondary sampling periods', as well as the conventional longer 'primary sampling periods' (Kendall et al. 1995). The population can be assumed to be closed over these short secondary sampling periods and this makes it possible to estimate capture probability based on just a single trapping session. With a Robust Design, it then becomes possible to estimate not just survival and capture probabilities but also the probability that a bird will undergo transitions to and from an unobservable state, perhaps by opting in different years to establish a territory which is inside or just outside the study area (Kendall et al. 1997; Kendall and Nichols 1995).

A number of terminology issues were recognized as being problematic in these areas. Firstly, multi-state models are sometimes referred to as 'multi-strata' models even though 'strata' usually refers to fixed states between which transition is not possible. Secondly, the probability of transition to an unobservable state outside the study area and the probability of remaining there have traditionally been referred to with the terms 'temporary emigration' γ'' and 'temporary immigration' γ' even though there has been some discomfort that these terms do not describe well meaning of the parameters estimated. Thirdly, in Robust Design models a distinction is made between the probability of capture for the first time within a trapping session, and the probability of subsequent captures within the trapping session. The probability of capture of an animal that has already been captured once within a trapping session has been given a separate name, 'resighting' probability, and denoted c. This same term 'resighting' probability is also used in the models of Barker (1997, 1999) with a different meaning and refers there to the probability that an animal marked with a field readable ring can be encountered live in the course of the conventional (primary) sampling periods.

The forum felt that these areas of confusion could be resolved as follows, and was unanimous in these recommendations:

- where it is possible to make transitions between states, we should use the term 'multi-state models' and should discontinue the use of the term 'multi-strata models' because strata are typically states between which transition is not possible (Lebreton and Pradel 2002). The use of only one term will avoid confusion, and 'multi-state models' is a better descriptor.
- in Robust Design models, transitions to and from unobservable states should be labeled with the terminology and notation Ψ of multi-state models, and since these parameters are normally nuisance parameters anyway, terms based on 'temporary emigration' need not normally be used. If the transition has biological meaning, for example when only breeding birds can be observed and where birds periodically take sabbatical years as non-breeders, then accurate descriptive terminology can be used but normally the notation will suffice. As a standard Greek letter, 'γ' will always be used widely by mathematicians in various contexts, but

within our field, discontinuation of the use of γ'' and γ' in Robust Design models should reduce confusion with the use of γ to denote seniority probability in the models of Pradel (1996)

- in Robust Design models, we see no need to create a new parameter 'resighting' probability or label it c; instead structure akin to modeling trap-dependence can be introduced whereby a distinction can be made between the capture probabilities of animals which have or have not previously been captured within the trapping session. This is more parsimonious and avoids all confusion with the 'resighting' probability of the Barker (1997) models.

In the context of resighting, the forum further suggested that rings which can be read in the field without capturing an animal should be referred to as 'field-readable rings'.

One further topic was raised during the discussions, namely the terminology which should be used when the exact age of trapped animals is unknown, but where the effects of age can crudely be built into the analysis by modeling the effect of 'time since marking'. Under some circumstances, this may be a good surrogate for age or it may otherwise have a clear biological meaning. For example, in cases where capture is impossible until animals recruit to the breeding population and where capture probabilities are high thereafter, 'time since marking' approximates time since recruitment, and this in turn approximates breeding experience. It was felt that some care should be exercised in using the term 'age' though, and under most circumstances it may be preferable to call these 'time since marking' models. In due course this issue may disappear if new models can be developed which estimate the effects of age on survival even when exact age of specific individuals is unknown.

Given that these recommendations have been established through open discussion and consensus, with the worthy goal of reducing confusion and simplifying communication and comprehension across our community, we hope very much that they will be adopted widely. We hope that these suggestions will not be blindly enforced or otherwise misused but that they will be taken up voluntarily and used intelligently to these ends. We urge authors of both manuscripts and software packages to be clear about what they mean, and we urge everyone not to invent new terms for established concepts when standard terminology and notation are already available.

Acknowledgments The authors wish to thank the organizers of the EURING conference in Radolfzell, particularly Wolfgang Fiedler and Juan Carlos Senar, who made this discussion forum possible. All authors acknowledge the support of their organizations. We thank Bill Link and Andy Royle for their helpful and entertaining comments on an earlier draft.

References

Baillie SR, North PM, Gosler AG (eds) (1999) Large-scale studies of marked birds. Proceedings of the EURING97 conference. Bird Study 46 supplement.

Barker RJ (1997) Joint modeling of live-recapture, tag-resight, and tag-recovery data. Biometrics 53:666–677.

Barker RJ (1999) Joint analysis of mark-recapture, resighting and ring-recovery data with age-dependence and marking-effect. Bird Study 46 Supplement:82–91.

Besbeas P, Freeman SN, Morgan BJT, Catchpole EA (2002) Integrating mark-recapture-recovery and census data to estimate animal abundance and demographic parameters. Biometrics 58:540–547.

Brownie C, Anderson DR, Burnham KP, Robson DS (1985) Statistical Inference from Band Recovery – A Handbook. United States Department of the Interior. Fish and Wildlife Service, Resource Publication No. 156, Washington, D.C.

Brownie C, Hines JE, Nichols JD, Pollock KH, Hestbeck JB (1993) Capture-recapture studies for multiple strata including non-Markovian transitions. Biometrics 49:1173–1187.

Burnham KP (1993) A theory for combined analysis of ring recovery and recapture data. pp 199–213 in J-D Lebreton and PM North (eds) Marked individuals in the study of bird population. Birkhauser Verlag, Basel, Switzerland.

Hestbeck JB, Nichols JD, Malecki RA (1991) Estimates of movement and site fidelity using mark-resight data of wintering Canada geese. Ecology 72:523–533.

Kendall WL, Nichols JD (1995) On the use of secondary capture-recapture samples to estimate temporary emigration and breeding proportions. Journal of Applied Statistics 22:751–762.

Kendall WL, Nichols JD, Hines JE (1997) Estimating temporary emigration using capture-recapture data with Pollock's robust design. Ecology 78:563–578.

Kendall WL, Pollock KH, Brownie C (1995) A likelihood-based approach to capture-recapture estimation of demographic parameters under the robust design. Biometrics 51:293–308.

Lebreton JD, North PM (eds) (1993) Marked Individuals in the Study of Bird Population. Birkhauser Verlag, Basel, Switzerland.

Lebreton JD, Pradel R (2002) Multistate recapture models: modelling incomplete individual histories. Journal of Applied Statistics 29:353–369.

Morgan BJT, North PM (1984) (eds) Statistics in Ornithology. Springer, New-York.

Morgan BJT, Thomson DL (eds) (2002) Statistical Analysis of Data from Marked Bird Populations. Journal of Applied Statistics 29(1–4).

North PM (ed) (1987) Ringing recovery analytical methods. Acta Ornithologica 23(1).

North PM, Nichols JD (eds) (1995) Statistics and Ornithology. Journal of Applied Statistics 22(5&6).

Pradel R (1996) Utilization of capture-mark-recapture for the study of recruitment and population growth rate. Biometrics 52:703–709.

Schwarz CJ (2002) Real and quasi-experiments in capture-recapture studies. Journal of Applied Statistics 29:459–473.

Seber GAF (1970) Estimating time-specific survival and reporting rates for adult birds from band returns. Biometrika 57:313–318.

Seber GAF (1971) Estimating age-specific survival rates from bird-band returns when the reporting rate is constant. Biometrika 58:491–497.

Senar JC, Dhondt A, Conroy MJ (eds) (2004) Proceedings of the International Conference EURING 2003, Deutschland. Animal Biodiversity and Conservation 27(1).

Estimating the Seasonal Distribution of Migrant Bird Species: Can Standard Ringing Data Be Used?

Kasper Thorup and Paul B. Conn

Abstract The true distribution of migrant species is rarely immediately apparent from the distribution of ring recoveries due to a heavy bias in regional recovery probabilities. For western Palearctic species, the recovery probability is especially low in Africa, but also varies within Europe. However, little work has been done to derive actual estimates of these recovery probabilities needed to infer the "true" underlying distribution. Here, we investigate the potential of using ringing data to estimate the seasonal distribution densities of migrant species. Using likelihoods based on a two point mixture distribution, the proportions of individuals wintering south of the Sahara are estimated using differences in recovery distributions among species in species groups where the location-specific probability of a ring recovery can be assumed to be essentially the same among species. We consider two such approaches. In the first, survival associated with a wintering area must be set constant across species. In the second, we assume the time series is long enough that a single binary response (recovered/not recovered) may be modeled independently of survival parameters. Under the first approach, we estimated the proportion of sub-Saharan migrants, together with 95% profile likelihood confidence intervals, for redstart as 0.84 [0.70,0.93], thrush nightingale 1.00 [0.49,1.00], garden warbler 0.95 [0.85;0.99], blackcap 0.60 [0.32;0.78], reed warbler 0.87 [0.72,0.95], and pied flycatcher 0.90 [0.76;0.97] using recovery data for birds ringed in Denmark and assuming that all robins winter north of Sahara. In the second approach, estimated proportions of sub-Saharan migrants were similar, but the confidence intervals were somewhat narrower. Although further work is required to examine the underlying assumptions, the models and analyses presented here provide a framework for making better use of existing ring recovery datasets to understand the "true" seasonal distribution patterns of European birds.

Keywords Capture-Recapture · Recovery Probability · Movement Rate · Winter Distribution · Sub-Saharan Africa

---

K. Thorup (✉)
Zoological Museum, University of Copenhagen. Universitetsparken 15, DK-2100 Copenhagen, Denmark
e-mail: kthorup@snm.ku.dk

D.L. Thomson et al. (eds.), *Modeling Demographic Processes in Marked Populations,* Environmental and Ecological Statistics 3, DOI 10.1007/978-0-387-78151-8_52,

1 Introduction

Birds are among the most mobile of all terrestrial organisms, and a few species travel up to more than 20 000 km annually between separate breeding and wintering areas. The task of defining the seasonal distribution of migratory species is considerably more complicated than for sedentary species. The recent demand for such data to assess risk of spread of diseases such as avian influenza (Delany et al. 2006) or effects of accelerating climate change on species of conservation concern (Walther et al. 2004) has created an urgent need to improve our ability to estimate the seasonal distribution and connectivity of migratory species.

From direct observations of birds in the field, we have acquired a general knowledge about seasonal species distributions (e.g. Cramp 1988). While different modeling approaches have been used to help explain variation in these observations (e.g. using climate variables, Walther et al. 2004), such approaches tell us little about the connectivity between populations and areas (Webster et al. 2002). While sophisticated approaches for tracking birds over long distances have been proposed using radio telemetry (Wikelski et al. 2007), these are not yet practical for most small passerines. As such, the most suitable available data bases for identifying connectivity are generally those based on ring recoveries. In Europe, very large ring recovery data bases exist. In Denmark alone, more than 4 million birds have been ringed and more than 180 000 of these have been recovered (Bønløkke et al. 2006). However, most analyses of these data to date have been qualitative or semi-quantitative (e.g. Zink 1973–1985, Zink and Bairlein 1995).

One obstacle to using such data for inference is the fact that spatial distributions of recoveries may provide biased estimates of the spatial distribution of birds due to differences in the regional recovery probabilities (Crissey 1955; Perdeck 1977). For western Palearctic bird migrants, the recovery probability is especially low in Africa, but it is also low in East Europe compared to West Europe. Because of difficulties in addressing spatial variation in recovery and survival probabilities, most analyses of ring recoveries from a larger spatial scale have not attempted any quantitative analysis and only presented the raw data for inferences on seasonal distributions (Bakken et al. 2003; Bønløkke et al. 2006; Fransson and Pettersson 2001; Wernham et al. 2002; Fig. 1; though see e.g. Kania 1981).

A number of studies have investigated movements using multi-site (or multi-state) capture-recapture methods. However, this modeling approach has seen only a few applications for estimating species densities/distributions. A few single-species studies have estimated seasonal distributions. For example, Sibert et al. (1999) used an advection-diffusion model to investigate skipjack tuna movement. Similarly, Skalski et al. (2002) estimated route-specific passage of smolt, in a framework similar to the migration system considered here. However, they used radio telemetry and the robust design (Pollock 1982; Kendall et al. 1995) at certain points to estimate detectability.

Ideally, we would like to model both the spatio-temporal distribution of recoveries as well as variation in survival probabilities for each population and age class. Apart from a general lack of extensive data sets necessary to estimate these many

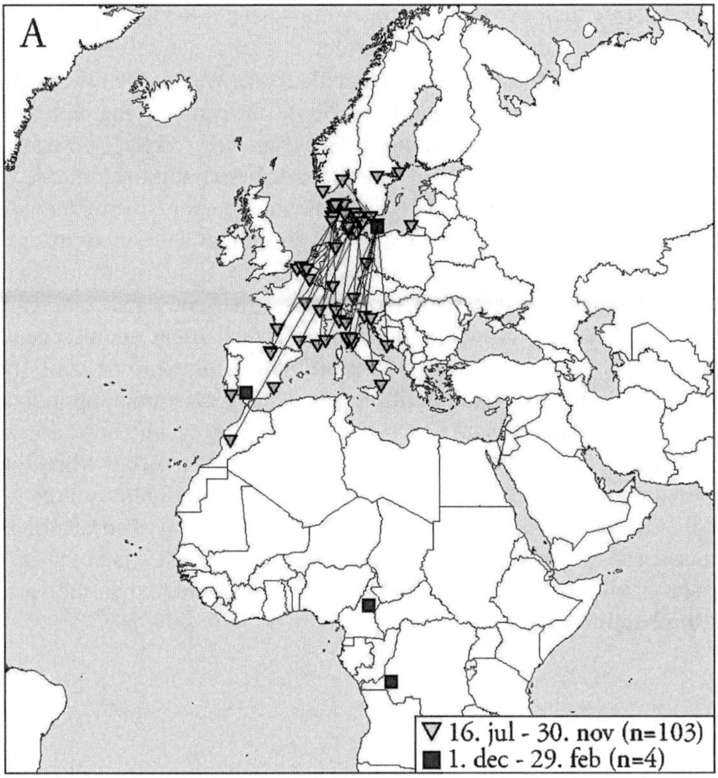

Fig. 1 Typical presentation of ring recovery data in national recovery atlases. Here garden warblers ringed in Denmark (from Bønløkke et al. 2006)

parameters, there is a general problem of parameter identifiability for such movement models. For example, Kendall et al. (2006) found that parameters are only identifiable if the number of "marking" areas with slightly different migration probabilities is greater or equal to the number of "recovery" areas.

Here, we present an approach for using ringing data to estimate the relative seasonal distribution (in winter) of migrant species based on a few simplifying assumptions. The rationale of this approach was already proposed by Busse and Kania (1977) and Kania and Busse (1987), who considered dividing the birds (of one species) into different groups. However, their model formulation only allowed for point estimation of parameters and thus statistical inference could not be based on their method.

The main focus will be estimating the species-specific proportions of songbirds wintering south of Sahara. In general, too little information is available in standard ringing data to separate survival, recovery and movement probabilities. However, standard ringing data generally includes many species, and the differences among species can provide the information necessary for estimating these parameters, if certain parameters such as survival and region-specific recovery probabilities can be treated as constant across species. In effect, this approach suffices to increase the number of "marking" areas (e.g., Kendall et al. 2006). We consider two methods for analyzing spatial ring recovery data. In the first, survival and ring recovery probabilities are assumed to be equal among species, but ring recovery probabilities are allowed to differ among regions. This approach utilizes traditional release-recovery arrays (e.g., Brownie et al. 1985). In the second method, the time series of recovery data are assumed to be long enough that survival is effectively zero (e.g., by censoring the last 15 years of ringing data), and only a binary response is modeled (recovered/not recovered). This second method requires that one throw out some of the data, but does not require the assumption that survival is equal among species. However, life-long ring reporting probabilities are still assumed to be constant across species within a region. Relaxation of the equal survival assumption in the second method will do little to alleviate bias if the reporting probabilities r differ between species. Because of the way in which r is defined, this can occur when sources of mortality within a region differ between species if each mortality source is associated with a different reporting rate. For both approaches, we restricted the species group considered to species where the variation among species in both survival and recovery probabilities are likely to be of smaller magnitude than the variation in recovery probabilities among regions.

2 Methods

We developed likelihoods based on a two point mixture distribution to estimate wintering distributions of songbirds. We assumed that individuals were ringed in the summer in Europe, and that the probability of capture for marking does not depend upon migration destination or individual attributes likely to affect survival or recovery probability. For simplicity, we consider the case where there are two wintering areas (A and B), and where interest focuses on estimating the proportion of birds that migrate to each area. We assume that an individual bird follows only one migration route; that is, an individual would not migrate to area A in one year and area B the next. As articulated in the introduction, we considered two possibilities for model construction.

2.1 *Likelihood 1: Survival and Recovery Probabilities Constant Across Species*

If we assume that survival is constant across species that use similar migration routes, we can specify a multinomial model for the spatial and temporal distribution of ring recoveries (Table 1). The parameters of the model include π_{gi}^{A}, the probability that an individual of group g, initially ringed in year i, is a migrant to wintering area A; S_i^A and S_i^B, the probability of annual survival from the time of ringing in

Table 1 Multinomial structure for the analysis of summer ringing and winter recovery data when migration occurs in the spring and fall and when individuals do not change migration route. A symbolic data description is provided in (A), where R_{gi} gives the number of individuals of group g that were ringed in year i, and m^{c}_{gij} gives the number of individuals recovered in wintering area c in year j, given that they were initially ringed as a member of group g in year i. Multinomial cell probabilities for each type of encounter are provided in (B). Remaining notation is defined in the text

A. Symbolic data description

Year of release	Group	Number released	Strata recovered	Year of recovery		
				1	2	3
1	1	R_{11}	A	m^{A}_{111}	m^{A}_{112}	m^{A}_{113}
	1		B	m^{B}_{111}	m^{B}_{112}	m^{B}_{113}
2	1	R_{12}	A		m^{A}_{122}	m^{A}_{123}
	1		B		m^{B}_{122}	m^{B}_{123}
3	1	R_{13}	A			m^{A}_{133}
	1		B			m^{B}_{133}
1	2	R_{21}	A	m^{A}_{211}	m^{A}_{212}	m^{A}_{213}
	2		B	m^{B}_{211}	m^{B}_{212}	m^{B}_{213}
2	2	R_{22}	A		m^{A}_{222}	m^{A}_{223}
	2		B		m^{B}_{222}	m^{B}_{223}
3	2	R_{23}	A			m^{A}_{233}
	2		B			m^{B}_{233}

B. Multinomial cell probabilities

Year of release	Group	Strata recovered	Year of recovery		
			1	2	3
1	1	A	$\pi^{A}_{11} f^{A}_{1}$	$\pi^{A}_{11} S^{A}_{1} f^{A}_{2}$	$\pi^{A}_{11} S^{A}_{1} S^{A}_{2} f^{A}_{3}$
	1	B	$(1-\pi^{A}_{11}) f^{B}_{1}$	$(1-\pi^{A}_{11}) S^{B}_{1} f^{B}_{2}$	$(1-\pi^{A}_{11}) S^{B}_{1} S^{B}_{2} f^{B}_{3}$
2	1	A		$\pi^{A}_{12} f^{A}_{2}$	$\pi^{A}_{12} S^{A}_{2} f^{A}_{3}$
	1	B		$(1-\pi^{A}_{12}) f^{B}_{2}$	$(1-\pi^{A}_{12}) S^{B}_{2} f^{B}_{3}$
3	1	A			$\pi^{A}_{13} f^{A}_{3}$
	1	B			$(1-\pi^{A}_{13}) f^{B}_{3}$
1	2	A	$\pi^{A}_{21} f^{A}_{1}$	$\pi^{A}_{21} S^{A}_{1} f^{A}_{2}$	$\pi^{A}_{21} S^{A}_{1} S^{A}_{2} f^{A}_{3}$
	2	B	$(1-\pi^{A}_{21}) f^{B}_{1}$	$(1-\pi^{A}_{21}) S^{B}_{1} f^{B}_{2}$	$(1-\pi^{A}_{21}) S^{B}_{1} S^{B}_{2} f^{B}_{3}$
2	2	A		$\pi^{A}_{22} f^{A}_{2}$	$\pi^{A}_{22} S^{A}_{2} f^{A}_{3}$
	2	B		$(1-\pi^{A}_{22}) f^{B}_{2}$	$(1-\pi^{A}_{22}) S^{B}_{2} f^{B}_{3}$
3	2	A			$\pi^{A}_{23} f^{A}_{3}$
	2	B			$(1-\pi^{A}_{23}) f^{B}_{3}$

year i for those birds migrating to wintering area A and B, respectively; and f_i^A and f_i^B, the probabilities that individual migrants to areas A and B are recovered in $(i,\ i+1)$, given that they are alive at time i. Groups in this case can include members of different species or different banding locations if only one species is under consideration. Whatever groups are considered, it is assumed that survival and recovery probabilities are the same for different groups.

Analytical and analytic-numeric (Burnham et al. 1987) methods were used to explore parameter identifiability with this model. As was found by Kendall et al. (2006) in the case of stochastic transitions to wintering areas, an argument based on the number of equations and number of unknowns suggested that parameters would only be identifiable if the number of groups (banding locations, number of species) were greater or equal to the number of wintering locations. Further, the proportion of individuals migrating to different wintering areas needs to differ among the groups under consideration; as a result, partitioning data from one species and banding location into two subsets will not remedy the parameter identifiability problem.

2.2 Likelihood 2: Long Term Survival Assumed to Be Zero

Eventually, all birds ringed will die. If ringing records are censored for a long enough duration at the end of the study (but recovery records are still compiled), one may eliminate one of the assumptions required by Model 1 by setting $S_*^A = S_*^B = 0$, where S_* is survival to the end of the study. Under this approach, each ringing event is accompanied by a binary response variable which equals 1 if the individual's ring is recovered later and 0 if not. Under this approach, we ignore temporal variation in ring recovery probabilities, and write a likelihood for ringing and recovery data as

$$L = \prod_{i=1}^{T}\prod_{g}\binom{R_{gi}}{m_{gi}^A,\, m_{gi}^B}\left[\pi_{gi}r^A\right]^{m_{gi}^A}\left[(1-\pi_{gi})r^B\right]^{m_{gi}^B}$$
$$\left[1-\pi_{gi}r^A-(1-\pi_{gi})r^B\right]^{R_{gi}-m_{gi}^A-m_{gi}^B},$$

where r^A and r^B give time constant ring reporting probabilities for wintering areas A and B in the sense of Seber (1982), R_{gi} gives the number of individuals ringed and released in group g at time i, m_{gi}^A and m_{gi}^B give the number of these individuals later recovered in wintering areas A and B, respectively, and T gives the number of years that ringing data are modeled. Requirements for parameter identification are similar in this approach as they were with Likelihood 1.

2.3 Example Analyses

We coded the preceding likelihoods into program SURVIV (White 1983) to estimate species-specific differences in the proportions of birds wintering in Europe/North

Africa and in sub-Saharan Africa, respectively, for 7 species of songbirds. We analyzed data on redstart *Phoenicurus phoenicurus*, thrush nightingale *Luscinia luscinia*, European robin *Erithacus rubecula*, reed warbler *Acrocephalus scirpaceus*, garden warbler *Sylvia borin*, blackcap *Sylvia atricapilla* and pied flycatcher *Ficedula hypoleuca* ringed in Denmark 1899–2002. These are all small passerines passing Denmark on migration in reasonable numbers. For these species, annual numbers ringed are known as well as detailed data on recoveries. For this analysis, only birds recovered dead were included. For the species included, live recaptures are rare and their associated probabilities are probably even more heterogeneous and have more complicated spatial and temporal variation than that for dead recoveries, as they depend on the distribution of ringers. For the species considered, most of the birds ringed are young birds but due to the low number of recoveries for several of the species, we pooled all age classes together.

All birds were considered to have been ringed in summer of year 0, as data sparseness precluded investigation of models with temporal variation in model parameters. The birds were primarily ringed in the migration seasons (spring and autumn), and very few were ringed in the winter (European robin only). The wintering period was defined for all species as December-February,, and no recoveries outside this season were included. For simplicity, we only modeled recoveries for a period of five years after ringing.

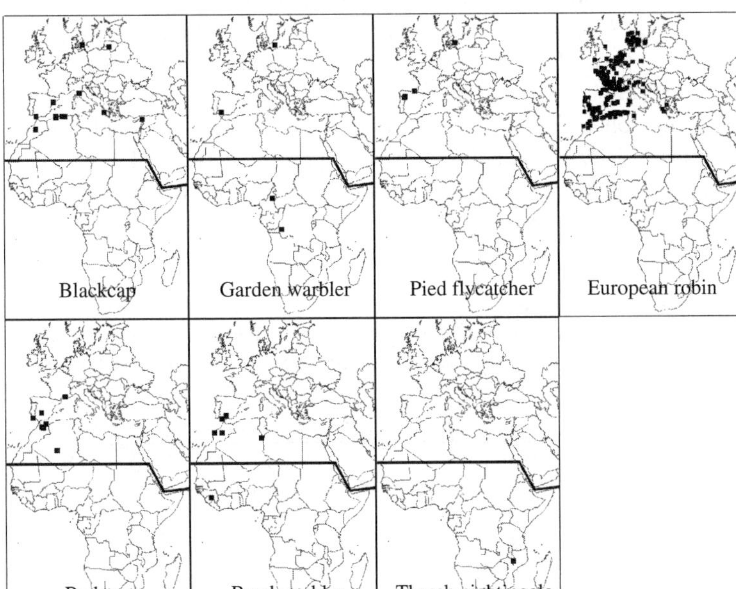

Fig. 2 Spatial distribution of ring recoveries of 7 species ringed in Denmark and recovered in winter (December–February)

Each recovery was either in Europe/North Africa or sub-Saharan Africa (Fig. 2). Due to data sparseness, we set mixture, survival, and recovery probabilities constant over time, and constrained survival to be independent of wintering area. There were only four recoveries in sub-Saharan Africa, which rendered parameters non-estimable if mixture parameters were included for all species. Thus, we also set the probability of migration to sub-Saharan Africa to zero for the European robin (which does not occur in sub-Saharan Africa; Keith et al. 1992), allowing estimation of mixture parameters for other species. For each likelihood formulation, two models were fit to the data. In the first, recovery probabilities were allowed to differ between Europe/North Africa and sub-Saharan Africa. In the second, recovery probability was set equal between the two wintering areas.

3 Results

Under likelihood 1, the model with differences in recovery probability between Europe/North Africa and sub-Saharan Africa was strongly favored by AIC_c ($\Delta AIC_c = 150.8$ relative to the model with a constant recovery probability; see Burnham and Anderson 2002). Under this model, the estimated ring recovery probability for Europe/North Africa was 4.6E-04, with a 95% profile likelihood interval of [4.1E-04; 5.2E-04] and that in sub-Saharan Africa 1.1E-05 [4.4E-06; 2.6E-05], thus around 40 times lower in sub-Saharan Africa. The common estimate of annual survival probability was 0.34 [0.30; 0.38]. Estimated probabilities of migrating to sub-Saharan Africa ranged from 0.60 in the blackcap to 1.0 in thrush nightingale (Table 2). These estimates can be compared with each species' actual recoveries from either region (Table 2 and Fig. 2).

Model selection criterion was similarly dismissive of the model with constant recovery probability when likelihood 2 was employed ($\Delta AIC_c = 178.6$). In this case, the estimated cumulative probability of a ring being reported over the lifetime of an individual was 1.4E-03 (95% profile CI: 1.3E-03, 1.6E-03) for birds wintering

Table 2 Numbers ringed (R) and recovered in Europe and North Africa (m^A) and sub-Saharan Africa (m^B) and estimated proportion of the population migrating to sub-Saharan Africa in autumn for likelihoods 1 and 2 in autumn (π_1^B and π_2^B, respectively). Also presented are 95% profile confidence intervals

Species	R	m^A	m^B	π_1^B (95% CI)	π_2^B (95% CI)
Redstart	66,387	7	0	0.84 (0.70–0.93)	0.86 (0.75–0.94)
Thrush nightingale	3,831	0	1	1.00 (0.49–1.00)	1.00 (0.63–1.00)
European robin	334,406	261	0	0.00†	0.00†
Garden warbler	57,057	2	2	0.95 (0.85–0.99)	0.96 (0.88–0.99)
Blackcap	44,155	12	0	0.60 (0.32–0.78)	0.63 (0.38–0.80)
Reed warbler	55,710	5	1	0.87 (0.72–0.95)	0.88 (0.75–0.96)
Pied flycatcher	57,384	4	0	0.90 (0.76–0.97)	0.93 (0.83–0.98)

† Fixed to zero for parameter estimability.

in Europe and North Africa, and 2.8E-05 (95% profile CI: 0.9E-05, 6.4E-05) for birds wintering in sub-Saharan Africa. Estimated probabilities of wintering in sub-Saharan Africa were similar to those from likelihood 1 (Table 2).

4 Discussion

The estimates of recovery probabilities and movement probabilities were obtained using standard ringing data with only limited trapping information. Only numbers of birds ringed each year are known. This is a common limitation for many of the ringing data bases in Europe, in that full information has only been computerized for birds recovered later in standard ringing data, but annual ringing totals for each species can generally be obtained with reasonable effort. The two approaches produced similar estimated movement probabilities that are in general similar to what one would expect based on biology of the species, and, despite the wide confidence limits, useful as quantitative estimates. For all species included, we estimated that the majority of individuals wintered in sub-Saharan Africa. These species are known to winter in large numbers in sub-Saharan Africa and they are comparatively rare in Europe in winter (Cramp 1988, 1992).

The rather strong assumptions made here are of course likely to bias the parameter estimates. Assuming equal survival probabilities, finding and reporting, and recovery probabilities among species and within regions are clearly not realistic. For example, mortality is not likely to be equal among short- and long-distance migrants. However, the variation in these parameters is likely to be on a smaller scale than the variation in the estimated parameters. The small differences between the parameter estimates under the two approaches provides some indication that our assumption about equal survival probabilities across species is not seriously violated. However, a full scale sensitivity analysis would likely be needed to confirm this assertion. Larger differences in parameter estimates between the two approaches are likely to arise if using less similar species.

Several possibilities exist to improve the estimates. First of all, the data set could be extended to include data from more countries, as these should be readily available, e.g. through the EURING data bank (http://www.euring.org/edb/). Another reasonable extension would be to include more species (e.g. wildfowl to assess spread of avian influenza). In that case, it would be necessary to model differences among species in survival and regional recovery probabilities. As long as these could be considered constant within groups this should still be possible. Alternatively, they could be modeled as relative differences, i.e. the recovery probability at site B could be constrained to be a constant proportion of the probability at site A. Estimates of the regional recovery probabilities could be further improved by using the seasonal changes in distribution for each species, since constant recovery probability within regions through seasons is probably a reasonable assumption. This does, however, cause some trouble due to the fact that many ringing schemes have not computerized detailed ringing information. Sensitivity to this can be modeled, but hopefully our approach may give some additional incentive for ringing organizations to digitize historical ringing records.

Extending the data set to include more species and more fine-scaled regions (and even time-specific probabilities from 100 years of ringing) raises the question of how to deal with high dimensional parameter sets. A custom-made framework is probably necessary for this type of analysis since complicated models are not easily accommodated in SURVIV.

The parameter estimates presented here were conditional on highly constrained models and few recoveries. As such, they are certainly not definitive. However, the analysis should serve as a template for how this framework could be extended to analyze larger, richer datasets. Even basic science may benefit from quantitative distribution maps, where previous studies on e.g. the migratory orientation program have used qualitative patterns only (Mouritsen 1998; Thorup and Rahbek 2004).

Acknowledgments We are grateful to Romaine Juillard for suggesting the use of a single binary response and to Fränzi Korner, Wojciech Kania and Charles Francis for constructive comments on and discussion about this paper.

References

Bønløkke J, Madsen JJ, Thorup K, Pedersen KT, Bjerrum M, Rahbek C (2006) The Danish Bird Migration Atlas. Rhodos, Humlebæk.

Bakken V, Runde O, Tjorve E (2003) Norwegian Bird Ringing Atlas. Stavanger, Stavanger Museum.

Brownie C, Anderson DR, Burnham KP, Robson DS (1985) Statistical Inference from Band-Recovery Data: A Handbook, 2nd edition. U. S. Fish and Wildlife Research Service, Resource Publication 156. U. S. Department of Interior, Washington D. C.

Burnham KP, Anderson DR (2002) Model Selection and Multimodel Inference: a Practical Information-Theoretic Approach, 2nd edn. Springer Verlag, New York.

Burnham KP, Anderson DR, White GC, Brownie C, Pollock KH (1987) Design and analysis for fish survival experiments based on release-recapture. American Fisheries Society Monograph 5. Bethesda, Maryland, USA.

Busse P, Kania W (1977) A quantitative estimation of distribution of ringed birds on the basis of recovery dispersal. Notatki Ornitologiczne 18:79–93. [Polish with English summary]

Cramp S (ed.) (1988) The Birds of the Western Palearctic. Vol. V. Oxford University Press, Oxford.

Cramp S (ed.) (1992) The Birds of the Western Palearctic. Vol. VI. Oxford University Press, Oxford.

Crissey WF (1955) The use of banding data in determining waterfowl migration and distribution. Journal of Wildlife Management 19:75–83.

Delany S, Veen J, Clarke J (Eds.) (2006) Urgent preliminary assessment of ornithological data relevant to the spread of Avian Influenza in Europe. Report to the European Commision.

Fransson T, Pettersson J (2001) Swedish Bird Ringing Atlas. Stockholm.

Kania W (1981) The autumn migration of the chaffinch *Fringilla coelebs* over the Baltic coast in Poland. Acta Ornithologica 18:371–414.

Kania W, Busse P (1987) An analysis of the recovery distribution based on finding probabilities. Acta Ornithologica 23:121–128.

Keith S, Urban EK, Fry CH (1992) The Birds of Africa. Vol. IV. Academic Press, London.
Kendall WL, Conn PB, Hines JE (2006) Combining multistate capture-recapture data with tag recoveries to estimate demographic parameters. Ecology 87:169–177.
Kendall WL, Pollock KH, Brownie C (1995) A likelihood-based approach to capture-recapture estimation of demographic parameters under the Robust Design. Biometrics 51:293–308.
Mouritsen H (1998) Modelling migration: the clock-and-compass model can explain the distribution of ringing recoveries. Animal Behaviour 56:899–907.
Perdeck AC (1977) The analysis of ringing data: Pitfalls and prospects. Vogelwarte 29 Sonderheft 33–44.
Pollock KH (1982) A capture-recapture design robust to unequal probability of capture. Journal of Wildlife Management 46:757–760.
Seber GAF (1982) The Estimation of Animal Abundance and Related Parameters, 2nd Edition. Griffin, London.
Sibert JR, Hampton J, Fournier DA, Bills PJ (1999) An advection-diffusion-reaction model for the estimation of fish movement parameters from tagging data, with application to skipjack tuna (*Katsuwonus pelamis*). Canadian Journal of Fisheries Aquatic Sciences 56:925–938.
Skalski JR, Townsend R, Lady J, Giorgi AE, Stevenson JR, McDonald RD (2002) Estimating route-specific passage and survival probabilities at a hydroelectric project from smolt radiotelemetry studies. Canadian Journal of Fisheries and Aquatic Sciences 59:1385–1393.
Thorup K, Rahbek C (2004) How do geometric constraints influence migration patterns. Animal biodiversity and conservation 27(1):319–329.
Walther BA, Wisz MS, Rahbek C (2004) Known and predicted African winter distributions and habitat use of the endangered Basra reed warbler (Acrocephalus griseldis) and the near-threatened cinereous bunting (Emberiza cineracea). Journal of Ornithology 145:287–299.
Webster MS, Marra PP, Haig SM, Bensch S, Holmes RT (2002) Links between worlds: unraveling migratory connectivity. Trends in Ecology & Evolution 17(2):76–83.
Wernham CV, Toms MP, Marchant JH, Clark JA, Siriwardena GM, Baillie SR (eds.) (2002) The Migration Atlas: movements of the birds of Britain and Ireland. Poyser, London.
White GC (1983) Numerical estimation of survival rates from band recovery and biotelemetry data. Journal of Wildlife Management 47:716–728.
Wikelski M, Kays R, Kasdin J, Thorup K, Smith JA, Cochran WW, Swenson GW Jr. (2007) Going wild: what a global small-animal tracking system could do for experimental biologists. Journal of Experimental Biology 210:181–186.
Zink G (1973–1985) Der Zug Europäischer Singvögel. Ein Atlas der Wiederfunde beringter Vögel. Teil 1–4. Radolfzell.
Zink G, Bairlein F (1995) Der Zug Europäischer Singvögel. Ein Atlas der Wiederfunde beringter Vögel. Teil 5. Radolfzell.

Evaluation of a Bayesian MCMC Random Effects Inference Methodology for Capture-Mark-Recapture Data

Gary C. White, Kenneth P. Burnham, and Richard J. Barker

Abstract Monte Carlo simulation was used to evaluate properties of a simple Bayesian MCMC analysis of the random effects model for single group Cormack-Jolly-Seber capture-recapture data. The MCMC method is applied to the model via a logit link, so parameters p, S are on a logit scale, where logit(S) is assumed to have, and is generated from, a normal distribution with mean μ and variance σ^2. Marginal prior distributions on logit(p) and μ were independent normal with mean zero and standard deviation 1.75 for logit(p) and 100 for μ; hence minimally informative. Marginal prior distribution on σ^2 was placed on $\tau^2 = 1/\sigma^2$ as a gamma distribution with $\alpha = \beta = 0.001$. The study design has 432 points spread over 5 factors: occasions (t), new releases per occasion (u), p, μ, and σ. At each design point 100 independent trials were completed (hence 43,200 trials in total), each with sample size $n = 10,000$ from the parameter posterior distribution. At 128 of these design points comparisons are made to previously reported results from a method of moments procedure. We looked at properties of point and interval inference on μ, and σ based on the posterior mean, median, and mode and equal-tailed 95% credibility interval. Bayesian inference did very well for the parameter μ, but under the conditions used here, MCMC inference performance for σ was mixed: poor for sparse data (i.e., only 7 occasions) or $\sigma = 0$, but good when there were sufficient data and not small σ.

Keywords Random effects · Variance components · MCMC · Program MARK · Bayesian estimation · Process variance · Process covariance

1 Introduction

Estimation of process variance across temporal and spatial scales of estimated survival parameters is a critical component of population viability analyses (White

G.C. White (✉)
Department of Fish, Wildlife, and Conservation Biology, Colorado State University, Fort Collins, CO, USA
e-mail: gwhite@cnr.colostate.edu

D.L. Thomson et al. (eds.), *Modeling Demographic Processes in Marked Populations*, Environmental and Ecological Statistics 3, DOI 10.1007/978-0-387-78151-8_53,

2000). Process variance is the variance of the true underlying parameter value across space or time. As an example, annual adult survival rates of a bird population likely vary annually because of differences in precipitation, temperature, and other weather effects that affect habitat quality. However, biologists have only estimates of annual survival available, and these estimates contain sampling variation (sometimes identified as measurement error). The variance of a set of annual survival estimates computed in the usual simple way provides an estimate of combined sampling and process variance. The separation of the sampling variance from the process variance is of immense interest to biologists because population viability analyses (PVA) should be constructed based on only the process variance, and not the combined variance. The estimation of process variance is often labeled as the analysis of random effects in the statistical literature.

Program MARK (White and Burnham 1999) provides a method of moments estimation procedure that allows estimation of a single random effect, and which has been shown by simulation to perform satisfactorily (Burnham and White 2002). Bayesian estimation procedures incorporating Markov chain Monte Carlo (MCMC) methods have been programmed into MARK, allowing an alternative method of estimating random effects for capture-recapture data. Specifically, the MCMC code in MARK allows for multiple hyperdistributions to create hierarchical models (Gelman et al. 2004:117–156), so that multiple random effects can be estimated simultaneously. So for example, a set of annual survival estimates, S_i, $i = 1, \ldots, k$, (specifically for MARK, the logit(S_i)) can be assumed to be observed from a hyperdistribution that is modeled as a normal distribution with mean μ and variance σ^2. Further enhancements include a design matrix capability to incorporate covariates into the hyperdistribution models, as well as estimation of the process correlation (hence covariance) across multiple random effects.

The objective of this paper is to evaluate by simulation the performance of the MCMC estimator for a single hyperdistribution.

2 Methods

2.1 Simulation Design

We simulated single-group live capture-recapture data of the CJS type (Lebreton et al. 1992) to evaluate the performance of the MCMC estimator. Random effects were simulated as normally distributed variables on the logit scale. The Monte Carlo simulations were done as a factorial treatment design using five factors:

Capture occasions ($t = k + 2$)	4 levels (7, 15, 23, 31)
Releases of new animals u on each occasion	3 levels (25, 100, 400)
Constant capture probability p on each occasion	3 levels (0.4, 0.6, 0.8)
Mean survival probability E(S)	3 levels (0.4, 0.6, 0.8)
Process variation σ on logit(S)	4 levels (0, 0.1, 0.25, 0.5)

All 432 combinations ($= 4 \times 3 \times 3 \times 3 \times 4$) of these levels defined the points used in the design space. At each design point we simulated 100 independent data sets. For cases of $\sigma > 0$ the $S_1, \ldots, S_k$ were generated as a random sample from a logit normal distribution as $\text{logit}(S_i) = \log[\text{E}(S)/(1 - \text{E}(S))] + \text{Normal}(0, \sigma)$. On each occasion a fixed number of new 'animals' ($u_i = 25, 100, 400$) were released into the population. Data sets were generated one at a time in SAS (SAS Institute, Inc 2003) and passed directly to MARK where the MCMC estimation was performed.

Processing time required to complete these simulations and ensuing MCMC estimation was 4 months on an Intel Pentium 4 dual processor machine with a 3.4 GHz clock speed and 2 GB of RAM.

2.2 MCMC Estimation

The CJS model was fitted to each of the simulated data sets with a logit link function. Prior distributions of the capture (p) parameters that were not included in a hyperdistribution were modeled on the logit scale with a normal distribution of mean zero and standard deviation 1.75. The value 1.75 was choosen to minimize the AIC discrepancy between the back-transformed logit variable and a uniform 0,1 variable. Survival parameters were included in a single hyperdistribution model, assumed to be a normal distribution, $N(\mu, \sigma)$, on the logit scale. The prior distribution of μ was taken as normal with mean zero and standard deviation 100, and the prior distribution of σ^2 was taken as an inverse gamma distribution with $\alpha = \beta = 0.001$.

The MCMC procedure was implemented sequentially in MARK using a Metropolis-Hastings sampler (Givens and Hoeting 2005, pages 183–188). Within each cycle, the p and S parameters were processed first, randomly ordered for each cycle (Givens and Hoeting 2005, page 198; Robert and Casella 1999). Then the μ and σ parameters of the hyperdistribution were processed to complete one cycle.

Proposal distributions were specified as random jumps from current parameter values. Jumps were normally distributed with mean zero and parameter-specific standard deviations adjusted during a tuning phase of 4,000 cycles to obtain an acceptance rate of approximately 45%. The proposal standard deviations were then set, and followed by a burn-in of 1,000 cycles. Summary statistics were computed from 10,000 sequential samples of the posterior distribution.

Convergence of each of the simulated chains was not evaluated directly. However, prior to starting these simulations, convergence of multiple test cases was evaluated through comparison of multiple chains (Gelman 1996) and graphically through plots of sequential values and histograms of the posterior distributions. In no case was a lack of convergence suggested.

2.3 Simulation Summaries

For each simulated data set, the mean, median, mode, and 2.5, 5, 10, 20, 80, 90, 95, and 97.5% percentiles were recorded from the posterior distributions of μ and

Table 1 Estimated mean of the hyperdistribution of survival rates (μ). The mean, median, and mode of the posterior distribution are reported, followed by the percent relative bias (PRB) computed for the mean. Each entry is based on 4,800 simulations summarized across the 3 remaining design factors. Average CV = SE/mean for table entries is approximately 0.008. Average SE for PRB was 0.97

	True $\log[E(S)/(1-E(S))]$ ($E(S)$ in parentheses)											
	(0.4) −0.405				(0.6) 0.405				(0.8) 1.386			
	Mean	Median	Mode	PRB	Mean	Median	Mode	PRB	Mean	Median	Mode	PRB
p												
0.4	−0.447	−0.452	−0.466	10.18	0.409	0.393	0.383	0.75	1.472	1.445	1.425	6.18
0.6	−0.407	−0.410	−0.413	0.39	0.408	0.403	0.398	0.56	1.399	1.390	1.383	0.92
0.8	−0.374	−0.376	−0.379	−7.70	0.420	0.417	0.413	3.52	1.397	1.392	1.386	0.74
Number of releases (u)												
25	−0.411	−0.419	−0.434	1.30	0.422	0.403	0.389	4.08	1.482	1.450	1.427	6.93
100	−0.410	−0.412	−0.414	1.02	0.408	0.405	0.402	0.70	1.396	1.390	1.384	0.73
400	−0.408	−0.408	−0.409	0.56	0.406	0.405	0.403	0.05	1.389	1.387	1.385	0.18

σ. Percent relative bias ($= (\hat{\theta} - \theta)/\theta$, except for $\sigma = 0$) and coverage of the 95% credibility intervals for true parameter values was computed.

Relative importance of the 5 design factors was assessed with ANOVA. Because of the number of simulations completed, trivially small effects could be detected, so only relative rankings were used. Results for the 3 most important design factors are presented.

3 Results

Bias in the estimates of μ was most affected by p, followed by the interaction of $E(S)$ and p and then the number of releases (Table 1). In general, there was only a small bias, and this bias seems somewhat inconsistent across 9 scenarios reported. Because the posterior distribution of μ is relatively symmetrical, the mean, median, and mode were approximately equivalent, with the median always intermediate between the mean and the mode.

Credibility interval coverage of the mean of the hyperdistribution of survival was close to the expected 95% for the 2.5 and 97.5 percentile summary statistics (Table 2). Estimated coverage was slightly below the expected value for the more sparse data sets, i.e., with only 25 releases per occasion or low capture probabilities combined with low survival. Otherwise, coverage tended to exceed slightly the expected 95%.

Bias of σ was most affected by the number of releases followed by the number of occasions (Table 3). As would be expected, the mean of the posterior distribution for scenarios with $\sigma = 0$ was biased high, because the lower bound of the posterior distribution is zero. Although the mode of the posterior was less biased for these scenarios, a bias persists. The median was always intermediate between the mean and mode, and the median was less biased than the mode in all cases except when $\sigma = 0$.

Credibility interval coverage was zero, as would be expected, for $\sigma = 0$. Coverage improved for positive values of σ (Table 4). However, credibility interval coverage

Table 2 Credibility interval coverage of the mean of the hyperdistribution of survival rates (μ). The proportion of the simulations with the 2.5 percentile $< \mu <$ 97.5 percentile are reported. Each entry is based on 4,800 simulations summarized across the 3 remaining design factors, giving a CV of approximately 0.0033, or 0.33%

	True $\log[E(S)/(1 - E(S))]$ ($E(S)$ in parentheses)		
	(0.4) −0.405	(0.6) 0.405	(0.8) 1.386
p			
0.4	0.923	0.949	0.948
0.6	0.948	0.958	0.959
0.8	0.952	0.958	0.958
Number of releases			
25	0.925	0.953	0.955
100	0.943	0.954	0.955
400	0.952	0.959	0.958

Table 3 Estimated standard deviation of the hyperdistribution of survival rates (σ). The mean, median, and mode of the posterior distribution are reported, followed by the percent relative bias (PRB) of the mean, except for $\sigma = 0$. Each entry for number of occasions is based on 2,700 simulations summarized across the 3 remaining design factors, and each entry for number of releases is based on 3,600 simulations summarized across the 3 remaining design factors. Average CV = SE/mean for table entries is approximately 0.022, or 2.2%. Average SE of PRB is 1.7

	True σ														
	0			0.1				0.25				0.5			
	Mean	Median	Mode	Mean	Median	Mode	PRB	Mean	Median	Mode	PRB	Mean	Median	Mode	PRB
Number of occasions															
7	0.262	0.175	0.076	0.255	0.170	0.069	155.1	0.333	0.238	0.116	33.2	0.568	0.435	0.278	13.7
15	0.123	0.101	0.055	0.142	0.120	0.072	41.9	0.236	0.213	0.160	−5.5	0.480	0.452	0.394	−4.00
23	0.102	0.087	0.054	0.123	0.109	0.075	22.9	0.231	0.217	0.183	−7.7	0.480	0.465	0.429	−3.9
31	0.091	0.079	0.052	0.115	0.104	0.077	15.0	0.228	0.219	0.191	−8.8	0.482	0.472	0.448	−3.6
Number of releases															
25	0.231	0.165	0.075	0.224	0.160	0.069	124.0	0.278	0.212	0.104	11.2	0.479	0.402	0.290	−4.2
100	0.125	0.101	0.056	0.140	0.116	0.071	40.0	0.239	0.215	0.169	−4.3	0.508	0.472	0.417	1.5
400	0.077	0.066	0.047	0.112	0.101	0.081	12.1	0.254	0.239	0.215	1.6	0.522	0.494	0.454	4.3

Table 4 Credibility interval coverage of the standard deviation of the hyperdistribution of survival rates (σ). The proportion of the simulations with the 2.5 percentile $< \sigma <$ 97.5 percentile are reported. Each entry for number of occasions is based on 2,700 simulations summarized across the 3 remaining design factors (CV of approximately 0.0044), and each entry for number of releases is based on 3,600 simulations summarized across the 3 remaining design factors (CV of approximately 0.0038, or 0.38%)

	True σ			
Number of occasions	0	0.1	0.25	0.5
7	0.000	0.990	0.987	0.960
15	0.000	0.993	0.957	0.940
23	0.000	0.983	0.960	0.930
31	0.000	0.979	0.941	0.924
Number of releases				
25	0.000	0.991	0.984	0.933
100	0.000	0.992	0.953	0.936
400	0.000	0.976	0.946	0.946

exceeded the expected 95% for $\sigma = 0.1$, but appears to be approaching the expected 95% for scenarios with ample data.

4 Discussion

Estimation of process variances on the logit scale may seem restrictive. However, often the logit scale is the biologically relevant scale at which to work. The logit scale is more likely to provide a linear scale to model the effects of environmental covariates, e.g., precipitation or temperature. Another consideration is modeling of correlations between parameters on the real scale, e.g., positively correlated survival rates such as young and adult survival rates. Positively correlated survival rates modeled on the real scale with a beta distribution cannot be generated with extensions of the beta distribution, whereas back-transformation of logit-normal variables makes this type of model easy to implement.

However, the back-transformation of an estimate of σ to the variance on the real scale depends on the mean of distribution. So, an estimate of $\sigma = 0.1$ with a mean of 0 on the logit scale results in a real variable with mean 0.5 and $\sigma = 0.025$. But, an estimate of $\sigma = 0.1$ with a mean of 4 on the logit scale results in a real variable with mean 0.982 and $\sigma = 0.0018$. Thus, interpretation of the estimates of process variance on the logit scale must consider the mean as well. Similarly, the correlation of 2 variables on the logit scale changes when back-transformed to the real scale.

A critical part of the Bayesian estimation procedure is to provide an appropriate prior distribution for each parameter in the model. A common approach is to provide non-informative or diffuse priors (i.e., a suitably-flat distribution that has little or no influence on the shape of the posterior distribution) so that nearly all the information contained in the posterior distribution is coming from the likelihood function

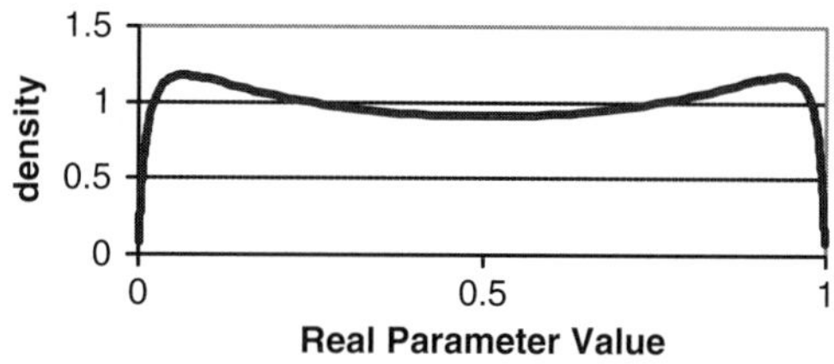

Fig. 1 Probability density function of the back-transformed $N(0, 1.75)$ density used as the default prior distribution for parameters not part of a hyperdistribution

and the data; for example, a normal distribution with a large standard deviation, e.g., $N(0, 100)$. The prior distribution for the parameters not in the random effects models are particularly problematic because these parameters are on the logit scale, and such a flat prior will result in a U-shaped distribution when back-transformed to the real scale. The default prior for parameters not included in the random effects model is $N(0, 1.75)$, which results in a reasonably flat distribution on the back-transformed scale (Fig. 1). Prior distributions for the hyperdistribution parameters are considered to be relatively uninformative distributions.

The Bayesian estimators of σ do not perform as well as the method of moments estimators (Burnham and White 2002) also available in Program MARK. Neither estimator performs particularly well with $\sigma = 0$, but because the method of moments estimator allows negative estimates (which get truncated back to zero), bias is reduced and confidence interval coverage is better than the performance of the Bayesian estimator. Most disappointing with the Bayesian estimators was the excessive coverage of the credibility intervals for $\sigma = 0.1$, and the apparent bias for sparse data, i.e., data with low capture probabilities and/or low numbers of animals released per occasion. This behavior is conservative in that coverage exceeds 95%, so it should not generally create problems.

The simulations reported in Burnham and White (2002) were for σ of parameters on the real scale (i.e., σ was estimated for parameter estimates constrained to the interval 0–1), whereas the simulations reported here are for σ estimated for parameters on the logit scale (i.e., parameter estimates are unconstrained). Thus, comparisons between these 2 sets of simulations must be made cautiously. However, Burnham and White (2002) (Table 2) showed negligible bias in estimates of σ, except for $\sigma = 0$. Likewise, they showed confidence interval coverage of 95% (Burnham and White 2002:254), with nearly equal misses both below and above the true σ. Thus, we conclude that the method of moments estimator of Burnham and White (2002) performs better than the Bayesian estimator reported here.

However, the major advantage of the Bayesian approach is the flexibility to handle multiple hyperdistributions, and to estimate the covariance or correlation across different random effects. The MCMC implementation in MARK provides for estimating process autocorrelation across time within a hyperdistribution, plus estimation of the process correlation between sets of parameters, each modeled with a separate hyperdistribution. This capability allows users to estimate the potentially positive process correlation between recruitment and survival in the Pradel (1996) model, although the correlation is between the $\log(f)$ and $\text{logit}(\varphi)$ estimates.

References

Burnham KP, White GC (2002) Evaluation of some random effects methodology applicable to bird ringing data. Journal of Applied Statistics 29:245–264.

Gelman A (1996) Inference and monitoring convergence. Pages 131–143 *in* Gilks WR, Richardson S, Spiegelhalter DJ, editors. Markov Chain Monto Carlo in Practice. Chapman and Hall/CRC, Boca Raton, Florida.

Gelman A, Carlin JB, Stern HS, Rubin DB (2004) Bayesian Data Analysis, 2nd Ed. Chapman & Hall/CRC, Boca Raton, Florida.

Givens GH, Hoeting JA (2005) Computational Statistics. Wiley & Sons, Hoboken, New Jersey, USA.

Lebreton JD, Burnham KP, Clobert J, Anderson DR (1992) Modeling survival and testing biological hypotheses using marked animals – a unified approach with case-studies. Ecological Monographs 62:67–118.

Pradel R (1996) Utilization of capture-mark-recapture for the study of recruitment and population growth rate. Biometrics 52:703–709.

Robert CP, Casella G (1999) Monte Carlo Statistical Methods. Springer-Verlag, New York, New York, USA.

SAS Institute, Inc. (2003) SAS System Help and SAS OnLineDoc. SAS Institute, Inc., Cary, North Carolina.

White GC (2000) Population viability analysis: data requirements and essential analyses. Pages 288–331 *in* Boitani L, Fuller TK, editors. Research techniques in animal ecology: controversies and consequences. Columbia University Press, New York, New York, USA.

White GC, Burnham KP (1999) Program MARK: survival estimation from populations of marked animals. Bird Study 46:120–139.

Index

Printed in the United States of America